Oxidative Stress in Applied Basic Research and Clinical Practice

Series Editor
Donald Armstrong

More information about this series at http://www.springer.com/series/8145

Ines Batinić-Haberle • Júlio S. Rebouças
Ivan Spasojević
Editors

Redox-Active Therapeutics

Humana Press

Editors
Ines Batinić-Haberle
Duke University School of Medicine
Durham, NC, USA

Júlio S. Rebouças
Universidade Federal da Paraíba
João Pessoa, PB, Brazil

Ivan Spasojević
Duke University School of Medicine
Durham, NC, USA

ISSN 2197-7224 ISSN 2197-7232 (electronic)
Oxidative Stress in Applied Basic Research and Clinical Practice
ISBN 978-3-319-30703-9 ISBN 978-3-319-30705-3 (eBook)
DOI 10.1007/978-3-319-30705-3

Library of Congress Control Number: 2016946368

Printed on acid-free paper

This Humana imprint is published by Springer Nature
The registered company is Springer International Publishing AG Switzerland

Foreword

Free Radical and Redox Biology in Disease States: The Era of Redox-Active Therapeutics

Free radicals and oxidant species have been associated with the initiation and progression of several disease states for the last three decades. Disruption of cell and tissue redox homeostasis occurs by a variety of molecular mechanisms including the alteration of mitochondrial function, overactivation of membrane-bound NAD(P)H oxidases, induction or uncoupling of nitric oxide synthases, redox cycling of xenobiotics, mobilization of iron pools, and enhanced activity of xanthine oxidase, myeloperoxidase, and other "redox" enzymes in specific cell types and tissue regions. These sources of reactive species have been identified as culprits of creating pro-oxidant environments that facilitate oxidative molecular damage. Additionally, oxidants, at controlled levels, have been more recently shown to participate in redox signaling pathways. Overall, significant changes in steady-state concentrations of reactive species can have a profound impact on the control of metabolism and gene expression. Two major challenges in the field have been to (a) dissect and prove the relevance of different oxidative processes in the pathophysiology of disease (in spite of its association) and (b) develop rational and effective redox-active therapeutics that can neutralize molecular events of disease. In one way, a + b are intertwined because current evidence is indicating that effective redox-active therapeutics should be targeted towards the main participating routes/reactive species and/or cell/tissue compartments. Additionally, a successful redox-based intervention serves as a "proof of concept" to connect oxidative processes to pathology. In the last decade a great deal of progress has been made on the sound development of such redox-active therapeutics. For example, in the case of transition metal-based synthetic compounds, modulation of the redox potential, structure, and hydrophobicity have largely improved the pharmacological action of these compounds by means of the efficient elimination of reactive species due to optimized kinetic capacity and biodistribution, together with a decrease of undesired interactions with biomolecules. Important findings through the years have been that many of the developed redox-active compounds are multifunctional in terms of both coping with a group of reactive species and catalytically participating in redox detoxification pathways at the expense of reducing equivalents obtained from endogenous reductants (such as uric acid, ascorbate, or glutathione) or even from redox enzymes or the mitochondrial electron transport chain. The high kinetic efficiency of some of the compounds together with catalytic character of these reactions makes it possible to have potent actions in the presence of micromolar or submicromolar amounts, with little or negligible associated toxicity.

Another important aspect, successfully developed lately to increase pharmacological actions, relates to the improved subcellular distribution. Indeed, targeting of redox-active compounds to specific cell compartments such as mitochondria by a variety of chemical modifications to the parent moiety has revealed to be a key breakthrough. Indeed, targeted compounds have been shown to increase their concentrations in specific intra- or extracellular compartment several hundreds of time, which has resulted in higher specificity and effectiveness.

Additionally, other important related conceptual demands (and challenges) have emerged lately in the development and testing of redox-active therapeutics. Firstly, many times the observed protective biological effects exceed what can be expected from carefully contrasting the actual concentrations of redox-active compound achieved intracellularly with the known rate constants of reaction with reactive species (important to say that the latter have been typically obtained in homogenous solutions, not in the crowded and compartmentalized cellular environment); indeed, it is not always obvious to visualize how some of the utilized compounds may largely outcompete the many parallel reactions of reactive species with critical intracellular targets or significantly augment pre-existing antioxidant capacities (e.g., how much could an "SOD mimic" increase SOD activity in vivo, when 10–20 μM SOD is present in different cellular compartments and react at near diffusion-controlled rates with superoxide radicals!). Part of the answer seems to rely on the capacity of these compounds to eliminate a variety of species, but *also* in their capacity to induce antioxidant responses at the transcriptional level (presumably via oxidation reactions involving protein thiols or maybe even iron-sulfur clusters) that ultimately results in the upregulation of endogenous antioxidant systems. Indeed, it is increasingly recognized that redox-active compounds may activate redox-sensitive transcription factors that, in turn, trigger cytoprotective gene responses. Thus, direct detoxification reactions may run simultaneously with more subtle and permanent changes in cell/tissue antioxidant capacities.

In the past, studies on different human pathologies administering both natural and synthetic compounds have been executed, suggesting positive pharmacological actions. Unfortunately, some of the promising studies carried out on small and controlled human populations were not confirmed later in studies with more demanding standards and/or on larger populations, which signified a large delay for the field of redox-active therapeutics and medical practice, and challenged the view on the role of disruption of redox processes on the basis of human disease conditions. But further scientific progress was made over the last decade both on the understanding of free radical/redox processes participating in molecular basis of disease and the design of synthetic redox-active compounds with improved chemical reactivity, selectivity, and biodistribution. Indeed, I firmly think that after the initial enthusiasm followed by a drawback experienced with redox-active therapeutics in humans, the field is now on a new and wide avenue of solid development, which is likely to become clinically successful. With the more profound understanding of the subtle redox reactions, mediators, and processes underlying the development of pathologies and the rationale design of more suitable synthetic compounds with enhanced selectivity, reactivity, and compartmentalization, the application of redox-active therapeutics to the clinical arena is getting very close. These new therapeutics are expected to be useful in inflammatory, cardiovascular, and neurodegenerative conditions and in radioprotection, among several other disease states. In addition to the large amount of existing preclinical data, some of the newer compounds are or will be promptly tested in clinical trials. Obviously, a positive pharmacological outcome will be of great impact to the field and to medical therapeutics.

All the key aspects of the field of redox-active therapeutics are nicely covered in this book edited by Drs. Batinić-Haberle, Rebouças, and Spasojević. A nice progression from very basic concepts and observations (starting with chapters from the discoverers of superoxide dismutase, Irwin Fridovich and Joe Mc Cord) to the application of redox-active therapeutics in large range of disease conditions is presented in a fully updated and coherent manner. With their vast experience in the field, the editors have been able to assemble an outstanding group of chapters written by leading investigators. This book represents a large and most welcome effort to bring together, and with a profound and solid view, the current developments and potential applications of redox-active therapeutics for the treatment of human pathologies. The work also leaves open the possibility for further research in the area including a deeper understanding of the pharmacological mechanism of redox-active drugs in vivo, identification of novel molecular targets and salutary mediators, development of tailored drugs for enhanced selectivity, and application of the more promising compounds for acute and chronic disease conditions in humans. I warmly congratulate the editors and authors for the outstanding work and I am positive this book, a very first dedicated to stress the key role of redox-active therapeutics in medicine, will constitute a reference material both for the free radical/redox biomedical community and for any biomedical and chemical researchers interested in the role of free radicals, oxidants, and antioxidant systems in human health and pathology.

Rafael Radi, MD, PhD
President, Society for Free Radical
Research International
Fellow, Society for Redox
Biology and Medicine
Foreign Associate, US National
Academy of Sciences
Universidad de la República
Montevideo, Uruguay

Preface

The inauguration of this book marks ~ half a century since the inception of the field of Free Radicals in Biology and Medicine with the seminal discovery of the major endogenous antioxidative defense, Cu,Zn superoxide dismutase enzyme (Cu,ZnSOD), by Irwin Fridovich and Joe McCord in 1969 [1].

The discovery of Cu,ZnSOD was followed by discoveries of mitochondrial (MnSOD) and extracellular (Cu,ZnSOD) isoforms. The subsequent studies of Babior et al. [2] supported the biological relevance of superoxide ($O_2^{\bullet-}$) demonstrating that this radical is formed, as a part of antibacterial strategy, by the action of NADPH oxidases in white blood cells. Over years it has been demonstrated that a number of enzymatic systems, oxidases, oxygenases, nitric oxide synthases, complexes I and III of mitochondrial respiration, and others produce $O_2^{\bullet-}$ (intentionally or not) and subsequently and rapidly H_2O_2 (enzymatically or not) under physiological and pathological conditions. Throughout half a century of research, immense knowledge has been collected on free radicals and other reactive species demonstrating their critical roles in redox biology of healthy, metabolically stressed and neoplastic cells. This in turn has inspired numerous studies that explore therapeutic approaches, many of which are covered in this book, to normalize physiological redox status in normal but diseased cells and induce apoptosis of cancer cells.

Ten years after the discovery of the SOD enzyme, the first study on an SOD mimic was reported by Pasternack and Halliwell [3]. The authors demonstrated the SOD-like activity of an Fe porphyrin. In the early 1980s Archibald and Fridovich [4] showed that Mn salts, such as Mn(II) lactate, possess high SOD-like activity thereby justifying the existence of organisms that accumulate mM levels of Mn to overcome their lack of an SOD enzyme. The first report on the Mn salen class of SOD mimics (EUK-8) appeared in 1993 [5]. One of those, EUK-134, is in use as an active ingredient in sunscreen products. Meanwhile Irwin Fridovich's group embarked on decades-long development of porphyrin-based SOD mimics. A highly efficacious Mn porphyrin-based SOD mimic was reported in 1997 by Ines Batinić-Haberle [6], setting the stage for the design of multiple redox-active metalloporphyrins on the basis of structure-activity relationships. In parallel with Mn porphyrin-based SOD mimics, works by Dennis Riley's group [7] gave rise to the

Mn(II) cyclic polyamine (aza-crown ethers) class of potent SOD mimics. Compound leads from both porphyrin and polyamine classes of SOD mimics are presently in clinical trials as radioprotectors of normal tissue. While not SOD mimics, in the early 1990s the redox-active (non-metal based) nitrones and nitroxides have been developed as therapeutics [8]. The OKN-007 nitrone (also known as NXY-059) went through clinical trials for stroke and is presently in a clinical trial as an anticancer therapeutic for recurrent malignant glioma. The redox-active quinone-based compound, MitoQ, has been in clinical trials for Parkinson's disease and for chronic hepatitis C and is presently used for skin care. In the late 2000s "shrinked" porphyrins, metallocorroles, emerged as prospective SOD mimics and therapeutics [9]. Numerous other redox-active metal complexes and SOD mimics were developed as therapeutics by different groups, some of which are addressed in this book. The wealth of data collected thus far demonstrates that structure-activity relationships, initially developed for Mn porphyrins, are valid for different classes of SOD mimics. The photosensitizing porphyrin ligand, $H_2TM\text{-}4\text{-}PyP^{4+}$, is being incorporated into nano-scaffolds to form nano-phototheranostics for diagnosis and treatment of various cancers. As alternatives to such photosensitizers, quantum dots made of semiconductor metal-containing materials (such as CdSe, CdTe, and InAs) conjugated or not to Zn(II) porphyrins have been explored [10].

Since the discovery of SOD enzymes, tremendous progress has been made on the chemistry of small endogenous reactive species and enzymes which maintain the balanced redox environment of a normal cell; the perturbed balance results in a pathological condition known as oxidative stress. Different classes of SOD mimics were initially developed and anticipated to be specific to $O_2^{\bullet-}$. It might have been obvious from the very beginning that such small molecules (relative to a protein-structured enzyme), with biologically compatible reduction potentials, would react with numerous reactive species. Yet, the wealth of knowledge on the chemistry of those species was not available to allow for such "free" thinking. Moreover the biological importance of numerous species such as nitric oxide, peroxynitrite, nitroxyl, carbon monoxide, reactive sulfur, and selenium species and their cross-talk has not yet fully emerged. In turn, not until the end of the 1990s and early 2000s did the rich reactivity of SOD mimics and other redox-active therapeutics towards species other than $O_2^{\bullet-}$ surfaced.

Researchers have often incorrectly assigned the effects of redox-active drugs to particular reactive species. The unselective chemistry of such compounds, the multitude of reactive species involved, and the biological milieu are too complex to define the mechanism of action with certainty. The use of genetically modified animals or microorganisms has allowed progress. While insight into the redox biology of a cell and redox-active compounds is expanding, the actions of compounds are still often incorrectly singularly attributed to the dismutation of either O_2^- or/and H_2O_2, when neither a true SOD mimic nor a functional catalase mimic is used. A recent comprehensive study pointed out that the majority of metal complexes (various Mn or Fe porphyrins, Mn salen EUK-8 or Mn(II) cyclic polyamines such as M40403) are not catalase mimics [11]. Only Fe(III) corroles have modest catalase-like activity, and its biological relevance awaits further exploration. Another important issue with any drug is its purity. Caution needs to be exercised, as commercial suppliers have frequently sold impure compounds, which in turn have hindered correct discussions of the effects observed.

It was not until the mid-2000s that evidence was provided to demonstrate the interaction between redox-active therapeutics and transcription factors, such as HIF-1α, AP-1, SP-1, and NF-κB. Effects on transcription factors have been found with different classes of therapeutics, both synthetic and natural, such as Mn porphyrins, nitroxides, Mn salen derivatives, sulforaphane, flavonoids, and polyphenols. Initially, it was speculated that the effects were due to the ability of Mn porphyrins to rapidly remove reactive species, produced upon oxidative stress, which would have otherwise activated one or more transcription factors and in turn transcription of a group of genes. It took nearly a decade before it became clearer that at least one of the major mechanisms involves protein thiols. This learning process required joint efforts of chemists, biochemists, pharmacologists, and biologists.

For years, an obvious fact was overlooked. Catalysis of $O_2^{\bullet -}$ dismutation is efficacious *ONLY* if a mimic (or SOD enzyme) is an equally good oxidant and antioxidant, i.e., it equally well oxidizes $O_2^{\bullet -}$ to oxygen and reduces it to H_2O_2. Thus an SOD mimic and/or SOD enzyme can act in vivo both as an antioxidant and (pro)oxidant. Moreover, the reduction of $O_2^{\bullet -}$ by an SOD mimic gives rise to an oxidant—H_2O_2. At that point it was a common understanding that H_2O_2 cannot accumulate in the cell as it is eliminated by numerous (redundant) enzymatic systems such as catalase, glutathione peroxidases (GPx), and peroxiredoxins. Only recently has it emerged that those systems may be downregulated and/or inactivated during disease, which would in turn lead to H_2O_2 accumulation—a frequent scenario in cancer. While the SOD enzyme is a tumor suppressor in healthy cells, under disease conditions, the SOD enzyme may become a tumor promoter. Such reports coincide with the conclusion that indeed an SOD mimic, with redox properties similar to SOD enzyme, can function as either a pro- or antioxidant depending upon the local environment.

Jon Piganelli was the first to suggest that, in diabetes models, Mn porphyrin can act as an oxidant, possibly oxidizing the p50 subunit of NF-κB in nucleus [12]. The notion was supported by a pharmacokinetic study on macrophages which showed that Mn porphyrin accumulates threefold more in the nucleus than in cytosol. A crucial study on lymphoma cells by Margaret Tome and her colleagues [13] furthered Piganelli's notion in helping understand which reactions are likely involved in the suppression of NF-κB transcriptional activity. Tome showed that H_2O_2 and GSH are indispensable in the actions of a redox-active Mn porphyrin in oxidatively modifying—*S*-glutathionylating—p65 and p50 subunits of NF-κB. Once glutathionylated, NF-κB cannot bind to DNA to initiate gene transcription. *S*-glutathionylation was then demonstrated by Tome's group to occur with other thiol-bearing proteins including mitochondrial complexes I, III, and IV. The inactivation of complexes I and III resulted in the suppression of ATP production. An effect of Mn porphyrin on glycolysis was also seen, possibly involving *S*-glutathionylation reactions. Importantly, no toxicity was observed with normal lymphocytes. Mitochondrial accumulation of Mn porphyrins and comprehensive aqueous chemistry on cysteine oxidase and/or GPx-like activity of Mn porphyrins by Batinić-Haberle and her colleagues supported such observations. Studies point to a major role of H_2O_2 in the actions of Mn porphyrins and agree well with a growing recognition of the critical role of this species in cell biology. A similar mechanism of action for other redox-active drugs is highly likely as several other compounds such as Mn salens, nitroxides, and polyphenols also inhibit NF-κB.

Studies as such have challenged how small molecule redox-active SOD mimics, originally considered as selective antioxidants, are viewed. Current data point to the activation of Nrf2 by different redox-active drugs including curcumin, sulforaphane, nitroxides, and Mn porphyrins, presumably via oxidation (or *S*-glutathionylation) of the cysteines of its Keap1 unit, reminiscent of the NF-κB story described above. Such action would result in upregulation of endogenous antioxidative defenses. The data provided by Thambi Dorai et al. [14] on rat kidney ischemia/reperfusion injury showed that Mn porphyrin, rather than acting as SOD mimic in its own right, may activate Nrf2. Activation of Nrf2 would upregulate numerous endogenous antioxidative defenses including mitochondrial and extracellular SOD enzymes. Recent preliminary data from Daret St. Clair's group suggest that the activation of Nrf2 by Mn porphyrin may, though, not occur in cancer. Indeed the activation of Nrf2 by Mn porphyrin was not seen with malignant hematopoietic stem cells derived from patients with myelodysplastic syndrome, but Nrf2 was activated in normal hematopoietic stem cells.

In addition to favorable redox properties, redox-active therapeutics MUST reach a site of action in the body without imposing significant toxicity. Significant efforts have been invested towards understanding how structural properties of Mn porphyrins affect their toxicity and biodistribution in organs, cells, and subcellular organelles. Similar studies on other redox-active drugs are needed.

This book provides a comprehensive, up-to-date source of information on the molecular design and mechanistic, pharmacological, and medicinal aspects of redox-active therapeutics. The first two sections of the book discuss the role of SOD enzymes under physiological and pathological conditions and address multiple classes of redox-active drugs. The basic aspects of the chemistry and biology of redox-active drugs and the brief overview of the redox-based pathways involved in cancer and the medical aspects of redox-active drugs are provided assuming little in the way of prior knowledge. The third and fourth sections of the book deal with the therapeutic approaches towards different diseases. Among therapeutic effects obtained with redox-active drugs, the radioprotection of normal tissues is of particular interest as there is no efficacious and nontoxic radioprotector of normal tissue available for human use. Such effects gave rise to ongoing clinical trials on the radioprotective effects of two efficacious SOD mimics (Mn porphyrin-based BMX-001 and Mn cyclic polyamine-based GC4419)—a critical step forward in the development of a redox-active SOD mimic as a human therapeutic. Up to 50 % of all cancer patients undergo some type of radiotherapy; thus there is a substantial need to protect normal tissue during tumor irradiation. While protecting normal tissue, it is essential that the radioprotector does not protect tumor tissue. Multiple studies now show that an SOD mimic can simultaneously function as a radioprotectant for normal tissue and as a radiosensitizer for tumor tissue. Only recently has a possible explanation for such apparent dichotomy emerged. It relies on differential redox environments of normal vs. cancer tissue and differential accumulation of Mn porphyrin in those tissues. This drives vast suppression of antiapoptotic NF-κB in the cancer cell while there is only moderate suppression in normal cells; in turn, tumor cells undergo apoptosis while survival pathways are upregulated in normal cell.

In recent years there has been a rapid advance in developing and exploiting redox-active therapeutics towards a wide range of diseases and pathophysiological states other than cancer, such as inflammation, diabetes, cardiovascular, and neurodegenerative diseases. Contrary to classical therapeutic approaches that target directly specific proteins, enzymes, or macromolecules, redox-active therapeutics constitute a novel strategy whose target is the organism redox network, the so-called *Redoxome*. Mimics of redox enzymes and compounds that are able to reestablish the physiological levels of reactive species in normal tissues (demonstrated as a suppression of oxidative stress-dependent immune responses and secondary inflammatory processes), or selectively increase the oxidative burden in some targeted cells and tissues (such as tumor), are being heavily sought as experimental therapeutics. The adventure into fully understanding the effects obtained with metal-based therapeutics in inducing subtle changes, which would be observed as the "healing" of a diseased or death of a cancerous tissue, is still in its infancy; yet remarkable therapeutic effects are driving research in this relatively new interdisciplinary area linking chemistry, biochemistry, biology, and medicine.

The abundance of new information and the paradigm shift in our understanding of the mechanism of how the redox-active drugs work in a wide variety of diseases impose a need for a book to synthesize the current state of knowledge regarding the role of redox-active compounds in healthy and diseased tissue. Thus, this appears to be the first book fully dedicated to addressing the critical role that redox processes play in the development of different classes of therapeutics from the bench to the clinic and stressing awareness of these concepts for the treatment of disease. The chapters in this book describe the progresses in defining the central role of redox biology in many disease processes including inflammation, immunology, and neoplasia. This sets the foundation for the development of new classes of small redox-active molecules with unique effects in control of redox biology through modulation of key transcription factors at the core of inflammation, immunity, and neoplasia. The next decade shows a promise for the translation of this body of knowledge into new, transformative human therapeutics.

Durham, NC, USA — Ines Batinić-Haberle
João Pessoa, PB, Brazil — Júlio S. Rebouças
Durham, NC, USA — Ivan Spasojević

References

1. McCord JM, Fridovich I. Superoxide dismutase. An enzymic function for erythrocuprein (hemocuprein). J Biol Chem. 1969;244:6049–55.
2. Curnutte JT, Whitten DM, Babior BM. Defective superoxide production by granulocytes from patients with chronic granulomatous disease. N Engl J Med. 1974;290:593–7.

3. Pasternack RF, Halliwell B. Superoxide dismutase activities of an iron porphyrin and other iron complexes. J Am Chem Soc. 1979;101:1026–31.
4. Archibald FS, Fridovich I. The scavenging of superoxide radical by manganous complexes: in vitro. Arch Biochem Biophys. 1982;214:452–63.
5. Baudry M, Etienne S, Bruce A, Palucki M, Jacobsen E, Malfroy B. Salen-manganese complexes are superoxide dismutase-mimics. Biochem Biophys Res Commun. 1993;192:964–8.
6. Batinic-Haberle I, Liochev SI, Spasojevic I, Fridovich I. A potent superoxide dismutase mimic: manganese beta-octabromo-meso-tetrakis-(N-methylpyridinium-4-yl) porphyrin. Arch Biochem Biophys. 1997;343:225–33.
7. Riley DP, Lennon PJ, Neumann WL, Weiss RH. Toward the rational design of superoxide dismutase mimics: mechanistic studies for the elucidation of substituent effects on the catalytic activity of macrocyclic manganese(II) complexes. J Am Chem Soc. 1997;119:6522–8.
8. Samuni A, Krishna CM, Riesz P, Finkelstein E, Russo A. A novel metal-free low molecular weight superoxide dismutase mimic. J Biol Chem. 1988;263:17921–4.
9. Eckshtain M, Zilbermann I, Mahammed A, Saltsman I, Okun Z, Maimon E, Cohen H, Meyerstein D, Gross Z. Superoxide dismutase activity of corrole metal complexes. Dalton Trans. 2009;7879–82.
10. Viana OS, Ribeiro MS, Rodas AC, Rebouças JS, Fontes A, Santos BS. Comparative study on the efficiency of the photodynamic inactivation of *Candida albicans* using CdTe quantum dots, Zn(II) porphyrin and their conjugates as photosensitizers. Molecules. 2015;20:8893–912.
11. Tovmasyan A, Maia CG, Weitner T, Carballal S, Sampaio RS, Lieb D, Ghazaryan R, Ivanovic-Burmazovic I, Ferrer-Sueta G, Radi R, Reboucas JS, Spasojevic I, Benov L, Batinic-Haberle I. A comprehensive evaluation of catalase-like activity of different classes of redox-active therapeutics. Free Radic Biol Med. 2015;86:308–21.
12. Tse HM, Milton MJ, Piganelli JD. Mechanistic analysis of the immunomodulatory effects of a catalytic antioxidant on antigen-presenting cells: implication for their use in targeting oxidation-reduction reactions in innate immunity. Free Radic Biol Med. 2004;36:233–47.
13. Jaramillo MC, Briehl MM, Crapo JD, Batinic-Haberle I, Tome ME. Manganese porphyrin, MnTE-2-PyP5+, acts as a pro-oxidant to potentiate glucocorticoid-induced apoptosis in lymphoma cells. Free Radic Biol Med. 2012;52:1272–84.
14. Dorai T, Fishman AI, Ding C, Batinic-Haberle I, Goldfarb DS, Grasso M. Amelioration of renal ischemia-reperfusion injury with a novel protective cocktail. J Urol. 2011;186:2448–54.

Acknowledgments

Special thanks go to my co-editors, long time collaborators, colleagues and dear friends, Júlio S. Rebouças, and Ivan Spasojević, without whom this Book would have never happened, as well as to the authors of the Chapters who helped this Book see the light of day. I greatly appreciate also the help of colleagues who, besides editors, served as reviewers of Chapters: Beatriz Alvarez, Alvaro Estevez, Miloš Filipović, Neil Hogg, Ivana Ivanović-Burmazović, Vesna Niketić, Katy Peters, Peter Reeh, Daret St. Clair, Artak Tovmasyan and David Warner.

In addition to numerous grants that have over decades supported the development of Mn porphyrin-based SOD mimics and other redox-active therapeutics, I wish to acknowledge all the past and present students, postdoctoral fellows and research associates, and all the colleagues within the USA and abroad whose contribution has been enormous in developing SOD mimics, addressing their aqueous chemistry, biochemistry, biology, and therapeutic effects, and who all became part of a porphyrin family. Among the Ph.D. students and postdoctoral fellows, who I had the immense pleasure to work with (and many of whom have since advanced in their own careers), special thanks go to: Artak Tovmasyan, Júlio S. Rebouças, Ivan Kos, Zrinka Rajić, Tin Weitner, Sumitra Miriyala, Aaron Holley, Luciana Hannibal, Yunfeng Zhao, Isabel Jackson, Zahid Rabbani, Jacqueline Bueno-Janice, Romulo Sampaio, Clarissa Maia, Gilson DeFreitas-Silva, Diane Fels, Kathleen Ashcraft, Mary-Keara Boss, Emily Roberts, Xiaodong Ye, Hubert Tse, Douglas Weitzel, Ayako Okado-Matsumoto, Sebastian Carballal, James Cobley, and Iva Vukelić. A number of other students were also associated with my Lab in some part of their careers. We have lost a fine young scientist and a wonderful colleague, Ivan Kos, who died of cancer. His stay in my Lab coincided with his stage IV lung cancer diagnosis, which was followed by immediate treatment at Duke affording him close to 5 years of life and enough time for two beautiful daughters to be born. His invaluable work on isocitrate dehydrogenase activity provided us with coauthorship in *New England Journal of Medicine*, still among the most cited works of ours. Zrinka Rajić has never worked on porphyrins before joining my Lab. Yet, with a strong determination and our support, she helped our lead clinical candidate (MnTnBuOE-2-PyP^{5+}, BMX-001) see the light of day. Artak Tovmasyan has been with me for

7 years. With his dedication, passion for science, and hard work, Artak has made a tremendous impact on the development of redox-active therapeutics and understanding of their mechanism(s) of action(s). While organic chemist by basic training, he excelled also in in vitro and in vivo experiments in support of our aqueous chemistry. With his exquisite artistic skills, the brilliant figures in bright colors became a growing part of our manuscripts.

I am also grateful to numerous colleagues who helped us on our way toward clinical trials, many of whom coauthored chapters in this book. We are most obliged to Irwin Fridovich (Duke University School of Medicine); Peter Hambright (Howard University); Ludmil Benov (Kuwait University School of Medicine); Kam Leong (Duke University School of Engineering, presently at Columbia University); Daret St. Clair (University of Kentucky at Lexington); Margaret Tome (University of Arizona at Tucson); Jon Piganelli (University of Pittsburgh School of Medicine); Christopher Lascola, Darell Bigner, Mark Cline, Huaxin Sheng, Bridget Koontz, and David Warner (Duke University School of Medicine); Željko Vujašković (University of Maryland School of Medicine, Baltimore); John Archambeau (Loma Linda University); and Ting-Ting Huang (Stanford University School of Medicine and Geriatric Research Education and Clinical Center, Veterans Affairs Palo Alto Health Care System). Shortly after I joined Irwin Fridovich's Lab, I was fortunate to get acquainted with Peter Hambright. While Irwin Fridovich is the father of *free radicals in biology and medicine*, Peter Hambright may be regarded as the father of *water-soluble porphyrins*. I could not have imagined my career in water-soluble porphyrins, whose structural modifications resulted in powerful SOD mimics and redox-active therapeutics, without the inputs of Peter Hambright. In 1995, when I joined Irwin Fridovich's Lab, I essentially knew nothing about superoxide and porphyrins. While Irwin Fridovich taught me the basics of free radical biology and medicine, Peter Hambright, a fascinating character in his own right, was the one who guided me through my first steps in porphyrin chemistry. Unfortunately he passed away 5 years afterwards. Ludmil Benov has been in charge of all experiments related to the bioavailability of different redox-active drugs in simple unicellular models of oxidative stress, *Escherichia coli* and *Saccharomyces cerevisiae*. Most recently his studies on cancer and normal mammalian cells helped in identifying the best Mn porphyrin candidates for clinical development. Also, Benov's experiments increased our insight into the bioavailability of Mn porphyrins and mechanisms of their in vitro and in vivo actions. Margaret Tome, a biochemist whose experiments are nothing short of art, contributed vastly to understanding the in vivo pro-oxidative mechanism of action of Mn porphyrins. Kam Leong occupies a very special place in my career, a role model of a scholar and mentor and a man of immense kindness and generosity, who, while on faculty at Duke University School of Engineering, supported my Lab with his own funds in times of great need. With Bridget Koontz, we recently launched our journey toward radioprotection of male sexuality during radiotherapy of prostate cancer patients. With Željko Vujašković, we performed the very first mice studies on the pulmonary radioprotective effects of Mn porphyrins at Duke University, which were followed by monkey studies in collaboration with

Mark Cline (Wake Forest School of Medicine). Christopher Lascola pioneered the use of Mn porphyrins for differential tumor vs. normal tissue MR imaging. Darell Bigner (co-leader of The Preston Robert Tisch Brain Tumor Center at Duke) was instrumental in supporting the initial mouse studies on the anticancer therapeutic potential of MnTnBuOE-2-PyP^{5+} in glioma, facilitating its progress toward clinical trials as a radioprotector of normal brain while radiosensitizer of brain tumor. David Warner and Huaxin Sheng were the first to point out the therapeutic potential of Mn porphyrins in central nervous system injuries. The unsurpassable expert surgical skills and the compassionate nature of Huaxin Sheng have been invaluable with CNS injuries and pharmacokinetic studies. Daret St. Clair pioneered the role of Mn porphyrins in mitochondrial actions thereby mimicking MnSOD; in collaboration with her, we showed that cationic Mn porphyrins favor mitochondrial over cytosolic accumulation. Our over-decade-long joint work has developed into gentle and highly supportive friendship. Jon Piganelli, spirited, enthusiastic, and a very dear colleague of mine, has always trusted the therapeutic potential of Mn porphyrins. He was the first to suggest they act at the level of NF-κB cysteines, supposedly as pro-oxidants challenging our views on Mn porphyrins as antioxidants. While provocative in the early 2000s, viewing the actions of so-called SOD mimics as pro-oxidative has been increasingly accepted by researchers nowadays. Importantly, Jon Piganelli was instrumental in getting MnTE-2-PyP^{5+} into clinical trials on islet transplants in Canada. A fine clinician, scientist, and a colleague, John Archambeau performed critical studies on proton radiation of rat rectum indicating the potential of Mn porphyrins as radioprotectants of the rectum and setting the stage for their development toward the clinic; related studies are in progress at Duke University. Christian Beausejour and Ting-Ting Huang conducted critical studies showing the therapeutic potential of Mn porphyrin in protecting hippocampal neurogenesis during irradiation, and substantiated similar studies conducted by Mark Dewhirst at Duke University. In collaboration with my Lab, Paula Lam, Emily Roberts/Xiaodong Ye/Kam Leong, Ludmil Benov and Margaret Tome showed that Mn porphyrins when combined with sources of reactive species (such as radio- and chemotherapy and ascorbate) induce differential effects in restoring the physiological environment of normal cells while killing cancer cells. We have just launched studies with Angeles Secord and Shara Reihani (Duke University School of Medicine) on the use of Mn porphyrins as single agents or combined with chemotherapy, radiotherapy, and ascorbate in ovarian and cervical cancer. The collaborations with Stephen Weber, University of Pittsburgh, and Robert Levy and Giovanni Ferrari, University of Pennsylvania School of Medicine, have just had an excellent start. An important place in addressing the therapeutic and mechanistic aspects of Mn porphyrins is occupied by the Uruguay group at Universidad de la República led by Rafael Radi and Gerardo Ferrer-Sueta; many of their students and colleagues have been involved in studies on Mn porphyrins during almost 20 years of collaboration, such as Homero Rubbo, Andres Trostchansky, and Madia Trujillo; they addressed peroxynitrite-related aspects of Mn porphyrins and are in a process of showing that Mn porphyrins also reduce hypochlorite. All of those studies paved the way toward clinical trials.

A special place in the development of electron-deficient metalloporphyrins occupies the information made available by Gordon M. Miskelly's group in *Richards et al., Inorg Chem 1996 (vol 35, 1940–1944)*. The manuscript described the complex of Li with octabrominated *para N*-methylpyridylporphyrin. The enclosed p*K*a values of the inner pyrrolic nitrogens indicated the exceptional acidity of the porphyrin, suggesting in turn the exceptional electron-withdrawing power of eight pyrrolic bromines. This report guided us towards the synthesis of the first powerful SOD mimic, $MnBr_8TM$-4-PyP^{4+}. That compound was unfortunately of insufficient metal/ligand stability *(Batinić-Haberle et al., Arch Biochem Biophys 1997)*. In addition, the information from that manuscript helped us with isolation and purification of different Mn porphyrins.

Many other researchers in the USA and abroad contributed also to the development of Mn porphyrins such as Dean Jones and Judith Fridovich (Emory University); Joan Valentine and Edith Gralla (University of California at Los Angeles); Lee-Ann MacMillan-Crow and John Crow (University of Arkansas for Medical Sciences at Little Rock); Daniela Salvemini (St. Louis University School of Medicine); Jeannette Vasquez-Vivar (University of Wisconsin); Bruce Freeman (University of Pittsburgh Medical Center); Thambi Dorai (New York Medical College); Sidhartha Tan (University of Chicago); Rebecca Oberley-Deegan (University of Nebraska Medical Center); Christian Beausejour (University of Montreal); Fabio Doctorovich (University of Buenos Aires); Candace Floyd (University of Alabama at Birmingham); Paula Lam (Duke-National University of Singapore Graduate Medical School, Singapore); Daohong Zhou (Winthrop P. Rockefeller Cancer Institute, University of Arkansas for Medical Sciences); Miloš Filipović, Jan Miljković, and Ivana Ivanović-Burmazović (Friedrich-Alexander University, Erlangen-Nürnberg, Germany); Won Park (Sungkyunkwan University School of Medicine, South Korea); Nuno Oliveira and Ana Fernandes (Faculty of Pharmacy, Universidade de Lisboa and Universidade Lusófona, Portugal); João Laranjinha (Universidade de Coimbra, Portugal); Peter Humphries (Trinity College of Dublin, Ireland); Ana Budimir and Mladen Biruš (University of Zagreb, School of Pharmacy and Biochemistry, Croatia); Dragica Bobinac (University of Rijeka School of Medicine, Croatia); and Mutay Aslan (Akdeniz University Medical School, Antalya, Turkey).

Upon return from my Lab to the Universidade Federal da Paraíba, at João Pessoa, Brazil, Júlio S. Rebouças has established collaborations with his computational colleagues (Silmar do Monte, Elizete Ventura and Otávio de Santana) who have since addressed different aspects of the development of Mn porphyrins. We are enjoying the spirit and enthusiasm with which many students at Universidade Federal da Paraíba, though with limited resources, are conducting porphyrin research. While relishing the tropical paradise of the Brazilian Atlantic coast during the 5th SILQCOM Latin American Symposium on *Coordination and Organometallic Chemistry* hosted by Eduardo N. dos Santos, I had the opportunity to meet many other passionate scientists in Brazil working on different aspects of drug development, some of whom coauthored chapters in this book. Insightful chatting with Henry Forman has been always greatly appreciated. The work on chapter "SOD Mimetics and Related Redox-Active Molecules" in the fifth edition of the *Free*

Radicals in Biology and Medicine book gave us the opportunity to offer our modest contribution to the epic efforts of Barry Halliwell and John Gutteridge to make such encyclopedia-style book an essential reading for all of us.

The story on SOD mimics started in the late 1970s with the seminal discovery of Irwin Fridovich and Joe McCord of the function of copper protein as superoxide dismutase enzyme. The very first steps toward commercialization were implemented by James Crapo. We are indebted to James Crapo (CEO, BioMimetix JVLLC), Mark Dewhirst, Katy Peters and David Brizel (Duke University School of Medicine) for invaluable efforts in enabling the Mn porphyrin lead candidate, BMX-001, to enter the two clinical trials at Duke University and University of Colorado. I am also much obliged to James Crapo, for continuous support of my and Ivan Spasojević's Labs. We have been very fortunate for having our administrator Dianne Young around; her smiling face has been there when different issues needed immediate attention; she exceeded her work duties in making the working environment cheerful and comforting, especially at stressful times of shortage in funding. The kind support and encouragement of Christopher Willett, the chair of the Department of Radiation Oncology, Duke University School of Medicine, has been always greatly appreciated.

I am also grateful to those supervisors of mine whose patience, kindness, and generosity were critical in early steps of my career: Blaženka Sebečić, Dubravka Barišin-Kahlina, Ranko Babić, and Mladen Biruš. The invaluable support of Ivan Spasojević in my career and life has always been more than collaborative: his ingenuity and creativity made everything happen smoothly. Together we went through numerous scientific ups and downs; not only did Ivan contribute his technical expertise to our joint research efforts, but his wonderful sense of humor made our work fun. We started our research endeavors with inorganic magnesium-containing mixtures in the early and mid 1980s and continued with iron and vanadium chemistry and biology in the late 1980s. By good fortune, Alvin Crumbliss, then the chair of the Department of Chemistry, Duke University, had an opening to work on an NSF project; in times of civil war in my country, Yugoslavia, such an opportunity was nothing short of a miracle. Two and a half years of my and Ivan's joint work on host–guest recognition chemistry of iron siderophores and crown ethers resulted in eight manuscripts, out of which one was published in *JACS*. In January 1995, we embarked on over a 2-decade long SOD epic journey, eventually succeeding in our major goal—seeing our compounds reach first-in-human clinical trial in 2016. Our son, Siniša Haberle, who joined our research while at Duke University as a radiology resident, has been our lifelong inspiration and joy; his newborn son, Luka, will continue to be a source of immense happiness beautifying the routine of our days. Last, but not the least, the free and adventurous spirit of my mother, Paula Sertić-Batinić, had nourished my curiosity. Supported by enormous kindness and generosity of my father, Branko Batinić, she directed me through my early childhood toward passionately exploring the knowledge and the world pushing me toward learning several languages that "would open my doors to the world" as if knowing where I would eventually find my place on this beautiful planet Earth.

Ines Batinić-Haberle

Contents

Contributors

Bryan G. Allen, M.D., Ph.D. Division of Free Radical and Radiation Biology, Department of Radiation Oncology, University of Iowa Hospitals and Clinics, Holden Comprehensive Cancer Center, Iowa City, IA, USA

Kathleen A. Ashcraft, Ph.D. Department of Radiation Oncology, Duke University Medical Center, Durham, NC, USA

Debashree Basudhar, Ph.D. Cancer Inflammation Program, NCI-Frederick, Frederick, MD, USA

Ines Batinić-Haberle, Ph.D. Department of Radiation Oncology, Duke University School of Medicine, Durham, NC, USA

Jayeeta Bhaumik, Ph.D. Department of Pharmaceutical Technology (Biotechnology), National Institute of Pharmaceutical Education and Research (NIPER), Punjab, India

Xandra O. Breakefield, Ph.D. Department of Neurology and Radiology, Program in Neuroscience, Massachusetts General Hospital, Harvard Medical School, Boston, MA, USA

Jacqueline C. Bueno-Janice, M.Sc. Departamento de Química, CCEN, Universidade Federal da Paraíba, Jardim Cidade Universitária, João Pessoa, PB, Brazil

Department of Radiation Oncology, Duke University Medical Center, Durham, NC, USA

Akshaya Chandrasekaran Nanobioscience Constellation, Colleges of Nanoscale Science and Engineering, SUNY Polytechnic Institute, State University of New York, Albany, NY, USA

Maria Rosa Ciriolo, Ph.D. Dipartimento di Biologia, Università di Roma "Tor Vergata", Rome, Italy

Gina M. Coudriet, Ph.D. Department of Surgery, University of Pittsburgh School of Medicine, Pittsburgh, PA, USA

James D. Crapo, M.D. Department of Medicine, National Jewish Health, Denver, CO, USA

Cynthia Demicheli, Ph.D. Departamento de Química, Instituto de Ciências Exatas, Universidade Federal de Minas Gerais, Belo Horizonte, MG, Brazil

Mark W. Dewhirst, D.V.M., Ph.D. Department of Radiation Oncology, Duke University Medical Center, Durham, NC, USA

Nivedh Dinesh, M.D. Laboratory of Cancer Gene Therapy, Humphrey Oei Institute of Cancer Research, National Cancer Centre of Singapore, Singapore, Singapore

Division of Neurosurgery, National University Hospital, Singapore, Singapore

Susan R. Doctrow, Ph.D. Department of Medicine, Pulmonary Center, Boston University School of Medicine, Boston, MA, USA

Alvaro G. Estevez, Ph.D. Burnett School of Biomedical Sciences, College of Medicine, University of Central Florida, Orlando, FL, USA

Nathália K.S.M. Falcão, M.Sc. Departamento de Química, CCEN, Universidade Federal da Paraíba, Jardim Cidade Universitária, João Pessoa, PB, Brazil

Nicholas P. Farrell, Ph.D. Department of Chemistry, Virginia Commonwealth University, Richmond, VA, USA

Ana S. Fernandes, Ph.D. CBiOS, Universidade Lusófona Research Center for Biosciences and Health Technologies, Lisbon, Portugal

Ana Maria Da Costa Ferreira, Ph.D. Departamento de Química, Instituto de Química, Universidade de São Paulo, São Paulo, SP, Brazil

Milos R. Filipovic, Ph.D. Department of Chemistry and Pharmacy, Friedrich-Alexander University of Erlangen-Nuremberg, Erlangen, Germany

Universite de Bordeaux, IBGC, UMR 5095, Bordeaux, France

CNRS, IBGC, UMR 5095, Bordeaux, France

Brian Fish Department of Radiation Oncology, Medical College of Wisconsin, Milwaukee, WI, USA

Robert A. Floyd, Ph.D. Experimental Therapeutics Research Laboratory, Oklahoma Medical Research Foundation, Oklahoma City, OK, USA

Adriana Fontes, Ph.D. Department of Biophysics and Radiobiology, Federal University of Pernambuco, Recife, Pernambuco, Brazil

Maria Clara Franco, Ph.D. Burnett School of Biomedical Sciences, College of Medicine, University of Central Florida, Orlando, FL, USA

Frédéric Frézard, Ph.D. Departamento de Fisiologia e Biofísica, Instituto de Ciências Biológicas, Universidade Federal de Minas Gerais, Belo Horizonte, MG, Brazil

Irwin Fridovich, Ph.D. Department of Biochemistry, Duke University Medical Center, Durham, NC, USA

L.P. Madhubhani P. Hemachandra, Ph.D. Nanobioscience Constellation, Colleges of Nanoscale Science and Engineering, SUNY Polytechnic Institute, State University of New York, Albany, NY, USA

Nadine Hempel, Ph.D. Department of Pharmacology, Penn State College of Medicine, Hershey, PA, USA

Aaron K. Holley, Ph.D. Department of Internal Medicine/Division of Gastroenterology, University of Kentucky, Lexington, KY, USA

Ting-Ting Huang, Ph.D. Department of Neurology and Neurological Sciences, Stanford University School of Medicine, Stanford, CA, USA

Geriatric Research, Education, and Clinical Center, Veterans Affairs Palo Alto Health Care System, Palo Alto, CA, USA

Wouter Hubens, M.Sc. Department of Biology, Faculty of Science, Toho University, Funabashi, Chiba, Japan

Karl D. Huffman Department of Medicine, Pulmonary Center, Boston University School of Medicine, Boston, MA, USA

Isabel L. Jackson, Ph.D. Division of Translational Radiation Sciences, Department of Radiation Oncology, University of Maryland School of Medicine, Baltimore, MD, USA

Kali Janes, Ph.D. Department of Pharmacological and Physiological Science, Saint Louis University School of Medicine, St. Louis, MO, USA

Seema Kirar, Ph.D. Department of Pharmaceutical Technology (Biotechnology), National Institute of Pharmaceutical Education and Research (NIPER), Punjab, India

Bridget F. Koontz, M.D. Department of Radiation Oncology, Duke University Medical Center, Durham, NC, USA

Joydev K. Laha, Ph.D. Department of Pharmaceutical Technology (Process Chemistry), National Institute of Pharmaceutical Education and Research (NIPER), Punjab, India

Paula Lam, Ph.D. Laboratory of Cancer Gene Therapy, Humphrey Oei Institute of Cancer Research, National Cancer Centre of Singapore, Singapore, Singapore

Department of Physiology, Yong Loo Lin School of Medicine, National University of Singapore, Singapore, Singapore

Duke-NUS Graduate Medical School, Singapore, Singapore

Zelmira Lazarova, MD. Department of Dermatology, Medical College of Wisconsin, Milwaukee, WI, USA

Joe M. McCord, Ph.D. Division of Pulmonary Sciences and Critical Care Medicine, Department of Medicine, University of Colorado Denver, Aurora, CO, USA

Meetha Medhora, Ph.D. Department of Radiation Oncology, Medical College of Wisconsin, Milwaukee, WI, USA

J. Andres Melendez, Ph.D. Nanobioscience Constellation, Colleges of Nanoscale Science and Engineering, SUNY Polytechnic Institute, State University of New York, Albany, NY, USA

Jan Lj. Miljkovic, Ph.D. Department of Chemistry and Pharmacy, Friedrich-Alexander University of Erlangen-Nuremberg, Erlangen, Germany

Katrina M. Miranda, Ph.D. Department of Chemistry and Biochemistry, University of Arizona, Tucson, AZ, USA

John E. Moulder, Ph.D. Department of Radiation Oncology, Medical College of Wisconsin, Milwaukee, WI, USA

William L. Neumann, Ph.D. Department of Pharmaceutical Sciences, School of Pharmacy, Southern Illinois University Edwardsville, Edwardsville, IL, USA

Huy Nguyen, Ph.D. Department of Neurology and Neurological Sciences, Stanford University School of Medicine, Stanford, CA, USA

Rebecca E. Oberley-Deegan, Ph.D. Department of Biochemistry and Molecular Biology, University of Nebraska Medical Center, Omaha, NE, USA

Ayako Okado-Matsumoto, Ph.D. Department of Biology, Faculty of Science, Toho University, Funabashi, Chiba, Japan

Nuno G. Oliveira, Ph.D. Research Institute for Medicines (iMed. ULisboa), Faculty of Pharmacy, Universidade de Lisboa, Lisbon, Portugal

Philippe Alexandre Divina Petersen, Ph.D. Departamento de Física dos Materiais, Instituto de Física, Universidade de São Paulo, São Paulo, SP, Brazil

Helena Maria Petrilli, Ph.D. Departamento de Física dos Materiais, Instituto de Física, Universidade de São Paulo, São Paulo, SP, Brazil

Jon D. Piganelli, Ph.D. Department of Surgery, University of Pittsburgh School of Medicine, Pittsburgh, PA, USA

Victor Hugo A. Pinto, Ph.D. Departamento de Química, CCEN, Universidade Federal da Paraíba, Jardim Cidade Universitária, João Pessoa, PB, Brazil

Clotilde Policar, Ph.D. Départementde Chimie, École Normale Supérieure-PSL Research University, Sorbonne Universités, UPMC Univ Paris 06, CNRS, Laboratoire des BioMolécules, UMR 7203, Paris, France

Dana M. Previte Department of Surgery, University of Pittsburgh School of Medicine, Pittsburgh, PA, USA

Kristina Ramdial Burnett School of Biomedical Sciences, College of Medicine, University of Central Florida, Orlando, FL, USA

Júlio S. Rebouças, Ph.D. Departamento de Química, CCEN, Universidade Federal da Paraíba, Jardim Cidade Universitária, João Pessoa, PB, Brazil

Martha S. Ribeiro, Ph.D. Center for Lasers and Applications, Nuclear and Energy Research Institute, National Nuclear Energy Commission, São Paulo, SP, Brazil

Lisa A. Ridnour, Ph.D. Cancer Inflammation Program, NCI-Frederick, Frederick, MD, USA

Timothy J. Robinson, M.D., Ph.D. Department of Radiation Oncology, Duke University Medical Center, Durham, NC, USA

Fabian H. Rossi, M.D. Orlando Veteran Affairs Medical Center, Orlando, FL, USA

Daniela Salvemini, Ph.D. Department of Pharmacological and Physiological Science, Saint Louis University School of Medicine, St. Louis, MO, USA

Beate S. Santos, Ph.D. Department of Pharmaceutical Sciences, Federal University of Pernambuco, Recife, Pernambuco, Brazil

Nuno Saraiva, Ph.D. CBiOS, Universidade Lusófona Research Center for Biosciences and Health Technologies, Lisbon, Portugal

Ulrich Schatzschneider, Ph.D. Institut für Anorganische Chemie, Julius-Maximilians-Universität Würzburg, Würzburg, Germany

Huaxin Sheng, M.D. Multidisciplinary Neuroprotection Laboratories, Department of Anesthesiology, Duke University Medical Center, Durham, NC, USA

Peter V. Simpson, Ph.D. Department of Chemistry, Curtin University, Bentley, WA, Australia

Ivan Spasojević, Ph.D. Department of Medicine and Duke Cancer Institute, Pharmaceutical Research Shared Resource, PK/PD Core Laboratory, Duke University School of Medicine, Durham, NC, USA

Douglas R. Spitz, Ph.D. Division of Free Radical and Radiation Biology, Department of Radiation Oncology, University of Iowa Hospitals and Clinics, Holden Comprehensive Cancer Center, Iowa City, IA, USA

Annapoorna Sreedhar Department of Pharmacology, Toxicology and Neurosciences, LSU Health Sciences Center, Shreveport, LA, USA

Chandra Srinivasan, Ph.D. Department of Graduate Studies and Research, California State University, Dominguez Hills, CA, USA

Daret K. St. Clair, Ph.D. Graduate Center for Toxicology, University of Kentucky, Lexington, KY, USA

Ashley M. Symons-Liguori, Ph.D. Department of Pharmacological and Physiological Science, Saint Louis University School of Medicine, St. Louis, MO, USA

Artak Tovmasyan, Ph.D. Department of Radiation Oncology, Duke University School of Medicine, Durham, NC, USA

Rheal A. Towner, Ph.D. Advanced Magnetic Resonance Center, Oklahoma City, OK, USA

Osnir S. Viana, Ph.D. Department of Pharmaceutical Sciences, Federal University of Pernambuco, Recife, Pernambuco, Brazil

Zeljko Vujaskovic, M.D., Ph.D. Division of Translational Radiation Sciences, Department of Radiation Oncology, University of Maryland School of Medicine, Baltimore, MD, USA

David S. Warner, M.D. Multidisciplinary Neuroprotection Laboratories, Department of Anesthesiology, Duke University Medical Center, Durham, NC, USA

Jacqueline P. Williams, Ph.D. Department of Environmental Medicine, University of Rochester Medical Center, Rochester, NY, USA

David A. Wink, Ph.D. Cancer Inflammation Program, NCI-Frederick, Frederick, MD, USA

Yunfeng Zhao, Ph.D. Department of Pharmacology, Toxicology and Neurosciences, LSU Health Sciences Center, Shreveport, LA, USA

Chapter 1
Superoxide and the Superoxide Dismutases: An Introduction by Irwin Fridovich

Irwin Fridovich

There is a family of enzymes devoted to catalyzing the conversion of univalently reduced molecular oxygen, called superoxide, into molecular oxygen plus hydrogen peroxide. A reaction that converts two identical substrates into two different products is called a dismutation reaction. Hence, these enzymes are called the superoxide dismutases (SODs).

At the time of their discovery, there was no thought that superoxide radical had any relevance to biology; much less that there could exist enzymes devoted to the elimination of this unstable free radical.

Finding an unsuspected enzyme that acts on an unsuspected substrate cannot be achieved by design. Nondirected research, in which the investigator follows one experimental lead after another, can lead to unsuspected discovery. That is how the enzymatic production of superoxide and the existence of the superoxide dismutases were discovered. The story of the trail that led to these discoveries has been told by myself and my former student and co-investigator, Joe McCord. I will, in the space allotted to me, try to explain why the formation of superoxide occurs so easily and how the dismutases speed a reaction that occurs rapidly even in the absence of catalysis.

Why is the univalent reduction of molecular oxygen to superoxide favored over its divalent reduction to hydrogen peroxide or its tetravalent reduction to two molecules of water? The short answer is the spin restriction. But that requires some elaboration.

Electrons can exist in two spin states and a spinning charge creates a magnetic field. The favored situation is for two electrons having opposite spin states to reside in the same orbital. In that way, their magnetic fields are oppositely oriented and mutually cancelling. Molecules or atoms in which all electrons are so arranged do not respond to magnetic fields and are said to be diamagnetic. Molecular oxygen

I. Fridovich (✉)
Department of Biochemistry, Duke University Medical Center, Durham, NC, USA
e-mail: irwin.fridovich@duke.edu

I. Batinić-Haberle et al. (eds.), *Redox-Active Therapeutics*,
Oxidative Stress in Applied Basic Research and Clinical Practice,
DOI 10.1007/978-3-319-30705-3_1

does respond to a magnetic field. It is paramagnetic and that property is exploited in designing oximeters that measure the level of oxygen in a mixture of gases.

The magnitude of the paramagnetism of oxygen indicates that it contains two parallel spinning electrons. These must reside in separate orbitals to avoid the impossibly high energy situation of two parallel spins in the same orbital, which is a statement of the Pauli exclusion principle.

Now imagine trying to add a spin-opposed pair of electrons, from some diamagnetic reductant, to molecular oxygen. One member of the spin-opposed pair would have a spin opposed to one of the parallel spins in the oxygen and so could join it in its orbital The other would have a spin that was parallel to the remaining unpaired electron in oxygen and could not join it. What we have described explains why molecular oxygen is less reactive than we might have expected and that hindrance to reaction is called the *spin restriction*. Were that not the case, organic matter would spontaneously combust in air.

There is a way around the spin restriction. Thus, electronic spin states can be inverted by interacting with nuclear spins. However, that is a relatively slow process. Indeed, it is many orders of magnitude slower than the lifetime of collisional complexes and so would be very unlikely to occur while the reacting oxygen and the reductant were in contact. However, it could occur during the longer times between collisions.

The foregoing explains why it is easier to add electrons to molecular oxygen one at a time, the so-called *univalent pathway*. Since it takes four electrons to reduce oxygen all the way to two molecules of water, there must be intermediates and these are, in the order of their formation, superoxide, hydrogen peroxide, and hydroxyl radical, all of which are capable of damaging certain classes of biomolecules.

Hence, defenses against these intermediates of oxygen reduction are essential to allow aerobic life. The first of these defenses is avoidance. Thus, there are enzymes that catalyze the divalent and even the tetravalent reduction of oxygen without releasing any of these intermediates. These enzymes accomplish such polyvalent reduction by binding the oxygen and keeping it bound while the electron transfers occur.

Additional defenses include the superoxide dismutases to eliminate superoxide; the catalases to dismute H_2O_2 onto oxygen plus water; the peroxidases to reduce H_2O_2 to water at the expense of some reductants; plus a plethora of enzymes to prevent and repair the oxidative damage that occurs in spite of these defensive enzymes.

Superoxide can engender the other intermediates of oxygen reduction by participating in reactions that will be described. Hence, its elimination is paramount.

Superoxide spontaneously dismutes to H_2O_2 plus O_2 in protic solvents, such as water. Since $O_2^{\bullet-}$ is the conjugate base of $HO_2^{\bullet}$ which has a pK_a of 4.8, we need to consider three dismutation reactions and they are: $HO_2^{\bullet}$ with $HO_2^{\bullet}$; $HO_2^{\bullet}$ with $O_2^{\bullet-}$; and $O_2^{\bullet-}$ with $O_2^{\bullet-}$.

The existence of $O_2^{\bullet-}$ was first proposed by Linus Pauling, and it was then found as the agent responsible for the transient oxygen-dependent conductivity of irradiated water. Radiolysis of water produces hydrogen atoms ($H^{\bullet}$), hydroxyl radicals ($OH^{\bullet}$), and hydrated electrons (e^{-}_{aq}). In the presence of O_2, the hydrogen atoms yield

$HO_2^{\bullet}$ which can ionize to $O_2^{\bullet-}$ and the hydrated electrons will add to O_2 yielding $O_2^{\bullet-}$. Radiation chemists then used pulse radiolysis to produce superoxide at will and to explore the dismutation reactions, the absorption spectra, and other properties of superoxide. When using this technique, they add formate to scavenge the hydroxyl radical that otherwise rapidly reform water by reacting with the hydrogen atoms and eliminate hydrated electrons by forming hydroxide. The formate also increases the yield of superoxide through the formation of the $CO_2^{\bullet-}$ anion radical that reduces O_2 to $O_2^{\bullet-}$.

$$HO_2^{\bullet-} + HO_2^{\bullet-} \leftrightarrow H_2O_2 + O_2 \quad k = 1\times10^5\,M^{-1}s^{-1}$$

$$HO_2^{\bullet} + O_2^{\bullet-} + H^+ \leftrightarrow H_2O_2 + O_2 \quad k = 1\times10^8\,M^{-1}s^{-1}$$

$$O_2^{\bullet-} + O_2^{\bullet-} \leftrightarrow O_2^{2-} + O_2 \quad k \approx 0$$

The $O_2^{\bullet-}$ plus $O_2^{\bullet-}$ dismutation reaction is so slow that it can hardly be measured because of the electrostatic repulsion of like changes, and this is the reaction that is catalyzed by the superoxide dismutases. We can be sure of this because the catalytic activity of SODs is maintained at pH values well above the pK_a of $HO_2^{\bullet}$.

The SODs avoid the electrostatic repulsion facing the close approach of one $O_2^{\bullet-}$ to another by taking one electron from one $O_2^{\bullet-}$ and then donating it to another $O_2^{\bullet-}$. The active sites of the SODs contain metals that mediate this electron transfer.

There are SODs based on copper plus zinc, manganese, iron, manganese or iron, and nickel. All of these SODs react with $O_2^{\bullet-}$ with rate constants that somewhat exceed 1×10^9 M^{-1} s^{-1}. When one considers that the spontaneous dismutation proceeds at neutrality with a rate constant of close to 1×10^5 M^{-1} s^{-1}, the advantage provided by the SOD seems less than impressive. Yet the SOD in *Escherichia coli* manages to intercept 95 % of the flux of $O_2^{\bullet-}$, and this is in competition with the sum of all the molecular targets for $O_2^{\bullet-}$ that exist in those cells.

The great advantage of the SOD catalyzed over the spontaneous dismutation is due to several factors. First, the rate constant of the enzyme-catalyzed reaction exceeds that of the spontaneous reaction, secondly the steady-state concentration of the $O_2^{\bullet-}$ is much less than the average concentration of the SOD; hence, an $O_2^{\bullet-}$ is far more likely to collide with SOD than it is to collide with another $O_2^{\bullet-}$ or $HO_2^{\bullet}$, and thirdly the spontaneous reaction is second order in $O_2^{\bullet-}/HO_2^{\bullet}$, while the catalyzed reaction is first order with respect to $O_2^{\bullet-}$. When reasonable numbers are assumed, we can calculate that the catalyzed reaction would exceed the spontaneous one by at least ten orders of magnitude.

Members of the SOD family of enzymes are found in different compartments. Thus, in mammals we find one Cu,ZnSOD (aka SOD1) in the cytosols, nuclei, and in the intermembrane space of mitochondria and a different one (SOD3) in the extracellular spaces while there is an MnSOD (SOD2) in the matrix of mitochondria. This is an indication that $O_2^{\bullet-}$ is produced in all of these compartments, that vital targets occur in all of them and that $O_2^{\bullet-}$ needs to be eliminated to the maximum degree possible, and that $O_2^{\bullet-}$ does not rapidly cross the membranes that separate these compartments.

This is the place to point out that since SODs compete with the sum of all targets for $O_2^{\bullet-}$, it follows that SODs need to be abundant in order to compete favorably, and also that 100 % protection cannot be achieved.

So what are the known targets susceptible to reaction with $O_2^{\bullet-}$? The first on this list should be the iron sulfur containing dehydratases, such as the aconitases and fumarases a and b of *E coli*. The (4Fe–4S) clusters are rapidly oxidized by $O_2^{\bullet-}$, and the oxidized cubane cluster is unstable and loses a ferrous ion. The ferrous ion can then react with H_2O_2, in what is called the Fenton reaction, yielding hydroxide plus hydroxyl radical. In this way, $O_2^{\bullet-}$ can engender hydroxyl radical, which is an extremely powerful oxidant, capable of attacking any biological molecule. Of course, oxidizing the cubane iron/sulfur cluster of dehydratases, such as aconitase, causes inactivation of that enzyme and thus inhibition of the citric acid cycle.

Another target for $O_2^{\bullet-}$ is the signaling molecule nitric oxide ($NO^{\bullet}$). Radical–radical reactions are usually rapid, and it is no surprise that the $O_2^{\bullet-}$ plus $NO^{\bullet}$ reaction proceeds with a rate constant of $1 \times 10^{10}\ M^{-1}\ s^{-1}$. That is the rate constant to be expected when every collision results in reaction. The product of the $O_2^{\bullet-}$ plus $NO^{\bullet}$ reaction is peroxynitrite ($ONOO^-$), which is the conjugate base of ONOOH, which can homolyze to $HO^{\bullet}$ plus NO_2, and that constitutes a potent nitrating mixture.

There are undoubtedly additional targets for $O_2^{\bullet-}$ such as sulfite, tetrahydrofolate, and other strong reductants. The special feature of the reaction of $O_2^{\bullet-}$ with a biological reductant is that the $O_2^{\bullet-}$ takes one electron from the reductant leaving a free radical of the reductant that can then initiate free radical chain reactions, thus amplifying the damage.

In summary, all aerobes all the time and anaerobes sometimes face the formation of $O_2^{\bullet-}$; that $O_2^{\bullet-}$ is a damaging species that can moreover engender other toxic species; that defenses are needed to minimize the damage and to repair that damage which nevertheless occurs; and that the superoxide dismutases, catalases, peroxidases, and the plethora of enzymes dedicated to repair or replacement of oxidatively damaged macromolecules are evolution's answer to this problem.

Chapter 2
The Discovery of Superoxide Dismutase and Its Role in Redox Biology

Joe M. McCord

2.1 The Discovery of Superoxide Dismutase

When I began as a graduate student under the direction of Irwin Fridovich, I was assigned a "back burner" project regarding the mechanism of cytochrome *c* reduction by xanthine oxidase and xanthine. This reaction required the presence of oxygen, but little oxygen was consumed by the reaction. Fridovich and Handler had proposed that a bound oxygen molecule might serve as an "electron bridge" to facilitate the transfer of an electron from a reduced active site (e.g., Fe^{+2}) on xanthine oxidase to the heme of the cytochrome [1]. Fridovich had also observed that certain preparations of myoglobin [2] or carbonic anhydrase [3] could inhibit the transfer of this electron to cytochrome *c*, presumably by competing with the cytochrome for a common binding site on xanthine oxidase. It was a logical hypothesis supported by classical adherence to Michaelis–Menten kinetic behavior, but lacking physical evidence. My job was to demonstrate the physical binding of these competing proteins to xanthine oxidase. To make a long story short, no such evidence could be acquired using a variety of techniques. I began to rethink the hypothesis. Was there any other way that these players could interact that would produce kinetic behavior that was so seemingly consistent with Michaelis–Menten enzyme kinetics of saturation and competitive inhibition? An alternative possibility occurred to me in what can only be described as a "eureka" moment. If the proposed bridging oxygen and its single extra electron were released from xanthine oxidase as a superoxide radical ($O_2^{\bullet-}$), an unstable entity already known to and studied by radiation chemists, it seemed to me that all could be explained via a competition between two

J.M. McCord (✉)
Division of Pulmonary Sciences and Critical Care Medicine, Department of Medicine, University of Colorado Denver, 12700 East 19th Avenue, Aurora, CO 80045, USA
e-mail: joe.mccord@ucdenver.edu

I. Batinić-Haberle et al. (eds.), *Redox-Active Therapeutics*,
Oxidative Stress in Applied Basic Research and Clinical Practice,
DOI 10.1007/978-3-319-30705-3_2

reactions, the first being the spontaneous dismutation reaction which would occur in the absence of cytochrome *c*:

$$HO_2^{\bullet} + O_2^{\bullet-} + H^+ \rightarrow H_2O_2 + O_2$$

and the second being the reduction of cytochrome *c* by the superoxide radical, if the cytochrome were present:

$$\text{cytochrome } c\text{-Fe}^{3+} + O_2^{\bullet-} \rightarrow \text{cytochrome } c\text{-Fe}^{2+} + O_2$$

The second reaction would obviously compete with the first, and would show saturation (i.e., the rate of cytochrome reduction would approach the rate of superoxide production) at very high concentrations of the cytochrome. If these reactions were taking place instead of the classical Michaelis–Menten enzyme-binding-substrate scenario, however, a very important distinction could be made. The Michaelis constant, K_m, reflects the binding constant between enzyme and substrate, and under conditions where the substrate concentration greatly exceeds the enzyme concentration, it is essentially constant. If, however, no binding were occurring then the concentration of cytochrome required to achieve half-maximal rate would be a function of enzyme concentration and steady-state concentration of superoxide achieved under those particular conditions–it would certainly not be a constant. The next day, I determined half-maximal rates of cytochrome reduction by xanthine oxidase over a 24-fold range of enzyme concentration, finding that the *apparent* K_m varied about 4.8 fold over this range. In fact, the half-maximal rates varied rather precisely with the square root of xanthine oxidase concentration, exactly as might have been predicted, given that the steady-state concentration of $O_2^{\bullet-}$ is determined by the spontaneous dismutation of the radical, a reaction that is second order in superoxide [4]. We concluded that superoxide was not an enzyme-bound intermediate but rather an actual product of the action of xanthine oxidase on xanthine, and was released into free solution where it could undergo spontaneous dismutation to produce hydrogen peroxide and oxygen, or where it could react with ferricytochrome *c*, reducing it to ferrocytochrome *c*. This realization required an immediate rethinking of how the inhibitory proteins (misidentified at this point as myoglobin and carbonic anhydrase) could be producing their effects. The most likely possibility was that they were eliminating superoxide by catalyzing the dismutation reaction–i.e., they were *superoxide dismutases*.

A subsequent study clarified things even more [5]. Superoxide dismutase (SOD) activity did not belong to carbonic anhydrase nor to myoglobin, but rather belonged to a minor impurity present in those preparations at a fraction of a percent. When isolated based on its activity from bovine erythrocytes, SOD was quickly identified as a previously studied copper-containing protein of unknown function that had been variously named hemocuprein, hepatocuprein, cerebrocuprein, erythrocuprein, and cytocuprein. We were able to demonstrate the activity of SOD in systems that involved neither xanthine oxidase nor cytochrome *c*, definitively ruling out protein–protein interactions of any kind. We also showed that hydrogen peroxide was the product of the SOD-catalyzed reaction, establishing that it was indeed a dismutation rather than something more complex, such as a four-electron reduction of oxygen to form water as the product, analogous to the reaction catalyzed by cytochrome oxidase.

2.2 The Existence of SOD Raised a Number of Questions

The discovery of SOD, a free radical-scavenging enzyme, was viewed by some as a biochemical curiosity—perhaps as "a solution in search of a problem." Why did this enzyme exist? Free radicals were known to be a generally reactive class of molecules, but superoxide seemed able to take care of its own destruction with a spontaneous dismutation rate of around $10^6\ M^{-1}\ s^{-1}$ at cytosolic pH. Is spontaneous dismutation not fast enough? Is superoxide toxic in biological systems? What are the biological targets of superoxide? Do all organisms living in oxygen require SOD? Are there pathological consequences to unscavenged superoxide?

2.3 SOD Was Everywhere

How widely was SOD distributed? It was the summer of 1968 and fellow graduate student Chuck Beauchamp had a vegetable garden, so every day it seemed there was a new vegetable in the lab blender. They all contained SOD. When microbiologist Bernie Keele joined the lab as a postdoc, he acquired microorganisms from the collections of all of his colleagues. They all had SOD—all except the strict anerobes. Unlike the facultative organisms that can live with or without oxygen, unlike the microaerophiles that require oxygen (just not too much), and unlike the aerotolerant anaerobes that can't utilize oxygen but can tolerate it, the strict anaerobes quickly die in oxygen, and are the ones without SOD [6]. Not only was SOD activity universally present in oxygen-metabolizing organisms, its concentration was found in a remarkably narrow range, whether in human brain or a tomato or *Escherichia coli*. This implied two things: that the metabolism of oxygen inevitably leads to superoxide production, and that the free radical, if left unscavenged, must be seriously toxic. It also quickly became apparent that not all SODs were the same. When we isolated the enzyme from *E. coli* it was pink, rather than the blue color of the Cu/Zn-SOD from bovine erythrocytes. While the native pink enzyme gave no EPR signal, the boiled enzyme showed the very characteristic signal of manganese [7]. This Mn-SOD was found to be structurally and genetically related to the avian and mammalian mitochondrial Mn-SODs [8], as well as to the iron-containing SODs isolated from bacteria such as *E. coli* [9] and from spinach [10]. The last genetically distinct family of SOD to be described in certain bacteria contains nickel at the active site [11].

2.4 Biological Toxicity and Chemical Reactivity Are Not Necessarily Related

While $O_2^{\bullet-}$ certainly meets the chemical definition of a free radical, it is rather mild mannered as free radicals go—a fact that chemists understood quite well. A few chemists offered the opinion that superoxide is so unreactive that it should not pose a problem [12], and that "superoxide dismutase" might be an incidental property of these metalloproteins whose real functions remained to be discovered [13]. At the same time, many biologists were struck by the novelty that biological systems could generate these dangerous-sounding "free radicals," and assumed them to be indiscriminately reactive and damaging to biological systems. Neither view proved to be correct because biological toxicity is not always related to broad chemical reactivity. Many lethal biological toxins act with surgical precision, sometimes reacting with a single specific target. Cyanide toxicity, e.g., is due to its ability to block mitochondrial electron transport by inhibition of the cytochrome *c* oxidase. Ricin catalytically depurinates ribosomes, halting protein synthesis. A number of specific biological targets have been shown to be inactivated by superoxide including the citric acid cycle enzyme aconitase, certain dehydratases in *E. coli* such as the α,β-dihydroxyisovalerate dehydratase and 6-phosphogluconate dehydratase, and many other important enzymes including catalase, glyceraldehyde-3-phosphate dehydrogenase, ornithine decarboxylase, glutathione peroxidase, myofibrillar ATPase, adenylate cyclase, creatine phosphokinase, and glutamine synthase.

The biological targets that lead to significant increases in lipid peroxidation deserve special mention due to the self-propagating nature of lipid peroxidation, once initiated, to the widespread nature of the damage that results when membrane integrity is breached, and to the large number of pathological states characterized by increased lipid peroxidation. Among the major classes of biological molecules (proteins, polysaccharides, nucleic acids, and lipids), the polyunsaturated fatty acid moieties of lipids are perhaps the most easily oxidized, and superoxide has been shown capable of initiating and promoting the propagation of the process. The perhydroxyl radical ($HO_2^{\bullet}$) is the conjugate acid of superoxide ($O_2^{\bullet-}$) and is present at lower concentrations whenever superoxide is generated in biological systems. Furthermore, it is uncharged and quite lipid soluble. It can initiate lipid peroxidation by abstraction of a *bis*-allyic hydrogen atom from a polyunsaturated fatty acid [14]. A second mechanism has been described wherein the perhydroxyl radical can also abstract a hydrogen atom from a lipid peroxide molecule (LOOH) to create a lipid peroxyl radical ($LOO^{\bullet}$), effectively creating a branch point in the otherwise linear propagation sequence [15]. Thus, it appears that unscavenged superoxide production does indeed pose a serious toxic threat to virtually all organisms.

2.5 It's Not All Bad: The Importance of Redox Balance and Bell-Shaped Curves

One of the characteristics of evolution is the ability to make the best of a bad situation, to make a silk purse from the proverbial sow's ear. Thus, there are clear examples of how superoxide can actually be put to constructive uses. A good example is the evolution of our phagocytic NADPH oxidase. When first recognized as a biologic metabolite, it appeared that the superoxide radical was simply a noxious cytotoxic by-product that served no good purpose. That view changed when Bernard Babior and colleagues [16] realized that superoxide plays a crucial role in our defense against invading microbes. Precisely because superoxide is cytotoxic, this NADPH oxidase of phagocytes has evolved to purposefully generate superoxide radical on the membrane surface lining the contained microenvironment of the phagolysosome, providing chemical destruction for the ingested microbes. In effect, superoxide acts as an extremely broad spectrum antibiotic. The neutrophil is also sacrificed in the process, in a kind of Kamikaze mission. Surrounding healthy host cells may be injured or even killed through collateral damage in this system that errs on the side of vigilance. This ability to fend off microbial invaders as injured tissues are repaired is what links superoxide production to the inflammatory/immune system.

Ironically, one of the "good deeds" that can be attributed to the superoxide radical goes back to its roles in lipid peroxidation, a free radical chain reaction process that requires a free radical to initiate the process, and another free radical to terminate the chain. It appears that $O_2^{\bullet-}$ (or its conjugate acid $HO_2^{\bullet}$) can do both [17]. At high concentrations of $O_2^{\bullet-}$ (little SOD) initiation events would be maximal, but so would termination events, resulting in a large number of short chains. At low concentrations of the radical (high SOD) initiation events would be far fewer but the chain length would be quite long (limited only by other terminators such as vitamin E, or by the mutual annihilation of propagating radicals). Somewhere in the middle is a "sweet spot" at which net lipid peroxidation is minimal at one optimal concentration of SOD. This behavior can be observed experimentally and can be predicted by mathematic modeling [17].

Considerable evidence suggests that our bodies regulate redox balance, just as we regulate pH, body temperature, rates of respiration and oxygen delivery, blood glucose levels, and numerous other factors to achieve homeostasis. SOD is by no means the sole determinant of physiological redox balance; rather, it results from a network of interactive antioxidant enzymes, regulatory kinases, and cascading transcription factors. The balance is readily upset by injury, disease, and by aging itself. The future of redox biology will, no doubt, have much more to tell us.

References

1. Fridovich I, Handler P. Detection of free radicals generated during enzymatic oxidations by the initiation of sulfite oxidation. J Biol Chem. 1961;236:1836–40.
2. Fridovich I. Competitive inhibition by myoglobin of the reduction of cytochrome c by xanthine oxidase. J Biol Chem. 1962;237:584–6.
3. Fridovich I. A reversible association of bovine carbonic anhydrase with milk xanthine oxidase. J Biol Chem. 1967;242:1445–9.
4. McCord JM, Fridovich I. The reduction of cytochrome c by milk xanthine oxidase. J Biol Chem. 1968;243:5753–60.
5. McCord JM, Fridovich I. Superoxide dismutase: an enzymic function for erythrocuprein (hemocuprein). J Biol Chem. 1969;244:6049–55.
6. McCord JM, Keele Jr BB, Fridovich I. An enzyme-based theory of obligate anaerobiosis: the physiological function of superoxide dismutase. Proc Natl Acad Sci U S A. 1971;68:1024–7.
7. Keele Jr BB, McCord JM, Fridovich I. Superoxide dismutase from *Escherichia coli B*: a new manganese-containing enzyme. J Biol Chem. 1970;245:6176–81.
8. Weisiger RA, Fridovich I. Superoxide dismutase. Organelle specificity. J Biol Chem. 1973; 248:3582–92.
9. Yost Jr FJ, Fridovich I. An iron-containing superoxide dismutase from *Escherichia coli*. J Biol Chem. 1973;248:4905–8.
10. Salin ML, Bridges SM. Isolation and characterization of an iron-containing superoxide dismutase from a eucaryote, *Brassica campestris*. Arch Biochem Biophys. 1980;201:369–74.
11. Youn HD, Kim EJ, Roe JH, Hah YC, Kang SO. A novel nickel-containing superoxide dismutase from *Streptomyces* spp. Biochem J. 1996;318:889–96.
12. Sawyer DT, Valentine JS. How super is superoxide? Accts Chem Res. 1981;14:393–400.
13. Fee JA. Is superoxide important in oxygen poisoning? Trends Biochem Sci. 1982;7:84–6.
14. Gebicki JM, Bielski BHJ. Comparison of the capacities of the perhydroxyl and the superoxide radicals to initiate chain oxidation of linoleic acid. J Am Chem Soc. 1981;103:7020–2.
15. Aikens J, Dix TA. Perhydroxyl radical (HOO•) initiated lipid peroxidation—the role of fatty acid hydroperoxides. J Biol Chem. 1991;266:15091–8.
16. Babior BM, Kipnes RS, Curnutte JT. Biological defense mechanisms. The production by leukocytes of superoxide, a potential bactericidal agent. J Clin Invest. 1973;52:741–4.
17. McCord JM. Superoxide dismutase, lipid peroxidation, and bell-shaped dose response curves. Dose Response. 2008;6:223–38.

Chapter 3
Manganese Superoxide Dismutase (MnSOD) and Its Importance in Mitochondrial Function and Cancer

Aaron K. Holley and Daret K. St. Clair

3.1 Introduction

Mitochondria are important centers of cellular metabolism, containing many components of myriad metabolic pathways. While mitochondria are best known for oxidative phosphorylation for ATP production [171], many other important metabolic pathways are found in mitochondria [126], including β-oxidation of fatty acids [128], amino acid metabolism [113, 134], Krebs cycle [30], heme synthesis [9], and the urea cycle [346]. Aberrant function of mitochondria is a cause and/or consequence of a number of diseases [145, 214], such as cardiovascular disease [1, 73, 83, 173, 179, 260, 315, 353], diabetes [145, 189, 302] and diabetes-associated pathologies [83, 98, 154, 255], and neurological disorders like Parkinson's disease [44, 52, 151, 311] and Alzheimer's disease [94, 213, 263, 366]. Altered mitochondrial function are also linked with cancer [27], with changes in the electron transport chain [331], mitochondrial DNA [130, 167], and mitochondria-associated metabolism [103, 348] all associated with development and progression of cancer.

Mitochondria are the primary oxygen-metabolizing organelles of cells because of oxidative phosphorylation. As a consequence, mitochondria are also the chief source of reactive oxygen species (ROS), by-products of oxygen metabolism, in

A.K. Holley (✉)
Department of Internal Medicine/Division of Gastroenterology, University of Kentucky,
800 Rose St., Lexington, KY 40536, USA
e-mail: aaron.holley@uky.edu

D.K. St. Clair (✉)
Graduate Center for Toxicology, University of Kentucky,
1095 VA Drive, Lexington, KY 40536, USA
e-mail: dstcl00@uky.edu

I. Batinić-Haberle et al. (eds.), *Redox-Active Therapeutics*,
Oxidative Stress in Applied Basic Research and Clinical Practice,
DOI 10.1007/978-3-319-30705-3_3

cells [91]. Superoxide radicals are the initial ROS produced in mitochondria [4, 129] and participate in the production of other radical and non-radical species, such as hydroxyl radical [277, 278] and the reactive nitrogen species (RNS) peroxynitrite [133]. ROS affect cellular homeostasis by altering the activity of numerous proteins, including transcription factors like p53 [84, 115, 328], HIF-1 [96], AP-1 [2], and NF-κB [144], as well as serine/threonine kinases [266], mitogen-activated protein kinases [297, 347], and tyrosine and serine/threonine phosphatases [360]. Proteins can undergo numerous oxidative modifications [36], either by direct reaction with ROS/RNS or by reaction with other oxidative products in cells, such as lipid peroxidation products [175]. Numerous amino acids are susceptible to oxidative modification, including histidine, tyrosine, tryptophan, and lysine [175], and these modifications can either be reversible or irreversible [36]. Of particular interest is modification of cysteine, which can undergo intra- and intermolecular disulfide bond formation, glutathionylation, S-nitrosation, and oxidation to sulfenic, sulfinic, or sulfonic acid [257]. These oxidative modifications can lead to changes in protein conformation, resulting in reduced enzyme activity and/or proteasomal degradation [175]. Metabolic enzymes found in mitochondria are especially susceptible to ROS-mediated damage due to their proximity to ROS-generating sites in mitochondria. Changes in metabolic enzyme activity due to ROS can have serious consequences on cell homeostasis and may contribute to cancer development and progression. This chapter will focus on the effects of ROS on these crucial metabolic pathways.

While believed to have only injurious effects on cells (damage to lipids, proteins, and DNA), careful work has revealed important effects of ROS in normal cellular function, such as cell growth and differentiation [28], cell adhesion, apoptosis, and the immune response [76]. ROS also act as vital second messengers in cellular signaling [87, 283]. A careful balance exists between ROS production and annihilation, with disturbances of this proportionality resulting in deviant ROS accrual that can contribute to the development of multiple diseases, such as myriad neurological disorders [349] and cancer [20, 107, 365]. Changes in basal ROS concentrations can occur by a number of mechanisms, including exogenous ROS-generating agents, increased production of ROS from endogenous sources, diminished antioxidant capacity, or a combination of the three.

Because of the injurious and signaling effects of ROS, the cell is equipped with numerous enzyme and non-enzyme systems to detoxify ROS [14, 160]. Superoxide radicals are particularly hazardous because they participate in the production of other damaging ROS. Superoxide dismutases (SODs) are the major ROS-eradicating enzymes of the cell [92], with manganese superoxide dismutase (MnSOD) being especially important because of its mitochondrial localization (the major site of ROS production). Changes in the expression or function of MnSOD can have significant consequences on mitochondrial function and cellular homeostasis due to oxidative damage to different mitochondria-localized proteins, resulting in disease development [205, 232]. This chapter will focus on different sources of mitochondrial ROS production and the importance of MnSOD in scavenging these reactive molecules, as well as the protective effects of MnSOD on mitochondrial function and the role of MnSOD in cancer development and progression.

3.2 Mitochondrial Production of ROS

3.2.1 Mitochondria Are a Major Source of ROS

Mitochondria are the primary source of ROS (especially superoxide radicals) due to oxygen metabolism [172, 219]. Several complexes in the electron transport chain produce superoxide, with complexes I (NADH-ubiquinone oxidoreductase) [112, 330] and III (ubiquinol-cytochrome *c* oxidoreductase) [340] as the major sites of production. The site of superoxide production in complex I is between the ferricyanide and ubiquinone reduction sites [122] and was pinpointed to iron-sulfur centers N1a [162] and N2 [104]. Another important source of superoxide is the proton-pumping activity of complex I. EIPA, a proton pump inhibitor, increases complex I superoxide production and enhances rotenone-mediated superoxide production [72]. Complex III superoxide production relies on the ubisemiquinone intermediate of the Q-cycle [340]. Superoxide from complex III can be released on both the matrix [218] and intermembrane space [47] sides of the inner mitochondrial membrane. Complex II also contributes to mitochondrial superoxide production, with the distal site of succinate oxidation [202], and later, either the reduced cytochrome b_{566} or ubisemiquinone of the Q_0 site of the cytochrome bc_1, identified as sites of superoxide production [372].

Other mitochondrial enzymes not directly involved in electron transport also contribute to mitochondrial ROS production. For example, α-ketoglutarate dehydrogenase (α-KGDH), a component of the citric acid cycle, generates ROS depending on the NADH/NAD^+ ratio [339], with the dihydrolipoyl dehydrogenase activity of α-KGDH generating ROS [323]. Dihydroorotic dehydrogenase (an important enzyme in pyrimidine synthesis) produces superoxide as a by-product of converting dihydroorotate to orotate [88, 89]. Cytochrome P450s [117, 118] and glycerophosphate dehydrogenase [75, 210] also contribute to mitochondrial ROS production.

Superoxide radicals participate in the production of other ROS that further damage mitochondria. Many iron-sulfur center-containing proteins are found in mitochondria, and these proteins are vulnerable to superoxide-induced damage, resulting in the release of free iron. The quintessential example of superoxide-induced iron release and deactivation is aconitase (discussed in more detail in Sect. 3.5.2). Aconitase is susceptible to superoxide-induced deactivation because it contains numerous FeS centers [37, 100, 120], with subsequent release of Fe(II) from the enzyme (reviewed in [99]). Iron cations participate in the Haber–Weiss reaction of hydroxyl radical production [31, 86, 277, 278]. Superoxide also reacts with nitric oxide to produce the reactive nitrogen species (RNS) peroxynitrite [319]. Peroxynitrite modifies numerous amino acids, including nitration of tyrosine [3] and oxidation of sulfhydryl groups on proteins [273]. Mitochondrial proteins are also susceptible to attack by peroxynitrite [274], with DNA polymerase gamma, MnSOD [187, 188, 333, 335], glutathione peroxidase [242], aconitase [42, 120], and complexes I [41, 220, 275, 287], II [41, 275], and V [275] of the electron transport chain as targets.

3.2.2 *Ways to Scavenge Mitochondrial ROS*

Cells contain several enzyme systems that detoxify ROS generated throughout the cell [14, 160]. Superoxide dismutases (SODs) are the primary ROS annihilating enzymes of the cell [92], and these enzymes catalyze the dismutation of superoxide one-electronically to hydrogen peroxide and molecular oxygen [93]. Catalase [228] catalyzes the dismutation (disproportionation) of H_2O_2 by reducing it two-electronically to water, H_2O, and oxidizing it to molecular oxygen, O_2. The peroxiredoxins [233], and glutathione peroxidases [81, 190], reduce hydrogen peroxide to water. Three forms of SOD are found in cells, each encoded by separate genes (reviewed in [371]). Copper- and zinc-containing SOD (CuZnSOD, SOD1) is a homodimer found mainly in the cytoplasm [312], although small amounts have also been found in the intermembrane space of mitochondria [240, 351]. Extracellular SOD (ECSOD, SOD3) has 40–60 % amino acid homology with CuZnSOD, has copper and zinc in its active site, but contrary to CuZnSOD, ECSOD is found in the extracellular region of the cell [85, 124]. MnSOD (SOD2) is a homotetramer found exclusively in the matrix of mitochondria. Expression of MnSOD is highly conserved, with reports of MnSOD being identified in prokaryotes [150], as well as lower [209, 280] and higher-order eukaryotes [351].

Hydrogen peroxide is another ROS that must be detoxified by cells, and there are several enzyme systems that remove this species either through catalysis of its H_2O_2 dismutation to H_2O and O_2 or through its reduction to water [14]. Glutathione peroxidase (GPX) is an important enzyme for the removal of hydrogen peroxide. Two forms of GPX are found in mitochondria: GPX1 [81] and phospholipid-hydroperoxide GPX (PHGPX) [190], with GPX1 in the matrix and PHGPX embedded in the inner membrane [244]. GPX uses glutathione (GSH) during hydrogen peroxide reduction to water, generating oxidized glutathione (GSSG). Glutathione reductase regenerates GSH by reducing GSSG [153]. Catalase is another enzyme that reduces hydrogen peroxide levels [46, 370], but there is some controversy concerning the mitochondrial localization of catalase. While some labs demonstrate localization of catalase in the nucleus, peroxisomes, and the sarcoplasm (but not in mitochondria) in mice overexpressing catalase [379], other laboratories report catalase in the mitochondria [293] and localization to the matrix [276]. Another important hydrogen peroxide-detoxifying enzyme is peroxiredoxin (PRX), with two forms discovered in mitochondria [233]: PRX III [45, 304] and PRX V [300]. PRX uses thioredoxin to reduce hydrogen peroxide to water and oxidized thioredoxin. Thioredoxin (Trx) is another important cellular antioxidant protein [182]. Two forms of thioredoxin (Trx) exist in cells, with Trx1 found in the cytoplasm and Trx2 found in mitochondria [318]. Thioredoxin reductase is used to regenerate reduced thioredoxin [53, 169].

3.3 MnSOD Is Essential for Aerobic Life

Myriad studies in model systems ranging from bacteria [110] to eukaryotes [111] have demonstrated the essential role for MnSOD in protecting aerobic life from the deleterious effects of oxygen. Studies in higher organisms have further validated the importance of MnSOD. Complete knockout of MnSOD results in death shortly after birth in mouse [178] and *Drosophila* [77] models due, in part, to decreased activity of numerous mitochondrial proteins [178, 256]. In MnSOD heterozygous knockout mice (with a 50 % reduction in MnSOD enzyme activity in all tissues) [343], there was an age-dependent, significant increase in oxidative DNA damage (8-oxodeoxyguanidine) in both nuclear and mitochondrial DNA compared to wild-type mice, but there were no increases in other markers of aging (immune response, cataract formation, etc.). Interestingly, there was no significant difference in the life span between wild-type and MnSOD heterozygous knockout mice, but there was a 100 % increase in the incidence of cancer in the MnSOD heterozygous knockout mice.

Interestingly, complete knockout of MnSOD does not affect embryonic development. Using a *Drosophila* model, Mukherjee et al. discovered that a lack of MnSOD does not affect embryogenesis and later development and differentiation [217], but only affects the viability of adult flies, which is consistent with mouse models demonstrating that some homozygous knockout of MnSOD are small size at birth but have no gross deformities at birth [178]. The increased death rate in MnSOD homozygous knockout neonates may be due to the failure of these animals to compensate for the higher oxygen levels in the atmosphere compared to oxygen levels in utero. Anatomical abnormalities contributing to early death in MnSOD homozygous knockout mice include several cardiovascular abnormalities, such as dilated left ventricular cavity and reduced left ventricular wall thickness, as well as myocardial hypertrophy leading to dilated cardiomyopathy [178]. Using a strain of MnSOD knockout mice with deletions of exon 1 and 2 in the *SOD2* gene, Lebovitz et al. found that these mice lived up to 3 weeks after birth, but suffered from anemia resulting from low levels of all hematopoietic cells, leading to hypocellular bone marrow. Approximately 10 % of the animals also had cardiac injury, demonstrated by balloon-like cardiac dilation and ventricular wall thinning [166]. Copin et al. found that simultaneous overexpression of CuZnSOD and knockout of MnSOD did not counteract the neonatal lethality resulting from decreased MnSOD expression, indicating that localization of antioxidant activity is imperative to oxidative stress-related cellular damage [60].

Conditional knockout of MnSOD using Cre-Lox technology is a valuable tool in ascertaining tissue-specific effects of diminished MnSOD, with several tissue-specific knockout models developed (recently reviewed in [196]). The first MnSOD floxed mouse was developed by Ikegami et al. [135] and used to develop a liver-specific MnSOD knockout model. Interestingly, liver-specific knockout of MnSOD did not affect morphology and there was no increase in oxidative damage, as measured by lipid peroxidation, indicating that the liver is either not susceptible to systemic oxidative stress or there are compensatory mechanisms to protect from oxidative stress. Kidney-specific knockout of MnSOD resulted in a decrease in body weight, altered kidney morphology, and increased kidney protein nitration, but with no change in overall renal function compared to Cre control mice [252]. In C57/BL6 mice with type IIB skeletal muscle-specific MnSOD knockout, there was a significant increase in mitochondrial superoxide production and oxidative damage, resulting in reduced aerobic exercise capacity of the gastrocnemius and extensor digitorum longus muscles to produce force over time [185]. However, muscle-specific knockout of MnSOD had no effect on age-dependent muscle atrophy [186].

In postnatal neurons, conditional knockout of MnSOD did not increase oxidative stress, but did lead to increased disorganization of distal nerve axons after injury [208]. In an attempt to study the effects of MnSOD knockout on diabetic neuropathy, Oh et al. endeavored to generate a nervous system-specific MnSOD knockout and crossing with the BKS.db/db mouse model of diabetes. Interestingly, the MnSOD knockout mice survived up to 3 weeks after birth. Since diabetes onset in BKS.db/db mice does not occur until 4 weeks after birth, the researchers revised their study to characterize the effects of nervous system-specific MnSOD knockout. These mice had an abnormal gait and seizures, had difficulty righting themselves between postnatal days 10–14 and became moribund between postnatal days 15–20. There was a complete knockout of MnSOD in the brain and substantial decrease in MnSOD in the lumbar and cervical spinal cord, a slight decrease of MnSOD in the sciatic nerve, and virtually no loss of MnSOD in the dorsal root ganglia. Further examination of the cortex revealed substantial extracellular vacuole and disruption of mitochondrial structure with MnSOD knockout [234].

If decreased MnSOD has harmful effects, then it is natural to deduce that increased MnSOD may be beneficial. While overexpression of MnSOD did not significantly affect life span or age-related pathologies compared to wild-type mice, there was an increase in aconitase activity, decreased lipid peroxidation, and a decrease in age-related decline in mitochondrial ATP production [140]. Overexpression of CuZnSOD, catalase, or a combination of CuZnSOD with MnSOD or catalase did not affect life span in mice [259], but MnSOD overexpression did increase life span in *Drosophila melanogaster* [326]. These studies suggest species-specific effects of MnSOD on longevity. MnSOD can have a dramatic impact on age-associated decline in cardiac function. Age-associated increases in oxidative stress markers in cardiac tissue (hydroperoxides, 8-isoprostanes, and 4-hydroxynonenal) were all reduced in mice overexpressing MnSOD compared to wild-type controls. MnSOD overexpression also significantly reduced age-associated cardiac remodeling compared to controls [163].

Several studies have demonstrated the importance of MnSOD in mitigating ischemia-mediated damage. Keller et al. studied the effects of MnSOD overexpression on ischemia-induced brain injury. In MnSOD-overexpressing mice, cortical infarct volume and levels of TBARS (thiobarbituric acid reactive substances) and nitrotyrosine immunoreactivity in brain sections were all significantly reduced compared to wild-type mice [152]. Comparable effects of MnSOD overexpression on ischemia/reperfusion-induced heart injury have been observed. There was a significant reduction in infarct size and lactate dehydrogenase release, as well as a significant improvement in post-ischemia recovery of cardiac function, in MnSOD-overexpressing transgenic mice compared to wild-type controls [51].

Overexpression of MnSOD can also ameliorate complications associated with diabetes. Kowluru et al., using a streptozotocin model of diabetes, discovered that MnSOD overexpression significantly reduced diabetes-induced oxidative DNA damage and significantly increased reduced glutathione and antioxidant capacity compared to non-transgenic diabetic mouse controls [161]. In a study on the effects of MnSOD on diabetic cardiomyopathy, Shen et al. generated cardiac-specific

MnSOD-overexpressing mice and crossed these mice with the OVE26 model of type 1 diabetes. Overexpression of MnSOD completely nullified the diabetes-induced changes in cardiac tissue morphology observed in OVE26 mice and increased the contractility of isolated cardiomyocytes. MnSOD/OVE26 mice also had improved mitochondrial function (increased respiratory control ratio and state 3 respiration) and prevented diabetes-induced increases in mitochondrial number, area, and protein and DNA content. MnSOD overexpression also increased the antioxidant capacity of cardiac tissue by increasing catalase enzyme activity and mitochondrial reduced glutathione content compared to OVE26 mice [303]. The beneficial effect in a streptozotocin model of diabetes was found with (Mn)SOD mimic, MnTM-2-PyP^{5+} [23]. Further, MnTE-2-PyP^{5+}, an (Mn)SOD mimic, also prevented adoptive transfer of autoimmune diabetes by diabetogenic T-cell clone when given Mn porphyrin at 10 mg/kg every second day starting 1 day before the adoptive transfer [264].

MnSOD can also dramatically impact cellular response to environmental insults and other toxins. Several studies have demonstrated neuroprotection by MnSOD overexpression. Klivenyi et al. found that overexpression of MnSOD protected mice from MPTP (1-methyl-4-phenyl-1,2,3,6-tetrahydropyridine)-induced neurotoxicity, in part, by attenuating MPTP-induced dopamine and dihydroxyphenylacetic acid (DOPAC) depletion and decreased 3-nitrotyrosine formation [159]. Overexpression of MnSOD suppressed paraquat-induced toxicity in C3H10 mouse embryonic fibroblasts [321]. This laboratory demonstrated that overexpression of MnSOD in C3H10T mouse embryonic fibroblasts protected the cells from death induced by the herbicide paraquat [321]. Mansouri et al. studied the effects of MnSOD overexpression on alcohol-induced liver injury. Damage to mitochondrial DNA and decreased mitochondrial function induced by acute alcohol exposure, but not chronic alcohol exposure, was suppressed by overexpression of MnSOD, with peroxynitrite mediating damage with acute alcohol exposure and iron contributing to damage with chronic alcohol exposure [195]. This laboratory was the first to demonstrate that MnSOD was important for protecting cardiac tissue from adriamycin-induced injury [369], in part, through protecting complex I from inactivation by adriamycin [368].

3.4 MnSOD and Mitochondrial Integrity/Function

3.4.1 Electron Transport Chain

Electron transport chain complexes I, II, and III are important sources of superoxide in mitochondria and potential targets of the superoxide they produce. This susceptibility to superoxide-induced damage is due to iron-sulfur centers present in major subunits in all three complexes [10, 11, 235, 236, 288, 337]. Oxidative modification of amino acids in these complexes can also affect their activity. Chen et al. identified the 51 kDa flavin mononucleotide-binding subunit of complex I (NADH dehydrogenase, NDH) as an important site of susceptibility to superoxide. Using the immunospin trapping technique, involving polyclonal antibodies raised against the nitrone adduct of the spin trap DMPO [50, 200], the researchers successfully identified superoxide-mediated oxidative modifications of NDH. This oxidative modification was significantly inhibited by the addition of SOD, but not catalase, confirming the importance of superoxide in this oxidative modification. Pretreatment of NDH with the thiol blocking agent *N*-ethyl maleimide (NEM) significantly inhibited the formation of the DMPO-nitrone adduct, suggesting the importance of thiyl radical formation in oxidative damage to NDH. Mass spectrometry identified Cys206 and Tyr177 of the 51 kDa subunit as the two amino acids vulnerable to oxidative damage. This oxidative modification of NDH results in reduced electron transport activity without affecting superoxide-generating activity [49]. This autooxidation of complex I by superoxide is thought to be involved in Parkinson's disease [151]. Peroxynitrite is another ROS that can affect the activities of complex I [32, 52, 258, 363] and complex III [258].

MnSOD is vital for scavenging superoxide radicals generated during oxidative phosphorylation and may inhibit ROS-induced inactivation of these complexes. MnSOD knockdown results in altered activities of complexes I, II, and III in numerous model systems. Homozygous knockout of MnSOD in mice results in a significant reduction in complex II levels in heart tissue [178], and mitochondria isolated from heterozygous MnSOD knockout mice have a reduced respiratory control ratio (RCR) compared to wild-type mice. Reduced RCR was greatest for complex I substrates glutamate/malate and complex III substrate duroquinol and was linked to decreased state III respiration. Complex I activity was reduced in heterozygous MnSOD knockout mice compared to wild-type mice due to oxidation of the FeS center of complex I [354]. Martin et al., using erythroblasts isolated from wild-type and MnSOD homozygous knockout mice, found that several nuclear gene-encoded subunits of all five complexes of oxidative phosphorylation were downregulated in MnSOD homozygous knockout cells compared to wild-type cells [197]. Upon exposure to ethanol, wild-type and MnSOD heterozygous knockout mice had a significant reduction in complex I and V activity, while these effects were blocked in MnSOD overexpressing mice. This decrease in activity was linked to an increase in iNOS expression and increased nitration of complexes I and V in the MnSOD heterozygous knockout mice, demonstrating the importance of MnSOD in modulating peroxynitrite formation and protecting mitochondrial proteins from nitration through the removal of excess superoxide contributing to peroxynitrite formation [165].

Work by this laboratory has focused on the off-target effects of cancer chemotherapeutic drugs, particularly adriamycin and its cardiac and neurological toxicities. Anthracycline chemotherapy can cause a dose-dependent cardiotoxicity in cancer patients [310], leading to dilated cardiomyopathy and congestive heart failure [206]. Mitochondria are important targets of adriamycin [295], and changes in ROS-scavenging capacity of cells can have a significant impact on the effects of excessive ROS production catalyzed by adriamycin in cardiac tissue [146, 306, 327]. This laboratory was the first to show mitochondrial ROS is vital for adriamycin-induced cardiac injury [369]. Adriamycin treatment resulted in a substantial reduction in RCR and state III respiration at complexes I and II in wild-type mice, but overexpression of MnSOD had a protective effect on complex I, with only complex II being affected by adriamycin treatment [368].

Another important side effect of cancer chemotherapy is a cognitive decline often referred to by patients as chemobrain [226, 350]. Characteristics of this chemotherapy-induced cognitive decline include decreased reaction time, diminished concentration, and memory loss [7, 336]. Adriamycin treatment can result in changes in the structure [33, 136] and activity [309] of different brain regions, with oxidative stress as a potential mechanism for this cognitive decline [143, 333]. This laboratory was the first to demonstrate a unique mechanism for adriamycin-induced neurotoxicity. Adriamycin does not cross the blood brain barrier [237], but is able to stimulate TNF-α levels in serum, whole brain, and the hippocampus and cortex of mouse brain. Increased TNF-α correlated with decreased state III respiration due to decreased complex I activity [334]. Changes in MnSOD enzyme activity is another important mechanism of adriamycin-induced neurotoxicity. MnSOD protein nitration increased with adriamycin treatment, correlating with decreased enzyme activity, and was not observed in iNOS knock-out mice, suggesting a role for iNOS in adriamycin-induced neurotoxicity and the importance of MnSOD in preventing chemotherapy-induced injury [333].

3.4.2 Krebs Cycle

The Krebs cycle is a vital metabolic pathway found in mitochondria. The Krebs cycle provides reducing equivalents that are used in oxidative phosphorylation for ATP production and generates substrates used in numerous cellular processes. Changes in the activity of different enzymes in the Krebs cycle have been tied to many neurological diseases and cancer [30, 289].

Aconitase catalyzes the conversion of aconitate to isocitrate [30] and is vulnerable to deactivation by superoxide because it contains FeS centers [37, 100, 120], resulting in release of Fe(II) from the enzyme (reviewed in [99]). Aconitase is also susceptible to deactivation by *S*-nitrosoglutathione [338] and peroxynitrite [42, 116, 338]. Peroxynitrite attacks two sites in aconitase in a concentration-dependent manner. Low levels of peroxynitrite inhibit aconitase activity, in the absence of citrate, by attacking the FeS center, converting the [4Fe-4S] center to a [3Fe-4S] center with a corresponding loss of Fe. Higher concentrations of peroxynitrite are needed to inactivate aconitase in the presence of citrate through oxidation of Cys385 (in the FeS center) to sulfonic acid, as well as nitration of important tyrosines and oxidation of cysteine residues near the active site of the enzyme [116].

Numerous studies in many model systems have established the importance of MnSOD in maintaining aconitase activity [77, 181, 215, 354]. In MnSOD homozygous knockout mouse erythroids, there is a significant decrease in the expression of nearly all proteins involved in the Krebs cycle compared to wild-type erythroids [197]. Liver mitochondria isolated from MnSOD heterozygous knockout mice have reduced aconitase activity, which was rescued by the addition of iron and dithiothreitol, suggesting that aconitase inactivation may be due to superoxide-mediated oxidation [354]. On the contrary, overexpression of MnSOD prevented hypoxia-reoxygenation-induced aconitase inactivation in A549 human lung adenocarcinoma cells [268], as well as electron transport chain inhibitors and the redox-cycling agent phenazine [100].

3.4.3 Iron Metabolism

Mitochondria are essential for proper iron utilization in cells and are the sites of two iron-consuming processes [174]: the synthesis of FeS centers [290] and heme [9]. Iron storage is another important task of mitochondria (reviewed in [284]). Improper iron handling can results in iron-induced oxidative damage [78] and has been linked with multiple disorders, such as hereditary myopathy, Friedreich ataxia, and X-linked sideroblastic anemia [225, 367].

Mutations in MnSOD, or altered expression of MnSOD, can have substantial effects on iron handling. Srinivasan et al. showed that knockdown of MnSOD, CuZnSOD, or both in yeast led to a marked increase in the amount of EPR-detectable iron compared to wild-type yeast [320]. The Ala16Val polymorphism of the *Sod2* gene (rs4880) was linked to changes in iron handling, with the Ala-MnSOD variant associated with increased intracellular iron in patients with alcohol-induced cirrhosis and an increased risk of hepatocellular carcinoma. In Huh7 human hepatoma cells, transfection with the Ala-MnSOD variant increased the expression of different iron-handling genes (frataxin, hepcidin, cytosolic ferritin, and transferrin receptors-1 and -2) [222], and complete knockout of MnSOD in mouse erythroblasts resulted in a significant increase in the expression of transferrin [197].

ABCb7 (ATP binding cassette subfamily B member-7) is also affected by MnSOD. ABCb7 is a transporter protein in the inner mitochondrial membrane [305], and is key for exporting FeS centers from mitochondria to the cytosol and the formation of cytosolic FeS-containing proteins [22, 267]. Changes in ABCb7 expression and mutations are associated with the mitochondrial accumulation of iron [43, 211, 296] and sideroblastic anemia with ataxia [12, 22, 296]. In erythroid cells, loss of MnSOD led to decreased expression of ABCb7 [197]. The studies imply a role for MnSOD in the regulation of iron levels in cells, particularly in mitochondria. Increased MnSOD expression or activity may be important for the treatment of iron toxicity associated with different diseases.

While MnSOD affects iron metabolism, the converse (iron metabolism affecting MnSOD expression) is also observed. Pinkham et al. found *Sod2* is a heme-responsive gene, with three *cis* elements that are bound by the heme-binding transcription factor Hap1p [265]. Heme oxygenase-1 (HO-1) is an enzyme that catalyzes the rate-limiting step in heme degradation. HO-1 is found in mitochondria, where it regulates mitochondrial heme content and the expression of multiple genes, such as mitochondrial nitric oxide synthase and cytochrome *c* oxidase subunit I [59]. Transfection of HO-1 in cultured rat astroglial cells results in increased MnSOD expression, which was abrogated by antioxidant treatment, indicating a role for HO-1-stimulated oxidative stress in the expression of MnSOD [90]. In fibroblasts isolated from Friedreich ataxia patients, iron-induced MnSOD expression is diminished compared to fibroblasts from healthy patients, and only high concentrations of iron were able to induce MnSOD expression in a NF-κb-dependent manner [142]. All together, these studies demonstrate a carefully regulated relationship between iron metabolism and MnSOD expression, with disease development resulting from a disruption of this balance.

3.4.4 Apoptosis

Apoptosis is a highly regulated type of cell death targeting single cells or small groups of cells and is characterized by cytoplasmic and nuclear condensation, the formation of apoptotic bodies (small membrane-bound fragments containing cellular components), and the absorption of these apoptotic bodies by surrounding healthy cells [6, 191]. Apoptosis occurs through either extrinsic (initiated by external stimuli) [281] or intrinsic (initiated by internal cellular stress) pathways [17, 291]. Apoptosis is vital for a number of cellular processes, such as embryonic development and the immune response [299].

Mitochondria are involved in the initiation and progression of apoptosis [114, 317]. Mitochondria-centered apoptosis can be triggered by ROS generation, permeabilization of the mitochondrial membrane, and changes in the mitochondrial membrane potential [109]. When mitochondria dysfunction, molecules such as apoptosis-inducing factor (AIF) and endonuclease G are released and translocate to the nucleus to participate in DNA degradation. Omi/Htr2 and Smac/DIABLO inhibit the activity of various members of the inhibitor of apoptosis (IAP) family. Cytochrome *c* is released into the cytoplasm and interacts with apoptotic protease activating factor-1 (Apaf-1) to form apoptosomes, protein complexes that participate in the cleavage and activation of caspase 9 [5].

Many mechanisms have been proposed for the protective effects of MnSOD from apoptosis. Overexpression of MnSOD protects mitochondria against loss of function and membrane potential caused by different apoptosis-inducing agents, such as ionizing radiation [79], tumor necrosis factor-related apoptosis-inducing ligand (TRAIL) [212], NO-generating agents, amyloid β-peptide, and Fe(II) [152], as well as numerous anticancer agents [123]. MnSOD also inhibits the release of inducers of apoptosis from mitochondria, like Smac/DIABLO [212] and cytochrome *c* [79].

Another important mechanism of apoptosis modulation by MnSOD is scavenging of ROS and suppression of ROS-mediated cellular damage. In FSa-II murine fibrosarcoma cells, MnSOD overexpression protected the cells from antimycin- and rotenone-induced apoptosis by inhibiting poly(ADP-ribose) polymerase cleavage and caspase-3 activation, in part, by changing mitochondrial ROS levels [157]. Overexpression of MnSOD in PC6 pheochromocytoma cells protected from Fe(II), amyloid β-peptide, and NO-generating agent-induced apoptosis by suppressing peroxynitrite production and lipid peroxidation, as well as preventing mitochondrial membrane potential collapse [152].

MnSOD also suppresses the apoptotic effects of inflammatory cytokines. Overexpression of MnSOD protected A375 human melanoma cells and Chinese hamster ovarian cells from inflammatory cytokine-induced toxicity [123]. Tumor necrosis factor (TNF) induces mitochondrial ROS production as part of its apoptosis stimulating effects [106]. TNF stimulates MnSOD expression, which may act as an adaptive response to further exposure to TNF [359], with overexpression of MnSOD conferring protection to multiple cell lines from TNF-stimulated apoptosis [358]. This adaptive response to TNF may be protective against other apoptosis-inducing agents, and is thought to occur through an increase in steady-state levels of hydrogen peroxide [65]. For example, hippocampal cells pretreated with TNF-α were protected from apoptosis induced by Fe(II) and amyloid β-peptide by stimulating MnSOD expression [201]. This laboratory found the antiestrogen tamoxifen enhanced TNF-α-induced MnSOD expression [64], and that this increase in MnSOD is an important part of tamoxifen-mediated protection from adriamycin-induced apoptosis in cardiac tissue [63]. MnSOD degradation is an important mechanism of Fas receptor-mediated apoptosis in human Jurkat T-cells, leading to increased superoxide production and apoptosis [253].

3.4.5 Mitochondrial Control of Innate and Adaptive Immunity

Innate immunity is the first line of protection for cells against attack by foreign microbes or cellular damage. Pattern recognition receptors (PRRs) carry out innate immunity by recognizing key components of microbial invaders (pathogen-associated molecular patterns (PAMPS)) or endogenous cellular components from damaged cells (danger-associated molecular patterns (DAMPS)). PRRs can be both membrane bound or cytosolic [56, 148, 199], and activation of PRRs results in activation of numerous transcription factors like interferon-regulator factor (IRF), NF-κB, and AP-1 [199]. PRR activation can also lead to the formation of inflammasomes, multiprotein complexes composed of a caspase (caspase-1), a sensing protein (NLR), and an adaptor protein (apoptosis-associated speck-like protein containing an CARD (ASC)) [308]. Inflammasomes activate numerous inflammatory cytokines, such as interleukin-1β, in response to PAMPs or DAMPS [199].

Mitochondria play a vital role for the initiation of immune response. In response to viral infection, mitochondrial antiviral signaling (MAVS) protein is important for interferon-β (IFN-β) expression regulated by NF-κB and IRF-3. Suppressing MAVS expression abolished IFN-β expression, while overexpression of MAVS stimulated IFN-β expression. Mitochondrial localization is essential for MAVS function, with endoplasmic reticulum or plasma membrane targeting of MAVS greatly reducing IFN-β expression [301]. Injury-induced release of mitochondrial DNA or formyl peptides activates polymorphonuclear neutrophils, leading to neutrophil-mediated organ injury [373]. Mitochondrial DNA is also important for innate immunity. Treatment of J774A.1 macrophages with ethidium bromide to deplete cells of mitochondrial DNA (ρ^0 cells) resulted in inhibition of LPS- and ATP-induced caspase-1 activation and IL-1β secretion. Cytosolic release of mitochondrial DNA is a key activator of inflammasomes. Treatment with DNase I inhibited LPS- and ATP-induced inflammasome formation due to digestion of mitochondrial DNA, while transfection with mitochondrial DNA enhanced inflammasome formation in bone marrow-derived macrophages [224].

ROS are another activator of inflammasome formation (reviewed in [198]). In human macrophages, NADPH oxidase (Nox)-generated ROS are involved in asbestos-induced inflammasome formation and IL-1β secretion. Asbestos-induced IL-1β production was hindered by inhibiting Nox activity by diphenylene iodonium chloride or apocynin, as well as scavenging ROS using *N*-acetylcysteine or (2*R*, 4*R*)-4-aminopyrrolidine-2,4-dicarboxylate (APDC) [74]. Sources of ROS other than Nox may also be involved in inflammasome activation, as Meissner et al. discovered that caspase-1 activation and IL-1β production can occur in mononuclear phagocytes that lack Nox [204].

Mitochondria are a major source of ROS involved in inflammasome formation. In THP1 macrophages, Zhou et al. discovered that chemical inhibition of different electron transport chain complexes increased mitochondrial ROS formation and correlated with increased IL-1β activation. NLRP3 and ASC localized to both the endoplasmic reticulum and mitochondria after treatment with inflammasome-stimulating agents MSU, nigericin, and alum. VDAC is vital for mitochondrial ROS-dependent inflammasome formation, with overexpression of Bcl-2 (which inhibits VDAC activity) or knockdown of VDAC1 or VDAC 2 inhibiting caspase-1 activation or the formation of mature IL-1β [71]. Rotenone treatment enhanced LPS- and ATP-stimulated caspase-1 activation and IL-1β secretion in macrophages, which was inhibited by treating with the antioxidant Mito-TEMPO [224].

Proinflammatory cytokines can also be activated by mitochondrial ROS independent of inflammasome formation. Bulua et al., using mouse embryonic fibroblasts expressing different mutants of type 1 TNF receptor (TNFR1), discovered that cells expressing mutant TNFR1 had increased basal levels of mitochondrial ROS. This mitochondrial ROS is important for lipopolysaccharide- (LPS-) stimulated production of proinflammatory cytokines TNF and IL-6, but not IL-1β, in the absence of inflammasome formation, and scavenging mitochondrial ROS suppressed LPS-induced cytokine production [35].

Autophagy has also been linked to inflammasome formation. Zhou et al. [71] and Nakahira et al. [224] found that inhibition of autophagy by 3-methyladenine (3-MA) treatment [71] or knockdown of autophagy proteins LC3 [224] or beclin-1 [71, 224] resulted in increased mitochondrial ROS and activation of inflammasomes, as evidenced by increased caspase-1 activation and IL-1β secretion. Autophagy inhibition prevented LPS- and ATP-stimulated mitochondrial DNA release and caspase-1 activation [224]. These studies imply a role for autophagy-mediated removal of mitochondria to prevent activation of innate immunity by inflammasome formation. Inhibition of autophagy leads to accumulation of damaged mitochondria, resulting in increased mitochondrial ROS and inflammasome formation. These studies suggest a potential role for mitochondrial antioxidant enzymes, especially MnSOD, in the regulation of innate immunity and inflammatory diseases through both inflammasome-dependent and -independent mechanisms.

ROS are also important for adaptive immunity, in part, by altering T-cell function. ROS play an important role in T-cell activation by stimulating lipid raft formation needed for T-cell activation complex assembly [184]. Decreased glutathione levels is a mechanism for 3-hydroxyanthranilic acid- (a tryptophan metabolite) mediated activated T-cell death [168]. ROS are also involved in hematopoietic progenitor cell differentiation in *Drosophila* [241]. Peroxynitrite formation in the thymus has been linked to thymocyte apoptosis [216]. Changes in the expression of MnSOD also appear to have an effect on adaptive immunity. Case et al. [40] found that increased superoxide levels have a negative impact on T-cell development, leading to alterations in adaptive immune system function due to increased apoptosis and developmental defects in T-cells. In thymus-specific MnSOD knockout mice, there is an increase in immunodeficiency and increased susceptibility to influenza A H1N1 infection compared to wild-type mice. Treatment with the superoxide scavengers Tempol or CTPO was able to rescue the thymus-specific MnSOD knockout mice, indicating the importance of MnSOD in adaptive immunity and identifying targets for potential therapies to treat T-cell dysfunction-related immunological disorders.

3.4.6 Mitochondrial DNA Stability

Mitochondrial DNA (mtDNA) comprises 16,569 base pairs encoding 37 genes for 22 transfer RNAs, 2 ribosomal RNAs, and important components of different complexes of the electron transport chain [13]. All other proteins found in mitochondria are encoded by nuclear DNA, synthesized in the cytoplasm, and imported into mitochondria by numerous mechanisms. mtDNA is organized into structures called nucleoids that contain 6–10 individual mtDNA molecules, as well as numerous proteins involved in the synthesis and transcription of mtDNA, such as mitochondrial single-stranded binding protein (mtSSB), mitochondrial DNA polymerase gamma (Polγ), and mitochondrial transcription factor A (mTFA) [48, 102, 170]. Mitochondrial function declines with age [26] and is linked to both decreased mtDNA content and increased oxidative damage [307]. mtDNA damage has been associated with a multitude of age-related disorders [332], such as Parkinson's disease, diabetes [145], and cancer [29, 183].

mtDNA is vulnerable to damage caused by myriad agents, such as ionizing [285] and ultraviolet radiation [329], as well as ROS [285, 329, 354], with the D-loop region of mtDNA as a particularly susceptible region to oxidative damage [194]. mtDNA sequences encoding different Complex I subunits are highly prone to damage, which may contribute to elevated superoxide production and age-related diseases [61]. Cells lacking mtDNA (ρ^0 cells) or cells expressing mtDNA containing a common 4977 bp deletion have significantly higher levels of ROS production compared to parental cells or cells which express wild-type mtDNA [137]. MnSOD expression increases as an adaptive response to mtDNA depletion [254]. A malicious cycle can occur in which mtDNA damage leads to altered mitochondrial function, which increases ROS production, resulting in more mtDNA damage [25]. In the GM00637E human fibroblast cell line, Yakes and van Houten discovered that mtDNA is more sensitive than nuclear DNA to hydrogen peroxide-induced damage. The mtDNA damage occurred more rapidly, and mtDNA damage is not repaired after prolonged hydrogen peroxide exposure. This mtDNA damage correlated with diminished mitochondrial function [362].

Polγ can undergo oxidative stress-mediated inactivation, which may affect mtDNA repair and replication. Graziewicz et al. discovered that treatment of the catalytic subunit of Polγ with hydrogen peroxide resulted in a 50 % reduction in polymerase activity, as well as diminished DNA binding activity. The researchers also found that Polγ was more sensitive to oxidative inactivation than Polβ or Polα, the two nuclear DNA polymerases [108].

MnSOD is vital for protecting mtDNA from ROS-induced damage. In *Escherichia coli* K-12 cells, MnSOD associates with DNA [324]. mtDNA oxidative damage is much greater in the livers of MnSOD heterozygous knockout mice than wild-type controls [343]. In type 2 diabetic patients, MnSOD expression in peripheral blood mononuclear cells is increased as an adaptive response to higher levels of oxidative mtDNA damage [98]. In bovine retina endothelial cells, MnSOD overexpression or treatment with MnSOD mimetics inhibited oxidative mtDNA damage induced by high glucose, leading to increased expression of different electron transport chain components compared to glucose alone [189]. MnSOD overexpression also protects mtDNA from UV-induced oxidative damage [329] and acute ethanol exposure [165, 195]. MnSOD is also part of the nucleoid complex in the mitochondrial matrix, interacting with mtDNA, Polγ, and glutathione peroxidase, and this nucleoid localization may be important in protecting mtDNA from oxidative stress-induced damage [156]. This laboratory has confirmed an interaction between MnSOD, Polγ, and mtDNA, and further demonstrated that MnSOD is vital for protecting mtDNA from UV-induced damage by preventing Polγ inactivation [16]. These studies imply a role for MnSOD in preventing or treating diseases associated with mtDNA damage.

3.4.7 Cellular Lipid Integrity

Lipids, particularly polyunsaturated lipids, are vulnerable to ROS-induced damage, leading to the production of different lipid peroxidation products like 4-hydroxynonenal (4HNE) and malondialdehyde and oxidative products of cholesterol, cholesterol esters, and sphingolipids [316]. RNS react with lipids to form both oxidation and nitration products [15, 230, 272, 292], and these RNS-derived lipid products may play an important role in functions as varied as ischemia/reperfusion injury [221] and inflammation [15].

Changes in MnSOD expression or activity can impact total cellular, and mitochondrial, lipid integrity. In heterozygous MnSOD knockout mice, there is a significant increase in lipid peroxidation (as measured by 8-isoprostane levels) at postnatal day 10 compared to wild-type littermates [325]. In PC6 pheochromocytoma cells, MnSOD overexpression diminished lipid peroxidation and 4HNE protein adduct formation caused by treatment with amyloid β-peptide, sodium nitroprusside, and Fe(II) [152]. In vivo, administration of MnSOD plasmid/liposome complexes intraesophageally or intraorally suppressed ionizing radiation-induced lipid peroxidation [80]. Ohtsuki et al. [238] found an age-related increase in lipid peroxidation of mitochondrial fractions of brain tissue in spontaneously hypertensive and deoxycorticosterone acetate (DOCA)-induced hypertensive rats compared to normotensive and vehicle-treated rats, respectively. MnSOD expression increased with age only in the brain tissue of normotensive rats, suggesting that altered superoxide metabolism may be a contributing factor of hypertension-associated neurological disorders [238].

Because MnSOD contains manganese in its active site [150], manganese availability may have a dramatic effect on lipid peroxidation. Manganese deficiency significantly reduced age-related increases in MnSOD activity compared to rats receiving adequate levels of manganese, correlating with a fivefold increase in liver lipid peroxidation in manganese-deficient rats [380]. In a study by Malecki and Greger, lipid peroxidation was increased in rat heart mitochondria of Mn-deficient rats compared to control rats, associating with a decrease in MnSOD enzyme activity [193]. Manganese deficiency is linked to myriad maladies, such as diabetes mellitus [149] and epilepsy [39].

Cardiolipin is another important lipid susceptible to oxidative damage. Cardiolipin is found almost exclusively in the inner mitochondrial membrane [57] and is a tetra-acylated glycerophospholipid made up of two phosphatidyl groups linked by glycerol, giving a lipid containing three chiral centers and four hydrocarbon chains [101]. Changes in the composition and levels of cardiolipin are associated with different mitochondria-linked diseases, such as aging [173, 247], heart failure, diabetes, ischemia/reperfusion injury [246, 315], Barth syndrome [119], and cancer [38, 139, 282].

Cardiolipin is essential for proper mitochondrial function. Cardiolipin is important for the initiation of apoptosis [82, 105, 203, 223, 260, 298, 314], with cardiolipin peroxidation resulting in cytochrome *c* release [223, 261] and deactivation of the adenine nucleotide translocator leading to opening of the mitochondrial permeability transition pore [223]. The activities of individual components of the electron transport chain [95], and the assembly of these components into supramolecular complexes [262, 357], are dependent on cardiolipin. Cardiolipin is susceptible to ROS-mediated damage [173, 342, 355], and oxidative damage of cardiolipin correlates with ROS-mediated inactivation of electron transport chain components. Different combinations of SOD and catalase, or supplementation with exogenous cardiolipin, inhibit the effects of ROS on the electron transport chain and cardiolipin peroxidation [155, 248–251]. In HaCaT human keratinocytes, UVB radiation suppresses electron transport chain activity concomitant with a decrease in cardiolipin content. MnSOD protein levels increase as an adaptive response to UVB exposure [356]. These studies suggest a potential mechanism for MnSOD in protecting mitochondria from ROS-induced damage by inhibiting peroxidation of cardiolipin.

3.5 The Role of MnSOD in Cancer Development

MnSOD is important in the development of cancer because of its ability to scavenge ROS. Many studies have shown that MnSOD expression is reduced in many types of cancer [54, 62, 132, 232, 313], while other studies show that MnSOD expression is greater in cancer compared to normal tissue [125, 131, 138, 141, 192, 239, 341]. These studies suggest that MnSOD may play a dual role in cancer development (reviewed in [121]) and has led to crucial studies to clarify the role of MnSOD in cancer development and progression.

3.5.1 MnSOD Accelerates Tumor Progression

Many studies have shown that an increase in MnSOD expression enhances cancer survival, growth, and aggressiveness. In HeLa cells, MnSOD overexpression protected the cells from serum starvation-induced cell death and growth inhibition. Serum starvation alone did not affect MnSOD expression in HeLa cells, but did stimulate MnSOD expression in HT29 colon carcinoma cells and was associated with resistance to serum starvation [243]. Overexpression of MnSOD protected HCT116 colon carcinoma cells from tumor necrosis factor-related apoptosis-inducing ligand (TRAIL)-induced apoptosis by suppressing the release of Smac/DIABLO and cytochrome *c* from mitochondria [212].

Metastasis is another behavior of cancer cells that is affected by MnSOD. In head and neck squamous cell carcinoma, Salzman et al. discovered that MnSOD enzyme activity was increased in patients with higher stages of cancer and patients with locoregional metastases [294]. In gastric cancer [192], as well as colorectal cancer [229], MnSOD expression was significantly higher in cancers that had lymph node metastases compared to cancers without metastasis. MnSOD expression is also linked to the metastatic behavior of the estrogen-independent MDA-MB-231 human breast cancer cell line [147]. In T47D human breast cancer cells, progestin treatment stimulated MnSOD expression, and MnSOD expression was linked to progestin stimulation of invasion, suggesting a possible mechanism by which progestins increase breast cancer aggressiveness [127]. One potential mechanism of increased metastasis by MnSOD expression may be an increase in hydrogen peroxide-dependent matrix metalloproteinase expression [227, 279].

In contrast to the numerous reports that MnSOD expression increases differentiation of cancer cells, others report that MnSOD expression actually leads to a less-differentiated state. Ladriscina et al. found an inverse correlation between

MnSOD expression and degree of differentiation in brain tumors of neuroepithelial origin, with the greatest amount of MnSOD in the most undifferentiated tumors [164]. In prostate cancer, neuroendocrine differentiation is a major step in the androgen-dependent to -independent progression of prostate cancer [344], with MnSOD being associated with neuroendocrine differentiation [271]. In LNCaP human prostate cancer cells, MnSOD overexpression led to androgen independence and neuroendocrine differentiation, as well as increased survival against TNF, docetaxel, and etoposide [270].

3.5.2 MnSOD Suppresses Tumorigenesis

While several studies have demonstrated a tumor-supporting role for MnSOD, other studies have revealed a tumor-suppressing function for MnSOD, with overexpression of MnSOD inhibiting many of the hallmarks of cancer such as invasiveness and anchorage-independent cell growth [21, 54, 180, 345, 352]. In the early 1990s, there was little known about the contribution of ROS generated in mitochondria in the neoplastic transformation process. To directly test the role of mitochondrial ROS in neoplastic transformation, this laboratory expressed MnSOD in mouse embryonic fibroblast C3H10T/2 cells because they are capable of being malignantly transformed by chemical and physical carcinogens and have low endogenous levels of MnSOD. Increased MnSOD expression reduced the frequency of in vitro neoplastic transformation caused by ionizing radiation. The formation of type 2 and 3 foci in parental C3H10T/2 cells and control cells transfected with a selectable vector alone was significantly higher than in C3H10T/2 cells overexpressing MnSOD (C3HSOD). This study was the first to link protection of mitochondria against ROS to reduction of neoplastic transformation [322]. Subsequently, this finding was substantiated by the introduction of a normal chromosome 6 into human melanoma cell lines, which results in suppression of tumorigenicity and suggests that gene(s) on chromosome 6 controls the malignant phenotype of human melanoma. Because MnSOD is localized to a region of chromosome 6 frequently lost in melanomas, the authors consequently examined the effect of transfecting sense and antisense human MnSOD cDNAs into melanoma cell lines. They found that cell lines expressing abundant (+)-sense MnSOD cDNAs significantly altered their phenotype in culture and lost their ability to form colonies in soft agar and tumors in nude mice. In contrast, the introduction of antisense MnSOD cDNA had no effect on melanoma tumorigenicity [55]. These studies have been followed in numerous publications confirming the tumor suppressing role of MnSOD in various cancers (reviewed in [67]).

Mechanisms of MnSOD-mediated tumor suppression are numerous, including sensitization of cancer cells to ROS-generating agents [176], as well as modulating carcinogen-induced ROS levels [231, 286, 374]. In HEK293 cells, overexpression of a MnSOD mutant lacking product inhibition led to a decrease in growth. This growth suppression was ROS-mediated, as overexpression of catalase blocked the effects of MnSOD overexpression [66]. MnSOD overexpression decreased metastatic potential, colony formation, and diminished tumor growth in vivo of XR23M transformed X-ray immortalized rat embryonic fibroblasts [286]. MnSOD overexpression also had substantial effects on PC-3 human prostate cancer cells, resulting in reduced cell growth concomitant with increased mitochondrial membrane potential and increased hydrogen peroxide levels [345].

Induction of differentiation by MnSOD is another important mechanism of cancer suppression. Church et al. found that overexpression of MnSOD by introducing a (+)-sense MnSOD cDNA resulted in a more differentiated morphology in UACC-903 human melanoma cells compared to empty vector controls [55]. In hepatocellular carcinoma, there is a direct correlation between relative MnSOD expression levels and degree of differentiation. Well differentiated cell lines and tumors have higher levels of MnSOD expression compared to poorly differentiated cell lines and tumors [8, 97, 364]. This laboratory discovered that overexpression of MnSOD in the FSa-II mouse fibrosarcoma cell line induced differentiation by inhibiting AP-1 DNA binding activity and AP-1 target gene expression [158]. MnSOD also increases differentiation by activating NF-κB and ERK MAP kinase [376].

A major emphasis of this laboratory has been to acquire a deeper knowledge of the effects of MnSOD on oxidative stress-induced tumor initiation/promotion. This laboratory was the first to report that MnSOD overexpression inhibited neoplastic transformation. In C3H 10 T1/2 mouse embryonic fibroblasts, MnSOD overexpression protected the cells from ionizing-radiation induced neoplastic transformation, but not DNA intercalating agent 3-methylcholanthrene, indicating a vital role for ROS in ionizing radiation-induced tumorigenesis and the potential for MnSOD in cancer prevention [322]. Overexpression of MnSOD suppressed 7,12-dimethylbenz(a)-anthracene/12-*O*-tetradecanoylphorbol-13-acetate (DMBA/TPA) induced papilloma incidence and multiplicity in C57BL/6 mice. A key mechanism of this tumor suppression was inhibition of TPA-induced oxidative stress [378].

Because MnSOD overexpression inhibited DMBA/TPA-induced tumor formation, logic suggests that decreased levels of MnSOD might enhance tumor formation. Interestingly, a similar number of papillomas were observed after DMBA/TPA treatment in both wild-type and MnSOD heterozygous knockout C57BL/6 mice due, in part, to increased proliferation and apoptosis in the basal layer of the epidermis. However, there was increased oxidative stress in MnSOD heterozygous knockout mice compared to wild-type mice, as measured by protein oxidation [377]. In a later study, this laboratory found that apoptosis occurred before proliferation in the basal layer of the epidermis. Apoptosis peaked 6 h after TPA treatment and mitosis peaked 24 h post-TPA. In wild-type mice, there was less proliferation compared to MnSOD heterozygous knockout mice. Treatment with MnTE-2-PyP^{5+} (a MnSOD mimetic, [18, 19, 207]) at 12 hrs after each TPA treatment led to a significant decrease in protein oxidation without affecting apoptosis, resulting in a 50 % reduction in tumor incidence. This study suggests that an important early event in cancer development is oxidative stress and exploiting this oxidative stress may be a vital mechanism for MnSOD inhibition of cancer development [375]. For more information on SOD mimetics, readers are directed to Forum Issue on "SOD therapeutics" in Antioxidant& Redox Signaling, 2014.

3.5.3 MnSOD Is a Double-Edged Sword in Cancer Development

Because of the inconsistent reports of MnSOD in cancer development, either as a tumor suppressor or as a supporter of tumor growth and progression, there are lingering questions of the exact role of MnSOD in cancer development. To address this seeming contradiction, one needs to consider how MnSOD affects hydrogen peroxide production in the cell. Buettner et al. found that MnSOD alters the steady-state level of hydrogen peroxide in systems where the equilibrium constant for the superoxide production (K) is below 1. In this instance, the rate constant for the conversion of superoxide back to molecular oxygen is greater than the rate constant for the conversion of molecular oxygen to superoxide. This condition is seen in the production of superoxide by the electron transport chain. When MnSOD is present, there is an increase in hydrogen peroxide production with increasing levels of MnSOD, with the greatest effects when MnSOD levels are low. When MnSOD levels are sufficiently high, however, there is only a small additional change in hydrogen peroxide production. The increase in hydrogen peroxide production is observed because MnSOD is drawing the equilibrium of the system toward the right, driving further superoxide production because a small amount of superoxide is being consumed and not participating in the reverse reaction, in keeping with Le Chatelier's principle [34].

Changes in hydrogen peroxide flux with changes in MnSOD expression can have broad repercussions in cancer progression. In the early stages of cancer development (when MnSOD levels are low [232]), MnSOD expression may inhibit cancer growth by multiple mechanisms because of increased hydrogen peroxide flux [177] and prevention of more reactive species including hydroxyl radicals. As cancer progresses (when cells display increased glucose and hydrogen peroxide metabolism and have chronic oxidative stress [24, 245, 269]), MnSOD expression may benefit cancer cells by promoting metastatic behavior [58, 121, 127, 227].

This laboratory has striven to better understand this nuanced role of MnSOD in the development and progression of cancer. This laboratory reported the dual nature of MnSOD in cancer using the DMBA/TPA two stage model of skin cancer development and a unique mouse model expressing a luciferase reporter gene driven by the MnSOD promoter [68]. DMBA treatment followed TPA treatments over 25 weeks resulted in a significant reduction in MnSOD-luciferase reporter activity, as well as decreased levels of MnSOD mRNA, protein, and enzyme activity in both DMBA/TPA-treated skin and papillomas compared to DMSO-treated controls. To allow for the formation of squamous cell carcinoma, the observation period was extended to 48–60 weeks. Interestingly, during the transition from papilloma to the more aggressive squamous cell carcinoma, MnSOD luciferase reporter activity, mRNA, protein, and enzyme activity were significantly increased.

The changes in MnSOD expression during the transition from papilloma to squamous cell carcinoma are due to changes in the transcription factors that bind the

Sod2 promoter. Basal expression of MnSOD is regulated by Sp1 [361], and Sp1 DNA binding activity was significantly reduced in DMBA/TPA-treated skin, with an additional decrease in papilloma. However, in squamous cell carcinoma, Sp1 DNA binding activity returned to nearly the same level as DMSO-treated skin control [68]. p53, a tumor-suppressing transcription factor, also regulates MnSOD expression through its interaction with Sp1 [69, 70]. p53 DNA binding activity was similar to DMSO treated skin in both DMBA/TPA-treated skin and papilloma but was significantly reduced in squamous cell carcinoma. Using the JB6 mouse epithelial cell line, knockdown of p53 or overexpression of Sp1 increased MnSOD luciferase reporter activity and MnSOD protein expression, with an additive effect with simultaneous p53 knockdown and Sp1 overexpression [68].

These results clearly demonstrate the magnitude of MnSOD in early cancer development and progression to a more aggressive state. MnSOD expression decreases in cells as they transition to cancer, indicative of the tumor-suppressing function of MnSOD, while MnSOD expression increases as the cancer transitions from a nonaggressive phenotype to a more aggressive phenotype, showing the tumor-supporting role of MnSOD in later cancer stages [207]. Because of the importance of MnSOD as a ROS-scavenging enzyme, and the role of MnSOD in maintaining the myriad functions of mitochondria (discussed above), changes in MnSOD expression and activity that impact mitochondrial function may play an important role in cancer development and progression.

3.6 Conclusions

Mitochondria are central players in metabolism in cells due to the presence of enzymes for myriad metabolic pathways, including oxidative phosphorylation, the urea cycle, fatty acid oxidation, heme synthesis, the Krebs cycle, iron metabolism, among other functions. Mitochondria consume oxygen to generate ATP, producing ROS as a by-product. ROS can be damaging to mitochondria, leading to diminished mitochondrial function and the development of many diseases, such as cardiovascular disease, diabetes, neurological disorders, and cancer. MnSOD is absolutely essential for the survival of aerobic life and for the conservation of mitochondrial function to maintain cellular homeostasis by detoxifying ROS and mitigating the harmful effects of ROS on mitochondrial metabolic enzymes and lipids. MnSOD is a two-edged sword in the development and progression of cancer: acting as a tumor suppressor during early stages of cancer and a tumor supporter during later stages of cancer. Because of the central role for MnSOD in the preservation of the numerous functions of mitochondria, strategies that seek to maintain, or even stimulate, MnSOD expression and/or enzyme activity by increasing expression of endogenous MnSOD, augmenting cells by the addition of exogenous MnSOD protein, or the use of pharmaceuticals that mimic the action of MnSOD will no doubt prove valuable in the treatment and prevention of various ROS-associated maladies.

References

1. Abaci N, Arikan M, Tansel T, Sahin N, Cakiris A, Pacal F, Sirma Ekmekci S, Gok E, Ustek D. Mitochondrial mutations in patients with congenital heart defects by next generation sequencing technology. Cardiol Young. 2015;25:705–11.
2. Abate C, Patel L, Rauscher III FJ, Curran T. Redox regulation of Fos and Jun DNA-binding activity in vitro. Science. 1990;249:1157–61.
3. Abello N, Kerstjens HAM, Postma DS, Bischoff R. Protein tyrosine nitration: selectivity, physicochemical and biological consequences, denitration, and proteomics methods for the identification of tyrosine-nitrated proteins. J Proteome Res. 2009;8:3222–38.
4. Adam-Vizi V, Chinopoulos C. Bioenergetics and the formation of mitochondrial reactive oxygen species. Trends Pharmacol Sci. 2006;27:639–45.
5. Adams JM. Ways of dying: multiple pathways to apoptosis. Genes Dev. 2003;17:2481–95.
6. Afford S, Randhawa S. Demystified…Apoptosis. J Clin Pathol Mol Pathol. 2000;53:55–63.
7. Ahles TA, Saykin AJ. Candidate mechanisms for chemotherapy-induced cognitive changes. Nat Rev Cancer. 2007;7:192–201.
8. Aida Y, Maeyama S, Takakuwa T, Uchikoshi T, Endo Y, Suzuki K, Taniguchi N. Immunohistochemical expression of manganese superoxide dismutase in hepatocellular carcinoma, using a specific monoclonal antibody. J Gastroenterol. 1994;29:443–9.
9. Ajioka RS, Phillips JD, Kushner JP. Biosynthesis of heme in mammals. Biochim Biophys Acta. 2006;1763:723–36.
10. Albracht SPJ. The prosthetic groups in succinate dehydrogenase number and stoichiometry. Biochim Biophys Acta. 1980;612:11–28.
11. Albracht SPJ, Subramanian J. The number of Fe atoms in the iron-sulfur centers of the respiratory chain. Biochim Biophys Acta. 1977;462:36–48.
12. Allikmets R, Raskind WH, Hutchinson A, Scheuck ND, Dean M, Koeller DM. Mutation of a putative mitochondrial iron transporter gene (*ABC7*) in X-linked sideroblastic anemia and ataxia (XLSA/A). Hum Mol Genet. 1999;8:743–9.
13. Anderson S, Bankier AT, Barrell BG, de Bruijn MHL, Coulson AR, Drouin J, Eperon IC, Nierlich DP, Roe BA, Sanger F, Schreier PH, Smith AJH, Staden R, Young IG. Sequence and organization of the human mitochondrial genome. Nature. 1981;290:457–65.
14. Andreyev AY, Kushnareva YE, Starkov AA. Mitochondrial metabolism of reactive oxygen species. Biochemistry (Mosc). 2005;70:246–64.
15. Baker PRS, Schopfer FJ, O'Donnell VB, Freeman BA. Convergence of nitric oxid and lipid signaling: anti-inflammatory nitro-fatty acids. Free Radic Biol Med. 2009;46:989–1003.
16. Bakthavatchalu V, Dey S, Xu Y, Noel T, Jungsuwadee P, Holley AK, Dhar SK, Batinic-Haberle I, St Clair DK. Manganese superoxide dismutase is a mitochondrial fidelity protein that protects Polgamma against UV-induced inactivation. Oncogene. 2012;31:2129–39.
17. Basu A, Castle VP, Bouziane M, Bhalla K, Haldar S. Crosstalk between extrinsic and intrainsic cell death pathways in pancreatic cancer: synergistic action of estrogen metabolite and ligands of death receptor family. Cancer Res. 2006;66:4309–18.
18. Batinic-Haberle I, Reboucas JS, Spasojevic I. Response to Rosenthal et al. Antioxid Redox Signal. 2011;14:1174–6.
19. Batinic-Haberle I, Tovmasyan A, Roberts ER, Vujaskovic Z, Leong KW, Spasojevic I. SOD therapeutics: latest insights into their structure-activity relationships and impact on the cellular redox-based signaling pathways. Antioxid Redox Signal. 2014;20:2372–415.
20. Bauer G. Targeting extracellular ROS signaling of tumor cells. Anticancer Res. 2014;34:1467–82.
21. Behrend L, Mohr A, Dick T, Zwacka RM. Manganese superoxide dismutase induces p53-dependent senescence in colorectal cancer cells. Mol Cell Biol. 2005;25:7758–69.
22. Bekri S, Kispal G, Lange H, Fitzsimons E, Tolmie J, Lill R, Bishop DF. Human ABC7 transporter: gene structure and mutation causing X-linked sideroblastic anemia with ataxia with disruption of cytosolic iron-sulfur protein maturation. Blood. 2000;96:3256–64.

23. Benov L, Batinic-Haberle I. A manganese porphyrin suppresses oxidative stress and extends the life span of streptozotocin-diabetic rats. Free Radic Res. 2005;39:81–8.
24. Biaglow JE, Miller RA. The thioredoxin reductase/thioredoxin system: novel redox targets for cancer therapy. Cancer Biol Ther. 2005;4:6–13.
25. Birch-Machin MA, Swalwell H. How mitochondria record the effects of UV exposure and oxidative stress using human skin as a model tissue. Mutagenesis. 2010;25:101–7.
26. Boffoli D, Scacco SC, Vergari R, Solarino G, Santacroce G, Papa S. Decline with age of the respiratory chain activity in human skeletal muscle. Biochim Biophys Acta. 1994;1226:73–82.
27. Boland ML, Chourasia AH, Macleod KF. Mitochondrial Dysfunction in Cancer. Front Oncol. 2013;3:292.
28. Boonstra J, Post JA. Molecular events associated with reactive oxygen species and cell cycle progression in mammalian cells. Gene. 2004;337:1–13.
29. Brandon M, Baldi P, Wallace DC. Mitochondrial mutations in cancer. Oncogene. 2006;25:4647–62.
30. Briere J-J, Favier J, Gimenez-Roqueplo A-P, Rustin P. Tricarboxylic acid cycle dysfunction as a cause of human diseases and tumor formation. Am J Physiol Cell Physiol. 2006;291:C1114–20.
31. Brookes PS, Yoon Y, Robotham JL, Anders MW, Sheu S-S. Calcium, ATP, and ROS: a mitochondrial love-hate triangle. Am J Physiol Cell Physiol. 2004;287:817–33.
32. Brown GC, Borutaite V. Inhibition of mitochondrial respiratory complex I by nitric oxide, peroxynitrite and S-nitrosothiols. Biochim Biophys Acta. 2004;1658:44–9.
33. Brown MS, Stemmer SM, Simon JH, Stears JC, Jones RB, Cagnoni PJ, Sheeder JL. White matter disease induced by high-dose chemotherapy: longitudinal study with MR imaging and proton spectroscopy. AJNR Am J Neuroradiol. 1998;19:217–21.
34. Buettner GR, Ng CF, Wang M, Rodgers VGJ, Schafer FQ. A new paradigm: manganese superoxide dismutase influences the production of H_2O_2 in cells and thereby their biological state. Free Radic Biol Med. 2006;41:1338–50.
35. Bulua AC, Simon A, Maddipati R, Pelletier M, Park H, Kim K-Y, Sack MN, Kastner DL, Siegel RM. Mitochondrial reactive oxygen species promote production of proinflammatory cytokines and are elevated in TNFR1-associated periodic syndrome (TRAPS). J Exp Med. 2011;208:519–33.
36. Cai Z, Yan LJ. Protein oxidative modifications: beneficial roles in disease and health. J Biochem Pharmacol Res. 2013;1:15–26.
37. Cantu D, Schaack J, Patel M. Oxidative inactivation of mitochondrial aconitase results in iron and H2O2-mediated neurotoxicity in rat primary mesencephalic cultures. PLoS One. 2009;4, e7095.
38. Canuto RA, Biocca ME, Muzio G, Dianzani MU. Fatty acid composition of phospholipids in mitochondria and microsomes during diethylnitrosamine carcinogenesis in rat liver. Cell Biochem Funct. 1989;7:11–9.
39. Carl GF, Keen CL, Gallagher BB, Clegg MS, Littleton WH, Flannery DB, Hurley LS. Association of low blood manganese concentrations with epilepsy. Neurology. 1986;36:1584–7.
40. Case AJ, McGill JL, Tygrett LT, Shirasawa T, Spitz DR, Waldschmidt TJ, Legge KL, Domann FE. Elevated mitochondrial superoxide disrupts normal T cell development, impairing adaptive immune response to an influenza challenge. Free Radic Biol Med. 2011;50:448–58.
41. Cassina A, Radi R. Differential inhibitory action of nitric oxide and peroxynitrite on mitochondrial electron transport. Arch Biochem Biophys. 1996;328:309–16.
42. Castro L, Rodriquez M, Radi R. Aconitase is readily inactivated by peroxynitrite, but not by its precursor, nitric oxide. J Biol Chem. 1994;269:29409–15.
43. Cavadini P, Biasiotto G, Poli M, Levi S, Verardi R, Zanella I, Derosas M, Ingrassia R, Corrado M, Arosio P. RNA silencing of the mitochondrial ABCB7 transporter in HeLa cells causes an iron-deficient phenotype with mitochondrial iron overload. Blood. 2007;109:3552–9.
44. Celardo I, Martins LM, Gandhi S. Unravelling mitochondrial pathways to Parkinson's disease. Br J Pharmacol. 2014;171:1943–57.

45. Chang T-S, Cho C-S, Park S, Yu S, Kang SW. Peroxiredoxin III, a mitochondrion-specific peroxidase, regulate apoptotic signaling by mitochondria. J Biol Chem. 2004;279:41975–84.
46. Chelikani P, Fita I, Loewen PC. Diversity of structures and properties among catalases. Cell Mol Life Sci. 2004;61:192–208.
47. Chen Q, Vazquez EJ, Moghaddas S, Hoppel CL, Lesnefsky EJ. Production of reactive oxygen species by mitochondria: central role of complex III. J Biol Chem. 2003;278:36027–31.
48. Chen XJ, Butow RA. The organization and inheritance of the mitochondrial genome. Nat Rev Genet. 2005;6:815–25.
49. Chen Y-R, Chen C-L, Zhang L, Green-Church KB, Zweier JL. Superoxide generation from mitochondrial NADH dehydrogenase induces self-inactivation with specific protein radical formation. J Biol Chem. 2005;280:37339–48.
50. Chen YR, Chen CL, Liu X, Li H, Zweier JL, Mason RP. Involvement of protein radical, protein aggregation, and effects on NO metabolism in the hypochlorite-mediated oxidation of mitochondrial cytochrome c. Free Radic Biol Med. 2004;37:1591–603.
51. Chen Z, Siu B, Ho YS, Vincent R, Chua CC, Hamdy RC, Chua BH. Overexpression of MnSOD protects against myocardial ischemia/reperfusion injury in transgenic mice. J Mol Cell Cardiol. 1998;30:2281–9.
52. Chinta SJ, Andersen JK. Nitrosylation and nitration of mitochondrial complex I in Parkinson's disease. Free Radic Res. 2011;45:53–8.
53. Choi JH, Kim TN, Kim S, Baek SH, Kim JH, Lee SR, Kim JR. Overexpression of mitochondrial thioredoxin reductase and peroxiredoxin III in hepatocellular carcinoma. Anticancer Res. 2002;22:3331–5.
54. Chuang T-C, Liu J-Y, Lin C-T, Tang Y-T, Yeh M-H, Chang S-C, Li J-W, Kao M-C. Human manganese superoxide dismutase suppresses HER2/neu-mediated breast cancer malignancy. FEBS Lett. 2007;581:4443–9.
55. Church SL, Grant JW, Ridnour LA, Oberley LW, Swanson PE, Meltzer PS, Trent JM. Increased manganese superoxide dismutase expression suppresses the malignant phenotype of human melanoma cells. Proc Natl Acad Sci U S A. 1993;90:3113–7.
56. Cinel I, Opal SM. Molecular biology of inflammation and sepsis: a primer. Crit Care Med. 2009;37:291–304.
57. Claypool SM. Cardiolipin, a critical determinant of mitochondrial carrier protein assembly and function. Biochim Biophys Acta. 2009;1788:2059–68.
58. Connor KM, Hempel N, Nelson KK, Dabiri G, Gamarra A, Balarmino J, van de Water L, Mian BM, Melendez JA. Manganese superoxide dismutase enhances the invasive and migratory activity of tumor cells. Cancer Res. 2007;67:10260–7.
59. Converso DP, Taille C, Carreras C, Jaitovich A, Poderoso JJ, Boczkowski J. HO-1 is located in liver mitochondria and modulates mitochondrial heme content and metabolism. FASEB J. 2006;20:E482–92.
60. Copin J-C, Gasche Y, Chan PH. Overexpression of copper/zinc superoxide dismutase does not prevent neonatal lethality in mutant mice that lack manganese superoxide dismutase. Free Radic Biol Med. 2000;28:1571–6.
61. Cortopassi G, Wang E. Modelling the effects of age-related mtDNA mutation accumulation; Complex I deficiency, superoxide and cell death. Biochim Biophys Acta. 1995;1271:171–6.
62. Cullen JJ, Weydert C, Hinkhouse MM, Ritchie J, Domann FE, Spitz D, Oberley LW. The role of manganese superoxide dismutase in the growth of pancreatic adenocarcinoma. Cancer Res. 2003;63:1297–303.
63. Daosukho C, Ittarat W, Lin S-M, Sawyer DB, Kiningham K, Lien Y-C, St. Clair DK. Induction of manganese superoxide dismutase (MnSOD) mediates cardioprotective effect of tamoxifen (TAM). J Mol Cell Cardiol. 2005;39:792–803.
64. Daosukho C, Kiningham K, Kasarskis EJ, Ittarat W, St. Clair DK. Tamoxifen enhancement of TNF-α induced MnSOD expression: modulation of NF-κB dimerization. Oncogene. 2002;21:3603–10.

65. Dasgupta J, Subbaram S, Connor KM, Rodriguez AM, Tirosh O, Beckman JS, Jourd'Heuil D, Melendez JA. Manganese superoxide dismutase protects from TNF-α-induced apoptosis by increasing the steady-state production of H_2O_2. Antioxid Redox Signal. 2006;8: 1295–305.
66. Davis CA, Hearn AS, Fletcher B, Bickford J, Garcia JE, Leveque V, Melendez JA, Silverman DN, Zucali J, Agarwal A, Nick HS. Potent anti-tumor effects of an active site mutant of human manganese-superoxide dismutase. Evolutionary conservation of product inhibition. J Biol Chem. 2004;279:12769–76.
67. Dhar SK, St Clair DK. Manganese superoxide dismutase regulation and cancer. Free Radic Biol Med. 2012;52:2209–22.
68. Dhar SK, Tangpong J, Chaiswing L, Oberley TD, St. Clair DK. Manganese superoxide dismutase is a *p53*-regulated gene that switches cancers between early and advanced stages. Cancer Res. 2011;71:6684–95.
69. Dhar SK, Xu Y, Chen Y, St. Clair DK. Specificity protein 1-dependent p53-mediated suppression of human manganese superoxide dismutase gene expression. J Biol Chem. 2006;281:21698–709.
70. Dhar SK, Xu Y, St. Clair DK. Nuclear factor κB- and specificity protein 1-dependent p53-mediated bi-directional regulation of the human manganese superoxide dismutase gene. J Biol Chem. 2010;285:9835–46.
71. Dhingra S, Feng W, Brown RE, Zhou Z, Khoury T, Zhang R, Tan D. Clinicopathologic significance of putative stem cell markers, CD44 and nestin, in gastric adenocarcinoma. Int J Clin Exp Pathol. 2011;4:733–41.
72. Dlaskova A, Hlavata L, Jezek P. Oxidative stress caused by blocking of mitochondrial complex I H^+ pumping as a link in aging/disease vicious cycle. Int J Biochem Cell Biol. 2008;40:1792–805.
73. Dominic EA, Ramezani A, Anker SD, Verma M, Mehta N, Rao M. Mitochondrial cytopathies and cardiovascular disease. Heart. 2014;100:611–8.
74. Dostert C, Petrilli V, van Bruggen R, Steele C, Mossman BT, Tschopp J. Innate immune activation through Nalp3 inflammasome sensing of asbestos and silica. Science. 2008;320:674–7.
75. Drahota Z, Chowdhury SKR, Floryk D, Mracek T, Wilhelm J, Rauchova H, Lenaz G, Houstek J. Glycerophosphate-dependent hydrogen peroxide production by brown adipose tissue mitochondria and its activation by ferricyanide. J Bioenerg Biomembr. 2002;34:105–13.
76. Droge W. Free radicals in the physiological control of cell function. Physiol Rev. 2002;82:47–95.
77. Duttaroy A, Paul A, Kundu M, Belton A. A *Sod2* null mutation confers severely reduced adult life span in drosophila. Genetics. 2003;165:2295–9.
78. Eaton JW, Qian M. Molecular bases of cellular iron toxicity. Free Radic Biol Med. 2002;32:833–40.
79. Epperly MW, Sikora CA, DeFilippi SJ, Gretton JE, Zhan Q, Kufe DW, Greenberger JS. Manganese superoxide dismutase (SOD2) inhibits radiation-induced apoptosis by stabilization of the mitochondrial membrane. Radiat Res. 2002;157:568–77.
80. Epperly MW, Tyurina YY, Nie S, Niu YY, Zhang X, Kagan VE, Greenberger JS. MnSOD-plasmid liposome gene therapy decreases ionizing irradiation-induced lipid peroxidation of the esophagus. In Vivo. 2005;19:997–1004.
81. Esworthy RS, Ho Y-S, Chu F-F. The Gpx1 gene encodes mitochondrial glutathione peroxidase in the mouse liver. Arch Biochem Biophys. 1997;340:59–63.
82. Fernandez MG, Troiano L, Moretti L, Nasi M, Pinti M, Salvioli S, Dobrucki J, Cossarizza A. Early changes in intramitochondrial cardiolipin distribution during apoptosis. Cell Growth Differ. 2002;13:449–55.
83. Fillmore N, Mori J, Lopaschuk GD. Mitochondrial fatty acid oxidation alterations in heart failure, ischaemic heart disease and diabetic cardiomyopathy. Br J Pharmacol. 2014;171:2080–90.

84. Fojta M, Kubicarova T, Vojtesek B, Palecek E. Effect of p53 protein redox states on binding to supercoiled and linear DNA. J Biol Chem. 1999;274:25749–55.
85. Folz RJ, Crapo JD. Extracellular superoxide dismutase (SOD3): tissue-specific expression, genomic characterization, and computer-assisted sequence analysis of the human EC SOD gene. Genomics. 1994;22:162–71.
86. Fong K-L, McCay PB, Poyer JL. Evidence for superoxide-dependent reduction of Fe^{3+} and its role in enzyme-generated hydroxyl radical formation. Chem Biol Interact. 1976;15:77–89.
87. Forman HJ, Fukuto JM, Torres M. Redox signaling: thiol chemistry defines which reactive oxygen and nitrogen species can act as second messengers. Am J Physiol Cell Physiol. 2004;287:C246–56.
88. Forman HJ, Kennedy J. Superoxide production and electron transport in mitochondrial oxidation of dihydroorotic acid. J Biol Chem. 1975;250:4322–6.
89. Forman HJ, Kennedy J. Dihydroorotate-dependent superoxide production in rat brain and liver: a function of the primary dehydrogenase. Arch Biochem Biophys. 1976;173:219–24.
90. Frankel D, Mehindate K, Schipper HM. Role of heme oxygenase-1 in the regulation of manganese superoxide dismutase gene expression in oxidatively-challenged astroglia. J Cell Physiol. 2000;185:80–6.
91. Fridovich I. The biology of oxygen radicals. Science. 1978;201:875–80.
92. Fridovich I. Superoxide dismutases. An adaptation to a paramagnetic gas. J Biol Chem. 1989;264:7761–4.
93. Fridovich I. Superoxide radical and superoxide dismutases. Annu Rev Biochem. 1995;64:97–112.
94. Friedland-Leuner K, Stockburger C, Denzer I, Eckert GP, Müller WE. Chapter Seven—Mitochondrial dysfunction: cause and consequence of Alzheimer's Disease. In: Heinz DO editor. Progress in molecular biology and translational science. Academic; 2014. pp. 183–210.
95. Fry M, Green DE. Cardiolipin requirement for electron transfer in complex I and III of the mitochondrial respiratory chain. J Biol Chem. 1981;256:1874–80.
96. Galanis A, Pappa A, Giannakakis A, Lanitis E, Dangaj D, Sandaltzopoulos R. Reactive oxygen species and HIF-1 signaling in cancer. Cancer Lett. 2008;266:12–20.
97. Galeotti T, Wohlrab H, Borrello S, De Leo ME. Messenger RNA for manganese and copper-zinc superoxide dismutases in hepatomas: correlation with degree of differentiation. Biochem Biophys Res Commun. 1989;165:581–9.
98. Garcia-Ramirez M, Francisco G, Garcia-Arumi E, Hernandez C, Martinez R, Andreu AL, Simo R. Mitochondrial DNA oxidation and manganese superoxide dismutase activity in peripheral blood mononuclear cells from type 2 diabetic patients. Diabetes Metab. 2008;34:117–24.
99. Gardner PR. Superoxide-driven aconitase FE-S center cycling. Biosci Rep. 1997;17:33–42.
100. Gardner PR, Raineri I, Epstein LB, White CW. Superoxide radical and iron modulate aconitase activity in mammalian cells. J Biol Chem. 1995;270:13399–405.
101. Garrett TA, Kordestani R, Raetz CRH. Quantification of cardiolipin by liquid chromatography-electrospray ionization mass spectrometry. In: Alen Brix H, editor. Methods in Enzymology. Amsterdam: Elsevier; 2007.
102. Garrido N, Griparic L, Jokitalo E, Wartiovaara J, van der Bliek AM, Spelbrink JN. Composition and dynamics of human mitochondrial nucleoids. Mol Biol Cell. 2003;14:1583–96.
103. Gaude E, Frezza C. Defects in mitochondrial metabolism and cancer. Cancer Metab. 2014;2:10.
104. Genova ML, Ventura B, Giuliano G, Bovina C, Formiggini G, Castelli GP, Lenaz G. The site of production of superoxide radical in mitochondrial complex I is not a bound ubisemiquinone but presumably iron-sulfur cluster N2. FEBS Lett. 2001;505:364–8.
105. Gonzalez F, Schug ZT, Houtkooper RH, MacKenzie ED, Brooks DG, Wanders RJA, Petit PX, Vaz FM, Gottlieb E. Cardiolipin provides an essential activating platform for caspase-8 on mitochondria. J Cell Biol. 2008;183:681–96.

106. Goossens V, Grooten J, De Vos K, Fiers W. Direct evidence for tumor necrosis factor-induced mitochondrial reactive oxygen intermediates and their involvement in cytotoxicity. Proc Natl Acad Sci U S A. 1995;92:8115–9.
107. Gorrini C, Harris IS, Mak TW. Modulation of oxidative stress as an anticancer strategy. Nat Rev Drug Discov. 2013;12:931–47.
108. Graziewicz MA, Day BJ, Copeland WC. The mitochondrial DNA polymerase as a target of oxidative damage. Nucleic Acids Res. 2002;30:2817–24.
109. Green DR, Reed JC. Mitochondria and apoptosis. Science. 1998;281:1309–12.
110. Gregory EM, Fridovich I. Oxygen toxicity and the superoxide dismutase. J Bacteriol. 1973;114:1193–7.
111. Gregory EM, Goscin SA, Fridovich I. Superoxide dismutase and oxygen toxicity in a eukaryote. J Bacteriol. 1974;117:456–60.
112. Grivennikova VG, Vinogradov AD. Generation of superoxide by the mitochondrial Complex I. Biochim Biophys Acta. 2006;1757:553–61.
113. Guda P, Guda C, Subramaniam S. Reconstruction of pathways associated with amino acid metabolism in human mitochondria. Genomics Proteomics Bioinformatics. 2007;5:166–76.
114. Gulbins E, Dreschers S, Bock J. Role of mitochondria in apoptosis. Exp Physiol. 2003;88:85–90.
115. Hainaut P, Milner J. Redox modulation of p53 conformation and sequence-specific DNA binding in vitro. Cancer Res. 1993;53:4469–73.
116. Han D, Canali R, Garcia J, Aguilera R, Gallaher TK, Cadenas E. Sites and mechanisms of aconitase inactivation by peroxynitrite: modulation by citrate and glutathione. Biochemistry. 2005;44:11986–96.
117. Hanukoglu I. Antioxidant protective mechanisms against reactive oxygen species (ROS) generated by mitochondrial P450 systems in steroidogenic cells. Drug Metab Rev. 2006;38:171–96.
118. Hanukoglu I, Rapoport R, Weiner L, Sklan D. Electron leakage from the mitochondrial NADPH-adrenodoxin reductase-adrenodoxin-P450scc (cholesterol side chain cleavage) system. Arch Biochem Biophys. 1993;305:489–98.
119. Hauff KD, Hatch GM. Cardiolipin metabolism and Barth syndrome. Prog Lipid Res. 2006;45:91–101.
120. Hausladen A, Fridovich I. Superoxide and peroxynitrite inactivate aconitases, but nitric oxide does not. J Biol Chem. 1994;269:29405–8.
121. Hempel N, Carrico PM, Melendez JA. Manganese superoxide dismutase (Sod2) and redox-control of signaling events that drive metastasis. Anticancer Agents Med Chem. 2011;11:191–201.
122. Herrero A, Barja G. Localization of the site of oxygen radical generation inside the complex I of heart and nonsynaptic brain mammalian mitochondria. J Bioenerg Biomembr. 2000;32:609–15.
123. Hirose K, Longo DI, Oppenheim JJ, Matsushima K. Overexpression of mitochondrial manganese superoxide dismutase promotes the survival of tumor cells exposed to interleukin-1, tumor necrosis factor, selected anticancer drugs, and ionizing radiation. FASEB J. 1993;7:361–8.
124. Hjalmarsson K, Marklund SL, Engstrom A, Edlund T. Isolation and sequence of complimentary DNA encoding human extracellular superoxide dismutase. Proc Natl Acad Sci U S A. 1987;84:6340–4.
125. Ho JC-m, Zheng S, Comhair SA, Farver C, Erzurum SC. Differential expression of manganese superoxide dismutase and catalase in lung cancer. Cancer Res. 2001;61:8578–85.
126. Holley AK, Bakthavatchalu V, Velez-Roman JM, St Clair DK. Manganese superoxide dismutase: guardian of the powerhouse. Int J Mol Sci. 2011;12:7114–62.
127. Holley AK, Kiningham KK, Spitz DR, Edwards DP, Jenkins JT, Moore MR. Progestin stimulation of manganese superoxide dismutase and invasive properties in T47D human breast cancer cells. J Steroid Biochem Mol Biol. 2009;117:23–30.

128. Houten SM, Wanders RJA. A general introduction to the biochemistry of mitochondrial fatty acid β-oxidation. J Inherit Metab Dis. 2010;33:469–77.
129. Hoye AT, Davoren JE, Wipf P, Fink MP, Kagan VE. Targeting mitochondria. Acc Chem Res. 2008;41:87–97.
130. Hsu CC, Lee HC, Wei YH. Mitochondrial DNA alterations and mitochondrial dysfunction in the progression of hepatocellular carcinoma. World J Gastroenterol. 2013;19:8880–6.
131. Hu H, Luo M-l, X-l D, Feng Y-B, Zhang Y, Shen X-M, Xu X, Cai Y, Y-l H, Wang M-R. Up-regulated manganese superoxide dismutase expression increases apoptosis resistance in human esophageal squamous cell carcinomas. Chin Med J. 2007;120:2092–8.
132. Hu Y, Rosen DG, Zhou Y, Feng L, Yang G, Liu J, Huang P. Mitochondrial manganese-superoxide dismutase expression in ovarian cancer: role in cell proliferation and response to oxidative stress. J Biol Chem. 2005;280:39485–92.
133. Huie RE, Padmaja S. The reaction of NO with superoxide. Free Radic Res Commun. 1993;18:195–9.
134. Hutson SM, Fenstermacher D, Mahar C. Role of mitochondrial transamination in branched chain amino acid metabolism. J Biol Chem. 1988;263:3618–25.
135. Ikegami T, Suzuki Y-I, Shimizu T, Isono K-I, Koseki H, Shirasawa T. Model mice for tissue-specific deletion of the manganese superoxide dismutase (MnSOD) gene. Biochem Biophys Res Commun. 2002;296:729–36.
136. Inagaki M, Yoshikawa E, Matsuoka Y, Sugawara Y, Nakano T, Akechi T, Wada N, Imoto S, Murakami K, Uchitomi Y, Group TBCSBMD. Smaller regional volumes of brain gray and white matter demonstrated in breast cancer survivors exposed to adjuvant chemotherapy. Cancer. 2007;109:146–56.
137. Indo HP, Davidson M, Yen H-C, Suenaga S, Tomita K, Nishii T, Higuchi M, Koga Y, Ozawa T, Majima HJ. Evidence of ROS generation by mitochondria in cells with impaired electron transport chain and mitochondrial DNA damage. Mitochondrion. 2007;7:106–18.
138. Izutani R, Asano S, Imano M, Kuroda D, Kato M, Ohyanagi H. Expression of manganese superoxide dismutase in esophageal and gastric cancers. J Gastroenterol. 1998;33:816–22.
139. Jahnke VE, Sabido O, Defour A, Castells J, Lefai E, Roussel D, Freyssenet D. Evidence for mitochondrial respiratory deficiency in rat rhabdomyosarcoma cells. PLoS One. 2010;5, e8637.
140. Jang YC, Perez VI, Song W, Lustgarten MS, Salmon AB, Mele J, Qi W, Liu Y, Liang H, Chaudhuri A, Ikeno Y, Epstein CJ, Van Remmen H, Richardson A. Overexpression of Mn superoxide dismutase does not increase life span in mice. J Gerontol A Biol Sci Med Sci. 2009;64A:1114–25.
141. Janssen AML, Bosman CB, van Duijn W, Oostendorp-van de Ruit MM, Kubben FJGM, Griffioen G, Lamers BBHW, van Krieken JHJM, van de Velde CJH, Verspaget HW. Superoxide dismutases in gastric and esophageal cancer and the prognostic impact in gastric cancer. Clin Cancer Res. 2000;6:3183–92.
142. Jiralerspong S, Ge B, Hudson TJ, Pandolfo M. Manganese superoxide dismutase induction by iron is impaired in Friedreich ataxia cells. FEBS Lett. 2001;509:101–5.
143. Joshi G, Sultana R, Tangpong J, Cole MP, St. Clair DK, Vore M, Estus S, Butterfield DA. Free radical mediated oxidative stress and toxic side effects in brain induced by the anticancer drug adriamycin: insight into chemobrain. Free Radic Res. 2005;29:1147–54.
144. Kabe Y, Ando K, Hirao S, Yoshida M, Handa H. Redox regulation of NF-kappaB activation: distinct redox regulation between the cytoplasm and the nucleus. Antioxid Redox Signal. 2005;7:395–403.
145. Kang D, Hamasaki N. Alterations of mitochondrial DNA in common diseases and disease states: aging, neurodegeneration, heart failure, diabetes, and cancer. Curr Med Chem. 2005;12:429–41.
146. Kang YJ, Sun X, Chen Y, Zhou Z. Inhibition of doxorubicin chronic toxicity in catalase-overexpressing transgenic mouse hearts. Chem Res Toxicol. 2002;15:1–6.
147. Kattan Z, Minig V, Leroy P, Dauca M, Becuwe P. Role of manganese superoxide dismtuase on growth and invasive properties of human estrogen-independent breast cancer cells. Breast Cancer Res Treat. 2008;108:203–15.

148. Kawai T, Akira S. The roles of TLRs, RLRs and NLRs in pathogen recognition. Int Immunol. 2009;21:317–37.
149. Kazi TG, Afridi HI, Kazi N, Jamali MK, Arain MB, Jalbani N, Kandhro GA. Copper, chromium, manganese, iron, nickel, and zinc levels in biological samples of diabetes mellitus patients. Biol Trace Elem Res. 2008;122:1–18.
150. Keele Jr BB, McCord JM, Fridovich I. Superoxide dismutase from *Escherichia coli* B. A new manganese-containing enzyme. J Biol Chem. 1970;245:6176–81.
151. Keeney PM, Xie J, Capaldi RA, Bennett Jr JP. Parkinson's disease brain mitochondrial complex I has oxidatively damaged subunits and is functionally impaired and misassembled. J Neurosci. 2006;26:5256–64.
152. Keller JN, Kindy MS, Holtsberg FW, St. Clair DK, Yen H-C, Germeyer A, Steiner SM, Bruce-Keller AJ, Hutchins JB, Mattson MP. Mitochondrial manganese superoxide dismutase prevents neural apoptosis and reduces ischemic brain injury: suppression of peroxynitrite production, lipid peroxidation, and mitochondrial dysfunction. J Neurosci. 1998;18:687–97.
153. Kelner MJ, Montoya MA. Structural organization of the human glutathione reductase gene: determination of correct cDNA sequence and identification of a mitochondrial leader sequence. Biochem Biophys Res Commun. 2000;269:366–8.
154. Khoshjou F, Dadras F. Mitochondrion and its role in diabetic nephropathy. Iran J Kidney Dis. 2014;8:355–8.
155. Kiebish MA, Han X, Cheng H, Chuang JH, Seyfried TN. Cardiolipin and electron transport chain abnormalities in mouse brain tumor mitochondria: lipidomic evidence supporting the Warburg theory of cancer. J Lipid Res. 2008;49:2545–56.
156. Kienhofer J, Haussler DJF, Ruckelshausen F, Muessig E, Weber K, Pimentel D, Ullrich V, Burkle A, Bachschmid MM. Association of mitochondrial antioxidant enzymes with mitochondrial DNA as integral nucleoid constituents. FASEB J. 2009;23:2034–44.
157. Kiningham KK, Oberley TD, Lin S-M, Mattingly CA, St. Clair DK. Overexpression of manganese superoxide dismutase protects against mitochondrial-initiated poly(ADP-ribose) polymerase-mediated cell death. FASEB J. 1999;13:1601–10.
158. Kiningham KK, St. Clair DK. Overexpression of manganese superoxide dismutase selectively modulates the activity of Jun-associated transcription factors in fibrosarcoma cells. Cancer Res. 1997;57:5265–71.
159. Klivenyi P, St Clair D, Wermer M, Yen HC, Oberley T, Yang L, Flint BM. Manganese superoxide dismutase overexpression attenuates MPTP toxicity. Neurobiol Dis. 1998;5:253–8.
160. Koehler CM, Beverley KN, Leverich EP. Redox pathways of the mitochondrion. Antioxid Redox Signal. 2006;8:813–22.
161. Kowluru RA, Kowluru V, Xiong Y, Ho YS. Overexpression of mitochondrial superoxide dismutase in mice protects the retina from diabetes-induced oxidative stress. Free Radic Biol Med. 2006;41:1191–6.
162. Kushnareva Y, Murphy AN, Andreyev A. Complex I-mediated reactive oxygen species generation: modulation by cytochrome *c* and $NAD(P)^+$ oxidation-reduction state. Biochem J. 2002;368:545–53.
163. Kwak HB, Lee Y, Kim JH, Van Remmen H, Richardson AG, Lawler JM. MnSOD Overexpression reduces fibrosis and pro-apoptotic signaling in the aging mouse heart. J Gerontol A Biol Sci Med Sci. 2015;70(5):533–44.
164. Landriscina M, Remiddi F, Ria F, Palazzotti B, De Leo ME, Iacoangeli M, Rosselli R, Scerrati M, Galeotti T. The level of MnSOD is directly correlated with grade of brain tumours of neuroepithelial origin. Br J Cancer. 1996;74:1877–85.
165. Larosche I, Letteron P, Berson A, Fromenty B, Huang T-T, Moreau R, Pessayre D, Mansouri A. Hepatic mitochondrial DNA depletion after an alcohol binge in mice: probable role of peroxynitrite and modulation by manganese superoxide dismutase. J Pharmacol Exp Ther. 2010;332:886–97.
166. Lebovitz RM, Zhang H, Vogel H, Cartwright Jr J, Dionne L, Lu N, Huang S, Matzuk MM. Neurodegeneration, mycardial injury, and perinatal death in mitochondrial superoxide dismutase-deficient mice. Proc Natl Acad Sci U S A. 1996;93:9782–7.

167. Lee HC, Huang KH, Yeh TS, Chi CW. Somatic alterations in mitochondrial DNA and mitochondrial dysfunction in gastric cancer progression. World J Gastroenterol. 2014;20:3950–9.
168. Lee S-M, Lee Y-S, Choi J-H, Park S-G, Choi I-W, Joo Y-D, Lee W-S, Lee J-N, Choi I, Seo S-K. Tryptophan metabolite 3-hydroxyanthranilic acid selectively induces activated T cell death via intracellular GSH depletion. Immunol Lett. 2010;132:53–60.
169. Lee S-R, Kim J-R, Kwon K-S, Yoon HW, Levine RL, Ginsburg A, Rhee SG. Molecular cloning and characterization of a mitochondrial selenocysteine-containing thioredoxin reductase from rat liver. J Biol Chem. 1999;274:4722–34.
170. Legros F, Malka F, Frachon P, Lombes A, Rojo M. Organization and dynamics of human mitochondrial DNA. J Cell Sci. 2004;117:2653–62.
171. Lemarie A, Grimm S. Mitochondrial respiratory chain complexes: apoptosis sensors mutated in cancer. Oncogene. 2011;30(38):3985–4003.
172. Lenaz G. The mitochondrial production of reactive oxygen species: mechanisms and implications in human pathology. IUBMB Life. 2001;52:159–64.
173. Lesnefsky EJ, Minkler P, Hoppel CL. Enhanced modification of cardiolipin during ischemia in the aged heart. J Mol Cell Cardiol. 2009;46:1008–15.
174. Levi S, Rovida E. The role of iron in mitochondrial function. Biochim Biophys Acta. 2009;1790:629–36.
175. Levine RL, Stadtman ER. Oxidative modification of proteins during aging. Exp Gerontol. 2001;36:1495–502.
176. Li N, Oberley TD, Oberley LW, Zhong W. Overexpression of manganese superoxide dismutase in DU145 human prostate carcinoma cells has multiple effects on cell phenotype. Prostate. 1998;35:221–33.
177. Li S, Yan T, Yang J-Q, Oberley TD, Oberley LW. The role of cellular glutathione peroxidase redox regulation in the suppression of tumor cell growth by manganese superoxide dismutase. Cancer Res. 2000;60:3927–39.
178. Li Y, Huang T-T, Carlson EJ, Melov S, Ursell PC, Olson JL, Noble LJ, Yoshimura MP, Berger C, Chan PH, Wallace DC, Epstein CJ. Dilated cardiomyopathy and neonatal lethality in mutant mice lacking manganese superoxide dismutase. Nat Genet. 1995;11:376–81.
179. Li YY, Maisch B, Rose ML, Hengstenberg C. Point Mutations in Mitochondrial DNA of Patients with Dilated Cardiomyopathy. J Mol Cell Cardiol. 1997;29:2699–709.
180. Liu R, Oberley TD, Oberley LW. Transfection and expression of MnSOD cDNA decreases tumor malignancy of human oral sqamous carcinoma SCC-25 cells. Hum Gene Ther. 1997;8:585–95.
181. Longo VD, Liou L-L, Valentine JS, Gralla EB. Mitochondrial superoxide decreases yeast survival in stationary phase. Arch Biochem Biophys. 1999;365:131–42.
182. Lu J, Holmgren A. The thioredoxin antioxidant system. Free Radic Biol Med. 2014;66:75–87.
183. Lu J, Sharma LK, Bai Y. Implications of mitochondrial DNA mutations and mitochondrial dysfunction in tumorigenesis. Cell Res. 2009;19:802–15.
184. Lu S-P, Feng M-HL, Huang H-L, Huang Y-C, Tsou W-I, Lai M-Z. Reactive oxygen species promote raft formation in T lymphocytes. Free Radic Biol Med. 2007;42:936–44.
185. Lustgarten MS, Jang YC, Liu Y, Muller FL, Qi W, Steinhelper M, Brooks SV, Larkin L, Shimizu T, Shirasawa T, McManus LM, Bhattacharya A, Richardson A, Van Remmen H. Conditional knockout of Mn-SOD targeted to type IIB skeletal muscle fibers increases oxidative stress and is sufficient to alter aerobic exercise capacity. Am J Physiol Cell Physiol. 2009;297:C1520–32.
186. Lustgarten MS, Jang YC, Liu Y, Qi W, Qin Y, Dahia PL, Shi Y, Bhattacharya A, Muller FL, Shimizu T, Shirasawa T, Richardson A, Van Remmen H. MnSOD deficiency results in elevated oxidative stress and decreased mitochondrial function but does not lead to muscle atrophy during aging. Aging Cell. 2011;10:493–505.
187. MacMillan-Crow LA, Cruthirds DL, Ahki KM, Sanders PW, Thompson JA. Mitochondrial tyrosine nitration precedes chronic allograft nephropathy. Free Radic Biol Med. 2001; 31:1603–8.

188. MacMillan-Crow LA, Thompson JA. Tyrosine modifications and inactivation of active site manganese superoxide dismutase mutant (Y34F) by peroxynitrite. Arch Biochem Biophys. 1999;366:82–8.
189. Madsen-Bouterse SA, Zhong Q, Mohammad G, Ho Y-S, Kowluru RA. Oxidative damage of mitochondrial DNA in diabetes and its protection by manganese superoxide dismutase. Free Radic Res. 2010;44:313–21.
190. Maiorino M, Scapin M, Ursini F, Biasolo M, Bosello V, Flohe L. Distinct promoters determine alternative transcription of gpx-4 into phospholipid-hydroperoxide glutathione peroxidase variants. J Biol Chem. 2003;278:34286–90.
191. Majno G, Joris I. Apoptosis, oncosis, and necrosis. An overview of cell death. Am J Pathol. 1995;146:3–15.
192. Malafa M, Margenthaler J, Webb B, Neitzel L, Christophersen M. MnSOD expression is increased in metastatic gastric cancer. J Surg Res. 2000;88:130–4.
193. Malecki EA, Greger JL. Manganese protects against heart mitochondrial lipid peroxidation in rats fed high levels of polyunsaturated fatty acids. J Nutr. 1996;126:27–33.
194. Mambo E, Gao X, Cohen Y, Guo Z, Talalay P, Sidransky D. Electrophile and oxidant damage of mitochondrial DNA leading to rapid evolution of homplasmic mutations. Proc Natl Acad Sci. 2003;100:1838–43.
195. Mansouri A, Tarhuni A, Larosche I, Reyl-Desmars F, Demeilliers C, Degoul F, Nahon P, Sutton A, Moreau R, Fromenty B, Ressayre D. MnSOD overexpression prevents liver mitochondrial DNA depletion after an alcohol binge but worsens this effect after prolonged alcohol consumption in mice. Dig Dis. 2010;28:756–75.
196. Marecki JC, Parajuli N, Crow JP, MacMillan-Crow LA. The use of the Cre/loxP system to study oxidative stress in tissue-specific manganese superoxide dismutase knockout models. Antioxid Redox Signal. 2014;20:1655–70.
197. Martin FM, Xu X, von Lohneysen K, Gilmartin TJ, Friedman J. SOD2 deficient erythroid cells up-regulate transferrin receptor and down-regulate mitochondrial biogenesis and metabolism. PLoS One. 2011;6, e16894.
198. Martinon F. Signaling by ROS drives inflammasome activation. Eur J Immunol. 2010;40:616–9.
199. Martinon F, Mayor A, Tschopp J. The inflammasomes: guardians of the body. Annu Rev Immunol. 2009;27:229–65.
200. Mason RP. Using anti-5,5-dimethyl-1-pyrroline N-oxide (anti-DMPO) to detect protein radicals in time and space with immuno-spin trapping. Free Radic Biol Med. 2004;36:1214–23.
201. Mattson MP, Goodman Y, Luo H, Fu W, Furukawa K. Activation of NF-κB protects hippocampal neurons against oxidative stress-induced apoptosis: evidence for induction of manganese superoxide dismutase and suppression of peroxynitrite production and protein tyrosine nitration. J Neurosci Res. 1997;49:681–97.
202. McLennan HR, Degli EM. The contribution fo mitochondrial respiratory complexes to the production of reactive oxygen species. J Bioenerg Biomembr. 2000;32:153–62.
203. McMillin JB, Dowhan W. Cardiolipin and apoptosis. Biochim Biophys Acta. 2002;1585:97–107.
204. Meissner F, Seger RA, Moshous S, Fischer A, Reichenbach J, Zychlinsky A. Inflammasome activation in NADPH oxidase defective mononuclear phagocytes from patients with chronic granulomatous disease. Blood. 2010;116:1570–3.
205. Miao L, St. Clair DK. Regulation of superoxide dismutase genes: implications in disease. Free Radic Biol Med. 2009;47(4):344–56.
206. Minotti G, Menna P, Salvatorelli E, Cairo G, Gianni L. Anthracyclines: molecular advances and pharmacologic developments in antitumor activity and cardiotoxicity. Pharmacol Rev. 2004;56:185–229.
207. Miriyala S, Spasojevic I, Tovmasyan A, Salvemini D, Vujaskovic Z, St Clair D, Batinic-Haberle I. Manganese superoxide dismutase, MnSOD and its mimics. Biochim Biophys Acta. 2012;1822:794–814.

208. Misawa H, Nakata K, Matsuura J, Moriwaki Y, Kawashima K, Shimizu T, Shirawawa T, Takahashi R. Conditional knockout of Mn superoxide dismutase in postnatal motor neurons reveals resistance to mitochondrial generated superoxide radicals. Neurobiol Dis. 2006;23:169–77.
209. Misra HP, Fridovich I. Purification and properties of superoxide dismutase from a red alga, Porphyridium cruentum. J Biol Chem. 1977;252:6421–3.
210. Miwa S, St.-Pierre J, Partridge L, Brand MD. Superoxide and hydrogen peroxide production by *Drosophila* mitochondria. Free Radic Biol Med. 2003;35:938–48.
211. Miyake A, Higashijima S-I, Kobayashi D, Narita T, Jindo T, Setiamarga DHE, Ohisa S, Orihara N, Hibiya K, Konno S, Sakaguchi S, Horie K, Imai Y, Naruse K, Kudo A, Takeda H. Mutation in the *abcb7* gene causes abnormal iron and fatty acid metabolism in developing medaka fish. Develop Growth Differ. 2008;50:703–16.
212. Mohr A, Buneker C, Gough RP, Zwacka RM. MnSOD protects colorectal cancer cells from TRAIL-induced apoptosis by inhibition of Smac/DIABLO release. Oncogene. 2008; 27(6):763–74.
213. Monteiro-Cardoso VF, Oliveira MM, Melo T, Domingues MR, Moreira PI, Ferreiro E, Peixoto F, Videira RA. Cardiolipin profile changes are associated to the early synaptic mitochondrial dysfunction in Alzheimer's disease. J Alzheimers Dis. 2015;43(4):1375–92.
214. Monteiro JP, Oliveira PJ, Jurado AS. Mitochondrial membrane lipid remodeling in pathophysiology: a new target for diet and therapeutic interventions. Prog Lipid Res. 2013; 52:513–28.
215. Morgan MJ, Lehmann M, Schwarzlander M, Baxter CJ, Sienkiewicz-Porzucek A, Williams TCR, Schauer N, Fernie AR, Fricker MD, Ratcliffe RG, Sweetlove LJ, Finkemeier I. Decrease in manganese superoxide dismutase leads to reduced root growth and affects tricarboxylic acid cycle flux and mitochondrial redox homeostasis. Plant Physiol. 2008;147:101–14.
216. Moulian N, Truffault F, Gaudry-Talarmain YM, Serraf A, Berrih-Aknin S. In vivo and in vitro apoptosis of human thymocytes are associated with nitrotyrosine formation. Blood. 2008;97:3521–30.
217. Mukherjee S, Forde R, Belton A, Duttaroy A. SOD2, the principal scavenger of mitochondrial superoxide, is dispensable for embryogenesis and imaginal tissue development but essential for adult survival. Fly. 2011;5:39–46.
218. Muller FL, Liu Y, Van Remmen H. Complex III releases superoxide to both sides of the inner mitochondrial membrane. J Biol Chem. 2004;279:49064–73.
219. Murphy MP. How mitochondria produce reactive oxygen species. Biochem J. 2009;417:1–13.
220. Murray J, Taylors SW, Zhang B, Ghosh SS, Capaldi RA. Oxidative damage to mitochondrial complex I due to peroxynitrite. Identification of reactive tyrosines by mass spectrometry. J Biol Chem. 2003;278:37223–30.
221. Nadtochiy SM, Baker PRS, Freeman BA, Brooks PS. Mitochondrial nitroalkene formation and mild uncoupling in ischaemic preconditioning: implications for cardioprotection. Cardiovasc Res. 2009;82:333–40.
222. Nahon P, Charnaux N, Friand V, Prost-Squarcioni C, Zoil M, Lievre N, Trinchet J-C, Beaugrand M, Gattegno L, Pessayre D, Sutton A. The manganese superoxide dismutase Ala16Val dimorphism modulates iron accumulation in human hepatoma cells. Free Radic Biol Med. 2008;45:1308–17.
223. Nakagawa Y. Initiation of apoptotic signal by the peroxidation of cardiolipin of mitochondria. Ann N Y Acad Sci. 2004;1011:177–84.
224. Nakahira K, Hapsel JA, Rathinam VAK, Lee S-J, Dolinay T, Lam HC, Englert JA, Rabinovitch M, Cernadas M, Kim HP, Fitzgerald KA, Ryter SW, Choi AMK. Autophagy proteins regulate innate immune responses by inhibiting the release of mitochondrial DNA mediated by the NALP3 inflammasome. Nat Immunol. 2011;8:222–30.
225. Napier I, Ponka P, Richardson DR. Iron trafficking in the mitochondrion: novel pathways revealed by disease. Blood. 2005;105:1867–74.

226. Nelson CJ, Nandy N, Roth AJ. Chemotherapy and cognitive deficits: mechanisms, findings, and potential interventions. Palliat Support Care. 2007;5:273–80.
227. Nelson KK, Ranganathan AC, Mansouri J, Rodriguez AM, Providence KM, Rutter JL, Pumiglia K, Bennett JA, Melendez JA. Elevated *Sod2* activity augments matrix metalloproteinase expression: evidence for the involvement of endogenous hydrogen peroxide in regulating metastasis. Clin Cancer Res. 2003;9:424–32.
228. Nohl H, Jordan W. The metabolic fate of mitochondrial hydrogen peroxide. Eur J Biochem. 1980;111:203–10.
229. Nozoe T, Honda M, Inutsuka S, Yasuda M, Korenaga D. Significance of immunohistochemical expression of manganese superoxide dismutase as a marker of malignant potential in colorectal carcinoma. Oncol Rep. 2003;10:39–43.
230. O'Donnell VB, Eiserich JP, Bloodsworth A, Chumley PH, Kirk M, Barnes S, Darley-Usmar VM, Freeman BA. Nitration of unsaturated fatty acids by nitric oxide-derived reactive species. Methods Enzymol. 1999;301:454–70.
231. Oberley LW. Mechanism of the tumor suppressive effect of MnSOD overexpression. Biomed Pharmacother. 2005;59:143–8.
232. Oberley LW, Buettner GR. Role of superoxide dismutase in cancer: a review. Cancer Res. 1979;39:1141–9.
233. Oberley TD, Verwiebe E, Zhong W, Kang SW, Rhee SG. Localization of the thioredoxin system in normal rat kidney. Free Radic Biol Med. 2001;30:412–24.
234. Oh SS, Sullivan KA, Wilkinson JE, Backus C, Hayes JM, Sakowski SA, Feldman EL. Neurodegeneration and early lethality in superoxide dismutase 2-deficient mice: a comprehensive analysis of the central and peripheral nervous systems. Neuroscience. 2012; 212:201–13.
235. Ohnishi T. Thermodynamic and EPR characterization of iron-sulfur centers in the NADH-ubiquinone segment of the mitochondrial respiratory chain in pigeon heart. Biochim Biophys Acta. 1975;387:475–90.
236. Ohnishi T. Iron-sulfur clusters/semiquinones in complex I. Biochim Biophys Acta. 1998;1364:186–206.
237. Ohnishi T, Tamai I, Sakanaka K, Sakata A, Yamashima T, Yamashita J, Tsuji A. *In vivo* and *in vitro* evidence for ATP-dependency of p-glycoprotein-mediated efflux of doxorubicin at the blood-brain barrier. Biochem Pharmacol. 1995;49:1541–4.
238. Ohtsuki T, Matsumoto M, Suzuki K, Taniguchi N, Kamada T. Mitochondrial lipid peroxidation and superoxide dismutase in rat hypertensive target organs. Am J Physiol Heart Circ Physiol. 1995;37:H1418–21.
239. Oida T, Suzuki K, Nanno M, Kanamori Y, Saito H, Kubota E, Kato S, Itoh M, Kaminogawa S, Ishikawa H. Role of gut cryptopatches in early extrathymic maturation of intestinal intraepithelial T cells. J Immunol. 2000;164:3616–26.
240. Okado-Matsumoto A, Fridovich I. Subcellular distribution of superoxide dismutases (SOD) in rat liver. Cu, Zn-SOD in mitochondria. J Biol Chem. 2001;276:38388–93.
241. Owusu-Ansah E, Banerjee U. Reactive oxygen species prime *Drosophila* haematopoietic progenitors for differentiation. Nature. 2009;461:537–41.
242. Padmaja S, Squadrito GL, Pryor WA. Inactivation of glutathione peroxidase by peroxynitrite. Arch Biochem Biophys. 1998;349:1–6.
243. Palazzotti B, Pani G, Colavitti R, de Leo ME, Bedogni B, Borrello S, Galeotti T. Increased growth capacity of cervical-carcinoma cells over-expressing manganous superoxide dismutase. Int J Cancer. 1999;82:145–50.
244. Panfili E, Sandri G, Ernster L. Distribution of glutathione peroxidases and glutathione reductase in rat brain mitochondria. FEBS Lett. 1991;290:35–7.
245. Pani G, Galeotti T, Chiarugi P. Metastasis: cancer cell's escape from oxidative stress. Cancer Metastasis Rev. 2010;29:351–78.
246. Paradies G, Petrosillo G, Paradies V, Ruggiero FM. Role of cardiolipin peroxidation and Ca^{2+} in mitochondrial dysfunction and disease. Cell Calcium. 2009;45:643–50.

247. Paradies G, Petrosillo G, Paradies V, Ruggiero FM. Oxidative stress, mitochondrial bioenergetics, and cardiolipin in aging. Free Radic Biol Med. 2010;48:1286–95.
248. Paradies G, Petrosillo G, Pistolese M, Di Venosa N, Federici A, Ruggiero FM. Decrease in mitochondrial complex I activity in ischemic/reperfused rat heart. Involvement of reactive oxygen species and cardiolipin. Circ Res. 2004;94:53–9.
249. Paradies G, Petrosillo G, Pistolese M, Ruggiero FM. The effect of reactive oxygen species generated from the mitochondrial electron transport chain on the cytochrome *c* oxidase activity and on the cardiolipin content in bovine heart submitochondrial particles. FEBS Lett. 2000;466:323–6.
250. Paradies G, Petrosillo G, Pistolese M, Ruggiero FM. Reactive oxygen species generated by the mitochondrial respiratory chain affect the complex III activity via cardiolipin peroxidation in beef-heart submitochondrial particles. Mitochondrion. 2001;1:151–9.
251. Paradies G, Petrosillo G, Pistolese M, Ruggiero FM. Reactive oxygen species affect mitochondrial electron transport complex I activity through oxidative cardiolipin damage. Gene. 2002;286:135–41.
252. Parajuli N, Marine A, Simmons S, Saba H, Mitchell T, Shimizu T, Shirasawa T, MacMillan-Crow LA. Generation and characterization of a novel kidney-specific manganese superoxide dismutase knockout mouse. Free Radic Biol Med. 2011;51:406–16.
253. Pardo M, Melendez JA, Tirosh O. Manganese superoxide dismutase inactivation during Fas (CD95)-mediated apoptosis in Jurkat T cells. Free Radic Biol Med. 2006;41:1795–806.
254. Park SY, Chang I, Kim JY, Kang SW, Park SH, Singh K, Lee MS. Resistance of mitochondrial DNA-depleted cells against cell death: role of mitochondrial superoxide dismutase. J Biol Chem. 2004;279:7512–20.
255. Patel HH, McDonough AA. Of mice and men: modeling cardiovascular complexity in diabetes. Focus on "Mitochondrial inefficiencies and anoxic ATP hydrolysis capacities in diabetic rat heart". Am J Physiol Cell Physiol. 2014;307:C497–8.
256. Paul A, Belton A, Nag S, Martin I, Grotewiel MS, Duttaroy A. Reduced mitochondrial SOD displays mortality characteristics reminiscent of natural aging. Mech Ageing Dev. 2007;128:706–16.
257. Paulsen CE, Carroll KS. Cysteine-mediated redox signaling: chemistry, biology, and tools for discovery. Chem Rev. 2013;113:4633–79.
258. Pearce LL, Epperly MW, Greenberger JS, Pitt BR, Peterson J. Identification of respiratory complexes I and III as mitochondrial sites of damage following exposure to ionizing radiation and nitric oxide. Nitric Oxide Biol Chem. 2001;5:128–36.
259. Perez VI, Van Remmen H, Bokov A, Epstein CJ, Vijg J, Richardson A. The overexpression of major antioxidant enzymes does not extend the lifespan of mice. Aging Cell. 2009;8:73–5.
260. Petrosillo G, Casanova G, Matera M, Ruggiero FM, Paradies G. Interaction of peroxidized cardiolipin with rat-heart mitochondrial membranes: induction of permeability transition and cytochrome *c* release. FEBS Lett. 2006;580:6311–6.
261. Petrosillo G, Ruggiero FM, Pistolese M, Paradies G. Reactive oxygen species generated from the mitochondrial electron transport chain induce cytochrome *c* dissociation from beef-heart submitochondrial particles via cardiolipin peroxidation. Possible role in the apoptosis. FEBS Lett. 2001;509:435–8.
262. Pfeiffer K, Gohil V, Stuart RA, Hunte C, Brandt U, Greenberg ML, Schagger H. Cardiolipin stabilizes respiratory chain supercomplexes. J Biol Chem. 2003;278:52873–80.
263. Picone P, Nuzzo D, Caruana L, Scafidi V, Di Carlo M. Mitochondrial dysfunction: different routes to Alzheimer's disease therapy. Oxid Med Cell Longev. 2014;2014:780179.
264. Piganelli JD, Flores SC, Cruz C, Koepp J, Batinic-Haberle I, Crapo J, Day B, Kachadourian R, Young R, Bradley B, Haskins K. A metalloporphyrin-based superoxide dismutase mimic inhibits adoptive transfer of autoimmune diabetes by a diabetogenic T-cell clone. Diabetes. 2002;51:347–55.
265. Pinkham JL, Wang Z, Alsina J. Heme regulates SOD2 transcription by activation and repression in Saccharomyces cerevisiae. Curr Genet. 1997;31:281–91.

266. Poli G, Leonarduzzi G, Biasi F, Chiarpotto E. Oxidative stress and cell signaling. Curr Med Chem. 2004;11:1163–82.
267. Pondarre C, Antiochos BB, Campagna DR, Clarke SL, Greer EL, Deck KM, McDonald A, Han A-P, Medlock A, Kutok JL, Anderson SA, Einsenstein RS, Fleming MD. The mitochondrial ATP-binding cassette transporter Abcb7 is essential in mice and participates in cytosolic iron-sulfur cluster biogenesis. Hum Mol Genet. 2006;15:953–64.
268. Powell CS, Jackson RM. Mitochondrial complex I, aconitase, and succinate dehydrogenase during hypoxia-reoxygenation: modulation of enzyme activities by MnSOD. Am J Physiol Lung Cell Mol Physiol. 2003;285:L189–98.
269. Powis G, Kirkpatrick DL. Thioredoxin signaling as a target for cancer therapy. Curr Opin Pharmacol. 2007;7:392–7.
270. Quiros-Gonzalez I, Sainz RM, Hevia D, Mayo JC. MnSOD drives neuroendocrine differentiation, androgen independence, and cell survival in prostate cancer cells. Free Radic Biol Med. 2011;50:525–36.
271. Quiros I, Sainz RM, Hevia D, Garcia-Suarez O, Astudillo A, Rivas M, Mayo JC. Upregulation of manganese superoxide dismutase (SOD2) is a common pathway for neuroendocrine differentiation in prostate cancer cells. Int J Cancer. 2009;125(7):1497–504.
272. Radi R, Beckman JS, Bush KM, Freeman BA. Peroxynitrite-induced membrane lipid peroxidation: the cytotoxic potential of superoxide and nitric oxide. Arch Biochem Biophys. 1991;288:481–7.
273. Radi R, Beckman JS, Bush KM, Freeman BA. Peroxynitrite oxidation of sulfhydryls. The cytotoxic potential of superoxide and nitric oxide. J Biol Chem. 1991;266:4244–50.
274. Radi R, Cassina A, Hodara R, Quijano C, Castro L. Peroxynitrite reactions and formation in mitochondria. Free Radic Biol Med. 2002;33:1451–64.
275. Radi R, Rodriquez M, Castro L, Telleri R. Inhibition of mitochondrial electron transport by peroxynitrite. Arch Biochem Biophys. 1994;308:89–95.
276. Radi R, Turrens JF, Chang LY, Bush KM, Crapo JD, Freeman BA. Detection of catalase in rat heart mitochondria. J Biol Chem. 1991;266:22028–34.
277. Raha S, Robinson BH. Mitochondria, oxygen free radicals, disease, and ageing. Trends Biochem Sci. 2000;25:502–8.
278. Raha S, Robinson BH. Mitochondria, oxygen free radicals, and apoptosis. Am J Med Genet. 2001;106:62–70.
279. Ranganathan AC, Nelson KK, Rodriguez AM, Kim K-H, Tower GB, Rutter JL, Brinckerhoff CE, Huang T-T, Epstein CJ, Jeffrey JJ, Melendez JA. Manganese superoxide dismutase signals matrix metalloproteinase expression via H_2O_2-dependent ERK1/2 activation. J Biol Chem. 2001;276:14264–70.
280. Ravindranath SD, Fridovich I. Isolation and characterization of a manganese-containing superoxide dismutase from yeast. J Biol Chem. 1975;250:6107–12.
281. Reed JC. Mechanisms of apoptosis. Am J Pathol. 2000;157:1415–30.
282. Reynier M, Sari H, d'Anglebermes M, Kye EA, Pasero L. Differences in lipid characteristics of undifferentiated and enterocytic-differentiated HT29 human colonic cells. Cancer Res. 1991;51:1270–7.
283. Rhee SG, Chang T-S, Bae YS, Lee S-R, Kang SW. Cellular regulation by hydrogen peroxide. J Am Soc Nephrol. 2003;14:S211–5.
284. Richardson DR, Lane DJR, Becker EM, Huang ML-H, Witnall M, Rahmanto YS, Sheftel AD, Ponka P. Mitochondrial iron trafficking and the integration of metabolism between the mitochondrion and cytosol. Proc Natl Acad Sci. 2010;107:10775–82.
285. Richter C, Park J-W, Ames BN. Normal oxidative damage to mitochondrial and nuclear DNA is extensive. Proc Natl Acad Sci U S A. 1988;85:6465–7.
286. Ridnour LA, Oberley TD, Oberley LW. Tumor suppressive effects of MnSOD overexpression may involve imbalance in peroxide generation versus peroxide removal. Antioxid Redox Signal. 2004;6:501–12.
287. Riobo NA, Clementi E, Melani M, Boveris A, Cadenas E, Moncada S, Poderoso JJ. Nitric oxide inhibits mitochondrial NADH:ubiquinone reductase activity through peroxynitrite formation. Biochem J. 2001;359:139–45.

288. Roessler MM, King MS, Robinson AJ, Armstrong FA, Harmer J, Hirst J. Direct assignment of EPR spectra to structurally defined iron-sulfur clusters in complex i by double electron-electron resonance. Proc Natl Acad Sci U S A. 2010;107:1930–5.
289. Rotig A, de Lonlay P, Chretien D, Foury F, Koenig M, Sidi D, Munnich A, Rustin P. Aconitase and mitochondrial iron-sulphur protein deficiency in Friedreich ataxia. Nat Genet. 1997;17:215–7.
290. Rouault TA, Tong W-H. iron-sulphur cluster biogenesis and mitochondrial iron homeostasis. Nat Rev Mol Cell Biol. 2005;6:345–51.
291. Roy S, Nicholson DW. Cross-talk in cell death signaling. J Exp Med. 2000;192:F21–5.
292. Rubbo H, Trostchansky A, O'Donnell VB. Peroxynitrite-mediated lipid oxidation and nitration: mechanisms and consequences. Arch Biochem Biophys. 2009;484:167–72.
293. Salvi M, Battaglia V, Brunati AM, La Rocca N, Tibaldi E, Pietrangeli P, Marcocci L, Mondovi B, Rossi CA, Toninello A. Catalase takes part in rat liver mitochondria oxidative stress defense. J Biol Chem. 2007;282:24407–15.
294. Salzman R, Kankova K, Pacal L, Tomandl J, Horakova Z, Kostrica R. Increased activity of superoxide dismutase in advanced stages of head and neck squamous cell carcinoma with locoregional metastases. Neoplasma. 2007;54:321–5.
295. Sarvazyan N. Visualization of doxorubicin-induced oxidative stress in isolated cardiac myocytes. Am J Physiol Heart Circ Physiol. 1996;271:H2079–85.
296. Sato K, Torimoto Y, Hosoki T, Ikuta K, Takahashi H, Yamamoto M, Ito S, Okamura N, Ichiki K, Tanaka H, Shindo M, Hirai K, Mizukami Y, Otake T, Fujiya M, Sasaki K, Kohgo Y. Loss of *ABCB7* gene: pathogenesis of mitochondrial iron accumulation in erythroblasts in refractory anemia with ringed siderblast with isodicentric (X)(q13). Int J Hematol. 2011;93: 311–8.
297. Schafer M, Schafer C, Ewald N, Piper HM, Noll T. Role of redox signaling in the autonomous proliferative response of endothelial cells to hypoxia. Circ Res. 2003;92:1010–5.
298. Schug ZT, Gottlieb E. Cardiolipin acts as a mitochondrial signalling platform to launch apoptosis. Biochim Biophys Acta. 2009;1788:2022–31.
299. Schwartzman RA, Cidlowski JA. Apoptosis: the biochemistry and molecular biology of programmed cell death. Endocr Rev. 1993;14:133–51.
300. Seo MS, Kang SW, Kim K, Baines IC, Lee TH, Rhee SG. Identification of a new type of mammalian peroxiredoxin that forms an intramolecular disulfide as a reaction intermediate. J Biol Chem. 2000;275:20346–54.
301. Seth RB, Sun L, Ea C-K, Chen ZJ. Identification and characterization of MAVS, a mitochondrial antiviral signaling protein that activates NF-κB and IRF3. Cell. 2005;122:669–82.
302. Sharoyko VV, Abels M, Sun J, Nicholas LM, Mollet IG, Stamenkovic JA, Gohring I, Malmgren S, Storm P, Fadista J, Spegel P, Metodiev MD, Larsson NG, Eliasson L, Wierup N, Mulder H. Loss of TFB1M results in mitochondrial dysfunction that leads to impaired insulin secretion and diabetes. Hum Mol Genet. 2014;23(21):5733–49.
303. Shen X, Zheng S, Metreveli NS, Epstein PN. Protection of cardiac mitochondria by overexpression of MnSOD reduces diabetic cardiomyopathy. Diabetes. 2006;55:798–805.
304. Shibata E, Nanri H, Ejima K, Araki M, Fukuda J, Yoshimura K, Toki N, Ikeda M, Kashimura M. Enhancement of mitochondrial oxidative stress and up-regulation of antioxidant protein peroxiredoxin III/SP-22 in the mitochondria of human pre-eclamptic placentae. Placenta. 2003;24:698–705.
305. Shimada Y, Okuno S, Kawai A, Shinomiya H, Saito A, Suzuki M, Omori Y, Nishino N, Kanemoto N, Fujiwara T, Horie M, Takahashi E-I. Cloning and chromosomal mapping of a novel ABC transporter gene (*hABC7*), a candidate for X-linked sideroblastic anemia with spinocerebellar ataxia. J Hum Genet. 1998;43:115–22.
306. Shioji K, Kishimoto C, Nakamura H, Masutani H, Yuan Z, Oka S-I, Yodoi J. Overexpression of thioredoxin-1 in transgenic mice attenuates adriamycin-induced cardiotoxicity. Circulation. 2002;106:1403–9.
307. Short KR, Bigelow ML, Kahl J, Singh R, Coenen-Schimke J, Raghavakaimal S, Nair KS. Decline in skeletal muscle mitochondrial function with aging in humans. Proc Natl Acad Sci. 2005;102:5618–23.

308. Sidiropoulos PI, Goulielmos G, Voloudakis GK, Petraki E, Boumpas DT. Inflammasomes and rheumatic diseases: evolving concepts. Ann Rheum Dis. 2008;67:1382–9.
309. Silverman DHS, Dy CJ, Castellon SA, Lai J, Pio BS, Abraham L, Waddell K, Petersen L, Phelps ME, Ganz PA. Altered frontocortical, cerebellar, and basal ganglia activity in adjuvant-treated breast cancer survivors 5–10 years after chemotherapy. Breast Cancer Res Treat. 2007;103:303–11.
310. Simbre II VC, Duffy SA, Dadlani GH, Miller TL, Lipshultz SE. Cardiotoxicity of cancer chemotherapy. Implications for children. Pediatr Drugs. 2005;7:187–202.
311. Siuda J, Jasinska-Myga B, Boczarska-Jedynak M, Opala G, Fiesel FC, Moussaud-Lamodiere EL, Scarffe LA, Dawson VL, Ross OA, Springer W, Dawson TM, Wszolek ZK. Early-onset Parkinson's disease due to PINK1 p.Q456X mutation—clinical and functional study. Parkinsonism Relat Disord. 2014;20(11):1274–8.
312. Slot JW, Geuze HJ, Freeman BA, Crapo JD. Intracellular localization of the copper-zinc and manganese superoxide dismutases in rat liver parenchymal cells. Lab Investig. 1986;55:363–71.
313. Soini Y, Vakkala M, Kahlos K, Paakko P, Kinnla V. MnSOD expression is less frequent in tumour cells of invasive breast carcinomas than in *in situ* carcinomas or non-neoplastic breast epithelial cells. J Pathol. 2001;195:156–62.
314. Sorice M, Manganelli V, Matarrese P, Tinari A, Misasi R, Malorni W, Garofalo T. Cardiolipin-enriched raft-like microdomains are essential activating platforms for apoptotic signals on mitochondria. FEBS Lett. 2009;583:2447–50.
315. Sparagna GC, Johnson CA, McCune SA, Moore RL, Murphy RC. Quantitation of cardiolipin molecular species in spontaneously hypertensive heart failure rats using electrospray ionization mass spectrometry. J Lipid Res. 2005;46:1196–204.
316. Spickett CM, Wiswedel I, Siems W, Zarkovic K, Zarkovic N. Advances in methods for the determination of biologically relevant lipid peroxidation products. Free Radic Res. 2010;44:1172–202.
317. Spierings D, McStay G, Saleh M, Bender C, Chipuk J, Maurer U, Green DR. Connected to death: the (unexpurgated) mitochondrial pathway of apoptosis. Science. 2005;310:66–7.
318. Spyrou G, Enmark E, Miranda-Vizuete A, Gustafsson J-Å. Cloning and expression of a novel mammalian thioredoxin. J Biol Chem. 1997;272:2936–41.
319. Squadrito GL, Pryor WA. The formation of peroxynitrite in vivo from nitric oxide and superoxide. Chem Biol Interact. 1995;96:203–6.
320. Srinivasan C, Liba A, Imlay JA, Valentine JS, Gralla EB. Yeast lacking superoxide dismutase(s) show elevated levels of "free iron" as measured by whole cell electron paramagnetic resonance. J Biol Chem. 2000;275:29187–92.
321. St Clair DK, Oberley TD, Ho YS. Overproduction of human Mn-superoxide dismutase modulates paraquat-mediated toxicity in mammalian cells. FEBS Lett. 1991;293:199–203.
322. St. Clair DK, Wang XS, Oberley TD, Muse KE, St. Clair WH. Suppression of radiation-induced neoplastic transformation by overexpression of mitochondrial superoxide dismutase. Mol Carcinog. 1992;6:238–42.
323. Starkov AA, Fiskum G, Chinopoulos C, Lorenzo BJ, Browne SE, Patel MS, Beal MF. Mitochondrial α-ketoglutarate dehydrogenase complex generates reactive oxygen species. J Neurosci. 2004;24:7779–88.
324. Steinman HM, Weinstein L, Brenowitz M. The manganese superoxide dismutase of *Escherichia coli* K-12 associates with DNA. J Biol Chem. 1994;269:28629–34.
325. Strassburger M, Bloch W, Sulyok S, Schuller J, Keist AF, Schmidt A, Wenk J, Peters T, Wlaschek M, Krieg T, Hafner M, Kumin A, Werner S, Muller W, Scharffetter-Kochanek K. Heterozygous deficiency of manganese superoxide dismutase results in severe lipid peroxidation and spontaneous apoptosis in murine myocardium in vivo. Free Radic Biol Med. 2005;38:1458–70.
326. Sun J, Folk D, Bradley TJ, Tower J. Induced overexpression of mitochondrial Mn-superoxide dismutase extends the life span of adult *Drosophila melanogaster*. Genetics. 2002;161: 661–72.

327. Sun X, Zhou Z, Kang YJ. Attenuation of doxorubicin chronic toxicity in metallothionein-overexpressing transgenic mouse heart. Cancer Res. 2001;61:3382–7.
328. Sun XZ, Vinci C, Makmura L, Han S, Tran D, Nguyen J, Hamann M, Grazziani S, Sheppard S, Gutova M, Zhou F, Thomas J, Momand J. Formation of disulfide bond in p53 correlates with inhibition of DNA binding and tetramerization. Antioxid Redox Signal. 2003; 5:655–65.
329. Takai D, Park S-H, Takada Y, Ichinose S, Kitagawa M, Akashi M. UV-irradiation induces oxidative damage to mitochondrial DNA primarily through hydrogen peroxide: analysis of 8-oxodGuo by HPLC. Free Radic Res. 2006;40:1138–48.
330. Takeshige K, Minakami S. NADH- and NADPH-dependent formation of superoxide anions by bovine heart submitochondrial particles and NADH-ubiquinone reductase preparation. Biochem J. 1979;180:129–35.
331. Tan AS, Baty JW, Berridge MV. The role of mitochondrial electron transport in tumorigenesis and metastasis. Biochim Biophys Acta. 2014;1840:1454–63.
332. Tanaka M, Kovalenko SA, Gong J-S, Borgeld H-JW, Katsumata K, Hayakawa M, Yoneda M, Ozawa T. Accumulation of deletions and point mutations in mitochondrial genome in degenerative diseases. Ann N Y Acad Sci. 1996;786:102–11.
333. Tangpong J, Cole MP, Sultana R, Estus S, Vore M, St. Clair W, Ratanachaiyavong S, St. Clair DK, Butterfield DA. Adriamycin-mediated nitration of manganese superoxide dismutase in the central nervous system: insight into the mechanism of chemobrain. J Neurochem. 2007;100:191–201.
334. Tangpong J, Cole MP, Sultana R, Joshi G, Estus S, Vore M, St. Clair W, Ratanachaiyavong S, St. Clair DK, Butterfield DA. Adriamycin-induced, TNF-α-mediated central nervous system toxicity. Neurobiol Dis. 2006;23:127–39.
335. Tangpong J, Sompol P, Vore M, St Clair W, Butterfield DA, St Clair DK. Tumor necrosis factor alpha-mediated nitric oxide production enhances manganese superoxide dismutase nitration and mitochondrial dysfunction in primary neurons: an insight into the role of glial cells. Neuroscience. 2008;151:622–9.
336. Tannock IF, Ahles TA, Ganz PA, van Dam FS. Cognitive impairment associated with chemotherapy for cancer: report of a workshop. J Clin Oncol. 2004;22:2233–9.
337. Teintze M, Slaughter M, Weiss H, Neupert W. Biogenesis of mitochondrial ubiquinol:cytochrome *c* reductase (cytochrome bc_1 complex). Precursor proteins and their transfer into mitochondria. J Biol Chem. 1982;257:10364–71.
338. Tortora V, Quijano C, Freeman B, Radi R, Castro L. Mitochondrial aconitase reaction with nitric oxide, *S*-nitrosoglutathione, and peroxynitrite: mechanisms and relative contributions to aconitase inactivation. Free Radic Biol Med. 2007;42:1075–88.
339. Tretter L, Adam-Vizi V. Generation of reactive oxygen species in the reaction catalyzed by α-ketoglutarate dehydrogenase. J Neurosci. 2004;24:7771–8.
340. Trumpower BL. The protonmotive Q cycle. Energy transduction by coupling of proton translocation to electron transfer by the cytochrome bc_1 complex. J Biol Chem. 1990;265: 11409–12.
341. Tsanou E, Ioachim E, Briasoulis E, Damala K, Charchanti A, Karavasilis V, Pavlidis N, Agnantis NJ. Immunohistochemical expression of superoxide dismutase (MnSOD) antioxidant enzyme in invasive breast carcinoma. Histol Histopathol. 2004;19:807–13.
342. Tyurina YY, Tyurin VA, Kaynar AM, Kapralova VI, Wasserloos K, Li J, Mosher M, Wright L, Wipf P, Watkins S, Pitt BR, Kagan VE. Oxidative lipidomics of hyperoxic acute lung injury: mass spectrometric characterization of cardiolipin and phosphatidylserine. Am J Physiol Lung Cell Mol Physiol. 2010;299:L73–85.
343. van Remmen H, Ikeno Y, Hamilton M, Pahlavani M, Wolf N, Thorpe SR, Alderson NL, Baynes JW, Epstein CJ, Huang T-T, Nelson J, Strong R, Richardson A. Life-long reduction in MnSOD activity results in increased DNA damage and higher incidence of cancer but does not accelerate aging. Physiol Genomics. 2003;16:29–37.
344. Vashchenko N, Abrahamsson PA. Neuroendocrine differentiation in prostate cancer: implications for new treatment modalities. Eur Urol. 2005;47:147–55.

345. Venkataraman S, Jiang X, Weydert C, Zhang Y, Zhang HJ, Goswami PC, Ritchie JM, Oberley LW, Buettner GR. Manganese superoxide dismutase overexpression inhibits the growth of androgen-independent prostate cancer cells. Oncogene. 2005;24:77–89.
346. Walker V. Ammonia toxicity and its prevention in inherited defects of the urea cycle. Diabetes Obes Metab. 2009;11:823–35.
347. Wang X, Martindale JL, Liu Y, Holbrook NJ. The cellular response to oxidative stress: influences of mitogen-activated protein kinase signaling pathways on cell survival. BioChem J. 1998;333:291–300.
348. Wang X, Peralta S, Moraes CT. Chapter four—mitochondrial alterations during carcinogenesis: a review of metabolic transformation and targets for anticancer treatments. In: Kenneth DT, Paul BF, editors. Advances in Cancer Research. Orlando: Academic; 2013. p. 127–60.
349. Waris G, Ahsan H. Reactive oxygen species: role in the development of cancer and various chronic conditions. J Carcinog. 2006;5:14–21.
350. Wefel JS, Lenzi R, Theriault R, Buzdar AU, Cruickshank S, Meyers CA. 'Chemobrain' in breast carcinoma? A prologue. Cancer. 2004;101:466–75.
351. Weisiger RA, Fridovich I. Mitochondrial superoxide dismutase. Site of synthesis and intramitochondrial localization. J Biol Chem. 1973;248:4793–6.
352. Weydert C, Roling B, Liu J, Hinkhouse MM, Ritchie JM, Oberley LW, Cullen JJ. Suppression of the malignant phenotype in human pancreatic cancer cells by the overexpression of manganese superoxide dismutase. Mol Cancer Ther. 2003;2:361–9.
353. Williams D, Venardos KM, Byrne M, Joshi M, Horlock D, Lam NT, Gregorevic P, McGee SL, Kaye DM. Abnormal mitochondrial L-arginine transport contributes to the pathogenesis of heart failure and rexoygenation injury. PLoS One. 2014;9:e104643.
354. Williams MD, Van Remmen H, Conrad CC, Huang T-T, Epstein CJ, Richardson A. Increased oxidative damage is correlated to altered mitochondrial function in heterozygous manganese superoxide dismutase knockout mice. J Biol Chem. 1998;273:28510–5.
355. Wiswedel I, Gardemann A, Storch A, Peter D, Schild L. Degradation of phospholipids by oxidative stress—exceptional significance of cardiolipin. Free Radic Res. 2010;44:135–45.
356. Wiswedel I, Keilhoff G, Dorner L, Navarro A, Bockelmann R, Bonnekoh B, Gardemann A, Gollnick H. UVB irradiation-induced impairment of keratinocytes and adaptive responses to oxidative stress. Free Radic Res. 2007;41:1017–27.
357. Wittig I, Schagger H. Supramolecular organization of ATP synthase and respiratory chain in mitochondrial membranes. Biochim Biophys Acta. 2009;1787:672–80.
358. Wong GHW, Elwell JH, Oberley LW, Goeddel DV. Manganese superoxide dismutase is essential for cellular resistance to cytotoxicity of tumor necrosis factor. Cell. 1989;58:923–31.
359. Wong GHW, Goeddel DV. Induction of manganous superoxide dismutase by tumor necrosis factor: possible protective role. Science. 1988;242:941–4.
360. Wright VP, Reiser PJ, Clanton TL. Redox modulation of global phosphatase activity and protein phosphorylation in intact skeletal muscle. J Physiol. 2009;587:5767–81.
361. Xu Y, Porntadavity S, St. Clair DK. Transcriptional regulation of the human manganese superoxide dismutase gene: the role of specificity protein 1 (Sp1) and activating protein-2 (AP-2). Biochem J. 2002;362:401–12.
362. Yakes FM, van Houten B. Mitochondrial DNA damage is more extensive and persists longer than nuclear DNA damage in human cells following oxidative stress. Proc Natl Acad Sci U S A. 1997;94:514–9.
363. Yamamoto T, Maruyama W, Kato Y, Yi H, Shamoto-Nagai M, Tanaka M, Sato Y, Naoi M. Selective nitration of mitochondrial complex I by peroxynitrite: involvement of mitochondria dysfunction and cell death of dopaminergic SH-SY5Y cells. J Neural Transm. 2002;109:1–13.
364. Yang L-Y, Chen W-L, Lin J-W, Lee S-F, Lee C-C, Hung TI, Wei Y-H, Shih C-M. Differential expression of antioxidant enzymes in various hepatocellular carcinoma cell lines. J Cell Biochem. 2005;96:622–31.
365. Yang Y, Karakhanova S, Werner J, Bazhin AV. Reactive oxygen species in cancer biology and anticancer therapy. Curr Med Chem. 2013;20:3677–92.

366. Yao J, Irwin RW, Zhao L, Nilsen J, Hamilton RT, Brinton RD. Mitochondrial bioenergetic deficit precedes Alzheimer's pathology in female mouse model of Alzheimer's disease. Proc Natl Acad Sci U S A. 2009;106:14670–5.
367. Ye H, Rouault TA. Human iron-sulfur cluster assembly, cellular iron homeostasis, and disease. Biochemistry. 2010;49:4945–56.
368. Yen H-C, Oberley TD, Gairola CG, Szweda LI, St. Clair DK. Manganese superoxide dismutase protects mitochondrial complex I against adriamycin-induced cardiomyopathy in transgenic mice. Arch Biochem Biophys. 1999;362:59–66.
369. Yen H-C, Oberley TD, Vichitbandha S, Ho Y-S, St. Clair DK. The protective role of manganese superoxide dismutase against adriamycin-induced acute cardiac toxicity in transgenic mice. J Clin Investig. 1996;98:1253–60.
370. Zamocky M, Furtmuller PG, Obinger C. Evolution of catalases from bacteria to humans. Antioxid Redox Signal. 2008;10:1527–47.
371. Zelko IN, Mariani TJ, Felz RJ. Superoxide dismutase multigene family: a comparison of CuZn-SOD (SOD1), Mn-SOD (SOD2), and EC-SOD (SOD3) gene structure, evolution, and expression. Free Radic Biol Med. 2002;33:337–49.
372. Zhang L, Yu L, Yu C-A. Generation of superoxide anion by succinate-cytochrome *c* reductase from bovine heart mitochondria. J Biol Chem. 1998;273:33972–6.
373. Zhang Q, Raoof M, Chen Y, Sumi Y, Sursal T, Junger W, Brohi K, Itagaki K, Hauser CJ. Circulating mitochondrial DAMPs cause inflammatory responses to injury. Nature. 2010;464:104–7.
374. Zhang Y, Smith BJ, Oberley LW. Enzymatic activity is necessary for the tumor-suppressive effects of MnSOD. Antioxid Redox Signal. 2006;8:1283–93.
375. Zhao Y, Chaiswing L, Oberley TD, Batinic-Haberle I, St. Clair W, Epstein CJ, St. Clair DK. A mechanism-based antioxidant approach for the reduction of skin carcinogenesis. Cancer Res. 2005;65:1401–5.
376. Zhao Y, Kiningham KK, Lin S-M, St. Clair DK. Overexpression of MnSOD protects murine fibrosarcoma cells (FSa-II) from apoptosis and promotes a differentiation program upon treatment with 5-azacytidine: involvement of MAPK and NFκB pathways. Antioxid Redox Signal. 2001;3:375–86.
377. Zhao Y, Oberley TD, Chaiswing L, Lin S-M, Epstein CJ, Huang T-T, St. Clair DK. Manganese superoxide dismutase deficiency enhances cell turnover via tumor promoter-induced alterations in AP-1 and p53-mediated pathways in a skin cancer model. Oncogene. 2002;21:3836–46.
378. Zhao Y, Xue Y, Oberley TD, Kiningham KK, Lin S-M, Yen H-C, Majima H, Hines J, St. Clair DK. Overexpression of manganese superoxide dismutase suppresses tumor formation by modulation of activator protein-1 signaling in a multistage skin carcinogenesis model. Cancer Res. 2001;61:6082–8.
379. Zhou Z, Kang YJ. Cellular and subcellular localization of catalase in the heart of transgenic mice. J Histochem Cytochem. 2000;48:585–94.
380. Zidenberg-Cherr S, Keen CL, Lonnerdal B, Hurley LS. Superoxide dismutase activity and lipid peroxidation in the rat: developmental correlations affected by manganese deficiency. J Nutr. 1983;113:2498–504.

Chapter 4
Regulation of the Cellular Redox Environment by Superoxide Dismutases, Catalase, and Glutathione Peroxidases During Tumor Metastasis

L.P. Madhubhani P. Hemachandra, Akshaya Chandrasekaran, J. Andres Melendez, and Nadine Hempel

Abbreviations

AhR	Aryl hydrocarbon receptor
AML	Acute myelogenous leukemia
AMPK	AMP activated-kinase A
Cat	Catalase
CRPC	Castration-resistant prostate cancer
CuZnSod/Sod1	Cytosolic/Copper-Zinc superoxide dismutase
EcSod/Sod3	Extracellular superoxide dismutase
EMT	Epithelial to mesenchymal transition
fALS	Familial amyotrophic lateral sclerosis
FAK	Focal adhesion kinase
Foxo	Forkhead box O
GAC	Gastric adenocarcinoma
GPx	Glutathione peroxidases
GR	Glutathione reductase
GSH	Glutathione
HCC	Hepatocellular carcinoma
H_2O_2	Hydrogen peroxide
iNOS	Inducible nitric oxide synthase

L.P.M.P. Hemachandra • A. Chandrasekaran • J.A. Melendez
Nanobioscience Constellation, Colleges of Nanoscale Science and Engineering, SUNY Polytechnic Institute, State University of New York, Albany, NY, USA

N. Hempel, Ph.D. (✉)
Department of Pharmacology, Penn State College of Medicine,
500 University Drive, P.O. Box 850, Hershey, PA 17033-0850, USA
e-mail: nhempel@hmc.psu.edu

I. Batinić-Haberle et al. (eds.), *Redox-Active Therapeutics*,
Oxidative Stress in Applied Basic Research and Clinical Practice,
DOI 10.1007/978-3-319-30705-3_4

IL-6	Interleukin-6
Keap1	Kelch-like erythroid cell-derived protein with CNC homology (ECH)-associated protein 1
MAPK	Mitogen-activated protein kinase
MMP-1	Matrix metalloproteinase 1
MnSod/Sod2	Mitochondrial/Manganese superoxide dismutase
mTOR	Mechanistic target of rapamycin
NADPH	Nicotinamide adenine dinucleotide phosphate
NF-κB	Nuclear factor κ B
Nox	NADPH oxidase
Nrf2	Nuclear factor erythroid 2 (NF-E2)-related factor 2
O_2	Oxygen
$O_2^{\bullet-}$	Superoxide
$OH^{\bullet}$	Hydroxyl radical
$ONOO^-$	Peroxynitrite
PI3K	Phosphoinositide 3-kinase
PKB	Protein kinase B (also known as Akt)
PHD	Prolyl hydroxylase
PPAR α	Peroxisome proliferator activated receptor α
PTP	Protein tyrosine phosphatases
SBP1	Selenium-binding protein 1
SNPs	Single nucleotide polymorphisms
Sod	Superoxide dismutase
RNS	Reactive nitrogen species
ROS	Reactive oxygen species
TM	Tetrathiomolybdate
TNF-α	Tumor necrosis factor α
TRX	Thioredoxin
UCP	Mitochondrial uncoupling proteins
WT1	Wilm's tumor suppressor 1

4.1 Introduction

The lethality of most cancers can be attributed to metastatic progression. This is primarily due to difficulties in identifying, targeting, and effectively treating metastatic lesions. The ability to survive the metastatic journey through hematological and lymphatic circulation, the homing to specific secondary sites, and the latency in appearance of metastatic spread many years after patients have been disease free illustrate the insidious nature of metastatic cells and their adaptability to cope with various stressful tumor microenvironments. Metastatic tumor cells also display clear differences from cells of the primary tumor as a result of distinct epigenetic, genetic, and cellular signaling that control their unique behavior. These alterations promote epithelial to mesenchymal transition (EMT); enhance invasion and traffic through the extracellular matrix; reprogram the cell cycle; and confer resistance to programmed cell death, nutrient deprivation, and oxidant production in the tumor microenvironment.

As depicted in Fig. 4.1, reactive oxygen and nitrogen species (ROS & RNS) are associated with both tumor initiation (carcinogenesis) and tumor progression (metastasis). These reactive species arise as a consequence of events such as deprivation of oxygen (hypoxia), immune cell infiltration and inflammation, and from chemo- and radiotherapeutic exposure. The effect of ROS and RNS on cancer cells essentially depends on the levels and type of reactive species [1]. Large surges of ROS, for example in response to ionizing radiation, will lead to oxidation of macromolecules and largely irreversible damage that may cause genomic instability and

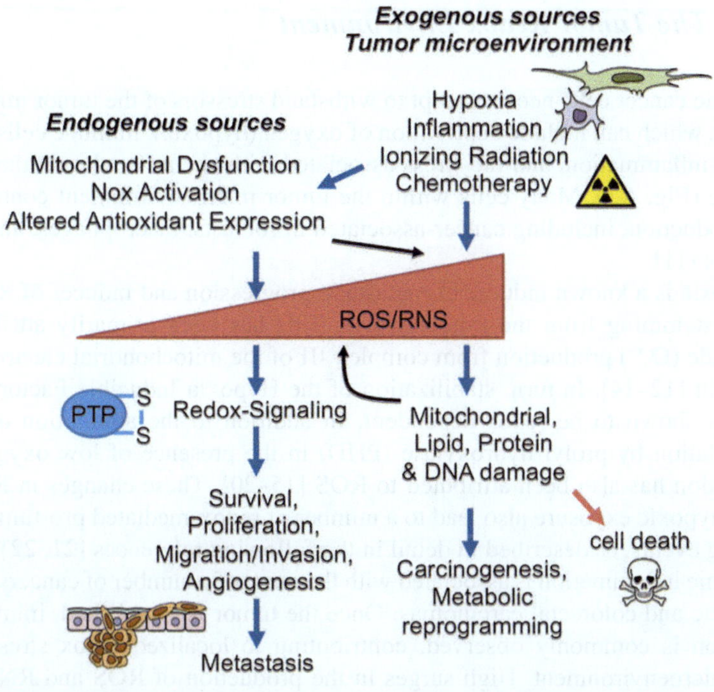

Fig. 4.1 Metastatic tumor cells are exposed to exogenous and endogenous sources of reactive oxygen (ROS) and nitrogen species (RNS). Examples of exogenous sources of ROS are those that emanate from local inflammation and radio- and chemotherapeutic exposure. Cancer-associated fibroblasts, macrophages, and senescent cells may also contribute to an altered redox environment. Further, hypoxia enhances production of ROS through mitochondria, and tumor microenvironment-derived growth factors and cytokines activate NADPH oxidase (Nox) enzymes to enhance intracellular production of ROS within tumor cells. Depending on the source, type, and amount of ROS/RNS produced, these changes in the tumor redox environment can either influence pro-metastatic signaling pathways or ROS-mediated cellular damage, which may be either tumor promoting or deleterious to the cancer cells

cancer initiation—carcinogenesis. Similarly, these may result in mitochondrial damage that can lead to alterations in cancer metabolism and drive glycolytic adaptation. In addition to exogenous sources of ROS, tumor cells themselves produce and are able to cope with elevated intracellular ROS [2–5]. These may be sublethal increases that can lead to ROS-mediated changes in cellular signaling, referred to as redox-signaling. Increases in the cellular redox steady state are often accompanied by changes in antioxidant enzyme expression, as an adaptation to survive under chronic redox stress. This inadvertently leads to enhanced ability of tumor cells to deal with excessive exogenous stress, and may be a contributing factor to chemoresistance. For example, high expression of Sod2 in cancer cells has been associated with eliciting radioresistance and chemoresistance to a number of agents, including Cisplatin [6–10].

4.1.1 The Tumor Redox Environment

Metastatic cancer cells need to adapt to withstand stressors of the tumor microenvironment, which can include deprivation of oxygen (hypoxia), immune cells infiltration and inflammation, and the stress associated with chemo- and radiotherapeutic exposure (Fig. 4.1). Many cells within the tumor microenvironment contribute to ROS production, including cancer-associated fibroblasts, macrophages, and senescent cells [11].

Hypoxia is a known inducer of metastatic progression and inducer of ROS production stemming from the mitochondria. This has been primarily attributed to superoxide ($O_2^{\bullet -}$) production from complex III of the mitochondrial electron transport chain [12–14]. In turn, stabilization of the Hypoxia Inducible Factor HIF-1α has been shown to be redox-dependent. In addition to the abrogation of HIF-α hydroxylation by prolyl hydroxylase (PHD) in the presence of low oxygen, HIF stabilization has also been attributed to ROS [15–20]. These changes in ROS following hypoxic exposure also lead to a number of redox-mediated pro-tumorigenic signaling events, as described in detail in the following references [21, 22].

Chronic inflammation is associated with the onset of a number of cancers, including gastric and colorectal carcinomas. Once the tumor is established, immune cell infiltration is commonly observed, contributing to localized redox stress in the tumor microenvironment. High surges in the production of ROS and RNS in the tumor environment are largely the result of NOX2 (NADPH oxidase) and iNOS (inducible Nitric oxide synthase) activation in phagocytic cells. These large doses of ROS/RNS are thought to primarily contribute to macromolecular damage. However, these may also contribute to alterations in cell signaling events that contribute to metastatic progression and provide a stimulus for the stress response pathways that initiate transcription of antioxidant defenses within the tumor cells. Inflammatory cytokines, such as Tumor Necrosis Factor α (TNF-α) and Interleukin-6 (IL-6) produced by tumor-associated macrophages, have also been shown to drive metastatic progression and can initiate ROS production in non-phagocytic cells [23, 24].

It has been proposed that tumor cells essentially become metastatic to escape from high ROS tumor microenvironments by enhancing their antioxidant defense and using redox-mediated signaling to migrate and invade. Further, enhanced redox damage to mitochondria in metastatic cells is thought to contribute to their metabolic adaptations towards glycolysis [25]. However, even in the absence of exogenous stressors, it has been known for some time that cancer cells exhibit intrinsic increases in steady-state levels of ROS [2–5]. Our own work suggests that this may be further enhanced as cells progress towards metastasis [26]. The origin of intracellular redox shifts within tumor cells has been demonstrated to lie at both the level of Nox enzymes and mitochondria. Membrane localized Nox enzymes reduce molecular oxygen (O_2) to $O_2^{\bullet-}$, which is then either spontaneously dismuted to hydrogen peroxide (H_2O_2) or via enzymatic action by superoxide dismutases (Sod). Nox-dependent ROS production has been aligned with cancer initiation, proliferation, angiogenesis, and metastasis and is directly engaged by oncogene activation and growth factor receptor signaling [27, 28]. The other major source of cellular ROS is the mitochondria, where increased electron leakage from the electron transport chain leads to the one-electron reduction of O_2 to $O_2^{\bullet-}$. Generation of ROS from mitochondria during tumorigenesis and metastasis has been attributed to mitochondrial dysfunction, hypoxia, and cellular signaling, such as the Phosphoinositide 3-kinase (PI3K)/Protein Kinase B (PKB/Akt) pathway [21, 25]. In addition, mitochondrial superoxide dismutases may contribute to conversion of $O_2^{\bullet-}$ to the primary signaling oxidant, H_2O_2.

4.1.2 Role of ROS and Redox-Signaling in Metastatic Disease

Increased ROS production outside and within metastatic tumor cells has been associated with a number of pro-metastatic events including angiogenesis, invasion, migration, survival, and anchorage-independent cell survival (anoikis resistance) [29–42]. Further, ROS-mediated damage to tumor cell mitochondria and DNA is thought to mediate shifts towards glycolytic metabolism and genetic alterations, respectively, that may further contribute to metastatic progression [25]. Examples of molecular mechanisms under redox control include the transcriptional regulation of matrix degradation enzymes, such as MMP-1, the activation of cellular signaling such as the PI3K/Akt and NFκB pathways, HIF-1α activation, and the redox-dependent regulation of the cytoskeleton by manipulating Rho/Rac activation. Redox-regulated pathways in cancer are described in more detail in a number of excellent reviews [21, 25, 43–45].

Redox-mediated signaling has largely been attributed to reversible oxidation of thiols, such as cysteine and methionine residues in phosphatases, kinases, and transcription factors, that either alter their kinetic activity or ability to bind other proteins or DNA [46]. An example of this is the oxidation of active site cysteine residues with phosphatases, which generally inhibits their catalytic activity. Conversely, oxidation of kinases can often result in their activation. This in turn results in the exacerbation of phosphorylation-mediated signaling pathways. The oxidation and inactivation of the phosphatase PTEN and subsequent activation of the Akt pathway is one such example, which regulates survival and angiogenesis [33, 47, 48]. While certain oxidation events can result in irreversible protein modifications, most redox-signaling relies on reversibility, which is generally achieved by the action of Glutathione reductase (GR) and thioredoxin (TRX) to reduce the oxidized protein, and which is dependent on reducing equivalents glutathione (GSH) and Nicotinamide adenine dinucleotide phosphate (NADPH) within the cells.

H_2O_2 has been suggested to be the major ROS responsible for cellular signaling, primarily due to its ability to traverse membranes and long half-life in comparison to other ROS such as $O_2^{\bullet-}$ and hydroxyl radical ($OH^{\bullet}$). While H_2O_2 is able to contribute directly to the oxidation of proteins, DNA damage in response to H_2O_2 can be largely attributed to $OH^{\bullet}$, which is generated in the Fenton reaction between H_2O_2 and ferrous iron Fe(II) [49, 50].

4.1.3 Acquisition of the Metastatic Antioxidant Phenotype

To cope with these changes in both intracellular and extracellular redox environments, tumor cells have uniquely evolved to alter their antioxidant enzyme expression. As described below many stress responsive signaling pathways, which are activated by increases in ROS from the extracellular and intracellular environment, induce the expression of antioxidant enzymes. This has explained some of the increases in antioxidant enzyme expression observed in metastatic disease. These pathways include the activation of the Nrf2/KEAP1 pathway (NF-E2-related factor 2/Kelch-like ECH-associated protein 1), NF-κB (Nuclear Factor κ B) signaling, and the Sirtuin-Foxo (Forkhead box O) transcription factor axis [51–54]. It is also thought that high expression of antioxidant enzymes is an adaptation of cancer stem cells and dormant tumor cells, which are thought to be the precursors of metastatic lesions. For example, cancer stem cells display low intracellular levels of ROS and high expression of antioxidant enzymes [55] and this has been suggested to be one of the causes for the enhanced radio- and chemoresistance of cancer stem cells. Analysis of gene expression array data suggests that expression of some antioxidant enzymes is further increased in metastatic cells compared to the primary tumor [56]. This enhanced antioxidant milieu provides a unique protection towards redox stress in the metastatic tumor niche, and might contribute to the enhanced chemoresistance of many metastatic tumor cells. As highlighted below it is increasingly becoming clear that many antioxidant enzymes are no longer considered simple tumor suppressor genes due to their ability to scavenge ROS. We have argued that certain antioxidant enzymes, such as Sod2, inadvertently manipulate the steady-state redox environment within cells. Their own actions on redox balance may further contribute to alterations in cellular signaling that may drive metastatic behavior.

The present review focuses on antioxidant enzymes important in the regulation of H_2O_2 balance within cells (Fig. 4.2). It aims to highlight some of the dichoto-

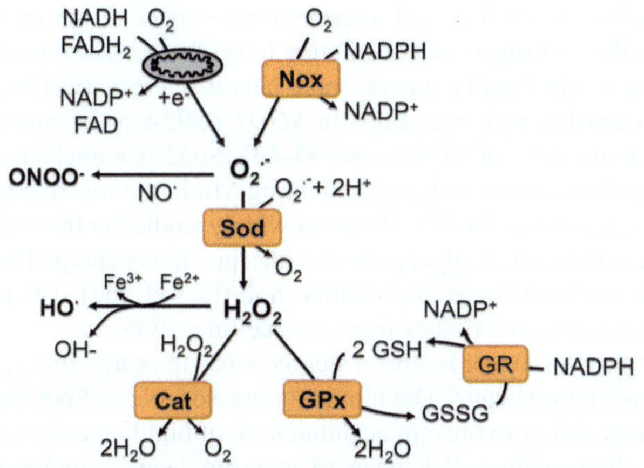

Fig. 4.2 The reactions catalyzed by the antioxidant enzymes Superoxide Dismutases (Sod), Catalase (Cat), and Glutathione Peroxidases (GPx). Sources of cellular superoxide most commonly stem from electron leakage of the mitochondrial electron transport chain or NADPH oxidases (Nox). Oxidized glutathione is reduced via Glutathione reductase (GR). Other reactive oxygen (ROS) or nitrogen species (RNS) may arise from the reaction with superoxide ($O_2^{\cdot-}$) or hydrogen peroxide (H_2O_2), including the generation of ($HO^\cdot$) and peroxynitrite ($ONOO^-$), respectively

mous roles demonstrated for these enzymes in cancer cells and how their enzymatic activity may influence the tumor redox environment and regulate carcinogenesis and metastasis. We focus on the role of superoxide dismutases, catalase, and glutathione peroxidases and give examples on their demonstrated roles as both tumor suppressors and promoters. Specifically, we discuss how metastatic cancer cells uniquely adapt to alter expression of these enzymes and how this may contribute to changes in the intracellular redox environment to drive certain metastatic phenotypes. We highlight a number of studies relating these enzymes to cancer and metastatic progression, but realize that this field is an ever-expanding research area and apologize for any omissions. While we will not discuss the changes in glutathione status, an important mediator of redox balance within cells, we want to point out that these reducing equivalents are of great importance in reversible redox-signaling.

4.2 Superoxide Dismutases

Superoxide dismutases (Sods) are important antioxidant enzymes in the family of oxidoreductases that catalyze the dismutation of superoxide anion ($O_2^{\cdot-}$) to O_2 and H_2O_2 [57] (Fig. 4.2). There are three forms of mammalian Sods: superoxide dismutase 1 (Sod1, Cu/ZnSod), superoxide dismutase 2 (Sod2, MnSod), and superoxide dismutase 3 (Sod3, EcSod). Sod1 is a homodimer with each monomer containing one Cu and one Zn ion in the center. Sod1 is localized primarily to the cytosol, but has been found in the nucleus and intermembrane space of the mitochondria [58, 59]. Sod1 was the first superoxide dismutase to be characterized in eukaryotic cells and is associated with familial amyotrophic lateral sclerosis (fALS), a fatal neurodegenerative disorder, with mutations in *SOD1* (G93A most commonly studied) connected to about 20 % of ALS cases [60–63]. Sod2 is a nuclear encoded, tetrameric mitochondrial matrix enzyme containing Mn in the reactive center [64–66]. Of the three Sod enzymes, Sod2 is the most widely studied in the context of cancer. Sod3 is a secreted tetrameric glycoprotein with a positively charged heparin-binding site, and is often referred to as extracellular Sod (EcSod) [67]. The primary role of Sod3 is to protect cells and tissues from extracellular ROS.

The role of Sods in cancer has been studied since the early 1970s, based on their role as important antioxidants. The tumor suppressor role of Sods has largely been attributed to their ability to prevent accumulation of highly reactive $O_2^{\bullet-}$, and therefore averting ROS-meditated damage to macromolecules, including DNA, that leads to cancer initiation. While a tumor suppressor function has been described for all Sod isoforms, it should be pointed out that Sod1 and Sod2 have been found to have dichotomous roles in cancer, having both oncogenic and anti-tumorigenic functions. Examples of these are described below.

Compared to Sod1 and Sod2, the extracellular superoxide dismutase Sod3 has largely been found to possess anti-tumorigenic properties and many studies demonstrate reduced expression of this extracellular Sod in a number of cancers, including pancreatic [68], prostate [69], thyroid [70], lung [71], and breast cancer [72, 73]. Re-expression or treatment with recombinant Sod3 has been shown to inhibit cancer cell proliferation [69], invasion [68, 69, 71, 72], clonogenic survival [71], and in vivo tumor growth and metastasis [68, 73]. Sod3 likely prevents carcinogenesis by scavenging extracellular ROS capable of initiating DNA damage and consequential genetic aberrations responsible for tumor initiation. In addition, Sod3 is thought to prevent the ROS-induced transcriptional induction of Heparanase, which is an important regulator of VEGF and heparan-sulfate, and therefore prevents the respective pro-angiogenic and pro-metastatic activity of these factors [72, 74, 75]. Interestingly, a recent study suggests that the tumor suppressor role of Sod3 may be due to Sod3-mediated increase in H_2O_2 levels, rather than $O_2^{\bullet-}$ scavenging. It was shown that enforced expression and recombinant Sod3 could inhibit cell proliferation, migration, and invasion of PC-3 prostate cancer cells, and that this was due to an accumulation of H_2O_2, leading to deleterious increases in DNA damage [69]. Loss of Sod3 appears to occur early in tumor development [71], hinting that Sod3 may be important in preventing carcinogenesis by exogenous ROS, emanating from the tumor environment or ROS-producing agents. Domann & coworkers have shown that epigenetic regulation may contribute to the decrease in Sod3 expression observed in a number of cancers. For example, hypermethylation of the *SOD3* promoter was observed in 8/10 adenocarcinomas compared with 0/5 normal lung samples and this epigenetic modification correlates with low expression of Sod3 in lung cancer [71]. Similarly, *SOD3* methylation and loss of Sod3 expression are associated with breast cancer [72, 73].

While the role of Sod1 and Sod2 in cancer depends on context and cancer type, a decrease in expression of these enzymes has been reported in a number of tumors [76]. Based on these observations and subsequent mechanistic studies, Sod2 in particular was initially classified as a tumor suppressor gene [77–80]. Like Sod3, the decreased expression of Sod2 in transformed cells is due to epigenetic control [80–82] and provides an explanation for its low level expression and putative tumor suppressor role in colorectal, breast, and prostate cancer [83–85]. Sod2 protects the mitochondria from $O_2^{\bullet -}$ radicals, emanating from electron leakage from the electron transport chain during oxidative phosphorylation. In the absence of Sod2, reduced clearance of $O_2^{\bullet -}$ leads to increased oxidative damage and mitochondrial dysfunction, which has been linked to several cancers [86]. In addition, it was recently suggested that Sod2 loss in cancer may precede the adaptation of cancer cells to aerobic glycolysis. Sod2 loss in skin tissue of heterozygous Sod2 (−/+) mice enhances expression of mitochondrial uncoupling proteins (UCPs), in a peroxisome proliferator activated receptor α (PPAR α) dependent manner, driving PI3K/Akt/mTOR (mechanistic Target of Rapamycin) pathway activation and the glycolytic phenotype [87].

While fewer studies have shown the role of Sod1 as a tumor suppressor, data from Sod1 −/− knockout mice demonstrate that a loss of Sod1 expression leads to the development of hepatocellular carcinoma in older mice, which is associated with increased cellular ROS accumulation and accumulation of 8-oxo-dG DNA modifications and genomic instability [88, 89]. The tumor suppresser activity of Sod1 was also observed in human glioma cells, where enforced overexpression of Sod1 significantly reduced the cell growth, clonogenicity in soft agar, and slowed the tumor doubling time [90]. Although a specific mechanism was not shown in that report, the authors suggested a possible mechanism of excess H_2O_2 accumulation and H_2O_2-mediated cell death in the Sod1 overexpressing cells.

In addition to preventing macromolecular redox damage, the manipulation of $O_2^{\bullet -}$ by Sods may contribute to cell cycle regulation. It has been shown that the intracellular redox environment fluctuates during the cell cycle and that a shift towards a more oxidizing environment correlates with cell cycle progression [91]. Spikes in $O_2^{\bullet -}$ are associated with mitosis, while H_2O_2 levels are linked with the G0/G1 phase of the cell cycle [92]. Since Sods play major roles in modulating cellular $O_2^{\bullet -}$, these enzymes may be critical regulators of the cell cycle. As such, it has been demonstrated that there is an inverse correlation between the periodic fluctuation of cellular $O_2^{\bullet -}$ levels, glucose and oxygen consumptions, and Sod2 activity during the cell cycle [93]. During carcinogenesis, the balance between $O_2^{\bullet -}$, which initiates proliferation, and H_2O_2, which triggers differentiation signals, is disrupted. It has therefore been proposed that increased $O_2^{\bullet -}$ accumulation due to reduced Sod expression observed in some tumor types may trigger increased cell proliferation, while differentiation is reduced due to abolished H_2O_2 generation [94, 95]. It is unclear whether these increases in $O_2^{\bullet -}$ result in immediate oxidation of macromolecules leading to this increase in proliferation, which presumably would need to be in close proximity to where $O_2^{\bullet -}$ is produced (i.e., mitochondria or Nox). Otherwise, due to the high rate constant of spontaneous dismutation of $O_2^{\bullet -}$ to H_2O_2, it is possible that increases in $O_2^{\bullet -}$ may lead to increases in the H_2O_2 pool.

The above studies demonstrate that the tumor suppressor functions of Sods are primarily centered around their ability to scavenge ROS that initiate carcinogenesis. Below we will describe how Sods may play a more complicated role during progression to metastatic disease. More recently, the dichotomous role of Sods, as both tumor suppressors and oncogenes, has come to light. The oncogenic role for Sods is supported by observations that expression of these enzymes is elevated in a number of cancers, and that high Sod levels can correlate with more advanced stage disease, implicating these enzymes with metastatic progression [56, 96]. This is particularly true for the intracellular-localized Sod1 and Sod2. Further, even though Sod2 expression can be decreased in many tumor cells, as eluded to above, total Sod activity is increased in many transformed cells compared to normal cells, suggesting that high Sod2 activity may be associated with tumorigenicity [79, 80].

The high expression of Sods in metastatic progression of cancer has been associated with a number of stress response pathways, initiated by exogenous and intracellular stress, including the NF-κB pathway [97–100], the aryl hydrocarbon receptor (AhR), and Nrf-2 transcription factors [101]. These stress responsive transcriptional pathways have also been implicated in driving Sod-dependent chemo- and radioresistance. In most cancers radiation sensitivity is suppressed by Sod expression, which can detoxify the highly reactive free radicals generated by radiotherapy [102]. For example, development of radioresistance in prostate cancers is mediated via the NF-κB-dependent regulation of Sod2 expression [103], and Sod1 expression in cisplatin-resistant human urothelial carcinoma cells is driven by the transcription factor by C/EBPδ [104]. Sod2 expression has also been associated with Cisplatin resistance, by Sod2-mediated NF-κB activation of the anti-apoptotic factor Bcl-2 [105].

The role of Sods in carcinogenesis and tumor progression appears to be context and disease stage dependent [30, 56, 106–109]. For example, Sod2 may play a protective role during tumor initiation by attenuating ROS-mediated DNA damage, yet it can mediate tumor metastasis by altering redox-dependent signaling during tumor progression [30]. High expression of Sod2 is associated with metastatic progression tested in many cancer cells in vitro, as evident from gene expression arrays comparing Sod2 mRNA levels between primary tumor and metastatic lesions [26, 30, 56, 96, 110–112]. The dichotomous role of Sod2 in cancer may be explained by the differential regulation of expression during tumor progression. It has been proposed that low Sod2 expression is associated with tumor initiation, while Sod2 levels increase as an adaptation during metastatic progression [56]. Given this observation, Dhar et al. investigated the timing of Sod2 transcriptional regulation in a multistage carcinogen-induced in vivo skin carcinogenesis model. In benign papillomas, it was shown that DMBA- and TPA-mediated activation of p53 results in Sod2 downregulation by inhibiting Sp1 binding to the *SOD2* promoter during early tumor initiation. Conversely, in squamous cell carcinomas, Sod2 expression increased, due to a loss in p53 expression in advanced stage disease [113].

It is now believed that Sods not only detoxify damaging $O_2^{\bullet -}$, but that these enzymes also alter intracellular redox homeostasis that contributes to redox damage and/or redox-signaling, which drive pro-tumorigenic and -metastatic events. The role of Sod enzymes as oncogenes has proven to be more difficult to rationalize based on the biochemical properties of the enzymatic reaction [114, 115]. For example, a number of labs have reported that increases in Sod expression are accompanied by increases in H_2O_2 levels in cancer cells [30, 116–120]. Theoretically, an increase in Sod should not result in higher levels of H_2O_2 production based on the enzyme's kinetics and spontaneous dismutation of $O_2^{\bullet-}$ [114, 115]. While it is beyond the scope of the present proposal to discuss these in detail, a number of plausible explanations have been put forth to explain changes in H_2O_2 steady state observed that correlate with increased Sod expression. These primarily pertain to altering reaction rates within the electron transport chain. For example, inhibition of cytochrome *c* oxidase by increased levels of nitric oxide that arise as a consequence of increased Sod2 expression may influence the reduction state of the mitochondrial electron transport chain and drive superoxide and H_2O_2 production [114, 121]. Alternatively, it has been proposed that Sod2 in the mitochondria may alter the flux of $O_2^{\bullet-}$ from some quinone/semiquinone/hydroquinone triads, such as that of coenzyme Q, thereby driving the reaction into the direction of $O_2^{\bullet-}$ production, potentially leading to enhanced localized dismutation to H_2O_2 [117, 122].

This increase in H_2O_2 has been suggested to lead to pro-tumorigenic and metastatic signaling in particular in cancers where levels of Sod1 and Sod2 are highly expressed. In metastatic cells Sod expression is often accompanied by shifts in the intracellular redox balance from $O_2^{\bullet-}$ to H_2O_2 and this is thought to contribute to pro-metastatic redox-signaling [29–33, 56, 112, 120, 123–126]. For example, a number of studies point to the role of H_2O_2 as the mediator of Sod1-dependent metastatic behavior. Enforced Sod1 expression enhances the formation of lung metastases by melanoma cells in a H_2O_2-dependent manner [126]. Further, Sod1 inhibition attenuates protein tyrosine phosphatases (PTPs) oxidation by H_2O_2 and subsequent inactivation, thereby leading to reduced EGF, IGF-1, and FGF-2 mediated phosphorylation of ERK1/2 and inhibition of angiogenesis, proliferation, and metastasis [10, 120, 127, 128]. Inhibition of Sod1 by the small molecule tetrathiomolybdate (TM) was also shown to enhance anchorage-independent cell death (anoikis) by activating the p38/MAPK (Mitogen-activated protein kinase) cell death pathway. These data suggest that anoikis resistance, a common feature of metastatic tumor cells, is in part attributed to high Sod1 levels [128].

Similarly, Sod2 has been associated with metastatic progression and expression of this enzyme often correlates with increasing stage and grade of tumors. Like Sod1, elevated expression of Sod2 promotes anoikis resistance and tumor metastasis in breast epithelial cells (MCF-10A), and it is thought that Sod2 may enhance scavenging of ROS that arise as a consequence of cell detachment from extracellular matrix [129]. Increased Sod2 expression is also accompanied by increases in H_2O_2 levels that influence redox-dependent pro-metastatic signaling. For example, we have shown that increases in H_2O_2, either endogenous or as a consequence of forced Sod2 expression, lead to inhibition of the phosphatase PTPN12 and enhanced p130cas-phosphorylation, Rac1 activation, and focal adhesion kinase (FAK) signaling to promote cell migration [31, 130]. In addition, increased Sod2 levels result in H_2O_2-dependent increases of Matrix metalloproteinase 1 (MMP-1) expression, promoting extracellular matrix degradation, migration, and invasion [32]. Migration and invasion of tongue squamous cell carcinoma (TSCC) were similarly mediated by Sod2-driven increases in intracellular H_2O_2 levels, leading to increased MMP-1, pERK1/2, and Snail protein expression and decreased E-cadherin levels [112]. Further, c-Myc-dependent Sod2 expression was associated with the migratory and invasive behavior of TSCC cancer stem cell subpopulations [131]. An increase in H_2O_2 as a consequence of enforced Sod2 expression has recently been demonstrated to contribute to AMP activated-kinase A (AMPK) activation and a consequential increase in glycolysis, demonstrating that Sod2 can influence metabolic reprogramming during metastatic progression [132]. Clearly, the effects of increased H_2O_2 levels as a consequence of Sod expression are dependent on the levels of H_2O_2 and its cellular or extracellular location. More studies are needed to understand the spatiotemporal role of H_2O_2 in redox-signaling during carcinogenesis and metastasis.

Sods appear to further alter the redox environment by reacting with H_2O_2. Spin trapping studies have shown that the reaction of Sod1 and H_2O_2 generates $OH^•$, which can in turn inactivate the enzyme by modification of histidine residue in the active site [133, 134]. In an in vivo model of atherosclerosis, the reaction of Sod3 with H_2O_2 was studied in the presence of bicarbonate ions and shown to produce $OH^•$ [135]. Like other Cu containing Sod, Sod3 was also inactivated after reacting with H_2O_2. This ability to catalyze the conversion of H_2O_2 to generate free $OH^•$ suggests that Sod enzymes also have the ability to be radical generators that can induce oxidative damage, and this may provide an alternate mechanism to explain the oncogenic role of Sods [136–139]. Further, studies have shown that apart from the ability to catalyze the conversion of $O_2^{•-}$ to H_2O_2, Sod2 also has peroxidase activity, as inorganic Mn (II) complexes display the ability to dismutate H_2O_2 [137, 140]. Similarly, Ansenberger-Fricano *et al.* reported that Sod2 has an intrinsic peroxidase activity when the enzyme is overexpressed in cancer cells, leading to mitochondrial damage and dysfunction [141]. Therefore, depending on its expression levels, Sod2 activity will differ and result in protection or sensitization of mitochondria to oxidative stress as a result of its dismutase or peroxidase activity, respectively.

4.3 Catalase

Catalase is a 60 kDa enzyme with four heme groups per tetramer which rapidly converts H_2O_2 into water and oxygen (Fig. 4.2) [142]. After the interaction with H_2O_2, catalase is converted to an oxidized heme intermediate (compound I) that is reduced by a second molecule of H_2O_2. Catalase is expressed in all organisms requiring oxygen for survival, and is almost exclusively found in peroxisomes [143]. Extended life spans have been observed in *Drosophila*, *Caenorhabditis elegans*, and transgenic mouse models that display catalase overexpression [144–146]. Most studies point to catalase as being a tumor suppressor, given that its expression is significantly reduced in a number of cancers, and that loss of expression results in H_2O_2 accumulation in cells. As discussed below, a decrease in catalase expression often leads to H_2O_2-mediated tumorigenesis and metastasis, but may in some instances lead to cancer cell death. Therefore, like Sod, enhanced expression of catalase may be a beneficial adaptation of certain tumor cells to protect against apoptotic cell death [147]. It has been reported that tumor cells may express a membrane localized form of catalase that specifically protects tumor cells from exogenous H_2O_2 [148, 149]. Similarly, catalase may protect tumor cells from autophagic cell death. It has been shown that catalase is selectively degraded during autophagy and that this is the prime cause for resulting ROS accumulation and non-apoptotic cell death [150]. It remains to be elucidated whether a loss of catalase expression is also associated with cytoprotective autophagy, a pathway often used by cancer cells to survive stress and nutrient deprivation. As eluded to previously, the consequential effects of altering steady-state H_2O_2 levels are likely dependent on the amounts and location of the ROS produced. Hence, a decrease in catalase may either drive protumorigenic signaling by increasing sublethal steady-state H_2O_2 levels, or promote the onset of more deleterious pathways, such as apoptosis and autophagy, when H_2O_2 levels rise above lethal thresholds.

It has been demonstrated that many tumor types exhibit decreased catalase expression, including lung [151, 152], cervical [153], liver [154], and breast cancers [155]. Low erythrocyte catalase levels have also been associated with patients suffering from lung cancer [152]. This decrease in catalase expression has been evaluated as a potential prognostic and predictive marker of disease outcome. For example, 72 % of hepatocellular carcinoma (HCC) specimens showed reduced catalase levels compared to surrounding non-tumor tissues, with further reduction observed in advanced stage IV cancers, suggesting that high catalase expression is a positive predictive marker of HCC patient survival and that a loss correlates with increasing tumor grade and metastatic spread [156]. Similarly, low catalase activity has been associated with enhanced aggressiveness of breast cancer [155]. Given that some tumors also display low glutathione peroxidase levels, it was proposed by Oberley and Oberley that most cancer cells lack the ability to detoxify H_2O_2 [76]. This increase in H_2O_2 is thought to contribute to ROS-mediated carcinogenesis, especially since a decrease in catalase expression is often seen in early cancerous lesions. An advantageous aspect to this loss of catalase expression in cancer cells is that we may take advantage of the cancer cells inability to effectively detoxify ROS for therapeutic purposes. For example, analyzing the pathological complete response of patients to anthracycline, a neoadjuvant chemotherapeutic used in advanced stage breast cancer, revealed that low catalase expression provided an advantage to the patient, presumably due to the tumor's inability to detoxify ROS emanating from the anthracycline [157]. This study suggests that cancers with low catalase expression may be targets for ROS-producing agents.

Besides a role for abrogated catalase expression during carcinogenesis, a number of studies have shown that this loss of H_2O_2 detoxification is also associated with metastasis, and that enforced expression or delivery of recombinant catalase to tumor cells can mitigate metastatic behavior. For example, decreased levels of catalase expression correlate with hepatocellular carcinoma grade and vascular invasion, which is suppressed by catalase overexpression [154]. In an established mouse model of human breast cancer, transgenic overexpression of mitochondria-targeted catalase drastically reduced metastatic tumor burden by >12-fold, emphasizing the role of mitochondrial oxidative stress in tumor progression and metastasis [158]. Similarly, enforced expression of catalase in MCF-7 breast cancer cell lines impaired their proliferative and migratory activity and increased their sensitivity to anticancer drugs [159]. We have shown that HT1080 fibrosarcoma cells engineered to overexpress catalase can attenuate metastatic capacity and invasive potential by blocking Sod2-dependent induction of the Ras/MAPK/AP-1 pathway that drives MMP-1 expression [32]. This study also demonstrated that H_2O_2-mediated redox-signaling as a consequence of altered catalase and Sod2 expression is an important pathway in the regulation of metastatic behavior. Interestingly, based on oncomine expression analysis, we have observed that many cancers which display enhanced Sod2 expression also have a concomitant reduction in catalase levels [26, 56]. This may be further exacerbated during metastatic progression. For example, we observed that an overall shift towards higher steady-state H_2O_2 levels in the highly metastatic bladder cancer cells is associated with low catalase and high Sod2 levels and that this contributes to the regulation of VEGF and MMP-9 production [26].

A number of mechanisms have been described that lead to the downregulation of catalase in cancer cells, including transcriptional and epigenetic regulation. Low catalase promoter activity has been observed in a number of cancer cell lines and was shown to be further repressed in advanced stage disease in a mouse skin tumor progression model [160, 161]. In this case, the transcriptional repression of catalase was shown to be mediated by Wilm's tumor suppressor 1 (WT1) binding within the proximal promoter region [161]. Catalase activity is also reduced by conditionally overexpressing C10orf10/DEPP, a transcriptional target of FOXO3, thereby sensitizing tumor cells to ROS-induced cell death [162]. In contrast, FOXO3 was not found to play a pivotal role in the regulation of catalase expression in parental mammary breast MCF-7 cancer cells. Conversely, in that study the authors reported an increase in catalase expression that was attributed to enhanced GSK3β and p70S6K signaling [163]. The tumor environment also appears to play a role in negatively regulating catalase expression in tumors. Sustained exposure to ROS has been found to induce methylation of CpG islands on the catalase promoter leading to its downregulation in hepatocellular carcinoma cell lines [164, 165]. The POU domain transcription factor Oct-1, which normally binds to the catalase promoter as an activator, is also downregulated by ROS-mediated CpG island promoter methylation in hepatocellular carcinoma [166]. Deacetylation of histone H4 has been shown to be responsible for the downregulation of catalase in the doxorubicin-resistant acute myelogenous leukemia (AML)-2/DX100 cells [167]. Thus, numerous epigenetic control mechanisms exist for the regulation of catalase gene expression.

Overall, it appears that catalase downregulation is involved in the etiology of a number of different cancers and that its loss during metastatic progression increases H_2O_2-dependent signaling. Restoring cellular H_2O_2 homeostasis has been suggested as a therapeutic strategy against metastatic cancers that display decreased catalase expression [35, 168]. At present, no specific catalase mimetic compounds exist; however, the delivery of recombinant catalase has been explored in therapeutic contexts [169]. Unfortunately, the feasibility of using recombinant proteins for therapeutics is associated with a number of challenges including delivery, stability, and immunoreactivity.

While the above work demonstrates that a loss in catalase expression is associated with a number of cancers and metastatic spread, there are some reports to show that enhanced catalase expression may also be associated with tumorigenesis and certain aspects of metastatic progression. High catalase levels are observed in malignant and highly invasive prostate cancer cells and were shown to contribute to reduced ROS levels [170]. Like Sods, catalase expression is also required for the survival of cancer cells in anchorage independence [171] and cytotoxic stress [172]. Catalase appears to be similarly important in preventing the ROS surge during extracellular matrix detachment of cells and contributing to anoikis resistance of cancer cells. This suggests that catalase expression may also be dynamically regulated in tumor cell progression and may provide a survival advantage to cancer cells under certain stress conditions.

4.4 Glutathione Peroxidases

The Glutathione peroxidase family comprises eight family members. Mammalian GPx 1–4 were the first well-defined GPxs and comprise four distinct mammalian selenoproteins including classical GPx1; gastrointestinal isoenzyme, GPx2; plasma, GPx3; and phospholipid hydroperoxide, GPx4. All individual isoenzymes function as efficient peroxidases to catalyze the degradation of peroxides and hydroperoxides to their corresponding water/alcohol, using reduced glutathione as the specific hydrogen donor and play varying tissue-specific roles in metabolic pathway regulation (Fig. 4.2) [173]. Gpx 5/6 and Gpx 7/8 are evolutionary variants of Gpx3 and Gpx4, respectively. Here we will focus on the many reported roles of the primary GPx 1–4 family members in distinct cancers. For a comprehensive review of the distinct biochemical activities of the various GPx family members, we refer the reader to review by Brigelius-Flohe & Maiorino [174].

The role of GPx1 in tumor progression and metastasis is often specific to the type and stage of cancer under study. For example, decreases in the immunoreactive selenium-binding protein 1 (SBP1), which limits GPx1 activity, lead to an increase in GPx1 levels. This increase in GPx1 expression is correlated with enhanced vascular invasion and HIF-1α downregulation and has been suggested to serve as a potential prognostic indicator for hepatocellular carcinoma [175, 176]. High GPx1 levels have also been correlated with resistance to chemotherapy in breast cancer patients [177]. Similar high level expression of Gpx1 can promote invasion, migration, proliferation, and cisplatin-resistance of esophageal cancer which is attenuated through a Vitamin D dependent inhibition of GPx1 [178]. In contrast, weakened

GPx1 expression has been strongly associated with tumor differentiation and progression in gastric adenocarcinoma (GAC). The loss of cytosolic expression of GPx1 has also been associated with poor differentiation and extensive lymph node involvement leading to aggressiveness and poor outcome in patients with GAC [179]. Promoter Hypermethylation is also responsible for the loss of expression of GPx1 and is associated with aggressiveness and poor gastric cancer patient outcome [180].

Similar to GPx1, the role of GPx2 is also tumor specific. GPx2 has been shown to be upregulated in response to the p53 family member p63 which serves to counter the apoptotic inducing properties of p53 and confer resistance to apoptotic cell death in MCF-7 cells [181]. GPx2 silencing can also significantly inhibit the subcutaneous growth of both rat and human castration-resistant prostate cancer (CRPC) cells. Furthermore GPx2 expression has been shown to be high in biopsy specimens and is correlated with low PSA recurrence-free survival and overall survival, suggesting GPx2 as a prognostic marker for CRPC [182]. In contrast, the homeobox gene Nkx3.1 which is essential for prostatic epithelial differentiation and suppressing prostate cancer protects against oxidative damage by maintaining GPx2 expression and loss-of-function of Nkx3.1 increases susceptibility to oxidative stress [183].

Reduced GPx3 expression is also significantly associated with lymph node metastasis and invasion of gallbladder cancers suggesting mechanistic involvement of oxidative stress in the progression of this tumor type [184]. High frequency of promoter hypermethylation and the subsequent loss of GPx3 expression have been implicated in Barrett's esophagus tumorigenesis that induces oxidative mucosal damage [185]. Hemizygous or homozygous deletion of GPx3 gene and hypermethylation of the GPx3 exon 1 region in prostate cancer samples leading to GPx3 inactivation are associated with pathogenesis of prostate cancer [186]. GPx3 expression is observed to be downregulated in multiple types of cancer, including gastric, cervical, thyroid, head and neck, lung cancer, melanoma, and inflammatory breast carcinogenesis when compared to healthy controls due to promoter hypermethylation [180, 187, 188]. It has been reported that promoter DNA hypermethylation in CpG islands around the transcription start sites of GPx3 and GPx7 represses their expression in Barrett's adenocarcinoma [189]. In endometrial adenocarcinoma, GPx3 expression is uniformly downregulated in rat and human tumor samples, regardless of tumor grade or histopathological subtype, as a result of promoter methylation [190]. A number of single nucleotide polymorphisms (SNPs) of GPx3 have been observed in differentiated thyroid cancer patients and allelic differences are associated with both decreased and increased risk for thyroid cancer [191].

The role of Gpx4 in tumor progression is more speculative. It has been shown that enforced expression of GPx4 results in the downregulation of irradiation-induced MMP-1 in dermal fibroblasts [192]. It is possible that GPx4 may play a similar role in restricting MMP-dependent migration and invasion of tumor cells. GPx4 also has a T/C SNP at position 718 within the 3′UTR near the selenocysteine amino acid site that is associated with increased gene reporter activity. Furthermore, comparison of SNP frequency indicates that patients with colorectal adenocarcinomas have a higher frequency of the CC genotype. Thus, GPx4 T/C 718 SNP variant is associated with higher gene activity and an increased risk for developing colorectal cancer [193]. A distinct GPx4 SNP (rs3746162) has also been associated with a 5.4-fold increase of recurrent disease following transurethral bladder cancer resection, but whether the rs3746162 (A;G) effects GPx4 levels has not been defined. However, the conclusion from this latter study strongly suggests that genetic variants in GSH pathway enzymes may influence recurrence in non-muscle invasive bladder cancer patients receiving curative treatment [194].

The diverse roles of the individual peroxidase enzymes render prediction of a general influence on carcinogenesis and tumor growth difficult. Most GPx enzymes tend to inhibit initiation and metastasis, suggesting that a loss of GPx enzymes is a phenotype of metastatic progression; however, as observed with catalase and Sods, the role of GPx in tumors is likely context dependent and this needs to be taken into account when designing GPx-targeted therapeutics.

4.5 Conclusion

It is evident that it is difficult to categorize antioxidant enzymes and specific isoforms as tumor suppressors or oncogenes, given their complex role in regulating redox balance in the tumor environment. Based on "common themes" associated with the role of the above antioxidant enzymes during metastatic progression, it appears that the H_2O_2 detoxifying GPx and Catalase enzymes are frequently downregulated in metastatic cancers. Forced expression of these can often abrogate metastatic behavior and their decrease may contribute to a higher cellular steady-state H_2O_2 status. Interestingly, a recent paper lays credence to this hypothesis, demonstrating that the ratios of highSod2:lowCat and highSod2:lowGPx expression are associated with progression of metastatic prostate, colon, and lung cancer [96]. Further, concomitant increases in H_2O_2 levels were observed in advanced stage prostate and lung cancers compared to normal tissue and low grade controls [96]. While the removal of H_2O_2 detoxifying enzymes may certainly increase the pool of cellular H_2O_2, it remains to be elucidated how the commonly observed increases in Sod1/2 contribute to this redox shift in metastatic tumor cells.

Expression analysis of proteins gives clues to their predicted role in a given tumor type; however, this provides little information on the spatial and temporal role of the antioxidant during tumor progression. The next challenges in determining the role of antioxidant enzymes in tumor progression are to understand their dynamic regulation during metastatic progression as these enzymes clearly provide protective advantages towards the survival of tumor cells in the hostile metastatic tumor environment (Fig. 4.3). Further, we will need to grasp the complexity of how different antioxidant enzymes and their altered expression and activity levels contribute to the redox balance of tumor cells, and how this influences redox-mediated signaling that either contributes or prevents tumorigenesis and metastasis. Monitoring specific ROS in real time remains difficult due to their highly reactive

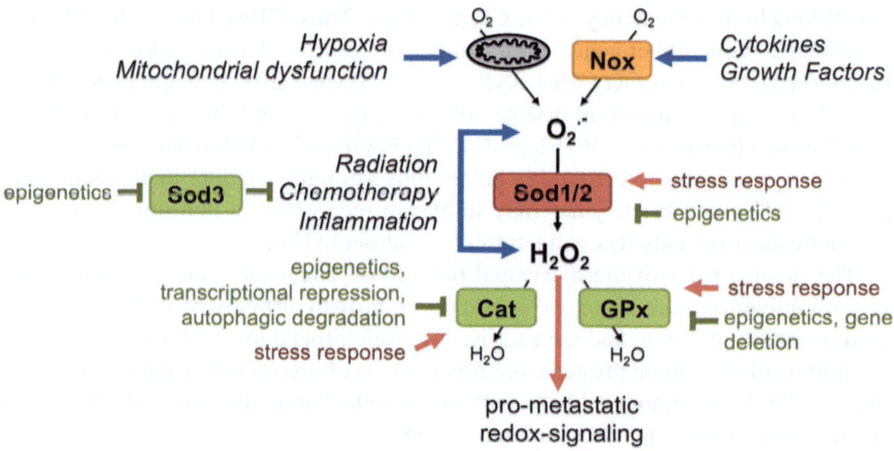

Fig. 4.3 Altered Superoxide Dismutase (Sod), Catalase (Cat), and Glutathione Peroxidase (GPx) expression can shift the redox balance within metastatic tumor cells. Expression of antioxidant enzymes is often tumor type and tumor stage-dependent. Epigenetics, such as promoter hypermethylation, has been described as a common mechanism for the downregulation of a number of antioxidant enzymes in cancer cells, while stress response signaling pathways such as the Nrf2/Keap1 and NF-κB contribute to high expression of antioxidant enzymes during tumor metastasis. Studies suggest that the antioxidant profile of metastatic cells may dictate an increase in steady-state H_2O_2, which can drive redox-mediated pro-metastatic signaling events. *Green* or *red* shading indicates that a majority of studies have found either decreased or increased expression in metastatic cancers, respectively

nature and lack of specific redox sensing dyes and molecular probes. There is a need for greater understanding of the actions of different ROS like $O_2^{\bullet-}$, H_2O_2, and $OH^{\bullet}$ and their interactions with cellular molecules. Tools that will allow monitoring of their temporal, spatial, and concentration dependent effects will enhance our ability to understand the dynamics of redox-signaling in tumor progression.

Given that the above antioxidant enzymes appear to be intricately associated with cancer and metastatic progression, and that ROS/RNS balance may play a significant role in altering cellular behavior of tumor cells, the use of redox-active therapeutics is an attractive strategy to explore for metastatic cancer. However, as discussed above, great care will need to be taken to ensure that the correct ROS are scavenged, and that reactions of these "scavengers" with reactive species do not yield other harmful species or result in the oxidation of macromolecules, that may lead to deleterious effects. The use of Sod and GPx mimetics, including metal containing porphyrins and ebselen, respectively, may prevent H_2O_2 buildup in cancer cells as a consequence of $O_2^{\bullet-}$ removal. However, like the observation of enhanced H_2O_2 levels as a consequence of Sod expression, it has been found that porphyrin-based Sod mimetics may similarly produce H_2O_2 following interaction with $O_2^{\bullet-}$ (reviewed in [195]). Also, many "antioxidant" compounds, such as the metal-based porphyrins, can act to oxidize macromolecules and may therefore themselves lead to thiol oxidation and the manipulation of redox-signaling (reviewed in [195]). We refer the reader to the chapter by Dr. Batinić-Haberle for further description on the reaction kinetics, mechanisms, and use of porphyrin compounds. It would appear that metastatic tumor cells with high expression of Sod1/2 and low Catalase/GPx, and concomitant increases in cellular H_2O_2 levels would benefit most from either H_2O_2 scavenging or by utilizing their high H_2O_2 threshold to effectively kill cells with exogenous H_2O_2. While specific scavenging of H_2O_2 by the catalase enzyme has been shown to inhibit metastatic behavior and decreases H_2O_2-mediated redox-signaling, there are no small molecules to date that exhibit the same specific activity as the enzyme [195, 196]. Research in this area may be a promising novel therapeutic approach to tackle metastatic cancer.

References

1. Sosa V, Moline T, Somoza R, Paciucci R, Kondoh H, LLeonart ME. Oxidative stress and cancer: an overview. Ageing Res Rev. 2013;12(1):376–90.
2. Dolado I, Swat A, Ajenjo N, De Vita G, Cuadrado A, Nebreda AR. p38alpha MAP kinase as a sensor of reactive oxygen species in tumorigenesis. Cancer Cell. 2007;11(2):191–205.
3. Zieba M, Suwalski M, Kwiatkowska S, Piasecka G, Grzelewska-Rzymowska I, Stolarek R, Nowak D. Comparison of hydrogen peroxide generation and the content of lipid peroxidation products in lung cancer tissue and pulmonary parenchyma. Respir Med. 2000;94(8):800–5.
4. Szatrowski TP, Nathan CF. Production of large amounts of hydrogen peroxide by human tumor cells. Cancer Res. 1991;51(3):794–8.
5. Lim SD, Sun C, Lambeth JD, Marshall F, Amin M, Chung L, Petros JA, Arnold RS. Increased Nox1 and hydrogen peroxide in prostate cancer. Prostate. 2005;62(2):200–7.
6. Xiong P, Li YX, Tang YT, Chen HG. Proteomic analyses of Sirt1-mediated cisplatin resistance in OSCC cell line. Protein J. 2011;30(7):499–508.
7. Benlloch M, Mena S, Ferrer P, Obrador E, Asensi M, Pellicer JA, Carretero J, Ortega A, Estrela JM. Bcl-2 and Mn-SOD antisense oligodeoxynucleotides and a glutamine-enriched diet facilitate elimination of highly resistant B16 melanoma cells by tumor necrosis factor-alpha and chemotherapy. J Biol Chem. 2006;281(1):69–79.
8. Hur GC, Cho SJ, Kim CH, Kim MK, Bae SI, Nam SY, Park JW, Kim WH, Lee BL. Manganese superoxide dismutase expression correlates with chemosensitivity in human gastric cancer cell lines. Clin Cancer Res. 2003;9(15):5768–75.
9. Suresh A, Guedez L, Moreb J, Zucali J. Overexpression of manganese superoxide dismutase promotes survival in cell lines after doxorubicin treatment. Br J Haematol. 2003; 120(3):457–63.
10. Izutani R, Kato M, Asano S, Imano M, Ohyanagi H. Expression of manganese superoxide dismutase influences chemosensitivity in esophageal and gastric cancers. Cancer Detect Prev. 2002;26(3):213–21.
11. Catalano V, Turdo A, Di Franco S, Dieli F, Todaro M, Stassi G. Tumor and its microenvironment: a synergistic interplay. Semin Cancer Biol. 2013;23(6 Pt B):522–32.
12. Chandel NS, McClintock DS, Feliciano CE, Wood TM, Melendez JA, Rodriguez AM, Schumacker PT. Reactive oxygen species generated at mitochondrial complex III stabilize hypoxia-inducible factor-1alpha during hypoxia: a mechanism of O2 sensing. J Biol Chem. 2000;275(33):25130–8.
13. Guzy RD, Hoyos B, Robin E, Chen H, Liu L, Mansfield KD, Simon MC, Hammerling U, Schumacker PT. Mitochondrial complex III is required for hypoxia-induced ROS production and cellular oxygen sensing. Cell Metab. 2005;1(6):401–8.

14. Bell EL, Klimova TA, Eisenbart J, Moraes CT, Murphy MP, Budinger GR, Chandel NS. The Qo site of the mitochondrial complex III is required for the transduction of hypoxic signaling via reactive oxygen species production. J Cell Biol. 2007;177(6):1029–36.
15. Emerling BM, Platanias LC, Black E, Nebreda AR, Davis RJ, Chandel NS. Mitochondrial reactive oxygen species activation of p38 mitogen-activated protein kinase is required for hypoxia signaling. Mol Cell Biol. 2005;25(12):4853–62.
16. Gerald D, Berra E, Frapart YM, Chan DA, Giaccia AJ, Mansuy D, Pouyssegur J, Yaniv M, Mechta-Grigoriou F. JunD reduces tumor angiogenesis by protecting cells from oxidative stress. Cell. 2004;118(6):781–94.
17. Pan Y, Mansfield KD, Bertozzi CC, Rudenko V, Chan DA, Giaccia AJ, Simon MC. Multiple factors affecting cellular redox status and energy metabolism modulate hypoxia-inducible factor prolyl hydroxylase activity in vivo and in vitro. Mol Cell Biol. 2007;27(3):912–25.
18. Yasinska IM, Sumbayev VV. S-nitrosation of Cys-800 of HIF-1alpha protein activates its interaction with p300 and stimulates its transcriptional activity. FEBS Lett. 2003;549(1–3):105–9.
19. Block K, Gorin Y, Hoover P, Williams P, Chelmicki T, Clark RA, Yoneda T, Abboud HE. NAD(P)H oxidases regulate HIF-2alpha protein expression. J Biol Chem. 2007;282(11):8019–26.
20. Dioum EM, Chen R, Alexander MS, Zhang Q, Hogg RT, Gerard RD, Garcia JA. Regulation of hypoxia-inducible factor 2alpha signaling by the stress-responsive deacetylase sirtuin 1. Science. 2009;324(5932):1289–93.
21. Hamanaka RB, Chandel NS. Mitochondrial reactive oxygen species regulate cellular signaling and dictate biological outcomes. Trends Biochem Sci. 2010;35(9):505–13.
22. Poyton RO, Ball KA, Castello PR. Mitochondrial generation of free radicals and hypoxic signaling. Trends Endocrinol Metab. 2009;20(7):332–40.
23. Kim S, Takahashi H, Lin WW, Descargues P, Grivennikov S, Kim Y, Luo JL, Karin M. Carcinoma-produced factors activate myeloid cells through TLR2 to stimulate metastasis. Nature. 2009;457(7225):102–6.
24. Landskron G, De la Fuente M, Thuwajit P, Thuwajit C, Hermoso MA. Chronic inflammation and cytokines in the tumor microenvironment. J Immunol Res. 2014;2014:149185.
25. Pani G, Galeotti T, Chiarugi P. Metastasis: cancer cell's escape from oxidative stress. Cancer Metastasis Rev. 2010;29(2):351–78.
26. Hempel N, Ye H, Abessi B, Mian B, Melendez JA. Altered redox status accompanies progression to metastatic human bladder cancer. Free Radic Biol Med. 2009;46(1):42–50.
27. Weyemi U, Redon CE, Parekh PR, Dupuy C, Bonner WM. NADPH Oxidases NOXs and DUOXs as putative targets for cancer therapy. Anticancer Agents Med Chem. 2013;13(3):502–14.
28. Block K, Gorin Y. Aiding and abetting roles of NOX oxidases in cellular transformation. Nat Rev Cancer. 2012;12(9):627–37.
29. Polytarchou C, Hatziapostolou M, Papadimitriou E. Hydrogen peroxide stimulates proliferation and migration of human prostate cancer cells through activation of activator protein-1 and up-regulation of the heparin affin regulatory peptide gene. J Biol Chem. 2005;280(49):40428–35.
30. Connor KM, Hempel N, Nelson KK, Dabiri G, Gamarra A, Belarmino J, Van De Water L, Mian BM, Melendez JA. Manganese superoxide dismutase enhances the invasive and migratory activity of tumor cells. Cancer Res. 2007;67(21):10260–7.
31. Hempel N, Bartling TR, Mian B, Melendez JA. Acquisition of the metastatic phenotype is accompanied by H2O2-dependent activation of the p130Cas signaling complex. Mol Cancer Res. 2013;11(3):303–12.
32. Nelson KK, Ranganathan AC, Mansouri J, Rodriguez AM, Providence KM, Rutter JL, Pumiglia K, Bennett JA, Melendez JA. Elevated Sod2 activity augments matrix metalloproteinase expression: Evidence for the involvement of endogenous hydrogen peroxide in regulating metastasis. Clin Cancer Res. 2003;9:424–32.

33. Connor KM, Subbaram S, Regan KJ, Nelson KK, Mazurkiewicz JE, Bartholomew PJ, Aplin AE, Tai YT, Aguirre-Ghiso J, Flores SC, Melendez JA. Mitochondrial H2O2 regulates the angiogenic phenotype via PTEN oxidation. J Biol Chem. 2005;280(17):16916–24.
34. Chiarugi P, Fiaschi T. Redox signalling in anchorage-dependent cell growth. Cell Signal. 2007;19(4):672–82.
35. Nishikawa M. Reactive oxygen species in tumor metastasis. Cancer Lett. 2008;266(1):53–9.
36. Nishikawa M, Tamada A, Hyoudou K, Umeyama Y, Takahashi Y, Kobayashi Y, Kumai H, Ishida E, Staud F, Yabe Y, Takakura Y, Yamashita F, Hashida M. Inhibition of experimental hepatic metastasis by targeted delivery of catalase in mice. Clin Exp Metastasis. 2004;21(3):213–21.
37. del Bello B, Paolicchi A, Comporti M, Pompella A, Maellaro E. Hydrogen peroxide produced during gamma-glutamyl transpeptidase activity is involved in prevention of apoptosis and maintenance of proliferation in U937 cells. FASEB J. 1999;13(1):69–79.
38. Chiarugi P, Pani G, Giannoni E, Taddei L, Colavitti R, Raugei G, Symons M, Borrello S, Galeotti T, Ramponi G. Reactive oxygen species as essential mediators of cell adhesion: the oxidative inhibition of a FAK tyrosine phosphatase is required for cell adhesion. J Cell Biol. 2003;161:933–44.
39. Chiarugi P. Reactive oxygen species as mediators of cell adhesion. Ital J Biochem. 2003;52(1):28–32.
40. Arbiser JL, Petros J, Klafter R, Govindajaran B, McLaughlin ER, Brown LF, Cohen C, Moses M, Kilroy S, Arnold RS, Lambeth JD. Reactive oxygen generated by Nox1 triggers the angiogenic switch. Proc Natl Acad Sci U S A. 2002;99(2):715–20.
41. Ushio-fukai M, Tang Y, Fukai T, Dikalov SI, Ma Y, Fujimoto M, Quinn MT, Pagano PJ, Johnson C, Alexander RW. Novel Role of gp91phox containing NAD(P)H oxidase in vascular endothelial factor-induced signaling and angiogenesis. Circ Res. 2014;91(12):1160–7.
42. Yoon SO, Park SJ, Yoon SY, Yun CH, Chung AS. Sustained production of H(2)O(2) activates pro-matrix metalloproteinase- 2 through receptor tyrosine kinases/phosphatidylinositol 3-kinase/NF- kappa B pathway. J Biol Chem. 2002;277(33):30271–82.
43. Ray PD, Huang BW, Tsuji Y. Reactive oxygen species (ROS) homeostasis and redox regulation in cellular signaling. Cell Signal. 2012;24(5):981–90.
44. Weinberg F, Chandel NS. Reactive oxygen species-dependent signaling regulates cancer. Cell Mol Life Sci. 2009;66(23):3663–73.
45. Wu W-S. The signaling mechanism of ROS in tumor progression. Cancer Metastasis Rev. 2006;25(4):695–705.
46. Brandes N, Schmitt S, Jakob U. Thiol-based redox switches in eukaryotic proteins. Antioxid Redox Signal. 2009;11(5):997–1014.
47. Kwon J, Lee SR, Yang KS, Ahn Y, Kim YJ, Stadtman ER, Rhee SG. Reversible oxidation and inactivation of the tumor suppressor PTEN in cells stimulated with peptide growth factors. Proc Natl Acad Sci U S A. 2004;101(47):16419–24.
48. Leslie NR, Bennett D, Lindsay YE, Stewart H, Gray A, Downes CP. Redox regulation of PI 3-kinase signalling via inactivation of PTEN. EMBO J. 2003;22(20):5501–10.
49. Henle ES, Linn S. Formation, prevention, and repair of DNA damage by iron/hydrogen peroxide. J Biol Chem. 1997;272(31):19095–8.
50. Imlay JA, Linn S. DNA damage and oxygen radical toxicity. Science. 1988; 240(4857):1302–9.
51. Jaramillo MC, Zhang DD. The emerging role of the Nrf2-Keap1 signaling pathway in cancer. Genes Dev. 2013;27(20):2179–91.
52. Oliveira-Marques V, Marinho HS, Cyrne L, Antunes F. Role of hydrogen peroxide in NF-kappaB activation: from inducer to modulator. Antioxid Redox Signal. 2009; 11(9):2223–43.
53. Kops GJ, Dansen TB, Polderman PE, Saarloos I, Wirtz KW, Coffer PJ, Huang TT, Bos JL, Medema RH, Burgering BM. Forkhead transcription factor FOXO3a protects quiescent cells from oxidative stress. Nature. 2002;419(6904):316–21.

54. Kansanen E, Kuosmanen SM, Leinonen H, Levonen AL. The Keap1-Nrf2 pathway: mechanisms of activation and dysregulation in cancer. Redox Biol. 2013;1(1):45–9.
55. Diehn M, Cho RW, Lobo NA, Kalisky T, Dorie MJ, Kulp AN, Qian D, Lam JS, Ailles LE, Wong M, Joshua B, Kaplan MJ, Wapnir I, Dirbas FM, Somlo G, Garberoglio C, Paz B, Shen J, Lau SK, Quake SR, Brown JM, Weissman IL, Clarke MF. Association of reactive oxygen species levels and radioresistance in cancer stem cells. Nature. 2009;458(7239):780–3.
56. Hempel N, Carrico PM, Melendez JA. Manganese superoxide dismutase (Sod2) and redox-control of signaling events that drive metastasis. Anticancer Agents Med Chem. 2011;11(2):191–201.
57. Fukai T, Ushio-Fukai M. Superoxide dismutases: role in redox signaling, vascular function, and diseases. Antioxid Redox Signal. 2011;15(6):1583–606.
58. Weisiger RA, Fridovich I. Superoxide dismutase. Organelle specificity. J Biol Chem. 1973;248(10):3582–92.
59. Tainer JA, Getzoff ED, Richardson JS, Richardson DC. Structure and mechanism of copper, zinc superoxide dismutase. Nature. 1983;306(5940):284–7.
60. Oh YK, Shin KS, Yuan J, Kang SJ. Superoxide dismutase 1 mutants related to amyotrophic lateral sclerosis induce endoplasmic stress in neuro2a cells. J Neurochem. 2008;104(4):993–1005.
61. Saccon RA, Bunton-Stasyshyn RK, Fisher EM, Fratta P. Is SOD1 loss of function involved in amyotrophic lateral sclerosis? Brain. 2013;136(Pt 8):2342–58.
62. Joyce PI, McGoldrick P, Saccon RA, Weber W, Fratta P, West SJ, Zhu N, Carter S, Phatak V, Stewart M, Simon M, Kumar S, Heise I, Bros-Facer V, Dick J, Corrochano S, Stanford MJ, Luong TV, Nolan PM, Meyer T, Brandner S, Bennett DL, Ozdinler PH, Greensmith L, Fisher EM, Acevedo-Arozena A. A novel SOD1-ALS mutation separates central and peripheral effects of mutant SOD1 toxicity. Hum Mol Genet. 2014;24(7):1883–97.
63. McCord JM, Fridovich I. Superoxide dismutase. An enzymic function for erythrocuprein (hemocuprein). J Biol Chem. 1969;244(22):6049–55.
64. Beck Y, Oren R, Amit B, Levanon A, Gorecki M, Hartman JR. Human Mn superoxide dismutase cDNA sequence. Nucleic Acids Res. 1987;15(21):9076.
65. Beckman G, Lundgren E, Tarnvik A. Superoxide dismutase isozymes in different human tissues, their genetic control and intracellular localization. Hum Hered. 1973;23(4):338–45.
66. Barra D, Schinina ME, Simmaco M, Bannister JV, Bannister WH, Rotilio G, Bossa F. The primary structure of human liver manganese superoxide dismutase. J Biol Chem. 1984;259(20):12595–601.
67. Fattman CL, Schaefer LM, Oury TD. Extracellular superoxide dismutase in biology and medicine. Free Radic Biol Med. 2003;35(3):236–56.
68. O'Leary BR, Fath MA, Bellizzi AM, Hrabe JE, Button AM, Allen BG, Case AJ, Altekruse SF, Wagner B, Buettner GR, Lynch CF, Hernandez BY, Cozen W, Beardsley RA, Keene J, Henry MD, Domann FE, Spitz DR, Mezhir JJ. Loss of SOD3 (EcSOD) expression promotes an aggressive phenotype in human pancreatic ductal adenocarcinoma. Clin Cancer Res. 2015;21:1741–51.
69. Kim J, Mizokami A, Shin M, Izumi K, Konaka H, Kadono Y, Kitagawa Y, Keller ET, Zhang J, Namiki M. SOD3 acts as a tumor suppressor in PC-3 prostate cancer cells via hydrogen peroxide accumulation. Anticancer Res. 2014;34(6):2821–31.
70. Laatikainen LE, Castellone MD, Hebrant A, Hoste C, Cantisani MC, Laurila JP, Salvatore G, Salerno P, Basolo F, Nasman J, Dumont JE, Santoro M, Laukkanen MO. Extracellular superoxide dismutase is a thyroid differentiation marker down-regulated in cancer. Endocr Relat Cancer. 2010;17(3):785–96.
71. Teoh-Fitzgerald ML, Fitzgerald MP, Jensen TJ, Futscher BW, Domann FE. Genetic and epigenetic inactivation of extracellular superoxide dismutase promotes an invasive phenotype in human lung cancer by disrupting ECM homeostasis. Mol Cancer Res. 2012;10(1):40–51.
72. Teoh ML, Fitzgerald MP, Oberley LW, Domann FE. Overexpression of extracellular superoxide dismutase attenuates heparanase expression and inhibits breast carcinoma cell growth and invasion. Cancer Res. 2009;69(15):6355–63.

73. Teoh-Fitzgerald ML, Fitzgerald MP, Zhong W, Askeland RW, Domann FE. Epigenetic reprogramming governs EcSOD expression during human mammary epithelial cell differentiation, tumorigenesis and metastasis. Oncogene. 2014;33(3):358–68.
74. Cohen I, Pappo O, Elkin M, San T, Bar-Shavit R, Hazan R, Peretz T, Vlodavsky I, Abramovitch R. Heparanase promotes growth, angiogenesis and survival of primary breast tumors. Int J Cancer. 2006;118(7):1609–17.
75. Gotte M, Yip GW. Heparanase, hyaluronan, and CD44 in cancers: a breast carcinoma perspective. Cancer Res. 2006;66(21):10233–7.
76. Oberley TD, Oberley LW. Antioxidant enzyme levels in cancer. Histol Histopathol. 1997;12(2):525–35.
77. Bravard A, Sabatier L, Hoffschir F, Ricoul M, Luccioni C, Dutrillaux B. SOD2: a new type of tumor-suppressor gene? Int J Cancer. 1992;51(3):476–80.
78. Church SL, Grant JW, Ridnour LA, Oberley LW, Swanson PE, Meltzer PS, Trent JM. Increased manganese superoxide dismutase expression suppresses the malignant phenotype of human melanoma cells. Proc Natl Acad Sci U S A. 1993;90(7):3113–7.
79. Oberley LW, Buettner GR. Role of superoxide dismutase in cancer: a review. Cancer Res. 1979;39(4):1141–9.
80. Huang Y, He T, Domann FE. Decreased expression of manganese superoxide dismutase in transformed cells is associated with increased cytosine methylation of the SOD2 gene. DNA Cell Biol. 1999;18(8):643–52.
81. Meng X, Wu J, Pan C, Wang H, Ying X, Zhou Y, Yu H, Zuo Y, Pan Z, Liu RY, Huang W. Genetic and epigenetic down-regulation of microRNA-212 promotes colorectal tumor metastasis via dysregulation of MnSOD. Gastroenterology. 2013;145(2):426–36.e1–6.
82. Cyr AR, Hitchler MJ, Domann FE. Regulation of SOD2 in cancer by histone modifications and CpG methylation: closing the loop between redox biology and epigenetics. Antioxid Redox Signal. 2013;18(15):1946–55.
83. Behrend L, Mohr A, Dick T, Zwacka RM. Manganese superoxide dismutase induces p53-dependent senescence in colorectal cancer cells. Mol Cell Biol. 2005;25(17):7758–69.
84. Plymate SR, Haugk KH, Sprenger CC, Nelson PS, Tennant MK, Zhang Y, Oberley LW, Zhong W, Drivdahl R, Oberley TD. Increased manganese superoxide dismutase (SOD-2) is part of the mechanism for prostate tumor suppression by Mac25/insulin-like growth factor binding-protein-related protein-1. Oncogene. 2003;22(7):1024–34.
85. Li JJ, Colburn NH, Oberley LW. Maspin gene expression in tumor suppression induced by overexpressing manganese-containing superoxide dismutase cDNA in human breast cancer cells. Carcinogenesis. 1998;19(5):833–9.
86. Boland ML, Chourasia AH, Macleod KF. Mitochondrial dysfunction in cancer. Front Oncol. 2013;3:292.
87. Xu Y, Miriyala S, Fang F, Bakthavatchalu V, Noel T, Schell DM, Wang C, St Clair WH, St Clair DK. Manganese superoxide dismutase deficiency triggers mitochondrial uncoupling and the Warburg effect. Oncogene. 2014;34(32):4229–37.
88. Elchuri S, Oberley TD, Qi W, Eisenstein RS, Jackson Roberts L, Van Remmen H, Epstein CJ, Huang TT. CuZnSOD deficiency leads to persistent and widespread oxidative damage and hepatocarcinogenesis later in life. Oncogene. 2005;24(3):367–80.
89. Busuttil RA, Garcia AM, Cabrera C, Rodriguez A, Suh Y, Kim WH, Huang TT, Vijg J. Organ-specific increase in mutation accumulation and apoptosis rate in CuZn-superoxide dismutase-deficient mice. Cancer Res. 2005;65(24):11271–5.
90. Zhang Y, Zhao W, Zhang HJ, Domann FE, Oberley LW. Overexpression of copper zinc superoxide dismutase suppresses human glioma cell growth. Cancer Res. 2002;62(4):1205–12.
91. Menon SG, Goswami PC. A redox cycle within the cell cycle: ring in the old with the new. Oncogene. 2007;26(8):1101–9.
92. Sarsour EH, Kumar MG, Chaudhuri L, Kalen AL, Goswami PC. Redox control of the cell cycle in health and disease. Antioxid Redox Signal. 2009;11(12):2985–3011.
93. Sarsour EH, Kalen AL, Goswami PC. Manganese superoxide dismutase regulates a redox cycle within the cell cycle. Antioxid Redox Signal. 2014;20(10):1618–27.

94. Oberley LW, Oberley TD, Buettner GR. Cell differentiation, aging and cancer: the possible roles of superoxide and superoxide dismutases. Med Hypotheses. 1980;6(3):249–68.
95. Oberley LW, Oberley TD, Buettner GR. Cell division in normal and transformed cells: the possible role of superoxide and hydrogen peroxide. Med Hypotheses. 1981;7(1):21–42.
96. Miar A, Hevia D, Munoz-Cimadevilla H, Astudillo A, Velasco J, Sainz RM, Mayo JC. Manganese superoxide dismutase (SOD2/MnSOD)/catalase and SOD2/GPx1 ratios as biomarkers for tumor progression and metastasis in prostate, colon, and lung cancer. Free Radic Biol Med. 2015;85:45–55.
97. Miao L, St Clair DK. Regulation of superoxide dismutase genes: implications in disease. Free Radic Biol Med. 2009;47(4):344–56.
98. Rojo AI, Salinas M, Martin D, Perona R, Cuadrado A. Regulation of Cu/Zn-superoxide dismutase expression via the phosphatidylinositol 3 kinase/Akt pathway and nuclear factor-kappaB. J Neurosci. 2004;24(33):7324–34.
99. Sompol P, Xu Y, Ittarat W, Daosukho C, St Clair D. NF-kappaB-associated MnSOD induction protects against beta-amyloid-induced neuronal apoptosis. J Mol Neurosci. 2006;29(3):279–88.
100. Delhalle S, Deregowski V, Benoit V, Merville MP, Bours V. NF-kappaB-dependent MnSOD expression protects adenocarcinoma cells from TNF-alpha-induced apoptosis. Oncogene. 2002;21(24):3917–24.
101. Park EY, Rho HM. The transcriptional activation of the human copper/zinc superoxide dismutase gene by 2,3,7,8-tetrachlorodibenzo-p-dioxin through two different regulator sites, the antioxidant responsive element and xenobiotic responsive element. Mol Cell Biochem. 2002;240(1–2):47–55.
102. Zhang Y, Martin SG. Redox proteins and radiotherapy. Clin Oncol (R Coll Radiol). 2014;26(5):289–300.
103. Holley AK, Xu Y, St Clair DK, St Clair WH. RelB regulates manganese superoxide dismutase gene and resistance to ionizing radiation of prostate cancer cells. Ann N Y Acad Sci. 2010;1201:129–36.
104. Hour TC, Lai YL, Kuan CI, Chou CK, Wang JM, Tu HY, Hu HT, Lin CS, Wu WJ, Pu YS, Sterneck E, Huang AM. Transcriptional up-regulation of SOD1 by CEBPD: a potential target for cisplatin resistant human urothelial carcinoma cells. Biochem Pharmacol. 2010;80(3):325–34.
105. Chen PM, Cheng YW, Wu TC, Chen CY, Lee H. MnSOD overexpression confers cisplatin resistance in lung adenocarcinoma via the NF-kappaB/Snail/Bcl-2 pathway. Free Radic Biol Med. 2015;79:127–37.
106. Kamarajugadda S, Cai Q, Chen H, Nayak S, Zhu J, He M, Jin Y, Zhang Y, Ai L, Martin SS, Tan M, Lu J. Manganese superoxide dismutase promotes anoikis resistance and tumor metastasis. Cell Death Dis. 2013;4:e504.
107. Hermann B, Li Y, Ray MB, Wo JM, Martin 2nd RC. Association of manganese superoxide dismutase expression with progression of carcinogenesis in Barrett esophagus. Arch Surg. 2005;140(12):1204–9; discussion 9.
108. Janssen AM, Bosman CB, Kruidenier L, Griffioen G, Lamers CB, van Krieken JH, van de Velde CJ, Verspaget HW. Superoxide dismutases in the human colorectal cancer sequence. J Cancer Res Clin Oncol. 1999;125(6):327–35.
109. Malafa M, Margenthaler J, Webb B, Neitzel L, Christophersen M. MnSOD expression is increased in metastatic gastric cancer. J Surg Res. 2000;88(2):130–4.
110. Nonaka Y, Iwagaki H, Kimura T, Fuchimoto S, Orita K. Effect of reactive oxygen intermediates on the in vitro invasive capacity of tumor cells and liver metastasis in mice. Int J Cancer. 1993;54(6):983–6.
111. Bur H, Haapasaari KM, Turpeenniemi-Hujanen T, Kuittinen O, Auvinen P, Marin K, Koivunen P, Sormunen R, Soini Y, Karihtala P. Oxidative stress markers and mitochondrial antioxidant enzyme expression are increased in aggressive Hodgkin lymphomas. Histopathology. 2014;65(3):319–27.
112. Liu Z, Li S, Cai Y, Wang A, He Q, Zheng C, Zhao T, Ding X, Zhou X. Manganese superoxide dismutase induces migration and invasion of tongue squamous cell carcinoma via H2O2-dependent Snail signaling. Free Radic Biol Med. 2012;53(1):44–50.

113. Dhar SK, Tangpong J, Chaiswing L, Oberley TD, St Clair DK. Manganese superoxide dismutase is a p53-regulated gene that switches cancers between early and advanced stages. Cancer Res. 2011;71(21):6684–95.
114. Liochev SI, Fridovich I. The effects of superoxide dismutase on H2O2 formation. Free Radic Biol Med. 2007;42(10):1465–9.
115. Fridovich I. Superoxide dismutases: anti- versus pro- oxidants? Anticancer Agents Med Chem. 2011;11(2):175–7.
116. Buettner GR. Superoxide dismutase in redox biology: the roles of superoxide and hydrogen peroxide. Anticancer Agents Med Chem. 2011;11(4):341–6.
117. Buettner GR, Ng CF, Wang M, Rodgers VG, Schafer FQ. A new paradigm: manganese superoxide dismutase influences the production of H2O2 in cells and thereby their biological state. Free Radic Biol Med. 2006;41(8):1338–50.
118. Dasgupta J, Subbaram S, Connor KM, Rodriguez AM, Tirosh O, Beckman JS, Jourd'Heuil D, Melendez JA. Manganese superoxide dismutase protects from TNF-alpha-induced apoptosis by increasing the steady-state production of H2O2. Antioxid Redox Signal. 2006;8(7–8):1295–305.
119. Zhang HJ, Zhao W, Venkataraman S, Robbins ME, Buettner GR, Kregel KC, Oberley LW. Activation of matrix metalloproteinase-2 by overexpression of manganese superoxide dismutase in human breast cancer MCF-7 cells involves reactive oxygen species. J Biol Chem. 2002;277(23):20919–26.
120. Juarez JC, Manuia M, Burnett ME, Betancourt O, Boivin B, Shaw DE, Tonks NK, Mazar AP, Donate F. Superoxide dismutase 1 (SOD1) is essential for H2O2-mediated oxidation and inactivation of phosphatases in growth factor signaling. Proc Natl Acad Sci U S A. 2008;105(20):7147–52.
121. Buerk DG, Lamkin-Kennard K, Jaron D. Modeling the influence of superoxide dismutase on superoxide and nitric oxide interactions, including reversible inhibition of oxygen consumption. Free Radic Biol Med. 2003;34(11):1488–503.
122. Song Y, Buettner GR. Thermodynamic and kinetic considerations for the reaction of semiquinone radicals to form superoxide and hydrogen peroxide. Free Radic Biol Med. 2010;49(6):919–62.
123. Liu J, Zhan X, Li M, Li G, Zhang P, Xiao Z, Shao M, Peng F, Hu R, Chen Z. Mitochondrial proteomics of nasopharyngeal carcinoma metastasis. BMC Med Genomics. 2012;5(1):62.
124. Ranganathan AC, Nelson KK, Rodriguez AM, Kim KH, Tower GB, Rutter JL, Brinckerhoff CE, Huang TT, Epstein CJ, Jeffrey JJ, Melendez JA. Manganese superoxide dismutase signals matrix metalloproteinase expression via H2O2-dependent ERK1/2 activation. J Biol Chem. 2001;276(17):14264–70.
125. Hurd TR, DeGennaro M, Lehmann R. Redox regulation of cell migration and adhesion. Trends Cell Biol. 2012;22(2):107–15.
126. Ferraro D, Corso S, Fasano E, Panieri E, Santangelo R, Borrello S, Giordano S, Pani G, Galeotti T. Pro-metastatic signaling by c-Met through RAC-1 and reactive oxygen species (ROS). Oncogene. 2006;25(26):3689–98.
127. Juarez JC, Betancourt Jr O, Pirie-Shepherd SR, Guan X, Price ML, Shaw DE, Mazar AP, Donate F. Copper binding by tetrathiomolybdate attenuates angiogenesis and tumor cell proliferation through the inhibition of superoxide dismutase 1. Clin Cancer Res. 2006;12(16):4974–82.
128. Kumar P, Yadav A, Patel SN, Islam M, Pan Q, Merajver SD, Teknos TN. Tetrathiomolybdate inhibits head and neck cancer metastasis by decreasing tumor cell motility, invasiveness and by promoting tumor cell anoikis. Mol Cancer. 2010;9:206.
129. Giannoni E, Buricchi F, Grimaldi G, Parri M, Cialdai F, Taddei ML, Raugei G, Ramponi G, Chiarugi P. Redox regulation of anoikis: reactive oxygen species as essential mediators of cell survival. Cell Death Differ. 2008;15(5):867–78.
130. Hempel N, Melendez JA. Intracellular redox status controls membrane localization of pro- and anti-migratory signaling molecules. Redox Biol. 2014;2:245–50.
131. Liu Z, He Q, Ding X, Zhao T, Zhao L, Wang A. SOD2 is a C-myc target gene that promotes the migration and invasion of tongue squamous cell carcinoma involving cancer stem-like cells. Int J Biochem Cell Biol. 2015;60:139–46.

132. Hart PC, Mao M, de Abreu AL, Ansenberger-Fricano K, Ekoue DN, Ganini D, Kajdacsy-Balla A, Diamond AM, Minshall RD, Consolaro ME, Santos JH, Bonini MG. MnSOD upregulation sustains the Warburg effect via mitochondrial ROS and AMPK-dependent signalling in cancer. Nat Commun. 2015;6:6053.
133. Cabelli DE, Allen D, Bielski BH, Holcman J. The interaction between Cu(I) superoxide dismutase and hydrogen peroxide. J Biol Chem. 1989;264(17):9967–71.
134. Viglino P, Scarpa M, Rotilio G, Rigo A. A kinetic study of the reactions between H2O2 and Cu, Zn superoxide dismutase; evidence for an electrostatic control of the reaction rate. Biochim Biophys Acta. 1988;952(1):77–82.
135. Hink HU, Santanam N, Dikalov S, McCann L, Nguyen AD, Parthasarathy S, Harrison DG, Fukai T. Peroxidase properties of extracellular superoxide dismutase: role of uric acid in modulating in vivo activity. Arterioscler Thromb Vasc Biol. 2002;22(9):1402–8.
136. Yim MB, Chock PB, Stadtman ER. Copper, zinc superoxide dismutase catalyzes hydroxyl radical production from hydrogen peroxide. Proc Natl Acad Sci U S A. 1990;87(13):5006–10.
137. Yim MB, Berlett BS, Chock PB, Stadtman ER. Manganese(II)-bicarbonate-mediated catalytic activity for hydrogen peroxide dismutation and amino acid oxidation: detection of free radical intermediates. Proc Natl Acad Sci U S A. 1990;87(1):394–8.
138. Yim MB, Chock PB, Stadtman ER. Enzyme function of copper, zinc superoxide dismutase as a free radical generator. J Biol Chem. 1993;268(6):4099–105.
139. Sankarapandi S, Zweier JL. Evidence against the generation of free hydroxyl radicals from the interaction of copper, zinc-superoxide dismutase and hydrogen peroxide. J Biol Chem. 1999;274(49):34576–83.
140. Stadtman ER, Berlett BS, Chock PB. Manganese-dependent disproportionation of hydrogen peroxide in bicarbonate buffer. Proc Natl Acad Sci U S A. 1990;87(1):384–8.
141. Ansenberger-Fricano K, Ganini D, Mao M, Chatterjee S, Dallas S, Mason RP, Stadler K, Santos JH, Bonini MG. The peroxidase activity of mitochondrial superoxide dismutase. Free Radic Biol Med. 2013;54:116–24.
142. Kirkman HN, Gaetani GF. Catalase: a tetrameric enzyme with four tightly bound molecules of NADPH. Proc Natl Acad Sci U S A. 1984;81(July):4343–7.
143. Chance B, Sies H, Boveris A. Hydroperoxide metabolism in mammalian organs. Physiol Rev. 1979;59(3):527–605.
144. Melov S, Ravenscroft J, Malik S, Gill MS, Walker DW, Clayton PE, Wallace DC, Malfroy B, Doctrow SR, Lithgow GJ. Extension of life-span with superoxide dismutase/catalase mimetics. Science. 2000;289(5484):1567–9.
145. Orr W, Sohal R. Extension of life-span by overexpression of superoxide dismutase and catalase in Drosophila melanogaster. Science. 1994;263(5150):1128–30.
146. Schriner SE, Linford NJ. Extension of mouse lifespan by overexpression of catalase. Age. 2006;28(December 2005):209–18.
147. Bauer G. Tumor cell-protective catalase as a novel target for rational therapeutic approaches based on specific intercellular ROS signaling. Anticancer Res. 2012;32(7):2599–624.
148. Heinzelmann S, Bauer G. Multiple protective functions of catalase against intercellular apoptosis-inducing ROS signaling of human tumor cells. Biol Chem. 2010;391(6):675–93.
149. Bauer G, Zarkovic N. Revealing mechanisms of selective, concentration-dependent potentials of 4-hydroxy-2-nonenal to induce apoptosis in cancer cells through inactivation of membrane-associated catalase. Free Radic Biol Med. 2015;81:128–44.
150. Yu L, Wan F, Dutta S, Welsh S, Liu Z, Freundt E, Baehrecke EH, Lenardo M. Autophagic programmed cell death by selective catalase degradation. Proc Natl Acad Sci U S A. 2006;103(13):4952–7.
151. Chung-man Ho J, Zheng S, Comhair SA, Farver C, Erzurum SC. Differential expression of manganese superoxide dismutase and catalase in lung cancer. Cancer Res. 2001;61(23):8578–85.

152. Cobanoglu U, Demir H, Duran M, Şehitogullari A, Mergan D, Demir C. Erythrocyte catalase and carbonic anhydrase activities in lung cancer. Asian Pac J Cancer Prev. 2010;11: 1377–82.
153. Jiang B, Xiao S, Khan MA, Xue M. Defective antioxidant systems in cervical cancer. Tumour Biol. 2013;34(4):2003–9.
154. Lim SO, Gu JM, Kim MS, Kim HS, Park YN, Park CK, Cho JW, Park YM, Jung G. Epigenetic changes induced by reactive oxygen species in hepatocellular carcinoma: methylation of the E-cadherin promoter. Gastroenterology. 2008;135(6):2128–40.e8.
155. Radenkovic S, Milosevic Z, Konjevic G, Karadzic K, Rovcanin B, Buta M, Gopcevic K, Jurisic V. Lactate dehydrogenase, catalase, and superoxide dismutase in tumor tissue of breast cancer patients in respect to mammographic findings. Cell Biochem Biophys. 2013;66(2):287–95.
156. M-y C, Cheong JY, Lim W, Jo S, Lee Y, Wang H-j, K-h H, Cho H. Prognostic significance of catalase expression and its regulatory effects on hepatitis B virus X protein (HBx) in HBV-related advanced hepatocellular carcinomas. Oncotarget. 2014;5(23):12233–46.
157. Woolston CM, Zhang L, Storr SJ, Al-Attar A, Shehata M, Ellis IO, Chan SY, Martin SG. The prognostic and predictive power of redox protein expression for anthracycline-based chemotherapy response in locally advanced breast cancer. Mod Pathol. 2012;25(8):1106–16.
158. Sotgia F, Martinez-Outschoorn UE, Lisanti MP. Mitochondrial oxidative stress drives tumor progression and metastasis: should we use antioxidants as a key component of cancer treatment and prevention? BMC Med. 2011;9(1):62.
159. Glorieux C, Dejeans N, Sid B, Beck R, Calderon PB, Verrax J. Catalase overexpression in mammary cancer cells leads to a less aggressive phenotype and an altered response to chemotherapy. Biochem Pharmacol. 2011;82(10):1384–90.
160. Sato K, Ito K, Kohara H, Yamaguchi Y, Adachi K, Endo H. Negative regulation of catalase gene expression in hepatoma cells. Mol Cell Biol. 1992;12(6):2525–33.
161. Kwei KA, Finch JS, Thompson EJ, Bowden GT. Transcriptional repression of catalase in mouse skin tumor progression. Neoplasia. 2004;6(5):440–8.
162. Salcher S, Hagenbuchner J, Geiger K, Seiter MA, Rainer J, Kofler R, Hermann M, Kiechl-Kohlendorfer U, Ausserlechner MJ, Obexer P. C10ORF10/DEPP, a transcriptional target of FOXO3, regulates ROS-sensitivity in human neuroblastoma. Mol Cancer. 2014;13:224.
163. Glorieux C, Auquier J, Dejeans N, Sid B, Demoulin JB, Bertrand L, Verrax J, Calderon PB. Catalase expression in MCF-7 breast cancer cells is mainly controlled by PI3K/Akt/mTor signaling pathway. Biochem Pharmacol. 2014;89(2):217–23.
164. Min JY, Lim SO, Jung G. Downregulation of catalase by reactive oxygen species via hypermethylation of CpG island II on the catalase promoter. FEBS Lett. 2010;584(11):2427–32.
165. Sun Y, Colburn NH, Oberley LW. Depression of catalase gene expression after immortalization and transformation of mouse liver cells. Carcinogenesis. 1993;14(8):1505–10.
166. Quan X, Lim SO, Jung G. Reactive oxygen species downregulate catalase expression via methylation of a CpG Island in the Oct-1 promoter. FEBS Lett. 2011;585(21):3436–41.
167. Lee T-B, Moon Y-S, Choi C-H. Histone H4 deacetylation down-regulates catalase gene expression in doxorubicin-resistant AML subline. Cell Biol Toxicol. 2012;28(1):11–8.
168. Pennington JD, Wang TJC, Nguyen P, Sun L, Bisht K, Smart D, Gius D. Redox-sensitive signaling factors as a novel molecular targets for cancer therapy. Drug Resist Updat. 2005;8:322–30.
169. Nishikawa M, Hashida M, Takakura Y. Catalase delivery for inhibiting ROS-mediated tissue injury and tumor metastasis. Adv Drug Deliv Rev. 2009;61(4):319–26.
170. Lim HW, Hong S, Jin W, Lim S, Kim SJ, Kang HJ, Park EH, Ahn K, Lim CJ. Up-regulation of defense enzymes is responsible for low reactive oxygen species in malignant prostate cancer cells. Exp Mol Med. 2005;37(5):497–506.
171. Ca D, Durbin SM, Thau MR, Zellmer VR, Chapman SE, Diener J, Wathen C, Leevy WM, Schafer ZT. Antioxidant enzymes mediate survival of breast cancer cells deprived of extracellular matrix. Cancer Res. 2013;73(12):3704–15.

172. Obrador E, Valles SL, Benlloch M, Sirerol JA, Pellicer JA, Alcácer J, Coronado JA, Estrela JM. Glucocorticoid receptor knockdown decreases the antioxidant protection of B16 melanoma cells: an endocrine system-related mechanism that compromises metastatic cell resistance to vascular endothelium-induced tumor cytotoxicity. PLoS One. 2014;9(5), e96466.
173. Brigelius-Flohé R. Tissue-specific functions of individual glutathione peroxidases. Free Radic Biol Med. 1999;27(99):951–65.
174. Brigelius-Flohe R, Maiorino M. Glutathione peroxidases. Biochim Biophys Acta. 2013;1830(5):3289–303.
175. Fang W, Goldberg ML, Pohl NM, Bi X, Tong C, Xiong B, Koh TJ, Diamond AM, Yang W. Functional and physical interaction between the selenium-binding protein 1 (SBP1) and the glutathione peroxidase 1 selenoprotein. Carcinogenesis. 2010;31(8):1360–6.
176. Huang C, Ding G, Gu C, Zhou J, Kuang M, Ji Y, He Y, Kondo T, Fan J. Decreased selenium-binding protein 1 enhances glutathione peroxidase 1 activity and downregulates HIF-1alpha to promote hepatocellular carcinoma invasiveness. Clin Cancer Res. 2012;18(11):3042–53.
177. Jardim BV, Moschetta MG, Leonel C, Gelaleti GB, Regiani VR, Ferreira LC, Lopes JR, Zuccari DA. Glutathione and glutathione peroxidase expression in breast cancer: an immunohistochemical and molecular study. Oncol Rep. 2013;30(3):1119–28.
178. Gan X, Chen B, Shen Z, Liu Y, Li H, Xie X, Xu X, Li H, Huang Z, Chen J. High GPX1 expression promotes esophageal squamous cell carcinoma invasion, migration, proliferation and cisplatin-resistance but can be reduced by vitamin D. Int J Clin Exp Med. 2014; 7(9):2530–40.
179. Han JJ, Xie DR, Wang LL, Liu YQ, Wu GF, Sun Q, Chen YX, Wei Y, Huang ZQ, Li HG. Significance of glutathione peroxidase 1 and caudal-related homeodomain transcription factor in human gastric adenocarcinoma. Gastroenterol Res Pract. 2013;2013:380193.
180. Min SY, Kim HS, Jung EJ, Jung EJ, Jee CD, Kim WH. Prognostic significance of glutathione peroxidase 1 (GPX1) down-regulation and correlation with aberrant promoter methylation in human gastric cancer. Anticancer Res. 2012;32(8):3169–75.
181. Yan W, Chen X. GPX2, a direct target of p63, inhibits oxidative stress-induced apoptosis in a p53-dependent manner. J Biol Chem. 2006;281:7856–62.
182. Naiki T, Naiki-Ito A, Asamoto M, Kawai N, Tozawa K, Etani T, Sato S, Suzuki S, Shirai T, Kohri K, Takahashi S. GPX2 overexpression is involved in cell proliferation and prognosis of castration-resistant prostate cancer. Carcinogenesis. 2014;35(9):1962–7.
183. Ouyang X, Deweese TL, Nelson WG, Abate-shen C. Loss-of-function of Nkx3.1 promotes increased oxidative damage in prostate carcinogenesis. Cancer Res. 2005;65(23):6773–9.
184. Yang Z-l, Yang L, Zou Q, Yuan Y, Li J, Liang L, Zeng G, Chen S. Positive ALDH1A3 and negative GPX3 expressions are biomarkers for poor prognosis of gallbladder cancer. Dis Markers. 2013;35(3):163–72.
185. Lee O-J, Schneider-Stock R, McChesney PA, Kuester D, Roessner A, Vieth M, Moskaluk CA, El-Rifai WE. Hypermethylation, loss of expression of glutathione peroxidase-3 in Barrett's tumorigenesis. Neoplasia. 2005;7(9):854–61.
186. Yu YP, Yu G, Tseng G, Cieply K, Nelson J, Defrances M, Zarnegar R, Michalopoulos G, Luo J-H. Glutathione peroxidase 3, deleted or methylated in prostate cancer, suppresses prostate cancer growth and metastasis. Cancer Res. 2007;67(17):8043–50.
187. Mohamed MM, Sabet S, Peng D-F, Nouh MA, El-Shinawi M, El-Rifai W. Promoter hypermethylation and suppression of glutathione peroxidase 3 are associated with inflammatory breast carcinogenesis. Oxid Med Cell Longev. 2014;2014:787195.
188. Zhang X, Yang JJ, Kim YSYSUN, Kim K-Y, Ahn WS, Yang S. An 8-gene signature, including methylated and down-regulated glutathione peroxidase 3, of gastric cancer. Int J Oncol. 2010;36(2):405–14.
189. Peng D, Razvi M, Chen H, Washington K, Roessner A, El-Rifai W. DNA hypermethylation regulates the expression of members of the Mu-class glutathione-S-transferases and glutathione peroxidases in Barrett's adenocarcinoma. Gut. 2010;58(1):5–15.

190. Falck E, Karlsson S, Carlsson J, Helenius G, Karlsson M, Klinga-Levan K. Loss of glutathione peroxidase 3 expression is correlated with epigenetic mechanisms in endometrial adenocarcinoma. Cancer Cell Int. 2010;10(1):46.
191. Lin J-C, Kuo W-R, Chiang F-Y, Hsiao P-J, Lee K-W, Wu C-W, Juo S-HH. Glutathione peroxidase 3 gene polymorphisms and risk of differentiated thyroid cancer. Surgery. 2009;145(5):508–13.
192. Wenk J, Schüller J, Hinrichs C, Syrovets T, Azoitei N, Podda M, Wlaschek M, Brenneisen P, Schneider L, Sabiwalsky A, Peters T, Sulyok S, Dissemond J, Schauen M, Krieg T, Wirth T, Simmet T, Scharffetter-Kochanek K. Overexpression of phospholipid-hydroperoxide glutathione peroxidase in human dermal fibroblasts abrogates UVA irradiation-induced expression of interstitial collagenase/matrix metalloproteinase-1 by suppression of phosphatidylcholine hydroperoxide-mediate. J Biol Chem. 2004;279(44):45634–42.
193. Bermano G, Pagmantidis V, Holloway N, Kadri S, Mowat NA, Shiel RS, Arthur JR, Mathers JC, Daly AK, Broom J, Hesketh JE. Evidence that a polymorphism within the 3′UTR of glutathione peroxidase 4 is functional and is associated with susceptibility to colorectal cancer. Genes Nutr. 2007;2:225–32.
194. Ke HL, Lin J, Ye Y, Wu WJ, Lin HH, Wei H, Huang M, Chang DW, Dinney CP, Wu X. Genetic variations in glutathione pathway genes predict cancer recurrence in patients treated with Transurethral resection and bacillus Calmette-Guerin instillation for non-muscle invasive bladder cancer. Ann Surg Oncol. 2015;22(12):4104–10.
195. Batinic-Haberle I, Tovmasyan A, Spasojevic I. An educational overview of the chemistry, biochemistry and therapeutic aspects of Mn porphyrins—from superoxide dismutation to HO-driven pathways. Redox Biol. 2015;5:43–65.
196. Tovmasyan A, Maia CG, Weitner T, Carballal S, Sampaio RS, Lieb D, Ghazaryan R, Ivanovic-Burmazovic I, Radi R, Reboucas JS, Spasojevic I, Benov L, Batinic-Haberle I. A comprehensive evaluation of catalase-like activity of different classes of redox-active therapeutics. Free Radic Biol Med. 2015;86:308–21.

Chapter 5
Superoxide Dismutase Family of Enzymes in Brain Neurogenesis and Radioprotection

Huy Nguyen, Chandra Srinivasan, and Ting-Ting Huang

5.1 Introduction

With the exception of a few extremely oxygen-sensitive anaerobic bacteria, superoxide dismutase (SOD, EC 1.15.1.1) is present ubiquitously in all organisms [1]. In the aerobic organisms, SOD is important for removing superoxide anions (O_2^-) generated from normal metabolism and certain enzymatic reactions wherein oxygen serves as the final electron acceptor. In anaerobic organisms, the presence of SOD leads to oxygen tolerance, and SOD is postulated as a virulent factor that enables survival of pathogenic anaerobes in an otherwise oxygenated tissue environment [2, 3].

There are three distinct classes of SODs, identified by their sequence homology and metal cofactors. The first class is the Cu and Zn containing SOD (CuZnSOD), which is commonly found in the cytosol of eukaryotic cells, in chloroplasts, and in some prokaryotes. The second class includes the Mn containing SOD (MnSOD) and the Fe containing SOD (FeSOD). Whereas MnSOD is found in the prokaryotes

H. Nguyen, Ph.D.
Department of Neurology and Neurological Sciences,
Stanford University School of Medicine, Stanford, CA, USA

C. Srinivasan, Ph.D.
Department of Graduate Studies and Research, California State University,
Dominguez Hills, CA, USA

T.-T. Huang, Ph.D. (✉)
Department of Neurology and Neurological Sciences, Stanford University School of Medicine, Stanford, CA, USA

Geriatric Research, Education, and Clinical Center, Veterans Affairs Palo Alto Health Care System, 3801 Miranda Avenue, Mail stop 154-I, Building 100, D3-101, Palo Alto, CA 94304, USA
e-mail: tthuang@stanford.edu

I. Batinić-Haberle et al. (eds.), *Redox-Active Therapeutics*,
Oxidative Stress in Applied Basic Research and Clinical Practice,
DOI 10.1007/978-3-319-30705-3_5

and in the mitochondria of eukaryotes, FeSOD can be found in prokaryotes, algae, and in some chloroplasts. The third class is the Ni containing SOD (NiSOD), which is found in Streptomyces and some marine cyanobacteria. With a few exceptions, CuZnSOD forms dimers, MnSOD forms tetramers, FeSOD forms dimers, and NiSOD forms hexamers for the active enzymes [1, 3–5].

5.2 Mammalian Superoxide Dismutases

In the mammalian system, there are three SODs encoded by three independent genes: CuZn containing SOD (CuZnSOD, SOD1), Mn containing SOD (MnSOD, SOD2), and extracellular SOD (EC-SOD, SOD3). Various biochemical and immunological studies have placed CuZnSOD in the cytosol, lysosomes, nucleus, and the intermembrane space of the mitochondria [6]; MnSOD in the matrix of mitochondria; and EC-SOD in the extracellular space with the majority of the protein attached to the extracellular matrix and a small fraction released to circulation after the heparin binding domain at the C terminus is removed enzymatically (Fig. 5.1). Whereas MnSOD requires Mn as the metal cofactor, CuZnSOD and EC-SOD both require Cu and Zn as cofactors for the enzymatic activities. In comparison of the monomeric form, CuZnSOD is the smallest protein among the three SODs at 17 kDa in size, while MnSOD is around 24 kDa. EC-SOD is the largest among the three, with glycosylation at the N terminus, which allows EC-SOD to bind with high affinity to heparin sulfate and collagen. The molecular weight of the bound EC-SOD is around 33 kDa, and the secreted form is around 30 kDa [7].

Naturally occurring genetic mutations and polymorphisms of human SODs have been identified and many are associated with disease susceptibilities. The best

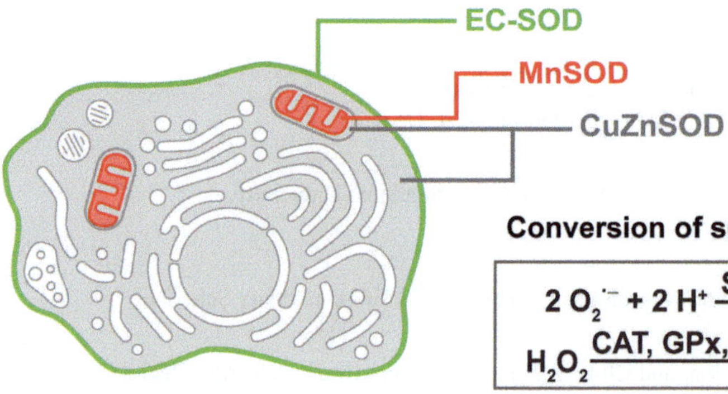

Conversion of superoxide to water

$$2\,O_2^{\cdot -} + 2\,H^+ \xrightarrow{SOD} H_2O_2 + O_2$$

$$H_2O_2 \xrightarrow{CAT,\ GPx,\ Prx} H_2O + \tfrac{1}{2}\,O_2$$

Fig. 5.1 Subcellular distribution of mammalian superoxide dismutases. The locations of CuZnSOD (*grey*) in the cytosol, nucleus, and the intermembrane space of mitochondria, MnSOD (*red*) in the matrix of mitochondria, and EC-SOD (*green*) on the extracellular matrix are shown. The circulating form of EC-SOD is not shown. CAT, catalase; GPx, glutathione peroxidase; Prx, peroxiredoxin known are mutations in SOD1 that lead to amyotrophic lateral sclerosis (ALS). To date, there are more than 100 ALS-associated mutations found in SOD1 since the initial study published in 1993. The mutations are detected in all 5 exons, and no specific hot spots have been identified. The underlying disease mechanism is more related to accumulation of the mutated protein, rather than a reduction of the CuZnSOD activity [8]. Two polymorphisms are known to exist in MnSOD. The Ala16Val polymorphism is located in the mitochondrial targeting sequence and existence of the Valine allele reduces mitochondrial transport by 30–40 % [9]. The Ile58Thr polymorphism is located in exon 3 and presence of the Threonine allele leads to a reduced stability in the tetrameric interface, and consequently a reduced MnSOD activity [10]. Both polymorphisms result in reduced MnSOD activities in the mitochondria. MnSOD plays an important role in cancer development and progression [11]. Consequently, associations between MnSOD polymorphisms and cancer susceptibilities have been extensively investigated [12–15]. A common polymorphism in EC-SOD Arg213Gly has been reported in 4–6 % Swedish and Japanese populations [16, 17]. A Glycine allele in the heparin binding domain significantly reduces the binding affinity to extracellular matrix and increases circulating level of EC-SOD by eight- to tenfolds [16]. Genetic epidemiological studies showed populations with the Arg213Gly polymorphism with increased risk for ischemic heart disease [18] and accelerated progression to renal failure in patients undergoing hemodialysis [19]. On the other hand, smokers with the Arg213Gly polymorphism were shown to have reduced risk for the development of chronic obstructive pulmonary disease [20].

5.3 Oxygen Free Radical Generation and Removal

Aerobic organisms, including animals, plants, and bacteria, require oxygen for efficient production of energy. These organisms use the electron transport chain (ETC), such as those in the mitochondria of eukaryotic cells, for the production of energy. Electrons collected from catabolism of glucose or fat at the end of the citric acid cycle are stored in the form of NADH or $FADH_2$. When these electrons are shuttled through the complexes of the ETC, O_2 serves as the final electron acceptor and is reduced to water in Complex IV. However, reduction of O_2 to H_2O does not occur in a single step, but in four sequential steps [21]. Each step includes one electron addition, which creates an opportunity for the existence of superoxide anions (O_2^-) as by-products. A small percentage of O_2^- is estimated to leak from the electron transport chain before they are completely reduced [21]. Outside of mitochondria, O_2^- can be produced from the enzymatic oxidation of NADPH by NADPH oxidases (NOXs). It is also generated from normal enzymatic reaction such as that of xanthine oxidase in the uric acid production pathway (Fig. 5.2).

O_2^- can be converted to hydrogen peroxide (H_2O_2) by compartment-specific SODs, or in the absence of SODs, by spontaneous dismutation. H_2O_2 is then converted to H_2O by peroxidases, such as catalase, peroxiredoxins (Prxs), glutathione

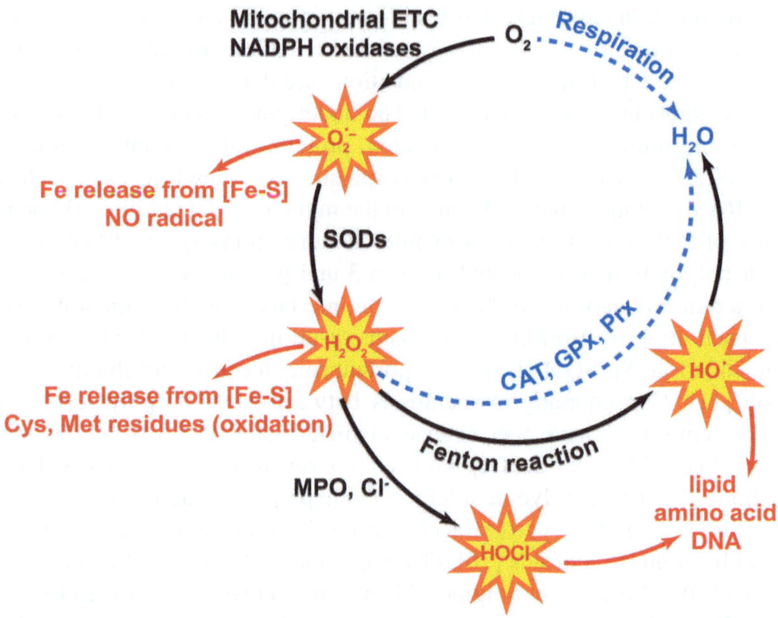

Fig. 5.2 The basics of reactive oxygen species and their effects on macromolecules. Reactive oxygen species (ROS) is a collective term that describes the highly reactive radicals formed upon incomplete reduction of oxygen. These products include superoxide radicals, hydroxyl radicals, hydrogen peroxide, and other related species. Intracellular superoxide ($^{\bullet}O_2^-$) is primarily produced from the enzymatic oxidation of NADPH by NADPH oxidases (NOXs) or by incomplete reduction of oxygen from mitochondria during oxidative phosphorylation. Superoxide is rapidly converted into hydrogen peroxide (H_2O_2) by compartment-specific superoxide dismutases (SODs), or by spontaneous dismutation. H_2O_2 is then converted to H_2O by peroxidases, such as peroxiredoxins (Prxs), glutathione peroxidases (GPxs), catalase (CAT), and glutaredoxins (Grxs). H_2O_2 can oxidize cysteine residues on redox-sensitive proteins and form disulfide bonds. Tight regulation of this process is thought to be involved in activation or termination of redox-sensitive signaling pathways. However, high level of H_2O_2 can have deleterious effects. In the presence of iron (Fe^{2+}), H_2O_2 can participate in Fenton reaction or Haber–Weiss reaction and generate the highly reactive hydroxyl radicals ($^{\bullet}OH$) and cause severe damage to cellular macromolecules. ECT, electron transport chain

peroxidases (GPxs), and glutaredoxins (Grxs). In the presence of Fe^{2+}, H_2O_2 may be converted to hydroxyl radicals ($^{\bullet}OH$) via Fenton reaction or Haber–Weiss reaction. Collectively, these highly reactive molecules are called reactive oxygen species (ROS). In addition to antioxidant enzymes, small molecule antioxidants, such as vitamin C, vitamin E, flavonoids, and glutathione (GSH), also contribute significantly to the overall antioxidant capacity in an organism. Oxidative stress occurs when production of free radical levels overwhelms antioxidant capacity in the cell. Oxidative stress is harmful to organisms as it can damage biomolecules. It is also thought to play a role in many diseases, and accelerates the process of aging [22–24].

5.4 Redox Balance and Cell Fate Decision

In a biological system, the reduction and oxidation potential is usually controlled by redox couples, such as GSH/GSSG and NADPH/NADP$^+$. Changes in the balance of these redox couples can significantly alter the intracellular and extracellular redox environment due to their abundance. Alterations in the redox environment in turn change the configuration of redox-sensitive proteins by changing primarily the inter- or intramolecular disulfide bounds in the cysteine residues [25]. A number of redox-sensitive signaling molecules critical for cell proliferation and survival have been identified [25, 26]. Consequently, changes in redox balance in the intracellular and the extracellular environment can impact cell fate decisions, including entering or exiting cell cycle [27], proliferation or differentiation [28, 29], and survival or cell death [30]. Studies with precise measurements of the cellular redox status indicate that under normal physiological conditions, a more reduced environment favors cell proliferation, whereas a more oxidized environment favors cell differentiation [30]. Under conditions of oxidative stress where oxidation potential exceeds the antioxidant capacity of the cells, apoptotic cell death or necrotic cell death ensues [30]. Production and regeneration of GSH and NADPH are tightly linked to cellular metabolism, and SODs and their downstream peroxidases play an important role in maintaining a normal balance in GSH/GSSH and NADPH/NADP$^+$ pairs to ensure normal cellular and tissue functions.

5.5 Redox Balance, Hippocampal Neurogenesis, and Learning and Memory

Hippocampal dentate gyrus (DG) and the *Cornu Ammonis* (CA) areas are the two separate structures in hippocampal formation (Fig. 5.3). Declarative memory, such as personal experience and facts that can be consciously recalled, is an important facet of learning and memory. In the medial temporal lobe of the brain, hippocampal formation and its surrounding areas, including perirhinal cortex and entorhinal cortex, are critical for the acquisition, consolidation, and retrieval of declarative memories. Hippocampus is also critical for spatial learning, the process of encoding the environment for spatial navigation, and the formation of spatial memory. The principal neurons in the dentate gyrus are granule cells. These are small excitatory neurons with apical dendrites that project into the molecular layer and receive input primarily from the entorhinal cortex. Axons from dentate granule cells then relay the information to the CA regions. Pyramidal neurons in the CA regions ultimately relay the information back to entorhinal cortex and thus complete the neural circuit in the hippocampal formation (Fig. 5.3).

Hippocampal neurogenesis, the production of new neurons in the hippocampal formation, plays an important role in hippocampal-dependent learning and memory. A positive correlation has been established in experimental animals between

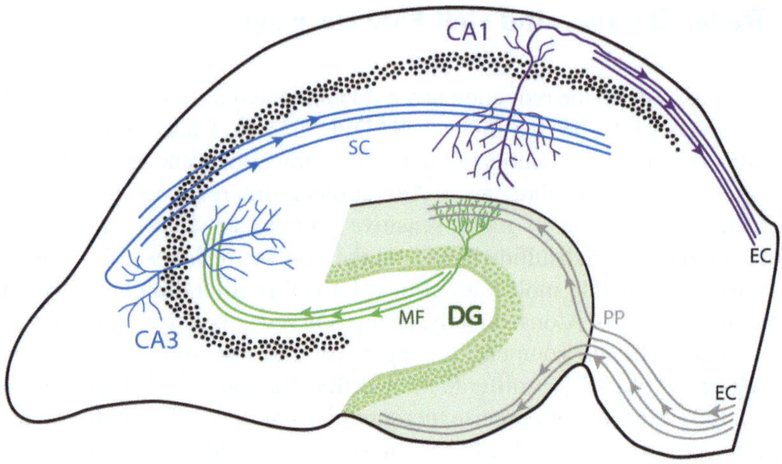

Fig. 5.3 Schematic illustration of synaptic pathways for hippocampal-dependent learning and memory. Hippocampus resides in the medial temporal lobe of the brain. It is important for declarative memory and spatial memory. The cellular organization and major circuitry in the hippocampal formation are depicted, showing the two separate structures, dentate gyrus (DG) and the *Cornu Ammonis* (CA) areas, that make up the hippocampal formation. Entorhinal cortex (EC) receives input from other cortical area. The information is transmitted by axonal projection of layer II neurons in the EC to DG neurons through the perforant pathway (PP). DG neurons then send projections to the pyramidal cells in CA3 through mossy fibers (MF). CA3 pyramidal neurons relay the information to CA1 pyramidal neurons through Schaffer collaterals (SC). CA1 pyramidal neurons then send projections back into deep layers of neurons in the EC to complete the hippocampal circuitry. The traditional excitatory trisynaptic pathway (EC → DG → CA3 → CA1 → EC) is depicted by *solid arrow heads*

hippocampal neurogenesis and cognitive performance that require encoding, consolidation, and retrieval of contextual or spatial memories [39–44]. Data from a recent study suggest that these are not just correlative observations. Cell lineage tracing and expression of learning-related immediate early genes show that adult-generated granule cells in the SGZ are preferentially incorporated into the spatial memory networks in the dentate gyrus of the hippocampus [45]. Consequently, reduction in hippocampal neurogenesis due to physiological changes or as a consequence of the normal aging process reduces the hippocampal function of learning and memory [46, 47].

Neurogenesis, beyond the early postnatal stage is limited to a few regions in the mammalian brain, including the subventricular zone (SVZ) of the lateral ventricle and the subgranular zone (SGZ) of the hippocampal dentate gyrus. Most of our understanding of hippocampal neurogenesis comes from studies in experimental animals. SGZ sits at the base of dentate granule cell layer. Neuronal progenitor cells in the SGZ possess the radial glial cell phenotype and go through asymmetrical replication to generate neuroblast, which then differentiate into immature neurons. In the span of 3–4 weeks, newborn immature neurons put out axons and complex dendritic trees as they mature and move further into the existing granule cell layer (Fig. 5.4).

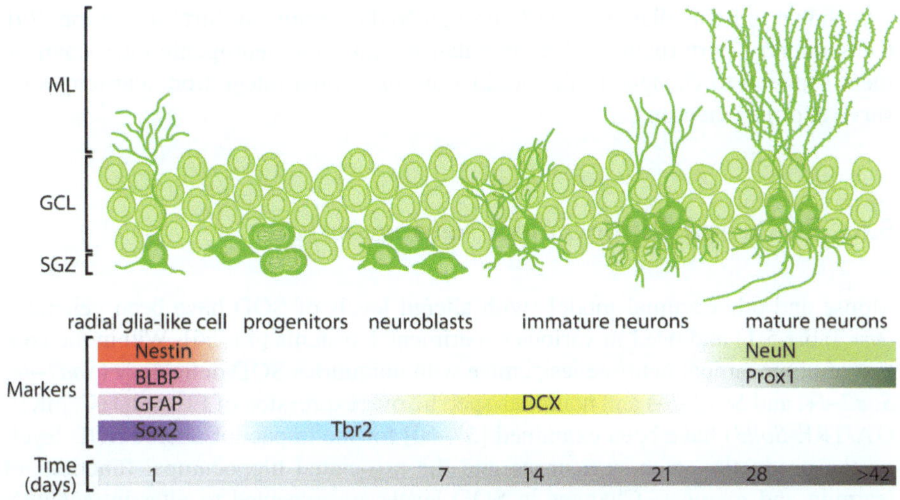

Fig. 5.4 Adult neurogenesis in the hippocampus. The developmental stages during adult hippocampal neurogenesis are shown with activation of quiescent radial glia-like cell in the subgranular zone (SGZ), followed by the proliferation of non-radial progenitor cells and intermediate progenitors, generation of neuroblasts, migration of immature neurons, and maturation and integration of adult-born dentate granule cells. Expression of stage-specific markers and the time scale are shown at the *bottom*. GCL, granule cell layer; ML, molecular layer

In the process, they establish synaptic connections with the existing neuronal network and become functionally integrated into the hippocampal circuitry. Recent studies show that newborn granule cells have passive membrane properties, action potentials, and functional synaptic inputs similar to those found in mature dentate granule cells [31]. Even though hippocampal neurogenesis continues through life, the rate of adult hippocampal neurogenesis decreases exponentially from 1 month of age [32]. The production of new neurons is balanced with apoptotic cell death in the granule cell layer. Consequently, the number of granule cells in the hippocampus stays relatively constant [32]. Similar observation has also been made in other species.

To achieve neurogenesis, i.e., to produce mature, functionally integrated neurons, a delicate redox balance needs to be achieved. Neural progenitor cells reside in an environment with low oxygen tension, and are known to maintain a low metabolic status. Therefore, the production of ROS is minimal [33]. With environmental cues, such as growth factors or other external stimuli, low levels of ROS are produced, which activate redox-sensitive signaling pathways that favor cell proliferation in stem cells and progenitor cells [33–38]. As newborn cells progress through the cell cycle, continuous increase in oxidation potential then leads to differentiation. A recent study suggests that, within the redox environment that favors neural progenitor cell differentiation, a more oxidized environment guides the differentiation towards the astroglia lineage, whereas a more reduced environment favors neuronal lineage [29]. The redox-dependent lineage decision in neural progenitor cells was shown to be mediated, in part, through NAD^+ dependent Sirt1 activation [29]. Consequently, perturbation of redox balance within the neurogenic microenvironment can lead to changes in the production, functional integration, and long-term survival of new neurons.

5.6 SODs and Hippocampal Neurogenesis

Mouse and other animal models with altered levels of SOD have been generated (see Table 5.1) and used in various experimental systems [48–56]. Within the context of hippocampal neurogenesis, mice with ubiquitous SOD deficiency (*Sod1*–/+, *Sod2*–/+, and *Sod3*–/–) and neuronal-specific overexpression of EC-SOD (CamKII-tTA/TRE-*Sod3*) have been examined [57–60] for the impact of altered SOD levels on the production of new neurons and the associated hippocampal functions of learning and memory. Changes in SOD levels are expected to alter intracellular (CuZnSOD and MnSOD) or extracellular (EC-SOD) redox balance and consequently, affecting redox-sensitive cellular processes such as the production and differentiation of newborn neurons in the hippocampus. Consistent with this idea, production of newborn cells and their lineage commitment in the hippocampal SGZ were altered in each of the three SOD deficient mouse strains examined.

While the overall new cell production in the SGZ was not significantly altered in *Sod1*–/+ and *Sod2*–/+ mice, the total number of newborn cells in *Sod3*–/– mice was significantly reduced [57–59]. The disparity may be due to the fact that *Sod1*–/+ and *Sod2*–/+ were only missing 50 % of CuZnSOD and MnSOD, respectively, while *Sod3*–/– mice were completely deprived of EC-SOD. Another possibility is that the extracellular redox environment played a more prominent role in hippocampal neurogenesis. Despite the relatively normal levels of new neuron production in *Sod1*–/+ and *Sod2*–/+ mice, the differentiation pattern was significantly altered with an increased preference for the astroglia lineage. Therefore, a 49 % reduction in the number of newborn neurons and a ninefold increase in newborn astroglia were observed in *Sod1*–/+ mice. In *Sod2*–/+ mice, the number of newborn neurons was reduced by 23–35 %, and the number of newborn astroglia ranged from no change to a 4.7-fold increase compared to that of *Sod2*+/+ controls [57, 58]. In contrast, the number of newborn neurons and astroglia in the SGZ were significantly reduced in *Sod3*–/– mice because of the overall reduction of newborn cells [59, 60]. Whereas there was a shift in the ratio of newborn neurons and astroglia in *Sod1*–/+ and *Sod2*–/+ mice, the ratio stayed relatively constant in *Sod3*–/– mice. The data suggested that lineage determination was perhaps more sensitive to changes in the intracellular redox balance. Consistent with the role of hippocampal neurogenesis in cognitive function, cognitive deficits in spatial memory and recognition memory were observed in *Sod3*–/– mice [60, 61]. On the other hand, the relatively normal hippocampal neurogenesis observed in *Sod2*–/+ mice correlated with normal cognitive performance [57].

Because EC-SOD deficiency had such a marked impact on the production of newborn neurons, a study was designed to restore EC-SOD to the principal neurons

Table 5.1 Rodent models with genetically altered levels of superoxide dismutases[a]

Strain ID	Species	Genetic alteration	References
[b]C57BL/6-Tg(*SOD1*)3Cje/J	Mouse	Human genomic CuZnSOD transgenes	Original designation: TgHS-218/3 [103, 104]
[b]C57BL/6-Tg(*SOD1*)10Cje/J	Mouse	Human genomic CuZnSOD transgenes	Original designation: TgHS-218/10 [103, 104]
[b]B6.Cg-Tg(*SOD1*)2Gur/J	Mouse	Human genomic CuZnSOD transgenes	Original designation: N1029 [105]
Tg(*SOD1*)66UCSF	Rat	Human genomic CuZnSOD transgenes	[106]
[b]B6SJL-Tg(Prnp-Immt/*SOD1*)1Gmnf/J	Mouse	Neuronal-specific, mitochondrial intermembrane space-targeted expression of human CuZnSOD cDNA transgenes	Original designation: *mitoSOD1* [107]
[c]B6.Cg-$Sod1^{tm1Cje}$/Mmmh	Mouse	Targeted deletion of CuZnSOD gene	Original designation: *Sod1* KO [108]
[b]B6.129S7-$Sod^{1tm1Leb}$/DnJ	Mouse	Targeted deletion of CuZnSOD gene	Original designation: *Sod1* KO [109]
ACTB-*Sod2*	Mouse	Actin-specific expression of human MnSOD cDNA transgenes	Tg-SOD-L, Tg-SOD-M, and Tg-SOD-H [55]
TgN(Sod2)274Cje	Mouse	Mouse genomic MnSOD transgenes	Original designation: Tg-274 [54]
TgE(Sod2)11Cje	Mouse	Mouse genomic MnSOD transgenes	Original designation: Tg-11 [54]
[b]NOD.FVB-Tg(INS-*SOD2*)3Pne/PneJ	Mouse	Insulin-specific expression of human MnSOD cDNA transgenes	Original designation: MnSOD3 (line 3) [110]
Sod2-TRE-*LacZ*	Mouse	Tetracycline-inducible mouse genomic MnSOD transgenes	[56]

(continued)

Table 5.1 (continued)

Strain ID	Species	Genetic alteration	References
[b]FVB-Tg(Myh6-*SOD2*,Tyr)3Pne/J	Mouse	Cardiac-specific expression of human genomic MnSOD transgenes	Original designation: MySOD [111]
[c]B6.Cg-*Sod2*^{tm1Cje}/Mmmh	Mouse	Targeted deletion of MnSOD gene	Original designation: *Sod2* KO [112]
[b]B6.129S7-*Sod2*^{tm1Leb}/J	Mouse	Targeted deletion of MnSOD gene	Original designation: *Sod2*^{m1BCM} [113]
Floxed *Sod2*	Mouse	Targeted inducible deletion of MnSOD gene	[114]
[b]B6N.129S6-*Sod2*$^{tm1(rtTA*M2,tetO)Jsg}$/J	Mouse	Targeted tetracycline-regulated MnSOD allele	Original designation: MnSOD^{tet} [115]
ACTB-*SOD3*	Mouse	Actin-specific expression of human EC-SOD cDNA transgenes	[116]
TRE-*Sod3*-GFP	Mouse	Tetracycline-inducible mouse EC-SOD genomic transgenes	[51]
[b]B6.129P2-*Sod3*^{tm1Mrkl}/J	Mouse	Targeted deletion of EC-SOD gene	Original designation: EC-SOD null mutants [52]

[a]Only transgenic animals with wild-type SOD transgenes are listed here. There are multiple strains of SOD1 transgenic mice and rats designed to model genetic alterations in amyotrophic lateral sclerosis (ALS) in humans and are available from the Jackson Laboratory or other repositories. With the constant addition of new rodent models, the list is not meant to be inclusive and only shows those that are available from different repositories

[b]Official designation from the Jackson Laboratory JAX® Mice repository

[c]Official designation from the Mutant Mouse Resource and Research Center (MMRRC) repository

(i.e., excitatory neurons) in hippocampus while maintaining the overall EC-SOD null environment in other cell populations. In this mouse model, overproduction of EC-SOD was limited to CamKIIα positive cells, which included all excitatory neurons in the hippocampus [51]. However, CamKIIα was not expressed in neuronal progenitor cells. Despite the sixfold increase in EC-SOD levels in CamKIIα positive neurons, and the short distance between the SGZ where neural progenitor cells resided and the granule cell layer, EC-SOD deficiency in the progenitor cell population led to a significant reduction in progenitor cell proliferation [60]. However, long-term survival of newborn cells was significantly improved [60], suggesting that EC-SOD was important for maintaining normal level of neural progenitor cell proliferation and long-term survival. On the other hand, high levels of EC-SOD had no influence on the lineage determination and the percentage of newborn neurons and astroglia in total newborn cells was not significantly changed in the EC-SOD transgenic mice.

5.7 Ionizing Radiation and Brain Injury

Ionizing radiation, even at low doses, leads to generation of oxygen free radicals and cellular injuries by direct interaction with macromolecules, or indirectly by radiolysis of water [62, 63]. The initial tissue injury then leads to additional oxygen free radical generation by several mechanisms (Fig. 5.6). First, proinflammatory mediators, such as prostaglandins, leukotrienes, and thromboxanes, are generated from the enzymatic metabolism of arachidonic acid (AA), which is the major cell membrane phospholipid in the nervous system and is released upon tissue injury [64]. Other proinflammatory mediators, including isoprostanes and 4-hydroxynonenal, can also be generated via nonenzymatic processing of AA. Second, inflammatory response of macrophage and microglia to tissue injury leads to activation of NADPH oxidase in these immune cells [65]. The family of NADPH oxidases 2 (NOX2) are membrane-bound proteins and generate superoxide radicals directly into the extracellular space or cytoplasm [66]. In the case of NOX2 activation, superoxide radicals are produced extracellularly. Third, excitatory toxicity most likely constitutes the other source of oxidative stress. Glutamate is the major excitatory neurotransmitter, and the most abundant neuron in the hippocampal formation is excitatory neurons. Damage to the synaptic membranes can lead to abnormal release of glutamate, and lead to overactivation of glutamate receptors and overproduction of ROS. These difference sources of ROS production following irradiation eventually lead to higher levels of lipid peroxidation and oxidative damage, and the positive feedback loop most likely contributes to the state of persistent oxidative stress following radiation exposure.

The brain is exposed to ionizing radiation in a number of clinical situations, predominantly in those involving cancer treatments. Recent statistics shows that approximately 69,000 people in the USA will be diagnosed with primary brain tumor and approximately 170,000 with secondary tumor each year (data from

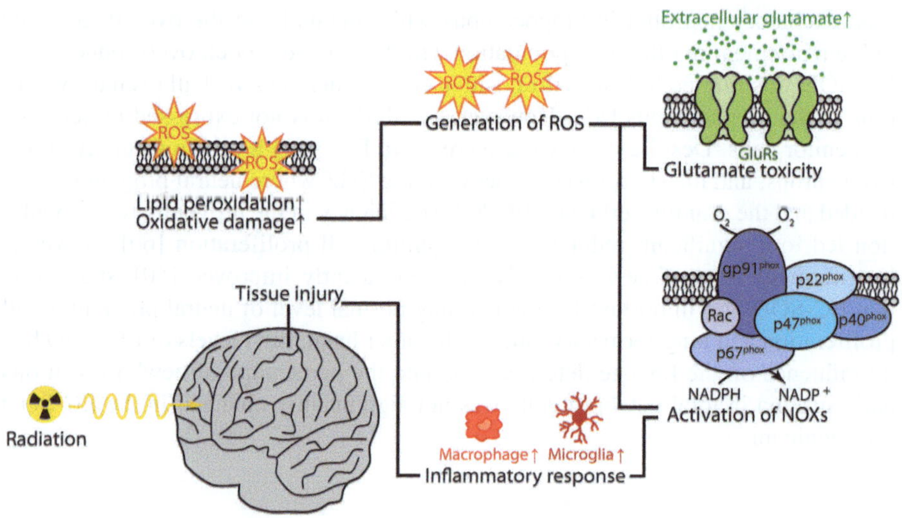

Fig. 5.5 Mechanism of radiation-induced brain damage. Ionizing radiation leads to generation of oxygen free radicals by several mechanisms, including inflammatory response of macrophage and microglia, activation of NADPH oxidases, and excitatory toxicity. These changes form a positive feedback loop to perpetuate tissue damage and continued production of oxygen free radicals. GluRs, glutamate receptors

National Brain Tumor Society and American Association of Neurological Surgeons). Brain tumors occur in all ages, with primary brain tumors occurring more frequently in children and metastatic brain tumors more common in adults. Although approximately 20–40 % of non-CNS tumors develop brain metastasis, metastatic breast cancers and small-cell lung cancers (SCLC) constitute the two most commonly observed metastases to the brain [67, 68]. Treatment of brain tumors includes surgical resection, radiation therapy, and chemotherapy. Radiation therapy is used to kill cancer cells or to shrink tumors; it is especially useful in cases where surgical resection cannot be used effectively to treat brain tumors. Therapeutic radiation includes high energy photons such as x-ray and gamma ray, and particle beams such as proton, neutron, and carbon ion.

Radiation brain injury could involve macroscopic tissue destruction after relatively high doses of irradiation [69]. With the improvement of technology and limited fraction size, overt tissue damage can be avoided. However, less severe morphologic injury can occur after radiotherapy and the injury often leads to neurocognitive dysfunctions manifested as deficits in hippocampal-dependent functions of learning, memory, and spatial information processing [70–72]. The generation of ROS is considered a main cause of radiation-mediated tissue damages. Ionizing radiation not only results in the acute generation of short-lived ROS [62], it also results in a persistent oxidative stress that extends up to several months after irradiation [59, 61, 73].

5.8 Radiation and Hippocampal Neurogenesis

The precise mechanism leading to radiation-induced cognitive dysfunction is not clear; however, experimental evidence suggests that damage to neural stem cells and suppression of neurogenesis after radiation therapy are the most immediate and persistent events [74, 75]. Hippocampal neurogenesis is exquisitely sensitive to irradiation, and irradiation in the CNS leads to a marked dose-dependent reduction in proliferating cells and immature neurons in the SGZ of hippocampal dentate gyrus within the first 48 h after irradiation [59]. Beyond the period immediately after radiation exposure, the effects of irradiation continue to manifest as suppression of hippocampal neurogenesis and reduction in dendritic arborization and spine densities [59, 60]. As a testament to the role of oxidative stress in radiation-mediated suppression of hippocampal neurogenesis, blockade of oxidative stress via antioxidant supplementation or inhibition of neuroinflammation have both minimized the negative impact of irradiation on hippocampal neurogenesis [60, 76–78].

5.9 EC-SOD and Radiation Protection

Given the impact of oxidative stress in radiation-mediated tissue damage and hippocampal neurogenesis, enhanced antioxidant capacity is expected to provide radiation protection. The concept of radiation protection via enhanced antioxidant capacity is not new. Induction of MnSOD is known to provide radiation protection in multiple cancers, and overexpression of MnSOD has been shown to provide organ-specific protection of normal tissues against ionizing radiation [79, 80]. High levels of CuZnSOD [81], as well as a number of antioxidant compounds, has also been shown to provide radiation protection [82, 83]. Furthermore, cancer stem cells are highly resistant to ionizing radiation, and recent studies correlate the resistance with high levels of antioxidant capacity [84].

Among the three mammalian SODs, EC-SOD has been tested in transgenic mice for its radiation protection of hippocampal neurogenesis and the associated cognitive functions [59–61]. As described earlier, the EC-SOD transgenic mice were designed as a "hybrid" system with overexpression of EC-SOD in CamKIIα positive neurons, and at the same time, were devoid of EC-SOD in all other cell types. The mouse model was used to test the effects of EC-SOD on the production, differentiation, and long-term survival of newborn neurons in the dentate SGZ. Consequently, the neural progenitor cells carried the EC-SOD null phenotype and showed significant reductions in proliferation. However, with a sixfold increase in EC-SOD levels in mature granule cells, EC-SOD transgenic mice had a significantly higher percentage of immature neurons advanced to the postmitotic stage and with a more complex dendritic structure (Fig. 5.6). Even though the immature neurons were not expected to express the EC-SOD transgene, their soma was in close proximity to mature granule cells, and their dendrites extended deep into the granule cell layers.

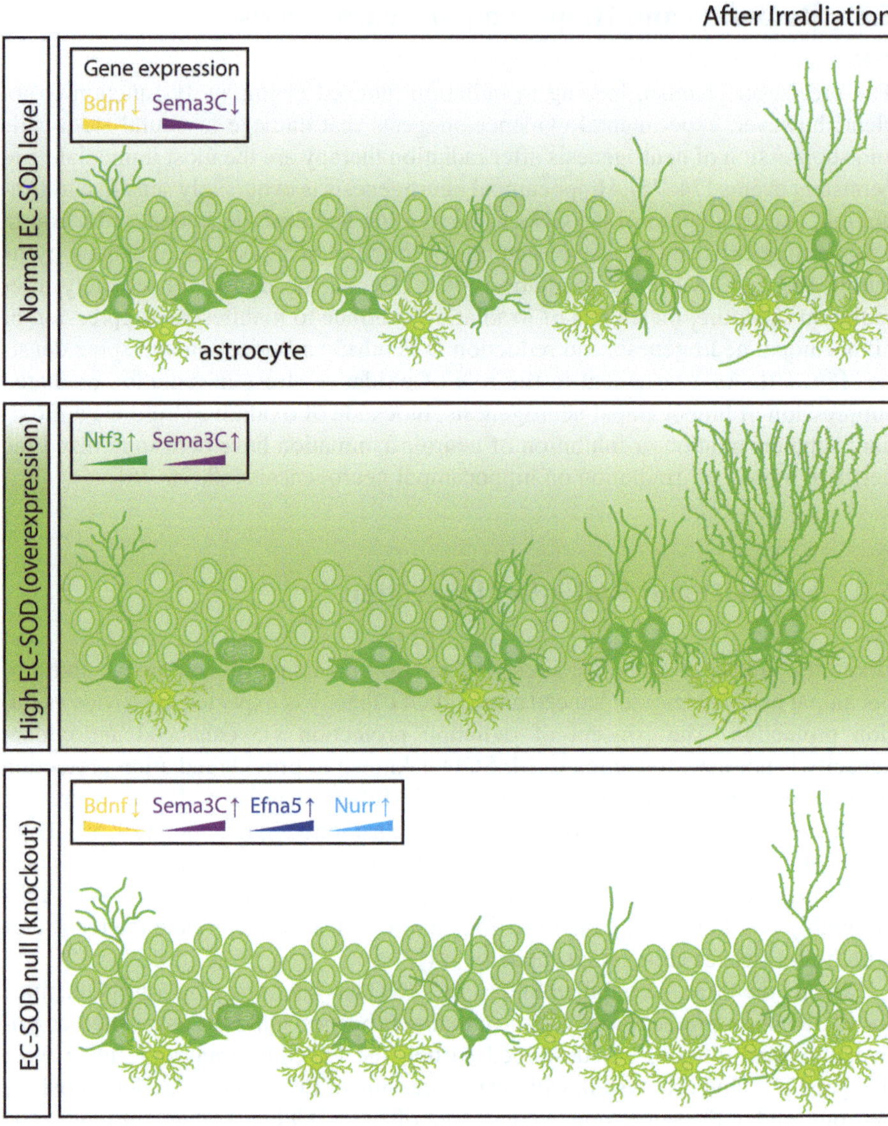

Fig. 5.6 The impact of EC-SOD on irradiation-induced changes in hippocampal neurogenesis. Under normal EC-SOD levels in the neurogenic environment, cranial irradiation leads to a significant reduction in hippocampal neurogenesis with reduced production of newborn neurons, reduced dendritic arborization, and reduced dendritic spine densities. The expression levels of brain-derived neurotrophic factor (BDNF) and the axon guidance molecule semaphorin 3C (sema3C), both important molecules for neuronal development and survival, are also significantly reduced following irradiation (*upper panel*). In transgenic mice expressing elevated levels of EC-SOD in CamKII positive neurons (*middle panel*), increased EC-SOD in the neurogenic environment mitigates irradiation effects and results in normalization of hippocampal neurogenesis, increased dendritic complexity, and increased dendritic spine densities. The expression level of BDNF remains normal in EC-SOD transgenic mice, while expressions of neurotrophin 3 (Ntf3) and Sema3C are significantly increased in the hippocampal formation. In contrast, without EC-SOD in the

It was possible that the EC-SOD-rich redox environment modified the dendrite and axon outgrowth, and encouraged the maturation of newborn neurons.

Irradiation reduced the number of immature neurons in wild-type and EC-SOD transgenic mice; however, the extent of reduction was statistically significant only in the wild-type mice. Parallel to the preservation of immature neurons following irradiation, the number of fully matured newborn neurons in EC-SOD transgenic mice was not reduced by irradiation. On the average, a similar percentage of immature neurons advanced to fully matured neurons in sham-irradiated wild-type and EC-SOD transgenic mice (6.0 % and 6.6 %, respectively). However, the outcome from irradiation was very different between these two genotypes. Whereas the percentage of immature neurons that advanced to fully matured neurons was decreased by 40 % in irradiated wild-type mice, a 39 % increase was observed in EC-SOD transgenic mice. The data suggested enhanced survival of newborn neurons in the EC-SOD-rich environment. Changes in the differentiation pattern following irradiation also played a role in the different neurogenic outcomes. As expected, the post-irradiation environment in wild-type mice became less favorable for neuronal lineage and consequently, a significant reduction in the number of fully matured newborn neurons was observed. However, no significant change was observed in EC-SOD transgenic mice following irradiation [60]. Collectively, the data suggested that constitutive overexpression of EC-SOD in mature hippocampal neurons was effective in providing radiation protection of hippocampal neurogenesis by safeguarding the maturation and differentiation process. Preserved hippocampal neurogenesis in EC-SOD transgenic mice also corresponded to preserved cognitive functions following irradiation.

Given the random nature of radiation targets and the extracellular location of EC-SOD, radiation protection by EC-SOD likely involves more complex mechanisms, such as modulation of the neurogenic microenvironment (Fig. 5.6). Expression of the brain-derived neurotrophic factor (BDNF) was significantly reduced in wild-type mice, but the level was not changed in EC-SOD transgenic mice following irradiation. In addition, expression of neurotrophin 3 (NTF3) was increased in EC-SOD transgenic mice following irradiation [60]. The combination of these two trophic factors in the postirradiation environment likely played an important role in the preservation of hippocampal neurogenesis. Similar to BDNF, expression of a guidance molecule, semaphorin 3C (Sema3C), was also reduced in wild-type mice following irradiation; however, no significant reduction was observed in EC-SOD transgenic mice. Sema3C is important for axon guidance and neurotigenesis [85, 86]. It has also been shown to promote survival and tumorigenicity of

Fig. 5.6 (continued) neurogenic environment, EC-SOD null mice (*bottom panel*) show a significant reduction in hippocampal neurogenesis and reduced dendritic complexity. However, these phenotypes are not affected by cranial irradiation, and the extent of hippocampal neurogenesis remains the same in EC-SOD knockout mice after irradiation. This may be due to the maintenance of normal Sema3C expression levels and increases in factors such as Efna5 and Nurr1 that are important for axon guidance, neuronal survival, or spatial learning and memory glioma stem cells when present at high levels [87]. Therefore, maintaining a normal level of Sema3C has the potential to promote survival of newborn neurons by making correct axonal connections and allowing functional integration of newborn neurons to the existing neuronal network.

One interesting observation in the role of EC-SOD in radiation-induced changes in hippocampal neurogenesis was the lack of further reduction in new neuron productions in EC-SOD null mice [59, 60]. This paradoxical observation at the cellular level correlated well with improved cognitive functions, especially in spatial learning and memory [60, 61], and led us to hypothesize in EC-SOD null tissues a condition similar to that in pre-conditioning where tissues or cells were more resistant to an insult if they were exposed to low levels of similar challenge beforehand. Results from the gene expression study corroborated this hypothesis and showed increased expression of ephrin A5 (Efna5) and nuclear receptor related 1 (Nurr1) in addition to a well-maintained Sema3 expression level in EC-SOD null mice after cranial irradiation (Fig. 5.6) [60]. Efna5 is a member of the membrane-bound ephrin A ligands that binds to ephrin receptors through cell–cell interactions [88]. It plays an important role in axon guidance, especially in migrating axons expressing ephrin A receptors. In the hippocampal dentate gyrus, Efna5 is abundantly expressed in mature neurons, neural progenitor cells, and astrocytes [89]. Efna5 deficient mice show decreased hippocampal neurogenesis due to a reduction in progenitor cell proliferation and decreased survival of newborn neurons. These mice also exhibit abnormal vascular formation in the dentate gyrus [89]. Increased Efna5 expression in EC-SOD null mice following cranial irradiation likely provided supports for hippocampal neurogenesis. Nurr1, on the other hand, is a transcription factor and is mainly known for its critical role in dopaminergic neuron development [90]. Animal studies showed Nurr1 to be upregulated in the hippocampus following spatial learning [91], and this upregulation enhanced memory retention [92]. Conversely, suppression of Nurr1 led to impaired spatial learning and memory [93]. Therefore, upregulation of Nurr1 in EC-SOD null mice after irradiation may have contributed towards improved hippocampal functions of learning and memory.

How irradiation changes transcriptional control of BDNF, NTF3, Sema3C, and a number of other critical factors is not fully understood. Epigenetic regulation of BDNF transcription via histone modification has been observed following higher doses of irradiation in rats [94]. Whether the same mechanism applies to lower doses is not clear. Not much is known about transcriptional regulation of NTF3 or Sema3C in the postirradiation environment at present time. It is also not clear how EC-SOD offset the negative impact of irradiation on the transcriptional control of BDNF and Sema3C. Mechanistic based investigation to a better understanding of how the transcriptional and translational landscape changes in the postirradiation environment and how the process impacts hippocampal neurogenesis will be crucial for rational designs of therapeutic strategies to provide protection of normal tissue from radiation damage or to enhance regeneration in the postirradiation stage.

5.10 Perspective of SOD-Mediated Radiation Protection for Cancer Treatment

Antioxidant treatment as a means of radiation protection of normal tissues can be a two-edged sword—it can provide protection to normal tissues, but can also protect cancer cells at the same time. In fact, high antioxidant capacity is a hallmark of radiation-resistant cancer cells. On the other hand, the redox balance in normal cells and cancer cells can be quite different due to differences in their metabolism. Therefore, differences in redox metabolism and sensitivity to changes in the redox environment between normal tissues and cancer cells may be harnessed for rational designs of SOD-based radiation protection of normal tissues in cancer treatment. Among the new generation of small molecular weight antioxidant compounds in the pipeline, Mn-containing porphyrins hold great promise to achieving the goal [95]. The Mn(III) *meso* tetrakis(*N*-alkylpyridinium-2-yl)porphyrins (such as MnTE-2-PyP^{5+}, BMX-010, where alkyl is ethyl) have been shown to protect normal eye tissues [96], bone marrow [97], lung tissue [98], salivary glands and mouth mucosa [99], and prostate tissues from radiation damage in animal models. Mn porphyrins that are designed to cross the blood brain barrier have been tested for tissue distribution in different brain regions and for CNS protections [100]. The efficacy of Mn porphyrins in mouse cranial irradiation models is currently under investigation, but preliminary study shows positive effects if the compound is administered before radiation therapy [101]. In mouse studies, the more lipophilic analog, Mn(III) *meso* tetrakis (*N*-n-butoxyethylpyridinium-2-yl)porphyrin, MnTnBuOE-2-PyP^{5+} (BMX-001) showed differential effects on normal vs. tumor tissue, i.e., the ability to radioprotect normal brain while radiosensitizing patient-derived glioblastoma multiforme D-245 MG [102]. Based on such data, MnTnBuOE-2-PyP^{5+} has entered Phase I/II Clinical Trials in May 2016, as a radioprotector of normal brain with glioma patients. The Clinical Trials on radioprotection of salivary glands and mouth mucosa with MnTnBuOE-2-PyP^{5+} will soon follow with head and neck cancer patients.

References

1. Fridovich I. Superoxide dismutases. Adv Enzymol Relat Areas Mol Biol. 1986;58:61–97.
2. Tally FP, et al. Superoxide dismutase in anaerobic bacteria of clinical significance. Infect Immun. 1977;16(1):20–5.
3. Lynch M, Kuramitsu H. Expression and role of superoxide dismutases (SOD) in pathogenic bacteria. Microbes Infect. 2000;2(10):1245–55.
4. Fink RC, Scandalios JG. Molecular evolution and structure—function relationships of the superoxide dismutase gene families in angiosperms and their relationship to other eukaryotic and prokaryotic superoxide dismutases. Arch Biochem Biophys. 2002;399(1):19–36.
5. Fukuhara R, Tezuka T, Kageyama T. Structure, molecular evolution, and gene expression of primate superoxide dismutases. Gene. 2002;296(1–2):99–109.
6. Slot JW, et al. Intracellular localization of the copper-zinc and manganese superoxide dismutases in rat liver parenchymal cells. Lab Invest. 1986;55(3):363–71.

7. Oury TD, et al. Human extracellular superoxide dismutase is a tetramer composed of two disulphide-linked dimers: a simplified, high-yield purification of extracellular superoxide dismutase. Biochem J. 1996;317(Pt 1):51–7.
8. Tafuri F, et al. SOD1 misplacing and mitochondrial dysfunction in amyotrophic lateral sclerosis pathogenesis. Front Cell Neurosci. 2015;9:336.
9. Sutton A, et al. The Ala16Val genetic dimorphism modulates the import of human manganese superoxide dismutase into rat liver mitochondria. Pharmacogenetics. 2003;13(3): 145–57.
10. Borgstahl GE, et al. Human mitochondrial manganese superoxide dismutase polymorphic variant Ile58Thr reduces activity by destabilizing the tetrameric interface. Biochemistry. 1996;35(14):4287–97.
11. Dhar SK, Clair DKS. Manganese superoxide dismutase regulation and cancer. Free Radic Biol Med. 2012;52(11–12):2209–22.
12. Iguchi T, et al. Association of MnSOD AA genotype with the progression of prostate cancer. PLoS One. 2015;10(7), e0131325.
13. Moradi MT, et al. Manganese superoxide dismutase (MnSOD Val-9Ala) gene polymorphism and susceptibility to gastric cancer. Asian Pac J Cancer Prev. 2015;16(2):485–8.
14. Atilgan D, et al. The relationship between ALA16VAL single gene polymorphism and renal cell carcinoma. Adv Urol. 2014;2014:932481.
15. Bresciani G, et al. The MnSOD Ala16Val SNP: relevance to human diseases and interaction with environmental factors. Free Radic Res. 2013;47(10):781–92.
16. Sandstrom J, et al. 10-fold increase in human plasma extracellular superoxide dismutase content caused by a mutation in heparin-binding domain. J Biol Chem. 1994;269(29): 19163–6.
17. Yamada H, et al. Molecular analysis of extracellular-superoxide dismutase gene associated with high level in serum. Jpn J Hum Genet. 1995;40(2):177–84.
18. Juul K, et al. Genetically reduced antioxidative protection and increased ischemic heart disease risk: The Copenhagen City Heart Study. Circulation. 2004;109(1):59–65.
19. Nakamura M, et al. Role of extracellular superoxide dismutase in patients under maintenance hemodialysis. Nephron Clin Pract. 2005;101(3):c109–15.
20. Juul K, et al. Genetically increased antioxidative protection and decreased chronic obstructive pulmonary disease. Am J Respir Crit Care Med. 2006;173(8):858–64.
21. Halliwell B, Gutteridge JMC. Free radicals in biology and medicine. New York: Oxford University Press; 1999.
22. Halliwell B, Cross CE. Oxygen-derived species: their relation to human disease and environmental stress. Environ Health Perspect. 1994;102 Suppl 10:5–12.
23. Knight JA. Free radicals: their history and current status in aging and disease. Ann Clin Lab Sci. 1998;28(6):331–46.
24. Wallace DC, Melov S. Radicals r'aging. Nat Genet. 1998;19(2):105–6.
25. Ray PD, Huang BW, Tsuji Y. Reactive oxygen species (ROS) homeostasis and redox regulation in cellular signaling. Cell Signal. 2012;24(5):981–90.
26. Rhee SG, Woo HA. Multiple functions of peroxiredoxins: peroxidases, sensors and regulators of the intracellular messenger H(2)O(2), and protein chaperones. Antioxid Redox Signal. 2011;15(3):781–94.
27. Sarsour EH, et al. Redox control of the cell cycle in health and disease. Antioxid Redox Signal. 2009;11(12):2985–3011.
28. Sarsour EH, et al. Manganese superoxide dismutase activity regulates transitions between quiescent and proliferative growth. Aging Cell. 2008;7(3):405–17.
29. Prozorovski T, et al. Sirt1 contributes critically to the redox-dependent fate of neural progenitors. Nat Cell Biol. 2008;10(4):385–94.
30. Schafer FQ, Buettner GR. Redox environment of the cell as viewed through the redox state of the glutathione disulfide/glutathione couple. Free Radic Biol Med. 2001;30(11): 1191–212.
31. van Praag H, et al. Functional neurogenesis in the adult hippocampus. Nature. 2002; 415(6875):1030–4.

32. Ben Abdallah NM, et al. Early age-related changes in adult hippocampal neurogenesis in C57 mice. Neurobiol Aging. 2010;31(1):151–61.
33. Wang K, et al. Redox homeostasis: the linchpin in stem cell self-renewal and differentiation. Cell Death Dis. 2013;4:e537.
34. Noble M, Mayer-Proschel M, Proschel C. Redox regulation of precursor cell function: insights and paradoxes. Antioxid Redox Signal. 2005;7(11–12):1456–67.
35. Noble M, et al. Redox state as a central modulator of precursor cell function. Ann N Y Acad Sci. 2003;991:251–71.
36. Smith J, et al. Redox state is a central modulator of the balance between self-renewal and differentiation in a dividing glial precursor cell. Proc Natl Acad Sci U S A. 2000;97(18): 10032–7.
37. Yoneyama M, et al. Endogenous reactive oxygen species are essential for proliferation of neural stem/progenitor cells. Neurochem Int. 2010;56(6–7):740–6.
38. Le Belle JE, et al. Proliferative neural stem cells have high endogenous ROS levels that regulate self-renewal and neurogenesis in a PI3K/Akt-dependant manner. Cell Stem Cell. 2011; 8(1):59–71.
39. Kempermann G, Kuhn HG, Gage FH. More hippocampal neurons in adult mice living in an enriched environment. Nature. 1997;386:493–5.
40. van Praag H, et al. Running enhances neurogenesis, learning, and long-term potentiation in mice. Proc Natl Acad Sci U S A. 1999;96(23):13427–31.
41. Kempermann G. Why new neurons? Possible functions for adult hippocampal neurogenesis. J Neurosci. 2002;22(3):635–8.
42. Kempermann G, Brandon EP, Gage FH. Environmental stimulation of 129/SvJ mice causes increased cell proliferation and neurogenesis in the adult dentate gyrus. Curr Biol. 1998; 8(16):939–42.
43. Kempermann G, Kuhn HG, Gage FH. Experience-induced neurogenesis in the senescent dentate gyrus. J Neurosci. 1998;18(9):3206–12.
44. Prickaerts J, et al. Learning and adult neurogenesis: survival with or without proliferation? Neurobiol Learn Mem. 2004;81(1):1–11.
45. Kee N, et al. Preferential incorporation of adult-generated granule cells into spatial memory networks in the dentate gyrus. Nat Neurosci. 2007;10(3):355–62.
46. Jinno S. Topographic differences in adult neurogenesis in the mouse hippocampus: a stereology-based study using endogenous markers. Hippocampus. 2011;21(5):467–80.
47. Burger C. Region-specific genetic alterations in the aging hippocampus: implications for cognitive aging. Front Aging Neurosci. 2010;2:140.
48. Huang TT, et al. Transgenic and mutant mice for oxygen free radical studies. Methods Enzymol. 2002;349:191–213.
49. Misawa H, et al. Conditional knockout of Mn superoxide dismutase in postnatal motor neurons reveals resistance to mitochondrial generated superoxide radicals. Neurobiol Dis. 2006;23(1):169–77.
50. Oury TD, Day BJ, Crapo JD. Extracellular superoxide dismutase in vessels and airways of humans and baboons. Free Radic Biol Med. 1996;20(7):957–65.
51. Zou Y, et al. A new mouse model for temporal- and tissue-specific control of extracellular superoxide dismutase. Genesis. 2009;47(3):142–54.
52. Carlsson LM, et al. Mice lacking extracellular superoxide dismutase are more sensitive to hyperoxia. Proc Natl Acad Sci U S A. 1995;92(14):6264–8.
53. Sakellariou GK, et al. Neuron-specific expression of CuZnSOD prevents the loss of muscle mass and function that occurs in homozygous CuZnSOD-knockout mice. FASEB J. 2014; 28(4):1666–81.
54. Raineri I, et al. Strain-dependent high-level expression of a transgene for manganese superoxide dismutase is associated with growth retardation and decreased fertility. Free Radic Biol Med. 2001;31(8):1018–30.
55. Yen HC, et al. The protective role of manganese superoxide dismutase against adriamycin-induced acute cardiac toxicity in transgenic mice. J Clin Invest. 1996;98(5):1253–60.

56. Kim A, et al. Enhanced expression of mitochondrial superoxide dismutase leads to prolonged in vivo cell cycle progression and up-regulation of mitochondrial thioredoxin. Free Radic Biol Med. 2010;48(11):1501–12.
57. Corniola R, et al. Paradoxical relationship between Mn superoxide dismutase deficiency and radiation-induced cognitive defects. PLoS One. 2012;7(11), e49367.
58. Fishman K, et al. Radiation-induced reductions in neurogenesis are ameliorated in mice deficient in CuZnSOD or MnSOD. Free Radic Biol Med. 2009;47(10):1459–67.
59. Rola R, et al. Lack of extracellular superoxide dismutase (EC-SOD) in the microenvironment impacts radiation-induced changes in neurogenesis. Free Radic Biol Med. 2007;42(8):1133–45; discussion 1131–2.
60. Zou Y, et al. Extracellular superoxide dismutase is important for hippocampal neurogenesis and preservation of cognitive functions after irradiation. Proc Natl Acad Sci U S A. 2012; 109(52):21522–7.
61. Raber J, et al. Irradiation enhances hippocampus-dependent cognition in mice deficient in extracellular superoxide dismutase. Hippocampus. 2011;21(1):72–80.
62. Riley PA. Free radicals in biology: oxidative stress and the effects of ionizing radiation. Int J Radiat Biol. 1994;65(1):27–33.
63. Azzam EI, Jay-Gerin JP, Pain D. Ionizing radiation-induced metabolic oxidative stress and prolonged cell injury. Cancer Lett. 2012;327(1–2):48–60.
64. Phillis JW, Horrocks LA, Farooqui AA. Cyclooxygenases, lipoxygenases, and epoxygenases in CNS: their role and involvement in neurological disorders. Brain Res Rev. 2006; 52(2):201–43.
65. Sorce S, Krause KH. NOX enzymes in the central nervous system: from signaling to disease. Antioxid Redox Signal. 2009;11(10):2481–504.
66. Bedard K, Krause KH. The NOX family of ROS-generating NADPH oxidases: physiology and pathophysiology. Physiol Rev. 2007;87(1):245–313.
67. Chargari C, et al. Whole-brain radiation therapy in breast cancer patients with brain metastases. Nat Rev Clin Oncol. 2010;7(11):632–40.
68. Quan AL, Videtic GM, Suh JH. Brain metastases in small cell lung cancer. Oncology (Williston Park). 2004;18(8):961–72; discussion 974, 979–80, 987.
69. Sheline GE, Wara WM, Smith V. Therapeutic irradiation and brain injury. Int J Radiat Oncol Biol Phys. 1980;6(9):1215–28.
70. Grill J, et al. Long-term intellectual outcome in children with posterior fossa tumors according to radiation doses and volumes. Int J Radiat Oncol Biol Phys. 1999;45(1):137–45.
71. Meyers CA, et al. Neurocognitive effects of therapeutic irradiation for base of skull tumors. Int J Radiat Oncol Biol Phys. 2000;46(1):51–5.
72. Surma-aho O, et al. Adverse long-term effects of brain radiotherapy in adult low-grade glioma patients. Neurology. 2001;56(10):1285–90.
73. Limoli CL, et al. Redox changes induced in hippocampal precursor cells by heavy ion irradiation. Radiat Environ Biophys. 2007;46(2):167–72.
74. Panagiotakos G, et al. Long-term impact of radiation on the stem cell and oligodendrocyte precursors in the brain. PLoS One. 2007;2(7), e588.
75. Abayomi OK. Pathogenesis of irradiation-induced cognitive dysfunction. Acta Oncol. 1996;35(6):659–63.
76. Belarbi K, et al. CCR2 deficiency prevents neuronal dysfunction and cognitive impairments induced by cranial irradiation. Cancer Res. 2013;73(3):1201–10.
77. Lee SW, et al. Absence of CCL2 is sufficient to restore hippocampal neurogenesis following cranial irradiation. Brain Behav Immun. 2013;30:33–44.
78. Monje ML, Toda H, Palmer TD. Inflammatory blockade restores adult hippocampal neurogenesis. Science. 2003;302(5651):1760–5.
79. Greenberger JS, Epperly MW. Review. Antioxidant gene therapeutic approaches to normal tissue radioprotection and tumor radiosensitization. In Vivo. 2007;21(2):141–6.
80. Eldridge A, et al. Manganese superoxide dismutase interacts with a large scale of cellular and mitochondrial proteins in low-dose radiation-induced adaptive radioprotection. Free Radic Biol Med. 2012;53(10):1838–47.

81. Breuer R, et al. Superoxide dismutase inhibits radiation-induced lung injury in hamsters. Lung. 1992;170(1):19–29.
82. Xiao X, et al. Catalase inhibits ionizing radiation-induced apoptosis in hematopoietic stem and progenitor cells. Stem Cells Dev. 2015;24(11):1342–51.
83. Jelveh S, et al. Investigations of antioxidant-mediated protection and mitigation of radiation-induced DNA damage and lipid peroxidation in murine skin. Int J Radiat Biol. 2013; 89(8):618–27.
84. Diehn M, et al. Association of reactive oxygen species levels and radioresistance in cancer stem cells. Nature. 2009;458(7239):780–3.
85. Moreno-Flores MT, et al. Semaphorin 3C preserves survival and induces neuritogenesis of cerebellar granule neurons in culture. J Neurochem. 2003;87(4):879–90.
86. Steup A, et al. Sema3C and netrin-1 differentially affect axon growth in the hippocampal formation. Mol Cell Neurosci. 2000;15(2):141–55.
87. Vaitkiene P, et al. High level of Sema3C is associated with glioma malignancy. Diagn Pathol. 2015;10:58.
88. Lisabeth EM, Falivelli G, Pasquale EB. Eph receptor signaling and ephrins. Cold Spring Harb Perspect Biol. 2013;5(9):a009159.
89. Hara Y, et al. Impaired hippocampal neurogenesis and vascular formation in ephrin-A5-deficient mice. Stem Cells. 2010;28(5):974–83.
90. Decressac M, et al. NURR1 in Parkinson disease—from pathogenesis to therapeutic potential. Nat Rev Neurol. 2013;9(11):629–36.
91. Pena de Ortiz S, Maldonado-Vlaar CS, Carrasquillo Y. Hippocampal expression of the orphan nuclear receptor gene hzf-3/nurr1 during spatial discrimination learning. Neurobiol Learn Mem. 2000;74(2):161–78.
92. Aldavert-Vera L, et al. Intracranial self-stimulation facilitates active-avoidance retention and induces expression of c-Fos and Nurr1 in rat brain memory systems. Behav Brain Res. 2013;250:46–57.
93. Colon-Cesario WI, et al. Knockdown of Nurr1 in the rat hippocampus: implications to spatial discrimination learning and memory. Learn Mem. 2006;13(6):734–44.
94. Ji S, et al. Irradiation-induced hippocampal neurogenesis impairment is associated with epigenetic regulation of bdnf gene transcription. Brain Res. 2014;1577:77–88.
95. Batinic-Haberle I, et al. Design of Mn porphyrins for treating oxidative stress injuries and their redox-based regulation of cellular transcriptional activities. Amino Acids. 2010;42(1):95–113.
96. Mao XW, et al. Radioprotective effect of a metalloporphyrin compound in rat eye model. Curr Eye Res. 2009;34(1):62–72.
97. Li H, et al. Mn(III) meso-tetrakis-(N-ethylpyridinium-2-yl) porphyrin mitigates total body irradiation-induced long-term bone marrow suppression. Free Radic Biol Med. 2011;51(1): 30–7.
98. Gauter-Fleckenstein B, et al. Early and late administration of MnTE-2-PyP5+ in mitigation and treatment of radiation-induced lung damage. Free Radic Biol Med. 2010;48(8): 1034–43.
99. Ashcraft KA, et al. Novel manganese-porphyrin superoxide dismutase-mimetic widens the therapeutic margin in a preclinical head and neck cancer model. Int J Radiat Oncol Biol Phys. 2015;93(4):892–900.
100. Batinic-Haberle I, et al. SOD therapeutics: latest insights into their structure-activity relationships and impact on the cellular redox-based signaling pathways. Antioxid Redox Signal. 2014;20(15):2372–415.
101. Yang P, et al. 229—Mitigation of radiation-mediated suppression of hippocampal neurogenesis. Free Radic Biol Med. 2014;76 Suppl 1:S97.
102. Weitzel DH, et al. Radioprotection of the brain white matter by Mn(III) n-Butoxyethylpyridylporphyrin-based superoxide dismutase mimic MnTnBuOE-2-PyP5+. Mol Cancer Ther. 2015;14(1):70–9.

103. Epstein CJ, et al. Transgenic mice with increased Cu/Zn-superoxide dismutase activity: animal model of dosage effects in Down syndrome. Proc Natl Acad Sci U S A. 1987;84(22):8044–8.
104. Shi YP, et al. The mapping of transgenes by fluorescence in situ hybridization on G-banded mouse chromosomes. Mamm Genome. 1994;5(6):337–41.
105. Gurney ME, et al. Motor neuron degeneration in mice that express a human Cu, Zn superoxide dismutase mutation. Science. 1994;264(5166):1772–5.
106. Chan PH, et al. Overexpression of SOD1 in transgenic rats protects vulnerable neurons against ischemic damage after global cerebral ischemia and reperfusion. J Neurosci. 1998;18(20):8292–9.
107. Fischer LR, et al. SOD1 targeted to the mitochondrial intermembrane space prevents motor neuropathy in the Sod1 knockout mouse. Brain. 2011;134(Pt 1):196–209.
108. Huang TT, et al. Superoxide-mediated cytotoxicity in superoxide dismutase-deficient fetal fibroblasts. Arch Biochem Biophys. 1997;344(2):424–32.
109. Matzuk MM, et al. Ovarian function in superoxide dismutase 1 and 2 knockout mice. Endocrinology. 1998;139(9):4008–11.
110. Chen H, Li X, Epstein PN. MnSOD and catalase transgenes demonstrate that protection of islets from oxidative stress does not alter cytokine toxicity. Diabetes. 2005;54(5):1437–46.
111. Shen X, et al. Protection of cardiac mitochondria by overexpression of MnSOD reduces diabetic cardiomyopathy. Diabetes. 2006;55(3):798–805.
112. Li Y, et al. Dilated cardiomyopathy and neonatal lethality in mutant mice lacking manganese superoxide dismutase. Nat Genet. 1995;11(4):376–81.
113. Lebovitz RM, et al. Neurodegeneration, myocardial injury, and perinatal death in mitochondrial superoxide dismutase-deficient mice. Proc Natl Acad Sci U S A. 1996;93(18):9782–7.
114. Ikegami T, et al. Model mice for tissue-specific deletion of the manganese superoxide dismutase (MnSOD) gene. Biochem Biophys Res Commun. 2002;296(3):729–36.
115. Epperly MW, et al. Conditional radioresistance of tet-inducible manganese superoxide dismutase bone marrow stromal cell lines. Radiat Res. 2013;180(2):189–204.
116. Oury TD, et al. Extracellular superoxide dismutase, nitric oxide, and central nervous system O2 toxicity. Proc Natl Acad Sci U S A. 1992;89(20):9715–9.

Chapter 6
Metabolic Production of H_2O_2 in Carcinogenesis and Cancer Treatment

Bryan G. Allen and Douglas R. Spitz

6.1 Introduction

In 1956, Otto Warburg proposed that cancer was fundamentally a metabolic disease where dysregulation of glycolysis and respiration was mechanistically related to carcinogenesis and maintenance of the malignant phenotype [1]. This hypothesis was based on 30 years of previous work with cancer cells showing increased rates of glycolysis and an inability to regulate the relationship between glucose and O_2 metabolism, relative to normal cells [1]. Warburg hypothesized that the dysregulation between glucose and O_2 metabolism in cancer cells was related to "damage" to the respiratory mechanisms (responsible for consuming O_2) requiring cancer cells to metabolize glucose uncontrollably to generate energy in the form of ATP for rapid proliferation. This hypothesis was initially challenged because cancer cells (relative to normal cells) were found to consume similar amounts of O_2 by respiration [2]. Recently, Warburg's original hypothesis has again been reformulated to propose that dysregulated metabolism of glucose by cancer cells occurs because of the need to meet the metabolic requirements of uncontrolled malignant cell proliferation leading to increased biomass [3]. The paradoxical data that continue to contradict hypotheses derived from Warburg's original proposal [1–3] are twofold. First, radiotracer and imaging techniques that allow for the monitoring of surrogate markers for both glucose metabolism (i.e., ^{18}FDG-PET imaging) and proliferation (i.e., ^{18}FLT-PET) show that normal tissues such as bone marrow and gut epithelium proliferate as well as creating biomass at much higher rates than solid tumor cells, and yet do not appear to undergo increases in glucose metabolism to match that of

B.G. Allen (✉) • D.R. Spitz
Division of Free Radical and Radiation Biology, Department of Radiation Oncology, University of Iowa Hospitals and Clinics, Holden Comprehensive Cancer Center, 200 Hawkins Drive, Iowa City, IA 52242-1009, USA
e-mail: bryan-allen@uiowa.edu

I. Batinić-Haberle et al. (eds.), *Redox-Active Therapeutics*, Oxidative Stress in Applied Basic Research and Clinical Practice, DOI 10.1007/978-3-319-30705-3_6

cancer tissues [2, 4, 5]. Secondly, cancer cells also appear to possess adequate capacity for metabolizing fat and amino acid substrates for the purpose of generating energy via oxidative phosphorylation, calling into question the assumption that cancer cells manifest any deficiency in mitochondrial oxidative energy metabolism [2, 6–8].

To reconcile these seemingly paradoxical findings, numerous studies have now demonstrated that cancer cells (relative to normal cells) have altered mitochondrial oxidative metabolism resulting in increased steady-state levels of reactive oxygen species (ROS) ($O_2^{\bullet-}$ and H_2O_2) and glucose metabolism is upregulated to provide reducing equivalents in the form of pyruvate and NADPH to detoxify hydroperoxides to compensate for these alterations in mitochondrial metabolism of O_2 [9–22]. Furthermore, there is now also strong evidence that NADPH oxidase enzymes, peroxisomal metabolism, and metabolism of xenobiotics through the cytochrome P450 monooxygenases can also generate $O_2^{\bullet-}$ and H_2O_2 that contribute to initiation, promotion, and progression of malignant phenotypes. [23–28] Finally, NADPH oxidase enzymes, peroxisomal metabolism, and cytochrome P450 metabolism can all also provide a potential feed forward mechanism by which membrane and cytosolic sources of $O_2^{\bullet-}/H_2O_2$ can dysregulate the communication between cytosolic and mitochondrial oxidative metabolism to drive the progression of malignancy as well as other degenerative diseases associated with aging [29, 30].

Given these new findings related to the relationship between glucose and O_2 metabolism in cancer cells as well as the theoretical construct relating electronic and cancer biology originally proposed by Szent Gyorgyi [31–34], mammalian organisms can now be conceptualized as complex higher order biological structures that generate their life force from the ability to extract electrons from food sources via oxidation reactions leading to the generation of ATP and reducing equivalents for biosynthesis, while consuming O_2 to dispose of electrons onto water molecules. In this regard the relationship between metabolic processes and gene expression pathways in mammals can be viewed as a "horse and a cart," where the "horse" is oxidative metabolism occurring in the cytoplasm and the "cart" is gene expression occurring in the nucleus (Fig. 6.1). The "horse" and the "cart" are connected by the driver of signal transduction, that is primarily influenced by the flows and fluxes of metabolic substrates as well as products of oxidation/reduction (redox) reactions between the cytoplasm and the nucleus. This fundamental relationship between metabolism and signal transduction directs the expression of genes that then coordinately control normal cellular functions that must be closely matched to the metabolic capabilities in the cell. Unfortunately for each individual living organism (but fortunately for Darwinian evolution) the reactive by-products of oxidative metabolism, such as $O_2^{\bullet-}/H_2O_2$ as well as oxidative damage products (i.e., aldehydes and hydroperoxides) produced by the "horse," can cause covalent changes in the genome, that if unrepaired or mis-repaired lead to the gradual deterioration of the "cart" (Fig. 6.1). The gradual deterioration of the integrity of the genome eventually results in stoichiometric mismatches in the expression of proteins necessary to assemble the metabolic machinery of the "horse" causing a more inefficient oxidative metabolism leading to the accelerated production of reactive species and

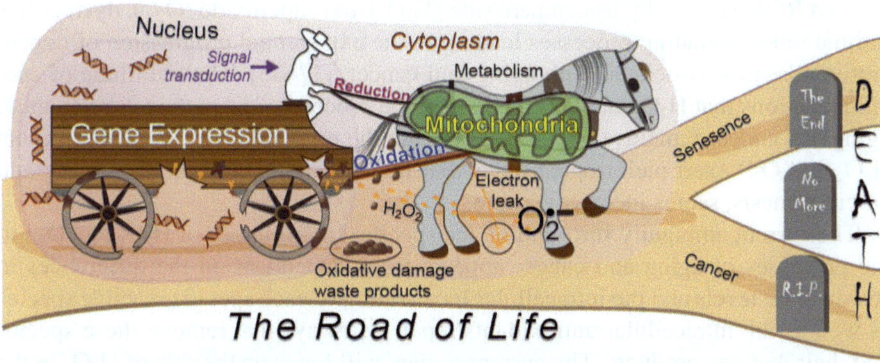

Fig. 6.1 The "Horse and Cart" model for understanding the relationship between metabolism, signal transduction, and gene expression in mammalian biology as well as degenerative diseases associated with aging. Oxidative metabolism (the horse) and gene expression (the cart) are tightly coupled by signal transduction via the nonequilibrium steady-state fluxes of reactive metabolic by-products and electron carriers that allow for the flow of electrons from reactions governing energy production and biosynthesis to gene expression pathways governing growth and development. In this way metabolic redox reactions provide the coordinated production of energy and reducing equivalents necessary for cell survival, proliferation, biosynthesis, and differentiated function. The model predicts, degenerative diseases associated with aging arise from the fact that metabolic oxidation reactions are not entirely efficient and by-products are formed via one-electron reductions of O_2 (electron leak) in electron transport chains leading to the formation of reactive oxygen species and oxidative damage products capable of disrupting gene structure and gene expression pathways. This inefficiency in metabolism slowly erodes the genetic integrity of the pathways governing the assembly and proper stoichiometry of the metabolic machinery necessary for the maintenance of normal cellular processes, causing the system to deteriorate as an exponential function of aging. Reactive oxygen species are therefore both necessary for living systems as redox signaling molecules and detrimental in mediating damage to the genome causing increased metabolic inefficiency accelerating the processes of senescence as well as onset of degenerative diseases associated with aging and cancer

subsequent exponential deterioration of the genome as a function of aging. In this fashion the deterioration of the mammalian organism driven by inefficiencies in O_2 metabolism as a function of age inevitably results in death either via degenerative diseases resulting from senescence or cancer induction (Fig. 6.1).

Since the original observations that free radicals and H_2O_2 produced by the radiolysis of H_2O were capable of contributing to the toxicity of radiation, hyperbaric O_2, and the onset of cancer induction during aging, [34–36] studies directed at testing the "free radical theory of aging" and the "free radical theory of cancer" have been accumulating significant momentum in the biomedical research community. The results to date are generally consistent with the concept of the "horse and cart" shown in Fig. 6.1, suggesting that alterations in the coordinate regulation of mitochondrial and cytosolic oxidative metabolism do indeed result in the accumulation of oxidative damage to genetic and epigenetic processes as a function of aging. This accumulation of damage results in a feed forward mechanism whereby alterations in oxidative metabolism leading to increased steady-state levels of metabolically derived ROS, such as hydrogen peroxide (H_2O_2) and superoxide ($O_2^{\bullet-}$), dysregulate normal redox signaling processes leading to the exponential progression of degenerative diseases associated with aging and cancer. [37–43]. The multitude of specific reactions that H_2O_2 and $O_2^{\bullet-}$ mediate in the processes and pathways relevant to aging and cancer remains to be elucidated, but the preponderance of evidence, that H_2O_2 and $O_2^{\bullet-}$ exert pleiotropic affects that fundamentally contribute to aging and carcinogenesis, seems incontrovertible.

The current ambiguity surrounding the specific contributions of H_2O_2 and $O_2^{\bullet-}$ to the processes of aging and cancer appears to be dependent on the differences in steady-state levels and the intracellular localization of each specific reactive species as well as the intracellular antioxidants/repair pathways that remove these species and their damage products. The current review will focus on the role of H_2O_2 in the process of carcinogenesis and aging with an emphasis on genomic instability and gene expression changes as well as altered cell function contributing to dysregulated cell growth, differentiation, and senescence. Finally, the possibility that this dysregulated oxidative metabolism could provide an "Achilles Heel" for selectively killing cancer vs. normal cells will be explored.

6.2 Hydroperoxides and Carcinogenesis

Oberley et al. originally proposed that $O_2^{\bullet-}$ and H_2O_2 produced during oxidative metabolism might represent signaling molecules that coordinately regulate cell division and differentiation that became dysregulated during induction of cancer (due to alterations in superoxide metabolism) leading to neoplastic transformation, uncontrolled growth, and the inability to differentiate which are hallmarks of malignancy [37, 38, 44]. This same group also proposed that these alterations in $O_2^{\bullet-}$ and H_2O_2 metabolism in cancer cells could be targeted with pharmacological and genetic approaches to improve cancer therapy outcomes [44–49]. These groundbreaking theoretical constructs as well as supporting data led to sustained and significant interest in the role of free radicals, ROS, and specifically H_2O_2 in cancer biology and therapy which persists to the present day.

As discussed previously, under normal steady-state conditions the generation of H_2O_2 and $O_2^{\bullet-}$ mediates many important signaling functions related to their reactivity and chemical biology. In mammalian cells hydroperoxides, including H_2O_2, reacting with redox active metal ions are believed to mediate two electron signaling including the oxidation of critical cysteine (and methionine) residues that activate kinases and inhibit phosphatases (reviewed in [14–16, 50–58]). In addition, hydroperoxides can also significantly impact thiol-mediated reactions governing both genetic and epigenetic expression of functional genes (reviewed in [14–16, 50–58]). H_2O_2 can also be chemotactic as part of the wound healing process initiating inflammation and recruitment of white blood cells to the site of injury [59]. In normal wound healing, the production of H_2O_2 is tightly controlled and is eventually downregulated during wound maturation. Intracellular H_2O_2 concentrations are also regulated by variety of enzyme mechanisms including catalase, cytosolic and mitochondrial glutathione peroxidases, and cytosolic as well as mitochondrial peroxiredoxins which all convert H_2O_2 to water [59–64].

In cancer cells, an increased flux of steady-state levels of H_2O_2 is often present [65] and appears to contribute to both pro-survival signaling processes and inducing genomic instability [14, 66]. Furthermore, H_2O_2 can diffuse freely across cell membranes or through specific aquaporin 3 channels and is relatively stable with a relatively long half-life compared to other ROS and thus can affect neighboring cells and surrounding cellular stroma [67–70]. Therefore, H_2O_2 metabolism may be considered to be highly dynamic as well as in steady-state equilibrium with the intracellular and extracellular environment of tumor and normal tissues reflecting the balance between cellular processes involving generation, transport, and removal.

In order for a normal cell capable of proliferation to undergo the full spectrum of phenotypic changes associated with malignant transformation, it must: (1) first become immortalized which is associated with the initiation phase of carcinogenesis, (2) lose control of normal growth regulatory and differentiation processes leading to uncontrolled cell proliferation, genomic instability, and apoptotic resistance that is associated with the promotion phase of carcinogenesis and, (3) as the transformed cell becomes increasingly genomically unstable, develop the ability to invade, metastasize, stimulate angiogenesis, and move through the circulatory system to seed distant normal tissue sites and evade the immune surveillance system during the progression phase of carcinogenesis.

H_2O_2 is known to be an excellent DNA damaging agent that with redox active metal ions (Fe and Cu) and reducing agents [71–74] can produce a broad spectrum of damage products including base damage, single and double strand breaks, sister chromatid exchanges, gene amplification, DNA protein cross-links, deletions, and DNA–DNA cross-links [73–81]. As DNA damaging agents as well as molecules capable of mediating changes in signal transduction and gene expression, $O_2^{\bullet-}$ and H_2O_2 as well as downstream reaction products such as organic hydroperoxides and other reactive lipid-derived mediators have been proposed to be involved in both the promotion and progression phases of carcinogenesis [65, 82–86]. Cancer cells have also commonly been shown to demonstrate increased steady-state levels of $O_2^{\bullet-}$ and H_2O_2 relative to their respective normal cell counterparts [11, 12, 65, 87–89]. Increased levels of $O_2^{\bullet-}$ and H_2O_2 in cancer cells have been suggested to be the result of mutagenesis, stoichiometric mismatches, improper assembly, and/or dysregulated electronic fluxes through mitochondrial electron transport chain proteins [11, 12, 14–16, 90, 91] as well as activation of NADPH oxidases [88, 89]. Increasing cellular concentrations of hydrogen peroxide has demonstrated an ability to stimulate cell proliferation, cell invasion, migration, apoptotic resistance, increased angiogenesis, inflammation, and metastasis [92–97]. H_2O_2 can also affect cell signaling of pro-survival pathways and stimulate pro-carcinogenic effects by reversibly inhibiting pathways involving phosphatases [98]. Specific pro-carcinogenic pathways affected by H_2O_2 include tumor necrosis factor-alpha (TNF-α), phosphatidylinositol 3-kinase (PI3K)/AKT, ERK 1 and 2, p38 MAPK, NFκB, and hypoxia-inducible factor 1 (HIF-1) pathways [98–103]. Furthermore, H_2O_2 has been suggested to be sufficient to convert normal fibroblasts to migratory cancer associated myofibroblasts thereby promoting the acceleration of cancer progression [104].

Increasing cellular concentrations of H_2O_2 has also been suggested to activate HIF-1 while overexpression of catalase and glutathione peroxidase prevents activation of HIF-1 and TNF-α [105–107]. Downstream targets of HIF-1α include pathways involving metabolism, cell growth, survival, invasion, migration, and angiogenesis [108]. Further support for H_2O_2 playing a significant role early in carcinogenesis includes the transformation of normal epithelial cells with oncogenes, such as Ras, that upregulate the production of H_2O_2 and superoxide, via stimulation of NADPH oxidase activity (NOX) [109–114]. Alternatively, overexpression of antioxidants, such as catalase, in transformed cells demonstrating a malignant phenotype can result in reduced steady-state levels of H_2O_2 and the reversion to a normal appearance with normal growth rates [114–116].

Catalase is highly specific for H_2O_2 scavenging as well as being a very catalytically active enzyme with a rate constant of 10^7 molecules of hydrogen peroxide converted to water and oxygen per second [117]. In addition, catalase-deficient mice spontaneously develop mammary tumors [118] suggesting a pro-carcinogenic function for H_2O_2. Similarly in humans, the CC allele of the catalase gene is associated with increased serum catalase activity and reduced incidence of breast cancer by 17 % compared to having at least one variant T allele [119]. Finally, treatment of murine tumor models with catalase has been demonstrated by several groups to inhibit tumor recurrence and metastasis [94, 120–123], again suggesting a pro-tumorigenic and pro-progression role for H_2O_2.

In addition to catalase, glutathione peroxidases [124] and peroxiredoxins [125] are also known to catalytically break down H_2O_2 as well as organic hydroperoxides (ROOH). Consistent with the findings of tumor suppression using catalase, overexpression of GPX1 (targeted to both the mitochondria and cytosol) as well as GPX4 (specifically targeted to the mitochondria) has been shown to reduce cancer cell proliferation, tumor growth, and metastasis [126–128]. Also, consistent with the cancer cell growth inhibition seen with GPX and catalase overexpression studies, knockdown of peroxiredoxins 1 and 4 promotes human lung cancer malignancy again supporting the hypothesis that increased steady-state levels of hydroperoxides can have pro-tumorigenic effects [125, 129].

In addition to the individual cancer cells, the tumor microenvironment plays a critical role in tumor development and again H_2O_2 originating from the cancer cells as well as the stromal fibroblasts is thought to play a pro-tumorigenic and pro-progression role [130]. Consistent with this idea, cancer cells have been shown to induce oxidative stress in adjacent stromal cells resulting in genomic instability including DNA damage and aneuploidy as well as inducing autophagy and mitophagy that appears to provide the cancer cells with nutrients (lactate, ketones, and glutamine) for continued growth [131, 132]. In addition, co-culture of fibroblasts with cancer cells results in fibroblasts beginning to demonstrate a transformed phenotype with increased expression of myofibroblast markers, extracellular matrix proteins, and activation of TGFβ/Smad2 signaling [133]. The addition of catalase prevented the phenotypic changes in fibroblasts identifying the predominant ROS signaling molecule as hydrogen peroxide to activate NFκB and HIF-1 signaling [132–135]. Similarly, the overexpression of either HIF-1α or NFκB in normal stromal fibroblasts induced mitophagy and stimulated the growth of human breast cancer xenografts by two to threefold [132]. Thus, in addition to creating genomic instability [80, 81] H_2O_2 may be considered as a driver mechanism for cancer cells to obtain nutrient support from adjacent cells and stroma through induction of autophagy and mitophagy.

Since cancer cells appear to demonstrate increased steady-state levels of H_2O_2, it has been postulated that H_2O_2 levels in vivo may be utilized to detect and diagnose cancer serving as an early marker as well as a marker of disease progression [117, 136–139]. While this methodology has not been widely tested or verified, there are recent reports that increases in exhaled H_2O_2 are detectable in breast and lung cancer patients and appear to correlate with clinical severity of disease [138, 139]. These findings also support the hypothesis discussed previously that, cancer cells may increase glucose metabolism to regenerate NADPH from the pentose cycle to fuel the detoxification of hydroperoxides via glutathione- and thioredoxin-dependent peroxidases as well as generate pyruvate from glycolysis which is capable of directly reacting with hydroperoxides in a deacetylation reaction leading to peroxide detoxification [11, 12, 16, 140–154]. While this increase in metabolism of glucose provides sufficient peroxide detoxification capacity to suppress redox stress-induced cell death pathways, the resulting disruption in normal redox signaling pathways inhibits the ability of cancer cells to differentiate, control cell division, and perform proper differentiated cellular functions. If this hypothesis was confirmed clinically, then ^{18}Fluoro-2-deoxy-D-Glucose (FDG) labeling followed by positron emission tomography (PET) imaging could be used as a predictor of increased steady-state levels of hydroperoxides in cancers to predict cancer cell sensitivity to interventions targeting dysregulated oxidative metabolism [11, 12, 143–154].

It is known that normal tissues experiencing chronic inflammation and oxidative stress during radiation-induced late normal tissue injury have also been suggested to have increased metabolic production of hydroperoxides as well as demonstrated increased uptake of ^{18}FDG on PET imaging that may be predictive of severity of normal tissue responses to radiation [150]. In this way, enhanced glucose uptake, monitored using FDG-PET imaging techniques, has also been proposed as a noninvasive marker in normal tissues of increased metabolic production of hydroperoxides that could be used to prospectively predict therapy outcomes to interventions involving manipulations of redox metabolism such as SOD mimics [150]. The mechanistic insights gained from these imaging technologies could also be used to guide therapy decisions based on the fundamental knowledge of how glucose metabolism may be regulating peroxide tone differentially in cancer vs. normal tissues in vivo [150, 151, 152].

6.3 Hydrogen Peroxide and Anticancer Treatments

The utility of hydrogen peroxide as an anticancer agent has a long and controversial history. In the 1950s, it was reported that rats bearing adenocarcinoma implants had a cure rate of 50–60 % when replacing drinking water with dilute solutions of H_2O_2 [155]. Unfortunately, consuming H_2O_2 as an adjuvant to cancer therapy was not successfully reproduced [156] but interest in hydrogen peroxide as an anticancer therapy continued. In the early 1980s, it was reported that glucose oxidase covalently coupled to polystyrene microspheres produced H_2O_2 and prolonged the survival of mice with implantable lymphoma cells with minimal toxicity towards the host mouse tissues [157]. However, in 1993, after reviewing the available literature, the American Cancer Society concluded that there was no evidence that treatment with H_2O_2 was safe or resulted in a therapeutic benefit for patients with cancer [158].

While delivery of external genuine H_2O_2 to cancer cells is difficult and has shown little benefit in cancer therapy, metabolic manipulations to increase intracellular steady-state levels of pro-oxidants such as H_2O_2 have been demonstrated to have significant anticancer effects [159]. Superoxide dismutase enzymes (SOD) are expressed in a variety of cellular compartments in addition to outside the cell and convert $2O_2^{\bullet-}$ into hydrogen peroxide and O_2. Several studies have shown that the overexpression of the SOD1 and SOD2 isoforms reduces tumor growth, metastasis potential, and malignant phenotype [46, 160–164]. Furthermore, overexpression of catalase which converts H_2O_2 in water and oxygen as well as the antioxidant enzyme glutathione peroxidase which breaks down lipid peroxides and hydrogen peroxide into their respective alcohols or water, reverses the antiproliferative effect of SOD overexpression indicating that hydroperoxides and specifically H_2O_2 are involved in the anticancer effects of SOD overexpression [164, 165]. This apparent differential SOD overexpression-induced H_2O_2 production in cancer vs. normal cells has been further studied using a kinetic model looking at the ability of SOD to pull electrons off electron transport chains when the reverse rate constant for the superoxide back reaction with the electron donor is more favorable than the forward rate constant [166]. This model was tested and fit very well with the observed H_2O_2 data and growth inhibition seen with MCF-7 human breast cancer cells overexpressing SOD2 [166]. Furthermore, increases in intracellular H_2O_2 can also modulate apoptotic pathways in cancer cells in seemingly paradoxical ways [167, 168]. Thus, it appears that in addition to its role in promoting carcinogenesis, hydrogen peroxide may also play an anticancer role at higher cellular concentrations that may explain why H_2O_2 can be both pro-tumorigenic as well as therapeutic [93, 94, 167, 168].

There have been numerous recent reports of using increased tumor cell metabolic production of hydrogen peroxide as a possible cancer therapy to kill and/or sensitize tumor cells to traditional therapies including chemotherapy and radiation therapy [169–176]. A rapidly emerging and readily clinically implementable way of selectively increasing H_2O_2 levels in cancer vs. normal cells is the use of very high dose intravenous pharmacological doses of ascorbate (Vitamin C) that raise plasma levels of ascorbate to ~20 mM [177–181]. Pharmacological ascorbate has been recently shown to selectively increase the sensitivity of both chemotherapy and radiation in a variety of cancer cell types relative to normal cells and appears to have this selective action via a mechanism that selectively increases cancer cell H_2O_2 [177–181].

Pharmacological levels of ascorbate being used as an antitumor agent also have an interesting and controversial history. In the 1970s, Drs. Cameron and Pauling showed that high dose ascorbate (10 g intravenous and 10 g orally) increased survival by a mean of 300 days in a variety of cancer types [182]. In contrast, in the late 1970s and early 1980s, two randomized controlled clinical trials demonstrated that there was no difference in survival between patients receiving placebo or 10 g of oral ascorbate [183, 184]. Unfortunately, after these two negative studies were published, interest in ascorbate as an anticancer agent fell from favor in the research community. However, the clear differences between the trials were that Cameron and Pauling's group dosed ascorbate intravenously for 10 days followed by oral administration, while the randomized clinical trials of the late 1970s–early 1980s dosed the ascorbate exclusively orally. Careful pharmacokinetic studies would later show that oral ascorbate dosing was only able to achieve plasma concentrations of 100–200 μM, while intravenous ascorbate dosing was able to achieve doses as high as 20–30 mM [185–188]. This difference in pharmacokinetics in the oral absorption of ascorbate appears to be due to tightly controlled uptake by gut and removal via kidney filtration [189]. These ascorbate dosing and plasma concentration differences likely explain the observed differences in between the two sets of clinical trials done prior to 1985.

While it is still not completely clear what mechanisms govern the ability of pharmacological ascorbate to selectively increase steady-state levels of hydrogen peroxide and clonogenic cell killing in cancer cells, basic research and clinical studies that are ongoing have provided new mechanistic insights. Ascorbic acid is water soluble and redox active with two ionizable hydroxyl groups (pK_1 of 4.2 and pK_2 of 11.6). Ascorbate auto-oxidation is pH dependent and can be accelerated by catalytic metal ions such as iron and copper [190, 191]. Ascorbate can be oxidized to the ascorbate radical in the process of reducing ferric (Fe^{3+}) iron to ferrous (Fe^{2+}) iron. The ascorbate radical can be detected in extracellular fluid but is not able to be detected in whole blood because of the presence of erythrocyte associated catalase, peroxidases, and peroxiredoxin [179]. Ferrous iron can be oxidized to ferric iron by reducing molecular oxygen to superoxide ($O_2^{\bullet -}$). Superoxide can then dismute either spontaneously or enzymatically driven by the SOD catalyzed reaction into H_2O_2 and molecular oxygen. Ferrous iron can also react with H_2O_2 to form the highly reactive hydroxyl radical ($OH^{\bullet}$) and ferric iron via the Fenton reaction [192].

In normal cells functioning under normal steady-state conditions, iron levels are tightly regulated by the iron regulatory proteins ferritin and transferrin [193]. Transferrin is a circulating iron regulatory glycoprotein with a high affinity for ferric iron making the iron redox inactive [194]. Iron bound transferrin is able to bind to transferrin receptors on cell surfaces with the complex being internalized to an endosome. Under acidic conditions, iron is released and ferric iron is reduced to labile ferrous iron where it is incorporated in iron containing proteins, such as iron-sulfur clusters in the electron transport chain, or stored in the Fe^{+3} state bound to ferritin. Ferric iron bound by ferritin can be released as Fe^{+2} by reducing agents including thiols, ascorbate, superoxide, and reduced flavins [195, 196]. Once released, iron has the potential to interact with H_2O_2 or organic hydroperoxides (ROOH) forming the highly reactive $OH^{\bullet}$ or $RO^{\bullet}$. Interestingly, unlike normal cells, cancer cells appear to exist under steady-state metabolic conditions where they have increased levels of superoxide and hydrogen peroxide [12]. Furthermore, cancer patients have been reported to have elevated serum and tissue ferritin concentrations [197–200]. Since high levels of ascorbate can induce the release of labile iron from proteins [201] and cancer cells may have increased labile iron as a function of having increased levels of superoxide and hydrogen peroxide, ascorbate might selectively auto-oxidize in cancer cells producing much greater [H_2O_2] and further increasing the labile iron pools and therefore selectively induce oxidative stress and radio-chemo-sensitivity in cancer cells, relative to normal cells. Furthermore, preincubation of cells with the iron chelator, deferoxamine, protects against ascorbate-induced toxicity [202]. An additional possible explanation as to why cancer cells are more sensitive to pharmacological doses of ascorbate is that normal epithelial cells have reduced endogenous generation of ROS and therefore normal cell H_2O_2 levels can be more closely regulated by glutathione peroxidase, catalase, and peroxiredoxins. In contrast, cancer cells have a high background rate of ROS production from electron transport chains and NADPH oxidase activity coupled with a potentially compromised ROS metabolism [12, 44–48] therefore existing closer to the threshold of cytotoxicity and can be more easily pushed to toxic levels by cellular insults including pharmacological ascorbate combined with chemotherapy and radiation therapy [203–205]. If metabolic sources of $O_2^{\bullet-}$ and H_2O_2 and downstream increases in redox active metal ion pools can be confirmed as being the causative agents for the selective toxicity of pharmacological ascorbate in cancer vs. normal cells, these mechanistic insights could be exploited using many different agents and biochemical approaches that take advantage of this metabolic frailty in cancer cell metabolism in order to develop new curative combined modality therapies.

To that end, several Phase I clinical trials have been published demonstrating the safety and tolerability of pharmacological doses of ascorbate with chemotherapy agents [187, 188, 206–208]. Contraindications of pharmacological ascorbate therapy include glucose-6-phosphate dehydrogenase (G-6-PD) deficiency. Removal of H_2O_2 in the blood requires NADPH produced from the pentose phosphate shunt via the G-6-PD enzyme [209]. Thus, G-6-PD deficiency could allow for the accumulation of excessive quantities of ROS resulting in a hemolytic anemia during treatment with pharmacological ascorbate [210]. Finally, it does not appear that ascorbate reduces the effectiveness of chemotherapy and/or radiation therapy and therefore should only contribute to improved clinical responses when combined with more traditional therapies [180, 181, 211–215].

6.4 Summary

In conclusion, metabolic production of H_2O_2 in the presence of redox active metal ions is known to contribute to several steps involved with carcinogenesis that can induce genomic instability, tumor promotion, and cancer progression. However, when selectively increased in cancer vs. normal cells, H_2O_2 can be utilized to enhance cancer therapy responses by increasing metabolic oxidative stress and sensitizing cancer cells to traditional chemotherapy and radiation therapy. In this way,

H_2O_2 can be considered as a double-edged sword that can exert both detrimental and beneficial effects in cancer biology and therapy. Understanding the critical mechanistic details of these pleiotropic effects of H_2O_2 could yield very exciting developments in the quest to develop new biochemical rationales for controlling cancer in human populations.

Acknowledgements The authors thank Drs. Larry Oberley, Garry Buettner, Frederick Domann, Prabhat Goswami, and Joseph Cullen for many helpful scientific discussions. The authors thank Gareth Smith for graphics and manuscript editorial assistance. The authors were supported in part by a grant from ASTRO JF2014-1, The Carver Research Program of Excellence in Redox Biology, P30CA086862, and NIH CA182804.

References

1. Warburg O. On the origin of cancer cells. Science. 1956;123(3191):309–14.
2. Weinhouse S, Warburg O, Burk D, Schade AL. On respiratory impairment in cancer cells. Science. 1956;124(3215):267–72.
3. Vander Heiden MG, Cantley LC, Thompson CB. Understanding the Warburg effect: the metabolic requirements of cell proliferation. Science. 2009;324(5930):1029–33. doi:10.1126/science.1160809. PMID: 19460998.
4. Salskov A, Tammisetti VS, Grierson J, Vesselle H. FLT: measuring tumor cell proliferation in vivo with positron emission tomography and 3′-deoxy-3′-[18F]fluorothymidine. Semin Nucl Med. 2007;37(6):429–39.
5. Shields AF, Grierson JR, Dohmen BM, Machulla HJ, Stayanoff JC, Lawhorn-Crews JM, Obradovich JE, Muzik O, Mangner TJ. Imaging proliferation in vivo with [F-18]FLT and positron emission tomography. Nat Med. 1998;4(11):1334–6. PMID: 9809561.
6. Maity A, Tuttle SW. 2-Deoxyglucose and radiosensitization: teaching an old DOG new tricks? Cancer Biol Ther. 2006;5(7):824–6. PMID: 16880734.
7. Weinhouse S. Isozyme alterations, gene regulation and the neoplastic transformation. Adv Enzyme Regul. 1983;21:369–86.
8. Fantin VR, St-Pierre J, Leder P. Attenuation of LDH-A expression uncovers a link between glycolysis, mitochondrial physiology, and tumor maintenance. Cancer Cell. 2006;9:425–34.
9. Springer EL. Comparative study of the cytoplasmic organelles of epithelial cell lines derived from human carcinomas and nonmalignant tissues. Cancer Res. 1980;40(3):803–17. PMID: 7193514.
10. Bize IB, Oberley LW, Morris HP. Superoxide dismutase and superoxide radical in Morris hepatomas. Cancer Res. 1980;40(10):3686–93. PMID: 62546387.
11. Ahmad IM, Aykin-Burns N, Sim JE, Walsh SA, Higashikubo R, Buettner GR, Venkataraman S, Mackey MA, Flanagan S, Oberley LW, Spitz DR. Mitochondrial $O_2^{\bullet-}$ and H_2O_2 mediate glucose deprivation-induced cytotoxicity and oxidative stress in human cancer cells. J Biol Chem. 2005;280(6):4254–63. PMID: 15561720.
12. Aykin-Burns N, Ahmad IM, Zhu Y, Oberley L, Spitz DR. Increased levels of superoxide and hydrogen peroxide mediate the differential susceptibility of cancer cells vs. normal cells to glucose deprivation. Biochem J. 2009;418:29–37. PMID: 18937644 PMCID: PMC2678564.
13. Graham NA, Tahmasian M, Kohli B, Komisopoulou E, Zhu M, Vivanco I, Teitell MA, Wu H, Ribas A, Lo RS, Mellinghoff IK, Mischel PS, Graeber TG. Glucose deprivation activates a metabolic and signaling amplification loop leading to cell death. Mol Syst Biol. 2012;8:589. doi:10.1038/msb.2012.20.
14. Zhou D, Shao L, Spitz DR. Reactive oxygen species in normal and tumor stem cells. Adv Cancer Res. 2014;122:1–67. PMID:24974178 PMCID: PMC4207279.

15. Spitz DR, Azzam EI, Li JJ, Gius D. Metabolic oxidation/reduction reactions and cellular responses to ionizing radiation: a unifying concept in stress response biology. Cancer Metastasis Rev. 2004;23:311–22. PMID: 15197331.
16. Spitz DR, Sim JE, Ridnour LA, Galoforo SS, Lee YJ. Glucose deprivation-induced oxidative stress in human tumor cells: a fundamental defect in metabolism? Ann NY Acad Sci. 2000;899:349–62. PMID: 10863552.
17. Wallace DC. Mitochondria and cancer: Warburg addressed. Cold Spring Harb Symp Quant Biol. 2005;70:363–74. PMID: 16869773.
18. Wallace DC. Why do we still have a maternally inherited mitochondrial DNA? Insights from evolutionary medicine. Annu Rev Biochem. 2007;76:781–821.
19. Ishikawa K, Takenaga K, Akimoto M, Koshikawa N, Yamaguchi A, Imanishi H, Nakada K, Honma Y, Hayashi J. ROS-generating mitochondrial DNA mutations can regulate tumor cell metastasis. Science. 2008;320(5876):661–4.
20. Polyak K, Li Y, Zhu H, Lengauer C, Willson JK, Markowitz SD, Trush MA, Kinzler KW, Vogelstein B. Somatic mutations of the mitochondrial genome in human colorectal tumours. Nat Genet. 1998;20(3):291–3. PMID: 9806551.
21. Liu VW, Shi HH, Cheung AN, Chiu PM, Leung TW, Nagley P, Wong LC, Ngan HY. High incidence of somatic mitochondrial DNA mutations in human ovarian carcinomas. Cancer Res. 2001;61(16):5998–6001. PMID: 11507041.
22. Dasgupta S, Hoque MO, Upadhyay S, Sidransky D. Mitochondrial cytochrome B gene mutation promotes tumor growth in bladder cancer. Cancer Res. 2008;68(3):700–6. PMID: 18245469.
23. Roy K, Wu Y, Meitzler JL, Juhasz A, Liu H, Jiang G, Lu J, Antony S, Doroshow JH. NADPH oxidases and cancer. Clin Sci (Lond). 2015;128(12):863–75.
24. Lambeth JD. Nox enzymes, ROS, and chronic disease: an example of antagonistic pleiotropy. Free Radic Biol Med. 2007;43(3):332–47.
25. Kamata T. Roles of Nox1 and other Nox isoforms in cancer development. Cancer Sci. 2009;100(8):1382–8. PMID: 19493276.
26. Heard JJ, Fong V, Bathaie SZ, Tamanoi F. Recent progress in the study of the Rheb family GTPases. Cell Signal. 2014;26(9):1950–7. PMID: 24863881.
27. Wen X, Wu J, Wang F, Liu B, Huang C, Wei Y. Deconvoluting the role of reactive oxygen species and autophagy in human diseases. Free Radic Biol Med. 2013;65:402–10.
28. Wells PG, McCallum GP, Chen CS, Henderson JT, Lee CJ, Perstin J, Preston TJ, Wiley MJ, Wong AW. Oxidative stress in developmental origins of disease: teratogenesis, neurodevelopmental deficits, and cancer. Toxicol Sci. 2009;108(1):4–18. PMID: 19126598.
29. Finkel T. Intracellular redox regulation by the family of small GTPases. Antioxid Redox Signal. 2006;8(9–10):1857–63. PMID: 16987038.
30. Bellot GL, Liu D, Pervaiz S. ROS, autophagy, mitochondria and cancer: Ras, the hidden master? Mitochondrion. 2013;13(3):155–62. doi:10.1016/j.mito.2012.06.007. PMID: 22750269.
31. Szent-Györgyi A. Towards a new biochemistry? Science. 1941;93(2426):609–11. PMID: 17841996.
32. Szent-Gyorgyi A. Molecules, electrons and biology. Trans N Y Acad Sci. 1969;31(4):334–40. PMID:5257509.
33. Szent-Györgyi A. Electronic biology and its relation to cancer. Life Sci. 1974;15(5):863–75. PMID: 4620969.
34. Szent-Györgyi A. Electronic biology and cancer. New York: Marcel Decker; 1976. p. 34–5.
35. Patt HM, Tyree EB, Straube RL, Smith DE. Cysteine protection against X irradiation. Science. 1949;110(2852):213–4. PMID: 17811258.
36. Gerschman R, Gilbert DL, Nye SW, Dwyer P, Fenn WO. Oxygen poisoning and x-irradiation: a mechanism in common. Science. 1954;119(3097):623–6. PMID: 13156638.
37. Harman D. Prolongation of the normal lifespan and inhibition of spontaneous cancer by antioxidants. J Gerontol. 1961;16:247–54. PMID: 13711616.

38. Oberley LW, Oberley TD, Buettner GR. Cell differentiation, aging and cancer: the possible roles of superoxide and superoxide dismutases. Med Hypotheses. 1980;6(3):249–68. PMID: 6253771.
39. Oberley LW, Oberley TD, Buettner GR. Cell division in normal and transformed cells: the possible role of superoxide and hydrogen peroxide. Med Hypotheses. 1981;7(1):21–42. PMID: 6259499.
40. Woo DK, Shadel GS. Mitochondrial stress signals revise an old aging theory. Cell. 2011;144(1):11–2. doi:10.1016/j.cell.2010.12.023.
41. Treiber N, Maity P, Singh K, Kohn M, Keist AF, Ferchiu F, Sante L, Frese S, Bloch W, Kreppel F, Kochanek S, Sindrilaru A, Iben S, Högel J, Ohnmacht M, Claes LE, Ignatius A, Chung JH, Lee MJ, Kamenisch Y, Berneburg M, Nikolaus T, Braunstein K, Sperfeld AD, Ludolph AC, Briviba K, Wlaschek M, Florin L, Angel P, Scharffetter-Kochanek K. Accelerated aging phenotype in mice with conditional deficiency for mitochondrial superoxide dismutase in the connective tissue. Aging Cell. 2011;10(2):239–54. PMID 21108731.
42. Schriner SE, Linford NJ, Martin GM, Treuting P, Ogburn CE, Emond M, Coskun PE, Ladiges W, Wolf N, Van Remmen H, Wallace DC, Rabinovitch PS. Extension of murine life span by overexpression of catalase targeted to mitochondria. Science. 2005;308(5730):1909–11. PMID: 15879174.
43. Holmström KM, Finkel T. Cellular mechanisms and physiological consequences of redox-dependent signaling. Nat Rev Mol Cell Biol. 2014;15(6):411–21. PMID: 24854789.
44. Lu T, Finkel T. Free radicals and senescence. Exp Cell Res. 2008;314(9):1918–22. doi:10.1016/j.yexcr.2008.01.011. PMID: 18282568.
45. Oberley LW, Buettner GR. Role of superoxide dismutase in cancer: a review. Cancer Res. 1979;39(4):1141–9. PMID: 217531.
46. Church SL, Grant JW, Ridnour LA, Oberley LW, Swanson PE, Meltzer PS, Trent JM. Increased manganese superoxide dismutase expression suppresses the malignant phenotype of human melanoma cells. Proc Natl Acad Sci U S A. 1993;90(7):3113–7. PMID: 8464931.
47. Urano M, Kuroda M, Reynolds R, Oberley TD, St Clair DK. Expression of manganese superoxide dismutase reduces tumor control radiation dose: gene-radiotherapy. Cancer Res. 1995;55(12):2490–3. PMID: 7780953.
48. Oberley LW, Rogers KL, Schutt L, Oberley TD, Leuthauser SW, Sorenson JR. Possible role of glutathione in the antitumor effect of a copper-containing synthetic superoxide dismutase in mice. J Natl Cancer Inst. 1983;71(5):1089–94. PMID: 6580486.
49. Leuthauser SW, Oberley LW, Oberley TD, Sorenson JR, Ramakrishna K. Antitumor effect of a copper coordination compound with superoxide dismutase-like activity. J Natl Cancer Inst. 1981;66(6):1077–81. PMID: 6941042.
50. Brown DI, Griendling KK. Nox proteins in signal transduction. Free Radic Biol Med. 2009;47(9):1239–53. PMID: 19628035.
51. Pervaiz S, Clement MV. Superoxide anion: oncogenic reactive oxygen species? Int J Biochem Cell Biol. 2007;39(7–8):1297–304. PMID: 17531522.
52. Hitchler MJ, Domann FE. Metabolic defects provide a spark for the epigenetic switch in cancer. Free Radic Biol Med. 2009;47(2):115–27. PMID: 19362589.
53. Hitchler MJ, Domann FE. Redox regulation of the epigenetic landscape in cancer: a role for metabolic reprogramming in remodeling the epigenome. Free Radic Biol Med. 2012;53(11):2178–87. PMID: 23022407.
54. Buettner GR. Superoxide dismutase in redox biology: the roles of superoxide and hydrogen peroxide. Anticancer Agents Med Chem. 2011;11(4):341–6. PMID: 21453242.
55. Gius D, Spitz DR. Redox signaling in cancer biology. Antioxid Redox Signal. 2006;8(7–8):1249–52. PMID: 16910772.
56. Sarsour EH, Kumar MG, Chaudhuri L, Kalen AL, Goswami PC. Redox control of the cell cycle in health and disease. Antioxid Redox Signal. 2009;11(12):2985–3011. PMID: 19505186.

57. Menon SG, Goswami PC. A redox cycle within the cell cycle: ring in the old with the new. Oncogene. 2007;26(8):1101–9. PMID: 16924237.
58. Schafer FQ, Buettner GR. Redox environment of the cell as viewed through the redox state of the glutathione disulfide/glutathione couple. Free Radic Biol Med. 2001;30(11):1191–212. PMID: 11368918.
59. Niethammer P, Grabher C, Look AT, Mitchison TJ. A tissue-scale gradient of hydrogen peroxide mediates rapid wound detection in zebrafish. Nature. 2009;459(7249):996–9. PMID: 19494811.
60. Chance B, Sies H, Boveris A. Hydroperoxide metabolism in mammalian organs. Physiol Rev. 1979;59(3):527–605. PMID: 37532.
61. Halliwell B. Reactive oxygen species and the central nervous system. J Neurochem. 1992;59(5):1609–23. PMID: 1402908.
62. Dringen R, Pawlowski PG, Hirrlinger J. Peroxide detoxification by brain cells. J Neurosci Res. 2005;79(1–2):157–65. PMID: 15573410.
63. Rhee SG. Cell signaling. H_2O_2, a necessary evil for cell signaling. Science. 2006;312(5782):1882–3. PMID: 16809515.
64. Mishina NM, Tyurin-Kuzmin PA, Markvicheva KN, Vorotnikov AV, Tkachuk VA, Laketa V, Schultz C, Lukyanov S, Belousov VV. Does cellular hydrogen peroxide diffuse or act locally? Antioxid Redox Signal. 2011;14(1):1–7. PMID: 20690882.
65. Szatrowski TP, Nathan CF. Production of large amounts of hydrogen peroxide by human tumor cells. Cancer Res. 1991;51(3):794–8. PMID: 1846317.
66. Zhu P, Tan MJ, Huang RL, Tan CK, Chong HC, Pal M, Lam CR, Boukamp P, Pan JY, Tan SH, Kersten S, Li HY, Ding JL, Tan NS. Angiopoietin-like 4 protein elevates the prosurvival intracellular $O_2^{\bullet-}$:H_2O_2 ratio and confers anoikis resistance to tumors. Cancer Cell. 2011;19(3):401–15. PMID: 21397862.
67. Bienert GP, Møller AL, Kristiansen KA, Schulz A, Møller IM, Schjoerring JK, Jahn TP. Specific aquaporins facilitate the diffusion of hydrogen peroxide across membranes. J Biol Chem. 2007;282(2):1183–92. PMID: 17105724.
68. Bienert GP, Schjoerring JK, Jahn TP. Membrane transport of hydrogen peroxide. Biochim Biophys Acta. 2006;1758(8):994–1003. PMID: 16566894.
69. Miller EW, Dickinson BC, Chang CJ. Aquaporin-3 mediates hydrogen peroxide uptake to regulate downstream intracellular signaling. Proc Natl Acad Sci U S A. 2010;107(36):15681–6. PMID: 20724658.
70. Bertolotti M, Bestetti S, García-Manteiga JM, Medraño-Fernandez I, Dal Mas A, Malosio ML, Sitia R. Tyrosine kinase signal modulation: a matter of H_2O_2 membrane permeability? Antioxid Redox Signal. 2013;19(13):1447–51. PMID: 23541115.
71. Freese E, Sklarow S, Freese EB. DNA damage caused by antidepressant hydrazines and related drugs. Mutat Res. 1968;5(3):343–8. PMID: 5727268.
72. Jonas SK, Riley PA, Willson RL. Hydrogen peroxide cytotoxicity. Low-temperature enhancement by ascorbate or reduced lipoate. Biochem J. 1989;264(3):651–5. PMID: 2515850.
73. Toyokuni S, Sagripanti JL. Iron-mediated DNA damage: sensitive detection of DNA strand breakage catalyzed by iron. J Inorg Biochem. 1992;47(3–4):241–8. PMID: 1431883.
74. Nyaga SG, Jaruga P, Lohani A, Dizdaroglu M, Evans MK. Accumulation of oxidatively induced DNA damage in human breast cancer cell lines following treatment with hydrogen peroxide. Cell Cycle. 2007;6(12):1472–8. PMID: 17568196.
75. Hartman PS, Eisenstark A. Killing of Escherichia coli K-12 by near-ultraviolet radiation in the presence of hydrogen peroxide: role of double-strand DNA breaks in absence of recombinational repair. Mutat Res. 1980;72(1):31–42. PMID: 7003364.
76. Meneghini R, Hoffmann ME. The damaging action of hydrogen peroxide on DNA of human fibroblasts is mediated by a non-dialyzable compound. Biochim Biophys Acta. 1980;608(1):167–73. PMID: 7388029.
77. Ward JF, Blakely WF, Joner EI. Mammalian cells are not killed by DNA single-strand breaks caused by hydroxyl radicals from hydrogen peroxide. Radiat Res. 1985;103(3):383–92. PMID: 2994167.

78. Aruoma OI, Halliwell B, Gajewski E, Dizdaroglu M. Copper-ion-dependent damage to the bases in DNA in the presence of hydrogen peroxide. Biochem J. 1991;273:601–4. PMID: 1899997.
79. Galloway SM, Painter RB. Vitamin C is positive in the DNA synthesis inhibition and sister-chromatid exchange tests. Mutat Res. 1979;60(3):321–7. PMID: 481430.
80. Hunt CR, Sim JE, Sullivan SJ, Featherstone T, Golden W, Von Kapp-Herr C, Hock RA, Gomez RA, Parsian AJ, Spitz DR. Genomic instability and catalase gene amplification induced by chronic exposure to oxidative stress. Cancer Res. 1998;58(17):3986–92. PMID: 9731512.
81. Dayal D, Martin SM, Limoli CL, Spitz DR. Hydrogen peroxide mediates the radiation-induced mutator phenotype in mammalian cells. Biochem J. 2008;413(1):185–91. PMID: 18352860.
82. Fridlich R, Annamalai D, Roy R, Bernheim G, Powell SN. BRCA1 and BRCA2 protect against oxidative DNA damage converted into double-strand breaks during DNA replication. DNA Repair (Amst). 2015;30:11–20. PMID: 25836596.
83. Kennedy AR, Troll W, Little JB. Role of free radicals in the initiation and promotion of radiation transformation in vitro. Carcinogenesis. 1984;5(10):1213–8. PMID: 6488446.
84. Zimmerman R, Cerutti P. Active oxygen acts as a promoter of transformation in mouse embryo C3H/10T1/2/C18 fibroblasts. Proc Natl Acad Sci U S A. 1984;81(7):2085–7. PMID: 6425826.
85. Nakamura Y, Colburn NH, Gindhart TD. Role of reactive oxygen in tumor promotion: implication of superoxide anion in promotion of neoplastic transformation in JB-6 cells by TPA. Carcinogenesis. 1985;6(2):229–35. PMID: 2982513.
86. Ames BN. Dietary carcinogens and anticarcinogens. Oxygen radicals and degenerative diseases. Science. 1983;221(4617):1256–64. PMID: 6351251.
87. Zieba M, Suwalski M, Kwiatkowska S, Piasecka G, Grzelewska-Rzymowska I, Stolarek R, Nowak D. Comparison of hydrogen peroxide generation and the content of lipid peroxidation products in lung cancer tissue and pulmonary parenchyma. Respir Med. 2000;94(8):800–5. PMID: 10955757.
88. Lim SD, Sun C, Lambeth JD, Marshall F, Amin M, Chung L, Petros JA, Arnold RS. Increased Nox1 and hydrogen peroxide in prostate cancer. Prostate. 2005;62(2):200–7. PMID: 15389790.
89. Arbiser JL, Petros J, Klafter R, Govindajaran B, McLaughlin ER, Brown LF, Cohen C, Moses M, Kilroy S, Arnold RS, Lambeth JD. Reactive oxygen generated by Nox1 triggers the angiogenic switch. Proc Natl Acad Sci U S A. 2002;99(2):715–20. PMID: 11805326.
90. Slane BG, Aykin-Burns N, Smith BJ, Kalen AL, Goswami PC, Domann FE, Spitz DR. Mutation of succinate dehydrogenase subunit C results in increased O2.-, oxidative stress, and genomic instability. Cancer Res. 2006;66(15):7615–20. PMID: 16885361.
91. Owens KM, Aykin-Burns N, Dayal D, Coleman MC, Domann FE, Spitz DR. Genomic instability induced by mutant succinate dehydrogenase subunit D (SDHD) is mediated by $O_2^{\bullet-}$and H_2O_2. Free Radic Biol Med. 2012;52(1):160–6. PMID: 22041456.
92. Polytarchou C, Hatziapostolou M, Papadimitriou E. Hydrogen peroxide stimulates proliferation and migration of human prostate cancer cells through activation of activator protein-1 and up-regulation of the heparin affin regulatory peptide gene. J Biol Chem. 2005; 280(49):40428–35. PMID: 16199533.
93. Nelson KK, Ranganathan AC, Mansouri J, Rodriguez AM, Providence KM, Rutter JL, Pumiglia K, Bennett JA, Melendez JA. Elevated sod2 activity augments matrix metalloproteinase expression: evidence for the involvement of endogenous hydrogen peroxide in regulating metastasis. Clin Cancer Res. 2003;9(1):424–32. PMID: 12538496.
94. Nishikawa M, Tamada A, Hyoudou K, Umeyama Y, Takahashi Y, Kobayashi Y, Kumai H, Ishida E, Staud F, Yabe Y, Takakura Y, Yamashita F, Hashida M. Inhibition of experimental hepatic metastasis by targeted delivery of catalase in mice. Clin Exp Metastasis. 2004; 21(3):213–21. PMID: 15387371.

95. del Bello B, Paolicchi A, Comporti M, Pompella A, Maellaro E. Hydrogen peroxide produced during gamma-glutamyl transpeptidase activity is involved in prevention of apoptosis and maintenance of proliferation in U937 cells. FASEB J. 1999;13(1):69–79. PMID: 9872931.
96. Qian Y, Luo J, Leonard SS, Harris GK, Millecchia L, Flynn DC, Shi X. Hydrogen peroxide formation and actin filament reorganization by Cdc42 are essential for ethanol-induced in vitro angiogenesis. J Biol Chem. 2003;278(18):16189–97. PMID:12598535.
97. Kobayashi Y, Nishikawa M, Hyoudou K, Yamashita F, Hashida M. Hydrogen peroxide-mediated nuclear factor kappaB activation in both liver and tumor cells during initial stages of hepatic metastasis. Cancer Sci. 2008;99(8):1546–52. PMID: 18754865.
98. Groeger G, Quiney C, Cotter TG. Hydrogen peroxide as a cell-survival signaling molecule. Antioxid Redox Signal. 2009;11(11):2655–71. PMID: 19558209.
99. Kaewpila S, Venkataraman S, Buettner GR, Oberley LW. Manganese superoxide dismutase modulates hypoxia-inducible factor-1 alpha induction via superoxide. Cancer Res. 2008;68(8):2781–8. PMID: 18413745.
100. Wang M, Kirk JS, Venkataraman S, Domann FE, Zhang HJ, Schafer FQ, Flanagan SW, Weydert CJ, Spitz DR, Buettner GR, Oberley LW. Manganese superoxide dismutase suppresses hypoxic induction of hypoxia-inducible factor-1alpha and vascular endothelial growth factor. Oncogene. 2005;24(55):8154–66. PMID: 16170370.
101. Giorgio M, Trinei M, Migliaccio E, Pelicci PG. Hydrogen peroxide: a metabolic by-product or a common mediator of ageing signals? Nat Rev Mol Cell Biol. 2007;8(9):722–8. PMID: 17700625.
102. Oliveira-Marques V, Marinho HS, Cyrne L, Antunes F. Role of hydrogen peroxide in NF-kappaB activation: from inducer to modulator. Antioxid Redox Signal. 2009;11(9):2223–43. PMID: 19496701.
103. Ma HP. Hydrogen peroxide stimulates the epithelial sodium channel through a phosphatidylinositide 3-kinase-dependent pathway. J Biol Chem. 2011;286(37):32444–53. PMID: 21795700.
104. Toullec A, Gerald D, Despouy G, Bourachot B, Cardon M, Lefort S, Richardson M, Rigaill G, Parrini MC, Lucchesi C, Bellanger D, Stern MH, Dubois T, Sastre-Garau X, Delattre O, Vincent-Salomon A, Mechta-Grigoriou F. Oxidative stress promotes myofibroblast differentiation and tumour spreading. EMBO Mol Med. 2010;2(6):211–30. PMID: 20535745.
105. Chandel NS, McClintock DS, Feliciano CE, Wood TM, Melendez JA, Rodriguez AM, Schumacker PT. Reactive oxygen species generated at mitochondrial complex III stabilize hypoxia-inducible factor-1alpha during hypoxia: a mechanism of O2 sensing. J Biol Chem. 2000;275(33):25130–8. PMID: 10833514.
106. Haddad JJ, Land SC. A non-hypoxic, ROS-sensitive pathway mediates TNF-alpha-dependent regulation of HIF-1alpha. FEBS Lett. 2001;505(2):269–74. PMID: 11566189.
107. Guzy RD, Hoyos B, Robin E, Chen H, Liu L, Mansfield KD, Simon MC, Hammerling U, Schumacker PT. Mitochondrial complex III is required for hypoxia-induced ROS production and cellular oxygen sensing. Cell Metab. 2005;1(6):401–8. PMID: 16054089.
108. Semenza GL. Targeting HIF-1 for cancer therapy. Nat Rev Cancer. 2003;3(10):721–32. PMID: 13130303.
109. Lee AC, Fenster BE, Ito H, Takeda K, Bae NS, Hirai T, Yu ZX, Ferrans VJ, Howard BH, Finkel T. Ras proteins induce senescence by altering the intracellular levels of reactive oxygen species. J Biol Chem. 1999;274(12):7936–40. PMID: 10075689.
110. Yang JQ, Buettner GR, Domann FE, Li Q, Engelhardt JF, Weydert CD, Oberley LW. v-Ha-ras mitogenic signaling through superoxide and derived reactive oxygen species. Mol Carcinog. 2002;33(4):206–18. PMID: 11933074.
111. Yang JQ, Li S, Domann FE, Buettner GR, Oberley LW. Superoxide generation in v-Ha-ras-transduced human keratinocyte HaCaT cells. Mol Carcinog. 1999;26(3):180–8. PMID: 10559793.
112. Hole PS, Pearn L, Tonks AJ, James PE, Burnett AK, Darley RL, Tonks A. Ras-induced reactive oxygen species promote growth factor-independent proliferation in human CD34+ hematopoietic progenitor cells. Blood. 2010;115(6):1238–46. PMID: 20007804.

113. Feng Y, Santoriello C, Mione M, Hurlstone A, Martin P. Live imaging of innate immune cell sensing of transformed cells in zebrafish larvae: parallels between tumor initiation and wound inflammation. PLoS Biol. 2010;8(12), e1000562. PMID: 21179501.
114. Arnold RS, Shi J, Murad E, Whalen AM, Sun CQ, Polavarapu R, Parthasarathy S, Petros JA, Lambeth JD. Hydrogen peroxide mediates the cell growth and transformation caused by the mitogenic oxidase Nox1. Proc Natl Acad Sci U S A. 2001;98(10):5550–5. PMID: 11331784.
115. Alexandrova AY, Kopnin PB, Vasiliev JM, Kopnin BP. ROS up-regulation mediates Ras-induced changes of cell morphology and motility. Exp Cell Res. 2006;312(11):2066–73. PMID: 16624288.
116. Hempel N, Ye H, Abessi B, Mian B, Melendez JA. Altered redox status accompanies progression to metastatic human bladder cancer. Free Radic Biol Med. 2009;46(1):42–50. PMID: 18930813.
117. Lisanti MP, Martinez-Outschoorn UE, Lin Z, Pavlides S, Whitaker-Menezes D, Pestell RG, Howell A, Sotgia F. Hydrogen peroxide fuels aging, inflammation, cancer metabolism and metastasis: the seed and soil also needs "fertilizer". Cell Cycle. 2011;10(15):2440–9. PMID: 21734470.
118. Ishii K, Zhen LX, Wang DH, Funamori Y, Ogawa K, Taketa K. Prevention of mammary tumorigenesis in acatalasemic mice by vitamin E supplementation. Jpn J Cancer Res. 1996;87(7):680–4. PMID: 8698615.
119. Ahn J, Gammon MD, Santella RM, Gaudet MM, Britton JA, Teitelbaum SL, Terry MB, Nowell S, Davis W, Garza C, Neugut AI, Ambrosone CB. Associations between breast cancer risk and the catalase genotype, fruit and vegetable consumption, and supplement use. Am J Epidemiol. 2005;162(10):943–52. PMID: 16192345.
120. Hyoudou K, Nishikawa M, Ikemura M, Kobayashi Y, Mendelsohn A, Miyazaki N, Tabata Y, Yamashita F, Hashida M. Prevention of pulmonary metastasis from subcutaneous tumors by binary system-based sustained delivery of catalase. J Control Release. 2009;137(2):110–5. PMID: 19361547.
121. Hyoudou K, Nishikawa M, Kobayashi Y, Umeyama Y, Yamashita F, Hashida M. PEGylated catalase prevents metastatic tumor growth aggravated by tumor removal. Free Radic Biol Med. 2006;41(9):1449–58. PMID: 17023272.
122. Nishikawa M, Tamada A, Kumai H, Yamashita F, Hashida M. Inhibition of experimental pulmonary metastasis by controlling biodistribution of catalase in mice. Int J Cancer. 2002;99(3):474–9. PMID: 11992420.
123. Goh J, Enns L, Fatemie S, Hopkins H, Morton J, Pettan-Brewer C, Ladiges W. Mitochondrial targeted catalase suppresses invasive breast cancer in mice. BMC Cancer. 2011;11:191. PMID: 21605372.
124. Brigelius-Flohé R, Maiorino M. Glutathione peroxidases. Biochim Biophys Acta. 2013;1830(5):3289–303. PMID: 23201771.
125. Mishra M, Jiang H, Wu L, Chawsheen HA, Wei Q. The sulfiredoxin-peroxiredoxin (Srx-Prx) axis in cell signal transduction and cancer development. Cancer Lett. 2015;366(2):150–9. PMID: 26170166.
126. Liu J, Du J, Zhang Y, Sun W, Smith BJ, Oberley LW, Cullen JJ. Suppression of the malignant phenotype in pancreatic cancer by overexpression of phospholipid hydroperoxide glutathione peroxidase. Hum Gene Ther. 2006;17(1):105–16. PMID: 16409129.
127. Liu J, Hinkhouse MM, Sun W, Weydert CJ, Ritchie JM, Oberley LW, Cullen JJ. Redox regulation of pancreatic cancer cell growth: role of glutathione peroxidase in the suppression of the malignant phenotype. Hum Gene Ther. 2004;15(3):239–50. PMID: 15018733.
128. Brigelius-Flohé R, Kipp A. Glutathione peroxidases in different stages of carcinogenesis. Biochim Biophys Acta. 2009;1790(11):1555–68. PMID: 19289149.
129. Jiang H, Wu L, Mishra M, Chawsheen HA, Wei Q. Expression of peroxiredoxin 1 and 4 promotes human lung cancer malignancy. Am J Cancer Res. 2014;4(5):445–60. PMID: 25232487.
130. Jezierska-Drutel A, Rosenzweig SA, Neumann CA. Role of oxidative stress and the microenvironment in breast cancer development and progression. Adv Cancer Res. 2013;119:107–25. PMID: 23870510.

131. Martinez-Outschoorn UE, Balliet RM, Rivadeneira DB, Chiavarina B, Pavlides S, Wang C, Whitaker-Menezes D, Daumer KM, Lin Z, Witkiewicz AK, Flomenberg N, Howell A, Pestell RG, Knudsen ES, Sotgia F, Lisanti MP. Oxidative stress in cancer associated fibroblasts drives tumor-stroma co-evolution: a new paradigm for understanding tumor metabolism, the field effect and genomic instability in cancer cells. Cell Cycle. 2010;9(16):3256–76. PMID: 20814239.
132. Martinez-Outschoorn UE, Trimmer C, Lin Z, Whitaker-Menezes D, Chiavarina B, Zhou J, Wang C, Pavlides S, Martinez-Cantarin MP, Capozza F, Witkiewicz AK, Flomenberg N, Howell A, Pestell RG, Caro J, Lisanti MP, Sotgia F. Autophagy in cancer associated fibroblasts promotes tumor cell survival: Role of hypoxia, HIF1 induction and NFκB activation in the tumor stromal microenvironment. Cell Cycle. 2010;9(17):3515–33. PMID: 20855962.
133. Martinez-Outschoorn UE, Pavlides S, Whitaker-Menezes D, Daumer KM, Milliman JN, Chiavarina B, Migneco G, Witkiewicz AK, Martinez-Cantarin MP, Flomenberg N, Howell A, Pestell RG, Lisanti MP, Sotgia F. Tumor cells induce the cancer associated fibroblast phenotype via caveolin-1 degradation: implications for breast cancer and DCIS therapy with autophagy inhibitors. Cell Cycle. 2010;9(12):2423–33. PMID: 20562526.
134. Martinez-Outschoorn UE, Lin Z, Trimmer C, Flomenberg N, Wang C, Pavlides S, Pestell RG, Howell A, Sotgia F, Lisanti MP. Cancer cells metabolically "fertilize" the tumor microenvironment with hydrogen peroxide, driving the Warburg effect: implications for PET imaging of human tumors. Cell Cycle. 2011;10(15):2504–20. PMID: 2177882.
135. Chiavarina B, Whitaker-Menezes D, Migneco G, Martinez-Outschoorn UE, Pavlides S, Howell A, Tanowitz HB, Casimiro MC, Wang C, Pestell RG, Grieshaber P, Caro J, Sotgia F, Lisanti MP. HIF1-alpha functions as a tumor promoter in cancer associated fibroblasts, and as a tumor suppressor in breast cancer cells: autophagy drives compartment-specific oncogenesis. Cell Cycle. 2010;9(17):3534–51. PMID: 20864819.
136. Khan N, Afaq F, Mukhtar H. Cancer chemoprevention through dietary antioxidants: progress and promise. Antioxid Redox Signal. 2008;10(3):475–510. PMID: 18154485.
137. Rui Q, Komori K, Tian Y, Liu H, Luo Y, Sakai Y. Electrochemical biosensor for the detection of H_2O_2 from living cancer cells based on ZnO nanosheets. Anal Chim Acta. 2010;670(1–2):57–62. PMID: 20685417.
138. Stolarek RA, Potargowicz E, Seklewska E, Jakubik J, Lewandowski M, Jeziorski A, Nowak D. Increased H_2O_2 level in exhaled breath condensate in primary breast cancer patients. J Cancer Res Clin Oncol. 2010;136(6):923–30. PMID: 19967414.
139. Chan HP, Tran V, Lewis C, Thomas PS. Elevated levels of oxidative stress markers in exhaled breath condensate. J Thorac Oncol. 2009;4(2):172–8. PMID: 19179892.
140. Averill-Bates DA, Przybytkowski E. The role of glucose in cellular defenses against cytotoxicity of hydrogen peroxide in Chinese hamster ovary cells. Arch Biochem Biophys. 1994;312(1):52–8. PMID: 8031146.
141. DeBoer LW, Bekx PA, Han L, Steinke L. Pyruvate enhances recovery of rat hearts after ischemia and reperfusion by preventing free radical generation. Am J Physiol. 1993;265(5 Pt 2):H1571–6. PMID: 8238569.
142. Salahudeen AK, Clark EC, Nath KA. Hydrogen peroxide-induced renal injury. A protective role for pyruvate. J Clin Invest. 1991;88(6):1886–93. PMID: 1752950.
143. Simons AL, Mattson DM, Dornfeld K, Spitz DR. Glucose deprivation-induced metabolic oxidative stress and cancer therapy. J Cancer Res Ther. 2009;5 Suppl 1:S2–6. PMID: 20009288.
144. Lee YJ, Galoforo SS, Berns CM, Chen JC, Davis BH, Sim JE, Corry PM, Spitz DR. Glucose deprivation-induced cytotoxicity and alterations in mitogen-activated protein kinase activation are mediated by oxidative stress in multidrug-resistant human breast carcinoma cells. J Biol Chem. 1998;273(9):5294–9. PMID: 9478987.
145. Scarbrough PM, Mapuskar KA, Mattson DM, Gius D, Watson WH, Spitz DR. Simultaneous inhibition of glutathione- and thioredoxin-dependent metabolism is necessary to potentiate 17AAG-induced cancer cell killing via oxidative stress. Free Radic Biol Med. 2012;52(2):436–43. PMID: 22100505.

146. Fath MA, Ahmad IM, Smith CJ, Spence J, Spitz DR. Enhancement of carboplatin-mediated lung cancer cell killing by simultaneous disruption of glutathione and thioredoxin metabolism. Clin Cancer Res. 2011;17(19):6206–17. PMID: 21844013.
147. Shutt DC, O'Dorisio MS, Aykin-Burns N, Spitz DR. 2-deoxy-D-glucose induces oxidative stress and cell killing in human neuroblastoma cells. Cancer Biol Ther. 2010;9(11):853–61. PMID: 20364116.
148. Hadzic T, Aykin-Burns N, Zhu Y, Coleman MC, Leick K, Jacobson GM, Spitz DR. Paclitaxel combined with inhibitors of glucose and hydroperoxide metabolism enhances breast cancer cell killing via H2O2-mediated oxidative stress. Free Radic Biol Med. 2010;48(8):1024–33. PMID: 20083194.
149. Coleman MC, Asbury CR, Daniels D, Du J, Aykin-Burns N, Smith BJ, Li L, Spitz DR, Cullen JJ. 2-deoxy-D-glucose causes cytotoxicity, oxidative stress, and radiosensitization in pancreatic cancer. Free Radic Biol Med. 2008;44(3):322–31. PMID: 18215740.
150. Dornfeld K, Hopkins S, Simmons J, Spitz DR, Menda Y, Graham M, Smith R, Funk G, Karnell L, Karnell M, Dornfeld M, Yao M, Buatti J. Posttreatment FDG-PET uptake in the supraglottic and glottic larynx correlates with decreased quality of life after chemoradiotherapy. Int J Radiat Oncol Biol Phys. 2008;71(2):386–92. PMID: 18164842.
151. Ahmad IM, Abdalla MY, Aykin-Burns N, Simons AL, Oberley LW, Domann FE, Spitz DR. 2-Deoxyglucose combined with wild-type p53 overexpression enhances cytotoxicity in human prostate cancer cells via oxidative stress. Free Radic Biol Med. 2008;44(5):826–34. PMID: 18155176.
152. Simons AL, Fath MA, Mattson DM, Smith BJ, Walsh SA, Graham MM, Hichwa RD, Buatti JM, Dornfeld K, Spitz DR. Enhanced response of human head and neck cancer xenograft tumors to cisplatin combined with 2-deoxy-D-glucose correlates with increased 18F-FDG uptake as determined by PET imaging. Int J Radiat Oncol Biol Phys. 2007;69(4):1222–30. PMID: 17967311.
153. Simons AL, Ahmad IM, Mattson DM, Dornfeld KJ, Spitz DR. 2-Deoxy-D-glucose combined with cisplatin enhances cytotoxicity via metabolic oxidative stress in human head and neck cancer cells. Cancer Res. 2007;67(7):3364–70. PMID: 17409446.
154. Lin X, Zhang F, Bradbury CM, Kaushal A, Li L, Spitz DR, Aft RL, Gius D. 2-Deoxy-D-glucose-induced cytotoxicity and radiosensitization in tumor cells is mediated via disruptions in thiol metabolism. Cancer Res. 2003;63(12):3413–7. PMID: 12810678.
155. Holman RA. Method of destroying a malignant rat tumour in vivo. Nature. 1957;179(4568):1033. PMID: 13430783.
156. Jeffree GM. Hydrogen peroxide and cancer. Nature. 1958;182(4639):892. PMID: 13590164.
157. Nathan CF, Cohn ZA. Antitumor effects of hydrogen peroxide in vivo. J Exp Med. 1981;154(5):1539–53. PMID: 7299347.
158. Questionable methods of cancer management: hydrogen peroxide and other 'hyperoxygenation' therapies. CA Cancer J Clin. 1993; 43(1):47–56. PMID: 8422605.
159. Wondrak GT. Redox-directed cancer therapeutics: molecular mechanisms and opportunities. Antioxid Redox Signal. 2009;11(12):3013–69. PMID: 19496700.
160. Safford SE, Oberley TD, Urano M, St Clair DK. Suppression of fibrosarcoma metastasis by elevated expression of manganese superoxide dismutase. Cancer Res. 1994;54(16):4261–5. PMID: 8044768.
161. Yan T, Oberley LW, Zhong W, St Clair DK. Manganese-containing superoxide dismutase overexpression causes phenotypic reversion in SV40-transformed human lung fibroblasts. Cancer Res. 1996;56(12):2864–71. PMID: 8665527.
162. Zhang Y, Zhao W, Zhang HJ, Domann FE, Oberley LW. Overexpression of copper zinc superoxide dismutase suppresses human glioma cell growth. Cancer Res. 2002;62(4):1205–12. PMID: 11861405.
163. Zhang HJ, Yan T, Oberley TD, Oberley LW. Comparison of effects of two polymorphic variants of manganese superoxide dismutase on human breast MCF-7 cancer cell phenotype. Cancer Res. 1999;59(24):6276–83. PMID: 10626823.

164. Rodríguez AM, Carrico PM, Mazurkiewicz JE, Meléndez JA. Mitochondrial or cytosolic catalase reverses the MnSOD-dependent inhibition of proliferation by enhancing respiratory chain activity, net ATP production, and decreasing the steady state levels of H_2O_2. Free Radic Biol Med. 2000;29(9):801–13. PMID: 11063906.
165. Li S, Yan T, Yang JQ, Oberley TD, Oberley LW. The role of cellular glutathione peroxidase redox regulation in the suppression of tumor cell growth by manganese superoxide dismutase. Cancer Res. 2000;60(14):3927–39. PMID: 10919671.
166. Buettner GR, Ng CF, Wang M, Rodgers VG, Schafer FQ. A new paradigm: manganese superoxide dismutase influences the production of H_2O_2 in cells and thereby their biological state. Free Radic Biol Med. 2006;41(8):1338–50. PMID: 17015180.
167. Ahmad KA, Iskandar KB, Hirpara JL, Clement MV, Pervaiz S. Hydrogen peroxide-mediated cytosolic acidification is a signal for mitochondrial translocation of Bax during drug-induced apoptosis of tumor cells. Cancer Res. 2004;64(21):7867–78. PMID: 15520193.
168. Ahmad KA, Clement MV, Hanif IM, Pervaiz S. Resveratrol inhibits drug-induced apoptosis in human leukemia cells by creating an intracellular milieu nonpermissive for death execution. Cancer Res. 2004;64(4):1452–9. PMID: 14973069.
169. Alexandre J, Nicco C, Chéreau C, Laurent A, Weill B, Goldwasser F, Batteux F. Improvement of the therapeutic index of anticancer drugs by the superoxide dismutase mimic mangafodipir. J Natl Cancer Inst. 2006;98(4):236–44. PMID: 16478742.
170. Mizutani H, Tada-Oikawa S, Hiraku Y, Kojima M, Kawanishi S. Mechanism of apoptosis induced by doxorubicin through the generation of hydrogen peroxide. Life Sci. 2005;76(13):1439–53. PMID: 15680309.
171. Wagner BA, Evig CB, Reszka KJ, Buettner GR, Burns CP. Doxorubicin increases intracellular hydrogen peroxide in PC3 prostate cancer cells. Arch Biochem Biophys. 2005; 440(2):181–90. PMID: 16054588.
172. Kajiwara K, Ikeda K, Kuroi R, Hashimoto R, Tokumaru S, Kojo S. Hydrogen peroxide and hydroxyl radical involvement in the activation of caspase-3 in chemically induced apoptosis of HL-60 cells. Cell Mol Life Sci. 2001;58(3):485–91. PMID: 11315194.
173. Tome ME, Jaramillo MC, Briehl MM. Hydrogen peroxide signaling is required for glucocorticoid-induced apoptosis in lymphoma cells. Free Radic Biol Med. 2011;51(11): 2048–59. PMID: 21964507.
174. Lee YS, Kang YS, Lee SH, Kim JA. Role of NAD(P)H oxidase in the tamoxifen-induced generation of reactive oxygen species and apoptosis in HepG2 human hepatoblastoma cells. Cell Death Differ. 2000;7(10):925–32. PMID: 11279538.
175. Cao L, Li LS, Spruell C, Xiao L, Chakrabarti G, Bey EA, Reinicke KE, Srougi MC, Moore Z, Dong Y, Vo P, Kabbani W, Yang CR, Wang X, Fattah F, Morales JC, Motea EA, Bornmann WG, Yordy JS, Boothman DA. Tumor-selective, futile redox cycle-induced bystander effects elicited by NQO1 bioactivatable radiosensitizing drugs in triple-negative breast cancers. Antioxid Redox Signal. 2014;21(2):237–50. PMID: 24512128.
176. Bey EA, Reinicke KE, Srougi MC, Varnes M, Anderson VE, Pink JJ, Li LS, Patel M, Cao L, Moore Z, Rommel A, Boatman M, Lewis C, Euhus DM, Bornmann WG, Buchsbaum DJ, Spitz DR, Gao J, Boothman DA. Catalase abrogates β-lapachone-induced PARP1 hyperactivation-directed programmed necrosis in NQO1-positive breast cancers. Mol Cancer Ther. 2013;12(10):2110–20. PMID: 23883585.
177. Chen Q, Espey MG, Krishna MC, Mitchell JB, Corpe CP, Buettner GR, Shacter E, Levine M. Pharmacologic ascorbic acid concentrations selectively kill cancer cells: action as a prodrug to deliver hydrogen peroxide to tissues. Proc Natl Acad Sci U S A. 2005;102(38):13604–9. PMID: 16157892.
178. Chen Q, Espey MG, Sun AY, Pooput C, Kirk KL, Krishna MC, Khosh DB, Drisko J, Levine M. Pharmacologic doses of ascorbate act as a prooxidant and decrease growth of aggressive tumor xenografts in mice. Proc Natl Acad Sci U S A. 2008;105(32):11105–9. PMID: 18678913.
179. Chen Q, Espey MG, Sun AY, Lee JH, Krishna MC, Shacter E, Choyke PL, Pooput C, Kirk KL, Buettner GR, Levine M. Ascorbate in pharmacologic concentrations selectively generates

ascorbate radical and hydrogen peroxide in extracellular fluid in vivo. Proc Natl Acad Sci U S A. 2007;104(21):8749–54. PMID: 17502596.

180. Du J, Martin SM, Levine M, Wagner BA, Buettner GR, Wang SH, Taghiyev AF, Du C, Knudson CM, Cullen JJ. Mechanisms of ascorbate-induced cytotoxicity in pancreatic cancer. Clin Cancer Res. 2010;16(2):509–20. PMID: 20068072.
181. Du J, Cieslak 3rd JA, Welsh JL, Sibenaller ZA, Allen BG, Wagner BA, Kalen AL, Doskey CM, Strother RK, Button AM, Mott SL, Smith B, Tsai S, Mezhir J, Goswami PC, Spitz DR, Buettner GR, Cullen JJ. Pharmacological ascorbate radiosensitizes pancreatic cancer. Cancer Res. 2015;75(16):3314–26. PMID: 26081808.
182. Cameron E, Pauling L. Supplemental ascorbate in the supportive treatment of cancer: reevaluation of prolongation of survival times in terminal human cancer. Proc Natl Acad Sci U S A. 1978;75(9):4538–42. PMID: 279931.
183. Creagan ET, Moertel CG, O'Fallon JR, Schutt AJ, O'Connell MJ, Rubin J, Frytak S. Failure of high-dose vitamin C (ascorbic acid) therapy to benefit patients with advanced cancer. A controlled trial. N Engl J Med. 1979;301(13):687–90. PMID: 384241.
184. Moertel CG, Fleming TR, Creagan ET, Rubin J, O'Connell MJ, Ames MM. High-dose vitamin C versus placebo in the treatment of patients with advanced cancer who have had no prior chemotherapy. A randomized double-blind comparison. N Engl J Med. 1985;312(3):137–41. PMID: 3880867.
185. Padayatty SJ, Sun H, Wang Y, Riordan HD, Hewitt SM, Katz A, Wesley RA, Levine M. Vitamin C pharmacokinetics: implications for oral and intravenous use. Ann Intern Med. 2004;140(7):533–7. PMID: 15068981.
186. Drisko JA, Chapman J, Hunter VJ. The use of antioxidants with first-line chemotherapy in two cases of ovarian cancer. J Am Coll Nutr. 2003;22(2):118–23. PMID: 12672707.
187. Riordan HD, Casciari JJ, González MJ, Riordan NH, Miranda-Massari JR, Taylor P, Jackson JA. A pilot clinical study of continuous intravenous ascorbate in terminal cancer patients. P R Health Sci J. 2005;24(4):269–76. PMID: 16570523.
188. Welsh JL, Wagner BA, van't Erve TJ, Zehr PS, Berg DJ, Halfdanarson TR, Yee NS, Bodeker KL, Du J, Roberts 2nd LJ, Drisko J, Levine M, Buettner GR, Cullen JJ. Pharmacological ascorbate with gemcitabine for the control of metastatic and node-positive pancreatic cancer (PACMAN): results from a phase I clinical trial. Cancer Chemother Pharmacol. 2013; 71(3):765–75. PMID: 23381814.
189. Graumlich JF, Ludden TM, Conry-Cantilena C, Cantilena Jr LR, Wang Y, Levine M. Pharmacokinetic model of ascorbic acid in healthy male volunteers during depletion and repletion. Pharm Res. 1997;14(9):1133–9.
190. Buettner GR, Jurkiewicz BA. Catalytic metals, ascorbate and free radicals: combinations to avoid. Radiat Res. 1996;145(5):532–41. PMID: 8619018.
191. Buettner GR. In the absence of catalytic metals ascorbate does not autoxidize at pH 7: ascorbate as a test for catalytic metals. J Biochem Biophys Methods. 1988;16(1):27–40. PMID: 3135299.
192. Winterbourn CC. Toxicity of iron and hydrogen peroxide: the Fenton reaction. Toxicol Lett. 1995;82–83:969–74. PMID: 8597169.
193. Richardson DR, Ponka P. The molecular mechanisms of the metabolism and transport of iron in normal and neoplastic cells. Biochim Biophys Acta. 1997;1331(1):1–40. PMID: 9325434.
194. Buettner GR. The reaction of superoxide, formate radical, and hydrated electron with transferrin and its model compound, Fe(III)-ethylenediamine-N, N′-bis[2-(2-hydroxyphenyl) acetic acid] as studied by pulse radiolysis. J Biol Chem. 1987;262(25):11995–8. PMID: 3040725.
195. Sirivech S, Frieden E, Osaki S. The release of iron from horse spleen ferritin by reduced flavins. Biochem J. 1974;143(2):311–5. PMID: 4462557.
196. Williams DM, Lee GR, Cartwright GE. The role of superoxide anion radical in the reduction of ferritin iron by xanthine oxidase. J Clin Invest. 1974;53(2):665–7. PMID: 11344583.
197. Güner G, Kirkali G, Yenisey C, Töre IR. Cytosol and serum ferritin in breast carcinoma. Cancer Lett. 1992;67(2–3):103–12. PMID: 1483258.

198. Miyata Y, Koga S, Nishikido M, Hayashi T, Kanetake H. Relationship between serum ferritin levels and tumour status in patients with renal cell carcinoma. BJU Int. 2001;88(9):974–7. PMID: 11851623.
199. Basso D, Fabris C, Del Favero G, Meggiato T, Panozzo MP, Vianello D, Plebani M, Naccarato R. Hepatic changes and serum ferritin in pancreatic cancer and other gastrointestinal diseases: the role of cholestasis. Ann Clin Biochem. 1991;28:34–8. PMID: 2024931.
200. Hann HW, Lange B, Stahlhut MW, McGlynn KA. Prognostic importance of serum transferrin and ferritin in childhood Hodgkin's disease. Cancer. 1990;66(2):313–6. PMID: 2369713.
201. Moser JC, Rawal M, Wagner BA, Du J, Cullen JJ, Buettner GR. Pharmacological ascorbate and ionizing radiation (IR) increase labile iron in pancreatic cancer. Redox Biol. 2013;2:22–7. PMID: 24396727.
202. Du J, Wagner BA, Buettner GR, Cullen JJ. Role of labile iron in the toxicity of pharmacological ascorbate. Free Radic Biol Med. 2015;84:289–95. PMID: 25857216.
203. Nicco C, Laurent A, Chereau C, Weill B, Batteux F. Differential modulation of normal and tumor cell proliferation by reactive oxygen species. Biomed Pharmacother. 2005;59(4):169–74. PMID: 15862711.
204. Oberley LW. Mechanism of the tumor suppressive effect of MnSOD overexpression. Biomed Pharmacother. 2005;59(4):143–8. PMID: 15862707.
205. Trachootham D, Alexandre J, Huang P. Targeting cancer cells by ROS-mediated mechanisms: a radical therapeutic approach? Nat Rev Drug Discov. 2009;8(7):579–91. PMID: 19478820.
206. Hoffer LJ, Levine M, Assouline S, Melnychuk D, Padayatty SJ, Rosadiuk K, Rousseau C, Robitaille L, Miller Jr WH. Phase I clinical trial of i.v. ascorbic acid in advanced malignancy. Ann Oncol. 2008;19(12):2095. PMID: 18544557.
207. Monti DA, Mitchell E, Bazzan AJ, Littman S, Zabrecky G, Yeo CJ, Pillai MV, Newberg AB, Deshmukh S, Levine M. Phase I evaluation of intravenous ascorbic acid in combination with gemcitabine and erlotinib in patients with metastatic pancreatic cancer. PLoS One. 2012;7(1), e29794. PMID: 22272248.
208. Ma Y, Chapman J, Levine M, Polireddy K, Drisko J, Chen Q. High-dose parenteral ascorbate enhanced chemosensitivity of ovarian cancer and reduced toxicity of chemotherapy. Sci Transl Med. 2014;6(222):222ra18. PMID: 24500406.
209. Tian WN, Braunstein LD, Apse K, Pang J, Rose M, Tian X, Stanton RC. Importance of glucose-6-phosphate dehydrogenase activity in cell death. Am J Physiol. 1999;276(5 Pt 1): C1121–31. PMID:10329961.
210. Rees DC, Kelsey H, Richards JD. Acute haemolysis induced by high dose ascorbic acid in glucose-6-phosphate dehydrogenase deficiency. BMJ. 1993;306(6881):841–2. PMID: 8490379.
211. Verrax J, Calderon PB. Pharmacologic concentrations of ascorbate are achieved by parenteral administration and exhibit antitumoral effects. Free Radic Biol Med. 2009;47(1):32–40. PMID: 19254759.
212. Prasad KN, Sinha PK, Ramanujam M, Sakamoto A. Sodium ascorbate potentiates the growth inhibitory effect of certain agents on neuroblastoma cells in culture. Proc Natl Acad Sci U S A. 1979;76(2):829–32. PMID: 284405.
213. Kurbacher CM, Wagner U, Kolster B, Andreotti PE, Krebs D, Bruckner HW. Ascorbic acid (vitamin C) improves the antineoplastic activity of doxorubicin, cisplatin, and paclitaxel in human breast carcinoma cells in vitro. Cancer Lett. 1996;103(2):183–9. PMID: 8635156.
214. Herst PM, Broadley KW, Harper JL, McConnell MJ. Pharmacological concentrations of ascorbate radiosensitize glioblastoma multiforme primary cells by increasing oxidative DNA damage and inhibiting G2/M arrest. Free Radic Biol Med. 2012;52(8):1486–93. PMID: 22342518.
215. Espey MG, Chen P, Chalmers B, Drisko J, Sun AY, Levine M, Chen Q. Pharmacologic ascorbate synergizes with gemcitabine in preclinical models of pancreatic cancer. Free Radic Biol Med. 2011;50(11):1610–9. PMID: 21402145.

Chapter 7
Mimicking SOD, Why and How: Bio-Inspired Manganese Complexes as SOD Mimic

Clotilde Policar

Abbreviations

ACN	Acetonitrile
ATP	Adenosine triphosphate
ALS	Amyotrophic lateral sclerosis
CYP450	Cytochrome P450
CD	Cyclodextrin
CN	Coordination number
DNA	Deoxyribonucleic acid
DMSO	Dimethylsulfoxide
DMF	*N,N*-Dimethylformamide
EPR	Electron paramagnetic resonance
EWG	Electron-withdrawing group
GSH	Glutathione
HAS	Human serum albumin
LFSE	Ligand-field stabilization energy
MMO	Methane monooxygenase
NBT	Nitroblue tetrazolium
PCET	Proton-coupled electron transfer
PROS	Partially reduced oxygen species
ROS	Reactive oxygen species
RNS	Reactive nitrogen species
SOD	Superoxide dismutase

C. Policar (✉)
Département de Chimie, École Normale Supérieure-PSL Research University, Sorbonne Universités, UPMC Univ Paris 06, CNRS, Laboratoire des BioMolécules, UMR 7203, 24, rue Lhomond, Paris 75005, France
e-mail: clotilde.policar@ens.fr

I. Batinić-Haberle et al. (eds.), *Redox-Active Therapeutics*,
Oxidative Stress in Applied Basic Research and Clinical Practice,
DOI 10.1007/978-3-319-30705-3_7

SOR	Superoxide reductase
UV-vis.	UV-visible spectroscopy
WT	Wild type

7.1 Mimicking Manganese Superoxide Dismutases: Why and How?

[Naturam] si sequemur ducem, numquam aberrabimus, Cicero, *De officiis*, I, 100
Let Nature guide us: we shall never get lost

Superoxide dismutases (SODs) are crucial proteins for protecting cells from superoxide, a reactive species derived from dioxygen by one-electron reduction. SODs catalyze the dismutation of superoxide, resulting in a tight control of its concentration in biological environments. Superoxide is a metastable anion endowed with both signaling functions and toxic properties, valuable for fighting against pathogens. Superoxide can be harmful leading to molecular modifications, which disturb endogenous functions of biomolecules. Oxidative stress reflects an imbalance between the continuous production of reactive oxygen species (ROS) and the antioxidant protective pathways, and is associated with a wide range of physiopathological conditions, including aging. SODs are remarkably efficient proteins that have elicited strong interests for many years. Indeed, besides the biological role of SOD in controlling oxidative stress conditions, understanding their physicochemical parameters, selected by evolution and responsible for their amazing efficiencies, is of fundamental interest and can also serve as a useful and effective guideline for chemists aiming at developing antioxidant derivatives for therapeutic applications against oxidative stress. This chapter focuses on the SOD physicochemical parameters that are useful for chemical design of low-molecular-weight complexes displaying superoxide scavenging activity. Nature is an endless source of both inspiration and challenges for chemists, and in turn, chemical modeling can help in deciphering underlying physicochemical characteristics determining bioactivity. Inspiration from natural systems and processes, providing their mechanisms have been dissected with care, can pave the way to efficient and original systems. The development of low-molecular-weight SOD mimics constitutes a shining example of this fertile interplay with Nature.

7.2 Dioxygen and Superoxide Dismutases

7.2.1 Dioxygen and Oxidative Stress

Dioxygen is necessary for aerobic life. Its concentration rose on Earth as a result of oxygenic photosynthesis that appeared in cyanobacteria about 2.5 billion years ago. This process enables the efficient retrieving of light energy for the synthesis of ATP, and ultimately organic sugars, and releases O_2 as a by-product. Rising concentration in dioxygen initiated a profound modification of the redox state on Earth, going from a reducing environment, still roughly found in the interior of cells, to an oxidant environment. In a certain sense, dioxygen can be considered as the first chemical pollutant and held responsible for an event called the *Oxygen Catastrophe* or the *Great Oxidation Event* [1, 2]. The increase in its concentration was associated with strong variations in the chemical environment on Earth: the greenhouse methane gas disappeared, with a fast decrease in temperature [2], oxidation of metal ions with drastic modifications in their solubility and bioavailability occurred [3], etc. The appearance of dioxygen triggered an increase in molecular diversity by opening the possibility of oxygen incorporation into organic molecules through new metabolic pathways. However, because of its triplet spin-state ($S=1$) resulting from its two-unpaired electron, diradical ground state nature, dioxygen is not very reactive with organic molecules ($S=0$, singlet, with all paired electrons): the incorporation of an oxygen atom into a C–H bond, leading to a C–OH moiety for instance, corresponds to the creation of new covalent bonds associated with a spin-flip from the $S=1$ diradical dioxygen species, which is energetically costly with a high activation barrier [4]. Hence, to be used by biological systems, dioxygen needs to be activated, which mainly occurs through reductive pathways as in cytochromes P450 (CYP450) or methane monooxygenase (MMO), etc. [5, 6]. Dioxygen can take up four electrons leading to water or can be sequentially reduced into superoxide ($O_2^{\bullet-}$), hydrogen peroxide (H_2O_2) or hydroxyl radical ($HO^{\bullet}$) and then H_2O (see Fig. 7.1). The partially reduced species—superoxide, hydrogen peroxide, and hydroxyl radical—are much more reactive than dioxygen itself, with no spin-state kinetic barrier for the reaction with singlet molecules, and hence are called *reactive oxygen species* (ROS) or sometimes, to point out their reduced nature, *partially reduced oxygen species* (PROS) [7]; in this chapter the more commonly used acronym ROS will be used. They are continuously produced in biological systems by endogenous mechanisms. Mitochondrial respiration, the reverse process of photosynthesis, consists of the four-electron reduction of dioxygen into water (Fig. 7.1) and results in a flow of ROS when the four-electron transfer is decoupled, that is, not performed in a single step. This decoupling leads to an incomplete reduction of dioxygen and about 1–3 % of the dioxygen processed by the mitochondria leaks as superoxide [8, 9].

$$2\,H_2O \underset{(2)}{\overset{(1)}{\rightleftharpoons}} O_2 + 4\,e^- + 4\,H^+$$

$$O_2 + e^- \longrightarrow O_2^{\bullet-} \qquad (3)$$

$$O_2^{\bullet-} + e^- + 2H^+ \longrightarrow H_2O_2 \qquad (4)$$

$$H_2O_2 + e^- \longrightarrow HO^- + HO^{\bullet} \qquad (5)$$

$$HO^{\bullet} + e^- \longrightarrow HO^- \qquad (6)$$

Fig. 7.1 Oxidation of water and reduction of dioxygen. Reduction cascade of O_2 leading to reactive oxygen species ROS. Note that superoxide can be labeled O_2^- or $O_2^{\bullet-}$ to emphasize its radical nature, as dioxygen could be labeled O_2 or $O_2^{\bullet\bullet}$ to indicate its diradical nature. In the main text, we have chosen to use $O_2^{\bullet-}$

Because of their reactivity, ROS can damage endogenous components. The hydroxyl radical $HO^•$ is the most reactive and deleterious species in the dioxygen-reduction cascade. $HO^•$ initiates radical production through hydrogen abstraction from any R–H molecule with a kinetics close to the diffusion limit, i.e., the first collision with another molecule being reactive-efficient. Superoxide ($O_2^{•-}$) and hydrogen peroxide (H_2O_2) are less reactive but superoxide is known to abstract hydrogen from DNA (DNA nicking) [10, 11], to deactivate radical proteins, such as ribonucleotide reductase, to initiate and also terminate lipid peroxidation [12]. One of the main biological effects of superoxide is the inhibition of iron-sulfur proteins, such as 6-phosphogluconate-dehydratase [13] or aconitase [14, 15], by the destruction of their metallic clusters. This process releases Fe^{II} that may in turn participate in the production of $HO^•$ through the Fenton reaction involving H_2O_2 (see Fig. 7.2). This reaction corresponds to the reduction of H_2O_2 into HO^- with the release of $HO^•$—note that the oxygen atoms in $HO^•$ and H_2O_2 are at the same redox state—and oxidation of Fe^{II} into Fe^{III}. Cu^I is able to perform a similar reaction, whereas most of Mn ions, having higher redox potential, are not [9, 16–18]. The concentration of Cu^I and Fe^{II} in biological media is tightly controlled and most of iron is coordinated by biomolecules or precipitated within ferritin [19]. This process would not be of much significance if Cu^I or Fe^{II} ions were not regenerated, which is made possible by superoxide or other cellular reductants, which may cycle Fe^{III} back to Fe^{II} or Cu^{II} to Cu^I (see reaction [8]) leading to the Haber–Weiss reaction (see Fig. 7.2).

But ROS are also useful species. Low levels of ROS are necessary for a variety of cellular processes, including cell adhesion, cell growth and differentiation, or intracellular signaling [20]. In addition, biological systems take advantage of ROS to fight against pathogens. Indeed, macrophages and other immune cells produce high extracellular flows and high intra-phagosome concentrations of ROS to kill infectious agents through oxidative damage [21]. Last, but not least, it has been proposed that the ROS-induced chemical stress on biomolecules, e.g., DNA and proteins, provided an essential source of chemical diversity that enabled adaptation and chemical evolution at the origins of life [22].

In living organisms, efficient pathways to finely tune the concentrations of superoxide and hydrogen peroxide have evolved to protect endogenous components against oxidative damage and to control ROS concentrations to appropriate levels. Such

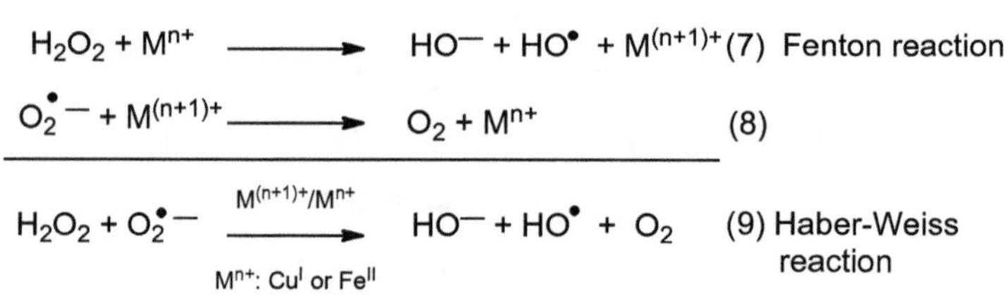

Fig. 7.2 Fenton and Haber Weiss reactions: oxidation of low-oxidation state metal ion by hydrogen peroxide and cycling back to low-oxidation state with superoxide. Note that in biological systems, other reductants can be involved in this reduction

pathways can involve stoichiometric antioxidants (e.g., vitamins E and C, and GSH) or redox enzymes, such as catalase, glutathione peroxidase, superoxide dismutases (SODs), and superoxide reductases (SORs); SODs are the focus of this chapter. Oxidative stress arises when these protective pathways are imbalanced by an increase in ROS flow or a decrease in protective enzymes expression or activity. This leads to a wide range of physiopathological processes, including aging, arthritis, stroke, neurodegenerative diseases (Parkinson and Alzheimer diseases), amyotrophic lateral sclerosis (ALS), cancer, or inflammation [22–26].

7.2.2 Superoxide Dismutases and Manganese: An Efficient Protection Against Oxidative Stress

SODs are metal-containing oxidoreductases that catalyze the dismutation of superoxide—oxidation to dioxygen and reduction to hydrogen peroxide (see Fig. 7.3). They maintain the concentration of superoxide inside cells at very low levels (less than 50 pM in *E. coli* or mammalian cells) [14, 15, 27]. They are crucial proteins in the protective antioxidant arsenal, since superoxide is the first step in the ROS cascade from dioxygen to water (see Fig. 7.1). Mammalian cells produce three different SODs, involving either copper-zinc or manganese metal centers: SOD1, CuZnSOD found in the cytosol, in the nuclear compartments, and in the mitochondrial intermembrane space; SOD2, a MnSOD found in the mitochondrial matrix, which is its exclusive location in human beings; and SOD3 (EC-SOD), an extracellular CuZnSOD, found in extracellular matrix of tissues and extracellular fluids (such as plasma) [1, 9]. SODs are compartmentalized, which means they are not able to cross membranes, move from one organelle to another, nor enter/leave cells. The mitochondrial MnSOD has a particular biological importance in mammals: its knockout is lethal to newborn mice [28, 29], whereas this is not the case for CuZnSOD [30].

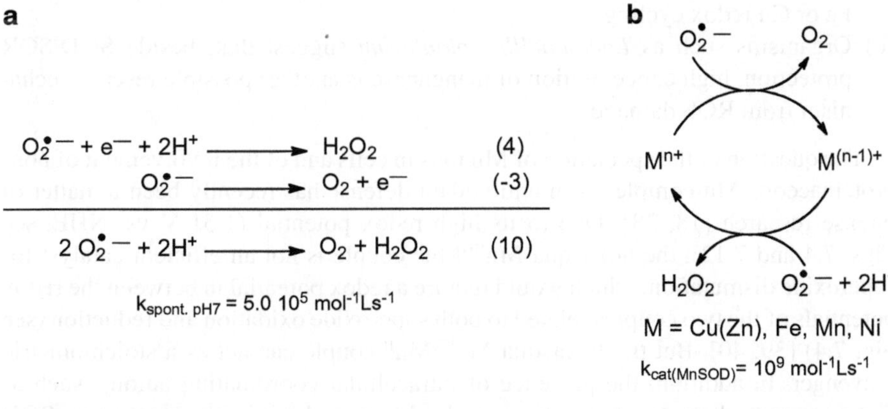

Fig. 7.3 (**a**) Spontaneous superoxide dismutation—or autodismutation, and (**b**) simplified ping-pong mechanism for the catalysis of superoxide dismutation. The kinetic data are from [61] and [58]

In addition, dysregulation of MnSOD or MnSOD natural mutations—an example of protein polymorphism—has been proposed to be involved in cancer [20]. Other metal ions can also be found as cofactors in non-mammalian SODs: iron in FeSOD, which coevolved with MnSOD from a common ancestor, shows high degrees of both sequence identity and structural homology with MnSOD and was discovered in 1973 [31]; and nickel in NiSOD, which belongs to a rarer SOD family discovered in 1996 [32].

SODs have been found in almost all aerobic, facultative anaerobic, and even in some anaerobic organisms, where they have been sought. An exception is the case of microaerophilic and anaerobic bacteria that contain a superoxide reductase (SOR), which detoxifies superoxide by its reduction into H_2O_2. Detoxification of superoxide by SORs avoids O_2 release which would be expected to be toxic in air-sensitive microorganisms [9, 33]. When SOD are not present or show a low activity (e.g., *Leptospira interrogans* serovars [34], *Lactobacillus plantarum* [35], and *Neisseria gonorrhea* [36]), cells are either very rich in catalase or show a high intracellular concentration of Mn^{II} (up to 15–30 mM in *Lactobacillus plantarum* [35]). These observations are interesting in several ways:

(a) Few organisms are found without any SOD or SOR: this observation emphasizes the importance of superoxide removal and the need for protection against superoxide.
(b) Viable organisms, although rare, are found with no SOD but with high concentration of catalase, which seems to rescue them in the absence of SOD or SOR. This could suggest that the key feature is to limit the time of co-residence of superoxide with hydrogen peroxide. $O_2^{\bullet -}$ and H_2O_2 are inevitably found at the same location in biological systems because superoxide self-dismutation produces H_2O_2. This leads potentially to Haber–Weiss chemistry (see Fig. 7.2) in the presence of soluble iron(II) or copper(I), redox states that can be found in cells where the environment is reducing [37]. Limiting the co-residence time of H_2O_2 and $O_2^{\bullet -}$ restricts possible Fenton/Haber–Weiss chemistry that would lead to the continuous production of the deleterious $HO^{\bullet}$ by the use of superoxide for Fe or Cu redox cycling.
(c) Organisms such as *Lactobacillus plantarum* suggest that, beside SOD/SOR protection, high concentration of manganese is another possible rescue mechanism from ROS damage.

The question of the speciation of Mn ions in cells and of the involvement of non-proteinaceous Mn-complexes in antioxidant defense has recently been a matter of intense research [18, 38]. Due to its high redox potential (1.51 V vs. NHE, see Figs. 7.4 and 7.12), the hexaaqua Mn^{III}/Mn^{II} couple is not an efficient catalyst for superoxide dismutation, which would require a redox potential in between the redox potentials of the two couples related to both superoxide oxidation and reduction (see Fig. 7.4) [39, 40]. But the hexaaqua Mn^{III}/Mn^{II} couple can act as a stoichiometric scavenger. In addition, the presence of intracellular coordinating anions, such as lactate or phosphate, has been shown to lead to complexes with an effective SOD activity, associated with cellular oxidative stress resistance [35, 39, 41, 42]. Even if

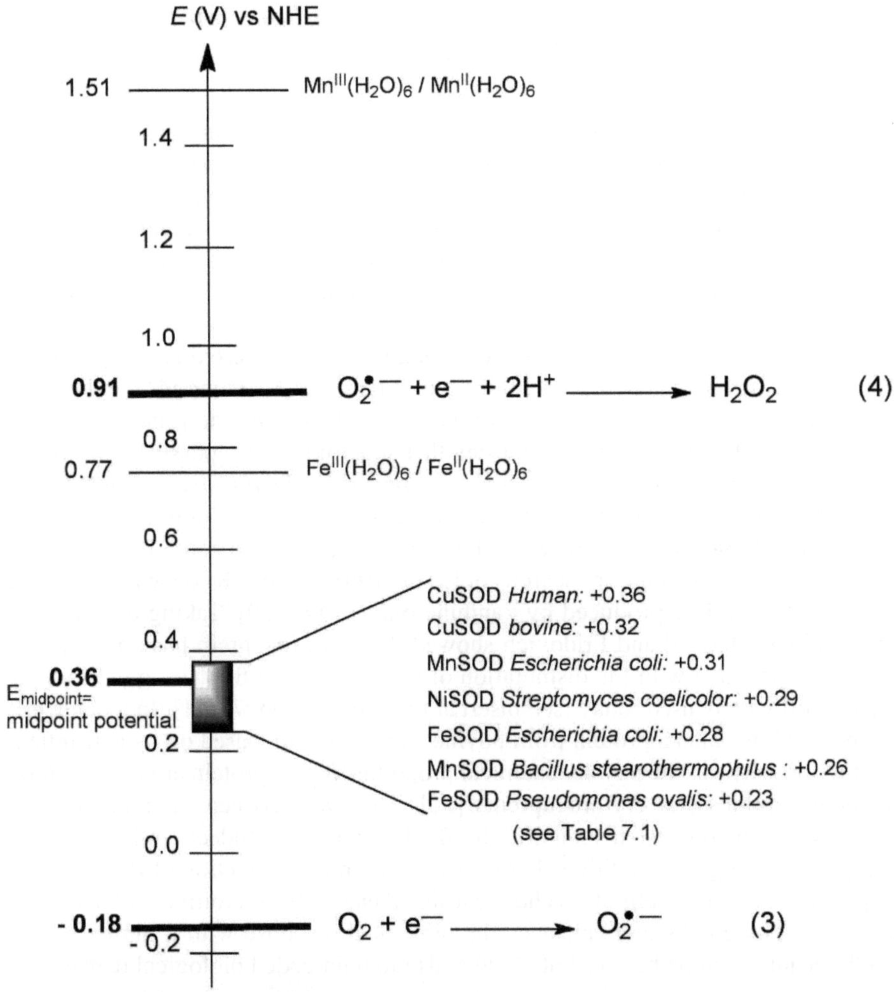

Fig. 7.4 Standard redox potentials E°′ V vs. NHE for superoxide at pH 7 [7], midpoint potential optimal for catalysis of the dismutation ($E_{midpoint}$) [60], redox potentials reported for several SOD (see also Table 7.1) and for hexa-aqua Mn^{III}/Mn^{II} and hexa-aqua Fe^{III}/Fe^{II}

their rate of superoxide dismutation is lower than that of SODs, the corresponding Mn complexes can efficiently protect cells if sufficiently concentrated. Mn ion can exchange its ligands, especially at the +2 oxidation state, and biological environments abound with coordinating molecules. It is thus possible that Mn complexes formed inside cells could act as efficient SOD mimics: they are possibly in dynamic exchange, and hence difficult to isolate and characterize, but some may show redox potential appropriate for SOD activity. Recently, it has been shown that this battery of Mn antioxidants was tightly regulated by specific pathways in *Saccharomyces cerevisiae* [43], indicating these non-proteinaceous Mn-antioxidants are not only passively formed in cells [18].

7.2.3 SOD Discovery: A History of Anti-Inflammatory Properties (See Also Chap. 1)

Through their anti-superoxide activity, SODs show interesting pharmacological properties. Indeed, the antioxidant and anti-inflammatory properties of SODs are closely associated with their discovery, which occurred, not once but three times [22]. In 1938, Mann and Keiling, interested in the speciation of copper ion in blood, isolated from horse serum a protein described as a copper-binding protein that they called hemocuprein [44]. Twenty years later, a similar copper-protein, erythrocuprein, was isolated from human erythrocytes [45]. In both occasions, several enzymatic activities for these proteins were sought: catalase, peroxidase, cytochrome oxidase, or carbonic anhydrase activities, by Mann and Keimann [44] and oxidase activities with various substrates by Markowitz et al. [45]. But none could be found.

In 1931–1932, Pauling had suggested that superoxide, as a possible "*intermediate between molecular oxygen and hydrogen peroxide, should have enough stability to exist*" [46] and superoxide had been characterized as potassium superoxide by Nauman in 1934 [46–48]. However, the idea that superoxide could be involved in biological environments appeared only in 1968, when Knowles found that superoxide could be produced by xanthine oxidase [49, 50]. Taking advantage of this finding, McCord and Fridovich showed that erythrocuprein has an enzymatic activity associated with the dismutation of superoxide and they designed an assay that is still widely used today (see insert at the end of Chap. 7) [51]. Independently, Huber et al. isolated a protein from bovine liver, orgotein, based on its anti-inflammatory properties. The physicochemical properties of this protein are close to those of hemocupreins and erythrocupreins [52]. The link between inflammation and superoxide was suggested soon after by Babior et al. [53]. Indeed, inflammation is a physiopathological condition characterized by a high production of ROS by macrophages or immune cells. This chemical signal causes the recruitment of additional ROS-producing cells and enhances the biological response with a higher flow in ROS meant to kill pathogens but leading also to unintended biological damage.

7.2.4 Mimicking SODs by Low-Molecular-Weight Complexes: SOD Mimics

Since this pioneering work, purified SODs have been successfully used in clinical trials [26, 54], and antioxidants in general, including SODs, are now quite well documented for their beneficial effects in oxidative stress situations [22, 24–26]. Nonetheless, major drawbacks limit the applications of the SOD enzymes in therapeutics: cost of production, immunogenicity, short half-life in plasma, inefficient per os delivery added to limited cellular accessibility [24], and clearly, stabilizing SODs in biological environment is not straightforward [55]. These limitations can be overcome by using low-molecular-weight complexes reproducing the *catalytic activity* of SODs, also called SOD mimics [17, 56]. In the literature, SOD mimics have been designed to react with superoxide outside of any cellular context [17, 56] and some are reportedly efficient in ameliorating oxidative stress, both in cells and in vivo [24–26, 54, 57]. Of note, we have used here a circumlocution to designate studies performed outside any cellular context, e.g., in UV-vis cuvette as in the McCord and Fridovich assay: in vitro, a term frequently used by chemists, is not fully appropriate here as for cell biologists it refers to studies in cell cultures as in contrast with studies performed in animal models. To avoid any misunderstanding, we refer herein to activities evaluated outside any cellular context as *intrinsic activities* in contrast with *in cell activities* and in vivo *activities*. In the case of SOD mimics, intrinsic activity will refer to the kinetics of the reaction with superoxide (see insert for methods of evaluation), possibly the catalytic constant in case of a true catalytic SOD mimic.

To obtain a complex acting as an SOD mimic with therapeutic efficacy against oxidative stress, a primary need is to design nontoxic complexes with a good intrinsic activity. Some excellent reviews have been previously published on the rational design of SOD mimics to delineate important features for an efficient intrinsic activity [17, 56], or for application in therapeutics [23–26, 54, 57]. In the following sections, we will focus on approaches aimed at designing SOD mimics bioinspired by SODs' main features, and more particularly those of MnSODs. We will draw the main characteristics of SODs on which chemists focused for the design of SOD mimics. We will then describe different strategies set up to enhance the activity of SOD mimics following the footsteps of Nature.

7.3 SODs Physicochemical Characteristics: A Guideline for Chemists

All SODs, whatever the metal ion involved at the active site, are very efficient enzymes, reacting with superoxide with kinetics close to the diffusion limit [58], and the parameters responsible for this efficient activity have been carefully analyzed (for an excellent recent review, see Sheng et al. [1]). As already mentioned, SODs can be classified into three subfamilies of different lineages: (a) CuSOD, with their bimetallic $Cu^{I/II}$-Zn^{II} active site, (b) Fe or MnSOD, with $Fe^{III/II}$ or $Mn^{III/II}$ active site, which are evolutionary related and share high degrees of sequence identity and structural homology, and (c) NiSOD, a rarer family of SOD with a $Ni^{III/II}$ active site. Interestingly, these families share some common features delineated below, a signature of a convergent evolution and indicating the strong efficiency of these particular characteristics and mechanisms selected by Nature [1]: (a) tuned redox potential (b) electrostatic guidance of the anionic superoxide, and (c) compartmentalization into specific organelles.

7.3.1 *Redox Potential: The Main Determinant for Catalytic Activity*

Redox Tuning at the Active Site of SODs

The superoxide anion is metastable and the reaction of dismutation is indicated in Fig. 7.3. The catalysis of this reaction is a redox process, involving both Mn^{II} and Mn^{III} oxidation states in the case of MnSOD, to respectively reduce superoxide into H_2O_2 in the presence of protons and to oxidize it into O_2. To be thermodynamically competent to carry out both steps, the redox potential of the Mn^{III}/Mn^{II} couple must lie between those of the two couples involving the superoxide anion, $O_2^{\bullet-}/H_2O_2$ and $O_2/O_2^{\bullet-}$, as indicated in Fig. 7.4, at pH 7 and with a concentration of dioxygen of 1.23 mM, which is that in water under standard conditions, 25 °C and 100 kPa [7]. In addition, for optimal kinetics, one should consider the fact that the closer the redox potential of two couples, the faster the reaction between them [59, 60]. Therefore, the value of the Mn^{III}/Mn^{II} redox potential optimizing the kinetics of the overall catalytic cycle is the mid-potential, at about 0.36 V vs. NHE (see Fig. 7.4) [60]. As a consequence of this redox tuning, the two half-reactions ((4) and (−3), see Figs. 7.1 and 7.3) proceed at the same rate, which is the overall optimal rate for the catalytic cycle.

Interestingly, all SODs indeed meet this criterion with the redox potential of Cu^{II}/Cu^{I}, Mn^{III}/Mn^{II}, Fe^{III}/Fe^{II}, and Ni^{III}/Ni^{II} falling within a very narrow range around 0.2–0.4 V vs. NHE (see Fig. 7.4 and Table 7.1). This observation is very informative as it clearly shows that the redox potential is a key parameter to be tuned for an efficient SOD activity. It should be further emphasized that this narrow potential range is very striking, since the intrinsic redox characteristics of the aqueous metal couples $M^{(n+1)}{}_{aq}/M^{n+}{}_{aq}$ involved in SODs are quite different. To take the example of Fe and MnSODs that has been extensively studied by Miller et al. [1, 65], the redox

Table 7.1 SODs redox potentials (see also Fig. 7.4)

SOD	Organism	pH	E_m (V vs. NHE)	References
CuSOD	Human	7.4	0.36	[62]
CuSOD	Bovine	7.4	0.32	[62]
MnSOD	*Escherichia coli*	7	0.31	[63]
MnSOD	*Escherichia coli*	9	0.18	[63]
NiSOD	*Streptomyces coelicolor*		0.29	[64]
FeSOD	*Escherichia coli*	7	0.28	[60]
FeSOD	*Escherichia coli*	9	0.12	[60]
MnSOD	*Bacillus stearothermophilus*	7	0.26	[63]
MnSOD	*Bacillus stearothermophilus*	9	0.12	[63]
FeSOD	*Pseudomonas ovalis*	7	0.23	[60]
FeSOD	*Pseudomonas ovalis*	9	0.19	[60]

Note that the E°s of SODs can be difficult to measure precisely due to slow equilibration and other slightly different values can be found in the literature potential of the hexaaqua complexes of $Mn^{III}_{aq}/Mn^{II}_{aq}$ in aqueous environment is 1.51 V vs. NHE and that of $Fe^{III}_{aq}/Fe^{II}_{aq}$ is 0.77 V vs. NHE (see Fig. 7.4). The role of the apoprotein is then to tightly control the metallic redox potential. The question of the tuning of redox potentials at the active site of redox proteins in general, including Fe/MnSOD and also copper proteins, has evoked some interest lately [1, 65–67]. Several parameters can be modulated for this tuning, including the nature of the coordinated Lewis bases, their protonation state, and the geometry of the coordination sphere. Tuning can occur in proteins, but also in low-molecular-weight complexes [40, 68, 69]. This is discussed in more detail below (see next section).

This redox tuning is key to the catalysis and the catalytic nature is most valuable for the design of a drug candidate. A catalytic drug is a compound displaying therapeutic properties based on catalysis [70]. To date, the conventional approaches to drug design consist of seeking for small organic or inorganic molecules that bind to the active sites of proteins and are thus stoichiometric reagents. SOD mimics belong to another class of therapeutic agents displaying a catalytic activity by performing both reduction of superoxide into H_2O_2 and its oxidation into O_2 (see Fig. 7.3). Catalytic complexes have been labeled *true SOD mimics*, in contrast to stoichiometric scavengers such as nitroxides [23]. Catalysis has many advantages, including the opportunity to lower dosage, which is important in a therapeutic perspective. The superoxide steady-state concentration is low in biological systems: 20–40 pM in aerobic log phase wild type (WT) *E. coli* (hence containing endogenous SOD) and 300 pM in SOD-deficient *E. coli* mutant [14]. As superoxide is continuously produced, an efficient control of its flow requires either an efficient catalyst or a stoichiometric scavenger in large amounts and/or of continuous renewal. In the case of a redox process in biological media, apparent catalysis is achievable through redox cycling using cell reductants, as in the superoxide reductase enzymes (SOR): these Fe^{II}-native-state enzymes are oxidized to Fe^{III} by superoxide and reduced back to their active Fe^{II} state by endogenous reductants rather than by superoxide itself [71, 72]. It should be kept in mind that it is not always easy to distinguish between highly concentrated scavengers and true catalytic species in assays used to characterize anti-superoxide activity: a true catalytic SOD mimic is able to perform several turnovers and is efficient even in a large excess of superoxide [73] (see insert at the end of Chap. 7).

Tuning of the Redox Potential of a Redox Metal Couple: What Tricks?

The redox potentials of $M^{(n+1)+}/M^{n+}$ can be tuned by exploiting the coordination sphere of the metal ion. This can be achieved in redox metalloproteins by the environment offered by the apoproteins, and can be translated at the level of low-molecular-weight complexes through the selection of ligands and second coordination sphere (see Sect. 7.4.2).

The redox potential can be controlled by the nature of the coordinated Lewis bases through electronic effects. For instance, increasing the number of the S or O-donors with regard to the N-donors will lower the potential by increasing the electron density onto the metal center. This can be easily achieved in proteins, by playing with coordinating amino acid side chains that can be N-donors as in histidine or histidinate, O-donors as in aspartate, glutamate, or tyrosinate, and S-donors as in cysteine, cysteinate, or methionine. Note that amidate from the terminal position or in the proteic chain are rarer, but can also be involved in metal coordination.

The protonation state of the coordinated ligand (water/hydroxide, histidine/histidinate, etc.) can be easily modulated in proteins within a network of hydrogen bonds. The more negative the ligand, the more easily the high oxidation and positive redox metallic states will be obtained and the lower the reduction potential. As an example, the redox potential of MnSOD or FeSOD was found to depend on the pH with a decrease in redox potential when pH is increased (see Table 7.1). This was rationalized by a deprotonation of the coordinated water [60], the higher oxidation state being more stabilized by the hydroxide OH^- ligand than by H_2O. It has been clearly demonstrated that the coordinated water molecule can switch from H_2O to OH^- depending on the oxidation state of the metal ion. This is associated with a process whereby proton and electron transfer are coupled (proton-coupled electron transfer, PCET), with a tight control of an H-bond network extending to the second coordination sphere [65]. The effect of protonation-deprotonation of coordinated imidazoles was also studied in low-molecular-weight complexes, in iron porphyrins with an axial imidazole [74] or nonporphyrinic complexes [75–78] with a ΔE of about—300 mV per lost proton, which is similar to protein systems [65]. In another study, PCET was observed in a low-molecular-weight Mn-complex, with coordinated water deprotonation occurring upon oxidation of Mn^{II} to Mn^{III} [79].

The geometry imposed upon the metal ion is also another possible source of redox-control, as geometric constraints may induce entatic states [80] and stabilize unusual redox states. Imposing a geometric environment on a metal cation can be achieved by selection of ligands/apoproteins in a constrained fashion motif or embedded in a rigid matrix, which organizes the resulting coordination sphere as in a template [69]. As described in any inorganic chemistry textbook, a specific redox state and d^n configuration of a metal ion is usually associated with a particular geometry. Interestingly, such relationship can be reversed: indeed, if a ligand imposes a specific geometry on a metal cation, this can, in turn, control the reduction potential of the redox couples related to the cation. Let us take the example of the Mn^{III}/Mn^{II} redox couple, of interest here. Hexaaqua Mn^{II} is rather stable and the Mn^{III}/Mn^{II} reduction potential of fully dissociated Mn salts in water is 1.51 V vs. NHE, consistent with a stable Mn^{II} state in aqueous environment. Water molecules as ligands organize themselves around the "free" Mn^{II} ion in an octahedral environment. Mn^{II} coordinated by flexible organic ligands will display a similar behavior, with a redox potential controlled by the electrodonating properties of the coordinated Lewis bases. But, in the case of a ligand imposing a specific geometry, there can be an effect of the constrained geometry on the redox potential. This can be understood as *geometric control of the redox potential* and was described in the case of Cu^{II}/Cu^{I} constrained complexes [69] but can also apply to Mn^{III}/Mn^{II} complexes [40, 68]. Mn^{III}, as a d^4 ion with an e (e_g) singly filled orbital, will prefer a geometry with an axial distortion from the octahedral structure, to lift the degeneracy of the e orbitals which is a stabilizing process known as the Jahn–Teller distortion [81, 82]. In contrast, Mn^{II}, as a d^5 ($S=5/2$) with half-filled t_2-e set (t_{2g}-e_g), shows no ligand-field stabilization energy (LFSE) and hence no geometric preference. Finally, a ligand imposing an axial distortion will stabilize the Mn^{III} state over the Mn^{II} state [40, 68], and this corresponds to a decrease in redox potential.

7.3.2 Electrostatic Guidance

Superoxide dismutases react with superoxide at a rate close to the diffusion limit. This efficiency is most surprising considering that the metal active site represents a surface that is less than 1 % of the enzyme [83–85], and that the net charge of SOD, which shows an isoelectric point (pI) ranging from 4.6 to 6.8 depending on the lineage, is negative at physiological pH [84, 85]. Calculations of electrostatic potential in CuSOD [83, 85] have led to the idea of an electrostatic guidance of the negatively charged superoxide by a cluster of positively charged amino acids within the substrate entrance channel [85], with an invariant arginine and the charge-stable Zn^{2+} ion in close proximity to the redox active Cu^{II}/Cu^{I} metal ion. A decrease in the k_{cat} was observed when ionic strength was increased for MnSOD, FeSOD [84, 86], and CuSOD [87, 88] in agreement with an electrostatic favorable interaction between the protein and its substrate, indicative of a positively charged target for the superoxide on the protein. Acylation of lysines eliminates this salt effect [84], and a network of positively charged amino acids, namely histidines, has been identified in X-ray structures [89] and by Brownian dynamics [86]. In some sense, we can consider that there is a double topographic effect, with the negatively charged protein surface repelling the superoxide and directing it towards the positively charged funnel, which then drives it to the active site (see Fig. 7.5). The electrostatic positively charged loops in the SOD proteins are thus important for long-range guidance from the protein exterior towards the active site, and attraction of superoxide by positive charges is a more general mechanism that can also be effective in low-molecular-weight molecules, as it will be described below (see Sect. 7.4.3).

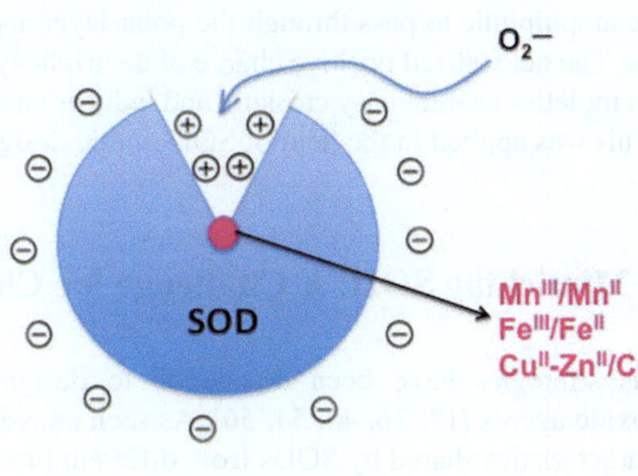

Fig. 7.5 Schematic representation of the electrostatic guidance: superoxide is repelled from the negatively charged surface towards the positively charge funnel leading to the active site

7.3.3 *Compartmentalization*

Another important feature of SODs is that they are compartmentalized: due to their size and negative net charge at physiological pH (see pI above), they are unable to cross cell membranes. MnSOD is synthesized in the cytosol, and imported into the mitochondria through an appropriate 24-amino acid targeting sequence which is then cleaved [90]. As mentioned above, mitochondrial MnSODs are key proteins in the protection of mammalian cells from oxidative stress and, indeed, MnSOD knockout is lethal to newborn mice [28, 29], whereas murine CuSOD is not crucial to survival [30]. Mitochondrial MnSOD expression is known to be altered in a wide range of diseases [20]. Mitochondria are the main cellular location at which superoxide is produced through uncoupling of the respiratory complexes involved in the reduction of dioxygen to water. The uncontrolled production of superoxide, associated with other ROS and RNS (reactive nitrogen species), induces oxidative damage leading to mitochondria dysfunction and contributes to several physiopathological processes (see a quick list in Sect. 7.2.1) or to cell death [20, 22, 24–26]. Targeting this organelle thus appears to be a key parameter to consider in the design of efficient anti-superoxide agents [24, 91–94].

The mitochondrion or the mitochondrial network is a dynamic structure with many shapes, which can take a form ranging from numerous individual capsules, most often depicted in textbooks, to a single large interconnected tubular network, as a net cast over the cell. They are key organelles: powerhouse of the cell, center of respiration and ATP production, and essential in lipid metabolism [95]. Their interior is protected by a double membrane, with a large membrane potential of up to 180 mV, negative in the interior. Smith and Murphy have set up a strategy to target small molecules to the mitochondria, taking advantage of this large membrane potential: lipophilic positively charged cations, such as triphenylphosphonium moieties [93], can be appended to various molecules in order to encourage the targeting of the mitochondria. Other groups have developed oligoguanidinium derivatives, showing a hydrophobic positively charged character as vectors to the mitochondria [96]. Cellular membranes contain negatively charged phospholipids, with polar and negatively charged groups pointing outward from the membrane and lipophilic chains inward. To cross them, a molecule must be neutral or positively charged, and must also be amphiphilic to pass through the polar layer and the hydrophobic membrane interior. The delocalized positive charge of the triphenylphosphonium or oligoguanidinium moieties enables easy crossing and induces an accumulation at the mitochondria. This was applied in the field of SOD mimic design as described in Sect. 7.4.4.

7.4 Mimicking SOD: A Challenge for Chemists

Various strategies have been developed to design efficient metal-based anti-superoxide agents [17, 26, 40, 54, 56]. As seen above, the common physicochemical characteristics shared by SODs from different lineages [1], namely, tuned redox potential, electrostatic attraction of superoxide, and mitochondria localization for MnSOD, constitute a guideline for the development of bioinspired SOD mimics. The main challenge for chemists in the field is the design of nontoxic stable complexes, with tuned redox potential and charge that would react quickly with superoxide, and show good cellular availability and possibly accumulation at the mitochondria. Of note, among the four metal ions responsible for the redox process at the active site of SODs (Cu, Fe, Mn, and Ni), manganese is most probably the least toxic [9, 16, 17], in particular in an oxidative stress context that could activate Haber–Weiss chemistry for copper and iron. This is why, although some neurotoxicity has been described [23], Mn ion is now the leading metal ion in the field of SOD mimic design for therapeutic use.

The stability and inertness of complexes are important parameters to keep in mind when designing metal-based drugs, as biological environments are rich in metal-coordinating molecules. Mn^{II} can be in fast exchange when coordinated to monodentate ligands, whereas more inert structures are obtained with Mn^{III}. The question of the speciation of manganese complexes in biological environments (see Sect. 7.2.2) is crucial to their bioactivity. Endogenous biologically available ligands have been shown to potentiate the antioxidant activity of Mn ions, most likely by modification of the Mn^{III}/Mn^{II} redox potential. This is the case with anions such as lactate or phosphate [35, 39, 41, 42]. Other ligands, such as citrate or EDTA, are thought to inhibit the activity [39, 40]. As Mn^{II} is a d^5 ion, the ligand field induces no stabilization and thus the thermodynamic constants for the association of Mn^{II} with most ligands are in a low range. Therefore, the requirement in association constants to avoid exchange with endogenous Lewis bases is less drastic for Mn^{II} than for other metal ions such as copper or iron. Polydentate ligands [40, 97, 98], particularly with a cyclic structure as in Mn^{II} cyclic polyamines (M40403) [25, 26, 57, 99] or porphyrins, stable at the Mn^{III} state [23], have been described to display association constants higher than 10^6 (or dissociation constants smaller than 10^{-6}), which should be enough considering that the archetypal bioligand, human serum albumin (HSA), display an association constant for Mn^{II} of about 8.4×10^3 (or 1.2×10^{-4} for the dissociation) [100].

The next criterion, after stability and inertness, is the kinetics of the reaction with superoxide. As stressed by Batinić-Haberle et al. [23], to be therapeutically active, a compound must be kinetically competent, which means it should react with superoxide with a kinetic constant higher than the autodismutation (see Fig. 7.3a) but, even more valuable, be a catalyst of its dismutation and, the faster, the better. There is a positive correlation between the catalytic kinetic constant for the superoxide dismutation, or $\log(k_{cat})$, and the therapeutic effects [23]. But other parameters are also of importance for in vivo efficacy. As shown in a study by Valentine et al., some derivatives can display a high intrinsic activity and be inactive in cellular models [101]. These authors assayed the ability of an assortment of SOD mimics to rescue *Escherichia coli* and *Saccharomyces cerevisiae* mutants lacking SOD. Only one of the compounds, a positively charged Mn^{III} porphyrin known to accumulate in the mitochondria [102], was found to be active in cells. The authors suggested that the absence of activity of the other derivatives was due to mislocation. In contrast, the active Mn^{III} porphyrin was able to reach the mitochondria, which is the appropriate location for an optimal rescue of these SOD-lacking strains. To be actually bioactive, a SOD mimic must reach its target, which can be the extracellular space, the cytosol or various organelles depending on the physiopathological conditions and origin of the oxidative stress. More generally speaking, the bioactivity of any kind of derivatives is in some sense the combination of its intrinsic activity, related to the redox potential in the case of SOD mimics, with the bioavailability and the cellular location [23, 103].

In the following section, we will describe (a) design of complexes directly inspired by the SOD active site; (b) strategies developed by chemists to tune redox potential to the $E_{midpoint}$ optimal value of 0.36 V vs. NHE, as one of the key parameters; (c) strategies developed by chemists to mimic efficient electrostatic attraction; and (d) strategies to enhance the bioavailability and control location within the cell. Structures discussed in this chapter are depicted in Figs. 7.8 and 7.11a, b.

7.4.1 Complexes Directly Inspired by SOD-Active Site (Mn-$N_{3/4}$O Complexes)

Policar et al. have designed a series of low-molecular-weight manganese complexes based on a tertiary amine or 1,2-diamino-ethane, directly inspired by the active site of SOD [40, 73, 97, 104–108], to reproduce the chemical environment of the Mn ion in SOD (see Figs. 7.6 and 7.8), with the initial goal of characterizing intermediates in the catalytic cycle of MnSOD. In native SOD, Mn^{III} is in a trigonal bipyramidal (TBP) geometry with an N_3O/water coordination core with one monodentate aspartate and two histidine moieties in the median triangle/equatorial plane (in green in Fig. 7.6), and a farther histidine with a water molecule (H_2O or HO^-) in the two apical positions.

The particular axial TBP geometry at the active site of SOD is important since it geometrically favors Mn^{III} over Mn^{II} and thus contributes to the tuning of the redox potential (see above) [40, 68], which must be lowered from 1.51 V vs. NHE down to *ca.* 0.36 V vs. NHE. In addition, the coordination number (CN) is lower than six, which is favorable for the direct coordination of the superoxide ion to the metal center. Indeed, superoxide is thought to coordinate to the active site of SOD as shown by experiments with the isostere and isoelectronic analog azide N_3^- [1, 110, 111]. Of note, superoxide can act either as an L-ligand or as an X-ligand leading to a non-redox coordination (L-ligand) or to an oxidative addition (X-ligand) onto the metal ion (see Fig. 7.7). Low CN complexes are difficult to prepare with low-molecular-weight ligands. In proteins, the polypeptide chain plays the role of a bulky matrix able to sterically impose low denticities. But with flexible open ligands, high-denticity structures are frequently obtained, with the coordination of a solvent molecule or Lewis bases acting as bidentate ligands possibly bridging two metal ions from two different molecular units. The corresponding complexes are sometimes said to be *coordinatively saturated*. Superoxide coordination will lead to the displacement of a ligand or

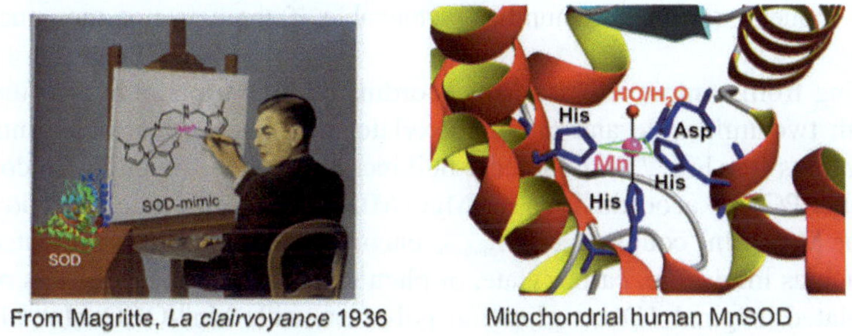

Fig. 7.6 Allegory of bioinspired chemistry: mimicking SOD from Nature—Active site of SOD, structure from [109]

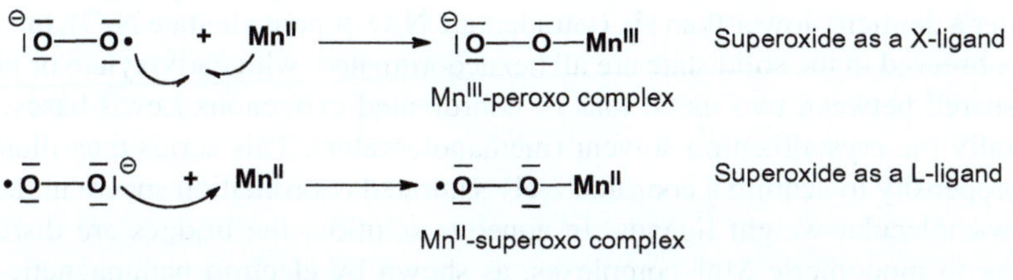

Fig. 7.7 Superoxide as a L or X ligand

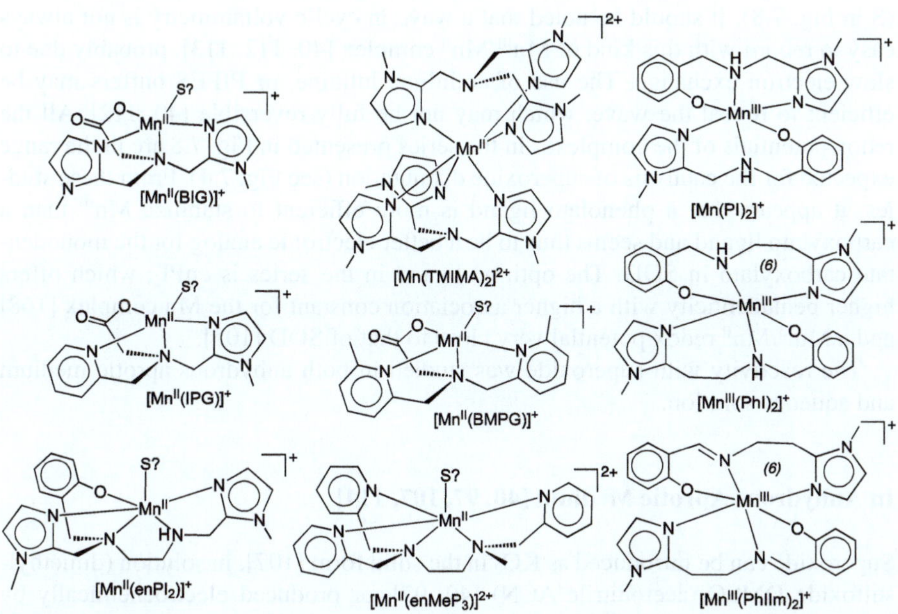

Fig. 7.8 Structure of the complexes discussed in Sect. 7.4.1: mononuclear units obtained after dissolution (see *text*). *S*? corresponds to possible solvent molecule. See: [40, 73, 97, 104–108]. *(6)* indicates a 6-membered metallacycle whereas all the others are 5-membered rings

to even higher coordination number, achievable if there is not too much steric crowding.

Starting from complexes with Mn^{II} coordinated by a tripodal amine functionalized with two imidazole and one carboxylato moieties, Policar et al. introduced modifications (see Fig. 7.8) to increase bulkiness around the metal ion (picolyl moieties in BMPG, hexacoordination in $[Mn(TMIMA)_2]^{2+}$) or to tune the redox potential of the Mn^{III}/Mn^{II} couple to the $E_{midpoint}$ encountered in SODs (coordinated Lewis bases such as imidazole, carboxylate, or phenolato). In the course of this research, they isolated original 1D coordination polymers with Mn^{II}-OCO-Mn^{II} along the polymeric chain, with bridging carboxylatos in an unusual *syn-anti* coordination mode [104]; dimeric structures with a bis-μ-carboxylato bridge [97] or a bis-μ-phenolato bridge [105]. Some other complexes were isolated in the solid state as monomers with two tridentate ligands [40, 106]. Although the ligands in the series all offer a denticity lower than six (tetradentate N_3O or pentadentate N_4O), the structures observed in the solid state are all hexacoordinated, with carboxylato or phenolato shared between two metal ions or coordinated exogenous Lewis bases, most generally the crystallization solvent (methanol, water). This series thus illustrates this propensity to acquire a coordinatively saturated coordination sphere in the case of low-molecular-weight ligands. In aqueous solution, the bridges are disrupted, leading to monomeric Mn^{II} complexes, as shown by electron paramagnetic resonance—characteristic 6-line signal for the Mn^{II} $I=5/2$ $S=5/2$ complexes [40, 97, 105]. Most probably, the coordination sphere is saturated with solvent molecule(s) (S in Fig. 7.8). It should be noted that a wave in cyclic voltammetry is not always easy to record with this kind of Mn^{III}/Mn^{II} complex [40, 112, 113], probably due to slow electron exchange. The use of collidine, lutidine, or PIPES buffers may be efficient to reveal the wave, which may not be fully reversible [40, 112]. All the redox potentials of the complexes in the series presented in Fig. 7.8 are in the range expected for the catalysis of superoxide dismutation (see Fig. 7.4). From these studies, it appears that a phenolato ligand is more efficient to stabilize Mn^{III} than a carboxylato ligand and seems thus to be a better electronic analog for the monodentate carboxylato in SOD. The optimal ligand in the series is $enPI_2$, which offers higher pentadenticity with a higher association constant for the Mn complex [168] and a Mn^{III}/Mn^{II} redox potential very close to that of SOD [105].

The reactivity with superoxide was studied in both anhydrous aprotic medium and aqueous solution.

In Anhydrous Aprotic Medium [40, 97, 107, 114]

Superoxide can be introduced as KO_2 in the solid form [107], in solution (dimethylsulfoxide DMSO, acetonitrile ACN) [40, 97], or produced electrochemically by reduction of dissolved O_2 [97, 114]. In anhydrous aprotic medium, where no protons are available for the reduction of superoxide into hydrogen peroxide (see Fig. 7.3), H_2O_2 cannot be released and intermediate adducts with the Mn complexes can be isolated: by reacting KO_2 in DMSO or ACN with Mn^{II}-complexes, MnOO adducts were characterized by low-temperature spectroscopies, including UV-visible spectroscopy and electron paramagnetic resonance (EPR). These adducts displayed a very nice blue color (see Fig. 7.9). For the first time, using parallel mode EPR, an Mn^{III} redox state ($S=2$) was unambiguously characterized, within a Mn-coordinated peroxo species [107]. Clearly, the reaction of Mn^{II} with superoxide in these complexes can be rationalized as an oxidative addition onto the Mn^{II} ion with superoxide being an X-ligand (see Fig. 7.7).

Similar Mn^{III}-peroxo adducts are expected when reacting Mn^{III} with H_2O_2, which is redox-equivalent to $Mn^{II}+O_2^{-}$. Other research groups have also characterized similar MnOO intermediates by reacting H_2O_2 with Mn^{II} at low temperature, which is clearly not redox-equivalent to $Mn^{II}+O_2^{\bullet-}$ or $Mn^{III}+H_2O_2$, but there is always an excess in H_2O_2 [115–119] and most probably the first step is a slow oxidation into Mn^{III}, which then reacts with H_2O_2. This kind of adduct can be labeled $\{MnOO\}^6$ [73] by analogy with the Enemark-Feltham nomenclature for Fe-nitrosyl [120]. These evolve at ambient temperature leading to Mn^{III}-Mn^{IV} di-μ-oxo derivatives

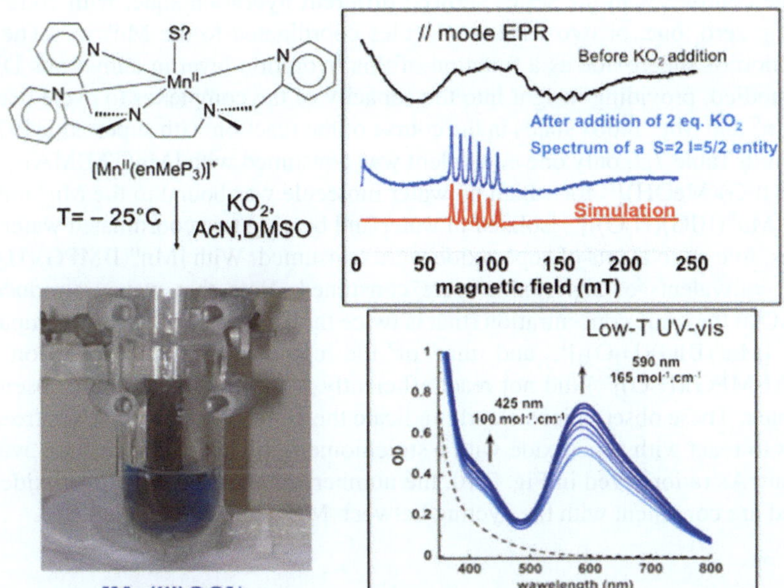

Fig. 7.9 Reaction in anhydrous medium of KO_2 with a Mn^{II}-tris-pyridyl-bis-amino complex. On the *left*: Photography showing the blue color of the Mn-OO adduct at –25 °C. Set-up to record low-temperature UV-vis spectra, with an immersion UV-vis probe (in *black*) within a double-jacket low-temperature cell. On the *right, top*: parallel mode EPR spectra showing that a Mn^{III} species was obtained. Recording conditions: microwave power 2 mW; modulation amplitude 5 G; modulation frequency 100 kHz; time constant 40.96 ms; $T=5$ K; parallel mode: $\upsilon=9.417$ GHz. On the *right, bottom*: UV-vis spectra recorded using the immersion probe (photo on the *left*) (**a**) *black dash line*: spectrum of the Mn^{II}-tris-pyridyl-bis-amino complex (**b**) *blue plain lines*: spectra of recorded at different times during the formation of the MnOO intermediate (*arrows* indicate growing absorption with time). Adapted from [107]

Table 7.2 Superoxide $O_2^{\bullet-}$ consumed by reaction of KO_2 with Mn-complexes in anhydrous DMSO (eq./complex). See [97]

	$[Mn^{II}(TMIMA)_2]^{2+}$	$[Mn^{II}(IPG)(MeOH)]^{2+}$	$[Mn^{II}(BMPG)(H_2O)]^{2+}$	$[Mn^{II}(BIG)(H_2O)_2]^{2+}$	H_2O
Bound H_2O	0	0	1	2	–
Consumed O_2^-	1 eq.	1 eq.	3 eq.	5 eq.	0 eq.

that were characterized by their UV-vis and EPR characteristic 16-lines spectra [97]. The pathway leading to Mn^{III}-Mn^{IV}-di-μ-oxo structures is a dead-end for the catalysis of the superoxide dismutation and must be avoided for efficient SOD mimics with a high turnover. This dimerization into $Mn^{III}Mn^{IV}$ derivatives can be sterically precluded and indeed, in the case of coordinatively saturated Mn^{II} complexes—hexacoordination in $[Mn(TMIMA)_2]^{2+}$—or bulky ligands—picolyl moieties in BMPG—no dimer Mn^{III}–Mn^{IV}-di-μ-oxo was obtained upon reaction with superoxide [40].

The complexes in the series showed different hydration state, with complexes bearing zero, one, or two water molecules coordinated to the Mn^{II} ion. The consumption of superoxide as a function of this hydration level in anhydrous DMSO was studied, providing insight into the capacity of the complexes to cycle between the Mn^{II} and Mn^{III} redox states in the course of the reaction with superoxide [97]. As shown in Table 7.2, only one equivalent was consumed with $[Mn(TMIMA)_2]^{2+}$ and $[Mn^{II}(IPG)(MeOH)]^{2+}$ for which no water molecule was bound to the Mn^{II} ion. But with $[Mn^{II}(BIG)(H_2O)_2]^+$, isolated in water and bearing two coordinated water molecules, five equivalents of superoxide were consumed. With $[Mn^{II}(BMPG)(H_2O)]^{2+}$ three equivalents of superoxide were consumed. Note that water introduced in DMSO at the same concentration (that is twice that of the complex for a comparison with $[Mn^{II}(BIG)(H_2O)_2]^+$, and that of the complex for a comparison with $[Mn^{II}(BMPG)(H_2O)]^{2+}$) did not react efficiently with superoxide in the absence of complex. These observations clearly indicate the ability of the complexes from this series to react with superoxide with a stoichiometry higher than one when water is present. As rationalized in Fig. 7.10, the numbers of equivalent of superoxide consumed are consistent with the cycling between Mn^{II} and Mn^{III}.

In Water [40, 73, 97, 105, 106]

The reactivity with superoxide was studied in water using the indirect assay developed by McCord and Fridovich (see insert), which produces a slow and continuous flow of superoxide reminiscent of what is found in biological environments. All the complexes were found to be active, with kinetic constant k_{McCF} (see insert) of the order of 10^6–10^7 mol^{-1} L s^{-1}. The study in anhydrous conditions with different hydration levels suggested the ability of the complexes to react with superoxide with several turnovers. But, to ascertain their ability to be true SOD mimics and to catalytically dismutate superoxide, which is not straightforward with the Fridovich

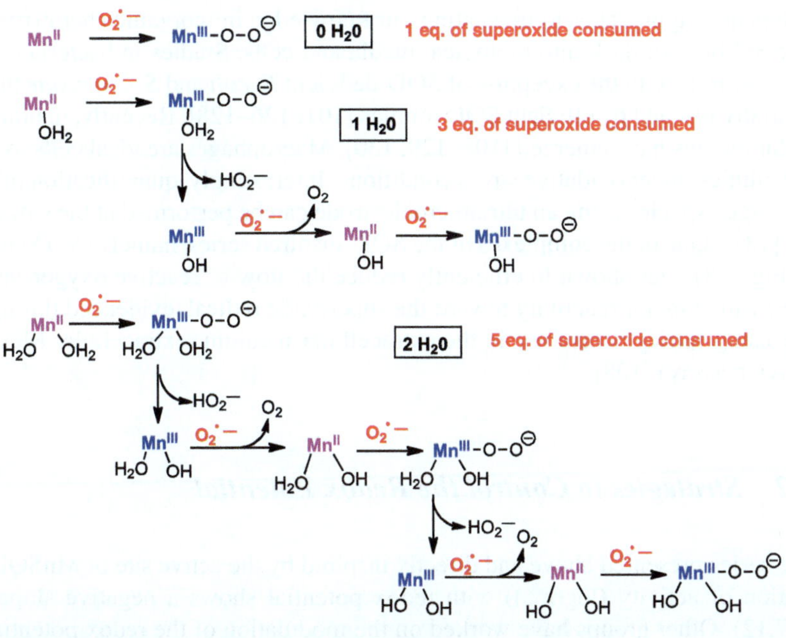

Fig. 7.10 Schematic rationale for the number of superoxide equivalents consumed in anhydrous DMSO upon the number of bound water molecules. See [97]

assay (see insert), reactivity was studied using pulsed radiolysis [40, 73]. A pulse of high concentration of superoxide can be produced in oxygenated aqueous solution in the presence of alcohol [121] upon irradiation. Superoxide disappearance in the presence of the SOD mimic at different concentrations can be followed at 270 nm ($\varepsilon = 1479$ L mol^{-1} cm^{-1}) [121–124]. Another way to investigate reactivity in large excess of superoxide is to use stopped-flow fast kinetic measurements with a highly concentrated solution of KO_2 in DMSO and rapid mixing in water [125]. Kinetic constants consistent with that from the McCord and Fridovich assay were obtained by pulsed radiolysis [40, 73] and more recently by stopped-flow [168].

Within this series, a correlation could be found between the kinetic constants derived from the McCord–Fridovich assay and the half-wave Mn^{III}/Mn^{II} potentials, with a negative slope (see Fig. 7.12, Δ markers). Within this series, as $E_{½}(Mn^{III}/Mn^{II}) > E_{midpoint}$, the SOD activity can be improved by lowering $E_{½}$ to $E_{midpoint}$, and hence by a stabilization of the Mn^{III} state [$E_{½}$ are defined as $(E_{anodic\ peak} + E_{cathodic\ peak})/2$]. Increasing the stabilization by imposing a Jahn–Teller distortion through a higher size metallacycle was efficient and led to the isolation of stable Mn^{III} complexes but poorer catalysts, probably because of an excessive Mn^{III} stabilization ($[Mn^{III}(PhI)_2]^+$ and $[Mn^{III}(PhIIm)_2]^+$, with (6) indicating a 6-membered metallacycle, see Fig. 7.8) [106].

These studies on the reactivity of these true SOD mimics were performed outside any cellular context and correspond to the intrinsic activity previously defined (see Sect. 7.2.4). As already mentioned, they are necessary, but not sufficient to ascertain bioactivity. Understanding and controlling inorganic compounds inside cells requires new approaches to translating our knowledge in inorganic chemistry from the round-bottom flask into biological media and cells. Studies in bacteria or cells are still scarce, with the exception of SOD-deficient *E. coli* and *S. cerevisiae* that are potentially rescued by efficient SOD mimics [101, 126–128]. Recently, mammalian cellular models have emerged [108, 129, 130]. Macrophages are ideal cells to study SOD mimics under oxidative stress conditions. Interestingly, quantification of reactive oxygen species using an ultramicroelectrode can be performed at the single cell level [131]. One of the complexes of the SOD-inspired series, namely $[Mn^{II}(enPI^2)]^+$ (see Fig. 7.8), was shown to efficiently reduce the flow of reactive oxygen species (ROS) with a major reactivity toward the superoxide radical, evidenced through an assay using ferricytochrome *c* in the extracellular medium (extracellular McCord–Fridovich assay) [108].

7.4.2 Strategies to Control the Redox Potential

In the series presented above and directly inspired by the active site of MnSOD, the variation of activity ($\log(k_{cat})$) with redox potential shows a negative slope (see Fig. 7.12). Other groups have worked on the modulation of the redox potentials of Mn^{III}/Mn^{II} redox couple. Walton et al. have developed a series of Mn^{III} complexes based on a 1,3,5-triaminocyclohexane central scaffold, with three amino moieties in *cis* configuration that were grafted with *para*-substituted phenols (see Fig 7.11b). Mn^{III} complexes were isolated with octahedral Mn coordinated to three amine and three phenolato moieties [132]. The redox potentials, measured in DMF, were shown to vary over a volt with *para*-substituents such as methoxy, methyl, chloro, nitro [132], and tri-methylammonium [133]. The complexes were reported to display an SOD-like activity [133].

A thorough study regarding the effects of the modulation of the redox potential variation was performed by the group of Batinić-Haberle et al. For more than 15 years, they have been developing a large family of porphyrins, stable in aerobic conditions at the Mn^{III} state. They show a redox potential smaller than the redox mid-point optimal for superoxide dismutation, $E_{midpoint}$. In this case, increasing the redox potential goes with an increase in activity (see Fig. 7.12). Metalloporphyrins are interesting molecules for a chemist aiming to mimic biological structures. First, porphyrins are valuable and versatile structural scaffolds, with a large range of possible modifications at the *meso*- or ß-positions. In porphyrins bearing bulky substituents—*ortho*-functionalized phenyl, 2-*N*-alkylpyridyl, or *N*-alkyl-2-imidazolyl on the *meso*- position, see Fig. 7.11a—the substituents are perpendicular to the average porphyrin plane, with atropisomerism and the possibility of constructing a molecular scaffold above and below the porphyrin. This atropisomery has been used since the mid-1970s to develop superstructured porphyrins as mimics of hemoproteins [134–137]. In addition, the 3D-structure of these functionalized porphyrins renders them less prone to interaction with DNA [126], known to occur with more planar porphyrins, such as the *para*-substituted-pyridinium (see $Mn^{III}TM\text{-}4\text{-}PyP^{5+}$ in

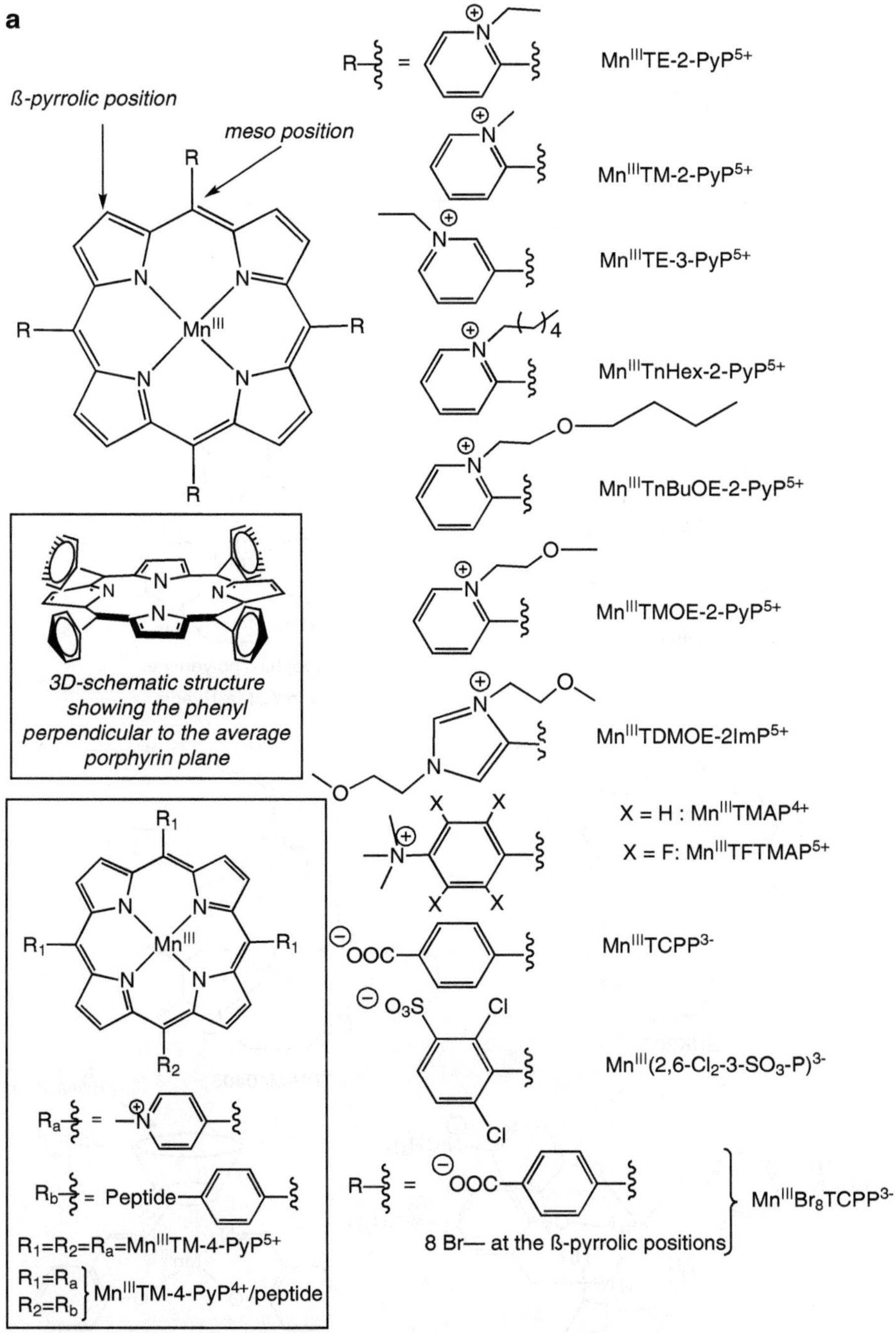

Fig. 7.11 (**a**) Structures of the Mn SOD mimics discussed in this chapter: structures of the Mn^{III} porphyrins. See: [23, 102, 126, 139, 141–144] and for Mn^{III}TM-4-PyP^{4+}/peptide see [126, 130, 142]. (**b**) Structure of the Mn SOD mimics discussed in this chapter: M40403 [99, 140], M40401, M40404 [140]; Mn^{II}CnMe$_2$Pyane [145]; EUK207 [23]; Mn^{III}*cis*-TACH-N3O3 [133]; Mn^{II}(ABCDSA) [146]. Note that some Cl^- ligands have been seen in the solid state, but they are not shown here as there is most probably dissociation in water medium

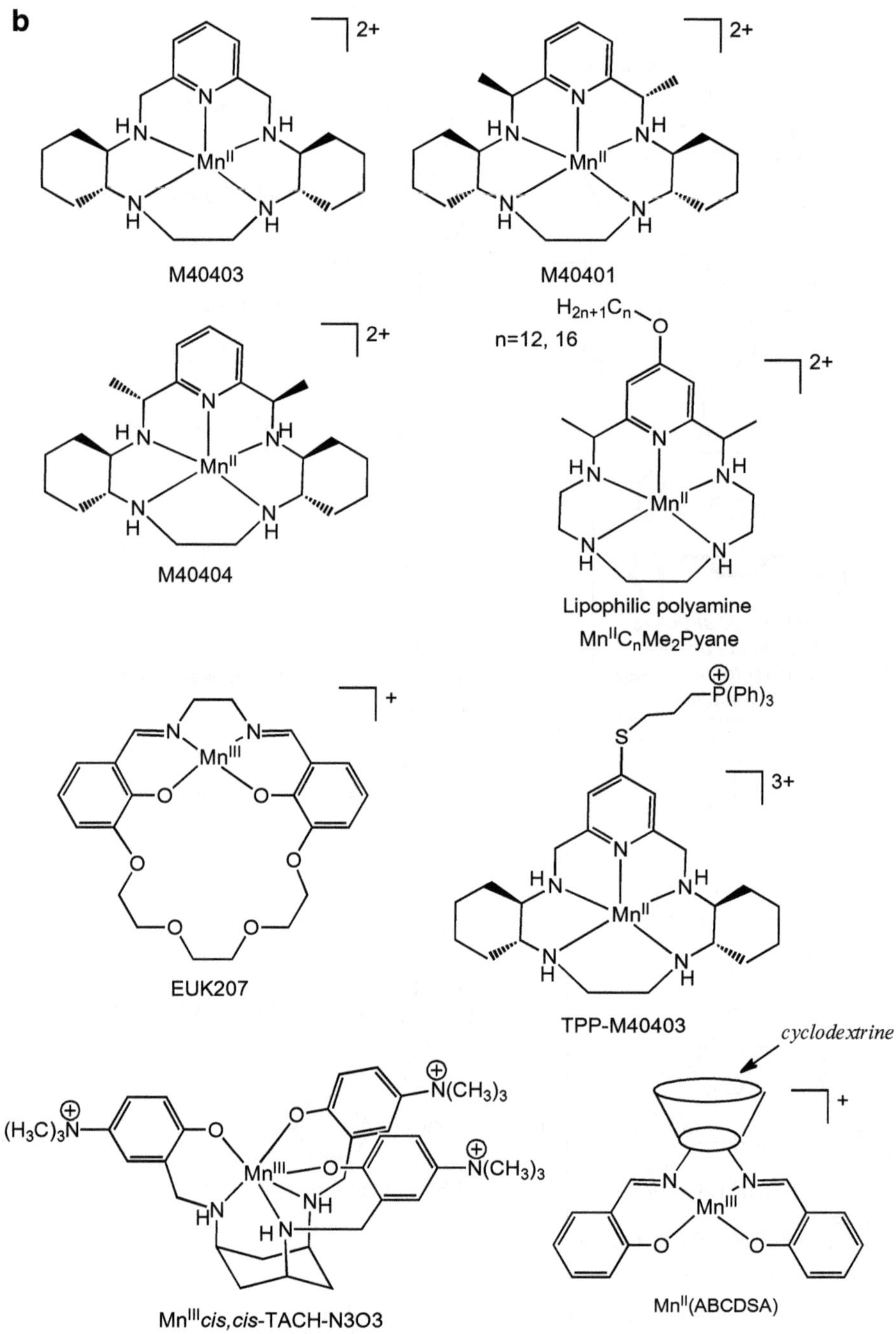

Fig. 7.11 (continued)

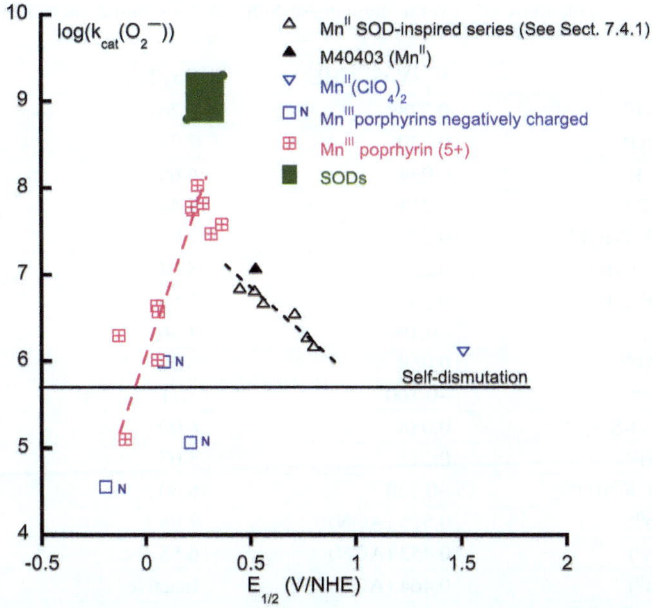

Fig. 7.12 Redox potentials (in water) and log$k_{cat}(O_2^-)$ for a series of Mn^{III}/Mn^{II} SOD mimics. *Triangles, up-ward*: Mn^{II} complexes directly inspired from SOD active site (see Sect. 7.4.1); *triangle, black*: M40403 (Mn^{II}); *triangle blue, down-ward*: $Mn^{II}(ClO_4)_2$ (Mn^{II}-hexa-aqua); *open-squares, purple N*: negatively charged porphyrins; *square with a cross, pink*: Mn^{III} porphyrins, 5+ charge. See Table 7.3 for the data and references. Inspired from [23, 40, 126, 139]

Fig. 7.11a, [138]). Porphyrins are efficient ligands, with high association constants and inertness for the metal complexes, although Mn has been shown to be lost in some rare cases for the reduced Mn^{II} form [23]. Electronic effects are easily propagated in these aromatic structures and redox potentials of the coordinated metal ions can be varied by functionalizing the porphyrin moiety with electroactive groups. Batinić-Haberle et al. prepared a wide series of Mn^{III} porphyrins showed to be true SOD mimics [23], using pulse radiolysis for instance [139]. Modulation of the redox potential over a wide range (800 mV) was achieved by functionalization with electron withdrawing groups (EWG) in *meso* or ß-positions, namely *N*-alkylpyridyl (PyP) or bis-*N*-alkylimidazolyl (ImP) groups (see Fig. 7.11a).

EWGs increase the electron deficiency on the metal center, which goes against oxidation into Mn^{III}. This favors low oxidation state and induces an increase in redox potential: the potential moves from negative values into the region of optimal $E_{½}$, i.e., to a redox potential which is closer to the $E_{midpoint}$. As expected, this variation was associated with an increase in activity, as shown in Fig. 7.12. Note that electrostatic effects are also important determinants of the activity and they will be discussed in Sect. 7.4.3.

Figure 7.12 shows dependence of log(k_{cat}) as a function of $E_{½}$ for a series of SOD mimics based on Mn^{III}/Mn^{II} couples (see Table 7.3 for the data and references).

Table 7.3 Redox potentials of several manganese-SOD mimics based on a Mn^{III}/Mn^{II} redox couple

SOD mimics	$E_{½}$ (V vs. NHE)	Log k_{cat}	References
Mn^{III}TE-2-PyP^{5+}	0.228	7.76	[126]
Mn^{III}TM-2-PyP^{5}	0.220	7.79	[126]
Mn^{III}TE-3-PyP^{5+}	0.054	6.65	[23]
Mn^{III}TnHex-2-PyP^{3+}	0.314	7.48	[23]
Mn^{III}TnBuOE-2-PyP^{5+}	0.277	7.83	[23]
Mn^{III}TMOE-2PyP^{5+}	0.251	8.04	[139]
Mn^{III}TDMOE-2-ImP^{5+}	0.365	7.59	[139]
Mn^{III}TCPP^{3-}	−0.194	4.56	[126]
Mn^{III}TFTMAP^{5+}	0.058	6.02	[126]
Mn^{III}TTMAP^{5+}	−0.100	5.11	[126]
$Mn^{III}(2,6\text{-}Cl_2\text{-}3\text{-}SO_3\text{-}P)^{3-}$	0.088	6.00	[126]
Mn^{III}Br8TCPP^{3-}	0.213	5.07	[144]
Mn^{III}salen(EUK207)$^{+}$	−0.130	6.30	[23]
M40403 (Mn^{II})	0.525 (ACN)	7.08	[140]
M40404 (Mn^{II})	0.452 (ACN)	6.55	[140]
M40401 (Mn^{II})	0.464 (ACN)	Inactive	[140]
Mn^{III}*cis*-TACH-N3O3	0.31	5.95	[133]
Mn^{II}(BIG)$^{+}$	0.80	6.18	[40]
Mn^{II}(IPG)$^{+}$	0.77	6.28	[40]
$Mn^{II}(TMIMA)_2^{2+}$	0.71	6.55	[40]
Mn^{II}(BMPG)$^{+}$	0.56	6.68	[40]
$Mn^{III}(PI)_2^{+}$	0.52	6.82	[40]
$Mn^{II}(enPI_2)^{+}$	0.45	6.85	[105]
Self-dismutation	–	5.7	[23]
SOD	0.2–0.40	8.8–9.3	See Table 7.1
$Mn^{II}(ClO_4)_2$	1.51	6.11	[23, 40]

$E_{½} = (E_{anodic\ peak} + E_{cathodic\ peak})/2$

Mn^{III}/Mn^{II} SOD mimics can be classified into two subgroups, those displaying a redox potential smaller than $E_{midpoint}$ (0.36 V vs. NHE), which is the case of most Mn^{III} porphyrins and for which an increase in potential is favorable to the SOD activity, and those with a redox potential higher than $E_{midpoint}$, which is the case of the Mn^{II}-N_3O complexes previously described. Interestingly, for Mn^{III} porphyrins, the correlation $\log(k_{cat}) = f(E_{½})$ was shown to reflect the Marcus equation for outer-sphere electron exchanges [126, 139] with a tenfold increase for each 120 mV increase in $E_{½}$, suggesting an outer-sphere electron transfer.

In the case of the series of the cyclic polyamines series, it was found that the redox potentials of the Mn^{III}/Mn^{II} couples determined in ACN do not correlate with the activity (see M4040x in Fig. 7.11b and Table 7.3), suggesting a mechanism that does not involve SOD-like dismutation, but does involve the interaction superoxide with a Mn^{III}-peroxo intermediate [140].

Although the compounds presented in Fig. 7.12 are of quite different molecular structures (see Fig. 7.8 and 7.11a, b), with porphyrinic and 1,2-amino-based ligands, they fall on the same bell curve with a maximum activity recorded for compounds with a redox potential close to that of SOD, as expected. The correlation also extends to corroles or billiverdins through their Mn^{IV}/Mn^{III} redox potential, Fe^{III}/Fe^{II} porphyrins, stoichiometric scavengers such as nitroxides oxidized into oxo-ammonium, and Ce^{IV}/Ce^{III} in nanoparticles of cerium oxide (CeO_2) [23, 126, 128, 138, 139, 141]. These observations strengthen, if necessary, the idea that activity can be optimized by tuning the redox potential.

7.4.3 Electrostatic Attraction and Other Effects Modulating the Intrinsic Activity

All the pairs ($E_{½}$, $\log(k_{cat})$) do not fall exactly on the correlation in Fig. 7.12. The scatter reflects factors other than the thermodynamic $E_{1/2}$ impacting the reactivity, such as electrostatic effects, shape, or local dipolar effects [23, 142, 144]. Electrostatic effects associated with superoxide attraction and interplay with the lipophilicity inducing solvation/desolvation effects come into play to modulate the redox potential [139, 144].

An exception in the previous correlation is seen in Fig. 7.12 with the negatively charged Mn^{III} porphyrins. For instance, the porphyrin $[Mn^{III}Br_8TCPP]^{3-}$ shows a redox potential for the Mn^{III}/Mn^{II} redox couple very close to that of $[Mn^{III}TE\text{-}2\text{-}PyP]^{5+}$ but with a kinetic constant 100-fold smaller. This was rationalized by an impaired interaction between the negatively charged porphyrin and the negatively charged superoxide [142].

In the case of the porphyrins series developed by Batinić-Haberle et al., the quaternarization of the pyridyl and imidazolyl in the ortho position, leading to positively charged pyridinium or imidazolium moieties conjugated with alkyl or polyether chains, was proven to be quite a valuable strategy with cumulative favorable effects [23]. In this series of positively charged Mn^{III} porphyrins with *N*-alkyl in *ortho*-position and for each individual Mn^{III} porphyrin, a decrease in k_{cat} upon increase in ionic strength (I) was observed with linear correlation between $\log(k_{cat})$ and $\frac{\sqrt{I}}{1+\sqrt{I}}$ [139, 142, 147]. This observation, similar to that observed for proteins themselves, clearly reveals a favorable electrostatic interaction between the negatively charged superoxide and the positive Mn^{III} porphyrins catalysts that is shielded when I increases. A very weak dependence of k_{cat} on I was recorded in the case of *para* and *meta*-substituted porphyrin [142]. This brings out the role of the charge topography: with *ortho*-alkylpyridinium ($Mn^{III}TE\text{-}2PyP^{5+}$), the charge is distributed over a donut above the metal center mimicking closely the topography of the funnel in the proteins. In contrast, with the *para*-alkylpyridinium ($Mn^{III}TM\text{-}4PyP^{5+}$), attraction is that of an overall charge distributed at the far periphery of the porphyrin, which was proven to be less efficient and a much weaker shielding is observed with increase in ionic strength [142]. It appears thus that it is necessary, in this biomimetic or bioinspired approach, to reproduce not only the charges but also their topology, with attraction along a funnel [147]. Recently, the SOD bio-inspired series (see Sect. 7.4.1) was conjugated to oligo-arginine moieties, with 1, 3, 6 and 9 Arg. Analyses of the dependence of the k_{cat} with ionic strenght showed that only the two or three first arginines were effective in the enhancement of k_{cat} associated with an electrostatic effect. This was rationalized by the fact that the long peptide chain was most probably topologically disordered [168]. These examples are reminiscent of the double topographic effect suggested above in the case of SODs: their overall negative charge repels the superoxide ion towards the attractive positively charged funnel (see Sect. 7.3.2 and Fig. 7.5).

Some interesting local effects have also been observed with porphyrins functionalized with alkyl and ether chains. More lipophilic porphyrins were found less sensitive to changes in ionic strength, revealing the importance of local dielectric environment [139]. This is an important consideration also for proteins, for which dielectric constants are not uniform but modulated by the amino acid side chains nearby. In addition, porphyrins functionalized with ether groups were found to display higher kinetic constants (see Table 7.3) with a stronger dependence on ionic strength, probably associated with a higher solvation [139]. As observed in the 1980s in the course of mimicking myoglobin with iron porphyrins [135], ether groups, able to accept hydrogen bonds, either from Fe-hydroperoxo adduct or from water, create a local hydration [139] that can play a role in superoxide dismutation especially in the reduction reaction, for which protons are necessary.

7.4.4 Bioavailability and Localization

Bioavailability and cellular distribution are also strong determinants of the bioactivity of the SOD mimics [23]. Water insolubility can be an issue preventing bioactivity from being revealed, and strategies have been developed to improve water solubility. In parallel, the complexes need to efficiently penetrate into cells, which requires amphiphilic properties.

Several groups have developed strategies to conjugate SOD mimics with moieties favoring water solubility. A Mn^{III} salen complex was built onto a cyclodextrin (CD) scaffold to efficiently improve the solubility and to take advantage of the CD $HO^{\bullet}$-scavenging properties (MnABCDSA, see Fig. 7.11b, [146]). Functionalization of porphyrins by negatively charged moieties, such as sulfonato in $Mn^{III}(2,6\text{-}Cl_2\text{-}3\text{-}SO_3\text{-}P)^{3-}$, or carboxylato in $Mn^{III}TCCP^{3-}$, meant to provide good water solubility, is detrimental for electrostatic attraction of superoxide, as discussed above [144]. In contrast, quaternarization of the pyridyl or imidazolyl moieties leads to efficient water-soluble positively charged SOD mimics, with redox potential tuned by the strong electrowithdrawing properties of pyridinium and imidazolium groups [23]. Increasing the amphiphilicity by long alkyl chains was shown to be favorable for the bioavailability and therapeutic efficacy, but some toxicity associated with detergent effect was observed at concentrations lower than that of lower-alkyl-chain analogs [128]. Interestingly, substituting the long alkyl chains with ether chains led to an increased intrinsic activity (k_{cat}), probably due to a solvation cavity and local polarity with an efficiency to restore normal growth in SOD-deficient yeast [143].

In a similar approach aimed at modulating lipophilicity, Mn^{II}-cyclic polyamine complexes were conjugated with long alkyl chains in $Mn^{II}C_nMe_2$Pyane [145]. The authors showed a tendency of the resulting compounds to form aggregates or micelles. Despite a lower thermodynamic stability and a lower reactivity toward superoxide, as determined by stopped-flow measurements, in comparison with the unsubstituted analog, cell studies revealed a beneficial effect on preventing lipid peroxidation at low concentration of the lipophilic complex [145].

Para-substituted pyridinium porphyrins (Mn^{III}TM-4-PyP^{5+}) were also associated with a catalase-polyethyleneglycol (cat-PEG) moiety [148, 149] with the double objective of enhancing the Mn^{III} porphyrin half-life in blood circulation and conjugating SOD with catalase activity for antioxidative stress applications in the plasma, such as ischemia-reperfusion injuries. In these physiopathological conditions, ROS are produced when blood circulation is re-established again in a de-perfused tissue, causing oxidative damages to proteins and lipid membranes.

Cellular recognition is also a key parameter of bioactivity. M40403 was conjugated to a galactose moiety [150] to encourage interaction with lectins, proteins that play an important role in recognition processes at the cell-membrane surface. Porphyrins functionalized with lactose were also developed with the same goal [151].

Intracellular targeting to mitochondria can be favorable for bioactivity, since superoxide potential overproduction occurs primarily in mitochondria. Asayama et al. have developed a porphyrin functionalized with a mitochondria-targeting peptide (Mn^{III}TM-4-PyP-peptide^{n+}), which showed a sixfold decrease in k_{cat} in comparison with the unsubstituted porphyrin (1.8×10^6 and 11×10^6 mol^{-1} L s^{-1}, respectively, measured by stopped-flow technique), but with mitochondrial accumulation. Protection against lipopolysaccharide toxicity on murine macrophages was more efficient than with the non-conjugated analog [130].

More generally, Mn^{III} porphyrins developed by Batinić-Haberle et al. are positively charged, exhibiting hydrophobic to amphiphilic characteristics and spontaneous accumulation at mitochondria [102], which was suggested as the reason for their efficiency in restoring growth in SOD-deficient *E. coli* and yeast mutants (see Sect. 7.4) [101].

Smith and Murphy have functionalized antioxidants like vitamin E with triphenylphosphonium moieties showing strong accumulation at mitochondria [93]. Recently, they have functionalized M40403 (TPP-M40403) with the same moiety [152]. The resulting derivative showed an intrinsic activity tenfold higher than the non-conjugated M40403, a 3000-fold accumulation in isolated mitochondria with an efficiency to protect mitochondrial aconitase from inactivation upon paraquat stress [152].

The subcellular distribution of MnSOD mimics are actually key to the antioxidant activity and new techniques are emerging to better characterize them [153]. These are important directions for future research in this field.

7.5 Conclusion

Superoxide dismutases are proteins that have been gradually carved by evolution to perform very effective catalysis of the superoxide dismutation aimed at protecting cells against damages from oxidative stress associated with superoxide. On the one hand, understanding the determinants for efficient activity at the molecular level is an important fundamental challenge. Tuning of redox potential, metal selectivity (not discussed here, see [1, 65]), and electrostatic attraction are crucial characteristics and mechanisms involved in SODs that have been analyzed along the years. On the other hand, in a bioinspired approach, chemists have taken advantage of what has been understood from these mechanisms to convey and apply them to the design of efficient catalysts for superoxide dismutation, with potential therapeutic perspective. This

Intrinsic Activity of SOD Mimics: Assays for SOD Activity in Water
The evaluation of the reaction kinetics of superoxide with a putative SOD mimic can be performed using direct or indirect tests. In the indirect methods, superoxide is provided at a constant rate and at concentrations close to what is encountered in biological systems under oxidative stress, providing a test for the ability of the putative SOD mimic to be useful in physiological conditions. However, in some cases it is difficult to distinguish between a catalyst and a scavenger (see below). In direct methods, since superoxide can be provided in large excess in comparison to the putative SOD mimic, unambiguous characterization of a catalyst versus a scavenger can be obtained.

Indirect assay: McCord and Fridovich test
The paradigmatic indirect assay is the McCord Fridovich test, which was originally set up to establish the activity of SOD [51]. In indirect assays, a continuous flow of superoxide is produced, usually by an enzymatic system, as in the McCord–Fridovich test: xanthine-oxidase oxidizes xanthine to urate, with specific conditions (pH, O_2) chosen in order to maximize the re-oxidation of the reduced enzyme with the release of superoxide [154]. The assay is based on a kinetic competition between the putative SOD mimic or SOD itself and a redox indicator that changes in color upon reacting with superoxide. The UV-vis. indicator is necessary since the flow in superoxide (1.2 μM min^{-1} [155]) is too low and the absorption coefficient of superoxide too weak to enable a direct spectrophotometric detection. The most frequently used indicators are cytochrome *c* Fe^{III} reduced in cytochrome *c* Fe^{II}, or nitro-blue tetrazolium (NBT)

reduced into formazan. Because NBT is not very soluble new tetrazolium salts leading to more soluble formazan forms have been developed [156]. A 50 % inhibition concentration (*IC50*) is determined, which corresponds to the concentration in the SOD mimic or SOD that reduces by 50 % the speed of the reduction of the indicator. *IC*50 values are dependent both on the nature of the indicator used and on its concentration: for a given putative SOD mimic, the smaller the detector concentration, the smaller the *IC*50 value. An important consequence is that *IC*50 values are not appropriate for comparisons across the literature. But from the measured *IC*50 values, an apparent kinetic constant value (k_{McCF}) can be calculated, which is independent of both the concentration and the nature of the detector [40].

At the *IC*50 concentration, superoxide reacts at the same speed with the detector and the putative SOD mimic:

$$k_{McCF} \cdot IC50 \cdot \left[O_2^{\bullet-}\right] = k_{detector} \cdot [\text{detector}] \cdot \left[O_2^{\bullet-}\right]$$

Then, $k_{McCF} = k_{detector} \cdot [\text{detector}]/IC50$ [40, 157].

In the case of cytochrome *c* Fe^{III} as the detector, $k_{Cyt\ c}$ (pH = 7.8; 21 °C) = 2.6×10^5 mol^{-1} L s^{-1} [158]. In the case of NBT, k_{NBT} (pH = 7.8) = 5.94 $\times 10^4$ mol^{-1} L s^{-1} [159].

Reliability of the McCord–Fridovich Assay [40, 160, 161]:
It is necessary to check that the putative SOD mimic does not inhibit the production of superoxide by xanthine oxidase. This can be performed by the determination of the rate of conversion of xanthine to urate. The putative SOD mimic should not react with ferri or ferrocytochrome *c*. Ideally, this should be checked for both redox states of the SOD mimic, but is not always possible.

A second important point to consider is the steady state in superoxide in this assay. Under canonical conditions, the test produces about 1.2 μM min^{-1} [155] and lasts about 10 min [162]. During this 10-min period, 12 μM of superoxide will have been produced and half of the superoxide has reacted with cytochrome *c* Fe^{III} and half with the putative SOD mimic. Hence, at *IC50* higher than 6 μM, the turnover of the SOD mimic in the experiment performed at the *IC50* concentration is not more than one. To ensure a catalytic nature would require high turnover number and this is obtained for *IC50* much smaller than μM. This is the case with SOD but generally not for most SOD mimics.

The cyt *c* assay has been shown to be free of artifacts in the large majority of cases and worked with a wide range of SOD mimics tested so far and its validity has been checked by stopped-flow and pulse radiolysis [73, 163]. The frequently used NBT (nitrobluetetrazolium) assay has artifacts as the assay also produces superoxide [164]. Moreover, NBT can be reduced to $NBT^{\bullet}$ radical (which disproportionates to blue formazan) by superoxide and numerous enzymes as well as by nonenzymatic reductants [165]. Thus, the assay with NBT is not superoxide specific. Once reduced, NBT• radical can be oxidized back with oxygen-producing superoxide and regenerating NBT, which would falsely suggest that NBT reduction is inhibited. Good controls as well as expressing the activity in terms of rate constants (see above) and not *IC50* is important to have meaningful data readily available. For comparison with literature data on available SOD mimics, one needs to express results on SOD-like activity in terms of rate constants and not in terms of *IC50* values.

Direct assays
Direct methods, which test the activity at a superoxide concentration that is much higher than the SOD mimic concentration, are important in ascertaining the catalytic nature of an SOD mimic. Two main techniques can be used. The first one consists of rapid mixing of solution in high superoxide concentration (typically KO_2 in DMSO or ACN) using a stopped-flow system [125, 166, 167]. The superoxide radical can also be produced in oxygenated aqueous solutions by pulsed radiolysis [122–124]. High-energy ionizing radiation generates primary radicals (H, e^-_{aq}, HO•) that are rapidly and quantitatively converted into either superoxide $O_2^{\bullet-}$ or its protonated form hydroperoxyl $HO_2^{\bullet}$, in the presence of either formate or alcohols, since formate might coordinate to manganese [61, 121]. See for example: [40, 73].

Concluding remark: In the indirect methods, superoxide is provided at a constant slow rate and at concentrations close to that encountered in biological systems. This provides a test for the ability of a putative SOD mimic to be useful under physiological situations. However, in most cases it is difficult to distinguish between a catalyst and a scavenger. With the direct methods, since superoxide can be used in large excess relative to the putative SOD mimic, unambiguous characterization of an SOD mimic as a catalyst or a scavenger can be obtained. Indirect and direct methods are thus complementary [56, 73].

is at the very heart of the bioinspired approach, with dual interest: to develop chemical models to improve the understanding of physicochemical processes in natural systems, and also to learn lessons from Nature to generate efficient artificial systems. SOD mimic designing is a paradigmatic field in realizing these goals.

References

1. Sheng Y, Abreu IA, Cabelli DE, Maroney MJ, Miller A-F, Teixeira M, Valentine JS. Superoxide dismutases and superoxide reductases. Chem Rev. 2014;114:3854–918.
2. Sessions AL, Doughty DM, Welander PV, Summons RE, Newman DK. The continuing puzzle of the great oxidation event. Cur Biol. 2009;19(14):R567–74.
3. Pierre J-L, Crichton RR. Old iron, young copper: from Mars to Venus. BioMetals. 2001;14:99–112.

4. Valentine JS. Dioxygen reactivity and toxicity. In: Bertini I, Gray SJ, Stiefel EI, Valentine JS, editors. Biological inorganic chemistry, structure and reactivity. Mill Valley: University Science Books; 2007. p. 319–31.
5. Koppenol WH. Oxygen activation by cytochrome P450: a thermodynamic analysis. J Am Chem Soc. 2008;129:9686–90.
6. Que LJ. Dioxygen activating enzymes. In: Bertini I, Gray SJ, Stiefel EI, Valentine JS, editors. Biological inorganic chemistry, structure and reactivity. Mill Valley: University Science Books; 2007. p. 388–413.
7. Koppenol WH, Stanbury DM, Bounds PL. Electrode potentials of partially reduced oxygen species, from dioxygen to water. Free Radical Biol Med. 2010;49:317–22.
8. Kirkinezosa IG, Moraesa CT. Reactive oxygen species and mitochondrial diseases. Semin Cell Dev Biol. 2001;12:449–57.
9. Halliwell B, Gutteridge JMC. Free radicals in biology and medicine. New York: Oxford University Press; 2007.
10. Goldberg IH. Mechanism of neocarzinostatin action: role of DNA microstructure in determination of chemistry of bistranded oxidative damage. Acc Chem Res. 1991;91(7):191–8.
11. Dix TA, Hess KM, Medina MA, Sullivan RW, Tilly SL, Webb TLL. Mechanism of site-selective DNA nicking by the hydrodioxyl (perhydroxyl) radical. Biochemistry. 1996;35(14):4578–83.
12. Nelson SK, Bose SK, McCord JM. The toxicity of high-dose superoxide dismutase suggests that superoxide can both initiate and terminate lipid peroxidation in the reperfused heart. Free Radical Biol Med. 1994;16:195–200.
13. Gardner PR, Fridovich I. Superoxide sensitivity of the Escherichia coli 6-phosphogluconate dehydratase. J Biol Chem. 1991;266:1478–83.
14. Gardner PR, Fridovich I. Inactivation-reactivation of aconitase in E. coli: a sensitive measured of superoxide radical. J Biol Chem. 1992;267:8757–63.
15. Gardner PR, Raineri I, Epstein LB, White CW. Superoxide radical and iron modulate aconitase activity in mammalian-cells. J Biol Chem. 1995;270:13399–405.
16. Charrier JG, Anastasio C. Impacts of antioxidants on hydroxyl radical production from individual and mixed transition metals in a surrogate lung fluid. Atmos Environ. 2011;45: 7555–62.
17. Iranzo O. Manganese complexes displaying superoxide dismutase activity: a balance between different factors. Bioorg Chem. 2011;39:73–87.
18. Aguirre JD, Culotta VC. Battles with iron: manganese in oxidative stress protection. J Biol Chem. 2012;287:13541–8.
19. Theil EC. Ferritin. In: Bertini I, Gray SJ, Stiefel EI, Valentine JS, editors. Biological inorganic chemistry, structure and reactivity. Mill Valley: University Science Books; 2007; p. 144–50.
20. Holley AK, Dhar SK, Xu Y, St. Clair DK. Manganese superoxide dismutase: beyond life and death. Amino Acids. 2012;42(1):139–58.
21. Inoue M, Sato EF, Nishikawa M, Park AM, Kira Y, Imada I, Utsumi K. Mitochondrial generation of reactive oxygen species and its role in aerobic life. Curr Med Chem. 2003;23:2495–505.
22. McCord JM, Edeas MA. SOD, oxidative stress and human pathologies: a brief history and a future vision. Biomed Pharmacother. 2005;59:139–42.
23. Batinic-Haberle I, Tovmasyan A, Roberts ER, Vujaskovic Z, Leong KW, Spasojevic I. SOD therapeutics: latest insight into their structure-activity relationships and impact on the cellular redox-based signaling pathways. Antioxid Redox Signal. 2014;20(15):2372–415.
24. Batinic-Haberle I, Reboucas Julio S, Spasojevic I. SOD mimic: chemistry, pharmacology and therapeutic potential. Antioxid Redox Signal. 2010;13(6):877–918.
25. Muscoli C, Cuzzocrea S, Riley DP, Zweier JL, Thiemermann C, Wang ZQ, Salvemini D. On the selectivity of SOD mimetics and its importance in pharmacological studies. Br J Pharmacol. 2003;140:445–60.

26. Salvemini D, Muscoli C, Riley DP, Cuzzocrea S. Superoxide dismutases mimetics. Pulm Pharmacol Ther. 2002;15:439–47.
27. Fridovich I, Imlay JA. Assay of metabolic superoxide production in Escherichia coli. J Biol Chem. 1991;266:6957–65.
28. Melov S, Coskun P, Patel M, Tuinstra R, Cottrell B, Jun AS, AZastawny TH, Dizdaroglu M, Goodman SI, Huang TT, Miziorko H, Epstein CJ, Wallace DC. Mitochondrial disease in superoxide dismutase 2 mutant mice. Proc Natl Acad Sci U S A. 1999;96:846–51.
29. Li Y, Huang T-T, Carlson EJ, Melov S, Ursell PC, Olson JL, Noble LJ, Yoshimura MP, Berger C, Chan PH, Wallace DC, Epstein CJ. Dilated carbiomyopathy and neonatal lethality in mutant mice lacking MnSOD. Nat Genet. 1995;11:376–81.
30. Reaume AG, Elliott JL, Hoffman EK, Kowall NW, Ferrante RJ, Siwek DR, Wilcox HM, Flood DG, Beal MF, Brown Jr RH, Scott RW, Snider WD. Motor neurons in Cu/Zn superoxide dismutase-deficient mice develop normally but exhibit enhanced cell death after axonal injury. Nat Genet. 1996;13:43–7.
31. Yost FJ, Fridovich I. An iron-containing superoxide dismutase from Escherichia coli. J Biol Chem. 1973;248(14):4905–8.
32. Youn H-D, Kim E-J, Roe J-H, Hah YC, Kang SO. A novel nickel-containing superoxide dismutase from Streptomyces spp. Biochem J. 1996;318:889–96.
33. Niviere V, Fontecave M. Discovery of superoxide reductase: an historical perspective. J Biol Inorg Chem. 2004;9:119–23.
34. Austin FE, Barbieri JT, Crorin RE, Grigas KE, Cox CD. Distribution of superoxide dismutase, catalase, and peroxidase activities among Treponema pallidum and other spirochetes. Infec Immun. 1981;33(2):372–9.
35. Archibald FS, Fridovich I. Mn and defenses against oxygen toxicity in Lactobacillus plantarum. J Bacteriol. 1981;445:442–51.
36. Seib KL, Tseng H-J, McEwan AG, Apicella MA, Jennings MP. Defenses against oxidative stress in Neisseria gonorrhoeae and Neisseria meningitidis: distinctive systems for different lifestyles. J Infect Disease. 2004;190:136–47.
37. Jones DP, Go YM, Anderson CL, Ziegler TR, Kinkade JM, Kirlin WG. Cysteine/cystine couple is a newly recognized node in the circuitry for biologic redox signaling and control. Faseb J. 2004;18(9):1246–8.
38. Latour J-M. Manganese, the stress reliever. Metallomics. 2015;7:25–8.
39. Archibald FS, Fridovich I. The scavenging of superoxide radical by manganous complexes: in vitro. Arch Biochem Biophys. 1982;214:452–63.
40. Durot S, Policar C, Cisnetti F, Lambert F, Renault J-P, Pelosi G, Blain G, Korri-Youssoufi H, Mahy J-P. Series of Mn complexes based on N-centered ligands and superoxide—reactivity in an anhydrous medium and SOD-like activity in an aqueous medium correlated to Mn^{II}/Mn^{III} redox potentials. Eur J Inorg Chem. 2005;3513–23.
41. Barnese K, Gralla EB, Cabelli DE, Valentine JS. Manganous phosphate acts as a SOD. J Am Chem Soc. 2008;130:4604–6.
42. McNaughton RL, Reddi AR, Clement MHS, Sharma A, Barnese K, Rosenfeld L, Butler Gralla E, Valentine JS, Culotta VC, Hoffman B. Probing in vivo Mn2+ speciation and oxidative stress resistance in yeast cells with electron-nuclear double resonance spectroscopy. Proc Natl Acad Sci U S A. 2010;107(35):15335–9.
43. Reddi AR, Culotta VC. Regulation of manganese antioxidants by nutrients sensing pathway in saccharomyces cerevisiae. Genetics. 2011;189:1261–70.
44. Mann T, Keilin D. Haemocuprein and hepatocuprein, copper protein compounds of blood and liver in mammals. Proc R Soc Ser B. 1938;126:303–15.
45. Markowitz H, Cartwright GE, Wintrobe MM. Studies on copper metabolism: XXVII. An Erythrocyte cuproprotein. The isolation and properties of an erythrocyte cuproprotein (erythrocuprein). J Biol Chem. 1959;234:40–5.
46. Pauling L. The discovery of superoxide radical. Trends Biochem Sci. 1979;4(11):N270–1.
47. Pauling L. The nature of the chemical bond. IV The energy of single bonds and the relative electronegativity of atoms. J Am Chem Soc. 1932;54:3570–82.

48. Neuman EW. Potassium superoxide and the three-electron bond. J Chem Phys. 1934;2:31–3.
49. Bray RC, Knowles PF. Electron spin resonance in enzyme chemistry: the mechanism of action of xanthine oxidase. Proc Roy Soc A. 1968;302:351–53.
50. Knowles PF, Gibson JF, Pick FM, Bray RC. Electron-spin-resonance evidence for enzymic reduction of oxygen to a free radical, the superoxide ion. Biochem J. 1969;111:53–8.
51. McCord JM, Fridovich I. Superoxide dismutase. An enzymatic function for erythrocuprein (hemocuprein). J Biol Chem. 1969;244(22):6049–55.
52. Carson S, Vogin EE, Huber W, Schulte TL. Safety tests of orgotein, an antiinflammatory protein. Toxicol Appl Pharmacol. 1973;26:184–202.
53. Babior BM, Kipnes RS, Curnutte JT. The production by leukocytes of superoxide radical, a potential bactericial agent. J Clin Invest. 1973;52:741–4.
54. Miriyala S, Spasojevic I, Tovmasyan A, Salvemini D, Vujaskovic Z, St. Clair D, Batinic-Haberle I. MnSOD and its mimic. Biochim Biophys Acta. 2012;1822(5):794–814.
55. Axthelm F, Casse O, Koppenol WH, Nauser T, Meier W, Palivan CG. Antioxidant nanoreactor based on SOD encapsulated in SOD permeable vesicles. J Phys Chem B. 2008;112(28):8211–6.
56. Riley DP. Functional mimics of superoxide dismutase enzymes as therapeutic agents. Chem Rev. 1999;99:2573–87.
57. Salvemini D, Riley DP, Cuzzocrea S. SOD mimetics are coming of age. Nat Rev Drug Discov. 2002;1(5):367–74.
58. Abreu IA, Cabelli DE. Superoxide dismutases—a review of the metal-associated mechanistic variations. Biochim Biophys Acta. 2010;1804:263–74.
59. Basolo F, Pearson RG. Mechanisms of inorganic reactions: a study of metal complexes in solution. New York: Wiley; 1967.
60. Barrette WCJ, Sawyer DT, Fee JA, Asada K. Potentiometric titration and oxidation-reduction potentials of several iron superoxide dismutases. Biochemistry. 1983;22(3):624–7.
61. Bielski BHJ, Cabelli DE, Arudi RL, Ross AB. Reactivity of HO_2/O_2^- radicals in aqueous solution. J Phys Chem Ref Data. 1985;14(4):1041–100.
62. Azab HA, Banci L, Borsari M, Luchinat C, Sola M, Viezzoli MS. Redox chemistry of superoxide-dismutase- Cyclic voltammetry of wild-type enzymes and mutants on functionally relevant residues. Inorg Chem. 1992;31(22):4649–55.
63. Lawrence GD, Sawyer DT. Potentiometric titrations and oxidation-reduction potentials of manganese and copper-zinc superoxide dismutases. Biochemistry. 1979;19:3045–50.
64. Herbst RW, Guce A, Bryngelson PA, Higgins KA, Ryan KC, Cabelli DE, Garman SC, Maroney MJ. Role of conserved tyrosine residues in NiSOD catalysis: a case of convergent evolution. Biochemistry. 2009;48(15):3354–69.
65. Miller A-F. Redox tuning over almost 1 V in a structurally conserved active site: lessons from Fe-containing superoxide dismutase. Acc Chem Res. 2008;41(4):501–10.
66. Marshall NM, Garner DK, Wilson TD, Gao Y-G, Robinson H, Nilges MJ. Rationally tuning the reduction potential of a single cupredoxin beyond the natural range. Nature. 2009;462:113–7.
67. New SY, Marshall NM, Hor TSA, Xue F, Lu Y. Redox tuning of biological copper centers through non-covalent interactions: same trends but different magnitude. Chem Commun. 2012;48:4217–9.
68. Drew MGB, Harding CJ, McKee V, Morgan GG, Nelson J. Geometric control of manganese redox state. J Chem Soc Chem Commun. 1995;10:1035–8.
69. Garcia L, Cisnetti F, Gillet N, Guillot R, Aumont-Nicaise M, Piquemal J-P, Desmadril M, Lambert F, Policar C. Entasis through Hook-and-Loop factening in a glycoligand with cumulative weak forces stabilizing Cu(I). J Am Chem Soc. 2015;137(3):1141–6.
70. Suh J, Chei WS. Metal complexes as artificial proteases: toward catalytic drugs. Curr Opin Chem Biol. 2008;12:207–13.
71. Lombard M, Fontecave M, Touati D, Nivière V. Reaction of the desulfoferrodoxin from Desulfoarculus baarsii with superoxide anion—Evidence for a superoxide reductase activity. J Biol Chem. 2000;275:115–21.

72. Pianzzola MJ, Soubes M, Touati D. Overproduction of the rbo gene product from Desulfovibrio species suppresses all deleterious effects of lack of superoxide dismutase in Escherichia coli. J Bacteriol. 1996;178:6736–42.
73. Durot S, Lambert F, Renault J-P, Policar C. A pulse radiolysis study of superoxide radical catalytic dismutation by a manganese(II) complex with a N-tripodal ligand. Eur J Inorg Chem. 2005;2789–93.
74. Quinn R, Mercer-Smith J, Burstyn JN, Valentine JS. Influence of hydrogen bonding on the properties of iron porphyrin imidazole complexes. An internally hydrogen bonded imidazole ligand. J Am Chem Soc. 1984;106:4136–44.
75. Haga M-A, Ano T-A, Kano K, Yamabe S. Proton-induced switching of metal-metal interactions in dinuclear ruthenium and osmium complexes bridged by 2,2′-bis(2-pyridyl)-bibenzimidazole. Inorg Chem. 1991;30:3843–9.
76. Carina RF, Verzegnassi L, Bernadinelli G, Williams AF. Modulation of iron reduction potential by deprotonation at a remote site. Chem Commun. 1998;24:2681–2.
77. Brewer C, Brewer G, Luckett C, Marbury GS, Viragh C, Beatty AM, Scheidt RW. Proton control of oxidation and spin state in a series of iron tripodal imidazole complexes. Inorg Chem. 2004;43:2402–15.
78. Lambert F, Policar C, Durot S, Cesario M, Yuwei L, Korri-Youssoufi H, Keita B, Nadjo L. Imidazole and imidazolate iron complexes: on the way for tuning 3D-structural characteristics and reactivity. Redox interconversions controlled by protonation state. Inorg Chem. 2004;43:4178–88.
79. Anxolabéhère-Mallart E, Costentin C, Policar C, Robert C, Savéant J-M, Teillout A-L. Proton-coupled electron transfers in biomimetic water bound metal complexes. The electrochemical approach. Faraday Discuss. 2011;48:83–95.
80. Vallee BL, Williams RJP. Metalloenzymes-entatic nature of their active sites. Proc Natl Acad Sci U S A. 1968;59(2):498–505.
81. Jahn HA, Teller E. Stability of polyatomic molecules in degenerate electronic states. I. Orbital degeneracy. Proc R Soc London Ser A. 1937;161(A905):220–35.
82. Deeth RJ. The ligand field molecular mechanics model and the stereoelectronic effects of d and s electrons. Coord Chem Rev. 2001;212:11–34.
83. Koppenol WH. The physiological role of charge distribution on SOD. In: Rodgers MAJ, editor. Oxygen and oxy-radicals in chemistry and biology. New York: Academic; 1981. p. 671–4.
84. Benovic J, Tillman T, Cudd A, Fridovich I. Electrostatic facilitation of the reaction catalyzed by the manganese-containing and iron-containing SOD. Arch Biochem Biophys. 1983;221:329–32.
85. Getzoff ED, Tainer JA, Weiner PK, Kollman PA, Richardson JS, Richardson DC. Electrostatic recognition between superoxide and copper-zinc SOD. Nature. 1983;306:287–90.
86. Sines J, Allison S, Wierzbicki A, McCammon JA. Brownian dynamics simulation of the superoxide-superoxide dismutase reaction: iron and manganese enzymes. J Phys Chem. 1990;94:959–61.
87. Cudd A, Fridovich I. Electrostatic interactions in the reaction mechanism of bovine Erythrocyte superoxide dismutase. J Biol Chem. 1982;257:11443–7.
88. Allison SA, Bacquet RJ, McCammon JA. Simulation of the diffusion-controlled reaction between superoxide and superoxide dismutase. II. Detailed models. Biopolymers. 1988;27:251–69.
89. Parker MW, Blake CCF. Crystal structure of MnSOD from Bacillus stearothermopilus at 2.4 Å resolution. J Mol Biol. 1988;199(2):649–61.
90. Wispe JR, Clark JC, Burhans MS, Kropp KE, Korfhagen TR, Whitsett JA. Synthesis and processing of the precursor for human mangano-superoxide dismutase. Biochim Biophys Acta. 1989;994:30–6.
91. Durand G, Polidori A, Pucci B. La vectorisation de pièges à radicaux libres. Nouvelle stratégie thérapeutique. Act Chimique. 2003;26–29.

92. Favier A. Le stress oxydant. Intérêt conceptuel et expérimental dans la compréhension des mécanismes des maladies et potentiel thérapeutique. Act Chimique. 2003;11–12:108–15.
93. Smith AJ, Porteous CM, Gane AM, Murphy MP. Delivery of active molecules to mitochondria *in vivo*. Proc Natl Acad Sci U S A. 2003;100:5407–12.
94. Dhanasekaran A, Kotamraju S, Karunakaran C, Kalivendi SV, Thomas S, Joseph J, Kalyanaraman B. Mitochondria superoxide dismutase mimetic inhibits peroxide-induced oxidative damage and apoptosis: role of mitochondrial superoxide. Free Radical Biol Med. 2005;39:567–83.
95. Rafelski S. Mitochondrial network morphology: building an integrative, geometrical view. BMC Biol. 2013;11:71.
96. Fernandez-Carneado J, van Gool M, MMartos V, Castel S, Prados P, de Mendoza J, Giralt E. Highly efficient, nonpeptidic oligoguanidinium vectors that selectively internalize into mitochondria. J Am Chem Soc. 2005;127:869–74.
97. Policar C, Durot S, Lambert F, Cesario M, Ramiandrasoa F, Morgenstern-Badarau I. New Mn(II) complexes with N/O coordination sphere from tripodal N-centered ligands. Characterization from solid state to solution and reaction with superoxide in non-aqueous and aqueous medium. Eur J Inorg Chem. 2001;1807–18.
98. Dees A, Zahl A, Puchta R, van Eikema Hommes NJR, Heinemann FW, Ivanovic-Burmazovic I. Water exchange on seven-coordinate Mn(II) complexes with macrocyclic pentadentate ligands: insight in the mechanism of Mn(II) SOD-mimetic. Inorg Chem. 2007;46:2459–70.
99. Salvemini D, Wang Z-Q, Zweier JL, Samouilov A, Macarthur H, Misko TP, Currie MG, Cuzzocrea S, Sikorski JA, Riley DP. A non-peptidyl mimic of SOD with therapeutic activity in rats. Science. 1999;286:304–5.
100. Xingcan S, Hong L, Zhiliang J, Xiwen H, Panwen P. Binding equilibrium study between Mn(II) and HSA or BSA. Chin J Chem. 2000;18(1):35–41.
101. Munroe W, Kingsley C, Durazo A, Butler Gralla E, Imlay JA, Srinivasan C, Valentine JS. Only one of a wide assortment of manganese-containing SOD mimicking compounds rescues the slow aerobic growth phenotypes of both Escherichia coli and Saccharomyces cerevisiae strains lacking superoxide dismutase enzymes. J Inorg Biochem. 2007; 101:1875–82.
102. Spasojevic I, Chen Y, Noel TJ, Yu Y, Cole MP, Zhang L, Zhao Y, St. Clair DK, Batinic-Haberle I. Mn-porphyrin-based superoxide dismutase mimic, [Mn(III)TE2PyP]5+ target mouse heart mitochondria. Free Radical Biol Med. 2007;42:1193–200.
103. Clède S, Lambert F, Saint-Fort R, Plamont MA, Bertrand H, Vessières A, Policar C. Influence of the side-chain length on the cellular uptake and the cytotoxicity of rhenium triscarbonyl derivatives: a bimodal infrared and luminescence quantitative study. Chem Eur J. 2014;20:8714–22.
104. Policar C, Lambert F, Cesario M, Morgenstern-Bararau I. An inorganic helix [Mn(IPG)(MeOH)]n[PF6]n: structural and magnetic properties of a syn-anti carboxylate-bridged manganese(II) chain involving a tetradentate ligand. Eur J Inorg Chem. 1999;(12):2201–7.
105. Cisnetti F, Lefevre AS, Guillot R, Lambert F, Blain G, Anxolabéhère-Mallart E, Policar C. A new pentadentate ligands forms both a di- and mononuclear Mn(II) complex: electrochemical, spectroscopic and SOD activity studies. Eur J Inorg Chem. 2007;4472–80.
106. Cisnetti F, Pelosi G, Policar C. Synthesis and superoxide dismutase-like activity of new manganese(III) complexes based on tridentate N2O ligands derived from histamine. Inorg Chim Acta. 2007;360:557–62.
107. Groni S, Blain G, Guillot R, Policar C, Anxolabéhère-Mallart E. Reactivity of Mn(II) with superoxide. Evidence for a $[Mn^{III}OO]^+$ unit by low-temperature spectroscopies. Inorg Chem. 2007;46:1951–3.
108. Bernard A-S, Giroud C, Ching HYV, Meunier A, Ambike V, Amatore C, Guille Collignon M, Lemaître F, Policar C. Evaluation of the anti-oxidant properties of a SOD-mimic Mn-complex in activated macrophages. Dalton Trans. 2012;41:6399–403.

109. Borgstahl GOE, Parge HE, Hickey MJ, Beyer WF, Hallewell RA, Tainer JA. The structure of human mitochondrial manganese superoxide dismutase reveals a novel tetrameric interface of two 4-helix bundles. Cell. 1992;71:107–18.
110. Tabares LC, Cortez N, Hiraoka BY, Yamakura F, Un S. Effects of substrate analogues and pH on manganese superoxide dismutases. Biochemistry. 2006;45(6):1919–29.
111. Whittaker MM, Whittaker JW. Low-temperature thermochromism marks a change in coordination for the metal ion in manganese superoxide dismutase. Biochemistry. 1996;35(21):6762–70.
112. Yamato K, Miyahara I, Ichimura A, Hirotsu K, Kojima Y, Sakurai H, Shiomi D, Sato K, Takui T. Superoxide dismutase mimetic complex of Mn(II)/N, N-bis(2-pyridylmethyl)-(S)-histidine. Chem Lett. 1999;28:295–6.
113. Nishida Y, Tanaka N, Yamazaki A, Tokii T, Hashimoto N, Ide K, Iwasawa K. Novel reactivity of dioxygen molecule in the presence of Mn(II) complex and reducing agents. Inorg Chem. 1995;34:3616–20.
114. Ching HYV, Anxolabehere-Mallart E, Colmer HE, Costentin C, Dorlet P, Jackson TA, Policar C, Robert M. Electrochemical formation and reactivity of a manganese peroxo complex: acid driven H_2O_2 generation vs. O-O bond cleavage. Chem Sci. 2014;5(6):2304–10.
115. Kitajima N, Komatsuzaki H, Hikichi S, Osawa M, Moro-Oka Y. A monomeric side-on peroxo Mn(III) complex: $Mn(O_2)(3,5\text{-}iPr_2pzH)(HB(3,5\text{-}iPr_2pz))$. J Am Chem Soc. 1994;116:11596–7.
116. Primus J-L, Grunenwald S, Hagedoorn P-L, Albrecht-Gary A-M, Mandon D, Veeger C. The nature of the intermediates in the reaction of Fe(III) and Mn(III)-microperoxidase 8 with H_2O_2: a rapid kinetics study. J Am Chem Soc. 2002;124(7):1214–21.
117. Seo MS, Kim JY, Annaraj J, Kim Y, Lee Y-M, Kim S-J, Kim J, Nam W. [Mn(tmc)(O2)]+: a side-on peroxido manganese(III) complex bearing a non-heme ligand. Angew Chem Int Ed Engl. 2007;46:377–80.
118. Groni S, Dorlet P, Blain G, Bourcier S, Guillot R, Anxolabehere-Mallart E. Reactivity of an aminopyridine LMn^{II} complex with H_2O_2. Detection of intermediates at low temperature. Inorg Chem. 2008;47(8):3166–72.
119. Leto DF, Chattopadhyay S, Daya VW, Jackson TA. Reaction landscape of a pentadentate N5-ligated MnII complex with $O_2^{\bullet -}$ and H_2O_2 includes conversion of a peroxomanganese(III) adduct to a bis(μ-oxo)-dimanganese(III, IV) species. Dalton Trans. 2013;42:13014–25.
120. Enemark JH, Feltham RD. Principles of structure, bonding, and reactivity for metal nitrosyl complexes. Coord Chem Rev. 1974;13(4):339–406.
121. Bielski BHJ, Arudi RL. Preparation and stabilization of aqueous/ethanolic superoxide solutions. Anal Biochem. 1983;133:170–8.
122. Bielski BHJ. Reevaluation of the spectral and kinetic properties of HO_2 and O_2^- free radicals. Photochem Photobiol. 1978;28:645–9.
123. Cabelli DE, Bielski BHJ. Pulse radiolysis study of the kinetics and mechanisms of the reaction between Mn(II) complex and HO_2/O_2- radicals. 1. sulfate, formate and pyrophosphate complexes. J Phys Chem. 1984;88(14):3111–5.
124. Cabelli D. Probing superoxide dismutases through radiation chemistry. Isr J Chem. 2014;54(3):272–8.
125. Friedel FC, Lieb D, Ivanovic-Burmazovic I. Comparative studies on manganese-based SOD mimetics, including the phosphate effect, by using global spectral analysis. J Inorg Biochem. 2012;109:26–32.
126. Batinic-Haberle I, Spasojevic I, Hambright P, Benov L, Crumbliss AL, Fridovich I. Relationship among redox potentials, proton dissociation constants of pyrrolic nitrogens and in vivo SOD activities of Mn(III) and iron(III) water soluble porphyrins. Inorg Chem. 1999;38:4011–22.
127. Okado-Matsumoto A, Batinic-Haberle I, Fridovich I. Complementation of SOD-deficient Escherichia coli by manganese porphyrin mimics of superoxide dismutase activity. Free Radical Biol Med. 2004;37(3):401–10.

128. Tovmasyan A, Carballal S, Ghazaryan R, Melikyan L, Weitner T, Maia CGC, Reboucas JS, Radi R, Spasojevic I, Benov L, Batinic-Haberle I. Rational design of superoxide dismutase (SOD) mimics: the evaluation of the therapeutic potential of New cationic Mn porphyrins with linear and cyclic substituents. Inorg Chem. 2014;53(21):11467–83.
129. Ndengele MM, Muscoli C, Zhi QW, Doyle TM, Matuschak GM, Salvemini D. Superoxide potentiates NF-kappa B activation and modulates endotoxin-induced cytokine production in alveolar macrophages. Shock. 2005;23(2):186–93.
130. Asayama S, Kawamura E, Nagaoaka S, Kawakami H. Design of Mn-porphyrin modified with a mitochondrial signal peptide for a new antioxidant. Mol Pharm. 2006;3:468–70.
131. Amatore C, Arbault S, Guille M, Lemaitre F. Electrochemical monitoring of single cell secretion: vesicular exocytosis and oxidative stress. Chem Rev. 2008;108:2585–621.
132. Lewis EA, Lindsay-Smith JR, Walton PH, Archibald SJ, Foxo SP, Giblin GMP. Tuning the metal-based redox potential of manganese cis, cis-1,3,5-triaminocyclohexane complexes. J Chem Soc Dalton Trans. 2001;1159–61.
133. Lewis EA, Khodr HH, Hider RC, Lindsay-Smith JR, Walton PH. A manganese superoxide dismutase mimic based on cis-cis-1,3,5-triaminocyclohexane. Dalton Trans. 2004;2:187–8.
134. Collman J. Synthetic models for the oxygen-binding hemoproteins. Acc Chem Res. 1977;10:265–72.
135. Momenteau M. Synthesis and coordination properties of superstructure iron-porphyrins. Pure Appl Chem. 1986;58(11):1493–502.
136. Collman J, Fu L. Synthetic models for hemoglobin and myoglobin. Acc Chem Res. 1999;32:455–63.
137. Collman J, Ghosh S. Recent applications of synthetic model of cytochrome c oxidase: beyond functional modeling. Inorg Chem. 2010;49:5798–810.
138. Batinic-Haberle I, Benov L, Spasojevic I, Fridovich I. The ortho effect makes Mn(III)porphyrins a powerful potentially used SOD mimic. J Biol Chem. 1998;273(38):24521–8.
139. Batinic-Haberle I, Spasojevic I, Stevens RD, Hambright P, Neta P, Okado-Matsumoto A, Fridovich I. New class of potent catalysts of superoxide dismutation. Mn(III) ortho-methoxyethylpyridyl- and di-ortho-methoxyethyl-imidazolylporphyrins. Dalton Trans. 2004; 11:1696–702.
140. Maroz A, Kelso GF, Smith RAJ, Ware DC, Anderson RF. Pulse radiolysis investigation on the mechanism of the catalytic action of Mn(II)—Pentaazamacrocycle compounds as superoxide dismutase mimetics. J Phys Chem A. 2008;112(22):4929–35.
141. Batinic-Haberle I, Liochev SI, Spasojevic I, Fridovich I. A potent SOD mimic: Mnßoctabromo meso tetrakis(Npyridinium4yl)porphyrin. Arch Biochem Biophys. 1997;343(2):225–33.
142. Spasojević I, Batinic-Haberle I, Rebouças JS, Idemori YM, Fridovich I. Electrostatic Contribution in the catalysis of superoxide dismutation by SOD mimics. J Biol Chem. 2003; 278:6831–7.
143. Rajic Z, Tovmasyan A, Spasojevic I, Sheng H, Lu M, Li A-M, Batinic-Haberle I, Benov LT. A new SOD mimic, Mn(III) ortho N-butoxyethylpyridylporphyrin, combines superb potency and lipophilicity with low toxicity. Free Radical Biol Med. 2012;52:1828–34.
144. Rebouças JS, DeFreitas-Silva G, Spasojević I, Idemori YM, Benov L, Batinić-Haberle I. Impact of electrostatics in redox modulation of oxidative stress by Mn porphyrins: Protection of SOD-deficient Escherichia coli via alternative mechanism where Mn porphyrin acts as a Mn carrier. Free Radical Biol Med. 2008;45:201–10.
145. Friedel FC, Kenkell I, Miljkovic JL, Moldenhauer D, Weber N, Filipovic ML, Gröhn F, Ivanovic-Burmazovic I. Amphiphilic pentaazamacrocyclic manganese superoxide dismutase mimetics. Inorg Chem. 2014;53:1009–20.
146. Puglisi A, Tabb G, Vecchio G. Bioconjugates of cyclodextrins of manganese salen-type with superoxide dismutase activity. J Inorg Biochem. 2004;98:969–76.
147. Rebouças JS, Spasojevic I, Tjahjono DH, Richaud A, Méndez F, Benov L, Batinic-Haberle I. Redox modulation of oxidative stress by Mn porphyrin-based therapeutics: the effect of charge distribution. Dalton Trans. 2008;9:1233–42.

148. Asayama S, Hanawa T, Nagaoka S, Kawakami H. Design of the complex between manganese porphyrins and catalase-poly(ethylene glycol) conjugates for a new antioxidant. Mol Pharm. 2007;4(3):484–6.
149. Hanawa T, Asayama S, Watanabe T, Owada S, Kawakami H. Protective effects of the complex between manganese porphyrins and catalase-poly(ethylene glycol) conjugates against hepatic ischemia/reperfusion injury in vivo. J Controlled Release. 2009;135(1):60–4.
150. D'agata R, Grasso G, Iacono G, Spoto G, Vecchio G. Lectin recognition of a new SOD mimic bioconjugate studied with surface plasmon resonance imaging. Org Biomol Chem. 2006;4:610–2.
151. Asayama S, Mizushima K, Nagaoka S, Kawakami H. Design of metalloporphyrin-carbohydrate conjugates for a New superoxide dismutase mimic with cellular recognition. Bioconjugate Chem. 2004;15:1360–3.
152. Kelso GF, Maroz A, Cocheme HM, Logan A, Prime TA, Peskin AV, Winterbourn CC, James AM, Ross MF, Brooker S, Porteous CM, Anderson RF, Murphy MP, Smith RAJ. A mitochondria-targeted macrocyclic Mn(II) superoxide dismutase mimetic. Chem Biol. 2012;19(10):1237–46.
153. Aitken JB, Shearer EL, Giles NM, Lai B, Vogt S, Reboucas JS, Batinic-Haberle I, Lay PA, Giles GI. Intracellular targeting and pharmacological activity of the SOD mimics MnTE-2-PyP(5+) and MnTnHex-2-PyP(5+) regulated by their porphyrin ring substituents. Inorg Chem. 2013;52:4121–3.
154. Fridovich I. Quantitative aspects of the production of superoxide anion radical by milk xanthine oxidase. J Biol Chem. 1970;245(16):4053–7.
155. Castello PR, Drechsel DA, Day BJ, Patel M. Inhibition of mitochondrial hydrogen peroxide production by lipophilic metalloporphyrins. J Pharmacol Exp Ther. 2008;324:970–6.
156. Okado-Matsumoto A, Fridovich I. Assay of superoxide dismutase: cautions relevant to the use of cytochrome c, a sulfonated tetrazolium, and cyanide. Anal Biochem. 2001; 298:337–42.
157. Pasternack RF, Halliwell B. Superoxide dismutase activities of an iron porphyrin and other iron complexes. J Am Chem Soc. 1979;101(4):1026–31.
158. Butler J, Koppenol WH, Margoliash E. Kinetics and mechanism of the reduction of ferricytochrome c by the superoxide anion. J Biol Chem. 1982;257(18):10747–50.
159. Liao Z-R, Zheng X-F, Luo B-S, Shen L-R, Li D-F, Liu H-L, Zhao W. Synthesis, characterization and SOD-like activities of manganese-containing complexes with N, N, N′, N′-tetrakis(2′-benzimidazolyl methyl)-1,2-ethanediamine (EDTB). Polyhedron. 2001;20:2813–21.
160. Weiss RH, Flickinger AG, Rivers WJ, Hardy MM, Aston KW, Ryan US, Riley DP. Evaluation of activity of putative superoxide dismutase mimics. Direct analysis by stopped-flow kinetics. J Biol Chem. 1993;268(31):23049–54.
161. Faulkner KM, Liochev SI, Fridovich I. Stable Mn(III) porphyrins mimic SOD in vitro and substitute for it in vivo. J Biol Chem. 1994;269(38):23471–6.
162. Note that after 10 min. the concentration in dioxygen drops too much and the OD as a function of time is no longer linear.
163. Spasojevic I, Batinic-Haberle I, Stevens RD, Hambright P, Thorpe AN, Grodkowski J, Neta P, Fridovich I. Mn(III) biliverdin IX dimethylester: a powerful catalytic scavenger of superoxide employing the Mn(III)/Mn(IV) redox couple. Inorg Chem. 2001;40:726–39.
164. Liochev SI, Fridovich I. Superoxide from glucose oxidase or from nitroblue tetrazolium? Arch Biochem Biophys. 1995;318:408–10.
165. Batinic-Haberle I, Rebouças Julio S, Spasojevic I. Response to Rosenthal et al. Antiox redox signal. 2011;14(6):1174–6.
166. Bull C, McClune GJ, Fee JA. The mechanism of Fe-EDTA catalysed superoxide dismutation. J Am Chem Soc. 1983;103:5290–300.
167. Riley DP, Rivers WJ, Weiss RH. Stopped-flow kinetic analysis for monitoring SO decay in aqueous systems. Anal Biochem. 1991;196:344–9.
168. Ching HYV, Kenkel I, Delsuc N, Mathieu E, Ivanović-Burmazovic I and Policar C, Bioinspired superoxide-dismutase mimics: The effects of functionalization with cationic polyarginine peptides. J Inorg Biochem. 2016;160:172–9.

Chapter 8
Mn Porphyrin-Based Redox-Active Therapeutics

Ines Batinić-Haberle, Artak Tovmasyan, and Ivan Spasojević

8.1 Introduction

The Mn porphyrin (MnP) story as superoxide dismutase (SOD) mimics started in the late 1960s when Irwin Fridovich and his then Ph.D. student, Joe McCord, discovered that an already known copper protein was able to catalyze $O_2^{\bullet-}$ dismutation at a diffusion-limited rate of ~10^9 M^{-1} s^{-1}, thus acting as superoxide dismutase enzyme [1]. While $O_2^{\bullet-}$ had been already known to the chemical audience, the seminal discovery of the superoxide dismutase enzyme marked the beginning of $O_2^{\bullet-}$-based biology and medicine. Irwin Fridovich got recognized as a Father of the field of *Free Radical Biology and Medicine* and became a co-founder of the related Society (originally *Oxygen Society* and presently *Society for Redox Biology and Medicine*) and of the famous yellow Journal of *Free Radical Biology and Medicine*. Soon upon the discovery of SOD enzyme, it became obvious that $O_2^{\bullet-}$ and its progenies, when beyond physiological nM concentrations, induce oxidative stress which subsequently impacts cellular metabolism giving rise to different pathological conditions. The discovery of different isoforms of SOD followed. The existential importance of mitochondrial isoform, MnSOD, became obvious with the studies on MnSOD knock-out mice. The Cu,ZnSOD knock-out mice, though, live longer than MnSOD knock-out mice but develop multiple cancers later in their lives. The experimental evidence collected over years demonstrated that nearly all diseases, including aging, have at least

I. Batinić-Haberle (✉) • A. Tovmasyan
Department of Radiation Oncology, Duke University School of Medicine,
Research Drive, 281b/285 MSRB I, Box 3455, Durham, NC 27710, USA
e-mail: ibatinic@duke.edu

I. Spasojević
Department of Medicine and Duke Cancer Institute, Pharmaceutical Research Shared Resource, PK/PD Core Laboratory, Duke University School of Medicine,
Duke Hospital South, 5317 Orange Zone, Trent Drive, Durham, NC 27710, USA

I. Batinić-Haberle et al. (eds.), *Redox-Active Therapeutics*,
Oxidative Stress in Applied Basic Research and Clinical Practice,
DOI 10.1007/978-3-319-30705-3_8

some component of oxidative stress in common. Very early on, with the growing awareness of the critical impact that the SOD enzymes have on our health, the idea that an SOD mimic, which is of small molecular size, able to enter the cell, may be an excellent therapeutic was inaugurated. The earliest work on SOD mimics was done on iron (Fe) porphyrin by Pasternack and Halliwell [2], followed by Fridovich et al. [3]. Why are metalloporphyrins the compounds frequently studied? Nature has used macrocyclic porphyrin ring as a ligand for Fe in different proteins and enzymes. Macrocyclic porphyrin ring affords vast stability to the metal complex, thereby assuring high integrity of the metal site where all actions of interest occur. The most critical bodily proteins, such as hemoglobin, myoglobin, cytochrome oxidase, cyt P450 family members, nitric oxide synthase, and guanylyl cyclase, have Fe porphyrin (Fe protoporphyrin IX) as active site. The first SOD mimics studied by Fridovich and collaborators were complexes of Mn with simple biological ligands (such as malate and citrate) [4] and with desferrioxamine B (which provides semicyclic structure around Mn). Porphyrin was introduced in the early 1990s as an SOD-based ligand for transition metals Mn, Fe, Cu, Ni, and Zn [3, 5].

It was not been before Ines Batinić-Haberle joined the Irwin Fridovich's Lab in early 1995 that the critical porphyrin features that gave rise to powerful SOD-like properties were identified [5–18]. The four cationic charges on pyridyl nitrogens, with their electron-withdrawing power, and their *ortho* positions close to the Mn site, appear to be necessary for the appropriate metal-centered reduction potential, $E_{1/2}$ (of ~ +200 to +300 mV vs. NHE) and in turn for the high SOD-like activity [7, 11]. *Ortho* cationic Mn(III) *N*-substituted pyridylporphyrins were established as the most powerful SOD mimics (Fig. 8.1). Such "*ortho*" effect provides nearly identical thermodynamics for the catalysis of $O_2^{\bullet-}$ dismutation by MnP as it does for SOD enzyme. On kinetic side, similar to SOD enzyme, cationic charges afford electrostatic facilitation for the approach of anionic $O_2^{\bullet-}$ towards the metal site. Thus the pentacationic MnTE-2-PyP^{5+} is about 100-fold more potent SOD mimic than monocationic MnBr$_8$T-2-PyP$^+$ which

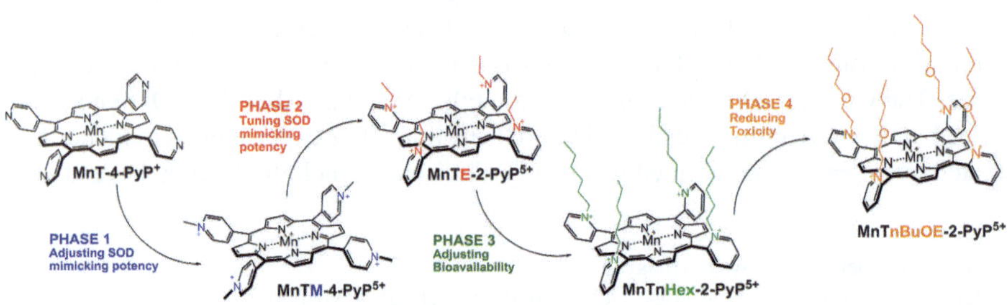

Fig. 8.1 Main stages of development of Mn porphyrins. The first lead drug, MnTE-2-PyP^{5+}, is presently in Canadian Clinical Trials on protection of islets during transplantation. The second lead, MnTnBuOE-2-PyP^{5+}, has entered First-in-Human Phase I/II Trials at Duke University as a radioprotector of normal tissue with cancer patients (NCT02655601). Based on "*ortho*" effect the di-*ortho* imidazolyl analog, MnTDE-2-ImP^{5+} (AEOL10150) was designed and is under aggressive development towards Clinical Trials as a radioprotector also. Lack of the intellectual property rights precludes the clinical development of MnTnHex-2-PyP^{5+}. M=methyl, E=ethyl, nHex=n-hexyl, nBuOE=n-butoxyethyl, 4=*para* position on pyridyl ring, 2=*ortho* position on pyridyl ring. Adapted from [14]

is neutral at the periphery, while both have the same $E_{1/2}$ for the $Mn^{III}P/Mn^{II}P$ redox couple [19]. Relative to small MnP molecule, the SOD protein structure affords the high specificity towards $O_2^{\bullet-}$. Both SOD and MnP can undergo the same reactions including the catalysis of $O_2^{\bullet-}$ dismutation as they have the same reduction potential (i.e., thermodynamics) of the metal site. Yet, the MnP reactions with species other than $O_2^{\bullet-}$ are few orders of magnitude faster than those of SOD enzyme due to the lack of steric hindrance towards the approach of such species. For example, MnTE-2-PyP^{5+} reduces $ONOO^-$ with $k_{red}(ONOO^-) > 10^7\ M^{-1}\ s^{-1}$, while MnSOD reduces $ONOO^-$ with $k_{red}(ONOO^-) \sim 10^4\ M^{-1}\ s^{-1}$. The diverse modifications of MnP structure aimed at finely tuning its bioavailability and, in particular, accumulation in mitochondria and brain. Also, the experiments were conducted to understand how the modification of MnP structure impacts its toxicity. MnP structure allowed for nearly limitless modifications at *meso* bridges and *beta* sites of pyrrolic rings. The progress on the design of Mn porphyrins as SOD mimics paralleled the growing insight into the key impact the small and large reactive species have in cellular redox biology. In addition to $ONOO^-$, and due to the biologically compatible $E_{1/2}$, we demonstrated that SOD mimics react with many different reactive species and with high rate constant such as O_2, $^{\bullet}NO$, $CO_3^{\bullet-}$, ClO^-, lipid radicals, H_2O_2, GS^-, RS^- (R being protein), ascorbate, and HNO [15]. We anticipate that many more reactions occur in vivo that have not yet been explored. We further demonstrated that on thermodynamic and kinetic grounds, the properties of MnPs that favor the SOD-like activity, favor all other reactions. In turn the rate constant for the catalysis of $O_2^{\bullet-}$, $k_{cat}(O_2^{\bullet-})$ parallels the rate constants for other reactions of MnPs [14, 20, 21]. The diverse redox reactivities of cationic MnP-based SOD mimics are in large part due to their ability to easily adopt four oxidation states in vivo, +2, +3, +4, and +5. The electron deficiency of the Mn site favors binding of the reactive species which frequently precedes the electron transfer. The mere fact that most of the reactive species are anionic (such as monodehydroascorbate, thiols, and peroxynitrite) further contributes to the high in vivo reactivity of cationic MnPs. The metal-centered reduction potential of cationic MnPs assures that they are only mild pro- and antioxidants. In summary, the possibilities of their in vivo interactions are limitless. While the type of the species, its size, and charge would control the outcome, only those reactions, where MnP and a particular reactive species are co-localized at high concentrations at the site of interest, have the highest likelihood to occur. In addition to appropriate redox properties, therefore, the bioavailability of MnP at the site of interest and at high concentration is the second major factor that controls their therapeutic effects. We are still far away from fully grasping the biology of MnPs. We however know, based on the existing data in animal models of diseases, that the reactivities of MnPs thus far identified and quantified parallel their therapeutic effects. Via four main phases of development, going from unsubstituted MnT-2-PyP$^+$, over our first lead compound, MnTE-2-PyP^{5+} (BMX-010, AEOl10133) over lipophilic MnTnHex-2-PyP^{5+} to MnTnBuOE-2-PyP^{5+} (BMX-001) (Fig. 8.1), we have reached our major milestone—First-in-Human Trial—with the ultimate goal of effectively treating patients.

While chemistry was primarily explored at Duke University, numerous groups of researchers, including those at Duke University, contributed to the growing insight into the biochemistry and biology of Mn porphyrins and their therapeutic effects. Several recent reviews addressed the chemistry, biology, and medical aspects of Mn porphyrins [6, 8–10, 13–18]. All studies combined helped identify compounds and conditions to be eventually tested in Clinical Trials. In addition to Mn porphyrins other redox-active drugs have also been addressed herein.

8.2 Design of an SOD Mimic

Dismutation of $O_2^{\bullet-}$ and its catalysis by SOD enzyme and its mimic. The best approach in mimicking the SOD-like activity is to mimic both the thermodynamics and kinetics of SOD-catalyzed $O_2^{\bullet-}$ dismutation. All isoforms of SODs, including Mn-, Fe-, Cu,Zn-, and NiSOD enzymes, operate at the same metal-centered reduction potential of ~+300 mV vs. NHE which is in between the potentials for the 1e⁻ reduction (+890 mV vs. NHE) and 1e⁻ oxidation of $O_2^{\bullet-}$(−180 mV vs. NHE). Such $E_{1/2}$ assures that both steps of dismutation process (oxidation of $O_2^{\bullet-}$ to O_2 and reduction of $O_2^{\bullet-}$ to H_2O_2) occur with similar rate constants of ~10^9 M^{-1} s^{-1} (Fig. 8.2) [11, 12, 14]. We started our design from an unsubstituted compound, MnT-4-PyP$^+$, which has unfavorable $E_{1/2}$ of −200 mV vs. NHE. Under such conditions Mn is stabilized in the +3 oxidation state and cannot be reduced with $O_2^{\bullet-}$ in a first step of dismutation process. The compound however possesses pyridyl nitrogens which allow quaternization with alkyl or modified alkyl substituents. Consequently the nitrogens become cationic and therefore impose electron-withdrawing effect upon the Mn site, reducing its electron density. In turn, Mn starts favoring the acceptance of electron from $O_2^{\bullet-}$.

Dismutation process involves oxidation and reduction of $O_2^{\bullet-}$. Like an SOD enzyme, MnP must be equally good anti- and prooxidant to be a good catalyst of $O_2^{\bullet-}$ dismutation. That in turn means that viewing SOD enzyme and its mimic solely as an antioxidant (reductant) [15] may be incorrect. Further, if H_2O_2 produced during reduction of $O_2^{\bullet-}$ is maintained at nM levels by enzymes such as catalase and glutathione peroxidase (GPx), the SOD enzyme and/or its mimic would be considered antioxidative defenses. If such conditions are not met, such as in cancer or grave conditions (such as exposure to high radiation), H_2O_2 gets accumulated and SOD enzyme does not function anymore as an antioxidative defense. In cancer H_2O_2-removing enzymes are frequently either downregulated or upregulated but inactivated ([16, 22, 23] and refs therein). For example, if three out of four thiols of thioredoxins/peroxiredoxins are oxidized, these enzymes lose their ability to reduce H_2O_2. Under such conditions, SOD enzyme contributes to the accumulation of H_2O_2 and may enhance the disease—in case of cancer SOD therefore may act as an oncogene.

Adjusting the thermodynamics and kinetics of the catalysis of $O_2^{\bullet-}$ dismutation. In Phase 1 the *para* MnT-4-PyP$^+$ was substituted with methyl groups in MnTM-4-PyP^{5+} (Fig. 8.1). The methylation shifted $E_{1/2}$ from −200 to +60 mV vs. NHE. Such shift allowed the compound to be a modest SOD mimic with log $k_{cat}(O_2^{\bullet-})$ of 6.58 [3]. When the nitrogens are moved from *para* (4) to *ortho* (2) positions, as in

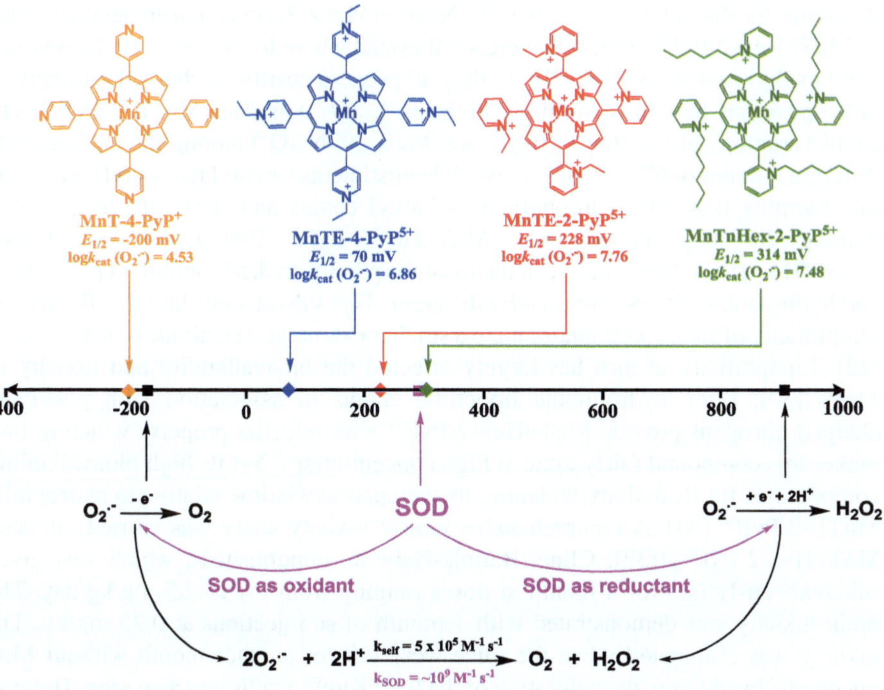

Fig. 8.2 Tuning of the thermodynamics of $O_2^{\bullet-}$ dismutation catalyzed by Mn porphyrins. Starting from MnT-4-PyP$^+$, over MnT-2-PyP$^+$, the derivatization (quaternization) of pyridyl nitrogens resulted in creation of the most powerful synthetic *ortho* isomeric cationic Mn porphyrin-based SOD mimics, MnTE-2-PyP^{5+} and MnTnHex-2-PyP^{5+}. Adapted from [15]

MnTM-2-PyP^{5+}, a stronger electron-withdrawing effect was imposed upon Mn site. In turn, $E_{1/2}$ is shifted from +60 to +220 mV vs. NHE and Mn starts favoring the acceptance of electron from $O_2^{\bullet-}$ more than it does in *para* MnTM-4-PyP^{5+}. Starting from unsubstituted MnT-2-PyP$^+$ to MnTM-2-PyP^{5+} a total shift in $E_{1/2}$ of 500 mV (from -280 mV to +220 mV vs. NHE) was achieved. The SOD-like activity was largely enhanced, log $k_{cat}(O_2^{\bullet-})=7.79$ [7]. In Phase 2, the ethyl analog, MnTE-2-PyP^{5+}, was synthesized [11] (log $k_{cat}(O_2^{\bullet-})=7.76$, $E_{1/2}=+228$ mV vs. NHE), and became the most studied Mn porphyrin in vitro and in vivo [6–18].

Further, *para* MnTM-4-PyP^{5+} is fairly planar and thus associates with nucleic acids, whereupon it loses SOD-like activity and becomes toxic [7]. Yet the bulkiness of *ortho* compound, where the methyl groups are located above or below the porphyrin plane, reduces such interactions; in turn *ortho* MnTM(and E)-2-PyP^{5+} is less toxic and has preserved in vitro and in vivo SOD-like activity relative to MnTM-4-PyP^{5+} [7]. Such "*ortho*" effect associated with four positive charges in the close vicinity of the Mn site electrostatically facilitates the approach of anionic $O_2^{\bullet-}$ to cationic MnP. It was soon demonstrated that such electrostatic facilitation could account for about two orders of magnitude increase in $k_{cat}(O_2^{\bullet-})$ going from a MnP that is neutral on periphery (MnBr$_8$T-2-PyP$^+$) to a compound that has four positive charges on periphery, MnTE-2-PyP^{5+}; importantly both compounds have same $E_{1/2}$, i.e., same thermodynamics for dismutation process [19]. Next, in Phase 3, several *ortho* analogs (such as $MnTnHex\text{-}2\text{-}PyP^{5+}$, Fig. 8.1) were synthesized where length of alkyl chain varied from methyl to nonenyl [28, 29]. As they all possess positively charged nitrogens in *ortho* positions they have similarly high log $k_{cat}(O_2^{\bullet-})$ as does $MnTE\text{-}2\text{-}PyP^{5+}$. The small variations in $E_{1/2}$, log $k_{cat}(O_2^{\bullet-})$ and log $k_{red}(ONOO^-)$ among the analogs with different number of CH_2 groups in pyridyl substituents (methyl to *n*-octyl) are due to the interplay between hydrophobicity of alkyl chains and steric effects [12, 20]. Among the long-chained ones, $MnTnHex\text{-}2\text{-}PyP^{5+}$ (log $k_{cat}(O_2^{\bullet-})=7.48$ and $E_{1/2}=+314$ mV vs. NHE) has been the most frequently studied compound [12]. Steric and hydrophobic effects, imposed by different alkyl substituents, largely affected the lipophilicity of such compounds and to a smaller extent the magnitude of log $k_{cat}(O_2^{\bullet-})$ [12]. Lipophilicity in turn has largely affected the bioavailability and toxicity of MnPs [30]. Long hydrophobic lipophilic chains in association with positively charged nitrogens provide $MnTnHex\text{-}2\text{-}PyP^{5+}$ with micellar property which in turn makes this compound fairly toxic at high concentrations. Yet its high bioavailability compensates for its toxicity widening its therapeutic window relative to hydrophilic $MnTE\text{-}2\text{-}PyP^{5+}$ [31]. A comprehensive mouse toxicity study was carried out with $MnTnHex\text{-}2\text{-}PyP^{5+}$ ([32], Cline, Batinić-Haberle, unpublished), which was given subcutaneously (sc) for a month at doses ranging from 0.1 to 2.5 mg/kg/day. The brain toxicity was demonstrated with 1 month of sc injections at 0.25 mg/kg. The toxicity was eliminated when the mice were left for another month without MnP injections. In addition, the coloration of the liver Kupffer cells was also seen (Batinić-Haberle, Vujaskovic, Cline et al., unpublished). To overcome the micellar-based toxicity of $MnTnHex\text{-}2\text{-}PyP^{5+}$, oxygen atoms were inserted deep into its alkyl chains in Phase 4 of the drug development (Fig. 8.1). Indeed, the butoxyethyl analog $MnTnBuOE\text{-}2\text{-}PyP^{5+}$ (BMX-001, log $k_{cat}(O_2^{\bullet-})=7.83$ and $E_{1/2}=+277$ mV vs. NHE) (Fig. 8.4) is ~4–5-fold less toxic than $MnTnHex\text{-}2\text{-}PyP^{5+}$, while equally lipophilic and equally powerful SOD mimic; it is now in aggressive development towards Clinic [35]. $MnTnHex\text{-}2\text{-}PyP^{5+}$, but not $MnTnBuOE\text{-}2\text{-}PyP^{5+}$ (Warner, Sheng et al., unpublished), is particularly suited for the suppression of stroke infarct volume as it crosses blood brain barrier to a large extent [36]. Yet, the inferior intellectual property rights of $MnTnHex\text{-}2\text{-}PyP^{5+}$ preclude its development towards Clinic. Numerous compounds were since synthesized by us and others where different *meso* and pyrrolic sites on porphyrin core were modified (Figs. 8.3 and 8.4).

Structure–activity relationship. Based on numerous compounds synthesized (Fig. 8.4) and many obtained from commercial sources, the structure–activity relationship (SAR) was established (Fig. 8.5); it is still the only available relationship of that kind for any class of SOD mimics [11, 14, 15, 19] The first SAR was established for Mn and Fe porphyrins of different properties [11]; it was later modified to account for the differences among those compounds with respect to their electrostatics, bulkiness and shape [19]. Most importantly we showed that such SAR is valid for different classes of redox active drugs and for different types of metal couples involved (Fig. 8.5) [14, 15]. What that basically means is that $O_2^{\bullet-}$ does not care with whom it exchanges electron as far as it happens at appropriate reduction potential. Thus $O_2^{\bullet-}$ can couple with Mn^{III}/Mn^{II} redox couple in porphyrins, salens,

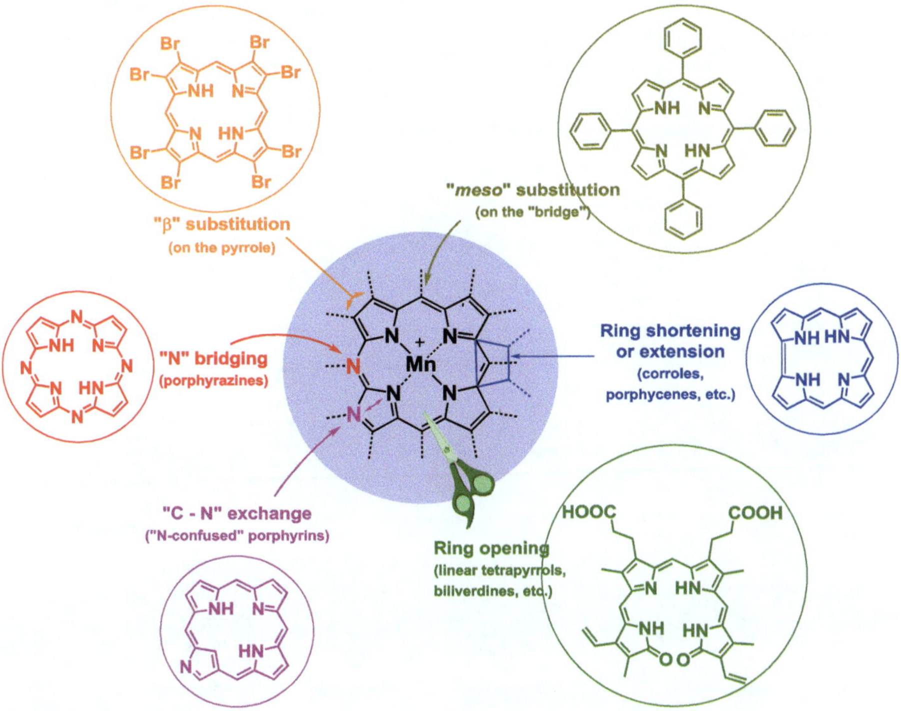

Fig. 8.3 Numerous compounds, bearing porphyrin core, were explored and used for mechanistic and therapeutic purposes. Over decades, the different sites on porphyrin core (*meso* and pyrrolic positions) were modified by us and others. Jonathan Sessler et al. worked on compounds with extended porphyrin core, texaphyrins [24], while Zeev Gross's group explored the compounds with shrinked core, corroles [25]. Our choice fell on porphyrins [14, 15]. We also studied linear tetrapyrroles, biliverdins and their analogs [26, 27]. For the sake of scientific pleasures we also briefly explored *N*-confused Mn porphyrins and Mn porphyrazines [27]

and cyclic polyamines, and with Mn^{IV}/Mn^{III} redox couple in biliverdins and corroles. It is further irrelevant whether the compound starts the dismutation process with reduction of $O_2^{\bullet -}$ to H_2O_2 in a first step while oxidizing Mn^{II} to Mn^{III}, such as with cyclic polyamines, or with oxidation of $O_2^{\bullet -}$ to O_2 in a first step while being reduced from Mn^{III} to Mn^{II}, as in MnPs.

Besides Mn(III) alkyl- and alkoxyalkylpyridylporphyrins, texaphyrins [24], porphyrazines, biliverdin and biliverdin analogs [26, 27]), several other types of compounds have been explored over the last 25 years (Fig. 8.4). Those are: Mn salen derivatives (EUK-8, EUK-132, EUK-207), Mn cyclic polyamines (M40403 and GC 4419), Mn and Fe corroles, Fe(III) polyethyleneglycolated porphyrin (FP-15), Fe(III) porphyrin with carboxylato substituents on pyridyl nitrogens (INO-4885) and metal oxides, such as CeO_2 [10, 14]. Non-metal based nitrones and nitroxides were often incorrectly described as SOD mimics since under physiological pH they lack SOD-like activity. However, when nitroxide is oxidized to oxoammonium

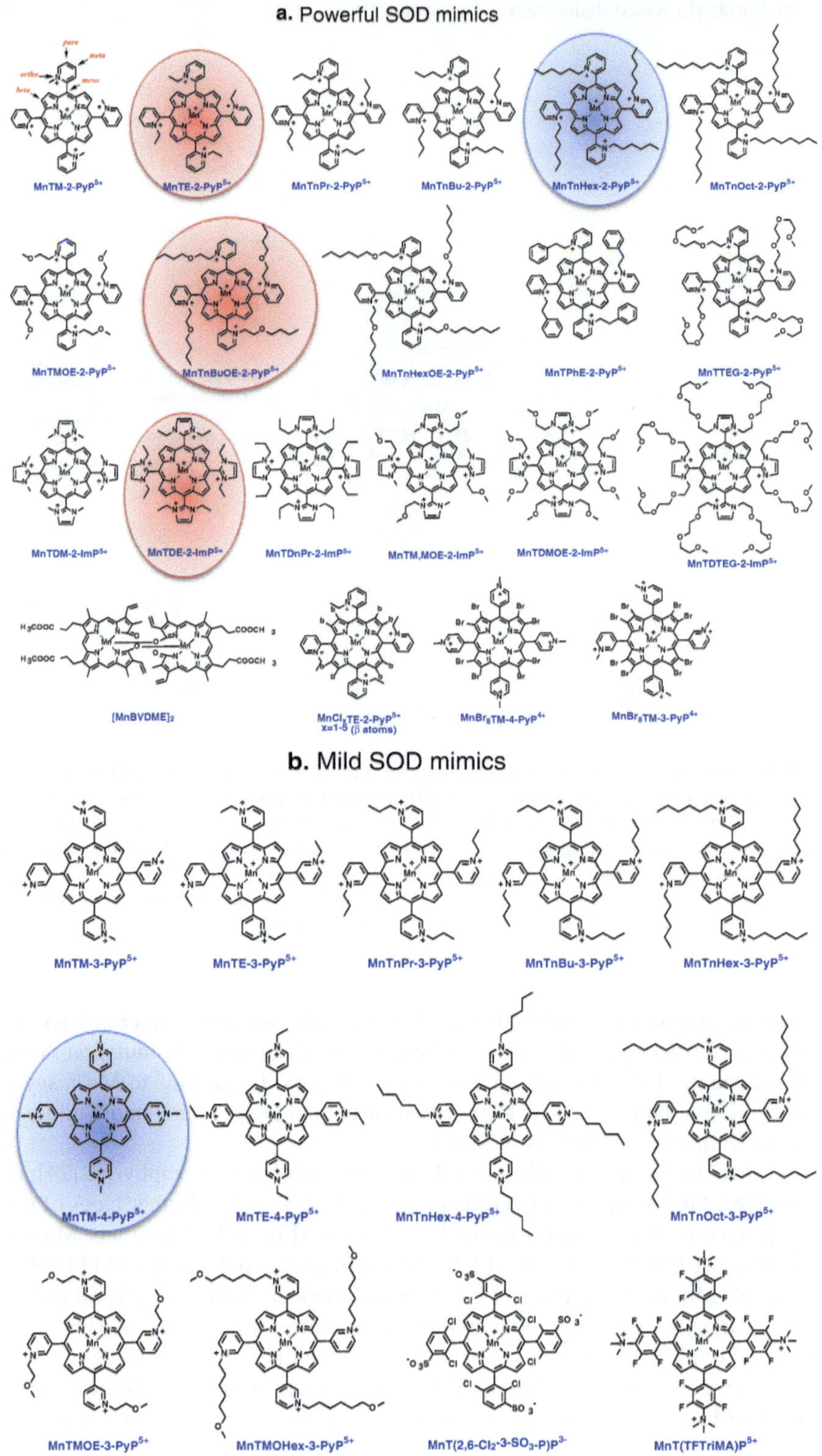

Fig. 8.4 Structures of compounds that have been tested in vitro and in vivo and produced therapeutic effects. Indicated in *circles* are those compounds which are the most studied ones (*blue*)

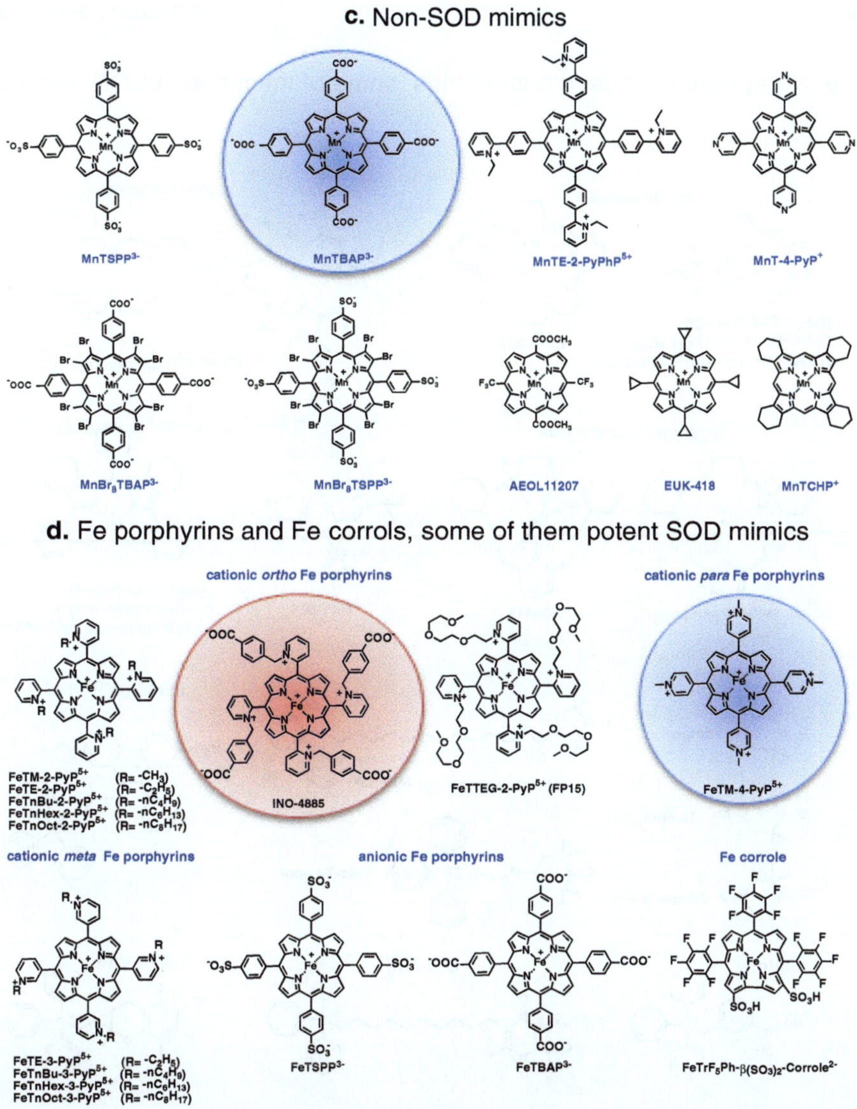

Fig. 8.4 (continued) and those which are in or are advancing towards Clinical Trials (*red*). Some of those ((**a**), (**b**), and (**e**)) are SOD mimics but the others are not (**c**). MnTE-2-PyP^{5+} is in Clinical Trials on islet transplants in Canada. MnTnBuOE-2-PyP^{5+} is entering two Phase I/II Clinical Trials at Duke University in spring of 2016 as a radioprotector of normal tissue with cancer patients: normal brain (with glioma patients) (NCT02655601) and salivary glands and mouth mucosa (with head and neck cancer patients) [33, 34]. Most metalloporphyrins are Mn complexes but few Fe porphyrins (**d**) have been tested also, and one of those, Fe(III) *ortho N*-substituted pyridyl analog, INO-4885 is in Clinical Trials. We have also listed the structures of Mn(III) salen derivatives and Mn(II) cyclic polyamines [GC4403 (also known as M40403, SC-72325 and KM40403), and GC4419] in (**e**). GC4419 is in Clinical Trials as a radioprotector of mouth mucosa with head and neck cancer patients (https://clinicaltrials.gov/ct2/show/NCT01921426). In (**e**) are also listed compounds which are not metal complexes such as nitrone, nitroxides, and MitoQ-based compounds. For other compounds please consult reviews in references [10, 14–18]. (**a**) Powerful SOD mimics. (**b**) Mild SOD mimics. (**c**) Non-SOD mimics. (**d**) Fe porphyrins and Fe corrroles, some of them potent SOD mimics. (**e**) Non-Mn porphyrin-based compounds, some of them potent SOD mimics

e. Non-porphyrin-based compounds, some of them potent SOD mimics

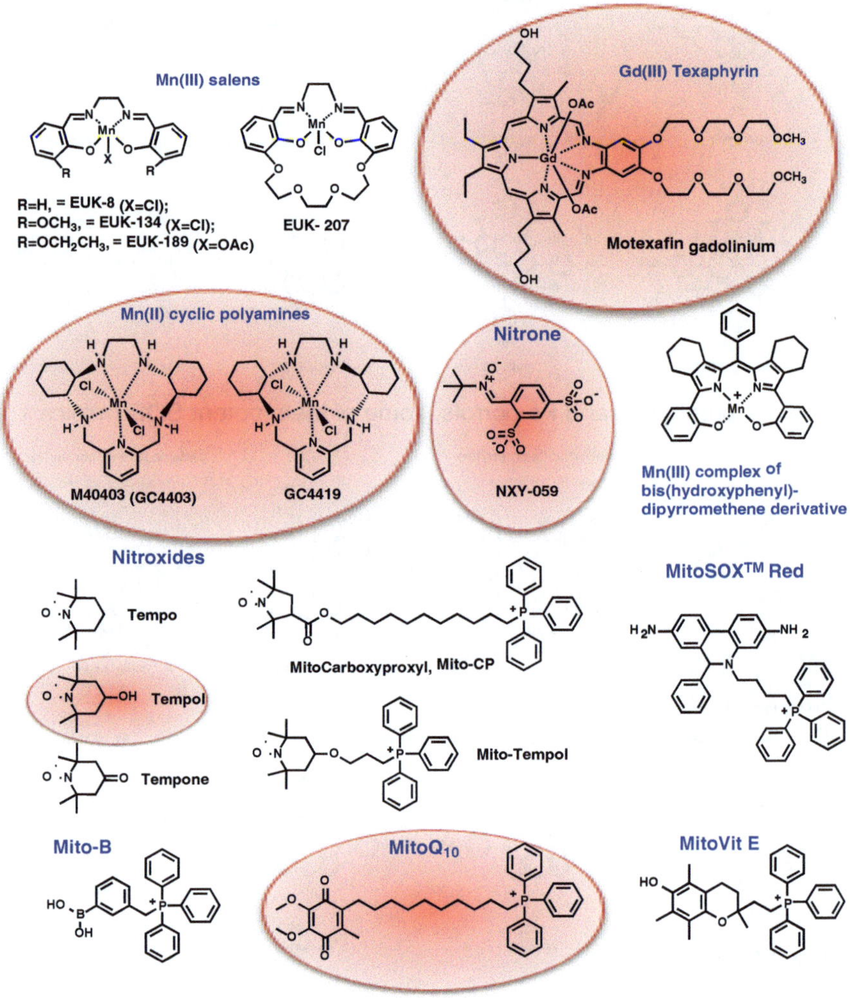

Fig. 8.4 (continued)

cation with reactive species, such as carbonate radical, it could be reduced back with $O_2^{\bullet-}$; in vivo this process could thus affect the levels of $O_2^{\bullet-}$. For review on SOD mimics and other redox-active therapeutics see also other reviews [6, 8–10, 13–18], the whole Issue of *Antioxid Redox Signal 2014* of "SOD therapeutics" dedicated to SOD mimics where chemistry and different therapeutic effects of Mn porphyrins are reviewed and other chapters of this "*Redox-Active Therapeutics*" book.

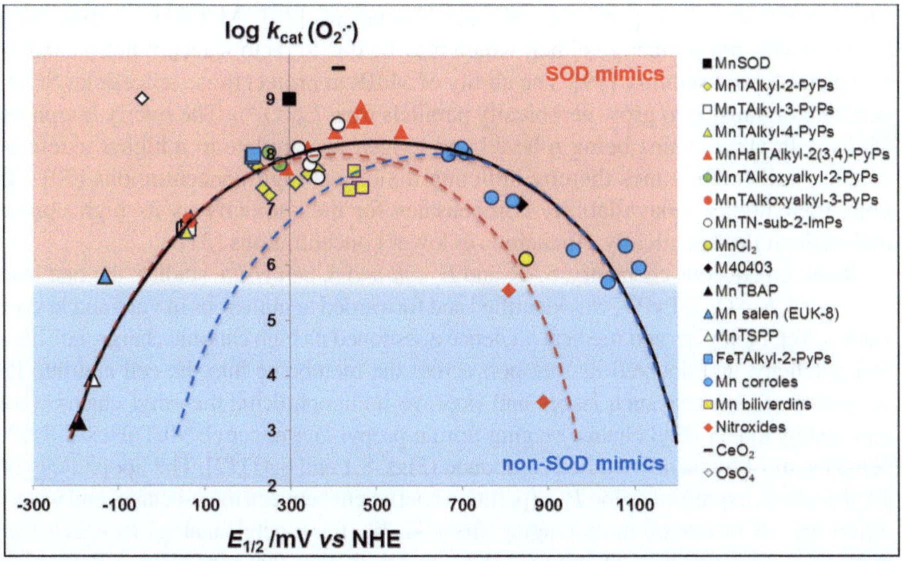

Fig. 8.5 Structure–activity relationship between the thermodynamic parameter, $E_{1/2}$ and the kinetic parameter $k_{cat}(O_2^{\cdot-})$ for $O_2^{\cdot-}$ dismutation. Listed are different classes of redox-active compounds. While some are aggressively developed towards clinic, the others are frequently used as mechanistic tools. At very negative values of $E_{1/2}$, compounds cannot be reduced with $O_2^{\cdot-}$ in a first step of dismutation process; such is the case of $MnTBAP^{3-}$. In turn they are not SOD mimics. At highly positive potentials the compounds are so electron-deficient, that they grabbed the electron ending in Mn^{2+} resting state, such as $MnTBr_8TM\text{-}4\text{-}PyP^{4+}$. While still very potent compound, its low stability constant of log $K = 8.08$ precludes its therapeutic development. Consequently at microM concentrations and at pH 7.8 the complex falls apart. At even higher positive potentials compounds are so stable in Mn +2 oxidation state that they cannot be oxidized with $O_2^{\cdot-}$ in a first step of dismutation process. It is generally true that the $E_{1/2}$ of redox-active compounds should be as close as possible to the potential which is in the middle between the potential for reduction (+890 mV vs. NHE) and oxidation of $O_2^{\cdot-}$(−180 mV vs. NHE). Yet, in certain cases the kinetic facilitation overcomes the inferior thermodynamics. Thus, M40403, a Mn(II) complex with aza crown ether (cyclic polyamine) operates at an inferior $E_{1/2}$ (close to the limit of +890 mV window); its favorable conformation though facilitates the dismutation process and compounds is among the most potent SOD mimics. Since Mn is in +2 oxidation state, the complex is not very stable; its log K is only 13.6, lower than the stability constant of Mn(II) with EDTA, log $K = 14.04$ [37, 38]. Adapted from [14]

8.3 Evaluation of Prospective Therapeutics in Unicellular $O_2^{\cdot-}$-Specific Models

Once the MnPs with good thermodynamic and kinetic properties were identified, they were tested in $O_2^{\cdot-}$-specific in vivo models, the aerobic growth of prokaryotic *Escherichia coli* and eukaryotic *Saccharomyces cerevisiae*. The SOD-deficient organisms cannot grow aerobically as they lack the protection against oxygen toxicity. As *E. coli* is more sensitive to the toxicity of lipophilic MnPs, the *S. cerevisiae* was introduced as an additional model for the evaluation of lipophilic long alkyl-chained Mn porphyrin-based SOD mimics [39]. Compounds that are not SOD mimics, such as

$MnTBAP^{3-}$, were not protective to such organisms [39]. M40403, despite high $k_{cat}(O_2^{\bullet -})$, was not protective either, which may be due to its low accumulation and/or low metal-ligand stability [39]. The ability of MnPs to protect those unicellular SOD-deficient organisms to grow aerobically parallels their $k_{cat}(O_2^{\bullet -})$. The overly lipophilic MnPs with alkyl chains being *n*-hexyl and longer, accumulate to a higher extent in cellular walls/membranes thereby inflicting toxicity at high concentrations [30]. Of note, their higher bioavailability compensates for their toxicity as its high uptake makes them therapeutically efficacious at lower concentrations [31].

Based on aqueous chemistry, SAR and *E. coli* and *S. cerevisiae* studies, the first lead compound, MnTE-2-PyP^{5+}, was identified and forwarded to numerous in vitro and in vivo studies. Yet, biologists and medical audience questioned its high cationic charge, i.e., high hydrophilicity, and doubted its transport across the membrane into the cell and into its organelles. To address such issues and improve its lipophilicity, the ethyl chains were replaced by longer alkyl chains, ranging from *n*-propyl to *n*-nonenyl; MnTnHex-2-PyP^{5+} being the most frequently studied compound (Figs. 8.1 and 8.4) [12]. The lipophilicity of alkyl analogs, expressed as log P_{OW} (partition coefficient between the *n*-octanol and water) differs by ~6 orders of units ranging from −2.32 (for *n*-octyl analog) to −8.16 (for methyl)—~1 log unit per CH_2 group [17]. It appears, though, that MnTnHex-2-PyP^{5+} with long hydrophobic alkyl chains and polar cationic pyridyl heads has strong micellar character and accumulates at high levels within the cells and exhibits thus fair toxicity [30]. Some toxicity, however, seems to be related to the factors other than lipophilicity and are not yet fully understood; fluorescent analog of identical lipophilicity, ZnTnHex-2-PyP^{4+}, but a potent photosensitizer, is not toxic to membranes if cells are grown in the dark [40]. To suppress the toxicity of MnTnHex-2-PyP^{5+}, oxygen atoms were introduced into alkyl chains. Oxygen atoms were firstly introduced at the end of the alkyl chains in *meta* (3) isomeric methoxyhexyl analog, MnTMOHex-3-PyP^{5+} [41]. However, the lipophilicity was largely reduced due to the interactions of polar oxygens with water molecules. In addition to the significant loss in lipophilicity, the synthetic problems were encountered due to the in situ reorganization of the alkylating agent during MnP synthesis. This resulted in an impure compound when the synthesis of *ortho* analog MnTMOHex-2-PyP^{5+} was attempted [Rajic et al., submitted]. However, when the oxygen atoms are buried deeper into the alkyl chain, such as in butoxyethyl derivative, MnTnBuOE-2-PyP^{5+} (Fig. 8.4), the lipophilicity was fully preserved and the micellar character suppressed. This was accompanied by ~4–5-fold reduced mouse toxicity of MnTnBuOE-2-PyP^{5+} in comparison with MnTnHex-2-PyP^{5+} [35]. Moreover a pure compound was isolated. The mechanism of the synthesis of alkoxyalkyl porphyrins including MnTnBuOE-2-PyP^{5+} has been studied [Rajic et al., submitted].

8.4 Exploring the Bioavailability of Mn Porphyrin-Based SOD Mimics

Mitochondrial accumulation. As soon as the excessive hydrophilicity of MnPs was questioned, HPLC/fluorescence and LC-MS/MS methods were developed to assess the bioavailability of different Mn and Fe porphyrins in different tissues and different cellular organelles [32, 42–44]. Given the critical role of mitochondria, most

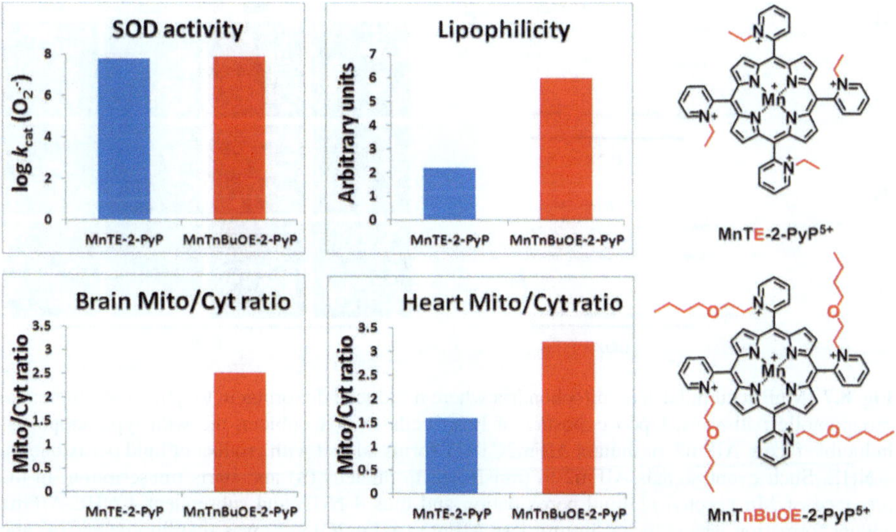

Fig. 8.6 The impact of lipophilicity of MnPs on their mitochondrial distribution. Two MnPs of nearly identical SOD-like activity [log $k_{cat}(O_2^{\bullet-})$ is 7.76 for ethyl (E) and 7.83 for butoxyethyl (BuOE) analog] have orders of magnitude different lipophilicities (expressed in distribution between *n*-octanol and water, P_{OW}) [28]. In turn only lipophilic MnTnBuOE-2-PyP^{5+} (log P_{OW}=−4.10) reaches brain mitochondria where it accumulates to a similar extent as in heart mitochondria. MnTE-2-PyP^{5+} (log P_{OW}=−7.79) was only found in heart but not in brain mitochondria [17, 44]

studies were aimed to quantify MnPs in these organelles. Independent studies of Skulachev and Murphy [45–47] showed that compounds must possess lipophilic character and positive charge(s) to accumulate in mitochondria, driven there by negative mitochondrial membrane potential. MnPs have excessive, pentacationic, charge; so we expected that such high cationic charge would compensate for the insufficient hydrophobicity and allow MnPs to accumulate in mitochondria. The ability of MnTE-2-PyP^{5+} and MnTBuOE-2-PyP^{5+} to accumulate in mitochondria was tested in mouse heart and brain. Mice were injected with MnPs for 5 days sc at 2 mg/kg/day [14, 44, 48]. Mice were perfused, organs extracted 6 h after last injection, and mitochondria isolated. Regardless of the type of organ (heart or brain), lipophilic MnTBuOE-2-PyP^{5+} was distributed between mitochondria and cytosol at similar ratio of ~3 (Fig. 8.6). The hydrophilic MnTE-2-PyP^{5+} accumulated 50 % more in mouse heart mitochondria than in cytosol—the same was true for its accumulation in the mitochondria of *S. cerevisiae* [49]. MnTE-2-PyP^{5+} was, however, not found in brain mitochondria (Fig. 8.6) [44, 50]. In an independent study on submitochondrial particles of bovine heart, Ferrer-Sueta et al. showed that MnTE-2-PyP^{5+} offers protection against ONOO$^-$-mediated damage [51].

Mn porphyrins mimic mitochondrial isoform MnSOD. Once it was shown that cationic MnPs are in mitochondria [17, 44] the question was asked: do they protect mitochondria? Several studies were performed, most of those in collaboration with St. Clair's group. The studies provided unambiguous evidence that cationic MnPs

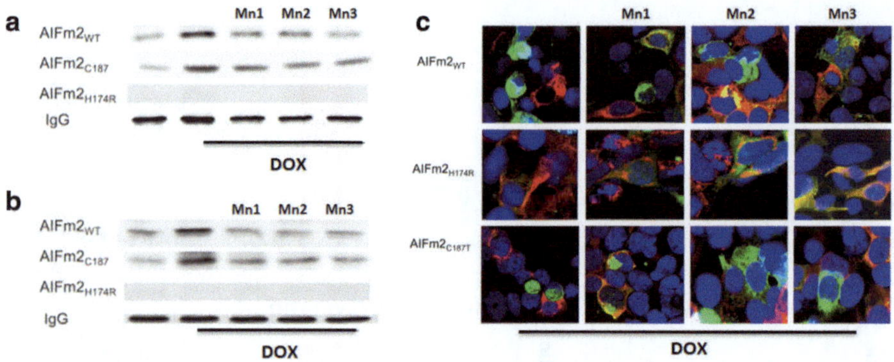

Fig. 8.7 MnP accumulates in mitochondria where it reduced doxorubicin toxicity and suppressed proapoptotic pathways. Upon exposure of heart cells to doxorubicin, the wild type apoptosis-inducible factor, AIFm2, or mutant AIFm2C187T forms adduct with product of lipid peroxidation, 4-NHE. Such event signals AIFm2 to translocate to nucleus (**a**) and starts transcription. In the presence of Mn porphyrins lipid peroxidation and thus 4-NHE and subsequent 4-NHE-AIFm2 adduct formation was reduced. In turn less AIFm2 (AIFm2C187T) was seen in nucleus (**a**) and cytosol (**b**); it stays in mitochondria. The effect is the largest with lipophilic MnTnBuOE-2-PyP^{5+}. Histidine 174 plays role in 4-HNE adduction of the AIFm2 protein; the H174R mutant did not bind to DNA even after exposure to 4-HNE [54]. (**c**) Immunofluorescence imaging of H9C2 cardiomyocytes transfected with lentiviral wild-type AIFm2 or H174 and C187 mutants treated with DOX and Mn porphyrins. AIFm2 proteins were linked to GFP (*green dye*). MitoTracker® was used to stain mitochondria (*red dye*), and DAPI to stain nuclei (*blue dye*). Mn1=MnTE-2-PyP^{5+}, M2=MnTnHex-2-PyP^{5+}, Mn3=MnTnBuOE-2-PyP^{5+}. Adapted from [54]

(MnTE-2-PyP^{5+}, MnTnBuOE-2-PyP^{5+}, and MnTnHex-2-PyP^{5+}) mimic MnSOD [52–56]. The cellular studies on cardiomyocytes showed that lipophilic MnTnBuOE-2-PyP^{5+} was somewhat more efficacious in protecting mitochondria than MnTE-2-PyP^{5+} [54].

The most fascinating stories on the ability of Mn porphyrins to mimic mitochondrial MnSOD enzyme are the ones reported by Miriyala et al. (Fig. 8.7) [54] and the one by Zhao et al. (Fig. 8.8) [55]. In a first study, cardiomyocytes were exposed to doxorubicin which resulted in lipid peroxidation and formation of 4-hydroxynonenal, 4-HNE. 4-HNE forms an adduct with mitochondrial apoptosis-inducible factor AIFm2, which then translocates from mitochondria into nucleus to start transcription and upregulate apoptosis. When the cells were treated with MnPs prior to exposure to doxorubicin, lipid peroxidation was reduced and AIFm2 remained to a large extent in mitochondria (Fig. 8.7).

In a second study, it was shown that MnTE-2-PyP^{5+} suppressed skin carcinogenesis to a larger extent than MnSOD in MnSOD overexpressor mice (Fig. 8.8) [55]. In an MnSOD-overexpressor mouse, the MnSOD enzyme suppressed both apoptosis and proliferation pathways. However, MnTE-2-PyP^{5+} was administered after cell underwent apoptosis but before proliferation started. Thus the effect of an MnSOD mimic occurred only at the level of proliferation. Tumor was induced by the mutagenic chemical initiator 7,12-dimethylbenz(α)-anthracene (DMBA), and

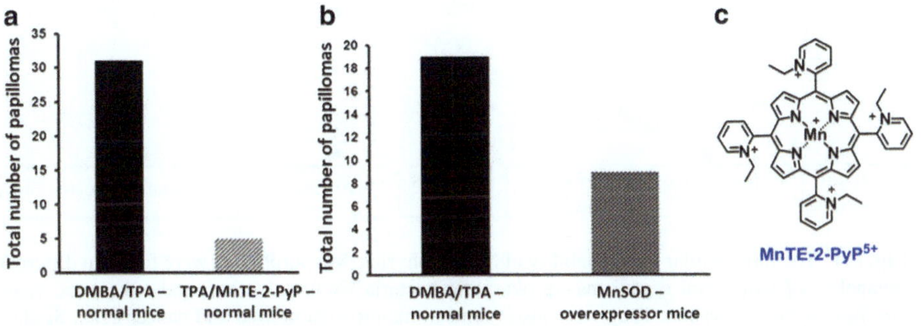

Fig. 8.8 MnSOD mimic, MnTE-2-PyP^{5+} (MnP) (**a** and **c**), and MnSOD enzyme (**b**) suppressed carcinogenesis as demonstrated by the reduced number of papillomas. The timing of the MnSOD expression in MnSOD overexpressor mice cannot be adjusted over the course of experiment. Thus MnSOD suppressed both apoptotic and proliferative pathways. Yet the timing of the exogenous administration of MnP was adjusted so that it did not affect apoptosis but prevented cell proliferation only. Thus the effect of MnP (**a**) was more profound and suppressed skin carcinogenesis to a larger extent than MnSOD enzyme (**b**). Tumor was induced by the mutagenic chemical initiator 7,12-dimethylbenz(α)-anthracene (DMBA), and was promoted by 12-O-tetradecanoylphorbol-13-acetate (TPA). Apoptosis peaks 6 h after TPA addition and proliferation at 24 h. The suppression of AP-1 activity was much more significant when MnTE-2-PyP^{5+} was given 12 h following TPA treatment (*not shown*). MnP also suppressed proliferating cellular nuclear antigen expression (*not shown*) [55]. MnTE-2-PyP^{5+} was applied to the skin (5 ng/mouse) either at 30 min prior to TPA (SOD/TPA) or at 12 h following the TPA application (TPA/SOD) 5 days per week for 14 weeks. Adapted from [55]

was promoted by 12-O-tetradecanoylphorbol-13-acetate (TPA). Apoptosis peaks 6 h after TPA addition while proliferation peaks at 24 h. MnTE-2-PyP^{5+} was applied to the skin (5 ng/mouse) either at 30 min prior to TPA (SOD/TPA) or at 12 h following the TPA application (TPA/SOD) 5 days per week for 14 weeks. The suppressive effect was much more significant when MnTE-2-PyP^{5+} was given 12 h following TPA treatment. TPA is known to cause the activation of activator protein-1 (AP-1), and activates the expression of several genes required for the cell proliferation in the skin, such as proliferating cellular nuclear antigen, PNCA [55].

Bioavailability of Mn porphyrins in different cellular organelles. The accumulation studies have been conducted on simple unicellular and complex organisms (Fig. 8.9) [30, 40, 57–60]. In addition to mitochondria, Mn porphyrins localize in other cellular organelles. Accumulation of Mn porphyrins in nucleus vs. cytosol was done on macrophages [13], while localization of Zn porphyrins was reported in different organelles of human colon adenocarcinoma, including plasma membrane, lysosomes, and endoplasmatic reticulum [40, 58]. Distribution of *ortho* isomeric Mn porphyrins, bearing alkyl chains of different length, in mitochondria, cytosol and cell wall were studied in unicellular eukaryotic and prokaryotic *S. cerevisiae* and *E. coli* [30, 49]. Finally, the levels of Mn porphyrins were also determined in mitochondria and cytosol of mouse heart and mouse brain [14, 42, 44]. In all studies, either cellular or animal, the increase in the length of alkyl chains increases the mitochondrial-to-cytosolic ratio. Few studies, carried out on Zn porphyrins,

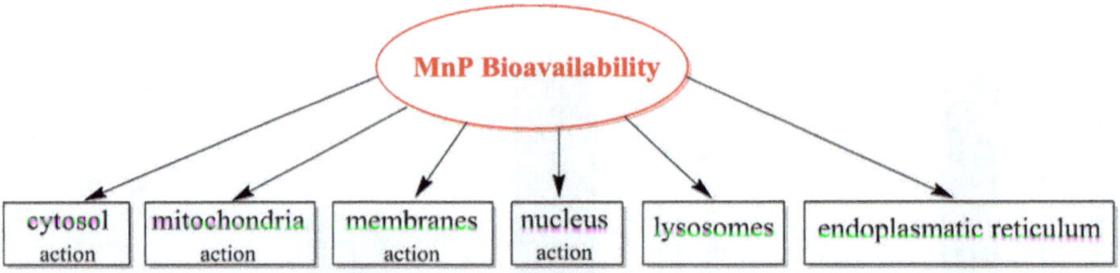

Fig. 8.9 The intracellular bioavailability of Mn porphyrins. Mn porphyrins were found in different organelles of single-cell organisms—prokaryotic bacteria *Escherichia coli* and eukaryotic yeast *Saccharomyces cerevisiae*. They were also found in mammalian normal and cancer cells. Studies demonstrated that MnPs act at the sites of their accumulation (noted as "action"). The fluorescent property of Zn analogs allows for the use of imaging techniques in order to confirm their accumulation in different cellular organelles. Such studies also provided insight into the accumulation of non-fluorescent Mn analogs. It is highly likely that MnPs accumulate and act at other cellular sites not assessed yet. The lipophilicity, charge and size control the site and magnitude of MnP accumulation

employed imaging techniques thereby taking the advantage of their fluorescent properties [40, 58]. LC-MS/MS technique was used to measure the bioavailability of Mn porphyrins in yeast [49] and mammalian systems [14, 32, 42, 44, 50], while uv/vis spectroscopy in *E. coli* [30]. Zn(II) and Mn(III) porphyrins differ with respect to a single charge on metal site. When the pyridyl substituents are long, such as is the case with *n*-hexyl chains, they hinder a single positive charge on metal site; in turn pentacationic MnPs are felt by their surrounding similarly to tetracationic ZnP as they are of identical lipophilicities. Moreover, Mn(III) porphyrins are in vivo likely reduced to their Mn(II) analogs, which have the same overall charge as Zn(II) porphyrins [59]. Therefore the knowledge obtained from the accumulation studies of fluorescent ZnPs may provide a good insight into the accumulation of Mn porphyrins.

As soon as the first in vitro and in vivo bioavailability studies were performed, the obvious question arose: does MnP act at the site of its accumulation? At this point we have substantial evidence that cationic MnPs mimic the mitochondrial matrix MnSOD enzyme, which in turn suggests that such action happens in mitochondria (see under *Mn porphyrins mimic mitochondrial isoform MnSOD*) [51, 53, 54]. We also have fair evidence that MnPs act in nucleus upon the cysteines of p50 subunit of NF-κB preventing its DNA binding [13]. The action upon the p65 subunit of NF-κB presumably happens in cytosol [61]. Accumulation within lipid membranes, and the toxicity associated with their damage, was also observed and to a much larger extent with lipophilic long-chained- than short methyl- and ethylpyridyl analogs [30, 62].

Bioavailability of MnPs in organs and tissues. MnPs are bioavailable when given subcutaneously (sc), intraperitoneally (ip), intravenously (iv), intramuscularly, transdermally, via inhalation but only slightly when given orally [32, 42, 44, 50]. The sc route will be used in Clinical Trials. The manuscript is in preparation to explain the artifacts that accounted for high oral availability reported [32]. The oral availabilities are re-evaluated to be 0.6 % for MnTE-2-PyP^{5+}, 2.9 % for MnTnHex-2-PyP^{5+}, and 3.9 % for MnTnBuOE-2-PyP^{5+} (Spasojevic et al., in preparation).

MnPs were found in all mouse and rat tissues thus far studied: liver, kidney, heart, lungs, spleen, colon, stomach, prostate, brain, spinal cord, tongue, and salivary glands [32, 42–44, 48]. The highest levels were found in liver and kidney, which are organs that serve as a depot for maintaining MnPs levels in other organs. The more lipophilic compounds accumulate in tissues more than the hydrophilic MnPs. Under same dosing regime, the highest liver concentration was found for MnTnHex-2-PyP^{5+}, and the lowest for MnTE-2-PyP^{5+}. While MnTnBuOE-2-PyP^{5+} is as lipophilic as MnTnHex-2-PyP^{5+}, it accumulates to a much lower extent in liver presumably due to the presence of four polar oxygen atoms in its pyridyl substituents (Huang, Spasojevic et al., unpublished). For the same reason, presumably, the brain levels of MnTnBuOE-2-PyP^{5+} are lower than those of MnTnHex-2-PyP^{5+} presumably limiting its efficacy in reducing infarct size ([36], Sheng, Warner, Spasojevic et al., unpublished). Yet, on positive side, less brain toxicity would be imposed by MnTnBuOE-2-PyP^{5+} than by MnTnHex-2-PyP^{5+}.

Bioavailability of MnPs in normal vs. tumor tissue. A mouse study on a 4T1 mammary tumor growing on a flank showed ~7-fold higher concentration of MnTE-2-PyP^{5+} in tumor than in normal tissue [63].

The redox-active drug must have proper thermodynamic of Mn site ($E_{1/2}$ for Mn^{III}P/Mn^{II}P redox couple) and kinetic properties (shape, size, charge) to undergo reactions of interest. *The bioavailability is the second major factor that controls the therapeutic efficacy of MnPs.* David Warner and Huaxin Sheng provided recently the clear evidence that those compounds, which have identical SOD-like properties, produce identical therapeutic effects in reducing (stroke) infarct volume ONLY if they reach the same target (brain) at similar concentrations (Sheng, Warner, unpublished). Thus, only when injected intracerebroventricularly (directly into the brain), hydrophilic MnTE-2-PyP^{5+} reduced stroke volume and improved neuroscore to a same degree as lipophilic MnTnHex-2-PyP^{5+} and MnTnBuOE-2-PyP^{5+}. Yet, when given subcutaneously, only MnTnHex-2-PyP^{5+} shows significant efficacy in suppressing stroke injury post middle cerebral artery occlusion (Warner, Sheng, unpublished). The LC-MS/MS analysis showed that, when those three MnPs were given subcutaneously, MnTnHex-2-PyP^{5+} accumulated in brain to the highest extent (Spasojevic, Sheng et al., unpublished).

8.5 MnPs and Reactive Species

Due to its evolved large protein structure, the SOD enzyme is $O_2^{\bullet -}$ specific, while small MnP molecule of ~1000 Da is not. As SOD and its mimic have the same thermodynamic properties, i.e., they operate at same metal-centered reduction potential, they can undergo same reactions yet at different rates. The reactivities of SOD enzymes toward species other than $O_2^{\bullet -}$ are orders of magnitude slower than those of MnPs. Thus, cationic MnPs reduce $ONOO^-$ with $k_{red}(ONOO^-) > 10^7$ M^{-1} s^{-1}, while the $k_{red}(ONOO^-)$ of MnSOD is ~10^4 M^{-1} s^{-1}. Listed in Fig. 8.10 are those

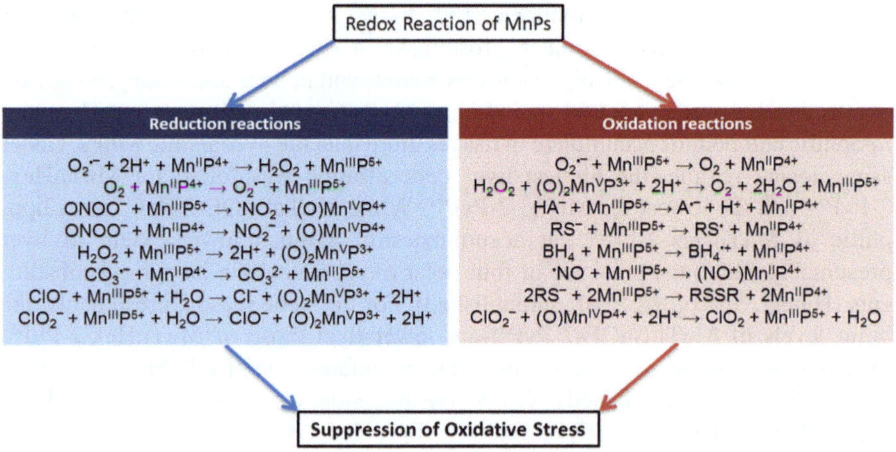

Fig. 8.10 MnPs undergo reduction and oxidation reactions with different species acting as either antioxidants (reductants) or prooxidants, respectively. With reactive species, listed in left column, MnPs act as antioxidants (reductants), while with those listed in right column as prooxidants. Such reactions result in suppression of oxidative stress presumably via affecting cellular transcription [14, 15, 63–66]. This is mostly true for the reaction of MnP with protein thiol (RS^-)—pro-oxidative in nature, resulting in their oxidation or *S*-glutathionylation. Subsequently the activation of signaling proteins such as NF-κB, and complexes I and III of mitochondrial respiration is suppressed, while Nrf2 gets activated. The magnitude of the suppression depends upon the levels of reactants: MnP, H_2O_2, thiols (GSH and protein thiols). The reaction of MnP with protein thiol was shown to require both H_2O_2 and glutathione. It demonstrates that in vivo MnP likely mimics the GPx enzyme [65]

reactions of MnPs that have been thus far addressed and quantified [15]. In Fig. 8.11 rate constants are listed for reactions of MnTE-2-PyP^{5+}. All *ortho* cationic Mn(III) *N*-substituted pyridyl porphyrins that bear positive charges on pyridyl nitrogens have similar $E_{1/2}$ for $Mn^{III}P/Mn^{II}P$ redox couple and similar electrostatics. The $E_{1/2}$ controls their SOD-like activity which in turn reportedly parallels all other activities thus far studied [14, 15, 21]. Thus such compounds react with the same reactive species at similar rates, which are also similar to those of MnTE-2-PyP^{5+} shown in Fig. 8.11. The differences in size and lipophilicity give rise to only minor differences in rate constants. MnPs *can* adopt in vivo four oxidation states, +2, +3, +4, and +5; all of those are apparently involved in their in vivo reactions. Given the complex biological milieu and complex reactivity of MnPs one may imagine that many more reactions, than those thus far explored, are possible. Many of the reactions listed have reportedly an impact on cellular transcription. The prominent roles in actions of MnPs occupy H_2O_2 and GSH. MnPs are not efficacious catalyst for H_2O_2 dismutation (i.e., they are not mimics of catalase): the 2-electron reduction of H_2O_2 to water and its 2-electron oxidation to oxygen [64]. However, MnPs employ H_2O_2 in oxidizing or *S*-glutathionylating protein thiols thereby exercising glutathione peroxidase, GPx-like activity; H_2O_2 gets reduced to H_2O in such a process [14, 15, 65, 66] ((8.10)–(8.16)). *S*-Glutathionylation was reported to occur with NF-κB and complexes I and III of mitochondrial respiration and lead to their inactivation. If the extent (yield) of *S*-glutathionylation of NF-κB is moderate (under

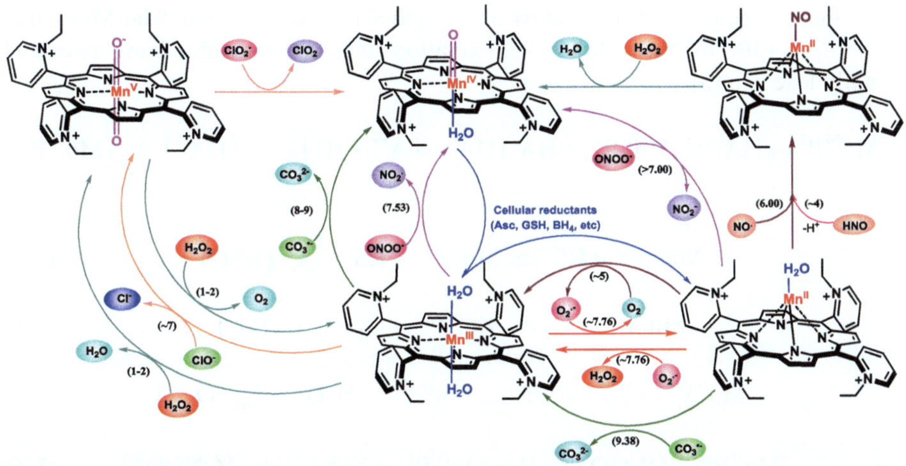

Fig. 8.11 The reactivity of MnTE-2-PyP^{5+} towards different reactive species. Mn in MnTE-2-PyP^{5+} exists in four different oxidation states (+2, +3, +4, +5) which are involved in its in vivo reactions with different reactive species. Such complex reactivity is enabled by biologically compatible reduction potential of Mn site which allowed MnP to act as either mild prooxidant or antioxidant depending upon the type of the species it encounters. Rate constants of all reactions of MnPs, that have *ortho* pyridyl nitrogens substituted with alkyl or alkoxyalkyl groups, are similar and thus similar to those of MnTE-2-PyP^{5+}. All are controlled by their (similar) metal-centered reduction potentials for Mn^{III}P/Mn^{II}P redox couple. The reactions involving high oxidation states often involve binding of reactive species (such as ONOO$^-$ and H_2O_2) to Mn site prior to electron transfer. The $E_{1/2}$ values for the couples involving Mn in high +4 and +5 oxidation states are very similar for all Mn porphyrins [14, 15, 21]. Thus the rates of such reactions are controlled by the binding of the species which is controlled by the electron density of the Mn site, which in turn is dependent upon the $E_{1/2}$ for Mn^{III}P/Mn^{II}P redox couple. Adapted from [15]

moderately high levels of H_2O_2 produced in the course of oxidative stress injury), the moderate suppression of NF-κB activation would result in a suppression of inflammation and would give rise to cell healing. The effect has been demonstrated in diabetes and stroke models [22, 36, 67, 68]. If the *S*-glutathionylation and subsequent inhibition of NF-κB is massive, the cell would undergo apoptosis (see under *Therapeutic effects of Mn porphyrins in cancer—radio- and chemosensitization of tumor while radioprotection of normal tissue*). MnPs can also react with thiols in the absence of H_2O_2, acting as cysteine oxidase giving rise to thiyl radical while reducing themselves from Mn^{III}P to Mn^{II}P [13].

The reactions of Mn porphyrins that are most likely to occur in vivo, are listed below. Those Mn^{III}Ps that are potent SOD mimics and have $E_{1/2}$ in the region of +100 to +400 mV vs. NHE get readily reduced in vivo to Mn^{II}P and re-oxidized with different species, (8.1)–(8.9). The species involved in reactions of Mn^{III}P/Mn^{II}P are oxygen (O_2), superoxide ($O_2^{\bullet -}$), ascorbate (at pH 7.8 it exists mostly as monodeprotonated species, HA$^-$), glutathione (GSH, partly deprotonated in vivo), nitric oxide ($^\bullet$NO, Mn^{III}P reacts also with reduced $^\bullet$NO, nitroxyl HNO, giving rise to the identical product as with $^\bullet$NO [69]), protein thiol (RSH, partly deprotonated in vivo), peroxynitrite (ONOO$^-$), carbonate anion radical ($CO_3^{\bullet -}$), hypochlorite (ClO$^-$), and H_2O_2 (likely the deprotonated peroxide is a reacting species). The reactions of MnPs with those species are listed below and in other reviews [14, 15]. The MnPs that have $E_{1/2}$>+400 mV have Mn in +2 oxidation state, are not stable complexes and readily lose Mn [5].

$$\mathrm{Mn^{III}P^{5+} + HA^- \leftrightarrows Mn^{II}P^{4+} + HA^{\bullet}\left(HA^{\bullet} \rightarrow A^{\bullet-} + H^+\right)} \quad k_{\mathrm{ox}}\left(\mathrm{HA^-}\right), k_{\mathrm{red}}\left(\mathrm{Mn^{III}P}\right) \tag{8.1}$$

$$\mathrm{Mn^{III}P^{5+} + RS^- \leftrightarrows Mn^{II}P^{4+} + RS^{\bullet}} \quad k_{\mathrm{ox}}\left(\mathrm{RSH}\right) \tag{8.2}$$

$$\mathrm{Mn^{II}P^{4+} + O_2 \leftrightarrows Mn^{III}P^{5+} + O_2^{\bullet-}} \quad k_{\mathrm{red}}\left(\mathrm{O_2}\right) \tag{8.3}$$

$$\mathrm{Mn^{II}P^{4+} + O_2^{\bullet-} + 2H^+ \leftrightarrows Mn^{III}P^{5+} + H_2O_2} \quad k_{\mathrm{red}}\left(\mathrm{O_2^{\bullet-}}\right) \tag{8.4}$$

$$\mathrm{Mn^{II}P^{4+} + ONOO^- \leftrightarrows O = Mn^{IV}P^{4+} + NO_2^-} \quad k_{\mathrm{red}}\left(\mathrm{ONOO^-}\right) \tag{8.5}$$

$$\mathrm{Mn^{II}P^{4+} + CO_3^{\bullet-} \leftrightarrows Mn^{III}P^{5+} + CO_3^{2-}} \quad k_{\mathrm{red}}\left(\mathrm{CO_3^{\bullet-}}\right) \tag{8.6}$$

$$\mathrm{Mn^{II}P^{4+} + {}^{\bullet}NO \leftrightarrows \left({}^{\bullet}NO\right)Mn^{II}P^{4+}} \quad k_{\mathrm{nitrosylation}}\left(\mathrm{Mn^{II}P}\right) \tag{8.7}$$

$$\mathrm{Mn^{II}P^{4+} + ClO^- \leftrightarrows O = Mn^{IV}P^{4+} + Cl^-} \quad k_{\mathrm{red}}\left(\mathrm{ClO^-}\right) \tag{8.8}$$

$$\mathrm{O = Mn^{IV}P^{4+} + HA^- + 2H^+ \rightleftarrows Mn^{III}P^{5+} + HA^{\bullet}\left(HA^{\bullet} \rightarrow A^{\bullet-} + H^+\right)} \\ \mathrm{+ H_2O} \quad k_{\mathrm{ox}}\left(\mathrm{HA^-}\right) \tag{8.9}$$

The catalase-like activity of MnPs is insignificant [64]. Yet MnPs would undergo critical in vivo reactions, involving H_2O_2 and GSH, giving rise to oxidative modifications of thiols of signaling molecules (presumably mimicking GPx), which in turn would affect cellular transcription. Further, when H_2O_2 oxidizes $Mn^{III}P$ to O = $Mn^{V}P$ = O, in order to act in a GPx-like fashion, the levels of H_2O_2 get decreased due to its reduction to water (see (8.10)–(8.16)). The possible reactions, starting from reduced $Mn^{II}P$ involved in the formation of *S*-glutathionylated protein, RSSG, are listed below, (8.10)–(8.16) [70–72]. Reactions starting from $Mn^{III}P$ with H_2O_2 resulting in $(O)_2Mn^{V}P^{3+}$ species are possible also [93], but somewhat less likely due to high in vivo reducibility of $Mn^{III}P$ in the presence of excessive levels of cellular reductants such as ascorbate and glutathione.

$$\mathrm{\left(H_2O\right)Mn^{II}P^{4+} + H_2O_2 \leftrightarrows \left(H_2O\right)\left(O\right)Mn^{IV}P^{4+} + H_2O} \quad k_{\mathrm{red}}\left(\mathrm{H_2O_2}\right) \tag{8.10}$$

$$\mathrm{\left(H_2O\right)\left(O\right)Mn^{IV}P^{4+} + 2H^+ + RS^- \leftrightarrows \left(H_2O\right)Mn^{III}P^{5+} + RS^{\bullet} + H_2O} \quad k_{\mathrm{ox}}\left(\mathrm{RS^-}\right) \tag{8.11}$$

$$\mathrm{RS^{\bullet} + RS^{\bullet} \leftrightarrows RSSR}, \text{disulfide formation} \tag{8.12}$$

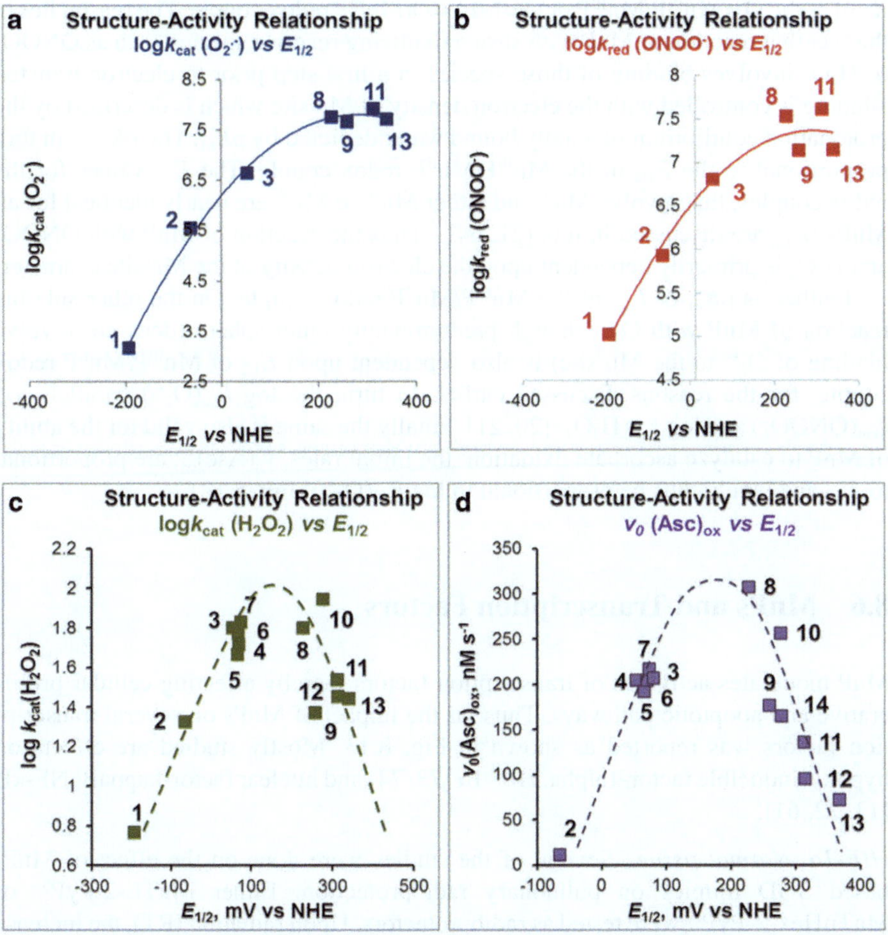

Fig. 8.12 The reactivities of MnPs thus far assessed correlate well with the thermodynamic property of Mn center, $E_{1/2}$ for $Mn^{III}P/Mn^{II}P$ redox couple. Because the log ($k_{cat}(O_2^{\bullet-})$) is linearly related to $E_{1/2}$, all other reactivities are proportional to the log $k_{cat}(O_2^{\bullet-})$ [21, 63]. Here are listed: (**a**) the log $k_{cat}(O_2^{\bullet-})$ vs. $E_{1/2}$; (**b**) the log $k_{red}(ONOO^-)$ vs. $E_{1/2}$; (**c**) the log $k_{cat}(H_2O_2)$ vs. $E_{1/2}$; (**d**) the $v_0(Asc)_{ox}$ vs. $E_{1/2}$ [15, 21, 63]. At pH 7.8 ascorbate is present predominantly as HA^- and peroxynitrite as $ONOO^-$

$$RSSR + GSH \leftrightarrows RSSG \tag{8.13}$$

$$RS^{\bullet} + GSH \leftrightarrows RSS^{\bullet-}G + H^+ \tag{8.14}$$

$$RSS^{\bullet-}G + O_2 \leftrightarrows RSSG + O_2^{\bullet-} \tag{8.15}$$

$$R-SOH + GS^- \leftrightarrows RSSG + OH^- \tag{8.16}$$

The rate constants for all the reactions thus far determined are dependent upon the $E_{1/2}$ of $Mn^{III}P/Mn^{II}P$ redox couple (Fig. 8.12). That is true even when reactions involve the $O{=}Mn^{IV}P/Mn^{III}P$ or $O{=}Mn^{V}{=}O/Mn^{III}P$ redox couple. The reason lies in the fact that reaction of MnP with strong oxidizing reactive species such as $ONOO^-$ or H_2O_2 involves binding of those species in a first step prior to electron transfer. Binding is controlled with the electron density of Mn site which is described by the protonation equilibrium of axially bound water depicted by pK_{a1}. The pK_{a1} is in turn proportional to the $E_{1/2}$ of the $Mn^{III}P/Mn^{II}P$ redox couple. The $E_{1/2}$ values for the redox couples, that involve Mn^{III} and either Mn^{IV} or Mn^{V}, are nearly identical for all MnPs ($E_{1/2}$ are discussed in refs [21, 64]). Thus the reaction of MnP with $ONOO^-$ and H_2O_2 is primarily dependent upon the electron density of the Mn site characterized either as pK_{a1} or $E_{1/2}$ of the $Mn^{III}P/Mn^{II}P$ redox couple. On the other side the reaction of MnP with $O_2^{\bullet-}$, though predominantly outer-sphere (does not involve binding of $O_2^{\bullet-}$ to the Mn site) is also dependent upon $E_{1/2}$ of $Mn^{III}P/Mn^{II}P$ redox couple, for the reasons discussed earlier. In turn, the log $k_{cat}(O_2^{\bullet-})$ parallels log $k_{red}(ONOO^-)$ and log $k_{cat}(H_2O_2)$ [20, 21]. Finally the same is also valid for the ability of MnP to catalyze ascorbate oxidation: the initial rates, $v_o(Asc)_{ox}$, are proportional to $E_{1/2}$ and would thus be proportional to log $k_{cat}(O_2^{\bullet-})$ (Fig. 8.12).

8.6 MnPs and Transcription Factors

MnP modulates activities of transcription factors thereby affecting cellular proliferative and apoptotic pathways. Thus far the impact of MnPs on several transcription factors was reported as shown in Fig. 8.13. Mostly studied are effects on hypoxia inducible factor-1alpha, HIF-1α [73, 74] and nuclear factor-kappaB, NF-κB [13, 22, 61].

HIF-1α. Normal tissue. Several of the studies were done on the effect of MnP-based SOD mimics on pulmonary radioprotection. Either MnTE-2-PyP^{5+} or MnTnHex-2-PyP^{5+} were tested as radioprotectors. Upon radiation (RT), the increase in reactive species resulted in the activation of HIF-1α, which upregulated VEGF. Also TGF-β and macrophage recruitments were activated, all of which favor pro-fibrotic processes. When MnP was administered either at 2, or 6 or 12 h after radiation and continued for different time periods (up to 2 weeks), HIF-1α, TGF-β, VEGF, and macrophage recruitments were all downregulated which resulted in normalized lung histology [73, 79, 80].

Tumor. In a 4T1 mammary cancer sc flank mouse model, MnP given at 15 mg/kg/day (but not at 2 mg/kg/day) for the duration of the study suppressed tumor growth. MnP reduced HIF-1α and VEGF levels, preventing tumor angiogenesis [81]. Also NADPH oxidase 4 (NOX4) expression was suppressed presumably via suppression of NF-κB transcription by MnP. HIF-1α could have been alternatively controlled via NF-κB pathway. Dewhirst's group studied the effects of radiation on tumor angiogenesis and demonstrated that HIF-1α is activated either via $H_2O_2^-$ or via radiation-induced oxidative stress [74]. The HIF-1α regulated cytokines- enhanced endothelial cell radioresistance. At concentrations as low as 2 μM, MnTE-2-PyP^{5+} fully inhibited HIF-1α-mediated secretion of the cytokines from tumor cells which

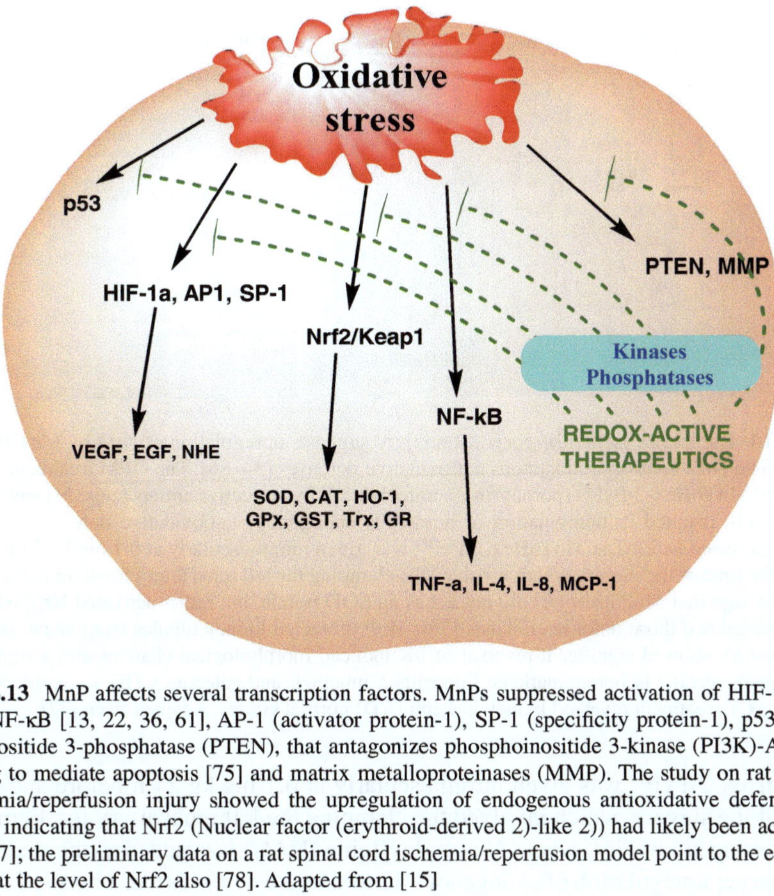

Fig. 8.13 MnP affects several transcription factors. MnPs suppressed activation of HIF-1α [73, 74], NF-κB [13, 22, 36, 61], AP-1 (activator protein-1), SP-1 (specificity protein-1), p53, phosphoinositide 3-phosphatase (PTEN), that antagonizes phosphoinositide 3-kinase (PI3K)-Akt signaling to mediate apoptosis [75] and matrix metalloproteinases (MMP). The study on rat kidney ischemia/reperfusion injury showed the upregulation of endogenous antioxidative defenses by MnPs indicating that Nrf2 (Nuclear factor (erythroid-derived 2)-like 2)) had likely been activated [76, 77]; the preliminary data on a rat spinal cord ischemia/reperfusion model point to the effect of MnP at the level of Nrf2 also [78]. Adapted from [15]

would have otherwise protected irradiated endothelial cells. In a mouse study, when combined with radiation, MnTE-2-PyP^{5+} (and not amifostine) reduced vascular density by 78.7 % at 72 h post-RT and delayed tumor growth by 9 days [82].

Based on our most recent understanding, MnP suppressed NF-κB transcription via H_2O_2/GSH-driven *S*-glutathionylation of protein cysteines. Tumor suppression could be driven thermodynamically with high concentrations of either or all of reactants: H_2O_2, MnP, protein thiol, or GSH. The 15 mg/kg/day dose was high enough for MnP to produce the anticancer effect in the presence of endogenous H_2O_2. In more recent studies [33, 63, 83] when MnP was given along with radiation and/or ascorbate, the antitumor effect was produced with either 0.2 or 2 mg/kg/day of MnP. Under such conditions of up to ~100-fold lower levels of MnP, the increased tumor H_2O_2 levels, via exogenous sources, accounted for the effects observed [63].

Nrf2. Normal tissue. In a rat kidney ischemia/reperfusion (I/R) model, the Dorai's group showed that GMP treatment, which contained growth factors (G), amino acids of Krebs cycle (M), and MnTnHex-2-PyP^{5+} (P), upregulated numerous endogenous antioxidative defenses (Fig. 8.14) - a suggestive of Nrf2 activation [77].

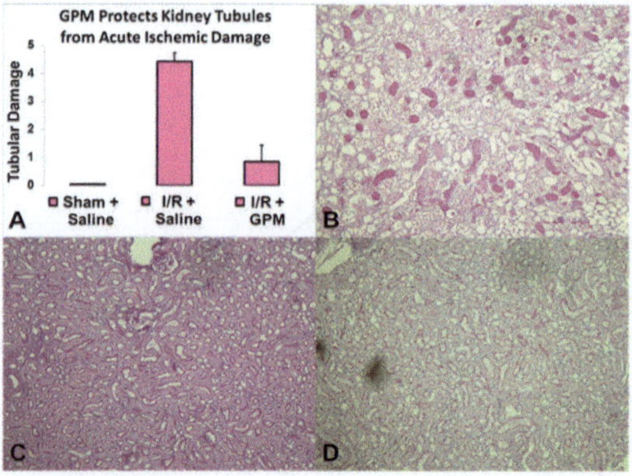

Genes	Fold-increase
Glutathione peroxidase 1	9.8
Glutathione peroxidase 3	23.9
Lactoperoxidase	16.7
NADPH oxidase 4	13.9
Peroxiredoxin	
2	9.6
5	9.8
3	6.6
Superoxide dismutase	
mitochondrial	10.2
extracellular	12.3
Inducible nitric oxide synthase	15.2
Thioredoxin	
Thioredoxin reductase	6.9
Interacting protein	7.3
Flavin monooxygenase	10.5
Catepsin B	12.2
Apoliprotein E	16.2
Prostaglandin peroxidase synthetase 1	11.1

Fig. 8.14 Rat kidney ischemia/reperfusion injury suggests upregulation of Nrf2 by MnTnHex-2-PyP^{5+}. The Nrf2 controls endogenous antioxidative defenses [84–86]. The GPM treatment, which included MnTnHex-2-PyP^{5+} (porphyrin P), mitochondrially protective amino acids (M) and growth factors (G), resulted in upregulation of numerous endogenous antioxidative defenses (some of which are listed here). The MnTnHex-2-PyP^{5+} was given intramuscularly at 0.1 mg/kg 24 h before and at the time of the surgery which involved the clamping the left renal artery for 40 min. Such data are clear sign that MnP itself (**A**) did not act as an SOD mimic, but rather activated Nrf2 which in turn upregulated those defense enzymes [76]. MnP protected kidney tubules from acute ischemic damage (**A**), showed significant reversal of I/R induced morphological changes and a significant decrease in specific ischemic markers: lipocalin-2, mucin-1, and galectin-3. (**B**), ischemic necrotic kidney, (**C**), treatment reversed kidney ischemia, (**D**) normal kidney. Adapted from [77]

MnTnHex-2-PyP^{5+} was given intramuscularly at 0.1 mg/kg 24 h before and at the time of the surgery, which involved the clamping the left renal artery for 40 min. In a following study [76], where *N*-acetylcysteine (NAC) was added to the treatment, even larger upregulation of endogenous antioxidative defenses was observed. While NAC in its own right acts as an antioxidant, when it redox cycles with MnP it gives rise to a H_2O_2 production. Subsequently, MnP/H_2O_2/GSH presumably activated Nrf2 by oxidizing/*S*-glutathionylating thiols of Keap1 [76, 77]. Study in a progress indicates the impact of MnPs on Nrf2 activation in a spinal cord ischemia/reperfusion model [78]. Carroll et al. reported that MnTnBuOE-2-PyP^{5+} upregulates Nrf2 in normal hematopoietic stem cells [87].

Tumor. However, Nrf2 is a double-edge sword. In numerous cancer studies Nrf2 upregulated antioxidative defenses thereby suppressing the effects of anticancer drugs. Studies are in progress exploring the expression of Nrf2 in 4T1 flank tumors of mice that were treated with MnP/H_2O_2/RT (see under *New anticancer strategies involving MnPs*) [63]. While MnTnBuOE-2-PyP^{5+} upregulates Nrf2 in normal hematopoietic stem cells, it does not activate Nrf2 but activates AP-1 in myelodysplastic cells from patients with myelodysplastic syndrome (MDS), a malignant blood disorder [87].

NF-κB. Normal tissue. Once the powerful MnP-based SOD mimic MnTE-2-PyP^{5+} was identified in the late 1990s, the first stroke study was performed to evaluate its efficacy; stroke was induced via middle cerebral artery occlusion. The assumption was made that MnP is solely an SOD mimic and scavenges superoxide produced

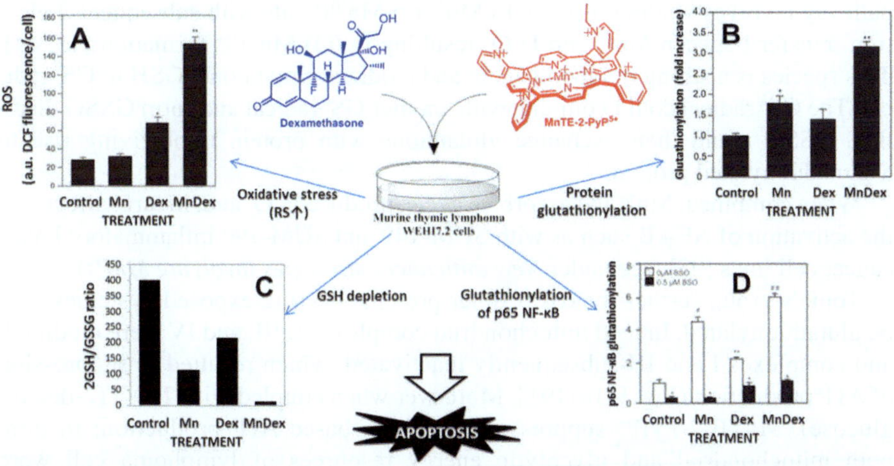

Fig. 8.15 The effect of MnTE-2-PyP^{5+} (indicated as Mn) on NF-κB *S*-glutathionylation. MnP *S*-glutathionylates p65 and to a smaller extent p50 subunit of NF-κB whereby suppressing anti-apoptotic pathways in cellular lymphoma model [61]. The effects are much larger if MnP is combined with dexamethasone (Dex)—chemosensitization was observed: higher oxidative stress, higher caspases-3 activity, and higher level of *S*-glutathionylation. When GSH synthesis is inhibited with buthionine sulphoximine, BSO, or if catalase was overexpressed (*not shown*), the *S*-glutathionylation was nearly fully eliminated. Adapted from [61]

during short period post-reperfusion. Indeed the protection of $O_2^{\bullet-}$-sensitive enzyme, aconitase, by MnTE-2-PyP^{5+} was demonstrated [36]. More comprehensive studies followed and demonstrated that MnP can reduce infarct size even when given at 6 h (but not at 12 h) post-reperfusion [88, 89]. Such effects questioned the exclusive role of MnP as an SOD mimic. The impact of MnP on secondary oxidative stress was hypothesized. To prove it, the activity of NF-κB as impacted by MnP was assessed; indeed MnP suppressed inhibition of NF-κB and in turn the inflammatory processes [68]. In a diabetes study MnP suppressed NF-κB activation reducing thereby the levels of inflammatory cytokines and chemokines such as TNF-α (tumor necrosis factor-alpha), IL-1β (interleukin-1 beta), IL-6 (interleukin-6), IFN-γ (interferon-gamma), and NADPH oxidase-dependent superoxide production [90]. The more recent studies provided the insight into the mechanism of NF-κB inhibition by MnP during an oxidative stress event [61, 91, 92]. Rather than SOD mimicking, the predominant action of MnP in suppressing the oxidative injury to normal tissue (such as stroke, I/R injury, radiation, and diabetes) seems to be the oxidation of signaling protein thiols by the combined action of MnP/H_2O_2/GSH in a GPx fashion [61, 65, 91, 92].

Tumor. The comprehensive mechanistic studies on NF-κB helped us understand the radio- and chemosensitizing effects of MnPs [61, 91, 92]. In combination with dexamethasone, MnP produced high levels of reactive species, induced *S*-glutathionylation of p65 subunit of NF-κB, and thereby inhibited NF-κB activation. Under such conditions, the activity of proapoptotic caspases-3 was largely enhanced (Fig. 8.15) [91]. None of this happened when either H_2O_2 or GSH had been eliminated (Fig. 8.15) [91]. The data provided the basis for understanding the mechanism of *S*-glutathionylation depicted in equations (8.10)–(8.16). Other possible pathway involves binding of H_2O_2 to Mn^{3+} (in $Mn^{III}P$) site with subsequent 2-electron transfer between Mn^{+3} and H_2O_2 resulting in $(O)_2Mn^VP^{3+}$ formation [15, 64]. This species is a strong oxidizing agent and oxidizes glutathione, GSH to GS$^•$ radical. The GS$^•$ radical could combine with another GS$^•$ radical and form GSSG disulfide. GSSG could then exchange glutathione with protein thiol giving rise to *S*-glutathionylated protein.

When combined, MnP and ascorbate cycle, produce H_2O_2 and, in turn, affect also the activation of NF-κB such as with SUM-149 and SUM-190 inflammatory breast cancer cell lines [93] (see under *New anticancer strategies involving MnPs*).

Tome's group further looked if other proteins bearing exposed cysteines may be glutathionylated. Indeed mitochondrial complexes I, III, and IV were modified, and complexes I and III subsequently inactivated, which resulted in suppression of ATP production (Fig. 8.16) [91]. Moreover when coupled with 2-DG (2-deoxyglucose), MnTE-2-PyP^{5+} suppressed glycolysis-based ATP production; in turn, both mitochondrial and glycolytic energy resources of lymphoma cell were reduced [Fig. 8.16]. *S*-Glutathionylation of glycolytic proteins might have also been involved. Importantly, MnP/dexamethasone was not cytotoxic to normal lymphocytes [91].

8.7 Therapeutic Effects of Mn Porphyrins in Cancer: Radio- and Chemosensitization of Tumor While Radioprotection of Normal Tissue

MnP-based SOD mimics demonstrated remarkable efficacy in numerous diseases that have oxidative stress in common. Many of those effects have been recently reviewed in Special Forum Issue of *Antioxidants and Redox Signaling* on "SOD therapeutics" and in several other reviews [6, 8–10, 13–18]. Herein we will concentrate on two effects—radioprotection of normal tissue while radio-, chemo-, and ascorbate-driven sensitization of tumor. Such differential effects enabled MnTnBuOE-2-PyP^{5+} to enter Phase I/II Clinical Trials. The other analog, MnTE-2-PyP^{5+}, is in Canadian Clinical Trials on the protection of beta cells during the transplant of islets. The related effects are reviewed in Special Forum Issue mentioned above [14, 39, 94–97] and in this Book on "*Redox-Active Therapeutics*" by Piganelli's group.

Prostate cancer. The remarkable therapeutic efficacy of MnPs has been demonstrated in nearly any model thus far studied. Efficacy and mechanistic studies were frequently done in parallel. Most recently due to the increase in aging population, we have concentrated our efforts on the treatment of cancer and protection of normal tissue exposed to radiation and chemotherapy. Among the most remarkable radioprotective effects thus far seen is the radioprotection of erectile function, protection of testes and preservation of prostate architecture during prostate radiation. This has been first demonstrated by Oberley-Deegan's and Crapo's groups using MnTE-2-PyP^{5+} (Fig. 8.17) [98] and then recapitulated by those groups [Oberley-Deegan

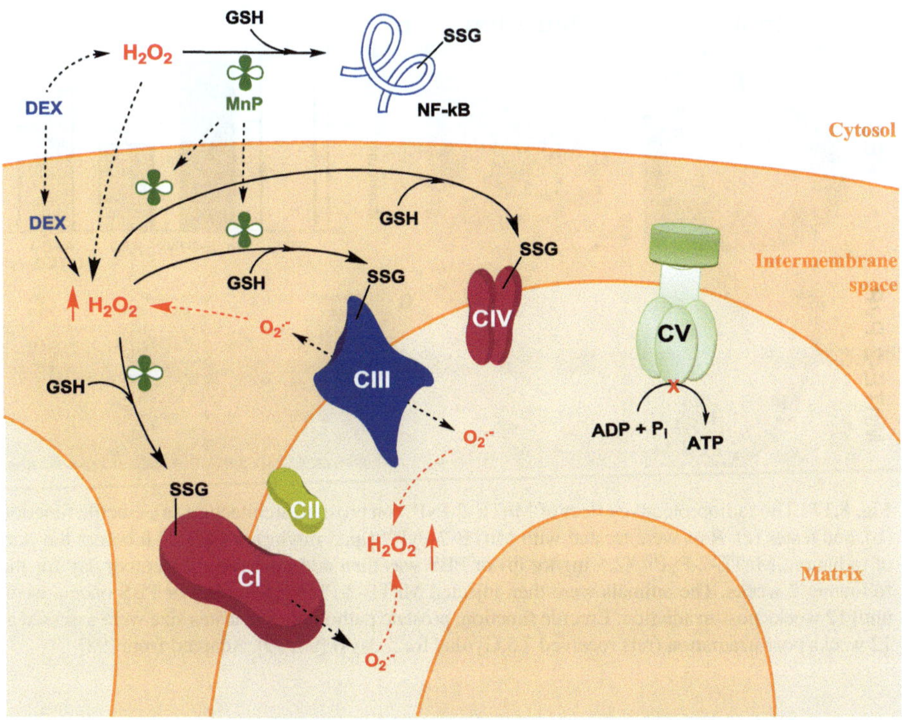

Fig. 8.16 *S*-Glutathionylation of NF-κB p65 subunit in cytosol and complexes I, III and IV of electron transport chain in mitochondria. The study was performed on MnTE-2-PyP^{5+} [61, 91, 92] and in part on MnTnBuOE-2-PyP^{5+} (Jaramillo et al., unpublished). Such modifications results in inactivation of thiol-containing proteins: NF-κB and complexes I and III (shown here). This resulted in suppression of cellular energetic and apoptotic pathways inducing lymphoma cell death. *No cytotoxicity was observed in normal lymphocytes*. Likely, the glycolytic proteins were modified by MnP/H_2O_2/GSH also. In combination with dexamethasone (DEX) and 2-deoxyglucose, MnP increases intracellular H_2O_2 which resulted in oxidation of either Mn(III) to Mn(V) or Mn(II) to Mn(IV) state. Subsequent reduction of Mn results in oxidation of glutathione to thiyl radical which couples with another thiyl radical to produce disulfide, GSSG. One possible way of protein thiol glutathionylation is the exchange of GSH between GSGG and protein thiol [91]. See (8.10)–(8.16) for the reactions that may be involved in *S*-glutathionylation of proteins, starting from reduced Mn^{II}P which due to high levels of cellular reductants is a more likely reactant in vivo than Mn^{III}P. Though reactions including Mn^{III}P [93] can not be excluded

et al., unpublished] and Koontz's and Batinić-Haberle's groups using MnTnBuOE-2-PyP^{5+} [98, 99].

Head and neck cancer. The radiation-induced injury to salivary glands and mouth mucosa are serious side effects of radiation therapy of head and neck cancer patients. In a non-tumor bearing mouse model, MnTnBuOE-2-PyP^{5+} showed efficacy as a radioprotector of those tissues (Fig. 8.18) [33]. Importantly, in a mouse sc xenograft model of head and neck FaDu cancer cell line, MnP radiosensitized tumor suppressing its growth [33]. Accompanying pharmacokinetic studies performed with a single sc injection of 10 mg/kg provided evidence of MnTnBuOE-2-PyP^{5+} accumulation

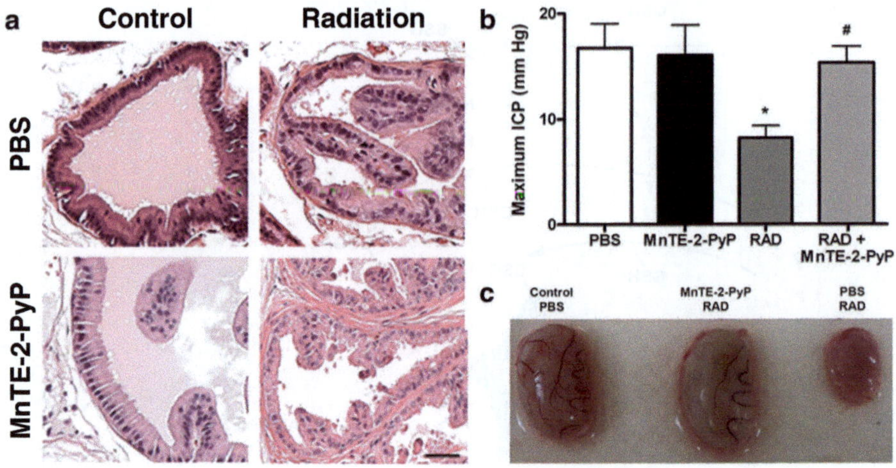

Fig. 8.17 The radioprotective effect of MnTE-2-PyP^{5+} on prostate architecture (**a**), erectile function (**b**), and testes (**c**). Rats were treated with MnTE-2-PyP^{5+} (ip, 5 mg/kg) or PBS 24 h before the start of radiation. MnTE-2-PyP^{5+} (2.5 mg/kg ip) or PBS was then administered every other day for the following 2 weeks. The animals were then injected MnTE-2-PyP^{5+} (5 mg/kg) or PBS once a week until 12 weeks post-irradiation. Erectile function, prostate pathology, and testes size were assessed at 12 weeks post-irradiation (rats received 7.5 Gy/day for 5 days) [98, 99]. Adapted from [98]

at sites of injury in tongue and salivary glands. In a dose-escalation study the dose-modifying factor was determined to be 0.77 for radioprotective and 1.30 for radiosensitizing action of MnP [33].

Glioma. Several groups at Duke University (Dewhirst and Batinić-Haberle), University of Montreal (Beausejour), and Stanford University (Huang) assessed the ability of MnPs to protect brain white matter, neurocognition, and hippocampal neurogenesis but not tumor (Fig. 8.19) [100–102]. In studies on hippocampal neurogenesis by Huang's group mice were injected sc with MnTnBuOE-2-PyP^{5+} starting a week before radiation and continuing for 4 weeks post-RT. Production of new neurons, including the immature neurons (Dcx+) and mature neurons (BrdU+/NeuN+), in the subgranular zone (SGZ) of hippocampal dentate gyrus was monitored. MnP also protected interneurons which are essential for neurogenesis [102]. The Duke University team dealt with different aspects of normal brain radioprotection. Mice were injected sc with MnTnBuOE-2-PyP^{5+} before and after RT and evaluated 3 months later. Mice treated with MnTnBuOE-2-PyP^{5+}+RT vs. mice treated with saline/RT have much higher preservation of myelin and lower loss of axons in corpus callosum (Fig. 8.19) [83]. The Beausejour's group demonstrated the radioprotection of neurogenesis and prevention of the loss of neurospheres by another lipophilic analog, MnTnHex-2-PyP^{5+} [100, 101]. The radioprotective effect occurred at least in part at the level of p16 pathway. The p16 is an apoptotic protein involved in senescence. With p16 knockout mice, neither RT nor MnP affected the number of neurospheres (Fig. 8.19).

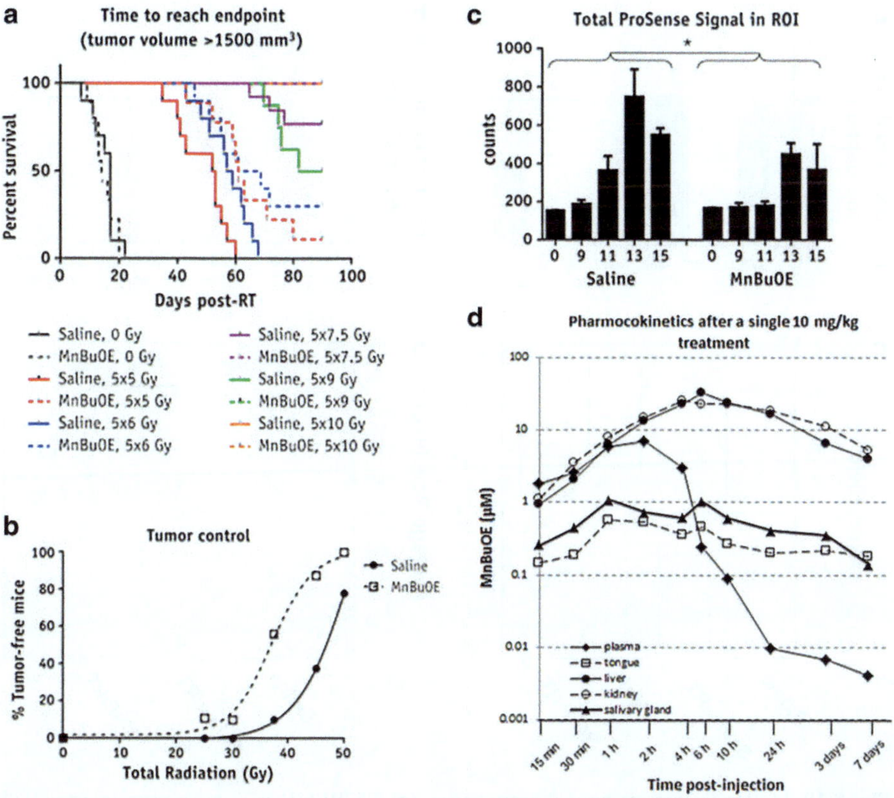

Fig. 8.18 Radiosensitization of head and neck FaDu tumor growth ((**a**) and (**b**)), radioprotection of salivary glands (**c**) by MnTnBuOE-2-PyP^{5+} and its pharmacokinetics with particular relation to its biodistribution to tissues of interest, tongue, and salivary glands (**d**). MnTnBuOE-2-PyP^{5+} was injected sc twice daily at 1.5 mg/kg in both radioprotection and tumor suppression studies [33]. The study indicated that the higher the radiation, the higher is the inflammation of salivary glands as measured by prosense marker; MnP suppressed inflammation at all RT doses (**c**). *ProSense 750 EX* is a fluorescent in vivo imaging agent and inflammatory marker that is activated by the protease Cathepsin which is elevated in lysosomes of tumors and inflammatory cells. The dose escalation study indicated that with MnP the radiation dose may be increased by ~30 % without producing radiation-induced damage—dose modifying factor was identified as 0.77. The dose-escalation study on sc xenograft mouse flank model of FaDu head and neck cancer showed that MnP radiosensitizes tumor—the tumor suppressing effect occurs at ~30 % lower RT dose if combined with MnP [33]. (**d**) The maximal tolerable dose (MTD) for sc dosing of MnP, 10 mg/kg, was used to study the 24-h single dose pharmacokinetics. The initial drug absorption in plasma (C_{max} = 7 µM at T_{max} = 2 h) was followed by an organ distribution phase. Liver and kidney profiles showed that their high concentrations were responsible for drug disappearance from the plasma early after injection. This fastest process consumed over 90 % of drug, and together with the distribution into other organs, left only 2 % of plasma MnP at 10 h postinjection. Very slow elimination of the remaining drug in plasma was observed between 1 and 7 days ($t_{1/2}$ = 4.8 days). Salivary glands and tongue levels followed the plasma profile initially (C_{max} = 0.5 µM at T_{max} = 2 h), and then the slow elimination profile observed in liver and kidney. Adapted from [33] and Spasojevic I et al, in preparation (Fig. 8.18d)

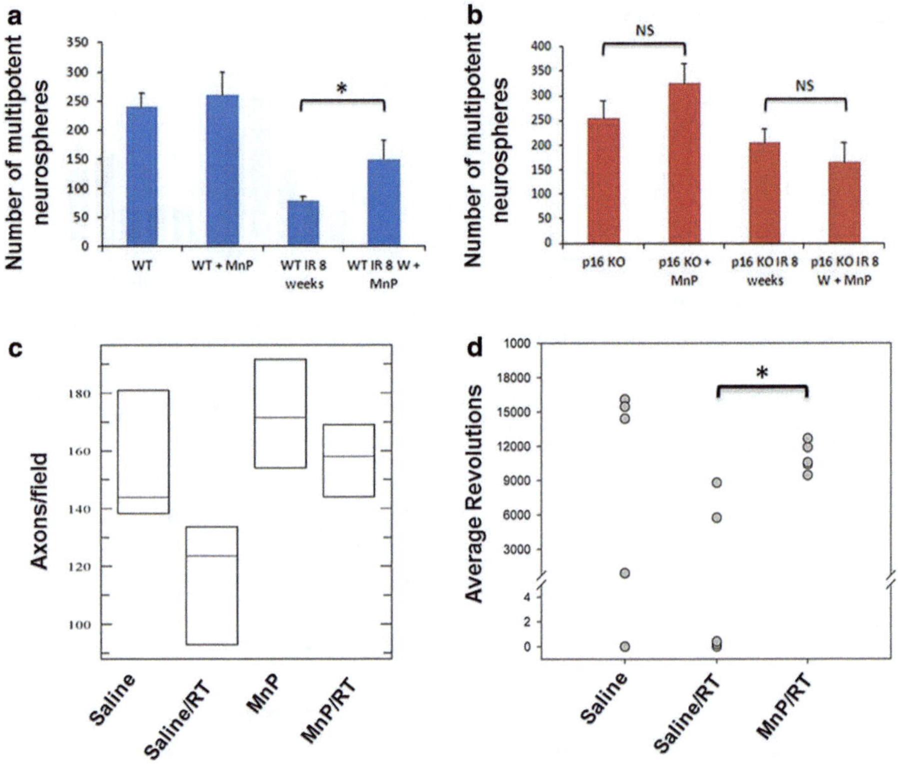

Fig. 8.19 MnP radioprotects normal brain tissue. (**a**) C57BL6/J (WT) mice were injected with MnTnHex-2-PyP at 1.6 mg/kg twice daily, starting 1 week before and continuing for 7 weeks after the 6 Gy whole brain RT. MnP protected neurogenesis; (**b**) Next experiment with p16 knock-out (KO) mice showed that the protection of neurogenesis by MnP is in large part dependent upon p16 pathway. The effect is eliminated when p16 knockout mice were radiated, the number of neurospheres was affected neither by RT nor with MnP [100, 101]; (**c**) Large preservation of myelinated axons in the corpus callosum of C57BL6/J mice was seen with MnBuOE-2-PyP^{5+}; (**d**) Improved neurobehavior (as measured by running wheel) was demonstrated with MnBuOE-2-PyP^{5+}. In (**c**) and (**d**) the twice daily sc injections of MnP at 1.5 mg/kg started 1 week before radiation and continued for the first 4 weeks and at 0.5 mg/kg/day for another 4 weeks [83]. At that time point the brain concentration of MnP was found to be 25 nM. Adapted from [83, 101]

The glioma radio- and chemosensitization (with temozolomide) by MnTnHex-2-PyP^{5+} and MnTnBuOE-2-PyP^{5+} was seen with D-245 MG glioblastoma multiforme in a sc patient-derived xenograft mouse model (Fig. 8.20) [14, 83]. The magnitude of tumor suppression was identical when MnP was combined with either radiation or temozolomide [103]. The following pathways were shown to be involved: metastatic (*ctss*, catepsin L, *becn1*, beclin1), antiapoptotic [NF-κB (*Nfkb1*, *Bcl211, Bcl2*) and PI3 kinase and mTOR (*Rsp6kb1*)] and protein translation (*EIF5b* and *Rsp6kb1*) [14].

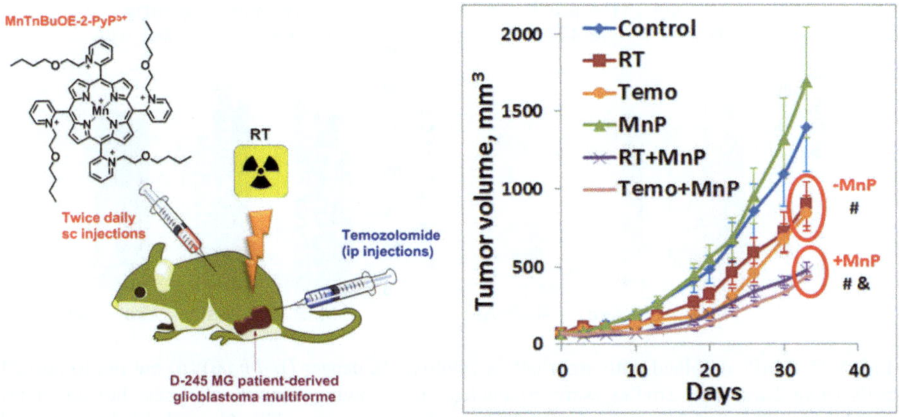

Fig. 8.20 MnTnBuOE-2-PyP^{5+} radio- and chemosensitizes tumor in a mouse sc xenograft model. BalbC nu/nu mice with sc patient-derived D-245 MG glioblastoma multiforme xenografts were injected sc with MnP at 1.6 mg/kg twice daily (throughout the study, starting 24 h before radiation) and/or treated with 1 Gy/day for 3 days, and/or temozolomide at 5 mg/kg/day injected ip for 5 days [103]. The animals were sacrificed at same time point to address mechanistic issues, while tumor regrowth was followed in a second study. The 10-day tumor growth delay was seen with MnP+Radiation vs. Radiation, the 17.5-day tumor growth delay with MnP+Radiation vs. control, and the 19.5-day tumor growth delay with MnP+Radiation vs. MnP

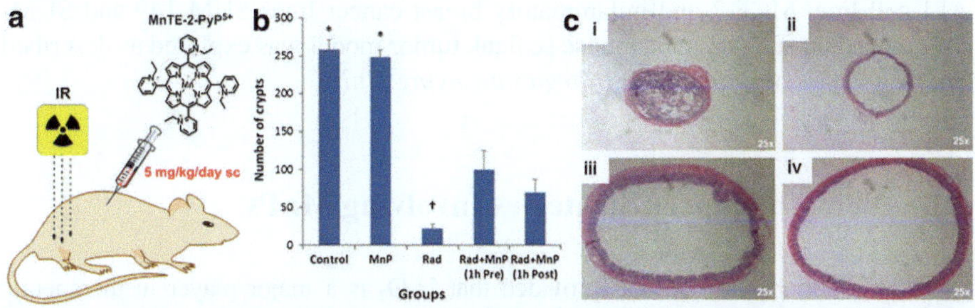

Fig. 8.21 MnTE-2-PyP^{5+} (**a**) protects rectum against radiation. Rats were radiated with single 21 Gy proton beam. (**b**) MnP reduced loss of crypts at 10 days post-RT. (**c**) MnP also suppressed chronic damage, reducing RT-induced rectal dilation at 150 days post-RT. (**i**) control; (**iii**) RT only; (**ii**) MnP given sc 1 h prior to RT; (**iv**) MnP given sc 1 h post-RT [104]. Adapted from [104]

Rectum. Remarkable radioprotection of rectum by MnTE-2-PyP^{5+} was demonstrated in a rat model. Rats were radiated with single 21 Gy proton beam. The suppression of radiation-based damage was demonstrated at both acute (reduction in loss of crypts at 10 days post-RT) and chronic level (suppression of rectal dilation at 150 days post-RT) (Fig. 8.21) [104].

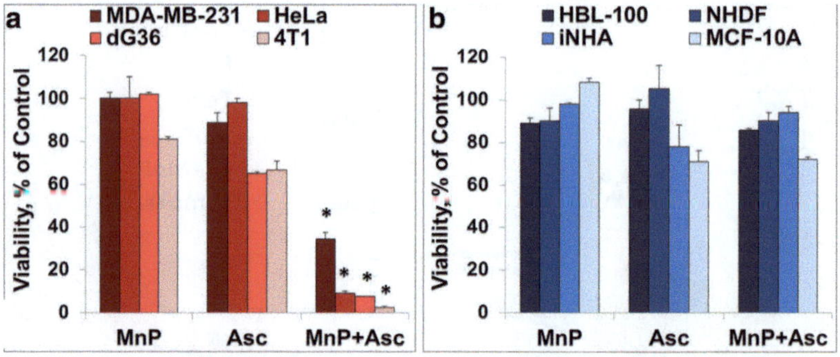

Fig. 8.22 MnP combined with ascorbate is cytotoxic to cancer (*red bars*) (**a**) but not to normal cells (*blue bars*) (**b**). Studies were conducted on following cancer cell lines: human breast MDA-MB-231, ovarian HeLa, glioma dG36, and mouse breast 4T1. Normal cell lines studied were: breast epithelial cancer cell line HBL-100, normal human astrocytes iNHA, dermal fibroblast NHDF, and breast cell line MCF-10A. Data are taken from [14, 63, 105, 114] and relate to either MnTE-2-PyP^{5+} or MnTnHex-2-PyP^{5+} or MnTnBuOE-2-PyP^{5+}

Breast. Several cellular studies confirmed the therapeutic potential of MnPs in the treatment of breast cancer. When combined with source of H_2O_2, such as is the presence of ascorbate, MnP was able to induce apoptosis of triple negative mammary 4T1 cell line, MCF-7 and inflammatory breast cancer lines SUM-149 and SUM-190. Based on such data the mouse sc flank tumor model was explored as described below under *New anticancer strategies involving MnPs.*

8.8 New Anticancer Strategies Involving MnPs

Once substantial evidence was provided that H_2O_2 is a major player in therapeutic actions of MnPs [91, 93, 105], an opportunity for another anticancer therapeutic approach has emerged. It was driven by the fact that Mn(III) *N*-substituted pyridylporphyrins readily cycle with ascorbate thereby producing ascorbyl radical and $Mn^{II}P$. Subsequently, $Mn^{II}P$ cycles back to $Mn^{III}P$ thereby reducing oxygen or superoxide to $O_2^{\bullet-}$ or H_2O_2, respectively [12, 106]. Due to higher concentration of oxygen than of $O_2^{\bullet-}$, the oxidation of $Mn^{II}P$ with oxygen is more likely than with $O_2^{\bullet-}$. In either case H_2O_2 will be eventually formed. Therefore, in addition to MnP and RT as a source of H_2O_2, the exogenous ascorbate (Asc) could be administered via intraperitoneal or intravenous routes to cycle with MnP giving rise to H_2O_2, thereby increasing the total intratumoral peroxide level [63, 93, 105, 107–113]. The impact of MnP/chemotherapy- and MnP/ascorbate-derived sources of H_2O_2 was explored in cellular studies on several breast, ovarian, lymphoma, glioma, and CaCo cell lines. Normal breast counterpart, fibroblasts, normal lymphocytes, and normal human astrocytes were also studied [14, 105, 114] (Fig. 8.22). Cytotoxicity was seen with cancer but not with normal cells [14, 63, 91, 93, 105, 114]. It was further shown that catalase eliminated toxicity of MnP/

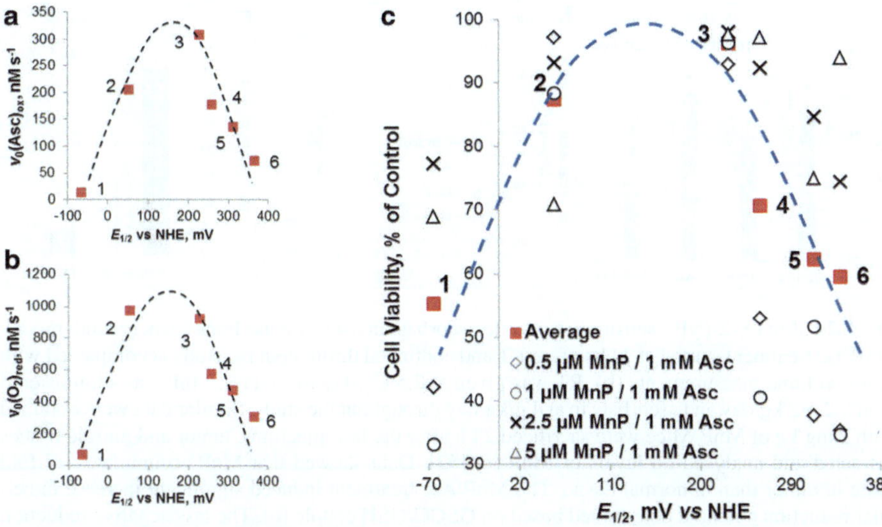

Fig. 8.23 The ability of MnP to catalyze ascorbate oxidation is described by the initial rates of ascorbate oxidation, $v_0(Asc)_{ox}$, and oxygen reduction, $v_0(O_2)_{red}$ ((**a**) and (**b**)). Such ability of MnP parallels its ability to exhibit cytotoxicity to 4T1 tumor cells when combined with ascorbate (**c**). Both are correlated in identical manner to the metal-centered $E_{1/2}$ of $Mn^{III}P/Mn^{II}P$ redox couple ((**a**), (**b**), and (**c**)) [63]. The study was done on 14 different MnPs; only the representative MnPs are plotted here. Among MnPs explored, MnTE-2-PyP^{5+} was identified as the best catalyst and was thus tested in a mouse model of tumor growth (Fig. 8.24). The MnPs plotted are: (#1) MnTE-2-PyPhP^{5+}, (#2) MnTE-3-PyP^{5+}, (#3) MnTE-2-PyP^{5+}, (#4) MnTPhE-2-PyP^{5+}, (#5) MnTnHexOE-2-PyP^{5+}, and (#6) MnTnOct-2-PyP^{5+}

ascorbate indicating that it is entirely due to the production of H_2O_2, predominantly at extracellular level [63, 93, 105, 115, 116]. GSH plays a major role in H_2O_2-driven actions of MnPs. MnP/Asc system was able to reduce cellular GSH levels and phosphorylation of NF-κB and ERK (extracellular signal-regulated kinase) in a study on inflammatory breast cancer cell lines, SUM-149 and SUM-190 [93]. The X-linked inhibitor of apoptosis protein, XIAP, was decreased and annexin V increased, indicating the apoptosis-related cytotoxicity imposed by MnP/Asc system [93].

In addition, the translocation of apoptosis-inducible factor, AIF, into nucleus implicates AIF-mediated and caspase-independent SUM 149 cell death pathway. Conversely (in normal cells), when cardiomyocytes were exposed to doxorubicin-mediated oxidative stress, MnTE-2-PyP^{5+}, MnTnHex-2-PyP^{5+}, MnTnBuOE-2-PyP^{5+} suppressed doxorubicin-induced lipid peroxidation and 4-HNE formation thereby preventing AIF_m translocation into nucleus and in turn apoptosis (see under *Mitochondrial accumulation*) [54]. Different redox environments of normal vs. cancer cell presumably contribute to such differential outcome. Even prior to anticancer therapy, the tumors are under higher oxidative stress than the normal tissue; oxidative stress is further enhanced with anticancer treatment. Moreover, MnP was found to accumulate in tumor much more than in normal tissue (Fig. 8.23). Higher

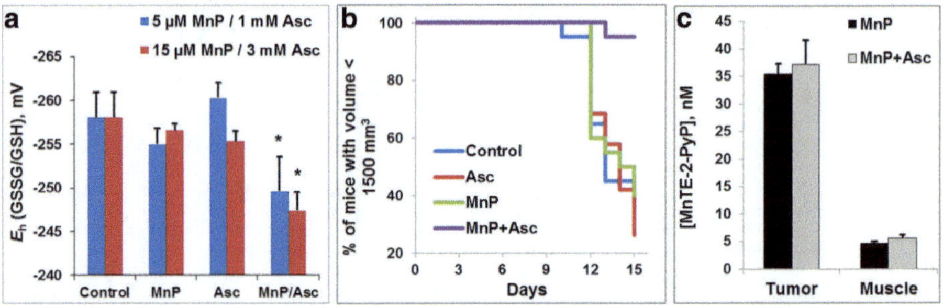

Fig. 8.24 MnTE-2-PyP^{5+} sensitizes tumor to ascorbate in a 4T1 mouse breast cancer flank model. MnP/asc treatment started at 24 h before RT and continued throughout the study accompanied with tumor volume measurements (**b**). RT was given at 2.5 Gy/day for 3 days. MnP was administered sc at 0.2 mg/kg/day and ascorbate ip at 4 g/kg/day throughout the study. Similar data were obtained with 2 mg/kg of MnP. Mice were sacrificed 24 h after the last injections, tumor and muscle tissues extracted and analyzed on levels of MnP (**c**) [63]. Data showed that MnP accumulates ~7-fold more in tumor than in normal tissue. The MnP/Asc treatment induced significant increase in cellular reduction potential determined based on GSGG/GSH couple (**a**). The less negative reduction potential obtained with MnP/Asc vs either control or MnP or Asc treatments suggests less reducing equivalent available for cellular metabolism when the cells were exposed to MnP/Asc. At day 15, all mice have tumor volumes higher than 1500 mm^3 except those from MnP/Asc treatment group (**b**). Adapted from [63]

H_2O_2 and MnP tumor levels jointly contribute to the differential effects (see below for further discussion). To substantiate such statements further, the redox environment of 4T1 cell, treated with MnP and ascorbate, was explored [63, 117]. The glutathione and cysteine redox couples were assessed (GSSG/GSH and Cys/CySS) and the related cellular reduction potentials, $E_{1/2}$ calculated. Indeed MnP/Asc treatment increased levels of oxidized thiols (GSSG and CySS) giving rise to less negative reduction potential and in turn less ability of a cell to provide reducing equivalents for its functioning; such data indicated that oxidative burden had indeed been imposed on cell via MnP/ascorbate redox cycling (Fig. 8.24) [63].

The ability of MnP to catalyze ascorbate oxidation in an aqueous solution was characterized in terms of initial rates of ascorbate oxidation, $v_o(Asc)_{ox}$. The $v_o(Asc)_{ox}$ relates to the metal-centered reduction potential, $E_{1/2}$ of $Mn^{III}P/Mn^{II}P$ redox couple, in the same manner as all other reactions of MnPs do (Fig. 8.12). Importantly, $v_o(Asc)_{ox}$ relates also in the same manner to the cytotoxicity exhibited upon the 4T1 cancer cell line (Fig. 8.23).

Ascorbate had been in Phase I Clinical Trials as an anticancer agent and showed no toxicity with marginal efficacy [112]. Its cytotoxicity was attributed to its oxidation catalyzed by endogenous metalloproteins. Yet those metalloproteins have neither the appropriate thermodynamics nor kinetics for the catalysis of ascorbate oxidation. Thus the effect had been only marginal. We believed that our MnPs which readily undergo cycling with ascorbate [12, 106] may be better catalysts of its oxidation. Our cellular and mouse studies supported the anticancer therapeutic potential of MnPs/ascorbate system [105]. Others subsequently also showed the benefit of combined MnP/ascorbate therapy [115, 118]. We then conducted a comprehensive study to identify which parameters control the catalysis. We showed that the efficacy of MnPs in terms of SOD mimicking parallels its ability to catalyze ascorbate oxidation; both are dependent upon $E_{1/2}$ of $Mn^{III}P/Mn^{II}P$ redox couple (Fig. 8.12) [63]. We then showed that those MnPs, which are the best catalysts for ascorbate oxidation, are the most cytotoxic in cellular studies where cells were threated with MnP/Ascorbate (Fig. 8.23). $MnTE\text{-}2\text{-}PyP^{5+}$ appeared to be the most successful among MnPs thus far studied and was therefore forwarded into a 4T1 sc mouse flank tumor model [63, 117]. This model is particularly useful in evaluation of drugs due to its short duration and the high success of tumor growth, approaching 100 %. The 4T1 cell line is a triple negative and an aggressive cell line and thus such studies are relevant to those breast tumors that are highly resistant to treatments. When the tumor volume reaches on average ~80 mm^3, the experiments are completed in a short period of ~2 weeks (Fig. 8.24). Once we showed that MnP/ascorbate treatment suppresses tumor growth, another study followed where RT was combined with MnP/ascorbate to account for an additional increase in tumor H_2O_2 levels. The triple MnP/ascorbate/RT treatment caused the most profound tumor suppression [117]. The study was terminated at 24 h after the last MnP injection to allow for the clearance of MnP from the blood not to interfere with determination of MnP tumor levels. At that time point mice were sacrificed, the tumors and muscles collected and MnP levels determined by LC-MS/MS method. MnP was found at ~7-fold higher concentration in tumor than in muscle (Fig. 8.22) [63]. Much higher levels of MnP in tumor than in normal tissue along with higher intratumoral levels of H_2O_2 (enhanced by RT or MnP/ascorbate cycling or chemotherapy) accounted for the higher yield of the reaction of MnPs with signaling protein thiols in tumor than in normal tissue. In turn, apoptotic processes are promoted in tumor and antiapoptotic processes in normal tissue. For additional discussion see under *Differential effects of MnPs on tumor vs. normal tissues.*

8.9 Differential Effects of MnPs on Tumor vs. Normal Tissues

Differential effects observed in a number of cellular and animal studies support therapeutic anticancer potential of MnPs and enable their development towards Clinical Trials. Our present understanding of such effect is as follows. It is highly likely that those tissues that are deficient in MnSOD would eventually develop tumor. Once normal cells undergo carcinogenesis, they may start upregulating MnSOD to fight oxidative stress. Yet production of H_2O_2 by MnSOD may not always be accompanied by its removal. Indeed studies showed that tumors have frequently upregulated MnSOD while catalase, GPx and peroxiredoxins were either downregulated or upregulated but inactive (Fig. 8.25) ([16, 63] and refs therein). Thus tumor is ill-equipped to handle additional increase in oxidative stress. Initially, tumor may take advantage of high levels of reactive species for activating signaling pathways that support its progression. Once overwhelmed with high H_2O_2 levels, tumor will undergo apoptosis and/or necrosis; meanwhile it will undergo metastases.

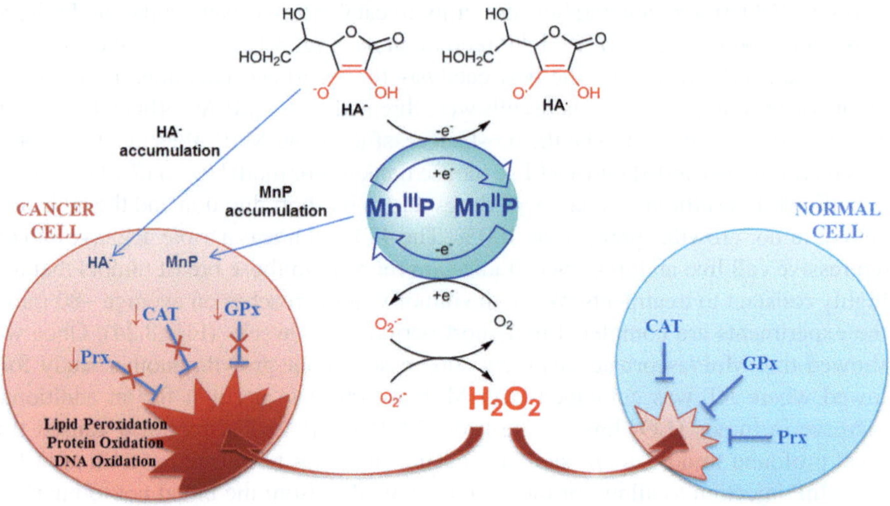

Fig. 8.25 Differential redox environment of cancer vs. normal cell affects the outcome of MnP action. Tumor has frequently overexpressed MnSOD in order to fight the perturbed cellular metabolism which gives rise to oxidative stress. If H_2O_2 produced during $O_2^{\bullet-}$ dismutation is not removed by H_2O_2-removing enzymes, such as catalase and GPx, its steady state levels start increasing. Such scenario has been frequently reported in tumors where catalase, GPx and other H_2O_2-removing enzymes are frequently downregulated or inactive ([16, 63] and refs therein). Initially, cancer uses the subsequent increase in oxidative stress for its own advantage, enhancing proliferative pathways. Yet cancer has only limited ability to handle oxidative stress and will eventually undergo apoptosis or necrosis while metastasizing meanwhile. Such scenario is used for therapeutic purposes where additional exogenous oxidative stress is imposed upon cancer via radiation and chemotherapy or quite recently via exogenously administered ascorbate. Monodeprotonated ascorbate, HA^-, predominant species at physiological pH 7.8, is oxidized by MnP to protonated ascorbyl radical, $HA^\bullet$. $HA^\bullet$ is readily deprotonated giving rise to $A^{\bullet-}$. During that process $Mn^{III}P$ undergoes reduction to $Mn^{II}P$ which is subsequently re-oxidized with either oxygen or superoxide; in either case H_2O_2 will be eventually produced. Mn site in MnP must have appropriate thermodynamics to undergo such reaction. We have reported that MnP accumulates ~7-fold more in tumor than in normal tissues [63] which along with high H_2O_2 and high ascorbate tumor levels drives tumor killing presumably via inducing massive inhibition of antiapoptotic NF-κB transcription. It was also reported that, when combined with sources of reactive species (radiation or chemotherapy or ascorbate), MnP (MnTnHex-2-PyP^{5+} and MnTnBuOE-2-PyP^{5+}) reduced cancer cell invasiveness and migration [119, 120] and lung metastases in orthotopically grown 4T1 mammary carcinoma model [119]. Conversely, lower degree of NF-κB inhibition results only in suppression of inflammation of normal cells. Adapted from [14]

Such compromised state of cancer redox environment has been taken advantage of in cancer therapy. Additional oxidative stress has been imposed by radiation therapy, chemotherapy, or MnP/ascorbate treatment, which are all good sources of reactive species, H_2O_2 being the most relevant due to its neutrality and long half-life [14, 15, 105]. Thus, if exposed to radiation, chemotherapy, or MnP/Asc, tumor will undergo death much sooner than normal tissue. *The co-localization of MnP, ascorbate, and H_2O_2 at much higher level in tumor than in normal tissue, resulted in higher yield of MnP reactions at the level of signaling protein thiols.* In other words, while rate constants for MnP reactions are identical regardless of the tissue in question, the thermodynamics (yield of reactions) rather than kinetics drives the differential effects in tumor vs normal tissue. Therefore, the massive oxidation/ *S*-glutathionylation of NF-κB thiols in tumor resulted in NF-κB inactivation and suppression of anti-apoptotic pathways. During tumor irradiation, normal tissue experiences a moderate increase in oxidative stress. That would allow for the inactivation of NF-κB to a modest extent so that inflammatory processes will be suppressed. We have reported supporting evidence for the inhibition of NF-κB activation in our stroke and diabetes studies resulting in lower levels of inflammatory cytokines [13, 22, 36, 68]. To clarify further the impact of the extent of NF-κB inhibition, let's look at the studies done by Takada and colleagues on the comparison of three anti-inflammatory drugs: aspirin, ibuprofen, and steroid. All three drugs are NF-κB inhibitors but are of different efficacies [121]. In the case of moderate inflammation, aspirin will be efficacious, while ibuprofen is needed to suppress larger inflammation. For massive inflammation, such as major low back pain or brain tumor, steroids would be the only drugs of choice. Due to the fact that NF-κB is our master transcription factor, if steroids are used at higher levels and/or for a prolonged period of time, the bodily metabolism would be greatly affected [13–15, 22, 36, 68]. Diabetes would often kick in. One can now draw a parallel concluding that MnP-driven inhibition of NF-κB in a normal tissue is of moderate extent such as is case with aspirin/ibuprofen; in turn the anti-apoptotic, anti-inflammatory pathways are promoted. Conversely, in cancer, the vast suppression of NF-κB by MnP results in massive suppression of antiapoptotic, survival pathways.

8.10 Toxicity

Among the FDA requirements for a drug aimed at clinical development is a good safety/toxicity profile. The safety/toxicity profile of our first lead, MnTE-2-PyP^{5+}, has been reported [122]. It is the safest Mn porphyrin thus far aimed at clinical development. Yet it is also the most hydrophilic compound among Mn porphyrins and is therefore not prospective for treatment of CNS injuries and neurodegenerative disorders. The lipophilic analogs, MnTnHex-2-PyP^{5+} and to a lesser extent the oxygen-modified MnTnBuOE-2-PyP^{5+}, have micellar properties and accumulate within tissues, cell and mitochondria to a high level. Such high bioavailability contributes to low toxicity of MnTnBuOE-2-PyP^{5+} which occurs at high doses; its mouse toxicity is 4–5-fold lower than that of MnTnHex-2-PyP^{5+}. The DNA comet assay was performed on rat and has demonstrated the lack of mutagenicity of MnTnBuOE-2-PyP^{5+} [Gad et al., unpublished]. Safety/toxicity studies of MnTnBuOE-2-PyP^{5+} have just been reported [129].

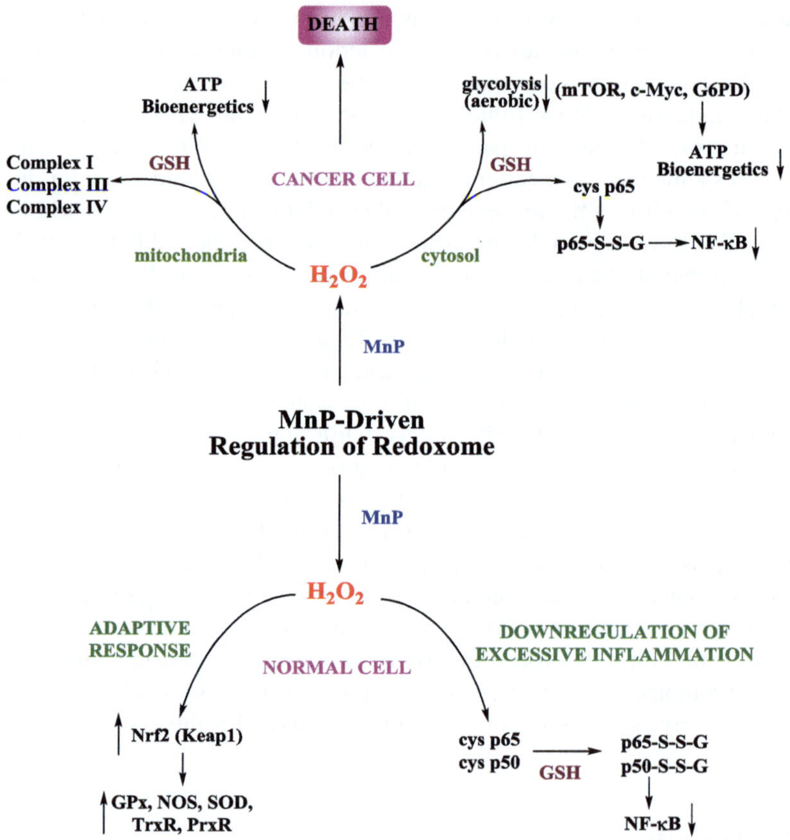

Fig. 8.26 The impact of H_2O_2 on MnP-driven pathways in cancer and normal cell. MnP couples with H_2O_2 and simple and protein thiols in both cancer and normal cell. Co-localized and higher tumor levels of MnP and H_2O_2 (both initial and enhanced by radiation and/or chemotherapy) contribute to the higher yield of reactions in cancer than in normal cell. Consequently cancer undergoes apoptosis/necrosis while redox environment of normal tissue gets restored after being exposed to some degree of radiation and chemotherapy which has been primarily targeted to cancer. The largest number of data has been collected on the impact of MnP on NF-κB and HIF-1α transcription factors, while effects may occur at the levels of other transcription factors (AP-1, SP-1, Nrf2, p53) and proteins, including those not yet studied

8.11 Summary

H_2O_2 occupies central role in in vivo actions/therapeutic effects of MnPs. The following Scheme (Fig. 8.26) summarizes our present knowledge on differential impact of MnP on the redox environment of cancer and normal cell when combined with H_2O_2 and thiols. MnPs use H_2O_2 and GSH for the critical modifications of transcriptions factors (TF) which result in either suppression of their activities, such as NF-κB and mitochondrial complexes I and III, or their activation such as Nrf2.

The co-localization of MnP, H_2O_2, and GSH and their levels control the fate of the cell—its death or survival. Indeed, the studies demonstrated that the differential impact on normal vs. tumor tissue primarily results from much higher levels of MnP and H_2O_2 in tumor than in normal tissue [63, 117].

Still little is known about the complex chemistry of MnPs within complex cellular milieu. We have extensively addressed the aqueous chemistry of MnPs and to a lesser extent their biology. The reason for the variety of reactions MnPs can undergo (thereby interfering with different pathways) lies in the fact that the Mn center can adopt four oxidation states (+2, +3, +4, and +5) in vivo, i.e., the related reduction potentials are biologically compatible. In addition, the pentacationic MnPs accumulate readily in different cellular organelles, being attracted to the negatively charged phosphates of cellular membranes and/or nucleic acids. Due to MnP complex chemistry and bioavailability and the also complex (and not fully grasped) biology of normal vs. cancer cell, it is very likely that we have only marginally addressed the MnP-based redox biology. What we know is that MnP has profound effect on restoration of the physiological environment of a diseased but normal cell while promoting the death of a tumor cell. We have also learned that the SOD-like potency of MnPs parallels (a) their ability to undergo all reactions thus far studied (in either stoichiometric or catalytic fashion) and (b) their therapeutic efficacy.

Cationic ortho Mn(III) N-substituted pyridylporphyrins vs. anionic para Mn(III) carboxylatophenyl-porphyrin. The story of an anionic compound, $MnTBAP^{3-}$, reminds us repeatedly that still little we know about the chemistry and biology of Mn porphyrins [123]. The log $k_{cat}(O_2^{\bullet-})$ of $MnTBAP^{3-}$ is 3.16, thus it is not an SOD mimic. Further, $MnTBAP^{3-}$ does not protect SOD-deficient *E. coli* when growing aerobically, another proof that it is not an SOD mimic [39]. Yet the researchers, including us, are repeatedly reporting that $MnTBAP^{3-}$ is therapeutically efficacious [78, 124]. We have dealt with this compound for over 20 years and still are not able to comprehend its biology. Often researchers use commercial samples which contain different impurities, ranging from a tiny fractions to as much as 25 % [123, 125]. While these commercial impurities complicate the discussion of the data, a pure $MnTBAP^{3-}$ sample was also efficacious in some models [78, 124, 126]. Most recently we undertook the 4T1 mammary carcinoma sc flank tumor mouse study where the anticancer effect of MnP/ascorbate was assessed. The fact that $MnTBAP^{3-}$ cannot be reduced with ascorbate in an aqueous solution agreed well with the lack of its anticancer effect in this mouse model [117]. However in an ongoing study on I/R spinal cord injury, $MnTnHex\text{-}2\text{-}PyP^{5+}$ performed as well as $MnTBAP^{3-}$ at both high and low doses of 1.5 and 0.15 mg/kg [78, 124]. Actions of $MnTBAP^{3-}$ at the level of peroxynitrite, $ONOO^-$, may be involved [126]). Such findings confirm that we are still far away from fully understanding the actions and effects of not only $MnTBAP^{3-}$ but of other MnPs also. Therefore, caution needs to be exercised when drawing the conclusions on the pathways involved in the actions of MnPs. Pharmacological and genetic studies need to be done in parallel to gain more accurate insight into the nature of the effects of MnPs.

On enthusiastic side, we almost ambiguously see favorable therapeutic outcomes of Mn porphyrins, even if we do not fully understand them, or misinterpret mechanistic data. Remarkable efficacies of MnPs in cellular and animal models motivate us to continue with our endeavors aimed at (a) understanding the underlying mechanisms of MnPs actions and (b) treating patients.

Antioxidative therapeutic effects vs. pro-oxidative actions of MnPs. As indicated here and elsewhere [15], the therapeutic effects of MnPs in normal tissue may appear antioxidative, yet may arise from pro-oxidative actions such as is the MnP/H_2O_2/GSH-driven *S*-glutathionylation of thiols of NF-κB and Nrf2/Keap1 [15, 65]. Modulation of the transcription activities of NF-κB and Nrf2 would result in suppression of inflammation and upregulation of endogenous antioxidative defenses, respectively. Thus the pro-oxidative actions at the levels of both transcription factors would translate into antioxidative effects. What we would detect in our in vitro and in vivo studies would be the reduction in the levels of $O_2^{\bullet-}$ and H_2O_2, yet not primarily as a consequence of MnP directly scavenging those species, though the direct removal of those species by MnP cannot be excluded. Further, when H_2O_2 oxidizes $Mn^{III}P$ to $(O)_2Mn^{V}P$, in order to act in a GPx-like fashion, the levels of H_2O_2 get decreased due to its reduction to water but not to its dismutation to water and oxygen in a catalase-like fashion (see equations (8.10)–(8.16)).

Development of MnPs towards Clinic. We have established the straightforward relationship between the SOD-like activity and therapeutic effects of MnPs. Our data teach us that designing compounds to be potent SOD mimics, while tuning their bioavailability and toxicity, is still the most valid drug development approach. Based on the "*ortho*" concept established with Mn(III) *N*-substituted pyridylporphyrins, the di-*ortho* imidazolyl analog, MnTDE-2-ImP^{5+} (AEOL10150), was synthesized. The Phase I Clinical Trial on ALS patients demonstrated the lack of the toxicity of MnTDE-2-ImP^{5+} (AEOL10150) at doses well above the therapeutic ones [127]. MnTDE-2-ImP^{5+} is now under aggressive development towards Clinical Trials as a radioprotector. The Phase I/II Canadian Clinical Trial on protection of beta cells during islet transplants with MnTE-2-PyP^{5+} (BMX-010, AEOL10113) is in progress (NCT02457858). The Phase I/II Clinical Trial on MnTnBuOE-2-PyP^{5+} (BMX-001), as a radioprotector of normal brain with glioma patients has started. Another Phase I/II Clinical Trial on radioprotection of salivary glands and mouth mucosa with head and neck cancer patients will commence later in 2016. The polyamine-based SOD mimic, GC4419, and Fe *ortho* substituted porphyrin, INO-4885, are also in Clinical Trials [15, 128]. First enthusiastic data on the efficacy of GC4419 (Fig. 8.4e) as a radioprotector of mouth mucosa in Clinical Trial have just emerged (Anderson CM et al, ASCO 2016). Therefore, we may soon gain further insight into the therapeutic effects of metal-based SOD mimics in humans and therefore justify their considerations both as therapeutics and mechanistic tools.

Acknowledgements Over years multiple funding mechanisms have supported development of SOD mimics; only few are listed here. IBH, AT, and IS acknowledge NIH 1R03-NS082704-01,

NIH U19AI067798, DTRI, Wallace H. Coulter Foundation, BioMimetix JVLLC (USA), and NC Biotechnology BIG Award (#2016-BIG-6518). AT acknowledges mini-fellowship award from SFRBM. IS is grateful for the support of NIH Core Grant, 5-P30-CA14236-29 that enabled the synthetic and pharmacokinetic studies. IBH also acknowledges the financial help from joint Benov/Batinić-Haberle Kuwait grants (MB02/12, YM04/14 and SRUL02/13), and joint Reboucas/ Batinić-Haberle Brazilian CNPq and CAPES grants. IBH and IS are consultants with BioMimetix JVLLC and hold equities in BioMimetix JVLLC. IBH, IS, and Duke University have patent rights and have licensed technologies to BioMimetix JVLLC.

References

1. McCord JM, Fridovich I. Superoxide dismutase. An enzymic function for erythrocuprein (hemocuprein). J Biol Chem. 1969;244:6049–55.
2. Pasternack RF, Halliwell B. Superoxide dismutase activities of an iron porphyrin and other iron complexes. J Am Chem Soc. 1979;101:1026–31.
3. Faulkner KM, Liochev SI, Fridovich I. Stable Mn(III) porphyrins mimic superoxide dismutase in vitro and substitute for it in vivo. J Biol Chem. 1994;269:23471–6.
4. Archibald FS, Fridovich I. The scavenging of superoxide radical by manganous complexes: in vitro. Arch Biochem Biophys. 1982;214:452–63.
5. Batinic-Haberle I, Liochev SI, Spasojevic I, Fridovich I. A potent superoxide dismutase mimic: manganese beta-octabromo-meso-tetrakis-(N-methylpyridinium-4-yl) porphyrin. Arch Biochem Biophys. 1997;343:225–33.
6. Batinic Haberle I, Tovmasyan A, Spasojevic I. The complex mechanistic aspects of redox-active compounds, commonly regarded as SOD mimics. BioInorg React Mech. 2013;9:35–58.
7. Batinic-Haberle I, Benov L, Spasojevic I, Fridovich I. The ortho effect makes manganese(III) meso-tetrakis(N-methylpyridinium-2-yl)porphyrin a powerful and potentially useful superoxide dismutase mimic. J Biol Chem. 1998;273:24521–8.
8. Batinic-Haberle I, Rajic Z, Tovmasyan A, Reboucas JS, Ye X, Leong KW, Dewhirst MW, Vujaskovic Z, Benov L, Spasojevic I. Diverse functions of cationic Mn(III) N-substituted pyridylporphyrins, recognized as SOD mimics. Free Radic Biol Med. 2011;51:1035–53.
9. Batinic-Haberle I, Reboucas JS, Benov L, Spasojevic I. Chemistry, biology and medical effects of water soluble metalloporphyrins. In: Kadish KM, Smith KM, Guillard R, editors. Handbook of porphyrin science. Singapore: World Scientific; 2011. p. 291–393.
10. Batinic-Haberle I, Reboucas JS, Spasojevic I. Superoxide dismutase mimics: chemistry, pharmacology, and therapeutic potential. Antioxid Redox Signal. 2010;13:877–918.
11. Batinić-Haberle I, Spasojević I, Hambright P, Benov L, Crumbliss AL, Fridovich I. Relationship among redox potentials, proton dissociation constants of pyrrolic nitrogens, and in vivo and in vitro superoxide dismutating activities of manganese(III) and iron(III) water-soluble porphyrins. Inorg Chem. 1999;38:4011–22.
12. Batinić-Haberle I, Spasojević I, Stevens RD, Hambright P, Fridovich I. Manganese(III) meso-tetrakis(ortho-N-alkylpyridyl)porphyrins. Synthesis, characterization, and catalysis of O2/—dismutation. Dalton Trans. 2002:2689–96.
13. Batinic-Haberle I, Spasojevic I, Tse HM, Tovmasyan A, Rajic Z, St. Clair DK, Vujaskovic Z, Dewhirst MW, Piganelli JD. Design of Mn porphyrins for treating oxidative stress injuries and their redox-based regulation of cellular transcriptional activities. Amino Acids. 2012;42:95–113.
14. Batinic-Haberle I, Tovmasyan A, Roberts ER, Vujaskovic Z, Leong KW, Spasojevic I. SOD therapeutics: latest insights into their structure-activity relationships and impact on the cellular redox-based signaling pathways. Antioxid Redox Signal. 2014;20:2372–415.

15. Batinic-Haberle I, Tovmasyan A, Spasojevic I. An educational overview of the chemistry, biochemistry and therapeutic aspects of Mn porphyrins—from superoxide dismutation to H_2O_2-driven pathways. Redox Biol. 2015;5:43–65.
16. Miriyala S, Spasojevic I, Tovmasyan A, Salvemini D, Vujaskovic Z, St. Clair D, Batinic-Haberle I. Manganese superoxide dismutase, MnSOD and its mimics. Biochim Biophys Acta. 2012;1822:794–814.
17. Tovmasyan A, Sheng H, Weitner T, Arulpragasam A, Lu M, Warner DS, Vujaskovic Z, Spasojevic I, Batinic-Haberle I. Design, mechanism of action, bioavailability and therapeutic effects of mn porphyrin-based redox modulators. Med Princ Pract. 2013;22:103–30.
18. Batinic-Haberle I, Tovmasyan A. Superoxide dismutase mimics and other redox-active therapeutics. In: Armstrong D, Stratton RD, editors. Oxidative stress and antioxidant protection: The science of free radical biology and disease. New York: Wiley; 2016. p. 415–70.
19. Reboucas JS, DeFreitas-Silva G, Spasojevic I, Idemori YM, Benov L, Batinic-Haberle I. Impact of electrostatics in redox modulation of oxidative stress by Mn porphyrins: protection of SOD-deficient Escherichia coli via alternative mechanism where Mn porphyrin acts as a Mn carrier. Free Radic Biol Med. 2008;45:201–10.
20. Ferrer-Sueta G, Vitturi D, Batinic-Haberle I, Fridovich I, Goldstein S, Czapski G, Radi R. Reactions of manganese porphyrins with peroxynitrite and carbonate radical anion. J Biol Chem. 2003;278:27432–8.
21. Tovmasyan A, Carballal S, Ghazaryan R, Melikyan L, Weitner T, Maia CG, Reboucas JS, Radi R, Spasojevic I, Benov L, Batinic-Haberle I. Rational design of superoxide dismutase (SOD) mimics: the evaluation of the therapeutic potential of new cationic Mn porphyrins with linear and cyclic substituents. Inorg Chem. 2014;53:11467–83.
22. Tse HM, Milton MJ, Piganelli JD. Mechanistic analysis of the immunomodulatory effects of a catalytic antioxidant on antigen-presenting cells: implication for their use in targeting oxidation-reduction reactions in innate immunity. Free Radic Biol Med. 2004;36:233–47.
23. Shan W, Zhong W, Zhao R, Oberley TD. Thioredoxin 1 as a subcellular biomarker of redox imbalance in human prostate cancer progression. Free Radic Biol Med. 2010;49:2078–87.
24. Arambula JF, Preihs C, Borthwick D, Magda D, Sessler JL. Texaphyrins: tumor localizing redox active expanded porphyrins. Anticancer Agents Med Chem. 2011;11:222–32.
25. Eckshtain M, Zilbermann I, Mahammed A, Saltsman I, Okun Z, Maimon E, Cohen H, Meyerstein D, Gross Z. Superoxide dismutase activity of corrole metal complexes. Dalton Trans. 2009;38:7879–82.
26. Spasojevic I, Batinic-Haberle I, Stevens RD, Hambright P, Thorpe AN, Grodkowski J, Neta P, Fridovich I. Manganese(III) biliverdin IX dimethyl ester: a powerful catalytic scavenger of superoxide employing the Mn(III)/Mn(IV) redox couple. Inorg Chem. 2001;40:726–39.
27. Spasojević I, Batinić-Haberle I. Manganese(III) complexes with porphyrins and related compounds as catalytic scavengers of superoxide. Inorg Chim Acta. 2001;317:230–42.
28. Kos I, Reboucas JS, DeFreitas-Silva G, Salvemini D, Vujaskovic Z, Dewhirst MW, Spasojevic I, Batinic-Haberle I. Lipophilicity of potent porphyrin-based antioxidants: comparison of ortho and meta isomers of Mn(III) N-alkylpyridylporphyrins. Free Radic Biol Med. 2009;47:72–8.
29. Tovmasyan A, Weitner T, Sheng H, Lu M, Rajic Z, Warner DS, Spasojevic I, Reboucas JS, Benov L, Batinic-Haberle I. Differential coordination demands in Fe versus Mn water-soluble cationic metalloporphyrins translate into remarkably different aqueous redox chemistry and biology. Inorg Chem. 2013;52:5677–91.
30. Kos I, Benov L, Spasojevic I, Reboucas JS, Batinic-Haberle I. High lipophilicity of meta Mn(III) N-alkylpyridylporphyrin-based superoxide dismutase mimics compensates for their lower antioxidant potency and makes them as effective as ortho analogues in protecting superoxide dismutase-deficient Escherichia coli. J Med Chem. 2009;52:7868–72.
31. Pollard JM, Reboucas JS, Durazo A, Kos I, Fike F, Panni M, Gralla EB, Valentine JS, Batinic-Haberle I, Gatti RA. Radioprotective effects of manganese-containing superoxide dismutase mimics on ataxia-telangiectasia cells. Free Radic Biol Med. 2009;47:250–60.
32. Weitner T, Kos I, Sheng H, Tovmasyan A, Reboucas JS, Fan P, Warner DS, Vujaskovic Z, Batinic-Haberle I, Spasojevic I. Comprehensive pharmacokinetic studies and oral bioavail-

ability of two Mn porphyrin-based SOD mimics, MnTE-2-PyP(5+) and MnTnHex-2-PyP(5+). Free Radic Biol Med. 2013;58:73–80.

33. Ashcraft KA, Boss MK, Tovmasyan A, Roy Choudhury K, Fontanella AN, Young KH, Palmer GM, Birer SR, Landon CD, Park W, Das SK, Weitner T, Sheng H, Warner DS, Brizel DM, Spasojevic I, Batinic-Haberle I, Dewhirst MW. Novel manganese-porphyrin superoxide dismutase-mimetic widens the therapeutic margin in a preclinical head and neck cancer model. Int J Radiat Oncol Biol Phys. 2015;93:892–900.
34. Prasanna PG, Narayanan D, Hallett K, Bernhard EJ, Ahmed MM, Evans G, Vikram B, Weingarten M, Coleman CN. Radioprotectors and radiomitigators for improving radiation therapy: the small business innovation research (SBIR) gateway for accelerating clinical translation. Radiat Res. 2015;184:235–48.
35. Rajic Z, Tovmasyan A, Spasojevic I, Sheng H, Lu M, Li AM, Gralla EB, Warner DS, Benov L, Batinic-Haberle I. A new SOD mimic, Mn(III) ortho N-butoxyethylpyridylporphyrin, combines superb potency and lipophilicity with low toxicity. Free Radic Biol Med. 2012;52:1828–34.
36. Sheng H, Spasojevic I, Tse HM, Jung JY, Hong J, Zhang Z, Piganelli JD, Batinic-Haberle I, Warner DS. Neuroprotective efficacy from a lipophilic redox-modulating Mn(III) *N*-hexylpyridylporphyrin, MnTnHex-2-PyP: rodent models of ischemic stroke and subarachnoid hemorrhage. J Pharmacol Exp Ther. 2011;338:906–16.
37. Martell AE, Motekaitis RJ. The determination and use of stability constants. New York: VCH; 1992.
38. Riley DP, Lennon PJ, Neumann WL, Weiss RH. Toward the rational design of superoxide dismutase mimics: mechanistic studies for the elucidation of substituent effects on the catalytic activity of macrocyclic manganese(II) complexes. J Am Chem Soc. 1997;119:6522–8.
39. Tovmasyan A, Reboucas JS, Benov L. Simple biological systems for assessing the activity of superoxide dismutase mimics. Antioxid Redox Signal. 2014;20:2416–36.
40. Ezzeddine R, Al-Banaw A, Tovmasyan A, Craik JD, Batinic-Haberle I, Benov LT. Effect of molecular characteristics on cellular uptake, subcellular localization, and phototoxicity of Zn(II) N-alkylpyridylporphyrins. J Biol Chem. 2013;288:36579–88.
41. Tovmasyan AG, Rajic Z, Spasojevic I, Reboucas JS, Chen X, Salvemini D, Sheng H, Warner DS, Benov L, Batinic-Haberle I. Methoxy-derivatization of alkyl chains increases the in vivo efficacy of cationic Mn porphyrins. Synthesis, characterization, SOD-like activity, and SOD-deficient E. coli study of meta Mn(III) N-methoxyalkylpyridylporphyrins. Dalton Trans. 2011;40:4111–21.
42. Spasojevic I, Chen Y, Noel TJ, Fan P, Zhang L, Reboucas JS, St. Clair DK, Batinic-Haberle I. Pharmacokinetics of the potent redox-modulating manganese porphyrin, MnTE-2-PyP(5+), in plasma and major organs of B6C3F1 mice. Free Radic Biol Med. 2008;45:943–9.
43. Spasojevic I, Li A, Tovmasyan A, Rajic Z, Salvemini D, St. Clair D, Valentine JS, Vujaskovic Z, Gralla EB, Batinic-Haberle I. Accumulation of porphyrin-based SOD mimics in mitochondria is proportional to their lipophilicity: S-cerevisiae study of ortho Mn(III) N-alkylpyridylporphyrins. Free Radic Biol Med. 2010;49:S199.
44. Spasojevic I, Weitner T, Tovmasyan A, Sheng H, Miriyala S, Leu D, Rajic Z, Warner DS, St. Clair D, Huang T-T, Batinic-Haberle I. Pharmacokinetics, brain hippocampus and cortex, and mitochondrial accumulation of a new generation of lipophilic redox-active therapeutic, Mn(III) meso tetrakis(N-n-butoxyethylpyridinium-2-yl)porphyrin, MnTnBuOE-2-PyP5+, in comparison with its ethyl and N-hexyl analogs, MnTE-2-PyP5+ and MnTnHex-2-PyP5+. Free Radic Biol Med. 2013;65:S132.
45. Liberman EA, Topaly VP, Tsofina LM, Jasaitis AA, Skulachev VP. Mechanism of coupling of oxidative phosphorylation and the membrane potential of mitochondria. Nature. 1969; 222:1076–8.
46. Murphy MP. Targeting lipophilic cations to mitochondria. Biochim Biophys Acta. 2008;1777:1028–31.
47. Murphy MP, Smith RA. Targeting antioxidants to mitochondria by conjugation to lipophilic cations. Annu Rev Pharmacol Toxicol. 2007;47:629–56.

48. Spasojevic I, Miriyala S, Tovmasyan A, Salvemini D, Vujaskovic Z, Batinic-Haberle I, St. Clair D. Lipophilicity of Mn(III) *N*-alkylpyridylporphyrins dominates their accumulation within mitochondria and therefore *in vivo* efficacy. A mouse study. Free Radic Biol Med. 2011;51:S98.
49. Li AM, Martins J, Tovmasyan A, Valentine JS, Batinic-Haberle I, Spasojevic I, Gralla EB. Differential localization and potency of manganese porphyrin superoxide dismutase-mimicking compounds in Saccharomyces cerevisiae. Redox Biol. 2014;3:1–6.
50. Spasojevic I, Chen Y, Noel TJ, Yu Y, Cole MP, Zhang L, Zhao Y, St. Clair DK, Batinic-Haberle I. Mn porphyrin-based superoxide dismutase (SOD) mimic, MnIIITE-2-PyP5+, targets mouse heart mitochondria. Free Radic Biol Med. 2007;42:1193–200.
51. Ferrer-Sueta G, Hannibal L, Batinic-Haberle I, Radi R. Reduction of manganese porphyrins by flavoenzymes and submitochondrial particles: a catalytic cycle for the reduction of peroxynitrite. Free Radic Biol Med. 2006;41:503–12.
52. Bakthavatchalu V, Dey S, Xu Y, Noel T, Jungsuwadee P, Holley AK, Dhar SK, Batinic-Haberle I, St. Clair DK. Manganese superoxide dismutase is a mitochondrial fidelity protein that protects Polγ against UV-induced inactivation. Oncogene. 2012;31:2129–39.
53. Holley AK, Xu Y, Noel T, Bakthavatchalu V, Batinic-Haberle I, St. Clair DK. Manganese superoxide dismutase-mediated inside-out signaling in HaCaT human keratinocytes and SKH-1 mouse skin. Antioxid Redox Signal. 2014;20:2347–60.
54. Miriyala S, Thippakorn C, Chaiswing L, Xu Y, Noel T, Tovmasyan A, Batinic-Haberle I, Vander Kooi CW, Chi W, Latif AA, Panchatcharam M, Prachayasittikul V, Allan Butterfield D, Vore M, Moscow J, St. Clair DK. Novel role of 4-hydroxy-2-nonenal in AIFm2-mediated mitochondrial stress signaling. Free Radic Biol Med. 2015;91:68–80.
55. Zhao Y, Chaiswing L, Oberley TD, Batinic-Haberle I, St. Clair W, Epstein CJ, St. Clair D. A mechanism-based antioxidant approach for the reduction of skin carcinogenesis. Cancer Res. 2005;65:1401–5.
56. Zhao Y, Miriyala S, Miao L, Mitov M, Schnell D, Dhar SK, Cai J, Klein JB, Sultana R, Butterfield DA, Vore M, Batinic-Haberle I, Bondada S, St. Clair DK. Redox proteomic identification of HNE-bound mitochondrial proteins in cardiac tissues reveals a systemic effect on energy metabolism after doxorubicin treatment. Free Radic Biol Med. 2014;72:55–65.
57. Aitken JB, Shearer EL, Giles NM, Lai B, Vogt S, Reboucas JS, Batinic-Haberle I, Lay PA, Giles GI. Intracellular targeting and pharmacological activity of the superoxide dismutase mimics MnTE-2-PyP5+ and MnTnHex-2-PyP5+ regulated by their porphyrin ring substituents. Inorg Chem. 2013;52:4121–3.
58. Odeh AM, Craik JD, Ezzeddine R, Tovmasyan A, Batinic-Haberle I, Benov LT. Targeting mitochondria by Zn(II)N-alkylpyridylporphyrins: the impact of compound sub-mitochondrial partition on cell respiration and overall photodynamic efficacy. PLoS One. 2014;9, e108238.
59. Spasojevic I, Kos I, Benov LT, Rajic Z, Fels D, Dedeugd C, Ye X, Vujaskovic Z, Reboucas JS, Leong KW, Dewhirst MW, Batinic-Haberle I. Bioavailability of metalloporphyrin-based SOD mimics is greatly influenced by a single charge residing on a Mn site. Free Radic Res. 2011;45:188–200.
60. Thomas M, Craik JD, Tovmasyan A, Batinic-Haberle I, Benov LT. Amphiphilic cationic Zn-porphyrins with high photodynamic antimicrobial activity. Future Microbiol. 2015;10:709–24.
61. Jaramillo MC, Briehl MM, Crapo JD, Batinic-Haberle I, Tome ME. Manganese porphyrin, MnTE-2-PyP5+, acts as a pro-oxidant to potentiate glucocorticoid-induced apoptosis in lymphoma cells. Free Radic Biol Med. 2012;52:1272–84.
62. Benov L, Craik J, Batinic-Haberle I. Protein damage by photo-activated Zn(II) N-alkylpyridylporphyrins. Amino Acids. 2012;42:117–28.
63. Tovmasyan A, Sampaio RS, Boss MK, Bueno-Janice JC, Bader BH, Thomas M, Reboucas JS, Orr M, Chandler JD, Go YM, Jones DP, Venkatraman TN, Haberle S, Kyui N, Lascola CD, Dewhirst MW, Spasojevic I, Benov L, Batinic-Haberle I. Anticancer therapeutic potential of Mn porphyrin/ascorbate system. Free Radic Biol Med. 2015;89:1231–47.
64. Tovmasyan A, Maia CG, Weitner T, Carballal S, Sampaio RS, Lieb D, Ghazaryan R, Ivanovic-Burmazovic I, Ferrer-Sueta G, Radi R, Reboucas JS, Spasojevic I, Benov L,

Batinic-Haberle I. A comprehensive evaluation of catalase-like activity of different classes of redox-active therapeutics. Free Radic Biol Med. 2015;86:308–21.
65. Bueno-Janice JC, Tovmasyan A, Batinic-Haberle I. Comprehensive study of GPx activity of different classes of redox-active therapeutics—implications for their therapeutic actions. Free Rad Biol Med. 2015;87:S86–7.
66. Tovmasyan A, Weitner T, Jaramillo M, Wedmann R, Roberts ERH, Leong KW, Filipovic M, Ivanovic-Burmazovic I, Benov L, Tome ME, Batinic-Haberle I. We have come a long way with Mn porphyrins: from superoxide dismutation to H2O2-driven pathways. Free Radic Biol Med. 2013;65:S133.
67. Delmastro-Greenwood MM, Votyakova T, Goetzman E, Marre ML, Previte DM, Tovmasyan A, Batinic-Haberle I, Trucco MM, Piganelli JD. Mn porphyrin regulation of aerobic glycolysis: implications on the activation of diabetogenic immune cells. Antioxid Redox Signal. 2013;19:1902–15.
68. Sheng H, Yang W, Fukuda S, Tse HM, Paschen W, Johnson K, Batinic-Haberle I, Crapo JD, Pearlstein RD, Piganelli J, Warner DS. Long-term neuroprotection from a potent redox-modulating metalloporphyrin in the rat. Free Radic Biol Med. 2009;47:917–23.
69. Alvarez L, Suarez SA, Bikiel DE, Reboucas JS, Batinic-Haberle I, Marti MA, Doctorovich F. Redox potential determines the reaction mechanism of HNO donors with Mn and Fe porphyrins: defining the better traps. Inorg Chem. 2014;53:7351–60.
70. Halliwell B, Gutteridge JMC. Free radicals in biology and medicine. New York: Oxford University Press; 2015.
71. Winterbourn CC. The biological chemistry of hydrogen peroxide. Methods Enzymol. 2013;528:3–25.
72. Winterbourn CC, Peskin AV, Parsons-Mair HN. Thiol oxidase activity of copper, zinc superoxide dismutase. J Biol Chem. 2002;277:1906–11.
73. Gauter-Fleckenstein B, Fleckenstein K, Owzar K, Jiang C, Reboucas JS, Batinic-Haberle I, Vujaskovic Z. Early and late administration of MnTE-2-PyP5+ in mitigation and treatment of radiation-induced lung damage. Free Radic Biol Med. 2010;48:1034–43.
74. Moeller BJ, Cao Y, Li CY, Dewhirst MW. Radiation activates HIF-1 to regulate vascular radiosensitivity in tumors: role of reoxygenation, free radicals, and stress granules. Cancer Cell. 2004;5:429–41.
75. Zhang Y, Zhang X, Rabbani ZN, Jackson IL, Vujaskovic Z. Oxidative stress mediates radiation lung injury by inducing apoptosis. Int J Radiat Oncol Biol Phys. 2012;83:740–8.
76. Cohen J, Dorai T, Ding C, Batinic-Haberle I, Grasso M. The administration of renoprotective agents extends warm ischemia in a rat model. J Endourol. 2013;27:343–8.
77. Dorai T, Fishman AI, Ding C, Batinic-Haberle I, Goldfarb DS, Grasso M. Amelioration of renal ischemia-reperfusion injury with a novel protective cocktail. J Urol. 2011;186:2448–54.
78. Vukelic I, Colic T, Rubinic N, Spanjol J, Bobinac M, Tovmasyan A, Oberley-Deegan R, Batinic-Haberle I, Bobinac D. Understanding redox biology behind the therapeutic effects of redox active Mn porphyrins in spinal cord ischemia/reperfusion injury. Free Radic Biol Med. 2015;87:S97–8.
79. Gauter-Fleckenstein B, Fleckenstein K, Owzar K, Jiang C, Batinic-Haberle I, Vujaskovic Z. Comparison of two Mn porphyrin-based mimics of superoxide dismutase in pulmonary radioprotection. Free Radic Biol Med. 2008;44:982–9.
80. Gauter-Fleckenstein B, Reboucas JS, Fleckenstein K, Tovmasyan A, Owzar K, Jiang C, Batinic-Haberle I, Vujaskovic Z. Robust rat pulmonary radioprotection by a lipophilic Mn N-alkylpyridylporphyrin, MnTnHex-2-PyP(5+). Redox Biol. 2014;2:400–10.
81. Rabbani ZN, Spasojevic I, Zhang X, Moeller BJ, Haberle S, Vasquez-Vivar J, Dewhirst MW, Vujaskovic Z, Batinic-Haberle I. Antiangiogenic action of redox-modulating Mn(III) meso-tetrakis(N-ethylpyridinium-2-yl)porphyrin, MnTE-2-PyP(5+), via suppression of oxidative stress in a mouse model of breast tumor. Free Radic Biol Med. 2009;47:992–1004.
82. Moeller BJ, Batinic-Haberle I, Spasojevic I, Rabbani ZN, Anscher MS, Vujaskovic Z, Dewhirst MW. A manganese porphyrin superoxide dismutase mimetic enhances tumor radioresponsiveness. Int J Radiat Oncol Biol Phys. 2005;63:545–52.

83. Weitzel DH, Tovmasyan A, Ashcraft KA, Rajic Z, Weitner T, Liu C, Li W, Buckley AF, Prasad MR, Young KH, Rodriguiz RM, Wetsel WC, Peters KB, Spasojevic I, Herndon II JE, Batinic-Haberle I, Dewhirst MW. Radioprotection of the brain white matter by Mn(III) N-butoxyethylpyridylporphyrin-based superoxide dismutase mimic MnTnBuOE-2-PyP5+. Mol Cancer Ther. 2015;14:70–9.
84. Hayes JD, Dinkova-Kostova AT. The Nrf2 regulatory network provides an interface between redox and intermediary metabolism. Trends Biochem Sci. 2014;39:199–218.
85. Jaramillo MC, Zhang DD. The emerging role of the Nrf2-Keap1 signaling pathway in cancer. Genes Dev. 2013;27:2179–91.
86. Ma Q. Role of nrf2 in oxidative stress and toxicity. Annu Rev Pharmacol Toxicol. 2013;53:401–26.
87. Carroll D, Zhao Y, Batinic-Haberle I, St. Clair D. A novel redox-based approach to myelodysplastic syndrome (MDS) therapy. Free Radic Biol Med. 2015;87:S87.
88. Sheng H, Enghild JJ, Bowler R, Patel M, Batinic-Haberle I, Calvi CL, Day BJ, Pearlstein RD, Crapo JD, Warner DS. Effects of metalloporphyrin catalytic antioxidants in experimental brain ischemia. Free Radic Biol Med. 2002;33:947–61.
89. Sheng H, Spasojevic I, Warner DS, Batinic-Haberle I. Mouse spinal cord compression injury is ameliorated by intrathecal cationic manganese(III) porphyrin catalytic antioxidant therapy. Neurosci Lett. 2004;366:220–5.
90. Piganelli JD, Flores SC, Cruz C, Koepp J, Batinic-Haberle I, Crapo J, Day B, Kachadourian R, Young R, Bradley B, Haskins K. A metalloporphyrin-based superoxide dismutase mimic inhibits adoptive transfer of autoimmune diabetes by a diabetogenic T-cell clone. Diabetes. 2002;51:347–55.
91. Jaramillo MC, Briehl MM, Batinic-Haberle I, Tome ME. Manganese (III) meso-tetrakis N-ethylpyridinium-2-yl porphyrin acts as a pro-oxidant to inhibit electron transport chain proteins, modulate bioenergetics, and enhance the response to chemotherapy in lymphoma cells. Free Radic Biol Med. 2015;83:89–100.
92. Jaramillo MC, Frye JB, Crapo JD, Briehl MM, Tome ME. Increased manganese superoxide dismutase expression or treatment with manganese porphyrin potentiates dexamethasone-induced apoptosis in lymphoma cells. Cancer Res. 2009;69:5450–7.
93. Evans MK, Tovmasyan A, Batinic-Haberle I, Devi GR. Mn porphyrin in combination with ascorbate acts as a pro-oxidant and mediates caspase-independent cancer cell death. Free Radic Biol Med. 2014;68:302–14.
94. Delmastro-Greenwood MM, Tse HM, Piganelli JD. Effects of metalloporphyrins on reducing inflammation and autoimmunity. Antioxid Redox Signal. 2014;20:2465–77.
95. Holley AK, Miao L, St. Clair DK, St. Clair WH. Redox-modulated phenomena and radiation therapy: the central role of superoxide dismutases. Antioxid Redox Signal. 2014;20:1567–89.
96. Jumbo-Lucioni PP, Ryan EL, Hopson ML, Bishop HM, Weitner T, Tovmasyan A, Spasojevic I, Batinic-Haberle I, Liang Y, Jones DP, Fridovich-Keil JL. Manganese-based superoxide dismutase mimics modify both acute and long-term outcome severity in a Drosophila melanogaster model of classic Galactosemia. Antioxid Redox Signal. 2014;20:2361–71.
97. Sheng H, Chaparro RE, Sasaki T, Izutsu M, Pearlstein RD, Tovmasyan A, Warner DS. Metalloporphyrins as therapeutic catalytic oxidoreductants in central nervous system disorders. Antioxid Redox Signal. 2014;20:2437–64.
98. Oberley-Deegan RE, Steffan JJ, Rove KO, Pate KM, Weaver MW, Spasojevic I, Frederick B, Raben D, Meacham RB, Crapo JD, Koul HK. The antioxidant, MnTE-2-PyP, prevents side-effects incurred by prostate cancer irradiation. PLoS One. 2012;7, e44178.
99. Granieri MA, Tovmasyan A, Yan H, Lu X, Mao L, Macias E, Spasojevic I, Batinic-Haberle I, Peterson A, Koontz BF. Radioprotection of erectile function using novel anti-oxidant in the rat. J Sex Med. 2015;12:178–9.
100. Batinic-Haberle I, Tovmasyan A, Weitner T, Rajic Z, Keir ST, Huang T-T, Leu D, Weitzel DH, Beausejour CM, Miriyala S, Roberts ERH, Dewhirst MW, St. Clair D, Leong KW, Spasojevic I, Piganelli J, Tome M. Mechanistic considerations of the therapeutic effects of

Mn porphyrins, commonly regarded as SOD mimics, in anticancer therapy: lessons from brain and lymphoma studies. Free Radic Biol Med. 2013;65:S120–1.
101. Beausejour CM, Palacio L, Le O, Batinic-Haberle I, Sharpless NE, Marcoux S, Laverdiere C. Decreased neurogenesis following exposure to ionizing radiation: A role for p16^{INK4a}-induced senescence. Cell Senescence in Cancer and Ageing. Hinxton: Wellcome Trust Genome Campus; 2013.
102. Leu D, Zou Y, Weitner T, Tovmasyan A, Spasojevic I, Batinic-Haberle I, Huang T-T. Radiation protection of hippocampal neurogenesis with Mn-containing porphyrins. The 60th annual meeting of the radiation research society, Las Vegas, Nevada; 2014.
103. Batinic-Haberle I, Keir ST, Rajic Z, Tovmasyan A, Spasojevic I, Dewhirst MW, Bigner DD. Glioma growth suppression via modulation of cellular redox status by a lipophilic Mn porphyrin. Mid-Winter SPORE Meeting. 2011;31–2.
104. Archambeau JO, Tovmasyan A, Pearlstein RD, Crapo JD, Batinic-Haberle I. Superoxide dismutase mimic, MnTE-2-PyP(5+) ameliorates acute and chronic proctitis following focal proton irradiation of the rat rectum. Redox Biol. 2013;1:599–607.
105. Ye X, Fels D, Tovmasyan A, Aird KM, Dedeugd C, Allensworth JL, Kos I, Park W, Spasojevic I, Devi GR, Dewhirst MW, Leong KW, Batinic-Haberle I. Cytotoxic effects of Mn(III) N-alkylpyridylporphyrins in the presence of cellular reductant, ascorbate. Free Radic Res. 2011;45:1289–306.
106. Batinic-Haberle I, Spasojevic I, Stevens RD, Hambright P, Neta P, Okado-Matsumoto A, Fridovich I. New class of potent catalysts of O2.-dismutation. Mn(III) ortho-methoxyethylpyridyl- and di-ortho-methoxyethylimidazolylporphyrins. Dalton Trans. 2004;(11):1696–702.
107. Chen Q, Espey MG, Krishna MC, Mitchell JB, Corpe CP, Buettner GR, Shacter E, Levine M. Pharmacologic ascorbic acid concentrations selectively kill cancer cells: action as a pro-drug to deliver hydrogen peroxide to tissues. Proc Natl Acad Sci U S A. 2005;102:13604–9.
108. Chen Q, Espey MG, Sun AY, Lee JH, Krishna MC, Shacter E, Choyke PL, Pooput C, Kirk KL, Buettner GR, Levine M. Ascorbate in pharmacologic concentrations selectively generates ascorbate radical and hydrogen peroxide in extracellular fluid in vivo. Proc Natl Acad Sci U S A. 2007;104:8749–54.
109. Chen Q, Espey MG, Sun AY, Pooput C, Kirk KL, Krishna MC, Khosh DB, Drisko J, Levine M. Pharmacologic doses of ascorbate act as a prooxidant and decrease growth of aggressive tumor xenografts in mice. Proc Natl Acad Sci U S A. 2008;105:11105–9.
110. Hoffer LJ, Levine M, Assouline S, Melnychuk D, Padayatty SJ, Rosadiuk K, Rousseau C, Robitaille L, Miller Jr WH. Phase I clinical trial of i.v. ascorbic acid in advanced malignancy. Ann Oncol. 2008;19:1969–74.
111. Levine M, Espey MG, Chen Q. Losing and finding a way at C: new promise for pharmacologic ascorbate in cancer treatment. Free Radic Biol Med. 2009;47:27–9.
112. Monti DA, Mitchell E, Bazzan AJ, Littman S, Zabrecky G, Yeo CJ, Pillai MV, Newberg AB, Deshmukh S, Levine M. Phase I evaluation of intravenous ascorbic acid in combination with gemcitabine and erlotinib in patients with metastatic pancreatic cancer. PLoS One. 2012;7, e29794.
113. Welsh JL, Wagner BA, van't Erve TJ, Zehr PS, Berg DJ, Halfdanarson TR, Yee NS, Bodeker KL, Du J, Roberts II LJ, Drisko J, Levine M, Buettner GR, Cullen JJ. Pharmacological ascorbate with gemcitabine for the control of metastatic and node-positive pancreatic cancer (PACMAN): results from a phase I clinical trial. Cancer Chemother Pharmacol. 2013;71:765–75.
114. Yulyana Y, Tovmasyan A, Ho IA, Sia KC, Newman JP, Ng WH, Guo CM, Hui KM, Batinic-Haberle I, Lam PY. Redox-active Mn porphyrin-based potent SOD mimic, MnTnBuOE-2-PyP, enhances carbenoxolone-mediated TRAIL-induced apoptosis in glioblastoma multiforme. Stem Cell Rev. 2016;12(1):140–55.
115. Rawal M, Schroeder SR, Wagner BA, Cushing CM, Welsh JL, Button AM, Du J, Sibenaller ZA, Buettner GR, Cullen JJ. Manganoporphyrins increase ascorbate-induced cytotoxicity by enhancing H2O2 generation. Cancer Res. 2013;73:5232–41.

116. Cieslak JA, Strother RK, Rawal M, Du J, Doskey CM, Schroeder SR, Button A, Wagner BA, Buettner GR, Cullen JJ. Manganoporphyrins and ascorbate enhance gemcitabine cytotoxicity in pancreatic cancer. Free Radic Biol Med. 2015;83:227–37.
117. Tovmasyan A, Bueno-Janice JC, Boss M-K, Orr M, Chandler JD, Weitzel DH, Sampaio RS, Reboucas JS, Go Y-M, Jones DP, Dewhirst MW, Spasojevic I, Benov L, Batinic-Haberle I. Mn porphyrin-based SOD mimic and vitamin C enhance radiation-induced tumor growth inhibition. Free Rad Biol Med. 2015;87:S97.
118. Tian J, Peehl DM, Knox SJ. Metalloporphyrin synergizes with ascorbic acid to inhibit cancer cell growth through fenton chemistry. Cancer Biother Radiopharm. 2010;25:439–48.
119. Boss MK, Tovmasyan A, Batinic-Haberle I, Malcolm J, Oldham M, Dewhirst MW. MnBuOE affects the development of pulmonary metastasis in an orthotopic mammary carcinoma model in a contrasting manner according to the tumor redox microenvironment. 61st annual meeting of the radiation research society. Weston; 2015: Submitted for publication.
120. Fernandes AS, Florido A, Cipriano M, Batinic-Haberle I, Miranda J, Saraiva N, Guerreiro PS, Castro M, Oliveira NG. Combined effect of the SOD mimic MnTnHex-2-PyP^{5+} and doxorubicin on the migration and invasiveness of breast cancer cells. Toxicol Lett. 2013;221:S59–256.
121. Takada Y, Bhardwaj A, Potdar P, Aggarwal BB. Nonsteroidal anti-inflammatory agents differ in their ability to suppress NF-kappaB activation, inhibition of expression of cyclooxygenase-2 and cyclin D1, and abrogation of tumor cell proliferation. Oncogene. 2004;23:9247–58.
122. Gad SC, Sullivan Jr DW, Crapo JD, Spainhour CB. A nonclinical safety assessment of MnTE-2-PyP, a manganese porphyrin. Int J Toxicol. 2013;32:274–87.
123. Reboucas JS, Spasojevic I, Batinic-Haberle I. Pure manganese(III) 5,10,15,20-tetrakis (4-benzoic acid)porphyrin (MnTBAP) is not a superoxide dismutase mimic in aqueous systems: a case of structure-activity relationship as a watchdog mechanism in experimental therapeutics and biology. J Biol Inorg Chem. 2008;13:289–302.
124. Celic T, Spanjol J, Bobinac M, Tovmasyan A, Vukelic I, Reboucas JS, Batinic-Haberle I, Bobinac D. Mn porphyrin-based SOD mimic, MnTnHex-2-PyP(5+), and non-SOD mimic, MnTBAP(3-), suppressed rat spinal cord ischemia/reperfusion injury via NF-kappaB pathways. Free Radic Res. 2014;48:1426–42.
125. Reboucas JS, Spasojevic I, Batinic-Haberle I. Quality of potent Mn porphyrin-based SOD mimics and peroxynitrite scavengers for pre-clinical mechanistic/therapeutic purposes. J Pharm Biomed Anal. 2008;48:1046–9.
126. Batinic-Haberle I, Ndengele MM, Cuzzocrea S, Reboucas JS, Spasojevic I, Salvemini D. Lipophilicity is a critical parameter that dominates the efficacy of metalloporphyrins in blocking the development of morphine antinociceptive tolerance through peroxynitrite-mediated pathways. Free Radic Biol Med. 2009;46:212–9.
127. Orrell RW. AEOL-10150 (Aeolus). Curr Opin Investig Drugs. 2006;7:70–80.
128. Slosky LM, Vanderah TW. Therapeutic potential of peroxynitrite decomposition catalysts: a patent review. Expert Opin Ther Pat. 2015;25(4):443–66.
129. Gad SC, Sullivan DW Jr, Spasojevic I, Mujer CV, Spainhour CB, Crapo JD. Nonclinical Safety and Toxicokinetics of MnTnBuOE-2-PyP5+ (BMX-001). Int J Toxicol. 2016;35:438–53.

Chapter 9
Cytochrome P450-Like Biomimetic Oxidation Catalysts Based on Mn Porphyrins as Redox Modulators

Victor Hugo A. Pinto, Nathália K.S.M. Falcão, Jacqueline C. Bueno-Janice, Ivan Spasojević, Ines Batinić-Haberle, and Júlio S. Rebouças

Abbreviations

AH	A generic 1-electron donor ($AH \rightarrow A^{\bullet} + H^{+} + 1e^{-}$)
AO	A generic oxygen donor such as PhIO, ClO^{-}, and H_2O_2 etc.
FeP	Iron porphyrin
$FeTPP^{+}$	Fe(III) *meso*-tetraphenylporphyrin
HRP	Horseradish peroxidase
LOOH	Lipid hydroperoxide or an alkyl hydroperoxide
MnP	Manganese porphyrin
$MnTBAP^{3-}$	Mn(III) *meso*-tetrakis(benzoic acid)porphyrin or Mn(III) *meso*-tetrakis(carboxyphenyl)porphyrin

V.H.A. Pinto • N.K.S.M. Falcão • J.S. Rebouças (✉)
Departamento de Química, CCEN, Universidade Federal da Paraíba, Jardim Cidade Universitária, João Pessoa, PB 58051-900, Brazil
e-mail: jsreboucas@quimica.ufpb.br; jsreboucas@gmail.com

J.C. Bueno-Janice
Departamento de Química, CCEN, Universidade Federal da Paraíba, Jardim Cidade Universitária, João Pessoa, PB 58051-900, Brazil

Department of Radiation Oncology, Duke University Medical Center, Durham, NC 27710, USA

I. Spasojević
Department of Medicine and Duke Cancer Institute, Pharmaceutical Research Shared Resource, PK/PD Core Laboratory, Duke University School of Medicine, Duke Hospital South, 5317 Orange Zone, Trent Drive, Durham, NC 27710, USA

I. Batinić-Haberle
Department of Radiation Oncology, Duke University School of Medicine, Research Drive, 281b/285 MSRB I, Box 3455, Durham, NC 27710, USA
e-mail: ibatinic@duke.edu

I. Batinić-Haberle et al. (eds.), *Redox-Active Therapeutics*, Oxidative Stress in Applied Basic Research and Clinical Practice, DOI 10.1007/978-3-319-30705-3_9

$MnTE\text{-}2\text{-}PyP^{5+}$	Mn(III) *meso*-tetrakis(*N*-ethylpyridinium-2-yl)porphyrin
$MnTM\text{-}2\text{-}PyP^{5+}$	Mn(III) *meso*-tetrakis(*N*-methylpyridinium-2-yl)porphyrin
$MnTM\text{-}3\text{-}PyP^{5+}$	Mn(III) *meso*-tetrakis(*N*-methylpyridinium-3-yl)porphyrin
$MnTM\text{-}4\text{-}PyP^{5+}$	Mn(III) *meso*-tetrakis(*N*-methylpyridinium-4-yl)porphyrin
$MnTnBuOE\text{-}2\text{-}PyP^{5+}$	Mn(III) *meso*-tetrakis(*N*-n-butoxyethylpyridinium-2-yl) porphyrin
$MnTnHex\text{-}2\text{-}PyP^{5+}$	Mn(III) *meso*-tetrakis(*N*-n-hexylpyridinium-2-yl) porphyrin
$MnTPP^{+}$	Mn(III) *meso*-tetraphenylporphyrin
$NADP^{+}$	Nicotinamide adenine dinucleotide phosphate
NADPH	Reduced form of nicotinamide adenine dinucleotide phosphate
P450	Cytochrome P450
PhIO	Iodosylbenzene
RH	A generic organic substrate usually an alkane
SOD	Superoxide dismutase

9.1 Highlights on Metalloporphyrin-Based Biomimetic Chemistry

Iron porphyrins (FePs) are ubiquitous in living organisms. Naturally occurring porphyrins, such as protoporphyrin IX (Fig. 9.1) or a closely related tetrapyrrole ligand, are able to bind Fe(II) or Fe(III) ions very tightly, yielding the well-known "heme" groups [1–3]. Heme is rarely free in biological systems [4, 5], being often bound to or associated with specific proteins collectively known as hemeproteins (or hemoproteins) [1–3]. These resulting metalloproteins play a variety of diverse and critical roles in many biological processes, in which the heme unit acts as the prosthetic group (e.g., hemoglobin, myoglobin, and cytochrome *c*) [1–3, 6–8]. Many important redox-based metalloenzymes, such as cytochromes P450, nitric oxide

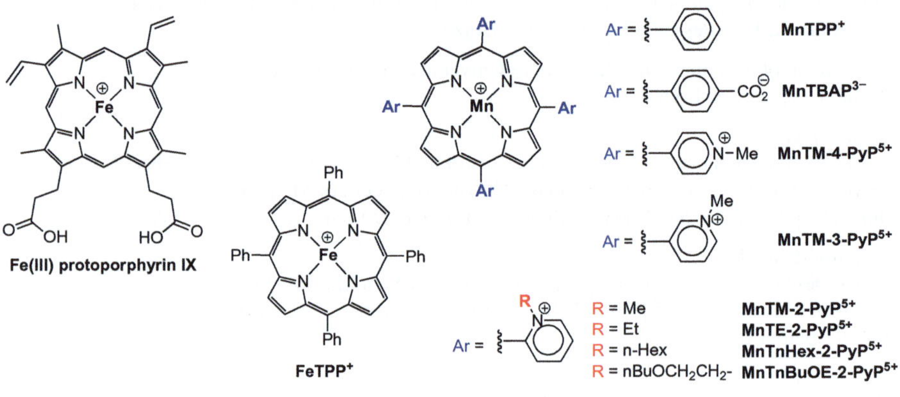

Fig. 9.1 Structure of metalloporphyrins relevant to this chapter

synthases, cytochrome *c* oxidase, and some peroxidases and catalases, are also hemeproteins [2, 3, 8–13]. The reactivity of the heme group is, thus, largely controlled by the apoprotein moiety, which ultimately dictates the heme biological function [6–11].

These hemeproteins have fascinated chemists and biochemists for more than a century [14]. The attempts to model their structural features, their electronic structure and corresponding spectroscopic signatures, and their reactivity have been under the scrutiny of biomimetic chemists before biomimetic chemistry was even termed as such. The term "biomimetic chemistry" was coined by Breslow [15, 16] in the early 1970s to describe a scientific investigation strategy that uses Nature as inspiration either to develop a chemical process related to those found in the natural systems, or to prepare a compound that imitates a biological material in its structure and/or function. The goal is not to merely reproduce Nature, but, nevertheless, to draw inspiration from it, which allows experimentations with ligands and metals not associated with biology, including non-naturally occurring ligands, and a full range of different metals, including those biologically uncommon ones (e.g., Ru [17–19]).

During the past 50 years or so, chemists have been quite busy using metalloporphyrins as biomimetic models of metalloproteins and/or metalloenzymes [6–14, 17–24]. Historically, most of the initial studies were devoted to the modeling of dioxygen transport and storage by hemoglobin and myoglobin, respectively [6, 7, 14]. In the 1970s, many groups (e.g., Collman, Baldwin, Momenteau, James, Dolphin, Traylor, and Basolo) carried out elegant studies on synthetic metalloporphyrins as models for describing the reversible coordination of O_2 to hemoglobin and myoglobin [6–8, 25–27]. The most promising systems as O_2 carriers were those of Fe(II); some success was also achieved with Co(II), but Mn(II) and Mn(III) were ineffective as O_2 carriers (of note, reversible O_2-coordination to Mn(II) porphyrins could be achieved under unsuitable conditions, i.e., in toluene at −78 °C [28]). The main challenges in these systems, in particular the Fe(II) ones, were [6, 7]: (a) to avoid the irreversible O_2-oxidation of $Fe^{II}P$ to $Fe^{III}P$, which may be achieved by preventing the approximation of water to the metal site; and (b) to prevent the irreversible formation of μ-oxo complexes, i.e., $PFe^{III}–O–Fe^{III}P$, which resulted from the reaction between the $(O_2)Fe^{II}P$ adduct and a second molecule of $Fe^{II}P$. In the hemeproteins, the bulky protein moiety, which is not present in the model systems, creates not only a hydrophobic region around the metal center but also prohibits the formation of μ-oxo species by sterically hindering the approximation of two heme groups [6, 7]. In this regard, the search for architecturally structured porphyrin macrocycles bearing bulky substituents with hydrophobic cavities around the metal center originated the classic "picket fence" [25], "capped" [26], and "basket handle" [27] porphyrins. Eventually, FePs were able to model hemoglobin and myoglobin with respect to reversible O_2-coordination, provided a good description of the spectroscopic characteristics of these systems, and revealed subtle details on the role played by the protein on modulating the reactivity of heme to yield reversible and selective O_2-coordination [6, 7]. The technological applications of these complexes as O_2-carriers or air fractioning devices, however, have been hindered by the complexity and fragility of the systems. Whereas the development of FePs as O_2-carriers for artificial blood formulations is still pursued [29], the drawbacks that need to be solved include concepts such as oxidative and nitrosative stress [30, 31]. Looking at these systems retrospectively, the early attempts to use simple FeP complexes as O_2-carriers in vivo in the 1970s were bound to fail, given that these "naked" systems are prone to side-reactions and side-effect complications associated with the implications of some reactivity patterns unforeseen then and still much of a challenge now: the reactivity of FePs toward NO [32], a molecule whose biology and biologically relevant chemistry blossomed from the 1980s on [33]. Coordination of O_2 to ferrous FePs yields the adduct $(O_2)Fe^{II}P$, which, by internal electron transfer, exists as a ferric superoxide-bound complex, $(O_2^-)Fe^{III}P$ [6]; superoxide may react with NO at diffusion limiting rates to yield the powerful oxidant peroxynitrite [34]. Despite the stress associated with peroxynitrite on its own [34], a high concentration, bolus dose of FePs (as artificial red blood cell substituent) would scavenge NO directly (via coordination to FePs) or indirectly (via FeP-derived superoxide pathways), leading to cardiovascular alterations, to the least.

The 1979 seminal paper by Groves and coworkers [35] on the use of a synthetic FeP as a biomimetic model of the cytochromes P450, which is a ferric heme-containing metalloenzyme, revealed that the same (or even simpler) FePs that used to be studied as hemoglobin/myoglobin models could also be investigated as P450 models [20]. The mechanism of the cytochromes P450 is rather complex [36] (*see Sect.* 9.2.1) and the use of model compounds appeared as an alternative to dissect the structural aspects, electronic features, and reactivity behavior of these enzymes [11, 37]. These possibilities were quickly taken by many groups worldwide and the results on these P450 model systems inaugurated the chemistry of metalloporphyrins as biomimetic oxidant catalysts, which remains quite an active and dynamic field of research since then [37]. Among the most studied metalloporphyrin-based catalysts are those of Mn(III), Fe(III), and Ru(VI) porphyrins [17, 20, 21, 38–40]. Upon the success in exploring metalloporphyrin complexes as biomimetic P450 models [37], the studies expanded rapidly to include other redox-active heme-based proteins, such as peroxidases, catalase, and nitric oxide synthase [11, 24, 38, 41].

Another very active area of research on using metalloporphyrins as biomimetic models is that of modeling of the superoxide dismutase (SOD) enzymes [22, 23, 42] (*see Chap.* 8 *by Batinić-Haberle et al.*), even though none of the four known isoforms of SOD (i.e., Cu,ZnSOD, MnSOD, FeSOD, and NiSOD) contains the porphyrin ring in their structure nor uses porphyrin as cofactor at all (*see Chap.* 7 *by Policar*). Despite SOD and metalloporphyrins being structurally unrelated, FePs and Mn porphyrins (MnPs) are among the most active models of SOD [22, 23, 43, 44], i.e., FePs and MnPs can model the SOD function, but, obviously, cannot account for any of the enzyme spectroscopic features. At this point, it is worth pointing out that along this chapter, the word "mimic" will be reserved for a model compound that is able to perform at rates meaningfully comparable to that of the modeled enzyme. For example, cationic MnPs, such as $MnTE\text{-}2\text{-}PyP^{5+}$ (Fig. 9.1), are SOD mimics, as their catalytic rate constant (log k_{cat}) are close to that of SOD; additionally, $MnTE\text{-}2\text{-}PyP^{5+}$ is able to substitute for the SOD enzyme in vivo in SOD-deficient microorganisms such as bacteria (*Escherichia coli*) and yeast (*Saccharomyces cerevisiae*) [23, 45]. Conversely, whereas FePs and MnPs are able to emulate the reactions of catalase, their catalase activities are rather low (as low as 0.006 % of that of the enzyme, in the best case scenario) [41]; these metalloporphyrins may, thus, be suitable biomimetic models of catalase, but are *not* catalase mimics [41].

The first studies on the SOD activity of metalloporphyrins appeared also in 1979 with the work by Pasternack and Halliwell [46], who showed that simple water-soluble metalloporphyrins could model the reactions of SOD on dismuting the superoxide radical ion to H_2O_2 and O_2. The metalloporphyrin-based SOD mimic field remained relatively dormant [47–52] as compared to the developments witnessed at P450 biomimetic systems. Of note, Meunier published in 1992 a classical review with already more than 600 references on P450-type reactions associated with metalloporphyrins [20]. The breakthrough in developing potent porphyrin-based SOD mimics appeared in 1998 with the work by Batinić-Haberle et al. [53] where the *ortho* substituted Mn(III) *meso*-tetrakis(*N*-methylpyridinium-2-yl)porphyrin, MnTM-2-PyP^{5+} (Fig. 9.1), was found able to dismutate superoxide at a, then, unusually high rate constant (log k_{cat}=7.72). The SOD-based biomimetic chemistry of metalloporphyrins flourished and soon became clear that a good SOD model should emulate the reduction potential of SOD enzymes (*ca.* +300 mV vs. NHE) [54–57] and bear positive charges in close vicinity to the metal center, in order to guide the superoxide anion to the reactive site via electrostatic facilitation [58, 59], in a similar fashion to the role that cationic amino acids residues play on SOD enzymes themselves. The first empirical structure–activity relationships on Mn-porphyrin-based SOD mimics accounting for *both* thermodynamic (reduction potential, $E_{½}$ [54]) *and* electrostatic facilitation (i.e., the impact the charges of cationic, anionic, and neutral porphyrins exert on SOD activity, log k_{cat} [58]) emerged [60] and have recently been extended to include many non-porphyrin complex models [42, 61]. Some MnP complexes are as SOD active as the SOD enzymes themselves [43], which represents a successful achievement in biomimetic chemistry: designed, nonstructurally related, small-molecule complexes mimicking closely the reactivity patterns and rates of Nature's own SOD enzymes.

Nowadays, water-soluble Mn porphyrins comprise the class of metalloporphyrins most investigated in biological systems [22, 23, 42, 45, 62, 63]. Their use in vivo spams a wide range of organisms, from microorganisms (e.g., bacteria, yeast) to animals, such as rodents, dogs, and primates, including humans [22, 23, 42, 45, 62, 63]. Historically, the development of MnPs for medicinal applications was initiated and had relied considerably on exploring the SOD mimic properties of the complexes. Their in vivo efficiency in a variety of oxidative stress-based injuries and pathological states has usually been ascribed to the SOD activity of the mimic [22, 23, 61–63]. As SOD enzymes are within the first lines of defense among the endogenous antioxidant enzymes against oxidative damage (*see Chap.* 8 *by Batinić-Haberle et al.*), these MnPs have, thus, been often referred to as catalytic antioxidants [64, 65].

In this chapter, we put forward an emerging concept on which the therapeutic action of some MnPs used in vivo might not be restricted to an SOD-type antioxidant catalytic role per se, but involve P450-type oxidation catalysis. In the next sections, a brief description of the general mechanism of the cytochromes P450 and related enzyme systems will be followed by the presentation of selected aspects of the 30+ years of biomimetic chemistry of MnPs as P450 models. It is not the goal of this chapter to provide a full account on the MnP-based biomimetic oxidation systems, but highlight some of the main features of P450-type MnP reactions that may become relevant on interpreting some of the biological results on the use of MnPs as redox-active therapeutics.

9.2 A Brief Overview on Cytochrome P450-Catalyzed Oxidations and Related Systems

9.2.1 Cytochrome P450 Family

The search for efficient catalysts that are able to promote the selective insertion of oxygen atoms (oxygenation reactions) in many organic substrates under mild conditions of temperature and pressure remains a challenge in chemical catalysis. This is particularly relevant when developing synthetic routes to the hydroxylation of the inert C–H bonds of alkanes or the selective oxidation of aromatic compounds [20, 21, 38–40, 66].

The direct conversion of an alkane into an alcohol using cheap and environmentally friendly oxidants, such as O_2, has been intensely sought after, as this would allow the development of novel synthetic routes to convert hydrocarbons, easily available and of low cost, in products to fine chemistry and pharmaceuticals [24, 67, 68]. On industrial settings, C–H bond activation is usually achieved using heterogeneous catalysis at high temperature, but polyoxygenation is often observed, which renders the systems of compromised selectivity [67].

In biological systems, however, the family of cytochrome P450 enzymes are able to catalyze the monooxygenation of a large variety of organic substrates (RH), under ambient temperature and pressure, using atmospheric O_2 and a biological reductant, such as the reduced form of nicotinamide adenine dinucleotide phosphate (NADPH), as represented in (9.1) [9, 36]. The cytochromes P450 belong to a superfamily of heme-containing monooxygenase enzymes with similar spectral properties and that share a distinct spectroscopic feature: they exhibit an intense absorption band (Soret band) at ~450 nm in the reduced form in the presence of CO [13, 36, 37, 69]. They all contain an Fe(III) protoporphyrin IX (Fig. 9.1) in the active site, whose binding to the protein is achieved by coordination of the thiolate moiety of a residual cysteine (the proximal ligand) to the heme Fe(III) center [11, 36]. The heme coordination site *trans* to the cysteine proximal ligand comprises the distal site, where binding of O_2 takes place. The distal site is often occupied by water molecules, which are easily replaced by O_2 in the presence of a suitable substrate [13, 36]. The polypeptide protein part has molecular weight of *ca.* 45,000 Da and is responsible for both rendering the catalytic site hydrophilic and directing the substrate toward the active metal site [20, 36, 37, 39].

$$RH + O_2 + NADPH + H^+ \rightarrow ROH + NADP^+ + H_2O \qquad (9.1)$$

The cytochromes P450 are membrane bound enzymes [70] widely spread among the living organisms, including unicellular microorganisms (e.g., bacteria and fungi) and complex organisms (such as humans) [71]. They exert important roles in vital processes of the organisms, being involved as catalysts for the biosynthesis of prostaglandins and steroids and for the oxidative biotransformation of xenobiotics, such as drugs, pesticides, carcinogens, polycyclic aromatic hydrocarbons, alkanes, and alkenes [12, 71].

The catalytic cycle of P450 is still a matter of intense research, with particular focus on the elucidation of the electronic structure and reactivity pattern of the high-valent intermediate species [12, 13, 72, 73]. Given the high molecular weight of the cytochromes P450, it is rather difficult to determine with accuracy the mechanistic details of the oxidation reactions, as well as the molecular structure of all intermediates involved in the processes. In this context, the use of biomimetic systems helped a great deal in shedding some light on the transformations associated with the enzymatic processes [37]. Even in simple model systems, the characterization of many short-lived intermediates is challenging. Two main issues remain of keen interest in the P450 reactions: the mechanism by which O_2 is activated to yield the active intermediate species and the mechanism by which the oxygen atom is transferred to the substrate [72, 74–78].

A proposed mechanism that accounts for the most likely intermediate species during P450-catalyzed oxidation reactions is illustrated in Fig. 9.2 [12, 36, 74–78]. The P450 resting state (**A**) is a hexacoordinate low-spin Fe(III) species with a water molecule as distal ligand. Upon the approximation of the substrate RH, the protein changes its conformation, releasing the axial water molecule to yield a pentacoordinate high-spin Fe(III) species (**B**), which accommodates the substrate in a pocket in the vicinity of the active site. The degree of low-to-high spin-state transition that accompanies iron coordination-sphere change may lead to *ca.* 80–130 mV increase in the P450 Fe(III)/Fe(II) reduction potential [79–82] (*see Sect.* 9.4), which is crucial for the Fe(III) → Fe(II) reduction that follows. Reduction of (**B**) by one electron leads to a pentacoordinate Fe(II) species (**C**) with the Fe atom slightly shifted toward the proximal cysteine ligand; coordination of O_2 to the pentacoordinate ferrous species (**C**) yields the hexacoordinate adduct $(O_2)Fe^{II}P$ (**D**). This species can also be described as a superoxide-bound ferric species (**D′**) resulting from an internal electron transfer from the ferrous iron to the coordinated O_2 in (**D**) [80, 81]. A second 1-electron reduction yields a Fe(III) species containing a peroxide moiety coordinated on an end-on (η^1) fashion (**E**) [77]. Protonation of the peroxide moiety leads to species (**F**, where X = H). Heterolytic cleavage of the O–OH bond followed by protonation of the resulting hydroxide leads to the release of water and concomitant formation of the high-valent Fe(IV)-oxo porphyrin π-cation species (**Compound I**), formally a Fe(V)-oxo species. Compound I, then, abstracts a hydrogen atom from the nearby substrate, yielding an organic radical $R^\bullet$ and Fe(IV)-hydroxo species (**G**). As the substrate-derived radical $R^\bullet$ is kept within the active site of the P450 enzyme, this allows a recombination of $R^\bullet$ with the coordinated

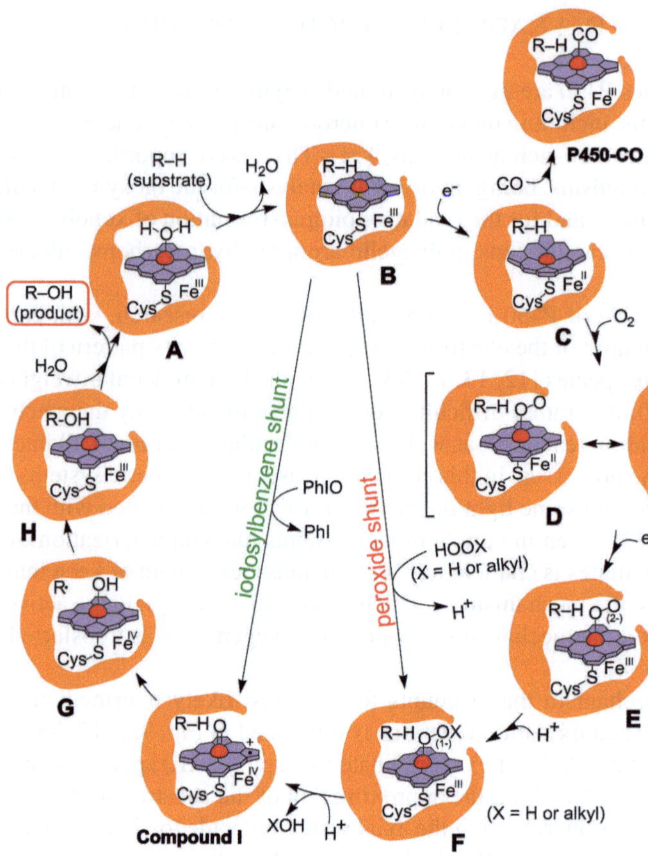

Fig. 9.2 Catalytic cycle of the cytochromes P450. The "long cycle" is associated with the following general steps, starting from the resting state species (**A**): exclusion of water from the active site upon substrate (RH) approach and accommodation to substrate pocket (**A** → **B**), iron reduction and O_2 coordination (**B** → **D**), heterolytic O–OH bond cleavage via proton-dependent reductive electron transfers to yield the high-valent Fe(IV)-oxo porphyrin π-cation-based active species, compound I (**D** → **Compound I**), substrate oxidation via a "oxygen rebound mechanism" (**Compound I** → **H**), followed by product release and water coordination restoring the enzyme resting state (**H** → **A**). The "short P450 cycles" are brought upon the use of peroxides XOOH, such as hydrogen peroxide (X = H) or alkyl hydroperoxides (X = alkyl), or active-oxygen donors, such as iodosylbenzene (PhIO), and involve the so-called peroxide shunt or iodosylbenzene shunt pathways, respectively. See text for additional details on the species along the cycle. Adapted from [2, 24, 77]

hydroxide moiety and an internal electron transfer to yield the Fe(III) species (**H**) with formation of the oxygenated product ROH. Release of the product and coordination of water restores the resting state species (**A**). The transformations from **Compound I** to species (**H**) are also referred to as the "oxygen rebound mechanism" [83, 84]. Of note, the reduction of the cytochromes P450 from the ferric to the ferrous form in the presence of CO leads to a catalytically dead carbonyl species (**P450-CO**) [13, 20, 36, 37, 69]. As noted previously, this species is the one responsible for the characteristic 450 nm absorption band of the cytochromes P450.

Shorter P450 catalytic cycles are available by using some oxygen donors other than O_2. The use of peroxides, such as hydrogen peroxide (XOOH, X = H) or alkyl hydroperoxides (XOOH, X = alkyl), shortcuts the cycle from species (**B**) into species (**F**) [77]. If oxidation is carried out with iodosylbenzene (PhIO) as oxygen donor, the cycle becomes even shorter and species (**B**) is converted directly into the high-valent active species (**Compound I**) [24]. These shortcuts are named "peroxide shunt" or "iodosylbenzene shunt" pathways, after the oxygen donor used, and have been the basis for justifying and validating the studies with such oxidants in biomimetic model systems [14].

This mechanism describes roughly most of the cytochromes P450 oxidation reactions, such as hydroxylations, epoxidations, oxidative dehydrogenation, sulfoxidations, *N*-oxidations, and *N*-demethylations [13, 24, 36]. These P450 reactions are usually 2-electron oxidation processes: the oxygen transfer from the Fe(IV)-oxo porphyrin π-cation radical active species Compound I (formally an Fe(V)-oxo species) to the substrate is accompanied by the reduction of the Fe center from a formal +5 oxidation state to a +3 oxidation state. This is however, not a universal behavior, as P450 enzymes may also be involved in two consecutive 1-electron oxidations in a similar fashion as peroxidases. Indeed, Hrycay and Bandiera [12] have recently reviewed the peroxidase-like activity of cytochromes P450.

9.2.2 Single-Electron Oxidation Systems

Heme-containing peroxidases, such as horseradish peroxidase (HRP), are usually associated with 1-electron oxidation of organic substrates (AH) using hydrogen peroxide as final electron acceptor (9.2) [85–87].

$$2\,AH + H_2O_2 \rightarrow 2A^{\bullet} + 2H_2O \qquad (9.2)$$

Analogously to the P450 systems, HRP also recruits a high-valent Fe(IV)-oxo porphyrin π-cation radical (Compound I) intermediate as active species. HRP-Compound I reactivity, however, is very different from that of P450-Compound I: HRP uses preferentially H_2O_2 as final electron acceptor for the oxidation of two equivalents of substrate by two independent 1-electron oxidations [86, 87]. The features that render the oxidation mode of HRP rather distinct from that of P450 start with the different amino acid residue used as heme proximal ligand: heme-Fe(III) in HRP is coordinated to the protein via an imidazole *N*-atom of a residual histidine [87], whereas P450s are Cys thiolate bound hemeproteins [11].

A simplified catalytic cycle for heme-containing peroxidase, such as HRP, is depicted in Fig. 9.3 [87]. The resting state in these systems is a pentacoordinate Fe(III) species that, upon coordination of a deprotonated hydrogen peroxide anion, yields a hexacoordinate Fe(III) species (**Compound 0**). Heterolytic cleavage of the coordinated hydroperoxide moiety in the presence of a proton yields the high-valent Fe(IV)-oxo porphyrin π-cation species (**Compound I**) and water. Compound I, reacts via a 1-electron process with a substrate AH (equivalent to A^- H^+) via a

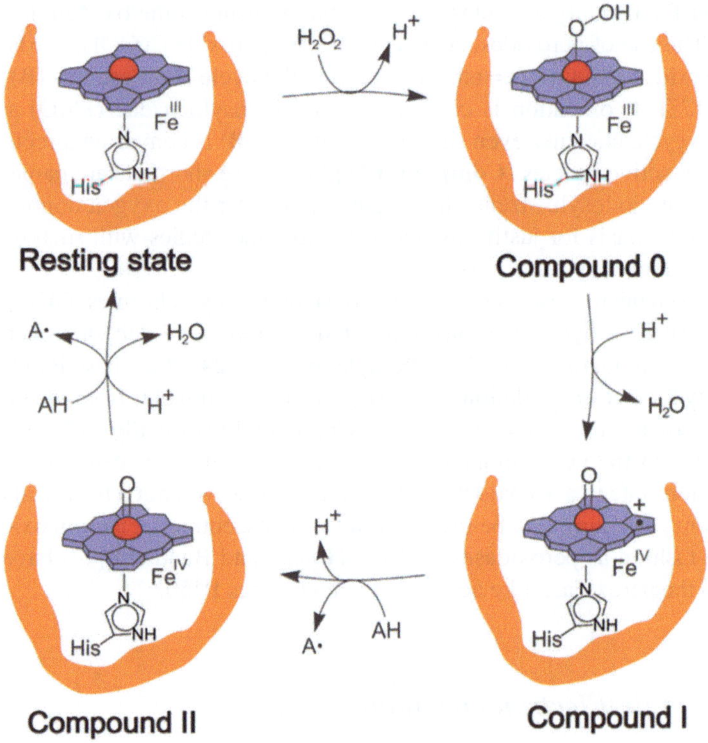

Fig. 9.3 Catalytic cycle of heme-containing peroxidases, such as horseradish peroxidase (HRP). Adapted from [87]

1-electron reduction at the porphyrin π-cation site, to yield a Fe(IV)-oxo porphyrin species (**Compound II**) and a 1-electron oxidized species $A^- \rightarrow A^\bullet$. Compound II reacts further with another AH molecule to restore the resting Fe(III) state via a 1-electron reduction at the meal site, producing a second equivalent of 1-eletron-oxidized $A^- \rightarrow A^\bullet$.

9.3 Selected Aspects of Metalloporphyrin-Based Biomimetic Oxidations

9.3.1 General Features

The development of synthetic porphyrin-based oxidation catalysts that are able to model the reactivity of cytochromes P450 has been the focus of many researchers for the past three decades [20, 24, 37, 38, 40, 66, 88–90]. During this period of time, the oxidation of a large variety of organic substrates (Fig. 9.4) using many oxygen

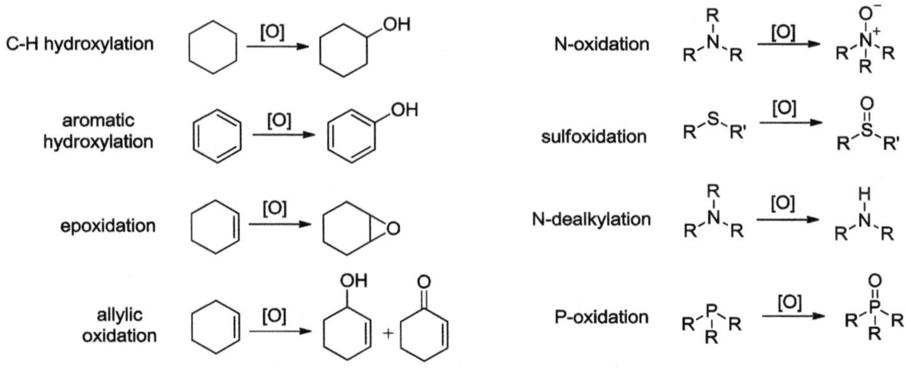

Fig. 9.4 Typical oxidation reactions catalyzed by metalloporphyrins

donors (such as iodosylbenzene, hypochlorite, hydrogen peroxide, organic peracids, monopersulfate, amine *N*-oxides, and O_2) have been accomplished [17, 20, 37]. Among the metalloporphyrins tested, those of Mn(III) and Fe(III) have yielded the most efficient oxidation catalysts [20, 24, 37, 40]. The mechanisms of such reactions are usually ascribed to a putative high-valent metal-oxo species, in close analogy with Compound I in the hemeprotein systems [21, 90]. The results accumulated on the use of metalloporphyrins as oxidation catalysts have been the subject of many reviews, books chapters, or entire books over those years. We highlight below a few selected aspects of these systems that could be useful for interpreting the biological effects observed with Mn porphyrins in vivo.

In natural enzyme systems, the wide range of reactivity, quimioselectivity, regioselectivity, stereoselectivity, oxidant preference (e.g., O_2 vs. H_2O_2), oxidant and substrate activation mode, and overall oxidation rate of enzymes that share the exact same heme cofactor, such as cytochromes P450 [12, 36, 74–78] and horseradish peroxidases [9, 85–87], cannot, thus, depend on the heme moiety by itself. In fact, the reactivity of the heme group in these systems is modulated by the proximal ligand (e.g., Cys vs. His on P450 vs. HRP, respectively), by the distal site amino acid envelope, and by the overall protein shape and charge distribution [11, 13, 36]. Conversely, biomimetic models lack the protein, which implies that the substrate and oxidant accessibility and the reactivity control must be modulated by the design of the metalloporphyrin itself, i.e., the metal selection, the nature of the porphyrin ring substituents, the choice of axial ligand and, finally, the microenvironment around the macrocycle [17, 20, 24, 37, 88, 91, 92].

Although O_2 would be the oxidant of choice for catalytic oxidations [24, 67], the activation of O_2 by the model systems poses some complications. Under aerobic conditions, most common Fe and Mn porphyrins exist in the +3 oxidation state. The coordination of O_2 to these metals is only favored, however, in lower oxidation states, such as Fe(II) and Mn(II) [6, 28]. Therefore, Fe(III) and Mn(III) porphyrins should first be reduced by some sacrificial reductant to yield their Fe(II) and Mn(II) analogues [17, 20]. Upon O_2-coordination, an internal 1-electron transfer from the

Fe(II) center to O_2 yields a Fe(III)-superoxo complex. A further 1-electron reduction at the superoxo site is required to yield a peroxo species, which upon O–O bond heterolytic cleavage oxidizes the metal(III) center by 2-electrons to give rise to the active high-valent Fe(V)-oxo or Mn(V)-oxo species. That is, in order to reach the formation of the putative metal-oxo active species, the complex needs to undergo two consecutive 1-electron reductions [17]. The highly oxidizing metal-oxo species needs, thus, to be formed in the presence of an excess of a sacrificial reductant. If the substrate is somewhat unreactive, such as alkanes, some significant O_2-oxidizing equivalents may be lost during the catalysis, by oxidation of the sacrificial reductant in detriment of substrate oxidation. Nevertheless, some O_2-oxidations have been accomplished by metalloporphyrins using various sacrificial reductants [93], such as aldehydes [20], Zn metal [20], and ascorbate [94–97]. The O_2-oxidation of a cancer pro-drug under biologically relevant conditions [97] is described below (*see Sect.* 9.3.2).

In most model systems, given the inherent impossibility to control the orchestrated access of oxidants and reductants to the metal center [17, 93], the majority of the biomimetic studies are carried out with oxygen donors, such as PhIO, sodium hypochlorite, alkyl hydroperoxides (LOOH), or hydrogen peroxide, to emulate either the peroxide-shunt or the PhIO-shunt pathways of P450 (Fig. 9.2) and yield the metal-oxo species equivalent to Compound I [14, 17, 20, 24, 90]. In the enzyme systems, the reactivity of Compound I is controlled by the protein moiety: whereas P450-Compound I usually carries out formally a single 2-electron substrate oxidation process to return to the resting state, HRP-Compound I carries out two 1-electron oxidations to reach the resting state [12]. It is worth noting that in the model systems, as model-Compound I lacks the protein moiety, its selectivity and mode of substrate oxidation cannot be accurately defined a priori. The prevalence of the 2-electron oxidation pathway over the two 1-electron oxidation pathway in the model-Compound I depends, thus, on a series of factor [20, 24, 90, 93], such as the chosen metal center, the porphyrin substituents, the nature of the oxygen donor system, and the medium conditions (e.g., pH [98]). It is worth noting that the organic transformations carried out by P450-type model reactions are typically of overall non-radical nature, whereas HRP-type model reactions are usually associated with radical species/products.

Although hemeproteins are naturally Fe-containing biomolecules, Mn(III) porphyrins have been considerably versatile oxidation catalysts [20, 37]. The hydroxylation of saturated hydrocarbons using the system Mn(III)-porphyrin/PhIO was simultaneously described by Groves and coworkers [99] and Hill and Schardt [100]. The advantages of the Mn(III) porphyrin systems over the corresponding Fe(III) counterparts are: longer life-time of the catalyst, and higher catalytic efficiency [20]. Mn porphyrin-based systems are, however, more prone to be involved in 1-electron processes (radical-type reactions) which compromise the selectivity of the Mn systems as compared to the corresponding Fe analogues [90, 101, 102]. Whereas the oxidation of alkanes by Fe(III) porphyrin/PhIO systems usually results in the formation of alcohol with little or no amount of the corresponding ketone, in the Mn(III) systems ketone is regularly found along with the desired alcohol product [20, 90]. In some systems, even alkyl halides are found as products resulting from the abstraction of halide of the MnP counter-ion, the porphyrin ring, or the solvent [101, 103]. The increased radical character of Mn(III) porphyrin systems is usually invoked to explain, for example, the decreased selectivity toward the alcohol [90]. A general reaction mechanism scheme for Mn(III) porphyrin systems is presented in Fig. 9.5.

A fundamental problem associated with metalloporphyrin catalyzed oxidations is the vulnerability of the macrocycle toward bimolecular oxidative destruction [24, 38, 104], in which a metal-oxo species generated in situ attacks another metalloporphyrin molecule. This is typically a situation where the macrocycle competes with the substrate for the active intermediate and is, therefore, accentuated when substrates of low reactivity are used [91, 103, 105, 106]. For example, the oxidation of alkanes catalyzed by MnPs is usually accompanied by partial-to-total destruction of the porphyrin, whereas if the substrate is an olefin instead, little destruction may be noticed under the same oxidizing conditions [105]. In the presence of an active oxidant, such as PhIO, ClO^-, or H_2O_2, but in the absence of suitable substrates, FePs and MnPs are prone to oxidative destruction [41, 103, 107]. The attempts to develop

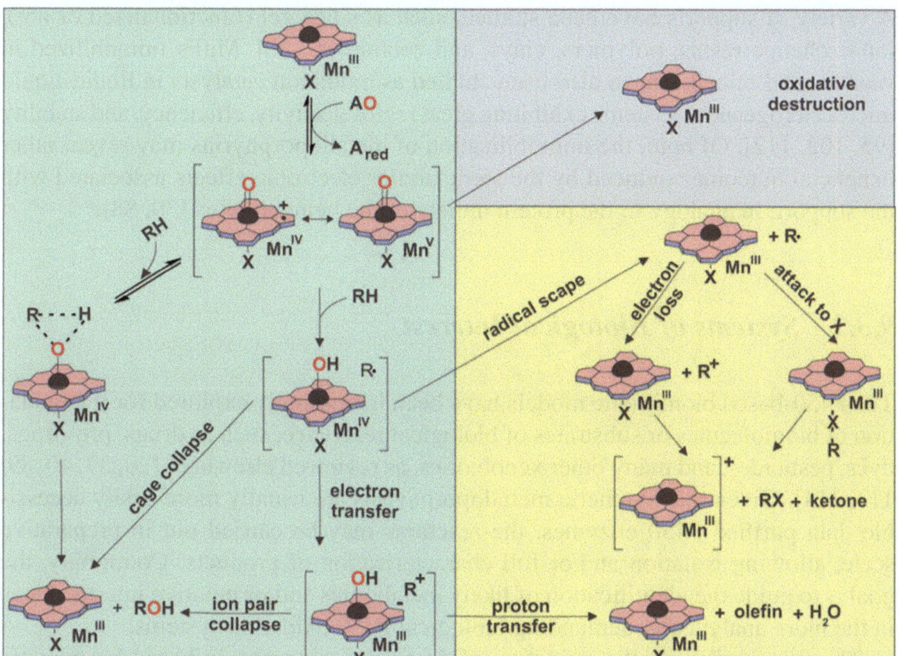

Fig. 9.5 Hypothetical mechanism scheme for the oxidations catalyzed by Mn(III) porphyrins using an oxygen donor AO. The box in the *left* highlights the major oxygenation pathways leading to the oxidized substrate ROH. The *right top* box presents the major catalyst deactivation pathway, which is associated with catalyst oxidative destruction. Other competing pathways, including those of radical nature, are presented in the *right bottom* box. Adapted from [100]

MnP- and FeP-based catalase models are particularly hindered by the destruction of the metalloporphyrins by H_2O_2 [41]. Of note, while H_2O_2 is a good oxygen donor to FePs and MnPs, leading to the corresponding metal-oxo species, H_2O_2 itself is not easily oxidized, being, thus, a very poor substrate to the metal-oxo species; in absence of a good substrate, the other metalloporphyrin molecules in the system become a suitable substrate to the electrophilic metal-oxo species [41, 104, 107]. The vulnerability of the metal-free macrocycle toward oxidation is increased with complexation of redox-active metals, such as Fe(III) and Mn(III): whereas bleaching of MnPs and FePs in the absence of a substrate is rapidly observed upon incubation with H_2O_2 [41], for example, the metal-free porphyrin is much less susceptible to destruction under the same conditions. In fact, incubation of free-base porphyrin mixtures with H_2O_2 [108] or other strong quinone-based oxidants [109, 110] is a common procedure in porphyrin synthesis.

A great deal of efforts has been dedicated toward the synthesis of new metalloporphyrins, introducing bulky and/or electron-withdrawing groups on the porphyrin ring, in an attempt to make the catalyst more efficient, selective, and, especially, more stable against oxidative destruction [24, 38, 111]. Another way to decrease the likelihood of the bimolecular oxidative destruction processes has been the immobilization of FePs and MnPs onto a solid surface or micelle interface [39, 88, 95, 96]. A variety of supports have been studied, such as silica gel (functionalized or not), ion-exchange resins, polymers, clays, and zeolite [20, 88]. MnPs immobilized in vesicles and micelles have also been studied as oxidation catalysts in liquid-liquid microheterogeneous systems exhibiting great regioselectivity, efficiency, and stability [95, 102, 112]. Of note, the immobilization of metalloporphyrins may reveal other beneficial outcomes induced by the steric and/or electronic effects associated with the support, in analogy to the protein moiety of the hemeproteins [39, 88].

9.3.2 Systems of Biological Interest

The P450-based biomimetic models have been increasingly explored for the oxidation of biomolecules or substrates of biological relevance, such as drugs, pro-drugs, dyes, pesticides, and many other xenobiotics, as reviewed elsewhere [20, 37, 40, 89, 113, 114]. Given that synthetic metalloporphyrins are usually more easily accessible than purified P450 enzymes, the reactions may be carried out in preparative scale, allowing isolation and/or full characterization of products. Commonly, the goal is to guide the identification of likely metabolites and/or putative intermediates in the more analytically demanding biological P450 oxidation systems.

The first studies on the oxidation of drugs under porphyrin-based biomimetic conditions were reported in the late 1980s and early 1990s, using the simplest, water-insoluble Fe(III) or Mn(III) porphyrins, such as $FeTPP^+$ and $MnTPP^+$ [115–118]. The oxidation of nicotine by $MnTPP^+$ using PhIO as oxidant yielded two products, 3-hydroxynicotine and cotinine, that were identical to the P450-oxidation metabolites observed in vivo [118]. The variety of drugs studied thus far increased

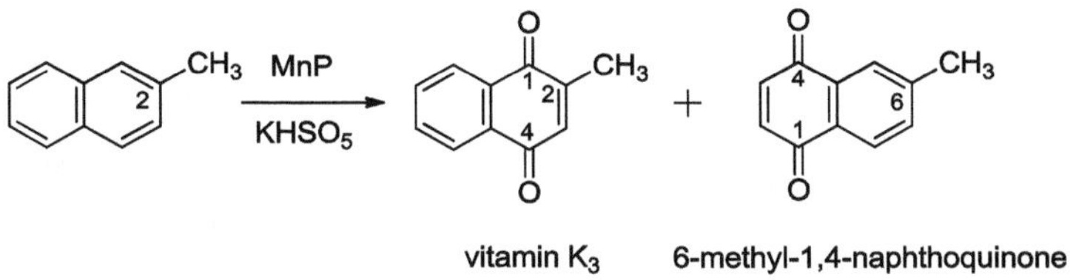

Fig. 9.6 MnP-based biomimetic oxidation of 2-methylnaphthalene to yield vitamin K_3. Adapted from [131]

considerably and included, for example, caffeine (the legal drug most universally consumed [119]), anti-inflammatory drugs, anesthetics, analgesics, steroids, natural products, antibiotics, anticancer drugs, anti-epileptic drugs, anti-psychotic drugs, antipyretic drugs, antiarrhythmic drugs, antihistaminic drugs, among others [113–130]. These studies are generally carried out under simple reaction conditions: the substrate (drug), the oxygen donor (e.g., H_2O_2, t-BuOOH, NaOCl, PhIO), with FePs or MnPs as catalyst, in a suitable solvent.

Vitamin K_3 was prepared in a single step via the oxidation of 2-methylnaphthalene with hydrogen persulfate (HSO_5^-) in aqueous solution at room temperature (Fig. 9.6) [131]. The reaction was catalyzed by anionic or cationic water-soluble metalloporphyrins (Fe and Mn). Although the substrate is not biologically available, it is worth noting that this demonstrates the potential of MnPs and FePs for the oxidation of aromatic compounds to biologically relevant quinones under ambient conditions. The biomimetic system makes use of HSO_5^-, which is prepared from sulfate and H_2O_2.

The rich chemistry of metalloporphyrin-based biomimetic catalysis on a naturally occurring substrate using a biologically ubiquitous oxidant can be illustrated by the H_2O_2-oxidation of all-(*E*)-retinol, a natural form of vitamin A, using the simplest Fe(III) porphyrin, $FeTPP^+$ under ambient conditions [132]. This biomimetic approach subjected the substrate to a variety of transformations, such as epoxidations, allylic hydroxylations, oxidative dehydrogenation, C–C bond cleavage, and isomerizations (Fig. 9.7). This is in direct contrast with the P450-catalyzed all-(*E*)-retinol oxidation, which yields primarily the alcohol dehydrogenation product, retinal. The unrelated nature of the products of the biomimetic versus the enzymatic systems is worth noting while using synthetic metalloporphyrin in vitro or in vivo: in the presence of both a suitable substrate and an active oxidant, metalloporphyrin-based oxidations could give rise to products of completely distinct nature from those usually expected as naturally occurring metabolites.

Naturally occurring reactive oxygen/nitrogen species, such as lipid hydroperoxides (LOOH) and peroxynitrite ($ONOO^-$), are suitable oxygen donors to MnPs under biological oxidative-stress conditions [133, 134]. Both oxidants can convert cationic MnPs (e.g., MnTE-2-PyP^{5+}) into the oxidizing Mn(IV)-oxo porphyrin species in aqueous solution. The high-valent species was further demonstrated to be able to carry out the oxidation of glutathione, ascorbate, and urate under biologically relevant conditions [134].

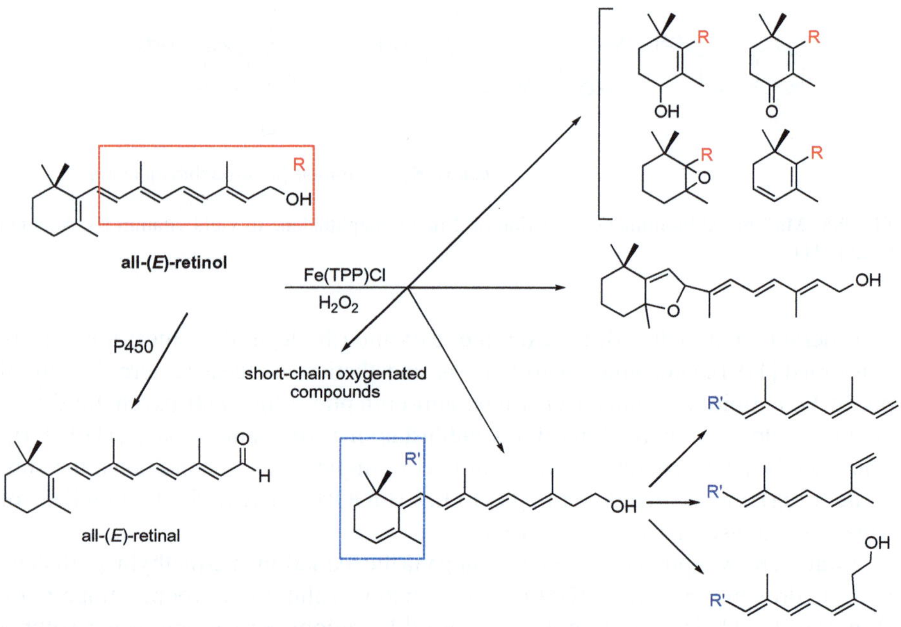

Fig. 9.7 H_2O_2-oxidation of naturally occurring all-(*E*)-retinol using FeP as biomimetic catalyst. Adapted from [132]

MnP-based biomimetic systems using O_2 to carry out oxidation reactions of biologically relevant substrates is uncommon. The activation of O_2 usually requires a sacrificial reductant. A biologically available reductant, present in cells at relative high concentrations (millimolar), is ascorbate. Mansuy and coworkers [94] used the O_2/ascorbate system for the $MnTPP^+$-catalyzed oxidations, epoxidations, and dehydrogenations of alkanes and olefins under biphasic (water/benzene) conditions. Groves and Neumann [95, 96] incorporated steroidal-appended MnPs and FePs into phospholipid bilayer vesicles (Fig. 9.8) and showed that the MnP-based catalytic constructs were able to carry out the selective hydroxylation of cholesterol at carbon C-25 using the O_2/ascorbate oxidant system under mild conditions in Tris buffer at pH 8.6; the FeP-based catalytic construct catalyzed the O_2/ascorbate-oxidation of desmosterol (Fig. 9.8). Of note, the corresponding homogeneous (non-vesicle) systems gave rise to small amounts of many products.

More recently, Spasojević and coworkers [97] showed that water-soluble Fe and Mn porphyrins are able to catalyze the hydroxylation of the anticancer pro-drug cyclophosphamide to active metabolite 4-hydroxycyclophosphamide using the O_2/ascorbate system (Fig. 9.9) in yields similar or higher than those typically obtained by the action of microsomal P450 enzymes in vivo. The cationic Mn porphyrins MnTM-2-PyP^{5+} and MnTM-3-PyP^{5+} used for the cyclophosphamide oxidation by the O_2/ascorbate system have regularly been investigated as SOD mimics for the development of redox-based experimental therapeutics. Of note, the O_2/ascorbate/MnP systems have recently been forwarded to in vivo studies as a prospective anticancer treatment [107].

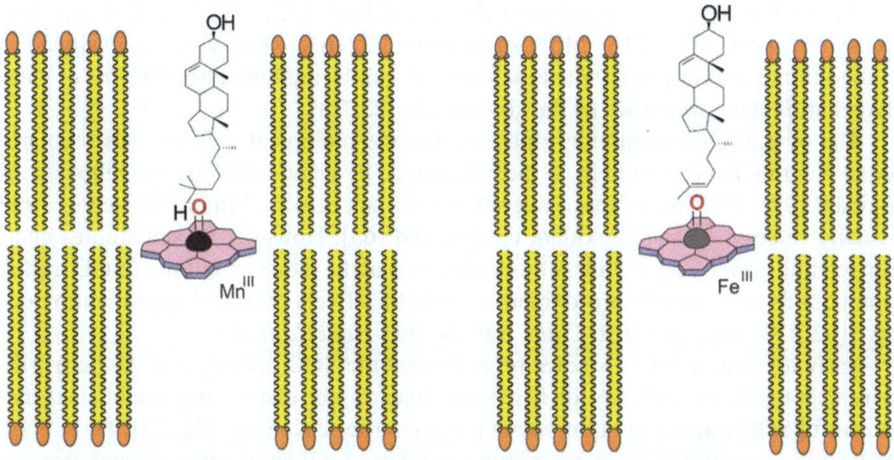

Fig. 9.8 Schematic representation of the incorporation of steroidal-appended MnPs (*left*) and FePs (*right*) into phospholipid bilayer vesicles for hydroxylation of cholesterol and epoxidation of desmosterol by the O_2/ascorbate system. Adapted from [95, 96]

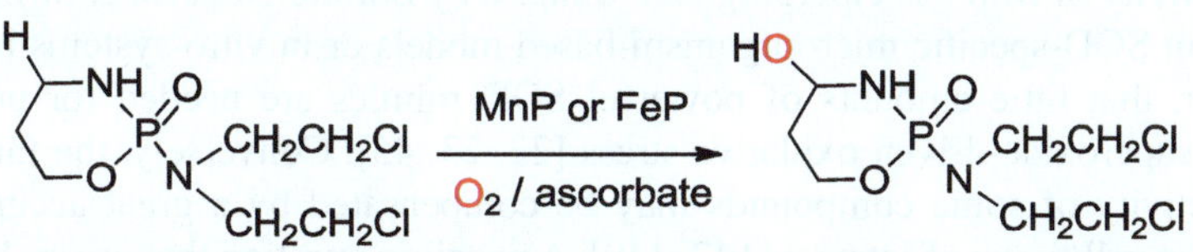

Fig. 9.9 The hydroxylation of the anticancer pro-drug cyclophosphamide to active metabolite 4-hydroxycyclophosphamide using the O_2/ascorbate system catalyzed by the MnP- and FeP-based biomimetic models. Adapted from [97]

9.4 Pro-oxidative Role of Mn-Porphyrins Under Oxidative Stress Conditions: Biomimetic Oxidation Catalysis in Biological Milieu?

The inorganic medicinal chemistry of MnPs is in its infancy as compared to classic metal-containing drugs, such as cisplatin. The design of MnP-based therapeutics did not follow the standard medicinal chemistry approach of defining a specific biomacromolecule target (e.g., protein, enzyme, nucleic acids), but aimed at ubiquitous reactive oxygen/nitrogen species (ROS/RNS) [22, 23, 135] and, thus, at the cell Redoxome [136].

The early studies on MnP in vivo were carried out in microorganism (e.g., *E. coli*) to verify their SOD activity in a complex biological matrix or to test them as an indirect mechanistic probe for ROS/RNS [45, 53, 137]. The reports on in vivo animal studies of MnPs as prospective experimental therapeutics began in the 2000s [22, 23]. Despite the short period of time, two of the designed and tested compounds (i.e., MnTE-2-PyP^{5+} [54] and MnTnBuOE-2-PyP^{5+} [138]) reached translational studies and are now entering phase I/II Clinical Trials [42, 62] (*see also Chap.* 8 *by Batinić-Haberle et al.*). The MnPs of medicinal interest were primarily designed as SOD mimics and/or peroxynitrite scavengers exploiting the radioprotector or catalytic antioxidant properties of the compounds [22, 23].

The overall effect commonly observed in the biological systems was regularly equivalent to that expected from the role played by an endogenous/exogenous antioxidant [22]. Because of these overall antioxidant effects, MnPs themselves were usually referred to as antioxidant catalyst [64, 65]. However, recent studies challenged this view by providing solid evidences that the mode of action of SOD mimic compounds is frequently pro-oxidative while the effects observed appear antioxidative [62, 139–146] (*see also Chap.* 8 *by Batinić-Haberle et al.*).

Early aqueous solution studies revealed that the SOD activity and the peroxynitrite decomposition efficacy of MnPs are linearly correlated, and established that powerful SOD mimics are also potent peroxynitrite scavengers [22, 23]. Thus, good SOD mimics would provide good drug candidates for treating ailments and diseases of oxidative-stress nature. Of course, the in vivo efficiency of any particular drug depends on its bioavailability, subcellular/cellular/tissue distribution, pharmacokinetics, dosing regimen, toxicity, etc. A correlation of animal in vivo efficacy and SOD activity of MnPs is emerging (*see Chap.* 8 *by Batinić-Haberle et al.*). A rough picture in SOD-specific microorganism-based models or in vitro systems indicates, however, that little amounts of powerful SOD mimics are needed for protection against superoxide-driven oxidative stress [22, 23, 45]. Conversely, the fair-to-low SOD activity of some compounds may be compensated by a great accumulation within the cell/tissue of interest [147–149]. A puzzling question that arises, however, is how to explain the overwhelming amount of biological data on MnPs that are not SOD mimics at all, such as $MnTBAP^{3-}$ [150]. Evidences associating $MnTBAP^{3-}$ with an oxidant role in biological systems have been recently reported [146].

The rational development of MnP-based SOD mimics have most commonly been guided by designing the porphyrin ligand so that the resulting MnP complex would exhibit Mn(III)/Mn(II) reduction potential close to that of the SOD enzymes [22, 23, 54, 55, 59, 60, 145, 151]. Although electrostatics plays also a major role in modulating the approach of superoxide to the metal center [58–60], the use of the enzyme reduction potential as a guide has been a rule of thumb. Indeed, cationic MnPs whose metal-centered reduction potentials are near *ca.* +300 mV vs. NHE have high SOD activity [22, 23, 42, 138] (Table 9.1). Conversely, MnPs (regardless of their overall charge) of moderate to negative reduction potential, such as $MnTM\text{-}4\text{-}PyP^{5+}$ [53] and $MnTBAP^{3-}$ [150], have moderate to negligible SOD activity, respectively [22, 23] (Table 9.1).

Cytochromes P450 exhibit a very rich electrochemical behavior [70, 79–82, 152]. Their Fe(III)/Fe(II) reduction potential show a remarkable substrate dependency giving that the substrate modulates the high- versus low-spin contributions on the Fe center spin state [80]. Table 9.2 exemplifies the dramatic changes in both human and bacterial P450 reduction potential with respect to the nature of the substrate [82, 152]. The Fe(III)/Fe(II) reduction potential of the resting state complex, a water-coordinated Fe(III) porphyrin thiolate species (Fig. 9.2, **A**) is very negative,

Table 9.1 Metal(III)/Metal(II) reduction potential ($E_{½}$) and SOD activity (log k_{cat}) of SOD enzymes and selected Mn(III) porphyrins

Compound	$E_{½}$, mV vs. NHE[a]	SOD activity (log k_{cat})[b]
SOD enzyme	*ca.* +300	8.84–9.30
MnTnBuOE-2-PyP^{5+}	+277	7.83
MnTnHex-2-PyP^{5+}	+314	7.48
MnTE-2-PyP^{5+}	+228	7.76
MnTM-2-PyP^{5+}	+220	7.79
MnTM-3-PyP^{5+}	+52	6.61
MnTM-4-PyP^{5+}	+60	6.58
MnTPP^{+}	−270	4.83
MnTBAP^{3-}	−194	3.16

Data compiled from [22, 23, 138]
[a]$E_{½}$ determined in 0.05 M phosphate buffer (pH 7.8, 0.1 M NaCl)
[b]k_{cat} determined by cytochrome *c* assay in 0.05 M phosphate buffer (pH 7.8, at 25 ± 1 °C)

Table 9.2 Substrate dependence of the Fe(III)/Fe(II) reduction potential of human cytochrome P450 CYP3A4 and bacterial cytochrome P450 CYP101 [81, 152]

Human cytochrome P450 CYP3A4		Bacterial cytochrome P450 CYP101	
Substrate	$E_{½}$, mV vs. NHE	Substrate	$E_{½}$, mV vs. NHE
–[a]	−220	–[a]	−303
Erythromycin	−210	TMCH[b]	−242
Testosterone	−140	Norcamphor	−206
Bromocriptine	−137	L-camphorquinone	−183
		Adamantanone	−175
		D-camphor	−173

[a]Corresponds to the water-bound, resting state of the enzyme
[b]TMCH = 3,3,5,5-tetramethylcyclohexanone

e.g., −220 mV vs. NHE for human cytochrome P450 CYP3A4 [81, 82]. This low potential is easily achieved even with the simplest *meso*-tetraphenylporphyrin derivatives [22, 23], such as MnTPP^{+} [153] or MnTBAP^{3-} [150]. The reaction of MnP-based model compounds with active oxygen donors (e.g., PhIO, H_2O_2) resembles, thus, that of the resting state of P450 to yield the active metal-oxo active species.

Although it is tempting to use the same reduction potential strategy used in the development of SOD mimics to guide the design of P450-model design, care must be exercised. In the SOD case, the Mn(III)/Mn(II) redox couple represents the exact Mn(III)P and Mn(II)P species involved in the dismutation process. In the P450 case, the rate-determining step is usually ascribed to the oxygen transfer from the formal Mn(V)-oxo species to the substrate. If the oxidation takes place via a typical P450-type reaction, the oxo-Mn(V)P species would be reduced directly to the Mn(III)P resting state in a single step. This implies, thus, that the redox couple of interest to describe this 2-electron process would be the Mn(V)/Mn(III) one, which is available for a few MnP systems only [154]. As the model compounds lack the protein moiety to favor a P450-type vs. peroxidase-type reaction, it is likely that 1-electron transfer processes may be also operative, depending on the nature of the substrate available, the porphyrin substituents, and the overall conditions. Irrespective of the mechanism being predominately either P450-like or peroxidase-like, if the catalytically relevant reactions are significantly dependent on the 2-electron cycling of the Mn(V)P/Mn(III)P species or the Mn(IV)P/Mn(II)P species, i.e., the rate determining step being associated with Mn(V)-oxo/Mn(III) or Mn(IV)-oxo/Mn(II) couples, then the electrochemically available Mn(III)/Mn(II) reduction potentials could be eventually a surrogate descriptor for the MnP reactivity under catalytic conditions. Comprehensive studies on correlating P450-type catalysis and Mn(III)/Mn(II) reduction potentials are still limited [105, 106, and references therein], but a bell-shape behavior is likely. However, recent data from our laboratories (Falcão, Pinto, Reboucas, unpublished) on the hydroxylation of cyclohexane catalyzed by isomeric *ortho*, *meta*, and *para* Mn(III) *N*-methylpyridinium porphyrins (MnTM-2-PyP^{5+}, MnTM-3-PyP^{5+}, and MnTM-4-PyP^{5+}, respectively, Fig. 9.1) under conditions of low catalyst degradation, resulted in approximately the same total yield of oxygenated products (~91 %) regardless of the isomer used. This suggests that the efficiency of the isomers in this particular oxidation system is essentially independent of the Mn(III)/Mn(II) reduction potential, which ranges from +52 mV (*meta* isomer [53]) to +220 mV vs. NHE (*ortho* isomer [53]): a 168 mV spam (Table 9.1). It is worth noting, however, that a more direct correlation between Mn(III)/Mn(II) reduction potential and the catalytic activity for metalloporphyrin-catalyzed oxidations should be likely found for radical O_2-autoxidation reactions in which the metal center cycles indeed between +3 and +2 oxidation states [67, 68, 155], resembling the behavior observed in the SOD systems.

Whether or not the Mn(III)/Mn(II) reduction potentials (or any another redox couple for that matter) would ever be a good descriptor for biomimetic oxidation catalysts is still uncertain. What is worth noting, however, is that even MnPs of very low Mn(III)/Mn(II) reduction potentials, which are unable to carry out SOD-like reactions in vivo, may give a suitable entry to Mn-oxo species in the presence of an oxidative stress-derived oxidant, such as H_2O_2 or ClO^- commonly used in the chemical biomimetic systems [20]. In MnP-loaded cells under oxidative stress, all ingredients needed for MnP-based biomimetic oxidation catalysis are present in the cell milieu: a good source of oxygen donor (ROS/RNS), the catalyst (MnP), and a plethora of substrates (e.g., proteins, lipids, nucleic acids). In such conditions of high substrate availability, vulnerability of the MnP toward bimolecular destruction should, thus, be minimized, preserving catalytic function [105]. An antioxidant overall effect could still be observed via biomimetic oxidation catalysis if ROS/RNS is intercepted by MnP species, which would then undergo oxidation to the corresponding Mn-oxo porphyrin species. Either Mn(V)-oxo or Mn(IV)-oxo porphyrin species are still very potent oxidants, being able to oxidize even inert aliphatic C-H bonds if better (more reducing) biomolecules are not present. Contrary to regular ROS/RNS (e.g., H_2O_2, $ONOO^-$, LOOH, etc) Mn-oxo porphyrin species are bulky oxidants. If the ROS/RNS-sensitive oxidation sites are either cofactors or active

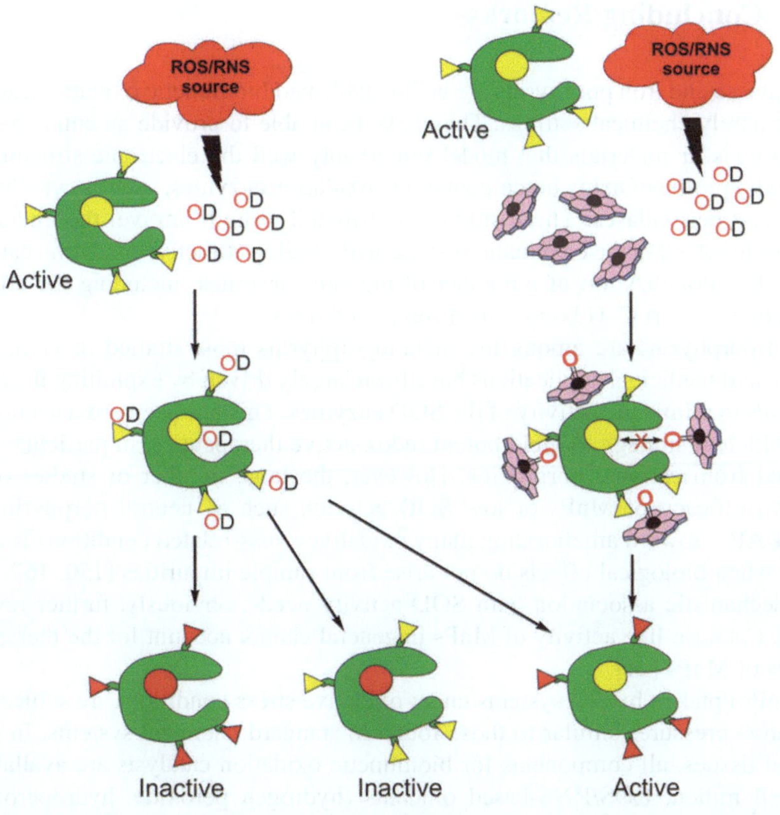

Fig. 9.10 Schematic cartoon illustrating the putative role of MnP-based biomimetic oxidation catalysts in protecting the active site (*central circle*) of a generic enzyme against oxidation by ROS/RNS-based oxygen donors (OD). MnP would intercept OD to form Mn-oxo porphyrin species, which could then direct the oxidation to sites of less physiological consequences, such as sacrificial residues [156–160] at protein surface (*triangles*). Oxidation of the active site by an OD-generated Mn-oxo porphyrin species would be hampered by steric demands. *Yellow* color indicates reduced state, whereas oxidizes sites are colored in *red*

centers buried within biomolecules such as enzymes, proteins, and transcription factors, the formation of the Mn-oxo porphyrin species would protect these ROS/RNS-sensitive sites against oxidation under oxidative stress conditions. Again, since Mn-oxo species are potent oxidants [154], they are likely to react within the cell milieu regenerating the MnP resting state. Given the size of the MnPs, the MnP-catalyzed oxidations in the cells is most likely to take place at sacrificial amino acid residues (e.g., Met or Cys) on protein surfaces [156–160], or other endogenous small molecular weight reductant, such as glutathione, ascorbate, or urate [107, 134, 161]. This mode of action where MnPs act as biomimetic oxidation catalysts of overall protective role is illustrated in Fig. 9.10.

9.5 Concluding Remarks

Manganese and iron porphyrins are well-established biomimetic oxidation catalysts under purely chemical settings. They have been able to provide an entry to single compounds or materials that model remarkably well the electronic structure and reactivity profile of many heme-containing oxidation enzymes, such as cytochromes P450 and peroxidases. These studies contributed to both unravel the underlying species involved in the enzymatic systems and develop attractive oxidation catalysts to the functionalization of a number of organic substrates, including the efficient activation of inert C–H bonds under mild conditions.

Mn porphyrins are among the metalloporphyrins most studied in vivo. Their design and medicinal applications have been largely driven by exploiting the ability of MnPs to mimic the activity of the SOD enzymes. This has been proven a remarkably efficient strategy to yield potent redox-active therapeutics, in particular those derived from cationic porphyrins. However, the large number of studies on the in vivo efficacy of MnPs of low SOD activity, such as neutral porphyrins and $MnTBAP^{3-}$, toward ameliorating many oxidative stress-related conditions is unsettling; when biological effects do not arise from sample impurities [150, 162, 163], the mechanistic association with SOD activity needs, obviously, further revision [137]. Catalase-like activity of MnPs in general cannot account for the therapeutic effects of MnPs [41].

MnPs uptaken by cell systems under oxidative stress conditions are subjected to oxidative pressures similar to those found in standard chemical systems. In MnP-treated tissues, all components for biomimetic oxidation catalysis are available in the cell milieu: ROS/RNS-based oxidants (hydrogen peroxide, hydroperoxides, hypochlorite, etc.), the MnP-based catalyst, and many biomolecules as substrates.

The transformation of ROS/RNS-derived oxygen donor to high-valent Mn-oxo species switches the oxidant size from a small reactive species into a bulkier and sterically demanding metal-oxo species of equally high oxidation power; such larger metal-based oxidant is most likely to react, however, with (sacrificial) surface protein-residues, while preserving untouched the buried, active sites of large biomolecules. The formation of MnP-based metal-oxo species may render the oxidant also less susceptible to diffusion, helping containing the oxidative burst to a restricted region of the cell. Conversely, as Mn-oxo porphyrin species react readily with thiol and thioether moieties [134, 161, 164], such as those of surface-active Cys and Met residues of transcription factors, Mn porphyrin action as biomimetic oxidation catalysts may play a significant role in affecting major cell signaling pathways (*see also Chap.* 8 *by Batinić-Haberle et al.*).

MnPs of low Mn(III)/Mn(II) reduction potential can barely work as SOD mimics (if at all) [150, 165], which increases the likelihood of the compounds being biologically active via a biomimetic oxidation catalysis pathway. This may be of particular relevance to negatively charged and/or neutral porphyrin-based compounds [166, 167], or other metal-based complexes of low reduction potential, such as Mn salens, whose biological data as redox-active therapeutics are clearly of undeniable significance [168] (*see also Chap.* 11 *by Doctrow and coworkers*). On the other hand, for cationic MnPs of high Mn(III)/Mn(II) reduction potentials, the contribution of the biomimetic oxidation catalysis versus the standard antioxidant catalysis to the overall biological effects remains uncertain, given that such compounds are also potent SOD mimics on their own, and should be of particular interest to research in the near future.

Acknowledgments The authors thank Drs. Ynara M. Idemori, Brian R. James, Artak Tovmasyan, Ludmil Benov, Margaret E. Tome, Dayse C. S. Martins, Gilson DeFreitas-Silva, and Shirley Nakagaki, for many helpful scientific discussions throughout these years. Financial support by The Brazilian Research Council (CNPq, Brazil), CAPES Foundation (Ministry of Education, Brazil), Financiadora de Estudos e Projetos (FINEP, Brazil), and the National Institutes of Health (NIH, USA) are greatly acknowledged.

References

1. Milgrom LR. The colours of life: an introduction to the chemistry of porphyrins and related compounds. Oxford: Oxford University Press; 1997.
2. Lippard SJ, Berg JM. Principles of bioinorganic chemistry. Mill Valley: University Science; 1994.
3. da Silva JJRF, Williams RJP. The biological chemistry of the elements: the inorganic chemistry of life. New York: Oxford University Press; 2001. p. 343–69.
4. Kumar S, Bandyopadhyay U. Free heme toxicity and its detoxification systems in human. Toxicol Lett. 2005;157:175–88.
5. Aich A, Freundlich M, Vekilov PG. The free heme concentration in healthy human erythrocytes. Blood Cells Mol Dis. 2015;55:402–9.
6. Jones RD, Summerville DA, Basolo F. Synthetic oxygen carriers related to biological systems. Chem Rev. 1979;79:139–79.
7. James BR. Interaction of dioxygen with metalloporphyrins. In: Dolphin D, editor. The porphyrins, vol. V. New York: Academic Press; 1977. p. 205–302.
8. Traylor TG. Synthetic model compounds for hemoproteins. Acc Chem Res. 1981;14:102–9.
9. de Montellano PRO. Catalytic sites of hemoprotein peroxidases. Annu Rev Pharmacol Toxicol. 1992;32:89–107.
10. Collman JP, Boulatov R, Sunderland CJ, Fu L. Functional analogues of cytochrome *c* oxidase, myoglobin, and hemoglobin. Chem Rev. 2004;104:561–88.
11. Mansuy D, Battioni P. Biochemistry and binding: activation of small molecules. In: Kadish KM, Smith KM, Guilard R, editors. The porphyrin handbook, vol. 4. New York: Academic Press; 2000. p. 1–15.
12. Hrycay EG, Bandiera SM. The monooxygenase, peroxidase, and peroxygenase properties of cytochrome P450. Archiv Biochem Biophys. 2012;522:71–89.
13. Meunier B, Visser SP, Shaik S. Mechanism of oxidation reactions catalyzed by cytochrome P450 enzymes. Chem Rev. 2004;104:3947–80.
14. Sheldon RA. Oxidation catalysis by metalloporphyrins: a historical perspective. In: Sheldon RA, editor. Metalloporphyrins in catalytic oxidations. New York: Marcel Dekker; 1994. p. 5–27.
15. Breslow R. Centenary lecture. Biomimetic chemistry. Chem Soc Rev. 1972;1:553–80.
16. Breslow R. Preface. In: Breslow R, editor. Artificial enzymes. Weinheim: Wiley-VCH; 2014. p. 9–10.

17. Ezhova MB, James BR. Catalytic oxidations using ruthenium porphyrins. In: Simándi LI, editor. Advances in catalytic activation of dioxygen by metal complexes. Dordrecht: Kluwer Academic; 2003. p. 1–77.
18. Rebouças JS, Patrick BO, James BR. Thiol, disulfide, and trisulfide complexes of Ru porphyrins: potential models for Iron-Sulfur bonds in heme-proteins. J Am Chem Soc. 2012;134:3555–70.
19. Rebouças JS, James BR. Molecular recognition using Ruthenium(II) porphyrin thiol complexes as probes. Inorg Chem. 2013;52:1084–98.
20. Meunier B. Metalloporphyrins as versatile catalysts for oxidation reactions and oxidative DNA cleavage. Chem Rev. 1992;92:1411–56.
21. Groves JT, Shalyaev K, Lee J. Oxometalloporphyrins in oxidative catalysis. In: Kadish KM, Smith KM, Guilard R, editors. The porphyrin handbook, vol. 4. San Diego: Academic Press; 2000. p. 17–40.
22. Batinić-Haberle I, Rebouças JS, Spasojević I. Superoxide dismutase mimics: chemistry, pharmacology and therapeutic potential. Antioxid Redox Signal. 2010;13:877–918.
23. Batinić-Haberle I, Rebouças JS, Benov B, Spasojević I. Chemistry, biology and medical effects of water-soluble metalloporphyrins. In: Kadish KM, Smith KM, Guilard R, editors. Handbook of porphyrin science, vol. 11. New York: World Scientific; 2011. p. 291–394.
24. Dolphin D, Traylor TG, Xie LY. Polyhaloporphyrins: unusual ligands for metals and metal-catalyzed oxidations. Acc Chem Res. 1997;30:251–9.
25. Collman JP, Gagne RR, Halbert TR, Marchon JC, Reed CA. Reversible oxygen adduct formation in ferrous complexes derived from a picket fence porphyrin. Model for oxymyoglobin. J Am Chem Soc. 1973;95:7868–70.
26. Almog J, Baldwin JE, Huff J. Reversible oxygenation and autoxidation of a capped porphyrin iron(II) complex. J Am Chem Soc. 1975;97:227–8.
27. Momenteau M, Loock B, Mispelter J, Bisagni E. "Basket handle" porphyrins and their ferrous complexes as stables oxygen carriers. Nouv J Chim. 1979;3:77–9.
28. Jones RD, Summerville DA, Basolo F. Manganese(II) porphyrin oxygen carriers. Equilibrium constants for the reaction of dioxygen with *para*-substituted *meso*-tetraphenylporphinatomanganese(II) complexes. J Am Chem Soc. 1978;100:4416–24.
29. Tsuchida E, Sou K, Nakagawa A, Sakai H, Komatsu T, Kobayashi K. Artificial oxygen carriers, hemoglobin vesicles and albumin-hemes, based on bioconjugate chemistry. Bioconjug Chem. 2009;20:1419–40.
30. Alayash AI. Blood substitutes: why haven't we been more successful? Trends Biotechnol. 2014;32:177–85.
31. Riess JG. Oxygen carriers ("blood substitutes")—raison d'etre, chemistry, and some physiology. Chem Rev. 2001;101:2797–920.
32. Cheng L, Richter-Addo GB. Binding and activation of nitric oxide by metalloporphyrins and heme. In: Kadish KM, Smith KM, Guilard R, editors. The porphyrin handbook, vol. 4. New York: Academic Press; 2000. p. 219–91.
33. Koshland Jr DE. The molecule of the year. Science. 1992;258:1861.
34. Szabó C, Ischiropoulos H, Radi R. Peroxynitrite: biochemistry, pathophysiology and development of therapeutics. Nat Rev Drug Discov. 2007;6:662–80.
35. Groves JT, Nemo TE, Myers RS. Hydroxylation and epoxidation catalyzed by iron-porphine complexes. Oxygen transfer from iodosylbenzene. J Am Chem Soc. 1979;101:1032–3.
36. de Montellano PRO. Hydrocarbon hydroxylation by cytochrome P450 enzymes. Chem Rev. 2010;110:932–48.
37. Mansuy D. A brief history of the contribution of metalloporphyrin models to cytochrome P450 chemistry and oxidation catalysis. C R Chimie. 2007;10:392–413.
38. Wijesekera TP, Dolphin D. Synthetic aspects of porphyrin and metalloporphyrin chemistry. In: Sheldon RA, editor. Metalloporphyrin in catalytic oxidations. New York: Marcel Dekker; 1994. p. 193–239.
39. Suslick KS. Shape-selective oxidations by metalloporphyrins. In: Kadish KM, Smith KM, Guilard R, editors. The porphyrin handbook, vol. 4. New York: Academic Press; 2000. p. 41–63.

40. Che CM, Lo VK, Zhou CY, Huang JS. Selective functionalisation of saturated C–H bonds with metalloporphyrin catalysts. Chem Soc Rev. 2011;40:1950–75.
41. Tovmasyan A, Maia CGC, Weitner T, Carballal S, Sampaio RS, Lieb D, Ghazaryan R, Ivanovic-Burmazovic I, Ferrer-Sueta G, Radi R, Rebouças JS, Spasojević I, Benov L, Batinić-Haberle I. A comprehensive evaluation of catalase-like activity of different classes of redox-active therapeutics. Free Radic Biol Med. 2015;86:308–21.
42. Batinić-Haberle I, Tovmasyan A, Roberts ER, Vujaskovic Z, Leong KW, Spasojević I. SOD therapeutics: latest insights into their structure-activity relationships and impact on the cellular redox-based signaling pathways. Antioxid Redox Signal. 2014;20:2372–415.
43. DeFreitas-Silva G, Rebouças JS, Spasojević I, Benov L, Idemori YM, Batinić-Haberle I. SOD-like activity of Mn(II) β-octabromo-meso-tetrakis(N-methylpyridinium-3-yl)porphyrin equals that of the enzyme itself. Arch Biochem Biophys. 2008;477:105–12.
44. Tovmasyan A, Weitner T, Sheng H, Lu M, Rajic Z, Warner DS, Spasojević I, Rebouças JS, Benov L, Batinić-Haberle I. Differential coordination demands in Fe versus Mn water-soluble cationic metalloporphyrins translate into remarkably different aqueous redox chemistry and biology. Inorg Chem. 2013;52:5677–91.
45. Tovmasyan A, Rebouças JS, Benov L. Simple biological systems for assessing the activity of superoxide dismutase mimics. Antioxid Redox Signal. 2014;20:2416–36.
46. Pasternack RF, Halliwell B. Superoxide dismutase activities of an iron porphyrin and other iron complexes. J Am Chem Soc. 1979;101:1026–31.
47. Faraggi M, Peretz P, Weinraub D. Chemical properties of water-soluble porphyrins. 4. The reaction of a 'picket-fence-like' iron (III) complex with the superoxide oxygen couple. Int J Radiat Biol Relat Stud Phys Chem Med. 1986;49:951–68.
48. Weinraub D, Peretz P, Faraggi M. Chemical properties of water-soluble porphyrins. 1. Equilibriums between some ligands and iron(III) tetrakis(4-N-methylpyridyl)porphyrin. J Phys Chem. 1982;86:1839–42.
49. Ilan Y, Rabani J, Fridovich I, Pasternack RF. Superoxide dismuting activity of an iron porphyrin. Inorg Nucl Chem Lett. 1981;17:93–6.
50. Pasternack RF, Banth A, Pasternack JM, Johnson CS. Catalysis of the disproportionation of superoxide by metalloporphyrins. III. J Inorg Biochem. 1981;15:261–7.
51. Peretz P, Solomon D, Weinraub D, Faraggi M. Chemical properties of water-soluble porphyrins 3. The reaction of superoxide radicals with some metalloporphyrins. Int J Radiat Biol Relat Stud Phys Chem Med. 1982;42:449–56.
52. Weinraub D, Levy P, Faraggi M. Chemical properties of water-soluble porphyrins. 5. Reactions of some manganese (III) porphyrins with the superoxide and other reducing radicals. Int J Radiat Biol Relat Stud Phys Chem Med. 1986;50:649–58.
53. Batinić-Haberle I, Benov L, Spasojević I, Fridovich I. The *ortho* effect makes manganese(III) *meso*-tetrakis(*N*-methylpyridinium-2-yl)porphyrin a powerful and potentially useful superoxide dismutase mimic. J Biol Chem. 1998;273:24521–8.
54. Batinić-Haberle I, Spasojević I, Hambright P, Benov L, Crumbliss AL, Fridovich I. Relationship among redox potentials, proton dissociation constants of pyrrolic nitrogens, and *in vivo* and *in vitro* superoxide dismutating activities of Manganese(III) and Iron(III) water-soluble porphyrins. Inorg Chem. 1999;38:4011–22.
55. Batinić-Haberle I, Liochev SI, Spasojević I, Fridovich I. A potent superoxide dismutase mimic: manganese beta-octabromo-*meso*-tetrakis-(N-methylpyridinium-4-yl)porphyrin. Arch Biochem Biophys. 1997;343:225–33.
56. Batinić-Haberle I, Spasojević I, Stevens RD, Hambright P, Fridovich I. Manganese(III) *meso*-tetrakis(*ortho*-*N*-alkylpyridyl)porphyrins. Synthesis, characterization, and catalysis of $O_2^{\bullet-}$ dismutation. J Chem Soc Dalton Trans. 2002:2689–96.
57. Batinić-Haberle I, Spasojević I, Stevens RD, Hambright P, Neta P, Okado-Matsumoto A, Fridovich I. New class of potent catalysts of O_2-dismutation. Mn(III) *ortho*-methoxyethylpyridyl- and di-*ortho*-methoxyethylimidazolylporphyrins. Dalton Trans. 2004:1696–702.
58. Spasojević I, Batinić-Haberle I, Rebouças JS, Idemori YM, Fridovich I. Electrostatic contribution in the catalysis of $O_2^{\bullet-}$ dismutation by superoxide dismutase mimics. $Mn^{III}TE\text{-}2\text{-}PyP^{5+}$ versus $Mn^{III}Br_8T\text{-}2\text{-}PyP^{+}$. J Biol Chem. 2003;278:6831–7.

59. Rebouças JS, Spasojević I, Tjahjono DH, Richaud A, Méndez F, Benov L, Batinić-Haberle I. Redox modulation of oxidative stress by Mn porphyrin-based therapeutics: the effect of charge distribution. Dalton Trans. 2008:1233–42.
60. Rebouças JS, DeFreitas-Silva G, Spasojević I, Idemori YM, Benov L, Batinić-Haberle I. Impact of electrostatics in redox modulation of oxidative stress by Mn porphyrins: protection of SOD-deficient Escherichia coli via alternative mechanism where Mn porphyrin acts as a Mn carrier. Free Radic Biol Med. 2008;45:201–10.
61. Batinić-Haberle I, Tovmasyan A, Spasojević I. The complex mechanistic aspects of redox-active compounds, commonly regarded as SOD mimics. BioInorg React Mech. 2013;9:35–58.
62. Batinić-Haberle I, Rajic Z, Tovmasyan A, Rebouças JS, Ye X, Leong KW, Dewhirst MW, Vujaskovic Z, Benov L, Spasojević I. Diverse functions of cationic Mn(III) *N*-substituted pyridyl porphyrins, recognized as SOD mimics. Free Radic Biol Med. 2011;51:1035–53.
63. Tovmasyan A, Sheng H, Weitner T, Arulpragasam A, Lu M, Warner DS, Vujaskovic Z, Spasojević I, Batinić-Haberle I. Design, mechanism of action, bioavailability and therapeutic effects of Mn porphyrin-based redox modulators. Med Princ Pract. 2013;22:103–30.
64. Sheng H, Spasojević I, Warner DS, Batinić-Haberle I. Mouse spinal cord compression injury is ameliorated by intrathecal cationic manganese(III) porphyrin catalytic antioxidant therapy. Neurosci Lett. 2004;366:220–5.
65. Mackensen GB, Patel M, Sheng H, Calvi CL, Batinić-Haberle I, Day BJ, Liang LP, Fridovich I, Crapo JD, Pearlstein RD, Warner DS. Neuroprotection from delayed post ischemic administration of a metalloporphyrin catalytic antioxidant. J Neurosci. 2001;21:4582–92.
66. Feiters MC, Rowan AE, Nolte RJM. From simple to supramolecular cytochrome P450 mimics. Chem Soc Rev. 2000;29:375–84.
67. Ellis PE, Lyons JE. Selective air oxidation of light alkanes catalyzed by activated metalloporphyrins—the search for a suprabiotic system. Coord Chem Rev. 1990;105:181–93.
68. Grinstaff MW, Hill MG, Labinger JA, Gray HB. Mechanism of catalytic oxygenation of alkanes by halogenated iron porphyrins. Science. 1994;264:1311–3.
69. Holm RH. Metal-centered oxygen atom transfer reactions. Chem Rev. 1987;87:1401–49.
70. Baylon JL, Lenov IL, Sligar SG, Tajkhorshid E. Characterizing the membrane-bound state of cytochrome P450 3A4: structure, depth of insertion, and orientation. J Am Chem Soc. 2013;135:8542–51.
71. Pochapsky TC, Kazanis S, Dang M. Conformational plasticity and structure/function relationships in cytochromes P450. Antioxid Redox Signal. 2010;13:1273–96.
72. Shaik S, Kumar D, Visser SP, Altun A, Thiel W. Theoretical perspective on the structure and mechanism of cytochrome P450 enzymes. Chem Rev. 2005;105:2279–328.
73. Nam W. High-valent iron(IV)—oxo complexes of heme and non-heme ligands in oxygenation reactions. Acc Chem Res. 2007;40:522–31.
74. Rittle J, Green MT. Cytochrome P450 compound I: capture, characterization, and C–H bond activation kinetics. Science. 2010;330:933–7.
75. Jung C. The mystery of cytochrome P450 Compound I: a mini-review dedicated to Klaus Ruckpaul. Biochim Biophys Acta. 1814;2011:46–57.
76. Krest CM, Onderko EL, Yosca TH, Calixto JC, Karp RF, Livada J, Rittle J, Green MT. Reactive intermediates in cytochrome p450 catalysis. J Biol Chem. 2013;288:17074–81.
77. Denisov IG, Mak PJ, Makris TM, Sligar SG, Kincaid JR. Resonance Raman characterization of the peroxo and hydroperoxo intermediates in cytochrome P450. J Phys Chem A. 2008;112:13172–9.
78. Cojocaru V, Balali-Mood K, Sansom MSP, Wade RC. Structure and dynamics of the membrane-bound cytochrome P450 2C9. PLoS Comput Biol. 2011;7, e1002152.
79. Das A, Grinkova YV, Sligar SG. Redox potential control by drug binding to cytochrome P450 3A4. J Am Chem Soc. 2007;129:13778–9.
80. Lewis DF, Hlavica P. Interactions between redox partners in various cytochrome P450 systems: functional and structural aspects. Biochim Biophys Acta. 2000;1460:353–74.

81. Fisher MT, Sligar SG. Control of heme protein redox potential and reduction rate: linear free energy relation between potential and ferric spin state equilibrium. J Am Chem Soc. 1985;107:5018–9.
82. Sligar SG. Coupling of spin, substrate, and redox equilibria in cytochrome P450. Biochemistry. 1976;15:5399–406.
83. Groves JT, McClusky GA. Aliphatic hydroxylation via oxygen rebound. Oxygen transfer catalyzed by iron. J Am Chem Soc. 1976;98:859–61.
84. Groves JT, McClusky GA. Aliphatic hydroxylation by highly purified liver microsomal cytochrome P-450. Evidence for a carbon radical intermediate. Biochem Biophys Res Commun. 1978;81:154–60.
85. Veitch NC. Horseradish peroxidase: a modern view of a classic enzyme. Phytochemistry. 2004;65:249–59.
86. Kobayashi S, Nakano M, Kimura T, Schaap AP. On the mechanism of the peroxidase-catalyzed oxygen-transfer reaction. Biochemistry. 1987;26:5019–22.
87. Fujii H. Model complexes of heme peroxidases. In: Raven E, Dunford B, editors. Heme peroxidases. Cambridge: RSC Metallobiology; 2016. p. 183–217.
88. Nakagaki S, Ferreira GKB, Marcalb AL, Ciuffi KJ. Metalloporphyrins immobilized on silica and modified silica as catalysts in heterogeneous processes. Curr Org Synth. 2014;11:67–88.
89. Hongjian L, Zhang XP. Catalytic C–H functionalization by metalloporphyrins: recent developments and future directions. Chem Soc Rev. 2011;40:1899–909.
90. Gunter MJ, Turner P. Metalloporphyrins as models for the cytochromes p-450. Coord Chem Rev. 1991;108:115–61.
91. Silva VS, Meireles AM, Martins DCS, Rebouças JS, DeFreitas-Silva G, Idemori YM. Effect of imidazole on biomimetic cyclohexane oxidation by first-, second-, and third-generation manganese porphyrins using PhIO and $PhI(OAc)_2$ as oxidants. Appl Catal A-Gen. 2015;491:17–27.
92. Iamamoto Y, Serra OA, Idemori YM. Iron(III) porphyrins atropisomers as catalysts for cyclohexane hydroxylations. A biomimetical system. J Inorg Biochem. 1994;54:55–66.
93. Tabushi I. Reductive dioxygen activation by use of artificial P-450 systems. Coord Chem Rev. 1988;86:1–42.
94. Mansuy D, Fontecave M, Bartoli JF. Mono-oxygenase-like dioxygen activation leading to alkane hydroxylation and olefin epoxidation by an Mn(porphyrin)-ascorbate biphasic system. J Chem Soc Chem Commun. 1983:253–4.
95. Groves JT, Neumann R. Enzymic regioselectivity in the hydroxylation of cholesterol catalyzed by a membrane-spanning metalloporphyrin. J Org Chem. 1988;53:3891–3.
96. Groves JT, Neumann R. Regioselective oxidation catalysis in synthetic phospholipid vesicles. Membrane-spanning steroidal metalloporphyrins. J Am Chem Soc. 1989;111:2900–9.
97. Spasojević I, Colvin OM, Warshany KR, Batinić-Haberle I. New approach to the activation of anti-cancer pro-drugs by metalloporphyrin-based cytochrome P450 mimics in all-aqueous biologically relevant system. J Inorg Biochem. 2006;100:1897–902.
98. Nam W, Choi HJ, Han HJ, Cho SH, Lee HJ, Han SY. Use of 2-methyl-1-phenylpropan-2-yl hydroperoxide (MPPH) as a mechanistic probe for the heterolytic versus homolytic O-O bond cleavage of tert-alkyl hydroperoxide by iron(III) porphyrin complex. Chem Commun. 1999;4:387–8.
99. Groves JT, Kruper Jr WJ, Haushalter RC. Hydrocarbon oxidations with oxometalloporphinates. Isolation and reactions of a (porphinato)manganese(V) complex. J Am Chem Soc. 1980;102:6375–7.
100. Hill CL, Schardt BC. Alkane activation and functionalization under mild conditions by a homogeneous manganese(III)porphyrin-iodosylbenzene oxidizing system. J Am Chem Soc. 1980;102:6374–5.
101. Smegal JA, Hill CL. Hydrocarbon functionalization by the (iodosylbenzene)manganese(IV) porphyrin complexes from the (tetraphenylporphinato)manganese(III)-iodosylbenzene cata-

lytic hydrocarbon oxidation system. Mechanism and reaction chemistry. J Am Chem Soc. 1983;105:3515–21.

102. Suslick KS, van Deusen-Jeffries S. Shape selective biomimetic oxidation catalysis. In: Atwood JL, Davies JED, MacNicol DD, Vögtle F, Lehn J-M, editors. Comprehensive supramolecular chemistry, vol. 5. Oxford: Elsevier; 1996. p. 141–70.
103. Rebouças JS, Carvalho MEMD, Idemori YM. Perhalogenated 2-pyridylporphyrin complexes: synthesis, self-coordinating aggregation properties, and catalytic studies. J Porphyrins Phthalocyanines. 2002;6:50–7.
104. Serra AC, Marçalo EC, Gonsalves AMd'AR. A view on the mechanism of metalloporphyrin degradation in hydrogen peroxide epoxidation reactions. J Mol Catal A-Chem. 2004;215:17–21.
105. do Nascimento E, de F Silva G, Caetano FA, Fernandes MA, da Silva DC, de Carvalho ME, Pernaut JM, Rebouças JS, Idemori YM. Partially and fully beta-brominated Mn-porphyrins in P450 biomimetic systems: effects of the degree of bromination on electrochemical and catalytic properties. J Inorg Biochem. 2005;99:1193–204.
106. da Silva DC, DeFreitas-Silva G, do Nascimento E, Rebouças JS, Barbeira PJ, de Carvalho ME, Idemori YM. Spectral, electrochemical, and catalytic properties of a homologous series of manganese porphyrins as cytochrome P450 model: the effect of the degree of beta-bromination. J Inorg Biochem. 2008;102:1932–41.
107. Tovmasyan A, Sampaio RS, Boss MK, Bueno-Janice JC, Bader BH, Thomas M, Rebouças JS, Orr M, Chandler JD, Go YM, Jones DP, Venkatraman TN, Haberle S, Kyui N, Lascola CD, Dewhirst MW, Spasojević I, Benov L, Batinić-Haberle I. Anticancer therapeutic potential of Mn porphyrin/ascorbate system. Free Radic Biol Med. 2015;89:1231–47.
108. Johnstone RAW, Nunes MLPG, Pereira MM, Gonsalves AMd'AR, Serra AC. Improved syntheses of 5, 10, 15, 20-tetrakisaryl- and tetrakisalkylporphyrins. Heterocycles. 1996;43: 1423–37.
109. Lindsey JS, Schreiman IC, Hsu HC, Kearney PC, Marguerettaz AM. Rothemund and Adler-Longo reactions revisited: synthesis of tetraphenylporphyrins under equilibrium conditions. J Org Chem. 1987;52:827–36.
110. Lindsey JS, Wagner RW. Investigation of the synthesis of *ortho*-substituted tetraphenylporphyrins. J Org Chem. 1989;54:828–36.
111. Atkinson ST, Brady SP, James JP, Nolan KB. Synthetic haems as mimics for high valent intermediates in haemoprotein catalysed oxidations. Synthesis and oxidation of chloro-7,8,17,18-tetracyano-5,10,15,20-tetraphenylporphyrinatoiron(III), a haem which contains strongly electron-withdrawing groups in the β-pyrrole positions. Pure Appl Chem. 1995;67:1109–16.
112. Borocci S, Marotti F, Mancini G, Monti D, Pastorini A. Selectivity in the oxidation of limonene by amphiphilized metalloporphyrins in micellar media. Langmuir. 2001;17:7198–203.
113. Lohmann W, Karst U. Biomimetic modeling of oxidative drug metabolism: strategies, advantages and limitations. Anal Bioanal Chem. 2008;391:79–96.
114. Bernadou J, Meunier B. Biomimetic chemical catalysts in the oxidative activation of drugs. Adv Synth Catal. 2004;346:171–84.
115. Masumoto H, Takeuchi K, Ohta S, Hirobe M. Application of chemical P450 model systems to studies on drug metabolism. I. Phencyclidine: a multi-functional model substrate. Chem Pharm Bull. 1989;37:1788–94.
116. Pautet F, Barret R, Daudon M. Optimization of biomimetic oxidation reactions achieved with the aid of an iodosylbenzene and meso-tetraphenylporphinatoiron (III): application to antergan. Pharm Acta Helv. 1988;63:140–4.
117. Salmeen IT, Foxall-VanAken S, Ball JC. A preliminary study of an iron porphyrin-iodosylbenzene system for activation of mutagens in the Ames assay. Mutat Res Lett. 1988;207:111–5.
118. Chauncey MA, Ninomiya SI. Metabolic studies with model cytochrome p-450 systems. Tetrahedron Lett. 1990;31:5901–4.

119. Neves CMB, Simões MMQ, Santos ICMS, Domingues FMJ, Neves MGPMS, Paz FAA, Silva AMS, Cavaleiro JAS. Oxidation of caffeine with hydrogen peroxide catalyzed by metalloporphyrins. Tetrahedron Lett. 2011;52:2898–902.
120. Schaab EH, Crotti AEM, Iamamoto Y, Kato MJ, Lotufo LVC, Lopes NP. Biomimetic oxidation of piperine and piplartine catalyzed by iron(III) and manganese(III) porphyrins. Biol Pharm Bull. 2010;33:912–6.
121. Breslow R, Zhang X, Huang Y. Selective catalytic hydroxylation of a steroid by an artificial cytochrome P-450 enzyme. J Am Chem Soc. 1997;119:4535–6.
122. Fang Z, Breslow R. A thiolate ligand on a cytochrome P-450 mimic permits the use of simple environmentally benign oxidants for biomimetic steroid hydroxylation in water. Bioorg Med Chem Lett. 2005;15:5463–6.
123. Melo AJB, Iamamoto Y, Maestrin APJ, Smith JRL, Santos MD, Lopes NP, Bonato PS. Biomimetic oxidation of praziquantel catalysed by metalloporphyrins. J Mol Catal A-Chem. 2005;226:23–31.
124. Hartmann J, Bartels P, Mau U, Witter M, Tümpling WV, Hofmann J, Nietzschmann E. Degradation of the drug diclofenac in water by sonolysis in presence of catalysts. Chemosphere. 2008;70:453–61.
125. Santos MD, Martins PR, Santos PA, Bortocan R, Iamamoto Y, Lopes NP. Oxidative metabolism of 5-o-caffeoylquinic acid (chlorogenic acid), a bioactive natural product, by metalloporphyrin and rat liver mitochondria. Eur J Pharm Sci. 2005;26:62–70.
126. Mac Leod TCO, Faria AL, Barros VP, Queiroz MEC, Assis MD. Primidone oxidation catalyzed by metalloporphyrins and Jacobsen catalyst. J Mol Catal A-Chem. 2008;296:54–60.
127. Neves CMB, Simões MMQ, Domíngues MRM, Santos ICMS, Neves MGPMS, Paz FAA, Silva AMS, Cavaleiro JAS. Oxidation of diclofenac catalyzed by manganese porphyrins: synthesis of novel diclofenac derivatives. RSC Adv. 2012;2:7427–38.
128. Simões MMQ, Neves CMB, Pires SMG, Neves MGPMS, Cavaleiro JAS. Mimicking P450 processes and the use of metalloporphyrins. Pure Appl Chem. 2013;85:1671–81.
129. Faria AL, Mac Leod TCO, Assis MD. Carbamazepine oxidation catalyzed by iron and manganese porphyrins supported on aminofunctionalized matrices. Catal Today. 2008; 133–135:863–9.
130. CarvalhoDa-Silva D, Mac Leod TCO, Faria AL, Santos JS, Carvalho MEMD, Rebouças JS, Idemori YM, Assis MD. Carbamazepine oxidation catalyzed by manganese porphyrins: effects of the β-bromination of the macrocycle and the choice of oxidant. Appl Catal A-Gen. 2011;408:25–30.
131. Song R, Sorokin A, Bernadou J, Meunier B. Metalloporphyrin-catalyzed oxidation of 2-methylnaphthalene to vitamin K3 and 6-methyl-1,4-naphthoquinone by potassium monopersulfate in aqueous solution. J Org Chem. 1997;62:673–8.
132. Waldmann D, König T, Schreier P. Iron(III)porphinate/H_2O_2-mediated conversion of all-(E)-retinol. Z Naturforsch B. 1995;50:589–94.
133. Bloodsworth A, O'Donnell VB, Batinić-Haberle I, Chumley PH, Hurt JB, Day BJ, Crow JP, Freeman BA. Manganese-porphyrin reactions with lipids and lipoproteins. Free Radic Biol Med. 2000;28:1017–29.
134. Ferrer-Sueta G, Batinić-Haberle I, Spasojević I, Fridovich I, Radi R. Catalytic scavenging of peroxynitrite by isomeric Mn(III) *N*-methylpyridylporphyrins in the presence of reductants. Chem Res Toxicol. 1999;12:442–9.
135. Batinić-Haberle I, Tovmasyan A, Spasojević I. An educational overview of the chemistry, biochemistry and therapeutic aspects of Mn porphyrins—from superoxide dismutation to H_2O_2-driven pathways. Redox Biol. 2015;5:43–65.
136. Buettner GR, Wagner BA, Rodgers VG. Quantitative redox biology: an approach to understand the role of reactive species in defining the cellular redox environment. Cell Biochem Biophys. 2013;67:477–83.
137. Batinić-Haberle I, Cuzzocrea S, Rebouças JS, Ferrer-Sueta G, Mazzon E, Di Paola R, Radi R, Spasojević I, Benov L, Salvemini D. Pure MnTBAP selectively scavenges peroxynitrite over

superoxide: comparison of pure and commercial MnTBAP samples to MnTE-2-PyP in two models of oxidative stress injury, an SOD-specific Escherichia coli model and carrageenan-induced pleurisy. Free Radic Biol Med. 2009;46:192–201.

138. Rajic Z, Tovmasyan A, Spasojević I, Sheng H, Lu M, Li AM, Gralla EB, Warner DS, Benov L, Batinić-Haberle I. A new SOD mimic, Mn(III) ortho N-butoxyethylpyridylporphyrin, combines superb potency and lipophilicity with low toxicity. Free Radic Biol Med. 2012;52:1828–34.
139. Jaramillo MC, Frye JB, Crapo JD, Briehl MM, Tome ME. Increased manganese superoxide dismutase expression or treatment with manganese porphyrin potentiates dexamethasone-induced apoptosis in lymphoma cells. Cancer Res. 2009;69:5450–7.
140. Jaramillo MC, Briehl MM, Crapo JD, Batinić-Haberle I, Tome ME. Manganese porphyrin, MnTE-2-PyP^{5+}, acts as a pro-oxidant to potentiate glucocorticoid-induced apoptosis in lymphoma cells. Free Radic Biol Med. 2012;52:1272–84.
141. Lee K, Briehl MM, Mazar AP, Batinić-Haberle I, Rebouças JS, Glinsmann-Gibson B, Rimsza LM, Tome ME. The copper chelator ATN-224 induces peroxynitrite-dependent cell death in hematological malignancies. Free Radic Biol Med. 2013;60:157–67.
142. Jaramillo MC, Briehl MM, Batinić-Haberle I, Tome ME. Manganese (III) *meso*-tetrakis-*N*-ethylpyridinium-2-yl porphyrin acts as a pro-oxidant to inhibit electron transport chain proteins, modulate bioenergetics, and enhance the response to chemotherapy in lymphoma cells. Free Radic Biol Med. 2015;83:89–100.
143. Tse HM, Milton MJ, Piganelli JD. Mechanistic analysis of the immunomodulatory effects of a catalytic antioxidant on antigen-presenting cells: implication for their use in targeting oxidation-reduction reactions in innate immunity. Free Radic Biol Med. 2004;36:233–47.
144. Delmastro-Greenwood MM, Tse HM, Piganelli JD. Effects of metalloporphyrins on reducing inflammation and autoimmunity. Antioxid Redox Signal. 2014;20:2465–77.
145. Batinić-Haberle I, Spasojević I, Tse HM, Tovmasyan A, Rajic Z, St Clair DK, Vujaskovic Z, Dewhirst MW, Piganelli JD. Design of Mn porphyrins for treating oxidative stress injuries and their redox-based regulation of cellular transcriptional activities. Amino Acids. 2012;42:95–113.
146. Celic T, Španjol J, Bobinac M, Tovmasyan A, Vukelic I, Rebouças JS, Batinić-Haberle I, Bobinac D. Mn porphyrin-based SOD mimic, MnTnHex-2-PyP^{5+}, and non-SOD mimic, $MnTBAP^{3-}$, suppressed rat spinal cord ischemia/reperfusion injury *via* NF-κB pathways. Free Radic Res. 2014;48:1426–42.
147. Kos I, Benov L, Spasojević I, Rebouças JS, Batinić-Haberle I. High lipophilicity of *meta* Mn(III) *N*-alkylpyridylporphyrin-based superoxide dismutase mimics compensates for their lower antioxidant potency and makes them as effective as *ortho* analogues in protecting superoxide dismutase-deficient Escherichia coli. J Med Chem. 2009;52:7868–72.
148. Kos I, Rebouças JS, DeFreitas-Silva G, Salvemini D, Vujaskovic Z, Dewhirst MW, Spasojević I, Batinić-Haberle I. Lipophilicity of potent porphyrin-based antioxidants: comparison of *ortho* and *meta* isomers of Mn(III) *N*-alkylpyridylporphyrins. Free Radic Biol Med. 2009;47:72–8.
149. Spasojević I, Kos I, Benov LT, Rajic Z, Fels D, Dedeugd C, Ye X, Vujaskovic Z, Rebouças JS, Leong KW, Dewhirst MW, Batinić-Haberle I. Bioavailability of metalloporphyrin-based SOD mimics is greatly influenced by a single charge residing on a Mn site. Free Radic Res. 2011;45:188–200.
150. Rebouças JS, Spasojević I, Batinić-Haberle I. Pure manganese(III) 5,10,15,20-tetrakis(4-benzoic acid)porphyrin (MnTBAP) is not a superoxide dismutase mimic in aqueous systems: a case of structure-activity relationship as a watchdog mechanism in experimental therapeutics and biology. J Biol Inorg Chem. 2008;13:289–302.
151. Tovmasyan A, Carballal S, Ghazaryan R, Melikyan L, Weitner T, Maia CGC, Rebouças JS, Radi R, Spasojević I, Benov L, Batinić-Haberle I. Rational design of superoxide dismutase (SOD) mimics: the evaluation of the therapeutic potential of new cationic Mn porphyrins with linear and cyclic substituents. Inorg Chem. 2014;53:11467–83.

152. Lewis DFV. The P450 catalytic cycle. In: Lewis DFV, editor. Guide to cytochromes P450: structure and function. New York: Taylor & Francis; 2001. p. 49–68.
153. Spasojević I, Batinić-Haberle I. Manganese(III) complexes with porphyrins and related compounds as catalytic scavengers of superoxide. Inorg Chim Acta. 2001;317:230–42.
154. Lahaye D, Groves JT. Modeling the haloperoxidases: reversible oxygen atom transfer between bromide ion and an oxo-Mn(V) porphyrin. J Inorg Biochem. 2007;101:1786–97.
155. Lyons JE, Ellis Jr PE. Reactions of alkanes with dioxygen: toward suprabiotic systems. In: Sheldon RA, editor. Metalloporphyrins in catalytic oxidations. New York: Marcel Dekker; 1994. p. 297–324.
156. Levine RL, Mosoni L, Berlett BS, Stadtman ER. Methionine residues as endogenous antioxidants in proteins. Proc Natl Acad Sci U S A. 1996;93:15036–40.
157. Stadtman ER, Moskovitz J, Levine RL. Oxidation of methionine residues of proteins: biological consequences. Antioxid Redox Signal. 2003;5:577–82.
158. Requejo R, Hurd TR, Costa NJ, Murphy MP. Cysteine residues exposed on protein surfaces are the dominant intramitochondrial thiol and may protect against oxidative damage. FEBS J. 2010;277:1465–80.
159. Pamplona R, Costantini D. Molecular and structural antioxidant defenses against oxidative stress in animals. Am J Physiol Regul Integr Comp Physiol. 2011;301:R843–63.
160. Bender A, Hajieva P, Moosmann B. Adaptive antioxidant methionine accumulation in respiratory chain complexes explains the use of a deviant genetic code in mitochondria. Proc Natl Acad Sci U S A. 2008;105:16496–501.
161. Bueno-Janice JC, Tovmasyan A, Batinić-Haberle I. Comprehensive study of GPx activity of different classes of redox-active therapeutics—implications for their therapeutic actions. Free Radic Biol Med. 2015;87:S86–7.
162. Rebouças JS, Spasojević I, Batinić-Haberle I. Quality of potent Mn porphyrin-based SOD mimics and peroxynitrite scavengers for pre-clinical mechanistic/therapeutic purposes. J Pharm Biomed Anal. 2008;48:1046–9.
163. Rebouças JS, Kos I, Vujasković Z, Batinić-Haberle I. Determination of residual manganese in Mn porphyrin-based superoxide dismutase (SOD) and peroxynitrite reductase mimics. J Pharm Biomed Anal. 2009;50:1088–91.
164. Marques A, Marin M, Ruasse MF. Hydrogen peroxide oxidation of mustard-model sulfides catalyzed by iron and manganese tetraarylporphyrins. Oxygen transfer to sulfides versus H_2O_2 dismutation and catalyst breakdown. J Org Chem. 2001;66:7588–95.
165. Lahaye D, Muthukumaran K, Hung CH, Gryko D, Rebouças JS, Spasojević I, Batinić-Haberle I, Lindsey JS. Design and synthesis of manganese porphyrins with tailored lipophilicity: investigation of redox properties and superoxide dismutase activity. Bioorg Med Chem. 2007;15:7066–86.
166. Rosenthal RA, Huffman KD, Fisette LW, Damphousse CA, Callaway WB, Malfroy B, Doctrow SR. Orally available Mn porphyrins with superoxide dismutase and catalase activities. J Biol Inorg Chem. 2009;14:979–91.
167. Rausaria S, Ghaffari MM, Kamadulski A, Rodgers K, Bryant L, Chen Z, Doyle T, Shaw MJ, Salvemini D, Neumann WL. Retooling manganese(III) porphyrin-based peroxynitrite decomposition catalysts for selectivity and oral activity: a potential new strategy for treating chronic pain. J Med Chem. 2011;54:8658–69.
168. Rosenthal RA, Fish B, Hill RP, Huffman KD, Lazarova Z, Mahmood J, Medhora M, Molthen R, Moulder JE, Sonis ST, Tofilon PJ, Doctrow SR. Salen Mn complexes mitigate radiation injury in normal tissues. Anticancer Agents Med Chem. 2011;11:359–72.

Chapter 10
Nitrones as Potent Anticancer Therapeutics

Rheal A. Towner and Robert A. Floyd

10.1 Brief History of Nitrones as Spin-Trapping Compounds

Nitrones have been used for over 50 years as spin-trapping compounds that allow the stabilization of reactive free radicals for further characterization as nitroxides with electron paramagnetic resonance (EPR) or electron spin resonance (ESR) spectroscopy. The term "spin trapping" was originally coined by Janzen in the 1960s [1, 2]. The three major nitrone spin traps that have been used in biological systems are *N-tert*-Butyl-α-phenyl nitrone (PBN) [2–15], α-(4-pyridyl-*N*-oxide)-*N-tert*-butyl nitrone (4-POBN) [14–16], and 5,5-dimethyl-1-pyrroline-*N*-oxide (DMPO) [2, 4, 7, 9, 10, 12–21]. The spin-trapping reaction regarding the coupling of a reactive free radical species (R•) to a PBN-type spin trap is depicted in Fig. 10.1. DMPO was found to be the least toxic of several spin traps tested for toxicity in Sprague–Dawley rats, whereas PBN was found to have an approximate lethal dose 100 mg (0.564 mmol)/100 g body weight (or 1000 mg/kg bw) [13]. Regarding the solubility of the three biological nitrone spin traps in 1-octanol/water, PBN is more hydrophobic and DMPO is the most hydrophilic [14]. When placed in biological media, such as micelles, vesicles or liver microsomes, it was found that PBN was intercalated in both hydrophobic and hydrophilic regions, whereas both 4-POBN and DMPO were mainly found in hydrophilic regions [22]. It is important to note that the selection of appropriate spin traps is dependent on the site of biological free radical production that is being studied. Other considerations regarding the

R.A. Towner (✉)
Advanced Magnetic Resonance Center, Oklahoma City, OK 73104, USA
e-mail: Rheal-Towner@omrf.org

R.A. Floyd
Experimental Therapeutics Research Laboratory, Oklahoma Medical Research Foundation, Oklahoma City, OK 73104, USA
e-mail: Robert-Floyd@omrf.org

I. Batinić-Haberle et al. (eds.), *Redox-Active Therapeutics*,
Oxidative Stress in Applied Basic Research and Clinical Practice,
DOI 10.1007/978-3-319-30705-3_10

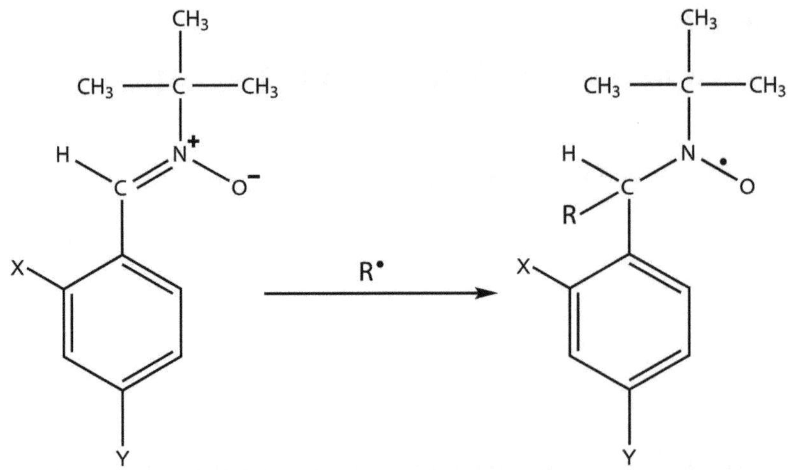

Fig. 10.1 General chemical structures of PBN-like nitrones, and their reaction to trap free radicals (R•) and form nitroxide spin adducts. For PBN, *X* and *Y* are hydrogen (H) atoms whereas for OKN-007, *X* and *Y* are sulfonato (SO_3^-) groups

metabolism of nitroxides in biological systems include the following principles: (1) the reduction of nitroxides by cells occurs mainly intracellularly; (2) the reduction by cells occurs usually via enzyme associated reactions; (3) the rate of reduction by cells is oxygen dependent; (4) the principle site of reduction is in the mitochondria; (5) the microsomal fractions of cells can metabolize nitroxides; (6) the structural characteristics (e.g., ring structure, substituents, hydrophilicity/lipophilicity) of the nitroxide affect the rate of reduction; (7) the principle products of the metabolism of nitroxides are corresponding hydroxylamines; (8) the rate of oxidation of hydroxylamines must be taken into consideration; (9) the oxidation of hydroxylamines mainly involves enzymes (e.g., cytochrome oxidase); (10) the enzymatic oxidation of hydroxylamines occurs mainly in membranes; (11) the rate of oxidation of hydroxylamines is dependent on the concentration of oxygen; and (12) the metabolism of nitroxides/hydroxylamines varies with the physiological state and environment of the cell [15, 16, 23–27]. Nucleophile addition to the β-position of nitrones (often called the Forrester-Hepburn mechanism), and oxidation of the spin trap and subsequent reaction with a nucleophile (inverted spin trapping) are sometimes sources of artifacts [25], but are not very likely to occur in biological systems [28]. Nitrone spin traps can not only react with free radicals, but also nonradicals such as peroxynitrite (ONOO–) via electrophilic and nucleophilic addition reactions [29].

Another more recent method that involves the fluorescence visualization of radical adducts trapped by the nitrone, DMPO, is immuno-spin-trapping (IST), which was developed by Mason et al. [30]. Briefly, free radicals generated as a result of oxidative stress processes can be trapped by 5,5-dimethyl-1-pyrroline *N*-oxide (DMPO) to form DMPO-radical adducts, which can then be further tagged by IST, a method that utilizes an antibody against DMPO-adducts [30–33]. Very recently, IST has been used in combination with molecular magnetic resonance imaging (mMRI) to visualize macromolecular radical adducts trapped with DMPO in vivo in various diseases [28, 34, 35], including a mouse model for gliomas [36]. The anti-DMPO-adduct antibody is coupled to an MRI contrast agent, rather than a fluorescent tag. The advantage of the IST approach coupled with mMRI is that heterogeneous tissues, such as tumors, can be studied to obtain spatial differences in the detection of free radical levels (detected by the presence of the anti-DMPO probe via coupling to DMPO-radical nitrone adducts) in vivo noninvasively. For example, in the glioma study, we were able to show that free radical adducts are mainly distributed in localized peripheral tumor regions near necrotic regions, emphasizing the heterogeneous nature of gliomas and the varied distribution of macromolecular free radicals in a tumor microenvironment [36]. This methodology can be applied to any preclinical models for various cancers to localize macromolecular free radical distribution in vivo which can be used to further characterize ex vivo oxidative stress-related molecular changes.

10.2 Brief Introduction of the Role of Free Radicals in Cancer

Numerous studies indicate that oxidative stress, a result of an imbalance in levels of reactive oxygen and/or nitrogen species (RONS) and antioxidative defense systems, plays a crucial role in cancer. Free radicals are involved and/or are the causal agents in several cancers. RONS may directly oxidize nucleic acids, proteins, carbohydrates, and lipids, causing intracellular and intercellular perturbations in homeostasis, including DNA mutations and interference with DNA repair [37]. High concentrations of lipid-derived electrophilic products resulting from the oxidation process readily react with proteins, DNA, and phospholipids, generating intra- and intermolecular toxic covalent adducts that lead to the propagation and amplification of oxidative stress [37]. The role of nitric oxide in the development of many cancers has also been well reported on [38].

10.2.1 Gliomas

Oxidative stress plays a major role in the growth of gliomas. For instance, the antioxidant status of glioma patients was found to be lowered [39]. It has also been suggested that there is a link between free radical generation and intra-mitochondrial cytochrome-c degradation, which could lead to the impairment of the apoptotic cytochrome-c-dependent cascade [40]. In addition, it has been found that manganese superoxide dismutase is over-expressed in most brain tumors, leading to the increased activation of mitogen-activated protein kinases and phosphatidylinositol-3-kinases, which promote migration and invasion in glioma cells [41]. Oxidative modification of cell lipids and proteins also has potential consequences for tumor cell proliferation [42]. It is also important to note that the distribution of free radicals in gliomas, as in many other tumors, is varied in different regions due to the heterogeneous nature of the tumor microenvironment, which Towner et al. (2013) demonstrated by using in vivo molecular imaging of DMPO-trapped macromolecular free radical adducts [36].

10.2.2 Hepatocarcinogenesis and Hepatocellular Carcinoma

It is well established that free radicals play an integral role in hepatocarcinogenesis [43], regardless of the induction mode (e.g., chronic hepatitis viral infections, carcinogens, toxins, steroid hormones, and dietary intervention) [44, 45]. The choline deficiency model, due to its relatively slow sequential changes in the induction of liver cell injury and neoplasia, was determined to be ideal to study free radical mechanism-of-action [46]. Free radical-initiated lipid peroxidation has been well documented to be associated with HCC [47, 48]. A study by Floyd and co-workers (2000) also found that hydrogen peroxide (H_2O_2) was significantly increased in mitochondria isolated from CD rat livers, compared with normal rats [49]. Additionally, oxidative DNA damage in the form of 8-hydroxydeoxyguanosine (8-OHdG) was demonstrated by Floyd et al. (1990) and Nakae et al. (1998) to be elevated in liver DNA of rats continuously fed a CDAA (choline deficient, L-amino acid-defined) diet [50, 51]. High levels of 8-OHdG were also found in liver DNA of transgenic mice that express hepatitis B virus (HBV) large envelope protein in liver [52]. Hepatic oxidative DNA damage is also considered an increased risk for HCC in chronic liver diseases, as established from chronic liver disease and HCC patient liver samples [53, 54]. It was also found that reactive oxygen species (ROS) damage human serum albumin (HAS) in HCC patients, and that carbonyl levels in these samples were quite elevated [55]. Other elevated molecular markers in HCC include hepatic Nox (NAPDH oxidase) proteins which sensitize mitochondria to generate ROS [56], and xCT (mammalian amino acid transporter also called SLC7A11) which is associated in maintaining glutathione (GSH) levels and its dysfunction has been shown to increase ROS levels [57]. Towner et al. (2002a, 2002b, 2003) were also able to demonstrate that mycotoxins that are highly associated with HCC incidence, such as aflatoxin [58, 59] or microcystin-LR [60], are involved in lipid peroxidation when administered to rats, which may play a key role in initiating hepatocarcinogenesis.

10.2.3 Colorectal Carcinoma

It is well established that oxidative stress plays an integral role in colorectal cancer [61, 62]. A recent study by Nedic et al. (2013) demonstrated that oxidized insulin-like growth factor-binding proteins and receptors are elevated in patients with colorectal carcinoma (CRC), and may be associated directly with tumor growth [61]. Another study by Kang et al. (2013) recently reported on the finding that ROS are involved in silencing the tumor suppressor RUNX3 (Runt-related transcription factor 3) which enhances the Akt (protein kinase B)-mediated signaling pathway and subsequently promotes the proliferation of colorectal cancer cells (SNU-407 cells exposed to H_2O_2) [62].

10.3 Brief Introduction of Nitrones as General Therapeutic Agents

Some of the molecular alterations associated with PBN have been provided by a series of papers by Kotake and co-workers regarding its possible therapeutic action. Kotake et al. (1998) initially showed in vitro, within macrophage cell cultures exposed to lipopolysaccharide (LPS), that PBN inhibits NF-κB, iNOS (inducible nitric oxide synthase) mRNA, COX2 (inducible cyclooxygenase), mRNA, and COX catalytic activity [63]. In endotoxin (LPS)-induced rodents, PBN was found to inhibit the induction of iNOS in mice [64] and also inhibited apoptosis-associated gene expression (e.g., Fas-A, Bax) [65] and multiple anti-inflammatory cytokines (tumor necrosis factor (TNF-α), interleukins (IL-1α and IL-1β), nuclear factor κB (NF-κB), activator protein-1 (AP-1)) [66] in rats. PBN was also found to provide a neuroprotective effect by reducing nitric oxide production in LPS-induced meningitis within rats [67], and inhibiting hypoxia-ischemia (HI)-induced up-regulation of IL-1β, TNF-α, and iNOS mRNA expression following HI [68]. For the HI study, it was also found that PBN had free radical scavenging activity [68]. In a streptozotocin (STZ)-induced type I diabetes model in mice, PBN was also found by the Kotake group to protect against diabetes in part by prevention of cytokine-induced nitric oxide (NO) generation by the pancreatic β-cells [69].

Other groups have also reported on the chemical properties and bioactivity of nitrone spin traps as therapeutic agents which elicit their therapeutic effects by altering cellular redox status via radical scavenging or nitric oxide donation [70]. The cytoprotective effects of DMPO and PBN against hydrogen peroxide-induced cytotoxicity and apoptosis in bovine aortic endothelial cells, investigated by Das et al. (2012), found that nitrone treatment-induced phase II antioxidant enzymes via nuclear translocation of NF-E2-related factor 2 (Nrf-2), and inhibited mitochondrial depolarization and subsequent activation of apoptosis via caspase-3 [71]. PBN and 2-sulfo-phenyl-*N*-tert-butyl nitrone (S-PBN) were found to attenuate cognitive disturbance and reduced tissue injury following traumatic brain injury (TBI) in rats, more than likely via free radical scavenging [72].

In 1988, Floyd and co-workers made the novel observation that PBN had neuroprotective activity in experimental ischemic stroke [73–76]. It was discovered that PBN-treated animals were protective from stroke even if the PBN was administered up to 1 h after the ischemia/reperfusion insult [73, 75]. These results were quickly confirmed and expanded upon by others [77, 78]. In subsequent studies, it was discovered that the disulfonyl derivative of PBN (2,4-disulfophenyl-*N*-tert-butylnitrone, also referred to as NXY-059, and which we have renamed OKN-007; see structure in Fig. 10.1) was much more effective than PBN in a stroke model [79]. NXY-059 was confirmed by others in rat stroke models (middle cerebral artery (MCA) occlusion, and transient focal cerebral ischemia) that it was protective against stroke-related complications [79–81]. In an in vitro study by Culot et al. (2009), they had established that the possible mechanism for the protective effect of NXY-059 in preclinical stroke models was due to cerebrovascular protection of the endothelial cells [82]. Once in clinical trials it was established that NXY-059 did not reveal any serious adverse events from initial pharmacokinetics Phase I and safety Phase II studies [79, 83–86]. In the first phase III clinical trial (SAINT-1), which enrolled about 1700 patients, the results showed significant efficacy, measured as a significant decrease in acute stroke-related injury in patients who received NXY-059 compared with placebo [86]. In patients, administered NXY-059 (0.5–0.8 mg/kg/h) intravenous infusion over 24 h resulted in a plasma concentration of 30 μmol/L [79, 85]. The initial phase III result subsequently led to a larger phase III trial (SAINT-2) that enrolled about 3200 patients [87]. The second phase III trial showed no beneficial effect or efficacy of using NXY-059 in stroke patients, and when the data from the SAINT-1 and SAINT-2 trials were pooled and analyzed, there was a nonsignificant effect when the window of administration was 6 h or less after the stroke [87–89].

10.4 Nitrones as Anticancer Agents

10.4.1 PBN and PBN-Analogues as Anticancer Agents in Hepatocellular Carcinoma

Hepatocellular carcinoma (HCC) is the most common primary malignant tumor of the liver, the fifth most common solid cancer worldwide, and the third leading cause of cancer-related deaths [90, 91]. Common etiological factors include persistent infections by hepatitis B or C viruses, aflatoxin exposure, heavy alcohol drinking, tobacco smoking, obesity, and diabetes to mention a few [91].

It has been known for some time by our groups (Towner and Janzen, Floyd and Nakae, Towner and Floyd) that nitrones are potent anticancer agents. Almost 25 years ago, it was first reported by Towner and Janzen (1990, 1991) that the nitrone, PBN, may be a potential anticancer agent when administered prior to hepatotoxin exposure (carbon tetrachloride (CCl_4) or CCl_4 + ethanol), known to initiate liver carcinogenesis, by preventing the free radical initiating process and early tissue injury [92, 93]. In a choline deficiency (CD) diet liver carcinogenesis model, it was initially reported by Nakae and Floyd (1998) over 15 years ago that PBN when administered orally (drinking water) was able to inhibit early phase carcinogenesis, such as the development of preneoplastic lesions, formation of 8-hydroxyguanine in DNA, and the activities of glutathione S-transferase and inducible cyclo-oxygenase (COX2) [94–97]. Hydroxylated derivatives of PBN (4- and 3-hydroxylated analogues of PBN) were also found to prevent hepatocarcinogenesis in a CD diet model, thought to be mediated by inducing apoptosis of the preneoplastic lesions [95]. Another report by Nakae and Floyd (2004) was able to further demonstrate that PBN could inhibit the formation of hepatocellular adenomas and carcinomas in CD diet fed rats when PBN was also administered, which was thought to occur via the exertion of pro-apoptotic and anti-proliferative effects of PBN towards the neoplastic hepatocytes [96, 98]. It was also shown by Inoue and Towner (2007) that PBN is anti-apoptotic (inhibition of Fas-induced apoptosis) in very early phases of CD-induced hepatocarcinogenesis [99], which apparently switches to a pro-apoptotic role when preneoplastic and neoplastic lesions are formed [95, 98]. See Fig. 10.1 for a summary of the anticancer effects of PBN and PBN-analogues in HCC studies.

10.4.2 PBN and PBN-Analogues as Anticancer Agents in Gliomas

Gliomas are the most common primary intracranial tumors that make up 81 % of all malignant brain tumors [100]. Gliomas comprise the majority of primary brain tumors diagnosed annually in the United States [100–102]. Gliomas are classified by the World Health Organization according to their morphologic characteristics into astrocytic, oligodendrogial, and mixed tumors [103]. Glioblastomas (GBM) which makes up to ~45 % of all gliomas have a ~5 % 5-year survival rate [100, 101]. Approximately 15,000 patients die with glioblastomas in the United States per year [102]. Due to the infiltrative nature of gliomas, surgery is rarely effective, i.e., after surgical removal tumors recur predominantly within 1 cm of the resection cavity [104]. Despite modern diagnostics and treatments the median survival time for patients with glioblastomas (GBM) does not exceed 15 months [104, 105]. Prognosis is related not only to tumor grade but to glioma subtype, i.e., oligodendrogliomas are characterized by a better prognosis [103]. Other important hallmarks of malignant gliomas are their invasive behavior and angiogenesis [104].

Towner and Floyd have reported on the anticancer activity of PBN in an orthotopic C6 rat glioma model [106–110], as well as the anticancer activity of the 2,4-disulfonyl derivative of PBN, (2,4-disulfonyl *tert-N*-butyl α-phenyl nitrone or disodium 4-[(tert-butyl-imino) methyl] benzene-1,3-disulfonate *N*-oxide or disufenton, previously called NXY-059, which we have renamed OKN-007 (*OK*lahoma *N*itrone 007)) in various rat glioma models (rat C6 and F98 gliomas, and human xenograft U87 in nude rats) [108–112]. In C6 rat gliomas, PBN was found to reduce tumor volumes and increase survival in ~50 % of animals treated once tumors were established, as measured by magnetic resonance imaging (MRI) [106]. In the study by Doblas et al. (2008), it was established that the anticancer effect of PBN was in part due to an anti-angiogenic activity, as determined from decreased tumor blood vessel volumes in treated animals, compared to untreated controls, as measured from magnetic resonance angiography (MRA) [106]. OKN was also assessed in a rat C6 glioma model and found to have a dramatic effect on regressing tumor formation and increasing survival [108, 111]. Rats treated with OKN-007 after tumors were visualized by MRI were found to have significantly decreased tumor volumes (~3-fold, $p<0.05$), decreased apparent diffusion coefficients (ADC) (~20 %, $p<0.05$), and increased tissue perfusion rates (~60 %, $p<0.05$) in tumors compared with non-treated rats [111]. OKN-007 was administered in the drinking water at a dose of 10 mg/kg/day starting when tumors had reached ~50 mm^3 in volume (about day 15 after intracerebral implantation of rat C6 glioma cells) and continued for a total of 10 days [111]. One group of rats was euthanized after the 10-day treatment period, and a second group was monitored for an additional 25 days after the treatment period [111]. In the cohort of animals that were treated for 10 days and then euthanized, percent survival was 100 % ($p<0.0001$), whereas for the rats that were monitored for an additional 25 days the percent survival was greater than 80 % ($p<0.001$) [111].

In a more recent study by Towner and co-workers, both the rat F98 orthotopic and the human U87 xenograft models were used to assess the effect of OKN-007 [112]. The F98 glioma model was used as it is more infiltrative and aggressive than the C6 model [113, 114] and is also synergetic in Fisher 344 rats [113, 115]. The F98 rat glioma model was considered since these gliomas have an infiltrative pattern of growth, have attributes associated with human GBM gliomas, and are classified as anaplastic malignant tumors [113, 116]. The F98 glioma cell line was obtained as a result of administering ethylnitrosourea (ENU) to pregnant rats, whose progeny developed brain tumors [113], and is commonly used as a preclinical rodent glioma model [112]. The human U87 xenograft model was used as it has some characteristics that are associated with GBM, such as high cellularity with atypia (e.g., mitotic figures and irregular nucleoli) and profuse neovascularization [115]. Pharmacokinetics data obtained from Fisher 344 rats indicated that OKN-007 reached a plasma level of 1.3 μmol/L in 1 h after a bolus dose of 10 mg/kg administered by gavage and was essentially cleared after ~4–5 h [112]. It was found by HPLC (high performance liquid chromatography) analysis of tissue samples following OKN-007 administration (oral via gavage; 18 mg/kg) that OKN-007 readily reaches the brain (despite its anionic form), lungs, heart, liver, spleen, and kidneys (see Fig. 10.2a). When OKN-007 levels were assessed in tumor versus normal brain tissues via HPLC, the compound was found to be equally taken up in both tissues, regardless of dose (see Fig. 10.2b). The route of administration, whether it was oral (via drinking water) or via i.v. injections (both daily), did not seem to alter the significant reduction in glioma tumor volumes (C6 rat glioma model) detected in both treatment regimes (see Fig. 10.3). Immunohistochemistry (IHC) assessment of commonly studied tumor markers

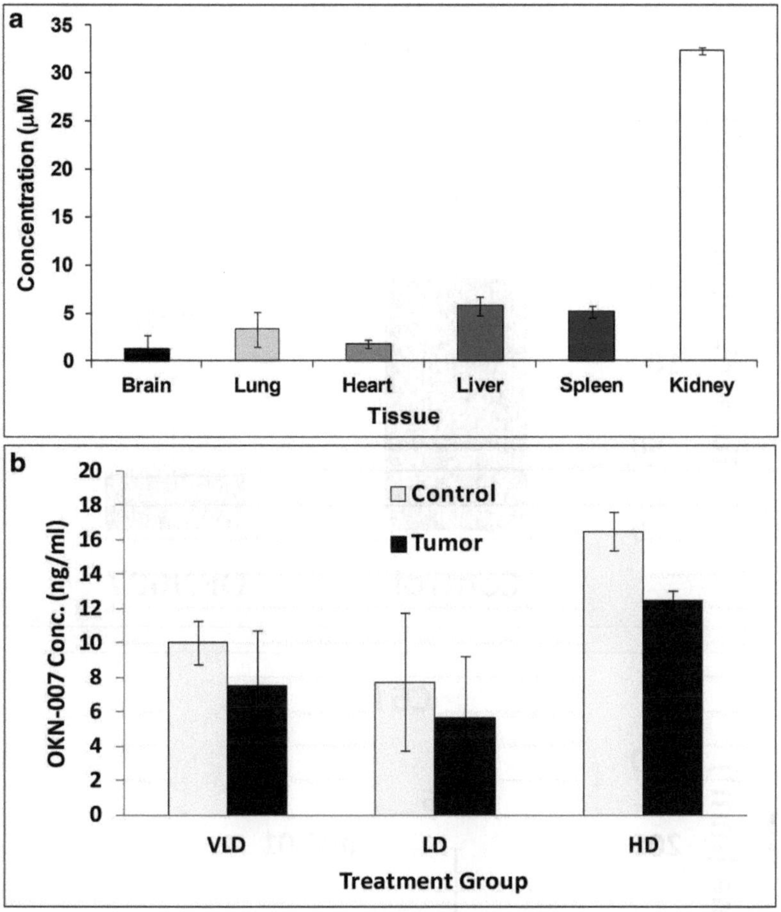

Fig. 10.2 (**a**) HPLC-derived tissue levels (concentration in µM) of OKN-007 following oral administration (via gavage; 18 mg/kg) in normal rats (male Fisher 344). (**b**) HPLC-derived tissue levels (concentration in ng/mL of tissue lysate) of OKN-007 following oral administration (via gavage; very-low dose (VLD): 18 mg/kg (0.015 % w/v), low-dose (LD): 35 mg/kg (0.035 w/v), or high-dose (HD): 75 mg/kg (0.065 %)) in tumor (Tumor) and contralateral normal brain (Control) tissues of C6 rat glioma-bearing rats

for cell proliferation or differentiation, hypoxia, angiogenesis, and apoptosis indicated that OKN-007 was able to significantly decrease cell proliferation (glucose transporter 1 (Glut-1) and the cell proliferation marker, MIB-1) but not cell differentiation (carbonate anhydrase IX), decrease angiogenesis (microvessel density (MVD; measured as levels of the endothelial marker, CD-31), but not the vascular endothelial growth factor (VEGF)), decrease hypoxia (hypoxia inducible factor 1 α (HIF-1α)), and increase apoptosis (cleaved caspase 3) compared with untreated controls [112]. OKN-007-induced decreases in Glut-1 and HIF-1α levels seemed to be similar in both F98 and U87 glioma models, whereas increased

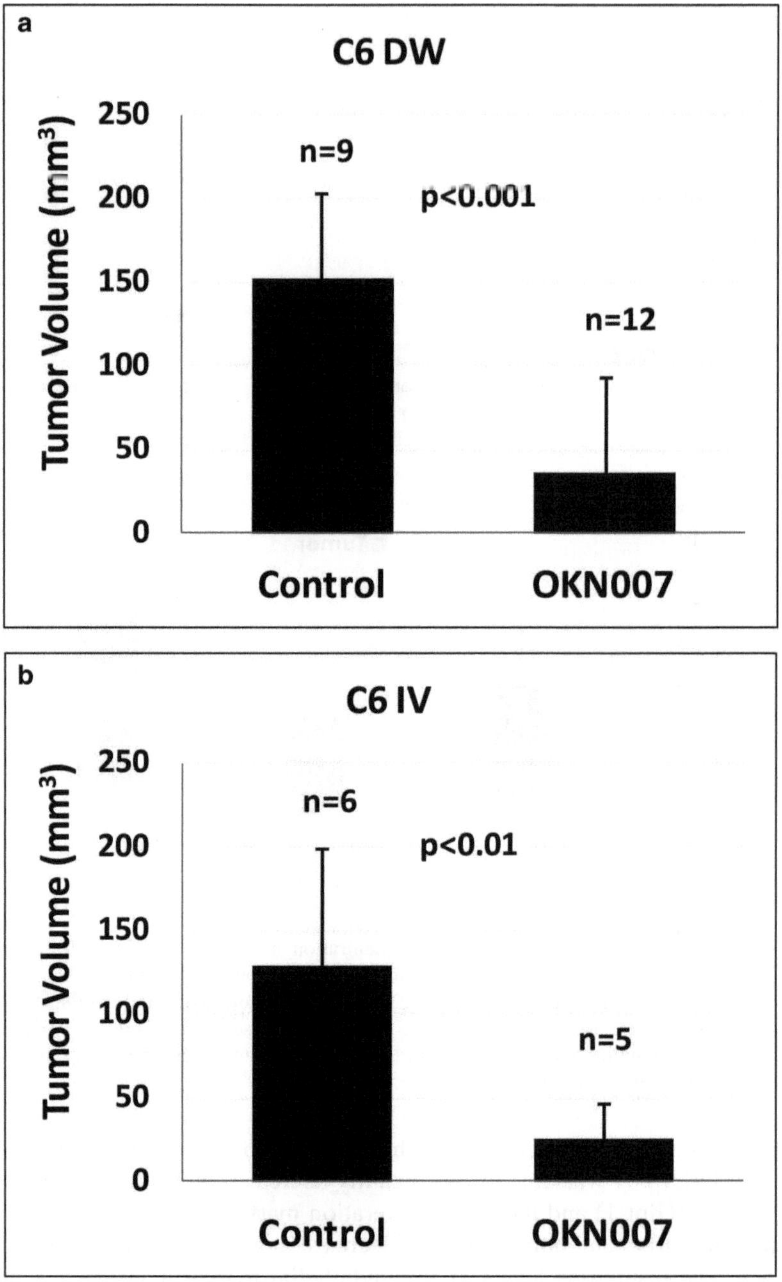

Fig. 10.3 Tumor volumes (mm^3; as measured by MRI) of C6 glioma-bearing rats either untreated (Control) or treated with OKN-007 (18 mg/kg daily) in drinking water (DW) or via IV (intravenous) administration. There was a significant decrease in OKN-treated gliomas for either DW- ($p<0.001$; $n=12$) or IV ($p<0.01$; $n=5$)-treated rats, when compared to untreated control rats ($n=9$ for the DW study, and $n=6$ for the IV study)

apoptosis seemed to be more elevated in the F98 gliomas compared to the U87 tumors [112]. We concluded that OKN-007 has the ability to cause glioma regression in aggressive rodent tumor models (F98 and U87), as well as in moderate gliomas (C6) [108, 111, 112]. The investigational drug, OKN-007, administered i.v., is currently in a Phase IIa (efficacy in a small cohort of patients) clinical trial in adult recurrent GBMs, with Phase Ib (pharmacokinetic study; three doses: 20, 40, and 60 mg/kg three times weekly in the first month, two times weekly in the second month, and once per week in subsequent months) recently completed. It is anticipated that the Phase IIa trial will be completed in the next 2 years. See Fig. 10.4 for a summary of the anticancer effects of PBN and PBN-analogues in glioma studies.

10.4.3 PBN as an Anticancer Agent Colorectal Polyps

Colorectal carcinoma (CRC) is one of the most prevalent cancers worldwide and has a very high mortality rate [117, 118]. Floyd and Towner have also reported on the anticancer activity of PBN in a transgenic mouse model, $Apc^{Min/+}$, for colorectal polyps [119]. After 3–4 months tumor size and numbers were examined at necropsy, and gadolinium-contrast-enhanced MRI was used to monitor colon tumors [119]. MRI data showed a time-dependent increase in total colonic signal intensity in sham-treated mice, which was noticeably decreased in the PBN-treated mice, which was confirmed by pathology at 10–17 weeks of age [119]. It is well known that iNOS expression is enhanced in colorectal carcinomas, including the $Apc^{Min/+}$ mouse model for colorectal polyps [120], and it was speculated that the antitumor effect of PBN was via inhibition of iNOS [119]. See Fig. 10.4 for a summary of the anticancer effects of PBN in CRC studies.

10.4.4 PBN and Other Nitrones as Anticancer Agents in Other Cancers

In addition to the anticancer effects of PBN in HCC, gliomas and colorectal polyps, there are other reports indicating that PBN may be effective in mammary tumors. A noticeable decline in tumor incidence ($p<0.0001$) in rat mammary tumors was observed in animals treated with PBN (160 mg/kg), compared to non-treated animals [121]. Another study by Ramadan et al. (2006) demonstrated that C-(2-chloroquinoline-3-yl)-*N*-phenyl nitrone (CQPN) strongly inhibited the growth of breast carcinoma MCF-7 cells via induction of apoptosis [122]. See Fig. 10.4 for a summary of the anticancer effects of PBN and PBN-analogues in breast cancer studies.

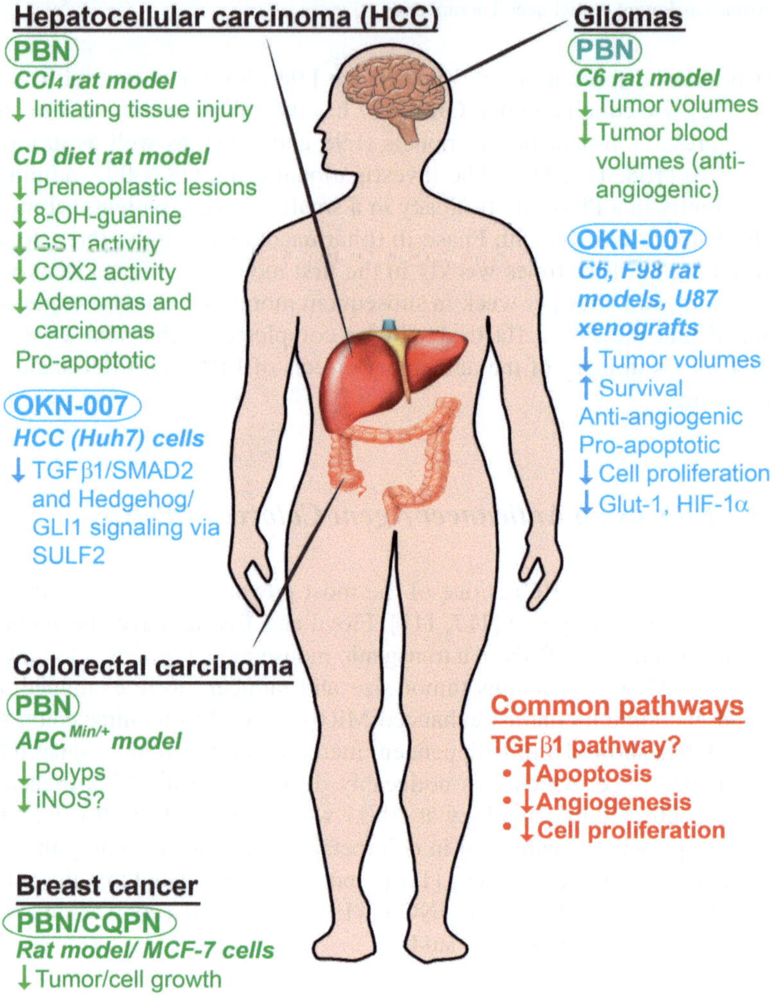

Fig. 10.4 Illustration and summary of the anticancer effect of PBN and PBN-analogues in cancers such as hepatocellular carcinomas (HCCs), gliomas, colorectal carcinomas (CRCs), and breast carcinomas. Based on the findings in specific tumor models, the most common anticancer effects of PBN/PBN-analogues included increasing apoptosis (HCC and glioma studies), decreasing angiogenesis (glioma studies), and decreasing cell proliferation (specifically in the glioma studies, and surmised in all other tumor studies due to decreased tumor/cell growth), which are all controlled by the TGFβ1 pathway. *Abbreviations*: *HCC* hepatocellular carcinoma, *PBN N-tert*-butyl α-phenyl nitrone, CCl_4 carbon tetrachloride, *CD* choline deficient, *8-OH-guanine* 8-hydroxy guanine, *GST* glutathione S-transferase, *COX2* cyclo-oxygenase 2, *OKN-007* OKlahoma Nitrone, *TGFβ1* transforming growth factor β1, *SMAD2* Sma- and Mad (Mother against DPP homolog)-related protein 2, *GLI1* effector of Hedgehog signaling, originally isolated in human glioblastoma, *SUF2* sulfatase 2, *iNOS* inducible nitric oxide synthase, *Glut-1* glucose transporter 1, *HIF-1α* hypoxia inducible factor 1α. For the HHC studies, the PBN: CCl_4 rat model results are summarized from references [85, 86], the PBN: CD diet rat model results are summarized from references [87–91], and the OKN-007: HCC cell results are summarized from reference [117]. For the glioma studies, the PBN: C6 rat model results are summarized from references [99–103], the OKN-007: C6 rat model results are summarized from references [101–104], and the OKN-007: F98/U87 rat/xenograft models results are summarized from reference [105]. For the CRC study, the PBN: $APC^{Min/+}$ mouse model results are summarized from reference [112]. For the breast cancer studies, the results are summarized from [114, 115]

10.5 Mechanisms of Action of PBN and PBN-Analogues

The parent nitrone compound, PBN, is known to inhibit the expression of cyclooxygenase-2 (COX-2), inducible nitric oxide synthase (iNOS), and nuclear factor kappaB (NF-κB) [100]. OKN-007 has been found to be neuroprotective in a rat transient middle cerebral artery (MCA) occlusion model for neurological ischemia/reperfusion injury [79]. The neuroprotective effect of OKN-007 has been suggested to be attributed in part to its ability to restore functionality of the brain endothelium [82]. The anticancer activity of PBN-nitrones has been attributed potentially to the suppression of nitric oxide production, suppression of iNOS expression, suppression of S-nitrosylation of critical proteins (caspases, Bcl-2, OGG1 DNA repair enzyme, and PTEN tumor suppressor protein, which are respectively involved in shutting down apoptosis, enhanced mutation events, and enhanced Akt-mediated signaling in oncogenesis), and inhibition of NF-κB activation [109]. The high trapping efficiency of the glutathiyl radical (GS•) by PBN that result in recycling of both nitrone spin trap and glutathione are also thought to contribute to the biological activities of nitrones [123]. OKN-007 was shown by the Towner group to decrease cell proliferation (MIB-1 and Glut-1), angiogenesis (MVD), and hypoxia (HIF-1α), as well as increase apoptosis (cleaved caspase 3) [112]. A common pathway that can effect cell proliferation, angiogenesis, and apoptosis during tumorigenesis is the transforming growth factor β1 (TGFβ1) pathway (also see Fig. 10.4), which plays an integral role in regulating the extracellular matrix [124]. It is also known that the NF-κB pathway, which has been previously shown to be inhibited by nitrones [109], is integrally linked to the TGFβ1 pathway [124]. Support for the anticancer effect of OKN-007 via the TGFβ1 pathway was recently reported by Zheng et al. (2013), where they found that OKN-007 mediates its antitumor effect in HCC cells (Huh7) via the suppression of TGFβ1/SMAD2 and Hedgehog/GLI1 signaling by inhibiting sulfatase 2 (SULF2) enzymatic activity [125]. Due to the inhibitory action of OKN-007 on SULF2, it is speculated this agent may decrease the migration of cancer cells in metastasis, which will need to be further investigated. In addition, the Towner research group is currently investigating the role of OKN-007 on the TGFβ1 pathway.

References

1. Janzen EG. Spin trapping. Methods Enzymol. 1984;105:188–98.
2. Janzen EG, Stronks HJ, Dubose CM, Poyer JL, McCay PB. Chemistry and biology of spin-trapping radicals associated with halocarbon metabolism in vitro and in vivo. Environ Health Perspect. 1985;64:151–70.
3. Kalyanaraman B, Mason RP, Perez-Reyes E, Chignell CF, Wolf CR, Philpot RM. Characterization of the free radical formed in aerobic microsomal incubations containing carbon tetrachloride and NADPH. Biochem Biophys Res Commun. 1979;89:1065–72.
4. Kalyanaraman B, Perez-Reyes E, Mason RP. Spin-trapping and direct electron spin resonance investigations of the redox metabolism of quinine anticancer drugs. Biochim Biophys Acta. 1980;630:119–30.

5. Albano E, Lott KA, Slater TF, Stier A, Symons MC, Tomasi A. Spin-trapping studies on the free-radical products formed by metabolic activation of carbon tetrachloride in rat liver microsomal fractions isolated hepatocytes and in vivo in the rat. Biochem J. 1982;204:593–603.
6. Lai EK, Crossley C, Sridhar R, Misra HP, Janzen EG, McCay PB. In vivo spin trapping of free radicals generated in brain, spleen, and liver during gamma radiation of mice. Arch Biochem Biophys. 1986;244:156–60.
7. Buettner GR. Spin trapping: ESR parameters of spin adducts. Free Radic Biol Med. 1987;3:259–303.
8. Janzen EG, Towner RA, Haire DL. Detection of free radicals generated from the in vitro metabolism of carbon tetrachloride using improved ESR spin trapping techniques. Free Radic Res Commun. 1987;3:357–64.
9. DuBose CM, Rehorek D, Oehler UM, Janzen EG. Spin trapping: ESR parameters of spin adducts. Free Radic Biol Med. 1988;5:55–6.
10. Janzen EG. Spin trapping and associated vocabulary. Free Radic Res Commun. 1990;9:163–7.
11. Janzen EG, Poyer JL, West MS, Crossley C, McCay PB. Study of reproducibility of spin trapping results in the use of C-phenyl-N-tert-butyl nitrone (PBN) for trichloromethyl radical detection in CCl4 metabolism by rat liver microsomal dispersions. Biological spin trapping I. J Biochem Biophys Methods. 1994;29:189–205.
12. Mason RP, Hanna PM, Burkitt MJ, Kadiiska MB. Detection of oxygen-derived radicals in biological systems using electron spin resonance. Environ Health Perspect. 1994;102 Suppl 10:33–6.
13. Janzen EG, Poyer JL, Schaefer CF, Downs PE, DuBose CM. Biological spin trapping. II. Toxicity of nitrone spin traps: dose-ranging in the rat. J Biochem Biophys Methods. 1995;30:239–47.
14. Janzen EG, West MS, Kotake Y, DuBose CM. Biological spin trapping methodology. III. Octanol-water partition coefficients of spin-trapping compounds. J Biochem Biophys Methods. 1996;32:183–90.
15. Towner RA. Chemistry of spin trapping. In: Rhodes C, editor. Toxicology of the human environment: the critical role of free radicals. London: Taylor and Francis; 2000. p. 7–24.
16. Rosen GM, Cohen MS, Britigan BE, Pou S. Application of spin traps to biological systems. Free Radic Res Commun. 1990;9:187–95.
17. Rosen GM, Pou S, Britigan BE, Cohen MS. Spin trapping of hydroxyl radicals in biological systems. Methods Enzymol. 1994;233:105–11.
18. Floyd RA, Soong LM. Spin trapping in biological systems. Oxidation of the spin trap 5,5-dimethyl-1-pyrroline-1-oxide by a hydroperoxide-hematin-system. Biochem Biophys Res Commun. 1977;74:79–84.
19. Buettner GR, Oberley LW. Considerations in the spin trapping of superoxide and hydroxyl radical in aqueous systems using 5,5-dimethyl-1-pyrroline-1-oxide. Biochem Biophys Res Commun. 1978;83:69–74.
20. Finkelstein E, Rosen GM, Rauckman EJ, Paxton J. Spin trapping of superoxide. Mol Pharmacol. 1979;16:676–85.
21. Floyd RA. Hydroxyl free-radical spin-adduct in rat brain synaptosomes. Observations on the reduction of the nitroxide. Biochim Biophys Acta. 1983;756:204–16.
22. Janzen EG, Haire DL, Coulter GA, Stronks HJ, Krygsman PH, Towner RA, Hilborn JW. Locating spin traps in heterogeneous media by carbon-13 NMR spectroscopy. Investigations in SDS micelles, DMPC vesicles, and rat liver microsomes. J Organic Chem. 1989;54:2915–20.
23. Swartz HS. Principles of the metabolism of nitroxides and their implications for spin trapping. Free Radic Res Commun. 1990;9:399–405.
24. Janzen EG, Zhdanov RI, Reinke LA. Metabolism of phenyl and alkyl spin adducts of PBN in rat hepatocytes. Rate dependence on size and type of addend group. Free Radic Res Commun. 1993;19:S157–62.

25. Hawkins CL, Davies MJ. Detection and characterization of radicals in biological materials using EPR methodology. Biochim Biophys Acta. 1840;2014:708–21.
26. Novakov CP, Feierman D, Cederbaum AI, Stoyanovsky DA. An ESR and HPLC-EC assay for the detection of alkyl radicals. Chem Res Toxicol. 2001;14:1239–46.
27. Novakov CP, Stoyanovsky DA. Comparative metabolism of N-tert-Butyl-N-[1-(1-oxy-pyridin-4-yl)-ethyl]- and N-tert-Butyl-N-(1-phenyl-ethyl)-nitroxide by the cytochrome P450 monooxygenase system. Chem Res Toxicol. 2002;15:749–53.
28. Towner RA, Smith N, Saunders D, Lupu F, Silasi-Mansat R, West M, Ramirez DC, Gomez-Mejiba SE, Bonini MG, Mason RP, Ehrenshaft M, Hensley K. *In vivo* detection of free radicals using molecular MRI and immuno-spin-trapping in a mouse model for amyotrophic lateral sclerosis (ALS). Free Radic Biol Med. 2013;63:351–60.
29. Nash KM, Rockenbauer A, Villamena FA. Reactive nitrogen species reactivities with nitrones: theoretical and experimental studies. Chem Res Toxicol. 2012;25:1581–97.
30. Mason RP. Using anti-5,5-dimethyl-1-pyrroline N-oxide (anti-DMPO) to detect protein radicals in time and space with immune-spin trapping. Free Radic Biol Med. 2004;36:1214–23.
31. Ramirez DC, Gomez-Mejiba SE, Mason RP. Mechanism of hydrogen peroxide-induced Cu, Zn-superoxide dismutase-centered radical formation as explored by immune-spin trapping: the role of copper- and carbonate radical anion-mediated oxidations. Free Radic Biol Med. 2005;38:201–14.
32. Ramirez DC, Gomez-Mejiba SE, Corbett JT, Deterding LJ, Tomer KB, Mason RP. Cu, Zn-superoxide dismutase-driven free radical modifications: copper- and carbonate radical anion-initiated protein radical chemistry. Biochem J. 2009;417:341–53.
33. Detweiler CD, Deterding LJ, Tomer KB, Chignell CF, Germolec D, Mason RP. Immunological identification of the heart myoglobin radical formed by hydrogen peroxide. Free Radic Biol Med. 2002;33:364–9.
34. Towner RA, Smith N, Saunders D, Henderson M, Downum K, Lupu F, Silasi-Mansat R, Ramirez DC, Gomez-Mejiba SE, Bonini MG, Ehrenshaft M, Mason RP. *In vivo* imaging of immune-spin trapped radicals with molecular MRI in a mouse diabetes model. Diabetes. 2012;61:2401–13.
35. Towner RA, Garteiser P, Bozza F, Smith N, Saunders D, d'Avila JCP, Magno F, Oliveira MF, Ehrenshaft M, Lupu F, Silasi-Mansat R, Ramirez DC, Gomez-Mejiba SE, Mason RP, Faria-Neto HCC. In vivo detection of free radicals in mouse septic encephalopathy using molecular MRI and immuno-spin-trapping. Free Radica Biol Med. 2013;65:828–37.
36. Towner RA, Smith N, Saunders D, De Souza PC, Henry L, Lupu F, Silasi-Mansat R, Ehrenshaft M, Mason RP, Gomez-Mejiba SE, Ramirez DC. Combined molecular MRI and immuno-spin-trapping for in vivo detection of free radicals in orthotopic mouse GL261 gliomas. Biochim Biophys Acta. 1832;2013:2153–61.
37. Federico A, Morgillo F, Tuccillo C, Ciardiello F, Loguercio C. Chronic inflammation and oxidative stress in human carcinogenesis. Int J Cancer. 2007;121:2381–6.
38. Floyd RA, Kotake Y, Towner RA, Guo WX, Nakae D, Konishi Y. Nitric oxide and cancer development. J Toxicol Pathol. 2007;20:77–92.
39. Kandavelu S, Vanisree AJ. A study on the biochemical and cytogenetic status in the blood of glioma patients. Pak J Biol Sci. 2011;14:511–8.
40. Macchioni L, Davidescu M, Sciaccaluga M, Marchetti G, Coaccioli S, Roberti R, Corazzi L, Castigli E. Mitochondrial dysfunction and effect of antiglycolytic bromopyruvic acid in GL15 glioblastoma cells. Lab Invest. 2011;91:1766–76.
41. Li F, Wang H, Huang C, Lin J, Zhu G, Hu R, Feng H. Hydrogen peroxide contributes to the manganese superoxide dismutase promotion of migration and invasion in glioma cells. Free Radic Res. 2011;45:1154–61.
42. Rice-Evans C, Burdon R. Free radical-lipid interactions and their pathological consequences. Prog Lipid Res. 1993;32:71–110.
43. Muriel P. Role of free radicals in liver diseases. Hepatol Int. 2009;3:526–36.
44. Tien Kou T, Savaraj N. Roles of reactive oxygen species in hepatocarcinogenesis and drug resistance gene expression in liver cancers. Mol Carcinog. 2006;45:701–9.

45. Sasaki Y. Does oxidative stress participate in the development of hepatocellular carcinoma? J Gastroenterol. 2006;41:1135–48.
46. Ghoshal AK, Farber E. Choline deficiency, lipotrope deficiency and the development of liver disease including liver cancer: a new perspective. Lab Invest. 1993;68:255–60.
47. Lalwani ND, Reddy MK, Qureshi SA, Reddy JK. Development of hepatocellular carcinomas and increased peroxisomal fatty acid beta-oxidation in rats fed [4-chloro-6-(2,3-xylidino)-2-pyrimidinylthio] acetic acid (Wy-14,643) in the semipurified diet. Carcinogenesis. 1981;2:645–50.
48. Iwagaki H, Hamazaki K, Matsubara N, Hiramatsu M, Orita K, Mori A. Lipid peroxidation in hepatocellular carcinoma. Acta Med Okayama. 1995;49:313–5.
49. Hensley K, Kotake Y, Sang H, Pye QN, Wallis GL, Kolker LM, Tabatabaie T, Stewart CA, Konishi Y, Nakae D, Floyd RA. Dietary choline restriction causes complex I dysfunction and increased H_2O_2 generation in liver mitochondria. Carcinogenesis. 2000;21:983–9.
50. Hinrichsen LI, Floyd RA, Sudilovsky O. Is 8-hydroxydeoxyguanosine a mediator of carcinogenesis by a choline-devoid diet in the rat liver? Carcinogenesis. 1990;11:1879–81.
51. Nakae D, Denda A, Kobayashi Y, Akai H, Kishida H, Tsujiuchi T, Konishi Y, Suzuki T, Muramatsu M. Inhibition of early-phase exogenous and endogenous liver carcinogenesis in transgenic rats harboring a rat glutathione S-transferase placental form gene. Jpn J Cancer Res. 1998;89:1118–25.
52. Hagen TM, Huang S, Curnutte J, Fowler P, Martinez V, Wehr CM, Ames BN, Chisari FV. Extensive oxidative DNA damage in hepatocytes of transgenic mice with chronic active hepatitis destined to develop hepatocellular carcinoma. Proc Natl Acad Sci USA. 1994;91:12808–12.
53. Ichiba M, Maeta Y, Mukoyama T, Saeki T, Yasui S, Kanbe T, Okano J, Tanabe Y, Hirooka Y, Yamada S, Kurimasa A, Murawaki Y, Shiota G. Expression of 8-hydroxy-2'-deoxyguanosine in chronic liver disease and hepatocellular carcinoma. Liver Int. 2003;23:338–45.
54. Tanake H, Fujita N, Sugimoto R, Urawa N, Horiike S, Kobayashi Y, Iwasa M, Ma N, Kawanishi S, Watanabe S, Kaito M, Takei Y. Hepatic oxidative DNA damage is associated with increased risk for hepatocellular carcinoma in chronic hepatitis C. Br J Cancer. 2008;98:580–6.
55. Rasheed Z, Ahmad R, Rasheed N, Ali R. Reactive oxygen species damaged human serum albumin in patients with hepatocellular carcinoma. J Exp Clin Cancer Res. 2007;26:395–404.
56. Choi J, Corder NL, Koduru B, Wang Y. Oxidative stress and hepatic Nox proteins in chronic hepatitis C and hepatocellular carcinoma. Free Radic Biol Med. 2014;72:267–84.
57. Guo W, Zhao Y, Zhang Z, Tan N, Zhao F, Ge C, Liang L, Jia D, Chen T, Yao M, Li J, He X. Disruption of xCT inhibits cell growth via the ROS/autophagy pathway in hepatocellular carcinoma. Cancer Lett. 2011;312:55–61.
58. Towner RA, Mason RP, Reinke LA. In vivo detection of aflatoxin-induced lipid free radicals in rat bile. Biochim Biophys Acta. 2002;1573:55–62.
59. Towner RA, Qian SY, Kadiiska MB, Mason RP. In vivo identification of aflatoxin-induced free radicals in rat bile. Free Radical Biol Med. 2003;35:1330–40.
60. Towner RA, Sturgeon SA, Hore KE. Assessment of in vivo oxidative lipid metabolism following acute microcystin-LR-induced hepatotoxicity in rats. Free Radic Res. 2002;36:63–71.
61. Nedic O, Robajac D, Sunderic M, Miljus G, Dukanovic B, Malenkovic V. Detection and identification of oxidized insulin-like growth factor-binding proteins and receptors in patients with colorectal carcinoma. Free Radic Biol Med. 2013;65:1195–200.
62. Kang KA, Kim KC, Bae SC, Hyun JW. Oxidative stress induces proliferation of colorectal cancer cells by inhibiting RUNX3 and activating the Akt signaling pathway. Int J Oncol. 2013;43:1511–6.
63. Kotake Y, Sang H, Miyajima T, Wallis GL. Inhibition of NF-kappaB, iNOS mRNA, COX2 mRNA, and COX catalytic activity by phenyl-N-tert-butylnitrone (PBN). Biochim Biophys Acta. 1998;1448:77–84.

64. Miyajima T, Kotake Y. Spin trapping agent, phenyl N-tert-butyl nitrone, inhibits induction of nitric oxide synthase in endotoxin-induced shock in mice. Biochem Biophys Res Commun. 1995;215:114–21.
65. Stewart CA, Hyam K, Wallis G, Sang H, Robinsin KA, Floyd RA, Kotake Y, Hensley K. Phenyl-N-tert-butylnitrone demonstrates broad-spectrum inhibition of apoptosis-associated gene expression in endotoxin-treated rats. Arch Biochem Biophys. 1999;365:71–4.
66. Sang H, Wallis GL, Stewart CA, Kotake Y. Expression of cytokines and activation of transcription factors in lipopolysaccharide-administered rats and their inhibition by phenyl N-tert-butylnitrone (PBN). Arch Biochem Biophys. 1999;363:341–8.
67. Endoh H, Kato N, Fujii S, Suzuki Y, Sato S, Kayama T, Kotake Y, Yoshimura T. Spin trapping agent, phenyl N-tert-butylnitrone, reduces nitric oxide production in the rat brain during experimental meningitis. Free Radic Res. 2001;35:583–91.
68. Lin S, Cox HJ, Rhodes PG, Cai Z. Neuroprotection of α-phenyl-n-tert-butyl-nitrone on the neonatal white matter is associated with anti-inflammation. Neurosci Lett. 2006;405:52–6.
69. Tabatabaie T, Graham KL, Vasquez AM, Floyd RA, Kotake Y. Inhibition of the cytokine-mediated inducible nitric oxide synthase expression in rat insulinoma cells by phenyl N-tert-butylnitrone. Nitric Oxide. 2000;4:157–67.
70. Villamena FA, Das A, Nash KM. Potential implication of the chemical properties and bioactivity of nitrone spin traps for therapeutics. Future Med Chem. 2012;4:1171–207.
71. Das A, Gopalakrishnan B, Voss OH, Doseff AI, Villamena FA. Inhibition of ROS-induced apoptosis in endothelial cells by nitrone spin traps via induction of phase II enzymes and suppression of mitochondrial-dependent pro-apoptotic signaling. Biochem Pharmacol. 2012;84:486–97.
72. Marklund N, Lewander T, Clausen F, Hillered L. Effects of the nitrone radical scavengers PBN and S-PBN on in vivo trapping of reactive oxygen species after traumatic brain injury in rats. J Cereb Blood Flow Metab. 2001;21:1259–67.
73. Floyd RA. Role of oxygen free radicals in carcinogenesis and brain ischemia. FASEB J. 1990;4:2587–97.
74. Carney JM, Starke-Reed PE, Oliver CN, Landrum RW, Chen MS, Wu JF, Floyd RA. Reversal of age-related increase in brain protein oxidation, decrease in enzyme activity, and loss in temporal and spacial memory by chronic administration of the spin-trapping compound N-*tert*-butyl-a-phenylnitrone. Proc Natl Acad Sci USA. 1991;88:3633–6.
75. Carney JM, Floyd RA. Phenyl butyl nitrone compositions and methods for treatment of oxidative tissue damage. [U.S. Patent # 5,025,032], 1–14. 6-18-1991.
76. Floyd RA. Serendipitous findings while researching oxygen free radicals. Free Radic Biol Med. 2009;46:1004–13.
77. Phillis JW, Clough-Helfman C. Protection from cerebral ischemic injury in gerbils with the spin trap agent N-tert-butyl-a-phenylnitrone (PBN). Neurosci Lett. 1990;116:315–9.
78. Cao X, Phillis JW. Alpha-phenyl-tert-butyl-nitrone reduces cortical infarct and edema in rats subjected to focal ischemia. Brain Res. 1994;644:267–72.
79. Maples KR, Green AR, Floyd RA. Nitrone-related therapeutics: potential of NXY-059 for the treatment of acute ischaemic stroke. CNS Drugs. 2004;18:1071–84.
80. Sydserff SG, Borelli AR, Green AR, Cross AJ. Effect of NXY-059 on infarct volume after transient or permanent middle cerebral artery occlusion in the rat: studies on dose, plasma concentration and therapeutic time window. Br J Pharmacol. 2002;135:103–12.
81. Kuroda S, Tsuchidate R, Smith M-L. Neuroprotective effects of a novel nitrone, NXY-059, after transient focal cerebral ischemia in the rat. J Cereb Blood Flow Metab. 1999;19:778–87.
82. Culot M, Mysiorek C, Renftel M, Roussel BD, Hommet Y, Vivien D, Cecchelli R, Fenart L, Berezowski V, Dehouck MP, Lundquist S. Cerebrovascular protection as a possible mechanism for the protective effects of NXY-059 in preclinical models: an in vitro study. Brain Res. 2009;1294:144–52.

83. Cheng Y-F, Jiang J, Hu P, Reinholdsson I, Guo W, Asenblad N, Nilsson D. Pharmacokinetics of 8-hour intravenous infusion of NXY-059: a phase I, randomized, double-blind (within dose panels), placebo-controlled study in healthy Chinese volunteers. Clin Therapeut. 2008;30:2342–53.
84. Lyden PD, Shuaib A, Lees KR, Davalos A, Davis SM, Diener H-C, Grotta JC, Ashwood TJ, Hardemark H-G, Svensson HH, Rodichok L, Wasiewski WW, Ahlberg G. Safety and tolerability of NXY-059 for acute intracerebral hemorrhage: the CHANT trial. Stroke. 2007;38:2262–9.
85. Wemer J, Cheng YF, Nilsson D, et al. Safety, tolerability and pharmacokinetics of escalating doses of NXY-059 in healthy young and elderly subjects. Curr Med Res Opin. 2006;22(9):1813–23.
86. Fong JJ, Rhoney DH. NXY-059: review of neuroprotective potential for acute stroke. Ann Pharmacother. 2006;40:461–71.
87. Shuaib A, Lees KR, Lyden P, Grotta J, Davalos A, Davis SM, Diener H-C, Ashwood T, Wasiewski WW, Emeribe U. NXY-059 for the treatment of acute ischemic stroke. N Engl J Med. 2007;357:562–71.
88. Wang CX, Shuaib A. Neuroprotective effects of free radical scavengers in stroke. Drugs Aging. 2007;24:537–46.
89. Ginsberg MD. Life after Cerovive: a personal perspective on ischemic neuroprotection in the post-NXY-059 era. Stroke. 2007;38:1967–72.
90. Mikhail S, Cosgrove D, Zeidan A. Hepatocellular carcinoma: systemic therapies and future perspectives. Expert Rev Anticancer Ther. 2014;14:1205–18.
91. Bosetti C, Turati F, La Vecchia C. Hepatocellular carcinoma epidemiology. Best Pract Res Clin Gastroenterol. 2014;28:753–70.
92. Janzen EG, Towner RA, Yamashiro S. The effect of phenyl tert-butyl nitrone (PBN) on CCl4-induced rat liver injury detected by proton magnetic resonance imaging (MRI) in vivo and electron microscopy (EM). Free Radic Res Commun. 1990;9:325–35.
93. Towner RA, Reinke LA, Janzen EG, Yamashiro S. Enhancement of carbon tetrachloride-induced liver injury by a single dose of ethanol: proton magnetic resonance imaging (MRI) studies in vivo. Biochim Biophys Acta. 1991;1096:222–30.
94. Nakae D, Kotake Y, Kishida H, Hensley KL, Denda A, Kobayashi Y, Kitayama W, Tsujiuchi T, Sang H, Stewart CA, Tabatabaie T, Floyd RA, Konishi Y. Inhibition by phenyl N-tert-butyl nitrone of early phase carcinogenesis in the livers of rats fed a choline-deficient, L-amino acid-defined diet. Cancer Res. 1998;58:4548–51.
95. Floyd RA, Kotake Y, Hensley K, Nakae D, Konishi Y. Reactive oxygen species in choline deficiency induced carcinogenesis and nitrone inhibition. Mol Cell Biochem. 2002;234–235:195–203.
96. Floyd RA. Nitrones as therapeutics in age-related diseases. Aging Cell. 2006;5:51–7.
97. Floyd RA, Towner RA, He T, Hensley K, Maples KR. Translational research involving oxidative stress and diseases of aging. Free Radic Biol Med. 2011;51:931–41.
98. Nakae D, Uematsu F, Kishida H, Kusuoka O, Katsuda S, Yoshida M, Takahashi M, Maekawa A, Denda A, Konishi Y, Kotake Y, Floyd RA. Inhibition of the development of hepatocellular carcinomas by phenyl N-tert-butyl nitrone in rats fed with a choline-deficient, L-amino acid-defined diet. Cancer Lett. 2004;206:1–13.
99. Inoue Y, Asanuma T, Smith N, Saunders D, Oblander J, Kotake Y, Floyd RA, Towner RA. Modulation of Fas-FasL related apoptosis by PBN in the early phases of choline deficient diet-mediated hepatocarcinogenesis in rats. Free Radic Res. 2007;41:972–80.
100. Ostrum QT, Gittleman H, Farah P, Ondracek A, Chen Y, Wolinsky Y, Stroup NE, Kruchko C, Barnholtz-Sloan JS. CBTRUS statistical report: primary brain and central nervous system tumors diagnosed in the United States in 2006–2010. Neuro-Oncol. 2013;15 Suppl 6:ii1–56.
101. Ostrum QT, Bauchet L, Davis FG, Deltour I, Fisher JL, Eastman Langer C, Pekmezci M, Schwartzbaum JA, Turner MC, Walsh KM, Wrensch MR, Barnholtz-Sloan JS. The epidemiology of glioma in adults: a "state of the science" review. Neuro-Oncol. 2014;16:896–913.

102. Central Brain Tumor Registry of the United States (CBTRUS). 2011 CBTRUS statistical report: primary brain and central nervous system tumors diagnosed in the United States in 2004–2007. 2011.
103. Sanson M, Thillet J, Hoang-Xuan K. Molecular changes in gliomas. Curr Opin Oncol. 2004;16:607–13.
104. Demuth T, Berens ME. Molecular mechanisms of glioma cell migration and invasion. J Neuro-Oncol. 2004;70:217–28.
105. Rao RD, James CD. Altered molecular pathways in gliomas: an overview of clinically relevant issues. Semin Oncol. 2004;31:595–604.
106. Doblas S, Saunders D, Kshirsagar P, Pye Q, Oblander J, Gordon B, Kosanke S, Floyd RA, Towner RA. Phenyl-tert-butylnitrone induces tumor regression and decreases angiogenesis in a C6 rat glioma model. Free Radic Biol Med. 2008;44:63–72.
107. Floyd RA, Kopke RD, Choi CH, Foster SB, Doblas S, Towner RA. Nitrones as therapeutics. Free Radic Biol Med. 2008;45:1361–74.
108. He T, Doblas S, Saunders D, Casteel R, Lerner M, Ritchey JW, Snider T, Floyd RA, Towner RA. Effects of PBN and OKN007 in rodent glioma models assessed by 1H MR spectroscopy. Free Radic Biol Med. 2011;51:490–502.
109. Floyd RA, Chandru HK, He T, Towner R. Anti-cancer activity of nitrones and observations on mechanism of action. Anticancer Agents Med Chem. 2011;11:373–9.
110. Floyd RA, Castro Faria Neto HC, Zimmerman GA, Hensley K, Towner RA. Nitrone-based therapeutics for neurodegenerative diseases: their use alone or in combination with lanthionines. Free Radic Biol Med. 2013;62:145–56.
111. Garteiser P, Doblas S, Watanabe Y, Saunders D, Hoyle J, Lerner M, He T, Floyd RA, Towner RA. Multiparametric assessment of the anti-glioma properties of OKN007 by magnetic resonance imaging. J Magn Reson Imaging. 2010;31:796–806.
112. Towner RA, Gillespie DL, Schwager A, Saunders DG, Smith N, Njoku CE, Krysiak 3rd RS, Larabee C, Iqbal H, Floyd RA, Bourne DW, Abdullah O, Hsu EW, Jensen RL. Regression of glioma tumor growth in F98 and U87 rat glioma models ny the nitrone OKN-007. Neuro Oncol. 2013;15:330–40.
113. Barth RF, Kaur B. Rat brain tumor models in experimental neuro-oncology: the C6, 9L, T9, RG2, F98, BT4C, RT-2 and CNS-1 gliomas. J Neurooncol. 2009;94(3):299–312.
114. Asanuma T, Doblas S, Tesiram YA, et al. Visualization of the protective ability of a free radical trapping compound against rat C6 and F98 gliomas with diffusion tensor fiber tractography. J Magn Reson Imaging. 2008;28(3):574–87.
115. Jacobs VL, Valdes PA, Hickey WF, De Leo JA. Current review of in vivo GBM rodent models: emphasis on the CNS-1 tumour model. ASN Neuro. 2011;3(3), e00063.
116. Barth RF. Rat brain tumor models in experimental neuro-oncology: the 9L, C6, T9, F98, RG2 (D74), RT-2 and CNS-1 gliomas. J Neurooncol. 1998;36(1):91–102.
117. Parkin DM, Bray F, Ferlay J, Pisani P. Global cancer statistics, 2002. CA Cancer J Clin. 2005;55:74–108.
118. Zhao YP, Ruan CP, Wang H, Hu ZQ, Fang M, Gu X, Ji J, Zhao JY, Gao CF. Identification and assessment of new biomarkers for colorectal cancer with serum N-glycan profiling. Cancer. 2012;118:639–50.
119. Floyd RA, Towner RA, Wu D, Abbott A, Cranford R, Branch D, Guo WX, Foster SB, Jones I, Alam R, Moore D, Allen T, Huycke M. Anti-cancer activity of nitrones in the Apc(Min/+) model of colorectal cancer. Free Radic Res. 2010;44:108–17.
120. Yagihashi N, Kasajima H, Sugai S, Matsumoto K, Ebina Y, Morita T, Murakami T, Yagihashi S. Increased in situ expression of nitric oxide synthase in human colorectal cancer. Virchows Arch. 2000;436:109–14.
121. Inano H, Onoda M. Nitric oxide produced by inducible nitric oxide synthase is associated with mammary tumorigenesis in irradiated rats. Nitric Oxide. 2005;12:15–20.
122. Ramadan M, Gamal-Eldeen AM, Abdel-Aziz M, Abuo-Rahma G-D, Abdel-Nabi H, Nagib AH. C-(2-chloroquinoline-3-yl)-N-phenyl nitrone: new synthetic antioxidant inhibits prolif-

eration and induces apoptosis of breast carcinoma MCF-7 cells. Arch Pharm (Weinheim). 2006;339:242–9.
123. Polovyanenko DN, Plyusnin VF, Reznikov VA, Khramtsov VV, Bagryanskaya EG. Mechanistic studies of the reactions of nitrone spin trap PBN with glutathiyl radical. J Phys Chem B. 2008;112:4841–7.
124. Tobar N, Villar V, Santibanez JF. ROS-NFκB mediates TGF-β1-induced expression of urokinase-type plasminogen activator, matrix metalloproteinase-9 and cell invasion. Mol Cell Biochem. 2010;340:195–202.
125. Zheng X, Gai X, Hans S, Moser CD, Hu C, Shire AM, Floyd RA, Roberts LR. The human sulfatase 2 inhibitor 2,4-disulfonylphenyl-tert-butylnitrone (OKN-007) has an antitumor effect in hepatocellular carcinoma mediated via suppression of TGFB1/SMAD2 and Hedgehog/GLI1 signaling. Genes Chromosomes Cancer. 2013;52:225–36.

Chapter 11
Salen Manganese Complexes Mitigate Radiation Injury in Normal Tissues Through Modulation of Tissue Environment, Including Through Redox Mechanisms

Susan R. Doctrow, Brian Fish, Karl D. Huffman, Zelmira Lazarova, Meetha Medhora, Jacqueline P. Williams, and John E. Moulder

11.1 Rationale for Testing Salen Mn Complexes as Radiation Mitigators

There is a need for therapeutic agents to reduce injury to normal tissues that can occur as a delayed consequence of not only radiation therapy, but also an accidental or terrorism-associated radiation exposure. While the mechanisms of delayed radiation injury are not well understood, there is evidence suggesting that oxidative stress is involved [1–3]. Therefore, we and others have studied synthetic agents that scavenge reactive oxygen species (ROS) such as superoxide and hydrogen peroxide as potential therapeutic agents to mitigate [4] normal tissue injury resulting from ionizing radiation. A wide range of antioxidant strategies has been studied for mitigation of radiation-induced normal tissue injury, including: nutraceuticals [5], free-radical scavengers [6], superoxide dismutase (SOD) mimetics [6], flavones [6–8], mitochondrial-targeted nitroxides [9], iron chelators [7], NADPH oxidase inhibitors [7], and Mn porphyrins [10–12]. This review focuses primarily on radiation mitigation studies employing one class of synthetic ROS scavenger, synthetic metal-containing compounds known as salen Mn complexes [13].

S.R. Doctrow (✉) • K.D. Huffman
Department of Medicine, Pulmonary Center, Boston University School of Medicine, Boston, MA, USA
e-mail: sdoctrow@bu.edu

B. Fish • M. Medhora • J.E. Moulder
Department of Radiation Oncology, Medical College of Wisconsin, Milwaukee, WI, USA

Z. Lazarova
Department of Dermatology, Medical College of Wisconsin, Milwaukee, WI, USA

J.P. Williams
Department of Environmental Medicine, University of Rochester Medical Center, Rochester, NY, USA

I. Batinić-Haberle et al. (eds.), *Redox-Active Therapeutics*,
Oxidative Stress in Applied Basic Research and Clinical Practice,
DOI 10.1007/978-3-319-30705-3_11

It is well known that exposure of biological materials to ionizing radiation produces a short-lived burst of ROS, damaging DNA and other cellular components [14]. Also important, however, is the possible contribution of chronically generated ROS to delayed radiation-induced damage [1–3]. Such outcomes, including fibrosis, necrosis, atrophy, and vascular damage, can occur months to years after exposure to radiation affecting such tissues as skin, kidney, lung, oral mucosa, and brain. Indicators of cumulative ROS and reactive nitrogen species (RNS)-mediated damage, for example, lipid peroxidation and protein and DNA oxidation, have been found following irradiation in vivo and in vitro [3]. There is also evidence for pro-inflammatory processes, namely acute activation of stress-sensitive transcriptional events and cytokine production [3, 15]. Infiltrating inflammatory cells are one likely source of damaging ROS, while another is damaged mitochondria. Evidence indicates that radiation leads to mitochondrial damage and, intriguingly, delivery of antioxidant enzymes to the mitochondria through gene therapy techniques can protect against delayed radiation damage in vivo [1] Furthermore, a mitochondrially targeted antioxidant reduces radiation injury in cell culture [9]. Therefore, both proinflammatory processes and mitochondrial dysfunction have been implicated in delayed radiation injury. These findings suggest that agents that interrupt these damaging subcellular processes might address the long-term effects of radiation exposure.

Such considerations led to our investigation of salen Mn complexes as potential mitigators of radiation-induced late effects. Earlier studies showed that treatment with these compounds could suppress proinflammatory processes, inhibit mitochondrial damage, and reduce oxidative changes in various experimental models of injury; and our 2011 review [16] summarized the ability of salen Mn complexes to mitigate radiation injury to lung, kidney, and oral mucosa, and mitochondrial injury in cell culture models. The current review builds on those previous findings, providing more recent data that continue to support the potential for mitigation of normal tissue injury by salen Mn complexes.

The only FDA-approved drug for preventing radiation injury is amifostine (Ethylol®), which is usually administered intravenously or subcutaneously and, according to current clinical guidelines, prior to radiation exposure [17]. It acts as a free radical scavenger, but may have additional mechanisms of action such as modulation of antioxidant enzymes [17, 18]. Only one agent, granulocyte colony stimulating factor (G-CSF) for hematological injury, has been FDA approved to diminish radiation injury when administered after radiation. In the radiation countermeasures area, mitigators refer to therapies begun after irradiation, but prior to clinical evidence of injury. This is to distinguish them from protectors, which are given prior to irradiation, and from treatment agents, given after clinical evidence of injury [4]. Salen Mn complexes, while not yet approved for human use, are interesting candidates for development with potential advantages over amifostine. One major advantage is their ability to mitigate delayed radiation injury when given *days to weeks after irradiation*.

Salen Mn complexes are synthetic low molecular weight agents that mimic the antioxidant enzymes superoxide dismutase (SOD), catalase, and peroxidases, scavenging superoxide and hydrogen peroxide, respectively [13]. Their hydrogen peroxide-scavenging properties, namely catalase and peroxidase activities, are more significant parameters than SOD activity in predicting certain cytoprotective effects of salen Mn complexes, though other factors such as pharmacokinetics and cytotoxicity are also important [13]. The in vitro ROS-scavenging activities of various salen Mn complexes, as well as of some other Mn compounds and native SOD and catalases, have been measured with a variety of in vitro and cell-based assay systems. Tabular and other data summarizing these findings for various analogs have been presented elsewhere [13, 16, 19]. It is important to note that synthetic metal-based complexes have specific activities that are orders of magnitude lower than those of the native enzymes [13]. However, when administered at sufficient quantities, they can be demonstrated to have comparable ROS-scavenging effects, for example, when compared to bovine liver catalase in protecting cells against exogenous enzymatically generated hydrogen peroxide [13]. Furthermore, evidence that salen Mn complexes act as peroxidases demonstrates that they scavenge hydrogen peroxide through mechanisms analogous to those of the heme- and Mn-based catalase enzymes [13, 20]. Salen Mn complexes can also scavenge RNS through mechanisms analogous to their catalase and peroxidase activities [21], and attenuate protein nitration in oxidative injury models [22, 23]. Overall, this combination of properties—low molecular weight, catalytic scavenging mechanism, and activity against multiple damaging species—gives salen Mn complexes potential advantages over noncatalytic ROS scavengers and antioxidant enzymes [24, 25].

Prototype salen Mn complexes EUK-8 and the improved catalase/peroxidases EUK-134 and EUK-189 (Fig. 11.1) are effective in a wide range of models for diseases involving oxidative stress [14, 26]. In many of these models, salen Mn complexes were not only functionally protective, but also suppressed biochemical indicators of oxidative stress, such as oxidative modification of protein, lipids, and nucleic acids [22, 27, 28]. EUK-207 (Fig. 11.1) is a "second generation" cyclized salen Mn complex with catalytic properties equivalent to EUK-134 and EUK-189, but with greater stability [19, 26, 27, 29], attaining much higher plasma levels than

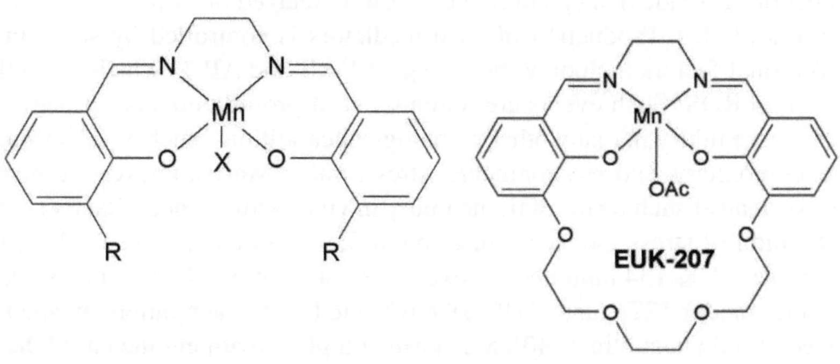

Fig. 11.1 Structures of salen Mn complexes: EUK-8, EUK-134, EUK-189, and EUK-207. SOD and catalase activities, cytoprotective, and other properties of EUK-207 [26, 29] and the non-cyclized salen Mn complexes [13] have been described. EUK-8, R=H, X=Cl; EUK-134, R=methoxy, X=acetoxy; EUK-189, R=ethoxy; X=acetoxy; EUK-207: OAc=acetoxy

EUK-189 when injected in vivo [16]. Both EUK-189 and EUK-207 improved cognitive performance and suppressed oxidative stress indicators (e.g., nucleic acid oxidation) in a mouse model for age-associated mild cognitive impairment [27, 30].

Many studies indicate that the mitochondria are particularly vulnerable to oxidative damage in various forms of injury, including by radiation. Endogenous antioxidant defense enzymes [14] include three types of SOD. These are the Cu- and Zn-containing cytosolic and extracellular forms of SOD (SOD1 and SOD3, respectively) and the Mn-containing mitochondrial form (SOD2). Mitochondria-targeted expression of either SOD2 or SOD1 protects against radiation injury in a mouse esophagitis model [1, 31] and mice deficient in SOD2 (*sod2*$^{-/+}$ mice) are more vulnerable to radiation lung injury [32]. Such findings suggest that a synthetic antioxidant reaching the mitochondria should have therapeutic benefits in radiation. Though salen Mn complexes were not designed specifically to target the mitochondria, several studies have shown that the compounds are "mito-protective." For example, in *sod2*$^{-/-}$ mice, an in vivo model for severe mitochondrial oxidative stress, salen Mn complexes prolong survival up to threefold, protect mitochondrial enzymes, and prevent oxidative pathologies [33, 34], more effectively than several other antioxidant agents [26]. EUK-207 preserved cardiac function and reversed mitochondrial impairments caused by ischemia/reperfusion in isolated hearts from ABCme$^{-/+}$ mice, whose mitochondria are particularly vulnerable to oxidative injury [35]. Finally, in rat astrocyte cultures, ionizing radiation caused mitochondrial injury, inducing changes in respiratory chain enzyme activity and reduced ATP production, and the salen Mn complex EUK-134, given up to 12 h after irradiation, significantly protected the mitochondria [16]. Since salen Mn complexes, unlike targeted agents such as MitoQ [36] and others [9] are *not* designed to exclusively reach the mitochondria, they might have broader therapeutic value because they will also address non-mitochondrial injury mechanisms. The "mito-protective" properties of salen Mn complexes have been reviewed previously [37].

Irradiation of late-responding tissues leads to acute activation of stress-responsive transcription leading to production of cytokines mediating an aberrant chronic inflammatory cascade that can ultimately lead to delayed outcomes such as fibrosis or necrosis [3, 15]. Production of such mediators is controlled by stress-induced transcriptional factors including those (e.g., NF-κB and AP-1) which are activated by increased ROS. Such events are reminiscent of proinflammatory processes that mediate tissue injury in many other pathological conditions, such as infection, ischemia/excitotoxicity, and environmental stress. Salen Mn complexes are markedly protective against such forms of tissue injury in vivo, while concomitantly suppressing activation of stress-induced transcription factors in target tissues [22, 38, 39]. For example, EUK-134 inhibited the excitotoxic activation of AP-1 and NF-κB in a rat seizure model [22], and EUK-189 inhibited AP-1 activation in aged mice subjected to mild heat-stress [40]. Salen Mn complex treatment increased the mean survival time of allogeneic C57BL/6 skin grafted into BALB/c mice, reduced proinflammatory type 1 alloresponse (IFN-γ) while promoting anti-inflammatory type 2 alloimmunity (IL-4 and IL-5) [41]. Recently, Kash et al. [42] showed that EUK-207 treatment improved survival of mice infected with the 1918 strain of influenza. The surviving mice showed greatly improved lung histopathology and reduced staining for oxidized nucleic acid. EUK-207 treatment did not affect viral titer, indicating that its benefits were related not to eliminating the virus, but instead to reducing the lethal host response that has been attributed to deaths caused by 1918 influenza. High-throughput sequencing and RNA expression analysis showed that the EUK-207 treated virus-infected mice had a remarkably different gene expression profile in the lung, as compared to vehicle-treated infected mice. Notably, with EUK-207 treatment, expression of inflammatory genes was suppressed and those associated with lung metabolic and repair processes were increased. Therefore, the suppressive effects of salen Mn complexes against proinflammatory, injury-associated transcriptional processes and, therefore, their ability to modulate the tissue microenvironment, provide a strong rationale for investigating their benefits in radiation injury models.

Taken together, such findings support the hypothesis that salen Mn complexes would mitigate radiation injury to normal tissues. Proposed mechanisms might involve preventing mitochondrial injury, suppressing "proinflammatory" processes, inhibiting other forms of oxidative injury to cellular constituents or most likely, complex combinations of the above that essentially modulate the tissue microenvironment to favor repair and remodeling over chronic injury.

11.2 Mitigation of Radiation Injury by Salen Mn Complexes

Salen Mn complexes mitigate radiation injury in several cell culture systems including, as described above, mitigating the mitochondrial injury induced by ionizing radiation in rat astrocytes [16]. These findings are not only relevant to long-term radiation-related damage to the brain, but also of potentially broader importance, given the hypothesized role of mitochondrial damage in other radiation late effects. Furthermore, several compounds with SOD and catalase/peroxidase activities, including the salen Mn complexes EUK-189 and EUK-207, inhibited radiation-induced apoptosis in bovine adrenal capillary endothelial cell cultures [43], and EUK-207 prevented several radiation-induced outcomes in human microvascular endothelial cell cultures [44]. These findings suggest that these agents might have broad applicability as radiation mitigators, since the vascular endothelial cell has long been regarded as being an important target in the progression to radiation late effects [45].

11.2.1 Oral Radiation Mucositis

Oral mucositis is a dose-limiting toxicity in patients receiving chemotherapy or localized radiation treatment for head and neck cancers [46, 47]. ROS have been an interventional target for the prevention and treatment of mucositis, with *N*-acetylcysteine as one agent under study (S. Sonis, personal communication). The ability of EUK-189 to mitigate radiation-induced mucositis was tested in a hamster model [48], using a single exposure of 35 Gy to induce severe to moderate mucositis by 15 days. EUK-189, given immediately after irradiation for 14 days, both topically (via a cheek pouch "wash" protocol, at 10 and 30 mg/mL [16]) and systemically (sc injection, 30 mg/kg-day) significantly reduced mucositis severity. The prolonged 15-day administration of EUK-189 was effective using both dosing protocols. In contrast, efficacy in attenuating oral mucositis by the SOD mimetic Mn cyclic polyamine M40403, of a compound class reported to lack catalase activity [49], had a very tight dose-schedule dependence [48]. For example, while M40403, given by ip injections (30 mg/kg) twice per day, from day 1 to day 3, mitigated oral mucositis, the mitigation effect was absent with a longer treatment, from day 1 to day 15. The investigators speculate that the longer treatment might impair healing. It is conceivable that excess hydrogen peroxide could prolong mucositis by impairing mucosal healing, similar to the complex role that this ROS is known to play in skin wound healing [50]. Possibly extended dosing with the SOD mimetic M40403 resulted in later increases in hydrogen peroxide, whereas EUK-189, as a catalase and peroxidase mimetic, lacks this effect. Both mucositis studies do support the potential value of an ROS-scavenging Mn complex, dosed appropriately, in mitigating radiation-induced oral mucositis. Given the commonality in underlying pathogenesis [48], it is also likely that both compounds may also be efficacious in attenuating chemotherapy-induced mucositis.

11.2.2 Pulmonary Radiation Injury

The lung is one of the most susceptible organs to potentially debilitating radiation injury. The functional effects seen following pulmonary irradiation are normally separated into two phases. Radiation pneumonitis generally occurs within 2–4 months and fibrosis tends to develop 4 months or later after irradiation [51] depending on dose. Acute pneumonitis is treated clinically with steroids, but can be life-threatening even with treatment. To date, radiation-induced lung fibrosis appears to be progressive and can ultimately lead to respiratory distress, pulmonary hypertension, right heart failure, and even death. Though the mechanisms are not well understood, extensive data suggest that pneumonitis and fibrosis may result from a cycle of chronic inflammation and oxidative damage initiated by radiation exposure to the lung [3, 15, 52, 53]. Several earlier studies examined the ability of salen Mn complexes to mitigate lung injury [16]. In Sprague–Dawley rats subjected to partial lung irradiation, EUK-189 given by a single sc injection suppressed micronucleus formation, an indicator of DNA damage in lung fibroblasts, measured 18 h later [54]. This suppression of micronuclei counts did not persist without subsequent EUK-189 injections, but it was observed even if the EUK-189 injection was delayed until 2 weeks after irradiation, consistent with micronucleus formation resulting from an ongoing cycle of injury and repair, that is, a continuous "cascade" of ROS formation after irradiation. This theory also implies that sustained treatment with an ROS scavenger such as a salen Mn complex would be required to have a significant impact on lung injury. Consequently, successful subsequent studies with salen Mn complexes have used longer periods of therapy, either by daily sc injection or by continuous sc infusion utilizing Alzet® osmotic pumps. One such study [55], also in Sprague–Dawley rats, administered EUK-207 by continuous sc infusion (~8 mg/kg/day) beginning 1 h after irradiation for 14 weeks. Rats received 10 Gy thoracic irradiation and were monitored for indicators of lung injury at various time points over 28 weeks. In the EUK-207 treated rats, a reduction in the radiation-induced increase in breathing rates indicated mitigation of pneumonitis. Furthermore, measurements of lung collagen synthesis at 28 weeks showed that EUK-207 treatment, although it had been discontinued after week 14, had also mitigated fibrosis. Importantly, the lungs from EUK-207 treated rats showed a reduction in oxidized nucleic acid, 8-OHdG, indicating that the compound, consistent with its hypothesized mechanism of action, inhibited oxidative stress in this model. More details of this study may be found in the original paper [55] as well as in the previous review [16].

In an alternative rat model for radiation pneumonitis, WAG/RijCmcr rats were exposed to 11 Gy total body irradiation (TBI) followed by syngeneic bone marrow transplant (BMT). This protocol (TBI/BMT) avoids death from hematopoietic injury, with rats developing pneumonitis by 6 weeks post-irradiation and those surviving the lung injury later developing radiation-induced renal injury. Among the consequences of this radiation protocol are increased breathing rate, increased right ventricular hypertrophy and pulmonary vascular resistance, and microvessel loss as indicated by imaging as well as measurements of the vascular marker angiotensin-converting enzyme (ACE) in perfused lung. A number of preliminary findings, reported in scientific abstracts and reviewed previously [16], indicated that EUK-207 mitigated such manifestations of radiation lung injury in this model. A subsequent publication [56] built upon these findings to demonstrate that EUK-207 mitigates pneumonitis and attenuates damage to the lung microvasculature.

Gao et al. [56] found that treatment with EUK-207 mitigated lung injury in rats given a single dose of total body or whole thoracic irradiation. When given TBI, the rats also received a BMT. EUK-207 (20 mg/kg, sc 5 days/week) from days 7 to 35 after 11 Gy TBI significantly decreased morbidity (Fig. 11.2). Examined at 42 days after irradiation, rats treated with EUK-207 exhibited several signs that vascular injury in the lungs was substantially attenuated. These included decreased pulmonary vascular resistance, increased angiotensin-converting enzyme (ACE) activity [16, 56] and, by analysis of planar angiograms, increased terminal arteriole counts (Fig. 11.3). In addition, EUK-207 treated rats had decreased arterial wall thickening and reduced numbers of foamy macrophages (Fig. 11.4).

To look at long-term fibrotic changes, rats received 13 Gy whole thoracic irradiation and collagen deposition was assessed at 210 days. In these animals, a similar EUK-207 treatment regimen reduced both pepsin-soluble (newly synthesized) and fibrotic collagen foci to normal, unirradiated levels [56]. Delivery of EUK-207 for these studies was started at 7 days after irradiation and stopped before symptoms of pneumonitis (30–35 days post-irradiation) were observed [56]. ACE inhibitors, that have successfully mitigated radiation pneumonitis, were not effective in a similar schedule. The ACE inhibitor enalapril failed to mitigate radiation pneumonitis or fibrosis when started at 7 days and terminated at 30 or 60 days [57]. The drug mitigated both endpoints

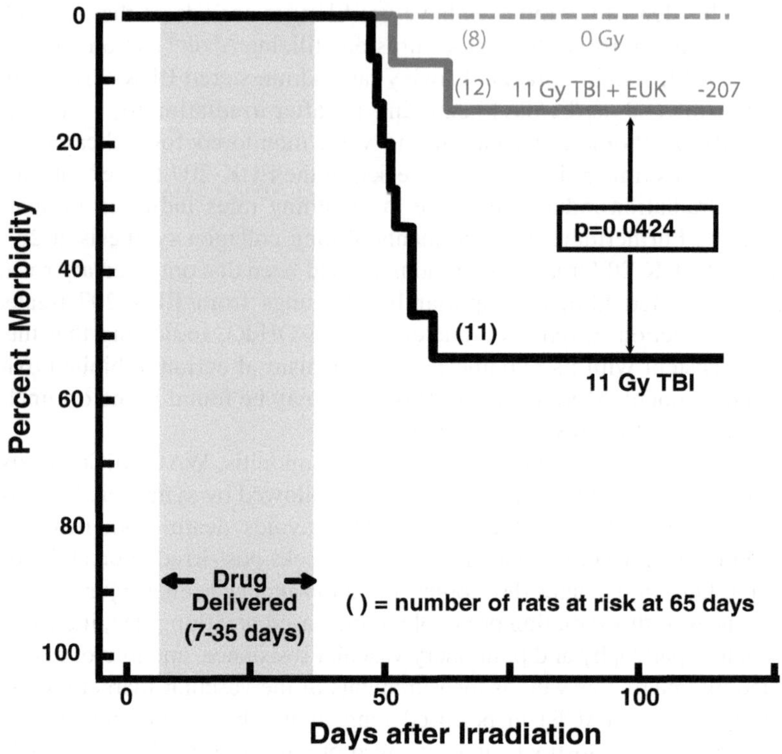

Fig. 11.2 EUK-207 treatment reduces morbidity in a rat radiation lung injury model. Rats received a total body irradiation and bone marrow transplantation (TBI/BMT) protocol followed by EUK-207 treatment, where indicated, as described in the text. Further details in [56]

when started at 7 days or even as late as 35 days after irradiation and continued. These results suggest that the mechanism of mitigation of the two drugs, EUK-207 and enalapril, are different; and further studies are needed to determine if serial mitigation with EUK-207 from 7 to 30 days and enalapril from 30 to 90 days will be more effective for mitigation than either drug alone. Recent data from a mouse model suggest that such would be the case (Williams et al., unpublished data). In that study, a mixed TBI/thoracic irradiation model in C57BL/J mice, EUK-207 or captopril were administered over an acute (3–10 days post-irradiation) and/or chronic (8–26 weeks post-irradiation) time course with the drugs being administered sc or po, respectively, every other day. Endpoints assessed included late survival and pulmonary structural changes (collagen accumulation and inflammatory cell infiltration). The acute EUK-207 schedule failed to show any mitigation activity although the chronic schedule, and both captopril protocols, demonstrated nonsignificant improvement trends in all indices. However, a combination of both agents led to 100 % survival and significantly reduced fibrosis. Alternative routes of administration, including by inhalation, and more targeted scheduling are currently being assessed.

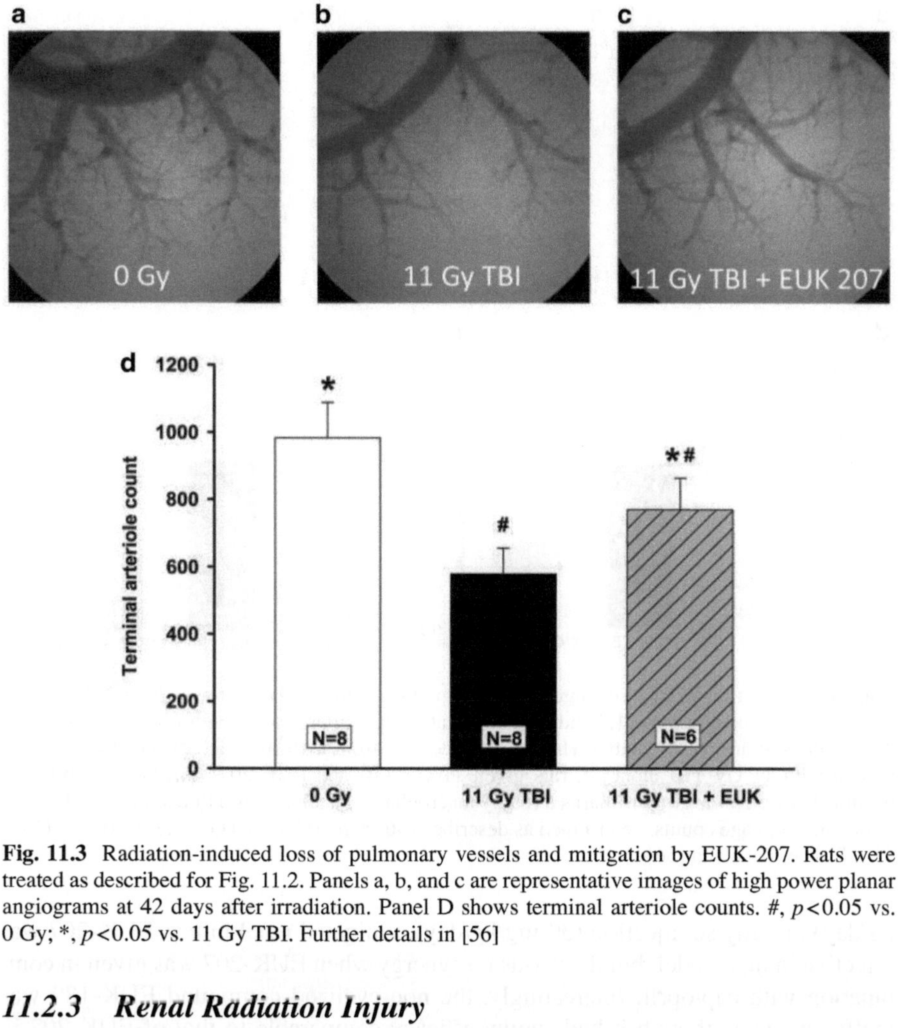

Fig. 11.3 Radiation-induced loss of pulmonary vessels and mitigation by EUK-207. Rats were treated as described for Fig. 11.2. Panels a, b, and c are representative images of high power planar angiograms at 42 days after irradiation. Panel D shows terminal arteriole counts. #, $p<0.05$ vs. 0 Gy; *, $p<0.05$ vs. 11 Gy TBI. Further details in [56]

11.2.3 Renal Radiation Injury

Renal radiation injury is a clinical concern months to several years after patients have undergone BMT when TBI is included in the preparatory regimen [58]. A TBI/BMT protocol induces similar radiation nephropathy in WAG/RijCmcr rats that is delayed in onset by weeks to months, with loss of renal function indicated by azotemia, proteinuria, and hypertension. Moulder and colleagues have reported that captopril and other modulators of the renin-angiotensin system mitigate renal injury [59, 60], while a variety of other agents, including the antioxidants deferiprone, genistein, and apocynin [7] are ineffective. Despite the lack of effect of other antioxidant approaches, including a genistein diet [7], salen Mn complexes were tested for mitigation of renal injury in this model [16]. EUK-207, given beginning at 3 weeks after TBI/BMT, either by continuous subcutaneous (sc) infusion (8–10 mg/

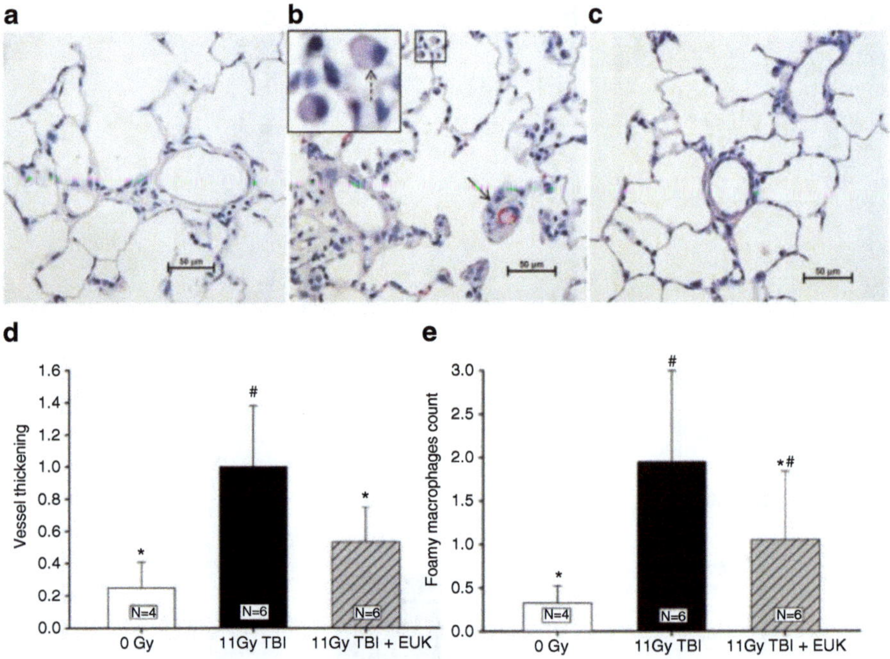

Fig. 11.4 Radiation-induced changes in lung structure with mitigation by EUK-207. Rats were treated as described for Fig. 11.2 and lungs fixed at 42 days after irradiation for histological analysis. Panels a–c are representative H/E stained sections from a, normal age matched rats; b, rats treated with 11 Gy TBI alone; c, rats given 11 Gy TBI and EUK-207. *Solid arrow* indicates occluded vessel, *dotted arrow* marks a foamy macrophage. Panel d, vessel thickening and Panel e, foamy macrophage counts, determined as described [56] #, $p<0.05$ vs. 0 Gy; *, $p<0.05$ vs. 11 Gy TBI. Further details in [56]

kg/day) or daily sc injection (30 mg/kg/day), mitigated renal injury. EUK-207 was effective in this model, but there was no synergy when EUK-207 was given in combination with captopril. Interestingly, the non-cyclized compound EUK-189 was ineffective even though it had shown efficacy comparable to that of EUK-207 in other animal models. Based on these and other observations, it was suggested that the beneficial effects of EUK-207 in the renal model might involve mechanisms other than its antioxidant properties [16].

11.2.4 Cutaneous Radiation Injury

Radiation dermatitis is a consequence of radiation therapy, for example, in breast and lung cancer patients, and cutaneous radiation exposure can impair wound healing. Cutaneous injury after radiation exposure involves inflammatory processes, alteration of cell proliferation and migration, and abnormal extracellular matrix synthesis (see review in [61, 62]). Oxidative stress is, in addition, implicated in the development of cutaneous radiation syndrome and delayed wound healing following radiation exposure [2, 15]. Such data imply that salen Mn complexes could have beneficial effects in models for cutaneous radiation injury, leading us to test EUK-207 in a rat model for combined radiation dermatitis and skin wound healing [61]. In this model, WAG/RijCmcr rats received a protocol delivering high doses (10–40 Gy, defined at the upper dermis) to the skin without significant exposure to internal organs, and then, within 1 h, received full thickness wounds by punch biopsy. Dermatitis was assessed weekly with a semiquantitative scoring for skin injury and wound closure was monitored by image analysis of traced wound areas. Irradiated rats developed, in a radiation dose-dependent manner, cutaneous reactions ranging in severity from transient erythema (onset 24 h, duration 1–2 days) to large deep erosions and non-healing ulcers (onset day 21, duration up to 90 days), plus alopecia (onset day 20, duration up to 90 days) and skin fibrosis (onset day 30, duration entire period of study). There was markedly impaired wound healing, also dependent on radiation dose. The peak of acute skin reactions was ~30 days. As reported [61] a number of other assessments demonstrated radiation-induced changes in extracellular matrix deposition, blood vessel proliferation, gene expression changes, and oxidative stress indicators in the skin.

EUK-207 was evaluated in this cutaneous combined injury model [62] using a 30 Gy radiation dose that, without drug treatment, induced severe cutaneous injury and wounds that failed to heal over the 90-day observation period. EUK-207 (~1.8 mg/kg/day) or vehicle (water) were given by sc Alzet® infusion pumps beginning 48 h after irradiation and continuing for up to 90 days. EUK-207 mitigated radiation dermatitis (Fig. 11.5) improved wound healing and increased angiogenesis at the wound edge (Fig. 11.6). Within 1 month, EUK-207 treated rats had reduced moist desquamation, reduced tissue inflammation, and increased wound contraction. Unlike the vehicle-treated groups, those treated with EUK-207 showed only slight tissue fibrosis at 90 days post-irradiation, and there was even growth of new hair within the center of the radiation field (Figs. 11.5 and 11.7). Unlike the vehicle-treated animals, those in the EUK-207 treatment groups had wounds that became fully healed. Skin microstructure evaluated by histopathology appeared substantially normalized with EUK-207, as compared with vehicle treatment (Fig. 11.7).

Data from this study indicated that the beneficial effects of EUK-207 appeared to be due, at least in part, to suppression of oxidative injury to the skin. Skin harvested 30 days after irradiation showed an increase in expression of a number of genes associated with oxidative stress, and these changes were abrogated in groups treated with EUK-207 [62]. In addition, 30 days after irradiation there was increased oxidative injury, as indicated by both protein and DNA oxidation markers, and both were inhibited by EUK-207 treatment (Fig. 11.8). In another model, examining radiation-induced skin damage in the mouse, EUK-207 treatment (sc beginning immediately after irradiation) caused suppression of skin lipid peroxidation detectable up to 4 weeks after irradiation [63]. In that study, several other agents with reported antioxidant properties, including curcumin, similarly suppressed lipid peroxidation, while

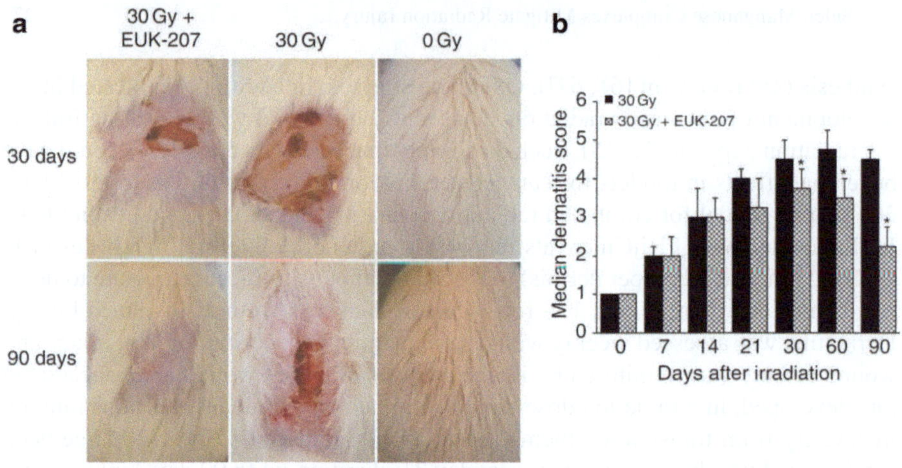

Fig. 11.5 Mitigation of radiation dermatitis by EUK-207. Rats received a combined skin irradiation exposure and full-thickness wound as described in the text. Radiation dermatitis in representative animals (**a**) and the entire groups scored semiquantitatively (**b**), receiving EUK-207 (1.8 mg/kg/day) or vehicle by sc infusion, beginning 48 h after irradiation. In (**b**) data shown are median scores, with error bars indicating the range. *, different from vehicle ($p<0.01$, Wilcoxon–Mann–Whitney rank sum test). Further details in [62]

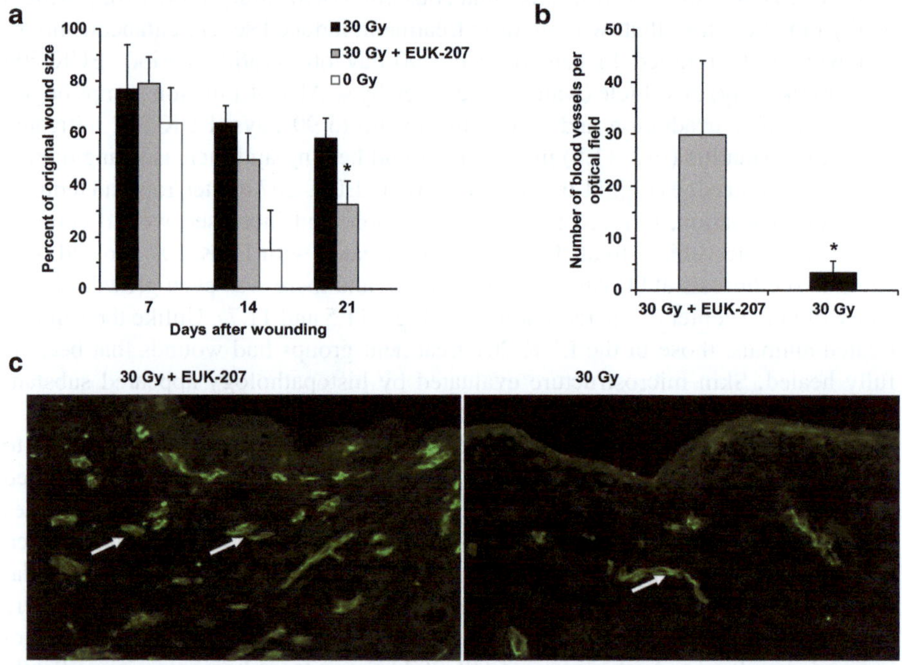

Fig. 11.6 Accelerated wound healing and enhanced angiogenesis in irradiated skin with EUK-207. Rats were treated as described for Fig. 11.5. (**a**) Wound areas and (**b**), (**c**) microvasculature visualized by staining endothelial cells (CD31) on the wound edge 30 days post-irradiation. Data are expressed as means ± SD, *, different from vehicle control "30 Gy," $p<0.01$. Bar = 100 μm. Further details in [62]

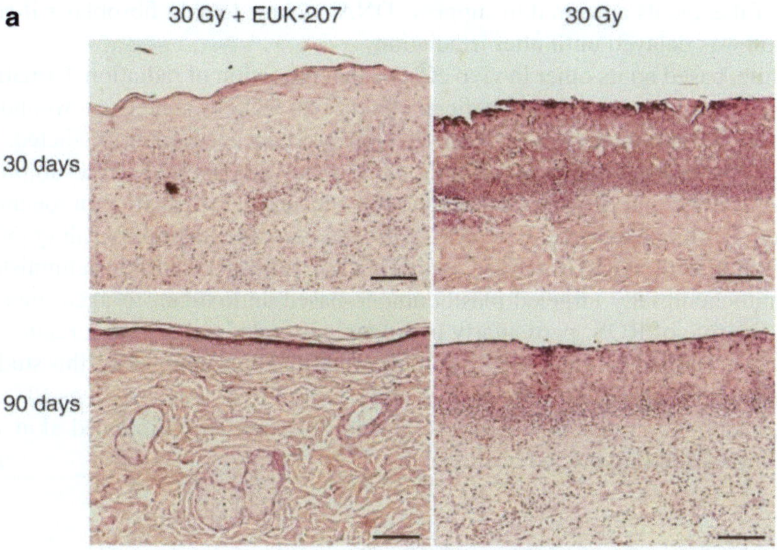

b

Experimental condition	Days after irradiation	Dermal thickness	Epidermal thickness	Hair follicles
30 Gy	30	608.4 ± 77.9	0 ± 0	0 ± 0
30 Gy + EUK-207	30	509.9 ± 75.2*	31.6 ± 30.0*	3.8 ± 3.9
30 Gy	90	711.7 ± 103.7	14.0 ± 14.6	1.3 ± 1.9
30 Gy + EUK-207	90	566.4 ± 70.1*	21.7 ± 5.10	6.2 ± 3.2**

Fig. 11.7 EUK-207 treatment improves skin architecture in irradiated rats. Rats were treated as described for Fig. 11.5 and histopathology was conducted 30 and 90 days after irradiation. Further details in [62]

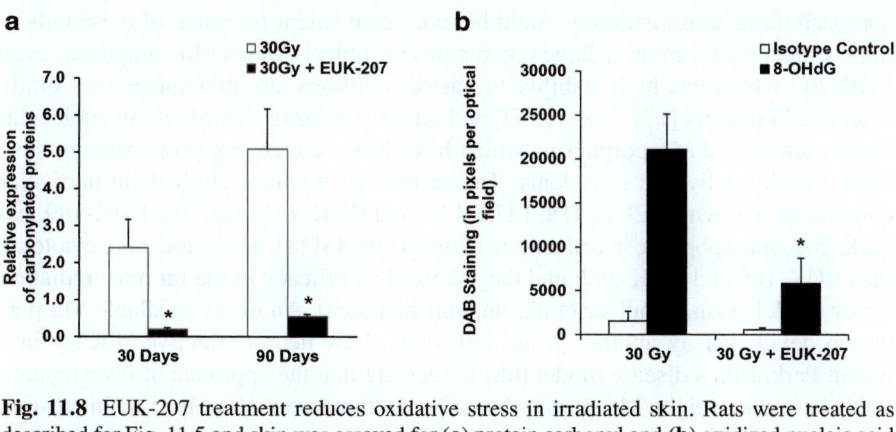

Fig. 11.8 EUK-207 treatment reduces oxidative stress in irradiated skin. Rats were treated as described for Fig. 11.5 and skin was assayed for (**a**) protein carbonyl and (**b**) oxidized nucleic acid (8-OHdG) content. Further details in [62]

none of the agents appeared to suppress DNA damage to skin fibroblasts if administration was delayed until after irradiation.

While, based on its other in vivo effects, the mitigation of radiation dermatitis by EUK-207 was expected, the beneficial effect of an ROS scavenger on wound healing in the rat combined-injury scenario [62] was not necessarily predicted. ROS, particularly hydrogen peroxide, are thought to play key signaling roles to promote wound healing, and localized transfection of catalase delays healing in rodents [50, 64]. Yet, in other experiments, excess hydrogen peroxide impaired healing [50] and transfection with SOD2 improved wound healing [65] as did chronic administration of a mitochondrially targeted plastoquinone-based antioxidant to aged mice [66]. Thus, the role of ROS, particularly hydrogen peroxide, and of redox regulation in cutaneous wound healing is highly complex. Despite this complexity, this study and others support the notion that selected ROS-scavenging agents such as EUK-207, given under the right circumstances, can mitigate radiation-induced skin injury including facilitating wound healing.

11.3 Future Directions and Challenges

Drug delivery methods: As discussed in this review, salen Mn complexes, particularly EUK-207, mitigate radiation injury to normal tissues. One important issue to be addressed in their further development as mitigating agents is the method of drug delivery. Evidence indicates that systemic use for a number of days, or even weeks, may be necessary. With the chronic, continuous nature of the oxidative stress that is implicated in delayed radiation injury [15], these findings are not at all surprising. It is promising, however, that mitigating effects of EUK-207 in pulmonary and renal models persist weeks after drug therapy has ended. Currently, daily injection or continuous sc infusion are the most effective means of delivery for salen Mn complexes. Such methods of delivery, even portable infusion pumps such as those delivering insulin, are tolerated by patients for certain chronic diseases, but are not the most desirable. Sustained release injectable formulations might be a more promising approach. Oral administration would be more convenient for some of these indications, especially for treating large populations of subjects. Salen Mn complexes, even EUK-207 which has high stability to gastric solutions, are nonetheless not orally available in rodents [19]. We [19, 67] and others [68] have described low molecular weight uncharged Mn porphyrins which have ROS-scavenging properties and also can be orally delivered to rodents. The series that we have studied (in particular compounds known as EUK-418, EUK-423, and EUK-451) are, like EUK-189 and EUK-207, anti-apoptotic in cell culture systems [19, 43] but, as a class, more cytotoxic than EUK-189 and EUK-207; and they showed no efficacy in the rat renal radiation injury model (Fish, Moulder et al., unpublished data). An orally available Mn porphyrin developed by another group [69] does show neuroprotective efficacy in a rodent Parkinson's disease model [68] suggesting that the approach of investigating low molecular weight Mn porphyrins still has the potential to lead to interesting orally bioavailable therapeutic agents.

For some normal tissue radiation mitigation indications, a topical means of delivering salen Mn complexes, if enabling sufficient drug distribution throughout the skin, might be therapeutically practical. Some studies with these compounds are promising in this regard. In the hamster mucositis model, a topical dosing regimen, consisting of direct application of EUK-189 to the cheek pouch mucosa, showed efficacy equivalent to that of sc injection. In patients, a lozenge or mouth rinse could be useful to mitigate oral mucositis. In the rat cutaneous injury study described here, only systemic EUK-207 was tested. However, EUK-189 and EUK-207 can be formulated into topical preparations that reduce ear inflammation in the mouse (Doctrow et al., unpublished data). Furthermore, topical application of an EUK-134 preparation to the skin of human volunteers prevented skin lipid peroxidation caused by UVA exposure [70]. Such data support the future testing of topical preparations of salen Mn complexes to mitigate cutaneous radiation injury. Inhaled methods of EUK-207 delivery are also under investigation, especially for pulmonary radiation injury. Preliminary findings (Williams et al., unpublished data) indicate that mice inhaling EUK-207 exhibited detectable drug levels, albeit with considerable variability, in lung tissue and plasma. Following a daily EUK-207 inhalation regimen, mice subjected to total body (5 Gy) plus thoracic (10 Gy) irradiation showed trends toward improvement in indicators of lung inflammation (macrophage infiltration) and fibrosis (collagen and airspace measurements), though they did not rise to the level of statistical significance. Studies are ongoing to determine whether inhaled administration of EUK-207 can be improved so as to produce results that are at least comparable to the lung mitigation effects of sc administered drug.

Mechanism of action: role of ROS scavenging in mitigation. As discussed throughout this review, the mechanisms of action of salen Mn complexes, as with any other mitigator of delayed radiation injury, are likely to be complex. Increased understanding of these mechanisms could help in future development of analogs, as well as improved drug delivery and dosing protocols, optimized for each specific radiation injury indication. Data in astrocyte cultures support the hypothesis that the compounds' "mito-protective" properties are relevant to delayed radiation injury [16] though this should be further studied in other cell types as well as in vivo. Findings in certain models support the hypothesis that salen Mn complexes are acting as ROS scavengers in vivo. For example, in rat lung injury models [55, 71], EUK-207 treatment mitigated signs of pneumonitis and fibrosis, while concomitantly decreasing lung levels of oxidized nucleic acid. Similar mitigation of oxidative stress-associated markers was reported in the cutaneous injury model [62]. While such associations do not prove causality, they are promising indicators that the ROS-scavenging properties of these agents measured in vitro extend to an ability to mitigate oxidative stress in vivo. In the renal injury model, however, the role of oxidative stress is in question and EUK-207 mitigated renal injury, while numerous antioxidant agents, including EUK-189, did not [16]. Although there is evidence for chronic oxidative stress after irradiation of lung, skin and brain, that has, thus far, not been found in radiation nephropathy [72–74]. Therefore, for the renal injury model, other mechanisms, including some hypothetical modulation of the renin-angiotensin system, should be addressed in future research. For all models, modulation of gene expression and other aspects of the tissue microenvironment might involve a combination of redox and other effects. More generally, however, it seems likely that combined therapies using two or more agents with distinct mechanisms of action are likely to be more effective than one agent alone for a complex indication such as delayed radiation injury [75]. In support of this, in their rat model for thoracic irradiation, Mahmood et al. [71] reported that a combination of EUK-207 and captopril was more effective than either agent alone in a number of indicators of radiation lung injury and lung oxidative stress. Captopril treatment also reduced indicators of oxidative stress in this study, not unexpected because of the redox implications of the renin angiotensin system (*e.g.*, angiotensin II stimulates ROS production by NADPH oxidase [76]). Whether the effects of the EUK-207 and captopril together were additive or synergistic has not yet been determined. Findings of Gao et al., as discussed earlier in this review, suggest that the ACE inhibitor enalapril [57] and EUK-207 [56] mitigate lung injury via mechanisms distinct from that of EUK-207, as evidenced by their different therapeutic windows in the rat TBI/BMT model.

Many other factors need to be considered when developing an antioxidant, metal-based compound as a potential therapeutic agent. These include, potentially toxic "pro-oxidant" effects, effects due to the metal itself versus the metal-ligand complex, and general questions of safety, tolerability, and biodistribution. While the doses of salen Mn complexes causing detectable toxicity in animals are at least 10–100-fold, depending on mode of administration, greater than those mitigating radiation injury, formal toxicity studies will be needed to assess the safety of these, or any other potential mitigating agent. Such factors, as they pertain to salen Mn complexes, are discussed in more detail in our previous review [16]. Overall, several factors such as lipophilicity, inherent stability, potential toxicity and, of course, in vivo efficacy must be weighed against one another in selecting a lead compound. The preponderance of data, in vitro as well as in vivo, indicates that EUK-207 exhibits a number of properties that justify its selection as a lead salen Mn complex for further study as a mitigator of radiation injury.

11.4 Conclusion

Synthetic compounds with SOD and catalase/peroxidase activities, particularly the salen Mn complexes exemplified by EUK-189 and EUK-207, mitigate radiation injury to normal tissues, including lung, kidney, skin, and oral mucosa. These novel agents could have potential value in reducing radiation injury, including as adjunct therapies after radiation therapy or mitigating agents given to exposed subjects after radiological accidents or terrorism. Future development of salen Mn complexes should focus on optimizing their methods of delivery, increased understanding of their mechanisms of action and potential side effects, and examination of their ability to work in combination with other radiation mitigators.

Acknowledgements Development of the salen Mn complexes, including EUK-207, was funded in part by GM57770 (SRD). Renal, cutaneous, and pulmonary studies and further compound development were supported by AI067734 (JEM). Some pulmonary studies were also supported by AI81294 (MM) and AI091036 (JPW). Funding for previously reviewed studies were acknowledged in [16] and the other cited primary journal articles. SRD is an inventor on patents describing salen Mn complexes. We thank Drs Eric Cohen and Richard Hill for helpful comments on the manuscript.

References

1. Greenberger JS, Epperly MW. Radioprotective antioxidant gene therapy: potential mechanisms of action. Gene Ther Mol Biol. 2004;8:31–44.
2. Zhao W, Diz DI, Robbins ME. Oxidative damage pathways in relation to normal tissue injury. Br J Radiol. 2007;80(Spec No 1):S23–31.
3. Zhao W, Robbins ME. Inflammation and chronic oxidative stress in radiation-induced late normal tissue injury: therapeutic implications. Curr Med Chem. 2009;16:130–43.
4. Stone HB, Moulder JE, Coleman CN, Ang KK, Anscher MS, Barcellos-Hoff MH, Dynan WS, Fike JR, Grdina DJ, Greenberger JS, Hauer-Jensen M, Hill RP, Kolesnick RN, Macvittie TJ, Marks C, McBride WH, Metting N, Pellmar T, Purucker M, Robbins ME, Schiestl RH, Seed TM, Tomaszewski JE, Travis EL, Wallner PE, Wolpert M, Zaharevitz D. Models for evaluating agents intended for the prophylaxis, mitigation and treatment of radiation injuries. Report of an NCI Workshop, December 3–4, 2003. Radiat Res. 2004;162:711–28.
5. Wambi CO, Sanzari JK, Sayers CM, Nuth M, Zhou Z, Davis J, Finnberg N, Lewis-Wambi JS, Ware JH, El-Deiry WS, Kennedy AR. Protective effects of dietary antioxidants on proton total-body irradiation-mediated hematopoietic cell and animal survival. Radiat Res. 2009;172:175–86.
6. Bourgier C, Levy A, Vozenin MC, Deutsch E. Pharmacological strategies to spare normal tissues from radiation damage: useless or overlooked therapeutics? Cancer Metastasis Rev. 2012;31:699–712.
7. Cohen EP, Fish BL, Irving AA, Rajapurkar MM, Shah SV, Moulder JE. Radiation nephropathy is not mitigated by antagonists of oxidative stress. Radiat Res. 2009;172:260–4.
8. Calveley VL, Jelveh S, Langan A, Mahmood J, Yeung IW, Van Dyk J, Hill RP. Genistein can mitigate the effect of radiation on rat lung tissue. Radiat Res. 2010;173:602–11.
9. Jiang J, Stoyanovsky DA, Belikova NA, Tyurina YY, Zhao Q, Tungekar MA, Kapralova V, Huang Z, Mintz AH, Greenberger JS, Kagan VE. A mitochondria-targeted triphenylphosphonium-conjugated nitroxide functions as a radioprotector/mitigator. Radiat Res. 2009;172:706–17.
10. Rabbani ZN, Batinic-Haberle I, Anscher MS, Huang J, Day BJ, Alexander E, Dewhirst MW, Vujaskovic Z. Long-term administration of a small molecular weight catalytic metalloporphyrin antioxidant, AEOL 10150, protects lungs from radiation-induced injury. Int J Radiat Oncol Biol Phys. 2007;67:573–80.
11. Gauter-Fleckenstein B, Fleckenstein K, Owzar K, Jiang C, Reboucas JS, Batinic-Haberle I, Vujaskovic Z. Early and late administration of MnTE-2-PyP5+ in mitigation and treatment of radiation-induced lung damage. Free Radic Biol Med. 2010;48:1034–43.
12. Batinic-Haberle I, Tovmasyan A, Roberts ER, Vujaskovic Z, Leong KW, Spasojevic I. SOD therapeutics: latest insights into their structure-activity relationships and impact on the cellular redox-based signaling pathways. Antioxid Redox Signal. 2014;20:2372–415.
13. Doctrow SR, Huffman K, Marcus CB, Tocco G, Malfroy E, Adinolfi CA, Kruk H, Baker K, Lazarowych N, Mascarenhas J, Malfroy B. Salen-manganese complexes as catalytic scavengers of hydrogen peroxide and cytoprotective agents: structure-activity relationship studies. J Med Chem. 2002;45:4549–58.

14. Halliwell B, Gutteridge JMC. Free radicals in biology and medicine. 4th ed. Oxford: Oxford University Press; 2007.
15. Robbins ME, Zhao W. Chronic oxidative stress and radiation-induced late normal tissue injury: a review. Int J Radiat Biol. 2004;80:251–9.
16. Rosenthal RA, Fish B, Hill RP, Huffman KD, Lazarova Z, Mahmood J, Medhora M, Molthen R, Moulder JE, Sonis ST, Tofilon PJ, Doctrow SR. Salen Mn complexes mitigate radiation injury in normal tissues. Anticancer Agents Med Chem. 2011;11:359–72.
17. Kouvaris JR, Kouloulias VE, Vlahos LJ. Amifostine: the first selective-target and broad-spectrum radioprotector. Oncologist. 2007;12:738–47.
18. Grdina DJ, Murley JS, Kataoka Y, Baker KL, Kunnavakkam R, Coleman MC, Spitz DR. Amifostine induces antioxidant enzymatic activities in normal tissues and a transplantable tumor that can affect radiation response. Int J Radiat Oncol Biol Phys. 2009;73:886–96.
19. Rosenthal RA, Huffman KD, Fisette LW, Damphousse CA, Callaway WB, Malfroy B, Doctrow SR. Orally available Mn porphyrins with superoxide dismutase and catalase activities. J Biol Inorg Chem. 2009;14:979–91.
20. Chance B, Sies H, Boveris A. Hydroperoxide metabolism in mammalian organs. Physiol Rev. 1979;59:527–605.
21. Sharpe MA, Ollosson R, Stewart VC, Clark JB. Oxidation of nitric oxide by oxomanganese salen complexes: a new mechanism for cellular protection by superoxide dismutase/catalase mimetics. Biochem J. 2002;366:97–107.
22. Rong Y, Doctrow SR, Tocco G, Baudry M. EUK-134, a synthetic superoxide dismutase and catalase mimetic, prevents oxidative stress and attenuates kainate-induced neuropathology. Proc Natl Acad Sci U S A. 1999;96:9897–902.
23. Pong K, Doctrow SR, Baudry M. Prevention of 1-methyl-4-phenylpyridinium- and 6-hydroxydopamine-induced nitration of tyrosine hydroxylase and neurotoxicity by EUK-134, a superoxide dismutase and catalase mimetic, in cultured dopaminergic neurons. Brain Res. 2000;881:182–9.
24. Doctrow SR, Huffman K, Marcus CB, Musleh W, Bruce A, Baudry M, Malfroy B. Salen-manganese complexes: combined superoxide dismutase/catalase mimics with broad pharmacological efficacy. In: Sies H, editor. Antioxidants in disease mechanisms and therapeutic strategies. New York: Academic Press; 1997. p. 247–70.
25. Doctrow SR, Adinolfi C, Baudry M, Huffman K, Malfroy B, Marcus CB, Melov S, Pong K, Rong Y, Smart J, Tocco G. Salen manganese complexes, combined superoxide dismutase/catalase mimetics demonstrate potential for treating neurodegenerative and other age-associated diseases. In: Rodriguez H, Cutler R, editors. Oxidative stress and aging: advances in basic science, diagnostics, and intervention. Singapore: World Scientific; 2003. p. 1324–42.
26. Doctrow SR, Baudry M, Huffman K, Malfroy B, Melov S. Salen Mn complexes: multifunctional catalytic antioxidants protective in models for neurodegenerative disease and aging. In: Doctrow SR, Sessler JL, McMurry TJ, Lippard SJ, editors. Medicinal inorganic chemistry. New York: American Chemical Society and Oxford University Press; 2005. p. 319–47.
27. Liu R, Liu IY, Bi X, Thompson RF, Doctrow SR, Malfroy B, Baudry M. Reversal of age-related learning deficits and brain oxidative stress in mice with superoxide dismutase/catalase mimetics. Proc Natl Acad Sci U S A. 2003;100:8526–31.
28. Jung C, Rong Y, Doctrow S, Baudry M, Malfroy B, Xu Z. Synthetic superoxide dismutase/catalase mimetics reduce oxidative stress and prolong survival in a mouse amyotrophic lateral sclerosis model. Neurosci Lett. 2001;304:157–60.
29. Malfroy-Camine B, Doctrow SR. Cyclic salen-manganese compounds: reactive oxygen species scavengers useful as antioxidants in the treatment and prevention of diseases. U.S. Patent Number 7,122,537, 2006.
30. Clausen A, Doctrow S, Baudry M. Prevention of cognitive deficits and brain oxidative stress with superoxide dismutase/catalase mimetics in aged mice. Neurobiol Aging (online 2008). 2010;31:425–33.
31. Rajagopalan MS, Stone B, Rwigema JC, Salimi U, Epperly MW, Goff J, Franicola D, Dixon T, Cao S, Zhang X, Buchholz BM, Bauer AJ, Choi S, Bakkenist C, Wang H, Greenberger JS.

Intraesophageal manganese superoxide dismutase-plasmid liposomes ameliorates novel total-body and thoracic radiation sensitivity of NOS1-/- mice. Radiat Res. 2010;174: 297–312.
32. Epperly MW, Epstein CJ, Travis EL, Greenberger JS. Decreased pulmonary radiation resistance of manganese superoxide dismutase (MnSOD)-deficient mice is corrected by human manganese superoxide dismutase-Plasmid/Liposome (SOD2-PL) intratracheal gene therapy. Radiat Res. 2000;154:365–74.
33. Hinerfeld D, Traini MD, Weinberger RP, Cochran B, Doctrow SR, Harry J, Melov S. Endogenous mitochondrial oxidative stress: neurodegeneration, proteomic analysis, specific respiratory chain defects, and efficacious antioxidant therapy in superoxide dismutase 2 null mice. J Neurochem. 2004;88:657–67.
34. Melov S, Doctrow SR, Schneider JA, Haberson J, Patel M, Coskun PE, Huffman K, Wallace DC, Malfroy B. Lifespan extension and rescue of spongiform encephalopathy in superoxide dismutase 2 nullizygous mice treated with superoxide dismutase-catalase mimetics. J Neurosci. 2001;21:8348–53.
35. Liesa M, Luptak I, Qin F, Hyde BB, Sahin E, Siwik DA, Zhu Z, Pimentel DR, Xu XJ, Ruderman NB, Huffman KD, Doctrow SR, Richey L, Colucci WS, Shirihai OS. Mitochondrial transporter ATP binding cassette mitochondrial erythroid is a novel gene required for cardiac recovery after ischemia/reperfusion. Circulation. 2011;124:806–13.
36. Smith RA, Murphy MP. Animal and human studies with the mitochondria-targeted antioxidant MitoQ. Ann N Y Acad Sci. 2010;1201:96–103.
37. Doctrow SR, Liesa M, Melov S, Shirihai OS, Tofilon PJ. Salen Mn complexes are superoxide dismutase and catalase mimetics that protect the mitochondria. Curr Inorg Chem. 2012;2:325–34.
38. Zhang HJ, Doctrow SR, Xu L, Oberley LW, Beecher B, Morrison J, Oberley TD, Kregel KC. Redox modulation of the liver with chronic antioxidant enzyme mimetic treatment prevents age-related oxidative damage associated with environmental stress. Faseb J. 2004;18:1547–9.
39. Peng J, Stevenson FF, Doctrow SR, Andersen JK. Superoxide dismutase/catalase mimetics are neuroprotective against selective paraquat-mediated dopaminergic neuron death in the substantial nigra: implications for Parkinson disease. J Biol Chem. 2005;280:29194–8.
40. Zhang HJ, Drake VJ, Xu L, Xie L, Oberley LW, Kregel KC. Heat-induced liver injury in old rats is associated with exaggerated oxidative stress and altered transcription factor activation. FASEB J. 2003;17:2293–5.
41. Tocco G, Illigens BM, Malfroy B, Benichou G. Prolongation of alloskin graft survival by catalytic scavengers of reactive oxygen species. Cell Immunol. 2006;241:59–65.
42. Kash JC, Xiao Y, Davis AS, Walters KA, Chertow DS, Easterbrook JD, Dunfee RL, Sandouk A, Jagger BW, Schwartzman LM, Kuestner RE, Wehr NB, Huffman K, Rosenthal RA, Ozinsky A, Levine RL, Doctrow SR, Taubenberger JK. Treatment with the reactive oxygen species scavenger EUK-207 reduces lung damage and increases survival during 1918 influenza virus infection in mice. Free Radic Biol Med. 2013;67:235–47.
43. Vorotnikova E, Rosenthal RA, Tries M, Doctrow SR, Braunhut SJ. Novel synthetic SOD/catalase mimetics mitigate capillary endothelial cell apoptosis. Radiation Res. 2010;173:748–59.
44. Otterson MF, Nie L, Link BJ, Schmidt JL, Rafiee P. Protective effect of EUK-207 on irradiated human intestinal microvascular endothelial cells (HIMEC). Abstracts, Experimental Biology 2011, 2011.
45. Thomas SR, Witting PK, Drummond GR. Redox control of endothelial function and dysfunction: molecular mechanisms and therapeutic opportunities. Antioxid Redox Signal. 2008;10:1713–65.
46. Sonis ST, Peterson RL, Edwards LJ, Lucey CA, Wang L, Mason L, Login G, Ymamkawa M, Moses G, Bouchard P, Hayes LL, Bedrosian C, Dorner AJ. Defining mechanisms of action of interleukin-11 on the progression of radiation-induced oral mucositis in hamsters. Oral Oncol. 2000;36:373–81.
47. Sonis ST, Lindquist L, Vugt AV, Stewart AA, Stam K, Qu GY, Iwata KK, Haley JD. Prevention of chemotherapy-induced ulcerative mucositis by TGF-beta. Cancer Res. 1994;54:1135–8.

48. Murphy CK, Fey EG, Watkins BA, Wong V, Rothstein D, Sonis ST. Efficacy of superoxide dismutase mimetic M40403 in attenuating radiation-induced oral mucositis in hamsters. Clin Cancer Res. 2008;14:4292–7.
49. Salvemini D, Wang ZQ, Zweier JL, Samouilov A, Macarther H, Misko TP, Currie MG, Cuzzocrea S, Sikorski JA, Riley DP. A nonpeptidyl mimic of superoxide dismutase with therapeutic activity in rats. Science. 1999;286:304–6.
50. Roy S, Khanna S, Nallu K, Hunt TK, Sen CK. Dermal wound healing is subject to redox control. Mol Ther. 2006;13:211–20.
51. Marks LB, Yu X, Vujaskovic Z, Small WJ, Folz R, Anscher MS. Radiation-induced lung injury. Sem Radiat Oncol. 2003;13:333–45.
52. Fleckenstein K, Zgonjanin L, Chen L, Rabbani Z, Jackson IL, Thrasher B, Kirkpatrick J, Foster WM, Vujaskovic Z. Temporal onset of hypoxia and oxidative stress after pulmonary irradiation. Int J Radiat Oncol Biol Phys. 2007;68:196–204.
53. Finkelstein JN, Johnston CJ, Baggs R, Rubin P. Early alterations in extracellular matrix and transforming growth factor beta gene expression in mouse lung indicative of late radiation fibrosis. Int J Radiat Oncol Biol Phys. 1994;28:621–31.
54. Langan AR, Khan MA, Yeung IW, Van Dyk J, Hill RP. Partial volume rat lung irradiation: the protective/mitigating effects of Eukarion-189, a superoxide dismutase-catalase mimetic. Radiother Oncol. 2006;79:231–8.
55. Mahmood J, Jelveh S, Calveley V, Zaidi A, Doctrow SR, Hill RP. Mitigation of radiation-induced lung injury by genistein and EUK-207. Int J Radiat Biol. 2011;87:889–901.
56. Gao F, Fish BL, Szabo A, Doctrow SR, Kma L, Molthen RC, Moulder JE, Jacobs ER, Medhora M. Short-term treatment with a SOD/catalase mimetic, EUK-207, mitigates pneumonitis and fibrosis after single-dose total-body or whole-thoracic irradiation. Radiat Res. 2012;178: 468–80.
57. Gao F, Fish BL, Moulder JE, Jacobs ER, Medhora M. Enalapril mitigates radiation-induced pneumonitis and pulmonary fibrosis if started 35 days after whole-thorax irradiation. Radiat Res. 2013;180:546–52.
58. Cohen E, Moulder J. Radiation nephropathy. In: Cohen E, editor. Cancer and the kidney. Oxford: Oxford University Press; 2011. p. 193–204.
59. Moulder JE, Fish BL, Cohen EP. ACE inhibitors and AII receptor antagonists in the treatment and prevention of bone marrow transplant nephropathy. Curr Pharm Des. 2003;9:737–49.
60. Cohen EP, Fish BL, Moulder JE. Mitigation of radiation injuries via suppression of the renin-angiotensin system: emphasis on radiation nephropathy. Curr Drug Targets. 2010;11:1423–9.
61. Jourdan MM, Lopez A, Olasz EB, Duncan NE, Demara M, Kittipongdaja W, Fish BL, Mader M, Schock A, Morrow NV, Semenenko VA, Baker JE, Moulder JE, Lazarova Z. Laminin 332 deposition is diminished in irradiated skin in an animal model of combined radiation and wound skin injury. Radiat Res. 2011;176:636–48.
62. Doctrow SR, Lopez A, Schock AM, Duncan NE, Jourdan MM, Olasz EB, Moulder JE, Fish BL, Mader M, Lazar J, Lazarova Z. A synthetic superoxide dismutase/catalase mimetic EUK-207 mitigates radiation dermatitis and promotes wound healing in irradiated rat skin. J Invest Dermatol. 2013;133:1088–96.
63. Jelveh S, Kaspler P, Bhogal N, Mahmood J, Lindsay PE, Okunieff P, Doctrow SR, Bristow RG, Hill RP. Investigations of antioxidant-mediated protection and mitigation of radiation-induced DNA damage and lipid peroxidation in murine skin. Int J Radiat Biol. 2013;89:618–27.
64. Sen CK, Roy S. Redox signals in wound healing. Biochim Biophys Acta. 1780;2008: 1348–61.
65. Luo JD, Wang YY, Fu WL, Wu J, Chen AF. Gene therapy of endothelial nitric oxide synthase and manganese superoxide dismutase restores delayed wound healing in type 1 diabetic mice. Circulation. 2004;110:2484–93.
66. Demianenko IA, Vasilieva TV, Domnina LV, Dugina VB, Egorov MV, Ivanova OY, Ilinskaya OP, Pletjushkina OY, Popova EN, Sakharov IY, Fedorov AV, Chernyak BV. Novel mitochondria-targeted antioxidants, "Skulachev-ion" derivatives, accelerate dermal wound healing in animals. Biochemistry (Mosc). 2010;75:274–80.

67. Meunier B, Cosledan F. Orally bioavailable low molecular weight metalloporphyrins. Eukarion, Inc., World Intellectual Property Organization (PCT application number 20060241095). 2005; p. 30.
68. Liang LP, Huang J, Fulton R, Day BJ, Patel M. An orally active catalytic metalloporphyrin protects against 1-methyl-4-phenyl-1,2,3,6-tetrahydropyridine neurotoxicity in vivo. J Neurosci. 2007;27:4326–33.
69. Trova MP, Gauuan PJ, Pechulis AD, Bubb SM, Bocckino SB, Crapo JD, Day BJ. Superoxide dismutase mimetics. Part 2: synthesis and structure-activity relationship of glyoxylate- and glyoxamide-derived metalloporphyrins. Bioorg Med Chem. 2003;11:2695–707.
70. Declercq L, Sente I, Hellemans L, Corstjens H, Maes D. Use of the synthetic superoxide dismutase/catalase mimetic EUK-134 to compensate for seasonal antioxidant deficiency by reducing pre-existing lipid peroxides at the human skin surface. Int J Cosmet Sci. 2004;26:255–63.
71. Mahmood J, Jelveh S, Zaidi A, Doctrow SR, Medhora M, Hill RP. Targeting the Renin-angiotensin system combined with an antioxidant is highly effective in mitigating radiation-induced lung damage. Int J Radiat Oncol Biol Phys. 2014;89:722–8.
72. Lenarczyk M, Cohen EP, Fish BL, Irving AA, Sharma M, Driscoll CD, Moulder JE. Chronic oxidative stress as a mechanism for radiation nephropathy. Radiat Res. 2009;171:164–72.
73. Cohen EP, Lenarczyk M, Fish BL, Jia S, Hessner MJ, Moulder JE. Evaluation of genomic evidence for oxidative stress in experimental radiation nephropathy. J Genet Disord Genet Rep. 2013;2: pii: 1000101.
74. Cohen SR, Cohen EP. Chronic oxidative stress after irradiation: an unproven hypothesis. Med Hypotheses. 2013;80:172–5.
75. Williams JP, Hill RP, Haston C, Johnston C, Miller J, Zimmerman C, Hernady E, Reed C, Finkelstein JN. A combined therapeutic approach to pulmonary mitigation following a radiological event. Abstract, 56th Annual Meeting of the Radiation Res. Society, 2010; PS1.61.
76. Jiang F, Zhang Y, Dusting GJ. NADPH oxidase-mediated redox signaling: roles in cellular stress response, stress tolerance, and tissue repair. Pharmacol Rev. 2011;63:218–42.

Chapter 12
Molecular Basis for Anticancer and Antiparasite Activities of Copper-Based Drugs

Ana Maria Da Costa Ferreira, Philippe Alexandre Divina Petersen, Helena Maria Petrilli, and Maria Rosa Ciriolo

12.1 Introduction

Different metal complexes have been described in last years as efficient alternative drugs for many diseases involving the organism redox network such as cancer, diabetes, and inflammation [1–4]. Complexes containing redox active metal ions that disrupt cell metabolism by binding to nucleic acids may also promote oxidative damage, influencing the redox imbalance in cells, tissues, or organs by the transfer of electrons, or by the formation of reactive oxygen and nitrogen species (ROS/RNS), especially hydroxyl radicals. By adequate design of the coordinated ligands, the resulting complexes can also act as inhibitors of specific proteins and enzymes, mostly by interactions of its particular ligands or the metal ion itself at the active site of such biomolecules. Therefore, a better understanding of the mechanism of action of such metal complexes could provide adequate settings for more controlled pharmacological and medicinal interventions in pathological conditions.

Among metal-based compounds recently developed for therapeutic purposes, copper complexes deserved special attention [5–9], usually due to their wide range of appropriate ligands, peculiar structural characteristics, and consequent redox potentials. As an essential ion present in two different oxidation states in biological medium, copper shows affinity for metal-binding sites with diverse features.

A.M.D.C. Ferreira (✉)
Departamento de Química, Instituto de Química, Universidade de São Paulo,
Av. Prof. Lineu Prestes 748, São Paulo, SP 05508-000, Brazil
e-mail: amdcferr@iq.usp.br

P.A.D. Petersen • H.M. Petrilli
Departamento de Física dos Materiais, Instituto de Física,
Universidade de São Paulo, São Paulo, SP, Brazil

M.R. Ciriolo
Dipartimento di Biologia, Università di Roma "Tor Vergata", Rome, Italy

I. Batinić-Haberle et al. (eds.), *Redox-Active Therapeutics*,
Oxidative Stress in Applied Basic Research and Clinical Practice,
DOI 10.1007/978-3-319-30705-3_12

Consequently, it can play different functions in many proteins including dismutation of reactive species, as in Cu, Zn superoxide dismutase; electron transfer, as in plastocyanin and laccases; catalytic reduction of nitrite, as in nitrite reductase; coordination and reduction of dioxygen, as in tyrosinases and cytochrome c oxidase; or oxidation of specific substrates, as in diverse oxidases. Moreover, copper has been lately considered as a key modulator of cell signal transduction pathways [10]. Enzymes such as kinases and phosphatases seem to be implicated in cellular mechanisms driven by a sequence of molecular interactions. Since changes in copper homeostasis are critical in many diseases, small molecules capable of modulating copper-protein interactions could therefore influence cell signaling associated to these diseases. Possible coordinating agents for copper ions include various types of compounds: carboxylates; thiosemicarbazones, thiosemicarbazides, and dithiocarbamates; flavones and naphtoquinones; and mostly nitrogen-containing ligands, such as imidazole, pyrazole, triazole, tetrazole, oxazole, and indole derivatives [11]. Most of such ligands have been specifically designed to provide adequate coordination environment to accommodate the two copper oxidation states usually found in biological systems, that is, Cu(I) and Cu(II), and result in complexes with proper structures, suitable for different targets [12, 13].

Herein, some studies on the reactivity of a particular class of copper complexes are discussed. Schiff base ligands derived from 2,3-dioxindoline (isatin), an oxindole formed as a metabolite of tryptophan, were designed and synthesized in order to further isolate bi-, tri-, or tetradentate coordinating agents, and the corresponding copper(II) species [14]. These complexes showed remarkable oxidative properties, catalyzing the oxidation of carbohydrates and proteins [15], in the presence of hydrogen peroxide, mainly by the formation of other more reactive ROS, and encouraged further studies about their influence in impairing the viability of tumor cells [16]. Analogous zinc(II) complexes, with no redox properties, were also prepared and investigated for comparison reasons. Different intracellular targets for these species were investigated, including DNA [17], mitochondria [18], and specific proteins [19] that could explain their remarkable proapoptotic properties. Based on these results, possible determining factors in the mechanisms of action of such copper and zinc complexes are presented, aiming at their potential application as alternative antitumor agents. Further, the influence of their redox properties were also investigated regarding their recently noticed antiparasite activity [20]; a controlled generation of reactive species that would induce an oxidative stress status with concomitant inhibition of certain proteins could be a suitable strategy to get rid of parasites, as in Trypanosomiasis or Leishmaniasis.

12.2 Oxindolimine Ligands and Corresponding Metal Complexes

Many naturally occurring or synthetic indole derivatives have been reported with diverse biological activities, and some of them revealed of medicinal importance [21]. Most of their biological activity is due to the presence of an indolin-2-one moiety in the molecule [22, 23].

Inspired by these results, different Schiff base derivatives of 2,3-dioxindoline (isatin) were designed and then synthesized by condensation reactions with suitable amines or diamines, usually in aqueous methanol or ethanol solutions. Some of these imine ligands were isolated, and characterized by NMR spectroscopy, and elemental analyses. In many cases, the corresponding metal complexes were prepared by in situ metallation of the imine ligand. In Fig. 12.1, some oxindolimine-based copper(II) and similar zinc(II) complexes are shown.

When dissolved in aqueous or buffer solutions at physiological pH, these complexes form cationic lipophilic species, a characteristic that is important for their application as potential pharmacological agents.

Similarly to the precursor isatin, all the compounds showed a keto-enol equilibrium, as illustrated for the complexes [Cu(isaepy)] and [Cu(isapn)] [24], in Fig. 12.2. Both complexes, containing the ligand in the corresponding keto- or enol-form can be isolated in solid state, depending on the pH used in the synthesis during the metallation step; in aqueous solution, however, all the possible tautomeric forms are present in different proportions.

By electron paramagnetic resonance (EPR) spectra, different structural features around the copper ion could be detected, as indicated by different hyperfine parameters, with complex geometry varying from typical tetragonal structure to more distorted tetrahedral species as shown in Table 12.1.

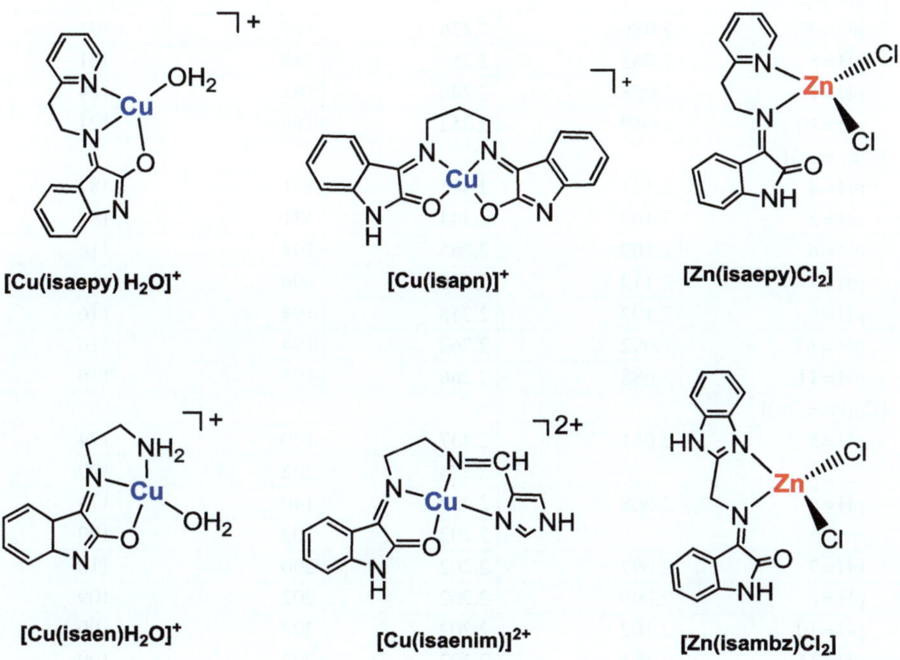

Fig. 12.1 Scheme of some oxindolimine–metal complexes, redox-active copper complexes and redox-inactive zinc complexes

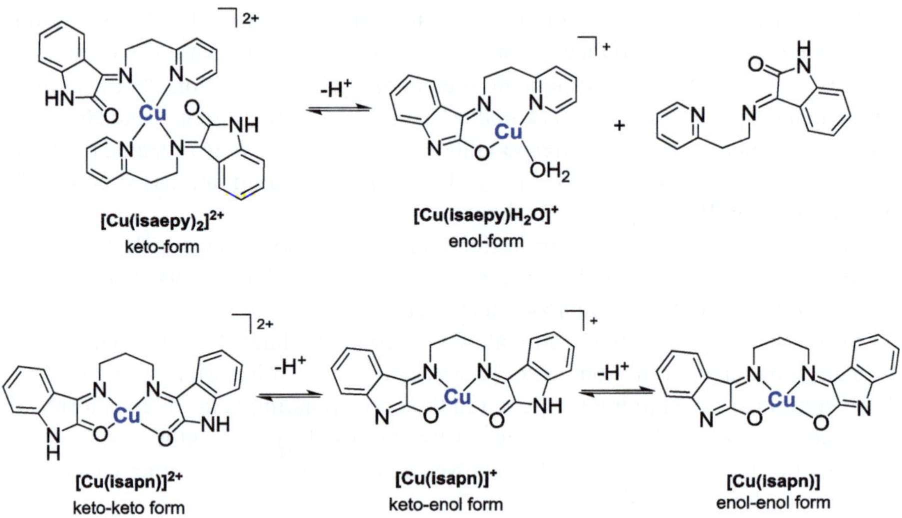

Fig. 12.2 Tautomeric equilibria for complexes [Cu(isaepy)] and [Cu(isapn)]

Table 12.1 EPR hyperfine parameters for some oxindolimine–copper(II) complexes

Compound	$g_{\perp}$	$g_{//}$	$A_{//}$, 10^{-4} cm	$g_{//}/A_{//}$, cm^{-1}
[Cu(isaepy)]				
pH = 3	2.086	2.426	127	191
pH = 5	2.065	2.259	186	121
pH = 7	2.058	2.246	181	124
pH = 10	2.059	2.252	186	121
[Cu(isapn)]				
pH = 4	2.101	2.445	131	187
pH = 5	2.102	2.444	131	187
pH = 6	2.102	2.255	194	116
pH = 7	2.112	2.256	196	115
pH = 9	2.103	2.258	194	116
pH = 10	2.092	2.262	194	116
pH = 11	2.088	2.266	195	116
[Cu(isaenim)]				
pH = 5	2.081	2.437	140	174
		2.204	202	109
pH = 6	2.068	2.437	140	174
		2.202	202	109
pH = 7	2.097	2.202	200	110
pH = 8	2.099	2.202	202	109
pH = 10	2.102	2.202	202	109
pH = 11	2.101	2.202	202	109

Note: $A_{//}$ (10^{-4} cm^{-1}) = $g_{//}$ β $A_{//}$(G) = 0.46686 × 10^{-4} $g_{//}$ $A_{//}$ (G); where β = 1.39969 MHz/G

Table 12.2 p*K* values for the tautomeric equilibria of some oxindolimine–copper(II) complexes

Complex	p*K*
[Cu(isapn)]	9.50[a]
[Cu(isaepy)]	8.94
[Cu(isaenim)]	7.86

[a]Value corresponding to two protons released

According to these data, enol forms tend to supply more tetragonal surroundings around the copper ion, while keto forms provide environments more strongly distorted toward tetrahedral geometry for the metal ion. The ratio of spectroscopic parameters $\mathbf{g}_{//}/\mathbf{A}_{//}$ is often used to estimate this distortion around the copper ion [25]. For all the complexes in Table 12.1, the parameters values indicate that the corresponding enol forms are predominant at pH 7.4. In the case of [Cu(isapn)], with two indole moieties, the mixed keto-enol form seems to be prevalent in aqueous solution at this pH. For [Cu(isaenim)] both forms are noticeable at pH 5–6.

Complementary data by UV/Vis spectrophotometric titrations of aqueous solutions of the copper(II) complexes with NaOH allowed the estimation of the corresponding p*K* values for the tautomeric equilibria (12.1) [26] as summarized in Table 12.2.

$$\begin{aligned}&\text{keto-form} + \text{OH}^- \leftrightarrows \text{enol-form} + \text{H}_2\text{O}\,\text{K}\\&= \left[\text{enol-form}\right] / \left[\text{keto-form}\right]\left[\text{OH}^-\right]\end{aligned} \quad (12.1)$$

12.3 Thermodynamic Stability of Oxindolimine–Copper(II) Complexes

In order to test the viability of these metal complexes in biological fluids, their relative thermodynamic stability was verified through circular dichroism (CD) spectra, monitoring the competitive equilibria of each complex with human serum albumin (HSA). Titrations of HSA (600 mM) with increasing amounts of each copper complex indicated a high stability of all of them, with respect to copper ion dissociation from the complexes and further insertion at the selective *N*-terminal site of the protein [27]. In these experiments, a standard curve with the aqua complex $[Cu(H_2O)_4]^{2+}$ was used to calculate the extension of [Cu(HSA)] species formation in each case monitored by the characteristic signal centered at 515 nm [(+)490 nm and (−)564 nm] [28]. The literature value for the association of copper at the terminal site of HSA, $\log K_{[Cu(HSA)]} = 12.0$ [29] was used to calculate the values for the relative stability constants ($\log K_{[CuL]}$) of the complexes (Table 12.3), considering the insertion of one copper ion per protein.

Similar experiments were carried out with the analogous zinc(II) complexes, indicating that the stability of the compounds is of the same order (Table 12.3) [19]

Table 12.3 Relative stability constants (log *K*) for some oxindolimine metal complexes determined by CD measurements using HSA as competitive metal ligand

Cu complex	log $K_{[CuL]}$	Zn complex	log $K_{[ZnL]}$
[Cu(isaepy)]	11.6	[Zn(isaepy)]	7.5
[Cu(isapn)]	12.8	[Zn(isapn)]	7.2
[Cu(isaenim)]	12.4	[Zn(HSA)]	7.1[a]
[Cu(HSA)]	12.0[b]		

Experimental conditions as indicated in cited references
[a][30]
[b][27, 29]

as that for the binding of zinc ion to the protein; for the [Zn(HSA)] species, log $K_{[Zn(HSA)]}=7.1$ [31]. In the zinc(II) case, at least two metal-binding sites are detected, one at Cys34. Therefore, these values attested that all the complexes in this series are of high thermodynamic stability and that they can be transported as metal–ligand species by blood influx until they reach intracellular targets.

Complementary experiments in the UV range indicated that some of the complexes were also able to modify the α helix content of the protein as verified by the decreasing of the band at 208 nm in CD spectra, after addition of increasing amount of the metal complex [28].

12.4 Pro-oxidant Properties

Previously, the pro-oxidant properties of the oxindolimine–copper(II) complexes have been evaluated toward oxidation of hexoses (glucose, fructose, and galactose), in alkaline media, in a process mediated by the formation of ROS, and showing first-order dependence on the catalyst, [Cu(isapn)] [15]. The proposed mechanism involved base-catalyzed substrate enolization and metal-catalyzed enediol oxidation as rate-determining steps.

This capability of the different copper complexes studied in generating ROS, in the presence of hydrogen peroxide, was further estimated by spin trapping EPR experiments, using DMPO as spin scavenger [16]. Spectra registered after 15 min of incubation with $[H_2O_2]=2.5$ mM in phosphate buffer 50 mM at pH 7.4 using [DMPO] = 100 mM and [CuL] = 100 μM indicated concentrations of formed DMPO-OH adduct (hyperfine constants, $a_N=a_H=14.9$ G) in the range 0.4–5.0 μM. This generation of hydroxyl radicals was verified to be very dependent on the nature of the ligand, being [Cu(isapn)] one of the most active complexes.

Also, regarding the oxidation of 2-deoxy-D-ribose, a series of oxindolimine–copper(II) complexes were tested, and some of them were more active than the aqua-complex such as [Cu(isaenim)], while others showed lower activity such as [Cu(isaepy)] [26]. In any case, both were able to cause oxidative damage to DNA.

12.5 DNA as One of the Preferential Targets

All the oxindolimine complexes in the series were able to displace ethidium bromide (EB) from DNA, as monitored by fluorescence measurements, in a process modulated by the ligand [32]. Addition of increasing amounts of each complex reduced the fluorescence of EB-DNA solutions, suggesting a probable competition for the EB-binding site at DNA. Based on these data, using plasmid DNA, values of the corresponding binding constant (K_a) in the range $3.64–8.85 \times 10^2\ M^{-1}$ and binding sites number (n) around 0.70 were obtained in the case of copper complexes. These values were comparable to those of [Cu(*o*-phen)] species ($K_a = 7.55 \times 10^2\ M^{-1}$, $n = 0.60$), a known DNA binder [33].

The copper(II) complexes in this series also showed good nuclease activity, being able to cause single- and double-strand DNA cleavage in the presence of hydrogen peroxide (120 μM). Typical results indicated significant single DNA cleavage, after 15 min incubation at complex concentrations 200 nM to 1 μM, and remarkable double cleavage in the concentration range 25–50 μM, after 30 min incubation [26, 32].

Further studies to verify the occurrence of hydrolytic steps, using DNA or RNA oligomers labeled at phosphorus 5′-end, showed no hydrolytic cleavage of the ^{32}P-phosphate groups [32], indicating that DNA damage occurs predominantly by an oxidative mechanism. However, the ability of damaging DNA seems not to be directly related to the facility on oxidizing carbohydrates. In spite of being more active in the oxidation of 2-deoxy-D-ribose, [Cu(isaenim)] was kinetically less efficient on forming linear DNA species than [Cu(isaepy)], which showed better nuclease properties [26].

12.6 Proapoptotic Activity

Apoptosis represents a cell death program designed to rapidly remove unwanted and potentially dangerous cells. In addition, the incorrect regulation of apoptosis is associated with a variety of pathologies including cancer, neurodegenerative diseases, and ischemic stroke [34]. The process is characterized by morphological changes typical of apoptotic cell (plasma membrane blebs, cell shrinkage, chromatin condensation, protein aggregates, and apoptotic bodies formation). A class of proteases, termed caspases (cysteine proteases that specifically cleave at an aspartate residue) [35], were fundamental in order to correctly accomplish the death program. Among the different classes of caspases, caspase-3, -6, and -7 (*executioners*) are directly involved in proteolysis of cellular macromolecules leading to cell death, whereas caspase-2, -8, -9, and -10 (*initiators*) are fundamental in the commitment of apoptosis, as they activate executer caspases. Three distinct pathways of induction of apoptosis have been characterized. The first is associated with perforin-mediated membrane alteration and granzyme B activation, which leads to direct activation of caspases during the activation of T killer lymphocytes on tumoral or viral-infected cells [34]. The second pathway is mediated by "death receptors" including tumor necrosis factor (TNF), Fas (APO1/CD95), and TNF-related apoptosis inducing ligand (TRAIL) [36]. The third pathway is the "mitochondrial pathway", which is considered the most important one because it embodies many apoptotic stimuli (e.g., chemical or physical stresses, growth factor withdrawal, and pro-oxidant molecules). Particularly, cytochrome *c* released into the cytosol is the primary event for the recruitment and formation of the apoptosome, a multimolecular complex in which procaspase-9 is activated and, in turn, activates the downstream executioner caspases [37]. Moreover, other proapoptotic factors are resident in mitochondria. Indeed, the permeabilization of the outer mitochondrial membrane, which is mainly controlled by members of the Bcl-2 family protein, is the crucial event that results in the release into the cytosol of a second mitochondria-derived activator of caspases/direct inhibitor-of-apoptosis (IAP) binding protein with low p*I* (SMAC/Diablo), Omi serine protease (Omi/HtrA2), apoptosis inducing factor (AIF), or endonuclease G [38, 39]. Mitochondria are also the place where reactive oxygen species (ROS) are mostly produced as by-product of oxygen consumption either under physiological conditions or during the execution of apoptosis. In fact, several apoptotic inducers are pro-oxidant molecules and some apoptotic pathways are inhibited by antioxidants [40].

Beside the key role in organs development, apoptosis serves as a natural obstacle to cancer development [41]. The ability of cancer cells to evade apoptosis is, therefore, an essential hallmark of cancer [42]. Moreover, tumor-related defects in apoptosis not only underpin tumorigenesis but also result in drug resistance [43]. Synthesis and characterization of novel antitumor compounds represented a forefront of research either for overcoming drug resistance or for yielding more specific and less toxic therapies. The principal and oldest targets of cancer treatment were DNA and cellular replication. These strategies have, in the last years, been enriched by developing drugs targeting other intracellular compartments and cell functions, as well as the microenvironment of cancer cells [44, 45]. For instance, cancer cells produce levels of ROS higher than normal cells do due to their more intense metabolic activity, lower levels of antioxidants, and higher mitochondrial defects. This intrinsic oxidative stress status selectively targets malignant cells for therapeutic strategies based on further ROS production and consequent irreversible oxidative insult finally inducing apoptosis [46–48]. Doxorubicin, daunorubicin, and bleomycin are among the most used and well-known examples of such chemotherapeutics [49, 50], which by easily undergoing one-electron redox cycling with oxygen are efficient mediators of oxidative insult. In this context, copper complexes could exert antitumor properties through their capability to efficiently react with molecular oxygen producing ROS. The molecular mechanisms underlying the activation of apoptosis upon treatment with two of the isatin–diimine copper(II) complexes, [Cu(isapn)] and [Cu(isaepy)], were investigated in SH-SY5Y neuroblastoma cells. The main results were outlined as follows:

12.6.1 [Cu(isapn)] and [Cu(isaepy)] Were Cell Permeable and Inducer of Oxidative Stress

These complexes showed preference for mitochondria as intracellular target in addition to DNA targeting [18]. This was demonstrated by following the copper uptake by atomic absorption analyses. Treatments with both compounds resulted in a rapid increase of intracellular copper content, with [Cu(isaepy)] being more efficiently incorporated within the cells than [Cu(isapn)] was. The extent of apoptosis was consistent with the kinetics of intracellular copper uptake, modulated by the ligand. For instance, *isaepy* ligand was able to deliver more copper inside neuroblastoma SH-SY5Y cells than *isapn*, after 12 or 24 h treatment with 50 μM [Cu(isapn)] or [Cu(isaepy)] complexes. The intracellular copper concentrations were of 7.5 and 4.0 nmol/mg protein, respectively, after 12 h incubation, in comparison to 1.2 nmols Cu/mg protein in the corresponding control experiments with simple copper salt [17].

Moreover, we demonstrated that the complex were chemically stable as no changes in the EPR parameters were evidenced under the experimental conditions. The most intriguing difference in the behavior of the two copper complexes was on their capability to produce ROS, [Cu(isapn)] being the fastest (ROS increase was detectable as early as 3 h after treatment) and most efficient (damages to both proteins and lipids was determined by increased carbonyls and malondialdehyde content). In contrast, [Cu(isaepy)] treatment resulted in oxidative protein damage at a later stage without compromising lipid structures [17].

12.6.2 [Cu(isapn)] and [Cu(isaepy)] Induce Cell Cycle Arrest and Caspase-Dependent Apoptosis

The cell cycle arrest was assessed by the fact that the p53/p21 pathway was induced at 12 h upon [Cu(isapn)] treatment and at 24 h upon treatment with [Cu(isaepy)]. Apoptosis was assayed by the identification of the active form of caspase-9 and caspase-3 and the proteolyzed PARP, as well as the release of cytochrome *c* into the cytosol. The caspase-dependent apoptotic response that occurred after treatment with either copper complex was efficiently inhibited by the pan-caspase inhibitor zVAD-fmk, confirming that caspase-mediated apoptosis was the principal mechanism underlying cell death [17] (Fig. 12.3).

12.6.3 [Cu(isapn)] and [Cu(isaepy)]-Mediated Toxicity Was a Copper-Mediated Event that Induces Nuclear and Mitochondrial Dysfunction

Due to their lipophilic behavior, [Cu(isapn)] and [Cu(isaepy)] could vehicle copper and let the catalysis of redox reactions and ROS production to take place also within specific intracellular organelles. We demonstrated, by atomic absorption analyses, a

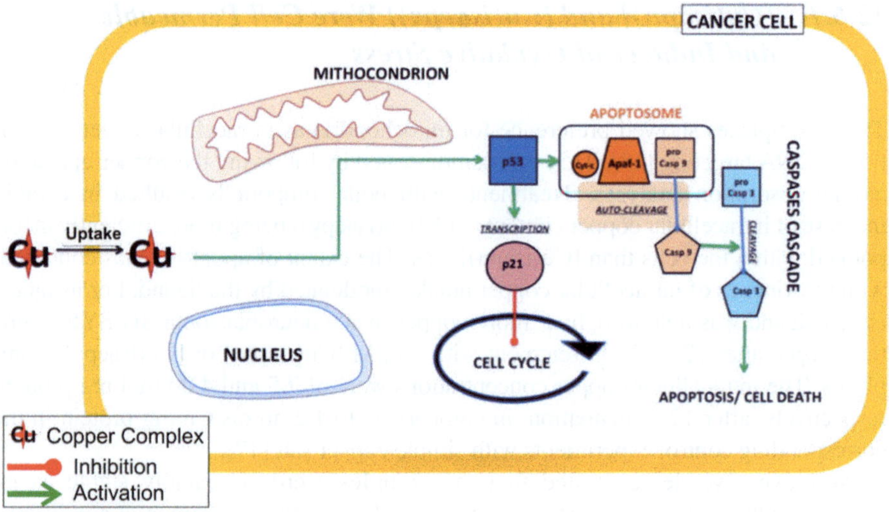

Fig. 12.3 [Cu(isapn)] and [Cu(isaepy)] induce cell cycle arrest and caspase-dependent apoptosis

significant increase of copper in nuclear and mitochondrial fractions upon treatment with both compounds. The degree and the occurrence of the oxidative insult was evidenced by staining nuclei with an antibody against the phospho-active histone H2A.x, which is phosphorylated on Ser139 after DNA double-strand break. In this case, the appearance of discrete *foci*, indicating the recruiting sites of the DNA repair machinery, revealed more DNA-specific damage mediated by [Cu(isaepy)]. Concomitantly, by the use of MitoTracker Red probe, a huge number of mitochondria resulted completely depolarized after treatment with [Cu(isaepy)]. These results give strength to the hypothesis that although [Cu(isaepy)] does produce less ROS, it was more efficient than [Cu(isapn)] in inducing apoptosis, because of its ability to specifically accumulate and damage fundamental organelles, such as the nucleus and mitochondria [17] (Fig. 12.4).

12.6.4 [Cu(isapn)] and [Cu(isaepy)] Induce Apoptosis Via the Mitochondrial Pathway in Different Tumor Cell Types

A general proapoptotic activity of the more efficient copper complex [Cu(isaepy)] was finally assessed by the results obtained with other human tumor cells such as the promonocytoma U937 and the melanoma M14. As previously done with SH-SY5Y cells, apoptotic analyses confirmed that the intrinsic mitochondrial pathway represents the preferential route for the induction of apoptosis.

The conclusion raised by these studies indicate that copper is fundamental in mediating [Cu(isapn)] and [Cu(isaepy)] cytotoxicity, because of its capability to

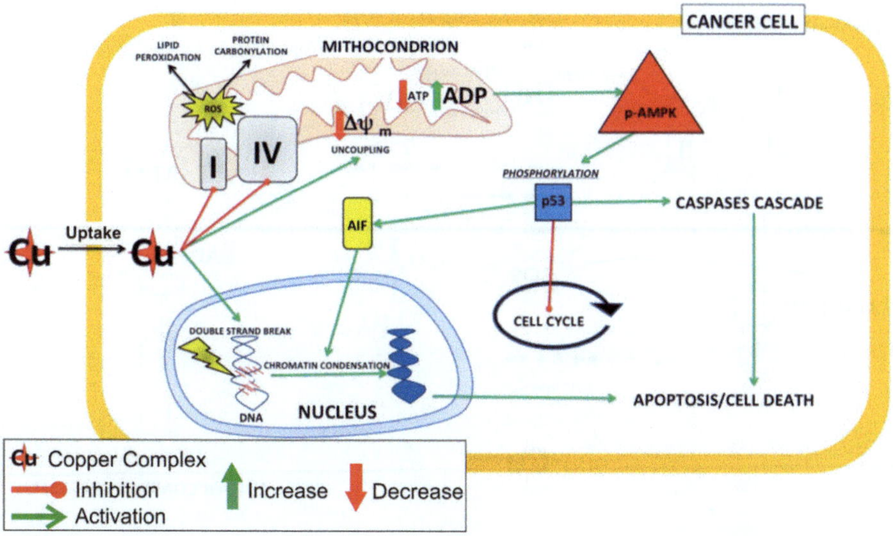

Fig. 12.4 [Cu(isapn)] and [Cu(isaepy)] induce nuclear and mitochondrial dysfunction

intracellular catalyze one-electron redox cycle reactions with oxygen thus producing ROS. Moreover, the combine of this property with the chemical structure of the organic ligand binding the metal ion seems to give the specificity of the cellular damage induced. We suggested that copper could function as iron in the presence of bleomycin [51, 52] in nuclear damage, whereas the isatin-imine ligand was fundamental to allow copper redox reactions at the mitochondrial level. This hypothesis was raised by observing a similarity in the chemical structure of [Cu(isapn)] and [Cu(isaepy)] with that of delocalized lipophilic cations (DLC), a class of charged molecules able to permeate mitochondria along their negative transmembrane potentials [53]. Therefore, taking into account that tumor cells are characterized by higher plasma and mitochondrial membrane potentials with respect to normal cells [54] by specifically changing the chemical characteristics of the isatin-imine ligand, we could modulate the cytotoxic effects induced. Exploiting the plasticity of this new class of compounds to improve the therapeutic selectivity to different tumor histotypes will be the challenge of next years.

12.6.5 [Cu(isaepy)] as a Delocalized Lipophilic Cations

Following research on the antitumor activity of copper complexes was centered on the mitochondrion toxicity exerted by [Cu(isaepy)] for further validating the role of mitochondria impairment in the cell commitment to apoptosis. We demonstrated that [Cu(isaepy)] increases oxygen consumption, mimicking uncoupling molecules, without a concomitant raise in ATP production. Moreover, carbonylated proteins, a

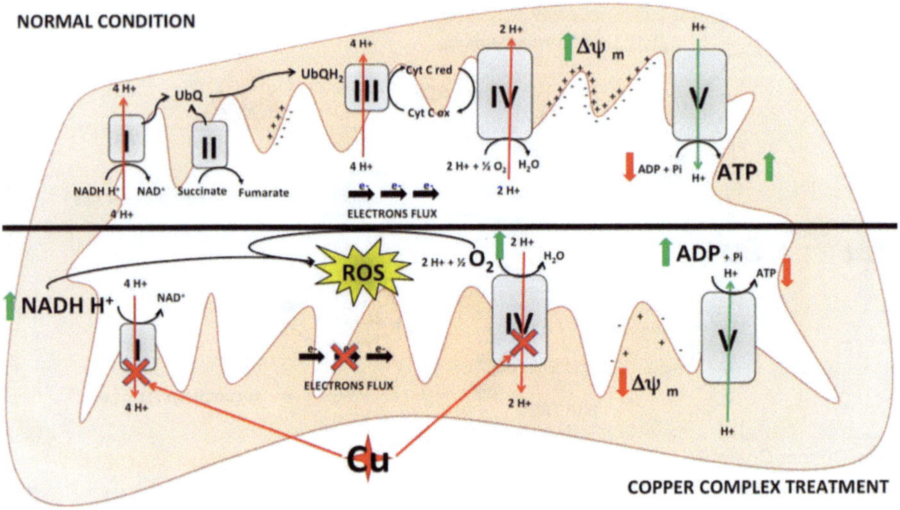

Fig. 12.5 [Cu(isaepy)] behaves as a delocalized lipophilic cations

byproduct of protein oxidation, were mostly occurring at the mitochondria level. These results suggested a specific damage to mitochondria that was confirmed by a decrease in the levels of respiratory complexes (Fig. 12.5). In fact, the decrease of complex I and IV, which we experimentally detected, represents a hallmark of oxidative damage to mitochondria and was previously reported to occur upon copper intoxication. Overall, the data indicate that [Cu(isaepy)] can be definitively included among DLCs and propose its putative use as a novel DLC compound useful for in vivo approaches in cancer treatment [18].

12.6.6 [Cu(isaepy)] in Combining Antitumor Therapy

Several studies demonstrated that cancer cells undergo a complex metabolic and bioenergetics reprogramming aimed at sustaining higher rate of proliferation. This biochemical reorganization involves mainly an increase in the glycolytic rate even under normal oxygen tension (aerobic glycolysis or the "Warburg effect"). Therefore, ATP-depleting molecules, such as glycolytic inhibitors (2-deoxyglucose) [55] and 3-bromopyruvate (3BrPA) [56], or compounds able to affect oxidative phosphorylation (e.g., rhodamine 123, dequalinium chloride, etc.) have been employed as death inducers of cancer cells [57]. Among chemotherapeutic strategies that of combining two or more drugs able to affect metabolic activity of cancer cells at low doses is envisaged as promising in order to enhance the drugs killing properties, reduce systemic toxicity, and overcome drug resistance.

We demonstrated that energy deficiency was profoundly implicated in [Cu(isaepy)]-mediated toxicity. Indeed, AMPK (AMP-activated protein kinase) was

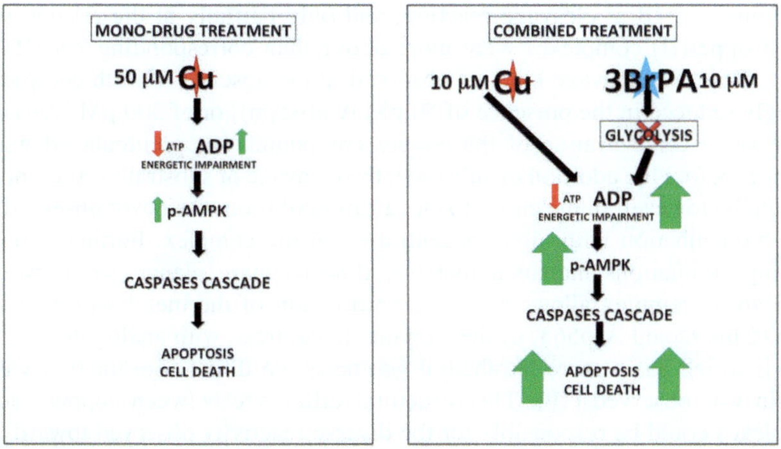

Fig. 12.6 [Cu(isaepy)] in combining antitumor therapy

efficiently activated under the treatment as upstream sensor of energy impairment. AMPK is a heterotrimeric serine/threonine protein kinase which acts as a fuel sensor in eukaryotic cells. It is activated physiologically when the AMP/ATP ratio increases and mediates the phosphorylation of a large number of metabolic enzymes aimed at the restoration of ATP levels [58]. Then, by using a combined treatment with low doses of [Cu(isaepy)] and 3BrPA, we demonstrated that it was effective and selective in inducing apoptosis in neuroblastoma cells, without any significant toxicity toward differentiated primary cortical neurons (Fig. 12.6).

Overall, the results obtained represent proof-of-principle evidence that anticancer effects can be potentiated by combining drugs that counteract the strategic biochemical adaptations of cancer cells and, also in this case, the copper complexes are of particular interest [59].

12.7 Inhibition of Specific Proteins

An extensively used strategy to target cancer is the development of new small molecules capable of inhibiting specific proteins, participating in vital processes, such as topological control of DNA strands or control of cell cycle. In this context, some specific proteins were verified as potential targets for the focused metal complexes.

Topoisomerases are enzymes that catalyze topological changes of DNA during replication, transcription, recombination, and repair processes [60]. Particularly, human topoisomerase IB has been a target for many drugs, classified as poisons or inhibitors, depending on its action upon the enzyme [61]. It has been verified through enzymatic kinetic assays and molecular docking simulations that the oxindolimine–metal complexes here described were able to efficiently inhibit topoisomerase IB at cleavage reaction, and only partially at the relegation step [19]. Copper(II) complexes were more active than corresponding zinc(II) complexes. The fast cleavage kinetics observed in the absence of each complex was strongly reduced in the presence of 50 μM [Cu(isapn)] or of 300 μM [Zn(isapn)]. If the same concentration of the copper compound was preincubated with the enzyme, before the addition of substrate, the cleavage of substrate was completely inhibited. However, in the case of zinc, a full inhibition was never observed, even after preincubation with high concentration of the complex. Further, molecular docking simulations indicated that the almost square planar geometry of the [Cu(isapn)] complex allows a direct coordination of the metal with two amino acids (Glu492 and Asp563) of the enzyme. In contrast, with analogous zinc compound, which has a more tetrahedral geometry, no direct coordination with the protein was observed [19]. This structural difference between copper and zinc complexes could be responsible for the diverse reactivity observed toward topoisomerase IB.

Kinases are a huge class of phosphate transferase proteins responsible for the regulation of cell survival and proliferation through phosphorylation reactions. There are approximately 518 kinases encoded in the human genome. Additionally, the inhibition of these proteins can often be tolerated by normal cells, and this opens a possibility for the selective killing of tumor cells. For this reason, these proteins are one of the most intensively pursued classes of drug targets, especially for cancer therapy [62]. Some inhibitors of kinases were developed as competitors at the ATP-binding site (type I inhibitors, acting at the hydrophobic pockets of activation loop of the protein and type II inhibitors, recognizing the inactive conformation of the protein) [22], and others were designed to act at an allosteric site [63]. Around 80 of those inhibitors advanced to some stage of clinical trial, including compounds with indole moieties [64, 65].

Having structural similarity to already tested inhibitors of kinases containing indole group, such as SU9516, SU6656 and SU6668 [66], shown in Fig. 12.7, the oxindolimine ligands provide similar features to copper(II) or zinc(II) complexes and are also expected to inhibit kinases. Theoretical studies were then performed to verify possible interactions of oxindolimine ligands at the ATP-binding site of kinase CDK2 and corroborate with the assumption of effectiveness of such ligands. Density Functional Theory (DFT) [67–70], molecular docking [71], and molecular dynamics (MD) [72, 73] were used to build a model that simulates the interaction of

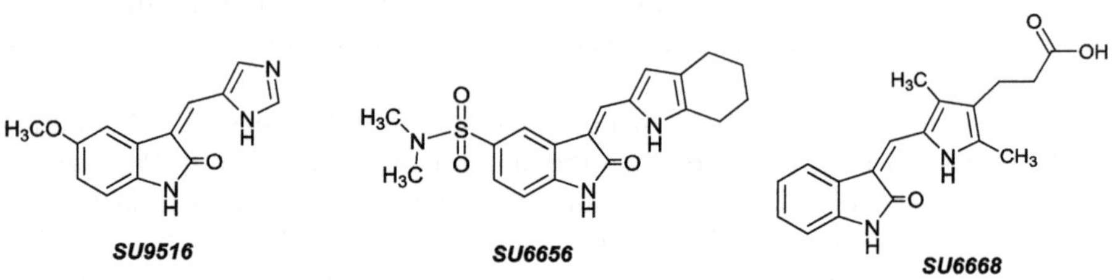

Fig. 12.7 Scheme of some indole derivatives capable of inhibiting kinases and already tested clinically as anticancer therapeutics

SU9516 and of *isapn* ligand inserted at the ATP-binding site. SU9516 was used as a reference model in the comparison with other ligands. In Fig. 12.8a, the docking of *isapn* ligand at the ATP site of CDK2 is shown as an example, including the amino acids (GLU81, LEU83, etc.) which are distant closer than 3 Å. The molecular docking of SU9516 was also carried out, in order to validate the procedure used, revealing that the theoretical conformation matches very well the X-ray determined structure as shown in Fig. 12.8b.

In Fig. 12.9, complementing the docking results, some relevant *isapn* interactions at CDK2 obtained by MD are shown. Figure 12.9a exhibits the number of

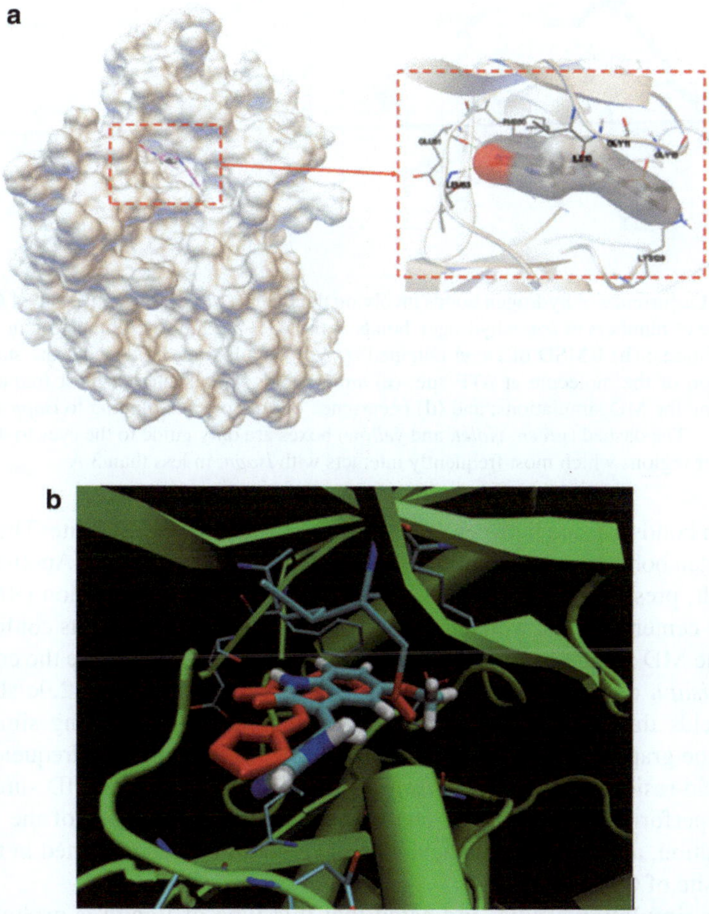

Fig. 12.8 Molecular docking of *isapn* ligand, in comparison to the SU9516 compound, at the ATP-binding site of CDK2. (**a**) Molecular docking of *isapn* at the ATP-binding site (*white* surface), exhibiting the potential surface of *isapn* (*grey*) and the amino acids localized less than 3 Å from it. (**b**) Comparison between the SU9516 structures obtained by X-ray crystal data (PDB ID: 3PY0), [22, 74] and by molecular docking at the ATP-binding site of CDK2. The determined X-ray structure is shown in *red*

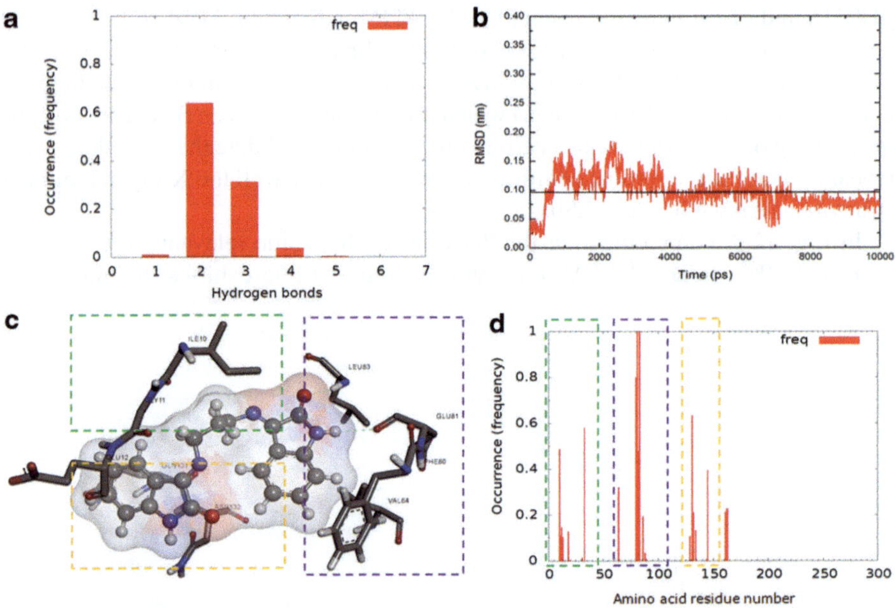

Fig. 12.9 Occurrence of hydrogen bonds involving the ligand *isapn* at the active site of CDK2. (**a**) Occurrence of numbers of *isapn* hydrogen bonds in the ATP-binding site of the protein during the MD simulations; (**b**) RMSD of *isapn* obtained by MD simulations, evidencing the stability and equilibration of the molecule at ATP site; (**c**) amino acids which interact most frequently with *isapn* during the MD simulations; and (**d**) occurrence of amino acids binding to *isapn* in the MD simulations. The dashed (*green*, *violet*, and *yellow*) boxes are only guide to the eyes to define residue number regions which most frequently interacts with *isapn*, in less than 3 Å

hydrogen bonds formed between *isapn* and amino acids at the ATP site. The number of hydrogen bonds suggests stability of the molecule at the ATP site. Another important result, presented in Fig. 12.9b, is the root-mean-square deviation (RMSD) of the *isapn* center of mass, which indicates that the ligand maintains its conformation during the MD simulations. The RMSD value is important to pursue the conformations of *isapn* during simulations as well as its stability. Figure 12.9c shows the amino acids that are most frequently in contact with *isapn* during simulations. Finally, the graph in Fig. 12.9d shows the number of occurrences (frequency) each amino acid residue interacts with *isapn* at the protein pocket. By MD simulations, one can perform all this kind of analysis supporting an estimate of the stability, conformation, and biological behavior of the oxindolic ligand inserted in the ATP-binding site of CDKs.

These simulation studies indicated that this type of ligand is probably very efficient on inserting itself at the ATP-active binding site of kinases. Experimental studies of kinase inhibition in the presence of oxindolimine metal complexes are presently under development in our lab, based on the above-reported simulation features, and preliminary results (data not shown) are very promising [75].

12.8 Antiparasite Activity

Chagas disease (American trypanosomiasis) is a neglected tropical disease [76] caused by protozoan parasites, whose currently available clinical drugs, *N*-benzyl-2-nitroimidazole-1-acetamide (benznidazole) or 4[(5-nitrofurfurylidene)amino]-3-methylthiomorpholine-1,1-dioxide (nifurtimox), have been developed many years ago [77, 78]. The efficacy of these traditional antiparasite drugs seems to be supported by reduction of nitro group to a nitroanion radical by nitroreductases, with subsequent formation of ROS in the presence of oxygen in the case of nifurtimox [79]. On the other hand, the main action of benznidazole seems to be based on interactions of nitroreduction intermediates with parasite components, or binding to proteins, lipids, and DNA [80–82], catalyzing double-stranded breaks in the parasite DNA [83]. Since these drugs are toxic and of limited efficacy, many natural and synthetic compounds have been reported in the literature as alternative trypanosomicidal agents [84], including some copper complexes [85, 86], based on different modes of action [87].

Based on its chemical properties and biological activities previously described, the oxindolimine complexes were tested against trypomastigote and amastigote forms of *T. cruzi,* a flagellate protozoa responsible for Chagas disease, and revealed good results, after 24 h incubation at 37 °C, as shown in Table 12.4. The corresponding selective indexes (SI) estimated were of the same order or better than those reported for benznidazole, after 24 h incubation of the parasites [90]. These results also showed that the activity of the complexes is modulated by the ligand. Further, similar values of SI were obtained for copper(II) and zinc(II) complexes, coordinated by the same ligand, indicating that redox activity was not crucial for its antichagasic action.

Topoisomerases have been already investigated as promising targets of antiparasite drugs [91], and very recently inhibition of kinases was claimed to be an essential factor in antitrypanosomiasis activity [92]. Therefore, both types of proteins can be important targets for these metal complexes and can explain their trypanosomicidal activity.

Table 12.4 Values of selectivity index (SI) for the tested oxindolimine complexes in comparison to benznidazole and other related copper or zinc-based compounds

Selective Index[a] (incubation time)		
[Cu(isapn)]	4.8 (24 h)	
[Cu(isaepy)]	3.7 (24 h)	
[Zn(isapn)]	5.6 (24 h)	
[Zn(isaepy)]	2.0 (24 h)	
Benznidazole	2.7[b] (24 h)	0.8[c] (72 h)
$[Cu(dmtp)_4(H_2O)_2](ClO_4)_2 \cdot 2dmtp \cdot 2H_2O$[c]		16.5 (72 h)
$[Zn(dmtp)_2(H_2O)_4](ClO_4)_2 \cdot 2H_2O$[c]		3.8 (72 h)

[a]SI = 50 % cytotoxic concentration (LD_{50}) obtained with macrophages divided by the 50 % inhibitory concentration (IC_{50}) obtained with *T. cruzi*
[b][88]
[c][89]; dmtp = 5,7-dimethyl-1,2,4-triazolo[1,5-a]pyrimidine

12.9 Concluding Remarks

Based on studies already carried out or under development in our lab, some determining aspects of possible mechanisms of action of oxindolimine–copper(II) complexes in biological systems were discussed. These metal-based compounds present tautomeric equilibria in aqueous solution, with predominance of corresponding keto or enol forms depending on the pH. This property is crucial for the modulation of biological activity of these metal complexes, defining the charge and the coordination sphere of the central metal. Keto forms, containing one oxindole moiety, act as bidentate ligands, whereas enol forms usually act as tridentate species, providing a more tetragonal or planar configuration to the complex.

These complexes showed good pro-oxidant properties, catalyzing the oxidation of carbohydrates via formation of reactive oxygen species, mainly hydroxyl radicals. They are able to bind to DNA, probably at major or minor grooves, depending on the ligand, causing remarkable oxidative damage with DNA double-strand cleavage in the presence of hydrogen peroxide. Inside the cell, they were able to oxidize proteins and lipids. Besides DNA, mitochondrion is also a preferential intracellular target, with consequent decreasing of membrane potential, and triggering of apoptosis. These copper complexes increase oxygen consumption, mimicking uncoupling molecules, without a concomitant raise in ATP production. Moreover, carbonylated proteins resulting from protein oxidation were mostly occurring at the mitochondria level, suggesting a specific damage to mitochondria. This was confirmed by a decrease in the levels of respiratory complexes. Further, it was also demonstrated that by using a combined treatment with low doses of an oxindolimine–copper(II) complex and a glycolytic inhibitor, such as 3-bromopyruvate (3BrPA), an effective and selective induction of apoptosis occurs in neuroblastoma cells, without any significant toxicity toward differentiated primary cortical neurons.

Additionally, interactions with certain proteins were verified, with molecular recognition of oxindole moieties present in such complexes at the active sites of kinases and topoisomerases, probably responsible for substantial inhibition of the protein functions. These results could also contribute significantly to elucidate antitumor and antiparasite activities exhibited by this class of metal complexes. Therefore, the coordinated ligand is important not only to vehicle copper ion into the cell and thus inducing oxidative stress but also to target organelles, and specific proteins, being responsible for the recognition of such metal complexes by key biomolecules.

Acknowledgments AMDCF is grateful to INCT—Redox Processes in Biomedicine, NAP/USP, and CEPID-Redoxoma (grant 2013/07937-8) for financial support. MRC is grateful to AIRC. PDP and HMP thank *Grupo Nanomol* for relevant discussions. The authors are also thankful to FAPESP (grant 2011/50318-1), CNPq, CAPES (Computational Biology Project), and to Marco Rosina for helping in drawing Figs. 12.3, 12.4, 12.5, and 12.6.

References

1. Ronconi L, Sadler PJ. Using coordination chemistry to design new medicines. Coord Chem Rev. 2007;251:1633–48.
2. Trondl R, Heffeter P, Kowol CR, Jakupec MA, Berger W, Keppler BK. NKP-1339, the first ruthenium-based anticancer drug on the edge to clinical application. Chem Sci. 2014;5:2925–32.
3. Groessl M, Hartinger CG. Anticancer metallodrug research analytically painting the "omics" picture—current developments and future trends. Anal Bioanal Chem. 2013;405:1791–808.
4. Gabbiani C, Magherini F, Modesti A, Messori L. Proteomic and metallomic strategies for understanding the mode of action of anticancer metallodrugs. Anti-Cancer Ag Med Chem. 2010;10:324–37.
5. Santini C, Pellei M, Gandin V, Porchia M, Tisato F, Marzano C. Advances in copper complexes as anticancer agents. Chem Rev. 2014;114:815–62.
6. Tardito S, Marchiò L. Copper compounds in anticancer strategies. Curr Med Chem. 2009;16:1325–48.
7. Weder JE, Dillon CT, Hambley TW, Kennedy BJ, Lay PA, Biffin JR, Regtop HL, Davies NM. Copper complexes of non-steroidal anti-inflammatory drugs: an opportunity yet to be realized. Coord Chem Rev. 2002;232:95–126.
8. Salas PF, Herrmann C, Orvig C. Metalloantimalarials. Chem Rev. 2013;113:3450–92.
9. Tamasi G, Serinelli F, Consumi M, Magnani A, Casolaro M, Cini R. Release studies from smart hydrogels as carriers for piroxicam and copper(II)-oxicam complexes as anti-inflammatory and anti-cancer drugs. X-ray structures of new copper(II)-piroxicam and -isoxicam complex molecules. J Inorg Biochem. 2008;102:1862–73.
10. Grubman A, White AR. Copper as a key regulator of cell signalling pathways. Expert Rev Mol Med. 2014;16:e11. doi:10.1017/erm.2014.11.
11. Marin-Hernandez A, Gracia-Mora I, Ruiz-Ramirez L, Moreno-Sanchez R. Toxic effects of copper-based antineoplastic drugs (Casiopeinas) on mitochondrial functions. Biochem Pharmacol. 2003;65:1979–89.
12. Hindo SS, Frezza M, Tomco D, Heeg MJ, Hryhorczuk L, McGarvey BR, Ping Dou Q, Verani CN. Metals in anticancer therapy: copper(II) complexes as inhibitors of the 20S proteasome. Eur J Med Chem. 2009;44:4353–61.
13. Zhou W, Wang X, Hu M, Zhu C, Guo Z. A mitochondrion-targeting copper complex exhibits potent cytotoxicity against cisplatin resistant tumor cells through multiple mechanisms of action. Chem Sci. 2014;5:2761–70.
14. Cerchiaro G, Da Costa Ferreira AM. Oxindoles and copper complexes with oxindole-derivatives as potential pharmacological agents. J Braz Chem Soc. 2006;17:1473–85.
15. Cerchiaro G, Saboya PL, Da Costa Ferreira AM, Tomazela DM, Eberlin MN. Keto-enolic equilibria of an isatin-Schiff base copper(II) complex and its reactivity towards carbohydrate oxidation. Transit Met Chem. 2004;29:495–504.
16. Cerchiaro G, Aquilano K, Filomeni G, Rotilio G, Ciriolo MR, Da Costa Ferreira AM. Isatin-Schiff base copper(II) complexes and their influence on cellular viability. J Inorg Biochem. 2005;99:1433–40.
17. Filomeni G, Cerchiaro G, Da Costa Ferreira AM, De Martino A, Pedersen JZ, Rotilio G, Ciriolo M. Pro-apoptotic activity of novel Isatin-Schiff base copper(II) complexes depends on oxidative stress induction and organelle-selective damage. J Biol Chem. 2007;282:12010–21.
18. Filomeni G, Piccirillo S, Graziani I, Cardaci S, Da Costa Ferreira AM, Rotilio G, Ciriolo MR. The isatin-Schiff base copper(II) complex [Cu(isaepy)$_2$] acts as delocalized lipophilic cation, yields widespread mitochondrial oxidative damage and induces AMP-activated protein kinase-dependent apoptosis. Carcinogenesis. 2009;30:1115–24.
19. Katkar P, Coletta A, Castelli S, Sabino GL, Alves Couto RA, Da Costa Ferreira AM, Desideri A. Effect of oxindolimine copper(II) and zinc(II) complexes on human topoisomerase I activity. Metallomics. 2014;6:117–25.

20. Sabino GL, de Paula QA, Couto RA, Dario BS, Vieira LQ, Ribeiro GA, Da Costa Ferreira AM. BR 10 2013 026558-6, deposit at INPI in October 15, 2013.
21. Kaushik NK, Kaushik N, Attri P, Kumar N, Kim CH, Verma AK, Choi EH. Biomedical importance of indoles. Molecules. 2013;18:6620–62.
22. García-Echeverría C, Traxler P, Evans DB. ATP site-directed competitive and irreversible inhibitors of protein kinases. Med Res Rev. 2000;20:28–57.
23. Mohammadi M, McMahon G, Sun L, Tang C, Hirth P, Yeh BK, Hubbard SR, Schlessinger J. Structures of the tyrosine kinase domain of fibroblast growth factor receptor in complex with inhibitors. Science. 1997;276:955–60.
24. For simplicity, the complexes will be referred here as [Cu(ligand)] species, with no explicit charge.
25. Sakaguchi U, Addison AW. Spectroscopic and redox studies of some copper(II) complexes with biomimetic donor atoms: implications for protein copper centres. J Chem Soc, Dalton Trans 1979:600–8. doi:10.1039/DT9790000600
26. da Silveira VC, Luz JS, Oliveira CC, Graziani I, Ciriolo MR, Da Costa Ferreira AM. Double-strand DNA cleavage induced by oxindole-Schiff base copper(II) complexes with potential antitumor activity. J Inorg Biochem. 2008;102:1090–103.
27. Bal W, Sokołowska M, Kurowska E, Faller P. Binding of transition metal ions to albumin: sites, affinities and rates. Biochim Biophys Acta. 2013;1830:5444–55.
28. da Silveira VC, Caramori GF, Abbott MP, Gonçalves MB, Petrilli HM, Da Costa Ferreira AM. Oxindole-Schiff base copper(II) complexes interactions with human serum albumin: spectroscopic, oxidative damage, and computational studies. J Inorg Biochem. 2009;103: 1331–41.
29. Rozga M, Sokołowska M, Protas AM, Bal W. Human serum albumin coordinates Cu(II) at its N-terminal binding site with 1 pM affinity. J Biol Inorg Chem. 2007;12:913–8.
30. Blindauer CA, Harvey I, Bunyan KE, Stewart AJ, Sleep D, Harrison DJ, Berezenko S, Sadler PJ. Structure, properties, and engineering of the major zinc binding site on human albumin. J Biol Chem. 2009;284:23116.
31. Ohyoshi E, Hamada Y, Nakata K, Kohata S. The interaction between human and bovine serum albumin and zinc studied by a competitive spectrophotometry. J Inorg Biochem. 1999;75: 213–8.
32. da Silveira VC, Benezra H, Luz JS, Georg RC, Oliveira CC, Da Costa Ferreira AM. Binding of oxindole-Schiff base copper(II) complexes to DNA and its modulation by the ligand. J Inorg Biochem. 2011;105:1692–703.
33. Thederahn TB, Kuwabara MD, Larsen TA, Sigman DS. Nuclease activity of 1,10-phenanthroline-copper: kinetic mechanism. J Am Chem Soc. 1989;111:4941–6.
34. Strasser A, O'Connor L, Dixit VM. Apoptosis signaling. Ann Rev Biochem. 2000;69: 217–45.
35. Thornberry NA, Lazebnik Y. Caspases: enemies within. Science. 1998;281:1312–6.
36. Chen M, Wang J. Initiator caspases in apoptosis signaling pathways. Apoptosis. 2002;7:313–9.
37. Adrain C, Martin SJ. The mitochondrial apoptosome: a killer unleashed by the cytochrome seas. Trends Biochem Sci. 2001;26:390–7.
38. Van Gurp M, Festjens N, Van Loo G, Saelens X, Vandenabeele P. Mitochondrial intermembrane proteins in cell death. Biochem Biophys Res Commun. 2003;30:487–97.
39. Penninger JM, Kroemer G. Mitochondria, AIF and caspases—rivaling for cell death execution. Nature Cell Biol. 2003;5:97–9.
40. Filomeni G, Ciriolo MR. Redox control of apoptosis: an update. Antioxid Redox Signal. 2006;8:2187–292.
41. Lowe SW, Cepero E, Evan G. Intrinsic tumour suppression. Nature. 2004;432:307–15.
42. Hanahan D, Weinberg RA. The hallmarks of cancer. Cell. 2000;100:57–70.
43. Xu RH, Pelicano H, Zhou Y, Carew JS, Feng L, Bhalla KN, Keating MJ, Huang P. Inhibition of glycolysis in cancer cells: a novel strategy to overcome drug resistance associated with mitochondrial respiratory defect and hypoxia. Cancer Res. 2005;65:613–21.

44. Blagosklonny MV, Pardee AB. Exploiting cancer cell cycling for selective protection of normal cells. Cancer Res. 2001;61:4301–5.
45. Michels J, Obrist F, Castedo M, Vitale I, Kroemer G. PARP and other prospective targets for poisoning cancer cell metabolism. Biochem Pharmacol. 2014;92:164–71.
46. Na HK, Surh YJ. Transcriptional regulation via cysteine thiol modification: a novel molecular strategy for chemoprevention and cytoprotection. Mol Carcinogen. 2006;45:368–80.
47. Hail Jr N, Cortes M, Drake EN, Spallholz JE. Cancer chemoprevention: a radical perspective. Free Radic Biol Med. 2008;45:97–110.
48. Cleveland JL, Kastan MB. Cancer. A radical approach to treatment. Nature. 2000;407: 309–11.
49. Stěrba M, Popelová O, Vávrová A, Jirkovský E, Kovaříková P, Geršl V, Simůnek T. Oxidative stress, redox signaling, and metal chelation in anthracycline cardiotoxicity and pharmacological cardioprotection. Antioxid Redox Signal. 2013;18:899–929.
50. Aouida M, Ramotar DA. New twist in cellular resistance to the anticancer drug bleomycin-A5. Curr Drug Metab. 2010;11:595–602.
51. Ciriolo MR, Magliozzo RS, Peisach J. Microsome-stimulated activation of ferrous bleomycin in the presence of DNA. J Biol Chem. 1987;262:6290–5.
52. Ciriolo MR, Peisach J, Magliozzo RS. A comparative study of the interactions of bleomycin with nuclei and purified DNA. J Biol Chem. 1989;264:1443–9.
53. Kurtoglu M, Lampidis TJ. From delocalized lipophilic cations to hypoxia: blocking tumor cell mitochondrial function leads to therapeutic gain with glycolytic inhibitors. Mol Nutr Food Res. 2009;53:68–75.
54. Solaini G, Sgarbi G, Baracca A. Oxidative phosphorylation in cancer cells. Biochim Biophys Acta. 2011;1807:534–42.
55. Dwarakanath BS. Cytotoxicity, radiosensitization, and chemosensitization of tumour cells by 2-deoxy-D-glucose in vitro. J Cancer Res Ther. 2009;5:S27–31.
56. Mathupala SP, Ko YH, Pedersen PL. Hexokinase-2 bound to mitochondria: cancer's stygian link to the "Warburg Effect" and a pivotal target for effective therapy. Semin Cancer Biol. 2009;19:17–24.
57. Modica-Napolitano JS, Aprille JR. Delocalized lipophilic cations selectively target the mitochondria of carcinoma cells. Adv Drug Delivery Rev. 2001;49:63–70.
58. Cardaci S, Filomeni G, Ciriolo MR. Redox implications of AMPK-mediated signal transduction beyond energetic clues. J Cell Sci. 2012;125:2115–25.
59. Filomeni G, Cardaci S, Da Costa Ferreira AM, Rotilio G, Ciriolo MR. Metabolic oxidative stress elicited by the copper(II) complex [Cu(isaepy)2] triggers apoptosis in SH-SY5Y cells through the induction of the AMP-activated protein kinase/p38MAPK/p53 signalling axis: evidence for a combined use with 3-bromopyruvate in neuroblastoma treatment. Biochem J. 2011;437:443–53.
60. Wang JC. DNA topoisomerases. Ann Rev Biochem. 1996;65:635–92.
61. Pommier Y. DNA topoisomerase I inhibitors: chemistry, biology and interfacial inhibition. Chem Rev. 2009;109:2894–902.
62. Zhang J, Yang PL, Gray NS. Targeting cancer with small molecule kinase inhibitors. Nat Rev Cancer. 2009;9:28–39.
63. Fang Z, Grütter C, Rauh D. Strategies for the selective regulation of kinases with allosteric modulators: exploiting exclusive structural features. Chem Biol. 2013;8:58–70.
64. Lapenna S, Giordano A. Cell cycle kinases as therapeutic targets for cancer. Nat Rev Drug Discov. 2009;8:547–66.
65. Hotte SJ, Oza A, Winquist EW, Moore M, Chen EX, Brown S, Pond GR, Dancey JE, Hirte HW. Phase I trial of UCN-01 in combination with topotecan in patients with advanced solid cancers: a Princess Margaret Hospital Phase II Consortium study. Ann Oncol. 2006;17:334–40.
66. Lane ME, Yu B, Rice A, Lipson KE, Liang C, Sun L, Tang C, McMahon G, Pestell RG, Wadler S. A novel cdk2-selective inhibitor, SU9516, induces apoptosis in colon carcinoma cells. Cancer Res. 2001;61:6170–7.

67. Laird AD, Vajkoczy P, Shawver LK, Thurnher A, et al. SU6668 is a potent antiangiogenic and antitumor agent that induces regression of established tumors. Cancer Res. 2000;60:4152–60.
68. Blake RA, Broome MA, Liu X, Wu J, Gishisky M, Sun L, Courtneidge SA. SU6656, a selective Src family kinase inhibitor, used to probe growth factor signaling. Mol Cell Biol. 2000;20:9018–27.
69. Hohenberg P, Kohn W. Inhomogeneous electron gas. Phys Rev. 1964;136:B864–71.
70. Kohn W, Sham LJ. Self-consistent equations including exchange and correlation effects. Phys Rev. 1965;140:A1133–8.
71. Trott O, Olson AJ. AutoDock Vina: improving the speed and accuracy of docking with a new scoring function, efficient optimization, and multithreading. J Comput Chem. 2010;31: 455–61.
72. Allen MP, Tildesley DJ. Computer simulations of liquids. Oxford: Oxford University Press; 1989.
73. Leach AR. Molecular modelling: principles and applications. 2nd ed. Englewood Cliffs: Prentice Hall; 2009.
74. Betzi S, Alam R, Martin M, Lubbers DJ, Han H, Jakkaraj SR, Georg GI, Schönbrunn E. Discovery of a potential allosteric ligand binding site in CDK2. ACS Chem Biol. 2011;6:492–501.
75. Bangwani NK, Petersen PD, Gonzalez F, Oliveira CC, Miguel RB, do Nascimento RR, Petrilli HM, Da Costa Ferreira AM. Inhibition of kinase protein CDK1 by oxindolimine ligands and corresponding metal complexes with antitumor and antiparasite activity, *manuscript in preparation.*
76. Coura JR. Chagas disease: what is known and what is needed—a background article. Mem Inst Oswaldo Cruz (Rio de Janeiro). 2007;102:113–22.
77. Coura JR, de Castro SL. A critical review on Chagas disease chemotherapy. Mem Inst Oswaldo Cruz (Rio de Janeiro). 2002;97:3–24.
78. McCall L-I, McKerrow JH. Determinants of disease phenotype in trypanosomatid parasites. Trends Parasitol. 2014;30:342–49.
79. DoCampo R, Moreno SNJ. Free radical metabolism of antiparasitic agents. Fed Proceed. 1986;45:2471–6.
80. Polak A, Richle R. Mode of action of 2-nitroimidazole derivative benznidazole. Ann Trop Med Parasitol. 1978;72:228–32.
81. DoCampo R. Sensitivity of parasites to free radical damage by antiparasitic drugs. Chem Biol Interact. 1990;73:1–27.
82. Diaz de Toranzo EG, Castro JA, Franke de Cazzulo BM, Cazzulo JJ. Interaction of benznidazole reactive metabolites with nuclear and kinetoplastic DNA, proteins and lipids from *T. cruzi*. Experientia. 1988;44:880–1.
83. Andrade Rajão M, Furtado C, Alves CL, Passos-Silva DG, de Moura MD, Schamber-Reis BL, Kunrath-Lima M, Zuma AA, Vieira-da-Rocha JP, Garcia JBF, Mendes IC, Pena SDJ, Macedo AM, Franco GR, de Souza-Pinto NC, de Medeiros MHG, Cruz AK, Motta MCM, Teixeira SMR, Machado CR. Unveiling benznidazole's mechanism of action through overexpression of DNA repair proteins in *T. cruzi*. Environ Mol Mutagen. 2014;55:309–21.
84. Duschak VG, Couto AS. An insight on targets and patented drugs for chemotherapy of Chagas disease. Recent Pat Anti-Infect Drug Discov. 2007;2:19–51.
85. Pérez-Rebolledo A, Teixeira LR, Batista AA, Mangrich AS, Aguirre G, Cerecetto H, González M, Hernández P, Ferreira AM, Speziali NL, Beraldo H. 4-Nitroacetophenone-derived thiosemicarbazones and their copper(II) complexes with significant in vitro anti-trypanosomal activity. Eur J Med Chem. 2008;43:939–48.
86. Navarro M, Cisneros-Fajardo EJ, Lehmann T, Sanchez-Delgado RA, Atencio R, Silva P, Lira R, Lira R. Toward a novel metal-based chemotherapy against tropical diseases. 6. Synthesis and characterization of new copper(II) and gold(I) clotrimazole and ketoconazole complexes and evaluation of their activity against Trypanosoma cruzi. Inorg Chem. 2001;40:6879–84.

87. Maya JD, Cassels BK, Iturriaga-Vásquez P, Ferreira J, Faúndeza M, Galanti N, Ferreira A, Morello A. Mode of action of natural and synthetic drugs against *T. cruzi* and their interaction with the mammalian host. Comp Biochem Physiol Part A. 2007;146:601–20.
88. Ciccarelli AB, Frank FM, Puente V, Malchiodi EL, Batle A, Lombardo ME. Antiparasitic effect of vitamin B12 on T. cruzi. Antimicrob Agents Chemother. 2012;56:5315–20.
89. Caballero AB, Rodríguez-Diéguez A, Quirós M, Salas JM, Huertas Ó, Ramírez-Macías I, Olmo F, Marín C, Chaves-Lemaur G, Gutierrez-Sánchez R, Sánchez-Moreno M. Triazolopyrimidine compounds containing first-row transition metals and their activity against the neglected infectious Chagas disease and Leishmaniasis. Eur J Med Chem. 2014;85:526–34.
90. Sabino GL, Couto RA, Ribeiro GA, Vieira LQ, Da Costa Ferreira AM. Oxindolimine-metal complexes showing remarkable antiparasite activity against *T. cruzi, manuscript in preparation.*
91. Fragoso SP, Mattei D, Hines JC, Ray D, Goldenberg S. Expression and cellular localization of T. cruzi type II DNA topoisomerase. Mol Biochem Parasitol. 1998;94:197–204.
92. Merritt C, Silva LE, Tanner AL, Stuart K, Pollastri MP. Kinases as druggable targets in trypanosomatid protozoan parasites. Chem Rev. 2014;114(22):11280–304. doi:10.1021/cr500197d

Chapter 13
Small Signaling Molecules and *CO-Releasing Molecules* (CORMs) for the Modulation of the Cellular Redox Metabolism

Peter V. Simpson and Ulrich Schatzschneider

13.1 Introduction

Together with nitric oxide (NO) and hydrogen sulfide (H_2S), carbon monoxide (CO) belongs to a class of small endogenously produced signaling molecules which has gained steadily increasing attention in the biomedical community over the last two decades due to their fascinating properties and potential therapeutic applications in human medicine. All three signaling systems are heavily intertwined and show not only common properties but also distinct differences [1]. Usually known to the general public as highly toxic gases only, they are sometimes grouped together under the term "gasotransmitters" [2], an unfortunate misnomer given that all three molecules are fully dissolved in body fluids at physiological concentrations (solubility in water at 20 °C: CO 1 mM, NO 2 mM, and H_2S 117 mM) [3] and therefore not present as gas bubbles [4].

In terms of their chemical properties, hydrogen sulfide is distinct from the other two since it can be deprotonated in the physiological pH range due to pK_a values for the H_2S/HS^- and HS^-/S^{2-} acid/base pairs at 6.8 and 14.1, respectively. Thus, hydrosulfide (HS^-) is the predominant species under these conditions together with H_2S and only traces of sulfide (S^{2-}) [4]. This strongly pH-dependent chemistry of H_2S contrasts with that of CO and NO and significantly complicates studies into H_2S bioactivity [1]. Both HS^- and S^{2-} can undergo one-electron redox reactions in a biologically accessible range, with the potentials for the $HS^{\cdot}/HS^-$ and $S^{\cdot-}$, H^+/HS^- redox couples in the 0.9–1.0 V range (vs. NHE). Although nitric oxide is a radical, its one-electron

P.V. Simpson
Department of Chemistry, Curtin University, Bentley, WA, Australia

U. Schatzschneider (✉)
Institut für Anorganische Chemie, Julius-Maximilians-Universität Würzburg, Würzburg D-97074, Germany
e-mail: ulrich.schatzschneider@uni-wuerzburg.de

I. Batinić-Haberle et al. (eds.), *Redox-Active Therapeutics*,
Oxidative Stress in Applied Basic Research and Clinical Practice,
DOI 10.1007/978-3-319-30705-3_13

reduction according to $NO + e^- + H^+ \rightarrow HNO$ occurs at only about −0.55 V (vs. NHE at pH 7). Thus, it is a poor one-electron oxidant that does not have much potential for hydrogen abstraction reactions from biological molecules but readily reacts with other radical species, in particular dioxygen, as well as metal centers, leading to a complicated follow-up chemistry [1]. In contrast to NO and H_2S, carbon monoxide is generally considered much more inert under physiological conditions, although enzymes such as CO dehydrogenase (CODH) can oxidize CO in a reaction reminiscent of the water-gas-shift reaction $CO + H_2O \leftrightarrows CO_2 + H_2$ with the redox potential for the CO_2/CO couple at −0.56 V (vs. NHE at pH 7) [1, 5].

Another important issue in small molecule signaling is the distance CO, NO, and H_2S can travel from the site of generation until they are scavenged by the constituents of biological systems. This "sphere of action" is also highly relevant in the design of artificial delivery systems for these molecules for therapeutic purposes, since it will determine how close to a biological target structure the release from the carrier system has to take place. While the diffusion coefficients of all three molecules in water are essentially identical and in the range of $1.8–2.1 \times 10^{-5}$ cm^2 s^{-1} at 20 °C [4], very little is known about the transport properties of these compounds in cells and tissues and to what extent they are able to cross biological membranes. However, in recent work by Alvarez and Möller, the partition coefficient of H_2S in models of biological membranes was determined and a rapid membrane permeability of 3 cm s^{-1} derived from this data. Further modeling of the diffusional spread of hydrogen sulfide originating from a single cell demonstrated that the sphere of action as defined by the distance at which the concentration is still 1/10 of that at the source is more than 200 neighboring cells at 1 min after formation [6]. In a much more detailed molecular model of a 1,2-dipalmitoyl-*sn*-glycero-3-phosphocholine (DPPC) bilayer surrounded by explicit water molecules and using molecular dynamics (MD) simulations, Riahi and Rowley very recently showed that H_2S experiences essentially no barrier to permeation and easily partitions into the hydrophobic interior of the membrane [7]. Unfortunately, no comparable data is available for CO and NO yet. Thus, it remains an open question of how far small signaling molecules can travel from the site of their enzymatic generation and how close synthetic carrier systems have to get to a biological target structure to deliver their "cargo."

This is illustrated in Fig. 13.1 for *CO-releasing molecules* (CORMs) as carrier systems for carbon monoxide, where the pathway at the top right has to be operative when the "sphere of action" is small and the CORM consequently must be internalized by the cell; while the one at the lower left would be sufficient if the extracellularly liberated CO can efficiently permeate membranes and travel considerable distances before getting trapped.

The mechanisms by which cells generate CO, NO, and H_2S also differ considerably. While the latter two originate from amino acid sources, namely L-arginine and L-methionine, respectively, carbon monoxide is generated by enzymatic degradation of heme by heme oxygenase (HO). NO generation is catalyzed by nitric oxide synthase (NOS) enzymes, which oxidize L-arginine (Arg) via intermediate N^{ω}-hydroxy-L-arginine (NOHA) to L-citrulline (Cit) and NO (Fig. 13.2) [8–10]. Three NOS isoforms are produced in mammals, endothelial NOS (eNOS) in the cardiovascular system, neuronal NOS (nNOS) in the nervous system, and inducible NOS (iNOS) by the immune system.

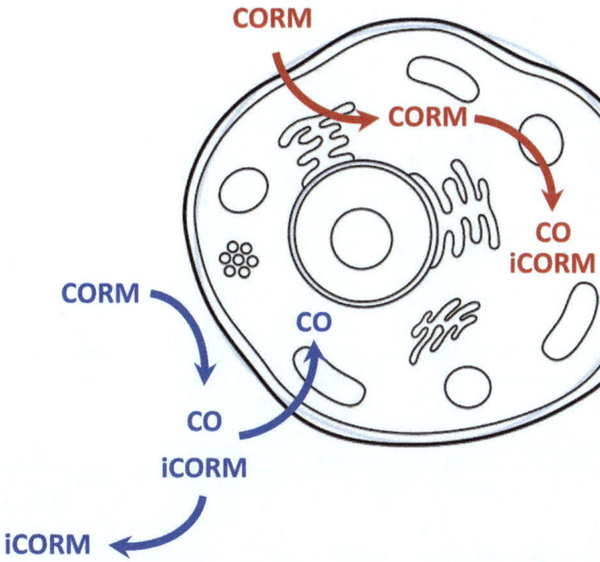

Fig. 13.1 Depending on the "sphere of action," which is the range a small signaling molecule can travel before getting scavenged, synthetic carrier systems have to be internalized by target cell in the case of a small sphere of action (*top right*) while it is sufficient to release the active molecule close to the cell but without uptake in the case of high membrane permeability and larger sphere of action (*bottom left*). The difference is illustrated for *CO-releasing molecules* (CORMs) as carrier systems for carbon monoxide and also shows the distinctive differences in the location of the *inactivated CORM* (iCORM) formed in addition to the small signaling molecule upon CO release from the carrier

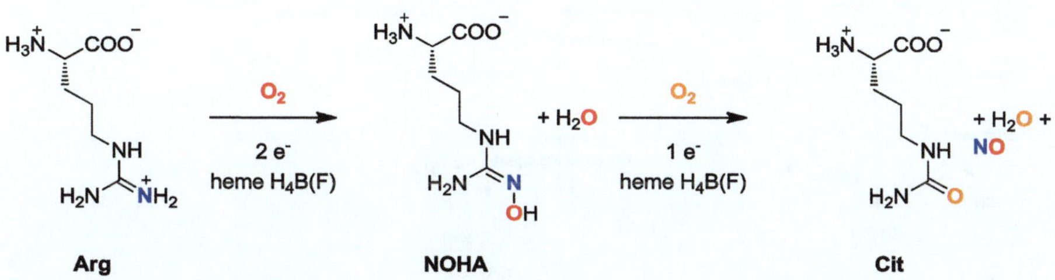

Fig. 13.2 Enzymatic generation of nitric oxide (NO) by oxidation of L-arginine via N^{ω}-hydroxy-L-arginine (NOHA) to L-citrulline (Cit) by activity of nitric oxide synthase (NOS) with $H_4B(F)$ = tetrahydrobiopterin (H_4B) or tetrahydrofolate (H_4F) [8, 9]

Hydrogen sulfide (H_2S) is also generated from amino acid sources, which can be traced back to L-methionine. However, its biosynthesis and degradation is much more complicated since a large number of different intermediates and enzymes are involved. A total of four pathways can be distinguished based on the enzymes involved (Fig. 13.3) [11–13, 15]. As the initial step of the methionine/transsulfuration pathway, L-methionine is converted to L-homocysteine via *S*-adenosylmethionine and *S*-adenosylhomo-cysteine (1). The β-elimination reaction of L-homocysteine with L-cysteine catalyzed by cystathionine β-synthase (CBS) leads to liberation of H_2S together with concomitant forma-

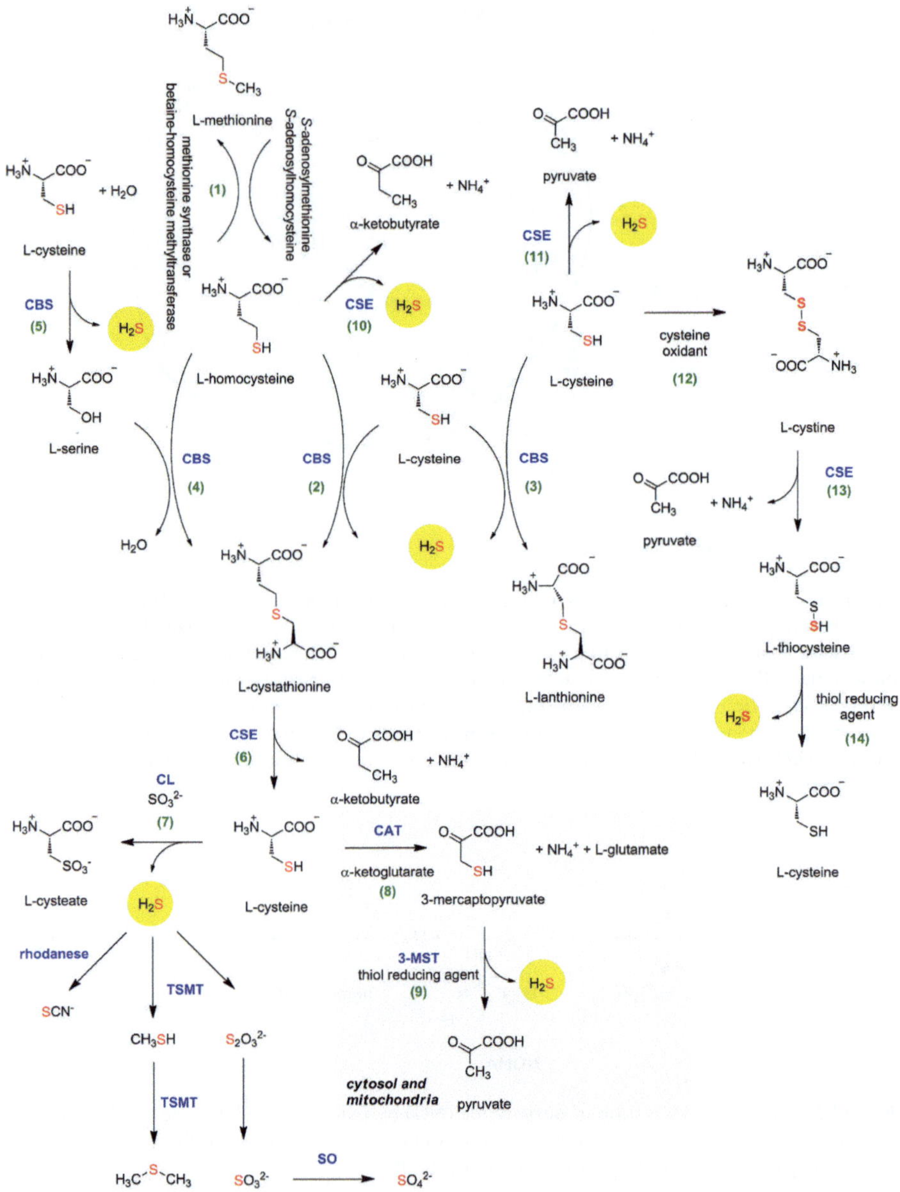

Fig. 13.3 Enzymatic generation of hydrogen sulfide (H_2S) from the methionine metabolism with involvement of different enzymatic systems, namely cystathionine β-synthase (CBS), cystathionine γ-lyase (CSE), cysteine aminotransferase (CAT), 3-mercaptopyruvate sulfurtransferase (3-MST), cysteine lyase (CL), thiol *S*-methyltransferase (TSMT), and sulfite oxidase (SO) [11–14]

tion of L-cystathionine (2). This enzyme can also utilize two molecules of L-cysteine as the substrate, leading to L-lanthionine and H_2S (3), while two molecules of L-homocysteine will give rise to L-homolanthionine and H_2S (not shown). Furthermore, L-homocysteine also reacts with L-serine to L-cystathionine but in this case, water instead of hydrogen sulfide is formed (4). Finally, reaction of L-cysteine with water leads to formation of L-serine and H_2S (5), which demonstrates the wide range of substrates utilized by this enzyme. L-cystathionine is further converted to L-cysteine, α-ketobutyrate, and ammonium by cystathionine γ-lyase (CSE) in an α,γ-elimination (6). L-Cysteine then undergoes a substitution reaction with inorganic sulfite catalyzed by cysteine lyase (CL) to form L-cysteate and H_2S (7). The latter is then further metabolized either to thiocyanate by action of rhodanese, converted to methyl mercaptan and then further on to dimethylsulfide by thiol *S*-methyltransferase (TSMT), or oxidized via thiosulfate and sulfite to sulfate. In an alternative pathway, the only one also taking place in the mitochondria, transamination of L-cysteine with α-ketoglutarate leads to 3-mercaptopyruvate, L-glutamate, and ammonia under the activity of cysteine aminotransferase (CAT) (8). The former is converted to pyruvate under release of H_2S by 3-mercaptopyruvate sulfur transferase (3-MST) (9). Alternatively, CSE can also utilize L-homocysteine instead of L-cystathionine, which produces α-ketobutyrate together with H_2S (10), and conversion of L-cysteine leads to pyruvate, ammonium, and H_2S (11).

Furthermore, oxidative coupling of two equivalents of L-cysteine leads to L-cystine (12). Under catalysis of cystathionine γ-lyase (CSE), this homodisulfide is cleaved to pyruvate, ammonia, and L-thiocysteine (13). In the presence of a thiol reducing agent, the latter is transformed to L-cysteine under release of hydrogen sulfide (14). While CBS, CSE, CAT, and CL are pyridoxal-5′-phosphate(PLP)-dependent enzymes [15], 3-MST contains a zinc active site. The intracellular localization of these enzymes also differs. While CAT and 3-MST are present both in mitochondria and the cytosol, CBS and CSE are exclusively found in the cytosol. Commonly used inhibitors of H_2S synthesis to explore these pathway have recently been reviewed by Whiteman and coworkers [16].

Carbon monoxide (CO) biosynthesis is catalyzed by conversion of heme (iron protoporphyrin IX) by heme oxygenase (HO) enzymes (Fig. 13.4) [17, 18]. The reaction sequence starts with regiospecific hydroxylation of heme at the α-meso carbon atom under consumption of one equivalent of dioxygen and two electrons provided by NADPH via an iron(III)-hydroperoxo intermediate. In the second step, Fe(III)-verdoheme is generated upon further oxidation with a second equivalent of dioxygen. It is this spontaneous autooxidation step in which a single carbon monoxide molecule, with the carbon atom originating from the heme α-meso carbon atom and the oxygen from dioxygen, is liberated. Interestingly, only this step is not inhibited by CO binding to the iron center. Under consumption of a third equivalent of dioxygen and another two reducing equivalents provided by NADPH, the iron is then released in the reduced ferrous form and the heme ring system cleaved to linear tetrapyrrole biliverdin, which is finally reduced to bilirubin by biliverdin reductase. In mammalian cells, two isoforms of heme oxygenase exists, the constitutively expressed HO-2 responsible for baseline production of CO and inducible HO-1, which is upregulated in response to stress conditions.

The three signaling molecules CO, NO, and H_2S also show important difference in their targeting ability. Nitric oxide is a radical species that readily reacts with metal centers, in particular iron, but can also act via *S*-nitrosylation of amino acid side chains [2]. On the other hand, much less is known about the relevancy of direct hydrogen sulfide binding to metalloproteins and the most important target of H_2S appears to be protein side-chain modification. The major covalent adducts result from cysteine thiol

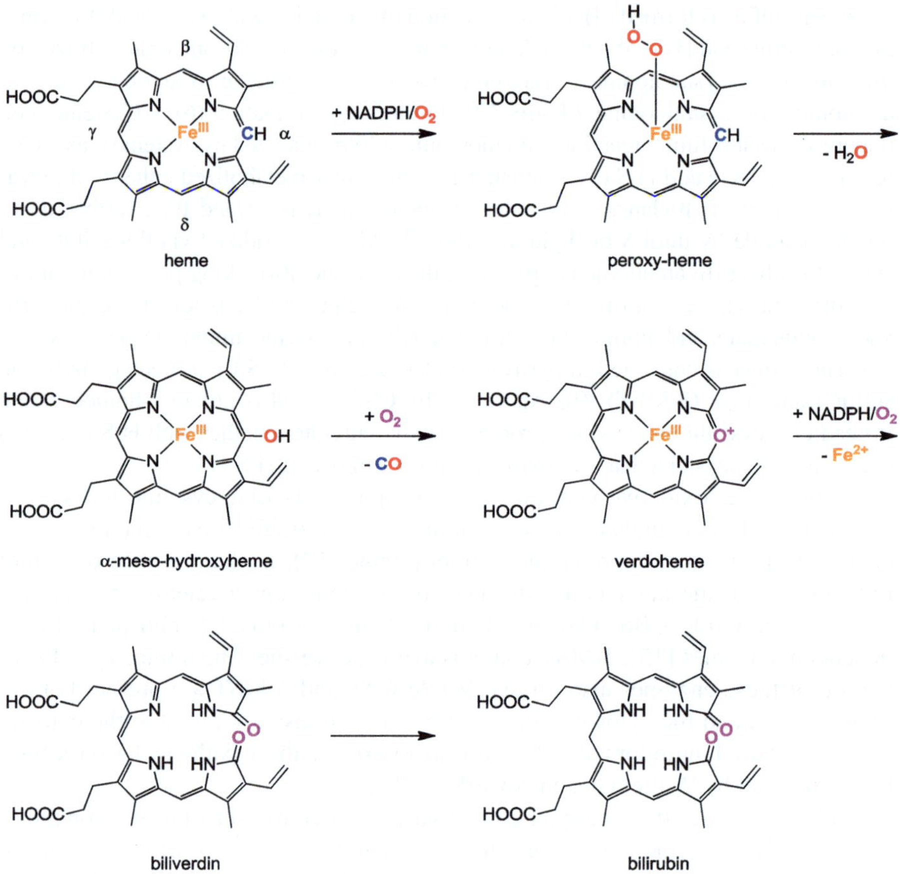

Fig. 13.4 Regiospecific oxidation of heme by heme oxygenase (HO) leading to formation of carbon monoxide (CO) together with ferrous iron and biliverdin, which is further reduced to bilirubin by biliverdin reductase [9, 17, 18]

S-sulfhydration, in which –SH is transformed to –S–SH groups, and the sulfane sulfur pool [11]. However, hydropersulfide –S–SH formation requires oxidation by two electrons and occurs only on low pK_a cysteine residues [19, 20]. This is probably also the molecular basis of hydrogen sulfide modulation of cellular signaling and metabolic processes, although a more direct link is yet to be elucidated. Another interesting aspect of H_2S bioactivity is the targeting of ion channels and receptors, which can either be activating (K^+ channel, cysteine/glutamate antiporter, TRPV) or inhibiting (Cl^- channel, L/T-type Ca^{2+} channel) [11]. In contrast, recognition of carbon monoxide is generally assumed to be restricted to metal centers and thus this signaling molecule is thought to be much more specific than the other two. Most work of CO binding to metalloproteins has focused on heme coordination while carbon monoxide modification of other metal centers, for example, iron–sulfur clusters, is more speculative. Although heme proteins, such as soluble guanylate cyclase (sGC), are often discussed as the primary target of CO, its activation by carbon monoxide is much lower compared to NO and requires additional mediators such as YC-1 (3-(5'-hydroxymethyl-2'-furyl)-1-benzyl indazole), although no natural analogue thereof has been discovered to date. More related to the cytotoxic than the signaling function of CO seems to be interaction with heme centers in the mitochondrial or bacterial electron transfer chain. A particularly fascinating emerging activity of carbon monoxide is its regulation of ion channel activity [21, 22], but the precise molecular nature of its mechanism has remained elusive to date.

In addition to potential modulation of these endogenous pathways of CO, NO, and H_2S generation by application of enzyme inhibitors or stimulators, a particular focus in recent years has also been on synthetic carrier systems for these small signaling molecules. Nitric oxide donors such as organic nitrates as well as metal nitrosyl complexes as simple as sodium nitroprusside (SNP) $Na_2[Fe(CN)_5(NO)]$ have been explored for nearly a century and thus will not be covered here [23]. The same also applies to hydrogen sulfide, where most of the work is still focused on the use of Na_2S or NaHS, two systems that present many drawbacks due to extremely fast and nonphysiological H_2S release [11, 16, 24, 25]. Thus, the following sections are restricted to CO delivery systems, commonly referred to as *CO-releasing molecules* (CORMs). Also excluded from the treatment will be technical solutions for the direct inhalative application of CO gas. A number of exogenous and endogenous trigger mechanisms have been developed over the last decade to initiate release of carbon monoxide from a stable prodrug that can be handled without the dangers associated with the use of CO gas.

13.2 Ligand-Exchange Triggered CORMs

The first two compounds explored as *CO-releasing molecules* (CORMs) by Motterlini and Mann in their landmark 2002 publication were $Mn_2(CO)_{10}$ and $Fe(CO)_5$ [26]. However, due to low bioavailability of these neutral metal carbonyls, the need for photoactivation to induce substantial CO transfer to myoglobin, and the formation of ill-defined decomposition products, these were quickly abandoned in favor of ruthenium carbonyl complexes. The first such compound evaluated was dimeric $[Ru(CO)_3Cl_2]_2$, commonly known as CORM-2 (Fig. 13.5). While being nontoxic to vascular smooth muscle cells at up to 420 μM, significant vasodilatory effects on precontracted rat aortic rings could be observed and considerable in vivo reduction of acute hypertension in rats demonstrated [26]. However, CORM-2 requires addition of dimethylsulfoxide (DMSO) to induce the liberation of carbon monoxide in a ligand-exchange triggered process leading to *fac*-$[RuCl_2(CO)_3(dmso)]$, resulting from dimer cleavage, and the CO-release product *cis, cis, trans*-$[RuCl_2(CO)_2(dmso)_2]$ which further slowly converts to the more stable all-*cis* isomer. Since further CO-release from the dicarbonyl complex does not occur even upon extended incubation [27], these final decomposition products are commonly called *inactivated CORMs* (iCORMs). More than 200 publications utilizing CORM-2 have appeared in the literature since then and it is also commercially available now. However, in a recent very thorough study by Romao and coworkers, no CO-release to the headspace of a solution of CORM-2 or other compounds of the general formula *fac*-$[RuCl_2(CO)_3(L)]$ could be observed while significant amounts of carbon dioxide were detected by gas chromatography (GC) [27]. In addition, a number of covalent protein adducts of Ru-carbonyl fragments to histidine residues of lysozyme could be identified by X-ray crystal structure analysis [28]. Since such ruthenium-protein adducts are also known to have independent biological activity from studies in the field of metal anticancer agents [29, 30], it remains highly questionable whether the biological activity induced by CORM-2 is indeed due to carbon monoxide release from the metal coordination sphere or rather results from protein modification by the iCORM fragment inevitably also generated. Any biological experiments using this compound will thus require very careful controls preferentially using both CO gas as well as alternative CORMs based on nonruthenium core structures.

To avoid the use of dimethylsulfoxide to trigger the CO-release from CORM-2 via dimer cleavage, a monomeric ruthenium(II) carbonyl species was introduced by glycinate coordination to the $Ru(CO)_3Cl$ core fragment, resulting in [RuCl(glycinate)$(CO)_3$], CORM-3. This compound, now also available for purchase, protects cardiac cells from hypoxia-reoxygenation damage and oxidative stress and has emerged as a major research tool to the CO bioactivity community [31]. However, similar to CORM-2, it has a very complicated and pH-dependent speciation in aqueous solution [32] and was also shown to form covalent protein adducts with surface-exposed histidine and to a lesser degree also aspartate side chains [33]. Furthermore, the half-life of CORM-3 is strongly solvent-dependent and ranges from 96 h in pure water to only 4 min in plasma [32]. Although very popular in CO research, CORM-3 thus should also be applied with great caution to exclude potential non-CO bioactivity via carefully designed control experiments.

One compound which avoids potential problems from CO-independent biological activity of metal-coligand fragments in iCORMs is transition metal-free sodium boranocarbonate $Na_2[H_3BCOO]$ CORM-A1 [34]. Initially prepared as an easy-to-handle carbon monoxide source for the preparation of SPECT imaging agents based on the $^{99m}Tc^{I}(CO)_3$ fragment [35], it not only is a rather slow and spontaneous CO-releaser with a half-life of about 20 min at a physiological temperature of 37 °C and pH 7.4, but also shows pronounced vasodilatory effects on rat aortic rings [34]. Although it can easily be functionalized by ester or amide formation [36, 37], only the parent compound was explored in a number of biological studies so far with about 45 publications up to early 2015. However, even this compound is not completely benign, since protonation and hydrolysis will lead to H_3B-CO, which quickly decomposes to carbon monoxide with concomitant formation of borane (BH_3). As a strongly electron-deficient compound, it will avidly react with Lewis bases L to form adducts H_3B-L. In aqueous medium, alternating addition and substitution steps will ultimately lead to formation of boric acid under release of three equivalents of dihydrogen. Although the toxicity of H_3BO_3 will probably not be much of a concern at the low doses required for CO delivery, the dihydrogen can act as a reducing agent and inject two electrons into electron transfer chains, a total of six electrons per equivalent of CORM-A1, and thus might alter the redox balance of a cell this way.

Since then, a large number of other ligand-exchange triggered CO-releasing molecules have also been explored, but evaluation of their therapeutic potential has usually remained restricted to a very few examples [38].

For example, Fairlamb and coworkers hypothesized that the combination of naturally occurring 2-pyrones with a proper metal carbonyl fragment would lead to promising new CORM lead structures. Thus, pyrone reaction with $Fe_2(CO)_9$ or $[Fe(coe)_2(CO)_3]$, with coe = *cis*-cyclooctene, lead to $[Fe(CO)_3(\eta^4\text{-pyrone})]$, in which the pyrone is coordinated to the iron(0) tricarbonyl moiety in a diene-like fashion. Spontaneous CO-release in solution under the conditions of the myoglobin assay and vasodilatory activity were demonstrated for the nonsubstituted pyrone complex as well as functionalized compounds with different 4,6-substituents [39–41].

Other core structures have also emerged over the last couple of years. For example, μ_2-alkyne-bridged dicobalt(0)hexacarbonyl complexes were also evaluated for their CO-release properties and allowed tuning of the kinetics via the alkyne substituents [42]. Also, simple metal salts of the general composition $[MX(CO)_5]^-$ with M=Cr, Mo, W and X=Cl, Br, I were explored as potential CORM lead structures [43–45]. Even some CO delivery systems based on the 5d transition metals have been reported. For example, different tetrachlorocarbonyliridates of general formula $[IrCl_5(CO)]^-$ and $[IrCl_5(CO)(L)]^-$ with L=H_2O, pyridine were investigated for their CO-release behavior and found to form stable protein adducts with lysozyme characterized by X-ray diffraction analysis [46, 47]. A particularly innovative strategy toward targeted CORMs was developed by the group of Zobi, who modified cobalamin (vitamin B_{12}) as a naturally occurring carrier system with a $ReBr_2(CO)_2$ CO-releasing moiety via a cyanide bridge [48–50].

A promising alternative to alleviate the problems with ruthenium-based CORMs described above might be CORM-401, with a functionalized dithiocarbamate coordinated to a manganese(I) tetracarbonyl moiety. An attached acetyl group makes the compound soluble in water upon deprotonation at physiological pH. The reversible dissociative CO loss results in relatively stable solutions of the compound, which will however be easily captured by heme proteins and other biomolecules that can act as CO scavengers [51]. Iron carbonyl complexes of the same ligand system were also explored, but beyond some cell viability studies, not much biological data seems to be reported to date on these compounds [52].

Interestingly, although iron is the most abundant transition metal in the human body, relatively few iron-based CORMs have been reported to date. Mascharak and coworkers presented a pentadentate N_5-chelator which coordinates an iron-monocarbonyl moiety of the general formula $[Fe(CO)(N_5)]^{2+}$. Rapid CO-release was observed in aqueous and nonaqueous media under oxygen as well as oxygen-free conditions and a vasodilatory effect on mouse aorta rings demonstrated, which was interestingly not influenced by a common inhibitor of guanylate cyclase (sGC) [53]. Another system with a mixed *N,S*-coordination sphere was reported by Mann, Motterlini and coworkers, who studied the CO-release behavior of [Fe(L-cysteinate)$_2$(CO)$_2$]. Due to the different relative orientation of the nitrogen and sulfur donor atoms relative to the carbonyl ligands, elucidation of the most stable isomeric form turned out to be quite difficult. Interestingly, the report speculates about the potential relevancy of this binding mode to CO recognition in nonheme proteins, in which two protein-derived cysteinate ligands could form a CO-binding site in situ by capture of iron(II) followed by carbon monoxide coordination [54].

Besides the wide range of CORM lead structures described above, very few of them have been studied for their biological activity beyond CO-release studies with the myoglobin assay, assessment of their cytotoxic potential, and the aortic ring contraction assay. Thus, the number of CORMs evaluated as viable drug candidates for a specific therapeutic application has remained very low. A notable exception is the research efforts of the researchers at Alfama, Inc., who have explored several hundred metal-

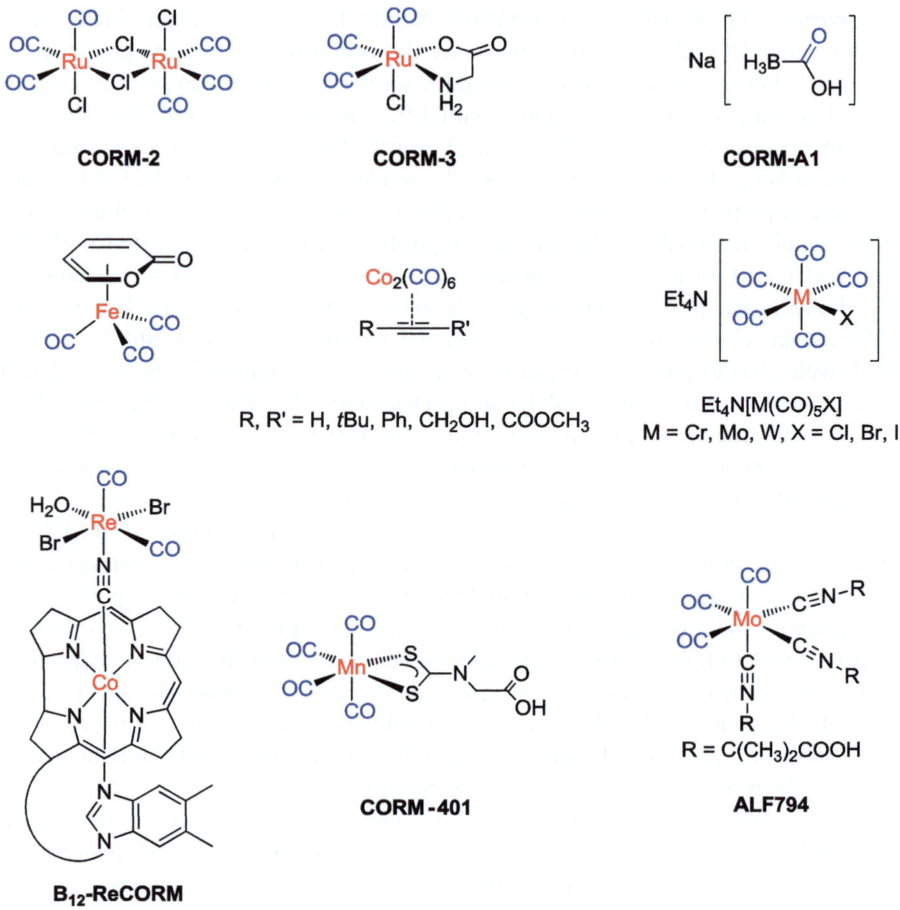

Fig. 13.5 Some important CORM lead structures activated by ligand exchange reactions

carbonyl complexes for a wide range of indications. An illustrative case study is that of molybdenum(0) tricarbonyl complexes with a range of isonitrile coligands for treatment of acetaminophen(APAP)-induced acute liver failure. Careful optimization of the outer ligand periphery, called the "drug sphere," allowed identification of complexes of the general formula $[Mo(CO)_3(CN\text{-}R)_3]$ with low general toxicity (LD_{50} in mice above 450 mg/kg) and high liver targeting capacity. Promising biological activity as quantified by a reduction in alanine aminotransferase (ALT) levels in serum as an indicator of hepatocellular death was observed which warrants further development [55]. This study also highlights the importance of optimizing not only the CO-release properties of the so-called CORM sphere, the direct metal-CO coordination environment, but also the outer parts of the molecule, the "drug sphere" [56]. Unfortunately, most CORMs reported to date do not pay proper attention to the drug-like features of the compounds explored, which will however be essential for further drug development. A particularly appealing feature of the molybdenum-based CORMs is final metabolization to heteropolyoxomolybdate clusters as demonstrated by X-ray crystallography, which are likely much more benign final products of the CO-release process than the ruthenium-coligand fragments discussed above [57]. Beyond these molecular carrier compounds for biological CO delivery, a number of more complex systems have also been developed in recent years. For example, crystals of the iron-based metal-organic framework (MOF) MIL-88B were loaded with carbon monoxide gas and the CO-release upon controlled degradation of the material studied under physiological conditions [58]. As an alternative strategy to modulate the targeting and carbon monoxide-release properties, some CORMs such as $[RuCl(glycinato)(CO)_3]$ have also been encapsulated in micelles and the hollow inner sphere of protein cages such as ferritin [59, 60]. In particular the latter system resulted in a pronounced biological effect, with NF-κB activation increased by a factor of about four compared to CORM-3 itself while the systems were not cytotoxic in the HEK293 cells studied. While most CORMs studied to date have only been applied in cell culture studies and likely will only be applicable by intravenous injection, very recently also a first system for oral CORM delivery has been reported. In the so-called *oral carbon monoxide-release system* (OCORS), CORM-2 is encapsulated in an oral tablet composed of sodium sulfite crystals and a filling material covered with by a cellulose acetate coating of variable water permeability. Diffusion of medium into the tablet leads to dissolution of the sodium sulfite that then triggers carbon monoxide release from CORM-2. A remaining problem is to ensure no ruthenium iCORM will leak from the tablet upon gastrointestinal passage to minimize detrimental non-CO mediated side-effects [61].

13.3 Photoactivated CORMs

13.3.1 Transition Metal-Based PhotoCORMs

Although the interaction of metal-carbonyl complexes with solvent or biological components such as proteins and enzymes may lead to CO-release, the use of external triggers is also being explored. In that context, photoactivation is a particularly promising way to achieve precise spatial and temporal control of biological activity due to the availability of focused and pulsed light sources (Fig. 13.6a) [62]. Absorption of light by a metal carbonyl complex leads to reactive excited states with a weakened M-CO bond, thus facilitating CO-release from the metal coordination sphere [63–67]. The known photochemistry of some simple metal-carbonyl complexes was initially exploited by Motterlini and Mann, who recognized that $Mn_2(CO)_{10}$ and $Fe(CO)_5$ can be used in this context [26, 68, 69]. Among a raft of other beneficial effects, studies revealed that $Mn_2(CO)_{10}$ could be used in the treatment of cardiovascular dysfunction in isolated rat hearts. These early compounds, however, displayed very poor bioavailability and required a "cold light source" for activation, which is why they were abandoned in favor of ligand-exchange triggered CORM-2 and CORM-3.

Despite the advent of metal-carbonyl complexes as *photoactivatable CO-releasing molecules* (PhotoCORMs), the inability to tune the properties of these early candidates limited their utility, and in fact it was not until several years later that the group of Schatzschneider reported the photoactivated release of carbon monoxide from manganese(I) tricarbonyl complex $[Mn(CO)_3(tpm)]^+$, bearing a facial tridentate tris(pyrazolyl)methane (tpm) ligand coordinated to a $Mn^I(CO)_3$ moiety. Upon illumination of the complex at 365 nm, two of the three carbon monoxide ligands were released from the metal coordination sphere [70]. The authors also showed passive uptake of the complex into HT-29 human colon cancer cells by atomic absorption spectroscopy, using the manganese as the intrinsic marker. The complex was completely inactive on HT-29 cell at a concentration of 100 μM over a 48 h incubation time in the dark while illumination at 365 nm for 10 min in the middle of a 48 h incubation cycle led to a significant reduction in cell biomass as shown by the crystal violet assay, with an activity comparable to that of organic standard 5-fluorouracil. These experiments demonstrated that high local concentrations of carbon monoxide could have a marked cytotoxic effect, beneficial in the eradication of aberrant cell populations from the body. Recent studies of the $[Mn(CO)_3(tpm)]^+$ PhotoCORM by UV-pump/IR-probe experiments on the femtosecond timescale revealed a rapid photodissociation of only one CO ligand [71]. The remaining CO, however, is liberated on a much slower timescale following oxidation of the metal center to give mononuclear manganese(II) tpm complexes, as evidenced by EPR studies, which finally dimerise to oxo-bridged manganese(III) dimers [72]. This raises the important questions of what remains of the CORM following CO-release, and what effect this fragment has on the overall biological activity. It is conceivable that the effect on the biological system is mediated not (or at least not exclusively) by the liberated CO, but instead by the action of the remaining metal-coligand fragment, commonly called *inactivated CORM* (iCORM). This alternative should, if possible, be excluded before any detailed analysis of the biological action, by control experiments using CO gas, or at least a CORM based on a different metal center.

Along those lines, Schatzschneider and coworkers recently reported the manganese tricarbonyl complex $[Mn(CO)_3(tpa\text{-}\kappa^3N)]^+$, containing a tris-(2-pyridylmethyl)amine (tpa) ligand bound to the metal center through the tertiary amine and only two of the three the pyridyl arms. Upon photorelease of CO, the remaining noncoordinated pyridyl group could fill one of the vacant coordination sites, thus forming a well-defined iCORM [73]. Internalization of this compound by *E. coli* cells was demonstrated by ICP-MS, with subsequent illumination at 365 nm leading to a pronounced and concentration-dependent decrease in the bacterial growth rate via inhibition of the terminal oxidases of the electron-transport chain from photo-released CO. Importantly, the compound had no detrimental effect on the bacterial growth rate in the dark. Another CORM with interesting antibacterial properties is the manganese complex $[Mn(CO)_3(CH_3CN)(trp)]$ very recently reported by Fairlamb and coworkers, in which trp is tryptophan coordinated via the amino and deprotonated carboxylate groups as a *N,O*-chelator reminiscent of glycinate in CORM-3, and thus named TryptoCORM [74]. The compound releases two equivalents of carbon monoxide upon illumination at 400 nm and is nontoxic toward mammalian RAW 264.7 cells in the dark or upon exposure to light. A bactericidal study using *E. coli* showed that, in the dark, the title compound had no effect of the bacterial growth rate, but following illumination, a 99.9 % loss of viability was observed. The authors tested whether the potent bactericidal effects were in fact due to the action of iCORM species formed following CO-release by preilluminating a solution of TryptoCORM and then performing the same antibacte-

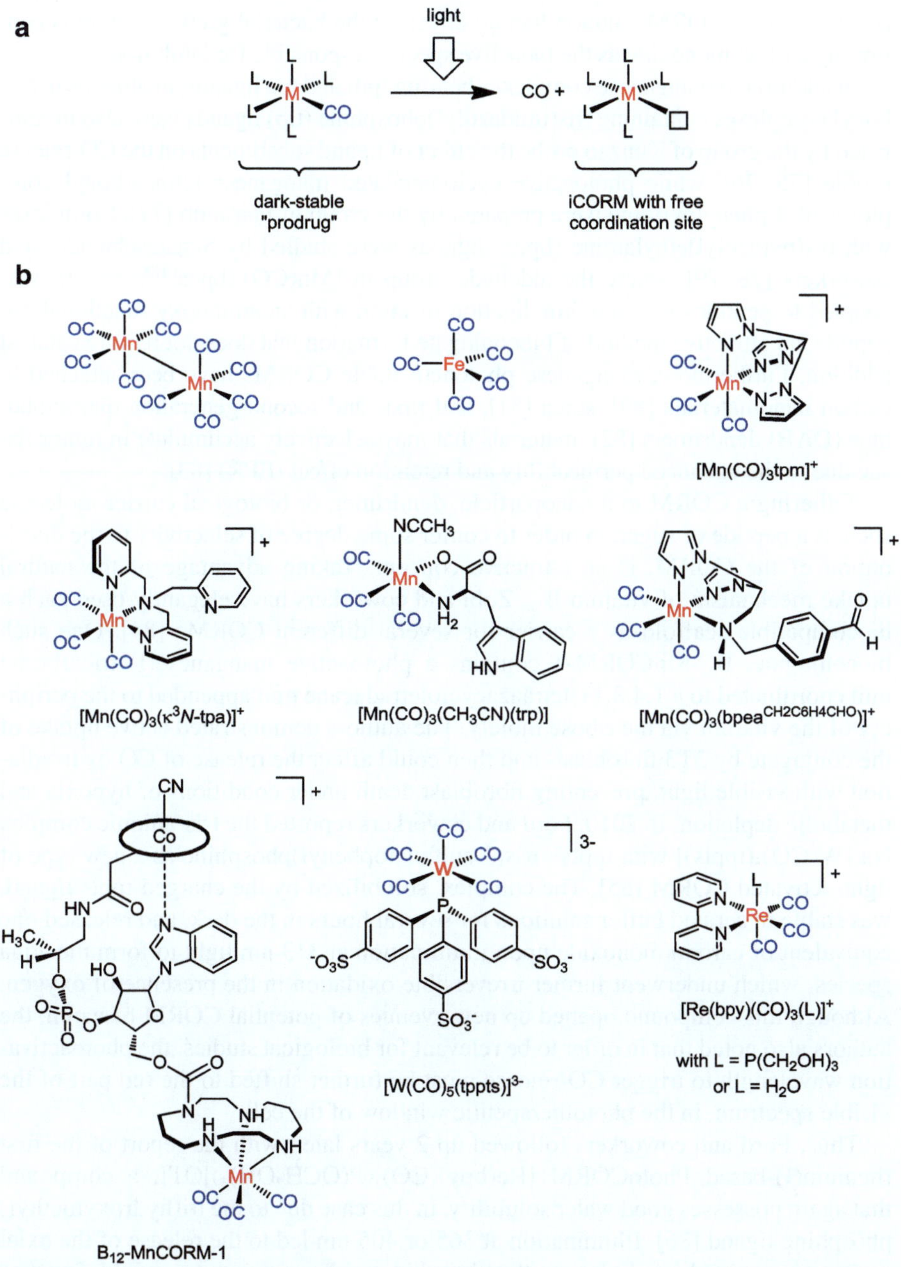

Fig. 13.6 (**a**) General concept of CO release by photoexcitation and (**b**) some recent examples of *photoactivatable CORMs* (PhotoCORMs)

rial assay. The iCORM solution had no effect on the bacterial growth-rate, thus confirming carbon monoxide as the bioactive species responsible for inhibition.

In addition to manganese complexes bearing tpm and tpa ligands, analogous tricarbonyl complexes containing tris(imidazolyl)phosphine (tip) ligands were also investigated by the group of Kunz to probe the effect of ligand substituents on the CO-release profile [75, 76], while photoactive cyclometalated manganese tetracarbonyl complexes of 2-phenylpyridine were prepared by the group of Fairlamb [77]. Complexes with bis(pyrazolyl)ethylamine (bpea) ligands were studied by Schatzschneider and coworkers [78, 79], where the aldehyde group in $[Mn(CO)_3(bpea^{CH2C6H4CHO})]^+$ was coupled to peptides via an oxime ligation reaction with an aminooxy functionalized peptide, an attractive method of bioconjugate formation that does not require catalyst addition. Furthermore, manganese photoactivatable CORMs have been attached to carbon nanomaterials [80], silica [81], and first- and second-generation diaminobutane (DAB) dendrimers [82], materials that may selectively accumulate in tumor tissue due to the enhanced permeability and retention effect (EPR) [83].

Tethering a CORM to a nanoparticle, dendrimer, or biological carrier molecule such as a peptide or sugar, in order to confer some degree of selectivity to the distribution of the CORM, is an attractive concept. Taking advantage of the natural uptake mechanism of vitamin B_{12}, Zobi and coworkers have elegantly used such a biocompatible scaffold as a carrier for several different CORMs [84]. One such bioconjugate, B_{12}-MnCORM-1 contains a photoactive manganese(I) tricarbonyl unit coordinated to a 1,4,8,11-tetraazacyclotetradecane unit appended to the periphery of the vitamin via the ribose moiety. The authors demonstrated active uptake of the conjugate by 3T3 fibroblasts and then could affect the release of CO by irradiation with visible light, preventing fibroblast death under conditions of hypoxia and metabolic depletion. In 2010, Ford and coworkers reported the tris-anionic complex $Na_3[W(CO)_5(tppts)]$ with tppts = tris(3-sulfanatophenyl)phosphine as a new type of light-activated CORM [85]. The complex, solubilized by the charged tppts ligand, was stable in aerated buffer solutions for several hours in the dark, and released one equivalent of carbon monoxide upon illumination at 313 nm light to form the aqua species, which underwent further irreversible oxidation in the presence of oxygen. Although this compound opened up new avenues of potential CORM research, the authors also noted that in order to be relevant for biological studies, the photoactivation wavelength to trigger CO-release must be further shifted to the red part of the visible spectrum, in the phototherapeutic window of the cell.

Thus, Ford and coworkers followed up 2 years later, with the report of the first rhenium(I)-based PhotoCORM $[Re(bpy)(CO)_3(P(OCH_2OH)_3)]OTf$, a compound that again possesses good water solubility, in this case due to the tri(hydroxymethyl) phosphine ligand [86]. Illumination at 365 or 405 nm led to the release of the axial carbon monoxide ligand *trans* to the phosphine with good quantum yield, forming the iCORM aqua complex $[Re(bpy)(CO)_2(P(OCH_2OH)_3)\text{-}(H_2O)]OTf$. Not surprisingly given the rich photophysical properties of rhenium-bis(imine) species, both complexes show a strong emission at 515 and 585 nm, respectively, upon excitation at 355 nm. The red-shift in the emission maxima in the aqua relative to the phosphine complex is due to the lower amount of $M \rightarrow CO$ back-donation, resulting in destabilization of the HOMO and a subsequent reduction of the HOMO-LUMO gap. In a novel study, the light from a confocal microscope was used to promote and also follow the formation of the iCORM from its CORM precursor inside living PPC-1 cells. The potential use of this compound in a biological context was strengthened by the nontoxic nature of both the CORM and iCORM, with no significant effect on PPC-1 cells up to concentrations 100 μM.

There is a significant need for materials that release carbon monoxide upon illumination with low energy, long wavelength light, which can penetrate deep into biological tissue [63]. In a major breakthrough in the field, Ford and coworkers very recently reported the release of carbon monoxide from water-soluble, lanthanide-doped upconversion nanoparticles (UCNPs) using 980 nm near-infrared (NIR) laser light [87]. The UCNPs were prepared by thermal decomposition of rare-earth oleate salts doped with the lanthanides ytterbium (20 %) and thulium (0.1 %), then encapsulated in an amphiphilic phospholipid functionalized poly(ethylene glycol) polymer. Incubation of UCNPs in the presence of the lipophilic manganese PhotoCORM $[Mn(bpy)(CO)_2(PPh_3)_2]^+$ led to the incorporation of the complex into the phospholipid layer of the nanoparticle with high affinity (Fig. 13.7). Illumination of the PhotoCORM-loaded UCNPs with 980 nm NIR light excites the Yb^{3+} present in the core, which then transfers the energy into higher excite-states of Tm^{3+}. Various deactivation pathways ensue, including emission at 450 and 475 nm, wavelengths that conveniently overlap with the absorption of the MLCT band of the complex. The authors show that upon illumination at 980 nm in PBS buffer, UCNPs loaded with $[Mn(bpy)(CO)_2(PPh_3)_2]^+$ release approximately two equivalents of carbon monoxide, while being inactive in the dark. This unique approach will allow the light-activated release of carbon monoxide at greater tissue depths and holds great promise in the delivery of CO to biological targets.

The group of Mascharak is also involved in the development of PhotoCORMs that can release carbon monoxide upon photoactivation with low energy light and have reported multiple examples of CORMs based on ruthenium and manganese [88–91]. One such example, $[MnBr(azpy)(CO)_3]$, containing the 2-phenylazopyridine (azpy) ligand, releases carbon monoxide upon exposure to low power 520 nm light due to a reduction of the M → CO back-bonding in the MLCT excited state (Fig. 13.8). Time-dependent density functional theory (TDDFT) showed a significant bromide → azpy charge transfer character of the visible transition, which also gives rise to a photoreactive excited state. To evaluate the anticancer properties of this compound, the PhotoCORM was added to HeLa and MDA-MB-231 cells and illuminated with visible light for 10 min. Following incubation and cell washing steps, a concentration-dependent decrease in the viabilities of the malignant cells was observed, with a 75 μM solution resulting in a 45 % reduction in viability. This is comparable to the effects of a 10 μM solution of the known anticancer agent 5-fluorouracil (5-FU) over a 74 h period. Interestingly, the morphological changes in the HeLa cells during the course of the experiment were typical of apoptosis. As mentioned earlier in this section, it is important to understand the fate of the iCORM remaining after release of carbon monoxide from the metal center, and whether or not this species remains intact or undergoes further reactions. In the case of manganese(I) complexes, the reactive nature of the iCORM often leads to dissociation of other ancillary ligands and oxidation of the metal to solvated manganese(II). Thus, there is potential release of two bioactive species, namely the carbon monoxide and also an ancillary ligand

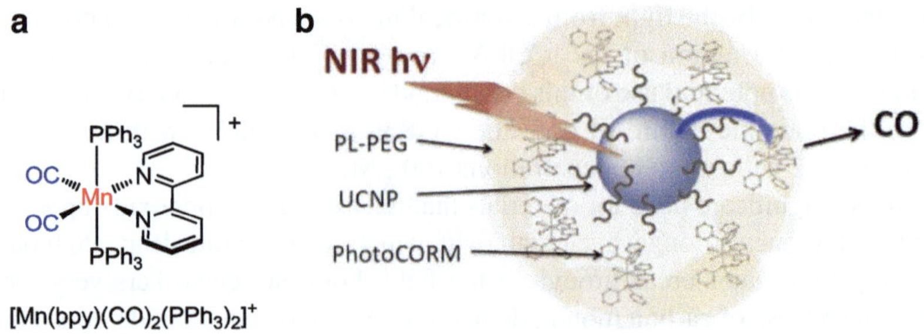

Fig. 13.7 (**a**) Manganese(I) PhotoCORM $[Mn(bpy)(CO)_2(PPh_3)_2]^+$ and (**b**) representation of CO-release mechanism from UCNPs. Adapted with permission from ref. [87] © 2014 Royal Society of Chemistry

such as a bidentate nitrogen donor. One such molecule in the field of CO-releasing molecules, $[MnBr(CO)_3(pbt)]$ containing the 2-phenylbenzothiazole (pbt) ligand, has very recently been reported by Mascharak and coworkers [92]. Upon illumination with broadband visible light, this remarkable complex releases all three carbon monoxide ligands and also the pbt. This latter process was detected by fluorescence spectroscopy through a growth of the band associated with the gradual formation of free pbt (Fig. 13.8). Therefore, $[MnBr(CO)_3(pbt)]$ is a "turn on" CO-donor, where release of CO is associated with an increase in luminescence due to free pbt. EPR studies also suggested that the manganese(I) undergoes oxidation to form a solvated manganese(II) species. MDA-MB-231 cells incubated with $[MnBr(CO)_3(pbt)]$ in the dark were nonemissive, but became highly fluorescent following a short exposure to visible light due to the free pbt in the cytosol, elegantly demonstrating the location of CO-release. A viability study of MDA-MB-231 cells in the presence of the complex showed a dose-dependent reduction in live cells following exposure to visible light, with 100 μM of $[MnBr(CO)_3(pbt)]$ killing around 50 % of the cells, comparable to around 25 μM for 5-fluorouracil. Repeating the viability study with the pbt ligand caused only a slight reduction in cell viability, consistent with the anticancer properties of this compound. Using this innovative approach, other carbonyl complexes could be applied in the eradication of malignant cells through the controlled dual-release of CO and other bioactive ligands.

Although most CORMs reported to date are based on manganese, and to a lesser extent ruthenium, there are also several examples of iron-based CORMs other than the earlier studied $Fe(CO)_5$ [93–96]. In fact, iron is an attractive metal in this context because its distribution is tightly regulated in biological systems. One such example is the iron(II)-based PhotoCORM $[Fe(CO)(N_4Py)](ClO_4)_2$, a mono-carbonyl complex that releases one equivalent of CO upon photoexcitation and in the process generates $[Fe(N_4Py)]^{2+}$, itself having growth inhibitory properties [93]. Treatment of PC-3 prostate cancer cells with the complex led to a loss in cell viability regardless of whether the experiment was carried out in the dark or with light irradiation, with a higher kill rate observed in the presence of light due to the cumulative effects of CO and $[Fe(N_4Py)]^{2+}$.

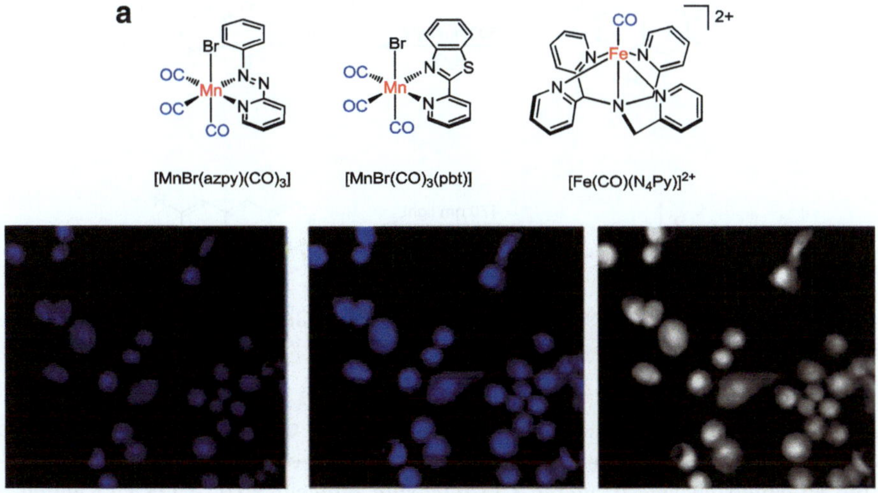

Fig. 13.8 (**a**) Structure of some PhotoCORM [MnBr(CO)$_3$(pbt)] and (**b**) Fluorescence images of MDA-MB-231 cells incubated with PhotoCORM. *Left*: without light exposure; *Center*: after three 10 s pulses of low power LED light, and *Right*: gray scale version of center image. Adapted with permission from ref. [92] © 2014 American Chemical Society

13.3.2 Organic PhotoCORMs

Although PhotoCORMs based on transition metals represent almost the entirety of those studied so far, there are two recently reported examples of purely organic PhotoCORMs [97, 98]. Proponents of such systems argue that organic molecules should be more easily taken up by cells compared to metal complexes, and that these molecules would not have the same toxicity on normal healthy cells, as is often assumed with metal complexes. The groups of Klán and Liao both reported in close succession the two organic molecules shown in Fig. 13.9 which release one and two equivalents of carbon monoxide upon illumination with 500 and 470 nm light, respectively (Fig. 13.9).

Furthermore, Liao and coworkers showed that the fluorescent iCORM of the anthracene-derived compound, formed upon CO-release, was present inside KG-1 leukaemia cells, while micelles thereof as well as the anthracene iCORM were non-toxic toward KG-1 cells up to concentrations of 40 μM. It seems certain that in the future more and more organic-based CORMs will be studied, but whether or not these systems become more prevalent then metal-based systems remains to be seen.

13.4 Other CO-Release Mechanisms

In contrast to conventional CORMs triggered by ligand-exchange reactions with medium and PhotoCORMs that require light as an externally applied stimulus, other mechanisms to induce the release of carbon monoxide from a molecular or

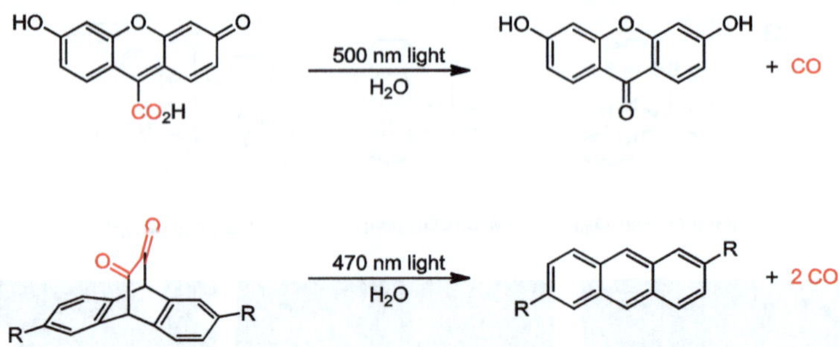

Fig. 13.9 Organic PhotoCORMs reported by the groups of Klán (*top*) and Liao (*bottom*)

supramolecular carrier system have been much less explored. However, a particularly interesting approach has recently been introduced by the group of Schmalz in the so-called *enzyme-triggered CO-releasing molecules* (ET-CORMs) which are based on an iron tricarbonyl moiety coordinated to a 1- or 2-substituted dienol [99]. In the inactive prodrug form, the molecule is "locked" in the dienol tautomeric form by *O*-acylation or *O*-phosphorylation and the $Fe(CO)_3$ group is η^4-coordinated to the diene, which is preferentially incorporated in a ring system such as 1,3-cyclohexadiene (Fig. 13.10). Cleavage of the ester or phosphate group by esterase or phosphatase enzymes leads to isomerization of the enolate to the keto form concomitant with a hapticity shift of the $Fe(CO)_3$ moiety from η^4 to η^2. The resulting $16e^-$-species is much less stable than the $18e^-$-prodrug and thus facile oxidation leads to release of ferric iron and three equivalents of carbon monoxide together with a cyclohexenone derivative [100–102]. The activity of ET-CORMs in hypothermic preservation damage and vascular cell adhesion protein 1 (VCAM-1) expression was recently explored and allowed structure–activity relationships to be derived based on the type of core structure (2-cyclohexenone vs. 1,3-cyclohexanedione vs. dimedone) and the ester substituent. VCAM-1 has an important physiological function in cell–cell interactions with the vascular endothelium and thus might be implicated in atherosclerosis and rheumatoid arthritis [103].

In addition to light, few other external stimuli have been explored for localized induction of CO-release. However, recently, Kunz and Janiak decorated maghemite-type iron oxide nanoparticles with a CORM-3 analog based on D/L-dihydroxyphenylalanine instead of glycine as the *N*,*O*-chelator to the $RuCl(CO)_3$ moiety [104]. When the CORM-functionalized particles are exposed to an external alternating magnetic field, local heating accelerates the rate of ligand-exchange-mediated CO-release from the $Ru(CO)_3$ moiety, with a significant field-dependency of the CO-release observed in the standard myoglobin assay (Fig. 13.10).

It remains to be seen whether other external stimuli as well as endogenous differences in parameters such as cellular pH and redox status can also be explored to control carbon monoxide release from CORM prodrugs.

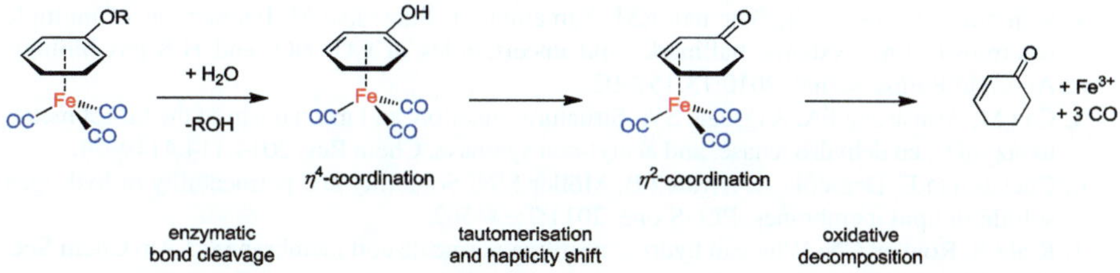

Fig. 13.10 Mechanism of enzymatic bond cleavage by esterases or phosphatases in ET-CORMs and sequence of tautomerization, hapticity shift and oxidative decomposition leading to release of carbon monoxide, ferric iron, and cyclohexenone

13.5 Conclusion

Carbon monoxide (CO), nitric oxide (NO), and hydrogen sulfide (H_2S) form a class of small signaling molecules endogenously produced in higher organisms, including human, by enzymatic activity. They are distinct in their redox and solubility properties, ability of bio(macro)molecule modification, and cellular precursors but nevertheless heavily intertwined in their biological activity. Their concentration-dependent balance between cytoprotective and cytotoxic activity is now increasingly explored for therapeutic applications in human medicine. Exogenous sources for these small signaling molecules have already been reviewed for NO and H_2S and thus, the present chapter is restricted to coverage of the use of metal carbonyl complexes as *CO-releasing molecules* (CORMs) only. Different trigger mechanism can be utilized to initiate carbon monoxide release from the metal carbonyl prodrug, with a particular focus on ligand-exchange mediated and photoinduced CO-release. Illustrative examples of important lead structures are presented and requirement for further CORM development discussed, in particular the need to carefully tune the outer ligand periphery or "drug sphere", for specific biomedical applications." Certainly, CO, NO, and H_2S have fascinating prospects for the study of fundamental biological processes and the development of new medicines.

Acknowledgments P.S. thanks the Alexander-von-Humboldt Foundation for a postdoctoral fellowship.

References

1. Fukuto JM, Carrington SJ, Tantillo DJ, Harrison JG, Ignarro LJ, Freeman BA, et al. Small molecule signaling agents: the integrated chemistry and biochemistry of nitrogen oxides, oxides of carbon, dioxygen, hydrogen sulfide, and their derived species. Chem Res Toxicol. 2012;25:769–93.
2. Szabo C. Gasotransmitters: new frontiers for translational science. Science Translat Med. 2010;2:59ps4.
3. Weast RC, editor. CRC handbook of chemistry and physics. Boca Raton: CRC Press; 1983.

4. Kajimura M, Fukuda R, Bateman RM, Yamamoto T, Suematsu M. Interactions of multiple gas-transducting systems: hallmarks and uncertainties of CO, NO, and H_2S gas biology. Antioxid Redox Signal. 2010;13:157–92.
5. Can M, Armstrong FA, Ragsdale SW. Structure, function, and mechanism of the nickel metalloenzymes, co dehydrogenase, and acetyl-coa synthase. Chem Rev. 2014;114:4149–74.
6. Cuevasanta E, Denicola A, Alvarez B, Möller MN. Solubility and permeability of hydrogen sulfide in lipid membranes. PLOS one. 2011;7:e34562.
7. Riahi S, Rowley CN. Why can hydrogen sulfide permeate cell membranes? J Am Chem Soc. 2014;136:15111–3.
8. Crane BR, Sudhamsu J, Patel BA. Bacterial nitric oxide synthases. Ann Rev Biochem. 2010;79:445–70.
9. Feng C. Mechanism of nitric oxide synthase regulation: electron transfer and interdomain interactions. Coord Chem Rev. 2012;256:393–411.
10. Poulos TL. Heme enzyme structure and function. Chem Rev. 2014;114:3919–62.
11. Li L, Rose P, Moore PK. Hydrogen sulfide and cell signaling. Ann Rev Pharmacol Toxicol. 2011;51:169–87.
12. Chiku T, Padovani D, Zhu W, Singh S, Vitvitsky V, Banerjee R. H_2S biogenesis by human cystathionine γ-lyase leads to the novel sulfur metabolites lanthionine and homolanthionine and is responsive to the grade of hyperhomocysteinemia. J Biol Chem. 2009;284:11601–12.
13. Wang R. Physiological implications of hydrogen sulfide: a whiff exploration that blossomed. Physiol Rev. 2012;92:791–896.
14. Singh S, Banerjee R. Plp-dependent H_2S biogenesis. Biochim Biophys Acta. 1814;2011:1518–27.
15. Whiteman M, Le Trionnair S, Chopra M, Fox B, Whatmore J. Emerging role of hydrogen sulfide in health and disease: critical appraisal of biomarkers and pharmacological tools. Clinical Sci. 2011;121:459–88.
16. Miller TW, Isenberg JS, Roberts DD. Molecular regulation of tumor angiogenesis and perfusion via redox signaling. Chem Rev. 2009;109:3099–124.
17. Maines MD. The heme oxygenase system: a regulator of second messenger gases. Ann Rev Pharmacol Toxicol. 1997;37:517–54.
18. Matsui T, Unno M, Ikeda-Saito M. Heme oxygenase reveals its strategy for catalyzing three successive oxygenation reactions. Acc Chem Res. 2010;43:240–7.
19. Predmore BL, Lefer DJ, Gojon G. Hydrogen sulfide in biochemistry and medicine. Antioxid Redox Signal. 2012;17:119–40.
20. Mustafa AK, Gadalla MM, Sen N, Kim S, Mu W, Gazi SK, et al. H2S signals through protein S-sulfhydration. Sci Signal. 2009;2:ra72 (1–8).
21. Wilkinson WJ, Kemp PJ. Carbon monoxide: an emerging regulator of ion channels. J Physiol. 2011;589:3055–62.
22. Hou S, Heinemann SH, Hoshi T. Modulation of Bkca channel gating by endogenous signaling molecules. Physiology. 2009;24:26–35.
23. Wang PG, Xian M, Tang X, Xu X, Wen Z, Cai T, et al. Nitric oxide donors: chemical activities and biological applications. Chem Rev. 2002;102:1091–134.
24. Whiteman M, Winyard PG. Hydrogen sulfide and inflammation: the good, the bad, the ugly and the promising. Expert Rev Clin Pharmacol. 2011;4:13–32.
25. Zhao Y, Wang H, Xian M. Cysteine-activated hydrogen sulfide (H_2S) donors. J Am Chem Soc. 2011;133:15–7.
26. Motterlini R, Clark JE, Foresti R, Sarathchandra P, Mann BE, Green CJ. Carbon-monoxide-releasing molecules - characterization of biochemical and vascular activity. Circ Res. 2002;90:e17–24.
27. Seixas JD, Santos MFA, Mukhopadhyay A, Coelho AC, Reis PM, Veiros LF, et al. A contribution to the rational design of $Ru(CO)_3Cl_2L$ complexes for *in vivo* delivery of CO. Dalton Trans. 2015;44:5058–75.
28. Santos MFA, Seixas JD, Coelho AC, Mukhopadhyay A, Reis PM, Romao MJ, et al. New insights into the chemistry of *fac*-$[Ru(CO)_3]^{2+}$ fragments in biologically relevant conditions:

the CO releasing activity of [$Ru(CO)_3Cl_2$(1,3-thiazole)], and the X-Ray crystal structure of its adduct with lysozyme. J Inorg Biochem. 2012;117:285–91.
29. Bergamo A, Sava G. Ruthenium anticancer compounds: myths and realities of the emerging metal-based drugs. Dalton Trans. 2011;40:7817–23.
30. Bergamo A, Gaiddon C, Schellens JHM, Beijnen JH, Sava G. Approaching tumor therapy beyond platinum drugs: status of the art and perspectives of ruthenium drug candidates. J Inorg Biochem. 2012;106:90–9.
31. Clark JE, Naughton P, Shurey S, Green CJ, Johnson TR, Mann BE, et al. Cardioprotective actions by a water-soluble carbon monoxide-releasing molecule. Circ Res. 2003;93:e2–8.
32. Johnson TR, Mann BE, Teasdale IP, Adams H, Foresti R, Green CJ, et al. Metal carbonyls as pharmaceuticals? [$Ru(CO)_3Cl$(glycinate)], a CO-releasing molecule with an extensive aqueous solution chemistry. Dalton Trans. 2007;2007:1500–8.
33. Santos-Silva T, Mukhopadhyay A, Seixas JD, Bernardes GJL, Romao CC, Romao MJ. Corm-3 reactivity towards proteins: the crystal structure of a Ru(II) dicarbonyl-lysozyme complex. J Am Chem Soc. 2011;133:1192–5.
34. Motterlini R, Sawle P, Bains S, Hammad J, Alberto R, Foresti R, et al. Corm-A1: a new pharmacologically active carbon monoxide-releasing molecule. FASEB J. 2004;18:284–6.
35. Alberto R, Ortner K, Wheatley N, Schibli R, Schubiger AP. Synthesis and properties of boranocarbonate: a convenient in Situ Co source for the aqueous preparation of $[^{99m}Tc(OH_2)_3(CO)_3]^+$. J Am Chem Soc. 2001;123:3135–6.
36. Pitchumony TS, Spingler B, Motterlini R, Alberto R. Derivatives of sodium boranocarbonate as novel CO-releasing molecules (CO-RMs). Chimia. 2008;62:277–9.
37. Pitchumony TS, Spingler B, Motterlini R, Alberto R. Syntheses, structural characterization and Co releasing properties of boranocarbonate $[H_3BCO_2H]^-$ derivatives. Org Biomol Chem. 2010;8:4849–954.
38. Motterlini R, Otterbein LE. The therapeutic potential of carbon monoxide. Nat Rev Drug Discov. 2010;9:728–43.
39. Fairlamb IJS, Duhme-Klair A-K, Lynam JM, Moulton BE, O'Brien CT, Sawle P, et al. H^4-Pyrone Iron(0)Carbonyl complexes as effective CO-releasing molecules (CO-RMs). Bioorg Med Chem Lett. 2006;16:995–8.
40. Sawle P, Hammad J, Fairlamb IJS, Moulton BE, O'Brien CT, Lynam JM, et al. Bioactive properties of iron-containing carbon monoxide-releasing molecules. J Pharmacol Exp Ther. 2006;318:403–10.
41. Fairlamb IJS, Lynam JM, Moulton BE, Taylor IE, Duhme-Klair A-K, Sawle P, et al. H1-2-pyrone metal carbonyl complexes as CO-releasing molecules (CO-RMs): a delicate balance between stability and CO liberation. Dalton Trans. 2007;2007:3603–5.
42. Atkin AJ, Williams S, Motterlini R, Lynam JM, Fairlamb IJS. M_2-Alkyne Dicobalt(0) Hexacarbonyl complexes as carbon monoxide-releasing molecules (CO-RMs): probing the release mechanism. Dalton Trans. 2009;2009:3653–6.
43. Zhang WQ, Atkin AJ, Thatcher RJ, Whitwood AC, Fairlamb IJS, Lynam JM. Diversity and design of metal-based carbon monoxide-releasing molecules (CO-RMs) in aqueous systems: revealing the essential trends. Dalton Trans. 2009;2009:4351–8.
44. Zhang W-Q, Whitwood AC, Fairlamb IJS, Lynam JM. Group 6 carbon monoxide-releasing metal complexes with biologically-compatible leaving groups. Inorg Chem. 2010;49:8941–52.
45. Liu H, Wang P, Zhao Q, Chen Y, Liu B, Zhang B, et al. Syntheses, toxicity and biodistribution of CO-releasing molecules containing M(CO)5 (M = Mo, W and Cr). Appl Organomet Chem. 2014;28:169–79.
46. Bikiel DE, Solveyra EG, Di Salvo F, Milagre HMS, Eberlin MN, Correa RS, et al. Tetrachlorocarbonyliridates: water-soluble carbon monoxide releasing molecules rate-modulated by the sixth ligand. Inorg Chem. 2011;50:2334–45.
47. Petruk AA, Vergara A, Marasco D, Bikiel DE, Doctorovich F, Estrin DA, et al. Interaction between proteins and Ir based CO releasing molecules: mechanism of adduct formation and Co release. Inorg Chem. 2014;53:10456–62.

48. Zobi F, Degonda A, Schaub MC, Bogdanova AY. Co releasing properties and cytoprotective effect of *cis-trans*-$[Re^{II}(CO)_2Br_2L_2]^n$ complexes. Inorg Chem. 2010;49:7313–22.
49. Zobi F, Blacque O. Reactivity of 17 e^- complex $[Re^{II}Br_4(CO)_2]^{2-}$ with bridging aromatic ligands. Characterization and CO-releasing properties. Dalton Trans. 2011;40:4994–5001.
50. Zobi F, Blacque O, Jacobs RA, Schaub MC, Bogdanova AY. 17 e^- rhenium dicarbonyl CO-releasing molecules on a cobalamin scaffold for biological application. Dalton Trans. 2012;41:370–8.
51. Crook SH, Mann BE, Meijer AJHM, Adams H, Sawle P, Scapens D, et al. $[Mn(CO)_4\{S_2CN Me(CH_2CO_2H)\}]$, a new water-soluble CO-releasing molecule. Dalton Trans. 2011;40:4230–5.
52. Hewison L, Crook SH, Mann BE, Meijer AJHM, Adams H, Sawle P, et al. New types of CO-releasing molecules (CO-RMs), based on iron dithiocarbamate complexes and $[Fe(CO)_3I(S_2COEt)]$. Organometallics. 2012;31:5823–34.
53. Gonzalez MA, Fry NL, Burt R, Davda R, Hobbs A, Mascharak PK. Designed iron carbonyls as carbon monoxide (CO) releasing molecules: rapid CO release and delivery to myoglobin in aqueous buffer, and vasorelaxation of mouse aorta. Inorg Chem. 2011;50:3127–34.
54. Hewison L, Johnson TR, Mann BE, Meijer AJHM, Sawle P, Motterlini R. A re-investigation of $[Fe(\text{L-cysteinate})_2(CO)_2]^{2-}$: an example of non-heme CO coordination of possible relevance to CO binding to ion channel receptors. Dalton Trans. 2011;40:8328–34.
55. Marques AR, Kromer L, Gallo DJ, Penacho N, Rodrigues SS, Seixas JD, et al. Generation of carbon monoxide releasing molecules (CO-RMs) as drug candidates for the treatment of acute liver injury: targeting of CO-RMs to the liver. Organometallics. 2012;31:5810–22.
56. Romao CC, Blättler WA, Seixas JD, Bernardes GJL. Developing drug molecules for therapy with carbon monoxide. Chem Soc Rev. 2012;41:3571–83.
57. Seixas JD, Mukhopadhyay A, Santos-Silva T, Otterbein LE, Gallo DJ, Rodrigues SS, et al. Characterization of a versatile organometallic Pro-Drug (CORM) for experimental CO based therapeutics. Dalton Trans. 2013;42:5985–98.
58. Ma M, Noei H, Mienert B, Niesel J, Bill E, Muhler M, et al. Iron metal–organic frameworks MIL-88B and NH_2-MIL-88B for the loading and delivery of the gasotransmitter carbon monoxide. Chem Eur J. 2013;19:6785–90.
59. Hasegawa U, van der Vlies AJ, Simeoni E, Wandrey C, Hubbell JA. Carbon monoxide-releasing micelles for immunotherapy. J Am Chem Soc. 2010;132:18273–80.
60. Fujita K, Tanaka Y, Sho T, Ozeki S, Abe S, Hikage T, et al. Intracellular CO release from composite of ferritin and ruthenium carbonyl complexes. J Am Chem Soc. 2014;136:16902–8.
61. Steiger C, Lühmann T, Meinel L. Oral drug delivery of therapeutic gases - carbon monoxide release for gastrointestinal diseases. J Controlled Release. 2014;189:46–53.
62. Simpson PV, Schatzschneider U, in *inorganic chemical biology: principles, technqiues and applications*, Gasser G. (Editor), Wiley, Chichester, 2014
63. Chakraborty I, Carrington SJ, Mascharak PK. Design strategies to improve the sensitivity of photoactive metal carbonyl complexes (PhotoCORMs) to visible light and their potential as CO-donors to biological targets. Acc Chem Res. 2014;47:2603–11.
64. García-Gallego S, Bernardes GJL. Carbon-monoxide-releasing molecules for the delivery of therapeutic CO *in vivo*. Angew Chem Int Ed. 2014;53:9712–21.
65. Rimmer RD, Pierri AE, Ford PC. Photochemically activated carbon monoxide release for biological targets. Toward developing air-stable PhotoCORMs labilized by visible light. Coord Chem Rev. 2012;256:1509–19.
66. Schatzschneider U. Photoactivated biological activity of transition-metal complexes. Eur J Inorg Chem. 2010;2010:1451–67.
67. Schatzschneider U. Photocorms: light-triggered release of carbon monoxide from the coordination sphere of transition metal complexes for biological applications. Inorg Chim Acta. 2011;374:19–23.
68. Motterlini R, Clark JE, Foresti R, Sarathchandra P, Mann BE, Green CJ. Carbon monoxide-releasing molecules: characterization of biochemical and vascular activities. Circ Res. 2002;90:e17–24.

69. Motterlini R, Foresti R, Green CJ, in *carbon monoxide and cardiovascular functions*, Wang R. (Editor) CRC Press, Boca Raton, Florida, USA, 2002, pp 249–71
70. Motterlini R, Mann BE, Foresti R. Therapeutic applications of carbon monoxide-releasing molecules. Expert Opin Investig Drugs. 2005;14:1305–18.
71. Niesel J, Pinto A, N'Dongo HWP, Merz K, Ott I, Gust R, et al. Photoinduced Co release, cellular uptake and cytotoxicity of a Tris(pyrazolyl)methane (tpm) manganese tricarbonyl complex. Chem Commun. 2008;2008:1798–800.
72. Rudolf P, Kanal F, Knorr J, Nagel C, Niesel J, Brixner T, et al. Ultrafast photochemistry of a manganese-tricarbonyl CO-releasing molecule (CORM) in aqueous solution. J Phys Chem Lett. 2013;4:596–602.
73. Berends H-M, Kurz P. Investigation of light-triggered carbon monoxide release from two manganese PhotoCORMs by IR. UV/Vis and EPR Spectroscopy. InorgChim Acta. 2012;380:141–7.
74. Nagel C, McLean S, Poole RK, Braunschweig H, Kramer T, Schatzschneider U. Introducing $[Mn(CO)_3(tpa\text{-}k^3N)]^+$ as a novel photoactivatable CO-releasing molecule with well-defined iCORM intermediates - synthesis, spectroscopy, and antibacterial activity. Dalton Trans. 2014;43:9986–97.
75. Ward JS, Lynam JM, Moir J, Fairlamb IJS. Visible-light-induced CO release from a therapeutically viable tryptophan-derived manganese(I) carbonyl (TryptoCORM) exhibiting potent inhibition against *E. coli*. Chem Eur J. 2014;20:15061–8.
76. Huber W, Linder R, Niesel J, Schatzschneider U, Spingler B, Kunz PC. A comparative study of tricarbonylmanganese photoactivatable CO releasing molecules (PhotoCORMs) by using the myoglobin assay and time-resolved IR spectroscopy. Eur J Inorg Chem. 2012;2012:3140–6.
77. Kunz PC, Huber W, Rojas A, Schatzschneider U, Spingler B. Tricarbonylmanganese(I) and rhenium(I) complexes of imidazol-based phosphane ligands: influence of the substitution pattern on the CO release properties. Eur J Inorg Chem. 2009;2009:5358–66.
78. Ward JS, Lynam JM, Moir JWB, Sanin DE, Mountford AP, Fairlamb IJS. A therapeutically viable photo-activated manganese-based CO-releasing molecule (Photo-CO-RM). Dalton Trans. 2012;41:10514–7.
79. Pai S, Hafftlang M, Atongo G, Nagel C, Niesel J, Botov S, et al. New modular manganese(I) tricarbonyl complexes as PhotoCORMs: *In vitro* detection of photoinduced carbon monoxide release using COP-1 as a fluorogenic switch-on probe. Dalton Trans. 2014;43:8664–78.
80. Pai S, Radacki K, Schatzschneider U. Sonogashira, CuAAC, and oxime ligations for the synthesis of Mn^I tricarbonyl PhotoCORM peptide conjugates. Eur J Inorg Chem. 2014;2014:2886–95.
81. Dördelmann G, Meinhardt T, Sowik T, Krueger A, Schatzschneider U. CuAAC Click functionalization of azide-modified nanodiamond with a photoactivatable CO-releasing molecule (PhotoCORM) based on $[Mn(CO)_3(tpm)]^+$. Chem Commun. 2012;48:11528–30.
82. Dördelmann G, Pfeiffer H, Birkner A, Schatzschneider U. Silicium dioxide nanoparticles as carriers for photoactivatable CO releasing molecules (PhotoCORMs). Inorg Chem. 2011;50:4362–7.
83. Govender P, Pai S, Schatzschneider U, Smith GS. Next generation PhotoCORMs: polynuclear tricarbonylmanganese(I)-functionalized polypyridyl metallodendrimers. Inorg Chemistry. 2013;52:5470–8.
84. Peer D, Karp JM, Hong S, Farokhzad OC, Margalit R, Langer R. Nanocarriers as an emerging platform for cancer therapy. Nat Nanotechnol. 2007;2:751–60.
85. Zobi F, Quaroni L, Santoro G, Zlateva T, Blacque O, Sarafimov B, et al. Live-fibroblast IR imaging of a cytoprotective photocorm activated with visible light. J Med Chem. 2013;56:6719–31.
86. Rimmer RD, Richter H, Ford PC. A photochemical precursor for carbon monoxide release in aerated aqueous media. Inorg Chem. 2010;49:1180–5.
87. Pierri AE, Pallaoro A, Wu G, Ford PC. A luminescent and biocompatible PhotoCORM. J Am Chem Soc. 2012;134:18197–200.

88. Pierri A, Huang P-J, Garcia JV, Stanfill JG, Chui M, Wu G, Zheng N, Ford PC. A PhotoCORM Nanocarrier for CO Release Using NIR Light. Chem Commun. 2014.
89. Gonzalez MA, Carrington SJ, Fry NL, Martinez JL, Mascharak PK. Syntheses, structures, and properties of new manganese carbonyls as photoactive CO-releasing molecules: design strategies that lead to CO photolability in the visible region. Inorg Chem. 2012;51:11930–40.
90. Gonzalez MA, Yim MA, Cheng S, Moyes A, Hobbs AJ, Mascharak PK. Manganese carbonyls bearing tripodal polypyridine ligands as photoactive carbon monoxide-releasing molecules. Inorg Chem. 2012;51:601–8.
91. Carrington SJ, Chakraborty I, Alvarado JR, Mascharak PK. Differences in the CO photolability of *cis*- and *trans*-[$RuCl_2$(azpy)$(CO)_2$] complexes: effect of metal-to-ligand back-bonding. Inorg Chim Acta. 2013;407:121–5.
92. Gonzalez MA, Carrington SJ, Chakraborty I, Olmstead MM, Mascharak PK. Photoactivity of mono- and dicarbonyl complexes of ruthenium(II) bearing an N, N, S-donor ligand: role of ancillary ligands on the capacity of CO photorelease. Inorg Chem. 2013;52:11320–31.
93. Carrington SJ, Chakraborty I, Bernard JML, Mascharak PK. Synthesis and Characterization of a "Turn-on" PhotoCORM for Trackable CO Delivery to Biological Targets. ACS Med Chem Lett. 2014.
94. Jackson CS, Schmitt S, Dou QP, Kodanko JJ. Synthesis, characterization, and reactivity of the stable iron carbonyl complex [Fe(CO)(N_4py)]$(ClO_4)_2$: photoactivated carbon monoxide release, growth inhibitory activity, and peptide ligation. Inorg Chem. 2011;50:5336–8.
95. Kretschmer R, Gessner G, Görls H, Heinemann SH, Westerhausen M. Dicarbonyl-Bis(cysteamine)Iron(II): a light induced carbon monoxide releasing molecule based on iron (CORM-S1). J Inorg Biochem. 2011;105:6–9.
96. Poh HT, Sim BT, Chwee TS, Leong WK, Fan WY. The dithiolate-bridged diiron hexacarbonyl complex $Na_2[(M\text{-}SCH_2CH_2COO)Fe(CO)_3]_2$ as a water-soluble photocorm. Organometallics. 2014;33:959–63.
97. Velásquez VPL, Jazzazi TMA, Malassa A, Görls H, Gessner G, Heinemann SH, et al. Derivatives of photosensitive CORM-S1 – CO complexes of iron and ruthenium with the $(OC)_2M(S\text{–}C\text{–}C\text{–}NH_2)_2$ fragment. Eur J Inorg Chem. 2012;2012:1072–8.
98. Antony LAP, Slanina T, Šebej P, Šolomek T, Klán P. Fluorescein analogue xanthene-9-carboxylic acid: a transition-metal-free CO releasing molecule activated by green light. Org Lett. 2013;15:4552–5.
99. Peng P, Wang C, Shi Z, Johns VK, Ma L, Oyer J, et al. Visible-light activatable organic CO-releasing molecules (PhotoCORMs) that simultaneously generate fluorophores. Org Biomol Chem. 2013;11:6671–4.
100. Romanski S, Kraus B, Schatzschneider U, Neudörfl J, Amslinger S, Schmalz H-G. Acyloxybutadiene iron tricarbonyl complexes as enzyme-triggered Co-releasing molecules (ET-CORMs). Angew Chem Int Ed. 2011;50:2392–6.
101. Romanski S, Kraus B, Guttentag M, Schlundt W, Rücker H, Adler A, et al. Acyloxybutadiene tricarbonyl iron complexes as enzyme-triggered CO-releasing molecules (ET-CORMs): a structure-activity relationship study. Dalton Trans. 2012;41:13862–75.
102. Romanski S, Rücker H, Stamellou E, Guttentag M, Neudörfl J, Alberto R, et al. Iron dienylphosphate tricarbonyl complexes as water-soluble enzyme-triggered CO-releasing molecules (ET-CORMs). Organometallics. 2012;31:5800–9.
103. Botov S, Stamellou E, Romanski S, Guttentag M, Alberto R, Neudörfl J, et al. Synthesis and performance of acyloxy-diene-$Fe(CO)_3$ complexes with variable chain lengths as enzyme-triggered carbon monoxide-releasing molecules. Organometallics. 2013;32:3587–94.
104. Romanski S, Stamellou E, Jaraba JT, Storz D, Kramer BK, Hafner M, et al. Enzyme-triggered CO-releasing molecules (ET-CORMs): evaluation of biological activity in relation to their structure. Free Rad Biol Med. 2013;65:78–88.
105. Kunz PC, Meyer H, Barthel J, Sollazzo S, Schmidt AM, Janiak C. Metal carbonyls supported on iron oxide nanoparticles to trigger the CO-gasotransmitter release by magnetic heating. Chem Commun. 2013;49:4896–8.

Chapter 14
HNO/Thiol Biology as a Therapeutic Target

Jan Lj. Miljkovic and Milos R. Filipovic

Abbreviations

AC	Adenylate cyclase
AS	Angeli's salt
ALDH	Aldehyde dehydrogenase
cAMP	Cyclic adenosine monophosphate
CBS	Cystathionine beta synthase
cGMP	Cyclic guanosine monophosphate
CGRP	Calcitonin gene-related peptide
CHF	Congestive heart failure
CLR	Calcitonin receptor-like receptor
CSE	Cystathionine gamma lyase
CuZnSOD	Cooper zinc superoxide dismutase
DTT	Dithiothreitol
EDHF	Endothelium derived hyperpolarizing factor
EDRF	Endothelium-derived relaxing factor
FeSOD	Iron superoxide dismutase
GAPDH	Glyceraldehyde 3-phosphate dehydrogenase

J.Lj. Miljkovic
Department of Chemistry and Pharmacy, Friedrich-Alexander University of Erlangen-Nuremberg, Egerlandstrasse 1, Erlangen 91058, Germany

M.R. Filipovic (✉)
Department of Chemistry and Pharmacy, Friedrich-Alexander University of Erlangen-Nuremberg, Egerlandstrasse 1, Erlangen 91058, Germany

Universite de Bordeaux, IBGC, UMR 5095, F-33077, Bordeaux, France

CNRS, IBGC, UMR 5095, F-33077, Bordeaux, France
e-mail: milos.filipovic@ibgc.cnrs.fr

I. Batinić-Haberle et al. (eds.), *Redox-Active Therapeutics*,
Oxidative Stress in Applied Basic Research and Clinical Practice,
DOI 10.1007/978-3-319-30705-3_14

GSH	Glutathione
IPA/NO	Isopropylamine diazeniumdiolate
IR	Ischemia/reperfusion
MHC	Myosin heavy chain
MLC1	Myosin light chain
MnSOD	Manganese superoxide dismutase
NMDA	*N*-methyl-D-aspartate receptor
NOS	Nitric oxide synthase
PKA	Protein kinase A
PLN	Phospholamban
RAMP	Receptor activity modifying protein
RyR	Ryanodine receptor
SERCA2a	Sarcoplasmic reticulum Ca^{2+} ATPase
sGC	Soluble guanylate cyclase
SNP	Sodium nitroprusside
TRPA1	Transient receptor potential ankyrin 1
VSMC	Vascular smooth muscle cells

14.1 Introduction

Signaling by gasotransmitters is an expanding field of research, which started with the seminal discovery that nitric oxide (NO), as a gas, is produced in the endothelial cells and that it is responsible for the endothelium-controlled relaxation [1, 2]. The advantage of the gaseous signaling molecules is that they can easily diffuse and cross cell membranes. They also have a specific "receptor" to which they bind to induce the desired biological action. In the case of NO that is guanylate cyclase, a hem-containing enzyme, which transforms guanosine triphosphate (GTP) to cyclic guanosine monophosphate (cGMP) [3]. cGMP is a powerful secondary messenger responsible for most of the "classical" NO effects, such as vasodilation or neuromodulation [4]. But not all actions of nitric oxide proceed via cGMP signaling. NO could be oxidized by one electron to form nitrosonium species (NO^+), which lead to an oxidative posttranslational modification of cysteine, called *S*-nitrosation. NO could be reduced by one electron to form nitroxyl (HNO). Nitrosyl hydride (hydrogen oxonitrate) or azanone, also commonly known as nitroxyl, has recently been recognized as a sibling of NO with beneficial effect similar to those ascribed to NO.

First documented indication of HNO dates more than 100 years ago [5, 6] where it was proposed by Angeli, that HNO is generated during the decomposition of sodium trioxodinitrate (Angeli's salt). Interestingly, HNO was also the first NO-bond containing interstellar molecule ever detected [7]. Its existence was later on confirmed in the Sagittarius Dwarf Galaxy, a satellite galaxy of our Milky Way [8]. The first reported attempt to isolate HNO was the one by Paul Harteck at Kaiser-Wilhelm-Institut für Physikalische Chemie und Elektrochemie in Berlin-Dahlem in 1933, where he managed to prepare HNO in the addition reaction of $H^•$ atom with $NO^•$ at cryogenic temperatures of liquid air (ca. −196 °C) [9]. This forgotten study actually suggests that HNO exists as a gas, which placed in the context of current knowledge about its biochemistry and pharmacology could even qualify it as a gasotransmitter.

In this chapter, we provide an overview of HNO pharmacology and discuss the new endogenous ways for HNO production. The special emphasis will be on the role of thiols and newly identified gasotransmitter hydrogen sulfide (H_2S) in the HNO generation and signaling.

14.2 HNO Donors

Due to the ephemeric nature of nitroxyl with respect to its rapid dimerization and dehydration to produce nitrous oxide (N_2O) (Sect. 14.3), nitroxyl solutions cannot be applied directly (with the exception of LiNO salt, [10]) for biological experiments or therapeutic treatment, and HNO must be generated in situ from donor compounds. Excellent short reviews that summarize information on commonly used as well as on promising novel HNO donors appeared recently [11–13].

The HNO donors can be classified into six major groups (Table 14.1): Angeli's salt, Piloty's acid and its derivatives, cyanamide, diazenium diolate-derived compounds, acyl nitroso compounds, and acyloxy nitroso compounds. To date, two major strategies have been used for HNO formation from organic and inorganic molecules (a) decomposition of hydroxylamine derivatives with good leaving groups attached to the nitrogen atom (such as Angeli's salt and Piloty's acid) and (b) decomposition of nitroso compounds [12].

Table 14.1 Structure and decomposition products of different HNO donors

HNO donor	Structure of HNO donor	Additional product released
Angeli's salt	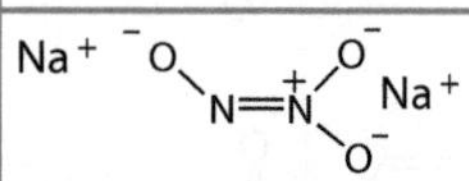	NO_2^-
Piloty's acid	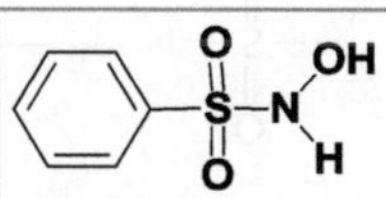	$PhSO_2H$
Cyanamide	$NH_2C\equiv N$	HCN
Diazeniumdiolate	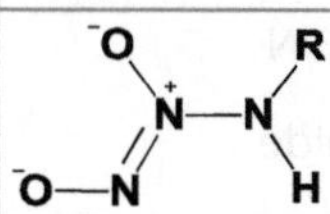	RNHNO
Acyloxy nitroso	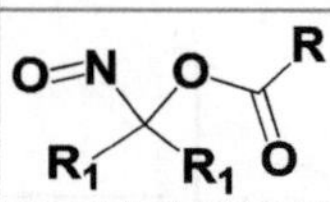	$RCOOH + R_1C(O)R_1$
Acyl nitroso	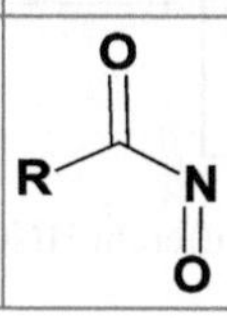	RCOOH

Angeli's salt (*sodium trioxodinitrate, $Na_2N_2O_3$*) is historically the first compound used for in situ generation of HNO [5, 6] and is currently the most popular HNO donor. In aqueous solutions at pH from 4 to 8 Angeli's salt decomposes to HNO and nitrite. A generally accepted mechanism for decomposition of Angeli's salt at this pH range is shown in Fig. 14.1a [14–17]. Briefly, the reaction starts with the

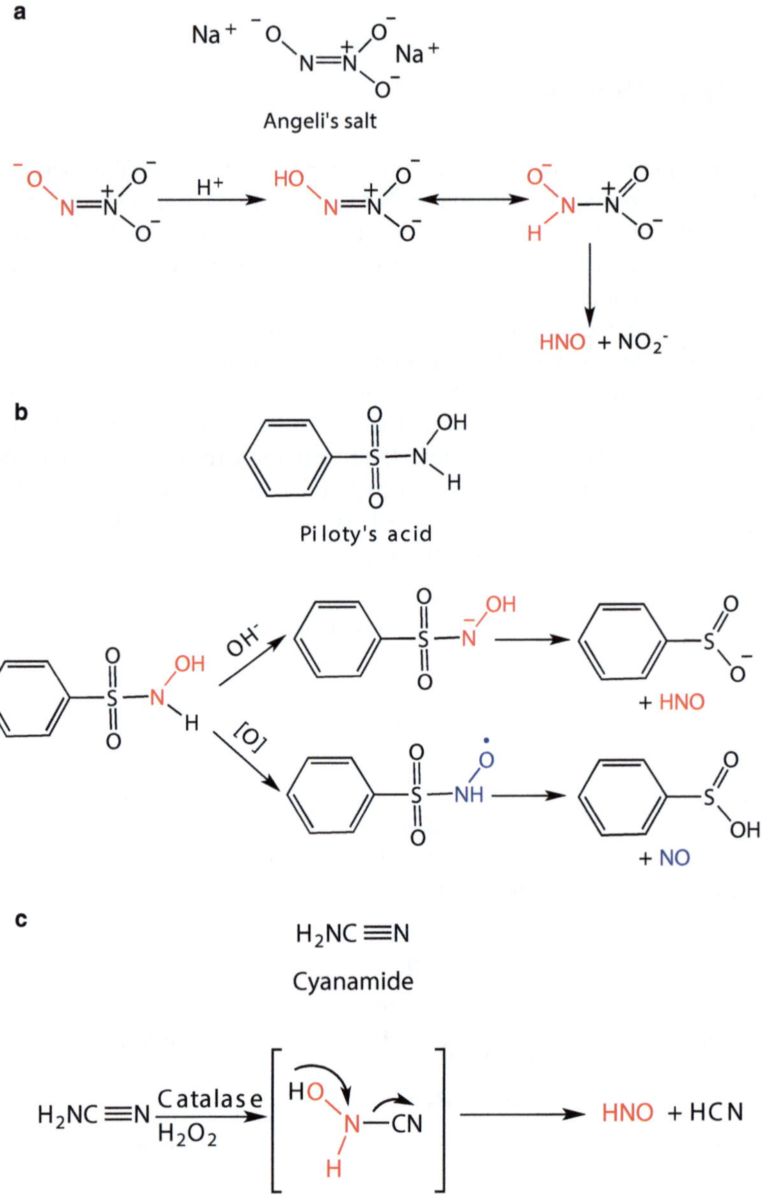

Fig. 14.1 Structures and decomposition pathways of different HNO donors (**a**) Angeli's salt, (**b**) Piloty's acid, and (**c**) cyanamide

protonation of oxygen in the nitroso group, followed by tautomerization to give a monoanion protonated at the nitroso nitrogen atom. By heterolytic cleavage of the N–N bond, this tautomer decomposes to give HNO and nitrite. It is worth noting that a recent study demonstrated that small amounts of NO are also generated upon Angeli's salt decomposition at neutral pH [18]. Below pH 3 Angeli's salt does not yield HNO and nitrite but generates exclusively NO [15, 16]. Since NO demonstrate distinct effects different from HNO and since reports on biological effects of nitrite are rapidly accumulating [19] formation of these two species should be taken into account when Angeli's salt is used in biological experiments.

From pH 4–8 Angeli's salt decomposes with the first-order rate constant of $4–5\times10^{-3}$ s^{-1} at 37 °C (half-life ~2 min) [17]. This rapid kinetics of decomposition leads to almost a "bolus dose" of HNO and nitrite, which should be taken into account when Angeli's salt is used for studying biological processes that are much slower.

Piloty's acid (*N-hydroxybenzenesulfonamide,* $C_6H_5SO_2NHOH$) represents another long known HNO donor. Its synthesis was first reported in 1896 [20] and the substance is commercially available at present. Piloty's acid releases HNO in strongly basic (pH 13) solutions under anaerobic conditions. Since pK_a of HNO is approximately 11.4 (Sect. 14.1) it is conceivable that nitroxyl anion (NO–) will be predominantly formed under these conditions. Under aerobic conditions Piloty's acid decomposes into NO rather than HNO [11].

Figure 14.1b shows generally accepted mechanisms for the decomposition of Piloty's acid at pH 13 under both anaerobic and aerobic conditions. Briefly, initial deprotonation occurs at the nitrogen atom, which under anaerobic conditions undergoes heterolytic S–N bond cleavage to produce HNO and benzenesulfinate [21–23]. The benzenesulfinate anion is a good nucleophile and a modest reducing agent [12], which should be taken into account when Piloty's acid is used in biological experiments. In the presence of oxygen, a nitroxide radical is formed which upon decomposition yields NO (Fig. 14.1b).

At pH 13, Piloty's acid decomposes with the first-order rate constant of 1.8×10^{-3} s^{-1} at 37 °C [22, 23] which is comparable to the rate constant for decomposition of Angeli's salt at neutral pH [22, 23]. At neutral pH, the rate of HNO release significantly decreases [11, 12], which together with the production of NO limit the use of Piloty's acid as an HNO donor in biological experiments. Considerable effort has been devoted to designing Piloty's acid derivatives, which are better suited for biological experiments and potential therapeutic treatments. These include syntheses of Piloty's acid derivatives with various aromatic or heterocyclic substituents at different positions, which release significant amounts of HNO at neutral pH under anaerobic conditions [24, 25] as well as syntheses of HNO prodrugs based on *N-*,*O*-diacylated or alkylated derivatives of Piloty's acid [13].

Cyanamide (NH_2CN), which is actually an HNO-releasing prodrug, is used therapeutically as an alcohol deterrent drug (under generic names Dipsan®, Abstem®, Temposil®, or Cyanamide Yoshitomi®) for the treatment of chronic alcoholism in Europe, Canada, and Japan. In liver, cyanamide is subjected to catalase-catalyzed oxidation by hydrogen peroxide to unstable *N*-hydroxycyanamide, which directly decomposes to HNO and HCN (Fig. 14.1c) [26–31]. Nitroxyl subsequently reacts with the thiol group of cysteine residue (Sect. 14.3.3) in the active site of aldehyde dehydrogenase (ALDH), which results in the enzyme inhibition. Although cyanide is formed upon metabolic activation of cyanamide, the drug shows very modest toxicity in humans [32].

Diazeniumdiolates (*NONOates*) are molecules which contain the N(O)NO– group. The secondary amine-derived NONOates release NO under neutral and acidic conditions and are traditionally the most commonly used NO donors in chemistry and biology (e.g., [33–35]). However, primary amine-derived NONOates (Table 14.1) release HNO in addition to NO. Of these, isopropylamine diazeniumdiolate (IPA/NO) is most extensively studied. IPA/NO decomposes in a pH-dependent manner to produce HNO above 7.8, and exclusively NO below pH 7 [36].

At pH 7.4, IPA/NO decomposes with the first-order rate constant of $5.1 \times 10^{-3}\ s^{-1}$ at 37 °C [12] which is comparable to the decomposition rate of Angeli's salt at neutral pH. Importantly, a number of studies showed that in biological milieu IPA/NO exhibits activities similar to those of Angeli's salt [36–39].

A new class of potential HNO donors is derived from *acyl nitroso [RC(=O)NO] compounds* [12]. Of these, the most promising for biological praxis seems to be cycloadducts prepared from acyl nitroso derivatives and 9,10-dimethylanthracene (Table 14.1) which release HNO via a retro-Diels-Alder reaction upon UV-A radiation [40]. The advantages of these compounds in comparison to other HNO donors include a possibility for the time control of HNO release as well as the fact that the only decomposition byproduct to consider is a primary amine.

Another novel class of HNO donors, bright brilliant blue *acyloxy nitroso compounds* (Table 14.1) that release HNO upon cleavage of the ester bond has been described recently [41]. The rate of HNO release depends on pH and the structure of the acyl group of the acyloxy nitroso compound [41]. Biological effects of these HNO donors, such as inhibition of platelet aggregation and cGMP-induced dilatation of preconstricted aortic rings were found to be similar to those of Angeli's salt [41]. Major advantages of these HNO donors include controlling the HNO release rate and lack of nitrite as a by-product which is formed upon decomposition of Angel's salt. Besides, the blue color of these compounds enables a very convenient monitoring of reaction kinetics of HNO release (Switzer et al., 2009). The drawback that limits the application of presently developed HNO donors from this class in biological experiments is the high reactivity with thiols. Since the biological activity of HNO is in the majority of cases associated with thiol reactions, the reactivity of "blue" HNO donors with thiols complicates the interpretation of biological experiments.

14.3 Biologically Relevant Chemistry of HNO

Nitroxyl is the simplest molecule with nitrogen in the +1 oxidation state and yet its aqueous chemistry is surprisingly complex. Even its deprotonation (dissociation) (14.1) is not typical.

$$^{1}HNO + H_2O \rightarrow {}^{3}NO^{-} + H_3O \qquad (14.1)$$

In variance to early studies, which reported pK_a values of 4.7 [42] and 7.2 [43], later studies documented that pK_a value for HNO is approximately 11.4 [44, 45]. Therefore, nitroxyl predominantly exists in the protonated, (HNO) form in aqueous solutions at physiological pH. Such low acidity of HNO is explained by the fact that HNO and NO^- have different spin states, ground state of HNO being singlet, whereas NO^- anion has a triplet ground state. Therefore, the transfer of proton requires a change of the spin state, which is forbidden and which makes this reaction extremely slow compared to a spin allowed process (e.g., $2H_2O \leftrightarrow H_3O^+ + OH^-$) [43, 45]. Consequently, both ^1HNO and $^3NO^-$ will take part in other reaction pathways (Sect. 14.3) rather than in the proton transfer. In addition to being vastly different chemical entities, HNO and NO^- also display very different chemical reactivity. While NO^- is a nucleophile, HNO can behave as either an electrophile or a nucleophile. The nitrogen center of HNO is electrophilic while the lone pair of electrons on the nitrogen atom can also be nucleophilic and capable of coordinating metals via donation of the nitrogen lone pair to the metal center. Both of these activities are highly relevant for the biological chemistry of HNO.

Estimated values of redox potential for a direct reduction of NO into $^1HNO/^3NO^-$ changed significantly as the understanding of HNO chemistry progressed. Based on experimental and computational data, the redox potentials of –0.81 and –0.55 V for the $NO/^3NO^-$ couple and for the proton coupled electron transfer at neutral pH (NO, $H^+/^1HNO$), respectively, were reported in recent studies [44, 46, 47]. Such high reduction potentials led to a suggestion that none of the biologically available reductants was capable of reducing NO to HNO directly. However, it was demonstrated that at pH 7 Angeli's salt (source of HNO, Sect. 14.2) did not reduce methyl viologen (redox potential –0.44 V) [44] which indicates that the redox potential for the reduction of NO into HNO should be more positive than –0.44 V. These findings question the view that NO cannot be reduced directly to HNO in biological milieu and open up a novel perspective for studying biologically relevant mechanisms of direct reduction of NO into HNO (Sects. 14.5 and 14.6).

14.3.1 *Dimerization and Reactions with Oxygen and NO*

This dual reactivity of HNO is responsible for a rapid self-dimerization reaction ($k = 8 \times 10^6\ M^{-1}\ s^{-1}$) whereby one HNO molecule acts as an electrophile and another acts as a nucleophile yielding unstable hyponitrous acid ($H_2N_2O_2$), which undergoes dehydration to form nitrous oxide (N_2O) (14.2) [11, 45, 48].

$$HNO + HNO \rightarrow [HONNOH] \rightarrow N_2O + H_2O \tag{14.2}$$

Besides fast dimerization of HNO, physiologically relevant reactions of HNO/NO^- species with other molecules (e.g., NO and O_2) additionally contribute to its short half-life under biological conditions. Auto-oxidation of HNO is relatively slow process compared with other biologically relative reactions (~$10^3\ M^{-1}\ s^{-1}$, [49, 50]); however, in the membranes where concentration of oxygen is higher and the other biological scavengers less abundant, the autooxidation could be of relevance.

$$NO^{-} + O_2 \rightarrow {}^{-}OONO \tag{14.3}$$

$$HNO + O_2 \rightarrow [HOONO]^{*} \tag{14.4}$$

The reaction product of NO^- with oxygen is peroxynitrite anion ($ONOO^-$) (14.3) [51, 52]. The structure of the reaction product of HNO with oxygen is, however, less clear. The extensive studies by Miranda and colleagues showed that the product of the reaction of HNO with oxygen is in fact a structural isomer of peroxynitrite but not peroxynitrite itself (14.4) [48, 53, 54]. Comparison with the synthetic peroxynitrite showed that although peroxynitrite reacts with CO_2 the presence of carbonate buffer does not affect the aerobic chemistry of HNO [55].

HNO could also react with NO [45, 56, 57]. The initial product of this reaction is $N_2O_2^-$ radical which reacts rapidly with another molecule of NO to form the closed shell $N_3O_3^-$ anion, which decays to the final products N_2O and NO_2^- with a rate constant of ~300 s^{-1} [11].

$$HNO + NO \rightarrow HN_2O_2 \tag{14.5}$$

$$HN_2O_2 + NO \rightarrow HN_3O_3 \tag{14.6}$$

$$HN_3O_3 \rightarrow N_2O + HNO_2 \tag{14.7}$$

Although NO is present in biological milieu at a much lower concentrations than oxygen, reaction of NO with HNO is considered biologically relevant since its reaction rate is significantly higher than that of the reaction of HNO with oxygen.

14.3.2 Reaction of HNO with Metalloproteins

In contrast to NO which reacts preferentially with ferrous heme proteins, HNO reacts preferentially with ferric heme proteins ($k \sim 10^6\ M^{-1}\ s^{-1}$), while its reaction with ferrous heme proteins is approximately 100 times slower ($k \sim 10^4\ M^{-1}\ s^{-1}$) [11, 48, 58]. The reaction of HNO with methemoglobin that yields Hb(Fe^{2+})–NO complex (14.8) [59, 60] is one of the earliest reactions applied for HNO detection and measurement [61].

$$\text{metHb}(Fe^{3+}) + HNO \rightarrow \text{Hb}(Fe^{2+}) - NO + H^{+} \tag{14.8}$$

Interestingly, HNO was found to form unusually stable complex with deoxymyoglobin. The stability of this complex was partially explained by hydrogen bonding between the proton from HNO and histidine (His64) residue from heme pocket of myoglobin [62, 63].

The most widely recognized and extensively studied heme-containing protein that interacts with HNO is the soluble guanylyl cyclase (sGC) [47]. Early studies demonstrated that HNO donors induce vasodilation as well as that this was associated with increase of cGMP level, which opened up the possibility that in addition to NO, which is well-known activator of sGC [4], HNO behaves as another activator of this enzyme [64–68]. However, it is not as yet quite clear whether HNO itself can directly activate sGC or if it is preceded by its oxidation to NO. Namely, a number of studies demonstrated that HNO is transformed by CuZnSOD catalyzed oxidation to NO, which then binds to ferrous heme in the enzyme's active site, which leads to enzyme activation and consequent production of cGMP [69–71]. However, [72] suggest that like NO, HNO may activate sGC by a direct interaction with the regulatory ferrous heme of the enzyme molecule. Interestingly, the same study has shown that at higher concentrations HNO causes enzyme inhibition by modifying enzyme thiols [72]. The authors propose that sGC activity can be regulated by HNO via interactions at both the regulatory heme and cysteine residues of the enzyme molecule.

14.3.3 Reactivity of HNO with Thiols

In variance to other biologically relevant electrophiles (such as hydrogen peroxide and peroxynitrite), which react with a variety of biological nucleophiles, HNO reacts significantly only with protein thiols. It is increasingly recognized that many of unique biological and pharmacological effects of HNO can be attributed to its ability to react with specific thiol protein targets in the cell.

That HNO reacts rapidly with thiols is clearly indicated by the rate constant for its reaction with glutathione (GSH) reported to be $2–8 \times 10^6\ M^{-1}\ s^{-1}$ [49, 73], while the reaction with active site cysteine residues (e.g., papain) was reported to be significantly greater ($k = 2 \times 10^7\ M^{-1}\ s^{-1}$) [74]. This makes HNO kinetically a much better thiol modifying agent compared to the well-known thiol oxidants/electrophiles such as hydrogen peroxide and peroxynitrite [47].

The reaction of HNO with thiols involves nucleophilic attack of the thiolate (deprotonated thiol anion RS^-) to the electrophilic nitrogen of HNO to give unstable *N*-hydroxysulfenamide (14.9). In the presence of excess (or vicinal) thiol, it reacts further with a second thiol to give the corresponding disulfide and hydroxylamine (14.10). In the absence of thiol, *N*-hydroxysulfenamide undergoes a rearrangement to give a sulfinamide (14.11) [47, 48, 60, 75]. The recent study which demonstrated that in BSA treated with Angelis's salt internal sulfinamide was formed (presumably through the reaction of *N*-hydroxysulfenamide with a nearby lysine residue) may indicate that the *N*-hydroxysulfenamide formed in protein molecule can possibly react with other nearby nucleophiles besides cysteine [76]. It was proposed that *N*-hydroxysulfenamide generated in protein active site may by stabilized by interactions with amino acid residues and thus have a prolonged life time [47].

$$RSH + HNO \rightarrow RSNHOH \quad (14.9)$$

$$RSNHOH + R'SH \rightarrow RSSR' + NH_2OH \quad (14.10)$$

$$RSNOH \rightarrow \left[RS^{+}NH + HO^{-}\right] \rightarrow RS(OH)NH \rightarrow RS(O)NH_2 \quad (14.11)$$

Of these, HNO-induced thiol modifications conversion of thiolate to a disulfide is considered to be a reversible process since there are cellular mechanisms which reduce disulfides back to thiols [77]. *N*-hydroxysulfenamide was also proposed to be readily reversible, while sulfinamide is considered to be an irreversible thiol modification (e.g., [46, 47]).

Since sulfinamide is a thiol modification exclusively produced by HNO, it is considered to be a specific marker of HNO activity (e.g., [46]). Indeed, by investigating thiol modifications in a variety of thiol proteins treated with Angeli's salt, [76] have confirmed that either sulfinamide or disulfide was formed depending on the protein microenvironment of the cysteine thiol. This was further confirmed in a recent proteomic study, which demonstrated the presence of sulfinic acid (presumably generated by hydrolysis of sulfinamide during the experimental procedures) in a number of proteins in platelet treated with Angeli's salt ex vivo [78]. Besides, disulfide bond (between Cys-152 in the enzyme active site and Cys-156) in glyceraldehyde-3-phosphate dehydrogenase (GAPDH) from HNO-treated platelets was identified in the cited study [78].

Many of the biological/pharmacological actions of HNO can be explained by HNO-induced modifications of crucial cysteine residues in specific thiol proteins in the cell. Just to mention that the earliest described example of pharmacological action of HNO was the inhibition of aldehyde dehydrogenase in liver through the modification of cysteine thiolate in the enzyme active site (presumably by the formation of sulfinamide) by HNO produced by catalase action on HNO-prodrug cyanamide (Sect. 14.1; [26–31]). Importantly, an inhibition of aldehyde dehydrogenase in cyanamide-treated hepatocytes did not affect intracellular GSH/GSSG levels [79] indicating that HNO can selectively modify certain protein thiols in the cell without affecting GSH and, presumably, other thiols in the cell. This was confirmed in the subsequent study which demonstrated that in yeast treated with Angeli's salt a significant inhibition of glyceraldehyde-3-phosphate dehydrogenase (by formation of a disulfide bond between Cys-152 in the enzyme active site and Cys-156, as mentioned above) was associated with minor changes in either GSH levels or GSH/GSSG ratios [80, 81]. Flores-Santana and colleagues reported recently an interesting calculation that based on the low concentration of GAPDH in the yeast (2 μM), the high concentration of intracellular GSH (1–10 mM) and relatively high rate for the reaction of HNO with GSH ($2–8 \times 10^6$ M^{-1} s^{-1}) it seems that the reaction of HNO with GAPDH in the cells proceeds in a diffusion-controlled rate [46].

Furthermore, we recently demonstrated that HNO selectively modifies critical cysteine residues on transient receptor potential channel A1 (TRPA1), leading to its activation and subsequent CGRP release ([82]; see Sect. 14.3.2). These studies provide an evidence for the ability of HNO to modify crucial cysteine residues in specific proteins leaving cellular glutathione levels unaltered. However, the reasons for such high selectivity of HNO, particularly with respect to high intracellular glutathione concentration are not fully clear.

The fact that a reaction of HNO with thiols involves, as mentioned above, the deprotonated thiolate anion (RS^-) makes the overall reactivity of thiol group strongly dependent on the pK_a value of the cysteine side chain. Cysteine residue in glutathione has a pK_a of 9.1, while in proteins pKa values of cysteine thiols are highly influenced by their local environment and values as low as 3.4 have been reported for some active site cysteines [83]. Low pK_a values appear to be primarily due to stabilizing interactions between the cysteine sulfur and the polypeptide backbone or nearby side chains. Therefore, it was speculated that that the rate of reaction of HNO with selected thiol proteins is significantly faster than, for example, with GSH (e.g., [46, 47]).

14.4 Pharmacological (Biological) Effects of HNO Donors: Thiol Modifications as the Main Mechanism of Action

As explained in previous sections, having pure HNO in the solution is not possible and the use of HNO donors is required in order to study its potential biological/pharmacological effects. Most studies done to date have been performed with Angeli's salt, but the recent development of new HNO donors has started contributing to an increased variety.

The observed pharmacological effects of HNO donors could be characterized as (a) vascular, (b) cardiovascular, (c) anticarcinogenic, and, eventually, (d) neuronal. The main mechanism causing all these effects is thiol modification of critical proteins, and they will be specifically covered in the sections that follow.

14.4.1 Vascular Effects of HNO

The seminal discovery by Moncada and colleagues showed that endothelium-derived relaxing factor (EDRF) is nitric oxide, followed by the later discovery that NO is in fact produced in the endothelium by the action of nitric oxide synthase (NOS), revolutionized our understanding of the vascular control [1, 2, 84–86]. Therefore, it was not surprising when HNO has been shown to do the same, albeit via different mechanisms.

First biological effects of Angeli's salt were related to its ability to induce vasodilation of blood vessels. AS caused vasodilation of large conduit and small resistance arteries, as well as of coronary and pulmonary blood vessels [64, 66, 68, 87–91]. More importantly, it seems that in vivo HNO causes predominantly venodilation [92, 93]. In fact, some studies suggested that endothelium-dependent vasorelaxation and nitrergic transmission are not only caused by NO, but also by HNO. In the pharmacological experiments using the model of isolated blood vessels, the group of Kemp-Harper and colleagues showed that EDRF in large conduit arteries of both mice and rats could be scavenged by L-cysteine, suggesting the role of HNO. The effects of NO were, conversely, enhanced by L-cysteine. These studied implied that HNO could be part of EDRF [87].

However, even when the NO production is inhibited and endothelial cells stimulated with acetylcholine (Ach), they produce yet unidentified molecule(s), which directly causes smooth muscle hyperpolarization characteristic of K^+ channel opening; the factor is therefore named endothelium-derived hyperpolarizing factor (EDHF) [94]. Andrews et al. [87] showed that HNO could be the part of EDHF produced by the Ach-caused stimulation of small resistance blood vessels. The group of Kemp-Harper and the colleagues showed that HNO targets Kv and ATP-sensitive K^+ channels (K_{ATP}) in the resistance vasculature [65, 67, 95]. 4-aminopyridine (Kv channel inhibitor) and glibenclamide (K_{ATP} channel inhibitor) attenuated Angeli's salt-stimulated relaxation of rat mesenteric and coronary arteries. Angeli's salt-induced hyperpolarization of vascular smooth muscle cells (VSMC), via cGMP-dependent activation of Kv channels, as ODQ, a selective sGC inhibitor, abolished Angeli's salt-induced VSMC hyperpolarization [67].

In addition to the beneficial hemodynamic effects of HNO donors, these compounds are resistant to the development of nitrate tolerance [67, 96], a major limitation of clinically used organic nitrates such as nitroglycerine. Nitrate tolerance is defined as a rapid attenuation of the hemodynamic and vasodilatory effects of nitrovasodilators following continuous exposure [97]. Studies from the Kemp-Harper group clearly showed that chronic infusion of Angeli's salt had no effect on subsequent vasorelaxation responses to themselves or to acetylcholine [95, 96].

14.4.2 HNO–TRPA1–CGRP Pathway for the Neurovascular Control

One of the main biomarkers of HNO-induced physiological effects, which distinguish it from the NO-induced effects, is the release of calcitonin gene-related peptide release (CGRP). CGRP also seems to contribute to coronary vasodilation in ex vivo models [98–100]. CGRP is a nonadrenergic/noncholinergic (NANC) type peptide, composed of 37-amino acids [100]. In humans, it exists in two forms: αCGRP (CGRPI), widely distributed in central and peripheral nervous system, and βCGRP (CGRPII), operational within enteric nervous system [101–105].

CGRP is stored in dense-core vesicles that are located in terminal endings of sensory nerve fibers [106]. Release of CGRP from terminal vesicles is triggered by Ca^{2+} influx as a consequence of neuronal depolarization. Released CGRP binds to its corresponding CGRP receptor, a transmembrane heterodimeric macromolecule that is composed of "calcitonin receptor-like receptor" (CLR) protein, one of the three peptide isoforms of "receptor activity modifying protein," (RAMPs; RAMP1, RAMP2, and RAMP3) and receptor component protein (RCP) [103, 107–110].

CGRP is the strongest known vasodilator produced in our body. When it binds to CGRP receptor, it may initiate several different signaling pathways (Fig. 14.2). Vasodilatory activity of CGRP was demonstrated in experiments where injection of picomolar concentration of CGRP in human skin [111] was able to provoke long-lasting erythema (5–6 h), while the injection of femtomolar CGRP concentration was followed by an increase of dermal microcirculation blood flow [99]. With its high potency (approx. tenfolds higher than prostaglandins), CGRP represent the most potent vasodilator known to date.

Main intracellular signaling molecule in CGRP-mediated cellular response is believed to be cyclic adenosine monophosphate (cAMP). Stimulation of CGRP

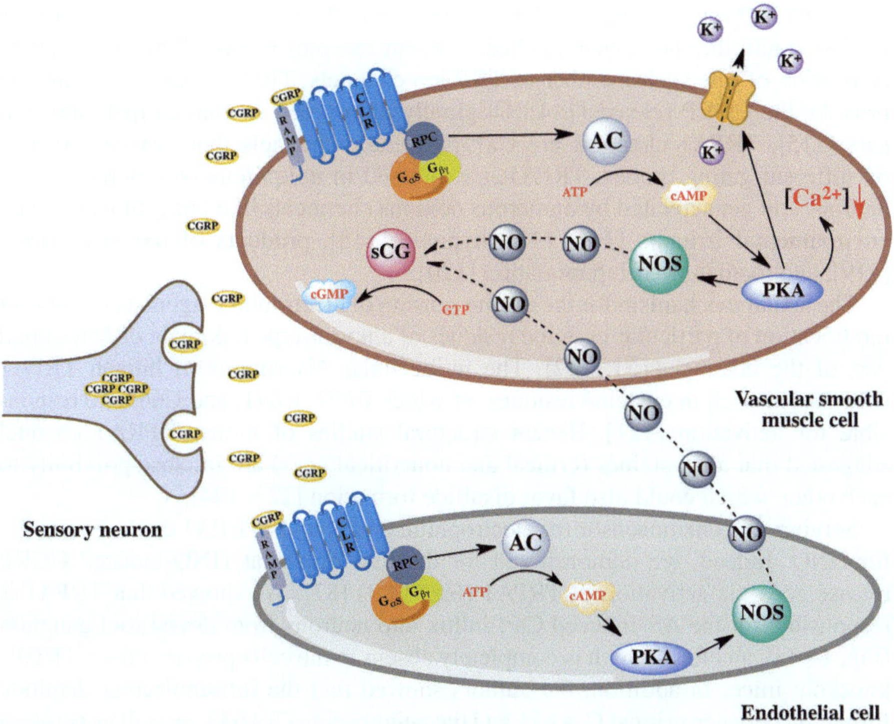

Fig. 14.2 Endothelium-dependent and -independent CGRP-induced vasodilation of blood vessels. *AC* adenylate cyclase, *PKA* protein kinase A, *NOS* nitric oxide synthase, *sGC* soluble guanylate cyclase, *CLR* and *RAMP* components of CGRP receptor. NO can be produced in both, endothelial cell and smooth muscle cells, mainly as a consequence of NOS phosphorylation caused by cAMP-activated PKA. Smooth muscle cell relaxation could be additionally caused by the K^+ channel activation and membrane hyperpolarization

receptor induces activation of adenylate cyclase (AC), an enzyme that generates cAMP. Generated cAMP subsequently activates protein kinase A (PKA), an enzyme that phosphorylates multiple intracellular molecules (Fig. 14.2) [101, 103].

However, CGRP is shown to induce the rise of cGMP as well, via stimulated NO production. CGRP-induced rise in cAMP activates PKA, which upregulates eNOS activity by phosphorylation. Consequently, NO generated by NOS diffuses from the endothelium to the smooth musculature, where it activates GC and via cGMP, additionally increasing the global vasodilator response [101, 103]. Furthermore, the same signaling could also occur, at smaller extent directly in the smooth muscle cells (Fig. 14.2). The essential role of NO/sGC/cGMP signaling cascade in CGRP-induced vasodilation was demonstrated in the early study by [112], and more recently, [113] examined the CGRP-induced vasodilating mechanism in humans to show that CGRP-induced vasodilation is dependent on the release of NO and does not involve activation of K^+_{Ca} or K^+_{ATP} channels.

The CGRP vesicles are stored in sensory nerve fibers, which also express a class of chemo and thermo sensors, called transient receptor potential channels (TRP). Activation of one such member of TRP ion channels, TRPA1, has been shown to provoke the CGRP release [114]. Originally discovered as cold temperature sensors [115], TRPA1 channels are Ca^{2+}-permeable channels that can be activated by different gating modes. TRPA1 is expressed in ubiquitous peripheral sensory neurons and get activated by numerous noxious chemicals like pungent natural and environmental irritants [116, 117], hypoxia [118], products of oxidative stress [119], and mediators of inflammation [120].

The actual mechanism for the chemo sensing of electrophilic agonists is covalent modification of particular cysteine residues of ankyrin-repeat domain of *N*-terminal part of the channel [121, 122]. The intracellular *N*-terminus of human TRPA1 (hTRPA1) is rich in cysteine residues, of which C621, C641, and C665 are responsible for activation [121]. Recent structural studies of mouse TPRA1 channel suggested that all cysteines (critical and noncritical ones) are in close proximity to each other, which could also favor disulfide formation [123, 124].

Serving as a chemosensor for electrophilic compounds TRPA1 could be a target for HNO. Indeed, we demonstrated for the first time that HNO-induced CGRP release goes via activation of TRPA1 (Fig. 14.3) [82]. We showed that TRPA1 is responsible for the AS-induced Ca^{2+} influx into neurons from dorsal root ganglion (Fig. 14.3a), an effect which is completely absent in the cells prepared from TRPA1 knockout mice. In addition, the authors showed that the intramolecular disulfide formation between critical Cys 621 and the neighboring Cys 633, as well as between Cys 651 and the critical Cys 665 is responsible for the channel activation. This opens the channel to allow Ca^{2+} influx, an effect that was long lasting and could only be reversed by the reducing agent DTT (Fig. 14.3b, c).

CGRP release from mice dura or sciatic nerve preparations was completely abolished in $TRPA1^{-/-}$ mice treated with AS, undoubtedly showing that CGRP is released only by the reaction between HNO and TRPA1 cysteine residues (Fig. 14.3d). This study finally answered a decade-long dilemma on how HNO actually stimulates CGRP release. More importantly, the HNO-induced blood pressure drop in mice was significantly reduced in TRPA1 knockout mice. The meningeal blood flow effects of AS were diminished by the application of TRPA1 antagonist or CGRP receptor antagonists, suggesting that this pathways plays important role in the control of meningeal blood flow. The control of the peripheral blood flow, in the sites, which are innervated with TRPA1 containing fibers, could represent the way by which HNO–TRPA1–CGRP signaling cascade contributes to the control of systemic blood pressure.

More interesting observation came from the psychophysics study. Subcutaneous injection of Angeli's salt lead to a rather painful (4/10) sensation followed by the axon-flare erythema, characteristic for the CGRP release (Fig. 14.3e). The effect was histamine-independent. This suggested that in certain pathologies HNO could be a pronociceptive agent.

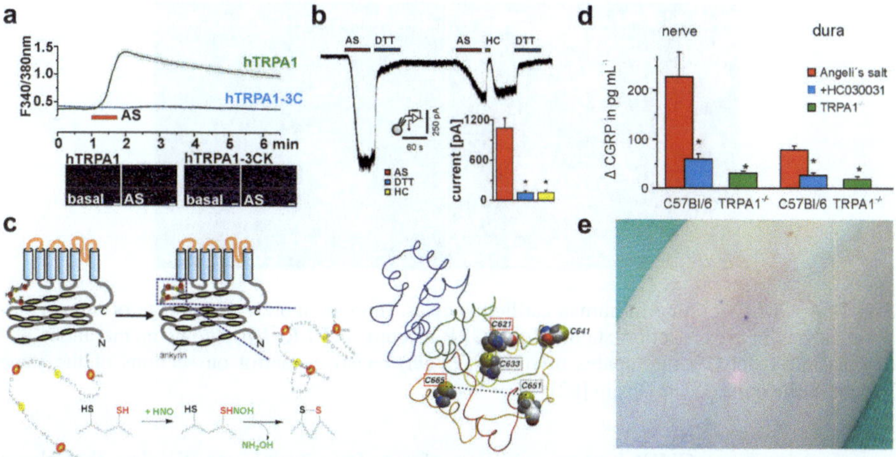

Fig. 14.3 Experimental dissection of HNO–TRPA1–CGRP pathway (adapted from [82]). (**a**) AS increases intracellular Ca^{2+} in hTRPA1 transfected HEK cells but not in cells expressing hTRPA1-3C; representative pseudocolor images (scale bar = 100 μm). (**b**) AS-evoked inward currents can be reversed by application of DTT, or temporarily blocked by HC030031. *Inset*: AS-evoked peak inward currents are significantly reduced by HC030031 or DTT. (**c**) Schematic model of TRPA1 with cysteine-rich region (*red dots*) and formation of disulfide bonds causing major conformational changes. Chemical structure (*bottom*) of two cysteine-SH residues reacting with HNO to form hydroxylamine (NH_2OH) and a disulfide bond and causing conformational change. Ab initio model of the 200 amino acid long polypeptide chain of the *N*-terminus ofhTRPA1 displaying five essential cysteine residues and two indicated disulfides (*dotted lines*). (**d**) Histogram showing the maximal CGRP released from C57Bl/6 and $TRPA1^{-/-}$ mouse sciatic nerve and dura mater upon stimulation with AS. AS effects were reduced by HC030031 and absent in $TRPA1^{-/-}$. (**e**) Photograph of a subject's volar forearm with noticeable axon reflex erythema upon intradermal injection of AS. Adapted from [82]

14.4.3 *Cardiovascular Effects of HNO: The Most Promising Therapeutic Target(s)*

The most promising of the beneficial pharmacological effects of HNO donors is certainly effect on heart. In the seminal work by Paolocci and colleagues, the authors showed previously unknown effect of HNO: an increase of myocardial contractility, known as positive inotropy, in combination with accelerated relaxation, known as positive lusitropy [93]. These effects could be of particular importance when treating congestive heart failure (CHF), a serious and life-threatening disease. Paolocci and colleagues also studied the pressure–volume relationship in a dog model of CHF [92]. They showed that Angeli's salt still has positive inotropic and lusitropic effects like it does in the control healthy dogs. Furthermore, coadministration of beta-agonist mimetics (beta-agonist are the most used therapeutic

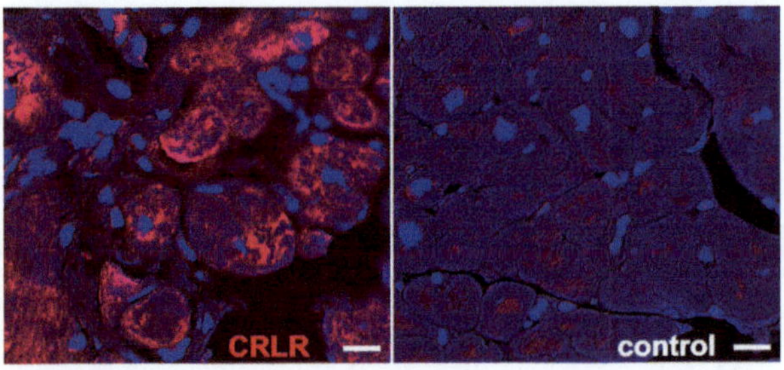

Fig. 14.4 CRLR-staining in human cardiomyocytes. Immunofluorescence image of human cardiomyocytes in cross section stained for CGRP receptor (CRLR, *left*) shows membranous and cytoplasmic reactivity that spares myofibers; (*right*) negative control on sections of the same patient. Scale bars: 25 μm. From [82]

approach to treat CHF) had an additive effect. It is worth mentioning that these effects of HNO are completely opposite to what has been reported for NO. This sparkled the research in the field of new HNO donor development for the treatment of heart failure. One such donor, named CXl-1020, a product of a company Cardioxyl, has been recently showed to have the beneficial effects in acute decompensated heart failure in humans, being the first reported example of the HNO donors tested in humans [125].

The mechanism by which HNO causes these effects is still a matter of intense research. Initial study by [93] showed that the positive inotropy could be completely abolished by the administration of CGRP antagonist, $CGRP_{8-37}$. However in an a follow-up study, the same group showed that CGRP-induced inotropic effects in dogs are indirect and independent of CGRP receptor activation, but rather caused by the release of noradrenaline from sympathetic efferent fibers [126]. Therefore, the whole paradigm of CGRP was disregarded in any of the follow-up studies. But the evidence accumulated to show that CGRP exerts direct inotropic and lusitropic effects on isolated rat cardiomyocytes, which are inhibited by CGRP receptor antagonists [127]. Previous experiments on porcine and human myocardial trabeculae have suggested the involvement of both functional CGRP1 and CGRP2 receptors in the mediation of CGRP-induced positive inotropic responses [128, 129]. Finally, in our recent study [82] we showed, for the first time, a massive expression of CRLR in the plasma membrane (and intracellularly) of the human heart tissue (Fig. 14.4). Therefore, it is still quite plausible that HNO-induced CGRP release is important for the positive inotropic and lusitropic effects, as originally observed in the study by [93]. In addition, an increase of the RAMP1 and RAMP3 has also been reported in rats with chronic heart failure [130], while in the rats with ischemic heart failure, elevations of myocardial adrenomedullin receptor and RAMP-2 mRNA levels were observed [131], offering the alternative CGRP-mediated explanation for the beneficial effects of HNO donors in failing hearts.

Several other non-CGRP-dependent mechanisms have been proposed to explain the HNO-induced positive inotropic and lusitropic effects, all of which involve the thiol modification on different proteins.

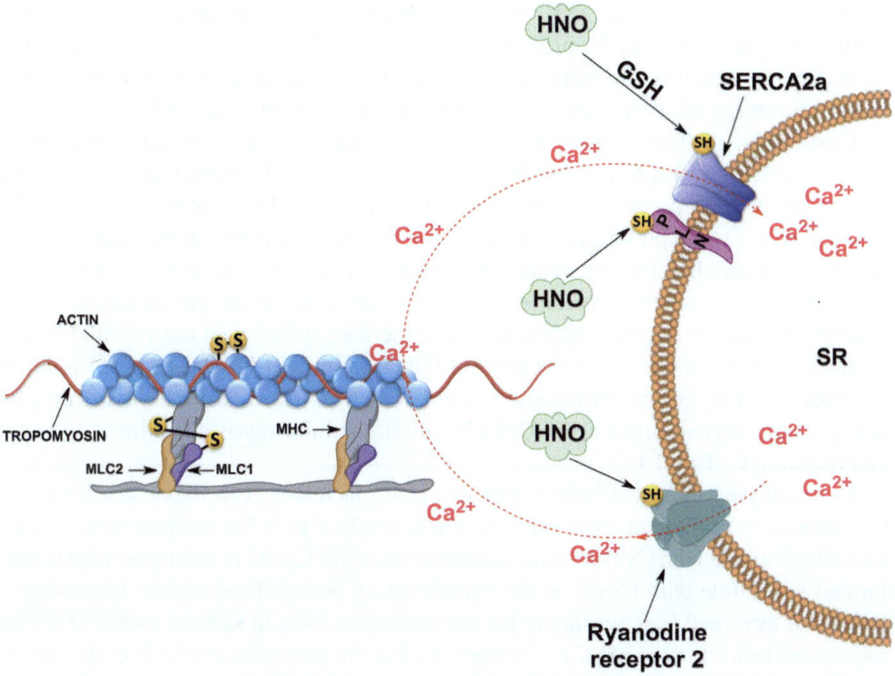

Fig. 14.5 Proposed mechanism(s) for the HNO-induced positive inotropic and lusitropic effects. HNO causes oxidation of cysteine residues on ryanodine receptor 2, opening the channel to pump the calcium (Ca^{2+}) out of SR. This makes more Ca^{2+} ions available for myofilaments. HNO also modifies the myofilament proteins: Cys257 in actin subdomain 4 forms a disulfide with Cys190 in tropomyosin while Cys81 in myosin light chain (MLC) formed a disulfide with Cys37 in the myosin heavy chain's (MHC) head region. This leads to the increase of both maximum force and Ca^{2+} sensitivity. Finally, HNO activates SERCA2a either directly, initially forming *N*-hydroxysulfenamide which further reacts with glutathione (GSH) to cause glutathionylation of C674, or by modifying cysteines of the SERCA's regulatory protein phospholamban (PLN). This leads to the pumping of the calcium (Ca^{2+}) back to the sarcoplasmic reticulum (SR)

The first reports indicated the role of cardiac ryanodine receptors (RyR) (Fig. 14.5) [132, 133]. RyRs are known to be redox sensors, suggesting that they could be easily modified by the reactive oxygen and nitrogen species [134]. Therefore, it came as no surprise that HNO-mediated Ca^{2+} release from the sarcoplasmic reticulum is caused by the RyR activation, which could be reversed by DTT [133]. Later studies broadened the view by introducing the direct interaction with the sarcoplasmic reticulum Ca^{2+} pump (SERCA). The authors suggested that HNO increases the opening probability of RyR and accelerates Ca^{2+} reuptake into sarcoplasmic reticulum by stimulating SERCA [133, 135]. Contraction therefore improves with no net rise in diastolic Ca^{2+}. The mechanism by which HNO activates SERCA has been proposed to be the oxidation of cysteine residues in the SERCA's regulatory protein phospholamban (PLN) (Fig. 14.5). Mutations of critical cysteines in PLN abolished the HNO effects, and the DTT treatment reversed them, confirming essential role of disulfide formation [136, 137]. Recently Paolocci's group observed that HNO enhances SERCA activity by promoting redox-dependent PLN oligomerization. The final effect is therefore the enhanced Ca^{2+} cycling which can explain both the HNO-induced increase of contractile force and myocardial relaxation [138].

Conversely, in their attempt to elucidate the mechanism by which HNO exerts inotropic effects, [139] found that HNO causes oxidation of a particular cysteine residue on SERCA. In particular, they observed *S*-glutathionylation of the channel at the cysteine 674. Overexpression of glutaredoxin-1 prevented HNO-stimulated oxidative modification and therefore activation of SERCA as it did the cysteine 674 replacements with serine. Although the actual mechanism has not been solved, the authors proposed that HNO causes *N*-hydroxysulfenamide intermediate, which then reacts with GSH.

In the most recent, extensive study, [140] have identified 12 sites of HNO-induced disulfides in the key myofilament proteins, such as actin, tropomyosin, myosin heavy chain, myosin light chain (MLC1), alpha-actinin, myosin-binding protein c, and troponin C (Fig. 14.5).

By modifying the myofilament proteins, HNO increases both maximum force and Ca^{2+} sensitivity of force in intact muscles. For example, Cys257 in the actin subdomain 4 formed a disulfide with Cys190 in the tropomyosin, while Cys81 in the myosin light chain formed a disulfide with Cys37 in the myosin heavy chain's head region. Interestingly, HNO also increased Ca^{2+} sensitivity but not maximum force in skeletal muscle. Skeletal muscle cell lack Cys81 in MLC1. This suggests that this particular residue is critical redox switch responsible for HNO-stimulated increase in cardiac force production. The study also compared two different HNO donors, AS with 1-nitrosocyclohexyl acetate (which release HNO slower and upon cleavage of the ester bond). A fewer modified proteins were observed in AS-treated cells suggesting that the rate of HNO release and its location could change the overall outcome of HNO donating drugs [140].

Despite all those in vitro studies on cells, the question remains which mechanism predominates under in vivo conditions. From a kinetic standpoint, it is unclear how HNO would survive all the thiols in the bloodstream and reach the cysteine residues of SERCA or myosin filaments, in a concentration sufficient to induce pronounced effect. On the other hand, even minor concentrations of HNO would be sufficient to modify the TRPA1 present in the membranes of nerve ending resulting in a massive Ca^{2+} influx and subsequent CGRP release.

14.4.4 HNO and Ischemia–Reperfusion Injury

It is worth mentioning that HNO also exhibited certain level of myocardial cardioprotection. When administrated at a low dose, just before ischemia, AS induced powerful cardioprotective actions which protected from ischemia–reperfusion (IR) injury [141]. However, when administered during or after the IR injury, HNO had detrimental effects [142]. These effects could be explained by the HNO-induced redox modulation of intracellular signaling, particularly mitochondrial signaling. [143, 144] for example showed that AS is able to induce expression of heme-oxygenase 1, which would be a desirable effects in patients experiencing IR injury and it is characteristic for the ischemic preconditioning. On the other hand, studies with IPA/NO showed that HNO could inhibit endothelial cell proliferation [145]. Antiangiogenic effects of HNO would certainly not be desirable in IR injury.

Shiva et al. [146] tried to unravel the effects of HNO on isolated mitochondria. The authors demonstrated that mitochondria are capable of reoxidizing HNO to NO, but that HNO exhibits inhibition of mitochondrial respiration through the inhibition of complexes I and II, most probably via modification of specific cysteine residues in the proteins. The proteomic approach identified extensive modifications of mitochondrial protein thiols. The protective effects of HNO in IR injury are reminiscent of those seen with diazoxide, a mitochondrial K_{ATP} channel opener, which the authors later proved to indeed be the case: HNO indeed caused the modifications of critical cysteines in mitochondrial K_{ATP} channels, leading to the channel opening, which is considered to be an important event for ischemic-preconditioning-induced cardioprotection [147].

Further studies on how HNO-induced (redox) modifications could cause changes in intracellular signaling (in cell types other than cardiomyocytes) might help in overall understanding of both beneficial and toxic effects of the HNO.

14.4.5 Anticarcinogenic Effects of HNO

One of the new therapeutic directions for the HNO drugs is their anticarcinogenic effect. Several early studies pointed out that in the cell culture HNO could induce double-strand DNA breaks [148, 149], albeit the high does of AS were used. However, more recent studies pointed out that HNO donors oxidize cysteine residues in GAPDH, which leads to the loss of its activity [80, 81]. Being dependent on glycolysis, cancer cells would be particularly sensitive to the loss of the activity of this critical enzyme [150]. More detailed overview of the anticancer effects of HNO donors will be covered in the next chapter of this book.

14.4.6 Neuronal Effects of HNO

The neuronal effects of HNO are still not well understood and there are only a few studies addressing this topic. One of the early studies showed that HNO, unlike NO, reacts mainly with Cys-399 in the NR2A subunit of the *N*-methyl-D-aspartate (NMDA) receptor to restrict excessive Ca^{2+} influx and thus provide neuroprotection from excitotoxic insults [151]. The fact that HNO can ameliorate excitotoxic neuronal damage raises the possibility that exogenous donor drugs may be used for therapeutic gain. Wide variety of neurologic disorders that are mediated, at least in part, by excessive stimulation of NMDA receptors could benefit from the HNO treatment.

On the other hand, injection of AS into cerebral spinal fluid led to motor neuron injury and it reduced striatal dopamine 7 h after treatment [152]. This may be due to the oxidation of intracellular GSH levels. In support of neurotoxic effects of HNO, another study recently showed that HNO is detrimental if Angeli's salt is infused before cerebral artery occlusion, mainly due to the oxidative damage [153].

Recent characterization of HNO–TRPA1–CGRP pathway opened up a new perspective on the potential role of HNO in both central and peripheral nervous system. Being a nociceptive transduction channel in the first place, TRPA1 contributes to pain/itch sensations [114]. Eberhardt et al. [82] showed the strong and persistent HNO-induced activation of TRPA1 due to the disulfide formation, which led to a strong pain sensation when injected subcutaneously. Furthermore, the role of CGRP release has never been considered outside the cardiovascular system, although CGRP is long known as important mediator of migraine attacks.

Neuropeptides that are released by trigeminal ganglion stimulation produce vasodilation of the meningeal vessels (mainly due to CGRP), plasma extravasation, and mast cell degranulation with secretion of other proinflammatory substances in the dura (Fig. 14.6). In humans suffering from migraine, increased level of CGRP is found in both the external and internal jugular blood during migraine attacks. The most recent development of the drugs to treat migraine goes in two directions (a) development of CGRP receptor antagonists and (b) development of anti-CGRP antibodies, the latter showing some promise and have moved further in clinical trial [154, 155].

All this shed a different light on the potential application of HNO donors and further studies are needed in order to clarify whether the drawback of HNO therapy could be strong pronociceptive stimulation [156].

14.5 Physiologically Relevant Reactions that Might Lead to Intracellular HNO Generation

Until recently there was no evidence for the endogenous HNO formation. This was mainly due to the short half-life that HNO is expected to have in the cells, which provides a methodological limitation for its intracellular detection. Several in vitro observations have suggested a possible endogenous sources of HNO: (a) decomposition of *S*-nitrosothiols (RSNO) [157], (b) L-arginine oxidation by nitric oxide synthase (NOS) under specific conditions [158], (c) oxidation of NH_2OH [159], and (d) reduction of NO by metalloproteins such as MnSOD [160] and cytochrome *c* [161].

Protein *S*-nitrosation is considered to be a very important posttranslational modification of proteins. The number of proteins found to be controlled by this modification is constantly increasing. The role of *S*-nitrosation in controlling the protein function has been covered in several extensive reviews [162–164]. To date, *S*-nitrosation has been implicated in the regulation of proteins involved in muscle contractility, neuronal

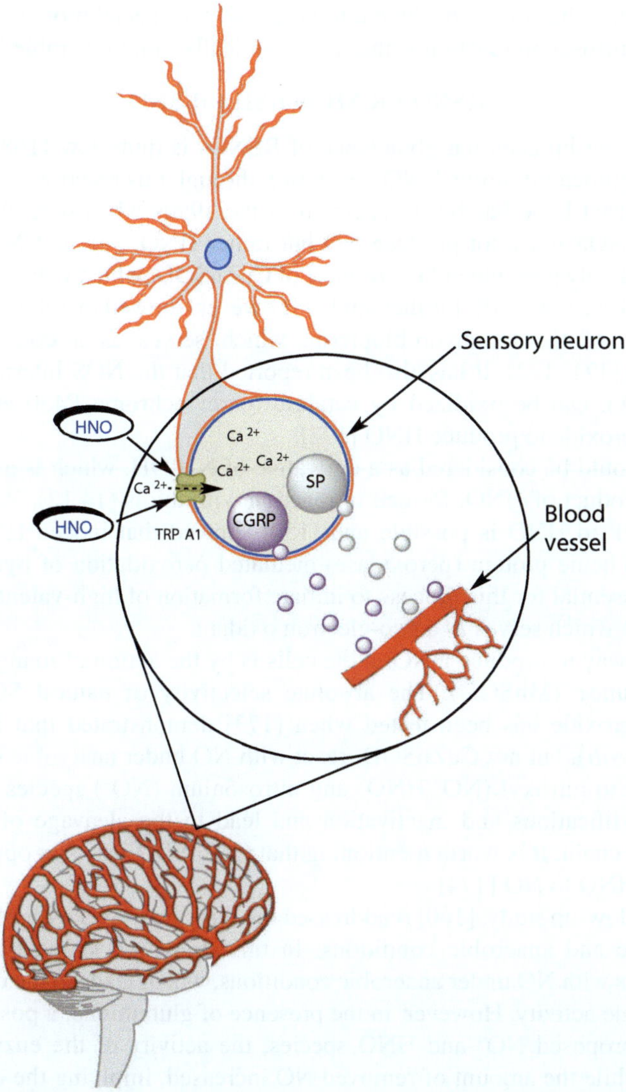

Fig. 14.6 Potential role of HNO in migraine attacks. HNO activates TRPA1 channels in trigeminal ganglion. Neuropeptides that are released (calcitonin gene-related peptide, CGRP, and substance P, SP) lead to vasodilation of the meningeal vessels, plasma extravasation, and mast cell degranulation with secretion of other proinflammatory substances in the dura (neurogenic inflammation)

transmission, host defense, cell trafficking, apoptosis, etc. [162–165]. One of the possible ways for the intracellular HNO generation is the reaction of *S*-nitrosothiols with some other thiol [157, 166]:

$$RSNO + RSH \rightarrow RSSR + HNO \quad (14.12)$$

This reaction is, however, considered to occur only in special protein environments as the transnitrosation reaction is thermodynamically more favorable [167]:

$$RSNO + R^1SH \rightarrow RSH + R^1SNO \tag{14.13}$$

In addition, the intracellular abundance of RSNOs is quite low [168, 169] which therefore significantly limits HNO generation through this reaction.

A significant hype has been generated in the 1990s when several studies suggested that NOS does not produce NO but rather HNO, and that NO production from NOS is only possible in the presence of CuZnSOD, which will reoxidize HNO back to NO [66, 158, 170]. Further studies, however, showed that this is true only in the absence of the tetrahydrobiopterin, which serves as a cofactor of NOS (Fig. 14.7a) [171, 172]. It has also been reported that the NOS intermediate, HO–Arg (NOHA), can be oxidized by catalase (or cytochrome P450 enzymes) and hydrogen peroxide to produce HNO [172].

NOHA could be considered as a derivative of NH_2OH, which is a two-electron reduction product of HNO, formed in reaction with thiols (14.12). Reoxidation of NH_2OH back to HNO is possible and [159] showed that indeed HNO could be formed in a heme protein (peroxidase)-mediated peroxidation of hydroxylamine. H_2O_2 was essential for this process to initiate formation of high-valent iron(IV)oxo intermediate which serves as a two-electron oxidant.

Another way to produce HNO in the cells is by the action of manganese superoxide dismutase (MnSOD). The absolute selectivity of natural SOD enzymes toward superoxide has been tested when [173] demonstrated that MnSOD and FeSOD (*E. coli*), but not CuZnSOD, react with NO under anaerobic conditions to transform it to nitroxyl (NO^-/HNO) and nitrosonium (NO^+) species which cause enzyme modifications and inactivation and lead to the cleavage of the enzyme polypeptide chain. It is worth mentioning that CuZnSOD reacts in opposite way, it reoxidizes HNO to NO [174].

In the follow-up study, [160] readdressed the reaction of NO with MnSOD under both aerobic and anaerobic conditions. In this study, we showed that MnSOD indeed reacts with NO under anaerobic conditions, which leads to the complete loss of the enzyme activity. However, in the presence of glutathione, a possible scavenger of the proposed NO^+ and HNO species, the activity of the enzyme remains preserved while the amount of removed NO increased, implying the catalytic process (Fig. 14.7b). Therefore we proposed that this reaction should be referred to as NO dismutation.

The reaction of MnSOD with NO under aerobic (and thus more physiological) conditions was also studied, and the second-order rate constant for the reaction of MnSOD (*E. coli*) with NO under aerobic conditions was determined to be 650 $M^{-1}\ s^{-1}$. It should be mentioned that this rate constant is not the catalytic rate constant for NO removal by MnSOD, since the experimental conditions used in the study were not catalytic. Further, it is possible that this constant is largely underestimated because of the relatively slow response time of the NO electrode used in the study to follow the reaction. Nonetheless, nitroxyl generation has been observed. We proposed that this reaction can represent a protective mechanism by which cell

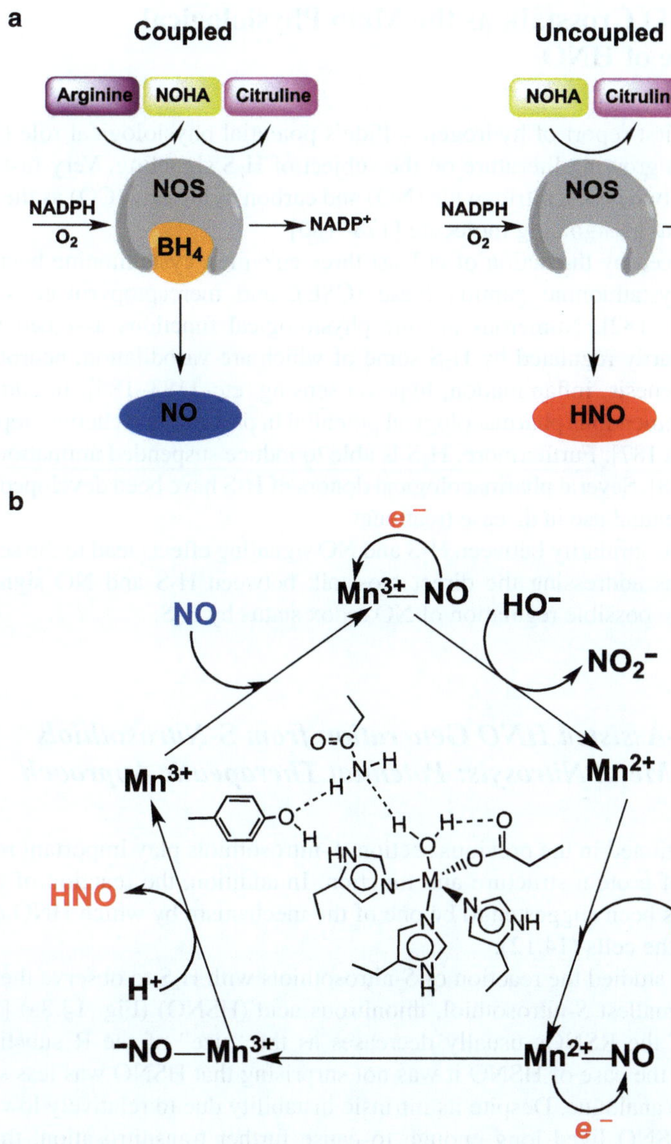

Fig. 14.7 Two potential enzymatic pathways for HNO generation. (**a**) Enzymatic formation of HNO, mediated by NOS, takes place only in the absence of tetrahydrobiopterin (BH_4). (**b**) MnSOD could use NO as a substrate and catalyze its dismutation to NO^+ and HNO

protects from the stress induced by overproduction of NO. Furthermore we also studied the reaction of pentaaza macrocyclic MnSOD mimics with NO in vitro [175] and on a cellular model [176] to further confirm the dismutation mechanism (Fig. 14.7b).

14.6 H_2S/NO Crosstalk as the Main Physiological Source of HNO

Ever since the first report of hydrogen sulfide's potential physiological role [177], there has been a growing literature on the subject of H_2S signaling. Very fast, H_2S joined the other two gases, nitric oxide (NO) and carbon monoxide (CO) as the third gaseous endogenous signaling molecule [178–180].

H_2S is produced by the action of at least three enzymes, cystathionine beta synthase (CBS), cystathionine gamma lyase (CSE), and mercaptopyruvate sulfurtransferase [181, 182]. Numerous are the physiological functions assigned to be exclusively or partly regulated by H_2S some of which are vasodilation, neurotransmission, angiogenesis, inflammation, hypoxia sensing, etc. [183–185]. In addition, H_2S showed a tremendous pharmacological potential in preventing ischemia–reperfusion injury [186, 187]. Furthermore, H_2S is able to induce suspended animation-like state in mice [188]. Several pharmacological donors of H_2S have been developed with hope of their eventual use in disease treatment.

The incredible similarity between H_2S and NO signaling effects lead to the several important studies addressing the direct crosstalk between H_2S and NO signaling pathways and the possible regulation of NO redox status by H_2S.

14.6.1 *H_2S-Assisted HNO Generation from S-Nitrosothiols and Metal Nitrosyls: Potential Therapeutic Approach*

As already mentioned in the previous section, *S*-nitrosothiols play important role in the regulation of protein structure and function. In addition, the reaction of thiols with RSNOs has been suggested to be one of the mechanism by which HNO could be produced in the cells (14.12).

We therefore studied the reaction of *S*-nitrosothiols with H_2S to observe the generation of the smallest *S*-nitrosothiol, thionitrous acid (HSNO) (Fig. 14.8a) [189]. The stability of the RSNOs usually decreases as the "size" of the R substituent decreases, so in the case of HSNO it was not surprising that HSNO was less stable than its cysteine analogue. Despite its intrinsic instability due to relatively low S–N bond energy, HSNO lived long enough to cause further transnitrosation, that is, transfer of NO^+ moiety, from one protein to another (Fig. 14.8a). More importantly, when HSNO reacted further with H_2S, nitroxyl was formed. The formation of HNO was observed by using methemoglobin as a scavenger and by monitoring the hydroxylamine, a final product of the reaction of nitroxyl with thiols.

We were able, using recently developed fluorescence sensor for nitroxyl (CuBOT1) [190], to detect HNO generation in the cells. An obvious increase of the CuBOT1 fluorescence was observed when the cells were treated with the combination of *S*-nitrosoglutathione and H_2S, but even stronger intracellular HNO generation was

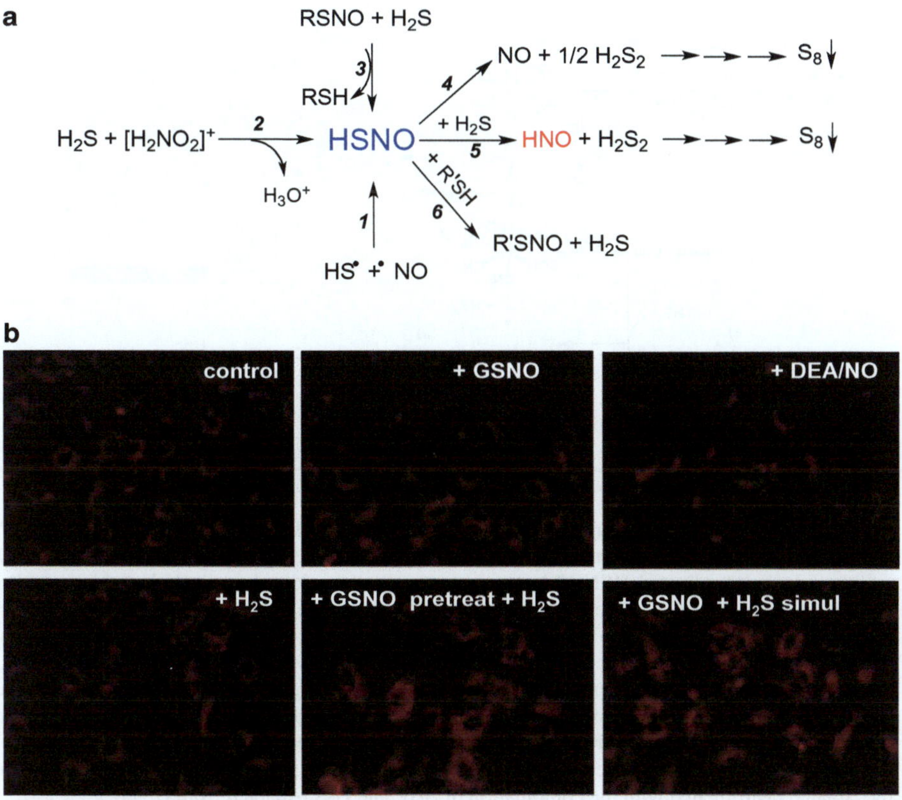

Fig. 14.8 HSNO-mediated HNO generation. (**a**) Reaction paths for HSNO generation (1–3) and subsequent biologically relevant reactions (4–6). HNO is generated in the reaction 5. (**b**) H_2S and GSNO react in cells to yield HNO. Human umbilical vein endothelial cells, loaded with 10 μM CuBOT1, were treated with either 100 μM Na_2S, 100 μM GSNO, or both for 20 min. Some cells were pretreated with 100 μM GSNO for 20 min to increase intracellular nitrosothiol content and then exposed to 100 μM Na_2S. A 100 μM DEA/NONOate solution served as a negative control. Adapted from [189]

observed in cells pretreated with *S*-nitrosoglutathione (in order to increase the intracellular content of *S*-nitrosothiols) and then washed and subsequently treated with H_2S (Fig. 14.8b). This was the first direct evidence for the intracellular generation of HNO, albeit not endogenously but rather pharmacologically. The fact that HSNO could freely diffuse from one cell to another (or between cellular compartments within one cell) suggests that HSNO could be a "carrier" of HNO. Aside its role in transnitrosation, if it would meet another H_2S molecule, HSNO would form HNO.

The similarity between metal-nitrosyls and *S*-nitrosothiols was used by in our follow-up study, which addressed the reaction of sodium nitroprusside with H_2S [191]. Sodium nitroprusside (SNP) is a vasodilator still used in acute hypertensive crises. In SNP, $Na_2[Fe(CN)_5(NO)]$, NO coordinated to iron has an NO^+ character

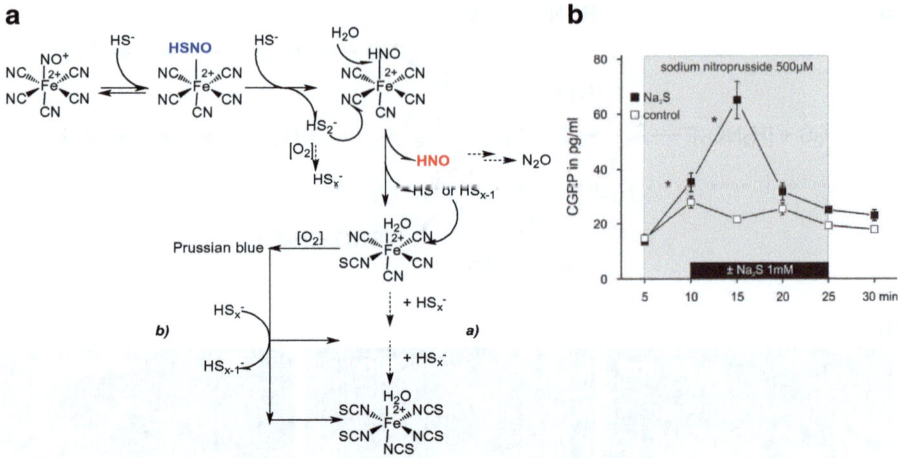

Fig. 14.9 Combination of SNP and H_2S is a good HNO donor. (**a**) Mechanistic depiction of the reaction steps behind sodium nitroprusside and H_2S reaction at pH 7.4 and aerobic conditions. SNP reacts very fast with H_2S to form intermediate $[(CN)_5FeN(O)SH]^{3-}$. In the second reaction step, sulfide reduces coordinated HSNO/SNO^- and gets initially oxidized to disulfide, which can be further oxidized to polysulfides along the reaction progress. As a result of this step, the $[(CN)_5Fe(HNO)]^{3-}$ intermediate is formed which is unreactive towards H_2S but does react with polysulfides forming thiocyanate adduct with elimination of HNO. HNO can dimerize to give N_2O, while the other coordinated cyanides are consecutively transformed to thiocyanate (main reaction path, *a*). Oxygen can induce oxidation of the iron(II) center that would lead to mixed-valent bridged complexes, known as Prussian blue, which are short-lived and react further with polysulfides to give the final thiocyanate product as well (minor reaction route, *b*). (**b**) CGRP release from the isolated heart treated with the combination of SNP and H_2S. Adapted from [191]

(similar like in RSNO) and undergoes chemistry that is best described as direct nitrosation biochemistry. SNP's effects are fast and efficient but the problem of cyanide toxicity exists, and this limits its application only to situations monitored by medical doctors.

Despite a long interest in the reactions of H_2S with SNP, the reaction mechanism has eluded researchers. The study by [191] is the first study at physiological pH and aerobic conditions, which combining the chemical tools with pharmacological/physiological experiments provided the overall reaction mechanism, depicted in Fig. 14.9.

SNP reacts very fast with H_2S to form the intermediate $[(CN)_5FeN(O)SH]^{3-}$. In the second reaction step H_2S actually has a catalytic role. By reducing coordinated HSNO/SNO^-, it gets initially oxidized to disulfide, which can be further oxidized to polysulfides along the reaction progress. As a result of this step the $[(CN)_5Fe(HNO)]^{3-}$ intermediate is formed, but like any cyanide species it could react with sulfane sulfur containing compounds (e.g., polysulfides), resulting in the conversion of CN^- into SCN^-. This process most probably labilizes HNO that can be released. The generation of HNO was so obvious in the cells treated with SNP and H_2S but more importantly, the isolated hearts treated with the combination of SNP and H_2S lead to the release of CGRP. CGRP was released from the primary afferent nerve fibers located in the epicardial surface of the heart suggesting the potential therapeutic effect of this combination. Combination of SNP with H_2S, which causes HNO-induced release of CGRP and transformation of toxic cyanide into thiocyanate, could be a new, more effective and less toxic therapeutic alternative to the SNP alone. Higher efficiency could be achieved even at lower SNP concentrations, but in addition, higher SNP doses could be used if needed due to the lower cytotoxicity.

14.6.2 H_2S-Assisted Iron Porphyrin-Catalyzed Nitrite Reduction as a Source of HNO

The past decade has witnessed the recognition of nitrite as an important pool of nitric oxide that acts as a vasodilator and intrinsic signaling molecule. Nitrite can be reduced in vivo, either nonenzymatically or enzymatically in reactions catalyzed by xanthine oxidase, deoxyhemoglobin, deoxymyoglobin, cytochrome *c*, or by thiols and metal center-assisted processes inside the cell [19, 192, 193]. However, H_2S, which could reach active sites of most of the metalloproteins, had not been considered as possible regulator of this reaction, until recently. In clinical praxis, nitrite is used as an antidote for hydrogen sulfide (H_2S) poisoning, being even superior to that of oxygen.

Although we observed that nitrite and sulfide do not react directly, in the cells they seem to generate NO [194]. More importantly, we observed very strong intracellular HNO generation, which was localized within mitochondria, and absent in Rho^0 cells, that is, cells depleted of mitochondria. This suggested essential role of mitochondria in catalyzing the reaction between NO and H_2S. Using iron–porphyrin model we proved at the structural level that the reaction proceeds with an intermediate formation of HSNO, which then probably ends up as HNO (Fig. 14.10).

On one side these data offer an additional explanation for the antidote effect of nitrite in H_2S poisoning: generation of all three, NO, HNO, and HSNO, would lead to an increase of blood flow, and finally, better oxygen delivery to the tissue. Furthermore, these results demonstrate that H_2S can transform biological metal-nitrosyl pools into *S*-nitrosothiols via HSNO and offer another mechanism for physiological generation of HNO. Finally, this study opens a new perspective for potential therapeutic application of iron–porphyrin drugs, which could, combined with H_2S donor and nitrite, serve as a good intracellular HNO generator.

14.6.3 H_2S Reacts Directly with NO to Form HNO: The Main Intracellular Source of HNO

Several groups have considered that H_2S effects could be linked to NO, including the first study that showed physiological effects of H_2S [177]. Additionally, Yong et al. [195] proposed that H_2S and NO together could give HNO and recently demonstrated positive inotropic effects of these combined gasotransmitters on the heart [196], which is one of the hallmarks of HNO physiology [92, 93].

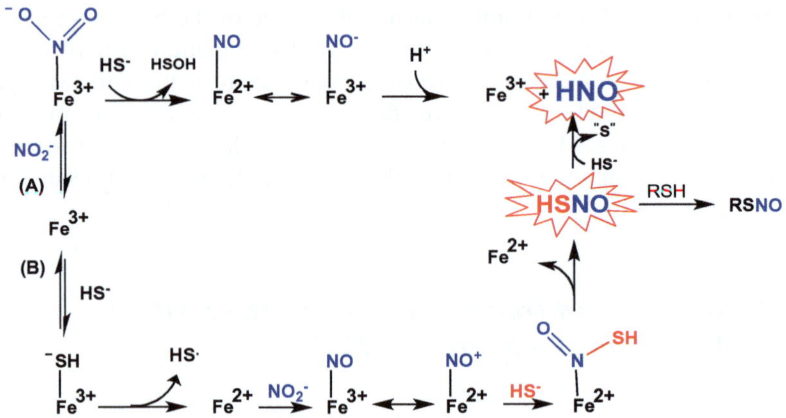

Fig. 14.10 Mechanism of iron-heme catalyzed, sulfide-assisted HNO generation from nitrite in mitochondria. Two paths could be envisioned and both could explain generation of HNO observed in cells, but under physiological conditions the excess of H_2S over nitrite would be more common situation meaning that intermediate formation of HSNO is probable (path B)

Reaction of NO with thiols is considered to be thermodynamically unfavorable and very slow to proceed at any detectable rate under physiological conditions [197]. The same could be extended to the reaction of NO with H_2S, if outer sphere electron transfer is to be the mechanism and the –0.81 V is a redox potential for the reduction of NO. However, in the study by Eberhardt and colleagues, we show the first chemical evidence for the HNO formation from a direct reaction of NO and H_2S [82]. Using recently developed HNO electrode [198], we demonstrated that 2 μM combination of NO and H_2S yields a peak HNO concentration of ca. 0.5 μM, similar to the effects of 1 mM AS concentration (Fig. 14.11a). The reaction proceeds even in the absence of oxygen and metal ions, therefore suggesting that it could be a major source of HNO in vivo. In fact, the intracellular levels of HNO, detected with fluorescence sensor, were diminished when cells were treated with NOS and CSE inhibitors. Furthermore, the combination, just like Angeli's salt, activated HNO–TRPA1–CGRP pathway [82] (Fig. 14.11b).

Unpublished data from our laboratory further support this mechanism. We managed to identify the reaction intermediate and study its decomposition to give HNO. The following reaction steps could lead to HNO formation:

$$HS^- + NO \rightarrow HSNO^{\bullet -} \tag{14.14}$$

$$2HSNO^{\bullet -} + 2H^+ \rightarrow 2\left[H_2SNO\right] \rightarrow 2HNO + HS^{\bullet} / \text{"S"} \tag{14.15}$$

$$HSNO^{\bullet -} + HS^- \rightarrow HNO + HS_2 \bullet^{2-} / \text{"S"} \tag{14.16}$$

$$HSNO^{\bullet -} + NO + H^+ \rightarrow HSNO + HNO \tag{14.17}$$

$$HSNO + H_2S \rightarrow HNO + H_2S_2 / \text{"S"} \tag{14.18}$$

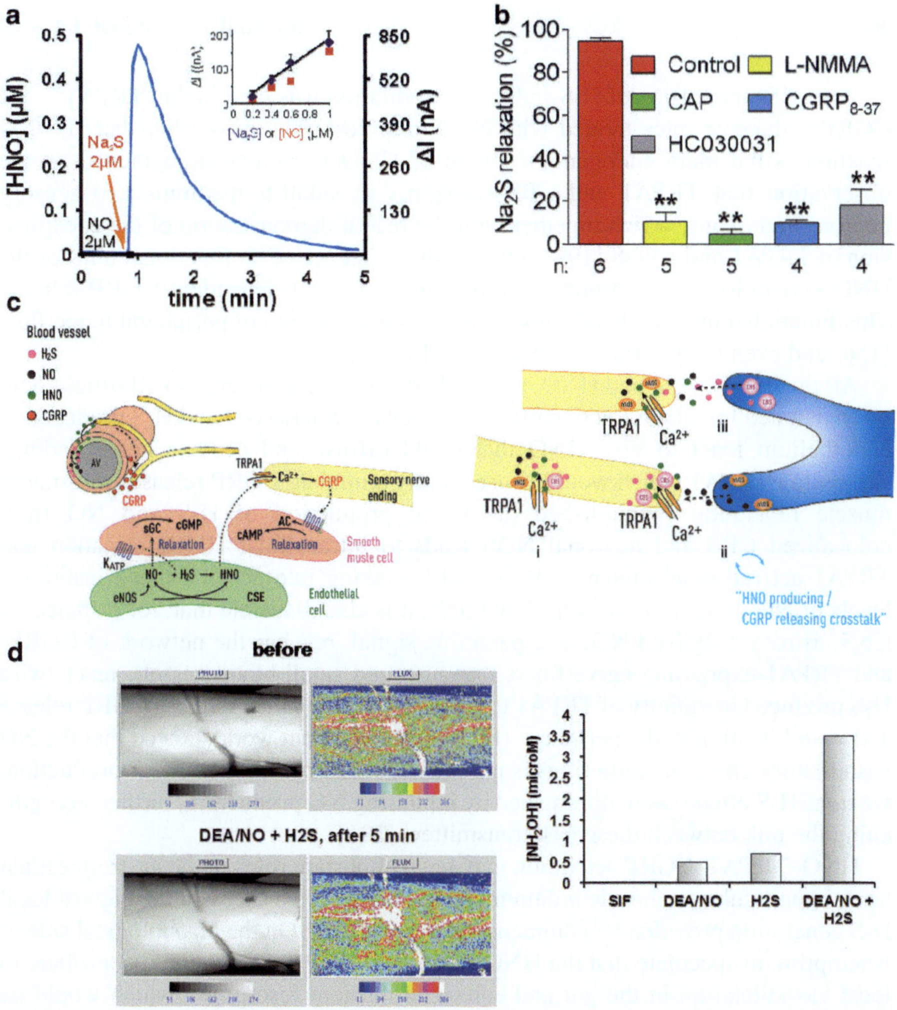

Fig. 14.11 H_2S and NO react directly to endogenously produce HNO and stimulate HNO–TRPA1–CGRP pathway. (**a**) Amperometric signal of the HNO-selective electrode after addition of H_2S (2 μM) to a solution of 2 μM NO (*left axis*: [HNO] after calibration, *right axis*: measured current). *Inset*: Signal peak vs. H_2S (*blue*) and NO (*red*) concentration, while the other reactant concentration is maintained constant and in excess. (**b**) Isometric tension recordings of phenylephrine- preconstricted ring segments of rat second-order mesenteric artery branches: Na_2S (10 μM)-induced vasodilation was reversed by $CGRP_{8-37}$ and abolished by $CGRP_{8-37}$. NOS inhibitor, L-NMMA, and TRPA1 antagonist, HC030031, pretreatment inhibited vasodilation induced by Na_2S. CGRP depletion by capsaicin (CAP) also abolished the Na_2S-induced blood vessel relaxation. (**c**) TRPA1/CGRP expressing nerve endings in the periphery communicate with the smooth muscle cells surrounding the endothelium of blood vessels. Endothelial cells are known to produce NO and H_2S, both of which freely diffuse and activate guanylyl cyclase and K_{ATP} channels, respectively, to induce vasodilatation. However, H_2S and NO also react with each other to give HNO, which could reach paravascular TRPA1-expressing sensory nerve fibers, inducing Ca^{2+} influx and CGRP release. Other potential sites of NO–H_2S interaction is in neurons (i) TRPA1 channels are coexpressed with nNOS and CBS in primary afferents forming a functional signaling complex that leads to confined HNO generation and TRPA1 gating upon activation of the gasotransmitter-generating enzymes. Additionally, NO (ii) or H_2S (iii) could originate from either side of a synaptic cleft (or from nearby axons of passage) and freely diffuse into adjacent neurons (or nerve fibers). There, they react with their counterpart producing HNO in vicinity of TRPA1. Adapted from [82]. (**d**) Experiment with intradermal injection of NO and H_2S through microdialysis tubings in parallel and 2 mm apart from each other: Laser Doppler images (*left*) and hydroxylamine levels (*right*) in effluate (unpublished data, a curtsey of Dr Barbara Namer)

A significant portion of H_2S-induced vasodilation was inhibited in TRPA1$^{-/-}$ and CGRP$^{-/-}$ mice or mice treated with NOS inhibitors. The possibility that H_2S/NO reaction is the main endogenous source of HNO is further strengthened by the observation that TRPA1 and CBS coexpress in small-to-medium-sized sensory neurons and axons, which together with the recent demonstration of the coexpression of TRPA1 and nNOS [199] suggests the existence of an functional unit for the HNO generation and subsequent activation of TRPA1-dependent CGRP release. This functional unit can be of importance in the regulation of peripheral blood flow [156] and even of systemic blood pressure [82].

All three, NO, H_2S, and HNO, are freely diffusible, so several possibilities could be envisioned for TRPA1 activation and CGRP release (a) NO and H_2S produced in endothelium react to give HNO that could diffuse and activate nerve endings expressing TRPA1 which would consequently stimulate CGRP release and smooth muscle relaxation (unpublished data); (b) production of H_2S and NO from colocalized CBS and neuronal NOS leads to intracellular HNO formation and TRPA1 activation as shown by [82]; and (c) taking into account that constitutive levels of NO in neurons are very low [200], it is also plausible that, for instance, in CNS, astrocyte-derived NO, as a paracrine signal, reaches the network of CGRP- and TRPA1-expressing nerve fibers that surround small blood vessels, reacts with H_2S produced in vicinity of TRPA1 to give HNO, which then induces CGRP release and vasodilatation in the periphery (Fig. 14.11c). Recent work showed that the NO vasodilatory effect on aorta rings is partially blocked by inhibiting H_2S production, whereas H_2S effects were diminished by inhibiting NO production, further strengthening the link between these gasotransmitters [201].

HNO–TRPA1–CGRP activation may have important roles in the gastrointestinal tract, in particular in chronic inflammation of the bowel, a site with the highest local H_2S generation provided by commensal bacteria [202]. On the physiological side, it is tempting to speculate that the HNO–TRPA1–CGRP pathway could contribute to local vasodilatation in the gut and enhanced nutrient resorption, which would be achieved by needs-based channel activation. However, in inflammation, characterized by high NO production, the pathway appears Janus-faced; it could contribute to the completion and resolution of the process by the protective effects of CGRP, or it can perpetuate the disease by TRPA1-controlled corelease of substance P that turns the immune system autoaggressive [203, 204].

Similar to Angeli's salt, combination of H_2S and NO, when injected subcutaneously, induced pain and itch, with an obvious axon-flare erythema, suggestive of CGRP signaling. When coinfused (but in close proximity) via separate microdialysis tubes, H_2S and NO also induced local vasodilation, but more importantly NH_2OH was detected in the effluate as a final product (Fig. 14.11d).

These recent data have several implications on the understanding of HNO signaling (a) they offer an endogenous source for HNO formation and complete the cycle spanning from HNO production to CGRP release; (b) they offer mechanistic explanation of H_2S-induced vasodilation and positive inotropic effect; and (c) they suggest a potential use of H_2S/NO combination as a pharmacological tool for HNO generation and treatment of cardiac failure. Interestingly, the simultaneous pharmacological NO/H_2S donors have been reported and they show anticancer affect, just like other HNO donors [205, 206].

14.7 Future Directions

Nitroxyl is a peculiar molecule with the remarkable (bio)chemical characteristics. In addition, its therapeutic potential in the treatment of the failing heart alone is worth of further investigation. Many questions of HNO biochemistry remain to be addressed. Besides the mechanisms described in this chapter, further studies on identifying the ways for its intracellular generation could be of great help in understanding how this molecule works and which (patho)physiological states are regulated by it. To do so, new and more selective tools for intracellular HNO detection are needed.

HNO also shows a peculiar reactivity. Its ability to specifically react with cysteine residues on a limited number of proteins, despite very high thiol content in the cells and around, still baffles the researchers. More detailed proteomic approach would help in identifying the potential cellular targets of HNO and maybe help in unraveling the nature of this selectivity.

Development of new slow releasing HNO drugs is certainly one of the major future directions. However, besides the development of the organic molecules, which could decompose and give HNO, new ways to modulate redox status of intracellular NO and convert it to HNO, might prove equally efficient. Metal-center based drugs could help in that approach.

Finally, finding the HNO targets could help in identifying some other disease states that could be improved by HNO treatment such as cancer. While keeping in mind the therapeutic potential, we should also not forget the potential shortcomings of HNO-based therapy. HNO–TRPA1–CGRP cascade could prove important for the development of migraine attacks. Nonetheless, this field of research will remain one of the hot topics for many years to come.

Acknowledgment The authors would like to thank professors Ivana Ivanovic-Burmazovic (FAU Erlangen-Nuremberg) and Vesna Niketic (University of Belgrade) for careful reading of the manuscript and helpful discussions. We are also grateful to the Friedrich-Alexander University Erlangen-Nuremberg (Emerging Field Initiative, MRIC) and to the French State in the frame of the "Investments for the future" Programme IdEx Bordeaux, reference ANR-10-IDEX-03-02.

References

1. Palmer RM, Ferrige AG, Moncada S. Nitric oxide release accounts for the biological activity of endothelium-derived relaxing factor. Nature. 1987;327(6122):524–6.
2. Palmer RMJ, Ashton DS, Moncada S. Vascular endothelial cells synthesize nitric oxide from L-arginine. Nature. 1988;333:664–6.
3. Moncada S, Palmer RM, Higgs EA. Nitric oxide: physiology, pathophysiology, and pharmacology. Pharmacol Rev. 1991;43(2):109–42.
4. Friebe A, Koesling D. Regulation of nitric oxide-sensitive guanylyl cyclase. Circ Res. 2003;93:96–105.
5. Angeli A, Angelico F. Reactions of nitroxyl [NOH]. Atti della Accademia Nazionale dei Lincei, Classe di Scienze Fisiche, Matematiche e Naturali, Rendiconti. 1901;10(v):164–8.
6. Angeli A, Angelico F. Nitrohydroxylaminic acid. Gazz Chim Ital. 1903;33:245.
7. Ulich BL, Hollis JM, Snyder LE. Radio detection of nitroxyl (HNO): the first interstellar NO bond. Astrophys J Lett. 1977;217:L105–8.

8. Turner BE. A molecular line survey of sagittarius B2 and orion-KL from 70 to 115 GHz. II—analysis of the data. Astrophys J Suppl Ser. 1991;76:617–86.
9. Harteck P. Die Darstellung von HNO bzw. [HNO]n (The preparation of HNO or [HNO]n). Berichte der Deutschen Chemischen Gesellschaft. 1933;66B:423–6.
10. Switzer CH, Miller TW, Farmer PJ, Fukuto JM. Synthesis and characterization of lithium oxonitrate (LiNO). J Inorg Biochem. 2013;118:128–33.
11. Doctorovich F, Bikiel D, Pellegrino J, Suárez SA, Larsen A, Martí MA. Nitroxyl (azanone) trapping by metalloporphyrins. Coordin Chem Rev. 2011;255(23–24):2764–84.
12. DuMond JF, King BS. The chemistry of nitroxyl-releasing compounds. Antioxid Redox Signal. 2011;9:1637–48.
13. Miranda KM, Nagasawa HT, Toscano JP. Donors of HNO. Curr Top Med Chem. 2005;5:647–64.
14. Bonner FT, Ravid B. Thermal-decomposition of oxyhyponitrite (sodium trioxodinitrate(II)) in aqueous-solution. Inorg Chem. 1975;14:558–63.
15. Dutton AS, Fukuto JM, Houk KN. Mechanisms of HNO and NO production from Angeli's salt: density functional and CBS-QB3 theory predictions. J Am Chem Soc. 2004;126:3795–800.
16. Hughes MN, Wimbledon PE. Chemistry of trioxodinitrates. 1. Decomposition of sodium trioxodinitrate (Angelis Salt) in aqueous-solution. J Chem Soc Dalton Trans. 1976;8:703–7.
17. Miranda KM, Dutton AS, Ridnour LA, Foreman CA, Ford E, Paolocci N, et al. Mechanism of aerobic decomposition of Angeli's salt (sodium trioxodinitrate) at physiological pH. J Am Chem Soc. 2005;127:722–31.
18. Amatore C, Arbault S, Ducrocq C, Hu S, Tapsoba I. Angeli's salt ($Na_2N_2O_3$) is a precursor of HNO and NO: a voltammetric study of the reactive intermediates released by Angeli's salt decomposition. ChemMedChem. 2007;2(6):898–903.
19. Gladwin MT, Schechter AN, Kim-Shapiro DB, Patel RP, Hogg N, Shiva S, et al. The emerging biology of the nitrite anion. Nat Chem Biol. 2005;1(6):308–14.
20. Piloty O. Ueber eine Oxydation des Hydroxylamins durch Benzolsulfochlorid. Ber Dtch Ges. 1896;29:1559–67.
21. Angeli A, Angelico F. New reactions of nitroxyl (dihydroxyammonia). Gazzetta Chimica Italiana. 1905;35(i):152–9.
22. Bonner FT, Ko YH. Kinetic, isotopic, and N15 NMR-study of *N*-hydroxybenzenesulfonamide decomposition—an HNO source reaction. Inorg Chem. 1992;31:2514–9.
23. Seel FBC. Mechanism of the decomposition of sodium benzenesulfohydroxamate in aqueous solution. Z Anorg Allg Chem. 1972;394:187–96.
24. Toscano JP, Brookfield FA, Cohen AD, Courtney SM, Frost LM, Kalish VJ. Preparation of *N*-hydroxylsulfonamide derivatives as nitroxyl (HNO) donors. Baltimore: Johns Hopkins University; 2007. p. 83.
25. Toscano JP, Brookfield FA, Cohen AD, Courtney SM, Frost LM, Kalish VJ. Preparation of aryl *N*-hydroxysulfonamides as physiologically useful nitroxyl donors. Baltimore: Johns Hopkins University; 2009. p. 24.
26. DeMaster EG, Kaplan E, Shirota FN, Nagasawa HT. Metabolic activation of cyanamide by liver mitochondria, a requirement for the inhibition of aldehyde dehydrogenase enzymes. Biochem Biophys Res Commun. 1982;107:1333–9.
27. Demaster EG, Nagasawa HT, Shirota FN. Metabolic activation of cyanamide to an inhibitor of aldehyde dehydrogenase in vitro. Pharmacol Biochem Be. 1983;18 Suppl 1:273–7.
28. DeMaster EG, Redfern B, Nagasawa HT. Mechanisms of inhibition of aldehyde dehydrogenase by nitroxyl, the active metabolite of the alcohol deterrent agent cyanamide. Biochem Pharmacol. 1998;55:2007–15.
29. DeMaster EG, Shirota FN, Nagasawa HT. Catalase mediated conversion of cyanamide to an inhibitor of aldehyde dehydrogenase. Alcohol. 1985;2:117–21.
30. DeMaster EG, Shirota FN, Nagasawa HT. The metabolic activation of cyanamide to an inhibitor of aldehyde dehydrogenase is catalyzed by catalase. Biochem Biophys Res Commun. 1984;122:358–65.

31. Nagasawa HT, Demaster EG, Redfern B, Shirota FN, Goon JW. Evidence for nitroxyl in the catalase-mediated bioactivation of the alcohol deterrent agent cyanamide. J Med Chem. 1990;33:3120–2.
32. Schep L, Temple W, Beasley M. The adverse effects of hydrogen cyanamide on human health: an evaluation of inquiries to the New Zealand National Poisons Centre. Clin Toxicol. 2009;472(1):58–60.
33. Keefer LK. Fifty years of diazeniumdiolate research. From laboratory curiosity to broad-spectrum biomedical advances. ACS Chem Biol. 2011;6(11):1147–55.
34. Keefer LK. Nitric oxide (NO)- and nitroxyl (HNO)-generating diazeniumdiolates (NONOates): emerging commercial opportunities. Curr Top Med Chem. 2005;5(7):625–36.
35. Keefer LK. Progress toward clinical application of the nitric oxide-releasing diazeniumdiolates. Annu Rev Pharmacol Toxicol. 2003;43:585–607.
36. Miranda KM, Katori T, Torres de Holding CL, Thomas L, Ridnour LA, McLendon WJ, et al. Comparison of the NO and HNO donating properties of diazeniumdiolates: primary amine adducts release HNO in Vivo. J Med Chem. 2005;48(26):8220–8.
37. Andrei D, Salmon DJ, Donzelli S, Wahab A, Klose JR, Citro ML, et al. Dual mechanisms of HNO generation by a nitroxyl prodrug of the diazeniumdiolate (NONOate) class. J Am Chem Soc. 2010;132(46):16526–32.
38. Holland RJ, Paulisch R, Cao Z, Keefer LK, Saavedra JE, Donzelli S. Enzymatic generation of the NO/HNO-releasing IPA/NO anion at controlled rates in physiological media using β-galactosidase. Nitric Oxide. 2013;35:131–6.
39. Salmon DJ, TorresdeHolding CL, Thomas L, Peterson KV, Goodman GP, Saavedra JE, et al. HNO and NO release from a primary amine-based diazeniumdiolate as a function of pH. Inorg Chem. 2011;50(8):3262–70.
40. Adachi Y, Nakagawa H, Matsuo K, Suzuki T, Miyata N. Photoactivatable HNO-releasing compounds using the retro-Diels-Alder reaction. Chem Commun (Camb). 2008;41:5149–51.
41. Sha X, Isbell TS, Patel RP, Day CS, King SB. Hydrolysis of acyloxy nitroso compounds yields nitroxyl (HNO). J Am Chem Soc. 2006;128:9687–92.
42. Graetzel M, Taniguchi S, Henglein A. Pulse radiolytic study of short-lived by-products of nitric oxide-reduction in aqueous solution. Berich Bunsen Ge. 1970;74:1003–10.
43. Bartberger MD, Fukuto JM, Houk KN. On the acidity and reactivity of HNO in aqueous solution and biological systems. Proc Natl Acad Sci USA. 2001;98:2194–8.
44. Bartberger MD, Liu W, Ford E, Miranda KM, Switzer C, Fukuto JM, et al. The reduction potential of nitric oxide (NO) and its importance to NO biochemistry. Proc Natl Acad Sci USA. 2002;99:10958–63.
45. Shafirovich V, Lymar SV. Nitroxyl and its anion in aqueous solutions: spin states, protic equilibria, and reactivities toward oxygen and nitric oxide. Proc Natl Acad Sci USA. 2002;99:7340–5.
46. Flores-Santana W, Salmon DJ, Donzelli S, Switzer CH, Basudhar D, Ridnour L, et al. The specificity of nitroxyl chemistry is unique among nitrogen oxides in biological systems. Antioxid Redox Signal. 2011;14(9):1659–74.
47. Fukuto JM, Carrington SJ. HNO signalling mechanisms. Antioxid Redox Signal. 2011;14:1649–57.
48. Miranda KM. The chemistry of nitroxyl (HNO) and implications in biology. Coord Chem Rev. 2005;249:433–55.
49. Jackson MI, Han TH, Dutton A, Ford E, Miranda KM, Houk KN, et al. Kinetic feasibility of nitroxyl (HNO) reduction by physiological reductants and biological implications. Free Radic Biol Med. 2009;47:1130–9.
50. Liochev SI, Fridovich I. The mode of decomposition of Angeli's salt (Na2N2O3) and the effects thereon of oxygen, nitrite, superoxide dismutase, and glutathione. Free Radical Biol Med. 2003;34:1399–404.
51. Donald CE, Hughes MN, Thompson JM, Bonner FT. Photolysis of the nitrogen-nitrogen double bond in trioxodinitrate: reaction between triplet oxonitrate(1-) and molecular oxygen to form peroxonitrite. Inorg Chem. 1986;25:2676–7.

52. Jorolan JH, Buttitta LA, Cheah C, Miranda KM. Comparison of the chemical reactivity of synthetic peroxynitrite with that of the autoxidation products of nitroxyl or its anion. Nitric Oxide. 2015;44:39–46.
53. Espey MG, Miranda KM, Thomas DD, Wink DA. Ingress and reactive chemistry of nitroxyl-derived species within human cells. Free Radical Biol Med. 2002;33:827–34.
54. Miranda KM, Espey MG, Yamada K, Krishna M, Ludwick N, Kim S, et al. Unique oxidative mechanisms for the reactive nitrogen oxide species, nitroxyl anion. J Biol Chem. 2001;276:1720–7.
55. Miranda KM, Yamada K, Espey MG, Thomas DD, DeGraff W, Mitchell JB, et al. Further evidence for distinct reactive intermediates from nitroxyl and peroxynitrite: effects of buffer composition on the chemistry of Angeli's salt and synthetic peroxynitrite. Arch Biochem Biophys. 2002;401(2):134–44.
56. Seddon WA, Young MJ. Pulse radiolysis of nitric oxide in aqueous solution. Can J Chem. 1970;48:393–4.
57. Seddon WA, Fletcher JW, Sopchyshyn FC. Pulse radiolysis of nitric oxide in aqueous solution. Can J Chem. 1973;51:1123–30.
58. Miranda KM, Nims RW, Thomas DD, Espey MG, Citrin D, Bartberger MD, et al. Comparison of the reactivity of nitric oxide and nitroxyl with heme proteins. A chemical discussion of the differential biological effects of these redox related products of NOS. J Inorg Biochem. 2003;93:52–60.
59. Bazylinski DA, Hollocher TC. Metmyoglobin and methemoglobin as efficient traps for nitrosyl hydride (nitroxyl) in neutral aqueous solution. J Am Chem Soc. 1985;107:7982–6.
60. Doyle MP, Mahapatro SN, Broene RD, Guy JK. Oxidation and reduction of hemoproteins by trioxodinitrate(II): the role of nitrosyl hydride and nitrite. J Am Chem Soc. 1988; 110:593–9.
61. Wink DA, Feelisch M. Formation and detection of nitroxyl and nitrous oxide. In: Feelisch M, Stamler JS, editors. Methods in nitric oxide research. London: Wiley; 1996.
62. Lin R, Farmer PJ. The HNO adduct of myoglobin: synthesis and characterization. J Am Chem Soc. 2000;122:2393–4.
63. Sulc F, Fleischer E, Farmer P, Ma D, La Mar G. 1H NMR structure of the heme pocket of HNO-myoglobin. J Biol Inorg Chem. 2003;8:348–52.
64. Ellis A, Li CG, Rand MJ. Differential actions of L-cysteine on responses to nitric oxide, nitroxyl anions and EDRF in the rat aorta. Br J Pharmacol. 2000;129:315–22.
65. Favaloro JL, Kemp-Harper BK. The nitroxyl anion (HNO) is a potent dilator of rat coronary vasculature. Cardiovasc Res. 2007;73:587–96.
66. Fukuto JM, Chiang K, Hszieh R, Wong P, Chaudhuri G. The pharmacological activity of nitroxyl: a potent vasodilator with activity similar to nitric oxide and/or endothelium-derived relaxing factor. J Pharmacol Exp Ther. 1992;263:546–51.
67. Irvine JC, Favaloro JL, Kemp-Harper BK. NO- activates soluble guanylate cyclase and Kv channels to vasodilate resistance arteries. Hypertension. 2003;41:1301–7.
68. Li CG, Karagiannis J, Rand MJ. Comparison of the redox forms of nitrogen monoxide with the nitrergic transmitter in the rat anococcygeus muscle. Br J Pharmacol. 1999;127:826–34.
69. Dierks EA, Burstyn JN. Nitric oxide (NO), the only nitrogen monoxide redox form capable of activating soluble guanylyl cyclase. Biochem Pharmacol. 1996;51:1593–600.
70. Zamora R, Grzesiok A, Weber H, Feelisch M. Oxidative release of nitric oxide accounts for guanylyl cyclase stimulating, vasodilator and antiplatelet activity of Piloty's acid: a comparison with Angeli's salt. Biochem J. 1995;312:333–9.
71. Zeller A, Wenzl MV, Beretta M, Stessel H, Russwurm M, Koesling D, et al. Mechanisms underlying activation of soluble guanylate cyclase by the HNO donor Angeli's Salt. Mol Pharmacol. 2009;76:1115–22.
72. Miller TW, Cherney MM, Lee AJ, Francoleon NE, Farmer PJ, King SB, et al. The effects of nitroxyl (HNO) on soluble guanylate cyclase activity interactions at ferrous heme and cysteine thiols. J Biol Chem. 2009;284:21788–96.

73. Miranda KM, Paolocci N, Katori T, Thomas DD, Ford E, Bartberger MD, et al. A biochemical rationale for the discrete behavior of nitroxyl and nitric oxide in the cardiovascular system. Proc Natl Acad Sci USA. 2003;100:9196–201.
74. Väänänen AJ, Salmenperä P, Hukkanen M, Miranda KM, Harjula A, Rauhala P, Kankuri E. Persistent susceptibility of cathepsin B to irreversible inhibition by nitroxyl (HNO) in the presence of endogenous nitric oxide. Free Radic Biol Med. 2008;45:749–55.
75. Sherman MP, Grither WR, McCulla RD. Computational investigation of the reaction mechanisms of nitroxyl and thiols. J Org Chem. 2010;75:4014–24.
76. Shen B, English AM. Mass spectrometric analysis of nitroxyl-mediated protein modification: comparison of products formed with free and protein-based cysteines. Biochemistry. 2005;44:14030–44.
77. Lu J, Holmgren A. The thioredoxin antioxidant system. Free Radic Biol Med. 2014;66:75–87.
78. Hoffman MD, Walsh GM, Rogalski JC, Kast J. Identification of nitroxyl-induced modifications in human platelet proteins using a novel mass spectrometric detection method. Mol Cell Proteom. 2009;8:887–903.
79. Hammond AH, Fry JR. Effect of cyanamide on toxicity and glutathione depletion in rat hepatocyte cultures: differences between two dichloropropanol isomers. Chem Biol Interact. 1999;122:107–15.
80. Lopez BE, Rodriguez CE, Pribadi M, Cook NM, Shinyashiki M, Fukuto JM. Inhibition of yeast glycolysis by nitroxyl (HNO): mechanism of HNO toxicity and implications to HNO biology. Arch Biochem Biophys. 2005;442:140–8.
81. Lopez BE, Wink DA, Fukuto JM. The inhibition of glyceraldehyde-3-phosphate dehydrogenase by nitroxyl (HNO). Arch Biochem Biophys. 2007;465:430–6.
82. Eberhardt M, Dux M, Namer B, Miljkovic J, Cordasic N, Will C, et al. H2S and NO cooperatively regulate vascular tone by activating a neuroendocrine HNO-TRPA1-CGRP signalling pathway. Nat Commun. 2014;5:4381.
83. Roos G, Foloppe N, Messens J. Understanding the pKa of redox cysteines: the key role of hydrogen bonding. Antioxid Redox Signal. 2013;18:94–127.
84. Furchgott RF, Zawadzki JV. The obligatory role of endothelial cells in the relaxation of arterial smooth muscle by acetylcholine. Nature. 1980;288:373–6.
85. Ignarro LJ, Buga GM, Wood KS, Byrns RE, Chaudhuri G. Endothelium-derived relaxing factor produced and released from artery and vein is nitric oxide. Proc Natl Acad Sci USA. 1987;84:9265–9.
86. Moncada S, Palmer RMJ, Higgs EA. Biosynthesis of nitric oxide from l-arginine: a pathway for the regulation of cell function and communication. Biochem Pharmacol. 1989;38: 1709–15.
87. Andrews KL, Irvine JC, Tare M, Apostolopoulos J, Favaloro JL, Triggle CR, et al. A role for nitroxyl (HNO) as an endothelium- derived relaxing and hyperpolarizing factor in resistance arteries. Br J Pharmacol. 2009;157:540–50.
88. De Witt BJ, Marrone JR, Kaye AD, Keefer LK, Kadowitz PJ. Comparison of responses to novel nitric oxide donors in the feline pulmonary vascular bed. Eur J Pharmacol. 2001;430(2–3):311–5.
89. Favaloro JL, Kemp-Harper B. Redox variants of NO (NO. and HNO) elict vasorelaxation of resistance arteries via distinct mechanisms. Am J Physiol Heart Circ Physiol. 2009; 296:H1274–80.
90. Irvine JC, Favaloro JL, Widdop RE, Kemp-Harper BK. Nitroxyl anion donor, Angeli's salt, does not develop tolerance in rat isolated aortae. Hypertension. 2007;49:885–92.
91. Wanstall JC, Jeffery TK, Gambino A, Lovren F. Triggle CR Vascular smooth muscle relaxation mediated by nitric oxide donors: a comparison with acetylcholine, nitric oxide and nitroxyl ion. Br J Pharmacol. 2001;134:463–72.
92. Paolocci N, Katori T, Champion HC, St John ME, Miranda KM, Fukuto JM, et al. Positive inotropic and lusitropic effects of HNO/NO- in failing hearts: independence from beta-adrenergic signaling. Proc Natl Acad Sci USA. 2003;100:5537–42.

93. Paolocci N, Saavedra WF, Miranda KM, Martignani C, Isoda T, Hare JM, et al. Nitroxyl anion exerts redox-sensitive positive cardiac inotropy in vivo by calcitonin gene-related peptide signaling. Proc Natl Acad Sci USA. 2001;98:10463–8.
94. Zygmunt PM, Edwards G, Weston AH, Larsson B, Högestätt ED. Involvement of voltage-dependent potassium channels in the EDHF-mediated relaxation of rat hepatic artery. Br J Pharmacol. 1997;121(1):141–9.
95. Bullen ML, Miller AA, Andrews KL, Irvine JC, Ritchie RH, Sobey CG, et al. Nitroxyl (HNO) as a vasoprotective signaling molecule. Antioxid Redox Signal. 2011;14(9):1675–86.
96. Irvine JC, Kemp-Harper BK, Widdop RE. Chronic administration of the HNO donor Angeli's salt does not lead to tolerance, cross-tolerance, or endothelial dysfunction: comparison with GTN and DEANO. Antioxid Redox Signal. 2011;14:1615–24.
97. Münzel T, Daiber A, Gori T. More answers to the still unresolved question of nitrate tolerance. Eur Heart J. 2013;34(34):2666–73.
98. Brain SD, Cambridge H. Calcitonin gene-related peptide: vasoactive effects and potential therapeutic role. Gen Pharmacol-Vasc S. 1996;27:607–11.
99. Brain SD, Williams TJ, Tippins JR, Morris HR, MacIntyre I. Calcitonin gene-related peptide is a potent vasodilator. Nature. 1985;313:54–6.
100. Morris HR, Panico M, Etienne T, Tippins J, Girgis SI, MacIntyre I. Isolation and characterization of human calcitonin gene-related peptide. Nature. 1984;308:746–8.
101. Brain SD, Grant AD. Vascular actions of calcitonin gene-related peptide and adrenomedullin. Physiol Rev. 2004;84:903–34.
102. Mulderry PK, Ghatei MA, Bishop AE, Allen YS, Polak JM, Bloom SR. Distribution and chromatographic characterisation of CGRP-like immunoreactivity in the brain and gut of the rat. Regul Peptides. 1985;12:133–43.
103. Russell FA, King R, Smillie SJ, Kodji X, Brain SD. Calcitonin gene-related peptide: physiology and pathophysiology. Physiol Rev. 2014;94(4):1099–142.
104. Sexton PM. Central nervous system binding sites for calcitonin and calcitonin gene-related peptide. Mol Neurobiol. 1991;5:251–73.
105. Wimalawansa SJ. Amylin, calcitonin gene-related peptide, calcitonin, and adrenomedullin. A peptide superfamily. Crit Rev Neurobiol. 1997;11:167–239.
106. Matteoli M, Haimann C, Torri-Tarelli F, Polak JM, Ceccarelli B, De Camilli P. Differential effect of alphalatrotoxin on exocytosis from small synaptic vesicles and from large dense-core vesicles containing calcitonin gene-related peptide at the frog neuromuscular junction. Proc Natl Acad Sci U S A. 1989;85:7366–70.
107. Chang CP, Pearse Ii RV, O'Connell S, Rosenfeld MG. Identification of a seven transmembrane helix receptor for corticotropin-releasing factor and sauvagine in mammalian brain. Neuron. 1993;11:1187–95.
108. Evans BN, Rosenblatt MI, Mnayer LO, Oliver KR, Dickerson IM. CGRP-RCP, a novel protein required for signal transduction at calcitonin gene-related peptide and adrenomedullin receptors. J Biol Chem. 2000;275:31438–43.
109. Luebke AE, Dahl GP, Roos BA, Dickerson IM. Identification of a protein that confers calcitonin gene-related peptide responsiveness to oocytes by using a cystic fibrosis transmembrane conductance regulator assay. Proc Natl Acad Sci USA. 1996;93:3455–60.
110. McLatchie LM, Fraser NJ, Main MJ, Wise A, Brown J, Thompson N, et al. RAMPs regulate the transport and ligand specificity of the calcitonin-receptor-like receptor. Nature. 1998;393:333–9.
111. Brain SD, Tippins JR, Morris HR, MacIntyre I, Williams TJ. Potent vasodilator activity of calcitonin gene-related peptide in human skin. J Invest Dermatol. 1986;87:533–6.
112. Gray DW, Marshall I. Human alpha-calcitonin gene-related peptide stimulates adenylate cyclase and guanylate cyclase and relaxes rat thoracic aorta by releasing nitric oxide. Br J Pharmacol. 1992;107:691–6.
113. de Hoon JN, Pickkers P, Smits P, Struijker-Boudier HA, Van Bortel LM. Calcitonin gene-related peptide: exploring its vasodilating mechanism of action in humans. Clin Pharmacol Ther. 2003;73:312–21.

114. Zygmunt PM, Högestätt ED. TRPA1. Handb Exp Pharmacol. 2014;222:583–630.
115. Story GM, Peier AM, Reeve AJ, Eid SR, Mosbacher J, Hricik TR, et al. ANKTM1, a TRP-like channel expressed in nociceptive neurons, is activated by cold temperatures. Cell. 2003;112:819–29.
116. Bautista DM, Jordt SE, Nikai T, Tsuruda PR, Read AJ, Poblete J, et al. TRPA1 mediates the inflammatory actions of environmental irritants and proalgesic agents. Cell. 2006;124:1269–82.
117. Bautista DM, Movahed P, Hinman A, Axelsson HE, Sterner O, Högestätt ED, et al. Pungent products from garlic activate the sensory ion channel TRPA1. Proc Natl Acad Sci USA. 2005;102:12248–52.
118. Takahashi N, Kuwaki T, Kiyonaka S, Numata T, Kozai D, Mizuno Y, et al. TRPA1 underlies a sensing mechanism for O2. Nat Chem Biol. 2011;7:701–11.
119. Andersson DA, Gentry C, Moss S, Bevan S. Transient receptor potential A1 is a sensory receptor for multiple products of oxidative stress. J Neurosci. 2008;28:2485–94.
120. Takahashi N, Mizuno Y, Kozai D, Yamamoto S, Kiyonaka S, Shibata T, et al. (2008) Molecular characterization of TRPA1 channel activation by cysteine-reactive inflammatory mediators. Channels. 2008;2:287–98.
121. Hinman A, Chuang HH, Bautista DM, Julius D. TRP channel activation by reversible covalent modification. Proc Natl Acad Sci USA. 2006;103:19564–8.
122. Macpherson LJ, Dubin AE, Evans MJ, Marr F, Schultz PG, Cravatt BF, et al. Noxious compounds activate TRPA1 ion channels through covalent modification of cysteines. Nature. 2007;445:541–5.
123. Cvetkov TL, Huynh KW, Cohen MR, Moiseenkova-Bell VY. Molecular architecture and subunit organization of TRPA1 channel revealed by electron microscopy. J Biol Chem. 2011;286:38168–76.
124. Wang L, Cvetkov TL, Chance MR, Moiseenkova-Bell VY. Identification of in vivo disulfide conformation of the TRPA1 ion channel. J Biol Chem. 2012;287:6169–76.
125. Sabbah HN, Tocchetti CG, Wang M, Daya S, Gupta RC, Tunin RS, et al. Nitroxyl (HNO): a novel approach for the acute treatment of heart failure. Circ Heart Fail. 2013;6(6):1250–8.
126. Katori T, Hoover DB, Ardell JL, Helm RH, Belardi DF, Tocchetti CG, Forfia PR, Kass DA, Paolocci N. Calcitonin gene-related peptide in vivo positive inotropy is attributable to regional sympatho-stimulation and is blunted in congestive heart failure. Circ Res. 2005;96:234–43.
127. Al-Rubaiee M, Gangula PR, Millis RM, Walker RK, Umoh NA, Cousins VM, et al. Inotropic and lusitropic effects of calcitonin gene-related peptide in the heart. Am J Physiol Heart Circ Physiol. 2013;304(11):H1525–37.
128. Saetrum Opgaard O, de Vries R, Tom B, Edvinsson L, Saxena PR. Positive inotropy of calcitonin gene-related peptide and amylin on porcine isolated myocardium. Eur J Pharmacol. 1999; 385(2–3):147–54.
129. Saetrum Opgaard O, Hasbak P, de Vries R, Saxena PR, Edvinsson L. Positive inotropy mediated via CGRP receptors in isolated human myocardial trabeculae. Eur J Pharmacol. 2000;397(2–3):373–82.
130. Cueille C, Pidoux E, de Vernejoul MC, Ventura-Clapier R, Garel JM. Increased myocardial expression of RAMP1 and RAMP3 in rats with chronic heart failure. Biochem Biophys Res Commun. 2002;294(2):340–6.
131. Øie E, Vinge LE, Yndestad A, Sandberg C, Grøgaard HK, Attramadal H. Induction of a myocardial adrenomedullin signaling system during ischemic heart failure in rats. Circulation. 2000;101:415–22.
132. Cheong E, Tumbev V, Abramson J, Salama G, Stoyanovsky DA. Nitroxyl triggers Ca2+ release from skeletal and cardiac sarcoplasmic reticulum by oxidizing ryanodine receptors. Cell Calcium. 2005;37:87–96.
133. Tocchetti CG, Wang W, Froehlich JP, Huke S, Aon MA, Wilson GM, et al. Nitroxyl improves cellular heart function by directly enhancing cardiac sarcoplasmic reticulum Ca2+ cycling. Circ Res. 2007;100:96–104.

134. Donoso P, Sanchez G, Bull R, Hidalgo C. Modulation of cardiac ryanodine receptor activity by ROS and RNS. Front Biosci (Landmark Ed). 2011;16:553–67.
135. Kohr MJ, Kaluderic N, Tocchetti CG, Gao WD, Kass DA, Janssen PML, et al. Nitroxyl enhances myocyte Ca2+ transients by exclusively targeting SR Ca^{2+}-cycling. Front Biosci. 2010;E2:614–26.
136. Froehlich JP, Mahaney JE, Keceli G, Pavlos CM, Goldstein R, Redwood AJ, et al. Phospholamban thiols play a central role in activation of the cardiac muscle sarcoplasmic reticulum calcium pump by nitroxyl. Biochemistry. 2008;47:13150–2.
137. Sivakumaran V, Stanley BA, Tocchetti CG, Ballin JD, Caceres V, Zhou L, et al. HNO enhances SERCA2a activity and cardiomyocyte function by promoting redox-dependent phospholamban oligomerization. Antioxid Redox Signal. 2013;19(11):1185–97.
138. Tocchetti CG, Stanley BA, Murray CI, Sivakumaran V, Donzelli S, Mancardi D, et al. Playing with cardiac "redox switches": the "HNO way" to modulate cardiac function. Antioxid Redox Signal. 2011;14(9):1687–98.
139. Lancel S, Zhang J, Evangelista A, Trucillo MP, Tong X, Siwik DA, et al. Nitroxyl activates SERCA in cardiac myocytes via glutathiolation of cysteine 674. Circ Res. 2009;104(6):720–3.
140. Gao WD, Murray CI, Tian Y, Zhong X, DuMond JF, Shen X, et al. Nitroxyl-mediated disulfide bond formation between cardiac myofilament cysteines enhances contractile function. Circ Res. 2012;111(8):1002–11.
141. Pagliaro P, Mancardi D, Rastaldo R, Penna C, Gattullo D, Miranda KM, et al. Nitroxyl affords thiol-sensitive myocardial protective effects akin to early preconditioning. Free Radic Biol Med. 2003;34:33–43.
142. Ma XL, Gao F, Liu GL, Lopez BL, Christopher TA, Fukuto JM, et al. Opposite effects of nitric oxide and nitroxyl on postischemic myocardial injury. Proc Natl Acad Sci U S A. 1999;96(25):14617–22.
143. Naughton P, Foresti R, Bains SK, Hoque M, Green CJ, Motterlini R. Induction of heme oxygenase 1 by nitrosative stress. A role for nitroxyl anion. J Biol Chem. 2002;277(43): 40666–74.
144. Naughton P, Hoque M, Green CJ, Foresti R, Motterlini R. Interaction of heme with nitroxyl or nitric oxide amplifies heme oxygenase-1 induction: involvement of the transcription factor Nrf2. Cell Mol Biol. 2002;48(8):885–94.
145. Tsihlis ND, Murar J, Kapadia MR, Ahanchi SS, Oustwani CS, Saavedra JE, et al. Isopropylamine NONOate (IPA/NO) moderates neointimal hyperplasia following vascular injury. J Vasc Surg. 2010;51:1248–59.
146. Shiva S, Crawford JH, Ramachandran A, Ceaser EK, Hillson T, Brookes PS, et al. Mechanisms of the interaction of nitroxyl with mitochondria. Biochem J. 2004;379(Pt 2):359–66.
147. Queliconi BB, Wojtovich AP, Nadtochiy SM, Kowaltowski AJ, Brookes PS. Redox regulation of the mitochondrial K(ATP) channel in cardioprotection. Biochim Biophys Acta. 2011;1813(7):1309–15.
148. Chazotte-Aubert L, Oikawa S, Gilibert I, Bianchini F, Kawanishi S, Ohshima H. Cytotoxicity and site-specific DNA damage induced by nitroxyl anion (NO-) in the presence of hydrogen peroxide—implications for various pathophysiological conditions. J Biol Chem. 1999;274:20909–15.
149. Ohshima H, Gilibert I, Bianchini F. Induction of DNA strand breakage and base oxidation by nitroxyl anion through hydroxyl radical production. Free Radic Biol Med. 1999;26:1305–13.
150. Norris AJ, Sartippour MR, Lu M, Park T, Rao JY, Jackson MI, et al. Nitroxyl inhibits breast tumor growth and angiogenesis. Int J Cancer. 2008;122:1905–10.
151. Kim WK, Choi YB, Rayudu PV, Das P, Asaad W, Arnelle DR, et al. Attenuation of NMDA receptor activity and neurotoxicity by nitroxyl anion, NO. Neuron. 1999;24(2):461–9.
152. Väänänen AJ, Moed M, Tuominen RK, Helkamaa TH, Wiksten M, Liesi P, Chiueh CC, Rauhala P. Angeli's salt induces neurotoxicity in dopaminergic neurons in vivo and in vitro. Free Radic Res. 2003;37(4):381–9.
153. Choe CU, Lewerenz J, Fischer G, Uliasz TF, Espey MG, Hummel FC, et al. Nitroxyl exacerbates ischemic cerebral injury and oxidative neurotoxicity. J Neurochem. 2009;110:1766–73.

154. Edvinsson L. Calcitonin gene-related peptide (CGRP) and the pathophysiology of headache: therapeutic implications. CNS Drugs. 2001;15:745–53.
155. Elvidge S. Anti-CGRP antibodies for migraine turn industry heads. Nat Biotechnol. 2014;32(8):707.
156. Dux M, Will C, Vogler B, Filipovic MR, Messlinger K. Meningeal blood flow is controlled by H2S-NO crosstalk activating a HNO-TRPA1-CGRP signalling pathway. Br J Pharmacol. 2016;173(3):431–45. doi:10.1111/bph.13164
157. Wong PSY, Hyun J, Fukuto JM, Shirota FN, DeMaster EG, Shoeman DW, Nagasawa HT. Reaction between S-nitrosothiols and thiols: generation of nitroxyl (HNO) and subsequent chemistry. Biochemistry. 1998;37:5362–71.
158. Schmidt HHHW, Hofmann H, Schindler U, Shutenko ZS, Cunningham DD, Feelisch M. No NO from NO synthase. Proc Natl Acad Sci USA. 1996;93:14492–7.
159. Donzelli S, Espey MG, Flores-Santana W, Switzer CH, Yeh GC, Huang J, et al. Generation of nitroxyl by heme protein-mediated peroxidation of hydroxylamine but not *N*-hydroxy-L-arginine. Free Radical Biol Med. 2008;45:578–84.
160. Filipovic MR, Stanic D, Raicevic S, Spasic M, Niketic V. Consequences of MnSOD interactions with •NO: •NO dismutation, peroxynitrite and H2O2 generation. Free Rad Res. 2007;41:62–72.
161. Sharpe MA, Cooper CE. Reactions of nitric oxide with mitochondrial cytochrome c: a novel mechanism for the formation of nitroxyl anion and peroxynitrite. Biochem J. 1998;332 (Pt 1):9–19.
162. Foster MW, Hess DT, Stamler JS. Protein S-nitrosylation in health and disease: a current perspective. Trends Mol Med. 2009;15:391–404.
163. Hess DT, Stamler JS. Regulation by S-nitrosylation of protein post-translational modification. J Biol Chem. 2012;287:4411–8.
164. Seth D, Stamler JS. The SNO-proteome: causation and classifications. Curr Opin Chem Biol. 2011;15:129–36.
165. Lima B, Forrester MT, Hess DT, Stamler JS. S-nitrosylation in cardiovascular signa. Circ Res. 2010;106:633–46.
166. Arnelle DR, Stamler JS. NO^+, NO•, and NO^- donation by *S*-Nitrosothiols: implications for regulation of physiological functions by *S*-Nitrosylation and acceleration of disulfide formation. Arch Biochem Biophys. 1995;318:279–85.
167. Talipov MR, Timerghazin QK. Protein control of S-nitrosothiol reactivity: interplay of antagonistic resonance structures. J Phys Chem B. 2013;117:1827–37.
168. Broniowska KA, Hogg N. The chemical biology of S-nitrosothiols. Antioxid Redox Signal. 2012;17:969–80.
169. Hogg N. Biological chemistry and clinical potential of S-nitrosothiols. Free Radic Biol Med. 2000;28:1478–86.
170. Hobbs AJ, Fukuto JM, Ignarro LJ. Formation of free nitric oxide from l-arginine by nitric oxide synthase: direct enhancement of generation by superoxide dismutase. Proc Natl Acad Sci USA. 1994;91:10992–6.
171. Adak S, Wang Q, Stuehr DJ. Arginine conversion to nitroxide by tetrahydrobiopterin-free neuronal nitric-oxide synthase: implications for mechanism. J Biol Chem. 2000;275:33554–61.
172. Rusche KM, Spiering MM, Marletta MA. Reactions catalyzed by tetrahydrobiopterin-free nitric oxide synthase. Biochemistry. 1998;37:15503–12.
173. Niketic V, Stojanovic S, Nikolic A, Spasic M, Michelson AM. Exposure of Mn and FeSODs, but not Cu/ZnSOD, to NO leads to nitrosonium and nitroxyl ions generation which cause enzyme modification and inactivation: an in vitro study. Free Radic Biol Med. 1999;27:992–6.
174. Liochev SI, Fridovich I. Nitroxyl (NO-): a substrate for superoxide dismutase. Arch Biochem Biophys. 2002;402(2):166–71.
175. Filipovic MR, Duerr K, Mojovic M, Simovic V, Zimmerman R, Niketic V, et al. NO dismutase activity of seven-coordinate manganese(II) pentaazamacrocyclic complexes. Angew Chem Int Edit. 2008;47:8735–9.

176. Filipovic MR, Koh ACW, Arbautlt S, Niketic V, Debus A, Schleicher U, et al. Striking inflammation from both sides: manganese(II) pentaazamacrocyclic SOD mimics act also as nitric oxide dismutases: a single-cell study. Angew Chem Int Edit. 2010;49:4228–32.
177. Abe K, Kimura H. The possible role of hydrogen sulfide as an endogenous neuromodulator. J Neurosci. 1996;16:1066–71.
178. Li L, Hsu A, Moore PK. Actions and interactions of nitric oxide, carbon monoxide and hydrogen sulphide in the cardiovascular system and in inflammation-a tale of three gases! Pharmacol Ther. 2009;123:386–400.
179. Mustafa AK, Gadalla MM, Sen N, et al. H2S signals through protein S-sulfhydration. Sci Signal. 2009;2:ra72.
180. Wang R. Two's company, three's a crowd: can H2S be the third endogenous gaseous transmitter? FASEB J. 2002;16:1792–808.
181. Kabil O, Banerjee R. Enzymology of H2S biogenesis, decay and signaling. Antioxid Redox Signal. 2014;20:770–82.
182. Kabil O, Motl N, Banerjee R. H2S and its role in redox signaling. Biochim Biophys Acta. 2014;1844(8):1355–66.
183. Kimura H, Nagai Y, Umemura K, Kimura Y. Physiological roles of hydrogen sulfide: synaptic modulation, neuroprotection, and smooth muscle relaxation. Antioxid Redox Signal. 2005;7:795–803.
184. Li L, Bhatia M, Zhu YZ, et al. Hydrogen sulfide is a novel mediator of lipopolysaccharide-induced inflammation in the mouse. FASEB J. 2005;19:1196–8.
185. Yang G, Wu L, Jiang B, et al. H2S as a physiologic vasorelaxant: hypertension in mice with deletion of cystathionine gamma-lyase. Science. 2008;322:587–90.
186. Calvert JW, Elston M, Nicholson CK, et al. Genetic and pharmacologic hydrogen sulfide therapy attenuates ischemia-induced heart failure in mice. Circulation. 2010;122:11–9.
187. Calvert JW, Jha S, Gundewar S, et al. Hydrogen sulfide mediates cardioprotection through Nrf2 signaling. Circ Res. 2009;105:365–74.
188. Blackstone E, Morrison M, Roth MB. H2S induces a suspended animation-like state in mice. Science. 2005;308:518.
189. Filipovic MR, Miljkovic JL, Nauser T, Royzen M, Klos K, Shubina T, et al. Chemical characterization of the smallest S-nitrosothiol, HSNO; cellular cross-talk of H2S and S-nitrosothiols. J Am Chem Soc. 2012;134:12016–27.
190. Rosenthal J, Lippard S. Direct detection of nitroxyl in aqueous solution using a TripodalCopper(II) BODIPY complex. J Am Chem Soc. 2010;132:5536–7.
191. Filipovic MR, Eberhardt M, Prokopovic V, Mijuskovic A, Orescanin Dusic O, et al. Beyond H2S and NO interplay: hydrogen sulfide and nitroprusside react directly to give nitroxyl (HNO). A new pharmacological source of HNO. J Med Chem. 2013;56:1499–508.
192. Lundberg J, Weitzberg E, Shiva S, Gladwin M. The nitrate-nitrite-nitric oxide pathway in mammals. In: Bryan NS, Loscalzo J, editors. Nitrite and nitrate in human health and disease. New York: Humana Press; 2011.
193. Lundberg JO, Weitzberg E, Gladwin MT. The nitrate-nitrite-nitric oxide pathway in physiology and therapeutics. Nat Rev Drug Discov. 2008;7:156–67.
194. Miljkovic JL, Kenkel I, Ivanovic-Burmazovic I, Filipovic MR. Generation of HNO and HSNO from nitrite by heme-iron-catalyzed metabolism with H2S. Angew Chem Int Edit. 2013;52:12061–4.
195. Yong QC, Hu LF, Wang S, Huang D, Bian JS. Hydrogen sulfide interacts with nitric oxide in the heart: possible involvement of nitroxyl. Cardiovasc Res. 2010;88:482–91.
196. Yong QC, Cheong JL, Hua F, Deng LW, Khoo YM, Lee HS, et al. Regulation of heart function by endogenous gaseous mediators-crosstalk between nitric oxide and hydrogen sulfide. Antioxid Redox Signal. 2011;14:2081–91.
197. Koppenol W. Nitrosation, thiols, and hemoglobin: energetics and kinetics. Inorg Chem. 2012;51(10):5637–41.
198. Suárez SA, Bikiel DA, Wetzler D, Martí MA, Doctorovich F. Time-resolved electrochemical quantification of azanone (HNO) at low nanomolar level. Anal Chem. 2013;85:10262–9.

199. Poole DP, Pelayo JC, Cattaruzza F, Kuo YM, Gai G, Chiu JV, et al. Transient receptor potential ankyrin 1 is expressed by inhibitory motoneurons of the mouse intestine. Gastroenterology. 2011;141:565–75.
200. Wood KC, Batchelor AM, Bartus K, Harris KL, Garthwaite G, Vernon J, Garthwaite J. Picomolar nitric oxide signals from central neurons recorded using ultrasensitive detector cells. J Biol Chem. 2011;286:43172–81.
201. Coletta C, Papapetropoulos A, Erdelyi K, Olah G, Modis K, Panopoulos P, et al. Hydrogen sulfide and nitric oxide are mutually dependent in the regulation of angiogenesis and endothelium-dependent vasorelaxation. Proc Natl Acad Sci USA. 2012;109:9161–6.
202. Medani M, Collins D, Docherty NG, Baird AW, O'Connell PR, Winter DC. Emerging role of hydrogen sulfide in colonic physiology and pathophysiology. Inflamm Bowel Dis. 2011;17:1620–5.
203. Engel MA, Leffler A, Niedermirtl F, Babes A, Zimmermann K, Filipović MR, et al. TRPA1 and substance P mediate colitis in mice. Gastroenterol. 2011;141:1346–58.
204. Engel MA, Khalil M, Siklosi N, Mueller-Tribbensee SM, Neuhuber WL, Neurath MF, et al. Opposite effects of substance P and calcitonin gene-related peptide in oxazolone colitis. Dig Liv Dis. 2012;44:24–9.
205. Chattopadhyay M, Kodela R, Olson KR, Kashfi K. NOSH-aspirin (NBS-1120), a novel nitric oxide- and hydrogen sulfide-releasing hybrid is a potent inhibitor of colon cancer cell growth in vitro and in a xenograft mouse model. Biochem Biophys Res Commun. 2012;419(3): 523–8.
206. Kodela R, Chattopadhyay M, Kashfi K. NOSH-aspirin: a novel nitric oxide-hydrogen sulfide-releasing hybrid: a new class of anti-inflammatory pharmaceuticals. ACS Med Chem Lett. 2012;3:257–62.

Chapter 15
Advances in Breast Cancer Therapy Using Nitric Oxide and Nitroxyl Donor Agents

Debashree Basudhar, Katrina M. Miranda, David A. Wink, and Lisa A. Ridnour

15.1 Introduction

The physiological importance of nitric oxide (NO), otherwise considered as an atmospheric pollutant [1], was firmly established in the early 1990s as endothelium-derived relaxing factor that revolutionized the whole field of NO biology. Ignarro, Furchgott, and Murad received the Nobel Prize in Medicine in 1998 for their contributions, thus changing the view of the scientific society on NO [2]. NO is an endogenous free radical messenger that plays a vital role in numerous physiological and pathophysiological processes like modulation of blood pressure, inhibition of platelet aggregation, neural transmission, and immune response [3–10]. Pharmacological actions of several well-known cardiovascular agents (sodium nitroprusside, nitroglycerin, isosorbide dinitrate, etc.) can be explained in terms of their capabilities to produce NO upon administration.

Biosynthesis of NO involves five-electron oxidation of L-arginine to L-citrulline and is catalyzed by NO synthase (NOS) in an NADPH-dependent process (Fig. 15.1) [11].

Three NOS isoforms have been identified: neuronal NOS (nNOS, NOS-1) [12], inducible NOS (iNOS, NOS-2) [13], and endothelial NOS (eNOS, NOS-3) [14].

The original version of this chapter was revised. An erratum to this chapter can be found at DOI 10.1007/978-3-319-30705-3_31

D. Basudhar • D.A. Wink • L.A. Ridnour (✉)
Cancer and Inflammation Program, NCI-Frederick, 1050 Boyles St, Bldg 567 Rm 253, Frederick, MD 21702, USA
e-mail: debashree.basudhar@nih.gov; wink@mail.nih.gov; ridnourl@mail.nih.gov

K.M. Miranda
Department of Chemistry and Biochemistry, University of Arizona, 1306 E. University Blvd, Tucson, AZ 85721, USA
e-mail: kmiranda@email.arizona.edu

I. Batinić-Haberle et al. (eds.), *Redox-Active Therapeutics*, Oxidative Stress in Applied Basic Research and Clinical Practice, DOI 10.1007/978-3-319-30705-3_15

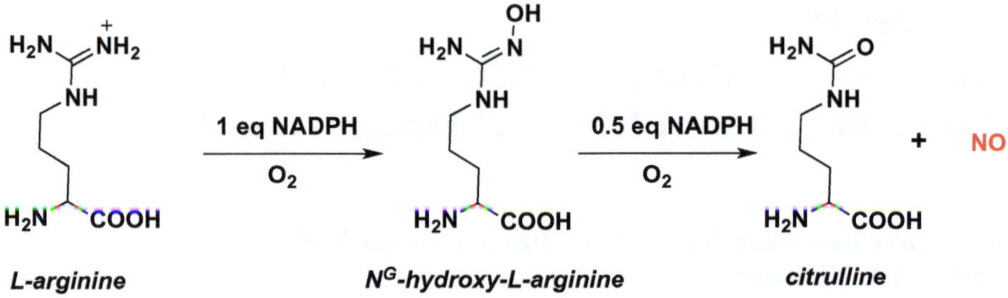

Fig. 15.1 Biosynthesis of NO by oxidation of L-arginine to L-citrulline catalyzed by NOS

While nNOS and eNOS are constitutively expressed and produce low NO flux in response to transient changes in calcium levels, iNOS is induced by pro-inflammatory cytokines [15]. NO production from dietary sources of nitrite and nitrate can also contribute to desired NO concentration and can play a crucial role under hypoxic conditions [16]. The biological properties of NO are decided by its chemistry [17]. NO can exert its physiological activities by direct activation of soluble guanylate cyclase (SGC) to generate cyclic guanosive monophospase (cGMP) followed by downstream signal transduction [18, 19] or by cGMP-independent mechanisms mediated by reactive nitrogen species (RNS) that are potent nitrating, nitrosating, and oxidizing agents [20–22]. The biological role of NO is complicated. It can play crucial roles in various physiological as well as disease processes in a concentration-dependent manner. Under pathophysiological settings, low levels of NO can lead to angina, hypertension, and impotency; while overproduction can lead to sepsis, stroke, inflammatory diseases, and cancer [15, 23, 24].

Nitroxyl (HNO), the neglected and less understood nitrogen oxide in the previous century, has emerged as a key player in the last decade. It is a promising pharmacological agent with distinct cardiovascular, tumoricidal, alcohol-deterrent properties among others [25–31]. Catalase-mediated generation of HNO from cyanamide is used clinically for the treatment of alcoholism [32–34]. Although mammalian biosynthesis of HNO has been postulated, the subject remains highly debatable, in part due to inefficient/nonspecific *in vivo* detection methods. Some of the pathways for possible endogenous production of HNO are: (1) N^{ω}-hydroxy-L-arginine (NOHA), a critical intermediate during NO biosynthesis, can be uncoupled from NOS and further oxidized by biological oxidants to release HNO, (2) absence of the cofactor tetrahydrobiopterin during NO biosynthesis, (3) reaction of thiols with *S*-nitrosothiols, (4) metalloproteins mediated reduction of NO [35–41]. Compared to NO, HNO has only recently generated interest as a promising chemotherapeutic agent [30, 31]. HNO can induce DNA double strand breaks under aerobic conditions and can irreversibly inhibit enzymes with critical thiols such as glyceraldehyde phosphate dehydrogenase (GAPDH) [42, 43]. In order to fully harness the potential of NO- and HNO-based therapeutics, it is important to understand their biological role in cancer. In this chapter, we will discuss developments in the field of NO- and HNO-based therapeutic approaches in breast cancer treatment illustrating some of their benefits and pitfalls.

15.2 Redox Biology of NO and HNO in Breast Cancer

Breast cancer is one of the most prevalent types of cancer among women. In 2015, 231,840 new cases of invasive and 60,290 new cases of noninvasive breast cancer are estimated in the USA alone with 40,290 projected cases of death [44]. Metastatic breast cancer has a high mortality rate among women. It is commonly associated with spreading of cancer from breast and lymph nodes to lung, liver, bone, and brain. Basal-like breast cancer associated with lack of expression of human epidermal growth factor (HER2), estrogen, and progesterone receptor is the most aggressive phenotype.

15.2.1 Nitric Oxide in Cancer

Dysfunction of redox homeostasis defined as the balance between oxidative/nitrosative stress and antioxidants, plays an important role in the development and progression of cancer. NOS2 is a pro-inflammatory mediator that generates intermediate flux of NO (>100 nM) leading to pro-growth signaling in cancer [45–49]. It has been established as a biomarker for poor patient survival in basal-like breast cancer [50]. Breast cancer is associated with overproduction of reactive oxygen species (ROS) and RNS. Apart from breast cancer, elevated NOS2 is also associated with pancreatic, ovarian, melanoma, prostate, oral, head and neck squamous cell cancer [51–56]. In addition, eNOS and nNOS are expressed in breast carcinoma [57], oligodendroglioma, and neuroblastoma cell lines [58].

The role of NO in cancer biology is much more complicated than previously thought. This is due in part to conflicting findings that have accumulated in the literature. NO is involved in multiple steps of tumor development with critical roles in hypoxia response, survival and metastasis, angiogenesis, and cancer immunology. However, it is also involved in cell cycle arrest, apoptosis, and enhances chemo- and radiotherapy effects [15, 59–61]. NO is involved in tumor suppression as well as progression and metastasis. Hibbs et al. showed tumoricidal effects of NO in activated murine macrophages [3, 62]. Leukemic cells are also susceptible to NO-mediated apoptosis [63]. This dichotomy has puzzled the scientific community over the decades.

A much clearer picture of the concentration-dependent role of NO in cancer (Fig. 15.2) has begun to emerge. In contrast to ligand-/receptor-dependent signaling, NO mechanisms in cancer are determined by its chemical properties. As such, signaling events generated from NO can be dramatically different depending on the concentration, duration of exposure, and tissue microenvironment. Most of its biological signaling is either due to its direct reaction with metal centers or nitrosation of biomolecules [20, 64–69]. Nitrosylation of metal centers occurs at a low concentration of NO and can be broadly classified as physiological response while protein nitrosation is an indicator of nitrosative stress, which is critical for understanding of various pathophysiological conditions. Thiol nitrosation is involved in inhibition as well as activation of key signaling pathways [67].

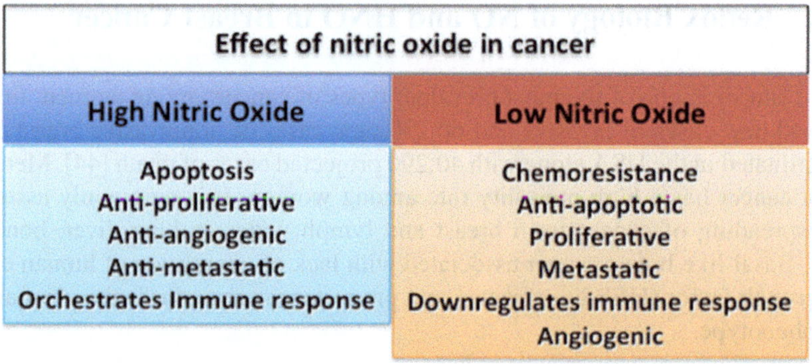

Fig. 15.2 Concentration-dependent effects of NO in cancer. Intermediate levels of NO >100 nM leads to pro-survival signaling and high levels in excess of 800 nM induces tumoricidal properties

One important factor, which is often overlooked, is that solid tumors often have hypoxic regions. So even with high expression of NOS2, the bioavailability of NO may be low due to the O_2 dependence of NOS for NO production as shown in Fig. 15.1. Hicock et al. showed that the maximum flux of NO generated from NOS2 occurred at O_2 concentrations ranging between 5 and 8 % [70]. Thus, availability of O_2 is a limiting factor in the rate of synthesis of NO. Moreover, for the same reason, the half-life of NO is longer under conditions of low O_2 as metabolism of NO is also O_2 dependent. Therefore, the O_2 dependence of NO synthesis and half-life is important during NO/RNS reaction with multiple cellular targets and downstream signaling [71]. Under normoxic conditions, NOS2 has the ability to produce a higher flux of NO. Moreover in hypoxic tumors, which typically have <3 % O_2, NO can inhibit mitochondrial function to control bioavailability of O_2 [70]. The interplay between O_2 and NO was recently reviewed [72]. In cancer biology under hypoxic conditions, the flux of NO is half of that under normoxic conditions suggesting that sGC signaling is a key mediator under hypoxic conditions.

Solid tumors often display hypoxic regions that are associated with more aggressive phenotypes and increased metastasis [73] and are harder to treat with both chemo and radiation therapy [74]. Hypoxia inducible factor 1α (HIF-1α), which plays a critical role in cell survival during hypoxia, is stabilized by NO-mediated S-nitrosylation at Cys 533 (Cys 520 in humans) even in the presence of oxygen [75]. A seminal study by Thomas et al., demonstrated the role of NO flux during the regulation of distinct signaling pathways in breast cancer cell lines; HIF-1α stabilization occurred at NO flux ranging from 100 to 500 nM using both macrophage induction and SPER/NO as NO donor [76]. Importantly, this study identified key pro-survival signaling pathways that were modulated by NO.

In addition to pro-survival signaling, low NO flux (50–100 nM) leads to increased angiogenesis and metastasis [76, 77]. Angiogenesis is a complex process that involves cellular proliferation, migration, and tube formation to replenish nutrient and oxygen requirements of growing tumor, thus playing a key role in shaping the tumor microenvironment. Several studies show NOS to be associated with tumor growth, invasion, and angiogenesis [77–79]. Elevated NOS2 tumor expression predicts poor disease-specific survival in ER-negative breast cancer patients [50]. Interestingly, elevated eNOS, which produces sustained low levels of NO (<10 nM) during angiogenesis induction, has also been shown to predict poor breast cancer survival [57, 80].

Thrombospondin-1 (TSP-1) is a glycoprotein known to exhibit anticancer and anti-angiogenic properties. The examination of NO regulation of TSP-1 during angiogenic response identified a novel cross-talk relationship between NO and TSP-1, where low flux NO suppressed TSP-1 levels during angiogenic response, which was ERK dependent. Importantly, exogenous TSP-1 blocked NO-induced angiogenic response [81, 82].

Dysregulation of matrix metalloproteinases (MMPs), a group of structurally similar endopeptidases that play a key role in controlling homeostasis of extracellular matrix (ECM) proteins, have long been implicated in various pathophysiological conditions like neurodegenerative diseases, stroke, arthritis, and cancer invasion and metastasis [83, 84]. MMP-9-deficient mice show reduced cancer-cell proliferation and metastasis compared to wild-type mice [85, 86]. Regulation of MMP-9 by NO during cerebral ischemia was shown to involve S-nitrosylation of Cys residue and further oxidation to stable sulfinic or sulfonic acid, which in turn led to neuronal apoptosis [87]. We have recently shown a dose-dependent, biphasic regulatory effect of NO on the activity of MMPs (MMP-9, -1, and -13) secreted from murine macrophages [88]. A concentration-dependent activation and inactivation of purified latent MMP-9 was observed on exposure to NO. Moreover, low flux NO-modulated MMP/tissue inhibitor of metalloproteinase (TIMP)-1 levels in cGMP-dependent manner by enhancing MMP activity and suppressing the endogenous inhibitor TIMP-1 [88]. Later studies demonstrated TIMP-1 nitration on specific tyrosine residues in hydrophobic pockets that abated the inhibitory function of TIMP-1 against active MMPs [89, 90]. In breast cancer, TIMP-1 predicted poor disease-specific breast cancer survival, which was restricted to patients with high tumor NOS2 expression. In contrast, TIMP-1 did not predict poor survival in patients with low tumor NOS2 expression. Importantly, elevated TIMP-1 increased the association between NOS2 and pAkt in these tumors, which suggests that elevated NOS2 and TIMP-1 promotes more aggressive tumor growth by Akt activation [90].

Higher NO concentrations similar to that produced by activated macrophages have been shown to induce p53 phosphorylation, which led to apoptosis in malignant cells (Fig. 15.2) [91, 92]. This elevated NO flux and downstream apoptotic signaling is key to the development of NO-releasing therapeutics and extended understanding of the role of high flux of NO in tumor cell death.

15.2.2 Nitroxyl and Cancer

Recently, nitroxyl (HNO) has received significant attention due to its distinct biological properties and potential as a pharmacological agent in the treatment of alcoholism, cardiovascular diseases, and cancer [25, 27, 28, 30, 31, 93]. While

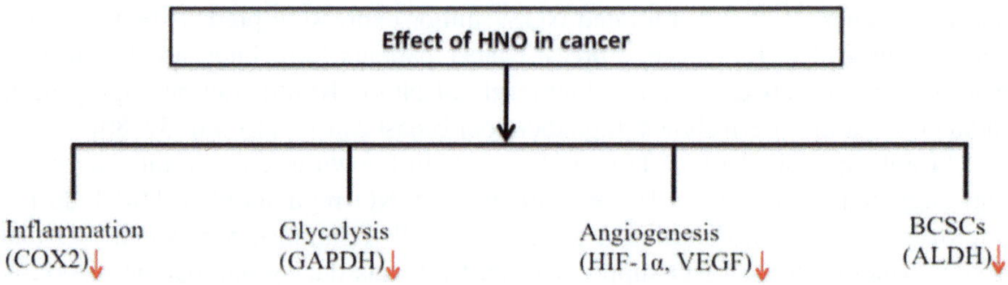

Fig. 15.3 Effect of HNO on different signaling pathways associated with cancer

the biological effect(s) of HNO is largely unexplored, recent studies have shown that HNO reduces angiogenesis by reducing HIF-1α and VEGF [30]. Toward this end, the chemotherapeutic properties of HNO have recently gained interest (Fig. 15.3) with additional studies showing cytotoxic effects [43] that are in part due to DNA double strand breaks caused by the formation of reactive nitrogen oxide species [94, 95]. HNO inhibited growth of breast cancer xenografts [30, 31] and neuroblastoma cells [96]. HNO also inhibited aldehyde dehydrogenase (ALDH), an enzyme key in metabolism of alcohol, and a marker of breast cancer stem cells (BCSC). While the effect of HNO on various signaling pathways is not well studied, much of its biological effects are based on its heme chemistry and the ability to modify critical thiol residues such as GAPDH [42], an enzyme involved in glycolysis, and other cellular process such as apoptosis and response to oxidative stress [97]. Cancer cells have a higher dependence on glycolysis than normal cells [98], which makes HNO an attractive chemotherapeutic candidate [99]. HNO also showed inhibition of poly(ADP)ribose polymerase (PARP), which is an important component of DNA repair mechanisms [100]. Thus, HNO-based pharmaceutical agents can not only be effective as single agent drugs but may also be useful in combination with chemo- and radiotherapy. In addition to the direct effects of HNO on cancer cells, HNO donors may be useful adjuvant agents to chemotherapy.

15.3 Development of Redox Therapeutics

The role of NO in tumor progression versus carcinogenesis is very complex and depends on factors including tumor type and response to NO and NO flux within the tumor microenvironment. Therefore, the choice of NOS inhibition or NO donors as effective therapeutic options must be based upon the tumor type and stage. Given the metastability of HNO, donor compounds have been developed to study its chemotherapeutic potential.

15.3.1 Inhibition of NOS

Recent clinical studies have identified NOS2 as a biomarker of prognosis, where elevated tumor NOS2 expression predicted poor outcome in ER(−) breast cancer patients (Fig. 15.2) [50, 51, 101–104]. Similarly, elevated tumor NOS2 (of ~300 nM NO) predicted poor survival in colon [105–107], lung cancers [108], as well as melanoma [109] and glioblastoma [110].

These observations generated the platform that launched the development of multiple classes of NOS2 inhibitors. Over the past few years, multiple NOS2 inhibitors have been patented and investigated [111–113]. As L-arginine is substrate of NOS, several structural analogs of arginine were developed with the idea of competitive inhibition of NOS. Well-known nonspecific NOS inhibitors that are most commonly used include L-N^{ω}-nitroarginine (L-NNA) and its methyl ester, L-N^{ω}-nitroarginine methyl ester (L-NAME). They have been extensively used *in vitro* and *in vivo* in part due to their chemical stability, water solubility, and low toxicity [114–116]. In a murine breast cancer model, NOS inhibition by L-NAME dramatically reduced the neovascular response [117]. In combination with tumor necrosis factor (TNF) and melphalan, administration of L-NAME inhibited growth of adenocarcinoma significantly in a renal subcapsular CC531 adenocarcinoma rat model [118]. Slight differences in the arginine binding domain for NOS isoforms can be exploited for developing therapeutically viable NOS-selective inhibitors and has attracted much attention [111–113]. In a non-obese diabetic/severe combined immunodeficient xenograft model, NOS2-selective small molecule antagonist *N*(6)-(1-iminoethyl)-L-lysine-dihydrochloride (L-nil) inhibited growth of human melanoma in vivo with an increased survival and was shown to synergize with cytotoxic cisplatin chemotherapy. NOS2 inhibition decreased the density of CD31+ microvessels and also enhanced number of apoptotic cells in tumor xenografts [119]. Chemopreventive properties of highly selective NOS2 inhibitors L-N(6)-(1-iminoethyl)lysine tetrazole-amide (SC-51) and aminoguanidine (AG) were assessed in a model of azoxymethane-induced colonic aberrant crypt foci (ACF), which showed dose-dependent inhibition of the incidence of colonic ACF by SC-51, while AG abated ACF at high concentrations [107]. In a separate study, AG significantly reduced tumor growth and metastases in MB-231 breast cancer xenografts [120]. Other NOS2 inhibitor including 2-cyano-3,12-dioxooleana-1,9(11)-dien-28-oic acid (CDDO) and its methyl ester (CDDO-ME), *S,S'*-1,4-phenylene-bis(1,2-ethanediyl)bis-isothiourea (PBIT), 7-nitroindazole,1-(2-trifluoromethylphenyl) imidazole (TRIM), *N*-(3-(aminomethyl)benzyl)acetamidine (1400 W) have shown promising inhibitory activity both *in vitro* and *in vivo* [121–126]. More recently, nanoparticle-based delivery vehicles have been explored in site-specific delivery of NOS inhibitors. Intravenous delivery of CDDO-Me using poly-lacticglycolic-acid nanoparticle in combination with subcutaneous tyrosinase-related protein 2 (Trp2) vaccine resulted in increased antitumor efficacy compared to Trp2 vaccine alone in B16F10 melanoma. CDDO-Me delivery resulted in remodeling of tumor-associated fibroblasts, collagen and vessel in tumor microenvironment, as well as enhanced Fas signaling, which can sensitize the tumor cells for cytotoxic T lymphocyte-mediated killing [127].

15.3.2 Donors of NO

NO donors have the potential to produce high intracellular concentrations of NO and act as tumor cytotoxic agents. Moreover, their potential as sensitizing agents to chemo-, radio-, and immunotherapies increases their utility as anticancer agents. Based on the potential pharmacological benefits that NO can offer, various classes of NO donors have been developed.

15.3.2.1 Organic Nitrates

Organic nitrates, the oldest class of NO donors, have long been used for clinical treatment of pain associated with angina [128]. Biochemical pathways leading to release of NO are not yet fully understood and several mechanisms have been postulated that can promote NO release [129–131]. Figure 15.4 shows some of the common examples of organic nitrates.

Glyceryl trinitrate (GTN) is the most studied nitrate and has been safely employed in the clinic for angina-related pain relief for more than a century, which makes GTN an attractive candidate for chemotherapeutic applications. Numerous studies have shown therapeutic potential of GTN in a variety of cancers and as an adjuvant; GTN can improve tumor response to other chemotherapeutic agents [132–136]. Reduced lung metastases of B16F10 murine melanoma cells were observed in mice treated with GTN [132]. Similarly, the therapeutic efficacy of doxorubicin was improved by GTN in prostate tumor xenografts [134]. A Phase II clinical study in

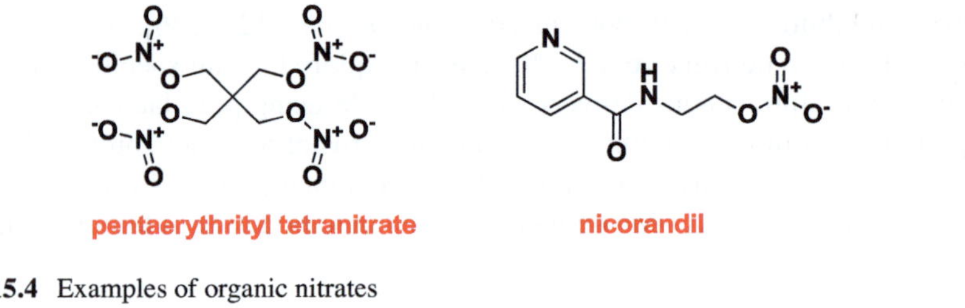

Fig. 15.4 Examples of organic nitrates

prostate cancer patients demonstrated increased time to relapse after surgery or radiotherapy, in patients that received low-dose GTN [137]. Also, small cell lung cancer patients that received GTN showed improved tumor response to chemotherapeutic drugs such as vinorelbine and cisplatin [138] or docetaxel and carboplatin regimen with no severe adverse effect [133].

Hypoxia-mediated acquisition of resistance to doxorubicin and 5-fluorouracil in human breast carcinoma and mouse melanoma cells was abated by low concentrations of GTN or diethylenetriamine-NO adduct (DETA/NO) [139]. Further studies demonstrated that hypoxic cancer cell overexpressed ADAM10, a key enzyme that limits immune surveillance [140]. However, treatment with NO donor agent, such as GTN, reduced tumor hypoxia, and abated resistance to immune system attack. The results implicate low-dose NO in boosting the body's natural immune response to cancer. Given the clinically established safety of GTN, additional clinical trials are required to establish anticancer applications of GTN.

15.3.2.2 S-Nitrosothiols

S-Nitrosothiols (RSNO) are the class of unstable organic compounds containing a nitroso group attached to thiol sulfur. They are potential vehicles for NO storage and delivery under physiological conditions [141, 142]. *S*-nitroso-*N*-acetylpenicillamine (SNAP) and *S*-nitrosoglutathione (GSNO) are two relatively stable RSNO agents (Fig. 15.5) that have been extensively utilized for studying the biological effects of RSNOs.

Decomposition of RSNO compounds yield one of the three species NO, NO^+, or NO^-, which can be triggered by light, thiol activation, heat, or metal ions [143]. It is now firmly established that a large number of proteins undergo S-nitrosylation *in vivo*, which plays an important role in various biological processes. Similar to GTN, SNAP and GSNO have demonstrated therapeutic potential in several cancer models [144–148]. For example, SNAP-induced p53-sensitive apoptosis in a Bcl-2/BAX-independent fashion in human neuroblastoma cells [145]. As a radiosensitizing agent, SNP-sensitized Fas-induced apoptotic cell death of HeLa human cervical cancers [146]. Pretreatment with GSNO suppressed doxorubicin resistance in MCF7/Dx breast cancer cells, which involved enhanced glutathionylation of histones [147]. GSNO treatment has also been shown to inhibit proliferation of several chemo-responsive and chemo-resistant ovarian cancer cell lines. Moreover, GSNO

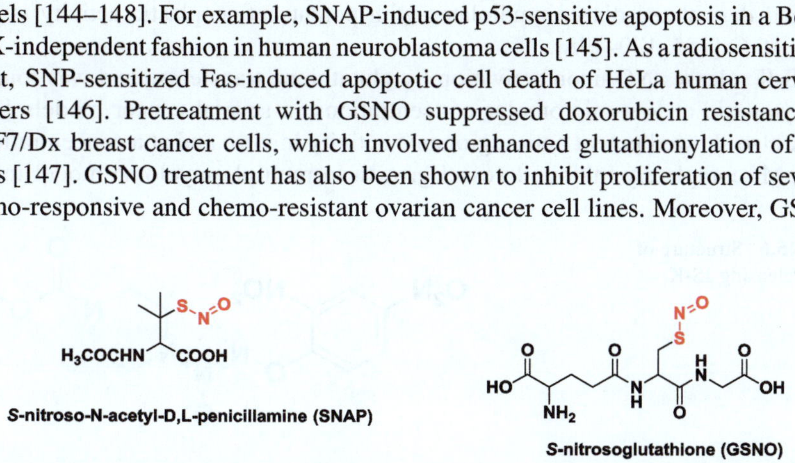

Fig. 15.5 Examples of stable *S*-nitrosothiols

not only attenuated tumor growth *in vivo*, but also potentiated cisplatin toxicity in an A2780 ovarian carcinoma nude mouse model [148]. These promising preclinical data findings warrant further investigation of *S*-nitrosothiols in cancer treatment.

15.3.2.3 Diazeniumdiolate

Diazeniumdiolates (also known as NONOates) are versatile NO donors with the basic structure: X-[N(O)NO]$^-$ where X can be C, N, O, or S-based nucleophilic center [149, 150]. These compounds dissociate spontaneously at neutral pH with half-life range from 2 s to 20 h and are widely used to study the time- and concentration-dependent biological effects of NO. Their popularity arises from the ability to produce NO over a broad timescale, and derivatization of the terminal oxygen to yield prodrugs that can be tuned to achieve site-specific NO delivery [143, 150–153]. Among the NO donors, diazeniumdiolates are the most extensively studied donors for cancer treatment. Keefer et al. pioneered the chemotherapeutic applications of diazeniumdiolates [154–158]. Several studies have shown diazeniumdiolates to enhance cytotoxicity of chemo drugs such as melphalan, doxorubicin, cisplatin, and carboplatin [159–163]. In prostate cancer cells, DETA/NO induced TRAIL-mediated apoptosis via inactivation of NF-κB and down-regulation of Bcl-xl expression [164]. While the cytotoxicity of ionic diazeniumdiolates in various cancer cells lines is low (typical $IC_{50}>1$ mM) [154, 165–167], several derivatized diazeniumdiolates have been shown to exert antitumor activity at low concentration [154–156]. DETA/NO, for instance, induced cytostasis and cell-cycle arrest in human breast cancer cells MDA-MB-231 at 1 mM concentration [168]. In contrast, esterase-sensitive acetoxymethyl protected DEA/NO (AcOM-DEA/NO) significantly decreased proliferation of leukemia cells at a concentration of 50 μM compared to >600 μM of parent donor in 24 h [154]. Also, AcOM-DEA/NO dose-dependently induced apoptosis in MB-231 breast cancer [169]. The most widely studied derivatized diazeniumdiolates with potent antineoplastic activity is JS-K (Fig. 15.6) [155, 170–178].

JS-K releases NO upon activation by glutathione transferase and was found to be cytotoxic at low micromolar concentrations in several cancer models [173]. JS-K also showed potent anti-angiogenic activity in vitro and at a concentration of 0.5 μM, completely inhibited vessel growth in multiple myeloma model in vivo

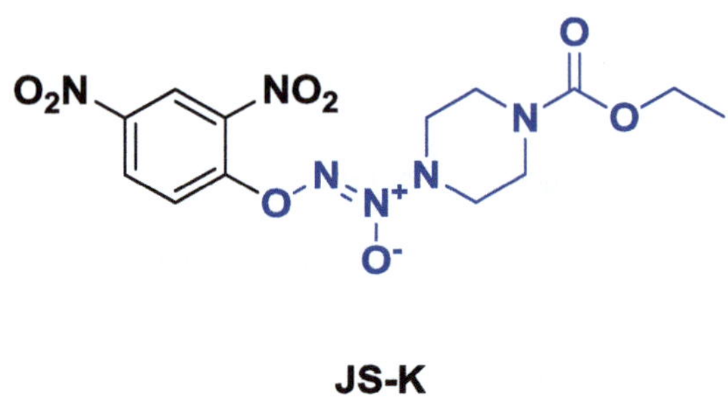

Fig. 15.6 Structure of NO-releasing JS-K

[174]. It also enhanced toxicity of existing cancer drugs such as cisplatin by increasing cellular accumulation [179]. JS-K thus represents a novel NO-based platform with therapeutic potential.

15.3.3 HNO Donors and Their Potential Usage

The role of HNO as a key pharmacological agent in the treatment of alcoholism, cardiovascular disease, and cancer has generated much interest. However, HNO chemical biology has been limited due to its irreversible dimerization, thus necessitating use of donor compounds for production of HNO in situ. Increasing interest in HNO has led to the development of various classes of HNO donors.

15.3.3.1 Diazeniumdiolate based HNO donors

Angeli's salt (AS, $Na_2N_2O_3$) was synthesized by Angeli in 1896 [180] and is one of the most commonly used HNO donors in the pH range of 4–8 with a first-order rate of decomposition of $4–5 \times 10^{-3}$ s^{-1} at 37 °C [34, 181, 182]. Dose-dependent antiproliferative effects of Angeli's salt were shown in human breast cancer cell lines [30]. While Angeli's salt has been used extensively in cancer cell models, its pharmacological utility as an anticancer agent is limited by its short half-life and millimolar concentrations required for cell killing. Moreover, being an inorganic salt, its structure cannot be modified to fine tune these properties. Toward this end, the structurally similar primary amine-based diazeniumdiolates have been developed [183] that release different ratios of NO and/or HNO depending upon the amine backbone and pH of the solution [165]. When compared to the large number of stable secondary amine-based diazeniumdiolates, only a few stable primary amine-based diazeniumdiolates have been isolated and studied including isopropylamine diazeniumdiolates (IPA/NO), cyclopentylamine diazeniumdiolates (CPA/NO), cyclohexylamine diazeniumdiolates (CHA/NO), cycloheptylamine diazeniumdiolates (CHPA/NO), and cyclooctylamine diazeniumdiolates (COA/NO), the structure of which are shown in Fig. 15.7 [183, 184]. Like Angeli's salt these donors have a short half-life in the range of 2.3–6.1 min [182, 184, 185] under physiological conditions and were shown to inhibit proliferation of MCF-7 breast cancer cells in millimolar concentrations [184].

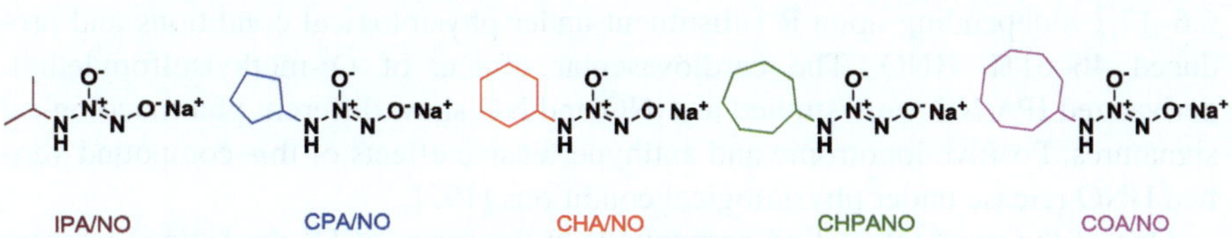

Fig. 15.7 Stable primary amine-based diazeniumdiolates

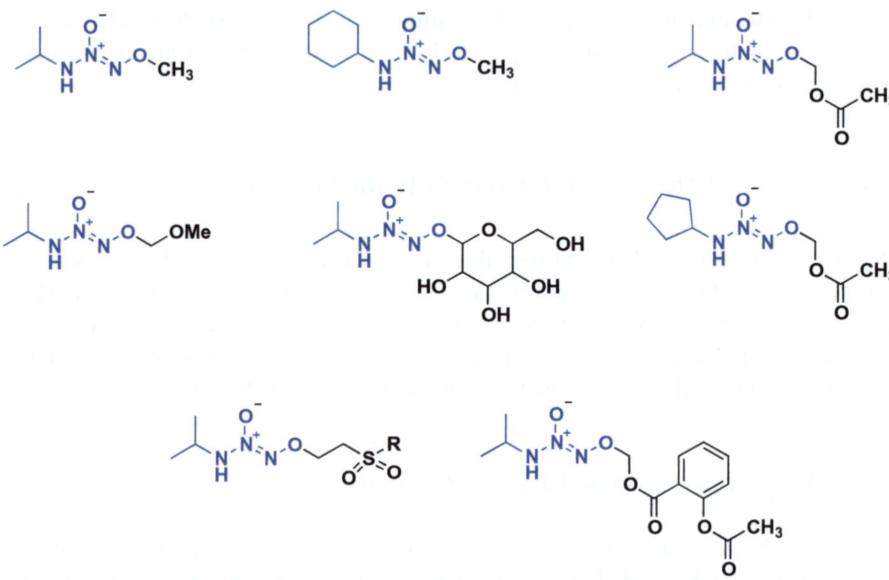

Fig. 15.8 Structures of O^2-protected primary amine diazeniumdiolates

A major challenge for the use of primary amine diazeniumdioate-based HNO donors for pharmaceutical purposes involved increasing their stability to achieve higher HNO delivery. Previously, O^2-alkylated secondary amine-based diazeniumdiolates were synthesized that exhibit increased half-lives and are activated enzymatically, photolytically or by hydrolysis [151, 154, 155, 186–189]. Saavedra et al. demonstrated that O^2-derivatization of primary amine diazeniumdiolates with alkyl, MOM, and glycosyl can also increase stability and purification (Fig. 15.8) [190].

The controlled release of NO and HNO/NO at pH 5 and pH 7.4, respectively, upon enzymatic cleavage of glycosylated IPA/NO demonstrated the ability to modulate IPA/NO release in a site-directed manner [190, 191]. However, care needs to be taken during the O^2-derivatization of primary amine diazeniumdiolates, as excess derivatizing agent can lead to the formation of N,O-derivatized product due to labile NH proton, where N is susceptible to electrophilic attack and this product cannot generate HNO [190].

Recently, Knaus et al. reported O^2-sulfonylethyl ($–CH_2CH_2SO_2R$ where R = OMe, NHOMe, NHOBn, Me) protected IPA/NO that can undergo a β-elimination reaction to release IPA/NO [192] (Scheme 15.1). These compounds exhibited half-lives of 6.6–17.1 h depending upon R substituent under physiological conditions and produced 46–61 % HNO. The cardiovascular effects of O^2-methylsulfonylethyl-derivatized IPA/NO were studied as HNO and NO show different pharmacological signatures. Positive ionotropic and antihypertensive effects of this compound verified HNO release under physiological conditions [192].

One of the most interesting compounds in the series of O^2-derivatized primary amine diazeniumdiolates was O^2-acetoxymethylated IPA/NO (Fig. 15.8) [193], which is a pure HNO donor with a half-life of 41 min at physiological temperature

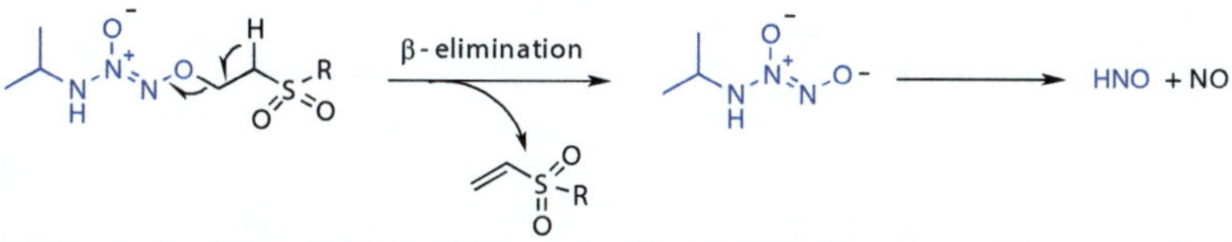

Scheme 15.1 Mechanism of decomposition of O^2-sulfonylethyl-protected IPA/NO

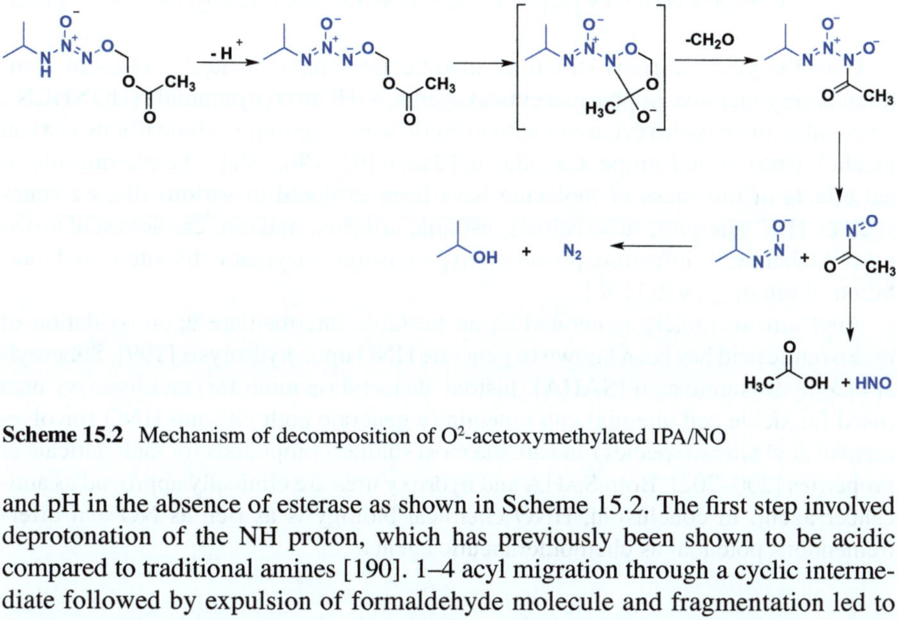

Scheme 15.2 Mechanism of decomposition of O^2-acetoxymethylated IPA/NO

and pH in the absence of esterase as shown in Scheme 15.2. The first step involved deprotonation of the NH proton, which has previously been shown to be acidic compared to traditional amines [190]. 1–4 acyl migration through a cyclic intermediate followed by expulsion of formaldehyde molecule and fragmentation led to the formation of an acylnitroso intermediate. Acylnitroso compounds are a well-studied class of HNO donors and intermediate in various HNO-releasing prodrugs. The increased half-life and the ability to release HNO without NO and nitrite production is an important step forward to investigate chemical and pharmaceutical properties of HNO.

O^2-acetoxymethylated CPA/NO (AcOM-CPA/NO) was also synthesized using a similar synthetic protocol and showed a half-life of 21 min and 96 % HNO release at pH 7.4 and 37 °C [184]. Thus, O^2-acetoxymethylation of primary amine diazeniumdiolates can be used to synthesize molecules with significantly increased half-lives [184, 193]. AcOM-CPA/NO was shown to inhibit proliferation of human breast cancer cells in micromolar concentration [184].

15.3.3.2 Hydroxamic Acid

Piloty's acid (benzenesulfohydroxamic acid, $C_6H_5SO_2NHOH$) produces HNO under basic condition (15.1) [194].

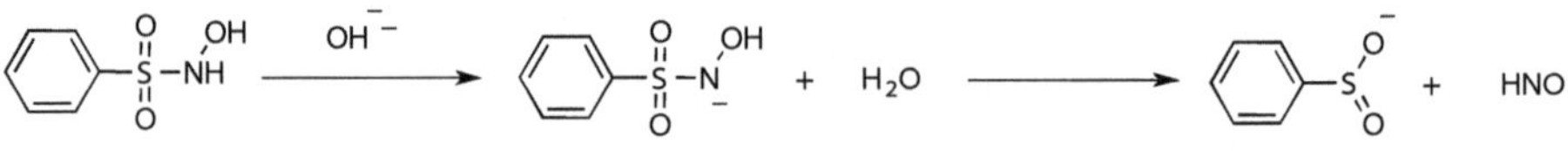

(15.1)

This limiting factor for biological use was compounded by its susceptibility to one-electron oxidation in air, which converts it to an NO donor (15.2) [195].

$$C_6H_5SO_2NHOH + [O] \rightarrow C_6H_5SO_2NHO \rightarrow C_6H_5SO_2H + NO \qquad (15.2)$$

Over the years, various structural modifications have made this class of compounds very lucrative as pharmaceutical agents. *N*-Hydroxycyanamide (HONHCN), an N-substituted hydroxylamine with cyanide leaving group is clinically used as an alcohol deterrent in Europe, Canada, and Japan [93, 196, 197]. The pharmacological effects of this class of molecule have been explored in various disease states such as HIV, allergies, tuberculosis, asthma, arthritis, malaria, cardiovascular disease, Alzheimer's, infection, psoriasis, hypertension, enzymatic disorders, and inhibition of tumor growth [198].

Acyl nitroso moiety generated as an unstable intermediate upon oxidation of hydroxamic acid has been known to generate HNO upon hydrolysis [199]. Suberoylanilide hydroxamic acid (SAHA) (histone deacetylase inhibitor) and hydroxy urea (used for sickle cell anemia) can potentially generate both NO and HNO (involvement of acyl nitroso species) and are the most studied compounds for their anticancer properties [200–202]. Both SAHA and hydroxy urea are clinically approved as anticancer agent. In conclusion, HNO chemical biology is as rich as NO and offers tremendous potential as chemotherapeutic agents.

15.3.4 NO/HNO Hybrid Drugs

The beneficial role of NO as a vasodilator, mucosal defense, and as an antineoplastic agent has led to the development of NO-hybrid drugs. This novel approach focuses on retaining the pharmacological activity of the parent compound along with synergistic activities with NO while minimizing the side effects that may be caused by the parent drug alone. NO donors have been covalently conjugated to anticancer drugs such as 5-fluorouracil (5-FU) [203], doxorubicin [204], and PARP inhibitor olaparib [157]. A 5-FU derivative was shown to have greater cytotoxicities than parent drug in human prostate and HeLa cancer cells [203], whereas doxorubicin derivative induced high toxicity in doxorubicin-resistant human colon cancer cells [204]. Olaparib derivative was based on JS-K analog thereby conjugating two active moieties into a single molecule. This derivative was shown to reduce growth rates of A549 human lung adenocarcinoma xenografts without any apparent systemic toxicity [157].

The most well-studied NO-conjugate drugs belong to the family of nonsteroidal anti-inflammatory drugs (NSAIDs) [205]. Several NO-NSAIDs have recently been developed and examined with organic nitrates [206–208] and diazeniumdiolates [209–211] being the frequent choice of NO donors. The NSAID Aspirin is extensively used worldwide to reduce pain, fever, and inflammation [212]. However, it is associated with serious side effects, including gastric ulceration and renal toxicity [213], resulting from nonspecific targeting of both COX-1 (constitutive isoform) and COX-2 (inducible isoform) [214]. NCX-4040, an NO-releasing aspirin analog was shown anti-proliferative activity against colon cancer [215]. NCX-4016, yet another NO-aspirin hybrid was shown not only to reverse sensitivity to cisplatin in recurrent human ovarian cancer cells [216] but also showed no toxicity when administered to healthy human subjects while retaining COX-1 and platelet inhibitory activity [217]. Given the tolerance issues related to NO donor nitrate esters, possibilities exist for the development of tolerance toward nitrate ester-based NO-NSAIDs [218]. Several other groups have developed other NO donor-based NO-NSAIDs [208, 211]. This concept was also used to synthesize diazeniumdiolates based HNO-releasing drug hybrid molecule IPA/NO-aspirin structure of which is shown in Fig. 15.8 [210]. IPA/NO-aspirin harnessed the beneficial effects of both HNO and aspirin. In the absence of serum, it showed a half-life of 7.5 h, which was reduced to 18 min when guinea pig serum was added. The protecting group facilitated cellular uptake as demonstrated by DAF-FM-2DA assay compared to IPA/NO, which mostly decomposes in the media followed by diffusion of HNO in the cell [210]. IPA/NO-aspirin and NO-releasing diazeniumdiolate-based NSAID analog DEA/NO-aspirin significantly inhibited the proliferation of A549 lung adenocarcinoma cells [210]. IPA/NO-aspirin also showed decreased angiogenesis, increased DNA damage, increased ROS production, and inhibition of GAPDH in various breast cancer cell lines [31]. Further investigation of the therapeutic applications of these prodrugs showed a decrease in growth and metastasis of breast cancer xenografts in mice with a high degree of cancer-specific sensitivity thereby suggesting significant pharmacological potential for these analogs [31]. IPA/NO-aspirin not only retained the anti-inflammatory properties of parent aspirin but was safer than aspirin in terms of gastrointestinal safety [210] and markedly reduced growth of breast cancer cells more effectively than IPA/NO while being nontoxic in a related non-tumorigenic cell line MCF-10A [31]. The high cytotoxicity of IPA/NO-aspirin can be explained in terms of increased ROS levels leading to DNA damage, inhibition of GAPDH, and reduced angiogenesis.

15.4 Conclusions

During the last decade, multitude of evidence suggests that elevated NOS levels are indicative of poor clinical prognosis. Role of NO in cancer have resulted in several controversies with groups claiming both pro- and anti-tumorogenic properties of NO. Only recently the importance of the level and duration of NO flux has been firmly established as a key player in explaining this dichotomy. Regardless of the dichotomy, there is a definite place for NO-based therapeutics in the field of oncology. Largely misunderstood and neglected redox sibling HNO has rapidly gained importance, and its potent chemotherapeutic potential has started to gain attention. A variety of versatile NO and HNO donors, as well as the rapid development of specific NOS inhibitors promises continued exploration and warrants further investigation as novel therapeutic agents.

Acknowledgments This work was supported in part by the Intramural Research Program of the NIH, Cancer and Inflammation Program.

References

1. Schwartz SE. Trace atmospheric constituents: properties, transformations and fates, Advances in environmental science and technology, vol. 12. New York: Wiley; 1983.
2. Bradbury J. Medicine Nobel Prize awarded to US pharmacologists. Lancet. 1998;352:1287.
3. Hibbs Jr JB, Taintor RR, Vavrin Z, Rachlin EM. Nitric oxide: a cytotoxic activated macrophage effector molecule. Biochem Biophys Res Commun. 1988;157:87–94.
4. Ignarro LJ, Buga GM, Wood KS, Byrns RE, Chaudhuri G. Endothelium-derived relaxing factor produced and released from artery and vein is nitric-oxide. Proc Natl Acad Sci U S A. 1987;84:9265–9.
5. Miwa M, Stuehr DJ, Marletta MA, Wishnok JS, Tannenbaum SR. Nitrosation of amines by stimulated macrophages. Carcinogenesis. 1987;8:955–8.
6. Ignarro LJ. Signal transduction mechanisms involving nitric oxide. Biochem Pharmacol. 1991;41:485–90.
7. Ignarro LJ. Nitric oxide-mediated vasorelaxation. Thromb Haemost. 1993;70:148–51.
8. Bredt DS, Snyder SH. Nitric-oxide, a novel neuronal messenger. Neuron. 1992;8:3–11.
9. Radomski MW, Palmer RMJ, Moncada S. Endogenous nitric-oxide inhibits human-platelet adhesion to vascular endothelium. Lancet. 1987;2:1057–8.
10. MacMicking J, Xie QW, Nathan C. Nitric oxide and macrophage function. Annu Rev Immunol. 1997;15:323–50.
11. Marletta MA. Nitric oxide synthase structure and mechanism. J Biol Chem. 1993;268: 12231–4.
12. Lamas S, Marsden PA, Li GK, Tempst P, Michel T. Endothelial nitric oxide synthase: molecular cloning and characterization of a distinct constitutive enzyme isoform. Proc Natl Acad Sci U S A. 1992;89:6348–52.
13. Xie QW, Cho HJ, Calaycay J, Mumford RA, Swiderek KM, Lee TD, Ding AH, Troso T, Nathan C. Cloning and characterization of inducible nitric oxide synthase from mouse macrophages. Science. 1992;256:225–8.
14. Hope BT, Michael GJ, Knigge KM, Vincent SR. Neuronal NADPH diaphorase is a nitric oxide synthase. Proc Natl Acad Sci U S A. 1991;88:2811–4.
15. Mocellin S, Bronte V, Nitti D. Nitric oxide, a double edged sword in cancer biology: searching for therapeutic opportunities. Med Res Rev. 2007;27:317–52.
16. Modin A, Bjorne H, Herulf M, Alving K, Weitzberg E, Lundberg JO. Nitrite-derived nitric oxide: a possible mediator of 'acidic-metabolic' vasodilation. Acta Physiol Scand. 2001;171:9–16.
17. Wink DA, Mitchell JB. Chemical biology of nitric oxide: insights into regulatory, cytotoxic, and cytoprotective mechanisms of nitric oxide. Free Radic Biol Med. 1998;25:434–56.
18. Stone JR, Marletta MA. Soluble guanylate cyclase from bovine lung: activation with nitric oxide and carbon monoxide and spectral characterization of the ferrous and ferric state. Biochemistry (Mosc). 1994;33:5636–40.

19. Moro MA, Russell RJ, Cellek S, Lizasoain I, Su YC, DarleyUsmar VM, Radomski MW, Moncada S. cGMP mediates the vascular and platelet actions of nitric oxide: confirmation using an inhibitor of the soluble guanylyl cyclase. Proc Natl Acad Sci U S A. 1996;93:1480–5.
20. Moncada S, Higgs EA. Endogenous nitric oxide: physiology, pathology and clinical relevance. Eur J Clin Invest. 1991;21:361–74.
21. Moncada S, Palmer RMJ, Higgs EA. Nitric-oxide—physiology, pathophysiology, and pharmacology. Pharmacol Rev. 1991;43:109–42.
22. Wink DA, Grisham MB, Mitchell JB, Ford PC. Direct and indirect effects of nitric oxide in chemical reactions relevant to biology. Methods Enzymol. 1996;268:12–31.
23. Thomas DD, Ridnour LA, Isenberg JS, Flores-Santana W, Switzer CH, Donzelli S, Hussain P, Vecoli C, Paolocci N, Ambs S, Colton CA, Harris CC, Roberts DD, Wink DA. The chemical biology of nitric oxide: Implications in cellular signaling. Free Radic Biol Med. 2008;45:18–31.
24. Pacher P, Beckman JS, Liaudet L. Nitric oxide and peroxynitrite in health and disease. Physiol Rev. 2007;87:315–424.
25. Paolocci N, Saavedra WF, Miranda KM, Martignani C, Isoda T, Hare JM, Espey MG, Fukuto JM, Feelisch M, Wink DA, Kass DA. Nitroxyl anion exerts redox-sensitive positive cardiac inotropy in vivo by calcitonin gene-related peptide signaling. Proc Natl Acad Sci U S A. 2001;98:10463–8.
26. Pagliaro P, Mancardi D, Rastaldo R, Penna C, Gattullo D, Miranda KM, Feelisch M, Wink DA, Kass DA, Paolocci N. Nitroxyl affords thiol-sensitive myocardial protective effects akin to early preconditioning. Free Radic Biol Med. 2003;34:33–43.
27. Miranda KM, Paolocci N, Katori T, Thomas DD, Ford E, Bartberger MD, Espey MG, Kass DA, Feelisch M, Fukuto JM, Wink DA. A biochemical rationale for the discrete behavior of nitroxyl and nitric oxide in the cardiovascular system. Proc Natl Acad Sci U S A. 2003;100:9196–201.
28. Paolocci N, Katori T, Champion HC, St. John ME, Miranda KM, Fukuto JM, Wink DA, Kass DA. Positive inotropic and lusitropic effects of HNO/NO-in failing hearts: independence from beta-adrenergic signaling. Proc Natl Acad Sci U S A. 2003;100:5537–42.
29. Nagasawa HT, DeMaster EG, Redfern B, Shirota FN, Goon DJ. Evidence for nitroxyl in the catalase-mediated bioactivation of the alcohol deterrent agent cyanamide. J Med Chem. 1990;33:3120–2.
30. Norris AJ, Sartippour MR, Lu M, Park T, Rao JY, Jackson MI, Fukuto JM, Brooks MN. Nitroxyl inhibits breast tumor growth and angiogenesis. Int J Cancer. 2008; 122:1905–10.
31. Basudhar D, Cheng RC, Bharadwaj G, Ridnour LA, Wink DA, Miranda KM. Chemotherapeutic potential of diazeniumdiolate-based aspirin prodrugs in breast cancer. Free Radic Biol Med. 2015;83:101–14.
32. Smith PAS, Hein GE. The alleged role of nitroxyl in certain reactions of aldehydes and alkyl halides. J Am Chem Soc. 1960;82:5731–40.
33. Kohout FC, Lampe FW. On role of nitroxyl molecule in reaction of hydrogen atoms with nitric oxide. J Am Chem Soc. 1965;87:5795.
34. Hughes MN, Wimbledon PE. Chemistry of trioxodinitrates. 1. Decomposition of sodium trioxodinitrate (angelis salt) in aqueous-solution. J Chem Soc Dalton 1976;(8):703–7.
35. Fukuto JM, Wallace GC, Hszieh R, Chaudhuri G. Chemical oxidation of N-hydroxyguanidine compounds. Release of nitric oxide, nitroxyl and possible relationship to the mechanism of biological nitric oxide generation. Biochem Pharmacol. 1992;43:607–13.
36. Schmidt HH, Hofmann H, Schindler U, Shutenko ZS, Cunningham DD, Feelisch M. No .NO from NO synthase. Proc Natl Acad Sci U S A. 1996;93:14492–7.
37. Adak S, Wang Q, Stuehr DJ. Arginine conversion to nitroxide by tetrahydrobiopterin-free neuronal nitric-oxide synthase. Implications for mechanism. J Biol Chem. 2000;275:33554–61.

38. Wong PS, Hyun J, Fukuto JM, Shirota FN, DeMaster EG, Shoeman DW, Nagasawa HT. Reaction between S-nitrosothiols and thiols: generation of nitroxyl (HNO) and subsequent chemistry. Biochemistry. 1998;37:5362–71.
39. Niketic V, Stojanovic S, Nikolic A, Spasic M, Michelson AM. Exposure of Mn and FeSODs, but not Cu/ZnSOD, to NO leads to nitrosonium and nitroxyl ions generation which cause enzyme modification and inactivation: an in vitro study. Free Radic Biol Med. 1999;27:992–6.
40. Saleem M, Ohshima H. Xanthine oxidase converts nitric oxide to nitroxyl that inactivates the enzyme. Biochem Biophys Res Commun. 2004;315:455–62.
41. Sharpe MA, Cooper CE. Reactions of nitric oxide with mitochondrial cytochrome c: a novel mechanism for the formation of nitroxyl anion and peroxynitrite. Biochem J. 1998; 332:9–19.
42. Lopez BE, Rodriguez CE, Pribadi M, Cook NM, Shinyashiki M, Fukuto JM. Inhibition of yeast glycolysis by nitroxyl (HNO): a mechanism of HNO toxicity and implications to HNO biology. Arch Biochem Biophys. 2005;442:140–8.
43. Wink DA, Feelisch M, Fukuto J, Chistodoulou D, Jourd'heuil D, Grisham MB, Vodovotz Y, Cook JA, Krishna M, DeGraff WG, Kim S, Gamson J, Mitchell JB. The cytotoxicity of nitroxyl: possible implications for the pathophysiological role of NO. Arch Biochem Biophys. 1998;351:66–74.
44. http://www.breastcancer.org/symptoms/understand_bc/statistics
45. Obermajer N, Wong JL, Edwards RP, Odunsi K, Moysich K, Kalinski P. PGE(2)-driven induction and maintenance of cancer-associated myeloid-derived suppressor cells. Immunol Invest. 2012;41:635–57.
46. Dominguez PM, Ardavin C. Differentiation and function of mouse monocyte-derived dendritic cells in steady state and inflammation. Immunol Rev. 2010;234:90–104.
47. Ikonomidis I, Michalakeas CA, Parissis J, Paraskevaidis I, Ntai K, Papadakis I, Anastasiou-Nana M, Lekakis J. Inflammatory markers in coronary artery disease. Biofactors. 2012;38:320–8.
48. Takahashi M, Mutoh M, Ishigamori R, Fujii G, Imai T. Involvement of inflammatory factors in pancreatic carcinogenesis and preventive effects of anti-inflammatory agents. Semin Immunopathol. 2013;35:203–27.
49. Sugiura H, Ichinose M. Nitrative stress in inflammatory lung diseases. Nitric Oxide. 2011;25:138–44.
50. Glynn SA, Boersma BJ, Dorsey TH, Yi M, Yfantis HG, Ridnour LA, Martin DN, Switzer CH, Hudson RS, Wink DA, Lee DH, Stephens RM, Ambs S. Increased NOS2 predicts poor survival in estrogen receptor-negative breast cancer patients. J Clin Invest. 2010;120:3843–54.
51. Loibl S, Buck A, Strank C, von Minckwitz G, Roller M, Sinn HP, Schini-Kerth V, Solbach C, Strebhardt K, Kaufmann M. The role of early expression of inducible nitric oxide synthase in human breast cancer. Eur J Cancer. 2005;41:265–71.
52. Ekmekcioglu S, Ellerhorst J, Smid CM, Prieto VG, Munsell M, Buzaid AC, Grimm EA. Inducible nitric oxide synthase and nitrotyrosine in human metastatic melanoma tumors correlate with poor survival. Clin Cancer Res. 2000;6:4768–75.
53. Raspollini MR, Amunni G, Villanucci A, Boddi V, Baroni G, Taddei A, Taddei GL. Expression of inducible nitric oxide synthase and cyclooxygenase-2 in ovarian cancer: correlation with clinical outcome. Gynecol Oncol. 2004;92:806–12.
54. Connelly ST, Macabeo-Ong M, Dekker N, Jordan RC, Schmidt BL. Increased nitric oxide levels and iNOS over-expression in oral squamous cell carcinoma. Oral Oncol. 2005;41:261–7.
55. Brennan PA, Dennis S, Poller D, Quintero M, Puxeddu R, Thomas GJ. Inducible nitric oxide synthase: correlation with extracapsular spread and enhancement of tumor cell invasion in head and neck squamous cell carcinoma. Head Neck. 2008;30:208–14.

56. Cronauer MV, Ince Y, Engers R, Rinnab L, Weidemann W, Suschek CV, Burchardt M, Kleinert H, Wiedenmann J, Sies H, Ackermann R, Kroncke KD. Nitric oxide-mediated inhibition of androgen receptor activity: possible implications for prostate cancer progression. Oncogene. 2007;26:1875–84.
57. Loibl S, von Minckwitz G, Weber S, Sinn HP, Schini-Kerth VB, Lobysheva I, Nepveu F, Wolf G, Strebhardt K, Kaufmann M. Expression of endothelial and inducible nitric oxide synthase in benign and malignant lesions of the breast and measurement of nitric oxide using electron paramagnetic resonance spectroscopy. Cancer. 2002;95:1191–8.
58. Cobbs CS, Brenman JE, Aldape KD, Bredt DS, Israel MA. Expression of nitric oxide synthase in human central nervous system tumors. Cancer Res. 1995;55:727–30.
59. Singh S, Gupta AK. Nitric oxide: role in tumour biology and iNOS/NO-based anticancer therapies. Cancer Chemother Pharmacol. 2011;67:1211–24.
60. Hickok JR, Thomas DD. Nitric oxide and cancer therapy: the emperor has NO clothes. Curr Pharm Des. 2010;16:381–91.
61. Wink DA, Ridnour LA, Hussain SP, Harris CC. The reemergence of nitric oxide and cancer. Nitric Oxide. 2008;19:65–7.
62. Hibbs Jr JB, Taintor RR, Vavrin Z. Macrophage cytotoxicity: role for L-arginine deiminase and imino nitrogen oxidation to nitrite. Science. 1987;235:473–6.
63. Ferry-Dumazet H, Mamani-Matsuda M, Dupouy M, Belloc F, Thiolat D, Marit G, Arock M, Reiffers J, Mossalayi MD. Nitric oxide induces the apoptosis of human BCR-ABL-positive myeloid leukemia cells: evidence for the chelation of intracellular iron. Leukemia. 2002;16:708–15.
64. Ford PC. Reactions of NO and nitrite with heme models and proteins. Inorg Chem. 2010;49:6226–39.
65. Flores-Santana W, Switzer C, Ridnour LA, Basudhar D, Mancardi D, Donzelli S, Thomas DD, Miranda KM, Fukuto JM, Wink DA. Comparing the chemical biology of NO and HNO. Arch Pharm Res. 2009;32:1139–53.
66. Marshall HE, Merchant K, Stamler JS. Nitrosation and oxidation in the regulation of gene expression. FASEB J. 2000;14:1889–900.
67. Anand P, Stamler JS. Enzymatic mechanisms regulating protein S-nitrosylation: implications in health and disease. J Mol Med. 2012;90:233–44.
68. Laval F, Wink DA, Laval J. A discussion of mechanisms of NO genotoxicity: implication of inhibition of DNA repair proteins. Rev Physiol Biochem Pharmacol. 1997;131:175–91.
69. Graziewicz M, Wink DA, Laval F. Nitric oxide inhibits DNA ligase activity: potential mechanisms for NO-mediated DNA damage. Carcinogenesis. 1996;17:2501–5.
70. Hickok JR, Vasudevan D, Jablonski K, Thomas DD. Oxygen dependence of nitric oxide-mediated signaling. Redox Biol. 2013;1:203–9.
71. Thomas DD, Liu X, Kantrow SP, Lancaster Jr JR. The biological lifetime of nitric oxide: implications for the perivascular dynamics of NO and O2. Proc Natl Acad Sci U S A. 2001;98:355–60.
72. Thomas DD. Breathing new life into nitric oxide signaling: a brief overview of the interplay between oxygen and nitric oxide. Redox Biol. 2015;5:225–33.
73. Lunt SJ, Chaudary N, Hill RP. The tumor microenvironment and metastatic disease. Clin Exp Metastasis. 2009;26:19–34.
74. Cosse JP, Michiels C. Tumour hypoxia affects the responsiveness of cancer cells to chemotherapy and promotes cancer progression. Anticancer Agents Med Chem. 2008;8:790–7.
75. Li F, Sonveaux P, Rabbani ZN, Liu S, Yan B, Huang Q, Vujaskovic Z, Dewhirst MW, Li CY. Regulation of HIF-1alpha stability through S-nitrosylation. Mol Cell. 2007;26:63–74.
76. Thomas DD, Espey MG, Ridnour LA, Hofseth LJ, Mancardi D, Harris CC, Wink DA. Hypoxic inducible factor 1 alpha, extracellular signal-regulated kinase, and p53 are regulated by distinct threshold concentrations of nitric oxide. Proc Natl Acad Sci U S A. 2004;101:8894–9.
77. Jadeski LC, Hum KO, Chakraborty C, Lala PK. Nitric oxide promotes murine mammary tumour growth and metastasis by stimulating tumour cell migration, invasiveness and angiogenesis. Int J Cancer. 2000;86:30–9.

78. Morbidelli L, Donnini S, Ziche M. Role of nitric oxide in tumor angiogenesis. Cancer Treat Res. 2004;117:155–67.
79. Bing RJ, Miyataka M, Rich KA, Hanson N, Wang X, Slosser HD, Shi SR. Nitric oxide, prostanoids, cyclooxygenase, and angiogenesis in colon and breast cancer. Clin Cancer Res. 2001;7:3385–92.
80. Lim KH, Ancrile BB, Kashatus DF, Counter CM. Tumour maintenance is mediated by eNOS. Nature. 2008;452:646–9.
81. Ridnour LA, Isenberg JS, Espey MG, Thomas DD, Roberts DD, Wink DA. Nitric oxide regulates angiogenesis through a functional switch involving thrombospondin-1. Proc Natl Acad Sci U S A. 2005;102:13147–52.
82. Isenberg JS, Martin-Manso G, Maxhimer JB, Roberts DD. Regulation of nitric oxide signalling by thrombospondin 1: implications for anti-angiogenic therapies. Nat Rev Cancer. 2009;9:182–94.
83. Loffek S, Schilling O, Franzke CW. Series "matrix metalloproteinases in lung health and disease": biological role of matrix metalloproteinases: a critical balance. Eur Respir J. 2011;38:191–208.
84. Tallant C, Marrero A, Gomis-Ruth FX. Matrix metalloproteinases: fold and function of their catalytic domains. Biochim Biophys Acta. 2010;1:20–8.
85. Itoh T, Tanioka M, Yoshida H, Yoshioka T, Nishimoto H, Itohara S. Reduced angiogenesis and tumor progression in gelatinase A-deficient mice. Cancer Res. 1998;58:1048–51.
86. Itoh T, Tanioka M, Matsuda H, Nishimoto H, Yoshioka T, Suzuki R, Uehira M. Experimental metastasis is suppressed in MMP-9-deficient mice. Clin Exp Metastasis. 1999;17:177–81.
87. Gu Z, Kaul M, Yan B, Kridel SJ, Cui J, Strongin A, Smith JW, Liddington RC, Lipton SA. S-nitrosylation of matrix metalloproteinases: signaling pathway to neuronal cell death. Science. 2002;297:1186–90.
88. Ridnour LA, Windhausen AN, Isenberg JS, Yeung N, Thomas DD, Vitek MP, Roberts DD, Wink DA. Nitric oxide regulates matrix metalloproteinase-9 activity by guanylyl-cyclase-dependent and -independent pathways. Proc Natl Acad Sci U S A. 2007;104:16898–903.
89. Patruno A, Pesce M, Marrone A, Speranza L, Grilli A, De Lutiis MA, Felaco M, Reale M. Activity of matrix metallo proteinases (MMPs) and the tissue inhibitor of MMP (TIMP)-1 in electromagnetic field-exposed THP-1 cells. J Cell Physiol. 2012;227:2767–74.
90. Ridnour LA, Barasch KM, Windhausen AN, Dorsey TH, Lizardo MM, Yfantis HG, Lee DH, Switzer CH, Cheng RY, Heinecke JL, Brueggemann E, Hines HB, Khanna C, Glynn SA, Ambs S, Wink DA. Nitric oxide synthase and breast cancer: role of TIMP-1 in NO-mediated Akt activation. PLoS One. 2012;7:e44081.
91. Klimp AH, de Vries EG, Scherphof GL, Daemen T. A potential role of macrophage activation in the treatment of cancer. Crit Rev Oncol Hematol. 2002;44:143–61.
92. Hofseth LJ, Saito S, Hussain SP, Espey MG, Miranda KM, Araki Y, Jhappan C, Higashimoto Y, He P, Linke SP, Quezado MM, Zurer I, Rotter V, Wink DA, Appella E, Harris CC. Nitric oxide-induced cellular stress and p53 activation in chronic inflammation. Proc Natl Acad Sci U S A. 2003;100:143–8.
93. DeMaster EG, Redfern B, Nagasawa HT. Mechanisms of inhibition of aldehyde dehydrogenase by nitroxyl, the active metabolite of the alcohol deterrent agent cyanamide. Biochem Pharmacol. 1998;55:2007–15.
94. Miranda KM, Yamada K, Espey MG, Thomas DD, DeGraff W, Mitchell JB, Krishna MC, Colton CA, Wink DA. Further evidence for distinct reactive intermediates from nitroxyl and peroxynitrite: effects of buffer composition on the chemistry of Angeli's salt and synthetic peroxynitrite. Arch Biochem Biophys. 2002;401:134–44.
95. Chazotte-Aubert L, Oikawa S, Gilibert I, Bianchini F, Kawanishi S, Ohshima H. Cytotoxicity and site-specific DNA damage induced by nitroxyl anion (NO^-) in the presence of hydrogen peroxide—implications for various pathophysiological conditions. J Biol Chem. 1999;274:20909–15.
96. Ivanova J, Salama G, Clancy RM, Schor NF, Nylander KD, Stoyanovsky DA. Formation of nitroxyl and hydroxyl radical in solutions of sodium trioxodinitrate: effects of pH and cytotoxicity. J Biol Chem. 2003;278:42761–8.

97. Dastoor Z, Dreyer JL. Potential role of nuclear translocation of glyceraldehyde-3-phosphate dehydrogenase in apoptosis and oxidative stress. J Cell Sci. 2001;114:1643–53.
98. Gatenby RA, Gillies RJ. Why do cancers have high aerobic glycolysis? Nat Rev Cancer. 2004;4:891–9.
99. Gatenby RA, Gillies RJ. Glycolysis in cancer: a potential target for therapy. Int J Biochem Cell Biol. 2007;39:1358–66.
100. Sidorkina O, Espey MG, Miranda KM, Wink DA, Laval J. Inhibition of poly(ADP-ribose) polymerase (PARP) by nitric oxide and reactive nitrogen oxide species. Free Radic Biol Med. 2003;35:1431–8.
101. Bulut AS, Erden E, Sak SD, Doruk H, Kursun N, Dincol D. Significance of inducible nitric oxide synthase expression in benign and malignant breast epithelium: an immunohistochemical study of 151 cases. Virchows Arch. 2005;447:24–30.
102. Vakkala M, Kahlos K, Lakari E, Paakko P, Kinnula V, Soini Y. Inducible nitric oxide synthase expression, apoptosis, and angiogenesis in in situ and invasive breast carcinomas. Clin Cancer Res. 2000;6:2408–16.
103. Tschugguel W, Schneeberger C, Unfried G, Czerwenka K, Weninger W, Mildner M, Gruber DM, Sator MO, Waldhor T, Huber JC. Expression of inducible nitric oxide synthase in human breast cancer depends on tumor grade. Breast Cancer Res Treat. 1999;56:145–51.
104. Thomsen LL, Miles DW, Happerfield L, Bobrow LG, Knowles RG, Moncada S. Nitric oxide synthase activity in human breast cancer. Br J Cancer. 1995;72:41–4.
105. Ambs S, Merriam WG, Bennett WP, Felley-Bosco E, Ogunfusika MO, Oser SM, Klein S, Shields PG, Billiar TR, Harris CC. Frequent nitric oxide synthase-2 expression in human colon adenomas: implication for tumor angiogenesis and colon cancer progression. Cancer Res. 1998;58:334–41.
106. Rao CV, Kawamori T, Hamid R, Reddy BS. Chemoprevention of colonic aberrant crypt foci by an inducible nitric oxide synthase-selective inhibitor. Carcinogenesis. 1999;20:641–4.
107. Rao CV, Indranie C, Simi B, Manning PT, Connor JR, Reddy BS. Chemopreventive properties of a selective inducible nitric oxide synthase inhibitor in colon carcinogenesis, administered alone or in combination with celecoxib, a selective cyclooxygenase-2 inhibitor. Cancer Res. 2002;62:165–70.
108. Okayama H, Saito M, Oue N, Weiss JM, Stauffer J, Takenoshita S, Wiltrout RH, Hussain SP, Harris CC. NOS2 enhances KRAS-induced lung carcinogenesis, inflammation and microRNA-21 expression. Int J Cancer. 2013;132:9–18.
109. Massi D, Franchi A, Sardi I, Magnelli L, Paglierani M, Borgognoni L, Maria Reali U, Santucci M. Inducible nitric oxide synthase expression in benign and malignant cutaneous melanocytic lesions. J Pathol. 2001;194:194–200.
110. Eyler CE, Wu Q, Yan K, MacSwords JM, Chandler-Militello D, Misuraca KL, Lathia JD, Forrester MT, Lee J, Stamler JS, Goldman SA, Bredel M, McLendon RE, Sloan AE, Hjelmeland AB, Rich JN. Glioma stem cell proliferation and tumor growth are promoted by nitric oxide synthase-2. Cell. 2011;146:53–66.
111. Maddaford S, Annedi SC, Ramnauth J, Rakhit S. Advancements in the development of nitric oxide synthase inhibitors (Chapter 2). In: John EM, editor. Annu. Rep. Med. Chem. New York: Academic; 2009. p. 27–50.
112. Tafi A, Angeli L, Venturini G, Travagli M, Corelli F, Botta M. Computational studies of competitive inhibitors of nitric oxide synthase (NOS) enzymes: towards the development of powerful and isoform-selective inhibitors. Curr Med Chem. 2006;13:1929–46.
113. Salerno L, Sorrenti V, Di Giacomo C, Romeo G, Siracusa MA. Progress in the development of selective nitric oxide synthase (NOS) inhibitors. Curr Pharm Des. 2002;8:177–200.
114. Shang ZJ, Li ZB, Li JR. In vitro effects of nitric oxide synthase inhibitor L-NAME on oral squamous cell carcinoma: a preliminary study. Int J Oral Maxillofac Surg. 2006;35:539–43.
115. Zhang Y, Bissing JW, Xu L, Ryan AJ, Martin SM, Miller Jr FJ, Kregel KC, Buettner GR, Kerber RE. Nitric oxide synthase inhibitors decrease coronary sinus-free radical concentration and ameliorate myocardial stunning in an ischemia-reperfusion model. J Am Coll Cardiol. 2001;38:546–54.

116. Boer R, Ulrich WR, Klein T, Mirau B, Haas S, Baur I. The inhibitory potency and selectivity of arginine substrate site nitric-oxide synthase inhibitors is solely determined by their affinity toward the different isoenzymes. Mol Pharmacol. 2000;58:1026–34.
117. Jadeski LC, Lala PK. Nitric oxide synthase inhibition by N(G)-nitro-L-arginine methyl ester inhibits tumor-induced angiogenesis in mammary tumors. Am J Pathol. 1999;155:1381–90.
118. de Wilt JH, Manusama ER, van Etten B, van Tiel ST, Jorna AS, Seynhaeve AL, ten Hagen TL, Eggermont AM. Nitric oxide synthase inhibition results in synergistic anti-tumour activity with melphalan and tumour necrosis factor alpha-based isolated limb perfusions. Br J Cancer. 2000;83:1176–82.
119. Sikora AG, Gelbard A, Davies MA, Sano D, Ekmekcioglu S, Kwon J, Hailemichael Y, Jayaraman P, Myers JN, Grimm EA, Overwijk WW. Targeted inhibition of inducible nitric oxide synthase inhibits growth of human melanoma in vivo and synergizes with chemotherapy. Clin Cancer Res. 2010;16:1834–44.
120. Heinecke JL, Ridnour LA, Cheng RY, Switzer CH, Lizardo MM, Khanna C, Glynn SA, Hussain SP, Young HA, Ambs S, Wink DA. Tumor microenvironment-based feed-forward regulation of NOS2 in breast cancer progression. Proc Natl Acad Sci U S A. 2014;111(17):6323–8.
121. Liby K, Royce DB, Williams CR, Risingsong R, Yore MM, Honda T, Gribble GW, Dmitrovsky E, Sporn TA, Sporn MB. The synthetic triterpenoids CDDO-methyl ester and CDDO-ethyl amide prevent lung cancer induced by vinyl carbamate in A/J mice. Cancer Res. 2007;67:2414–9.
122. Chen T, Nines RG, Peschke SM, Kresty LA, Stoner GD. Chemopreventive effects of a selective nitric oxide synthase inhibitor on carcinogen-induced rat esophageal tumorigenesis. Cancer Res. 2004;64:3714–7.
123. Janakiram NB, Rao CV. iNOS-selective inhibitors for cancer prevention: promise and progress. Future Med Chem. 2012;4:2193–204.
124. Southan GJ, Szabo C. Selective pharmacological inhibition of distinct nitric oxide synthase isoforms. Biochem Pharmacol. 1996;51:383–94.
125. Volke V, Wegener G, Bourin M, Vasar E. Antidepressant- and anxiolytic-like effects of selective neuronal NOS inhibitor 1-(2-trifluoromethylphenyl)-imidazole in mice. Behav Brain Res. 2003;140:141–7.
126. Garvey EP, Oplinger JA, Furfine ES, Kiff RJ, Laszlo F, Whittle BJ, Knowles RG. 1400W is a slow, tight binding, and highly selective inhibitor of inducible nitric-oxide synthase in vitro and in vivo. J Biol Chem. 1997;272:4959–63.
127. Zhao Y, Huo M, Xu Z, Wang Y, Huang L. Nanoparticle delivery of CDDO-Me remodels the tumor microenvironment and enhances vaccine therapy for melanoma. Biomaterials. 2015;68:54–66.
128. Parker JO. Nitrate therapy in stable angina pectoris. N Engl J Med. 1987;316:1635–42.
129. Torfgard KE, Ahlner J. Mechanisms of action of nitrates. Cardiovasc Drugs Ther. 1994;8:701–17.
130. Horowitz JD, Antman EM, Lorell BH, Barry WH, Smith TW. Potentiation of the cardiovascular effects of nitroglycerin by N-acetylcysteine. Circulation. 1983;68:1247–53.
131. Thatcher RGJ, Weldon H. NO problem for nitroglycerin: organic nitrate chemistry and therapy. Chem Soc Rev. 1998;27:331–7.
132. Postovit LM, Adams MA, Lash GE, Heaton JP, Graham CH. Nitric oxide-mediated regulation of hypoxia-induced B16F10 melanoma metastasis. Int J Cancer. 2004;108:47–53.
133. Yasuda H, Nakayama K, Watanabe M, Suzuki S, Fuji H, Okinaga S, Kanda A, Zayasu K, Sasaki T, Asada M, Suzuki T, Yoshida M, Yamanda S, Inoue D, Kaneta T, Kondo T, Takai Y, Sasaki H, Yanagihara K, Yamaya M. Nitroglycerin treatment may enhance chemosensitivity to docetaxel and carboplatin in patients with lung adenocarcinoma. Clin Cancer Res. 2006;12:6748–57.
134. Frederiksen LJ, Sullivan R, Maxwell LR, Macdonald-Goodfellow SK, Adams MA, Bennett BM, Siemens DR, Graham CH. Chemosensitization of cancer in vitro and in vivo by nitric oxide signaling. Clin Cancer Res. 2007;13:2199–206.

135. Frederiksen LJ, Siemens DR, Heaton JP, Maxwell LR, Adams MA, Graham CH. Hypoxia induced resistance to doxorubicin in prostate cancer cells is inhibited by low concentrations of glyceryl trinitrate. J Urol. 2003;170:1003–7.
136. Millet A, Bettaieb A, Renaud F, Prevotat L, Hammann A, Solary E, Mignotte B, Jeannin JF. Influence of the nitric oxide donor glyceryl trinitrate on apoptotic pathways in human colon cancer cells. Gastroenterology. 2002;123:235–46.
137. Siemens DR, Heaton JP, Adams MA, Kawakami J, Graham CH. Phase II study of nitric oxide donor for men with increasing prostate-specific antigen level after surgery or radiotherapy for prostate cancer. Urology. 2009;74:878–83.
138. Yasuda H, Yamaya M, Nakayama K, Sasaki T, Ebihara S, Kanda A, Asada M, Inoue D, Suzuki T, Okazaki T, Takahashi H, Yoshida M, Kaneta T, Ishizawa K, Yamanda S, Tomita N, Yamasaki M, Kikuchi A, Kubo H, Sasaki H. Randomized phase II trial comparing nitroglycerin plus vinorelbine and cisplatin with vinorelbine and cisplatin alone in previously untreated stage IIIB/IV non-small-cell lung cancer. J Clin Oncol. 2006;24:688–94.
139. Matthews NE, Adams MA, Maxwell LR, Gofton TE, Graham CH. Nitric oxide-mediated regulation of chemosensitivity in cancer cells. J Natl Cancer Inst. 2001;93:1879–85.
140. Barsoum IB, Hamilton TK, Li X, Cotechini T, Miles EA, Siemens DR, Graham CH. Hypoxia induces escape from innate immunity in cancer cells via increased expression of ADAM10: role of nitric oxide. Cancer Res. 2011;71:7433–41.
141. Ng ESM, Cheng ZJ, Ellis A, Ding H, Jiang YF, Li Y, Hollenberg MD, Triggle CR. Nitrosothiol stores in vascular tissue: modulation by ultraviolet light, acetylcholine and ionomycin. Eur J Pharmacol. 2007;560:183–92.
142. Huerta S, Chilka S, Bonavida B. Nitric oxide donors: novel cancer therapeutics (review). Int J Oncol. 2008;33:909–27.
143. Wang PG, Xian M, Tang XP, Wu XJ, Wen Z, Cai TW, Janczuk AJ. Nitric oxide donors: chemical activities and biological applications. Chem Rev. 2002;102:1091–134.
144. Kamoshima W, Kitamura Y, Nomura Y, Taniguchi T. Possible involvement of ADP-ribosylation of particular enzymes in cell death induced by nitric oxide-donors in human neuroblastoma cells. Neurochem Int. 1997;30:305–11.
145. Kitamura Y, Kamoshima W, Shimohama S, Nomura Y, Taniguchi T. Nitric oxide donor-induced p53-sensitive cell death is enhanced by Bcl-2 reduction in human neuroblastoma cells. Neurochem Int. 1998;32:93–102.
146. Park IC, Woo SH, Park MJ, Lee HC, Lee SJ, Hong YJ, Lee SH, Hong SI, Rhee CH. Ionizing radiation and nitric oxide donor sensitize Fas-induced apoptosis via up-regulation of Fas in human cervical cancer cells. Oncol Rep. 2003;10:629–33.
147. de Luca A, Moroni N, Serafino A, Primavera A, Pastore A, Pedersen JZ, Petruzzelli R, Farrace MG, Pierimarchi P, Moroni G, Federici G, Sinibaldi Vallebona P, Lo Bello M. Treatment of doxorubicin-resistant MCF7/Dx cells with nitric oxide causes histone glutathionylation and reversal of drug resistance. Biochem J. 2011;440:175–83.
148. Giri S, Rattan R, Deshpande M, Maguire JL, Johnson Z, Graham RP, Shridhar V. Preclinical therapeutic potential of a nitrosylating agent in the treatment of ovarian cancer. PLoS One. 2014;9(6):e97897.
149. Keefer LK. Fifty years of diazeniumdiolate research. From laboratory curiosity to broad-spectrum biomedical advances. ACS Chem Biol. 2011;6:1147–55.
150. Hrabie JA, Keefer LK. Chemistry of the nitric oxide-releasing diazeniumdiolate ("nitrosohydroxylamine") functional group and its oxygen-substituted derivatives. Chem Rev. 2002;102:1135–54.
151. Davies KM, Wink DA, Saavedra JE, Keefer LK. Chemistry of the diazeniumdiolates. 2. Kinetics and mechanism of dissociation to nitric oxide in aqueous solution. J Am Chem Soc. 2001;123:5473–81.
152. Keefer LK, Flippen-Anderson JL, George C, Shanklin AP, Dunams TA, Christodoulou D, Saavedra JE, Sagan ES, Bohle DS. Chemistry of the diazeniumdiolates. 1. Structural and spectral characteristics of the [N(O)NO]-functional group. Nitric Oxide. 2001;5:377–94.

153. Srinivasan A, Kebede N, Saavedra JE, Nikolaitchik AV, Brady DA, Yourd E, Davies KM, Keefer LK, Toscano JP. Chemistry of the diazeniumdiolates. 3. Photoreactivity. J Am Chem Soc. 2001;123:5465–72.
154. Saavedra JE, Shami PJ, Wang LY, Davies KM, Booth MN, Citro ML, Keefer LK. Esterase-sensitive nitric oxide donors of the diazeniumdiolate family: in vitro antileukemic activity. J Med Chem. 2000;43:261–9.
155. Shami PJ, Saavedra JE, Wang LY, Bonifant CL, Diwan BA, Singh SV, Gu YJ, Fox SD, Buzard GS, Citro ML, Waterhouse DJ, Davies KM, Ji XH, Keefer LK. JS-K, a glutathione/glutathione S-transferase-activated nitric oxide donor of the diazeniumdiolate class with potent antineoplastic activity. Mol Cancer Ther. 2003;2:409–17.
156. Saavedra JE, Srinivasan A, Buzard GS, Davies KM, Waterhouse DJ, Inami K, Wilde TC, Citro ML, Cuellar M, Deschamps JR, Parrish D, Shami PJ, Findlay VJ, Townsend DM, Tew KD, Singh S, Jia L, Ji XH, Keefer LK. PABA/NO as an anticancer lead: analogue synthesis, structure revision, solution chemistry, reactivity toward glutathione, and in vitro activity. J Med Chem. 2006;49:1157–64.
157. Maciag AE, Holland RJ, Kim Y, Kumari V, Luthers CE, Sehareen WS, Biswas D, Morris NL, Ji X, Anderson LM, Saavedra JE, Keefer LK. Nitric oxide (NO) releasing poly ADP-ribose polymerase 1 (PARP-1) inhibitors targeted to glutathione S-transferase P1-overexpressing cancer cells. J Med Chem. 2014;57:2292–302.
158. Kiziltepe T, Hideshima T, Ishitsuka K, Ocio EM, Raje N, Catley L, Li CQ, Trudel LJ, Yasui H, Vallet S, Kutok JL, Chauhan D, Mitsiades CS, Saavedra JE, Wogan GN, Keefer LK, Shami PJ, Anderson KC. JS-K, a GST-activated nitric oxide generator, induces DNA double-strand breaks, activates DNA damage response pathways, and induces apoptosis in vitro and in vivo in human multiple myeloma cells. Blood. 2007;110:709–18.
159. Weyerbrock A, Walbridge S, Pluta RM, Saavedra JE, Keefer LK, Oldfield EH. Selective opening of the blood-tumor barrier by a nitric oxide donor and long-term survival in rats with C6 gliomas. J Neurosurg. 2003;99:728–37.
160. Cook JA, Krishna MC, Pacelli R, DeGraff W, Liebmann J, Mitchell JB, Russo A, Wink DA. Nitric oxide enhancement of melphalan-induced cytotoxicity. Br J Cancer. 1997;76:325–34.
161. Muir CP, Adams MA, Graham CH. Nitric oxide attenuates resistance to doxorubicin in three-dimensional aggregates of human breast carcinoma cells. Breast Cancer Res Treat. 2006;96:169–76.
162. Deng L, Zhang E, Chen C. Synergistic interaction of beta-galactosyl-pyrrolidinyl diazeniumdiolate with cisplatin against three tumor cells. Arch Pharm Res. 2013;36:619–25.
163. Wink DA, Cook JA, Christodoulou D, Krishna MC, Pacelli R, Kim S, DeGraff W, Gamson J, Vodovotz Y, Russo A, Mitchell JB. Nitric oxide and some nitric oxide donor compounds enhance the cytotoxicity of cisplatin. Nitric Oxide. 1997;1:88–94.
164. Huerta-Yepez S, Vega M, Jazirehi A, Garban H, Hongo F, Cheng G, Bonavida B. Nitric oxide sensitizes prostate carcinoma cell lines to TRAIL-mediated apoptosis via inactivation of NF-kappa B and inhibition of Bcl-xl expression. Oncogene. 2004;23:4993–5003.
165. Miranda KM, Katori T, Torres de Holding CL, Thomas L, Ridnour LA, McLendon WJ, Cologna SM, Dutton AS, Champion HC, Mancardi D, Tocchetti CG, Saavedra JE, Keefer LK, Houk KN, Fukuto JM, Kass DA, Paolocci N, Wink DA. Comparison of the NO and HNO donating properties of diazeniumdiolates: primary amine adducts release HNO in vivo. J Med Chem. 2005;48:8220–8.
166. Jarry A, Charrier L, Bou-Hanna C, Devilder MC, Crussaire V, Denis MG, Vallette G, Laboisse CL. Position in cell cycle controls the sensitivity of colon cancer cells to nitric oxide-dependent programmed cell death. Cancer Res. 2004;64:4227–34.
167. Shao C, Furusawa Y, Aoki M. Sper/NO-induced reversible proliferation inhibition and cycle arrests associated with a micronucleus induction in HSG cells. Nitric Oxide. 2003;8:83–8.
168. Pervin S, Singh R, Chaudhuri G. Nitric oxide-induced cytostasis and cell cycle arrest of a human breast cancer cell line (MDA-MB-231): potential role of cyclin D1. Proc Natl Acad Sci U S A. 2001;98:3583–8.

169. Simeone AM, Colella S, Krahe R, Johnson MM, Mora E, Tari AM. N-(4-Hydroxyphenyl) retinamide and nitric oxide pro-drugs exhibit apoptotic and anti-invasive effects against bone metastatic breast cancer cells. Carcinogenesis. 2006;27:568–77.
170. Shami PJ, Kaur G, Thillainathan J, Jia L, Saavedra JE, Keefer LK. JS-K, a novel nitric oxide (NO) generator, shows potent anti-angiogenic activity. Blood. 2004;104:931A.
171. Shami PJ, Saavedra JE, Bonifant CL, Chu J, Udupi V, Malaviya S, Carr BI, Kar S, Wang M, Jia L, Ji X, Keefer LK. Antitumor activity of JS-K [O2-(2,4-dinitrophenyl) 1-[(4-ethoxycarbonyl)piperazin-1-yl]diazen-1-ium-1,2-diolate] and related O2-aryl diazeniumdiolates in vitro and in vivo. J Med Chem. 2006;49:4356–66.
172. Kitagaki J, Yang Y, Saavedra JE, Colburn NH, Keefer LK, Perantoni AO. Nitric oxide prodrug JS-K inhibits ubiquitin E1 and kills tumor cells retaining wild-type p53. Oncogene. 2009;28:619–24.
173. Maciag AE, Saavedra JE, Chakrapani H. The nitric oxide prodrug JS-K and its structural analogues as cancer therapeutic agents. Anticancer Agents Med Chem. 2009;9:798–803.
174. Kiziltepe T, Anderson KC, Kutok JL, Jia L, Boucher KM, Saavedra JE, Keefer LK, Shami PJ. JS-K has potent anti-angiogenic activity in vitro and inhibits tumour angiogenesis in a multiple myeloma model in vivo. J Pharm Pharmacol. 2010;62:145–51.
175. Maciag AE, Chakrapani H, Saavedra JE, Morris NL, Holland RJ, Kosak KM, Shami PJ, Anderson LM, Keefer LK. The nitric oxide prodrug JS-K is effective against non-small-cell lung cancer cells in vitro and in vivo: involvement of reactive oxygen species. J Pharmacol Exp Ther. 2011;336:313–20.
176. McMurtry V, Saavedra JE, Nieves-Alicea R, Simeone AM, Keefer LK, Tari AM. JS-K, a nitric oxide-releasing prodrug, induces breast cancer cell death while sparing normal mammary epithelial cells. Int J Oncol. 2011;38:963–71.
177. Duan SF, Cai S, Yang QH, Forrest ML. Multi-arm polymeric nanocarrier as a nitric oxide delivery platform for chemotherapy of head and neck squamous cell carcinoma. Biomaterials. 2012;33:3243–53.
178. Maciag AE, Holland RJ, Robert Cheng YS, Rodriguez LG, Saavedra JE, Anderson LM, Keefer LK. Nitric oxide-releasing prodrug triggers cancer cell death through deregulation of cellular redox balance. Redox Biol. 2013;1:115–24.
179. Liu J, Li C, Qu W, Leslie E, Bonifant CL, Buzard GS, Saavedra JE, Keefer LK, Waalkes MP. Nitric oxide prodrugs and metallochemotherapeutics: JS-K and CB-3-100 enhance arsenic and cisplatin cytolethality by increasing cellular accumulation. Mol Cancer Ther. 2004;3:709–14.
180. Angeli A. Nitrohydroxylamine. Gazz Chim Ital. 1896;26:17–25.
181. Dutton AS, Fukuto JM, Houk KN. Mechanisms of HNO and NO production from Angeli's salt: density functional and CBS-QB3 theory predictions. J Am Chem Soc. 2004;126: 3795–800.
182. Salmon DJ, Torres de Holding CL, Thomas L, Peterson KV, Goodman GP, Saavedra JE, Srinivasan A, Davies KM, Keefer LK, Miranda KM. HNO and NO release from a primary amine-based diazeniumdiolate as a function of pH. Inorg Chem. 2011;50:3262–70.
183. Drago RS, Karstett BR. Reaction of nitrogen(II) oxide with various primary and secondary amines. J Am Chem Soc. 1961;83:1819–22.
184. Bharadwaj G, Benini PG, Basudhar D, Ramos-Colon CN, Johnson GM, Larriva MM, Keefer LK, Andrei D, Miranda KM. Analysis of the HNO and NO donating properties of alicyclic amine diazeniumdiolates. Nitric Oxide. 2014;2:70–8.
185. Maragos CM, Morley D, Wink DA, Dunams TM, Saavedra JE, Hoffman A, Bove AA, Isaac L, Hrabie JA, Keefer LK. Complexes of NO with nucleophiles as agents for the controlled biological release of nitric-oxide-vasorelaxant effects. J Med Chem. 1991;34:3242–7.
186. Saavedra JE, Billiar TR, Williams DL, Kim Y-M, Watkins SC, Keefer LK. Targeting nitric oxide (NO) delivery in vivo. Design of a liver-selective NO donor prodrug that blocks tumor necrosis factor-Î±-induced apoptosis and toxicity in the liver. J Med Chem. 1997;40:1947–54.

187. Tang X, Xian M, Trikha M, Honn KV, Wang PG. Synthesis of peptide-diazeniumdiolate conjugates: towards enzyme activated antitumor agents. Tetrahedron Lett. 2001;42:2625–9.
188. Makings LR, Tsien RY. Caged nitric oxide. Stable organic molecules from which nitric oxide can be photoreleased. J Biol Chem. 1994;269:6282–5.
189. Saavedra JE, Srinivasan A, Bonifant CL, Chu J, Shanklin AP, Flippen-Anderson JL, Rice WG, Turpin JA, Davies KM, Keefer LK. The secondary amine/nitric oxide complex ion R2N[N(O)NO]$^{-}$ as nucleophile and leaving group in SNAr reactions. J Org Chem. 2001;66:3090–8.
190. Saavedra JE, Bohle DS, Smith KN, George C, Deschamps JR, Parrish D, Ivanic J, Wang YN, Citro ML, Keefer LK. Chemistry of the diazeniumdiolates. O- versus N-alkylation of the RNH[N(O)NO](-) ion. J Am Chem Soc. 2004;126:12880–7.
191. Holland RJ, Paulisch R, Cao Z, Keefer LK, Saavedra JE, Donzelli S. Enzymatic generation of the NO/HNO-releasing IPA/NO anion at controlled rates in physiological media using beta-galactosidase. Nitric Oxide. 2013;35:131–6.
192. Huang Z, Kaur J, Bhardwaj A, Alsaleh N, Reisz JA, DuMond JF, King SB, Seubert JM, Zhang Y, Knaus EE. O2-sulfonylethyl protected isopropylamine diazen-1-ium-1,2-diolates as nitroxyl (HNO) donors: synthesis, β-elimination fragmentation, HNO release, positive inotropic properties, and blood pressure lowering studies. J Med Chem. 2012;55(22): 10262–71.
193. Andrei D, Salmon DJ, Donzelli S, Wahab A, Klose JR, Citro ML, Saavedra JE, Wink DA, Miranda KM, Keefer LK. Dual mechanisms of HNO generation by a nitroxyl prodrug of the diazeniumdiolate (NONOate) class. J Am Chem Soc. 2010;132:16526–32.
194. Bonner FT, Ko YH. Kinetic, isotopic, and nitrogen-15 NMR study of N-hydroxybenzenesulfonamide decomposition: an nitrosyl hydride (HNO) source reaction. Inorg Chem. 1992;31:2514–9.
195. Zamora R, Grzesiok A, Weber H, Feelisch M. Oxidative release of nitric oxide accounts for guanylyl cyclase stimulating, vasodilator and anti-platelet activity of Piloty's acid: a comparison with Angeli's salt. Biochem J. 1995;312:333–9.
196. Nagasawa HT, Lee MJ, Kwon CH, Shirota FN, DeMaster EG. An N-hydroxylated derivative of cyanamide that inhibits yeast aldehyde dehydrogenase. Alcohol. 1992;9:349–53.
197. Shirota FN, Goon DJ, DeMaster EG, Nagasawa HT. Nitrosyl cyanide, a putative metabolic oxidation product of the alcohol-deterrent agent cyanamide. Biochem Pharmacol. 1996;52:141–7.
198. Muri EM, Nieto MJ, Sindelar RD, Williamson JS. Hydroxamic acids as pharmacological agents. Curr Med Chem. 2002;9:1631–53.
199. King SB. N-hydroxyurea and acyl nitroso compounds as nitroxyl (HNO) and nitric oxide (NO) donors. Curr Top Med Chem. 2005;5:665–73.
200. Huang J, Sommers EM, Kim-Shapiro DB, King SB. Horseradish peroxidase catalyzed nitric oxide formation from hydroxyurea. J Am Chem Soc. 2002;124:3473–80.
201. Samuni Y, Wink DA, Krishna MC, Mitchell JB, Goldstein S. Suberoylanilide hydroxamic acid radiosensitizes tumor hypoxic cells in vitro through the oxidation of nitroxyl to nitric oxide. Free Radic Biol Med. 2014;73:291–8.
202. Xu YP, Mull CD, Bonifant CL, Yasaki G, Palmer EC, Shields H, Ballas SK, Kim-Shapiro DB, King SB. Nitrosylation of sickle cell hemoglobin by hydroxyurea. J Org Chem. 1998;63:6452–3.
203. Cai TWB, Tang XP, Nagorski J, Brauschweiger PG, Wang PG. Synthesis and cytotoxicity of 5-fluorouracil/diazeniumdiolate conjugates. Biorg Med Chem. 2003;11:4971–5.
204. Chegaev K, Riganti C, Lazzarato L, Rolando B, Guglielmo S, Campia I, Fruttero R, Bosia A, Gasco A. Nitric oxide donor doxorubicins accumulate into Doxorubicin-resistant human colon cancer cells inducing cytotoxicity. ACS Med Chem Lett. 2011;2:494–7.
205. Rigas B, Kashfi K. Nitric-oxide-donating NSAIDs as agents for cancer prevention. Trends Mol Med. 2004;10:324–30.
206. Rao CV, Reddy BS, Steele VE, Wang CX, Liu XP, Ouyang NT, Patlolla JMR, Simi B, Kopelovich L, Rigas B. Nitric oxide-releasing aspirin and indomethacin are potent inhibitors

against colon cancer in azoxymethane-treated rats: effects on molecular targets. Mol Cancer Ther. 2006;5:1530–8.
207. Hundley TR, Rigas B. Nitric oxide-donating aspirin inhibits colon cancer cell growth via mitogen-activated protein kinase activation. J Pharmacol Exp Ther. 2006;316:25–34.
208. Nortcliffe A, Ekstrom AG, Black JR, Ross JA, Habib FK, Botting NP, O'Hagan D. Synthesis and biological evaluation of nitric oxide-donating analogues of sulindac for prostate cancer treatment. Bioorg Med Chem. 2014;22:756–61.
209. Velazquez C, Rao PNP, Knaus EE. Novel nonsteroidal antiinflammatory drugs possessing a nitric oxide donor diazen-1-ium-1,2-diolate moiety: design, synthesis, biological evaluation, and nitric oxide release studies. J Med Chem. 2005;48:4061–7.
210. Basudhar D, Bharadwaj G, Cheng RY, Jain S, Shi S, Heinecke JL, Holland RJ, Ridnour LA, Caceres VM, Spadari-Bratfisch RC, Paolocci N, Velázquez-Martínez CA, Wink DA, Miranda KM. Synthesis and chemical and biological comparison of nitroxyl- and nitric oxide-releasing diazeniumdiolate-based aspirin derivatives. J Med Chem. 2013;56:7804–20.
211. Abdellatif KR, Chowdhury MA, Velazquez CA, Huang Z, Dong Y, Das D, Yu G, Suresh MR, Knaus EE. Celecoxib prodrugs possessing a diazen-1-ium-1,2-diolate nitric oxide donor moiety: synthesis, biological evaluation and nitric oxide release studies. Bioorg Med Chem Lett. 2010;20:4544–9.
212. Vane JR. Inhibition of prostaglandin synthesis as a mechanism of action for aspirin-like drugs. Nat New Biol. 1971;231:232–5.
213. Wallace JL. Nonsteroidal anti-inflammatory drugs and gastroenteropathy: the second hundred years. Gastroenterology. 1997;112:1000–16.
214. Seibert K, Zhang Y, Leahy K, Hauser S, Masferrer J, Perkins W, Lee L, Isakson P. Pharmacological and biochemical demonstration of the role of cyclooxygenase 2 in inflammation and pain. Proc Natl Acad Sci U S A. 1994;91:12013–7.
215. Tesei A, Ulivi P, Fabbri F, Rosetti M, Leonetti C, Scarsella M, Zupi G, Amadori D, Bolla M, Zoli W. In vitro and in vivo evaluation of NCX 4040 cytotoxic activity in human colon cancer cell lines. J Transl Med. 2005;3:7.
216. Bratasz A, Weir NM, Parinandi NL, Zweier JL, Sridhar R, Ignarro LJ, Kuppusamy P. Reversal to cisplatin sensitivity in recurrent human ovarian cancer cells by NCX-4016, a nitro derivative of aspirin. Proc Natl Acad Sci U S A. 2006;103:3914–9.
217. Fiorucci S, Santucci L, Gresele P, Faccino RM, Del Soldato P, Morelli A. Gastrointestinal safety of NO-aspirin (NCX-4016) in healthy human volunteers: a proof of concept endoscopic study. Gastroenterology. 2003;124:600–7.
218. Gori T, Parker JD. Nitrate tolerance—a unifying hypothesis. Circulation. 2002;106:2510–3.

Chapter 16
Mechanisms by Which Manganese Porphyrins Affect Signaling in Cancer Cells

Rebecca E. Oberley-Deegan and James D. Crapo

16.1 Introduction

This chapter will explore the role of metalloporphyrins in cancer growth and progression. It will focus on the effect of metalloporphyrins on signaling events in cancer cells. In order to understand the mechanisms by which these porphyrins affect signaling in cancer cells, it is critical to understand how the redox environment modulates signaling in cancer cells.

16.2 Cancer Cells Are Under Metabolic Oxidative Stress

Cancer cells, as compared to normal cells, display increased levels of oxidative stress. In general, cancer cells have dysfunctional mitochondria, which results in metabolic changes and leakage of reactive oxygen species (ROS) [1]. To compensate for the dysfunctional mitochondria, cancer cells commonly rely on glucose metabolism to produce pyruvate and NADPH to detoxify ROS [2] as well as other

R.E. Oberley-Deegan (✉)
Department of Biochemistry and Molecular Biology, University of Nebraska Medical Center, Omaha, NE 68198-5870, USA
e-mail: becky.deegan@unmc.edu

J.D. Crapo
Department of Medicine, National Jewish Health,
K701 National Jewish Health 1400 Jackson St, Denver, CO 80206, USA
e-mail: crapoj@njhealth.org

I. Batinić-Haberle et al. (eds.), *Redox-Active Therapeutics*,
Oxidative Stress in Applied Basic Research and Clinical Practice,
DOI 10.1007/978-3-319-30705-3_16

types of reactive species [2]. Cancer cells also use the intermediates of glycolysis for nucleotide and amino acid synthesis to support the rapidly dividing cells [3]. Besides mitochondrial leakage, there appears to be many sources of oxidative stress, which vary across cancer types. Endogenous antioxidant enzymes can be reduced in cancer cells and, therefore, shift the redox balance to a more oxidized state [4]. It has been shown that cancer cells overexpress NADPH oxidases (NOX) enzymes, which generate superoxide [5]. Levels of other antioxidant enzymes, such as glutathione peroxidases, peroxyredoxins, and catalase, that detoxify hydrogen peroxide, can be increased or decreased in cancer cells [6, 7]. This oxidative environment is instrumental in shaping multiple aspects of the cancer cell such as altering cell signaling, supporting uncontrolled cancer growth, and cancer invasion.

There are multiple studies demonstrating that reducing the oxidative environment of a cancer cell results in a more normal phenotype. Superoxide dismutase (SOD) enzymes, enzymes that scavenge the free radical superoxide and produce hydrogen peroxide, are downregulated in many primary cancers. Accordingly, when SOD activities are artificially enhanced in cancer cells, tumor growth is inhibited. This has been documented in cell and animal models of prostate, sarcoma, breast, liver, melanoma, skin, head and neck, leukemia, and lymphoma cancers by a variety of labs across the world [8–26]. There are also some cancers, especially highly metastatic cancers, that have very high levels SOD activities and usually very low levels of the hydrogen peroxide scavenger, catalase [27–32]. Thus, even though these tumors have high levels of SOD activity, the tumor redox environment may be oxidizing due to the buildup of hydrogen peroxides [27–29, 32, 33]. Accordingly, when these tumors are treated with catalase, which removes hydrogen peroxide, the tumors become less oxidatively stressed and return to a more normal phenotype [27–29, 32, 33]. Thus, reducing the oxidative environment of a cancer cell results in a cell that behaves more normally by reducing proliferation, invasion, and migration of the cancer. This large body of data suggests that targeting the oxidative environment of a cancer cell should provide an effective means to inhibit cancer progression while not causing damage to the normal cells in the human body.

16.3 Redox Regulation of Cell Signaling

In normal cells, low levels of ROS are required for signaling. The redox environment affects everything from gene transcription to post-translational modification [34]. For example, many transcription factors are redox-sensitive primarily through cysteine residues present in DNA binding domains. Nuclear factor-κB (NF-κB) and activating protein-1 (AP-1) demonstrate increased activity inside cells under oxidative stress [35]. Other transcription factors, such as Sp-1, glucocorticoid receptor (GR) and early growth response-1 (Egr-1), have reduced affinity for DNA binding under oxidizing conditions [35]. Under oxidizing conditions, in general, protein kinases are activated and protein phosphatases are inhibited. Protein modifications, such as phosphorylation, methylation, and glutathionylation, can all be affected by the redox environment. Epigenetic modifications that regulate gene transcription are also sensitive to oxidative stress [36]. Sirtuins are redox-sensitive histone deacetylases (HDACs) that remove acetyl groups from histones, which allows the chromatin to have a high affinity for histones, resulting in an overall reduction in gene transcription [37]. Conversely, the histone acetyl transferase (HAT) enzyme, p300, is also redox-sensitive and enhances gene transcription. Besides p300 acting as a HAT enzyme, p300 also acts as an enhancer for other redox-sensitive transcription factors such as NF-κB and hypoxia inducible factor-1α (HIF-1α) [38, 39]. Thus, normal cell signaling involves multiple redox-sensitive pathways and given that cancer cells have increased oxidative stress, it is logical to surmise that changes in the redox environment directly affect signaling in cancer cells. The next several sections provide greater detail as to how oxidative stress affects tumor progression through alterations of redox-sensitive signaling pathways in cancer cells.

16.4 Redox Environment Alters Cell Cycle in Cancer Cells

Normal cell growth requires a burst of ROS to progress through the cell cycle. For example, in mammalian cells, a burst in cellular pro-oxidant levels in G1 are required to transition to the S phase in the cell cycle [40]. Also, growth factors can stimulate cell division by increasing ROS production by activating NOX enzymes [41]. In addition, there are several redox-sensitive cell-cycle proteins. Cyclin D1 drives the cell from G1 to S phase. Oxidative stress promotes cyclin D1 levels and conversely, manganese superoxide dismutase (MnSOD), a potent scavenger of superoxide, inhibits cyclin D1 levels [42]. Overexpression of MnSOD in many cancer cells inhibits proliferation as well as induces quiescence [25, 43]. Another cell cycle protein that is sensitive to ROS is the cell division cycle 25 (CDC25) phosphatase. CDC25 specifically dephosphorylates cyclin-dependent kinases (CDKs) and activates the cyclin-CDK activity, which allows the cell to progress through the cell cycle [40]. CDC25 has two cysteine sites that are able to undergo thiol redox reactions [44]. The reduced form of CDC25 has phosphatase activity, while the oxidized form loses activity.

When cells are exposed to high levels of ROS to the point of near cell death, cell cycle arrest has been observed [45]. Extremely high ROS levels can activate the p38/JNK/MAPK signaling cascade and cause the downregulation of cyclins and activation of CDK inhibitors, which result in cell cycle arrest [46].

In conclusion, ROS levels produce a biphasic response in cell cycle. Moderate levels of ROS drive the cell cycle while scavenging of ROS results in inhibition of cell division, and extremely high levels of ROS also can inhibit the cell cycle.

16.5 Oxidative Environment Promotes Angiogenesis

The tumor needs to establish and maintain a sufficient blood supply to support the quickly growing mass. In areas of the tumor that are hypoxic, ROS are produced [47]. ROS, mainly produced by NOX enzymes, serve as a signal that more blood supply is needed at the hypoxic region of the tumor. Increased ROS levels, in particular superoxide levels, help stabilize the transcription factor, HIF-1α, which regulates the gene expression of vascular endothelial growth factor (VEGF). VEGF is the major mediator of blood vessel formation and can also induce proliferation, migration, and rearrangement of the extracellular matrix of endothelial cells. Accordingly, overexpression of MnSOD or ECSOD results in decreased VEGF levels [23, 48, 49]. However, a biphasic response is observed with SOD expression [49, 50]. When SOD is expressed in physiological concentrations, inhibition of HIF-1α is observed, but when SOD is expressed at very high levels, HIF-1α is stabilized. HIF-1α stabilization was caused by a buildup of hydrogen peroxide in the cells with high SOD expression [49]. When the high SOD expressing cells were treated with hydrogen peroxide-removing proteins, HIF-1α stabilization was significantly reduced. These data indicate that both superoxide and hydrogen peroxide levels can affect HIF-1α activity.

Another component of the redox environment, the thioredoxin (Trx) system, also affects VEGF signaling in tumors [47]. Trx-1 proteins act as antioxidants by facilitating the reduction of signaling proteins by cysteine thiol-disulfide exchange and are not direct ROS scavengers. Overexpression of Trx-1 enhances VEGF expression and tumor revascularization while inhibiting Trx-1 expression decreases VEGF, angiogenesis, and tumor growth [51, 52]. Presumably, Trx-1 is regulating VEGF by directly reducing proteins involved in the HIF-1α signaling pathway resulting in enhanced VEGF expression. Trx-1 is not an ROS scavenger; in contrast, SOD enzymes and catalase inhibit the stabilization of HIF-1α and VEGF expression through superoxide and hydrogen peroxide scavenging, respectively [53]. Thus, ROS serves as direct signaling molecules to enhance angiogenesis in cancers and inhibition of the ROS signal inhibits angiogenic pathways related to tumors.

16.6 Redox Environment Affects Cancer Cell Migration and Invasion

In order for a tumor to grow and invade, cancer cells need to alter their extracellular environment. Hydrogen peroxide is a key signaling molecule that affects many aspects of cancer cell migration and invasion [33]. Hydrogen peroxide is a highly diffusible and relatively stable ROS, which makes it a good signaling molecule both inside and outside of the cell. Extracellular ROS affect a variety of components on the outside of the cell: cell surface receptors, actin cytoskeleton rearrangement, cell adhesion, and extracellular matrix degradation.

Hydrogen peroxide activates cell surface receptors on cancer cells, such as epidermal growth factor (EGF) receptors and platelet-derived growth factor (PDGF) receptors [54]. When these growth factor receptors are stimulated they in turn release intracellular ROS (through either NOX, lipoxygenases (LOX), or the electron transport chain), which result in signal amplification inside the tumor cell that promotes tumor progression. An example of this is the activation of protein kinase C (PKC) by ROS released by the growth factor, PDGF, binding to its receptor embedded in the cell membrane [55]. The activated PKC then activates the MAPK kinase signaling cascade, which leads to the activation of matrix metalloproteinases (MMPs) and ultimately tumor progression. When cell adhesion molecules, such as integrins, are activated by changes in the extracellular matrix, ROS are also released inside the cancer cell. The ROS result in phosphorylation of the focal adhesion kinase (FAK), which activates other signaling molecules downstream, resulting in membrane ruffling and cell migration [56]. Thus, ROS can act both as an extracellular and intracellular signaling molecule to promote tumor progression.

Another key component in cell migration is cytoskeleton alternation. Many different types of cellular protrusions have been documented while cancer cells undergo migration, such as membrane ruffling, blebs, and lamellipodia [54]. These cellular protrusions are largely controlled by the signaling molecule, Rac, which ultimately affects actin remodeling through ROS. Rac activates NOX enzymes that are responsible for producing intracellular superoxide. The increased superoxide levels promote actin polymerization [57]. ROS can also directly oxidize actin, which promotes actin-myosin disassembly and results in cell spreading [58].

ROS are also involved in promoting the invasion of cancer cells. In order for a cancer cell to invade, the cell has to degrade the surrounding extracellular matrix [54]. Hydrogen peroxide promotes aberrant activation of extracellular matrix proteases: MMPs and urokinase plasminogen activators (uPAs) [59, 60]. ROS trigger MAPK signaling, which leads to activation of transcription factors (AP-1, Ets-1 and NF-κB) that regulate gene expression of MMPs and uPAs [61, 62]. Not only do ROS enhance extracellular matrix proteases, they also inhibit the activity of tissue inhibitors of metalloproteinases (TIMPs) [63]. ROS can also directly control the activity of MMP through the oxidation of critical cysteine residues on MMPs, which causes the autocatalytic cleavage of the MMP protein and results in an active MMP [64]. Thus, ROS are involved in multiple metabolic processes that influence migration and invasion of cancer cells. This is of critical importance since a metastasized tumor is more difficult to treat than a localized tumor.

16.7 Enhancing Antioxidant Activity Inhibits Tumor Progression

As demonstrated by the previous sections in this chapter, there is a wealth of data, both in cultured cells and in animal models, showing that oxidative stress helps to drive cancer progression. There is an imbalance in the production of ROS and the ability of these cells to detoxify ROS as compared to normal cells, which results in a more oxidized cellular redox state and facilitates proliferation and metastasis of cancer cells. Many tumor types have reduced levels of antioxidant activity coupled with enhanced pro-oxidant enzyme activity. Cancer types that have low levels of antioxidant enzyme activity include: prostate, breast, and oral cancers [23, 65–67]. In most cases, the addition of antioxidant enzymes reduces oxidative stress in cancers and inhibits the pro-oxidant signaling that the tumor depends on to survive, proliferate, and metastasize (Fig. 16.1). Over-expression of antioxidants such as SOD enzymes, catalase, and glutathione peroxidases causes many forms of cancer to grow more slowly and behave less aggressively both in vitro and in vivo [23, 25, 48].

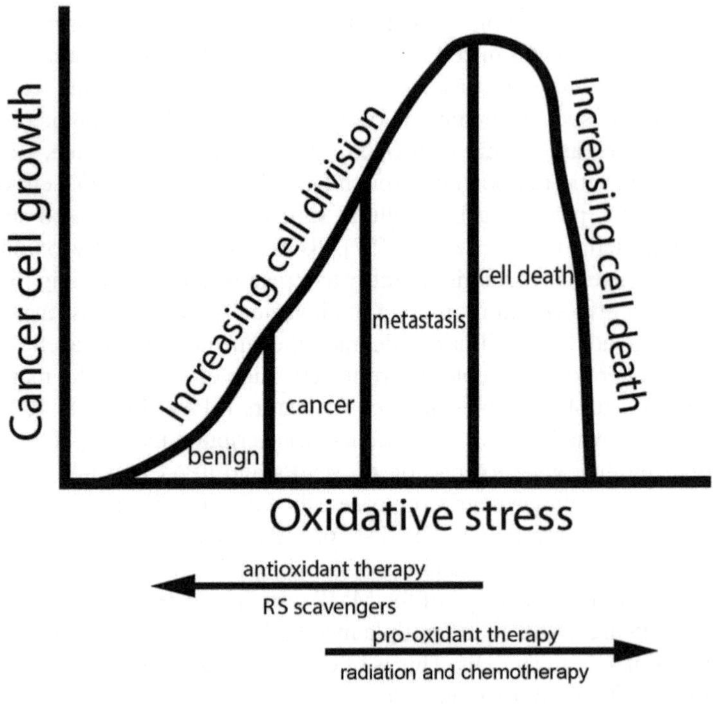

Fig. 16.1 Schematic illustrating how antioxidant and pro-oxidant therapies can reduce cancer cell growth. All cancer cells are under oxidative stress and the redox imbalance in these cells helps to promote the cancerous phenotype. The goal of antioxidant therapy is to change the redox state of the cancer cell from oxidizing to a more reduced state. The addition of antioxidants (superoxide and/or hydrogen peroxide scavengers) to an oxidatively stressed cancer cell will make these cells proliferate more slowly and be less metastatic. It is postulated that antioxidant enzymes alter cell signaling in cancer cells which leads to a more normal phenotype. Standard treatments for cancer include radiation and chemotherapy, which kill cancer cells by releasing extremely high levels of ROS in cancer cells over a short period of time and/or through other mechanisms not related to redox state. The major limitation of radiation and chemotherapy is that the therapies also injure or kill normal cells. The use of antioxidants in combination with radiation and/or chemotherapy should produce additive antitumor effects and protect normal cells from radiation and chemotherapeutic damage

Clinical trials using dietary antioxidants have displayed little protection against various types of cancers [45]. There are two main reasons that likely explain why these clinical trials failed. First, the dietary antioxidants used in these trials may not effectively target the critical sources of ROS, i.e., mitochondria and NAPDH oxidases. Second, vitamin E, vitamin C, and *N*-acetyl-L-cysteine (NAC) are nonenzymatic antioxidants, thus high concentrations are needed for these antioxidants to adequately scavenge ROS. Oral bioavailability of these antioxidants is exceedingly low [68, 69]. Thus, low concentrations of these dietary antioxidants would not be expected to effectively reduce the oxidative environment in cancers. In support of this rationale, a recent phase I clinical trial using high doses of vitamin C administered intravenously showed that patient survival was enhanced while pancreatic tumor progression was reduced [70]. This study also demonstrated that the levels of vitamin C were elevated and isoprostane levels were reduced in the plasma of patients receiving the high doses of vitamin C [70]. Therefore, more work needs to be done using high doses of dietary antioxidants to truly show whether or not these compounds have antitumor capabilities. In addition, catalytically active antioxidants that can target critical subcellular sources of ROS with enhanced bioavailability need to be tested clinically to effectively test the hypothesis that adding antioxidants to certain cancer cells will reduce pro-oxidative signaling and inhibit cancer cell growth and metastasis.

16.8 Very High Levels of ROS Are Anti-tumorigenic

It is well known that standard cancer therapies, such as ionizing irradiation and chemotherapeutic agents, kill cancer cells, at least in part, by release of high levels of ROS. Ionizing radiation causes DNA damage in cancer cells, at least in part, by the creation of the hydroxyl radical. Radiation can cause water radiolysis that directly produces either the hydroxyl radical or forms superoxide which may damage iron bearing proteins, such as aconitase, giving rise to free Fe2+ and in turn higher levels of the hydroxyl radical. Chemotherapeutic drugs, such as taxanes, promote mitochondrial cell death and disrupt the electron transport chain resulting in large bursts of ROS [71]. The enhancement of ROS to promote cancer cell death has been used clinically for the past century. However, there are two main drawbacks of enhancing ROS production to kill cancer cells. First, the cancer cell can become resistant to the increased ROS levels and grow back more aggressively. Second, the enhanced ROS production also damages and kills normal cells. It is a sledgehammer approach to treatment of cancer.

Although there is a lot of data indicating that reducing oxidative stress slows tumor growth and inhibits tumor progression, paradoxically, there is also data indicating that enhancing oxidative stress can also be toxic to cancer cells. Some cancer cell types, such as lung and ovarian cancer cells, are able to adapt to high levels of oxidative stress by elevating endogenous antioxidant enzymes including peroxyredoxins (Prxs), thioredoxins (Trxs), and SOD enzymes [30, 72]. The transcription factor, nuclear factor (erythroid derived 2)-like 2 (NRF2), which is a master regulator of antioxidant gene expression is elevated in some cancers to combat high oxidative stress [73, 74]. In many cases, SOD enzyme activities are increased but catalase activity is decreased resulting in high concentrations of hydrogen peroxide. Thus, even though SOD activity may be elevated in certain tumors, this does not mean that the redox state of the tumor environment is reduced. In fact, it has been shown that in some tumors with high SOD levels, the overall redox state is highly oxidized due to increased hydrogen peroxide production. These high hydrogen peroxide levels allow the cancer cells to maintain pro-oxidant signaling though elevated ROS [45]. Extremely high levels of ROS, produced by treatments such as radiation therapy or chemotherapy, can induce cancer cell apoptosis and inhibit proliferation [45] (Fig. 16.1). Alternatively, one could speculate that inhibiting SOD activity in cancer cells with high levels of SOD activity and concurrently low hydrogen peroxide scavenging capabilities may reduce the oxidative environment by inhibiting hydrogen peroxide levels and thereby result in a more normal phenotype (Fig. 16.1).

There have been recent clinical trials that support this hypothesis. Lung and ovarian cancer cells generally have high levels of ROS with correspondingly high levels of Trx activity [75]. Several inhibitors of Trx, such as PX-12, motexafin gadolinium, and the gold compound auranofin, have produced antitumor effects in clinical trials [52, 76, 77]. Cu/ZnSOD is also elevated in some lung adenocarcinoma cells [78]. Methoxyestradiol (2-ME) increases superoxide in cancer cells and preclinical data indicate that this compound inhibits tumor growth and appears to selectively target the cancerous cells to undergo apoptosis while not affecting normal cells [79]. Clinical trials are now ongoing for 2-ME [80, 81]. It should be noted that 2-ME and PX-12 also inhibit major signaling pathways in tumors that appear to be separate from their ability to enhance ROS levels [82, 83]. Both PX-12 and 2-ME can enhance oxidative stress but are also HIF-1α inhibitors and 2-ME has been demonstrated to inhibit AKT, ERK, and activate JNK pathways [82–86]. Therefore, the antitumor effects observed may be not be caused solely by inhibiting antioxidant enzymes, but also through alteration of key signaling pathways in cancer cells.

16.9 The Paradox of Antioxidant Enzyme Activity Levels and Cancer Progression

There are conflicting results in the literature as to whether antioxidants are good or bad for cancer therapy. Most of the confusion is a result of not fully understanding the redox state of cancer cells. We hypothesize that therapies which change the overall oxidizing environment in cancer cells, i.e., superoxide and or hydrogen peroxide scavengers to a more reduced state will produce a more normal cell phenotype and result in tumor growth inhibition and inhibition of metastasis. A cancer cell can be in an oxidizing redox state even though a particular antioxidant enzyme, such as SOD, is highly active. Most, if not all cancers are in a state of oxidative stress. Cancer cells prosper in high oxidative stress. Oxidative stress occurs by either an increased rate of ROS generation or reduced ability to efficiently remove ROS or a combination of these events. The increase of ROS in cancer cells is due, at least in part, to dysfunctional mitochondria, which results in electron leakage, and elevated activity of NOX enzymes. In most cancer cells, there is also a decrease in antioxidant enzyme activity. Generally, MnSOD, Cu/ZnSOD, ECSOD, catalase, and other antioxidant enzyme activities are reduced and/or inhibited in cancer cells. The addition of catalytically active antioxidant enzymes reduces cancer growth and metastasis. In some cases, there is an opposite effect observed where SOD, NRF2, Trx and/or other antioxidant activities show increased intracellular levels in cancer cells as compared to normal cells. It is postulated that these cancer cells have induced antioxidant enzyme activities that enable them to survive in a highly oxidized environment. However, even though these cells have high antioxidant activity, these cells are still under oxidative stress.

Melendez's laboratory has elegantly demonstrated that certain cancers have high levels of MnSOD expression but are also oxidatively stressed [27, 28, 32, 33]. These cancers generally have concurrently low levels of catalase and are often highly metastatic. When MnSOD is administered to these cells, the tumor becomes even more aggressive associated with an increased production of hydrogen peroxide [27, 28, 32, 33]. In these cases, the addition of SOD results in a more oxidizing environment, which results in a more metastatic cancer cell. When catalase was added along with MnSOD, the excess hydrogen peroxide was removed. This resulted in an overall reduced redox environment in the cancer cells and as a consequence the cancer cells reverted back to a more normal phenotype [27, 28, 32, 33]. Therefore, in certain cancers, the addition of SOD may not be enough to produce a reducing tumor environment and the addition of hydrogen peroxide scavengers are needed to effectively reduce the redox state of the cancer cell. We postulate that the addition of appropriate levels of SOD and/or hydrogen peroxide removing enzymes will effectively inhibit growth and progression of any cancer cell by producing a more reduced redox state.

16.10 Manganese Porphyrins as Potent ROS Scavengers

Redox-active manganese porphyrins have been developed with characteristics that mimic the active center of naturally occurring metalloenzymes, which catalytically breakdown superoxide and inhibit peroxynitrite formation. Although there are a variety of redox-active manganese porphyrins, we will focus on three cationic N-substituted pyridylporphyrins that have been well characterized and most frequently studied: MnTE-2-PyP, MnTnHex-2-PyP, and MnTnBuOE-2-PyP [87] (Fig. 16.2). These small molecular weight and catalytically active porphyrins have a rate constant for the dismutation of superoxide that is close to the native SOD enzyme activity [88].

All three compounds are able to cross cell membranes and enter cells [87]. MnTE-2-PyP was the first of these three compounds to be synthesized by Batinić-Haberle and colleagues [89]. Despite being highly positively charged, MnTE-2-PyP enters cells readily [87]. MnTE-2-PyP is found in concentrations about three times

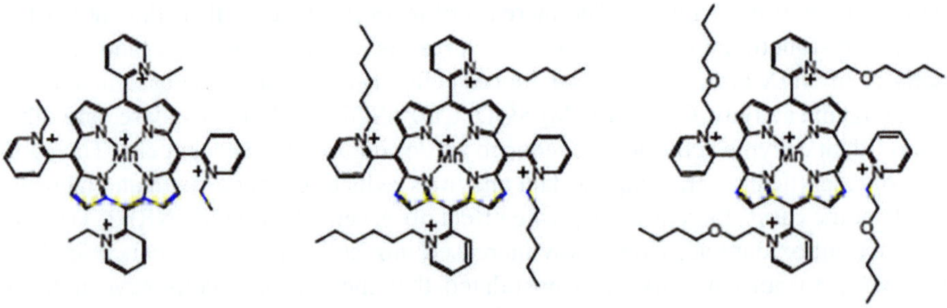

Fig. 16.2 Structures of the manganese porphyrins discussed in this chapter. MnTE-2-PyP is the most studied of the three compounds. Placing the positively charged pyridyl nitrogens closer to the porphyrin ring enhances both the electron-withdrawing effects and the attraction of superoxide and other anionic species to the molecule. MnTE-2-PyP is a potent superoxide scavenger. MnTE-2-PyP is hydrophilic, which limits its ability to cross the blood–brain barrier. MnTnHex-2-PyP was created to enhance the lipophilicity of manganese porphyrins. Longer alkyl chains were added to the pyridyl nitrogens to enhance lipophilicity. MnTnHex-2-PyP readily enters cells and crosses the blood–brain barrier, but due to its high micellar properties, it can damage lipid membranes and can be toxic at higher doses. Therefore, a polar oxygen atom was added to each of the four alkyl chains to interrupt its micellar property; in turn, MnTnBuOE-2-PyP was created. MnTnBuOE-2-PyP is similarly lipophilic to MnTNHex-2-PyP and crosses the blood–brain barrier but due to the oxygen molecules is less toxic than MnTnHex-2-PyP. All three compounds have antitumor effects in vitro and in vivo alone and in combination with other cancer therapies [118]

higher in the nucleus as compared to the cytoplasm [90]. In *E. coli*, MnTE-2-PyP accumulates more in membranes than in other parts of the cell [91]. In yeast and mouse tissues, higher levels of MnTE-2-PyP are found in mitochondria as compared to the cytosol [92, 93]. This compound does not readily cross the blood–brain barrier due to its hydrophilic nature. In order to make a more lipophilic redox-active manganese porphyrin, hydrophilic ethyl chains were replaced with hydrophobic long alkyl chains (MnTnHex-2-PyP) [94]. MnTnHex-2-PyP is ~8500-fold more lipophilic than MnTE-2-PyP and is found throughout the cell but accumulates in higher levels in mitochondria and in cell membranes relative to the cytosol. This compound enters cells more effectively than MnTE-2-PyP and, thus, up to 120-fold lower concentrations of MnTnHex-2-PyP are needed for targeted therapy [95]. MnTnHex-2-PyP also crosses the blood–brain barrier and can be used for treatment of the brain and spinal cord tissues [96]. However, due to its high lipophilicity, MnTnHex-2-PyP is more toxic at higher concentrations compared to MnTE-2-PyP [97]. To reduce toxicity, oxygen was added to the alkyl chains to disrupt the micellar properties, MnTnBuOE-2-PyP emerged as a result of such design strategy. MnTnBuOE-2-PyP has similarly high lipophilicity, but reduced toxicity, as compared to MnTnHex-2-PyP [97]. MnTnBuOE-2-PyP readily enters cells, accumulates in mitochondria and cell membranes, and crosses the blood–brain barrier [94]. Taken together, all three compounds have high SOD activity and target areas of the cell that are at high risk of ROS damage and, thus, bear therapeutic potential.

16.11 Manganese Porphyrins Inhibit Cancer Growth

MnTE-2-PyP as a single therapy inhibits the ability of human prostate cancer cells to form colonies in vitro at high concentrations (10 μM, Fig. 16.3). This porphyrin has also been shown to inhibit migration and invasion of human prostate cancer cells at doses of 10–30 μM in cell culture [98]. Similarly, MnTnHex-2-PyP has also been shown to inhibit migration of breast cancer cells [99]. In vivo, when given at high doses, MnTE-2-PyP (15 mg/kg daily), inhibits mouse breast cancer cell growth [100]. Topical administration can result in localized high concentration in skin cancer cells, and topical administration of MnTE-2-PyP dramatically inhibits skin tumor growth [101]. However, when administered systemically and at less frequent dosing intervals (2.5 mg/kg, three times a week), the tumor growth inhibition ability of MnTE-2-PyP was less robust [102–105]. However, even though tumor inhibition is not drastic with this MnTE-2-PyP dosing scheme, substantial protection of normal tissue injury from radiation exposure is observed [106]. Metalloporphyrins that enter cells more readily can inhibit tumor growth at lower concentrations. For example, the more lipophilic porphyrin, MnTnHex-2-PyP, significantly inhibited brain tumor xenografts in nude mice at concentrations of 1.6 mg/kg given systemically twice daily [107]. Specifically, MnTnHex-2-PyP alone inhibited the growth of both adult and pediatric glioblastomas and pediatric medulloblastoma cancers [107]. Medulloblastoma tumor bearing animals treated with MnTnHex-2-PyP had a 171 % increase in survival as compared to controls [107]. In comparison, to observe the same amount of tumor growth inhibition, MnTE-2-PyP needs to be administered at 15 mg/kg daily [100]. MnTnBuOE-2-PyP has been shown to inhibit brain tumor growth in vitro (Fig. 16.4). MnTnBuOE-2-PyP at (0.5–1 μM) effectively inhibited the growth of the medulloblastoma ONS-76 cell line (Fig. 16.4). Thus, both lipophilic compounds, MnTnHex-2-PyP and MnTnBuOE-2-PyP, inhibit tumor growth at a lower concentration than MnTE-2-PyP, presumably because the more lipophilic compounds enter the tumor more readily and can achieve higher activity in the tumor. In summary, manganese porphyrins are able to inhibit tumor growth and migration when high localized levels are achieved in the tumor.

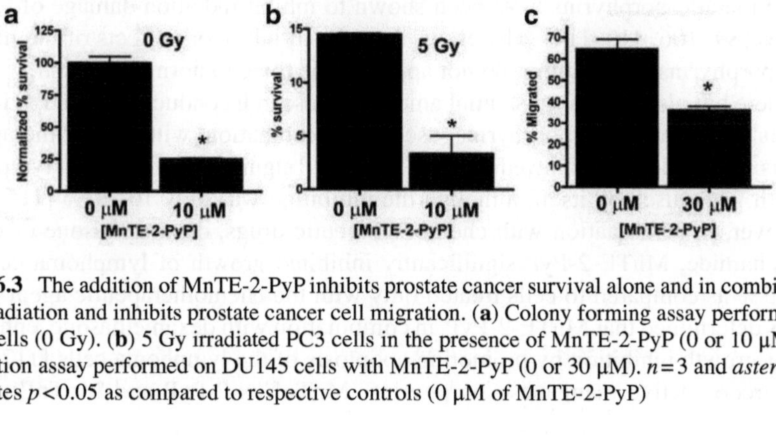

Fig. 16.3 The addition of MnTE-2-PyP inhibits prostate cancer survival alone and in combination with radiation and inhibits prostate cancer cell migration. (**a**) Colony forming assay performed on PC3 cells (0 Gy). (**b**) 5 Gy irradiated PC3 cells in the presence of MnTE-2-PyP (0 or 10 μM). (**c**) Migration assay performed on DU145 cells with MnTE-2-PyP (0 or 30 μM). $n=3$ and *asterisk* (*) indicates $p<0.05$ as compared to respective controls (0 μM of MnTE-2-PyP)

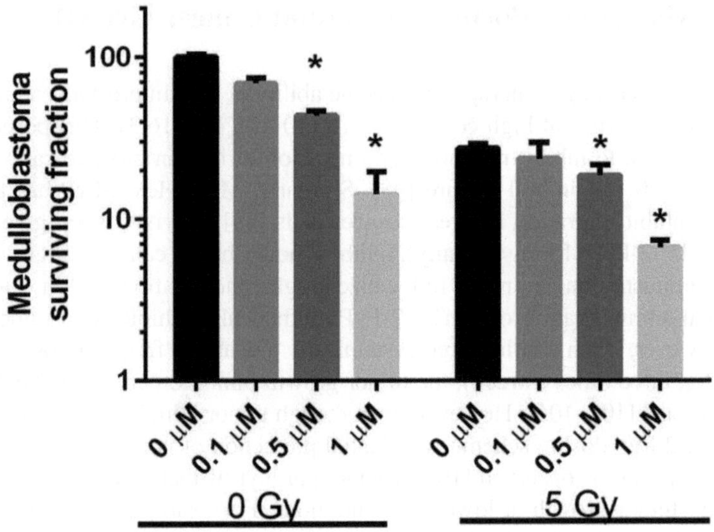

Fig. 16.4 MnTnBuOE-2-PyP inhibits medulloblastoma cell survival alone and in combination with radiation. Colony-forming assays performed on ONS-76 cells with either 0 or 5 Gy of irradiation in combination with MnTnBuOE-2-PyP (0–1 μM). $n = 3$ and *asterisk* (*) indicates $p < 0.05$ as compared to respective controls (0 μM of MnTnBuOE-2-PyP)

The ability of manganese porphyrins to inhibit tumor growth is enhanced when combined with traditional cancer therapies. In combination with radiation, MnTE-2-PyP inhibits both tumor growth and tumor angiogenesis in mouse mammary cancer cells [100]. MnTE-2-PyP also sensitizes human prostate cells (DU145, PC3, and LNCaP) to irradiation in vitro and further inhibits mouse prostate cancer growth in combination with radiation in vivo [98]. MnTnBuOE-2-PyP radiosensitizes human head and neck cancer cells, FaDu, with a dose-modifying factor of 1.3 [108]. Both MnTnBuOE-2-PyP and MnTnHex-2-PyP were found to promote the anticancer effect of radiation in brain tumors [94, 109, 110]. At the same time, the redox-active metallo-porphyrins have been shown to inhibit radiation damage of normal tissues [89, 106, 110–114]. Therefore, the growth inhibitory effects of the manganese porphyrins on the tumor do not appear to be toxic to normal tissues.

There has also been a substantial amount of research conducted investigating the effects of manganese porphyrins used in combination with chemotherapeutic agents. Jaramillo et al. showed that MnTE-2-PyP significantly inhibited lymphoma growth when used by itself, although the inhibition was only 10–20 % [115–117]. However, in combination with chemotherapeutic drugs, dexamethasone or cyclophosphamide, MnTE-2-PyP significantly inhibited growth of lymphoma cells by 80–95 % as compared to cells treated only with the chemotherapeutic agent [116]. It was determined that MnTE-2-PyP, in combination with dexamethasone, enhanced tumor growth inhibition by promoting apoptosis of the lymphoma cells [117]. The other redox-active manganese porphyrins, MnTnHex-2-PyP and MnTnBuOE-2-

PyP, have also been shown to act as chemosensitizers [109, 118]. In a subcutaneous xenograft model, both of these porphyrins sensitized brain tumors to treatment with the chemotherapy agent, temozolomide [109, 118]. All three manganese porphyrins have been shown to enhance the anticancer potency of ascorbate [119–121]. The porphyrins enhanced the cytotoxicity of ascorbate and decreased clonogenic survival of several different cancer cell lines, but are nontoxic to normal cells [94, 119, 120]. The manganese porphyrins are able to enhance the effectiveness of a variety of chemotherapeutic agents in tumors, but they are also able to protect normal cells from cytotoxicity. Doxorubicin is an effective chemotherapeutic agent, but its use is limited by its acute and chronic cardiotoxicity. While MnTE-2-PyP did not affect the toxicity of doxorubicin to lymphoma cells, MnTE-2-PyP did significantly enhance the viability of cardiomyocytes when treated with doxorubicin [117]. Thus, the redox-active manganese porphyrins protect normal cells from the injury that occurs with chemotherapy while at the same time not protecting the tumor and in many cases enhance tumor killing.

16.12 Manganese Porphyrins Alter Signaling in Cancer Cells

There are multiple mechanisms by which redox-active manganese porphyrins can inhibit cancer growth. There is evidence showing that these porphyrins are able to affect transcription of genes involved in tumor survival and cause post-translational modifications of key signaling proteins. These changes in signaling affect the way the cancer grows and how the body responds to the tumor.

16.13 Manganese Porphyrins Effects on Tumor Immunology

Makinde et al. investigated the role that MnTE-2-PyP plays on the immune system of a tumor bearing mouse [103, 122]. Splenocytes of prostate tumor bearing animals that underwent radiation therapy were altered in the presence of MnTE-2-PyP. MnTE-2-PyP inhibited IL-4 and IL-13 cytokines, Th2 type cytokines, as well as transforming growth factor-β1 (TGF-β1) expression in splenocytes [122]. At the same time, MnTE-2-PyP treatment enhanced Th1 type cytokine production in the spleen. Treatment with MnTE-2-PyP significantly elevated CD4+ T helper cells and CD8+ T cytotoxic cells [103]. In the tumor, MnTE-2-PyP treatment caused a significant reduction in Th2 cytokine expression as well as TGF-β1 and VEGF expression in prostate cancer cells [122]. From these studies, it was concluded that MnTE-2-PyP likely reduced tumor growth by inhibiting tumor vasculature while enhancing the Th1 immune response in tumor bearing animals. MnTE-2-PyP has also been shown to suppress the Th2 immune response in a completely different disease model, asthma [123]. The mechanism as to how MnTE-2-PyP inhibits the Th2 response is not well understood. It is speculated that MnTE-2-PyP may be scavenging the ROS burst used to differentiate a naive T cell into a Th2 type cell. Another hypothesis is that MnTE-2-PyP may inhibit a transcription factor that is required to express the appropriate genes for differentiation of a naive T cell into a Th2 type cell.

16.14 Manganese Porphyrins Regulate Aerobic Glycolysis

In T cells from diabetic mice, it was demonstrated that MnTE-2-PyP promotes metabolic quiescence and impedes cell proliferation by restricting aerobic glycolysis [124]. Specifically, MnTE-2-PyP had no effect on mitochondrial numbers, but enhanced glucose oxidation and reduced fatty acid oxidation [124]. MnTE-2-PyP also reduced lactate production and reduced the expression of the glucose transporter 1 (Glut 1) [124]. Taken together, these data indicate that the addition of MnTE-2-PyP restricts glycolysis and overall metabolism of T cells. The increased metabolism required by activated immune cells is similar to the increased demand observed in cancer cells [3]. Both cell types rely on glycolysis to proliferate rapidly [3]. Accordingly, Jaramillo et al., have also found that MnTE-2-PyP inhibits glycolysis of lymphoma cells [115].

16.15 Manganese Porphyrins Effects on Transcription in Cancer Cells

MnTE-2-PyP has been shown to affect the activity of several transcription factors that promote tumor progression. Radiation causes tumor hypoxia, which promotes HIF-1α, to enhance VEGF expression and revascularize the tumor [105]. MnTE-2-PyP inhibits HIF-1α activity in hypoxic areas of the tumor after irradiation [104]. AP-1 controls genes involved in proliferation of skin cancer cells and MnTE-2-PyP inhibits skin tumor growth, in part, by inactivating AP-1 [101]. NF-κB is another transcription factor that is upregulated in cancer cells and which promotes tumor survival, tumor migration, and tumor invasion [125, 126]. MnTE-2-PyP inhibits the ability of NF-κB to bind to DNA in a variety of disease models, including cancer [116, 119, 127, 128]. MnTE-2-PyP was also found to inhibit SP-1 activity [128, 129]. Finally, MnTE-2-PyP inhibits the translocation of the transcription factor, p53, a tumor suppressor, to the nucleus from the cytoplasm in skin cancer cells [130].

A gene array was conducted on tumors isolated from a subcutaneous mouse glioblastoma cancer xenograft that was irradiated and treated with or without manganese porphyrins [131]. It was shown that the majority of gene transcription is downregulated in tumors that were irradiated in the presence of redox-active metallo-porphyrins (Table 16.1). This observation led researchers to investigate whether the manganese porphyrins are affecting transcription at a broader level.

Table 16.1 Differences in gene expression in irradiated brain tumors isolated from the hind flank of mice treated with MnTnBuOE-2-PyP as compared to irradiated alone tumors

Gene name	Fold difference	Gene name	Fold difference
Rik	18.96	Eif5b	−408.81
Abl1	3.4	Fas	−15.35
Akt1	−6.53	Fasl	−11.88
Apaf1	−12.87	Foxi1	−3.65
App	−8.89	Grb2	−45.54
Atg12	−20.73	Hspbap1	−28.25
Atg16/1	−16.32	Htt	−7.75
Atg3	−13.31	Igf1	−25.77
Atg5	−43.26	Irgm1	−6.71
Bax	−4.19	Mag	−5.88
Bclsl1	−59.14	Map 1lc3a	−10.15
Bclsl11	−6.2	Mapk8	−7.78
Becn1	−15.56	Mcl1	−8.75
Birc2	−3.37	Nfkb1	−13.88
Birc3	−20.11	Olfr1404	−9.98
Bmf	−16.37	Parp1	−5.39
Casp1	−37.32	Parp2	−3.31
Casp2	−8.65	Pvr	−17.44
Casp3	−3.27	Rps6kb1	−176.6
Cd40	−21.09	Snca	−3.83
Cflar	−38.66	Sqstm1	−10.06
Commd4	−17.8	Tmem57	−3.19
Ctss	−15.91	Tnfrsf10b	−3.19
Dennd4a	−11.14	Txn14b	5.61
Dpysl4	−3	Ulk1	−12.69
		Cyld	−3.46

Mice were irradiated three times with 1 Gy of irradiation, treated with MnTnBuOE-2-PyP (1.6 mg/kg/b.i.d.), and harvested 1 month after irradiation. A cell death gene array through Qiagen was performed on the extracted tumors. Genes shown in *blue* were downregulated and the genes in *red* were upregulated. This experiment was repeated in three irradiated alone tumors and three irradiated tumors treated with MnTnBuOE-2-PyP with similar results. Overall, there was a significant decrease in gene expression in irradiated tumors treated with MnTnBuOE-2-PyP [94, 131]

The HAT enzyme, p300, acetylates histones and increases overall transcription [98]. p300 also acts as an enhancer for redox-sensitive transcription factors such as HIF-1α and NF-κB [38]. MnTE-2-PyP significantly reduces overall HAT activity after irradiation in prostate tumor cells and specifically inhibits p300 activity [98]. The mechanism as to how MnTE-2-PyP alters the activity of the p300 enzyme remains unknown at this time. Thus, MnTE-2-PyP may not only be altering activity of transcription factors directly, but also affecting transcription globally through modifying the redox transcription factor enhancer, p300.

16.16 Manganese Porphyrins Enhance Post-translation Modifications

Tse et al. were the first to show that MnTE-2-PyP alters activity of proteins through the oxidation of cysteine residues [128]. Using lipopolysaccharide (LPS)-stimulated macrophages, it was shown that NF-κB cannot bind to DNA in the presence of MnTE-2-PyP because the cysteines on the p50 subunit of NF-κB were oxidized [128]. In subsequent studies, NF-κB activity has also been shown to be inhibited in cancer cells [116]. Many other redox-sensitive transcription factors have critical cysteine residues that regulate activity, and it is speculated that MnTE-2-PyP alters the activity of these other transcription factors in a similar manner [118]. Recently, another study showed that MnTE-2-PyP can alter the activity of NF-κB through an entirely different mechanism. The combination of dexamethasone and MnTE-2-PyP causes the p65 subunit of NF-κB to be glutathionylated [116]. Glutathionylation of NF-κB inactivates the transcription factor, which results in increased apoptosis in lymphoma cells [116]. In this same study, it was shown that MnTE-2-PyP, in combination with dexamethasone treatment, resulted in a 2.5-fold enhanced glutathionylation of all intracellular proteins [116]. Jaramillo et al. have also recently shown that MnTE-2-PyP in combination with dexamethasone results in the glutathionylation of mitochondrial proteins, specifically complex I and III [115]. The glutathionylation of these electron transport chain proteins causes decreased activity and leads to apoptosis of lymphoma cells but not normal lymphocytes [115]. Thus, cells treated with MnTE-2-PyP and dexamethasone likely alter the activity of many proteins through glutathionylation. Finally, in a recent study conducted by Evans et al., it was shown that MnTE-2-PyP in combination with ascorbate caused a decrease in the phosphorylation of key signaling pathways regulating apoptosis in breast cancer cells [119]. Specifically, NF-κB and ERK1/2 phosphorylation (p44/42) are reduced, which leads to the inhibition of pro-survival factor signaling and promotes apoptosis of the breast cancer cells [119]. It is possible that there are other, as yet undiscovered, ways in which the manganese porphyrins are able to alter activity of proteins through post-translational modifications.

16.17 Mechanisms as to How Manganese Porphyrins Change Cell Signaling

Currently, there is lack of solid data demonstrating exactly how manganese porphyrins alter signaling in a cancer cell. Therefore, based on the current data, we hypothesize that the manganese porphyrins may work in three ways: (1) By acting as a direct scavenger of superoxide and/or peroxynitrite. (2) By producing localized levels of H_2O_2, which inhibits pro-cancerous signaling. (3) By directly oxidizing or reducing critical cysteine residues of signaling proteins [53] (Fig. 16.5). In reaction with peroxynitrite, manganese porphyrins act as antioxidants/reductants by reducing

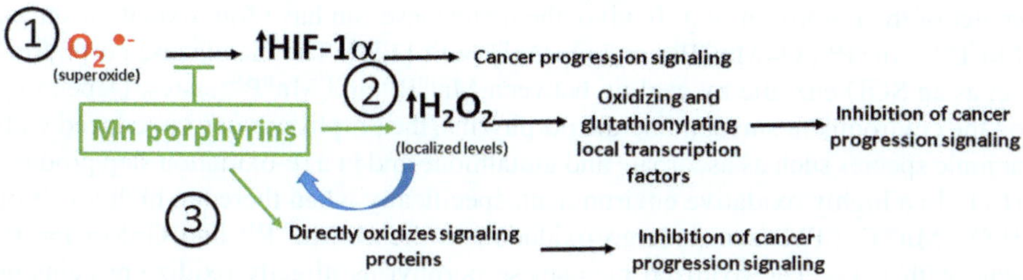

Fig. 16.5 Proposed model illustrating the three potential mechanisms by which manganese porphyrins can alter signaling in cancer cells. Cancer cells are under oxidative stress and have elevated intracellular levels of ROS such as superoxide and H_2O_2. Aberrant superoxide and H_2O_2 production is thought to drive cancer progression signaling. By scavenging superoxide, manganese porphyrins both inhibit HIF-1α signaling and produce localized high levels of H_2O_2 (only if hydrogen peroxide removing enzymes are suppressed). Oxidized manganese (Mn) porphyrins (MnP/H_2O_2) can oxidize thiols of local transcription factors which leads to inhibition of gene expression and inhibition of pro-cancer signaling

peroxynitrite to either nitrite (in the presence of cellular reductants) or NO_2 radical [118]. Dismutation of superoxide by manganese porphyrins, however, gives rise to both O_2 (oxidized form) or H_2O_2 (reduced form). Thus, during superoxide dismutation, manganese porphyrins act as both anti- and pro-oxidants. When production of H_2O_2 is coupled with its removal by catalases or peroxidases, manganese porphyrins, like SOD enzymes, may be viewed as antioxidants. Superoxide directly activates the HIF-1α signaling pathway, which promotes tumor progression. Therefore, the catalytically active manganese porphyrins remove superoxide and inhibit HIF-1α stability and signaling pathways [53]. Manganese porphyrins may also be oxidized with oxygen and give rise to superoxide and eventually H_2O_2 (Fig. 16.6). Hydrogen peroxide, produced during dismutation, can directly oxidize cysteine residues in key signaling proteins [118] (Fig. 16.6). In a normal cell, the production of both superoxide and hydrogen peroxide are tightly regulated. In cancer cells, this regulation is lost and it is postulated that the addition of manganese porphyrins in cancer cells helps to restore normal signaling in the cancer cell. Thus, the addition of manganese porphyrins is postulated to have little effect on signaling in normal, healthy cells, but major effects in cancer cells. Producing differential effects in normal vs. cancer cells is a critical element in the treatment of cancer. In summary, manganese porphyrins can act as both antioxidants and pro-oxidants.

Although the structure of the redox-active manganese porphyrins was designed based on the active sites of SOD enzymes, these porphyrins can interact with other molecules besides superoxide. This is likely due to the lack of tertiary structure surrounding the active site of the manganese porphyrins as compared to SOD enzymes, which allows the porphyrins to interact with a variety of molecules [87, 90, 121, 132]. For instance, the redox-active manganese porphyrins not only have excellent affinity for superoxide, but also for other anionic species such as ascorbate, glutathione, cysteine, and protein thiols [90, 121, 132]. The ability of manganese porphyrins to act as SOD mimetics depends on the oxidation state of the manganese at the center of the porphyrin ring. In vivo, the manganese can have four oxidation states: $Mn^{II}P^{4+}$, $Mn^{III}P^{5+}$, $O{=}Mn^{IV}P^{4+}$, and $O{=}Mn^{V}P{=}O^{3+}$ [133]. The manganese porphyrins act as an SOD enzyme by cycling between $Mn^{III}P^{5+}$ and $Mn^{II}P^{4+}$ states. Depending on the environment surrounding the porphyrins, the porphyrins can be reduced with anionic species such as ascorbate and glutathione and in a re-oxidation step produce H_2O_2. In a highly oxidative environment, specifically when there are high levels of H_2O_2, MnTE-2-PyP can undergo oxidation to the $O{=}Mn^{IV}P^{4+}$ and $O{=}Mn^{V}P{=}O^{3+}$ state with H_2O_2. The oxidized manganese porphyrins directly oxidize proteins or lipids if not reduced by cellular reductants (Fig. 16.6). Batinić-Haberle et al. have recently published a review article with more details on biologically relevant manganese porphyrin redox-chemistry [118].

Therefore, the cellular reductant levels (ascorbate, glutathione, etc.) and the localized hydrogen peroxide levels surrounding the manganese porphyrin can cause the porphyrin to act as an antioxidant or a pro-oxidant. It has been shown that in a mildly oxidative environment (a cancer cell or a normal cell undergoing stress) the

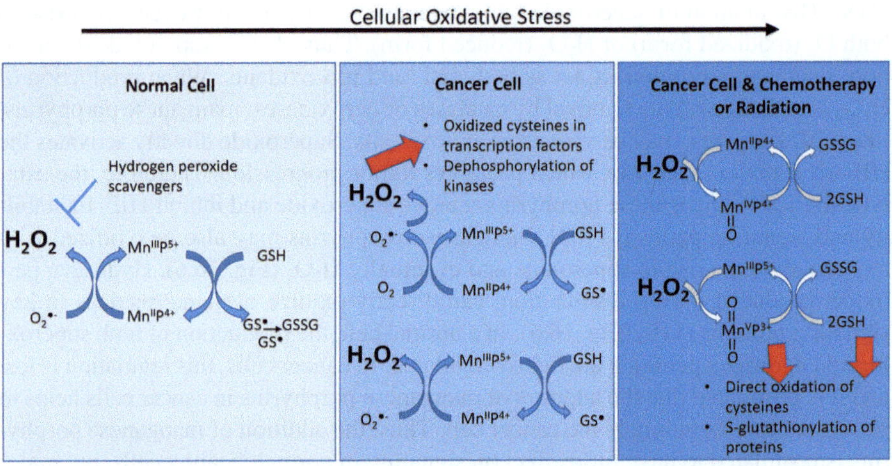

Fig. 16.6 Schematic illustrating the mechanisms by which the manganese porphyrins can be both antioxidants and pro-oxidants depending on the cellular environment. Manganese porphyrins use small anionic molecules such as GSH and ascorbate to cycle between oxidation states. In normal cells, the redox-active metallo-porphyrins behave as mild antioxidants or mild pro-oxidants depending, in part, upon the activity of hydrogen peroxide scavengers. In cancer cells, where the environment is more oxidizing, the porphyrins behave as pro-oxidants by additionally increasing the oxidative burden due to already existing localized levels of hydrogen peroxide. These localized increases in hydrogen peroxide affect signaling through oxidation of cysteines and dephosphorylation of kinases. In highly oxidized tumor cells, such as cells treated with chemotherapy, the manganese porphyrins in combination with H_2O_2 can directly oxidize proteins and lipids and inhibit protein activity through glutathionylation. It should be noted that in every cell there is a potential for any of the above reactions to take place. However, the cellular environment of the particular cells (normal vs. cancer cells) will differentially favor the magnitude and yield of certain reactions. For simplicity, hydrogen atoms have been left off the diagrams

manganese porphyrins are cycling between $Mn^{III}P^{5+}$ and $Mn^{II}P^{4+}$ states [118]. Depending on the local concentration of anionic species, the porphyrin is either scavenging superoxide and producing H_2O_2 and O_2 or cycling with anionic species and producing H_2O_2 and O_2 (Fig. 16.6). The cell's ability to effectively scavenge H_2O_2 will also determine the local concentration of H_2O_2. These localized increases in H_2O_2 can lead to the oxidation of cysteine residues and can also change the phosphorylation state of proteins and ultimately change the signaling pathways that affect the entire cell. In a cell that has a highly oxidizing environment (a cancer cell treated with chemotherapy or radiation), the manganese porphyrins will become oxidized with H_2O_2 to either the $O{=}Mn^{IV}P^{4+}$ or $O{=}Mn^{V}P{=}O^{3+}$ states, which can in turn oxidize or S-glutathionylated cysteines [118]. The oxidized porphyrins require H_2O_2 to act as a catalyst to oxidize and glutathionylate thiols on signaling proteins. Therefore, the redox-active manganese porphyrins act differently in a normal cell as compared to a cancer cell due to differences in oxidative stress, the abundance of cellular reductants and in the ability to effectively remove H_2O_2 in these two different cell types (Fig. 16.6).

16.18 Conclusions

There has been 30 years of research showing that redox biology plays an important part in both normal cell homeostasis and in disease progression, in particular cancer [1, 33, 40, 45, 47]. Almost every tumor type has an imbalanced redox system. Some tumors have low levels of antioxidant defense systems while other tumors have high levels of antioxidants, but in both cases the cancer cells are under oxidative stress. On the other hand, normal cells have tight regulation of their redox balance, are in a reduced state and can effectively combat momentarily high levels of oxidative stress and return to redox homeostasis. In cancer therapy, the redox environment is a logical process to modulate because oxidative stress promotes cancer cell progression and the redox environments are markedly different between cancer cells and normal cells.

There has been a significant amount of work showing that altering the tumor redox environment reduces tumor growth. It is postulated that manganese porphyrins are scavenging superoxide and inhibiting signaling that is directly affected by superoxide, i.e., HIF-1α signaling. In addition by scavenging superoxide, an increased rate or burst of H_2O_2 may occur in localized intracellular or extracellular domains. These changes in concentration and duration of H_2O_2 exposure can change signaling in and between cells. Thus, it is hypothesized that the manganese porphyrins can contribute to regulation of H_2O_2 production [134], which results in H_2O_2 signaling that occurs in normal cells [118].

Over the past 15 years, a wealth of data shows that manganese porphyrins protect normal tissues from oxidative stress and inflammation in a variety of disease models, including cancer therapy [87]. The porphyrins enhance the killing of cancer cells both alone and in combination with other cancer therapeutic agents [116, 117, 119].

These compounds are effective in minimizing damage to normal cells while enhancing damage to cancer cells.

These redox-active metallo-porphyrins hold promise as therapeutic agents because they are able to affect cell signaling in a normal cell differently than a cancer cell. Clinical trials need to investigate the ability of these porphyrins to protect normal tissues from the adverse effects of radiation and chemotherapy and assess their simultaneous effects on tumor growth. The redox-active metalloporphyrins offer a new approach in the treatment of human cancers.

Acknowledgment Rebecca E. Oberley-Deegan acknowledges financial support from RO1CA178888.

References

1. Gius D, Spitz DR. Redox signaling in cancer biology. Antioxid Redox Signal. 2006;8(7–8):1249–52.
2. Simons AL, Mattson DM, Dornfeld K, Spitz DR. Glucose deprivation-induced metabolic oxidative stress and cancer therapy. J Cancer Res Ther. 2009;5 Suppl 1:S2–6.
3. Palsson-McDermott EM, O'Neill LA. The Warburg effect then and now: from cancer to inflammatory diseases. Bioessays. 2013;35(11):965–73.
4. Kinnula VL, Crapo JD. Superoxide dismutases in malignant cells and human tumors. Free Radic Biol Med. 2004;36(6):718–44.
5. Weyemi U, Redon CE, Parekh PR, Dupuy C, Bonner WM. NADPH oxidases NOXs and DUOXs as putative targets for cancer therapy. Anticancer Agents Med Chem. 2013; 13(3):502–14.
6. Bauer G. Targeting extracellular ROS signaling of tumor cells. Anticancer Res. 2014;34(4):1467–82.
7. Nystrom T, Yang J, Molin M. Peroxiredoxins, gerontogenes linking aging to genome instability and cancer. Genes Dev. 2012;26(18):2001–8.
8. Bai J, Zhu X, Zheng X, Wu Y. Overexpression of CuZnSOD gene suppresses the growth of hepatocellular cancer cell line HepG2. Chin Med J (Engl). 1998;111(9):789–92.
9. Echiburu-Chau C, Roy D, Calaf GM. Deleterious MnSOD signals lead to abnormal breast cell proliferation by radiation and estrogen exposure. Int J Oncol. 2011;38(6):1703–11.
10. Feng W, Mei S, Wenjie Y, Luyuan H. High-level soluble expression of recombinant human manganese superoxide dismutase in Escherichia coli, and its effects on proliferation of the leukemia cell. Protein Expr Purif. 2011;77(1):46–52.
11. Huang F, Ma B, Wang Y, Xiao R, Kong Y, Zhou X, Xia D. Targeting gene-virus-mediated manganese superoxide dismutase effectively suppresses tumor growth in hepatocellular carcinoma in vitro and in vivo. Cancer Biother Radiopharm. 2014;29(10):403–11.
12. Huang WC, Kuroiwa K, Serio AM, Bianco Jr FJ, Fine SW, Shayegan B, Scardino PT, Eastham JA. The anatomical and pathological characteristics of irradiated prostate cancers may influence the oncological efficacy of salvage ablative therapies. J Urol. 2007;177(4):1324–9; quiz 1591.
13. Jeon YJ, Yoo H, Kim BH, Lee YS, Jeon B, Kim SS, Kim TY. IFNgamma-mediated inhibition of cell proliferation through increased PKCdelta-induced overexpression of EC-SOD. BMB Rep. 2012;45(11):659–64.
14. Kim SH, Kim MO, Gao P, Youm CA, Park HR, Lee TS, Kim KS, Suh JG, Lee HT, Park BJ, Ryoo ZY, Lee TH. Overexpression of extracellular superoxide dismutase (EC-SOD) in mouse skin plays a protective role in DMBA/TPA-induced tumor formation. Oncol Res. 2005;15(7–8):333–41.

15. Liu D, Liu A. Superoxide dismutase induces G1-phase cell cycle arrest by down-regulated expression of Cdk-2 and cyclin-E in murine sarcoma S180 tumor cells. Cell Biochem Funct. 2013;31(4):352–9.
16. Oberley LW, Oberley TD. Role of antioxidant enzymes in cell immortalization and transformation. Mol Cell Biochem. 1988;84(2):147–53.
17. Piazuelo E, Cebrian C, Escartin A, Jimenez P, Soteras F, Ortego J, Lanas A. Superoxide dismutase prevents development of adenocarcinoma in a rat model of Barrett's esophagus. World J Gastroenterol. 2005;11(47):7436–43.
18. Plymate SR, Haugk KH, Sprenger CC, Nelson PS, Tennant MK, Zhang Y, Oberley LW, Zhong W, Drivdahl R, Oberley TD. Increased manganese superoxide dismutase (SOD-2) is part of the mechanism for prostate tumor suppression by Mac25/insulin-like growth factor binding-protein-related protein-1. Oncogene. 2003;22(7):1024–34.
19. Rausalu K, Karo-Astover L, Kilk A, Ustav M. CuZn-SOD suppresses the bovine papillomavirus-induced proliferation of fibroblasts. APMIS. 2007;115(12):1415–21.
20. Robbins D, Zhao Y. The role of manganese superoxide dismutase in skin cancer. Enzyme Res. 2011;2011:409295.
21. Tanaka M, Kogawa K, Nishihori Y, Kuribayashi K, Nakamura K, Muramatsu H, Koike K, Sakamaki S, Niitsu Y. Suppression of intracellular Cu-Zn SOD results in enhanced motility and metastasis of Meth A sarcoma cells. Int J Cancer. 1997;73(2):187–92.
22. Tanaka M, Kuribayashi K, Kogawa K, Nakamura K, Watanabe N. Intracellular superoxide dismutase activity defines invasiveness of the murine T-lymphoma cell line L5187Y-ML25 in vitro and in vivo. Leuk Res. 2013;37(1):89–92.
23. Teoh ML, Fitzgerald MP, Oberley LW, Domann FE. Overexpression of extracellular superoxide dismutase attenuates heparanase expression and inhibits breast carcinoma cell growth and invasion. Cancer Res. 2009;69(15):6355–63.
24. Thongphasuk J, Oberley LW, Oberley TD. Induction of superoxide dismutase and cytotoxicity by manganese in human breast cancer cells. Arch Biochem Biophys. 1999;365(2):317–27.
25. Weydert C, Roling B, Liu J, Hinkhouse MM, Ritchie JM, Oberley LW, Cullen JJ. Suppression of the malignant phenotype in human pancreatic cancer cells by the overexpression of manganese superoxide dismutase. Mol Cancer Ther. 2003;2(4):361–9.
26. Wheeler MD, Smutney OM, Samulski RJ. Secretion of extracellular superoxide dismutase from muscle transduced with recombinant adenovirus inhibits the growth of B16 melanomas in mice. Mol Cancer Res. 2003;1(12):871–81.
27. Connor KM, Hempel N, Nelson KK, Dabiri G, Gamarra A, Belarmino J, Van De Water L, Mian BM, Melendez JA. Manganese superoxide dismutase enhances the invasive and migratory activity of tumor cells. Cancer Res. 2007;67(21):10260–7.
28. Hempel N, Melendez JA. Intracellular redox status controls membrane localization of pro- and anti-migratory signaling molecules. Redox Biol. 2014;2:245–50.
29. Hempel N, Ye H, Abessi B, Mian B, Melendez JA. Altered redox status accompanies progression to metastatic human bladder cancer. Free Radic Biol Med. 2009;46(1):42–50.
30. Hu Y, Rosen DG, Zhou Y, Feng L, Yang G, Liu J, Huang P. Mitochondrial manganese-superoxide dismutase expression in ovarian cancer: role in cell proliferation and response to oxidative stress. J Biol Chem. 2005;280(47):39485–92.
31. Kuninaka S, Ichinose Y, Koja K, Toh Y. Suppression of manganese superoxide dismutase augments sensitivity to radiation, hyperthermia and doxorubicin in colon cancer cell lines by inducing apoptosis. Br J Cancer. 2000;83(7):928–34.
32. Nelson KK, Ranganathan AC, Mansouri J, Rodriguez AM, Providence KM, Rutter JL, Pumiglia K, Bennett JA, Melendez JA. Elevated sod2 activity augments matrix metalloproteinase expression: evidence for the involvement of endogenous hydrogen peroxide in regulating metastasis. Clin Cancer Res. 2003;9(1):424–32.
33. Hempel N, Carrico PM, Melendez JA. Manganese superoxide dismutase (Sod2) and redox-control of signaling events that drive metastasis. Anticancer Agents Med Chem. 2011;11(2):191–201.

34. Circu ML, Aw TY. Reactive oxygen species, cellular redox systems, and apoptosis. Free Radic Biol Med. 2010;48(6):749–62.
35. Sun Y, Oberley LW. Redox regulation of transcriptional activators. Free Radic Biol Med. 1996;21(3):335–48.
36. Cencioni C, Spallotta F, Martelli F, Valente S, Mai A, Zeiher AM, Gaetano C. Oxidative stress and epigenetic regulation in ageing and age-related diseases. Int J Mol Sci. 2013;14(9):17643–63.
37. Merksamer PI, Liu Y, He W, Hirschey MD, Chen D, Verdin E. The sirtuins, oxidative stress and aging: an emerging link. Aging (Albany NY). 2013;5(3):144–50.
38. Rahman I, Marwick J, Kirkham P. Redox modulation of chromatin remodeling: impact on histone acetylation and deacetylation, NF-kappaB and pro-inflammatory gene expression. Biochem Pharmacol. 2004;68(6):1255–67.
39. Reece KM, Richardson ED, Cook KM, Campbell TJ, Pisle ST, Holly AJ, Venzon DJ, Liewehr DJ, Chau CH, Price DK, Figg WD. Epidithiodiketopiperazines (ETPs) exhibit in vitro anti-angiogenic and in vivo antitumor activity by disrupting the HIF-1alpha/p300 complex in a preclinical model of prostate cancer. Mol Cancer. 2014;13:91.
40. Sarsour EH, Kumar MG, Chaudhuri L, Kalen AL, Goswami PC. Redox control of the cell cycle in health and disease. Antioxid Redox Signal. 2009;11(12):2985–3011.
41. Cucoranu I, Clempus R, Dikalova A, Phelan PJ, Ariyan S, Dikalov S, Sorescu D. NAD(P)H oxidase 4 mediates transforming growth factor-beta1-induced differentiation of cardiac fibroblasts into myofibroblasts. Circ Res. 2005;97(9):900–7.
42. Menon SG, Sarsour EH, Kalen AL, Venkataraman S, Hitchler MJ, Domann FE, Oberley LW, Goswami PC. Superoxide signaling mediates N-acetyl-L-cysteine-induced G1 arrest: regulatory role of cyclin D1 and manganese superoxide dismutase. Cancer Res. 2007;67(13):6392–9.
43. Church SL, Grant JW, Ridnour LA, Oberley LW, Swanson PE, Meltzer PS, Trent JM. Increased manganese superoxide dismutase expression suppresses the malignant phenotype of human melanoma cells. Proc Natl Acad Sci U S A. 1993;90(7):3113–7.
44. Savitsky PA, Finkel T. Redox regulation of Cdc25C. J Biol Chem. 2002;277(23):20535–40.
45. Glasauer A, Chandel NS. Targeting antioxidants for cancer therapy. Biochem Pharmacol. 2014;92(1):90–101.
46. Ichijo H, Nishida E, Irie K, ten Dijke P, Saitoh M, Moriguchi T, Takagi M, Matsumoto K, Miyazono K, Gotoh Y. Induction of apoptosis by ASK1, a mammalian MAPKKK that activates SAPK/JNK and p38 signaling pathways. Science. 1997;275(5296):90–4.
47. Tertil M, Jozkowicz A, Dulak J. Oxidative stress in tumor angiogenesis—therapeutic targets. Curr Pharm Des. 2010;16(35):3877–94.
48. Venkataraman S, Jiang X, Weydert C, Zhang Y, Zhang HJ, Goswami PC, Ritchie JM, Oberley LW, Buettner GR. Manganese superoxide dismutase overexpression inhibits the growth of androgen-independent prostate cancer cells. Oncogene. 2005;24(1):77–89.
49. Wang M, Kirk JS, Venkataraman S, Domann FE, Zhang HJ, Schafer FQ, Flanagan SW, Weydert CJ, Spitz DR, Buettner GR, Oberley LW. Manganese superoxide dismutase suppresses hypoxic induction of hypoxia-inducible factor-1alpha and vascular endothelial growth factor. Oncogene. 2005;24(55):8154–66.
50. Kaewpila S, Venkataraman S, Buettner GR, Oberley LW. Manganese superoxide dismutase modulates hypoxia-inducible factor-1 alpha induction via superoxide. Cancer Res. 2008;68(8):2781–8.
51. Welsh SJ, Bellamy WT, Briehl MM, Powis G. The redox protein thioredoxin-1 (Trx-1) increases hypoxia-inducible factor 1alpha protein expression: Trx-1 overexpression results in increased vascular endothelial growth factor production and enhanced tumor angiogenesis. Cancer Res. 2002;62(17):5089–95.
52. Welsh SJ, Williams RR, Birmingham A, Newman DJ, Kirkpatrick DL, Powis G. The thioredoxin redox inhibitors 1-methylpropyl 2-imidazolyl disulfide and pleurotin inhibit hypoxia-induced factor 1alpha and vascular endothelial growth factor formation. Mol Cancer Ther. 2003;2(3):235–43.

53. Buettner GR. Superoxide dismutase in redox biology: the roles of superoxide and hydrogen peroxide. Anticancer Agents Med Chem. 2011;11(4):341–6.
54. Tochhawng L, Deng S, Pervaiz S, Yap CT. Redox regulation of cancer cell migration and invasion. Mitochondrion. 2013;13(3):246–53.
55. Wu WS, Tsai RK, Chang CH, Wang S, Wu JR, Chang YX. Reactive oxygen species mediated sustained activation of protein kinase C alpha and extracellular signal-regulated kinase for migration of human hepatoma cell Hepg2. Mol Cancer Res. 2006;4(10):747–58.
56. Ben Mahdi MH, Andrieu V, Pasquier C. Focal adhesion kinase regulation by oxidative stress in different cell types. IUBMB Life. 2000;50(4–5):291–9.
57. Moldovan L, Mythreye K, Goldschmidt-Clermont PJ, Satterwhite LL. Reactive oxygen species in vascular endothelial cell motility. Roles of NAD(P)H oxidase and Rac1. Cardiovasc Res. 2006;71(2):236–46.
58. Hung RJ, Pak CW, Terman JR. Direct redox regulation of F-actin assembly and disassembly by Mical. Science. 2011;334(6063):1710–3.
59. Brenneisen P, Sies H, Scharffetter-Kochanek K. Ultraviolet-B irradiation and matrix metalloproteinases: from induction via signaling to initial events. Ann N Y Acad Sci. 2002;973:31–43.
60. Kim MH, Yoo HS, Kim MY, Jang HJ, Baek MK, Kim HR, Kim KK, Shin BA, Ahn BW, Jung YD. Helicobacter pylori stimulates urokinase plasminogen activator receptor expression and cell invasiveness through reactive oxygen species and NF-kappaB signaling in human gastric carcinoma cells. Int J Mol Med. 2007;19(4):689–97.
61. Ho BY, Wu YM, Chang KJ, Pan TM. Dimerumic acid inhibits SW620 cell invasion by attenuating H(2)O(2)-mediated MMP-7 expression via JNK/C-Jun and ERK/C-Fos activation in an AP-1-dependent manner. Int J Biol Sci. 2011;7(6):869–80.
62. Reddy SP, Mossman BT. Role and regulation of activator protein-1 in toxicant-induced responses of the lung. Am J Physiol Lung Cell Mol Physiol. 2002;283(6):L1161–78.
63. Li WQ, Qureshi HY, Liacini A, Dehnade F, Zafarullah M. Transforming growth factor Beta1 induction of tissue inhibitor of metalloproteinases 3 in articular chondrocytes is mediated by reactive oxygen species. Free Radic Biol Med. 2004;37(2):196–207.
64. Nelson KK, Melendez JA. Mitochondrial redox control of matrix metalloproteinases. Free Radic Biol Med. 2004;37(6):768–84.
65. Bostwick DG, Alexander EE, Singh R, Shan A, Qian J, Santella RM, Oberley LW, Yan T, Zhong W, Jiang X, Oberley TD. Antioxidant enzyme expression and reactive oxygen species damage in prostatic intraepithelial neoplasia and cancer. Cancer. 2000;89(1):123–34.
66. Lam EW, Zwacka R, Seftor EA, Nieva DR, Davidson BL, Engelhardt JF, Hendrix MJ, Oberley LW. Effects of antioxidant enzyme overexpression on the invasive phenotype of hamster cheek pouch carcinoma cells. Free Radic Biol Med. 1999;27(5–6):572–9.
67. Saito T, Kurasaki M, Kaji H, Saito K. Deficiency of erythrocyte superoxide dismutase and catalase activities in patients with malignant lymphoma and acute myeloid leukemia. Cancer Lett. 1984;24(2):141–6.
68. Buur JL, Diniz PP, Roderick KV, KuKanich B, Tegzes JH. Pharmacokinetics of N-acetylcysteine after oral and intravenous administration to healthy cats. Am J Vet Res. 2013;74(2):290–3.
69. Duconge J, Miranda-Massari JR, Gonzalez MJ, Jackson JA, Warnock W, Riordan NH. Pharmacokinetics of vitamin C: insights into the oral and intravenous administration of ascorbate. P R Health Sci J. 2008;27(1):7–19.
70. Welsh JL, Wagner BA, van't Erve TJ, Zehr PS, Berg DJ, Halfdanarson TR, Yee NS, Bodeker KL, Du J, Roberts 2nd LJ, Drisko J, Levine M, Buettner GR, Cullen JJ. Pharmacological ascorbate with gemcitabine for the control of metastatic and node-positive pancreatic cancer (PACMAN): results from a phase I clinical trial. Cancer Chemother Pharmacol. 2013;71(3):765–75.
71. Kaufmann SH, Earnshaw WC. Induction of apoptosis by cancer chemotherapy. Exp Cell Res. 2000;256(1):42–9.
72. Saydam N, Kirb A, Demir O, Hazan E, Oto O, Saydam O, Guner G. Determination of glutathione, glutathione reductase, glutathione peroxidase and glutathione S-transferase levels in human lung cancer tissues. Cancer Lett. 1997;119(1):13–9.

73. Hayes JD, McMahon M. NRF2 and KEAP1 mutations: permanent activation of an adaptive response in cancer. Trends Biochem Sci. 2009;34(4):176–88.
74. Satoh H, Moriguchi T, Takai J, Ebina M, Yamamoto M. Nrf2 prevents initiation but accelerates progression through the Kras signaling pathway during lung carcinogenesis. Cancer Res. 2013;73(13):4158–68.
75. Young TW, Mei FC, Yang G, Thompson-Lanza JA, Liu J, Cheng X. Activation of antioxidant pathways in ras-mediated oncogenic transformation of human surface ovarian epithelial cells revealed by functional proteomics and mass spectrometry. Cancer Res. 2004;64(13): 4577–84.
76. Magda D, Miller RA. Motexafin gadolinium: a novel redox active drug for cancer therapy. Semin Cancer Biol. 2006;16(6):466–76.
77. Sobhakumari A, Love-Homan L, Fletcher EV, Martin SM, Parsons AD, Spitz DR, Knudson CM, Simons AL. Susceptibility of human head and neck cancer cells to combined inhibition of glutathione and thioredoxin metabolism. PLoS One. 2012;7(10):e48175.
78. Somwar R, Erdjument-Bromage H, Larsson E, Shum D, Lockwood WW, Yang G, Sander C, Ouerfelli O, Tempst PJ, Djaballah H, Varmus HE. Superoxide dismutase 1 (SOD1) is a target for a small molecule identified in a screen for inhibitors of the growth of lung adenocarcinoma cell lines. Proc Natl Acad Sci U S A. 2011;108(39):16375–80.
79. Huang P, Feng L, Oldham EA, Keating MJ, Plunkett W. Superoxide dismutase as a target for the selective killing of cancer cells. Nature. 2000;407(6802):390–5.
80. James J, Murry DJ, Treston AM, Storniolo AM, Sledge GW, Sidor C, Miller KD. Phase I safety, pharmacokinetic and pharmacodynamic studies of 2-methoxyestradiol alone or in combination with docetaxel in patients with locally recurrent or metastatic breast cancer. Invest New Drugs. 2007;25(1):41–8.
81. Sweeney C, Liu G, Yiannoutsos C, Kolesar J, Horvath D, Staab MJ, Fife K, Armstrong V, Treston A, Sidor C, Wilding G. A phase II multicenter, randomized, double-blind, safety trial assessing the pharmacokinetics, pharmacodynamics, and efficacy of oral 2-methoxyestradiol capsules in hormone-refractory prostate cancer. Clin Cancer Res. 2005;11(18):6625–33.
82. Florczak U, Toulany M, Kehlbach R, Peter Rodemann H. 2-Methoxyestradiol-induced radiosensitization is independent of SOD but depends on inhibition of Akt and DNA-PKcs activities. Radiother Oncol. 2009;92(3):334–8.
83. Kim YH, Coon A, Baker AF, Powis G. Antitumor agent PX-12 inhibits HIF-1alpha protein levels through an Nrf2/PMF-1-mediated increase in spermidine/spermine acetyl transferase. Cancer Chemother Pharmacol. 2011;68(2):405–13.
84. Baker AF, Dragovich T, Tate WR, Ramanathan RK, Roe D, Hsu CH, Kirkpatrick DL, Powis G. The antitumor thioredoxin-1 inhibitor PX-12 (1-methylpropyl 2-imidazolyl disulfide) decreases thioredoxin-1 and VEGF levels in cancer patient plasma. J Lab Clin Med. 2006;147(2):83–90.
85. Kachadourian R, Liochev SI, Cabelli DE, Patel MN, Fridovich I, Day BJ. 2-methoxyestradiol does not inhibit superoxide dismutase. Arch Biochem Biophys. 2001;392(2):349–53.
86. Lee YM, Ting CM, Cheng YK, Fan TP, Wong RN, Lung ML, Mak NK. Mechanisms of 2-methoxyestradiol-induced apoptosis and G2/M cell-cycle arrest of nasopharyngeal carcinoma cells. Cancer Lett. 2008;268(2):295–307.
87. Tovmasyan A, Sheng H, Weitner T, Arulpragasam A, Lu M, Warner DS, Vujaskovic Z, Spasojevic I, Batinic-Haberle I. Design, mechanism of action, bioavailability and therapeutic effects of mn porphyrin-based redox modulators. Med Princ Pract. 2013;22(2):103–30.
88. Batinic-Haberle I, Spasojevic I, Stevens RD, Hambright P, Neta P, Okado-Matsumoto A, Fridovich I. New class of potent catalysts of O2.-dismutation. Mn(III) ortho-methoxyethylpyridyl- and di-ortho-methoxyethylimidazolylporphyrins. Dalton Trans. 2004;11:1696–702.
89. Vujaskovic Z, Batinic-Haberle I, Rabbani ZN, Feng QF, Kang SK, Spasojevic I, Samulski TV, Fridovich I, Dewhirst MW, Anscher MS. A small molecular weight catalytic metalloporphyrin antioxidant with superoxide dismutase (SOD) mimetic properties protects lungs from radiation-induced injury. Free Radic Biol Med. 2002;33(6):857–63.

90. Batinic-Haberle I, Spasojevic I, Tse HM, Tovmasyan A, Rajic Z, Clair DKS, Vujaskovic Z, Dewhirst MW, Piganelli JD. Design of Mn porphyrins for treating oxidative stress injuries and their redox-based regulation of cellular transcriptional activities. Amino Acids. 2012;42(1):95–113.
91. Kos I, Benov L, Spasojevic I, Reboucas JS, Batinic-Haberle I. High lipophilicity of meta Mn(III) N-alkylpyridylporphyrin-based superoxide dismutase mimics compensates for their lower antioxidant potency and makes them as effective as ortho analogues in protecting superoxide dismutase-deficient Escherichia coli. J Med Chem. 2009;52(23):7868–72.
92. Li AM, Martins J, Tovmasyan A, Valentine JS, Batinic-Haberle I, Spasojevic I, Gralla EB. Differential localization and potency of manganese porphyrin superoxide dismutase-mimicking compounds in Saccharomyces cerevisiae. Redox Biol. 2014;3:1–6.
93. Spasojevic I, Chen Y, Noel TJ, Fan P, Zhang L, Reboucas JS, Clair DKS, Batinic-Haberle I. Pharmacokinetics of the potent redox-modulating manganese porphyrin, MnTE-2-PyP(5+), in plasma and major organs of B6C3F1 mice. Free Radic Biol Med. 2008;45(7):943–9.
94. Batinic-Haberle I, Tovmasyan A, Roberts ER, Vujaskovic Z, Leong KW, Spasojevic I. SOD therapeutics: latest insights into their structure-activity relationships and impact on the cellular redox-based signaling pathways. Antioxid Redox Signal. 2014;20(15):2372–415.
95. Miriyala S, Spasojevic I, Tovmasyan A, Salvemini D, Vujaskovic Z, Clair DS, Batinic-Haberle I. Manganese superoxide dismutase, MnSOD and its mimics. Biochim Biophys Acta. 2012;1822(5):794–814.
96. Sheng H, Spasojevic I, Tse HM, Jung JY, Hong J, Zhang Z, Piganelli JD, Batinic-Haberle I, Warner DS. Neuroprotective efficacy from a lipophilic redox-modulating Mn(III) N-hexylpyridylporphyrin, MnTnHex-2-PyP: rodent models of ischemic stroke and subarachnoid hemorrhage. J Pharmacol Exp Ther. 2011;338(3):906–16.
97. Rajic Z, Tovmasyan A, Spasojevic I, Sheng H, Lu M, Li AM, Gralla EB, Warner DS, Benov L, Batinic-Haberle I. A new SOD mimic, Mn(III) ortho N-butoxyethylpyridylporphyrin, combines superb potency and lipophilicity with low toxicity. Free Radic Biol Med. 2012;52(9):1828–34.
98. Tong, Q, Weaver MR, Kosmacek EA, O'Connor B, Harmacek L, Venkataraman S, and Oberley-Deegan RE, MnTE-2-PyP reduces prostate cancer growth and metastasis by suppressing p300 activity and p300/HIF-1/CREB binding to the promoter region of the PAI-1 gene. Free Radic Biol Med. 2016;94.185–94.
99. Fernandes AS, Florido A, Cipriano M, Batinic-Haberle I, Miranda J, Saraiva N, Guerreiro PS, Castro M, Oliveira NG. Combined effect of the SOD mimic MnTnHex-2-PyP5+ and doxorubicin on the migration and invasiveness of breast cancer cells. Toxicol Lett. 2013;221:S70–1.
100. Rabbani ZN, Spasojevic I, Zhang X, Moeller BJ, Haberle S, Vasquez-Vivar J, Dewhirst MW, Vujaskovic Z, Batinic-Haberle I. Antiangiogenic action of redox-modulating Mn(III) meso-tetrakis(N-ethylpyridinium-2-yl)porphyrin, MnTE-2-PyP(5+), via suppression of oxidative stress in a mouse model of breast tumor. Free Radic Biol Med. 2009;47(7):992–1004.
101. Zhao Y, Chaiswing L, Oberley TD, Batinic-Haberle I, Clair WS, Epstein CJ, Clair DS. A mechanism-based antioxidant approach for the reduction of skin carcinogenesis. Cancer Res. 2005;65(4):1401–5.
102. Gridley DS, Makinde AY, Luo X, Rizvi A, Crapo JD, Dewhirst MW, Moeller BJ, Pearlstein RD, Slater JM. Radiation and a metalloporphyrin radioprotectant in a mouse prostate tumor model. Anticancer Res. 2007;27(5A):3101–9.
103. Makinde AY, Luo-Owen X, Rizvi A, Crapo JD, Pearlstein RD, Slater JM, Gridley DS. Effect of a metalloporphyrin antioxidant (MnTE-2-PyP) on the response of a mouse prostate cancer model to radiation. Anticancer Res. 2009;29(1):107–18.
104. Moeller BJ, Batinic-Haberle I, Spasojevic I, Rabbani ZN, Anscher MS, Vujaskovic Z, Dewhirst MW. A manganese porphyrin superoxide dismutase mimetic enhances tumor radioresponsiveness. Int J Radiat Oncol Biol Phys. 2005;63(2):545–52.

105. Moeller BJ, Cao Y, Li CY, Dewhirst MW. Radiation activates HIF-1 to regulate vascular radiosensitivity in tumors: role of reoxygenation, free radicals, and stress granules. Cancer Cell. 2004;5(5):429–41.
106. Oberley-Deegan RE, Steffan JJ, Rove KO, Pate KM, Weaver MW, Spasojevic I, Frederick B, Raben D, Meacham RB, Crapo JD, Koul HK. The antioxidant, MnTE-2-PyP, prevents side-effects incurred by prostate cancer irradiation. PLoS One. 2012;7(9):e44178.
107. Keir ST, Dewhirst MW, Kirkpatrick JP, Bigner DD, Batinic-Haberle I. Cellular redox modulator, ortho Mn(III) meso-tetrakis(N-n-hexylpyridinium-2-yl)porphyrin, MnTnHex-2-PyP(5+) in the treatment of brain tumors. Anticancer Agents Med Chem. 2011;11(2):202–12.
108. Dewhirst MWAK, Batinic-Haberle I, Spasojevic I, Brizel DM. A novel MnSOD mimetic widens the therapeutic margin by simultaneously radioprotecting normal tissue and radiosensitizing tumor. In: International journal of radiation oncology, San Fransisco; 2014.
109. Batinic-Haberle I, Kerr ST, Rajic Z, Tovmasyan A, Bigner DD. Lipophilic Mn porphyrins in the treatment of brain tumors. In: Society for free radical biology and medicine, Atlanta; 2011.
110. Weitzel DH, Tovmasyan A, Ashcraft KA, Rajic Z, Weitner T, Liu C, Li W, Buckley AF, Prasad MR, Young KH, Rodriguiz RM, Wetsel WC, Peters KB, Spasojevic I, Herndon 2nd JE, Batinic-Haberle I, Dewhirst MW. Radioprotection of the brain white matter by Mn(III) n-butoxyethylpyridylporphyrin-based superoxide dismutase mimic MnTnBuOE-2-PyP5+. Mol Cancer Ther. 2015;14(1):70–9.
111. Archambeau JO, Tovmasyan A, Pearlstein RD, Crapo JD, Batinic-Haberle I. Superoxide dismutase mimic, MnTE-2-PyP(5+) ameliorates acute and chronic proctitis following focal proton irradiation of the rat rectum. Redox Biol. 2013;1(1):599–607.
112. Gauter-Fleckenstein B, Fleckenstein K, Owzar K, Jiang C, Reboucas JS, Batinic-Haberle I, Vujaskovic Z. Early and late administration of MnTE-2-PyP5+ in mitigation and treatment of radiation-induced lung damage. Free Radic Biol Med. 2010;48(8):1034–43.
113. Mao XW, Crapo JD, Mekonnen T, Lindsey N, Martinez P, Gridley DS, Slater JM. Radioprotective effect of a metalloporphyrin compound in rat eye model. Curr Eye Res. 2009;34(1):62–72.
114. Pearlstein RD, Higuchi Y, Moldovan M, Johnson K, Fukuda S, Gridley DS, Crapo JD, Warner DS, Slater JM. Metalloporphyrin antioxidants ameliorate normal tissue radiation damage in rat brain. Int J Radiat Biol. 2010;86(2):145–63.
115. Jaramillo MC, Briehl MM, Batinic-Haberle I, Tome ME. Manganese (III) meso-tetrakis N-ethylpyridinium-2-yl porphyrin acts as a pro-oxidant to inhibit electron transport chain proteins, modulate bioenergetics, and enhance the response to chemotherapy in lymphoma cells. Free Radic Biol Med. 2015;83:89–100.
116. Jaramillo MC, Briehl MM, Crapo JD, Batinic-Haberle I, Tome ME. Manganese porphyrin, MnTE-2-PyP5+, acts as a pro-oxidant to potentiate glucocorticoid-induced apoptosis in lymphoma cells. Free Radic Biol Med. 2012;52(8):1272–84.
117. Jaramillo MC, Frye JB, Crapo JD, Briehl MM, Tome ME. Increased manganese superoxide dismutase expression or treatment with manganese porphyrin potentiates dexamethasone-induced apoptosis in lymphoma cells. Cancer Res. 2009;69(13):5450–7.
118. Batinic-Haberle I, Tovmasyan A, Spasojevic I. An educational overview of the chemistry, biochemistry and therapeutic aspects of Mn porphyrins—from superoxide dismutation to HO-driven pathways. Redox Biol. 2015;5:43–65.
119. Evans MK, Tovmasyan A, Batinic-Haberle I, Devi GR. Mn porphyrin in combination with ascorbate acts as a pro-oxidant and mediates caspase-independent cancer cell death. Free Radic Biol Med. 2014;68:302–14.
120. Rawal M, Schroeder SR, Wagner BA, Cushing CM, Welsh JL, Button AM, Du J, Sibenaller ZA, Buettner GR, Cullen JJ. Manganoporphyrins increase ascorbate-induced cytotoxicity by enhancing H2O2 generation. Cancer Res. 2013;73(16):5232–41.
121. Ye X, Fels D, Tovmasyan A, Aird KM, Dedeugd C, Allensworth JL, Kos I, Park W, Spasojevic I, Devi GR, Dewhirst MW, Leong KW, Batinic-Haberle I. Cytotoxic effects of Mn(III) N-alkylpyridylporphyrins in the presence of cellular reductant, ascorbate. Free Radic Res. 2011;45(11–12):1289–306.

122. Makinde AY, Rizvi A, Crapo JD, Pearlstein RD, Slater JM, Gridley DS. A metalloporphyrin antioxidant alters cytokine responses after irradiation in a prostate tumor model. Radiat Res. 2010;173(4):441–52.
123. Jungsuwadee P, Weaver MR, Gally F, Oberley-Deegan RE. The metalloporphyrin antioxidant, MnTE-2-PyP, Inhibits Th2 cell immune responses in an asthma model. Int J Mol Sci. 2012;13(8):9785–97.
124. Delmastro-Greenwood MM, Votyakova T, Goetzman E, Marre ML, Previte DM, Tovmasyan A, Batinic-Haberle I, Trucco MM, Piganelli JD. Mn porphyrin regulation of aerobic glycolysis: implications on the activation of diabetogenic immune cells. Antioxid Redox Signal. 2013;19(16):1902–15.
125. Wu Y, Zhou BP. TNF-alpha/NF-kappaB/Snail pathway in cancer cell migration and invasion. Br J Cancer. 2010;102(4):639–44.
126. Zhao X, Laver T, Hong SW, Twitty Jr GB, Devos A, Devos M, Benveniste EN, Nozell SE. An NF-kappaB p65-cIAP2 link is necessary for mediating resistance to TNF-alpha induced cell death in gliomas. J Neurooncol. 2011;102(3):367–81.
127. Oberley-Deegan RE, Lee YM, Morey GE, Cook DM, Chan ED, Crapo JD. The antioxidant mimetic, MnTE-2-PyP, reduces intracellular growth of Mycobacterium abscessus. Am J Respir Cell Mol Biol. 2009;41(2):170–8.
128. Tse HM, Milton MJ, Piganelli JD. Mechanistic analysis of the immunomodulatory effects of a catalytic antioxidant on antigen-presenting cells: implication for their use in targeting oxidation-reduction reactions in innate immunity. Free Radic Biol Med. 2004;36(2): 233–47.
129. Khan I, Batinic-Haberle I, Benov LT. Effect of potent redox-modulating manganese porphyrin, MnTM-2-PyP, on the Na(+)/H(+) exchangers NHE-1 and NHE-3 in the diabetic rat. Redox Rep. 2009;14(6):236–42.
130. Zhao Y, Chaiswing L, Velez JM, Batinic-Haberle I, Colburn NH, Oberley TD, Clair DKS. p53 translocation to mitochondria precedes its nuclear translocation and targets mitochondrial oxidative defense protein-manganese superoxide dismutase. Cancer Res. 2005;65(9): 3745–50.
131. Batinic-Haberle I, Tovmasyan A, Weitner T, Rajic Z, Keir ST, Huang TT, Leu D, Weitzel DH, Beausejour CM, Miriyala S, Roberts ERH, Dewhirst MW, St. Clair D, Leong KW, Spasojevic I, Piganelli J, Tome M. Mechanistic considerations of the therapeutic effects of Mn porphyrins, commonly regarded as SOD mimics in anticancer therapy. Lessons from brain and lymphoma studies. In: Society for free radical biology and medicine; 2013. p. S2.
132. Batinic-Haberle I, Spasojevic I, Fridovich I. Tetrahydrobiopterin rapidly reduces the SOD mimic Mn(III) ortho-tetrakis(N-ethylpyridinium-2-yl)porphyrin. Free Radic Biol Med. 2004;37(3):367–74.
133. Batinic-Haberle I, Reboucas JS, Spasojevic I. Superoxide dismutase mimics: chemistry, pharmacology, and therapeutic potential. Antioxid Redox Signal. 2010;13(6):877–918.
134. Tovmasyan A, Maia CG, Weitner T, Caraballal S, Sampaio RS, Lieb D, Ghazaryan R, Ivanovic-Burmazovic I, Radi R, Reboucas JS, Spasojevic I, Benov L, Batinic-Haberle I. A comprehensive evaluation of catalase-like activity of different classes of redox-active therapeutics. Free Radic Biol Med. 2015;86:308–21.

Chapter 17
Targeted Therapy for Malignant Brain Tumors

Paula Lam, Nivedh Dinesh, and Xandra O. Breakefield

17.1 Introduction

Glioblastoma (WHO grade IV) (GBM) is among the most lethal of all cancers. The prognosis for patients diagnosed with these tumors remains dismal with a median survival of less than 15 months, and the 5-year median survival of less than 3 % [1]. In recent years, improvements have been targeted at improving the precision of tumor debulking during surgery. For example, fluorescence-guided surgery of malignant gliomas based on 5-aminolevulinic acid has enhanced the visibility of the tumor border region, thereby making the radical resection of tumors easier [2]. Other strategies exploit the electrical fields or heat, with or without hydrogel, for increase tumor cell kill. These include the NovoTTF-100A system which is a local treatment that produces low-intensity, intermediate-frequency alternating electric

P. Lam (✉)
Laboratory of Cancer Gene Therapy, Humphrey Oei Institute of Cancer Research, National Cancer Centre of Singapore, Singapore, Singapore

Department of Physiology, Yong Loo Lin School of Medicine, National University of Singapore, Singapore 117597, Singapore

Duke-NUS Graduate Medical School, Singapore 169547, Singapore
e-mail: cmrlyp@nccs.com.sg

N. Dinesh, M.D.
Laboratory of Cancer Gene Therapy, Humphrey Oei Institute of Cancer Research, National Cancer Centre of Singapore, Singapore, Singapore

Division of Neurosurgery, National University Hospital, Singapore 119074, Singapore

X.O. Breakefield
Department of Neurology and Radiology, Program in Neuroscience, Massachusetts General Hospital, Harvard Medical School, Boston, MA 02114, USA

I. Batinić-Haberle et al. (eds.), *Redox-Active Therapeutics*,
Oxidative Stress in Applied Basic Research and Clinical Practice,
DOI 10.1007/978-3-319-30705-3_17

fields that can interfere with the proper formation of mitotic spindles and affects polar molecules in telophase, thus preventing cell division [3]. The Nano-Therm uses "thermotherapy," which involves surgery to insert a liquid containing 15 nm-wide magnetic particles directly into the brain tumor of the patient. The heat generated from the oscillating magnetic nanoparticles can prime the tumors cells for increased sensitivity when combined with fractionated stereotactic radiotherapy [4]. Of interest, bioengineers have also contributed to innovative therapeutic platforms. The MRI-compatible long-term therapeutic hydrogel system consists of a thermosensitive/magnetic hydrogel. When the hydrogel is mixed with drug, sustained drug release and long-term MRI monitoring is feasible due to the properties of the biocompatible hydrogel. Paclitaxel-loaded biodegradable microsphere entrapped in an alginate gel matrix is one good example [5]. In this chapter however, we focus on innovative-targeted therapy for malignant brain tumors.

17.2 Current Challenges in GBM Treatment

17.2.1 Infiltrative Nature of GBM

The lack of long-term survivors is attributed to the highly infiltrating and heterogeneous nature of these neoplasms. These malignant cells are often seen surrounding the neurons and blood vessels (also known as perineuronal and perivascular satellitosis) [6] and migrate through the white matter tracts and along blood vessels to regions distant from the original tumor mass within the brain [7]. The lack of defined tumor margins makes complete surgical resection challenging. Thus, glioma recurrence is a common problem that typically occurs within a few centermeter of the margins of the resected cavity owing to the invasive nature of these tumors [8, 9]. Some of these migratory cells may temporarily exit the cell cycle during migration, thus making them resistant to therapies that target dividing cells [10]. Unfortunately, very few chemotherapeutic drugs can access the brain tumor from the body due to the presence of the blood–brain barrier, which is a highly selective barrier formed by the capillary endothelial cells connected by epithelial-like tight junctions [11]. Even if the drugs can pass through the blood–brain barrier, drug resistance and tumor recurrence remain as challenges to the treatment of GBM.

17.2.2 Resistance to Temozolomide (TMZ)

The initial management of patients with glioblastoma includes maximal surgical resection, followed by radiotherapy and concomitant and adjuvant chemotherapy with TMZ. The median survival with the combination of all three of these is 15 months and the median time to recurrence after standard therapy is 6.9 months

(95 % confidence interval, 5.8–8.2 months) [12]. Subsequently, the same group reported that survival of patients who received the combination therapy of TMZ and radiation exceeded that of radiation alone [13].

TMZ is a second-generation imidazotetrazine prodrug that undergoes spontaneous conversion to the active alkylating agent 3-methyl-(trizaen-1-yl)imidazole-4-carboxamide (MTIC) at physiological pH [14]. MTIC react with water to produce 5-aminoimidazole-4-carboxamine (AIC) and the highly reactive methyldiazonium cation which transfers a methyl group to DNA, thereby damaging it and causing TMZ to be cytotoxic to replicating glioma cells. The methylation occurs at purine bases of the DNA at the O^6 and the N^7 positions of guanine and the N^3 position of adenine [15, 16]. TMZ does not require hepatic metabolism for activation [17]. In addition, its small molecular weight allows efficient penetration across the blood–brain barrier [18]. However, resistance to TMZ is a common problem, and understanding these resistance mechanisms are keys to developing therapies that provide either synergistic or additive effects to the standard therapeutic protocols, in order to improve the management of patients and kill malignant glioblastoma cells. DNA repair mechanisms that act by the removal of TMZ-induced DNA adduct can contribute to the resistance of glioblastoma cells to TMZ.

There are three major pathways mediated by O6-methylguanine DNA methyltransferase (MGMT), DNA mismatch repair (MMR) and DNA base excision repair (BER). MGMT has alternatively been denoted as O6-alkyguanine DNA alkyltransferase (AGT) in a number of publications. In essence, the MGMT protein is capable of removing the methyl groups attached to the O^6 position in guanine (O6-meG; [19]). Approximately 45 % of newly diagnosed GBM patients have hypermethylation of the MGMT promoter [20]. As a result, the expression of MGMT is silenced and these patients tend to respond better to TMZ [21]. When TMZ-induced O6-meG adducts are left unrepaired by MGMT, they will pair with thymine. The resulting O6-meG/T is recognized by the MMR repair system which initiates a futile repair process leading to DNA strand breakage and cell death. However, alkylating agents can also induce mutations leading to inactivation of MMR. Under such a scenario, MMR deficiency can promote TMZ resistance despite low levels of MGMT [22]. Another possible mechanism of resistance for TMZ is the BER system which is involved in the repair of DNA damage caused by oxidizing agents, ionizing radiation, or alkylating agents [23]. Among the proteins and enzymes that constitute the BER pathway is poly(ADP-ribose) polymerase-1 (PARP-1) which is capable of repairing the N^3 and N^7 methylations induced by TMZ. Thus, pharmacological inhibition of PARP activity increases sensitivity to TMZ in gliomas and glioma stem cells (GSCs) [24, 25]. Interestingly, inactivating mutation of the MSH6 mismatch repair gene was identified in recurrent GBM tumors only after treatment with the alkylating agent TMZ but not in untreated tumors, suggesting that TMZ may have led to positive selection of the MSH6-mutant cells, thus leading to tumor resistance [26]. This clonal enrichment for MSH6-mutant cells in tumors leads to a transient reduction in tumor heterogeneity, followed by a rapid evolution of clones with highest survival fitness, which contribute to tumor progression and resistance to subsequent lines of treatment.

17.2.3 Tumor Heterogeneity

Recently, surgical multisampling scheme has been performed to collect spatially distinct tumor fragments from 11 GBM patients [27]. The results uncovered different GBM subtypes even within the same tumor. In one patient, the founder clone displayed amplification/gain of epidermal growth factor receptor (EGFR), CDK6, and MET, and loss/deletion of CDKN2a/B, PTEN, and PARK2. This clone then gave rise to other subclones. One of these clones inherited a gain of chromosome 3 which contains the PIK3CA gene and gradually expanded to form a clone including the partial loss of chromosome 17 (containing the NF1 and TP53 genes). Based on the presentation of an integrated analysis of tumor heterogeneity at the levels of genotype (copy number), cellular phenotype (gene expression profile), and single-molecule mitosis (molecular clocks), the authors proposed an expanded view on the clonal evolution of these tumors. In this view, resistance to treatment is associated with a heterogeneous, rather than a single population of malignant cells, with various genetic aberrations coexisting within the same cancer. The survival of these cells is restricted only by the spatial structure of the neoplasm. This increases the challenge of brain tumor treatment. A subsequent study by Patel and colleagues showed extensive heterogeneity in transcripts predictive of therapeutic treatment in single cells within GBM tumors [28]. The study revealed intracellular variability in the mosaic distribution of mutant and wild-type (wt) EGFR within the same tumor harboring cells with exclusive wt EGFR (7 %), mutant EGFRvIII (19 %), or an exon4-deletion variant (25 %). Several cells that lacked EGFR express other tyrosine kinase receptors including PDGFRA, EPHA3, or EPHA4. The increase in heterogeneity is associated with a decrease in overall survival. Along the same notion, Endaya and colleagues recently reported that the extensive heterogeneity in transcripts in GBM was shown to arisen from the proliferating tumor cells enriched in transcripts belonging to the cancer and neuronal stem cells [29].

17.2.4 Glioma Stem Cells

A possible mechanism underlying resistance to therapy and the renewable source of GBM tumors is thought to be attributed to the presence of GSCs [30]. These cells are defined to exhibit self-renewal ability ex vivo and in vivo to have the capacity to generate nontumorigenic and tumorigenic offspring, and to exhibit altered karyotype and multilineage differentiation abilities [31]. CD133 (also known as prominin-1) and other markers such as nestin, SOX2, Bmi1, Musashi, CD44, CD15, and A2B5 have been used as markers for GSCs [32]. Although there has been some controversy about the cell type of origin from which GSCs arise, their existence as "the fountain of youth" for gliomas has gained credence [33–35].

17.3 Recent Developments in Gliomas Therapies

17.3.1 Modulating the MGMT Activities in Overcoming TMZ Resistance

TMZ resistance can be resolved in several ways. One strategy is to reduce the MGMT activity through the use of a small nucleoside inhibitor, O6-benzylguanine (O6BG) capable of reducing MGMT activity by mimicking the methylated guanine nucleotide base targeted by MGMT, thereby binding MGMT protein and causing structural changes that mark the protein–nucleotide complex for degradation [36]. However, concomitant administration of O6BG and TMZ can produce severe off-target myelosuppression [37]. In a subsequent Phase II trial against recurrent, TMZ-resistant malignant glioma, the addition of O6BG to a 1-day dosing regimen of TMZ was able to restore TMZ sensitivity in anaplastic gliomas, but not in GBM patients [38]. Further, the most commonly reported adverse events were grade 4 hematologic events which were experienced in 32 of the 66 patients enrolled. The hematopoietic-specific toxicity was attributed to low-to-nonexistent levels of MGMT detected in hematopoietic stem cells (HSCs) and progenitor cells. Research has shown that expression of the O6BG-resistant MGMT mutant P140K by hematopoietic cells provides significant chemoprotection against hematopoietic toxicity from O6BG/alkylator chemotherapy [39]. By combining this information, the Fred Hutchinson Group formulated a method of transplanting patients with autologous P140K gene-modified hematopoietic CD34+ cells to prevent hematopoietic toxicity during combination O6BG/TMZ chemotherapy. Phase I and II clinical studies showed that patients tolerated an increased number of cycles of chemotherapy with delayed tumor growth [40]. Preparation is underway for Phase III trials. Other research groups are focusing on engineering new TMZ derivatives, such that the modified DNA at O6-meG is no longer recognized or repaired by MGMT [41]. Two novel imidazotetrazine analogues, D68 and D86, have also been developed to overcome TMZ resistance. The efficacy of these compounds was found to be independent of MGMT and MMR functions [42]. Bifunctional D68 can crosslink DNA with significant gains in potency when compared to monofunctional D86. Thus, it is more potent when compared to D86 and is capable of a distinct type of cell cycle arrest. Other mechanism of TMZ resistance in GBM includes elevated expression of an extracellular matrix protein (EFEMP1) and Chitinase 3-like 1 (Chi3l1, also known as YKL-40/BRP-39) ([43, 44] respectively). Strategies that target the downregulation of these gene products could result in sensitization of glioblastoma cells to TMZ.

17.3.2 Antiangiogenic Therapies

Copper chelation is known to inhibit the angiogenic response stimulated by implants of human brain tumors [45]. However, a clinical study showed that an oral agent used to treat intracerebral copper overload did not significantly improve the survival of patients with GBM [46]. The discord between preclinical and clinical practice is likely due to the suboptimal copper chelator used in the clinical trial, and yet, high dose of membrane permeable metal chelators could potentially interfere with intracellular processes with pathological consequences [47]. A copper chelator, known as disulfiram (tetraethylthiuram disulfide, SDF), has good safety profiles, can penetrate the blood–brain barrier and can override resistance to TMZ [48]. This effect is partly mediated by inhibition of a polo-like kinase 1, a brain cancer overexpressed serine/threonine kinase involved in cell cycle regulation.

Bevacizumab is currently the most prominent antiangiogenic agent in the field of recurrent glioblastoma. It is a humanized monoclonal antibody which binds to circulating vascular endothelial growth factor (VEGF) and prevents its interaction with the VEGF receptors, thereby suppressing VEGF signaling [49, 50]. The de novo formation of blood vessels is defined as vasculogenesis, a process mediated by the differentiation of precursor cells into endothelial cells, whereas angiogenesis represents the development of new vessels from a pre-existing vascular network [51]. Bevacizumab has both antivascular properties by causing regression of existing tumor vasculature and antiangiogenesis properties by causing the inhibition of new and recurrent tumor vessel growth [52]. Disappointingly, initial trials with bevacizumab, radiation and TMZ did not improve survival in patients with newly diagnosed glioblastoma [53, 54], although it did improve progression-free survival in patients with recurrent GBM as evidenced from the BRAIN study [55, 56]. In a recently completed European Organization for Research and Treatment of Cancer (EORTC) 26101 randomized phase III trial that compare lomustine against the combination of bevacizumab and lomustine in recurrent GBM patients, there was also no improvement in overall survival but progression-free survival (Wick et al, Neuro-Oncology 17; v1-2015 from The 20th Annual Scientific Meeting of the Society for Neuro-Oncology, Neuro-Oncology, 19-22 November, 2015). It remains to be determined whether in this trial, patients with proneural GBM may derive overall survival benefit from the addition of bevacizumab, a finding from the AVAglioa Trial [135]. Regardless, improved progression-free survival and maintenance of baseline quality of life and performance status were observed (for review see [49]). Another anti-VEGF agent, cediranib, is a potent inhibitor of VEGFR tyrosine kinases with additional activity against platelet-derived growth factor β and c-kit [57]. Treatment with cediranib can transiently normalize tumor vasculature and alleviate tumor-induced cerebral edema [58]. Edema alleviation can result in prolonged survival even without inhibition of tumor growth [59]. In addition, normalization of tumor vasculature can also reduce tumor hypoxia and enhance sensitivity to concurrently administered cytotoxic therapies. Unfortunately, there was no significant difference in the overall survival or progression-free survival of patients with recurrent glioblastoma when they are treated with either cediranib alone or in combination with lomustine versus lomustine alone [60]. Similarly, the combination of vandetanib (an inhibitor VEGFR2, EGFR, and RET) with standard chemoradiation also did not prolong the overall survival despite being well tolerated [61].

17.3.3 Anti-EGFR Therapies

The EGFR signaling system is an attractive target for therapeutic intervention. EGFR gene amplification and overexpression account for approximately 40–60 % of GBMs [62] In one report, 42.6 % of GBM patients failed to express detectable EGFR, 25.9 % had an overexpression of wtEGFR, and 31.5 % expressed a specific EGFR mutant (EGFRvIII, also known as EGFR type III, de2-7, ΔEGFR) [63]. Others have reported 46 % of GBMs lacking EGFR [64]. Of note, EGFRvIII expression in GBM is frequently associated with amplification and coexpression of the wtEGFR [65] Elevated levels of EGFR or EGFRvIII expression confer enhanced cell proliferation and invasion of glioblastomas [66]. Ligand-binding by EGF results in the activation of the receptor tyrosine kinase/RAS/PI3K pathway [67]. Phosphorylation of PI3K leads to activation of AKT [68]. This in turn induces activation of mammalian target of rapamycin (mTOR) which plays a critical role in controlling mRNA translation, ribosome biogenesis, autophagy, and metabolism [69]. The EGFRvIII mutant is generated from an in-frame deletion of 267 amino acids from the extracellular domain of the wtEGFR [70]. As a consequence, its tyrosine kinase is constitutively activated, which accounts for its oncogenic potential. Dimerization of wtEGFR and EGFRvIII leads to gliomagenesis in a STAT3/5 dependent manner [71]. Preclinical results using tyrosine kinase inhibitors (TKIs) against EGFR, gefitinib, and erlotinib have been disappointing [72]. The lack of efficacy may be attributed to several factors including the presence of efflux transporters on the endothelial cells that inhibit drug permeability across the blood–brain barrier [73]. Furthermore, these TKIs are known to be more efficacious when targeting tumor cells that express mutations in exons 19 and 21 of the EGFR kinase domain which has yet to be identified in GBM [74]. The most successful antibodies against EGFR activity used for GBM include cetuximab, panitumumab, and nimotuzumab [75, 76]. However, in a phase II study, cetuximab had no significant correlation between EGFR status and response or overall survival [77]. A recent in vitro study has reported that nimotuzumab is effective against human glioma cell lines regardless of their EGFR status [78]. This finding is in an agreement with a phase III study where no obvious correlation of nimotuzumab efficacy was found with regard to the EGFR status [79]. This could be explained by off-target effects. Overall, the inefficient antibody-based therapies have been attributed to the large molecular weight of the antibodies which reduce their ability to traverse the blood–brain barrier [80]. Importantly, even though the phase III nimotuzumab trial was statistically insignificant, the study indicated that targeting EGFR with monoclonal antibody therapy may be more promising in a stratified group of patients who are MGMT nonmethylated and have EGFR amplification [81]. Rindopepimut is emerging as a safe and potentially effective drug for the treatment of GBM. It consists of a 14-mer peptide that spans the EGFRvIII epitope which is conjugated with keyhole limpet hemocyanin (KLH) (Swartz AM, Li QJ, Sampson JH, Rindopepimut: a promising immunotherapeutic for the treatment of glioblastoma multiforme). KLH is a high-molecular-weight carrier protein that could enhance the production of an immunological response against the conjugated low-molecular weight peptide [82]. The administration of rindopepimut with granulocyte-macrophage colony-stimulating factor (GM-CSF) further increases the tumor-specific immune responses [83]. However, the effectiveness awaits further studies as the current findings are limited to a small number of EGFRvIII positive GBM patients.

17.4 Virus-Based Strategies

Oncolytic viruses (OVs) typically are mutated or deleted in viral genes that encode proteins essential for viral replication. The absence of these viral proteins leads to an abortive replication cycle in normal cells. However, in tumor cells, the mutant virus can selectively replicate because the missing viral protein function is provided by the tumor, but is expressed at low levels in normal cells. Oncolytic virotherapy is emerging as a promising strategy for the treatment of human cancers including gliomas [84]. First, the therapeutic index of a viral therapy is estimated as high as 100,000:1 which means that for every 100,000 tumor cells killed by viral therapy, one normal cell is killed. In contrast, therapeutic index of chemotherapy is reported to be only 6:1 [85]. Second, OVs are less prone to tumor resistance and tumor heterogeneity due to their direct and rapid killing of tumor cells. Third, OVs can induce an antitumor response and its efficacy is augmented in the presence of radiation, chemotherapy, and other agents [86]. Furthermore, these viruses can be engineered to encode therapeutic transgenes for enhanced efficacy [87]. To date, the herpes simplex virus-based viruses, HSV1716 and G207, have completed phase I trials in patients with recurrent gliomas in the United Kingdom and the United States, respectively [88, 89]. Both trials used a 34.5-null virus with the difference of HSV1716 being strain 17+ while G207 belongs to strain F. In addition, G207 has a further inactivation of UL39 gene that encodes the large subunit of ribonucleotide reductase (ICP6), which is required for the synthesis of DNA precursors. At present, G207 is under re-evaluation in combination with radiation for recurrent GBM in a phase I trial [90]. G47delta was created by the deletion of the α47 gene and the overlapping US11 promoter region in the G207 [91]. These deletions inhibit a transporter associated with antigen presentation (also known as TAP) which translocates peptides across the endoplasmic reticulum and can down-regulate MHC class I.

Another two OVs T-VEC and JX-549 have moved into clinical trials for other solid malignancies. T-VEC (also known as talimogene laherparepvec or oncoVEX$^{GM\text{-}CSF}$) is a second-generation herpes simplex virus based on the JS1 strain and expresses the immune stimulatory factor, human granulocyte-monocyte colony-stimulating factor (hGM-CSF). A randomized open-label phase III trial using T-VEC has completed with success where T-VEC is well tolerated and resulted in higher durable response rate and longer median overall survival in advanced metastatic melanoma [92]. JX-594 is poxvirus engineered for replication, transgene expression and amplification in cancer cells harboring activation of the EGFR/Ras pathway [93–95]. A randomized phase II trial in advanced hepatocellular carcinoma showed a significant increased medium overall survival in advanced hepatocellular carcinoma in a small number of patients. The human poliovirus receptor CD155 is commonly overexpressed in GBM [96] and PVS-RIPO, a prototype nonpathogenic poliovirus recombinant, is in an ongoing Phase I trial at Preston Robert Tisch Brain Tumor Center at Duke University Medical Center, The trial appears to be safe with 12 of the first 20 patients treated remaining alive, including the first and second patients who have survived more than 31-month posttreatment [97].

Toca 511 is a retroviral replicating vector, which is currently in clinical development [98]. Toca 511 spreads with high efficiency throughout solid tumors in a nonlytic manner, resulting in widespread integration of the yeast cytosine deaminase prodrug activator gene into the tumor cells' own genome, with tumor death being achieved by administration of 5-fluorocytosine, which crosses the blood–brain barrier and is converted within tumors to the chemotherapeutic agent, 5-fluorouridine [99]. By mid-2011, FDA reviewed the safety data from the first cohort of three patients and concluded that safety was acceptable, and in 2012, approved a second clinical trial of Toca 511 gene therapy administered into the postresection tumor bed, also in recurrent high-grade glioma patients. To date, multiple neuro-oncology centers in the United States have treated more than 30 patients with Toca 511, with dose escalation up to four dose cohorts. All patients have tolerated the treatment well, without dose-limiting toxicity, and promising signs of therapeutic efficacy have been observed, including symptomatic improvement, radiographic evidence of tumor stabilization, or shrinkage and progression-free survival. Recently, the efficacy of combining Toca511 with lomustine, an alkylating agent, was shown to prolong survival in syngeneic murine glioma models which provided further support for initiating clinical trials for patients with high-grade glioma [100].

Last but not least, a novel idea has been to create a "zone of resistance" using adeno-associated virus (AAV)-mediated expression of human interferon-β (hIFN) in normal brain tissue [101]. This therapeutic approach could potentially be translated into clinical trials by multiple injections of the AAV vector into the brain cavity during debulking, thus creating a zone of resistance to tumor recurrence.

17.5 MicroRNA-Based Strategies

MicroRNAs (miRNAs) are a family of small noncoding RNAs that regulate gene expression at the posttranscriptional level [102]. They are conserved across species and have a length of 20–25 nucleotides. Mature miRNAs are incorporate into the RNA-induced silencing complex (RISC) and directed to complementary sequences of target messenger RNAs. The complementary pairing usually occurs at the 3′ untranslated region resulting in target mRNA degradation or translational inhibition [103]. miRNAs play critical roles in multiple biological processes in GBM and GSCs including cell growth, migration, invasion, angiogenesis, and stem cells behavior [104]. They are also involved in the normal development of the central nervous system [105]. Highly expressed miR-9 has been correlated with glioblastoma progression [106]. MiR-9 has been shown to stimulate the migration of glioma cells in part through targeting NF1 [107]. Other microRNAs such as miR-21 promote proliferation of glioma cells, in part via downregulation of IGFBP3 [108]. Differential microRNA expression profiling revealed that miR-100 expression is downregulated in GBM compared to normal controls [109]. Further analysis showed that miR-100 inhibits proliferation of glioma cells by targeting the "silencing mediator of retinoid or thyroid hormone receptor-2" (SMRT/NCOR2) gene which is essential for histone deacetylase 3 (HDAC3) activity [110]. A recent study has identified miR-138 as a molecular signature of GSCs (GSCs) and demonstrates a vital role for miR-138 in promoting growth and survival of *bona fide* tumor-initiating cells with self-renewal potential. Further, its correlation to tumor recurrence and survival highlights the clinical significance of miR-138 as a prognostic biomarker and a therapeutic target for treatment of malignant gliomas [134]. Silencing of Mir-210 could reduce hypoxic GSCs stemness and radioresistance [111]. Further, it has been shown to mediate its effect on glioma cell apoptosis and proliferation in part through regulating the SIN3A expression [112]. All of these findings suggest that miRNAs possess clinical potential for GBM therapy.

17.6 Other Innovative Strategies

Metformin is a guanidine derivative that is widely used to treat type II diabetes [113]. It suppresses the actions of insulin and insulin-like growth factors. Insulin and insulin-like growth factors activate a signaling kinase pathway that depends on PI3-kinase and stimulates growth. With the inhibition of the insulin and insulin-like growth factors, there is less mitogenic stimulation on the cancer cells and this is a postulated theory as to how metformin suppresses cancer growth. Metformin also directly inhibits the mTOR pathway by promoting the interaction of two upstream molecules that stop the pathway's function-through enhanced PRAS40 binding to RAPTOR. Inhibition of mTOR has been shown to induce cancer cell death by stimulating autophagy or apoptosis. This is due to the key role of mTOR in the control of cell growth, proliferation, and metabolism through regulation of the phosphoinositide 3-kinase (PI3K)/Akt signaling pathway [114].

An interesting avenue which is currently being studied is the administration of a novel mimic, Mn(III) *meso*-tetrakis(*N*-n-butoxyethylpyridinium-2-yl)porphyrin (MnTnBuOE-2-PyP^{5+}) of the superoxide dismutase enzyme (SOD) [115]. This drug has been shown to provide long-term neuroprotection against oxidative stress damage caused by radiation exposure [116] and is currently entering phase I/II clinical trials as radioprotector for normal brain in patients with high-grade gliomas. While acting as a radioprotector of normal tissue MnTnBuOE-2-PyP^{5+} proved to be powerful glioma radio- and chemosensitizer (temozolomide) (Batinić-Haberle et al, unpublished) in a patient-derived D-245 MG GBM xenograft mouse model [116]. Differential effects arise from differential redox environment of normal versus tumor tissue [115]. Cancer therapeutics that increase the production of reactive oxygen species (ROS) or decrease the intracellular ROS scavengers could also potentially provide a powerful treatment strategy. An example of a natural compound with selective killing of tumor cells in an ROS-dependent mechanism is obtusaquinone (OBT) [117]. The novel antioxidant thiol, *N*-acetylcysteine amide could also augment the therapeutic effect of neural stem cell-based antiglioma oncolytic virotherapy [118].

In the field of cancer therapy, the most attractive feature of stem cells is their inherent ability to migrate toward tumor cells or "a wound that never heals." Multiple types of stem cells, including neural stem cells and mesenchymal stem cells (MSCs), have been shown to exhibit inherent tumor tropism [119, 120]. Further, studies have shown that naïve stem cells can exhibit an antitumor effect [121, 122] while others have shown that different sources of stem cells exhibit different effects on proliferation of malignant glioma cells [123]. We have shown that the antitumor effect of MSCs may be mediated through a paracrine system via the downregulation of the PDGF/PDGFR axis, which is known to play a key role in glioma angiogenesis [121]. Alternatively, exosomes (extracellular vesicles) from MSCs have been shown to contain miR-146b which can inhibit the growth of gliomas [124]. While ongoing studies are probing further into the precise role of MSCs, others have focused on modifying stem cells for cancer gene therapy. Systemic injection of neural stem cells expressing a secreted form of TRAIL increased glioma regression when administered together with the cardiac glycoside lanatoside C [125]. Likewise, MSCs have proven effective when modified with a number of agents, including cytokines, such as interleukin-12 [126]; interferon-β [127]; proapoptotic protein, TRAIL [128, 129]; prodrug activation platforms, such as herpes simplex kinase thymidine kinase [130]; antiangiogenic agents, such as endostatin and carboxylesterase [131], OV [132]; and nanoparticles [133].

17.7 Future Directions

With the advent of new discoveries and new technology platforms, we have summarized briefly the effective agents/vectors that could effectively inhibit (1) growth of cancer cells; (2) angiogenesis; (3) invasion; (4) tumor-associated inflammation; and (5) proliferative signal, as well as induces (6) specific apoptosis and immune responses to tumors (Fig. 17.1). In the future, it will be important to improve the match between the molecular subtype of tumors with the therapeutic strategies and to correlate them with clinical pathological outcomes of GBM for better therapeutic efficacies.

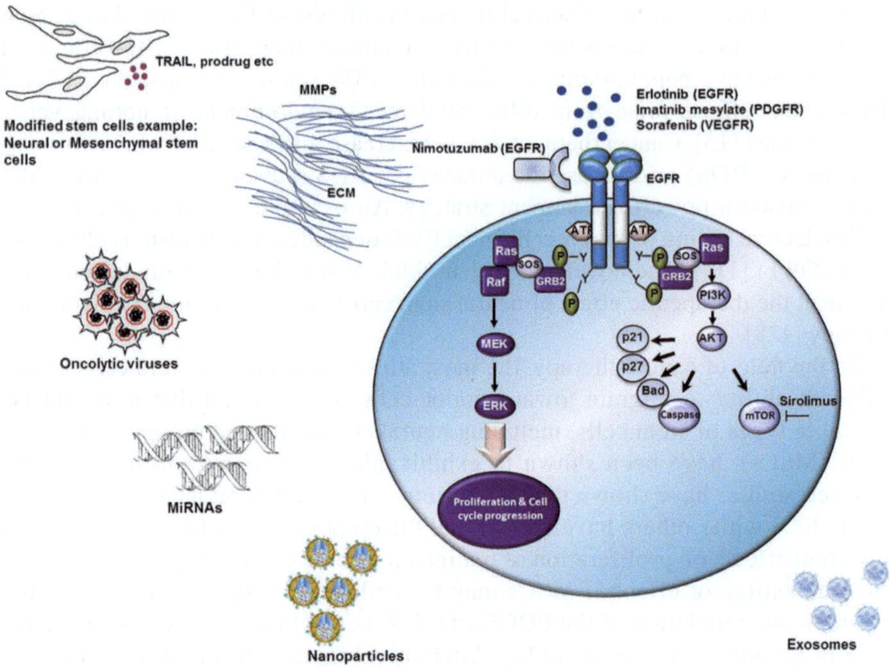

Fig. 17.1 Illustration of various strategies to target GBM

Acknowledgments We would like to thank Assistant Prof. Katherine Peters (Duke University Medical Center, USA) for her valuable inputs and for proof reading this chapter, Ms. Suzanne McDavitt for her skilled editorial assistance. Special thanks to A/Prof. Toh HC (National Cancer Center, Singapore) and A/Prof Yeo Tseng Tsai (National University Hospital, Singapore) for their support. Last but not least, we would also like to express our gratitude to funding agencies, National Medical Research Council of Singapore, Singhealth Research Grant Foundation and National Cancer Center Research Fund for their supports. X.O.B. is supported by the NIH Common Fund through the Office of Strategic Coordination/Office of the NIH Director, NCI U19 CA179563 and NIN NCI P01 CA069246.

References

1. Stupp R, et al. Promising survival for patients with newly diagnosed glioblastoma multiforme treated with concomitant radiation plus temozolomide followed by adjuvant temozolomide. J Clin Oncol. 2002;20(5):1375–82.
2. Stummer W, et al. Fluorescence-guided resection of glioblastoma multiforme by using 5-aminolevulinic acid-induced porphyrins: a prospective study in 52 consecutive patients. J Neurosurg. 2000;93(6):1003–13.
3. Turner SG, et al. The effect of field strength on glioblastoma multiforme response in patients treated with the NovoTTF-100A system. World J Surg Oncol. 2014;12(1):162.

4. Maier-Hauff K, et al. Efficacy and safety of intratumoral thermotherapy using magnetic iron-oxide nanoparticles combined with external beam radiotherapy on patients with recurrent glioblastoma multiforme. J Neurooncol. 2011;103(2):317–24.
5. Ranganath SH, et al. Hydrogel matrix entrapping PLGA-paclitaxel microspheres: drug delivery with near zero-order release and implantability advantages for malignant brain tumour chemotherapy. Pharm Res. 2009;26(9):2101–14.
6. Holland EC. Glioblastoma multiforme: the terminator. Proc Natl Acad Sci. 2000;97(12):6242–4.
7. Silbergeld DL, Chicoine MR. Isolation and characterization of human malignant glioma cells from histologically normal brain. J Neurosurg. 1997;86(3):525–31.
8. Aydın H, et al. Patterns of failure following CT-based 3-D irradiation for malignant glioma. Strahlenther Onkol. 2001;177(8):424–31.
9. Wallner KE, et al. Patterns of failure following treatment for glioblastoma multiforme and anaplastic astrocytoma. Int J Radiat Oncol Biol Phys. 1989;16(6):1405–9.
10. Berens ME, Giese A. "… those left behind". Biology and oncology of invasive glioma cells. Neoplasia. 1999;1(3):208–19.
11. Pardridge WM. Drug transport across the blood–brain barrier. J Cereb Blood Flow Metab. 2012;32(11):1959–72.
12. Stupp R, et al. Radiotherapy plus concomitant and adjuvant temozolomide for glioblastoma. N Engl J Med. 2005;352(10):987–96.
13. Stupp R, et al. Effects of radiotherapy with concomitant and adjuvant temozolomide versus radiotherapy alone on survival in glioblastoma in a randomised phase III study: 5-year analysis of the EORTC-NCIC trial. Lancet Oncol. 2009;10(5):459–66.
14. Stevens MF, et al. Antitumor activity and pharmacokinetics in mice of 8-carbamoyl-3-methyl-imidazo [5, 1-d]-1, 2, 3, 5-tetrazin-4 (3H)-one (CCRG 81045; M & B 39831), a novel drug with potential as an alternative to dacarbazine. Cancer Res. 1987;47(22):5846–52.
15. Drabløs F, et al. Alkylation damage in DNA and RNA—repair mechanisms and medical significance. DNA Repair. 2004;3(11):1389–407.
16. Zhang J, et al. Temozolomide: mechanisms of action, repair and resistance. Curr Mol Pharmacol. 2012;5(1):102–14.
17. Clark A, et al. Antitumor imidazotetrazines. 32.1 synthesis of novel imidazotetrazinones and related bicyclic heterocycles to probe the mode of action of the antitumor drug temozolomide. J Med Chem. 1995;38(9):1493–504.
18. Patel M, et al. Plasma and cerebrospinal fluid pharmacokinetics of temozolomide. Proc Am Soc Clin Oncol 1995;14:461a.
19. Kaina B, et al. MGMT: key node in the battle against genotoxicity, carcinogenicity and apoptosis induced by alkylating agents. DNA Repair. 2007;6(8):1079–99.
20. Park C-K, et al. The changes in MGMT promoter methylation status in initial and recurrent glioblastomas. Transl Oncol. 2012;5(5):393–7.
21. Hegi ME, et al. MGMT gene silencing and benefit from temozolomide in glioblastoma. N Engl J Med. 2005;352(10):997–1003.
22. Johannessen T-CA, et al. DNA repair and cancer stem-like cells–potential partners in glioma drug resistance? Cancer Treat Rev. 2008;34(6):558–67.
23. Wood RD, et al. Human DNA repair genes. Science. 2001;291(5507):1284–9.
24. Tentori L, et al. Pharmacological inhibition of poly (ADP-ribose) polymerase (PARP) activity in PARP-1 silenced tumour cells increases chemosensitivity to temozolomide and to a N3-adenine selective methylating agent. Curr Cancer Drug Targets. 2010;10(4):368–83.
25. Tentori L, et al. Pharmacological inhibition of poly (ADP-ribose) polymerase-1 modulates resistance of human glioblastoma stem cells to temozolomide. BMC Cancer. 2014;14(1):151.
26. Hunter C, et al. A hypermutation phenotype and somatic MSH6 mutations in recurrent human malignant gliomas after alkylator chemotherapy. Cancer Res. 2006;66(8):3987–91.
27. Sottoriva A, et al. Intratumor heterogeneity in human glioblastoma reflects cancer evolutionary dynamics. Proc Natl Acad Sci. 2013;110(10):4009–14.

28. Patel AP, et al. Single-cell RNA-seq highlights intratumoral heterogeneity in primary glioblastoma. Science. 2014;344(6190):1396–401.
29. Endaya BB, et al. Transcriptional profiling of dividing tumor cells detects intratumor heterogeneity linked to cell proliferation in a brain tumor model. Mol Oncol. 2016;10(1):126–37.
30. Eramo A, et al. Chemotherapy resistance of glioblastoma stem cells. Cell Death Differ. 2006;13(7):1238–41.
31. Vescovi AL, et al. Brain tumour stem cells. Nat Rev Cancer. 2006;6(6):425–36.
32. Gilbert CA, Ross AH. Cancer stem cells: cell culture, markers, and targets for new therapies. J Cell Biochem. 2009;108(5):1031–8.
33. Heywood RM, et al. A review of the role of stem cells in the development and treatment of glioma. Acta Neurochir. 2012;154(6):951–69.
34. Seymour T, et al. Targeting aggressive cancer stem cells in glioblastoma. Front Oncol. 2015;5(159):1–9.
35. Suva ML, et al. Reconstructing and reprogramming the tumor-propagating potential of glioblastoma stem-like cells. Cell. 2014;157(3):580–94.
36. Gerson SL. Clinical relevance of MGMT in the treatment of cancer. J Clin Oncol. 2002;20(9):2388–99.
37. Quinn JA, et al. Phase 1 trial of temozolomide plus O6-benzylguanine for patients with recurrent or progressive malignant glioma. J Clin Oncol. 2005;23(28):7178–87.
38. Quinn JA, et al. Phase II trial of temozolomide plus O6-benzylguanine in adults with recurrent, temozolomide-resistant malignant glioma. J Clin Oncol. 2009;27(8):1262–7.
39. Beard BC, et al. Efficient and stable MGMT-mediated selection of long-term repopulating stem cells in nonhuman primates. J Clin Invest. 2010;120(7):2345–54.
40. Adair JE, et al. Gene therapy enhances chemotherapy tolerance and efficacy in glioblastoma patients. J Clin Invest. 2014;124(9):4082.
41. Pletsas D, et al. Polar, functionalized guanine-O6 derivatives resistant to repair by O6-alkylguanine-DNA alkyltransferase: implications for the design of DNA-modifying drugs. Eur J Med Chem. 2006;41(3):330–9.
42. Ramirez YP, et al. Evaluation of novel imidazotetrazine analogues designed to overcome temozolomide resistance and glioblastoma regrowth. Mol Cancer Ther. 2015;14(1):111–9.
43. Akiyama Y, et al. YKL-40 downregulation is a key factor to overcome temozolomide resistance in a glioblastoma cell line. Oncol Rep. 2014;32(1):159–66.
44. Hiddingh L, et al. EFEMP1 induces γ-secretase/notch-mediated temozolomide resistance in glioblastoma. Oncotarget. 2014;5(2):363–74.
45. Alpern-Elran H, Brem S. Angiogenesis in human brain tumors: inhibition by copper depletion. Surg Forum. 1985;36:498–500.
46. Brem S, et al. Phase 2 trial of copper depletion and penicillamine as antiangiogenesis therapy of glioblastoma. Neuro Oncol. 2005;7(3):246–53.
47. Cuajungco MP, Lees GJ. Nitric oxide generators produce accumulation of chelatable zinc in hippocampal neuronal perikarya. Brain Res. 1998;799(1):118–29.
48. Triscott J, et al. Disulfiram, a drug widely used to control alcoholism, suppresses the self-renewal of glioblastoma and over-rides resistance to temozolomide. Oncotarget. 2012;3(10):1112–23.
49. Castro BA, Aghi MK. Bevacizimab for glioblastoma: current indications, surgical implications, and future directions. Neurosurg Focus. 2014;37(6):E9.
50. Curry RC, et al. Bevacizumab in high-grade gliomas: past, present, and future. Expert Rev Anticancer Ther. 2015;15(4):387–97.
51. Risau W, Flamme I. Vasculogenesis. Annu Rev Cell Dev Biol. 1995;11(1):73–91.
52. Ellis LM. Mechanisms of action of bevacizumab as a component of therapy for metastatic colorectal cancer. Semin Oncol. 2006;33(5 Suppl 10):S1–7.
53. Chinot OL, et al. Bevacizumab plus radiotherapy–temozolomide for newly diagnosed glioblastoma. N Engl J Med. 2014;370(8):709–22.
54. Lai A, et al. Phase II study of bevacizumab plus temozolomide during and after radiation therapy for patients with newly diagnosed glioblastoma multiforme. J Clin Oncol. 2011;29(2):142–8.

55. Chamberlain MC. Radiographic patterns of relapse in glioblastoma. J Neurooncol. 2011;101(2):319–23.
56. Pope W, et al. Patterns of progression in patients with recurrent glioblastoma treated with bevacizumab. Neurology. 2011;76(5):432–7.
57. Sahade M, et al. Cediranib: a VEGF receptor tyrosine kinase inhibitor. Future Oncol. 2012;8(7):775–81.
58. Batchelor TT, et al. AZD2171, a pan-VEGF receptor tyrosine kinase inhibitor, normalizes tumor vasculature and alleviates edema in glioblastoma patients. Cancer Cell. 2007;11(1):83–95.
59. Kamoun WS, et al. Edema control by cediranib, a vascular endothelial growth factor receptor–targeted kinase inhibitor, prolongs survival despite persistent brain tumor growth in mice. J Clin Oncol. 2009;27(15):2542–52.
60. Batchelor TT, et al. Antiangiogenic therapy for glioblastoma: current status and future prospects. Clin Cancer Res. 2014;20(22):5612–9.
61. Lee EQ, et al. A multicenter, phase II, randomized, noncomparative clinical trial of radiatioand temozolomide with or without vandetanibin newly diagnosed glioblastoma patients. Clin Cancer Res. 2015;21(16):3610–8.
62. Smith JS, et al. PTEN mutation, EGFR amplification, and outcome in patients with anaplastic astrocytoma and glioblastoma multiforme. J Natl Cancer Inst. 2001;93(16):1246–56.
63. Heimberger AB, et al. The natural history of EGFR and EGFRvIII in glioblastoma patients. J Transl Med. 2005;3:38.
64. Heimberger AB, et al. Prognostic effect of epidermal growth factor receptor and EGFRvIII in glioblastoma multiforme patients. Clin Cancer Res. 2005;11(4):1462–6.
65. Viana-Pereira M, et al. Analysis of EGFR overexpression, EGFR gene amplification and the EGFRvIII mutation in Portuguese high-grade gliomas. Anticancer Res. 2008;28(2A):913–20.
66. Pines G, et al. Oncogenic mutant forms of EGFR: lessons in signal transduction and targets for cancer therapy. FEBS Lett. 2010;584(12):2699–706.
67. Patel R, Leung HY. Targeting the EGFR-family for therapy: biological challenges and clinical perspective. Curr Pharm Des. 2012;18(19):2672–9.
68. Chakravarti A, et al. The prognostic significance of phosphatidylinositol 3-kinase pathway activation in human gliomas. J Clin Oncol. 2004;22(10):1926–33.
69. Guertin DA, Sabatini DM. Defining the role of mTOR in cancer. Cancer Cell. 2007;12(1):9–22.
70. Huang H-JS, et al. The enhanced tumorigenic activity of a mutant epidermal growth factor receptor common in human cancers is mediated by threshold levels of constitutive tyrosine phosphorylation and unattenuated signaling. J Biol Chem. 1997;272(5):2927–35.
71. Fan Q-W, et al. EGFR phosphorylates tumor-derived EGFRvIII driving STAT3/5 and progression in glioblastoma. Cancer Cell. 2013;24(4):438–49.
72. Gan HK, et al. The EGFRvIII variant in glioblastoma multiforme. J Clin Neurosci. 2009;16(6):748–54.
73. Sampson JH, et al. Clinical utility of a patient-specific algorithm for simulating intracerebral drug infusions. Neuro Oncol. 2007;9(3):343–53.
74. Karpel-Massler G, et al. Therapeutic inhibition of the epidermal growth factor receptor in high-grade gliomas: where do we stand? Mol Cancer Res. 2009;7(7):1000–12.
75. Liang W, et al. Multi-targeted antiangiogenic tyrosine kinase inhibitors in advanced non-small cell lung cancer: meta-analyses of 20 randomized controlled trials and subgroup analyses. PLoS One. 2014;9(10):e109757.
76. Wu C, et al. Gefitinib as palliative therapy for lung adenocarcinoma metastatic to the brain. Lung Cancer. 2007;57(3):359–64.
77. Neyns B, et al. Stratified phase II trial of cetuximab in patients with recurrent high-grade glioma. Ann Oncol. 2009;20(9):1596–603.
78. Chong DQ, et al. Combined treatment of nimotuzumab and rapamycin is effective against temozolomide-resistant human gliomas regardless of the EGFR mutation status. BMC Cancer. 2015;15(1):255.

79. Westphal M, et al. A randomised, open label phase III trial with nimotuzumab, an anti-epidermal growth factor receptor monoclonal antibody in the treatment of newly diagnosed adult glioblastoma. Eur J Cancer. 2015;51(4):522–32.
80. Chen KS, Mitchell DA. Monoclonal antibody therapy for malignant glioma. New York: Springer; 2012. p. 121–41.
81. Bode U, et al. Nimotuzumab treatment of malignant gliomas. Expert Opin Biol Ther. 2012;12(12):1649–59.
82. Harris JR, Mark IJ. Keyhole limpet hemocyanin: molecular structure of a potent marine immunoactivator. Eur Urol. 2000;37 Suppl 3:24–33.
83. Swartz AM, et al. Rindopepimut: a promising immunotherapeutic for the treatment of glioblastoma multiforme. Immunotherapy. 2014;6(6):679–90.
84. Kaufmann JK, Chiocca EA. Glioma virus therapies between bench and bedside. Neuro Oncol. 2014;16(3):334–51.
85. Ring JAM. Cytolytic viruses as potential anti-cancer agents. J Gen Virol. 2002;83:491–502.
86. Kumar S, et al. Virus combinations and chemotherapy for the treatment of human cancers. Curr Opin Mol Ther. 2008;10(4):371–9.
87. Russell SJ, et al. Oncolytic virotherapy. Nat Biotechnol. 2012;30(7):658–70.
88. Markert J, et al. Conditionally replicating herpes simplex virus mutant, G207 for the treatment of malignant glioma: results of a phase I trial. Gene Ther. 2000;7(10):867–74.
89. Rampling R, et al. Toxicity evaluation of replication-competent herpes simplex virus (ICP 34.5 null mutant 1716) in patients with recurrent malignant glioma. Gene Ther. 2000;7(10):859–66.
90. Markert JM, et al. A phase 1 trial of oncolytic HSV-1, G207, given in combination with radiation for recurrent GBM demonstrates safety and radiographic responses. Mol Ther. 2014;22(5):1048–55.
91. Todo T, et al. Oncolytic herpes simplex virus vector with enhanced MHC class I presentation and tumor cell killing. Proc Natl Acad Sci. 2001;98(11):6396–401.
92. Andtbacka RH, et al. Talimogene laherparepvec improves durable response rate in patients with advanced melanoma. J Clin Oncol. 2015;33(25):2780–8.
93. Brown MC, et al. Mitogen-activated protein kinase-interacting kinase regulates mTOR/AKT signaling and controls the serine/arginine-rich protein kinase-responsive type 1 internal ribosome entry site-mediated translation and viral oncolysis. J Virol. 2014;88(22):13149–60.
94. Kim JH, et al. Systemic armed oncolytic and immunologic therapy for cancer with JX-594, a targeted poxvirus expressing GM-CSF. Mol Ther. 2006;14(3):361–70.
95. Mastrangelo MJ, et al. Intratumoral recombinant GM-CSF-encoding virus as gene therapy in patients with cutaneous melanoma. Cancer Gene Ther. 1999;6(5):409–22.
96. Merrill MK, et al. Poliovirus receptor CD155-targeted oncolysis of glioma. Neuro Oncol. 2004;6(3):208–17.
97. Desjardins A, et al. Oncolytic polio/rhinovirus recombinant (PCSRIPO) against recurrent glioblastoma (GBM): optional dose determination. In: ASCO annual meeting, J Clin Oncol. 2015;33(15):2068.
98. Perez OD, et al. Design and selection of Toca 511 for clinical use: modified retroviral replicating vector with improved stability and gene expression. Mol Ther. 2012;20(9):1689–98.
99. Miller CR, et al. Intratumoral 5-fluorouracil produced by cytosine deaminase/5-fluorocytosine gene therapy is effective for experimental human glioblastomas. Cancer Res. 2002;62(3):773–80.
100. Robbins JM, et al. Additive therapeutic effect of a non-lytic retroviral replicating vector (Toca 511) and 5-fluorocytosine in combination with lomustine in experimental glioma models. In: ASCO annual meeting, J Clin Oncol 2015;33(15)_suppl e13033.
101. Maquire CA. Preventing growth of brain tumors by creating a zone of resistance. Mol Ther. 2008;16(10):1695–702.
102. Bartel DP. MicroRNAs: genomics, biogenesis, mechanism, and function. Cell. 2004;116(2):281–97.

103. Filipowicz W, et al. Mechanisms of post-transcriptional regulation by microRNAs: are the answers in sight? Nat Rev Genet. 2008;9:102–14.
104. Godlewski J, et al. MicroRNAs and glioblastoma; the stem cell connection. Cell Death Differ. 2010;17(2):221–8.
105. Leucht C, et al. MicroRNA-9 directs late organizer activity of the midbrain-hindbrain boundary. Nat Neurosci. 2008;11(6):641–8.
106. Malzkorn B, et al. Identification and functional characterization of microRNAs involved in the malignant progression of gliomas. Brain Pathol. 2010;20(3):539–50.
107. Tan X, et al. The CREB-miR-9 negative feedback minicircuitry coordinates the migration and proliferation of glioma cells. PLoS One. 2012;7(11):e49570.
108. Yang CH, et al. MicroRNA-21 promotes glioblastoma tumorigenesis by down-regulating insulin-like growth factor-binding protein-3 (IGFBP3). J Biol Chem. 2014;289(36):25079–87.
109. Alrfaei BM, et al. microRNA-100 targets SMRT/NCOR2, reduces proliferation, and improves survival in glioblastoma animal models. PLoS One. 2013;8(11):e80865.
110. You SH, et al. Nuclear receptor co-repressors are required for the histone-deacetylase activity of HDAC3 in vivo. Nat Struct Mol Biol. 2013;20(2):182–7.
111. Yang W, et al. Knockdown of miR-210 decreases hypoxic glioma stem cells stemness and radioresistance. Exp Cell Res. 2014;326(1):22–35.
112. Shang C, et al. MiR-210 up-regulation inhibits proliferation and induces apoptosis in glioma cells by targeting SIN3A. Med Sci Monit. 2014;20:2571–7.
113. Ucbek A, et al. Effect of metformin on the human T98G glioblastoma multiforme cell line. Exp Ther Med. 2014;7(5):1285–90.
114. Liu X, et al. Discrete mechanisms of mTOR and cell cycle regulation by AMPK agonists independent of AMPK. Proc Natl Acad Sci. 2014;111(4):E435–44.
115. Batinic-Haberle I, et al. Latest insights into their structure-activity relationships and impact on the cellular redox-based signaling pathways. Antioxid Redox Signal. 2014;20:2372–415.
116. Weitzel DH, et al. Radioprotection of the brain white matter by Mn (III) N-butoxyethylpyridylporphyrin–based superoxide dismutase mimic MnTnBuOE-2-PyP5+. Mol Cancer Ther. 2015;14(1):70–9.
117. Badr CE, et al. Targeting cancer cells with the natural compound obtusaquinone. J Natl Cancer Inst. 2013;105(9):643–53.
118. Kim CK, et al. N-acetylcysteine amide augments the therapeutic effect of neural stem cell-based antiglioma oncolytic virotherapy. Mol Ther. 2013;21(11):2063–73.
119. Aboody KS, et al. Neural stem cells display extensive tropism for pathology in adult brain: evidence from intracranial gliomas. Proc Natl Acad Sci U S A. 2000;97(23):12846–51.
120. Stuckey DW, Shah K. Stem cell-based therapies for cancer treatment: separating hope from hype". Nat Rev Cancer. 2014;14(10):683–91.
121. Ho IAW, et al. Human bone marrow-derived mesenchymal stem cells suppress human glioma growth through inhibition of angiogenesis. Stem Cells. 2013;31:146–55.
122. Motaln H, et al. Human mesenchymal stem cells exploit the immune response mediating chemokines to impact the phenotype of glioblastoma. Cell Transplant. 2012;21(7):1529–45.
123. Akimoto K, et al. Umbilical cord blood-derived mesenchymal stem cells inhibit, but adipose tissue-derived mesenchymal stem cells promote, glioblastoma multiforme proliferation. Stem Cells Dev. 2013;22(9):1370–86.
124. Katakowski M, et al. Exosomes from marrow stromal cells expressing miR-146b inhibit glioma growth. Cancer Lett. 2013;335(1):201–4.
125. Teng J, et al. Systemic anticancer neural stem cells in combination with a cardiac glycoside for glioblastoma therapy. Stem Cells. 2014;32(8):2021–32.
126. Ryu CH, et al. Gene therapy of intracranial glioma using interleukin 12-secreting human umbilical cord blood-derived mesenchymal stem cells. Hum Gene Ther. 2011;22(6):733–43.
127. Nakamizo A, et al. Human bone marrow-derived mesenchymal stem cells in the treatment of gliomas". Cancer Res. 2005;65(8):3307–18.

128. Sasportas LS, et al. Assessment of therapeutic efficacy and fate of engineered human mesenchymal stem cells for cancer therapy. Proc Natl Acad Sci U S A. 2009;106(12):4822–7.
129. Yulyana Y, et al. carbenoxolone enhances TRAIL-induced apoptosis through the upregulation of death receptor 5 and inhibition of gap junction intercellular communication in human glioma. Stem Cells Dev. 2013;22(13):1870–82.
130. Matuskova M, et al. HSV-tk expressing mesenchymal stem cells exert bystander effect on human glioblastoma cells. Cancer Lett. 2010;290(1):58–67.
131. Yin J, et al. hMSC-mediated concurrent delivery of endostatin and carboxylesterase to mouse xenografts suppresses glioma initiation and recurrence". Mol Ther. 2011;19(6):1161–9.
132. Duebgen M. Stem cells loaded with multimechanistic oncolytic herpes simplex virus variants for brain tumor therapy. J Natl Cancer Inst. 2014;106(6):1–9.
133. Roger M, et al. Ferrociphenol lipid nanocapsule delivery by mesenchymal stromal cells in brain tumor therapy. Int J Pharm. 2012;423(1):63–8.
134. Chan XH, et al. Targeting glioma stem cells by functional inhibition of a prosurvival oncomiR-138 in malignant gliomas. Cell Rep. 2012;2(3):591–602.
135. Sandmann T, et al., Patients With Proneural Glioblastoma May Derive Overall Survival Benefit From the Addition of Bevacizumab to First-Line Radiotherapy and Temozolomide: Retrospective Analysis of the AVAglio Trial. J Clin Oncol. 2015;33(25):2735–44.

Chapter 18
Redox Therapeutics in Breast Cancer: Role of SOD Mimics

Ana S. Fernandes, Nuno Saraiva, and Nuno G. Oliveira

18.1 Introduction

Breast cancer is the most commonly diagnosed cancer among women in the USA (excluding skin cancers) and the second cause of cancer death [1]. The estimated new breast cancer cases and deaths for women in the USA in 2015 are 231,840 and 40,290, respectively [2]. Moreover, according to global estimates, female breast cancer is the most prevalent cancer worldwide [3]. It is therefore unquestionable that the impact of this disease clearly justifies the research that is currently ongoing in this field, mainly in metastatic disease. In fact, metastasis of breast cancer is a key hallmark of this type of cancer, being life-threatening and the major cause of treatment failure [4]. Many efforts have been made in order to understand the processes of invasion and subsequent formation of metastasis.

Reactive species (RS), especially reactive oxygen species, have been greatly implicated in cancer at different levels, including cell proliferation, invasiveness, and metastases. RS production leading to cellular oxidative stress is considered a risk factor for cancer development and an important aspect in disease progression [5, 6]. In order to survive, cancer cells need to adapt to a new environment. The level of adaptation may determine the cancer cell ability to respond to radio- and chemotherapy, to invade, and to generate metastasis. Many studies report that tumor cells

A.S. Fernandes (✉) • N. Saraiva
CBiOS, Universidade Lusófona Research Center for Biosciences and Health Technologies, Campo Grande, 376, ECTS, Lisbon 1749-024, Portugal
e-mail: ana.fernandes@ulusofona.pt; nuno.saraiva@ulusofona.pt

N.G. Oliveira
Research Institute for Medicines (iMed. ULisboa), Faculty of Pharmacy, Universidade de Lisboa, Av. Professor Gama Pinto, Lisbon 1649-003, Portugal
e-mail: ngoliveira@ff.ulisboa.pt

I. Batinić-Haberle et al. (eds.), *Redox-Active Therapeutics*,
Oxidative Stress in Applied Basic Research and Clinical Practice,
DOI 10.1007/978-3-319-30705-3_18

display an increased production of RS and significant changes in antioxidant enzyme expression profiles [5–7].

The cellular and tissue redox homeostasis is partially achieved by the actions of superoxide dismutases (SOD): SOD1 (cytoplasmic and mitochondrial intermembrane space, Cu,Zn-SOD) [8, 9], SOD2 (mitochondrial matrix, MnSOD) [10], and SOD3 (extracellular, EC-SOD) [11]. The role of SOD, especially SOD2, in cancer is a controversial topic, where conflicting data exist [12–14]. A substantial body of evidence for an anticancer role of MnSOD has emerged through the years [15]. Moreover, there are several reports on the role of SOD in breast cancer. The present book chapter will summarize some of the relevant existing data, aiming to foster interest in this topic. Notably, redox modulation in breast cancer can be achieved using chemicals with SOD-like activity (SOD mimics—SODm). This book chapter will address the effects of SODm and other chemicals with superoxide scavenger activity in the context of breast cancer.

18.2 Role of SOD in Breast Cancer

SOD2 has emerged as a key enzyme with a dual role in tumorigenic progression. The expression level of SOD2 is altered in several tumors, including thyroid [16], brain and CNS [17, 18], gastric [19, 20], ovarian [21], and breast [22, 23]. In breast cancer cells, the expression of SOD2 is commonly up-regulated in advanced breast cancers whereas it can be often down-regulated in early cancer stages (reviewed in [24]). The altered expression levels of SOD2 affect cell proliferation and invasion properties of cancer cells and tumor angiogenesis, hence impacting disease progression. SOD2 mRNA levels determined from microarrays analysis of human breast tumors specimens are contradictory: one study supports increased SOD2 levels, whereas two other found reduced SOD2 levels in tumor specimens [25].

The *sod2* gene expression regulation occurs mainly at the transcriptional stage. The increased levels of repressive transcription factors or epigenetic modifications are the two major mechanisms of SOD2 downregulation [22, 26–28]. Sp1 and NF-κB are the two main transcription factors responsible for high expression of SOD2 in invasive cells [22, 27, 28]. Tumor microenvironment conditions such as hypoxia and hypoglycemia can induce SOD2 expression in breast cancer cell lines [29]. Isolated migrating cancer cells at the invasive front of primary tumors exhibit higher SOD2 expression than those inside the tumor mass. Moreover, SOD2 levels in primary tumors appear to be lower than the matched lymph-node metastases [30]. Increased SOD2 levels have been shown to directly correlate with breast cancer histological grading and invasive properties [23], reinforcing the importance of SOD2 in breast cancer progression.

The levels of SOD2 in non-tumor breast cells (MCF-10A, HMEC) are significantly higher than in noninvasive, estrogen-sensitive breast cancer cell lines such as MCF7 or T47D [22, 31, 32]. Highly invasive and metastatic breast cancer cell lines

(MDA-MB-231 or SKBR3) present higher SOD2 protein levels when compared to MCF7 or T47D [22, 31, 32]. Importantly, upregulation of SOD1 or SOD2 can inhibit breast cancer growth in a mouse model [33].

The increase in SOD expression in some cancers is proposed to be mainly associated with a cellular response to intracellular oxidative stress. In this context, the scavenging of superoxide by SOD may attenuate the reactive species (RS)-induced stress, thus reducing the simulating effect of RS on cell proliferation [21]. Overexpression of SOD2 in human cancer cell lines such as MCF-7 (breast) [34], UACC-903 (melanoma) [35], SCC-25 (oral squamous carcinoma) [36], U118 (glioma) [37], and DUI45 (prostate carcinoma) [38] significantly inhibited their malignant phenotypes. The impact of SOD2 on cell proliferation is highlighted by the variations in SOD2 expression during the cell cycle. SOD2 levels are lower during S and G2/M phases, leading to RS accumulation [38, 39]. Because SOD2 levels are reduced in certain cancer cells and induced overexpression of SOD2 is able to suppress malignant phenotypes in certain experimental models, *sod2* has been proposed as a tumor suppressor gene [13, 35, 37, 40]. Additionally, SOD2 overexpression suppresses the tumor promoting effects of cancer-associated fibroblasts with caveolin-1 reduced expression in MDA-MB-231 derived tumor xenografts [41]. The overexpression of SOD1 and SOD3 is also related to reduced cell growth in human cancers [33, 42]. SOD3 levels are lower in human breast cancer samples and SOD3 mRNA levels are inversely correlated with breast cancer clinical stage [43].

The overall activity of SOD2 can be altered by certain *sod2* genetic polymorphisms [44] and by the deacetylase SIRT3 [45]. The *sod2* $Ala^{16}Val$ allele produces a protein that is not efficiently transported to the mitochondria, resulting in a reduced antioxidant activity [44]. The presence of the Ala/Ala genotype has been shown to correlate with an increased breast cancer risk in premenopausal women [46], especially when in combination with the glutathione peroxidase-1 $Pro^{198}Leu$ polymorphism [47]. A recent study supporting the opposite conclusion demonstrated an association between the SOD2 Val/Val genotype and the appearance of axillary lymph node metastasis [48]. The primary mitochondrial deacetylase SIRT3 activates SOD2 in response to cellular stress, by deacetylating the dismutase [45]. The importance of this regulatory mechanism in tumorigenesis is highlighted by the development of invasive ductal mammary tumors *Sirt3* KO mice [49] and by SIRT3 low expression in human cancers, including breast cancer [49].

Although much conflicting data still exists, a correlation between SOD2 expression/activity levels and metastatic breast carcinoma is apparent. This correlation has also been reported for brain [18], gastric and colorectal cancers [20, 50]. Conversely, low SOD2 expression correlates with an increased cell proliferation during early stages of tumorigenesis. This has also been observed for mesothelioma and prostate cancers [51, 52]. Additionally, SOD2 overexpression in many tumor-derived cell lines is able to attenuate their malignant phenotype and reduce their metastatic potential, thus justifying the aforementioned proposal of SOD2 as a tumor suppressor gene (reviewed in [53, 54]).

18.3 SOD Mimics and Breast Cancer

Despite the potential benefits of SOD in breast cancer, the clinical use of the native enzyme presents several limitations, including antigenicity, high-manufacturing costs, low cell permeability, and short circulating half-life [55]. To overcome these restrictions, small synthetic compounds with the ability to mimic the functional properties of SOD (SODm) have been developed [55, 56]. These compounds have shown remarkable effects in experimental models of several diseases, including cancer. Some of these are currently being evaluated in clinical trials [55, 56]. It is important to mention that many of the compounds known as SODm are catalytical polyfunctional antioxidants. In addition to the disproportionation of superoxide anion, SODm may react with other reactive oxygen, nitrogen and sulfur species, as well as with redox units of cellular signaling proteins involved in transcription [57, 58]. Although many compounds with SOD-like activity with distinct chemical structures have been reported, most studies regarding their role in breast cancer use manganese(III) porphyrins, copper(II) complexes, or nitroxides. In the subsequent sections, the studies of SODm in the context of breast cancer are reviewed, shedding light on the usefulness of these agents either as therapeutic drugs or as mechanistic tools.

18.4 Antiproliferative Effects of SODm

Several compounds with SOD-like activity have been reported to have anticancer activity, which has been mostly ascribed to the ability of these compounds to push the level of endogenous RS (namely H_2O_2) to the threshold of toxicity. The disproportionation of superoxide anion by an SODm leads to H_2O_2 production. In a normal cell, the physiological levels of H_2O_2 may be maintained due to the abundance of catalase, peroxidases and other peroxide-removing enzymes [59]. However, a cancer cell has generally a higher RS content and compromised antioxidant defenses [59, 60]. In such case, the additional H_2O_2 burden would force the cell to die [59–62] (Fig. 18.1a). Moreover, in a lymphoma cellular model, Mn porphyrin-based SODm were reported to take advantage of high H_2O_2 levels and use it to oxidize thiols of signaling proteins such as NF-κB, resulting in potentiation of glucocorticoid-induced apoptosis [63]. SODm can thus act both as anti- or pro-oxidants, depending on cell type, cellular levels of RS and antioxidants, cellular ratio of superoxide- to peroxide-removing systems, levels of oxygen, and cellular and subcellular distribution of SODm [58, 60]. In turn, they could produce therapeutic effects of different nature—protection of normal tissue (demonstrated as antioxidative effects) or cancer cell killing (demonstrated as pro-oxidative therapeutic effects) [60, 64] (Fig. 18.1b). Based on data reported on pro-oxidative actions in cancer cell, compounds with SOD-like activity have anticancer therapeutic potential as single agents or may be useful in combination regimens to boost chemo- and radiotherapy [58].

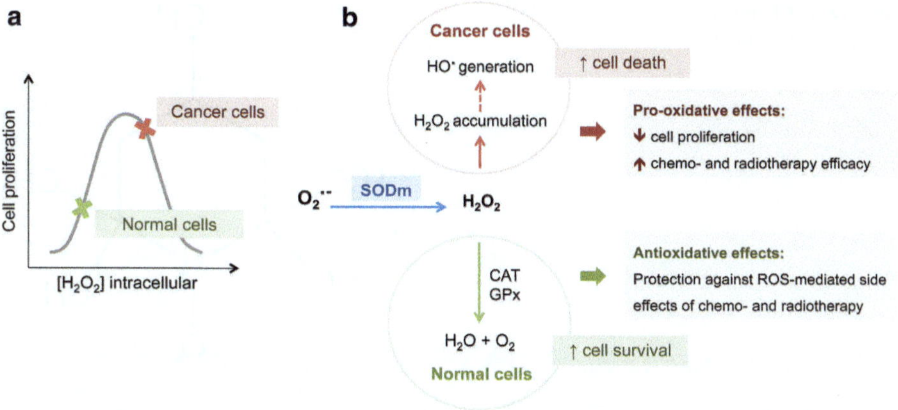

Fig. 18.1 Differential role of SOD mimics in cancer and normal cells. (**a**) Model proposed to explain the opposing effects of intracellular H_2O_2 concentrations on the proliferation of cancer and normal cells (adapted from [62]). (**b**) The redox status of cancer and normal cells differs considerably, controlling their differential sensitivity to additional increase in H_2O_2. In cancer cells, SODm may decrease cell proliferation and boost chemo- and radiotherapy treatments. Conversely, SODm may increase the survival of normal cells, protecting against RS-mediated adverse effects of chemo- and radiotherapy

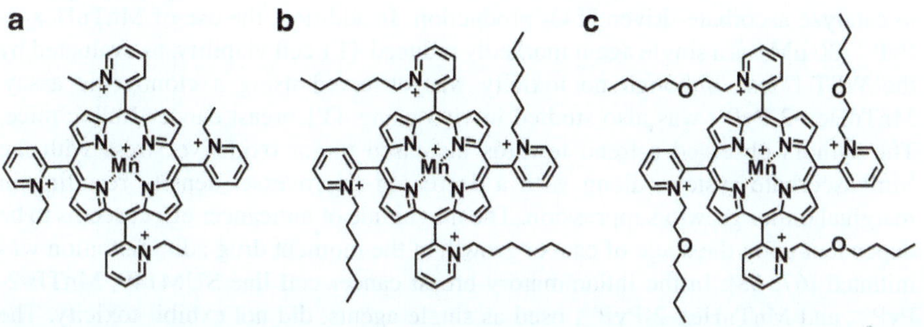

Fig. 18.2 Chemical structures of manganese(III) N-substituted pyridylporphyrins. (**a**) MnTE-2-PyP^{5+}, (**b**) MnTnHex-2-PyP^{5+}, (**c**) MnTnBuOE-2-PyP^{5+}

The in vitro studies using mammary carcinoma cell lines from mouse (4T1) and rat (R3230) showed that a member of *N*-substituted pyridylporphyrins (Fig. 18.2), the manganese(III) porphyrin MnTE-2-PyP^{5+} (1, 2 and 5 μM) had significant growth inhibitory effects during the first 24 h of exposure, although there was no effect with longer incubation times [65]. A study carried out in human breast adenocarcinoma cells (MCF-7 cells) cultured with 1 % O_2, showed that the exposure to the nitroxide Tempol (Fig. 18.3) decreased the clonogenic survival, but only at concentrations ≥10 mM. Although less pronounced, an effect was also observed for MnTE-2-PyP^{5+} (200 μM) under the same conditions [66].

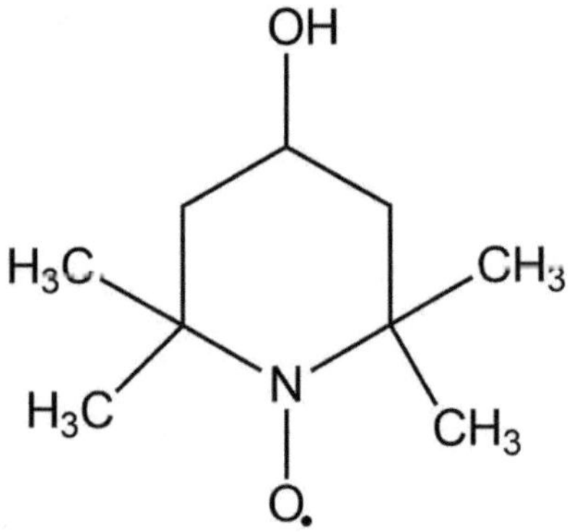

Fig. 18.3 Chemical structure of Tempol

Recently, combinations of MnPs with ascorbate have been explored for their anticancer potential. Ye et al. [60] studied the effects of *ortho(2)* MnTE-2-PyP^{5+}, *ortho(2)* MnTnHex-2-PyP^{5+} and the *meta(3)* isomer MnTnHex-3-PyP^{5+}, alone or combined with ascorbate, in different cell lines, including mouse mammary cancer 4T1 and human inflammatory breast cancer SUM149. These MnPs were shown to potentiate the toxic effects of ascorbate in 4T1 cells treated with cytotoxic concentrations of ascorbate. The increased efficacy could be ascribed to the ability of MnPs to catalyze ascorbate-driven H_2O_2 production. In addition, the use of MnTnHex-2-PyP^{5+} (30 μM) as a single agent markedly reduced 4T1 cell viability, as evaluated by the WST-1 test, although no toxicity was observed using a clonogenic assay. MnTnHex-2-PyP^{5+} was also studied in vivo using 4T1 breast cancer Balb/c mice. The authors observed a trend towards increased tumor oxidative stress with the MnP/ascorbate system, along with a decreased microvessel density resulting in marginal tumor growth suppression. The magnitude of anticancer effect seems to be dependent upon the stage of cancer growth at the moment drug administration was initiated [67, 68]. In the inflammatory breast cancer cell line SUM149, MnTE-2-PyP^{5+}, and MnTnHex-2-PyP^{5+}, used as single agents, did not exhibit toxicity. The combination of MnTE-2-PyP^{5+} and ascorbate decreased cell viability as measured by the mitochondrial membrane potential marker TMRE. However, the combination of MnTnHex-2-PyP^{5+} and ascorbate was not efficacious under the same conditions [60]. Evans et al. [69] further explored the anticancer properties of MnP/ascorbate systems in models of inflammatory breast cancer (SUM149, rSUM149, and SUM190 cell lines). The study was focused on MnTE-2-PyP^{5+} and MnTnBuOE-2-PyP^{5+}. These MnPs alone had no significant effect on viability in any of the cell lines. However, in combination with ascorbate both MnPs decreased viability in all three cell types. The authors showed that the observed cell death was dependent on H_2O_2. The production of H_2O_2 led to decreased prosurvival ERK and NF-κB signaling and decreased mitochondrial integrity [69].

Many copper(II) complexes with superoxide scavenging activity exhibited cytotoxic effects in in vitro models of breast cancer. In the case of copper(II) complexes, an additional mechanism underlying the anticancer effect has been reported. Due to the high levels of H_2O_2 in cancer cells, and copper Fenton chemistry, hydroxyl radical may be generated [61] and induce protein oxidation, lipid peroxidation, and oxidative DNA cleavage [70, 71]. O'Connor et al. [71] described three copper(II) complexes derived from salicylic acid combining superoxide dismutase mimicking properties with DNA binding and cleaving properties. These compounds exhibited cytotoxicity against MCF-7 cells at micromolar levels. Another study reported two square-planar copper(II) complexes with dicarboxylate *o*-phthalate and 1,10-phenanthroline or 2,20-dipyridyl ligands with potent SOD activity and the ability to induce DNA strand breaks. They exhibited anticancer potential against MCF-7 cells [72]. Different mixed-ligand copper(II) complexes with the general composition [Cu(qui)(L)] $BF_4 \cdot xH_2O$ were also cytotoxic to MCF-7 cells. Along with SOD-like activity, these compounds have DNA cleavage properties [70]. In addition, Kovala-Demertzi et al. [73] studied different complexes of meclofenamic acid. These authors found that the complexes [Mn(meclo)$_2$] and [Cu(meclo)$_2$(H_2O)$_2$] show superoxide dismutase activity and a mild cytotoxic effect in MCF-7 cells. Another complex, dichloro[1-(4,5-dihydro-1*H*-imidazol-2-yl)-1*H*-pyrazole-*N*,*N′*] copper(II), with SOD-like activity and cytotoxicity to MCF-7 cells, was described by Saczewski et al. [74]. In another report [75], three pyridazolato-bridged copper complexes showed potent superoxide radical scavenging activity and remarkable antiproliferative activity against estrogen independent breast cancer cells (BT-20 cell line). A similar finding was reported for a copper-zinc complex of a catecholamine Schiff base ligand [76]. In the studies with copper(II) complexes, the correlation between cytotoxic potency and superoxide-scavenging activity is not always clear. In many cases, it is not clear if those agents are true SOD mimetics, i.e., disproportionating superoxide anion in a catalytic way, or just act as stoichiometric scavengers of O_2^-. Nevertheless, those studies justify the search for new copper(II) complexes with potential application in breast cancer therapy. Furthermore, the in vitro results warrant studies with more sophisticated models.

18.5 Beneficial Effects of SODm Combined with Cytotoxic Drugs

There are interesting reports suggesting an important role for SODm when combined with standard anticancer drugs. For example, Tempol was suggested as a potential agent that may improve the clinical effect of doxorubicin (dox) in tumors exhibiting a multidrug resistance (MDR) phenotype [77]. Cytotoxicity studies in MCF-7 wildtype and their MDR variant MCF-7 AdrR cells showed that Tempol potentiates dox cytotoxicity in both cell lines, with significant synergy in MDR cells. Such effect seems to be at least in part due to Tempol ability to increase peroxide levels and to deplete cellular glutathione pools. In addition, Tempol increased dox accumulation in MCF-7 AdrR cells by interfering with P-glycoprotein, and also affected the expression of proteins involved in response to drug treatment (e.g., p53, bcl2, bax, p21) [77].

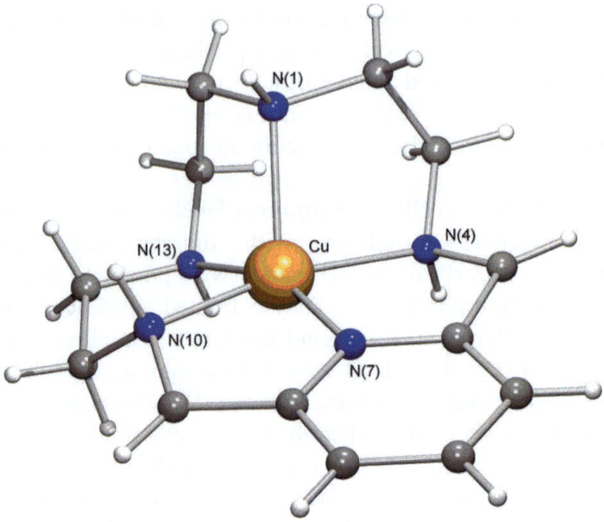

Fig. 18.4 Chemical structure of [Cu[15]pyN$_5$] complex [79]

As aforementioned, normal and cancer cells are generally distinct in terms of RS content and antioxidant defenses. These differences associated with the capacity of SODm to produce differential effects in normal vs. cancer cells or in cancer cells at earlier and later stage of progression may explain the differential behavior of these compounds reported in the literature. Mito-CP$_{11}$ is a mitochondrially targeted SODm formed by conjugating a triphenylphosphonium cation to the nitroxide carboxyproxyl. Mito-CP11 inhibited MCF-7 cells proliferation and exacerbated the cytotoxic effects of fluvastatin in breast cancer cells (MCF-7 and MDA-MB-231 cells). Although fluvastatin is clinically used as a cholesterol-lowering drug, it was previously shown to inhibit proliferation of breast cancer cells. Mito-CP$_{11}$ significantly inhibited NF-κB activity in MCF-7 cells, which might contribute to the antiproliferative effects observed. Interestingly, despite the antiproliferative effects in breast cancer cells, only a minimal effect was observed in MCF-10A normal-like mammary epithelial cells [78]. This dual effect was also observed for a pyridine-containing macrocyclic copper(II) complex, Cu[15]pyN$_5$ (Fig. 18.4), with superoxide scavenging activity developed by Fernandes et al. [61, 79]. While this complex is devoid of a marked cytotoxic effect as a single agent, it potentiated the antiproliferative activity of the anticancer drug oxaliplatin in MCF-7 cells. Conversely, this complex reduced the cytotoxicity of oxaliplatin in MCF-10A cells [61].

The ability to act differently in cancer and normal cells opens another interesting field of therapeutic opportunities for SODm in the context of breast cancer. RS are associated with side effects of the drugs commonly used in breast cancer chemotherapy, such as anthracyclines-induced cardiotoxicity [80, 81] or neuropathic pain induced by paclitaxel [82, 83]. SODm may therefore lessen the chemotherapy side effects. For example, the SODm MnTM-4-PyP^{5+} was shown to protect neonatal rat ventricular myocytes from daunorubicin-induced cell death [80]. Fidanboylu et al.

[83] assessed the ability of Tempol to inhibit established paclitaxel-induced (neuropathic) pain, as demonstrated in a rat model. The authors found a protective effect of Tempol on established paclitaxel-induced mechanical hypersensitivity at 1 h postinjection. However, the protective effects were not consistent under all experimental conditions tested. Doyle et al. [82] demonstrated that the administration of MnTE-2-PyP^{5+} or FeTM-4-PyP^{5+} in a rat model attenuated the development of neuropathic pain in a dose-dependent manner and reversed established pain. The authors attribute this protection to the ability of MnTE-2-PyP^{5+} and FeTM-4-PyP^{5+} to decompose peroxynitrite. Interestingly, this study also showed that the antitumor activity of paclitaxel in human breast cancer cells SKBR3 was enhanced by MnTE-2-PyP^{5+} [82]. A dual behavior was observed for the pyrroline nitroxyl derivative, pirolin, in a rat model with 7,12-dimethylbenz(*a*)anthracene-induced mammary tumors [84]. As a single agent, pirolin displayed both antioxidant and pro-oxidant properties depending on the cumulative dose of this nitroxide and the regimen of treatment. Combined chemotherapy with doxorubicin and docetaxel generated oxidative damage to plasma proteins. The inclusion of pirolin in dox-docetaxel chemotherapy resulted in a partial protection of blood plasma proteins and lipids against oxidative stress generated by these drugs [84]. The ability of SODm to protect non-cancer cells from toxic side effects induced by chemotherapy, along with their capacity to boost the efficacy of anticancer drugs, may lead to an increase in the therapeutic index of breast cancer treatments.

18.6 Benefits of SODm Combined with Radiotherapy

The ability of SODm to protect normal tissues while sensitizing tumors was also previously described in the context of radiotherapy. Moeller et al. [65] have shown that MnTE-2-PyP^{5+} may improve tumor radioresponsiveness. Conditioned medium from mouse breast 4T1 cells protected endothelial cells from ionizing radiation toxicity. However, medium that was conditioned in the presence of MnTE-2-PyP^{5+} inhibited the ability of tumor cells to secrete endothelial cell-rescuing factors into the medium. This effect may be explained by the ability of MnTE-2-PyP^{5+} to disrupt RS-induced cytokine secretion from tumor cells. This finding was corroborated in an in vivo model of Balb/C mice with 4T1 tumors. While MnTE-2-PyP^{5+} alone had no effect on tumor vascularity, the combined treatment with radiation led to a striking tumor devascularization, resulting in a significant tumor growth delay [65]. Another report using a slightly different treatment protocol also demonstrated that the combination of MnTE-2-PyP^{5+} with ionizing radiation led to a significant tumor growth delay in mice [85]. In addition, the ability of SODm to act as radiosensitizers was demonstrated in other cancer types (reviewed in [58]).

It has also been pointed out that, as the Warburg effect is implicated in resistance to cytotoxic stress, treatment methods that reduce glycolytic metabolism after radiotherapy may increase tumor cell sensitivity to radiation and chemotherapeutic killing. Zhong et al. [86] showed that the concurrent use of MnTnBuOE-2-PyP^{5+} and radiation prevented the deleterious post-radiation glycolytic changes in mice with

4T1 mammary carcinomas, which may constitute another advantage of the combination of SODm with radiotherapy.

To our knowledge, the use of SODm for the protection against radiotherapy-induced side effects has not yet been reported in breast cancer models. Fibrosis, following breast radiotherapy for mammary cancer, is a frequent undesired effect. Clinical tests were previously carried out with Cu,Zn-SOD locally applied as ointment on the fibrotic palpable area. The results showed attenuation of fibrosis size and pain after SOD treatment [87]. In addition, SODm were shown to play a radioprotective role in different models (reviewed in [57, 58, 88]). Considering such data, we may anticipate that SODm are likely to be useful in relieving the fibrosis and pain associated with breast radiotherapy.

18.7 Role of SODm in Angiogenesis and Cell Migration

An additional mechanism by which SODm may be beneficial in cancer treatment was recently suggested. The complex Cu[15]pyN$_5$ decreased MCF-7 cell migration although it did not significantly change the migration of MDA-MB-231 cells. A combined treatment with this complex and dox reduced the invasion of MDA-MB-231 cells [89]. Moreover, the co-exposure of breast cancer cells to MnTnHex-2-PyP^{5+} and dox reduced cell migration and invasion [90]. These data suggest that the potential of such complexes in breast cancer therapy may not be limited to effects on tumor proliferation. These compounds can also impact processes implicated in tumor metastases, offering a broader spectrum of therapeutic opportunities. Another phenomenon crucial for cancer progression that may be modulated by SODm is angiogenesis. Rabbani et al. [91] have demonstrated that MnTE-2-PyP^{5+} was able to suppress tumor growth in Balb/C mice with 4T1 cells sc xenografts. The antitumor effect of the SODm was suggested not to be due to a direct effect on cancer cells proliferation, but rather due to a decrease in angiogenesis. The subcutaneous administration of MnTE-2-PyP^{5+} (twice daily) over the course of the experiment markedly decreased the microvessel density and the endothelial cell proliferation by acting on HIF/VEGF pathways [91]. A reduction in microvascular density was also reported by Ye et al. [60] for MnTnHex-2-PyP^{5+}, either given alone or in combination with ascorbate, using the same experimental model.

18.8 SODm as a Mechanistic Tool in Breast Cancer Research

Besides the potential benefits of SODm as therapeutic agents, these compounds are also valuable mechanistic tools for breast cancer research. In some reports, SODm were used to clarify the role of RS in carcinogenesis. Kumari Kanchan et al. [92] published a study that aimed to investigate whether superoxide anion acts as a mediator of breast carcinogenesis by activating mTORC2, and whether this phenomenon is regulated by estrogen receptor. They found that treatment with $MnTBAP^{3-}$ impaired estrogen-induced mTORC2 activation in MCF-7 cells. To assess the role of RS on HIF-1α modulation, studies with Tempol or $MnTE\text{-}2\text{-}PyP^{5+}$ were carried out on MCF-7 cells. These studies demonstrated that superoxide anion is an important molecular effector underlying hypoxic HIF-1α stabilization [66]. $MnTM\text{-}4\text{-}PyP^{5+}$ was used to explore the involvement of mitochondrially produced H_2O_2 in the regulation of MDA-MB-231 cell growth [32]. Bandarra et al. [93] addressed the role of RS in the toxicity of glycidamide (GA) in MCF-10A cells using $MnTE\text{-}2\text{-}PyP^{5+}$ and other redox modulators. This SODm did not alter GA-induced cytotoxicity. The ultimate goal of this work was to further the understanding of the mechanisms of GA toxicity due to the pivotal role of this metabolite on the carcinogenesis induced by acrylamide, an industrial and food toxicant.

SODm have also been used to elucidate the involvement of RS in the mechanism of action of potential novel therapeutics for breast cancer. For example, an SODm was used by Hsieh et al. [94] as a tool to study the anticancer effect of the natural lignan arctigenin in MDA-MB-231 cells. The SODm EUK-8 antagonized arctigenin-mediated p38 activation and Bcl-2 protein downregulation, suggesting the elevation of RS as an upstream event of arctigenin-induced cell death. Different studies have also been carried out in in vitro models of inflammatory breast cancer, such as SUM149 and SUM190 cell lines, using SODm as tools to understand the impact of RS on the efficacy of therapeutic strategies. Allensworth et al. [95] showed that $MnTnHex\text{-}2\text{-}PyP^{5+}$ protected against RS accumulation and reversed the decrease in cell viability caused by embelin + TRAIL treatment. In addition, Aird et al. [96] used $MnTnHex\text{-}2\text{-}PyP^{5+}$ to explore the influence of RS on the induction of apoptosis by the lapatinib analog GW583340. Regarding radiotherapy, Moeller et al. [85] used $MnTE\text{-}2\text{-}PyP^{5+}$ in 4T1 breast cancer models (in vitro and in vivo) to clarify the role of HIF-1 regulation in determining tumor radiosensitivity.

Due to exceedingly complex chemistry of SODm, and considering the lack of specificity to a single reactive species, caution needs to be exercised when identifying the reactive species involved in particular pathways. The use of SODm combined with other pharmacological approaches should be complemented with genetic studies to corroborate conclusions.

18.9 Conclusion

In summary, SODm could impact different aspects of breast cancer therapy. This could be due to their inherent anticancer activity, or to their ability to boost chemotherapy and radiotherapy efficacy, ameliorate chemo- and radiotherapy side effects, hamper metastases formation or reduce angiogenesis. However, despite the amount of encouraging data from in vitro and in vivo studies, none of the SODm was already approved for clinical use in cancer. An aspect that needs further clarification is the timing for the use of SODm as an anticancer strategy. The effect of SOD depends on the stage of breast cancer, but so far only few studies addressed the impact of

SODm in models representative of different cancer stages. More studies are required to duly define the circumstances in which SODm may be beneficial. SODm are also useful tools to unravel drugable targets or to characterize the mechanisms of action of novel therapeutics. One important feature of most SODm is their ability to scavenge several RS and impact redox-dependent cellular transcriptional activity. Future work in this field should benefit from a thorough comprehension of the redox basis of breast carcinogenesis. The growing understanding of redox mechanisms, along with the continuous quest for optimized SODm, will impact the progress of those redox-active drugs towards clinic.

References

1. DeSantis C, Ma J, Bryan L, Jemal A. Breast cancer statistics, 2013. CA Cancer J Clin. 2014;64(1):52–62.
2. Siegel RL, Miller KD, Jemal A. Cancer statistics, 2015. CA Cancer J Clin. 2015;65(1):5–29.
3. Bray F, Ren JS, Masuyer E, Ferlay J. Global estimates of cancer prevalence for 27 sites in the adult population in 2008. Int J Cancer. 2013;132(5):1133–45.
4. Eccles SA, Aboagye EO, Ali S, Anderson AS, Armes J, Berditchevski F, et al. Critical research gaps and translational priorities for the successful prevention and treatment of breast cancer. Breast Cancer Res. 2013;15(5):R92. Pubmed Central PMCID: 3907091.
5. Evans MD, Dizdaroglu M, Cooke MS. Oxidative DNA damage and disease: induction, repair and significance. Mutat Res. 2004;567(1):1–61.
6. Pelicano H, Carney D, Huang P. ROS stress in cancer cells and therapeutic implications. Drug Resist Updat. 2004;7(2):97–110.
7. Szatrowski TP, Nathan CF. Production of large amounts of hydrogen peroxide by human tumor cells. Cancer Res. 1991;51(3):794–8.
8. Crapo JD, Oury T, Rabouille C, Slot JW, Chang LY. Copper, zinc superoxide dismutase is primarily a cytosolic protein in human cells. Proc Natl Acad Sci U S A 1992;89(21):10405–9. Pubmed Central PMCID: 50347.
9. Okado-Matsumoto A, Fridovich I. Subcellular distribution of superoxide dismutases (SOD) in rat liver: Cu, Zn-SOD in mitochondria. J Biol Chem. 2001;276(42):38388–93.
10. Weisiger RA, Fridovich I. Mitochondrial superoxide simutase. Site of synthesis and intramitochondrial localization. J Biol Chem. 1973;248(13):4793–6.
11. Marklund SL, Holme E, Hellner L. Superoxide dismutase in extracellular fluids. Clin Chim Acta. 1982;126(1):41–51.
12. Kinnula VL, Crapo JD. Superoxide dismutases in malignant cells and human tumors. Free Radic Biol Med. 2004;36(6):718–44.
13. Li JJ, Colburn NH, Oberley LW. Maspin gene expression in tumor suppression induced by overexpressing manganese-containing superoxide dismutase cDNA in human breast cancer cells. Carcinogenesis. 1998;19(5):833–9.
14. Connor KM, Hempel N, Nelson KK, Dabiri G, Gamarra A, Belarmino J, et al. Manganese superoxide dismutase enhances the invasive and migratory activity of tumor cells. Cancer Res. 2007;67(21):10260–7.
15. Batinic-Haberle I. SOD enzymes and their mimics in cancer: pro vs anti-oxidative mode of action. Part I. Anticancer Agents Med Chem. 2011;11(2):172–4.
16. Nishida S, Akai F, Iwasaki H, Hosokawa K, Kusunoki T, Suzuki K, et al. Manganese superoxide dismutase content and localization in human thyroid tumours. J Pathol. 1993;169(3):341–5.
17. Cobbs CS, Levi DS, Aldape K, Israel MA. Manganese superoxide dismutase expression in human central nervous system tumors. Cancer Res. 1996;56(14):3192–5.

18. Landriscina M, Remiddi F, Ria F, Palazzotti B, De Leo ME, Iacoangeli M, et al. The level of MnSOD is directly correlated with grade of brain tumours of neuroepithelial origin. Br J Cancer. 1996;74(12):1877–85. Pubmed Central PMCID: 2074812.
19. Janssen AM, Bosman CB, van Duijn W, Oostendorp-van de Ruit MM, Kubben FJ, Griffioen G, et al. Superoxide dismutases in gastric and esophageal cancer and the prognostic impact in gastric cancer. Clin Cancer Res. 2000;6(8):3183–92.
20. Malafa M, Margenthaler J, Webb B, Neitzel L, Christophersen M. MnSOD expression is increased in metastatic gastric cancer. J Surg Res. 2000;88(2):130–4.
21. Hu Y, Rosen DG, Zhou Y, Feng L, Yang G, Liu J, et al. Mitochondrial manganese-superoxide dismutase expression in ovarian cancer: role in cell proliferation and response to oxidative stress. J Biol Chem. 2005;280(47):39485–92.
22. Ennen M, Minig V, Grandemange S, Touche N, Merlin JL, Besancenot V, et al. Regulation of the high basal expression of the manganese superoxide dismutase gene in aggressive breast cancer cells. Free Radic Biol Med. 2011;50(12):1771–9.
23. Tsanou E, Ioachim E, Briasoulis E, Damala K, Charchanti A, Karavasilis V, et al. Immunohistochemical expression of superoxide dismutase (MnSOD) anti-oxidant enzyme in invasive breast carcinoma. Histol Histopathol. 2004;19(3):807–13.
24. Becuwe P, Ennen M, Klotz R, Barbieux C, Grandemange S. Manganese superoxide dismutase in breast cancer: from molecular mechanisms of gene regulation to biological and clinical significance. Free Radic Biol Med. 2014;77:139–51.
25. Hempel N, Carrico PM, Melendez JA. Manganese superoxide dismutase (Sod2) and redox-control of signaling events that drive metastasis. Anticancer Agents Med Chem. 2011;11(2):191–201. Pubmed Central PMCID: 3902825.
26. Cyr AR, Hitchler MJ, Domann FE. Regulation of SOD2 in cancer by histone modifications and CpG methylation: closing the loop between redox biology and epigenetics. Antioxid Redox Signal. 2013;18(15):1946–55. Pubmed Central PMCID: 3624766.
27. Bianchi A, Becuwe P, Franck P, Dauca M. Induction of MnSOD gene by arachidonic acid is mediated by reactive oxygen species and p38 MAPK signaling pathway in human HepG2 hepatoma cells. Free Radic Biol Med. 2002;32(11):1132–42.
28. Zhao Y, St Clair DK. Detection of the content and activity of the transcription factor AP-1 in a multistage skin carcinogenesis model. Methods Mol Biol. 2003;218:177–84.
29. Rieber M, Strasberg-Rieber M. Hypoxia, Mn-SOD and H(2)O(2) regulate p53 reactivation and PRIMA-1 toxicity irrespective of p53 status in human breast cancer cells. Biochem Pharmacol. 2012;84(12):1563–70.
30. Kamarajugadda S, Cai Q, Chen H, Nayak S, Zhu J, He M, et al. Manganese superoxide dismutase promotes anoikis resistance and tumor metastasis. Cell Death Dis. 2013;4:e504. Pubmed Central PMCID: 3734828.
31. Li JJ, Oberley LW, St Clair DK, Ridnour LA, Oberley TD. Phenotypic changes induced in human breast cancer cells by overexpression of manganese-containing superoxide dismutase. Oncogene. 1995;10(10):1989–2000.
32. Kattan Z, Minig V, Leroy P, Dauca M, Becuwe P. Role of manganese superoxide dismutase on growth and invasive properties of human estrogen-independent breast cancer cells. Breast Cancer Res Treat. 2008;108(2):203–15.
33. Weydert CJ, Waugh TA, Ritchie JM, Iyer KS, Smith JL, Li L, et al. Overexpression of manganese or copper-zinc superoxide dismutase inhibits breast cancer growth. Free Radic Biol Med. 2006;41(2):226–37.
34. Zhang HJ, Yan T, Oberley TD, Oberley LW. Comparison of effects of two polymorphic variants of manganese superoxide dismutase on human breast MCF-7 cancer cell phenotype. Cancer Res. 1999;59(24):6276–83.
35. Church SL, Grant JW, Ridnour LA, Oberley LW, Swanson PE, Meltzer PS, et al. Increased manganese superoxide dismutase expression suppresses the malignant phenotype of human melanoma cells. Proc Natl Acad Sci U S A 1993;90(7):3113–7. Pubmed Central PMCID: 46247.
36. Liu R, Oberley TD, Oberley LW. Transfection and expression of MnSOD cDNA decreases tumor malignancy of human oral squamous carcinoma SCC-25 cells. Hum Gene Ther. 1997;8(5):585–95.

37. Zhong W, Oberley LW, Oberley TD, St Clair DK. Suppression of the malignant phenotype of human glioma cells by overexpression of manganese superoxide dismutase. Oncogene. 1997;14(4):481–90.
38. Li N, Oberley TD, Oberley LW, Zhong W. Overexpression of manganese superoxide dismutase in DU145 human prostate carcinoma cells has multiple effects on cell phenotype. Prostate. 1998;35(3):221–33.
39. Oberley LW. Anticancer therapy by overexpression of superoxide dismutase. Antioxid Redox Signal. 2001;3(3):461–72.
40. Plymate SR, Haugk KH, Sprenger CC, Nelson PS, Tennant MK, Zhang Y, et al. Increased manganese superoxide dismutase (SOD-2) is part of the mechanism for prostate tumor suppression by Mac25/insulin-like growth factor binding-protein-related protein-1. Oncogene. 2003;22(7):1024–34.
41. Trimmer C, Sotgia F, Whitaker-Menezes D, Balliet RM, Eaton G, Martinez-Outschoorn UE, et al. Caveolin-1 and mitochondrial SOD2 (MnSOD) function as tumor suppressors in the stromal microenvironment: a new genetically tractable model for human cancer associated fibroblasts. Cancer Biol Ther. 2011;11(4):383–94. Pubmed Central PMCID: 3047109.
42. Teoh ML, Fitzgerald MP, Oberley LW, Domann FE. Overexpression of extracellular superoxide dismutase attenuates heparanase expression and inhibits breast carcinoma cell growth and invasion. Cancer Res. 2009;69(15):6355–63. Pubmed Central PMCID: 2728021.
43. Teoh-Fitzgerald ML, Fitzgerald MP, Zhong W, Askeland RW, Domann FE. Epigenetic reprogramming governs EcSOD expression during human mammary epithelial cell differentiation, tumorigenesis and metastasis. Oncogene. 2014;33(3):358–68. Pubmed Central PMCID: 3711965.
44. Shimoda-Matsubayashi S, Matsumine H, Kobayashi T, Nakagawa-Hattori Y, Shimizu Y, Mizuno Y. Structural dimorphism in the mitochondrial targeting sequence in the human manganese superoxide dismutase gene. A predictive evidence for conformational change to influence mitochondrial transport and a study of allelic association in Parkinson's disease. Biochem Biophys Res Commun. 1996;226(2):561–5.
45. Tao R, Coleman MC, Pennington JD, Ozden O, Park SH, Jiang H, et al. Sirt3-mediated deacetylation of evolutionarily conserved lysine 122 regulates MnSOD activity in response to stress. Mol Cell. 2010;40(6):893–904. Pubmed Central PMCID: 3266626.
46. Liu G, Sun G, Wang Y, Wang D, Hu W, Zhang J. Association between manganese superoxide dismutase gene polymorphism and breast cancer risk: a meta-analysis of 17,842 subjects. Mol Med Rep. 2012;6(4):797–804.
47. Cox DG, Tamimi RM, Hunter DJ. Gene x Gene interaction between MnSOD and GPX-1 and breast cancer risk: a nested case-control study. BMC Cancer. 2006;6:217. Pubmed Central PMCID: 1569434.
48. Bica CG, da Silva LL, Toscani NV, Zettler CG, Gottlieb MG, Alexandre CO, et al. Polymorphism (ALA16VAL) correlates with regional lymph node status in breast cancer. Cancer Genet Cytogenet. 2010;196(2):153–8.
49. Kim HS, Patel K, Muldoon-Jacobs K, Bisht KS, Aykin-Burns N, Pennington JD, et al. SIRT3 is a mitochondria-localized tumor suppressor required for maintenance of mitochondrial integrity and metabolism during stress. Cancer Cell. 2010;17(1):41–52. Pubmed Central PMCID: 3711519.
50. Toh Y, Kuninaka S, Oshiro T, Ikeda Y, Nakashima H, Baba H, et al. Overexpression of manganese superoxide dismutase mRNA may correlate with aggressiveness in gastric and colorectal adenocarcinomas. Int J Oncol. 2000;17(1):107–12.
51. Kahlos K, Soini Y, Paakko P, Saily M, Linnainmaa K, Kinnula VL. Proliferation, apoptosis, and manganese superoxide dismutase in malignant mesothelioma. Int J Cancer. 2000;88(1): 37–43.
52. Venkataraman S, Jiang X, Weydert C, Zhang Y, Zhang HJ, Goswami PC, et al. Manganese superoxide dismutase overexpression inhibits the growth of androgen-independent prostate cancer cells. Oncogene. 2005;24(1):77–89.
53. Bravard A, Sabatier L, Hoffschir F, Ricoul M, Luccioni C, Dutrillaux B. SOD2: a new type of tumor-suppressor gene? Int J Cancer. 1992;51(3):476–80.

54. Oberley LW. Mechanism of the tumor suppressive effect of MnSOD overexpression. Biomed Pharmacother. 2005;59(4):143–8.
55. Day BJ. Catalytic antioxidants: a radical approach to new therapeutics. Drug Discov Today. 2004;9(13):557–66.
56. Batinic-Haberle I, Reboucas JS, Spasojevic I. Superoxide dismutase mimics: chemistry, pharmacology, and therapeutic potential. Antioxid Redox Signal. 2010;13(6):877–918. Pubmed Central PMCID: 2935339.
57. Tovmasyan A, Sheng H, Weitner T, Arulpragasam A, Lu M, Warner DS, et al. Design, mechanism of action, bioavailability and therapeutic effects of mn porphyrin-based redox modulators. Med Princ Pract. 2013;22(2):103–30. Pubmed Central PMCID: 3640855.
58. Batinic-Haberle I, Tovmasyan A, Roberts ER, Vujaskovic Z, Leong KW, Spasojevic I. SOD therapeutics: latest insights into their structure-activity relationships and impact on the cellular redox-based signaling pathways. Antioxid Redox Signal. 2014;20(15):2372–415. Pubmed Central PMCID: 4005498.
59. Nicco C, Laurent A, Chereau C, Weill B, Batteux F. Differential modulation of normal and tumor cell proliferation by reactive oxygen species. Biomed Pharmacother. 2005;59(4):169–74.
60. Ye X, Fels D, Tovmasyan A, Aird KM, Dedeugd C, Allensworth JL, et al. Cytotoxic effects of Mn(III) N-alkylpyridylporphyrins in the presence of cellular reductant, ascorbate. Free Radic Res. 2011;45(11–12):1289–306. Pubmed Central PMCID: 3500602.
61. Fernandes AS, Costa J, Gaspar J, Rueff J, Cabral MF, Cipriano M, et al. Development of pyridine-containing macrocyclic copper(II) complexes: potential role in the redox modulation of oxaliplatin toxicity in human breast cells. Free Radic Res. 2012;46(9):1157–66.
62. Laurent A, Nicco C, Chereau C, Goulvestre C, Alexandre J, Alves A, et al. Controlling tumor growth by modulating endogenous production of reactive oxygen species. Cancer Res. 2005;65(3):948–56.
63. Jaramillo MC, Briehl MM, Crapo JD, Batinic-Haberle I, Tome ME. Manganese porphyrin, MnTE-2-PyP5+, Acts as a pro-oxidant to potentiate glucocorticoid-induced apoptosis in lymphoma cells. Free Radic Biol Med. 2012;52(8):1272–84. Pubmed Central PMCID: 3331723.
64. Batinic-Haberle I, Tovmasyan A, Spasojevic I. An educational overview of the chemistry, biochemistry and therapeutic aspects of Mn porphyrins—from superoxide dismutation to H2O2-driven pathways. Redox Biol. 2015;5:43–65.
65. Moeller BJ, Batinic-Haberle I, Spasojevic I, Rabbani ZN, Anscher MS, Vujaskovic Z, et al. A manganese porphyrin superoxide dismutase mimetic enhances tumor radioresponsiveness. Int J Radiat Oncol Biol Phys. 2005;63(2):545–52.
66. Kaewpila S, Venkataraman S, Buettner GR, Oberley LW. Manganese superoxide dismutase modulates hypoxia-inducible factor-1 alpha induction via superoxide. Cancer Res. 2008;68(8):2781–8. Pubmed Central PMCID: 2633869.
67. Tovmasyan A, Carballal S, Ghazaryan R, Melikyan L, Weitner T, Maia CG, et al. Rational design of superoxide dismutase (SOD) mimics: the evaluation of the therapeutic potential of new cationic Mn porphyrins with linear and cyclic substituents. Inorg Chem. 2014;53(21):11467–83. Pubmed Central PMCID: 4220860.
68. Tovmasyan A, Roberts ER, Yuliana Y, Haberle S, Boss M, Venkatraman T, et al. The role of ascorbate in therapeutic actions of cationic Mn porphyrin-based SOD mimics. Free Radic Biol Med. 2014;76:S94–5.
69. Evans MK, Tovmasyan A, Batinic-Haberle I, Devi GR. Mn porphyrin in combination with ascorbate acts as a pro-oxidant and mediates caspase-independent cancer cell death. Free Radic Biol Med. 2014;68:302–14.
70. Buchtik R, Travnicek Z, Vanco J. In vitro cytotoxicity, DNA cleavage and SOD-mimic activity of copper(II) mixed-ligand quinolinonato complexes. J Inorg Biochem. 2012;116:163–71.
71. O'Connor M, Kellett A, McCann M, Rosair G, McNamara M, Howe O, et al. Copper(II) complexes of salicylic acid combining superoxide dismutase mimetic properties with DNA binding and cleaving capabilities display promising chemotherapeutic potential with fast acting in vitro cytotoxicity against cisplatin sensitive and resistant cancer cell lines. J Med Chem. 2012;55(5):1957–68.

72. Kellett A, Howe O, O'Connor M, McCann M, Creaven BS, McClean S, et al. Radical-induced DNA damage by cytotoxic square-planar copper(II) complexes incorporating o-phthalate and 1,10-phenanthroline or 2,2′-dipyridyl. Free Radic Biol Med. 2012;53(3):564–76.
73. Kovala-Demertzi D, Staninska M, Garcia-Santos I, Castineiras A, Demertzis MA. Synthesis, crystal structures and spectroscopy of meclofenamic acid and its metal complexes with manganese(II), copper(II), zinc(II) and cadmium(II). Antiproliferative and superoxide dismutase activity. J Inorg Biochem. 2011;105(9):1187–95.
74. Saczewski F, Dziemidowicz-Borys E, Bednarski PJ, Gdaniec M. Synthesis, crystal structure, cytotoxic and superoxide dismutase activities of copper(II) complexes of N-(4,5-dihydroimidazol-2-yl)azoles. Arch Pharm. 2007;340(7):333–8.
75. Dutta S, Padhye S, Ahmed F, Sarkar F. Pyridazolate-bridged dicopper (II) SOD mimics with enhanced antiproliferative activities against estrogen and androgen independent cancer cell lines. Inorg Chim Acta. 2005;358(13):3617–24. PubMed PMID: WOS:000232384400009.
76. Dutta S, Bendre R, Padhye S, Ahmed F, Sarkar F. Synthesis, antioxidant properties and antiproliferative activities of tetrameric copper and copper-zinc metal complexes of catecholamine Schiff base ligand. Synth React Inorg Metal Org Nano-Met Chem. 2005;35(1):3–10. PubMed PMID: WOS:000228531500002.
77. Gariboldi MB, Terni F, Ravizza R, Meschini S, Marra M, Condello M, et al. The nitroxide Tempol modulates anthracycline resistance in breast cancer cells. Free Radic Biol Med. 2006;40(8):1409–18.
78. Cheng G, Lopez M, Zielonka J, Hauser AD, Joseph J, McAllister D, et al. Mitochondria-targeted nitroxides exacerbate fluvastatin-mediated cytostatic and cytotoxic effects in breast cancer cells. Cancer Biol Ther. 2011;12(8):707–17. Pubmed Central PMCID: 3218525.
79. Fernandes AS, Cabral MF, Costa J, Castro M, Delgado R, Drew MG, et al. Two macrocyclic pentaaza compounds containing pyridine evaluated as novel chelating agents in copper(II) and nickel(II) overload. J Inorg Biochem. 2011;105(3):410–9.
80. Ikegami E, Fukazawa R, Kanbe M, Watanabe M, Abe M, Watanabe M, et al. Edaravone, a potent free radical scavenger, prevents anthracycline-induced myocardial cell death. Circ J. 2007;71(11):1815–20.
81. Takemura G, Fujiwara H. Doxorubicin-induced cardiomyopathy from the cardiotoxic mechanisms to management. Prog Cardiovasc Dis. 2007;49(5):330–52.
82. Doyle T, Chen Z, Muscoli C, Bryant L, Esposito E, Cuzzocrea S, et al. Targeting the overproduction of peroxynitrite for the prevention and reversal of paclitaxel-induced neuropathic pain. J Neurosci. 2012;32(18):6149–60. Pubmed Central PMCID: 3752044.
83. Fidanboylu M, Griffiths LA, Flatters SJ. Global inhibition of reactive oxygen species (ROS) inhibits paclitaxel-induced painful peripheral neuropathy. PLoS One. 2011;6(9):e25212. Pubmed Central PMCID: 3180385.
84. Tabaczar S, Koceva-Chyla A, Czepas J, Pieniazek A, Piasecka-Zelga J, Gwozdzinski K. Nitroxide pirolin reduces oxidative stress generated by doxorubicin and docetaxel in blood plasma of rats bearing mammary tumor. J Physiol Pharmacol. 2012;63(2):153–63.
85. Moeller BJ, Cao Y, Li CY, Dewhirst MW. Radiation activates HIF-1 to regulate vascular radiosensitivity in tumors: role of reoxygenation, free radicals, and stress granules. Cancer Cell. 2004;5(5):429–41.
86. Zhong J, Rajaram N, Brizel DM, Frees AE, Ramanujam N, Batinic-Haberle I, et al. Radiation induces aerobic glycolysis through reactive oxygen species. Radiother Oncol. 2013;106(3):390–6. Pubmed Central PMCID: 3770265.
87. Campana F, Zervoudis S, Perdereau B, Gez E, Fourquet A, Badiu C, et al. Topical superoxide dismutase reduces post-irradiation breast cancer fibrosis. J Cell Mol Med. 2004;8(1):109–16.
88. Soule BP, Hyodo F, Matsumoto K, Simone NL, Cook JA, Krishna MC, et al. The chemistry and biology of nitroxide compounds. Free Radic Biol Med. 2007;42(11):1632–50. Pubmed Central PMCID: 1991293.
89. Fernandes AS, Flórido A, Saraiva N, Cerqueira S, Ramalhete S, Cipriano M, et al. Role of the copper(II) complex Cu[15]pyN5 in intracellular ROS and breast cancer cell motility and invasion. Chem Biol Drug Des. 2015;86(4):578–88.

90. Fernandes AS, Florido A, Cipriano M, Batinic-Haberle I, Miranda J, Saraiva N, et al. Combined effect of the SOD mimic MnTnHex-2-PyP5+ and doxorubicin on the migration and invasiveness of breast cancer cells. Toxicol Lett. 2013;221:S70–1. PubMed PMID: WOS:000323865800213.
91. Rabbani ZN, Spasojevic I, Zhang X, Moeller BJ, Haberle S, Vasquez-Vivar J, et al. Antiangiogenic action of redox-modulating Mn(III) meso-tetrakis(N-ethylpyridinium-2-yl) porphyrin, MnTE-2-PyP(5+), via suppression of oxidative stress in a mouse model of breast tumor. Free Radic Biol Med. 2009;47(7):992–1004. Pubmed Central PMCID: 2749298.
92. Kumari Kanchan R, Tripathi C, Singh Baghel K, Kumar Dwivedi S, Kumar B, Sanyal S, et al. Estrogen receptor potentiates mTORC2 signaling in breast cancer cells by upregulating superoxide anions. Free Radic Biol Med. 2012;53(10):1929–41.
93. Bandarra S, Fernandes AS, Magro I, Guerreiro PS, Pingarilho M, Churchwell MI, et al. Mechanistic insights into the cytotoxicity and genotoxicity induced by glycidamide in human mammary cells. Mutagenesis. 2013;28(6):721–9. PubMed PMID: WOS:000326380000012.
94. Hsieh CJ, Kuo PL, Hsu YC, Huang YF, Tsai EM, Hsu YL. Arctigenin, a dietary phytoestrogen, induces apoptosis of estrogen receptor-negative breast cancer cells through the ROS/p38 MAPK pathway and epigenetic regulation. Free Radic Biol Med. 2014;67:159–70.
95. Allensworth JL, Aird KM, Aldrich AJ, Batinic-Haberle I, Devi GR. XIAP inhibition and generation of reactive oxygen species enhances TRAIL sensitivity in inflammatory breast cancer cells. Mol Cancer Ther. 2012;11(7):1518–27.
96. Aird KM, Allensworth JL, Batinic-Haberle I, Lyerly HK, Dewhirst MW, Devi GR. ErbB1/2 tyrosine kinase inhibitor mediates oxidative stress-induced apoptosis in inflammatory breast cancer cells. Breast Cancer Res Treat. 2012;132(1):109–19. Pubmed Central PMCID: 3734382.

Chapter 19
Anticancer Action of Mn Porphyrins in Head and Neck Cancer

Kathleen A. Ashcraft and Mark W. Dewhirst

19.1 Introduction

Manganese porphyrins have been successfully used preclinically as adjuvant therapies to radiotherapy and/or chemotherapy for tumors of multiple histologies, including brain, breast, prostate, and melanoma [1–6]. In contrast, literature regarding the use of Mn porphyrins in head and neck cancer (HNC) is conspicuously scant.

Theoretically, HNC is an ideal disease to target with Mn porphyrins. First, the pathologies and risk factors underlying HNC are linked to oxidative stress. In addition, there are several major sources of oxidative stress in HNC (i.e., products of smoking, hypoxia, virally induced gene expression, and immune cell infiltration) that drive tumor promotion and progression by contributing to genetic instability. Further, oxidative stress promotes tumorigenesis through induction of multiple signal transduction pathways and cellular and mitochondrial reactive oxygen species (reviewed in [7]). In fact, tumor hypoxia is a clear prognostic factor for poor outcome in HNC patients (Fig. 19.1).

Although Mn porphyrins have not been used clinically in HNC to date, there are several published reports using manganese superoxide dismutase (MnSOD)-based methods in preclinical HNC models that suggest that Mn porphyrins could be efficacious in the treatment of HNC. Thus, the focus of this chapter will be on historical use of MnSOD therapies to provide a rationale for future work using Mn porphyrins.

K.A. Ashcraft • M.W. Dewhirst (✉)
Department of Radiation Oncology, Duke University Medical Center,
Box 3455, Medical Sciences Research Building 1, Room 201, Durham, NC 27710, USA
e-mail: kathleen.ashcraft@duke.edu; mark.dewhirst@dm.duke.edu

I. Batinić-Haberle et al. (eds.), *Redox-Active Therapeutics*,
Oxidative Stress in Applied Basic Research and Clinical Practice,
DOI 10.1007/978-3-319-30705-3_19

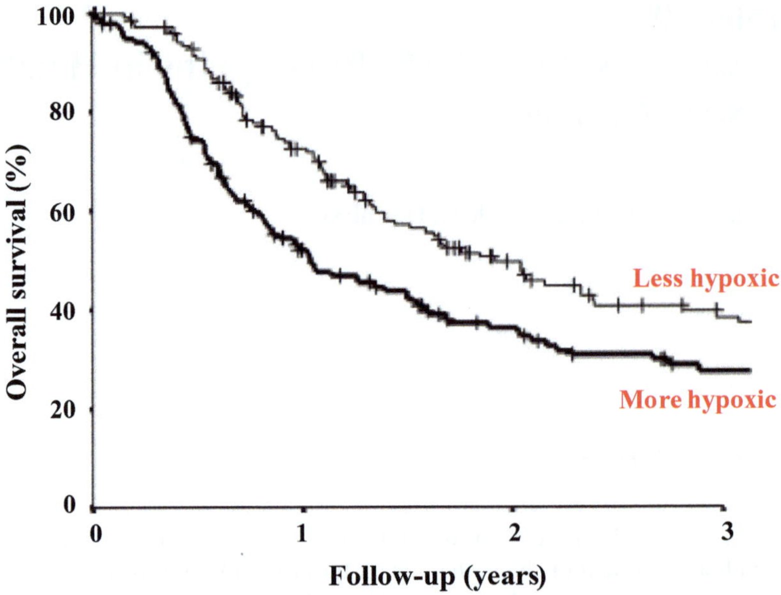

Fig. 19.1 Overall survival curves for HNC patients stratified by tumor hypoxia. Patients with less hypoxic tumors showed a significant increase in survival compared to patients with more hypoxic tumors. In this study, which was done using the Eppendorf oxygen electrode, hypoxia was defined as a tumor in which at least 19 % of the measurements were less than 2.5 mmHg. $p = 0.006$. Taken from Nordsmark et al., Radiotherapy and Oncology, 77(1)18–24 (2005). Used with permission

19.2 Head and Neck Cancers: Types, Etiology, and Underlying Pathologies

HNC comprise multiple tumor types and locations. The most common sites of origin are the oral cavity, pharynx/tonsils, paranasal sinuses and nasal cavity, larynx, and salivary glands. Though technically not included as a HNC, esophageal cancers will be considered within this chapter as well. The types, incidences, and underlying causes of HNC are described in Table 19.1. HNC accounts for 3 % of all cancers in the United States.

HNC are commonly associated with environmental and/or lifestyle choices although the primary risk factors have shifted in recent decades. For instance, 75 % of oral and pharyngeal cancers were associated with tobacco and alcohol use in 1988 [8]. More recently, a 2011 report estimated that nearly 50 % of oropharynx tumors resulted from human papilloma virus (HPV) infection [9]. Three lifestyle/environmental risk factors will be discussed briefly below. Although superficially they seem disparate, all three have been linked to cellular and DNA damage following oxidative stress.

The first risk factor, and the one that is most directly under the patient's control, is tobacco and alcohol use. Smoking tobacco exposes the head and neck tissues to

Table 19.1 Head and neck cancers: Incidence and underlying pathologies

Tumor type	Estimated cases in the United States for 2014	Estimated deaths in the United States in 2014	Underlying pathologies/risk factors
Laryngeal	12,630	3,610	History of smoking and excessive alcohol use [53]
Pharyngeal	14,410	2,540	~98 % of nasopharyngeal cancers are associated with EBV [54]. Hypopharyngeal cancers are associated with loss of chromosome 18 [55]
Oral and oropharyngeal	41,440	8,390	History of smoking, HPV16 infection [56]
Paranasal and nasal cavity			~20 % are associated with HPV [57]
Esophageal	18,170	15,540	Associated with GERD [58]

Numbers taken from cancer.gov, December 2014

smoke-borne ROS, as well as tobacco smoke-induced injury, which increases endogenous ROS. Rats exposed to cigarette smoke for 8 weeks (six cigarettes worth of smoke per day) had increased free radicals in their laryngeal tissues compared to control rats [10]. Because the combination of alcohol and tobacco raises cancer risk, one hypothesis is that alcohol acts as a solvent to facilitate the entry of tobacco/smoke-derived chemicals into cells. A second cause of tobacco-related cancers is smokeless tobacco. In vitro studies with human dermal fibroblasts demonstrated that exposure to smokeless tobacco extract increased intracellular reactive oxygen and nitrogen species [11].

A second causative agent of oral cancers that must be mentioned is HPV and to a lesser extent, Epstein–Barr virus (EBV). In the United States, the epidemiology of HPV has historically been focused on cervical cancer; however, the incidence of oral-based cancers caused by HPV is increasing at such a rate that there will be more oral tumors than cervical tumors caused by HPV by the year 2020. EBV infection is linked to nasopharyngeal cancer though its incidence is more concentrated in Southeast Asia. The oncogenic nature of HPV and EBV results from expression of proteins E6 and E7 in HPV and EBNA-1 in EBV. However, studies from the past decade have revealed that E6 and EBNA-1 may also promote tumor development by increasing cellular and DNA damage caused by increased cellular oxidative stress (Fig. 19.2). Williams et al. reported that expression of truncated E6 isoforms (referred to as E6*) increase oxidative stress even in the absence of viral infection or full E6 genome insertion [12]. Normal oral keratinocytes transfected with E6* showed reduced SOD2 (MnSOD) and glutathione peroxidase (GPx1/2), and increased oxidative stress-dependent DNA damage. EBNA-1 has also been linked to increased oxidative stress [13, 14]. These studies are exciting because by implicating oxidative stress as an additional mechanism of tumor development following oncogenic viral infection, they may pave the way for new therapeutics to slow or prevent tumor growth, including the use MnSOD mimics as chemopreventive agents.

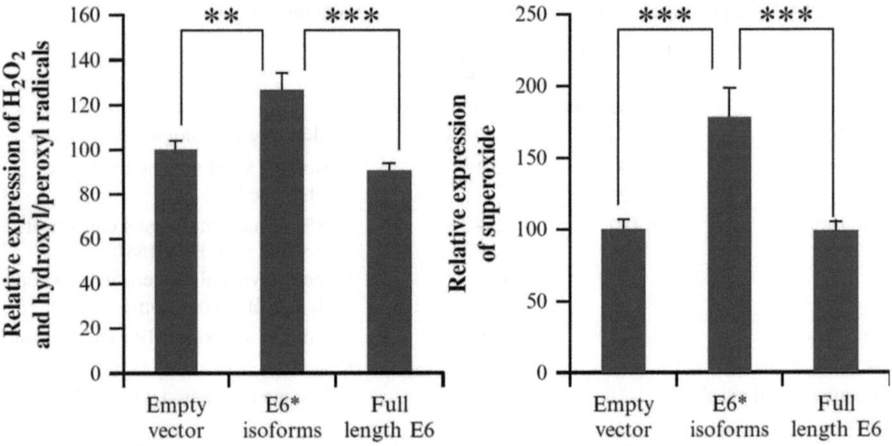

Fig. 19.2 Expression of the truncated isoforms of HPV16 E6 protein (E6*) induce cellular oxidative stress independent of viral infection, by increasing relative expression of H_2O_2, and hydroxyl/peroxyl radicals (*left panel*) and superoxide (*right panel*). **indicates a 95 % level of confidence; ***indicates a 99 % level of confidence. Modified from Williams et al., J. Virology, 88(12): 6751–6761, 2014

Finally, a risk factor for esophageal cancer is gastroesophageal reflux disease (GERD). Like tobacco use and HPV infection, GERD is associated with lifestyle choices, as its risk and severity is linked to obesity and consumption of trans fats and possibly alcohol. GERD is implicated in esophagitis, Barrett's esophagus, and progression to esophageal adenocarcinoma. However, the tissue damage seen in GERD and Barrett's esophagus patients is greater than would be expected from exposure to bile salts and gastric acid alone. Tissue analysis revealed that exposure to bile salts and gastric acid causes oxidative stress in the esophageal mucosa, which is responsible for the severe cellular damage observed, and progression to cancer. Tissues from esophagitis and Barrett's esophagus patients have increased superoxide-related markers and peroxynitrite-related nitrotyrosine compared to controls. These patients' tumors also had increased expression of reduced glutathione (GSH), catalase, and MnSOD, but decreased activity of all SOD isoforms [15]. Recently, gastric reflux was found to be predictive of laryngopharyngeal carcinoma [16]. Taken together, these results indicate that oxidative stress confers sufficient cellular damage to cause GERD-induced mucosal damage to progress to esophagitis, Barrett's esophagus, and potentially esophageal adenocarcinoma.

Salivary gland cancers (SGC) are quite different from other HNC carcinomas. The ten different types of cancers that can develop in the salivary glands are often benign and constitute roughly 5 % of all HNC. SGC have not been linked to any strong risk factor although various reports have suggested tumor growth can be predicated by: (1) being blood group A [17], (2) having HPV or EBV infection [18], (3) having occupational exposure to rubber or nickel [19], or (4) having been treated with radiotherapy for prior cancers [20]. There has been no link between SGC hypoxia and patient prognosis [21], and the hypoxia markers (hypoxia-inducible factors HIF-1α, HIF-2α, carbonic anhydrase [CAIX], glucose transporter

[GLUT-1], and erythropoietin receptor) are not predictive for local control or survival [22]. Elevating MnSOD in the salivary gland may be useful as a radioprotectant against xerostomia [23, 24]. However, there have been no studies investigating the use of MnSOD as a therapy for SGC.

19.3 Phenotypes of HNC Tumors That Make MnSOD-Based Therapies Attractive

The notoriously hypoxic microenvironment of HNC tumors has been the central theme of many studies (reviewed in [25]). Hypoxia can contribute to oxidative stress because mitochondria over produce superoxide during periods of hypoxia (reviewed in [7]). Thus, alleviating oxidative stress that emanates from hypoxia may improve treatment response. Multiple studies with HNC patients have shown that the clinical ramifications of tumor hypoxia, measured prior to radiotherapy or chemotherapy, include decreased overall survival [26–28]. Patients with median tumor $pO_2 \leq 10$ mmHg show significantly worse local control after radiotherapy and reduced disease-free survival [27, 28], which is especially sobering as one study reported that 70 % of oral squamous cell carcinomas expressed high levels of HIF-1α [29]. Relatedly, oxidative stress can push tumors towards a Warburg phenotype, resulting in lactate production in hypoxic areas. High concentrations of lactate are predictive for metastatic HNC [30] and lower overall survival (Fig. 19.3). Therefore, even if MnSOD treatments do not possess intrinsic anticancer properties, using them to help the cell cope with oxidative stress could improve sensitivity to conventional radiotherapy. Such research strengthens our opinion that MnSOD-based therapies are a worthy pursuit. This has been exemplified by studies focusing on MnSOD levels and oxygenation status within the tumors.

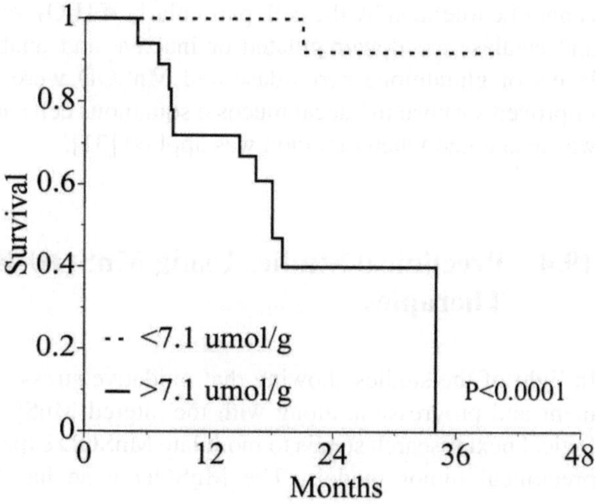

Fig. 19.3 Overall survival curves for HNC patients stratified by tumor lactate concentrations. Patients with low levels of tumor lactate (<7.1 μmol/g) showed significantly higher overall survival. Taken from Brizel et al., Int. J. Radiation Oncology Biol. Phys., 51(2): 349–353, 2001. Used with permission

Firstly, although exceptions exist, the majority of studies have shown that HNC tumors have decreased MnSOD expression; this is in contrast to what has been seen in other types of cancers (i.e., gastric, colorectal). Expression levels of MnSOD are prognostically important. For example, elevated MnSOD in buccal mucosal squamous cell carcinoma showed a trend for improved outcome [31] compared to patients with lower expression, particularly when radiation was applied.

MnSOD levels exhibit variable expression in esophageal cancers. Toh et al. showed an inverse correlation between MnSOD expression and venous and lymphatic invasion of esophageal carcinoma [32]. For 22/34 untreated tumors, MnSOD expression was lower in tumor than surrounding normal tissue. Notably, MnSOD expression was even further reduced in tumors that were poorly differentiated [32] or in tissue from Barrett's esophagus patients whose pathology has progressed to esophageal adenocarcinoma [33, 34]. Schiffman et al. reported on the relationship between bile salts and oxidative stress [35]. A human esophageal squamous carcinoma cell line (OC-3) showed a significant increase in MnSOD 4 h after in vitro exposure to bile salts, but an esophageal adenocarcinoma cell line did not. This would suggest that the esophageal adenocarcinoma cell line was unable to cope with the bile salt exposure—in an in vivo setting this may allow for accumulation of DNA and cellular damage.

In contrast, it has also been reported that larynx and oral cavity squamous cell cancers have increased MnSOD expression compared to healthy tissue either adjacent to or distant from the tumor [36]. It should be noted that this study comprised a broad sampling of oral cavity cancers with noted differences in terms of differentiation, nodal involvement, stage and metastasis, and the authors did not explore whether MnSOD varied between those subsets. It is possible that collapsing all the data together masked a decrease in MnSOD expression in more advanced tumors.

Regardless of whether MnSOD is up- or downregulated relative to healthy tissue, a significant difference in its expression is indicative of oxidative stress. Reduced MnSOD levels may reflect an inability to dismute mitochondrial $O_2^{\bullet-}$ giving rise to H_2O_2. Elevated MnSOD could result in excessive H_2O_2 accumulation at rates that cannot be tolerated by the cell, particularly if H_2O_2 removing enzymes such as GPx and catalase are downregulated or inactive and unable to compensate. Increased levels of glutathione peroxidase and MnSOD were significantly associated with improved survival in buccal mucosal squamous cell carcinoma, and again, the effect was also noted when radiation was applied [31].

19.4 Preclinical Studies Using MnSOD-Based Gene Therapies

In light of the studies showing that oxidative stress plays a role in HNC development and progression, along with the altered MnSOD expression within HNC, a logical next research step is to modulate MnSOD expression, either in vitro or using preclinical tumor models. The MnSOD gene has been introduced via several mechanisms, including stable transfection, plasmid liposome delivery, and adenoviral vector. This section will summarize the knowledge gleaned using these approaches.

Liu et al. linked MnSOD expression level with a reduction in tumor growth rate. This study used stable transfection of MnSOD at different expression levels to generate a dose-response curve in SCC-25 (squamous cell carcinoma of the tongue) [37]. Six different SCC-25 clones were prepared, with varying degrees of SOD activity. When transplanted into nude mice, the clones transfected with MnSOD grew slower compared to the parental and vector control cells. Tumors were excised after 40 days and examined for MnSOD activity; an inverse correlation between SOD enzymatic activity and growth was found (Fig. 19.4). In vitro work from another laboratory showed that MnSOD overexpression in SCC-25 cells increased their radioresistance 4–5-fold by causing cells to accumulate in the G_2 phase of the cell cycle [38]. It is important to note that this study tested only a single dose of radiation (6 Gy), rather than performing a complete radiation response curve, and they did not further investigate their findings using an in vivo model.

Epperly and Greenberger increased MnSOD expression in multiple oral squamous cell carcinoma cell lines using a plasmid liposome (MnSOD-PL) that provided a transfection efficiency of ~80 %. Increased MnSOD expression sensitized cells to radiation in vitro [39, 40]. Intratumoral injection of MnSOD-PL prior to irradiation rendered the tumors less hypoxic, as determined by PET imaging 24 h post-radiation [40]. The authors speculated that the effect of SOD depends on the cellular concentrations of catalase and GPx. If catalase and glutathione peroxidase levels are not sufficient to convert H_2O_2 produced by SOD, then the cell will be further damaged by H_2O_2 accumulation and either die or slow its replication cycle.

In a separate study, Darby Weydert et al. overexpressed MnSOD using an intratumorally injected adenovirus [41]. In vitro, AdMnSOD alone did not affect SCC-25 or HCPC-1 cell viability, but when combined with BCNU (a GPx and catalase inhibitor) there was a synergistic increase in cell death. Growth delay studies using SCC-25 and HCPC-1 xenografts gave conflicting results and were difficult to interpret due

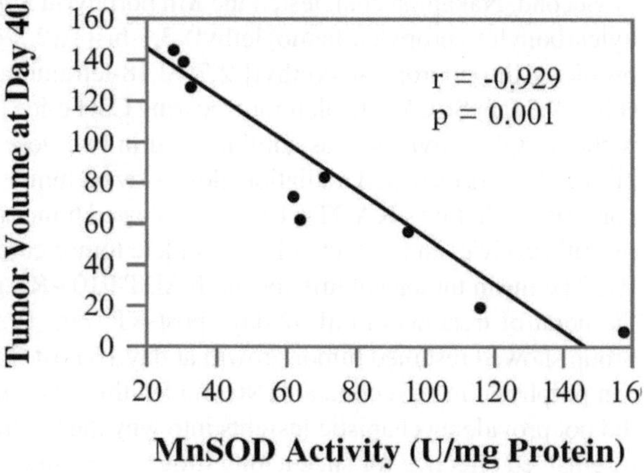

Fig. 19.4 SCC-25 xenografts expressing different levels of transfected MnSOD show disparate growth rates that inversely correlate with MnSOD expression. $r=-0.929$, $p=0.001$. Modified from Liu et al., Human Gene Therapy, 8:585–595, 1997

to the small sample size of groups (*N*=4). The authors hypothesized that increased MnSOD leads to increased levels of H_2O_2, and if catalase and GPx are inhibited by BCNU, the cell cannot handle the extra H_2O_2 and dies.

Finally, Lam et al. intratumorally injected AdMnSOD into HCPC-1 (hamster cheek pouch carcinoma) flank xenografts every 5 days. Western blots of tumor lysates confirmed that the tumors expressed increased MnSOD levels following the injections; uniformity of expression throughout the tumor was not reported. Initially, tumor growth was slowed in the AdMnSOD-treated mice, but growth accelerated after the final treatment (day 21 post-transplant). The authors attributed the biphasic tumor growth curve to tumor cell turnover and loss of the transgene [42].

In summary, these data show the promise of MnSOD-based gene therapies, but highlight that it may be difficult to maintain gene expression over prolonged periods, which would compromise therapy.

19.5 Early Signs of Mn Porphyrin Efficacy in HNC

The use of Mn porphyrins as anticancer agents in HNC is clearly in its infancy. At present, most of the work has been preclinical in vitro work, but recent murine studies have highlighted the promise of Mn porphyrin anticancer effects.

First, in vitro work suggested that the Mn porphyrin (Mn(III) *meso* tetrakis (4-benzoic acid); MnTBAP, a non-SOD mimic) could prevent oxidative stress in GERD patients. Schiffman et al. showed that treating the Barrett's esophageal line, Bar-T, and the human esophageal line (Het-1A) in vitro with MnTBAP increased the cellular expression of MnSOD in vitro [43]. Bile treatment reduced MnSOD activity in both cell lines, but MnTBAP pretreatment restored it in Het-1A cells. If the hypothesis that Barrett's esophagus progresses to esophageal adenocarcinoma due to bile salt-induced oxidative stress that cannot be handled by cellular MnSOD is true, than augmenting MnSOD expression by MnTBAP treatment may impede progression and reduce the risk of cancer.

Second, Nakajima et al. tested the Mn porphyrin KADT-F10 (13,17-bis(carboxyethylcarbonyloxypropylcarbamoylethyl)-3,8-bis[1-[2,2-difluoro-3-(2′-nitroimidazoyl)propionyl]iminopropyl-oxyethyl]-2,7,12,18-tetramethylporphynatomanganese(III)) (Fig. 19.5a) in SCCVII flank tumor-bearing C3H mice [44]. KADT-F10, which is also without SOD activities, was administered in two doses of 25 mg/kg at 24 and 1.5 h prior to RT. As expected, radiation alone slowed tumor growth compared to untreated controls. Radiation + KADT-F10 further slowed tumor growth, significantly improved overall survival, and in fact led to complete tumor cure in 3/7 mice by day 120 post-RT. The mean tumor volume for the KADT-F10 + RT group remained stagnant from the point of treatment until 32 days post-RT (Fig. 19.5b). In contrast, the RT alone group showed resumed tumor growth at day 16 post-RT. As the first in vivo use of an Mn porphyrin in a preclinical HNC model, this study provided proof-of-concept, but did not provide mechanistic insights into why the treatment was so effective, and how it remained effective for such a long time posttreatment.

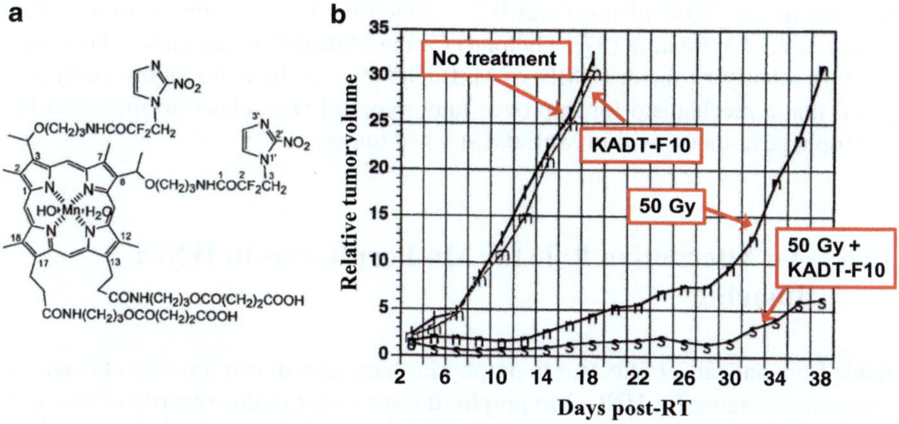

Fig. 19.5 The Mn porphyrin KADT-F10 (panel **a**) was used as an adjuvant therapy to radiation of SCCVII tumors in mice. The relative tumor volume for SCCVII tumor-bearing mice following no treatment, KADT-F10, 50 Gy radiation, or KADT-F10 + 50 Gy radiation is shown (panel **b**). Tumor volumes for the 50 Gy and 50 Gy KADT-F10 groups were significantly different at all time points from Day 18 on ($p < 0.01$). Modified from Nakajima et al., Cancer Letters 181:173–178, 2002. Used with permission

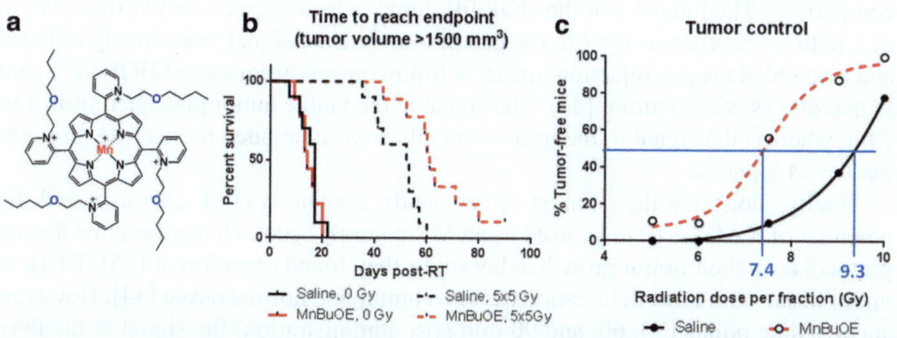

Fig. 19.6 Although the Mn porphyrin MnBuOE (shown in panel **a**) did not affect tumor growth by itself, it radiosensitized FaDu xenografts to fractionated radiation. This led to increased survival times (panel **b**) and improved tumor control (panel **c**). Modified from Ashcraft et al., IJROBP 93(4): 892–900, 2015

The most recent efforts towards Mn porphyrin treatments in preclinical HNC were reported by Ashcraft et al. in 2015 [45]. This study treated human FaDu xenograft-bearing mice with fractionated RT and the SOD-mimic Mn porphyrin Mn(III) *meso*-tetrakis(*N*-n-butoxyethylpyridinium-2-yl)porphyrin (MnBuOE) (Fig. 19.6a) in a TCD_{50} study. Similar to the KADT-F10 study by Kato et al., MnBuOE alone (1.5 mg/kg, b.i.d.) did not affect tumor growth. However, the addition of fractionated RT slowed tumor growth (Fig. 19.6b) and increased local control rates (Fig. 19.6c). As indicated by the blue lines in Fig. 19.6c, the radiation dose required to cure 50 % of mice was 9.3 Gy/fraction (46.5 Gy total) for the control mice and 7.4 Gy/fraction (37.1 Gy total) for the MnBuOE-treated mice. Therefore, the dose modifying factor (DMF) of MnBuOE was 1.3. In addition, this study proposed that a mechanism for improved tumor control was related to increased M1 macrophage numbers within MnBuOE-treated tumors.

19.6 An Alternative Role for Mn Porphyrins in HNC: Imaging

Aside from anticancer effects, Mn porphyrins have also shown promise in magnetic resonance imaging for HNC. Mn porphyrins are useful in this regard because they form stable complexes and are retained within tumor cells.

Takehara successfully imaged tumors using MRI with the Mn porphyrin HOP-9P (13,17-bis (1-carboxypropionyl) carbamoylethyl-3,8-bis (1-phenylpropyloxyethyl)-2,7,12,18-tetramethyl-porphyrinato manganese(III)) [46]. HOP-9P was selectively retained in syngeneic SCCVII tumors in mice and remained higher in the tumors than in the muscles at 2 and 24 h postinjection. The signal intensity in the tumor was higher in the mice that received HOP-9P than a gadolinium chelate, included for comparison. The intensity of the HOP-9P signal in the tumor increased from 5 min, to 2 h, to 24 h, whereas by 2 h, the gadolinium chelate signal was already reduced and resembled the pre-injection image. A follow-up study compared HOP-9P signal in necrotic vs. viable tumor [47]. The signal in the viable tumor plateaued from 1 to 24 h, whereas the signal in the necrotic/viable areas continued to increase throughout the 24 h period.

Finally, alongside their tumor control study, Nakajima et al. also explored the potential of KADT-F10 to be used as an MR contrast agent. Using the same dosing protocol as in their tumor growth delay study, they found retention of KADT-F10 in tumor tissue with clear delineation between tumor and normal tissue [44]. However, at early time points (30, 60, and 90 min post-administration) the signal in the liver and kidneys was comparable to that of the tumor. Later time points were not investigated though from the HOP-9P studies one would expect enhanced retention in the tumor compared to other organs.

19.7 Don't Forget the Normal Tissue: Can Mn Porphyrins Act as Radioprotectants?

Finally, the success of curing HNC with radiation must be balanced with concern for the delicate normal tissues at risk of inclusion in the radiation field. HNC patients are at risk of developing mucositis (a short-term effect of RT, which may be severe enough to disrupt treatment plans and necessitate feeding tubes) and xerostomia

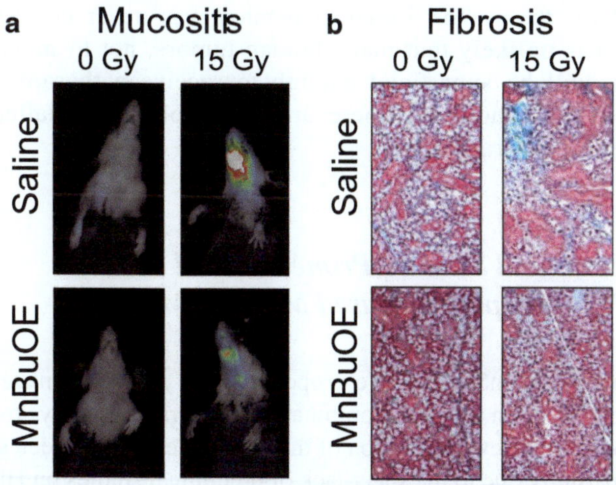

Fig. 19.7 In addition to radiosensitizing tumors, the Mn porphyrin MnBuOE protected normal head and neck tissues following irradiation with 15 Gy. MnBuOE reduced oral mucositis (**a**) at 10 days post-RT. At 6 weeks post-RT, Masson's trichrome staining revealed fibrosis in the salivary glands, as indicated by the bright turquoise staining (**b**). However, MnBuOE treatment reduced fibrosis following RT. Modified from Ashcraft et al., IJROBP 93(4): 892–900, 2015

(a long-term loss of salivation, which may lead to dysphagia, dental caries, malnutrition, and can severely impact quality of life). Currently, there are no FDA-approved treatments for mucositis, and the only approved radioprotectant for xerostomia, amfostine, causes side effects that are serious enough to preclude its use. As Mn porphyrins have been shown to protect normal tissue from radiation injury in other models, Ashcraft et al. utilized a model of head and neck RT and reported that the Mn porphyrin MnBuOE administered at 1.5 mg/kg b.i.d. reduced mucositis (at 10 days post-RT) and fibrosis (at 6 weeks post-RT) (Fig. 19.7). This was the first report of an Mn porphyrin widening the therapeutic margin by radiosensitizing a tumor while radioprotecting normal tissue [45].

19.8 What's Next?

19.8.1 MnSOD Gene Therapy

MnSOD is clearly implicated in HNC. Various studies have found it to be up- or downregulated [32, 36] in such tumors, and a single nucleotide polymorphism, Val16Ala, has been linked to laryngeal cancer risk in smokers [48]. Yet, correcting the problem of altered MnSOD expression and redox status will be trickier in humans than in preclinical mouse studies. Delivering a functional MnSOD to a human tumor presents challenges not seen in mouse models. The AdMnSOD studies described here used intratumoral injections throughout the tumor to deliver the vector. Equal dispersion was never confirmed in those studies, and it is unlikely that many human tumors, not to mention head and neck tumors, will be superficial enough to receive a therapy that must be directly injected. In addition, there are immunological challenges to gene therapy using adenoviral vectors.

19.8.2 Preclinical Work to Prime Clinical Use of Mn-Porphyrin-Based MnSOD Mimics

Mn porphyrins with MnSOD mimic properties have shown promise as anticancer agents, especially when used in combination with radiotherapy, for a variety of tumor histologies. However, adoption of this novel therapy has been slow in HNCs. This may be partially due to the sheer lack of preclinical studies on HNC, compared to mammary and brain tumors. However, there is enough evidence from MnSOD-based therapies [37–42] to warrant expansion of Mn porphyrin studies to models of HNC. Mn porphyrins are an attractive choice because that they can be delivered systemically and be detected in tissues throughout the body, they are nearly as efficacious as native MnSOD, and they remain stable and have not been shown to elicit an immune response.

Researchers would be wise to take several factors into consideration when designing studies to test anticancer potential of Mn porphyrins in HNC. First, Nakajima et al. and Ashcraft et al. both showed that Mn porphyrins increased the efficacy of radiation therapy, but had no effect when given alone [44, 45]. Other studies have reported similar findings [1]. Researchers should bear in mind that it is unlikely Mn porphyrins will ever replace current therapies, and should test them in concert with clinically relevant therapies, such as radiotherapy and chemotherapy. A proposed mechanism is that Mn porphyrins with MnSOD properties employ H_2O_2 produced by either RT or chemotherapy to oxidize cysteine residues on signal cascade proteins to affect transcription and tumor growth. Tumor cells frequently have upregulated MnSOD and reduced expression of catalase and GPx, which are needed to eliminate H_2O_2. Under such conditions, H_2O_2 produced during O_2 dismutation accumulates. Consequently, tumors have elevated H_2O_2, which is further increased by RT or chemotherapy. Mn porphyrin subsequently uses H_2O_2 to oxidatively modify thiols of signaling proteins, such as NF-κB, suppressing its activation, which in turn induces apoptotic processes [49]. However, in normal tissue, where abundant peroxide-removing enzymes ensure lower physiological levels of H_2O_2, signaling cascades may be transiently inactivated only, which will suppress inflammation but not result in cell death [49]. Therefore, logic suggests that Mn porphyrins will be effective adjuvants to any therapeutic strategy that raises levels of H_2O_2. Future studies should investigate whether comparable tumor control can be achieved using lower radiation doses in the presence of Mn porphyrin.

19.8.3 A Shift in Focus Towards Viral-Mediated Cancers

Research at the preclinical level should reflect the increased prevalence of HPV-related oral tumors. HPV-positive oral cancer patients typically have smaller primary tumors that have progressed to advanced neck disease [50]. Treatment protocols have shifted to reflect the change in epidemiology; HPV-positive oral cancers tout high cure rates, and there are investigational treatment protocols exploring the effectiveness of modified radiation treatment plans [51, 52]. Investigators considering Mn porphyrins as adjuvant therapy for HNC would be wise to consider this paradigm shift when designing their preclinical studies. An Mn porphyrin therapy that is effective in one patient population may have little effect on tumors in another.

In summary, the use of Mn porphyrins in HNC is relatively unexplored. Studies using MnSOD-based therapies have shown promise, but much work remains to be done before this type of therapy can be considered for HNC patients.

References

1. Weitzel DH, Tovmasyan A, Ashcraft KA, Rajic Z, Weitner T, Liu C, et al. Radioprotection of the brain white matter by Mn(III) N-butoxyethylpyridylporphyrin-based superoxide dismutase mimic, MnTnBuOE-2-PyP5+. Mol Cancer Ther. 2014;14(9):70–9.
2. Keir ST, Dewhirst MW, Kirkpatrick JP, Bigner DD, Batinic-Haberle I. Cellular redox modulator, ortho Mn(III) meso-tetrakis(N-n-hexylpyridinium-2-yl)porphyrin, MnTnHex-2-PyP(5+) in the treatment of brain tumors. Anticancer Agents Med Chem. 2011;11(2):202–12.
3. Rabbani ZN, Spasojevic I, Zhang X, Moeller BJ, Haberle S, Vasquez-Vivar J, et al. Antiangiogenic action of redox-modulating Mn(III) meso-tetrakis(N-ethylpyridinium-2-yl) porphyrin, MnTE-2-PyP(5+), via suppression of oxidative stress in a mouse model of breast tumor. Free Radic Biol Med. 2009;47(7):992–1004.
4. Makinde AY, Luo-Owen X, Rizvi A, Crapo JD, Pearlstein RD, Slater JM, et al. Effect of a metalloporphyrin antioxidant (MnTE-2-PyP) on the response of a mouse prostate cancer model to radiation. Anticancer Res. 2009;29(1):107–18.
5. Gridley DS, Makinde AY, Luo X, Rizvi A, Crapo JD, Dewhirst MW, et al. Radiation and a metalloporphyrin radioprotectant in a mouse prostate tumor model. Anticancer Res. 2007;27(5A):3101–9.
6. Moeller BJ, Batinic-Haberle I, Spasojevic I, Rabbani ZN, Anscher MS, Vujaskovic Z, et al. A manganese porphyrin superoxide dismutase mimetic enhances tumor radioresponsiveness. Int J Radiat Oncol Biol Phys. 2005;63(2):545–52.
7. Sabharwal SS, Schumacker PT. Mitochondrial ROS in cancer: initiators, amplifiers or an Achilles' heel? Nat Rev Cancer. 2014;14(11):709–21.
8. Blot WJ, McLaughlin JK, Winn DM, Austin DF, Greenberg RS, Preston-Martin S, et al. Smoking and drinking in relation to oral and pharyngeal cancer. Cancer Res. 1988; 48(11):3282–7.
9. Jayaprakash V, Reid M, Hatton E, Merzianu M, Rigual N, Marshall J, et al. Human papillomavirus types 16 and 18 in epithelial dysplasia of oral cavity and oropharynx: a meta-analysis, 1985-2010. Oral Oncol. 2011;47(11):1048–54.
10. Uneri C, Sari M, Baglam T, Polat S, Yuksel M. Effects of vitamin E on cigarette smoke induced oxidative damage in larynx and lung. Laryngoscope. 2006;116(1):97–100.

11. Malpass GE, Arimilli S, Prasad GL, Howlett AC. Regulation of gene expression by tobacco product preparations in cultured human dermal fibroblasts. Toxicol Appl Pharmacol. 2014; 279(2):211–9.
12. Williams VM, Filippova M, Filippov V, Payne KJ, Duerksen-Hughes P. Human papillomavirus type 16 E6* induces oxidative stress and DNA damage. J Virol. 2014;88(12):6751–61.
13. Kamranvar SA, Masucci MG. The Epstein-Barr virus nuclear antigen-1 promotes telomere dysfunction via induction of oxidative stress. Leukemia. 2011;25(6):1017–25.
14. Ma N, Thanan R, Kobayashi H, Hammam O, Wishahi M, El Leithy T, et al. Nitrative DNA damage and Oct3/4 expression in urinary bladder cancer with Schistosoma haematobium infection. Biochem Biophys Res Commun. 2011;414(2):344–9.
15. Jimenez P, Piazuelo E, Sanchez MT, Ortego J, Soteras F, Lanas A. Free radicals and antioxidant systems in reflux esophagitis and Barrett's esophagus. World J Gastroenterol. 2005;11(18):2697–703.
16. Langevin SM, Michaud DS, Marsit CJ, Nelson HH, Birnbaum AE, Eliot M, et al. Gastric reflux is an independent risk factor for laryngopharyngeal carcinoma. Cancer Epidemiol Biomarkers Prev. 2013;22(6):1061–8.
17. Singh K, Kote S, Patthi B, Singla A, Singh S, Kundu H, et al. Relative risk of various head and neck cancers among different blood groups: an analytical study. J Clin Diagn Res. 2014;8(4):ZC25–8.
18. Lin FC, Chen PL, Tsao TY, Li CR, Jeng KC, Tsai SC. Prevalence of human papillomavirus and Epstein-Barr virus in salivary gland diseases. J Int Med Res. 2014;42(5):1093–101.
19. Horn-Ross PL, Ljung BM, Morrow M. Environmental factors and the risk of salivary gland cancer. Epidemiology. 1997;8(4):414–9.
20. Boukheris H, Ron E, Dores GM, Stovall M, Smith SA, Curtis RE. Risk of radiation-related salivary gland carcinomas among survivors of Hodgkin lymphoma: a population-based analysis. Cancer. 2008;113(11):3153–9.
21. Wijffels KI, Hoogsteen IJ, Lok J, Rijken PF, Marres HA, de Wilde PC, et al. No detectable hypoxia in malignant salivary gland tumors: preliminary results. Int J Radiat Oncol Biol Phys. 2009;73(5):1319–25.
22. Roh JL, Cho KJ, Kwon GY, Choi SH, Nam SY, Kim SY. Prognostic values of pathologic findings and hypoxia markers in 21 patients with salivary duct carcinoma. J Surg Oncol. 2008;97(7):596–600.
23. Epperly MW, Carpenter M, Agarwal A, Mitra P, Nie S, Greenberger JS. Intraoral manganese superoxide dismutase-plasmid/liposome (MnSOD-PL) radioprotective gene therapy decreases ionizing irradiation-induced murine mucosal cell cycling and apoptosis. In Vivo. 2004; 18(4):401–10.
24. Nagler RM, Reznick AZ, Slavin S, Nagler A. Partial protection of rat parotid glands from irradiation-induced hyposalivation by manganese superoxide dismutase. Arch Oral Biol. 2000;45(9):741–7.
25. Perez-Sayans M, Suarez-Penaranda JM, Pilar GD, Barros-Angueira F, Gandara-Rey JM, Garcia-Garcia A. Hypoxia-inducible factors in OSCC. Cancer Lett. 2011;313(1):1–8.
26. Nordsmark M, Bentzen SM, Rudat V, Brizel D, Lartigau E, Stadler P, et al. Prognostic value of tumor oxygenation in 397 head and neck tumors after primary radiation therapy. An international multi-center study. Radiother Oncol. 2005;77(1):18–24.
27. Brizel DM, Dodge RK, Clough RW, Dewhirst MW. Oxygenation of head and neck cancer: changes during radiotherapy and impact on treatment outcome. Radiother Oncol. 1999;53(2):113–7.
28. Brizel DM, Sibley GS, Prosnitz LR, Scher RL, Dewhirst MW. Tumor hypoxia adversely affects the prognosis of carcinoma of the head and neck. Int J Radiat Oncol Biol Phys. 1997;38(2):285–9.
29. Liu SY, Chang LC, Pan LF, Hung YJ, Lee CH, Shieh YS. Clinicopathologic significance of tumor cell-lined vessel and microenvironment in oral squamous cell carcinoma. Oral Oncol. 2008;44(3):277–85.

30. Brizel DM, Schroeder T, Scher RL, Walenta S, Clough RW, Dewhirst MW, et al. Elevated tumor lactate concentrations predict for an increased risk of metastases in head-and-neck cancer. Int J Radiat Oncol Biol Phys. 2001;51(2):349–53.
31. Fu TY, Hou YY, Chu ST, Liu CF, Huang CH, Chen HC, et al. Manganese superoxide dismutase and glutathione peroxidase as prognostic markers in patients with buccal mucosal squamous cell carcinomas. Head Neck. 2011;33(11):1606–15.
32. Toh Y, Kuninaka S, Mori M, Oshiro T, Ikeda Y, Nakashima H, et al. Reduced expression of manganese superoxide dismutase mRNA may correlate with invasiveness in esophageal carcinoma. Oncology. 2000;59(3):223–8.
33. Hermann B, Li Y, Ray MB, Wo JM, Martin 2nd RC. Association of manganese superoxide dismutase expression with progression of carcinogenesis in Barrett esophagus. Arch Surg. 2005;140(12):1204–9; discussion 9.
34. Martin RC, Liu Q, Wo JM, Ray MB, Li Y. Chemoprevention of carcinogenic progression to esophageal adenocarcinoma by the manganese superoxide dismutase supplementation. Clin Cancer Res. 2007;13(17):5176–82.
35. Schiffman SC, Li Y, Xiao D, Li X, Aiyer HS, Martin RC. The resistance of esophageal adenocarcinoma to bile salt insult is associated with manganese superoxide dismutase expression. J Surg Res. 2011;171(2):623–30.
36. Piyathilake CJ, Bell WC, Oelschlager DK, Heimburger DC, Grizzle WE. The pattern of expression of Mn and Cu-Zn superoxide dismutase varies among squamous cell cancers of the lung, larynx, and oral cavity. Head Neck. 2002;24(9):859–67.
37. Liu R, Oberley TD, Oberley LW. Transfection and expression of MnSOD cDNA decreases tumor malignancy of human oral squamous carcinoma SCC-25 cells. Hum Gene Ther. 1997;8(5):585–95.
38. Kalen AL, Sarsour EH, Venkataraman S, Goswami PC. Mn-superoxide dismutase overexpression enhances G2 accumulation and radioresistance in human oral squamous carcinoma cells. Antioxid Redox Signal. 2006;8(7–8):1273–81.
39. Epperly MW, Franicola D, Zhang X, Nie S, Greenberger JS. Effect of EGFR antagonists gefitinib (Iressa) and C225 (Cetuximab) on MnSOD-plasmid liposome transgene radiosensitization of a murine squamous cell carcinoma cell line. In Vivo. 2006;20(6B):791–6.
40. Epperly MW, Lai SY, Kanai AJ, Mason N, Lopresi B, Dixon T, et al. Effectiveness of combined modality radiotherapy of orthotopic human squamous cell carcinomas in Nu/Nu mice using cetuximab, tirapazamine and MnSOD-plasmid liposome gene therapy. In Vivo. 2010;24(1):1–8.
41. Darby Weydert CJ, Smith BB, Xu L, Kregel KC, Ritchie JM, Davis CS, et al. Inhibition of oral cancer cell growth by adenovirusMnSOD plus BCNU treatment. Free Radic Biol Med. 2003;34(3):316–29.
42. Lam EW, Hammad HM, Zwacka R, Darby CJ, Baumgardner KR, Davidson BL, et al. Immunolocalization and adenoviral vector-mediated manganese superoxide dismutase gene transfer to experimental oral tumors. J Dent Res. 2000;79(6):1410–7.
43. Schiffman SC, Li Y, Martin RC. The association of manganese superoxide dismutase expression in Barrett's esophageal progression with MnTBAP and curcumin oil therapy. J Surg Res. 2012;176(2):535–41.
44. Nakajima S, Fujii T, Murakami N, Aburano T, Sakata I, Nakae Y, et al. Therapeutic and imaging capacity of tumor-localizing radiosensitive Mn-porphyrin KADT-F10 for SCCVII tumors in C3H/He mice. Cancer Lett. 2002;181(2):173–8.
45. Ashcraft KA, Boss M-K, Tovmasyan A, Roy Choudhury K, Fontanella AN, Young KH, et al. Novel manganese-porphyrin superoxide dismutase-mimetic widens the therapeutic margin in a preclinical head and neck cancer model. Int J Radiat Oncol Biol Phys. 2015;93(4): 892–900.
46. Takehara Y, Sakahara H, Masunaga H, Isogai S, Kodaira N, Takeda H, et al. Tumour enhancement with newly developed Mn-metalloporphyrin (HOP-9P) in magnetic resonance imaging of mice. Br J Cancer. 2001;84(12):1681–5.

47. Nasu H, Takehara Y, Isogai S, Kodaira N, Takeda H, Saga T, et al. Tumor enhancement using Mn-metalloporphyrin in mice: magnetic resonance imaging and histopathologic correlation. J Magn Reson Imaging. 2004;20(2):294–9.
48. Aynali G, Dogan M, Sutcu R, Yuksel O, Yariktas M, Unal F, et al. Polymorphic variants of MnSOD Val16Ala, CAT-262 C<T and GPx1 Pro198Leu genotypes and the risk of laryngeal cancer in a smoking population. J Laryngol Otol. 2013;127(10):997–1000.
49. Batinic-Haberle I, Tovmasyan A, Spasojevic I. An educational overview of the chemistry, biochemistry and therapeutic aspects of Mn porphyrins—from superoxide dismutation to HO-driven pathways. Redox Biol. 2015;5:43–65.
50. Deschler DG, Richmon JD, Khariwala SS, Ferris RL, Wang MB. The "new" head and neck cancer patient-young, nonsmoker, nondrinker, and HPV positive: evaluation. Otolaryngology Head Neck Surg. 2014;151(3):375–80.
51. Shaw RJ, Holsinger FC, Paleri V, Evans M, Tudur-Smith C, Ferris RL. Surgical trials in head & neck oncology: renaissance and revolution? Head Neck. 2015;37(7):927–30.
52. Masterson L, Moualed D, Liu ZW, Howard JE, Dwivedi RC, Tysome JR, et al. De-escalation treatment protocols for human papillomavirus-associated oropharyngeal squamous cell carcinoma: a systematic review and meta-analysis of current clinical trials. Eur J Cancer. 2014;50(15):2636–48.
53. Spitz MR. Epidemiology and risk factors for head and neck cancer. Semin Oncol. 1994;21(3):281–8.
54. Tsao SW, Tsang CM, To KF, Lo KW. The role of Epstein-Barr virus in epithelial malignancies. J Pathol. 2015;235(2):323–33.
55. Poetsch M, Kleist B, Lorenz G, Herrmann FH. Different numerical chromosomal aberrations detected by FISH in oropharyngeal, hypopharyngeal and laryngeal squamous cell carcinoma. Histopathology. 1999;34(3):234–40.
56. Chaturvedi AK, Engels EA, Pfeiffer RM, Hernandez BY, Xiao W, Kim E, et al. Human papillomavirus and rising oropharyngeal cancer incidence in the United States. J Clin Oncol. 2011;29(32):4294–301.
57. Alos L, Moyano S, Nadal A, Alobid I, Blanch JL, Ayala E, et al. Human papillomaviruses are identified in a subgroup of sinonasal squamous cell carcinomas with favorable outcome. Cancer. 2009;115(12):2701–9.
58. Cook MB, Corley DA, Murray LJ, Liao LM, Kamangar F, Ye W, et al. Gastroesophageal reflux in relation to adenocarcinomas of the esophagus: a pooled analysis from the Barrett's and Esophageal Adenocarcinoma Consortium (BEACON). PLoS One. 2014;9(7):e103508.

Chapter 20
Redox-Based Therapeutic Strategies in the Treatment of Skin Cancers

Annapoorna Sreedhar, Ines Batinić-Haberle, and Yunfeng Zhao

20.1 Introduction

Cancer is a second leading cause of death in the United States. In 2013, according to the National Cancer Institute, approximately 1,660,290 new cases of cancer were diagnosed and 550,350 Americans died of cancer, approximately 1600 people per day. Cancer has one of the highest mortality and morbidity rates, making it the second most prevalent disease worldwide.

Skin, the largest organ of the body, is prone to cancer. Its epithelial cells are the fastest dividing cells and therefore are under the highest risk of becoming malignant. Skin cancer is the most common cancer and accounts for 50 % of all cancers in the United States. More than 3.5 million skin cancers in more than two million people are diagnosed in the United States annually. Current estimates are such that one in five Americans will develop skin cancer in their lifetimes. From the year 2006 to 2012, there has been an average of 2.6 % increase in the number of skin cancer cases per year.

Basal cell carcinoma, squamous cell carcinoma, and melanoma are the three most common forms of skin cancer. Basal and squamous cell carcinomas, together called non-melanoma skin cancers (NMSC), arise from the basal and the squamous cells, respectively. NMSC are curable and treatable if detected at an early stage of cancer progression. Melanoma is a more severe type of skin cancer originating from

A. Sreedhar • Y. Zhao (✉)
Department of Pharmacology, Toxicology and Neurosciences, LSU Health Sciences Center, 1501 Kings Highway, Shreveport, LA 71130, USA
e-mail: yzhao1@lsuhsc.edu

I. Batinić-Haberle
Department of Radiation Oncology, Duke University School of Medicine, Durham, NC 27710, USA

I. Batinić-Haberle et al. (eds.), *Redox-Active Therapeutics*, Oxidative Stress in Applied Basic Research and Clinical Practice, DOI 10.1007/978-3-319-30705-3_20

the melanocytes—cells that are responsible for skin coloring or pigmentation. Though it accounts for only about 2 % of all skin cancers in the United States, melanoma causes 75 % of skin cancer-related deaths.

20.1.1 Causes of Skin Cancer

What are the causes of skin cancer? A short answer is, almost anything and everything. The sun itself is one of the biggest risk factors for skin cancer. Anyone who spends a substantial amount of time outside and has been exposed to the radiation of the sun can potentially develop skin cancer. The UV radiation from the sun is thought to be carcinogenic in nature (Fig. 20.1). Regions of the skin exposed to the UV rays respond by generating more melanin pigment that imparts the color to the skin. Melanin moves towards the outer layer of the skin surface and develops into a tan, which helps protect against UV-induced skin cancer. Being fair skinned can also be a risk for developing skin cancer since fair or pale skin can easily be tanned or

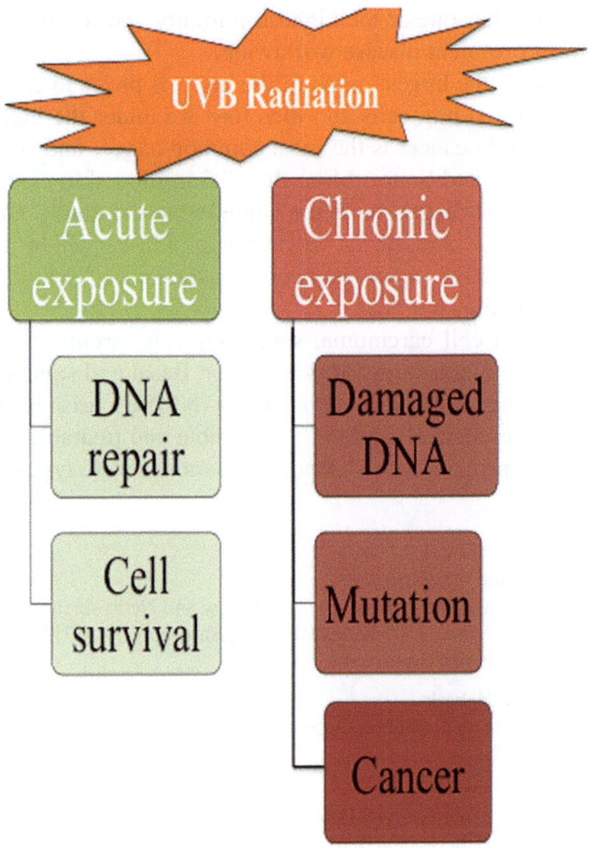

Fig. 20.1 UV radiation is one of the biggest risk factors for skin cancer. The DNA readily absorbs UVB photons and by targeting DNA, these photons can cause direct damage. Under acute exposure, the DNA repair mechanisms come into action. The DNA repair is closely associated with the cell cycle; the checkpoints in the cell cycle can arrest the growth till repair is completed. However, under chronic exposure, these checkpoints could be damaged. Failures in the cell cycle arrest can lead to accumulation of the damaged DNA leading to mutations and ultimately to skin cancer

burnt. Family or personal history of skin cancer, precancerous lesions or moles, exposure to radiation, or exposure to certain substances such as arsenic can increase the lifetime risk of skin cancer.

20.1.2 Current Available Treatments

If left untreated, skin cancers can result in penetration of tumors deep within the skin and can even metastasize to distant organs. However, NMSC, if detected early, is almost always curable. So the earlier the skin cancer is detected, the greater is the possibility of successful treatment. Hence, it is always advisable to do a self-assessment of one's own body. This is called the ABCD of detection—Appearance, Border, Color, and Depth of tumor penetration. Once identified, one or more of the following treatments are needed.

Radiation—Destruction of the tumor by directing a beam of X-ray. This method is advantageous as it is a noninvasive treatment and requires no anesthesia.

Chemotherapy—Most widely used cancer treatment that results in destruction of tumor cells either by directly killing the tumor cells or by inhibiting their division. Chemotherapy can be administered orally or injected into a vein (called the systemic chemotherapy). The US Food and Drug Administration has approved various drugs for the treatment of skin cancers. Aldara, Efudex, and Carac are useful in treating basal carcinoma. Erbitux and Blenoxane are drugs associated with squamous cell carcinoma. Zelboraf, Yervoy, Tafinlar, and Mekinist are the common drugs for the treatment of melanomas.

Surgery—this treatment involves one or more of the following surgical procedures:

Excisional surgery—the removal of the entire tumor growth along with a surrounding tissue layer using a scalpel.
Electrosurgery—burning heat produced by an electric needle destroys the tumor.
Cryosurgery—destruction of a tumor cell by freezing the tumor with liquid nitrogen using a spray device.

Photodynamic therapy—a more selective treatment for destroying only the cancer cells while causing the least or minimal damage to surrounding cells. A topical agent such as 5-aminolevulinic acid (5-ALA) is applied to the tumor region and after a 24-h interval the medicated areas are activated by strong light.

Several topical medications such as 5-fluorouracil (5-FU) are FDA approved treatments for basal cell carcinoma.

20.2 Redox-Based Therapy in Skin Cancer

Cancer is neither a single disease nor is it just a genetic mutation. It is a term used to describe a group of complex diseases. Hence, targeting a specific gene mutation may not be the best strategy to treat cancers. Moreover, these mutations are more diverse in nature; they may differ between individuals or even within a single tumor. Over a period of time, cancer cells will become resistant to anticancer drugs that are used to target specific mutations that previously lead to cancer. Therefore, new concepts and new approaches are needed for skin cancer therapy.

20.2.1 Concept of Redox-Based Therapy

Oxidative stress is a prime "ingredient" for the proliferation of cancer cells. Because of the metabolic dysfunction that exists in cancer cells, they show increased levels of reactive oxygen species (ROS). ROS can in turn lead to damage of DNA, proteins, and lipids, which leads to tumor initiation and progression. Therefore, higher levels of ROS can provide growth advantage to cancer cells and thus can be targeted in skin cancer therapy.

Eleni Papadopulos was the first to propose the oxidative stress theory of cancer in 1982 [1]. "Free radicals" play a major role in the manifestation of the disease. Free radicals or simply radicals are unstable molecules containing unpaired electrons, which are susceptible to either donating their unpaired electrons or extracting electrons from a neighboring molecule. Mitochondria are the biggest source of intracellular ROS. Approximately 2 % of the consumed oxygen is utilized to form ROS [2]. Superoxide is generated via incomplete one-electron reduction of oxygen, which can give rise to hydrogen peroxide (in dismutation or peroxidation reactions). $O_2^{\bullet-}$ and H_2O_2, and reactive species derived from them, can alter the structure and function of biological entities. To fight against reactive species, nature has developed antioxidant defenses. Major antioxidant enzymes include families of superoxide dismutases, glutathione peroxidases, catalases, peroxiredoxins, etc. In addition, small molecules, such as glutathione, ascorbate, tocopherol, and NAD(P)H, are donors of reducing equivalents/electrons. These antioxidants have been used in clinical trials to treat human cancers including skin cancer. Phytochemicals are another class of naturally existing small molecules with antioxidant properties and are also widely explored for skin cancer treatment.

20.2.2 Phytochemicals in Redox Therapy

Phytochemicals as medicinal herbs first appeared in the bible of Indian medicine—Ayurveda [3]. Nowadays, numerous studies report the use of these chemicals as drugs.

Turmeric, a natural herb that originated in India, is a potent anti-inflammatory, antibacterial, antiseptic, and antitumor agent. It is also an effective antioxidant. Around 5000 years ago, the efficacy of turmeric as a natural healing agent was already established. It is gaining increased attention as a treatment option. Turmeric has been effective in the treatment of cancer, diabetes, bowel disorders, and other gastrointestinal (GI) conditions. The application of turmeric cream as a possible cure for skin cancers is under investigation.

Curcumin, as an active component of turmeric, is also an effective antioxidant. It is a flavonoid with anti-inflammatory and anticancerous properties. Curcumin could inhibit cutaneous squamous cell carcinoma in a xenograft mouse model [4]. The mice were treated with curcumin either orally or topically and then exposed to ultraviolet B radiation (UVB). The group of mice that received curcumin have delayed tumor onset and developed fewer tumors compared to the control mice, indicating a potential of curcumin for the treatment of skin carcinogenesis. Mechanistically, curcumin suppresses the activation of nuclear factor-kappa B (NF-κB). NF-κB is a redox-sensitive transcription factor that gets activated in response to oxidative stress or in response to an oncogene, leading to cell proliferation, angiogenesis, and resistance to chemotherapy [5]. Curcumin has also been shown to protect against inorganic arsenic-induced cytotoxicity in human keratinocytes through the activation of nuclear factor erythroid 2-related factor 2 (NRF2) [6]. The Phase I clinical trial conducted by Cheng et al. demonstrated that up to 8000 mg/day per person of curcumin was nontoxic to humans [7].

Aeglemarmelos, commonly called the bael, is another medicinal plant with potential in skin cancer therapy. Bael fruit extract was effective in a mouse model of skin carcinogenesis. Research conducted by Agarwal et al. showed a significant reduction in the skin tumors induced by 7,12-dimethylbenz[*a*]anthracene (DMBA) [8]. Mice were treated orally at 100 mg/kg body weight with the bael fruit extract starting at the pre-initiation phase of tumor development and continued throughout the study. Analysis of skin and liver tissues showed increased levels of antioxidant enzymes such as SOD and catalase as well as low-molecular weight cellular reductants including glutathione and ascorbate (vitamin C). Thus, bael shows a potential to reduce skin carcinogenesis by inducing the upregulation of endogenous antioxidative defenses.

Pomegranate is an excellent source of retinoid, ascorbic acid, tocopherols, and folic acid. It exhibits potent antioxidant activities. A study published in 2005 [9] revealed that treatment with pomegranate fruit extract (PFE) inhibited the tumors that were induced by UVB radiations in a mouse model. The animals were fed with PFE via drinking water prior to UVB exposure. Data show that oral feeding of PFE inhibited the edema, hyperplasia, cyclooxygenase and ornithine decarboxylase. PFE treatment of NHEK resulted in a dose- and time-dependent inhibition of nuclear translocation and phosphorylation of NF-κB/p65 at the serine residues with subsequent NF-κB inactivation.

Garlic, another household ingredient, is a natural spice widely used as a food flavoring agent. It has also been well established that it exhibits medicinal activities. According to a study conducted in 2013 [10] at the Jiangsu Center for Disease Control and Prevention in China, volunteers who consumed garlic at least twice a week at 8.4 g/week developed fewer lung tumors. They also had a 44 % lower risk of developing cancer in general. Garlic contains water soluble derivatives of alkyl amino acids, lipid soluble allyl sulfides such as allicin, flavonoids, and other essential nutrients. Allicin is the prime component that is responsible for the biological activity of garlic. Another study conducted by Das et al. revealed the effects of aqueous garlic preparations on skin carcinogenesis [11]. Garlic extracts are shown to inhibit early and late skin carcinogenesis due to the reduction in lipid peroxidation (LPO). Daily consumption of 250 mg/kg body weight of garlic helped in reducing the number of skin papillomas. Reduction in tumors and LPO is shown to be due to the induction of phase II detoxifying enzymes such as glutathione *S*-transferase (GST) as well as antioxidant enzymes including superoxide dismutase, catalase, and glutathione peroxidase. The protective effect of garlic might be therefore at least in part due to the induction of endogenous cellular defense systems.

Henna leaf, commonly called Mehndi, is a natural dye used to color hair and hands. The important chemicals found in henna leaf are sugars, tannins, gallic acids, coumarins, and resins. Coumarins are the principal components responsible for the coloring properties of henna leaf. Tannins and gallic acids are found to have other beneficial aspects. Studies have reported the effects of henna leaf on the drug metabolism enzymes (e.g., cytochrome P450, glutathione-*S*-transferase, and DT-diaphorase), antioxidant enzymes, lactate dehydrogenase, and LPO in Swiss albino mice [12]. At doses of 200 and 400 mg/kg bodyweight, henna extracts were effective in increasing the SOD and catalase activities. Moreover, skin carcinogenesis studies revealed that significant reduction of papilloma formation was demonstrated with both doses, proving that henna leaf has potential as an antioxidant and anticarcinogenic agent.

Grapes are one of the world's healthiest fruits. The health benefits of grapes are innumerable, ranging from the ability to treat constipation, indigestion, and fatigue to that of treating heart diseases, kidney disorder, diabetes, and even cancer. Grapes are rich in calcium, potassium, iron, retinoic acid, and pyridoxamine. They consist of 70–80 % water and are a rich source of flavonoids. Flavonoids are phenolic compounds, which contribute to the color and the taste of wine. Major flavonoids present in grapes are anthocyanins and tannins, which are powerful antioxidants.

Proanthocyanidines belong to the class of polyphenolic compounds naturally present in grapes. Katiyar showed the effectiveness of grape seed proanthocyanidines (GSP) in suppression of oxidative stress [13]. Dietary supplementation of GSP was associated with a reduction in tumor incidence and multiplicity in SHK-1 hairless mice, where the tumor development was induced via UVB. Administration of these supplements at concentrations of 0.2 and 0.5 % (w/w) resulted in a dose-dependent decrease in photocarcinogenesis. The chemopreventive antioxidative effects of grape seed extract are thought to be due to the inhibition of oxidative stress that is associated with UVB radiation, through the inhibition of mitogen-activated protein kinase (MAPK) and NF-κB signaling pathways, which are usually activated in response to oxidative stress.

Studies show muscadine grapes to be one of the best sources of antioxidants among all fruits, as well as the richest source of polyphenols. They have been extensively cultivated throughout North America since ancient times and are used for production of commercial wines due to their unique aroma and flavor. A more recent study conducted by Burton et al. demonstrated the effectiveness of muscadine grape skin extract (MSKE) in the inhibition of prostate cancer, by inducing apoptosis without causing any deleterious effects to the normal prostate epithelial cell [14]. Androgen-repressed human prostate cancer cells, ARCaP, and androgen-sensitive human prostate adenocarcinoma LnCaP cells, overexpressing the Snail superfamily of zinc-finger transcription factors, were used in this study. Overexpression of Snail causes a decrease in E-cadherin and an increase in mesenchymal markers, leading to increased invasion and metastasis thus inducing epithelial–mesenchymal transition. Snail-overexpressed ARCaP and LnCap cells exhibited elevated levels of mitochondrial superoxide. MSKE treatment decreased the superoxide levels in both cell lines.

Ginger, commonly called the healthiest spice, is an aromatic or pungent smelling spicy root that adds a special flavor to foods. Historically, it is known to be effective in alleviating many disorders including constipation, diarrhea, and GI discomfort. It has been used as an antiinflammatory, antiemetic, antibacterial, anticancer, antipyretic, analgesic, and anticoagulant. In clinical trials, ginger has been used in doses ranging from 250 mg to 1 g, given either three or four times per day. Ginger consists of zingerone, shagaols, and gingerols which are responsible for its characteristic flavor. Other volatile oils comprise 3 % of the mass. In 2013, ginger made headlines with an article entitled "Ginger destroys cancer more effectively than the death-linked cancer drugs." Various studies have been done to show the anticancer effects of ginger. It is known to be effective against ovarian, skin, prostate, and breast cancer. A study by Kim et al. demonstrated the protective effect of ginger against UVB-induced skin disorders [15]. Topical administration of 98 % pure gingerol (30 μM) could effectively inhibit the expression and activation of cyclooxygenase (COX-2) in 8–10-week-old male SKH-1 hairless mice prior to irradiation by UVB light. Those authors also showed that pretreatment with ginger extract could inhibit the ROS production and the activation of caspases as a result of UVB radiation.

Ginger is also found to be effective in the treatment of ovarian cancers. A study by Rhode et al. [16] showed that ginger extract at a dose of 50 μg/mL in cell culture could successfully inhibit tumor growth and modulate angiogenic factors in epithelial human-derived ovarian cancer SKOV3, CaOV3, and ES-2 cell lines. Dried whole ginger root powder extract standardized to 5 % gingerols was used in this study. Ginger root extract could effectively block NF-κB activation in ovarian cancer thereby decreasing the expression of interleukin-8 (IL-8) and secretion of vascular endothelial growth factor (VEGF) involved in cell proliferation and angiogenesis.

In conclusion, as summarized in Table 20.1, phytochemicals have long been tested in skin cancer prevention and treatment. It is highly likely that next approved drug would belong to this class of compounds.

20.2.3 MnSOD in Skin Cancer

Superoxide is produced as a by-product of energy metabolism and at the levels beyond nM is toxic to the cell [17]. Under physiological conditions, the family of superoxide dismutases (SOD), the body's first line of defense, catalyzes the dismutation of superoxide to molecular oxygen and hydrogen peroxide, maintaining nM

Table 20.1 Phytochemicals with antioxidant activities tested in skin carcinogenesis

Scientific names	Structure	Phytochemicals	Mechanism of action	Application in cancer
Curcuma longa	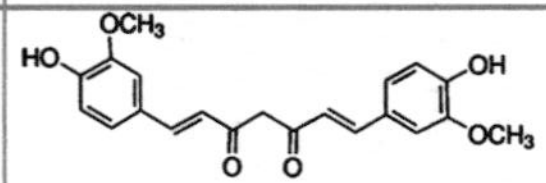	Turmeric	Downregulates pro-oncogenes through c-Jun NH2 terminal kinase (JNK), extracellular signal-regulated kinase (ERK) and p38 kinase. Also inhibits MAPK pathway	Skin, ovarian, prostate, and gastrointestinal tumors
Lawsonia inermis	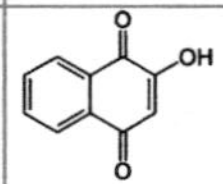	Henna leaf	Activates p38, MAPK, c-Jun and NH2 terminal kinase (JNK) stress signals Affects epidermal growth factor receptor (EGFR) phosphorylation and activation of PI3K/Akt pathway	Skin, ovarian
Aegle marmelos	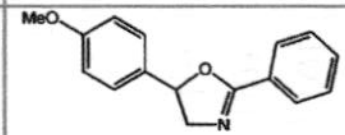	Bael fruit	Inhibits PI3K/Akt, MAPK proteins p38 and Erk1/2 and phosphorylation of IkB alpha and NF-kB	Skin, prostate, breast, and GI cancer
Punica granatum	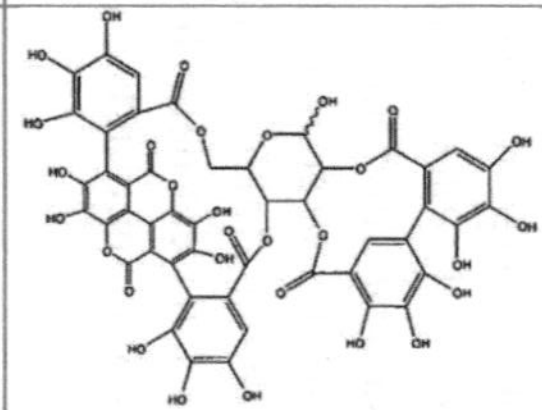	Pomegranate	Inhibits the phosphorylation of ERK1/2, p38 & JNK1/2 and the activation of NF-kB & IKK alpha Inhibition of cyclin D1 expression	Prostate, skin, and breast
Allium sativum	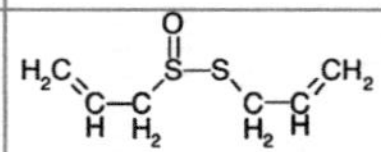	Garlic	Inhibits NF-kB activation and AP-1. Decreases oxidative stress and inflammation by suppressing Cox2 and downregulation of TF	Skin, rectal, lung, prostate, and breast
Zingiber officinale	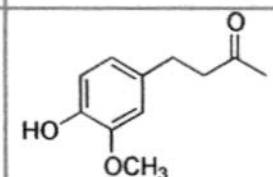	Ginger	Suppresses NF-kB activation and IKK/MAPK signaling pathway and hence decreases ROS	Lung, ovarian, hepatocarcinoma, and skin disorders
Vitis vinifera	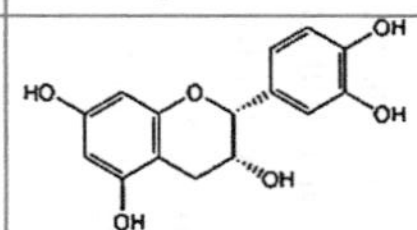	Grapes	Inhibits the IL-17 induced mRNA expression and decreased IL-17 induced phosphorylation of IKKalpha Decreases ERK1/2, p38 & MAPK signaling and also attenuates TNF-alpha and IL-6 production	Melanoma, skin, and prostate

levels of superoxide. Hydrogen peroxide is subsequently either dismuted to oxygen and water by catalases or reduced to water by peroxidases. Antioxidants can play diverse roles in tumor initiation, promotion, progression, or in cancer suppression. An increase in the levels of reactive species and a decrease in the levels of antioxidant enzymes is often a hallmark of a cancerous cell. Therefore, overexpression of antioxidant enzymes has been one of the strategies to suppress cancer growth via reducing the levels of RS.

Superoxide dismutases, when coupled with catalases and peroxidases, are powerful tools for the detoxification of cells. In humans, there are three isoforms of SOD—CuZnSOD, MnSOD, and ECSOD. CuZnSOD is primarily a cytosolic protein in human cells; MnSOD is a mitochondrial protein having manganese in its center; and ECSOD is found extracellularly and possesses similarities with CuZnSOD. There is also CuZnSOD (isoform identical to cytosolic) found in intermembrane space of mitochondria. In terms of skin carcinogenesis, animal studies indicated that MnSOD plays a major role in skin cancer [18].

MnSOD is considered a tumor suppressor in several human and murine tumor cell lines. An earlier study conducted by Zhao et al. demonstrated that overexpression of MnSOD could reduce the tumor incidence and multiplicity in a DMBA/TPA (12-*O*-tetradecanoylphorbol-13-acetate) treated multi-stage skin carcinogenesis model [19]. The expression levels of activator protein-1 (AP-1) were significantly reduced in MnSOD transgenic mice, in addition to suppression of oxidative stress. MnSOD heterozygous knock-out mice showed increased levels of oxidative damage and higher expression levels of oncogenes during skin carcinogenesis. Cell proliferation was also accelerated in MnSOD knock-out mice, further suggesting that MnSOD could suppress cell growth. MnSOD also promoted cell differentiation [20]. In FSa-II fibrosarcoma cells, 10 μM azacytidine induced apoptosis in the control cells whereas in MnSOD overexpressed cells, the transcription factor MyoD (a bHLH transcription factor regulating myogenic differentiation) was highly induced, leading to inhibition of apoptosis and induction of differentiation. Most recent data indicate that in cancer, H_2O_2-removing enzymes are frequently downregulated allowing for H_2O_2 accumulation which cancer cell uses for its progression. In turn, MnSOD switches its role from tumor suppressor to oncogene [21].

In the same skin carcinogenesis mouse model, Mn porphyrin-based SOD mimic, MnTE-2-PyP^{5+}, was able to mimic MnSOD when applied to skin at 5 ng daily, 4 days per week for 14 weeks. Moreover, since it presents a pharmacological approach, the dosing regimen of MnTE-2-PyP^{5+} could be adjusted to be given after cell undergo apoptosis and before their proliferation in order to afford much larger effect than that seen with MnSOD overexpression [22–24].

In another skin model, MnTE-2-PyP^{5+} mimicked MnSOD in protecting polγ from inactivation due to the nitrotyrosine formation as a consequence of UVB radiation. The polγ is a polymerase enzyme responsible for replication and repair of mtDNA. The mtDNA is organized in mitochondrial inner membrane as nucleoid. The MnSOD wild type (MnSOD$^{+/+}$) and MnSOD-knock-out mice (MnSOD$^{+/-}$) were used. MnTE-2-PyP^{5+} was given at 5 mg/kg intraperitoneally twice daily for 2 days before radiation. Quantification of polγ was done via co-immunoprecipitation by anti-3-nitrotyrosine antibody in MnTE-2-PyP^{5+}- and saline-pre-treated MnSOD$^{+/-}$

and MnSOD$^{+/+}$ mice skin exposed to 5 kJ/m^2 of UVB radiation. When the data on protein nitration were compared at 1 and 24 h after UVB treatment, MnTE-2-PyP^{5+} was able to fully protect polγ from oxidative damage; the levels of protein nitration in MnSOD$^{+/+}$ and MnTE-2-PyP^{5+}-treated MnSOD$^{+/-}$ mice were identical [25].

The therapeutic potential of another, more lipophilic but equally redox-active Mn porphyrin, Mn(III) *meso*-tetrakis(*N*-n-butoxyethylpyridinium-2-yl)porphyrin, MnTnBuOE-2-PyP^{5+} [26], in skin carcinogenesis has been reported [27]. The MnSOD heterozygous knock-out and transgenic SKH-1 hairless, albino mice and MnSOD knockdown and overexpressing HaCaT human keratinocytes were used to study the effects of MnSOD and MnTnBuOE-2-PyP^{5+} on ultraviolet (UV) radiation-induced inside-out signaling. The decreased MnSOD expression in MnSOD$^{+/-}$ knock-out mice enhnanced UV-induced activation of different oncogenic signaling pathways through an inside-out signaling-mediated mechanism. The epidermal growth factor receptor, EGFR, is an important contributor to ultraviolet (UV)-induced skin cancer development. EGFR activation is dependent on NADPH oxidase isoform, Nox4, expression and Src kinase activation, with Src activation upstream of Nox4 in regulation of EGFR activation. Inhibition of Src kinase activity or knockdown of Nox4 by lentiviral transduction of Nox4-specific shRNA blocked EGFR activation [27]. The inverse correlation exists between MnSOD expression and UV-induced activation of EGFR, induced by phosphorylation of tyrosine Tyr1068 both in vitro and in vivo, which correlates with increased ROS production. Pretreatment of MnSOD-knockdown HaCaT cells with the SOD mimetic MnTnBuOE-2-PyP^{5+} prevents UV-induced EGFR activation in a concentration-dependent manner [27].

All above said indicates impressive therapeutic potential of redox-active class of cationic Mn(III) N-substituted pyridylporphyrins [23, 28].

20.2.4 Glutathione

Glutathione (GSH) is a tri-peptide—a thiol consisting of cysteine, glutamate, and glycine. It is also a potent reducing agent and due to its high mM levels it maintains physiological cellular redox status [29]. It is involved in cell protection directly or as a cofactor of different peroxidases, including glutathione peroxidase and thioredoxin [17]. GSH may play different roles at different stages of cancer development. Increased levels of intracellular GSH are associated with conferring resistance to chemotherapeutic drugs, while depletion in GSH levels is associated with an increased sensitivity to carcinogens. Hence, the strategies to decrease intracellular GSH levels for chemotherapy are under exploration [17]. A novel glutathione transferase P1-1 inhibitor NBDHEX (7-nitro-2,1,3-benzoxadiazol-4-ylthio) showed efficacy against melanoma. NBDHEX along with TMZ (temozolomide) was effective in reducing the expression levels of cyclin D1 and cyclin D3 through the activation of c-Jun N terminal kinase in murine B16 and human A375 melanoma cells [30]. Thus, it has a potential for combinational therapies as an antiproliferative agent, especially in the treatment of melanomas.

20.3 Future Directions

20.3.1 Cancer Stem Cells

Stem cells (SC) are a group of undifferentiated cells that have the potential to differentiate into specialized cell types. These cells are highly proliferative and pluripotent in nature.

In mammals there are two types of stem cells—embryonic stem cells and adult stem cells. As the name suggests, embryonic stem cells are derived from embryos, i.e., from the inner cell mass of a 4–5-day-old blastocyst, whereas adult stem cells are obtained from adult tissues such as the bone marrow, blood, brain, liver, skin, and other tissues. These cells exhibit a self-renewal property making them different from normal cells.

Cancer stem cells (CSC) are a rare cell population of cancers. CSC are capable of self-renewal and can differentiate into specialized organs just like stem cells. Moreover, they are considered to be a main drive for cancer progression, metastasis, cancer recurrence, and the adaptation of tumor cells in response to oxidative stress. One could hypothesize that CSC which are responsible for drug resistance likely utilize cellular redox regulatory machinery. An interaction between the CSC and cellular redox-based pathways could contribute to cancer cell progression, survival, and tolerance to anticancer drugs.

Another important feature is that both SC and CSC share the same signaling pathway, PTEN/PI3K/mTOR, which is responsible for maintaining a population of CSC [31]. All these features justify further exploration of CSC for redox therapy of skin cancer.

20.3.2 Stem Cell Therapy

Although radiation has been an effective treatment, it results in severe damage to the surrounding normal cells. Stem cell therapy can help restore and repair radiation-induced damage. A recent review by Benderitter et al. [32] showed the therapeutic potential of stem cells in counteracting the side effects caused by radiotherapy. Their early clinical studies showed promising results in lung-, bone-, and skin cancer complications associated with radiation. Although these results are promising, the stem cell safety issues regarding tumor formation, tumor progression, and immune rejection, as well as concerns related to the selection of appropriate stem cells need to be addressed.

20.3.3 Antimalarial Agent

Although tremendous progress in recent years has been achieved with respect to cancer therapy, the complex nature of cancer poses a continuous challenge. Due to the difficulties encountered, researchers realize that single therapy is not sufficient in treating cancers. Combinational chemotherapies have been the best bet in preventing proliferation and resistance of tumor cells. Co-treatment of conventional chemo drugs with antimalarial drugs (e.g., artemisinin, chloroquine, and hydroxychloroquine) is one of the new combinational strategies for human cancers including skin cancer.

β-dihydroartemisinin (DHA), which is a major component of artemisinin, is shown to be effective as an adjunct to enhance tumor cell cytotoxicity, thereby improving the efficacy of anticancer treatment [33]. Since tumor cells can become resistant to oxidative stress via upregulation of their antioxidant defenses, Merendino et al. showed the ability of DHA to induce the GSH extrusion in cancer cell lines [34]. After 6 h of treatment with DHA the intracellular GSH level was reduced by 60 % due to the active extrusion of GSH. Another study by Ding et al. demonstrated that DHA treatment could reduce glutathione peroxidase-4 protein levels by at least 50 % in various human cancer cell lines, such as ovarian carcinoma, B-cell lymphoma, T-cell leukemia, and myeloma [35].

20.3.4 Amitriptyline

As one of the major tricyclic antidepressants (TCA), amitriptyline has a potential as an anticancer agent. It is frequently used to stimulate the mood especially in cancer patients. It recently became appreciated that this TCA could be effective in the treatment of cancer. Kulaksiz-Erkmen et al. demonstrated the effectiveness of amitriptyline in the inhibition of GST and hence hypothesized that it could potentially play a role as an anticancer drug as well [36]. Another study by Cordero et al. demonstrated the antitumor effect of amitriptyline against a large number of human cancer cells [37]. This team provided substantial evidence that amitriptyline can cause cellular damages, thereby leading to elevated levels of RS and decreased levels of antioxidant enzymes when compared to traditional chemotherapeutic drugs—Doxorubicin, Methotrexate, and Camptothecin. In conclusion, amitriptyline has a capacity to cause an increase in RS production of tumor cell, leading to its apoptosis and thus may be useful in the treatment of cancer.

References

1. Papadopulos-Eleopulos E, Turner VF, Papadimitriou JM. Oxidative stress, HIV and AIDS. Res Immunol. 1992;143:145–8.
2. Chance B, Sies H, Boveris A. Hydroperoxide metabolism in mammalian organisms. Physiol Rev. 1979;59:527–605.
3. Bagde AB, Sawant RS, Sawai RV, Muley SK, Dhimdhime RS. Charaka Samhita—complete encyclopedia of ayurvedic science. Int Natl J Ayur Altern Med. 2013;1(1):12–20.
4. Phillips JM, Clark C, Herman-Ferdinandez L, Moore-Medlin T, Rong X, Gill JR, Clifford JL, Abreo F, Nathan CO. Curcumin inhibits skin squamous cell carcinoma tumor growth in vivo. Otolaryngol Head Neck Surg. 2011;145:58–63.

5. Chen YR, Tan TH. Inhibition of the c-Jun N-terminal kinase (JNK) signaling pathway by curcumin. Oncogene. 1998;17:173–8.
6. Zhao R, Yang B, Wang L, Xue P, Deng B, Zhang G, Liu M, Pi J, Guan D. Curcumin protects human keratinocytes against inorganic arsenite-induced acute cytotoxicity through an NRF2-dependent mechanism. Oxid Med Cell Longev. 2013;2013.
7. Cheng AL, Hsu CH, Lin JK, Hsu MM, Ho YF, Shen TS, Ko JY, Lin JT, Lin BR, Ming-Shiang W, Yu HS, Jee SH, Chen GS, Chen TM, Chen CA, Lai MK, Pu YS, Pan MH, Wang YJ, Tsai CC, Hsieh CY. Phase I clinical trial of curcumin, a chemopreventive agent, in patients with high-risk or pre-malignant lesions. Anticancer Res. 2001;21:2895–900.
8. Agrawal A, Jahan S, Soyal D, Goyal E, Goyal PK. Amelioration of chemical-induced skin carcinogenesis by Aegle marmelos, an Indian medicinal plant, fruit extract. Integr Cancer Ther. 2012;11:257–66.
9. Afaq F, Malik A, Syed D, Maes D, Matsui MS, Mukhtar H. Pomegranate fruit extract modulates UV-B-mediated phosphorylation of mitogen-activated protein kinases and activation of nuclear factor kappa B in normal human epidermal keratinocytes paragraph sign. Photochem Photobiol. 2005;81:38–45.
10. Jin ZY, Wu M, Han RQ, Zhang XF, Wang XS, Liu AM, Zhou JY, Lu QY, Zhang ZF, Zhao JK. Raw garlic consumption as a protective factor for lung cancer, a population-based case-control study in a Chinese population. Cancer Prev Res (Phila). 2013;6:711–8.
11. Das I, Saha T. Effect of garlic on lipid peroxidation and antioxidation enzymes in DMBA-induced skin carcinoma. Nutrition. 2009;25:459–71.
12. Dasgupta T, Rao AR, Yadava PK. Modulatory effect of henna leaf (Lawsoniainermis) on drug metabolizing phase I and phase II enzymes, antioxidant enzymes, lipid peroxidation and chemically induced skin and forestomachpapillomagenesis in mice. Mol Cell Biochem. 2003;245:11–22.
13. Katiyar SK. Grape seed proanthocyanidines and skin cancer prevention: inhibition of oxidative stress and protection of immune system. Mol Nutr Food Res. 2008;52 Suppl 1:S71–6.
14. Burton LJ, Barnett P, Smith B, Arnold RS, Hudson T, Kundu K, Murthy N, Odero-Marah VA. Muscadine grape skin extract reverts snail-mediated epithelial mesenchymal transition via superoxide species in human prostate cancer cells. BMC Complement Altern Med. 2014;14:97.
15. Kim JK, Kim Y, Na KM, Surh YJ, Kim TY. Gingerol prevents UVB-induced ROS production and COX-2 expression in vitro and in vivo. Free Radic Res. 2007;41:603–14.
16. Rhode J, Fogoros S, Zick S, Wahl H, Griffith KA, Huang J, Liu JR. Ginger inhibits cell growth and modulates angiogenic factors in ovarian cancer cells. BMC Complement Altern Med. 2007;7:44.
17. Halliwell B, Gutteridge JMC. Free radicals in biology and medicine. New York: Oxford University Press; 2007.
18. Robbins D, Zhao Y. The role of manganese superoxide dismutase in skin cancer. Enzyme Res. 2011;2011:409295. doi:10.4061/2011/409295.
19. Zhao Y, Oberley TD, Chaiswing L, Lin SM, Epstein CJ, Huang TT, St Clair D. Manganese superoxide dismutase deficiency enhances cell turnover via tumor promoter-induced alterations in AP-1 and p53-mediated pathways in a skin cancer model. Oncogene. 2002;21: 3836–46.
20. Zhao Y, Kiningham KK, Lin SM, St Clair DK. Overexpression of MnSOD protects murine fibrosarcoma cells (FSa-II) from apoptosis and promotes a differentiation program upon treatment with 5-azacytidine: involvement of MAPK and NFκB pathway. Antioxid Redox Signal. 2001;3:375–86.
21. Miriyala S, Spasojevic I, Tovmasyan A, Salvemini D, Vujaskovic Z, St Clair D, Batinic-Haberle I. Manganese superoxide dismutase, MnSOD and its mimics. Biochim Biophys Acta. 2012;1822:794–814.
22. Zhao Y, Chaiswing L, Oberley TD, Batinic-Haberle I, St. Clair W, Epstein CJ, St. Clair D. A mechanism-based antioxidant approach for the reduction of skin carcinogenesis. Cancer Res. 2005;65:1401–5.

23. Batinic-Haberle I, Tovmasyan A, Roberts ER, Vujaskovic Z, Leong KW, Spasojevic I. SOD therapeutics: latest insights into their structure-activity relationships and impact on the cellular redox-based signaling pathways. Antioxid Redox Signal. 2014;20:2372–415.
24. Batinic-Haberle I, Reboucas JS, Spasojevic I. Superoxide dismutase mimics: chemistry, pharmacology, and therapeutic potential. Antioxid Redox Signal. 2010;13:877–918.
25. Bakthavatchalu V, Dey S, Xu Y, Noel T, Jungsuwadee P, Holley AK, Dhar SK, Batinic-Haberle I, St Clair DK. Manganese superoxide dismutase is a mitochondrial fidelity protein that protects Polγ against UV-induced inactivation. Oncogene. 2012;31:2129–39.
26. Rajic Z, Tovmasyan A, Spasojevic I, Sheng H, Lu M, Li AM, Gralla EB, Warner DS, Benov L, Batinic-Haberle I. A new SOD mimic, Mn(III) ortho N-butoxyethylpyridylporphyrin, combines superb potency and lipophilicity with low toxicity. Free Radic Biol Med. 2012;52:1828–34.
27. Holley AK, Xu Y, Noel T, Bakthavatchalu V, Batinic-Haberle I, St Clair DK. Manganese superoxide dismutase-mediated inside-out signaling in HaCaT human keratinocytes and SKH-1 mouse skin. Antioxid Redox Signal. 2014;20:2347–60.
28. Batinic-Haberle I, Tovmasyan A, Spasojevic I. An educational overview of the chemistry, biochemistry and therapeutic aspects of Mn porphyrins—from superoxide dismutation to H_2O_2-driven pathways. Redox Biol. 2015;5:43–65.
29. Traverso N, Ricciarelli R, Nitti M, Marengo B, Furfaro AL, Pronzato MA, Marinari UM, Domenicotti C. Role of glutathione in cancer progression and chemoresistance. Oxid Med Cell Longev. 2013;2013:972913.
30. Tentori L, Dorio AS, Mazzon E, Muzi A, Sau A, Cuzzocrea S, Vernole P, Federici G, Caccuri AM, Graziani G. The glutathione transferase inhibitor 6-(7-nitro-2,1,3-benzoxadiazol-4-ylthio)hexanol (NBDHEX) increases temozolomide efficacy against malignant melanoma. Eur J Cancer. 2011;47(8):1219–30.
31. Dayem AA, Choi HY, Kim JH, Cho SG. Role of oxidative stress in stem, cancer, and cancer stem cells. Cancers (Basel). 2010;2:859–84.
32. Benderitter M, Caviggioli F, Chapel A, Coppes RP, Guha C, Klinger M, Malard O, Stewart F, Tamarat R, Luijk PV, Limoli CL. Stem cell therapies for the treatment of radiation-induced normal tissue side effects. Antioxid Redox Signal. 2014;21:338–55.
33. Chen T, Li M, Zhang R, Wang H. Dihydroartemisinin induces apoptosis and sensitizes human ovarian cancer cells to carboplatin therapy. J Cell Mol Med. 2009;13:1358–70.
34. Merendino N, Loppi B, D'Aquino M, Molinari R, Pessina G, Romano C, Velotti F. Docosahexaenoic acid induces apoptosis in the human PaCa-44 pancreatic cancer cell line by active reduced glutathione extrusion and lipid peroxidation. Nutr Cancer. 2005;52:225–33.
35. Ding WQ, Lind SE. Phospholipid hydroperoxide glutathione peroxidase plays a role in protecting cancer cells from docosahexaenoic acid-induced cytotoxicity. Mol Cancer Ther. 2007;6:1467–74.
36. Kulaksiz-Erkmen G, Dalmizrak O, Dincsoy-Tuna G, Dogan A, Ogus IH, Ozer N. Amitriptyline may have a supportive role in cancer treatment by inhibiting glutathione S-transferase pi (GST-π) and alpha (GST-α). J Enzyme Inhib Med Chem. 2013;28:131–6.
37. Cordero MD, Sánchez-Alcázar JA, Bautista-Ferrufino MR, Carmona-López MI, Illanes M, Ríos MJ, Garrido-Maraver J, Alcudia A, Navas P, de Miguel M. Acute oxidant damage promoted on cancer cells by amitriptyline in comparison with some common chemotherapeutic drugs. Drugs. 2010;21:932–44.

Chapter 21
Role of Oxidative Stress in Erectile Dysfunction After Prostate Cancer Therapy

Timothy J. Robinson and Bridget F. Koontz

Abbreviations

4HNE	4-Hydroxynonenal
8-OHdG	8-Hydroxy-2′-deoxyguanosine
ED	Erectile dysfunction
eNOS	Endothelial nitric oxide synthase
H_2O_2	Hydrogen peroxide
ICP	Intracavernous pressure
IMRT	Intensity modulated radiotherapy
MAP	Mean arterial pressure
$^{\bullet}NO$	Nitric oxide
$ONOO^-$	Peroxynitrite
nNOS	Neuronal nitric oxide synthase
$O_2^{\bullet -}$	Superoxide anion
$^{\bullet}OH$	Hydroxyl radical
PDE-5	Phosphodiesterase-5
PIN	Protein inhibitor of NOS
RNOS	Reactive nitrogen species
ROS	Reactive oxygen species
sGC	Soluble guanylyl cyclase
SOD	Superoxide

T.J. Robinson • B.F. Koontz (✉)
Department of Radiation Oncology, Duke University Medical Center,
Morris Bldg Rm 05151, Box 3085, Durham, NC 27707, USA
e-mail: Timothy.robinson@duke.edu; Bridget.koontz@duke.edu

I. Batinić-Haberle et al. (eds.), *Redox-Active Therapeutics*,
Oxidative Stress in Applied Basic Research and Clinical Practice,
DOI 10.1007/978-3-319-30705-3_21

Prostate cancer is the most frequently diagnosed cancer in men in the developed world, with an estimated 240,000 new cases but only 28,000 deaths caused by the disease each year within the United States alone [1]. The vast majority of prostate cancers are detected at an early stage and are amenable to curative treatment using surgery or radiation. The use of radiation therapy to treat prostate cancer, via either external beam or brachytherapy, has been in use now for half a century [2] and continues to be a major treatment modality, particularly among men who are either older, poor surgical candidates, or present with high-risk disease. The most common side effect of radiation therapy is erectile dysfunction (ED) [3], which has been a known side effect of radiation therapy since the 1970s [4] that continues to incur significant financial and psychosocial costs. Although the delivery of prostate radiation therapy has changed significantly over the past 50 years, radiation-induced ED remains the leading toxicity of prostate cancer radiation therapy. Observational studies of newer, intensity modulated radiation therapy (IMRT), have observed increased diagnoses of ED in patients treated with IMRT as compared to conventional 3D conformal treatment [5]. Despite awareness of radiation-induced ED for over 40 years, as of 2014 there are still no proven strategies or treatments to reduce the risk of radiation-induced ED.

Radiation-induced ED, by definition, impairs normal erectile function, which requires engorgement of intact vascular erectile tissues with an adequate blood supply driven by neurogenic stimuli. It has been long known that neurogenic stimuli of erection involves a complex integration of both voluntary and involuntary psychological, tactile, and autonomic stimuli [6]. However, despite the likely inherently complex integration of these signals, it has been shown over a century and a half ago using a canine model that direct autonomic nerve stimulation can result in penile erection [7]. From a vascular perspective, it was observed 30 years ago that direct vascular damage of the corpus cavernosum from radiation damage can also cause ED [8], with current estimates suggest that 80 % of all causes of ED are caused at least in part by vascular disease of the penile arteries [9]. Taken together, the most basic model of physiologic erectile dysfunction can be simplified as a result of damage to nerves and/or vessels involved in erectile function.

Advances in molecular biology have provided an appreciation for the role of nitric oxide (•NO) as the principle stimulator of cavernosal vasodilation and penile erection, which stimulates smooth muscle relaxation to increase cavernosal arteriole blood flow (Fig. 21.1) [10–12]. •NO is generated by neuronal or endothelial sources and causes activation of guanylyl cyclase, which increases cGMP levels and in turn activates a cGMP-specific protein kinase, resulting in decreased free intracellular calcium. Low levels of cytoplasmic calcium results in smooth muscle relaxation, which allows engorgement of the cavernosal sinusoids [13]. Nitric oxide can be generated by neuronal nitric oxide synthase (nNOS) in cavernosal nerve endings or by endothelial nitric oxide synthase (eNOS) in the endothelial cells lining the sinusoids of the cavernosa. Thus damage to either the cavernosal nerve or pudendal vessels serving the penis can result in reduced NO production and a poor erection. The cGMP nitric oxide pathway is the mechanism by which PDE-5 inhibitors are beneficial in treatment of ED, as they prevent cGMP hydrolysis, thus increasing the effect of the •NO present [14]. Radiation-induced ED has been recently extended to rat models of erectile

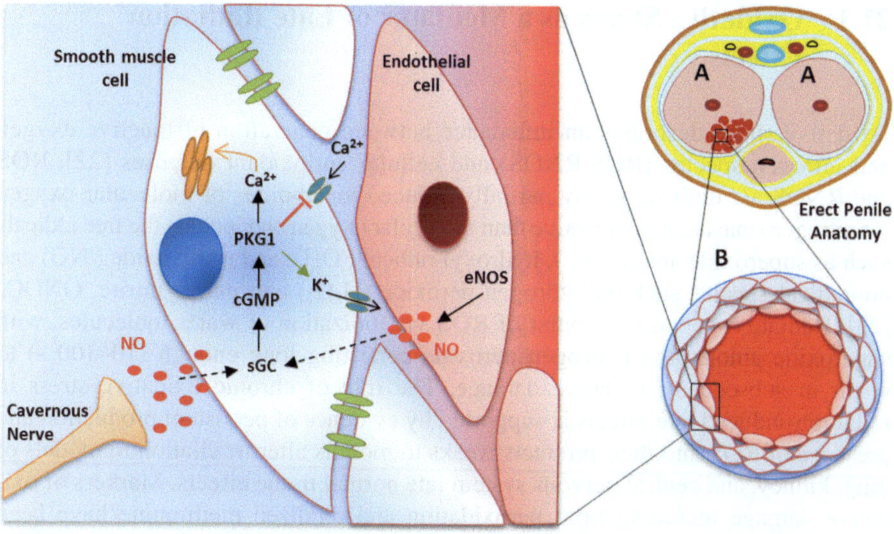

Fig. 21.1 Penis anatomy includes two penile cavernosa (**a**), which fill with blood to achieve erection. Each cavernosa is filled with many sinusoids (**b**), which is a capillary surrounded by smooth muscle. When smooth muscle relaxes, the sinusoid fills and expands, compressing venous outflow and causing increased cavernosal pressure and an erect penis. To stimulate an erection, the cavernous nerve and the endothelium release nitric oxide (•NO) produced by nNOS and eNOS, respectively. The •NO causes activation of cGMP-specific kinases. GTP then activates K channel pumps, inhibits Ca channel pumps in the cell membrane, and promotes Ca uptake by endoplasmic reticulum, resulting in reduced intracellular calcium and smooth muscle relaxation. *cGMP* cyclic guanosine monophosphate, *eNOS* endothelial nitric oxide synthase, *nNOS* neuronal nitric oxide synthase, *NO* nitric oxide, *PKG1* cGMP-dependent protein kinase G1 (figure courtesy of Artak Tovmasyan Ph.D.)

dysfunction at the molecular level, where radiation has been shown to decrease nitric oxide synthase in the nerve endings of the penis (nNOS) and the endothelial lining of the sinusoids (eNOS) [15], thereby reducing overall levels of •NO [16].

Radiation-induced ED typically manifests within the first 2 years of radiation therapy and is considered a late effect of radiation therapy [17]. Acute radiation toxicities are generally attributable to clonogenic death of a specific cell lineage with a well-defined temporal course specified on the order of hours to days. In contrast, late effects of radiation therapy involve a more complicated interaction between stromal tissues, vasculature, and cytokines and may manifest over a much broader, poorly defined, and unpredictable time frame that have not been adequately explained by classic models of clonogenic death used to explain acute toxicities [17, 18]. Following irradiation of bystander normal tissues, late effects of radiation develop over the course of months to years. These late effects are thought to be mediated by a chronic inflammatory and wound healing response that results in chronic vascular and parenchymal dysfunction and cell loss. Importantly, recent evidence supports the notion that radiation-induced injury can be mitigated by targeting this series of events [19–21]. Among these, chronic oxidative stress has emerged as a potential driver of late normal tissue injury [21].

21.1 Oxidative Stress as a Mediator of Late Radiation Effects

Oxidative stress describes an imbalance between generation of reactive oxygen and nitrogen species (ROS/RNOS) and cellular antioxidant defenses [22]. ROS and RNOS are defined as any partially reduced metabolites of molecular oxygen (or nitrogen) that are more reactive than molecular oxygen and include the free radicals such as superoxide anion ($O_2^{\bullet-}$), hydroxyl radical ($^{\bullet}OH$), and nitric oxide ($^{\bullet}NO$) and nonradical species such as hydrogen peroxide (H_2O_2) and peroxynitrite, $ONOO^-$ [23], Radiation generates a burst of ROS via ionization of water molecules, with superoxide anions and hydrogen peroxide persisting long enough (10–100 s) to result in subsequent oxidative damage. The role of chronic oxidative stress in radiation-induced late effects is supported by evidence of persistent production and presence of ROS and their products weeks to months after irradiation in models of lung, kidney, and central nervous system late normal tissue effects. Markers of oxidative damage including lipid peroxidation and oxidized methionine have been observed in bronchoalveolar lavage fluid in lung cancer patients after the completion of radiation therapy [24–26] and in mouse models of lung irradiation 15–20 weeks following radiation. In these mouse models, overexpression of superoxide dismutase (SOD), which catalyzes the dismutation of the superoxide anion, that is, its reduction to hydrogen peroxide and its oxidation to oxygen, was able to mitigate radiation-induced late effects and fibrosis [27]. In neuronal tissues, lipid peroxidation has been observed in the hippocampi of adult mice 2 weeks after radiation [28] and the expression of HMOX-1, a marker of oxidative stress, has been observed in rat spinal cords prior to the onset of myelopathy [29]. In kidney, radiation has been shown to increase 8-hydroxy-2′-deoxyguanosine (8-OHdG), a marker of oxidative DNA damage, evident by 4 weeks following radiation that persisted up to 24 weeks later [30]. In addition to direct evidence of oxidative stress in late normal tissue, production of ROS can also be increased by hypoxia [31, 32] which has been observed to precede histologic evidence of late effects in both lung [33] and spinal cord. [34] The production of ROS, at least in the period immediately following radiation, may be mitigated not only by stoichiometric antioxidants such as *N*-acetyl cysteine, but also by more general mediators of anti-inflammatory signaling such as the angiotensin receptor inhibitor losartan [22].

21.2 Oxidative Stress in Erectile Dysfunction

Although there is now a substantial body of work describing the role of ROS in promoting and maintaining late toxicities of radiation therapy in lung, kidney, and nervous tissue, most of our current knowledge for an explicit role of oxidative stress in erectile dysfunction (ED) comes from studies of nonradiation-related sources of oxidative stress including smoking [35], diabetes [36], sleep apnea [37], and aging

[38, 39]. A number of these and other stimuli, including nicotine and mechanical injury, are known to impair erectile function and simultaneously result in generation of superoxide anions in cavernosal smooth muscle cells [40–42]. In conditions such as hyperlipidemia or diabetes, increased superoxide anion production by NADPH oxidase in cavernosal tissue appears to mediate ED [43–45]. This appears to be related to interference with neuronal •NO transmission, supported by a study of bovine penile strips where inhibition of endogenous SOD and subsequent increase in superoxide anions would enhance removal of •NO by superoxide (resulting in $ONOO^-$ production) and would in turn result in near complete inhibition of nerve-mediated vascular relaxation [46]. In animal models, the use of catalytic and stoichiometric antioxidants has been shown to improve erectile function, increase circulating •NO, and enhance the efficacy of phosphodiesterase-5 (PDE-5) inhibitors [36, 45]. In humans, a high level of systemic oxidative stress (measured by erythrocyte glycohydrolases) is associated with the presence ED as compared to non-ED controls [47].

In addition to more historical physiologic markers of erectile function, there is emerging support of the role of oxidative stress impacting ED-related molecular signaling. Following nerve activation, normal erectile function requires intact •NO signaling in order to induce vascular smooth muscle relaxation. Nitric oxide signaling, in turn, appears to act directly through only a single known receptor known as soluble guanylyl cyclase (sGC), which is encoded by the genes GUCY1A2, GUCY1A3, GUCY1B2, and GUCY1B3. It is a cytosolic, soluble heterodimer consisting of an alpha and beta subunit, which together bind a heme moiety. These heme moieties are critical to the sensing of NO signaling and are capable in other proteins of interacting with diatomic gases including •NO, O_2, and CO [48, 49]. In the sGC enzyme, the iron contained within heme moiety of the functional protein exists in a reduced ferrous, Fe^{2+} state. Oxidization of this heme moiety results in the ferric, Fe^{3+}, state, which results in the form of the enzyme that is no longer responsive to •NO signaling, providing a direct means by which oxidative stress can impact physiologic erectile signaling [50]. In addition, reactive oxygen species directly scavenge •NO, reducing its availability to signal vascular smooth muscle relaxation [51]. In models of aging associated ED, siRNA targeting of the protein inhibitor of NOS (PIN) has restored erectile function in rat models, providing additional evidence that intact •NO signaling is required for normal erectile function [39].

Direct evidence of a role for oxidative stress in ED following prostate irradiation is now emerging. One of the first studies to provide support for a unifying role of oxidative stress in radiation-induced ED observed that among men who experienced ED following prostate irradiation, that the vast majority of those who developed ED (93 %) were smokers, as compared to only 25 % of patients who did not develop ED [8]. More direct evidence of the role of oxidative stress in radiation-induced ED has come from recently developed and validated models of radiation-induced ED in the rat [15, 52, 53]. In this model, erectile function in rats is reproducibly quantified using the ratio of intracavernous pressure (ICP) to mean arterial pressure (MAP) following cavernous nerve electrical stimulation. ED is a late-occurring event, requiring 9–12 weeks to be evident [15, 54, 55]. Using image-guidance to provide prostate-confined radiation, *unirradiated* penile corpora of rats treated with a single 20 Gy fraction conformal to the prostate were found to have increased expression of a number of oxidative stress markers, including the NADPH oxidase subunit gp91, 8-OHdG, and 4-hydroxynonenal (4HNE). It is notable that these oxidative stress markers were normal shortly after radiation, indicating that the oxidative stress is not directly radiation-induced (in fact the cavernosal tissue were out of the radiation field in this particular experiment) but more likely secondary to nerve and endothelial injury. Normal erectile function, as measured using ICP/MAP ratios, was inversely correlated with the expression of oxidative stress markers [53]. This evidence for the role of oxidative stress in radiation-induced ED as well as more common causes of ED including diabetes, smoking, and sleep apnea have raised significant interest in oxidative stress as a potential therapeutic target to prevent ED.

21.3 Strategies to Mitigate Oxidative Stress in ED

Several nonradiation-based models of ED have now demonstrated potential in vitro and in vivo efficacy of antioxidant approaches for the treatment of ED in settings such as obesity/diabetes and aging. Work by Silva et al. [56] have established a model of diabetes-induced ED in mice fed a high-fat diet. In this model, they were able to quantify ED in mice as measured by decreased ICP following cavernosal nerve stimulation. Diabetes-induced ED was fully reversible following two weeks of therapy with BAY 60-2770, an activator of sGC that was capable of reactivating heme-oxidized sGC and decreasing ROS levels [56]. The role of antioxidant-based interventions has also been examined in the role of aging induced-ED in a second model described [38]. In this model, male Wistar rats were observed to have differential corpus cavernosum relaxation depending on whether rats were young (3.5 months) vs. middle-aged (10 months). They observed that treatment with apocynin an NADPH oxidase inhibitor (a primary producer of superoxide and H_2O_2) either ex (10^{-4} M) or in vivo (85 mg per day orally for 4 weeks), nearly normalized age-related declines in function, including response to treatment with the PDE-5 inhibitor sildenafil. Ex vivo treatment of the corpus cavernosum with SOD (75 U/mL) showed similar improvements in smooth muscle relaxation (required for engorgement) [38]. In another model of diabetes-induced ED in rats, Huan et al. used the traditional Chinese medicine Panax notoginseng saponins (PNS), selected for its antioxidant properties. Treatment with 100–150 mg/kg of PNS was able to increase ICP, increase levels of SOD, decrease levels of the ROS marker malondialdehyde, and prevent rupture of endothelium and generation of thrombi within the penis [57].

Testing strategies to prevent radiation-induced ED is becoming more prevalent now that the role of oxidative stress is known. Oberley-Deegan et al. used a rat model of whole pelvis radiation-induced ED in rats treated with 7.5 Gy × 5 to investigate the impact of the SOD mimetic MnTE-2-PyP^{5+} (Mn(III) *ortho meso*-tetrakis (*N*-ethylpyridinium-2-yl)porphyrin) [54, 58]. They showed that MnTE-2-PyP^{5+} was able to mitigate damage to the skin, prostate, testes, and maintain erectile function 12 weeks after 37.5 Gy in 5 daily fractions. The porphyrin was given 5 mg/kg 24 h prior to radiation therapy followed by 2.5 mg/kg every other day for 2 weeks and then 5 mg/kg throughout the observation period. Bassett et al. [59] tested exercise as a mediator of global oxidative stress and vascular function; unfortunately this was found not to have a protective role. Adipose-derived stem cells can have antioxidant properties [60] and have been successfully applied in a rat model of radiation-induced ED to protect nNOS expression in corpus cavernosa and maintain erectile function [61]. The use of metalloporphyrin SOD-mimetics is particularly exciting in the realm of treating prostate cancer, with evidence that the use of these therapies may not only spare normal tissue toxicity, but may also provide potential antitumor activity in models of irradiated prostate cancer xenografts [62].

21.4 Conclusion

Prostate cancer remains a significant public health concern for which thousands of men are treated with radiation each year. Despite our knowledge of radiation-induced ED in men treated with prostate cancer over the past 40 years, there remain no proven strategies to mitigate the risk of radiation-induced ED. Progress in understanding the molecular underpinnings of ED have elucidated a prominent role of oxidative stress in impairing normal erectile physiology and molecular signaling within a number of disease entities including diabetes, ageing, and more recently radiation therapy. Models of radiation-induced ED in rats are now becoming available, including models of previously unavailable, organ-confined external beam radiation therapy. These preclinical models will hopefully provide the data needed to progress to human trials targeting oxidative stress for the prevention of radiation-induced ED in the near future.

References

1. Siegel R, Naishadham D, Jemal A. Cancer statistics, 2013. CA Cancer J Clin. 2013;63(1): 11–30.
2. Budhraja SN, Anderson JC. An assessment of the value of radiotherapy in the management of carcinoma of the prostate. Br J Urol. 1964;36:535–40.
3. Meltzer D, Egleston B, Abdalla I. Patterns of prostate cancer treatment by clinical stage and age. Am J Public Health. 2001;91(1):126–8.
4. Bagshaw MA, Ray GR, Pistenma DA, Castellino RA, Meares EM. External beam radiation therapy of primary carcinoma of the prostate. Cancer. 1975;36(2):723–8.
5. Bekelman JE, Mitra N, Efstathiou J, et al. Outcomes after intensity-modulated versus conformal radiotherapy in older men with nonmetastatic prostate cancer. Int J Radiat Oncol Biol Phys. 2011;81(4):e325–34.
6. Weiss HD. The physiology of human penile erection. Ann Intern Med. 1972;76(5):793–9.
7. Eckhardt C. Untersuchungen uber die Erection des Penis beim Hunde. Beitr Anat Physiol. 1863;3(123).

8. Goldstein I, Feldman MI, Deckers PJ, Babayan RK, Krane RJ. Radiation-associated impotence. A clinical study of its mechanism. JAMA. 1984;251(7):903–10.
9. Sullivan ME, Keoghane SR, Miller MA. Vascular risk factors and erectile dysfunction. BJU Int. 2001;87(9):838–45.
10. Andersson KE. Pharmacology of penile erection. Pharmacol Rev. 2001;53(3):417–50.
11. Andersson KE. Mechanisms of penile erection and basis for pharmacological treatment of erectile dysfunction. Pharmacol Rev. 2011;63(4):811–59.
12. Burnett AL. Nitric oxide in the penis—science and therapeutic implications from erectile dysfunction to priapism. J Sex Med. 2006;3(4):578–82.
13. Lue TF. Erectile dysfunction. N Engl J Med. 2000;342(24):1802–13.
14. Corbin JD. Mechanisms of action of PDE5 inhibition in erectile dysfunction. Int J Impot Res. 2004;16 Suppl 1:S4–7.
15. Kimura M, Rabbani ZN, Zodda AR, et al. Role of oxidative stress in a rat model of radiation-induced erectile dysfunction. J Sex Med. 2012;9(6):1535–49.
16. Carrier S, Hricak H, Lee SS, et al. Radiation-induced decrease in nitric oxide synthase—containing nerves in the rat penis. Radiology. 1995;195(1):95–9.
17. Withers HR, Peters LJ, Kogelnik HS. The pathobiology of late effects in irradiation. New York: Raven; 1980.
18. Rubin P, Casarett GW. Clinical radiation pathology as applied to curative radiotherapy. Cancer. 1968;22(4):767–78.
19. Moulder JE, Robbins ME, Cohen EP, Hopewell JW, Ward WF. Pharmacologic modification of radiation-induced late normal tissue injury. Cancer Treat Res. 1998;93:129–51.
20. Moulder JE. Post-irradiation approaches to treatment of radiation injuries in the context of radiological terrorism and radiation accidents: a review. Int J Radiat Biol. 2004;80(1):3–10.
21. Robbins ME, Zhao W. Chronic oxidative stress and radiation-induced late normal tissue injury: a review. Int J Radiat Biol. 2004;80(4):251–9.
22. Zhao W, Diz DI, Robbins ME. Oxidative damage pathways in relation to normal tissue injury. Br J Radiol. 2007;80(Spec No. 1):S23–31.
23. Fridovich I. Fundamental aspects of reactive oxygen species, or what's the matter with oxygen? Ann N Y Acad Sci. 1999;893:13–8.
24. Jack CI, Cottier B, Jackson MJ, Cassapi L, Fraser WD, Hind CR. Indicators of free radical activity in patients developing radiation pneumonitis. Int J Radiat Oncol Biol Phys. 1996;34(1):149–54.
25. Beinert T, Binder D, Stuschke M, et al. Oxidant-induced lung injury in anticancer therapy. Eur J Med Res. 1999;4(2):43–53.
26. Beinert T, Binder D, Oehm C, et al. Further evidence for oxidant-induced vascular endothelial growth factor up-regulation in the bronchoalveolar lavage fluid of lung cancer patients undergoing radio-chemotherapy. J Cancer Res Clin Oncol. 2000;126(6):352–6.
27. Kang SK, Rabbani ZN, Folz RJ, et al. Overexpression of extracellular superoxide dismutase protects mice from radiation-induced lung injury. Int J Radiat Oncol Biol Phys. 2003; 57(4):1056–66.
28. Limoli CL, Rola R, Giedzinski E, Mantha S, Huang TT, Fike JR. Cell-density-dependent regulation of neural precursor cell function. Proc Natl Acad Sci U S A. 2004;101(45):16052–7.
29. Tofilon PJ, Fike JR. The radioresponse of the central nervous system: a dynamic process. Radiat Res. 2000;153(4):357–70.
30. Robbins ME, Zhao W, Davis CS, Toyokuni S, Bonsib SM. Radiation-induced kidney injury: a role for chronic oxidative stress? Micron. 2002;33(2):133–41.
31. Duranteau J, Chandel NS, Kulisz A, Shao Z, Schumacker PT. Intracellular signaling by reactive oxygen species during hypoxia in cardiomyocytes. J Biol Chem. 1998;273(19):11619–24.
32. Killilea DW, Hester R, Balczon R, Babal P, Gillespie MN. Free radical production in hypoxic pulmonary artery smooth muscle cells. Am J Physiol Lung Cell Mol Physiol. 2000;279(2): L408–12.
33. Vujaskovic Z, Anscher MS, Feng QF, et al. Radiation-induced hypoxia may perpetuate late normal tissue injury. Int J Radiat Oncol Biol Phys. 2001;50(4):851–5.

34. Sawaya R, Yamamoto M, Ramo OJ, Shi ML, Rayford A, Rao JS. Plasminogen activator inhibitor-1 in brain tumors: relation to malignancy and necrosis. Neurosurgery. 1995;36(2): 375–80. discussion 380-371.
35. Tostes RC, Carneiro FS, Lee AJ, et al. Cigarette smoking and erectile dysfunction: focus on NO bioavailability and ROS generation. J Sex Med. 2008;5(6):1284–95.
36. De Young L, Yu D, Bateman RM, Brock GB. Oxidative stress and antioxidant therapy: their impact in diabetes-associated erectile dysfunction. J Androl. 2004;25(5):830–6.
37. Liu K, Liu XS, Xiao L, et al. NADPH oxidase activation: a mechanism of erectile dysfunction in a rat model of sleep apnea. J Androl. 2012;33(6):1186–98.
38. Silva FH, Monica FZ, Bau FR, et al. Superoxide anion production by NADPH oxidase plays a major role in erectile dysfunction in middle-aged rats: prevention by antioxidant therapy. J Sex Med. 2013;10(4):960–71.
39. Magee TR, Kovanecz I, Davila HH, et al. Antisense and short hairpin RNA (shRNA) constructs targeting PIN (Protein Inhibitor of NOS) ameliorate aging-related erectile dysfunction in the rat. J Sex Med. 2007;4(3):633–43.
40. Koupparis AJ, Jeremy JY, Muzaffar S, Persad R, Shukla N. Sildenafil inhibits the formation of superoxide and the expression of gp47 NAD[P]H oxidase induced by the thromboxane A2 mimetic, U46619, in corpus cavernosal smooth muscle cells. BJU Int. 2005; 96(3):423–7.
41. Hotston M, Jeremy JY, Persad R, Bloor J, Shukla N. 8-isoprostane F2alpha up-regulates the expression of type 5 phosphodiesterase in cavernosal vascular smooth muscle cells: inhibition with sildenafil, iloprost, nitric oxide and picotamide. BJU Int. 2010;106(11):1794–8.
42. Ozkara H, Alan C, Atukeren P, et al. Changes of nitric oxide synthase-containing nerve fibers and parameters for oxidative stress after unilateral cavernous nerve resection or manuplation in rat penis. Chin J Physiol. 2006;49(3):160–6.
43. Shukla N, Jones R, Persad R, Angelini GD, Jeremy JY. Effect of sildenafil citrate and a nitric oxide donating sildenafil derivative, NCX 911, on cavernosal relaxation and superoxide formation in hypercholesterolaemic rabbits. Eur J Pharmacol. 2005;517(3):224–31.
44. Kovanecz I, Ferrini MG, Vernet D, Nolazco G, Rajfer J, Gonzalez-Cadavid NF. Pioglitazone prevents corporal veno-occlusive dysfunction in a rat model of type 2 diabetes mellitus. BJU Int. 2006;98(1):116–24.
45. Musicki B, Burnett AL. Endothelial dysfunction in diabetic erectile dysfunction. Int J Impot Res. 2007;19(2):129–38.
46. Mok JS, Paisley K, Martin W. Inhibition of nitrergic neurotransmission in the bovine retractor penis muscle by an oxidant stress: effects of superoxide dismutase mimetics. Br J Pharmacol. 1998;124(1):111–8.
47. Massaccesi L, Melzi d'Eril GV, Colpi GM, Tettamanti G, Goi G, Barassi A. Levels of human erythrocyte membrane-bound and cytosolic glycohydrolases are associated with oxidative stress in erectile dysfunction patients. Dis Markers. 2014;2014:485917.
48. Poulos TL. Soluble guanylate cyclase. Curr Opin Struct Biol. 2006;16(6):736–43.
49. Cary SP, Winger JA, Derbyshire ER, Marletta MA. Nitric oxide signaling: no longer simply on or off. Trends Biochem Sci. 2006;31(4):231–9.
50. Stasch JP, Schmidt PM, Nedvetsky PI, et al. Targeting the heme-oxidized nitric oxide receptor for selective vasodilatation of diseased blood vessels. J Clin Invest. 2006; 116(9):2552–61.
51. Fernandez-Sanchez A, Madrigal-Santillan E, Bautista M, et al. Inflammation, oxidative stress, and obesity. Int J Mol Sci. 2011;12(5):3117–32.
52. Koontz BF, Yan H, Kimura M, Vujaskovic Z, Donatucci C, Yin FF. Feasibility study of an intensity-modulated radiation model for the study of erectile dysfunction. J Sex Med. 2011; 8(2):411–8.
53. Kimura M, Yan H, Rabbani Z, et al. Radiation-induced erectile dysfunction using prostate-confined modern radiotherapy in a rat model. J Sex Med. 2011;8(8):2215–26.
54. Oberley-Deegan RE, Steffan JJ, Rove KO, et al. The antioxidant, MnTE-2-PyP, prevents side-effects incurred by prostate cancer irradiation. PLoS One. 2012;7(9), e44178.

55. van der Wielen GJ, Mulhall JP, Incrocci L. Erectile dysfunction after radiotherapy for prostate cancer and radiation dose to the penile structures: a critical review. Radiother Oncol. 2007;84(2):107–13.
56. Silva FH, Leiria LO, Alexandre EC, et al. Prolonged therapy with the soluble guanylyl cyclase activator BAY 60-2770 restores the erectile function in obese mice. J Sex Med. 2014;11(11):2661–70.
57. Li H, He WY, Lin F, Gou X. Panax notoginseng saponins improve erectile function through attenuation of oxidative stress, restoration of Akt activity and protection of endothelial and smooth muscle cells in diabetic rats with erectile dysfunction. Urol Int. 2014;93(1):92–9.
58. Batinic-Haberle I, Tovmasyan A, Roberts ER, Vujaskovic Z, Leong KW, Spasojevic I. SOD therapeutics: latest insights into their structure-activity relationships and impact on the cellular redox-based signaling pathways. Antioxid Redox Signal. 2014;20(15):2372–415.
59. Bassett M, Lu X, Yan H, et al. Exercise as a radioprotector of erectile function after prostate radiotherapy. In: Moderated poster discussion at AUA Annual Meeting San Diego, CA, 2013.
60. Kim WS, Park BS, Sung JH. The wound-healing and antioxidant effects of adipose-derived stem cells. Expert Opin Biol Ther. 2009;9(7):879–87.
61. Qiu X, Villalta J, Ferretti L, et al. Effects of intravenous injection of adipose-derived stem cells in a rat model of radiation therapy-induced erectile dysfunction. J Sex Med. 2012;9(7):1834–41.
62. Gridley DS, Makinde AY, Luo X, et al. Radiation and a metalloporphyrin radioprotectant in a mouse prostate tumor model. Anticancer Res. 2007;27(5A):3101–9.

Chapter 22
Theranostic Nanoconjugates of Tetrapyrrolic Macrocycles and Their Applications in Photodynamic Therapy

Jayeeta Bhaumik, Seema Kirar, and Joydev K. Laha

22.1 Introduction

Nanoparticles (NPs) that fall in the broad size range of 1–100 nm [1] and represent a suitable scaffold to incorporate both therapeutic and imaging moieties can be defined as nanotheranostic agents (NTNs) [2]. The skeleton on NPs can be composed of quantum dots, metal nanoparticles [3], liposomes, dendrimers [4], carbon nanotubes [5], silica, or iron oxide [6]. The cytotoxicity of the NPs are mainly dependent on their composition, whereas their in vivo performance is highly adjusted by parameters such as size and surface groups. Upon systemic injection, nanotherapeutics travel to different parts of the body via blood circulation. Since the pore size of endothelium is <5 nm, NPs with small (<5 nm) hydrodynamic diameter can easily extravasate through it. Importantly, the clearance pathway is highly dependent on the size of the NPs [2]. In nanotheranostics, nanometer sized particles are used in combination with diagnostic and therapeutic agents.

For the production of this book chapter, some excerpts were reproduced with permission from Bhaumik, J. *et al. Nano Research* (**2015**, *8*, 1373–1394), DOI 10.1007/s12274-014-0628-3, Copyright (2014) Springer.

J. Bhaumik (✉) • S. Kirar
Department of Pharmaceutical Technology (Biotechnology), National Institute of Pharmaceutical Education and Research (NIPER),
Sector-67, S.A.S. Nagar 160062, Punjab, India
e-mail: jbhaumi@gmail.com

J.K. Laha
Department of Pharmaceutical Technology (Process Chemistry), National Institute of Pharmaceutical Education and Research (NIPER),
Sector-67, S.A.S. Nagar 160062, Punjab, India

I. Batinić-Haberle et al. (eds.), *Redox-Active Therapeutics*,
Oxidative Stress in Applied Basic Research and Clinical Practice,
DOI 10.1007/978-3-319-30705-3_22

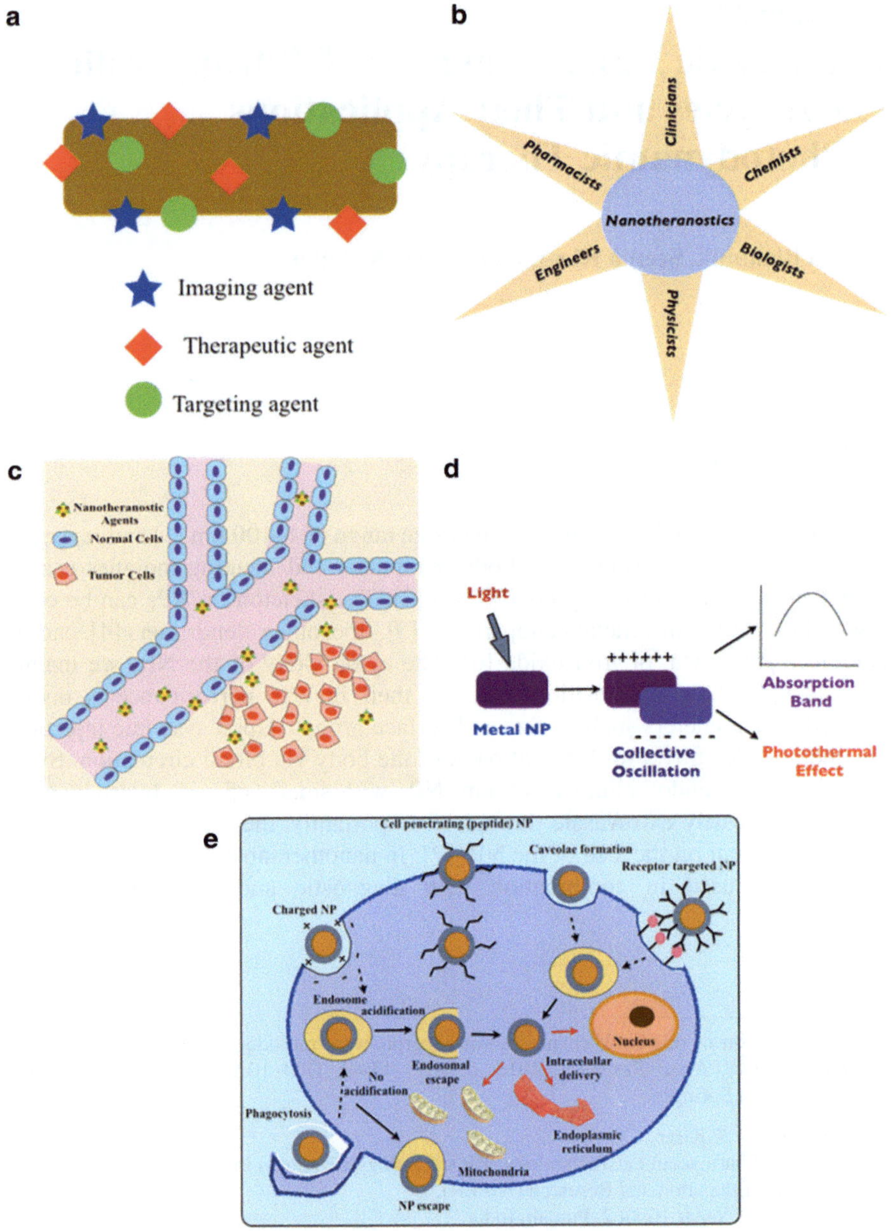

Fig. 22.1 (**a**) Diagram of a nanotheranostic agent as a combination of drugs along with imaging and targeting agents. (**b**) Nanotheranostic development and application can be successfully achieved by collaborative work of researchers from various backgrounds. (**c**) Delivery of nanotheranostic agents are achieved through the pores of blood vessels located near the tumor area via enhanced permeability and retention (EPR) effect. (**d**) Surface plasmon resonance (SPR) and photothermal effect in metal nanoparticles. (**e**) Various mechanisms of cellular internalization of nanoparticles (NP). Adapted with permission from Kievit, F. M. *et al. Adv. Mater.* 23 (2011), H217-H247, Copyright (2011) John Wiley and Sons

Incorporation of imaging modalities to nanotherapeutics affords NTNs, which have strong potential for wholesome management or cure for a complicated disease such as cancer [7] (Fig. 22.1a).

Due to large size and large surface area nanobodies have high capabilities to carry several therapeutically relevant molecules on their surface. These physical properties of the nanocarriers assist them to be composed of medicines, diagnostic agents, targeting moieties, and so on. Additionally, biocompatibility can be achieved with these nanoscaffolds by coating them with biomolecules thereby enhancing their in vivo circulation time [6]. The main advantage of this novel tool is that, both therapeutic and diagnostic moieties will be presented within the same nanoscaffold and made concomitantly available to the diseased site. This contrasts with conventional agents for diagnosis and therapy, which by being administered independently into the body, need further developments in order to target and/or treat a particular disease site. Taking the advantage of relatively large surface area of nanoparticles, NTNs can act as multimodal agents with functionalities ranging from targeting, diagnosis, and treatment.

The term "theranostic" was first introduced by Funkhauser [2]. Theranostic nanoagents are a package which can simultaneously "see" and "treat" a diseased state [8, 9]. Additionally, due to the added advantage to attach targeting moieties, high level of selectivity and biodistribution is observed. The main objective of this field is to acquire capacity to simultaneously image and monitor the diseased state, to follow the extent of drug delivery, and to determine the efficacy of the treatment. By the aid of NTN, personalized medicine can be adopted instead of "one size fits all" strategy [2].

Despite the young nature of nanotheranostics field, its development is progressing at a fast pace. Scientists from various disciplines (chemistry, physics, biology, engineering, and many other fields) from all over the world are working to find appropriate nanotherapeutics for the treatment of life-threatening diseases (Fig. 22.1b). In parallel, many researchers are developing techniques on how to optimize imaging of diseased states [10].

22.2 Nanotheranostics in Biological Systems

22.2.1 Enhanced Permeability and Retention Effect

Due to the dilation caused by highly metabolizing cancer cells, tumor blood vessels tend to be very leaky. Thus, nanomaterials can extravasate from the blood stream and reach tumor tissues. On the other hand, due to poor lymphatic drainage, the NPs can be retained in the tumor tissues. This entire phenomenon of NP drainage and retention is known as enhanced permeability and retention (EPR) effect (Fig. 22.1c) [11, 12]. Taking advantage of this phenomenon, drugs and imaging agents conjugated to nanoparticles can be successfully delivered to tumor site, thereby making diagnosis and treatment highly feasible.

22.2.2 Surface Plasmon Resonance

Metal NPs, including gold and silver ones, show optical activity at nanoscale associated with surface plasmon resonance (SPR) [13]. Noble metals in nanosize range can therefore be used for optical imaging. Upon the modification of size and shape, their optical properties can be adjusted. Upon shining light on noble metals, collective oscillation is induced to the electrons present at the metal surface. This results in the generation of a sharp and intense absorption band. Nonradiative decay of surface electron oscillation converts the absorbed light energy into heat, thus causing photothermal effect (Fig. 22.1d). This unique property of metal nanoparticles assist them in playing a key role in nanotheranostics upon combination to diagnostic agents.

Once delivered to the body, NTNs may be attacked and destroyed by numerous components of the biological system. This imparts major challenges in delivering nanomaterials to diseased tissues. Thus, good understanding of biological system is crucial while designing nanotherapeutics. Blood, which is composed of various sugars, salts, amino acids, and other chemical molecules, can be responsible for the instability of nanoagents. Immune cells, such as monocytes, present in human blood can also force removal of nanoparticles from the biological system. Hence, the use of biocompatible coatings (such as polymers) on nanoparticle surface can improve their availability to their final target (e.g., tumor tissues) [14]. Nanoparticles with high surface-area-to-volume ratio possess the capability to incorporate large number of therapeutic molecules as well as imaging agents. Additionally, the large surface area can be derivatized with targeting agents (e.g., antibodies) and coated with stabilizing agents (e.g., polyethylene glycol) for increasing in vivo circulation time [10].

There are various mechanisms for the uptake of theranostic nanoagents by cells (Fig. 22.1e). Nanoagents may contain positively charged molecules on the surface, which can interact with the negative charges of the molecules present on the cell surface. Alternatively, cell-penetrating peptides or antibodies used for targeting via conjugation on the surface of nanoparticles can also assist them in entering into cells. Upon cellular internalization, endosomes are formed, which then deliver nanoagents to either nucleus or mitochondria after acidification.

There are various methodologies for developing biocompatible theranostic nanoagents. Both imaging and therapeutic moieties can be incorporated into nanoparticles during NP synthesis through encapsulation; alternatively, these agents can be conjugated on nanoparticle surface after NP synthesis via bioconjugation (Fig. 22.2a). There are both advantages and disadvantages for either techniques and each setup has different mechanism of action [15].

Theranostic photosensitizers or nano-phototheranostics can be majorly used in the treatment of cancer, as well as in the treatment of microbial infection and cardiovascular disease (CVD). The use of photosensitizers (PS) along with nanoparticles (NPs) has improved the efficacy of these agents to a great extent. Many reports have been published recently on the use of PS in photodynamic therapy (PDT) to treat various diseases, including cancer and CVD. Overall, nano-phototheranostic agents are robust platforms for finding cure for deadly diseases [16].

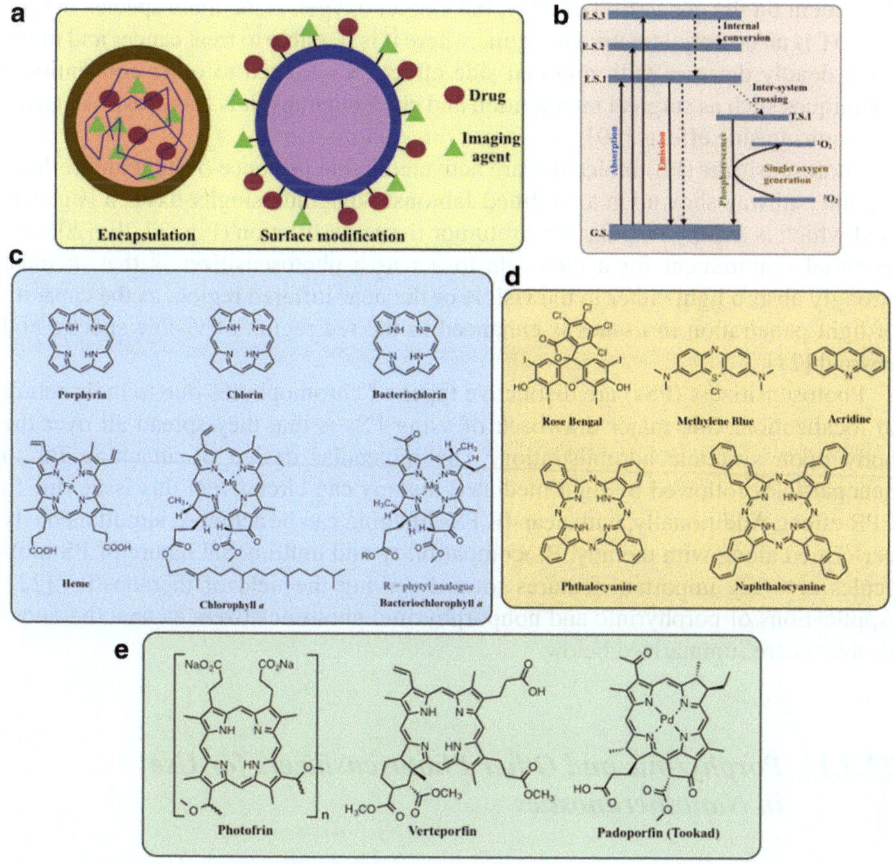

Fig. 22.2 (**a**) Two different strategies to prepare nanotheranostics: encapsulation (left) and surface modification (right). (**b**) Modified Jablonski Diagram, which shows the mechanism for photogeneration of singlet oxygen ($^1O_2$). (**c**) Types of tetrapyrrolic macrocycles (photosensitizers) (*top panel*) and their presence in naturally occurring molecules (*bottom panel*). (**d**) More examples of macrocycles with potential to act as photosensitizers in PDT: nonpyrrolic PS (*top panel*) and pyrrolic macrocycles (*bottom panel*). (**e**) Commercially available tetrapyrrolic photosensitizers [photofrin® (porfimer sodium), verteporfin (Visudyne®), and padoporfin (TOOKAD)] (Fig. 22.2c–e: adapted with permission from Bhaumik, J. *et al. Nano Research* (2014), DOI 10.1007/s12274-014-0628-3, Copyright (2014) Springer)

22.3 Photodynamic Therapy

In PDT, upon shining light, a chromophore captures energy, which upon release is taken up by the tissue oxygen in order to be converted into the singlet oxygen ($^1O_2$) species. Singlet oxygen thus generated can be a reactive oxygen species (ROS) to cause destruction of cancer cells [17, 18]. There are two mechanisms by which PDT agents may promote photo-killing: type I and type II. Both mechanisms are dependent on the generation of ROS, but singlet oxygen is the main species in type II. PDT is an essential treatment regime where it is possible to treat cancer and many other deadly diseases with minimal side effects. Compared to other conventional techniques such as surgical intervention and chemotherapy, it is less drastic in terms of acquiring side effects [19].

Photosensitizer (PS) molecules are activated in the presence of light and following the pathway shown via a modified Jablonski diagram, singlet oxygen is generated which is mainly responsible for tumor tissue destruction (Fig. 22.2b) [20]. An essential requirement for a molecule to act as a photosensitizer is that, it must strongly absorb light either in the visible or the near-infrared region, as the capacity of light penetration in tissues is enhanced in the red region of visible spectra and beyond [21].

Photosensitizers (PSs) are distinctive types of chromophores due to their nature of localization. One major drawback of using PSs is that they spread all over the body upon systemic administration. The molecular design of attaching PS to nanoparticles followed by light mediated therapy can circumvent this issue due to EPR effect. Additionally, with near-IR PSs imaging can be achieved simultaneously performed along with therapy. Biocompatibility and multimodal nature of PS molecules are very important features toward serving the field of theranostics [22]. Applications of porphyrinic and nonporphyrinic photosensitizers as nanotheranostic agents are summarized below.

22.3.1 *Porphyrinic and Other Photosensitizers for Use in Nanotheranostics*

Porphyrinic macrocycles namely, porphyrins, chlorins, and bacteriochlorins, are tetrapyrrolic aromatic systems (Fig. 22.2c, top panel). Due to the presence of a large aromatic core, they are capable of absorbing visible and near-IR light. Naturally occurring porphyrinic molecules include heme, chlorophylls, and bacteriochlorophylls (Fig. 22.2c, bottom panel). Examples of some nonporphyrinic macrocycles, which have PDT potential, include rose bengal [23], methylene blue, and acridine (Fig. 22.2d, top panel). Phthalocyanines and naphthalocyanines are examples of extended tetrapyrrolic macrocycles, which absorb at the near-IR region of the electromagnetic spectra (Fig. 22.2d, bottom panel) [24]. The members of porphyrin family and other nonporphyrinic photosensitizers are potential candidates for PDT due to their strong light absorption, high triplet quantum yield, and long excited triplet state lifetime [25].

Porphyrinic macrocycles encompass significant classes of chromophores for their characteristic absorption properties on the visible region. These tetrapyrrolic compounds take an essential part in delivering PDT to treat cancer and several microbial infections. There are several commercially available tetrapyrrolic macrocycles (photofrin, verteporfin, padoporfin, etc.), which are being used in clinical settings (Fig. 22.2e) [21]. However, most of these drugs (e.g., photofrin) possess different drawbacks including the presence of a mixture of components, which are difficult to separate. Hydrophobicity is another drawback, which makes pharmaceutical formulations a challenge. There is an urgent need to fulfill these drawbacks for using them in biological settings. In order to circumvent this problem, these macrocycles are nowadays being incorporated into nanoparticles. The bioavailability of those hydrophobic photosensitizers can be significantly improved via conjugation with nanomaterials. Several tetrapyrrolic macrocycles (either from natural or synthetic sources) are therefore frequently loaded on biocompatible nanoparticles for both in vitro and in vivo experiments [25].

22.3.2 How Can Photosensitizers and Nanotheranostics Be Combined to Treat Life-Threatening Diseases?

Photosensitizers, which are generally hydrophobic, with the aid of nanotechnology, can be successfully utilized in biological applications [26]. PS can be associated with biocompatible materials via the formation of a covalent bond on nanoparticle surface or through encapsulation into polymeric nanoparticles. Recent literature reveals that nanotheranostics have already become an important class of delivery system in biomedical research [27]. When both therapeutic and diagnostic tools work from the same scaffold, the outcome of the treatment of any complicated disease can be superior. First-generation hydrophobic photosensitizers (e.g., photofrin and eosin) lacked water solubility and, hence, their in vitro and in vivo administration was very challenging [25]. In the second-generation PSs (e.g., verteporfin and photosensitivity), hydrophilic moieties were attached to them [25]. Though hydrophilicity was achieved, targetability still remained an issue resulting poor in vivo delivery. In third-generation PS (e.g., glycosylated porphyrin), ideally a PS is attached to a targeting moiety for better penetration into tumor cell membrane [25]. Introduction of multimodal photosensitizer–nanoparticle (PS–NP) system would further enhance PS delivery taking the advantage of EPR effect [28, 29]. Due to the presence of both therapeutic efficacy and imaging properties, near-IR absorbing photosensitizers are ideal candidates to form nano-phototheranostics in combination with nanomaterials [30].

22.4 Role of Tetrapyrrolic Nanoconjugates in Treating Cancer

Cancer is the third principal cause of mortality worldwide after heart and infectious diseases [31]. Major limitations to cancer therapeutics include (a) poor bypass through biological membranes, (b) low biodistribution of drugs, (c) inability to treat metastasis, (d) drug resistance, and (e) lack of imaging. The field of cancer nanotheranostics is a combination of therapy and imaging of tumor via the aid of nanomaterials. Nano-phototheranostics can be taken into consideration for the treatment of metastatic and drug-resistant cancers as it can combine targeting, imaging, and therapeutic entities [14]. Typical reason for the failure of cancer treatment is due to the fact that the majority of patients are diagnosed at tumor metastasis stage, thus making the treatment almost impossible. With the aid of nanotherapeutics even cancer can be treated efficiently as compared to the use of conventional therapeutics [28].

Recently, photosensitizers loaded on nanoscaffolds are widely being used as theranostic nanoagents for their capability to serve as both therapeutic and imaging module through a single moiety. Near-IR absorbing photosensitizers can take part in performing both detection and treatment of various diseases. An advantage of using nano-photosensitizers as theranostic agents is that they tend to possess lower toxicity and provide sophisticated treatment module thus minimizing lengthy recovery time for patients [32].

Recently published literature reports show the evidence of substantial use of nano-phototheranostics for the diagnosis and treatment of cancer and other diseases. Remarkable systems in this topic are summarized in Table 22.1. For example, lipoprotein nanoparticles in combination with bacteriopheophorbide (used as both imaging and therapeutic agent) formed nano-phototheranostics and used in the treatment of oral cancer (Table 22.1, entry 1) [33]. Chlorin e6 is a photosensitizer, which can act as both diagnostic and therapeutic agent due to its strong absorption in the red region of visible spectra along with its singlet oxygen generation (SOG) capacity. Chlorin e6 was conjugated to human serum albumin NPs by Jeong et al. to construct nano-photosensitizers and administered in vivo to image and treat colon carcinoma (Table 22.1, entry 6) [34]. In another instance, chlorin e6 was incorporated into hyaluronic acid NPs via conjugation technique and in vivo diagnosis and therapy was carried out using a colon cancer model (Table 22.1, entry 7) [35]. Huang et al. reported the use of chlorin e6 in combination with iron oxide NPs as nano-photosensitizers for dual mode near-IR fluorescence imaging and MRI to treat gastric cancer via PDT (Table 22.1, entry 8) [36]. Upconversion nanoparticles (UCNPs) possess the capability of converting near-infrared radiations (lower energy) into visible radiations (higher energy) through a nonlinear optical process; chlorin e6 was used in combination with UCNPs to image and treat breast and cervical cancer [37, 38]. In another photosensitizer-nanoparticle combination, bacteriopheophorbide–lipoprotein conjugate was used for the treatment of oral cancer (Table 22.1, entry 1) [33]. A nanoconjugate consisting of pyropheophorbide-a (as both imaging and diagnostic agent) and polyacrylamide (as a nanoscaffold) was successfully used in the treatment of colon cancer (Table 22.1, entry 4) [39]. Pyropheophorbide-a was also utilized in combination with gold nanorods (GNR) to treat fibrosarcoma (Table 22.1, entry 5) [40]. Lung cancer could be treated by using a cluster of protoporphyrin IX and PEG-PLA copoymer (Table 22.1, entry 11) [41]. Protoporphyrin IX was also applied in the treatment of squamous cell carcinoma along with glycol chitosan nanoparticles. Methylene blue was effective in the treatment of breast cancer, glioma, and melanoma when combined with polyacrylamide nanogel (Table 22.1, entry 17) [43]. 5,10,15,20-tetrakis(*N*-methylpyridinium-4-yl)porphyrin ($H_2TM\text{-}4\text{-}PyP^{4+}$) was incorporated into different nanoscaffolds to form nano-phototheranostics and

Table 22.1 List of nano-phototheranostic agents for the treatment of various diseases[a]

Entry	Photosensitizer	Imaging agent	Nanomaterial	Disease	RReference
1	Bacteriopheophorbide	Bacteriopheophorbide	Lipoprotein	Oral cancer	[33]
2	Al(III) phthalocyanine chloride tetrasulfonic acid	Al(III) phthalocyanine chloride tetrasulfonic acid	Gold nanorods	Squamous cell carcinoma	[46]
3	Pyropheophorbide-a	Cyanine dye	Polyacrylamide NPs	Breast cancer	[47]
4	Pyropheophorbide-a	Pyropheophorbide-a	Polyacrylamide NPs	Colon cancer	[39]
5	Pyropheophorbide-a	Pyropheophorbide-a	Gold nanorods	Fibrosarcoma	[40]
6	Chlorin e6	Chlorin e6	Human serum albumin NPs	Colon cancer	[34]
7	Chlorin e6	Chlorin e6	Hyaluronic acid NPs	Colon cancer	[35]
8	Chlorin e6	Chlorin e6	Iron oxide NP	Gastric cancer	[36]
9	Chlorin e6	FITC[b]	Glycol chitosan NPs	Squamous cell carcinoma	[48]
10	UCNP-Chlorin e6	UCNP	UCNP[c]	Breast and cervical cancer	[38]
11	Protoporphyrin IX	Protoporphyrin IX	PEG-PLA co-polymer[d]	Lung cancer	[41]
12	Protoporphyrin IX	FITC[b]	Glycol chitosan NPs	Squamous cell carcinoma	[42]
13	Bacteriochlorin e6 bisoleate	Bacteriochlorin e6 bisoleate	LDL[e]	Hepatoblastoma	[49]
14	Pd-meso-tetra(4-carboxyphenyl) porphyrin	Pd-meso-tetra(4-carboxyphenyl) porphyrin	Mesoporous silica NPs	Breast cancer	[50]
15	Pheophorbide-a	Pheophorbide-a	Bovine serum albumin NPs	Breast cancer	[51]
16	Porphyrin–brucine conjugate	Porphyrin–brucine conjugate	Gold NPs	Breast cancer	[52]
17	Methylene blue	Methylene blue	Polyacrylamide nanogel	Breast cancer, glioma, melanoma	[43]
18	FITC[b]	FITC[b]	Liposomes	Ovarian cancer	[53]

(continued)

Table 22.1 (continued)

Entry	Photosensitizer	Imaging agent	Nanomaterial	Disease	RReference
19	TM-4-PyP^{4+f}	Ru(bpy)$_3$^{g}	Iron oxide-silica core shell NPs	Breast cancer	[46]
20	Phthalocyanines	Phthalocyanines	Gold NPs	Colon cancer	[12]

[a]Adapted with permission from Bhaumik, J. *et al. Nano Research* (2014), DOI 10.1007/s12274-014-0628-3, Copyright (2014) Springer
[b]Fluorescein isothiocyanate
[c]Upconversion nanoparticles
[d]Poly(ethylene glycol)-b-poly(D,L-lactide)
[e]Low-density lipoprotein
[f]5,10,15,20-tetrakis(*N*-methylpyridinium-4-yl)porphyrin
[g]Tris(2,20-bipyridyl)-dichlororuthenium(II) hexahydrate

applied in the diagnosis and treatment of various cancers [44]. Use of encapsulated photosensitizers, such as methylene blue [45] and tosylate salt of H_2TM-4-PyP^{4+} [26], was reported by few research groups who developed nano-phototheranostics for cancer diagnosis and treatment.

A photosensitizer-gold nanoconstruct was developed by Jang et al. using Al(III) phthalocyanine chloride tetrasulfonic acid (AlPcS4) for near-IR imaging and treatment of cancer [46]. Upon complex formation with GNR, both fluorescence and SOG from AlPcS4 were quenched. When AlPcS4–-GNR was compared with free AlPcS4 in terms of intracellular uptake, the nanomaterial seemed to be four times more efficient than the free PS. After systemic administration of AlPcS4–GNR complex the tumor site could be easily located by near-IR fluorescence imaging within 1 h of injection. Interestingly, the tumor-to-background ratio reached up to 3.7 after 24 h of probe administration. Taking the advantage of the presence of both photosensitizer (AlPcS4) and metal nanoparticles, (gold) a combination of PDT and photothermal therapy (PTT) could be performed simultaneously: when PDT was applied alone tumor growth regression was 79 %, whereas when the combination of PDT and PTT was applied the regression reached up to 95 %. This research work is a good example of combining photosensitizers to nanoparticles for dual imaging and therapy of various cancers.

Recently, a multifunctional nano-phototheranostic carrier was developed by Wang et al. [47]. The probe was based on biodegradable polyacrylamides, which included an imaging agent (a cyanine dye), a photosensitizer drug (HPPH) and a targeting agent (F3-Cys) [47]. This novel multimodal theranostic system was spherical in shape with ~44 nm in size and consisted of a biocompatible PEG surface coating. The nanoplatform was based on "see" and "treat" strategy, which combined PDT with fluorescence imaging to efficiently destroy cancer. The conjugation of targeting peptide to the NPs significantly improved accumulation in cancer cells. Additionally, presence of imaging probes on the nanocarrier assisted in the detection of tumor cells. Most importantly, PSs attached to the polymeric nanomaterial generated singlet oxygen upon shining light thereby destructing cancer cells without showing any dark toxicity. This unique combination of biodegradable, targeting, imaging, and therapeutic nanotheranostic platform is certainly an immense contribution to cancer research.

Jang et al. described the development of gold nanorods–photosensitizer (MMP2P-GNR) conjugates for cancer treatment [40]. In this nano-platform, a protease cleavable peptide linker was used to conjugate photosensitizers on the surface of GNR. They have demonstrated that the fluorescence and phototoxicity properties of the nano-photosensitizers is kept suppressed until protease cleavage takes place. Thus, the photosensitizer-decorated GNR remained inactive until delivered into the tumor cells, which are rich in matrix metalloprotease-2 (MMP2). The peptide linkages in MMP2P–GNR were cleaved by the protease and the photosensitizer gained both fluorescence and phototoxicity, thereby damaging cancer cell lines. The in vitro studies with phototheranostic nanoagent MMP2P–GNR showed that the probe was highly effective in SOG when protease rich HT1080 cell lines were used. Interestingly, with BT20 cell lines, which lack MMP2 (protease), the efficacy of the probe was rather low. Additionally, while comparing the nanoprobe MMP2P-GNR with a free photosensitizer (pyropheophorbide-a), the nanoprobe showed much higher efficacy in damaging the cells lines. This study indicates the importance and efficacy of conjugating photosensitizers to nanomaterials. This innovative activatable nano-phototheranostic agent can be applied efficiently for simultaneous detection and treatment of various cancers.

22.5 Concluding Remarks and Future Considerations

Nano-phototheranostics field is very young compared to the area of nanotherapeutics. However, this multifunctional imaging plus drug delivery nanoplatform has the full potential to eradicate many life-threatening diseases due to its targetability along with minimal side effects. Given the potential of these entities to incorporate versatile therapeutic and diagnostic moieties, they can definitely be beneficial for the treatment of various diseases. Upon incorporation of targeting moieties such as antibodies, small molecules, or peptides onto nanosensitizers, their output will certainly be enhanced.

Phototheranostic nanoagents are attracting much attention for imaging and therapy of certain cancers and other diseases. Imaging and therapy based on phototheranostics is minimally invasive, rapid, inexpensive, and safer than the conventional alternatives [54–67]. Novel combinations of photosensitizers, nanoparticles, and targeting agents are improving the theranostic capabilities, reducing side effects, and broadening the applications of this emerging technology. Further research is needed in broadening the applicability of nano-phototheranostics. This emerging imaging and drug delivery platform has the potential to be more broadly used than at present. Highly specific targeting agents including antibodies can be combined with more effective photosensitizers for enhanced usefulness. Developing new treatments based on phototheranostic nanotechnology requires a collaborative effort of a range of specialists including physicians, chemists, biologists, physicists, and engineers.

Acknowledgment J.B. would like to acknowledge a generous funding support under Women Scientists Scheme A from the Department of Science and Technology (DST WOS-A), Govt. of India. S. K. thanks Department of Biotechnology (DBT), Govt. of India for providing graduate fellowship. Authors thank Prof. U. C. Banerjee, NIPER, Mohali for his support during the preparation of this book chapter.

References

1. Mittal AK, Chisti Y, Banerjee UC. Synthesis of metallic nanoparticles using plant extracts. Biotechnol Adv. 2013;31:346–56.
2. Kelkar SS, Reineke TM. Theranostics: combining imaging and therapy. Bioconjug Chem. 2011;22:1879–903.

3. Kuo WS, Chang YT, Cho KC, Chiu KC, Lien CH, Yeh CS, Chen SJ. Gold nanomaterials conjugated with indocyanine green for dual-modality photodynamic and photothermal therapy. Biomaterials. 2012;33:3270–8.
4. Nishiyama N, Morimoto Y, Jang W-D, Kataoka K. Design and development of dendrimer photosensitizer-incorporated polymeric micelles for enhanced photodynamic therapy. Adv Drug Deliv Rev. 2009;61:327–38.
5. Paszko E, Ehrhardt C, Senge MO, Kelleher DP, Reynolds JV. Nanodrug applications in photodynamic therapy. Photodiagnosis Photodyn Ther. 2011;8:14–29.
6. Lammers T, Aime S, Hennink WE, Storm G, Kiessling F. Theranostic nanomedicine. Acc Chem Res. 2011;44:1029–38.
7. Melancon MP, Stafford RJ, Li C. Challenges to effective cancer nanotheranostics. J Control Release. 2012;164:177–82.
8. Mura S, Couvreur P. Nanotheranostics for personalized medicine. Adv Drug Deliv Rev. 2012;64:1394–416.
9. Caldorera-Moore ME, Liechty WB, Peppas NA. Responsive theranostic systems: integration of diagnostic imaging agents and responsive controlled release drug delivery carriers. Acc Chem Res. 2011;44:1061–70.
10. Wang LS, Chuang MC, Ho JA. Nanotheranostics—a review of recent publications. Int J Nanomed. 2012;7:4679.
11. McCarthy JR, Bhaumik J, Karver MR, Erdem SS, Weissleder R. Targeted nanoagents for the detection of cancers. Mol Oncol. 2010;4:511–28.
12. Obaid G, Chambrier I, Cook MJ, Russell D. Targeting the oncofetal Thomsen–Friedenreich disaccharide using Jacalin-PEG phthalocyanine gold nanoparticles for photodynamic cancer therapy. Angew Chem Int Ed. 2012;51:6158–62.
13. Tong R, Kohane DS. Shedding light on nanomedicine. Wiley Interdiscip Rev Nanomed Nanobiotechnol. 2012;4:638–62.
14. Kievit FM, Zhang M. Cancer nanotheranostics: improving imaging and therapy by targeted delivery across biological barriers. Adv Mater. 2011;23:H217–47.
15. Geng J, Li K, Pu K-Y, Ding D, Liu B. Conjugated polymer and gold nanoparticle co-loaded PLGA nanocomposites with eccentric internal nanostructure for dual-modal targeted cellular imaging. Small. 2012;8:2421–9.
16. Khdair A, Handa H, Mao G, Panyam J. Nanoparticle-mediated combination chemotherapy and photodynamic therapy overcomes tumor drug resistance in vitro. Eur J Pharm Biopharm. 2009;71:214–22.
17. Mroz P, Bhaumik J, Dogutan DK, Aly Z, Kamal Z, Hamblin MR. Imidazole metalloporphyrins as photosensitizers for photodynamic therapy: role of molecular charge, central metal and hydroxyl radical production. Cancer Lett. 2009;282:63–76.
18. Nann T. Nanoparticles in photodynamic therapy. Nano Biomed Eng. 2011;3:137–43.
19. Bhaumik J, Mittal AK, Banerjee A, Chisti Y, Banerjee UC. Applications of phototheranostic nanoagents in photodynamic therapy. Nano Res. 2015;8:1373–94. doi:10.1007/s12274-014-0628-3.
20. Shibu ES, Hamada M, Murase N, Biju V. Nanomaterials formulations for photothermal and photodynamic therapy of cancer. J Photochem Photobiol C. 2012;15:53–72.
21. Ethirajan M, Chen Y, Joshi P, Pandey RK. The role of porphyrin chemistry in tumor imaging and photodynamic therapy. Chem Soc Rev. 2011;40:340–62.
22. Zhang Y, Lovell JF. Porphyrins as theranostic agents from prehistoric to modern times. Theranostics. 2012;2:905–15.
23. Uppal A, Jain B, Gupta PK, Das K. Photodynamic action of Rose Bengal silica nanoparticle complex on breast and oral cancer cell lines. Photochem Photobiol. 2011;87:1146–51.
24. Giribabu L, Sudhakar K, Velkannan KV. Phthalocyanines: potential alternative sensitizers to Ru(II) polypyridyl complexes for dye-sensitized solar cells. Curr Sci. 2012;102:991–1000.
25. Josefsen LB, Boyle RW. Unique diagnostic and therapeutic roles of porphyrins and phthalocyanines in photodynamic therapy, imaging and theranostics. Theranostics. 2012;2:916.

26. Abdelghany SM, Schmid D, Deacon J, Jaworski J, Fay F, McLaughlin KM, Gormley J, et al. Enhanced anti-tumor activity of the photosensitizer Meso-tetra (N-methyl-4-pyridyl) porphine tetra tosylate (TMP) through encapsulation in antibody targeted chitosan/alginate nanoparticles. Biomacromolecules. 2013;14:302–10.
27. Shi J, Liu TWB, Chen J, Green D, Jaffray D, Wilson BC, Wang F, et al. Transforming a targeted porphyrin theranostic agent into a PET imaging probe for cancer. Theranostics. 2011;1:363.
28. Lim C-K, Heo J, Shin S, Jeong K, Seo YH, Jang W-D, Park CR, et al. Nanophotosensitizers toward advanced photodynamic therapy of cancer. Cancer Lett. 2012;334:176–87.
29. Rai P, Mallidi S, Zheng X, Rahmanzadeh R, Mir Y, Elrington S, Khurshid A, et al. Development and applications of photo-triggered theranostic agents. Adv Drug Deliv Rev. 2010;62:1094–124.
30. Liu TWB, Chen J, Burgess L, Cao W, Shi J, Wilson BC, Zheng G. Multimodal bacteriochlorophyll theranostic agent. Theranostics. 2011;1:354.
31. Narayan KV, Ali MK, Koplan JP. Global noncommunicable diseases-where worlds meet. N Engl J Med. 2010;363(2011):1196–8.
32. Qian J, Wang D, Cai F, Zhan Q, Wang Y, He S. Photosensitizer encapsulated organically modified silica nanoparticles for direct two-photon photodynamic therapy and In vivo functional imaging. Biomaterials. 2012;33:4851–60.
33. Cao W, Ng KK, Corbin I, Zhang Z, Ding L, Chen J, Zheng G. Synthesis and evaluation of a stable bacteriochlorophyll-analog and its incorporation into high-density lipoprotein nanoparticles for tumor imaging. Bioconjug Chem. 2009;20:2023–31.
34. Jeong H, Huh M, Lee SJ, Koo H, Kwon IC, Jeong SY, Kim K. Photosensitizer-conjugated human serum albumin nanoparticles for effective photodynamic therapy. Theranostics. 2011;1:230.
35. Yoon HY, Koo H, Choi KY, Lee SJ, Kim K, Kwon IC, Leary JF, et al. Tumor-targeting hyaluronic acid nanoparticles for photodynamic imaging and therapy. Biomaterials. 2012;33:3980–9.
36. Huang P, Li Z, Lin J, Yang D, Gao G, Xu C, Bao L, et al. Photosensitizer-conjugated magnetic nanoparticles for *in vivo* simultaneous magnetofluorescent imaging and targeting therapy. Biomaterials. 2011;32:3447–58.
37. Shan J, Budijono SJ, Hu G, Yao N, Kang Y, Ju Y, Prud'homme RK. Pegylated composite nanoparticles containing upconverting phosphors and meso-tetraphenyl porphine (TPP) for Photodynamic Therapy. Adv Funct Mater. 2011;21:2488–95.
38. Wang C, Tao H, Cheng L, Liu Z. Near-infrared light induced in vivo photodynamic therapy of cancer based on upconversion nanoparticles. Biomaterials. 2011;32:6145–54.
39. Wang S, Fan W, Kim G, Hah HJ, Lee Y-E, Kopelman R, Ethirajan M, et al. Novel methods to incorporate photosensitizers into nanocarriers for cancer treatment by photodynamic therapy. Lasers Surg Med. 2011;43:686–95.
40. Jang B, Choi Y. Photosensitizer-conjugated gold nanorods for enzyme-activatable fluorescence imaging and photodynamic therapy. Theranostics. 2012;2:190.
41. Ding H, Sumer BD, Kessinger CW, Dong Y, Huang G, Boothman DA, Gao J. Nanoscopic micelle delivery improves the photophysical properties and efficacy of photodynamic therapy of protoporphyrin IX. J Control Release. 2011;151:271–7.
42. Lee SJ, Koo H, Lee HD-E, Min S, Lee S, Chen X, Choi Y, et al. Tumor-homing photosensitizer-conjugated glycol chitosan nanoparticles for synchronous photodynamic imaging and therapy based on cellular on/off system. Biomaterials. 2011;32:4021–9.
43. Hah HJ, Kim G, Lee Y-EK, Orringer DA, Sagher O, Philbert MA, Kopelman R. Methylene blue-conjugated hydrogel nanoparticles and tumor-cell targeted photodynamic therapy. Macromol Biosci. 2011;11:90–9.
44. Yin M, Li Z, Liu Z, Ren J, Yang X, Qu X. Photosensitizer-incorporated G-quadruplex DNA-functionalized magnetofluorescent nanoparticles for targeted magnetic resonance/fluorescence multimodal imaging and subsequent photodynamic therapy of cancer. Chem Commun (Camb). 2012;48:6556–8.

45. He X, Wu X, Wang K, Shi B, Hai L. Methylene blue-encapsulated phosphonate-terminated silica nanoparticles for simultaneous in vivo imaging and photodynamic therapy. Biomaterials. 2009;30:5601–9.
46. Jang B, Park J, Tung C, Kim I, Choi Y. Gold nanorod–photosensitizer complex for near-infrared fluorescence imaging and photodynamic/photothermal therapy in vivo. ACS Nano. 2011;5:1086–94.
47. Wang S, Kim G, Lee Y-EK, Hah HJ, Ethirajan M, Pandey RK, Kopelman R. Multifunctional biodegradable polyacrylamide nanocarriers for cancer theranostics a "see and treat" strategy. ACS Nano. 2012;6:6843–51.
48. Lee SJ, Koo H, Jeong H, Huh MS, Choi Y, Jeong SY, Byun Y, et al. Comparative study of photosensitizer loaded and conjugated glycol chitosan nanoparticles for cancer therapy. J Control Release. 2011;152:21–9.
49. Marotta DE, Cao W, Wileyto EP, Li H, Corbin I, Rickter E, Glickson JD, et al. Evaluation of bacteriochlorophyll-reconstituted low-density lipoprotein nanoparticles for photodynamic therapy efficacy in vivo. Nanomedicine (Lond). 2011;6:475–87.
50. Cheng SH, Lee C-H, Yang C-S, Tseng F-G, Mou C-Y, Lo L-W. Mesoporous silica nanoparticles functionalized with an oxygen-sensing probe for cell photodynamic therapy: potential cancer theranostics. J Mater Chem. 2009;19:1252–7.
51. Ozgur A, Lambrecht FY, Ocakoglu K, Gunduz C, Yucebas M. Synthesis and biological evaluation of radiolabeled photosensitizer linked bovine serum albumin nanoparticles as a tumor imaging agent. Int J Pharm. 2012;422:472–8.
52. Záruba K, Králová J, Rezanka P, Poucková P, Veverková L, Král V. Modified porphyrin–brucine conjugated to gold nanoparticles and their application in photodynamic therapy. Org Biomol Chem. 2010;8:3202–6.
53. Rahmanzadeh R, Rai P, Celli JP, Rizvi I, Baron-Lühr B, Gerdes J, Hasan T. Ki-67 as a molecular target for therapy in an in vitro three-dimensional model for ovarian cancer. Cancer Res. 2010;70:9234–42.
54. Schwiertz J, Wiehe A, Gräfe S, Gitter B, Epple M. Calcium phosphate nanoparticles as efficient carriers for photodynamic therapy against cells and bacteria. Biomaterials. 2009;70:3324–31.
55. Huang P, Pandoli O, Wang X, Wang Z, Li Z, Zhang C, et al. Chiral guanosine 5′-monophosphate-capped gold nanoflowers: Controllable synthesis, characterization, surface-enhanced Raman scattering activity, cellular imaging and photothermal therapy. Nano Res. 2012;5:630–9.
56. Johnson NJJ, van Veggel FCJM. Sodium lanthanide fluoride core-shell nanocrystals: A general perspective on epitaxial shell growth. Nano Res. 2013;6:547–61.
57. Wu H, Wang P, He H, Jin Y. Controlled synthesis of porous Ag/Au bimetallic hollow nanoshells with tunable plasmonic and catalytic properties. Nano Res. 2012;5:135–44.
58. Wu L, Cai X, Nelson K, Xing W, Xia J, Zhang R, et al. A green synthesis of carbon nanoparticles from honey and their use in real-time photoacoustic imaging. Nano Res. 2013;6:312–25.
59. Yu J, Hao R, Sheng F, Xu L, Lii G, Hou Y. Hollow manganese phosphate nanoparticles as smart multifunctional probes for cancer cell targeted magnetic resonance imaging and drug delivery. Nano Res. 2012;5:679–94.
60. Chen J, Zhang R, Han L, Tu B, Zhao D. One-pot synthesis of thermally stable gold@mesoporous silica core-shell nanospheres with catalytic activity. Nano Res. 2013;6:871–9.
61. Panfilova E, Shirokov A, Khlebtsov B, Matora L, Khlebtsov N. Multiplexed dot immunoassay using Ag nanocubes, Au/Ag alloy nanoparticles, and Au/Ag nanocages. Nano Res. 2012;5:124–34.
62. Gilroy KD, Sundar A, Farzinour P, Hughes RA, Neretina S. Mechanistic study of substrate-based galvanic replacement reactions. Nano Res. 2014;7:365–79.
63. Li J, He Y, Sun W, Luo Y, Cai H, Pan Y, et al. Hyaluronic acid-modified hydrothermally synthesized iron oxide nanoparticles for targeted tumor MR imaging. Biomaterials. 2014;35:3666–77.

64. Liu Y, Li LL, Qi GB, Chen XG, Wang H. Dynamic disordering of liposomal cocktails and the spatio-temporal favorable release of cargoes to circumvent drug resistance. Biomaterials. 2014;35:340–15.
65. Chen J, Liu H, Zhao C, Qin G, Xi G, Li T, et al. One-step reduction and PEGylation of graphene oxide for photothermally controlled drug delivery. Biomaterials. 2014;35:4986–95.
66. Gao FP, Lin YX, Li LL, Liu Y, Mayerhöffer U, Spenst P, et al. Supramolecular adducts of squaraine and protein for noninvasive tumor imaging and photothermal therapy in vivo. Biomaterials. 2014;35:1004–14.
67. Zha Z, Wang J, Zhang S, Wang S, Qu E, Zhang Y, et al. Engineering of perfluorooctylbromide polypyrrole nano-microcapsules for simultaneous contrast enhanced ultrasound imaging and photothermal treatment of cancer. Biomaterials. 2014;35:287–93.

Chapter 23
Quantum Dots in Photodynamic Therapy

Osnir S. Viana, Martha S. Ribeiro, Adriana Fontes, and Beate S. Santos

23.1 Introduction

Since the first applications of nanostructured semiconductor materials in heterogeneous photocatalysis in the 1980s, their unique optical and electrical properties have open up several research fields. These semiconductor nanoparticles were denominated "quantum dots" (QDs) when their optical and electrical properties could be controlled by the dimensions of the nanocrystals. Soon, their unique properties attracted scientists at the biomedical and health interfaces and the first accomplishments of these QDs included the concept of a new class of fluorescent labels. A deeper knowledge of the preparation, stabilization, and characterization of these systems allowed their recent applications in Photodynamic Therapy. Photodynamic Therapy (PDT) is a well-established clinical treatment modality for various diseases, including especially malign tumors (e.g., melanoma, and esophagus and bladder cancers) and antimicrobial inactivation. This treatment is based on the administration of a photosensitizer, which induces the production of oxygen

O.S. Viana (✉) • B.S. Santos
Department of Pharmaceutical Sciences, Federal University of Pernambuco,
Av. Prof Artur de Sá, s/n, Recife, Pernambuco 50740-520, Brazil
e-mail: osnirviana@yahoo.com.br; beate_santos@yahoo.com.br

M.S. Ribeiro
Center for Lasers and Applications, Nuclear and Energy Research Institute, National Nuclear Energy Commission, Av. Lineu Prestes, 2242, São Paulo, São Paulo 05508-000, Brazil
e-mail: marthasr@usp.br

A. Fontes
Department of Biophysics and Radiobiology, Federal University of Pernambuco,
Av. Prof. Moraes Rego S/N, Recife, Pernambuco 50670-901, Brazil
e-mail: adriana.fontes.biofisica@gmail.com

I. Batinić-Haberle et al. (eds.), *Redox-Active Therapeutics*,
Oxidative Stress in Applied Basic Research and Clinical Practice,
DOI 10.1007/978-3-319-30705-3_23

reactive species upon excitation by light within the targeted tissue, causing cellular death by oxidative stress. Only in the last decade (>2000s) QDs have emerged as new photosensitizer candidates in PDT processes [1–3] and, although a deeper scenario must still be drawn, a great number of studies already delineate the state of the art of their applications in this treatment modality. In the following sections of this Chapter, we will present the fundaments of Photodynamic Therapy and some features related to QDs as active redox materials. The last section in the Chapter summarizes the QD applications in PDT and shows that more studies are still needed to completely understand the therapeutic role of these nanoparticles.

23.2 Photodynamic Therapy

Photodynamic therapy (PDT) is the application of light-activated photosensitizers (PSs) in the presence of oxygen for either host cell destruction, as desired in cancer PDT, or antimicrobial purposes (also known as photodynamic inactivation, PDI). This process results in light-activated transfer of energy or electrons from the PS molecule to the surrounding environment. The first step in this process is to administer the PS to the target, which may be achieved by topical (skin lesions and cancer) or systemic (mainly in cancer PDT) routes. After a certain period of time (pre-irradiation time) sufficient to allow biodistribution of PS, the treatment site is irradiated. The absorption of light by PS induces a cascade of photochemical reactions producing reactive oxygen species (ROS). Subsequently, there is the generation of multiple redox processes that install oxidative stress in the site of action leading to irreversible damage to the target cell and/or tissue.

The first experiences with PDT date from over 115 years ago, were reported by Raab, who observed that the exposure of paramecium to acridine dye and light could be lethal to the protozoan that causes malaria disease. He postulated that this effect was caused by the transfer of the luminous energy to the chemical environment within the cell, similarly to the process that occurs in plants by the absorption of light by chlorophyll. Neither the dye nor the light alone had any apparent effect on the paramecium organisms, but together they were highly cytotoxic [4]. In 1903, Von Tappeiner and his colleague Jesionek attempted to apply PDT in humans, for treatment of cancerous, syphilitic, and tuberculosis conditions [5].

Since then, several examples of photodynamic activity were observed for a variety of photosensitizers, both in vitro and in vivo. The research then took a different turn, concentrating the studies on the diagnosis and treatment of cancer and also the development of new photosensitizers and new light sources. In 1972, Diamond and colleagues tried to destroy tumor cells with an hematoporphyrin association, currently one of the most accepted photosensitizing agents, with X-ray exposure. However, experiments were unsuccessful, demonstrating that the light source (i.e., its electromagnetic radiation) should be in resonance with the absorption spectrum of the dye photosensitizer [6].

Diseases, such as macular degeneration of the retina, several types of cancer, psoriasis, fungal and bacterial infections, warts, which are characterized by an abnormal growth of cells, are also being treated by photodynamic therapy [2, 7, 8].

The photodynamic effect depends on the combined action of drugs that are photoactivated (photosensitizers) by a specific wavelength and energy of light, to destroy pathogenic organisms or tissues that have rapid multiplication. The photodynamic process is based on the photo-oxidation of organic matter. The presence of PS, oxygen, and light irradiation at the site to be treated is needed. When PS absorbs light, it goes from the ground state to an excited state. The return to the ground state may take place through a triplet state, which then reacts with either the substrate (mechanism type I) or with molecular oxygen (mechanism type II) (Fig. 23.1).

Representative type I and type II reactions are expressed in Eqs. (23.1) and (23.2), respectively:

$$\text{Type I}: \begin{array}{l} {}^{3}\text{PS}^{*} + \text{sub} \rightarrow \text{PS}^{\bullet +} + \text{sub}^{\bullet -} \left(e^{-} \text{ transfer}\right) \\ \text{sub}^{\bullet -} + {}^{3}\text{O}_2 \rightarrow \text{sub} + \text{O}_2^{\bullet -} \rightarrow \text{HO}^{-} + \text{HO}^{\bullet} \end{array} \tag{23.1}$$

$$\text{Type II}: \begin{array}{l} {}^{3}\text{PS}^{*} + {}^{3}\text{O}_2 \rightarrow {}^{1}\text{PS} + {}^{1}\text{O}_2 \text{ (energy transfer)} \\ {}^{1}\text{O}_2 + \text{sub} \rightarrow \text{sub} - \text{OOH} \end{array} \tag{23.2}$$

where sub is substrate.

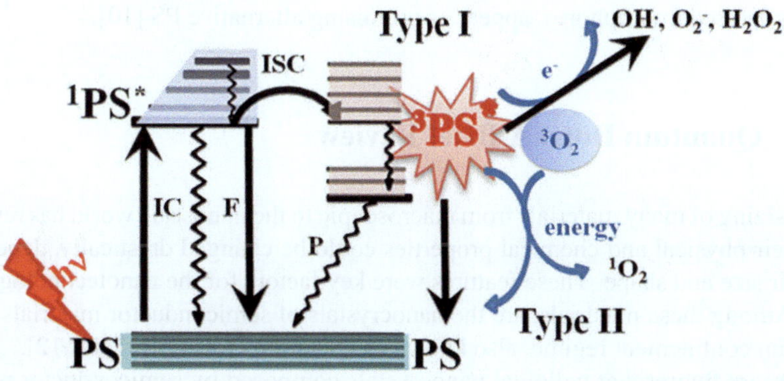

Fig. 23.1 Modified Jablonski diagram adapted from personal communication of Michael R. Hamblin. The photosensitizer in the ground state (^{0}PS) absorbs luminous energy ($E = h\nu$), resulting in an excited singlet state (^{1}PS*). Returning to the ground state, the excited photosensitizer may undergo a transition to its triplet excited state (^{3}PS*), where it may have two types of reaction. In the type I reaction, it can react directly with the substrate, such as the cell membrane or a molecule, transferring electrons (e^{-}) to form a radical or radical ion in both PS and substrate. These radicals can further interact with oxygen to produce oxidized products. Otherwise, in the type II reaction, the excited photosensitizer transfers its energy directly to molecular oxygen ($^{3}O_2$), to produce singlet oxygen ($^{1}O_2$), a highly reactive oxygen species, which may oxidize the substrate. *IC* internal conversion, *F* fluorescence, *ISC* intersystem crossing, *P* phosphorescence

Thus, reactive oxygen species (hydroxyl radical, superoxide anion, hydrogen peroxide in reaction type I, and singlet oxygen in reaction type II) are generated locally, resulting in cell death in tumor tissue or inactivation of infectious agents by oxidative stress.

Literature strongly supports the hypothesis that PDT can represent a viable alternative to treat cancer and infectious diseases. For cancer, the major favorable features of PDT comprise the avoidance of systemic treatment, selectivity for malignant cells, and repeatability. Regarding antimicrobial PDT (PDI), the main advantages are (i) the broad spectrum of action (bacteria, yeasts, and parasitic protozoa), (ii) outcome is independent of the antibiotic resistance pattern of the microbial strain, (iii) wide decrease in pathogens with minimal damage to the host tissue, (iv) absence of selection of photoresistant strains after multiple treatments, and (v) lack of mutagenicity. In addition, PDT is a low cost and minimally invasive localized therapy.

For successful PDT, biological, physical, and chemical factors should be considered. The selectivity and retention of the photosensitizer by the target cell is of outmost importance. Additionally, the intensity of light reaching the treatment region should be adequate to promote oxidizing effects on PS. Different types of organic molecules and complexes have been synthesized and studied as PS agents for use in PDT, such as phenothiazines, porphyrins and their derivatives, chlorins, and phthalocyanines [9]. Although literature reports successful uses of PDT for treatment of some types of cancer and resistant infections in clinical trials, few photosensitizing agents have been produced commercially, due to the difficulty in obtaining stable and highly efficient photosensitizers. Therefore, the search for new alternatives has driven the use of nanoparticles for this type of therapy and QDs, as a unique class of nanostructured fluorophores, appear as promising alternative PS [10].

23.3 Quantum Dots: A Brief Review

The resizing of many materials from macroscopic to the nanoscale world has revealed that their physical and chemical properties could be changed drastically, depending on their size and shape. These features were key factors for the nanotechnology success. Among these materials, are the nanocrystals of semiconductor materials in the quantum confinement regime, also known as Quantum Dots (QDs) [11, 12].

QDs are fluorescent colloidal nanocrystals composed by semiconductor materials mainly comprised of II-VI or III-V chemical elements in binary or ternary combinations. With dimensions ranging from 2 to 10 nm, they present unique optical properties, such as their size-dependent fluorescence wavelength and great photostability. Due to these special properties, QDs have been applied as fluorescent probes for molecular, cellular, and in vivo imaging, cytometry and fluoroimmunoassays, for biosensors and also, more recently, as photosensitizer in photodynamic therapy [13–24]. As fluorescent probes, QDs present the following advantages:

1. Broad absorption spectrum band, which allows the fluorescence excitation of different QDs sizes at the same time (Fig. 23.2a) [14].

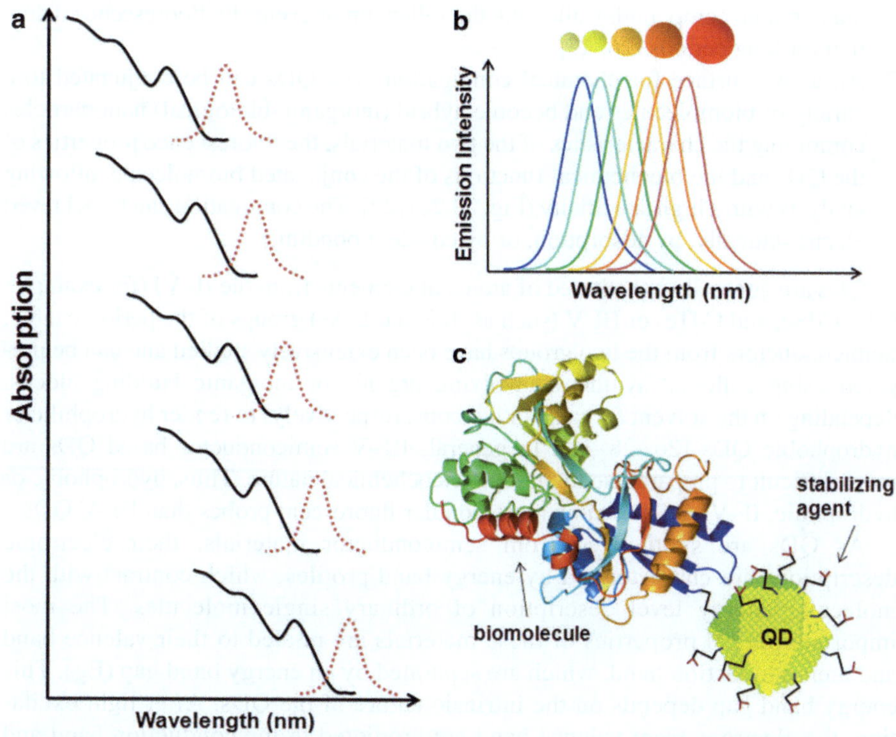

Fig. 23.2 Some interesting optical and structural properties of II–VI semiconductor quantum dots. (**a**) Size-dependent behaviour of absorption and emission spectra. (**b**) Characteristic size-dependent narrow emission band profile of the nanocrystals. (**c**) Chemical active surface of the quantum dots, allowing the bioconjugation of targeting molecules

2. Fluorescence emission tuned according only to the nanoparticle size, for QDs synthesized from the same semiconductor materials, which provides access to a wide range of fluorophores colours tuned by the particle size (Fig. 23.2b).
3. Narrower and symmetric emission bands (full width at half-maximum ~25–40 nm), from the blue to the near infrared region, which allow multicolour analyses avoiding spectral crosstalk.
4. Large Stokes shift (Fig. 23.2a), interesting for experiments with QDs as FRET (Forster resonance energy transfer) donors for minimizing the possibility of direct excitation of the acceptors [25].
5. Long emission lifetimes (dozens to hundreds of nanoseconds), allowing elimination of cell autofluorescence (which has considerably shorter lifetime) by fluorescence lifetime analysis.
6. Higher brightness and photostability, especially for core/shell QDs, which presents an even greater resistance against photobleaching (1000-fold more photostable

than organic compounds), allowing the follow up of events by fluorescence detection for long periods [26, 27].

7. An active surface for chemical conjugation. The QDs can be conjugated to a variety of biomolecules and become hybrid (inorganic-biological) nanoparticles combining the characteristics of the two materials, the fluorescence properties of the QDs and the biochemical functions of the conjugated biomolecule, allowing analysis with a high specificity (Fig. 23.2c) [22]. The conjugation can be achieved electrostatically, by adsorption, or by covalent bonding.

QDs are generally constituted of atoms of elements from the II–VI (for example: CdS, CdSe, and CdTe) or III-V (such as, InP and InAs) groups of the periodic table. Semiconductors from the two groups have been extensively studied and can be prepared using colloidal synthesis involving organic or inorganic building blocks, depending on the solvent (organic or aqueous, respectively) to render hydrophilic or hydrophobic QDs [26, 28–31]. In general, III–V semiconductor based QDs are more difficult to prepare due to their physicochemical nature. Thus, hydrophobic or hydrophilic, II–VI QDs became more popular fluorescent probes than III–V QDs.

As QDs are synthetized from semiconductor materials, their electronic descriptions are characterized by energy band profiles, which contrast with the molecular energy level description of ordinary single molecules. The most important optical properties of these materials are related to their valence band and their conduction band, which are separated by an energy band gap (Eg). This energy band gap depends on the intrinsic nature of the QDs. After light excitation, the electrons from valence band are promoted to the conduction band and this transient charge separation within the particle is called "exciton". After being excited to the conduction band, electrons return to the valence band producing the excitonic recombination and producing fluorescence. The colour of the fluorescence is proportionally related to the QD band gap. The higher the QD band gap, the more energetic is the observed emission band. The exciton is described as an independent electronic species, which describes an excited electron (*e*) interacting with the hole (*h*) that was left in the valence band. The e-h pair may be described with an electronic level configuration distinct from that of the energy band of the semiconductor material and also by its radius (named exciton Bohr radius, $\mathbf{a}_B$). When the nanocrystals reach the size dimension of the exciton Bohr radius of that semiconductor material, their excitons sense the size confinement and as a consequence the nanocrystals' energy levels modify. This is called the quantum confinement of the exciton and it is possible to tune the band gap energy and consequently the optical properties in this size regime (as depicted in Fig. 23.2a, b). The size regime in which the quantum-confinement effect is observed is dependent on the chemical nature of QD composition. QDs are nanocrystals quantum-confined in all three dimensions (3D), and as a consequence, smaller QDs present higher band gaps than larger QDs. Thus, smaller QDs present fluorescence towards the blue spectral region and, in the other hand, larger QDs emit light toward the red spectral region reaching, occasionally, the infrared spectral region (as in the case of PbS QDs).

The QDs absorption and emission spectra can offer important information about these nanoparticles. The position (wavelength) and width of the maximum of the first absorption band (Fig. 23.2a) is related to the nanoparticles size and size dispersion, respectively. The particle size of CdS and CdTe QDs has been correlated experimentally to their absorption band maximum wavelengths; the absorbance value can also provide the nanoparticles` concentrations for CdTe QDs [26, 32, 33]. Additionally, the width of the emission spectrum reflects the combination of the excitonic emission and the presence of crystal surface defects. The presence of surface dangling bonds results in new electronic state traps between the valence and conduction bands. These new electronic states are responsible for the red shift and an enlargement of the emission band [14, 31, 34]. Figure 23.3 depicts the radiative pathways usually observed for light-excited QDs.

The emission quantum yield of these systems is related to the energy conversion output in excitonic recombination processes. In order to decrease the possibility of energy losses by traps, the dangling bonds of the QDs may be compensated by growing a shell on the semiconductor core. QDs semiconductor core can be capped with organic or inorganic materials. For emission enhancement purposes, a shell of another semiconductor material of greater band gap is usually applied. This chemical shell is called passivation shell and is formed by depositing some monolayers of a second semiconductor onto the QD core to guarantee a physical separation of the optically active core from the surrounding medium. This removes the trapping electronic stages and reduces the QDs sensitivity to chemical changes of the neighbouring environment, which results in improved optical properties (Fig. 23.3) [35].

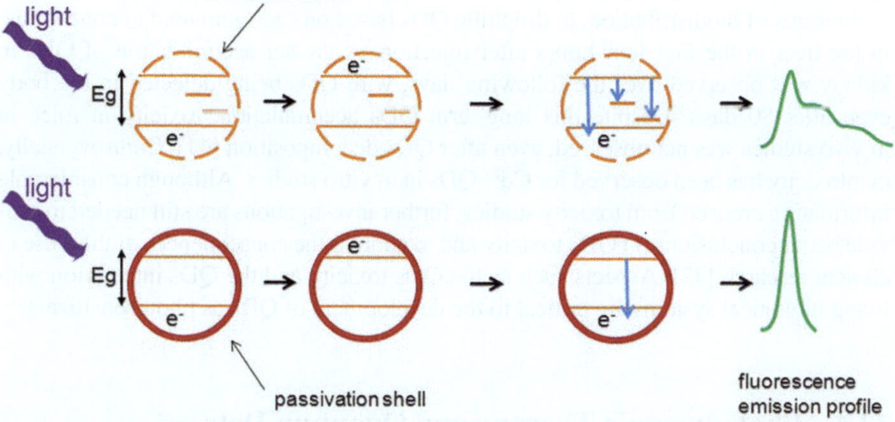

Fig. 23.3 Light deactivation pathways observed in QDs. For surfaces where defects are in great concentration, the emission band is composed of wider bands of lower energy than that observed for the exciton recombination band

QDs can label living biological systems by non-specific and specific interactions. Usually, the labeling of living cells by QDs is limited to the membranes. The use of QDs for intracellular study is still restricted because their passage through the cell membrane does not occur passively. QDs usually are internalized by cells and become trapped in endocytic vesicles, requiring the development of methods to deliver them freely into the cytosol. The internalization process depends either on the QDs properties (such as size, shape, surface functionalization, surface charge, and the combination of these properties) or on the cell type, for example: cell lineage, metabolic state, cell cycle stage, normal or cancerous cells [31, 36, 37].

As the stages of preparation, conjugation, and application of QDs in biology are relatively well-established, there is a growing interest in testing these nanocrystals in clinical research. Although promising, there is a noteworthy concern in using QDs in humans due to the potential risk of toxicity, especially when they are based on heavy metals such as cadmium [38]. In this context, QDs of different compositions and coatings are being evaluated not only in in vitro cell culture assays, but also in in vivo small animal and primates experiments [39]. Although there are QDs based on InP, ZnSe, and silicon that can be considered less toxic, QDs based on cadmium stand for their optical quality and good fluorescence intensity.

One of the alternatives explored to minimize the effect of QD toxicity is to modify their surface with materials that form a protective shell around the core of the nanocrystal. Among the types of coatings, there are: (1) polyethylene glycol, which is able to circumvent the immune system responses and (2) zinc-based shells, as this is an essential element for humans. These coatings can be important to avoid degradation of cadmium-based QD cores and consequently release of toxic Cd^{2+} ions. A study using mass spectrometry showed that the release rate of Cd^{2+} ions resulting from the deterioration of the core was greater in CdTe QDs without a passivation layer. Conversely, the release of Cd^{2+} ions from coated QDs was lower, which resulted thus in lower toxicity, especially when QD were coated with a zinc shell [40].

In terms of biodistribution, hydrophilic QDs based on cadmium tend to concentrate in the liver in the first four hours after injection. A greater accumulation of QDs in kidney was observed over the following days, with QDs being detected in the body even after 80 days. Despite this long-term QDs accumulation, toxicity in mice in in vivo studies was not observed, even after QDs decomposition [41]. Controversially, cytotoxicity has been observed for Cd^{2+} QDs in in vitro studies. Although considerable information aroused from toxicity studies, further investigations are still needed to provide better conclusions on QDs toxicity and to address the consequences of their use in clinical research [42]. Aspects such as the QDs toxicity and the QDs interaction with living biological systems are critical to the development of QDs as photosensitizers.

23.4 Photodynamic Therapy and Quantum Dots

Almost a decade after QDs were successfully applied as fluorescent nanolabels in biological systems, their potential as ROS inductors for PDT started to be tested [43]. What attracted scientists to apply these nanocrystals as PSs was their known capability of selectively transferring energy and/or charge to the surrounding species when excited by light. Unlike most organic photosensitizers, QDs do not have metastable excited states, which can promote effective energy transfer, and distinct processes are expected to occur in these systems to yield reactive oxygen species [44]. Two processes, namely type I and type II (depicted in Fig. 23.4), are described in the literature as possible mechanisms for QD-promoted ROS production.

In the so-called type I-induced photoreaction, QDs absorb a photon and produce electron-hole pairs, named excitons. These excited electrons normally return to the ground state emitting light, but they may also migrate to electronic traps present on the surface of the QDs. These trapped electrons may eventually be scavenged by O_2 or O-containing molecules present in the surrounding medium, resulting in ROS production. This effect is even more expected for unpassivated QDs or those that were partially corroded by chemical processes exposing unshared bonds and increasing electronic traps. Some authors observed that up to a certain amount of corrosion of CdTe QD surface in water, the particles increased their ROS production, indicating a successful trapping of electrons at the surface electronic levels [44, 45].

The type II photoreactions induce the excitation of the surrounding species after the primary excitation of QDs followed by relaxation via electron-hole recombination. This transferred energy may be sufficient to promote singlet oxygen ($^1O_2$) species. For these processes, efficient luminescence is a determining factor for ROS production efficiency.

In the late 1980s, some authors reported on the energy transfer occurring between QDs and other species (molecules, proteins, and nanoparticles). The first attempts observed in the literature of producing ROS using QDs were summarized by Samia and coworkers in 2003 [43]. They found that direct photoactivation of QDs produces singlet oxygen ($^1O_2$) due to the excitonic energy transfer from QD to molecular oxygen ($^3O_2$) (Fig. 23.5); this process is marked, however, by an energy transfer

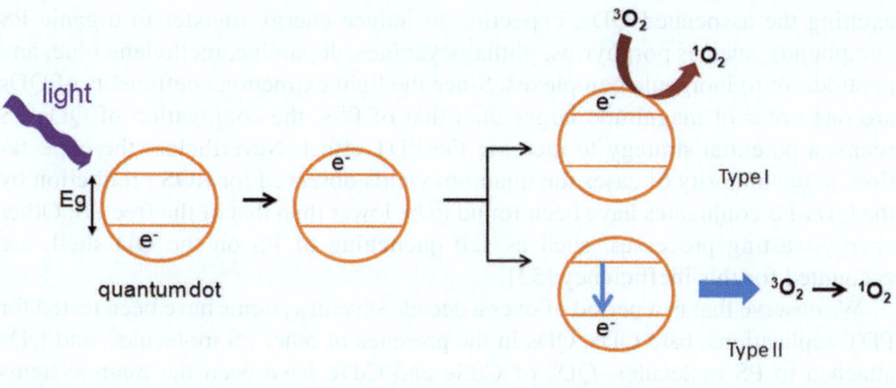

Fig. 23.4 Light excitation of QDs followed by their possible mechanisms (type I and II) of producing ROS

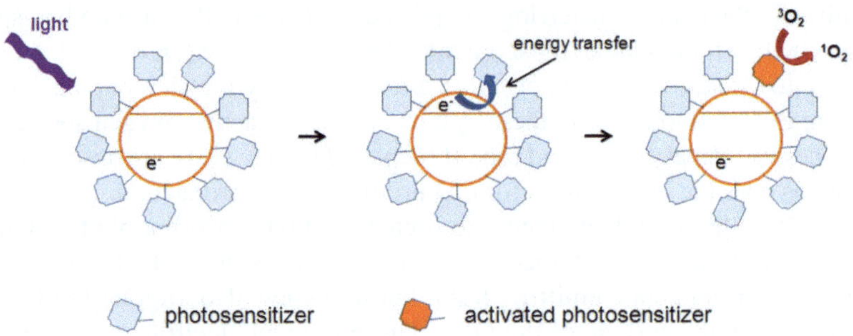

Fig. 23.5 ROS production process of QDs associated to other PS molecules

of low efficiency. Some years later, Lovric et al. [46] were studying the cytotoxicity of QDs and they observed that one of the mechanisms related to their toxicity involved the production of reactive oxygen species within the cell cytoplasm. They concluded that applying bare QDs to biological systems caused a greater cell injury including damage to the plasma membrane, mitochondrion, and nucleus. Moreover, they suggested that QDs induced ROS formation and demonstrated that they avoided the oxidative processes by treating the cells with molecular antioxidants such as *N*-acetylcysteine or by surface modifications of QDs.

Alternatively, there is also a third indirect process by which QDs may induce ROS production. This process involves the QD energy transfer (via radiative or non-radiative manner by FRET) to other PS species present nearby the nanoparticles (Fig. 23.5). In this case, the PS molecules attached to the particles or present in the colloidal medium can be activated by radiative or non-radiative energy transfers.

The use of QDs associated with photosensitizers has attracted the attention of many researchers particularly due to the ability of QDs to produce light over a broad range of wavelengths, which can be tuned to the correct wavelength to excite the PS [47–52]. Several authors have tried to increase the PS efficiency by exciting the associated QDs, expecting to induce energy transfer to organic PS compounds, such as porphyrins, phthalocyanines, dopamine, methylene blue, and peptides, or to inorganic complexes. Since the light extinction coefficients of QDs are one order of magnitude larger than that of PSs, the conjugation of QDs-PS seems a potential strategy to increase the PDT effect. Nevertheless the expectation, in the majority of cases the quantum yields observed for ROS production by the QDs-PS conjugates have been found to be lower than that of the free PS. Other energy-wasting processes, such as self-quenching of PS on the QD shell, are accounted for this inefficiency [53].

We observe that in a period of over a decade several systems have been tested for PDT applications: bare QDs, QDs in the presence of other PS molecules, and QDs attached to PS molecules. QDs of CdSe and CdTe have been the main systems tested. Some examples of systems tested in PDT are described in Table 23.1.

Table 23.1 Studies exemplifying the application of QDs as photosensitizers in PDT

QD	PS	Main results	References
CdSe	Phthalocyanine	The system produced $^1O_2$ via FRET by the system. ROS was also produced by the bare QDs but they were less effective	[43]
CdSe/CdS/ZnS	Rose Bengal and Chlorine e_6	PS produced $^1O_2$ via FRET with a quantum yield in the range of 0.1–0.3 depending on their excitation conditions and of the applied PS	[51]
CdTe-Thiol	Sulphonated aluminum phthalocyanine	The conjugation of QDs to the PS increased ~15 % the $^1O_2$ quantum yield	[54]
CdTe	Methylene blue (MB)	QDs within the PS solution contributed to a greater $^1O_2$ production compared to the MB alone via FRET and charge transference	[48]
CdSe	Porphyrin	The $^1O_2$ production was twofold greater for the conjugated system than for the porphyrin alone	[55]
CdSe/CdS/ZnS	Sulphonated aluminum phthalocyanine	The conjugate was capable of penetrating nasopharyngeal carcinome cells and the authors detected the $^1O_2$ production via FRET (84 % efficiency)	[47]
CdTe-Thiol	Phthalocyanine derivatives (AlPc, GaPc, and InPc)	The presence of QDs increased the $^1O_2$ quantum yield via FRET processes	[56]
CdSe/ZnS	Chlorine e_6	There was a more efficient $^1O_2$ production when the QD-PS conjugates were excited than for the non-conjugated PS	[57]
CdTe	Zn(II) Porphyrin	Bare QDs and conjugates were able to produce ROS. However, bare QDs and conjugates showed reduced photodynamic activity against *C. albicans* compared to Zn(II) Porphyrin alone	[53]

Biological systems are, however, much more complicated as compared to aqueous solution chemistry. Up to the present date, a great number of studies showed that the QDs are indeed good candidates for PDT purposes [13, 20, 52, 55, 58]. So far, the (energy transfer)-mediated PDT by PS-QD conjugates has not been achieved in living systems, leaving a doubt whether the PS-QD conjugates can really work in PDT [47]. The mechanisms by which QDs are internalized within the cell moiety, for instance, also play an important role in the PDT action. We have recently reported that the internalization mechanism of QDs by cells can control the PDT performance [53]. For cells without cell wall such as parasites and human cells, QDs end up internalized more easily and the overall PDT action usually can be greater. For those cells that present a cell wall, such as the *Candida albicans* yeast cells, QDs are not internalized by the common uptake mechanisms. This challenge may be overcome by bioconjugating the QDs with specific targeting molecules.

23.5 Conclusions and Perspectives

There are several advantages and limitations for both conventional PS drugs and QDs when individually applied for PDT. For example, some properties such as NIR absorption, broad absorption bands, and photostability are desirable for PDT. In contrast, the current PSs still present some disadvantages when compared to the unique optical properties of QDs, such as narrow absorption band, poor photostability, and visible light absorption. The need to meet most of these properties together has led the search for new photosensitizers for PDT [62]. Besides the QDs' advantageous properties such as efficient light outputs and the capacity of tuning their absorption and emission bands, they present an exceptional photoresistance, when compared to the known organic photosensitizers. This last characteristic allows light excitation for continuous ROS production either via type I or type II processes enabling prolonged phototherapies. Some photoresistance studies were performed with different organic PS molecules and QDs, confirming that PSs are more prone to photodegradation (e.g. Kunz et al. [61]).

Although the overall reports [59] show a potential application of QDs and their conjugates with other PSs in PDT, the Φ_{Δ} (quantum yield of singlet oxygen) of different systems (mainly CdSe and CdTe QDs) determined by distinct authors in different colloidal media lie within the $\Phi_{\Delta}=0.01–0.05$ range [20, 43, 54, 60]. These values are considered low when compared to the most common photosensitizers and several key points must be also taken into account when choosing the systems such as: (1) absorption and emission profiles of both QDs and PS species, (2) resilient time decay of the excited electron on the surface of the nanocrystals, which models transfer time interval of the excited light, and (3) photoresistance of the QDs.

Looking at the big picture, after a decade since the first results on application of QDs as PSs, there is still no definite answers, but rather some nice clues on the real potential of using QDs in PDT.

Acknowledgment The authors are grateful for discussions with the members of Nanotecnologia Biomédica/UFPE research group. Funding from Fundação de Amparo a Ciência e Tecnologia do Estado de Pernambuco (FACEPE), Conselho Nacional de Desenvolvimento Científico e Tecnológico (CNPq), Coordenação de Aperfeiçoamento de Pessoal de Nível Superior (CAPES), Academia Brasileira de Ciências, Instituto Nacional de Ciência e Tecnologia de Fotônica (INCT-INFo), and the Universidade Federal de Pernambuco (UFPE) is greatly acknowledged.

References

1. Wieder ME, et al. Intracellular photodynamic therapy with photosensitizer-nanoparticle conjugates: cancer therapy using a 'Trojan horse'. Photochem Photobiol Sci. 2006;5(8):727–34.
2. Allison RR, Bagnato VS, Sibata CH. Future of oncologic photodynamic therapy. Future Oncol. 2010;6(6):929–40.
3. Paszko E, et al. Nanodrug applications in photodynamic therapy. Photodiagnosis Photodyn Ther. 2011;8(1):14–29.
4. Raab O. Uber die Wirkung fluoreszierender Stoffe auf Infusorien. Z Biol. 1900;39(5):524–46.
5. Von Tappeiner H, Jesionek A. Therapeutische versuche mit fluoreszierenden stoffen. Munch Med Wochenschr. 1903;47:2042–4.
6. Diamond I, et al. Photodynamic therapy of malignant tumours. Lancet. 1972;300(7788):1175–7.
7. Hanout M, et al. Therapies for neovascular age-related macular degeneration: current approaches and pharmacologic agents in development. Bio Med Res Int. 2013;2013:8.
8. Kharkwal GB, et al. Photodynamic therapy for infections: clinical applications. Lasers Surg Med. 2011;43(7):755–67.
9. Wainwright M. Photodynamic therapy: the development of new photosensitisers. Anticancer Agents Med Chem. 2008;8(3):280–91.
10. Geszke-Moritz M, Moritz M. Quantum dots as versatile probes in medical sciences: synthesis, modification and properties. Mater Sci Eng C. 2013;33(3):1008–21.
11. Bruchez M, et al. Semiconductor nanocrystals as fluorescent biological labels. Science. 1998;281(5385):2013–6.
12. Chan WCW, Nie S. Quantum dot bioconjugates for ultrasensitive nonisotopic detection. Science. 1998;281(5385):2016–8.
13. Brannon-Peppas L, Blanchette JO. Nanoparticle and targeted systems for cancer therapy. Adv Drug Deliv Rev. 2012;64:S206–12.
14. Fontes A, Santos BS, editors. Quantum dots: applications in biology. In: Methods in molecular biology. 2nd ed. New York: Springer; 2014. p. 258.
15. Law W-C, et al. Aqueous-phase synthesis of highly luminescent CdTe/ZnTe core/shell quantum dots optimized for targeted bioimaging. Small. 2009;5(11):1302–10.
16. Tian J, et al. Synthesis of CdTe/CdS/ZnS quantum dots and their application in imaging of hepatocellular carcinoma cells and immunoassay for alpha fetoprotein. Nanotechnology. 2010;21(30):305101.
17. Vannoy CH, et al. Biosensing with quantum dots: a microfluidic approach. Sensors. 2011;11(10):9732.
18. Lira RB, et al. Non-specific interactions of CdTe/Cds quantum dots with human blood mononuclear cells. Micron. 2012;43(5):621–6.
19. Andrade CG, et al. Evaluation of glycophenotype in breast cancer by quantum dot-lectin histochemistry. Int J Nanomed. 2013;8:4623.
20. Samia ACS, Dayal S, Burda C. Quantum dot-based energy transfer: perspectives and potential for applications in photodynamic therapy. Photochem Photobiol. 2006;82(3):617–25.
21. Li Y, et al. Electrochemiluminescence biosensor based on CdSe quantum dots for the detection of thrombin. Electrochim Acta. 2012;65:1–6.

22. Carvalho KHG, et al. Fluorescence plate reader for quantum dot-protein bioconjugation analysis. J Nanosci Nanotechnol. 2014;14(5):3320–7.
23. Tenório DPLA, et al. CdTe quantum dots conjugated to concanavalin A as potential fluorescent molecular probes for saccharides detection in Candida albicans. J Photochem Photobiol B Biol. 2015;142:237–43.
24. Ma Q, Su X. Recent advances and applications in QDs-based sensors. Analyst. 2011;136(23):4883–93.
25. Medintz IL, Mattoussi H. Quantum dot-based resonance energy transfer and its growing application in biology. Phys Chem Chem Phys. 2009;11(1):17–45.
26. Santos BS, Farias PMA, Fontes A. Semiconductor quantum dots for biological applications. In: Henini M, editor. Handbook of self assembled semiconductor nanostructures for novel devices in photonics and electronics. Amsterdam: Elsevier; 2008. p. 773–98.
27. Choi AO, Maysinger D. Applications of quantum dots in biomedicine. In: Rogach AL, editor. Semiconductor nanocrystal quantum dots. Vienna: Springer; 2008. p. 349–65.
28. Chan WCW, et al. Luminescent quantum dots for multiplexed biological detection and imaging. Curr Opin Biotechnol. 2002;13(1):40–6.
29. Mićić OI, Nozik AJ. Colloidal quantum dots of III–V semiconductors (Chapter 5). In: Nalwa HS, editor. Nanostructured materials and nanotechnology. San Diego: Academic Press; 2002. p. 183–205.
30. Gaponik N, et al. Efficient phase transfer of luminescent thiol-capped nanocrystals: from water to nonpolar organic solvents. Nano Lett. 2002;2(8):803–6.
31. Fontes A, et al. Quantum dots in biomedical research. In: Hudak R, editor. Biomedical engineering—technical applications in medicine. Rijeka: InTech; 2012.
32. Yu WW, et al. experimental determination of the extinction coefficient of CdTe, CdSe, and CdS nanocrystals. Chem Mater. 2003;15(14):2854–60.
33. Dagtepe P, et al. Quantized growth of CdTe quantum dots; observation of magic-sized CdTe quantum dots. J Phys Chem C. 2007;111(41):14977–83.
34. Dabbousi BO, et al. (CdSe)ZnS core–shell quantum dots: synthesis and characterization of a size series of highly luminescent nanocrystallites. J Phys Chem B. 1997;101(46):9463–75.
35. Dai T, et al. Concepts and principles of photodynamic therapy as an alternative antifungal discovery platform. Front Microbiol. 2012;3:120.
36. Lira RB, et al. Studies on intracellular delivery of carboxyl-coated CdTe quantum dots mediated by fusogenic liposomes. J Mater Chem B. 2013;1(34):4297–305.
37. Jaiswal JK, et al. Long-term multiple color imaging of live cells using quantum dot bioconjugates. Nat Biotechnol. 2003;21(1):47–51.
38. Nawrot T, et al. Cadmium-related mortality and long-term secular trends in the cadmium body burden of an environmentally exposed population. Environ Health Perspect. 2008;116(12): 1620–8.
39. Yong K-T, et al. Nanotoxicity assessment of quantum dots: from cellular to primate studies. Chem Soc Rev. 2013;42(3):1236–50.
40. Xu M, et al. Free cadmium ions released from CdTe-based nanoparticles and their cytotoxicity on Phaeodactylum tricornutum. Metallomics. 2010;2(7):469–73.
41. Su Y, et al. In vivo distribution, pharmacokinetics, and toxicity of aqueous synthesized cadmium-containing quantum dots. Biomaterials. 2011;32(25):5855–62.
42. Hauck TS, et al. In vivo quantum-dot toxicity assessment. Small. 2010;6(1):138–44.
43. Samia ACS, Chen X, Burda C. Semiconductor quantum dots for photodynamic therapy. J Am Chem Soc. 2003;125(51):15736–7.
44. Chen J-Y, et al. Quantum dot-mediated photoproduction of reactive oxygen species for cancer cell annihilation. Photochem Photobiol. 2010;86(2):431–7.
45. Cho SJ, et al. Long-term exposure to CdTe quantum dots causes functional impairments in live cells. Langmuir. 2007;23(4):1974–80.
46. Lovrić J, et al. Unmodified cadmium telluride quantum dots induce reactive oxygen species formation leading to multiple organelle damage and cell death. Chem Biol. 2005;12(11): 1227–34.

47. Li L, et al. Quantum dot-aluminum phthalocyanine conjugates perform photodynamic reactions to kill cancer cells via fluorescence resonance energy transfer. Nanoscale Res Lett. 2012;7(1):1–8.
48. Rakovich A, et al. CdTe quantum dot/dye hybrid system as photosensitizer for photodynamic therapy. Nanoscale Res Lett. 2010;5(4):753–60.
49. Jhonsi MA, Renganathan R. Investigations on the photoinduced interaction of water soluble thioglycolic acid (TGA) capped CdTe quantum dots with certain porphyrins. J Colloid Interface Sci. 2010;344(2):596–602.
50. Wen Y, et al. Activation of porphyrin photosensitizers by semiconductor quantum dots via two-photon excitation. Appl Phys Lett. 2009;95(14):143702.
51. Tsay JM, et al. Singlet oxygen production by peptide-coated quantum dot—photosensitizer conjugates. J Am Chem Soc. 2007;129(21):6865–71.
52. Clarke SJ, et al. Photophysics of dopamine-modified quantum dots and effects on biological systems. Nat Mater. 2006;5(5):409–17.
53. Viana O, et al. Comparative study on the efficiency of the photodynamic inactivation of Candida albicans using CdTe quantum dots, Zn(II) Porphyrin and their conjugates as photosensitizers. Molecules. 2015;20(5):8893.
54. Ma J, et al. generation of singlet oxygen via the composites of water-soluble thiol-capped CdTe quantum dotssulfonated aluminum phthalocyanines. J Phys Chem B. 2008;112(15): 4465–9.
55. Qi Z-D, et al. Biocompatible CdSe quantum dot-based photosensitizer under two-photon excitation for photodynamic therapy. J Mater Chem. 2011;21(8):2455–8.
56. Tekdaş DA, et al. Photodynamic therapy potential of thiol-stabilized CdTe quantum dot-group 3A phthalocyanine conjugates (QD-Pc). Spectrochim Acta A Mol Biomol Spectrosc. 2012;93:313–20.
57. Rotomskis R, et al. Complexes of functionalized quantum dots and chlorin e6 in photodynamic therapy. Lithuanian J Phys. 2013;53(1).
58. Narband N, et al. Quantum dots as enhancers of the efficacy of bacterial lethal photosensitization. Nanotechnology. 2008;19(44):445102.
59. Planas O, et al. Newest approaches to singlet oxygen photosensitisation in biological media (Chapter 9). In: Fasani E, Albini A, editors. Photochemistry, vol. 42. London: Royal Society of Chemistry; 2015. p. 233–78.
60. Zenkevich EI, et al. Quantitative analysis of singlet oxygen ($^1O_2$) generation via energy transfer in nanocomposites based on semiconductor quantum dots and porphyrin ligands. J Phys Chem C. 2011;115(44):21535–45.
61. Kunz L, et al. Photodynamic modification of disulfonated aluminium phthalocyanine fluorescence in a macrophage cell line. Photochem Photobiol Sci. 2007;6(9):940–8.
62. Biju V, Itoh T, Ishikawa M. Delivering quantum dots to cells: bioconjugated quantum dots for targeted and nonspecific extracellular and intracellular imaging. Chem Soc Rev. 2010;39(8): 3031–56.

Chapter 24
Metalloporphyrin in CNS Injuries

Huaxin Sheng and David S. Warner

Abbreviations

$CO_3^{\bullet-}$	Carbonate anion radical
CNS	Central nervous system
ECSOD	Extracellular superoxide dismutase
EMSA	Electromobility shift essay
FeP	Fe(III) porphyrin
FeTE-2-PyP^{5+}	Fe(III) *meso*-tetrakis (*N*-ethylpyridinium-2-yl) porphyrin
FeTM-4-PyP^{5+}	Fe(III) *meso*-tetrakis(*N*-methylpyridinium-4-yl)porphyrin
FeTnHex-2-PyP^{5+}	Fe(III) *meso*-tetrakis (*N*-*N*-hexylpyridinium-2-yl) porphyrin
FeTSPP^{3-}	Fe(III) *meso*-tetrakis(4-sulfonatophenyl)porphyrin
$^{\bullet}OH$	Hydroxyl radicals
8-OHdG	8-Hydroxy-2′-deoxyguanosine
H_2O_2	Hydrogen peroxide
HClO	Hypochlorous acid
ICV	Intracerebroventricular
IL-6	Interleukin-6
IP	Intraperitoneal
IT	Intrathecal
IV	Intravenous
LDH	Lactate dehydrogenase
MAP	Mean arterial blood pressure
MCAO	Middle cerebral artery occlusion

H. Sheng (✉) • D.S. Warner
Multidisciplinary Neuroprotection Laboratories, Department of Anesthesiology,
Duke University Medical Center, Durham, NC 27710, USA
e-mail: sheng001@mc.duke.edu

I. Batinić-Haberle et al. (eds.), *Redox-Active Therapeutics*,
Oxidative Stress in Applied Basic Research and Clinical Practice,
DOI 10.1007/978-3-319-30705-3_24

MnP	Mn(III)-substituted *N*-pyridyl- and *N,N'*-imidazolylporphyrin
MnTDE-2-ImP^{5+}	Mn(III) *meso*-tetrakis (*N,N'*-diethylimidazolium-2-yl) porphyrin
MnTE-2-PyP^{5+}	Mn(III) *meso*-tetrakis (*N*-ethylpyridinium-2-yl) porphyrin
MnTnHex-2-PyP^{5+}	Mn(III) *meso*-tetrakis (*N*-*n*-hexylpyridinium-2-yl) porphyrin
MnTnOct-2-PyP^{5+}	Mn(III) *meso*-tetrakis (*N*-*n*-octylpyridinium-2-yl) porphyrin
NADPH	Nicotinamide adenine dinucleotide phosphate
NF-κB	Nuclear factor -kappa-B
$^{\bullet}NO$	Nitric oxide radical
$^{\bullet}NO_2$	Nitrogen dioxide radical
$O_2^{\bullet-}$	Superoxide anion radical
OGD	Oxygen glucose deprivation
$ONOO^-$	Peroxynitrite
PBS	Phosphate-buffered saline
RNS	Reactive nitrogen species
ROS	Reactive oxygen species
SAH	Subarachnoid hemorrhage
SOD	Superoxide dismutase
Sq	Subcutaneous
TNF-α	Tumor necrosis factor α
tPA	Tissue plasminogen activator

24.1 Introduction

Metalloporphyrins belong to a class of porphyrin-based redox-active compounds. In the center of a macrocyclic porphyrin ligand sits the metal, such as manganese, iron, or zinc. All reactions of interest occur at the metal site. The type and magnitude of the reactivity of metalloporphyrins are largely impacted by the structure of the porphyrin ligand. The macrocyclic character of porphyrin ligand results in vast stability of the metal complex towards the loss of metal. Such immense stability is the reason why nature has used porphyrin ring for essential metabolic functions such as transport of oxygen and cytochrome P450-mediated metabolism of thousands of endogenous and exogenous chemicals.

Most preclinical studies have focused on cationic Mn(III)-substituted *N*-pyridyl- and *N,N'*-imidazolylyl porphyrins (MnPs), which were optimized to exhibit high SOD-like activities [11]. Few studies examined Fe(III) porphyrins (FePs) also [67]. Initially, MnPs were described as mimics of SOD because their SOD-like activity, similar in magnitude to that of enzymes, was demonstrated [8, 15]. Animal studies have shown that they protect tissue and organs against ischemic brain injury [45, 68, 72], spinal cord injury [71], radiation-induced lung injury [31, 57, 58, 87], breast cancer [1, 2], and diabetes [55]. While MnPs possess same thermodynamic and kinetic properties as SOD enzymes, they are small molecules and thus no steric hindrance to reactive species larger than $O_2^{\bullet-}$ is imposed. In turn, MnPs were shown to very efficiently scavenge other reactive species such as peroxynitrite and hypochlorite [11, 13]. MnPs, due to a biologically compatible reduction potential, oxidize signaling thiols (presumably acting in thiol oxidase or glutathione peroxidase fashion) and thereby modulate redox signaling and activation of cellular transcription factors such as NF-κB [10, 40, 59, 60], which has opened new avenues for treating CNS injury [13].

FePs have been also synthesized and proved effective in promoting the growth of SOD-deficient *E. coli* [86]. The protective effects are seen only at low <1 μM levels and are toxic at concentrations of ~20 μM where Mn analogs are efficacious [86]. The effects of FePs were reported in animal models, and one compound, INO-4885, is in clinical trials [67]. The data on FePs vs. MnPs suggest different mechanisms of action and toxicity; further studies on animal models are needed to understand chemistry and biology of FePs. Redox-inactive Zn porphyrins are used predominantly as photosensitizers [53] or as redox-inactive controls to MnPs or FePs. The largest database for metalloporphyrins in CNS injury has been obtained from the studies of MnPs, and it will be the focus of this chapter.

24.2 Role of Oxidative Stress in CNS Injury

Numerous studies have reported that oxidative stress is increased following CNS injury including cerebral ischemia [17, 21, 34, 37, 76], traumatic brain injury [3, 5, 24, 54, 61, 74], spinal cord injury [38, 43, 44, 49, 81], and hemorrhagic stroke [6, 27, 33, 95]. Oxidative stress plays a major role in the progression of injury and contributes to the severity of the outcome, not only in animal models, but also in humans [14, 18, 25, 28]. The key mechanism of action of MnPs involves restoring the balance of reactive oxygen and nitrogen species via redox-based activity [7, 10]. Therefore, it is important to learn how injury induces oxidative stress and to determine the status of those reactive species developed at different stages in the progression of injury. Such findings will allow for the best management plan for preventing tissue damage resulting from oxidative stress in CNS injury.

Superoxide ($O_2^{\bullet-}$) is a primary free radical that is generated during the injury process [20, 75, 89]. It reacts with nitric oxide ($^{\bullet}NO$) to form peroxynitrite ($ONOO^-$), a powerful nitrating agent, resulting in nitration of tyrosine in cellular proteins and initiation of cell dysfunction and death. The human body has three known SOD isoforms: MnSOD, which is localized in the mitochondrial matrix [41], Cu,ZnSOD, which is located in cytoplasm and intermembrane space of mitochondria [23, 39], and a Cu,ZnSOD isoform that is localized in the extracellular space (ECSOD) [65]. These enzymes present our first line of antioxidative defense and prevent the adverse effects of O_2^- by dismuting it to hydrogen peroxide (H_2O_2) and oxygen, thereby maintaining physiological nM levels of $O_2^{\bullet-}$. SODs operate as an antioxidant system only if coupled with removal of H_2O_2. H_2O_2 is readily removed by a number of redundant systems. Such are catalases, which dismute H_2O_2, i.e., reduce it to water and oxidize it to oxygen. Glutathione peroxidases

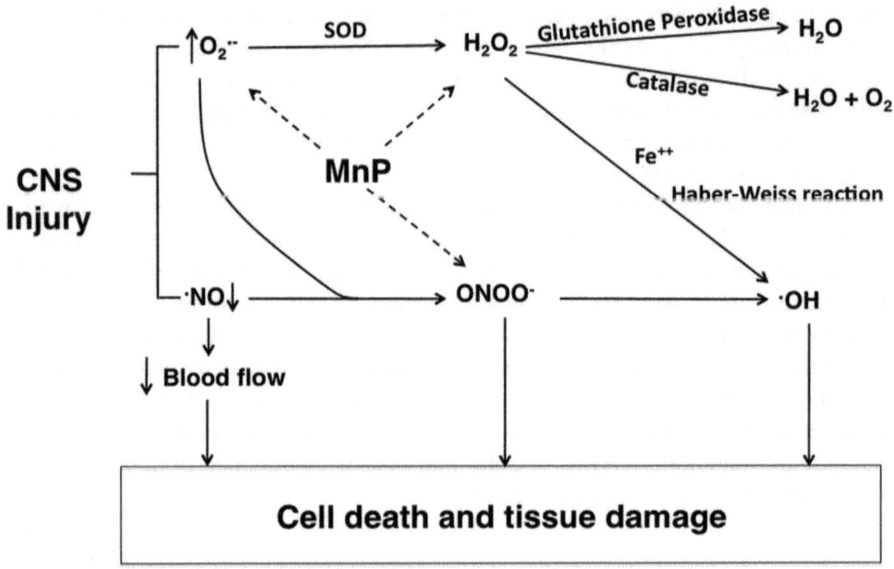

Fig. 24.1 Schematic showing formation of reactive oxygen and nitrogen species, and the action targets of MnPs in CNS injuries. While the ability of MnP to catalyze H_2O_2 dismutation in catalase-like fashion is negligible [83], its reaction with H_2O_2 in a glutathione peroxidase fashion, leading to oxidative modifications of signaling protein cysteines, is likely to affect cellular signaling pathways [13]

and peroxyredoxins reduce H_2O_2 to H_2O (Fig. 24.1). If these enzymes are deficient or inactivated during injury, damage can occur. H_2O_2 reacts with reduced iron Fe^{2+} to form highly damaging but shortly lived hydroxyl radicals ($^{\bullet}OH^-$) via the Haber–Weiss reaction. Another source of $^{\bullet}OH^-$ in tissue is spontaneous degradation of $ONOO^-$ to form $^{\bullet}OH^-$ and highly oxidizing nitrogen dioxide radical ($^{\bullet}NO_2^{\bullet}$) [20]. Yet, due to high in vivo levels of carbonate, $ONOO^-$ readily forms an adduct with CO_2 producing $ONOOCO_2^-$, which then gives rise to $CO_3^{\bullet-}$ and $^{\bullet}NO_2$ radicals. The $^{\bullet}OH$ radical is able, at sites of its production, to initiate lipid peroxidation by hydrogen abstraction and cause protein oxidation and DNA damage in cells. The endogenuous antioxidant enzymes may be impaired/inactivated due to oxidative modifications of their amino acid residues. Additionally, their expression may be suppressed due to modified cellular transcriptional activity associated with CNS injury. Cumulatively, oxidative stress is induced and perpetuated, resulting in continuous/cycling damage to cells and tissues. In addition, $^{\bullet}NO$, a critical molecule that regulates local blood flow, may be removed in the reaction with $O_2^{\bullet-}$ producing $ONOO^-$. This results in decreased local $^{\bullet}NO$ levels, which indirectly decreases tissue perfusion and thereby worsens the tissue damage. Additionally, nitric oxide synthase activity may be suppressed.

Cationic MnPs are powerful redox-active compounds which catalytically remove $O_2^{\bullet-}$ and reduce $ONOO^-$ [10], as well as other reactive species including carbonate anion ($CO_3^{\bullet-}$), hypochlorous acid (HClO), lipid peroxyl, and alkoxyl radicals [85].

In vivo Mn porphyrins also readily couple with cellular reductants (ascorbate, and thiols) and therefore remove reactive species in catalytic manner [10, 11]. Such has been shown with $ONOO^-$ and $O_2^{\bullet-}$. Further, MnPs, in a thiol oxidase or glutathione peroxidase fashion, can couple with H_2O_2 to oxidize and/or glutathionylate signaling thiols of p50 and p65 subunits of NF-κB subunits, which would result in inhibition of NF-κB and subsequent suppression of excessive inflammation [35]. This unique activity has been demonstrated repeatedly in in vitro and in vivo experiments [9, 45]. Reactive oxygen and nitrogen species (ROS/RNS) also participate in regulating cellular transcriptional signaling [19, 63, 64]. It has been shown in middle cerebral artery occlusion (MCAO) models that MnPs protect tissue in part, by inhibiting NF-κB transcription activity [12, 70] and in turn suppression of inflammatory responses. These findings form a scientific basis for developing MnP-based therapy for CNS injury.

The efficacy of treatment depends on dosing regime: administration during the optimal therapeutic window and for a sufficient duration, in the absence of adverse drug effect. Thus, it is critical to define the status of oxidative stress and the dynamic changes that occur during injury progression. $O_2^{\bullet-}$, the initial free radical generated during injury, comes from several sources. In stroke or trauma, the interruption of blood flow leads to energy depletion and impaired metabolism. Electrons that are normally transferred along the respiratory chain of the mitochondria to oxygen at the site of cytochrome oxidase begin to accumulate at different sites of electron transport chain (e.g., complexes I and III). When blood flow is restored, oxygen, O_2, is reduced by accepting an electron, giving rise to superoxide $O_2^{\bullet-}$. This is believed to be the major source of reperfusion-derived $O_2^{\bullet-}$. Other biological chemicals including epinephrine, norepinephrine, ascorbate, semiquinones, and hemoglobin can also autoxidize to form $O_2^{\bullet-}$. NADPH oxidases are another source of $O_2^{\bullet-}$. The NADPH oxidase complex is a cluster of proteins that donate an electron from NADPH to molecular oxygen to produce $O_2^{\bullet-}$. Although NADPH oxidases present a natural defense system to kill bacteria, they can also damage tissue at the injury site. Studies have shown that suppression of NADPH oxidases improves functional and histological recovery after cerebral ischemia and reperfusion injury [32, 48, 51].

Oxidative stress, through activation of redox-regulated transcription factors, leads to expression of inflammatory cytokines and chemokines, further disruption of the blood-brain barrier, recruitment of neutrophils, and activation of macrophage and microglia, all of which in turn amplify and sustain the inflammatory response. This response may persist for several weeks. While inflammation causes direct tissue injury, it also serves a role in repair and neurogenesis. Hence, the interaction between Mn porphyrins and recovery may be complex and defined, at least in part, by timing of treatment onset and duration of treatment. Better definition of these interactions will likely lead to improved efficacy from the MnPs. We know from existing data (see below) that MnP treatment must be sustained for at least 1 week to provide long-term efficacy, presumably because this interval represents the acute phase of oxidative/inflammatory injury. Effects of longer treatment intervals on the biology of recovery are not yet defined.

24.3 Therapeutic Potential of MnPs After CNS Injury

Transgenic and knockout mice have served as novel tools in understanding the role of endogenous antioxidant enzymes in the progression of CNS injury. Overexpression of Cu, ZnSOD is neuroprotective in multiple injury models including cerebral ischemia [23, 29], spinal cord injury [80, 94], intracerebral hemorrhage [88], and cardiac arrest and resuscitation [93]. MnSOD overexpression is also neuroprotective in models of cerebral ischemia [30, 41, 47, 50, 52] and subarachnoid hemorrhage [46]. Our group examined the role of EC-SOD in brain injury and confirmed that it is protective as well [56, 65, 66, 69]. These studies provide a rationale for investigation of exogenous SOD mimics in treating brain and spinal cord injury.

MnPs are more effective SOD mimics in SOD-deficient *E. coli* and *S. cerevisiae* models when compared to other compounds such as Mn(II) macrocyclic polyamines, nitroxide, nitrone, and Mn(III) salen complexes [84]. While Mn(II) macrocyclic polyamines are as efficient in catalyzing $O_2^{\bullet-}$ dismutation as MnPs, they are less stable towards the loss of Mn. The log value of their stability constant log *K* is 13.6 for M40403, which is similar to the stability constant for complex of Mn(II) with EDTA, log *K* = 14.04; in turn biological chelators can easily extract Mn from M40403 [7, 10]. MnPs are of such high stability that detection of their stability constant is not possible. The log K is likely much higher than 30. Most efficacy studies of MnPs have focused on three compounds: manganese(III) *meso*-tetrakis (*N*-ethylpyridinium-2-yl) porphyrin (MnTE-2-PyP^{5+}) [45, 68], manganese(III) *meso*-tetrakis (*N,N′*-diethylimidazolium-2-yl) porphyrin (MnTDE-2-ImP^{5+}) [68, 72], and manganese(III) *meso*-tetrakis (*N-N*-hexylpyridinium-2-yl) porphyrin (MnTnHex-2-PyP^{5+}) [70] (Fig. 24.2). These compounds have similar SOD-like activities, but have different redox properties and different lipophilicities and bulkiness, which affect their pharmacokinetic profiles, ability to cross the blood-brain barrier, and accumulation/distribution within cells and mitochondria [13]. All three of these MnPs have been demonstrated to be efficacious [11, 77–79, 90], albeit with differential impacts on outcome, most likely due to bioavailability.

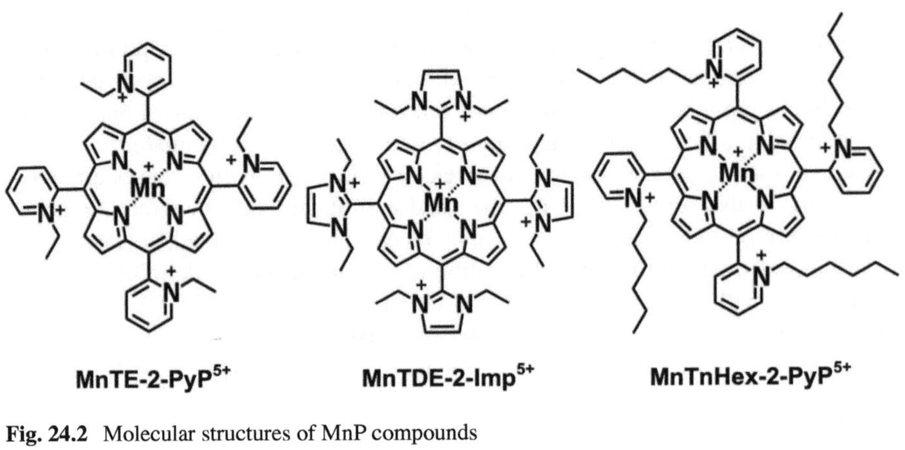

Fig. 24.2 Molecular structures of MnP compounds

24.3.1 *Manganese(III) Meso-Tetrakis (N-Ethylpyridinium-2-yl)Porphyrin (MnTE-2-PyP^{5+})*

MnTE-2-PyP^{5+} was the first MnP demonstrated to have high SOD-like activity and was thus the first examined in animal CNS injury models [13]. The MCAO is the most studied model of CNS injuries. It is performed by inserting a nylon monofilament coated with silicone into the right internal carotid artery until it reached to the anterior cerebral artery, largely blocking blood flow to the MCA. Cerebral blood flow was restored after 90 min of occlusion and injured rats were left to survive for 1 week. To assure bioavailability to brain, 150 or 300 ng MnTE-2-PyP^{5+} was directly injected into the right lateral ventricle 60 min before ischemia. Both groups treated with MnTE-2-PyP^{5+} lost less body weight ($P=0.004$) and had improved neuroscores ($P<0.01$) and reduced infarct volumes ($P<0.01$) compared to the phosphate buffered saline (PBS)-treated vehicle group. These findings confirmed that MnTE-2-PyP^{5+} is neuroprotective in ischemic brain injury [45].

Treatment following stroke is often delayed several hours since patients are often slow to present for medical care. An effective compound for treating stroke, therefore, must have a wide therapeutic window. To assess the therapeutic window for MnTE-2-PyP^{5+}, rats were subjected to 90 min MCAO and received treatment 60 min before ischemia, or at 5 min, 90 min, 6 h, or 12 h after ischemia [45]. PBS or 300 ng MnTE-2-PyP^{5+} was injected directly into the right lateral ventricle. Injured rats were left to survive for 1 week. Neurological deficits and infarct volumes were then measured. In rats treated with MnTE-2-PyP^{5+} before ischemia, or at 5 or 90 min post-injury, the infarct volume was reduced to approximately 25 % of the infarct volume of PBS-treated vehicle group. In the group that received delayed treatment at 6 h post-ischemia, the infarct volume was ~50 % of infarct volume of PBS-treated animals. However, in rats treated at 12 h post-ischemia, the infarct volume was not significantly different from the one of the vehicle group (Fig. 24.3). These findings indicate that MnTE-2-PyP^{5+} offers a sufficiently wide therapeutic window of 6 h.

To determine the efficacy in other species, MCAO was performed in mice [45]. Ischemia was induced for 90 min. PBS, or 50 or 100 ng MnTE-2-PyP^{5+}, was injected into the right lateral ventricle at 5 min following reperfusion. Neuroscores and infarct volumes were assessed at 24 h post-injury. Again, both MnTE-2-PyP^{5+} treatment groups had improved neuroscores and reduced infarct volumes, indicating that efficacy is reproducible across species.

In humans, intravenous administration is preferred. To test efficacy following intravenous administration, mice were subjected to 90 min MCAO [45]. PBS, or 1 or 2 mg/kg MnTE-2-PyP^{5+}, was given through a jugular vein catheter at 5 min following reperfusion, and neurological deficits and infarct volumes were examined at 24 h. Compared to vehicle, both treatment groups had improved neuroscores ($p=0.04$, vehicle 3 ± 0.75, 1 mg/kg MnTE-2-PyP^{5} 3 ± 2, 2 mg/kg MnTE-2-PyP^{5} 2 ± 2) and reduced infarct volumes ($p=0.009$, vehicle 77 ± 24 mm^3, 1 mg/kg MnTE-2-PyP^{5} 52 ± 33 mm^3, 2 mg/kg MnTE-2-PyP^{5} 42 ± 33 mm^3).

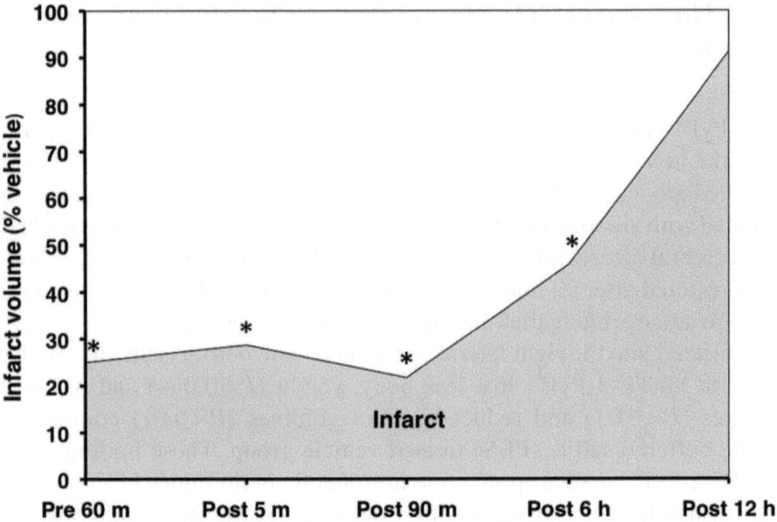

Fig. 24.3 Therapeutic window of MnTE-2-PyP^{5+} in a rat model of middle cerebral artery occlusion. PBS or 300 ng MnTE-2-PyP^{5+} was injected into the right lateral cerebral ventricle at 60 min before ischemia, or 5 or 90 min, or 6 or 12 h following reperfusion. Infarct volume measured at 7 days post-injury is reported as a percentage change from the mean infarct volume in the respective vehicle-treated groups ($n = 13$–16 per group). $*P < 0.05$

With MCAO treatments, we are targeting oxidative stress. Thus, in the rat model of transient MCAO, the extent to which MnTE-2-PyP^{5+} decreases oxidative stress in ischemic brain injury was evaluated [45]. Aconitase activity, an enzyme that is sensitive to superoxide-driven inactivation, and 8-hydroxy-2′-deoxyguanosine (8-OHdG), a biomarker of oxidative stress-related DNA damage, were measured in brain tissue samples harvested at 4 h posttreatment. Ischemic injury caused decreased aconitase activity, while 8-OHdG increased. However, in rats treated with 300 ng MnTE-2-PyP^{5+}, brain aconitase activity was 47 % greater than in the vehicle group, and 8-OHdG levels were decreased compared to the vehicle group. These findings demonstrated that MnTE-2-PyP^{5+} attenuates oxidative stress, which is consistent with its beneficial therapeutic effects.

MnTE-2-PyP^{5+} showed no dose-dependent effect in mouse MCAO studies even when the dose was doubled, regardless of administration route. One explanation is that a dose higher than the minimal dose required for efficacy does not enhance the protective effects due to the catalytic nature of its action. Hence, doses examined to date may have exceeded the dose necessary to provide maximal efficacy. More detailed studies are needed to determine the minimally efficacious dose and associate outcome effects with impact on oxidative stress.

The efficacy of MnTE-2-PyP^{5+} in other CNS injury models remains unclear. Two studies performed in forebrain ischemia and closed-head injury models yielded negative results [42, 45]. A rat model of forebrain ischemia mimics brain dysfunction simulates some elements of cardiac arrest. In this model, blood flow to the cortex and hippocampus is transiently interrupted by ligating both common carotid arteries for 10 min in a setting of hypotension, which was induced by withdrawing blood from the right jugular vein to maintain mean arterial blood pressure (MAP) at 30 mmHg. This low MAP prevents the collateral blood flow from the basilar artery, thereby causing near-complete ischemia in the cortex, basal ganglia, and hippocampus; however, the thalamus and brain stem retain sufficient flow to prevent the injury [91].

PBS or a single 300 ng dose of MnTE-2-PyP^{5+} was injected directly into the right lateral cerebral ventricle 60 min before injury. Neurological function and cell death of region 1 of hippocampus proper (CA1) were assessed at 5 days post-injury. No difference in neurological function or percentage of dead cells in CA1 was detected between MnTE-2-PyP^{5+}- and PBS-treated groups. This suggests that this compound does not protect CA1 neurons against ischemic insult, even though it was present in the ventricle adjacent to the hippocampus. Yet, only single 300 ng dose was tested. Plausibly, the sustained treatment would allow for a different result (see below).

Similar lack of effect was reported in a mouse traumatic brain injury model [42]. Closed-head injury was induced by placing a concave 4.2-mm metallic disc on the midline of the skull surface just caudal to bregma and then using a pneumatic cylinder containing a metal piston (3 mm diameter) to impact the disc at a velocity of 6.8 ± 0.2 m/s, with a skull depression of 3 mm. PBS or 3 mg/kg MnTE-2-PyP^{5+} was given intravenously 15 min after the injury. Neurological function and histology were assessed at 5 days post-injury. No effect of MnTE-2-PyP^{5+} was found in this acute outcome experiment. Because aconitase activity was significantly preserved in the treatment group at 2 h after injury, but not at 6 or 24 h, a single dose and its level might not have been sufficient for outcome. Therefore, another experiment was performed using 6 mg/kg MnTE-2-PyP^{5+}. Motor function was tested daily on a rotarod for the first 5 days after injury, but the treatment group showed no improvement. The Morris water maze test was performed at 3 weeks post-injury. Again, the treatment group showed no improvement in learning memory deficit. A skull depression of 3 mm normally causes severe brain damage. Again, MnTE-2-PyP^{5+} failed to rescue tissue from this insult. A single dose may not have been adequate to cause sustained improvement in a long-term survival outcome study [77]. Thus, while evidence of efficacy was absent, a sustained treatment paradigm might yield a different result.

MnTE-2-PyP^{5+} has two major adverse side effects in rodents, behavioral change and arterial hypotension. Normal mice tolerate up to 5 mg/kg intravenous MnTE-2-PyP^{5+}. However, when this dose was given 5 min after 90 min MCAO, severe behavioral changes including proptosis, ataxia, and hypersensitivity to sound appeared at approximately 2 h after injection [45]. This adverse effect was sometimes lethal and was also observed in injured rats given an intracerebroventricular injection of 1 μg MnTE-2-PyP^{5+}. Mice with cerebral ischemia tolerated up to 2 mg/kg intravenously without behavioral changes. Thus, 2 mg/kg served as the upper limit for the study. Hypotension is an adverse effect dependent upon the specific MnP, its dose, and the species and appears to be mediated by an intravascular mechanism [62]. Transient, but significant, hypotension was induced after intravenous injection of 25 μg/kg

$MnTE\text{-}2\text{-}PyP^{5+}$ in normal rats. However, 5 mg/kg intravenous $MnTE\text{-}2\text{-}PyP^{5+}$ did not induce hypotension in mice [62]. When injected into the lateral cerebral ventricle, however, hypotension was not induced.

Beneficial effects of $MnTE\text{-}2\text{-}PyP^{5+}$ have not yet been investigated in cases when blood flow to the brain is not restored, as in a permanent focal ischemia state, which is commonly seen in clinical practice. Overexpression of Cu,ZnSOD does not protect brain in permanent cerebral ischemia [22]. The extent to which this predicts efficacy of $MnTE\text{-}2\text{-}PyP^{5+}$ is not known. Stroke patients who present for care early and have a thrombotic vascular occlusion are treated with tissue plasminogen activator (tPA) or endovascular thrombectomy. $MnTE\text{-}2\text{-}PyP^{5+}$ combined with tPA or endovascular intervention may increase the efficacy of current stoke treatment interventions. Preclinical testing is needed to explore this possibility.

24.3.2 *Manganese(III) Tetrakis (N,N′-Diethylimidazolium-2-yl)Porphyrin (MnTDE-2-ImP^{5+})*

$MnTDE\text{-}2\text{-}ImP^{5+}$ is less toxic due to its bulky structure, which suppresses the unfavorable interaction with biological targets. Thus, it accumulates less in *E. coli* and is significantly less protective to SOD-deficient *E. coli* [84]. Its SOD-like activity is similar to that of $MnTE\text{-}2\text{-}PyP^{5+}$ [11]. Yet it has 118 mV higher $E_{1/2}$ for $Mn^{III}P/Mn^{II}P$ redox couple, the implication of which on therapeutic effects has not yet been assessed [68]. Behavioral side effects were compared in normal rats using multiple graded doses of these two compounds. The maximal intracerebroventricular doses that induced only slightly noticeable behavioral effects were 600 ng $MnTE\text{-}2\text{-}PyP^{5+}$ and 9 μg $MnTDE\text{-}2\text{-}ImP^{5+}$ [68], a reduction in toxicity to 1/15 of $MnTE\text{-}2\text{-}PyP^{5+}$.

The differential efficacy of these two compounds was assessed in rat focal cerebral ischemia model [68]. Rats were subjected to 90 min MCAO, and given one dose of PBS, or either 300 or 4500 ng of $MnTDE\text{-}2\text{-}ImP^{5+}$, or 300 ng $MnTE\text{-}2\text{-}PyP^{5+}$ via intracerebroventricular injection at 90 min after reperfusion. Neurological deficit and infarct volume were assessed at 7 days post-injury. Surprisingly, these compounds improved functional and histological outcomes equally at 300 ng doses. The 4500 ng dose of $MnTDE\text{-}2\text{-}ImP^{5+}$did not provide more protection compared to 300 ng. The compound is catalytic and perhaps 300 ng may have produced a ceiling effect due to its efficiency. To test this, another experiment was performed. The same experimental design was used except that $MnTDE\text{-}2\text{-}ImP^{5+}$ was administered in doses of 0, 100, 300, or 900 ng. Both sensorimotor function ($P=0.006$) and total infarct volume ($P=0.03$) were improved in a dose-dependent fashion (Fig. 24.4). Thus, a ceiling effect is yet to be defined, but likely is near 900 ng as this dose produced outcome benefit similar to that previously observed at 4500 ng. If so, this is important because limited $MnTDE\text{-}2\text{-}ImP^{5+}$ bioavailability, when given parenterally, may still provide near-maximal therapeutic efficacy.

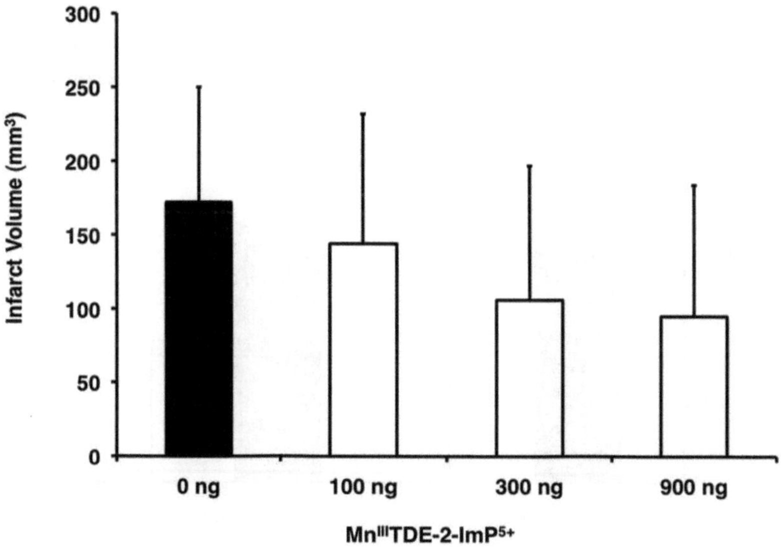

Fig. 24.4 Dose-dependent effect of MnTDE-2-ImP^{5+} in a rat model of middle cerebral artery occlusion. At 90 min following reperfusion, 0, or 100, or 300, or 900 ng MnTDE-2-ImP^{5+} was injected into the right lateral cerebral ventricle. Infarct volume was measured at 7 days post-injury. A dose-dependent decrease in total infarct volume (mean ± SD) was observed ($P=0.03$)

To define the brain elimination half-life of MnTDE-2-ImP^{5+}, 300 ng was injected into the right lateral ventricle in normal rats, and brain samples were harvested at 0.5, 1, 3, 6, 24, or 48 h after injection. The brain half-life was determined to be ~10 h [68].

To determine the therapeutic window, rats were subjected to 90 min MCAO, and 300 ng MnTDE-2-ImP^{5+} was given 6 h following reperfusion. The infarct volume was 160 ± 74 mm^3 in the vehicle group, and 92 ± 80 mm^3 in the treatment group ($P=0.015$) [68]. The data indicate that the therapeutic window is similar to MnTE-2-PyP^{5+}.

All rats in experiments described above survived for 7 days. To assess the long-term efficacy of MnTDE-2-ImP^{5+}, another experiment was performed. The protective effect of a single intracerebroventricular dose of 900 ng was compared to the effects of an intracerebroventricular loading dose of 900 ng followed by a continuous infusion of 56 ng/h for 1 week [72]. Sensorimotor function ($P=0.03$) and total infarct volume ($P=0.003$) were improved at 8 weeks post-injury in rats treated for 1 week (Fig. 24.5), but not in rats treated with a single dose. This suggests ongoing molecular events in the post-ischemic brain amenable to MnP treatment. However, should that treatment be discontinued prior to resolution of such events, a rebound may occur resulting in loss of therapeutic benefit.

To test the protective effects of MnTDE-2-ImP^{5+} in mice, 60 min of transient MCAO was performed [68]. An intravenous bolus dose of PBS, or 125 or 500 μg/kg MnTDE-2-ImP^{5+}, was given at 5 min after reperfusion onset followed by PBS, or

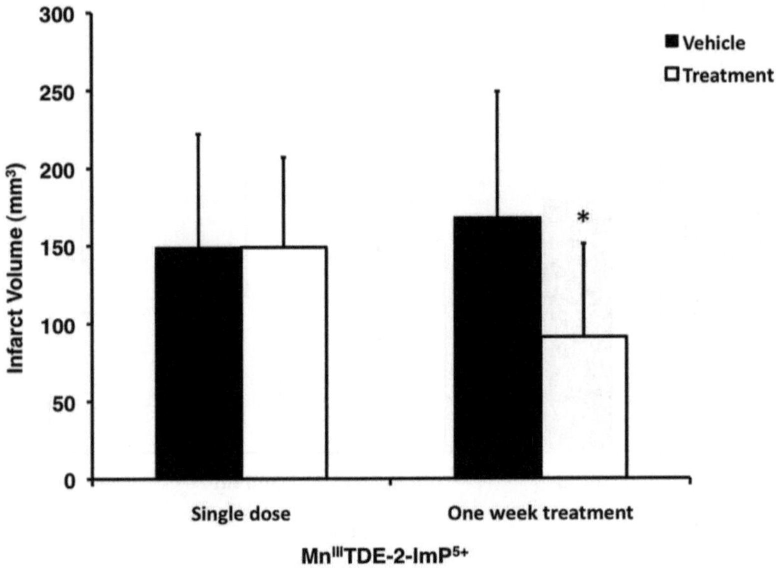

Fig. 24.5 MnTDE-2-ImP^{5+} efficacy in a long-term survival study in a rat model of middle cerebral artery occlusion. At 90 min following reperfusion, rats were treated with vehicle (PBS) or a single dose of 900 ng MnTDE-2-ImP^{5+}. A separate cohort received vehicle or a 900 ng bolus of MnTDE-2-ImP^{5+} followed by 56 ng/h continuous infusion for 1 week ($n = 13–15$ per group). Infarct volume (mean ± SD) was measured at 8 weeks post-injury. *$P = 0.003$

250 μg/kg/h or 1 mg/kg/h for 24 h. Infarct volume at 24 h post-injury was reduced by 28 % in treated groups compared to vehicle; however, no difference was found between the two MnTDE-2-ImP^{5+} doses. Neurologic scores followed the same pattern. A pharmacokinetic study showed that, in the high-dose group, plasma MnTDE-2-ImP^{5+} concentration was 2475 ± 471 ng/mL at 1 h post-treatment and slightly increased at 5 h. At 24 h, the concentration increased to 5353 ± 4770 ng/mL. In the ischemic hemisphere, brain concentrations in the high-dose group were 78 ± 22, 151 ± 97, and 582 ± 645 ng/mL at 1, 5, and 24 h post-treatment, respectively, which was two- to threefold higher than in the contralateral (uninjured) hemisphere. At 5 h following the low-dose infusion, the MnTDE-2-ImP^{5+} concentration in brain tissue was 34 ± 21 ng/mL and was identified as the brain protective concentration. Efficacy was further demonstrated in an in vitro experiment in which MnTDE-2-ImP^{5+} attenuated oxygen/glucose/deprivation (OGD)-induced lactate dehydrogenase (LDH) release and preserved aconitase activity in primary neuronal cell cultures ($P < 0.01$) [68, 92].

A proteomic analysis was performed in mouse brain samples harvested from sham surgery and ischemic injury at 6 h following vehicle or high-dose MnTDE-2-ImP^{5+} given via intravenous infusion as described above [68]. The 2D gel electrophoresis showed that 40 proteins were upregulated and 6 proteins were downregulated twofold or more in the treatment compared to the vehicle group. Ischemia alone upregulated 41 proteins and downregulated 9 proteins compared to sham surgery. Intravenous infusion of high-dose MnTDE-2-ImP^{5+} upregulated 37 proteins in ischemic brain and downregulated no proteins compared to vehicle. The same dose of MnTDE-2-ImP^{5+} also attenuated mRNA expression of inflammatory genes defined by reverse transcriptase polymerase chain reactions [16].

The efficacy of MnTDE-2-ImP^{5+} was also tested in a different mouse injury model, transient spinal cord compression injury. It was induced by transversely placing a flexible silicone tube into the vertebral canal and removing it after 60 min. Such procedure generates a moderate degree of traumatic and ischemic tissue damage. Intrathecal injection of PBS, or 2.5 or 5.0 μg MnTDE-2-ImP^{5+} at the onset of compression, improved rotarod performance and total histology damage score in a 3-week survival study [71]. However, intravenous continuous infusion of 0.5 mg/kg followed by 1 mg/kg/h for 24 h did not provide protection, indicating that the efficacy in this model was dependent upon the route of administration, and that bioavailability is critical in defining therapeutic efficacy.

24.3.3 *Manganese(III) Meso-Tetrakis(*N*-Hexylpyridinium-2-yl)Porphyrin (MnTnHex-2-PyP^{5+})*

Both MnTE-2-PyP^{5+} and MnTDE-2-ImP^{5+} are hydrophilic compounds and have demonstrated efficacy in treating cerebral ischemia, but exhibit poor ability to penetrate blood-brain barrier. MnTnHex-2-PyP^{5+} is ~4 orders of magnitude more lipophilic than MnTE-2-PyP^{5+} [11]. In normal mice, brain MnTnHex-2-PyP^{5+} accumulation is approximately ninefold greater than MnTE-2-PyP^{5+} at 24 h after treatment with a single intraperitoneal dose [70].

The maximal intravenous dose of MnTnHex-2-PyP^{5+} in rats was determined based on arterial blood pressure changes. In 300 g Wistar rats, 75 μg MnTnHex-2-PyP^{5+} was the maximal dose at which arterial blood pressure was not substantively affected, compared to a maximal dose of only 10 μg/kg MnTE-2-PyP^{5+} [62]. Thus, MnTnHex-2-PyP^{5+} has less effect on arterial blood pressure than MnTE-2-PyP^{5+} and can be administered to rats via a systemic route. The reason may lie in fact that cationic charges are less exposed to interactions with biotargets when they are surrounded by a long hexyl chain when compared to short ethyl chains.

Based on this data, we performed MCAO outcome studies in rats treated with 75 μg MnTnHex-2-PyP^{5+} bolus intravenous infusion followed by 225 μg MnTnHex-2-PyP^{5+} injected subcutaneously twice a day [70]. To define the dynamic change in brain concentration, brain and plasma samples were harvested at 1 or 4 h, and at 1, 3, or 7 days after start of treatment. Pharmacokinetic analysis revealed that up to 25 nM MnTnHex-2-PyP^{5+} was accumulated in the brain at the end of 7 days [70]. The dose is similar to ~35 nM identified as protective dose of MnTDE-2-ImP^{5+} when given intracerebroventricular injection. After 90 min MCAO, treatment began at 5 min or 6 h after reperfusion and improved both neurological function and infarct volume at 7 days post-injury. In a second confirmatory experiment, the efficacy of treatment beginning at 6 h post-reperfusion was assessed. Cerebral blood flow was monitored with laser Doppler during the surgical procedure and for the duration of ischemia to evaluate the severity of ischemia insult for all rats. Intra-ischemia laser Doppler red blood cell flow velocity was reduced to 20–30 % of baseline in both vehicle and MnTnHex-2-PyP^{5+} treatment groups. Given this documented similarity of ischemic insult, MnTDE-2-ImP^{5+} treatment continued to improve both neurological ($P=0.002$) and histological outcome ($P=0.01$), as measured at 7 days post-ischemia [70].

Since MnTDE-2-ImP^{5+} modulates inflammatory responses at the transcriptional level [16, 68], an experiment was performed to determine whether MnTnHex-2-PyP^{5+} operates through a similar mechanism of action. Rats were subjected to 90 min MCAO and received 225 μg MnTnHex-2-PyP^{5+} at 5 min after reperfusion onset, followed by another subcutaneous injection 12 h later. MnTnHex-2-PyP^{5+} inhibited TNF-α and IL-6 expression in the ischemic brain at 24 h post-injury. To determine whether this decrease in pro-inflammatory cytokines was modulated by NF-κB, electromobility-shift assays (EMSA) were performed on nuclear extracts obtained at 6 h post-MCAO in rats treated with MnTnHex-2-PyP^{5+} or vehicle. NF-κB activation was increased in the vehicle group. Treatment with MnTnHex-2-PyP^{5+} inhibited nuclear NF-κB binding, which corresponded to a decrease in nuclear translocation of NF-κB, as shown by NF-κB p65 supershift analysis. Thus, MnTnHex-2-PyP^{5+} suppressed inflammatory responses by inhibiting NF-κB activation.

Cerebral artery aneurysm rupture is a hemorrhagic stroke that causes subarachnoid hemorrhage (SAH) leading to cerebral vasospasm and delayed neurological deficit. Lysis of red blood cells in the blood clot releases $O_2^{\bullet-}$, which reacts with $^{\bullet}NO$ to form $ONOO^-$. Decreased $^{\bullet}NO$ and vascular wall damage by $ONOO^-$ contribute to cerebral artery vasospasm. In a mouse model of subarachnoid hemorrhage, the hypothesis was tested whether MnTnHex-2-PyP^{5+} scavenges $O_2^{\bullet-}$ and $ONOO^-$ and preserves $^{\bullet}NO$ [70]. A blunt 6-0 filament was inserted into the internal carotid artery (ICA) and allowed to penetrate through the bifurcation of the anterior and middle cerebral arteries (ACA and MCA) to induce subarachnoid hemorrhage. Sixty minutes after injury, 225 μg MnTnHex-2-PyP^{5+} or vehicle was given intraperitoneally twice per day for 3 days. Neurological function was assessed at 72 h, and cerebral arteries were casted by ink-gel perfusion. The internal diameters of the cerebral arteries were measured and the hemorrhagic size was scored. No differences in hemorrhagic scores were found between groups. However, treatment improved neurological function score at 3 days post-injury ($P=0.02$). Further, the diameters of all major cerebral arteries ipsilateral to the perforation were greater in the treatment group compared to vehicle ($P=0.015$ in ACA, 0.0005 in MCA, and 0.003 in ICA). Basilar artery diameter was similar for both groups ($P=0.73$), indicating that the ink-gel perfusion technique was consistent. Hypotension is considered as a side effect of Mn porphyrin treatment, but, here, the vasodilation effect of MnTnHex-2-PyP^{5+} prevents SAH-induced cerebral artery vasospasm and improves tissue perfusion and subsequently attenuated the neurological deficit. Importantly, this vascular effect is not selective; therefore, patients who have already developed vasospasm may be adversely affected by MnTnHex-2-PyP^{5+}, although this has not yet been examined at the preclinical level.

24.3.4 Other Metalloporphyrins in CNS Injury

Two iron porphyrins—Fe(III) *meso*-tetrakis(4-sulfonatophenyl)porphyrin ($FeTSPP^{3-}$) and Fe(III) *meso*-tetrakis(*N*-methylpyridinium-4-yl)porphyrin (FeTM-4-PyP^{5+})—can catalytically decompose peroxynitrite [36, 73]. Intravenous administration of 3 mg/kg $FeTSPP^{3-}$ or FeTM-4-PyP^{5+} at 2 and 6 h post-MCAO attenuated neurological deficit and decreased infarct volume in a rat focal cerebral ischemia 3-day survival study [82]. Intraperitoneal administration of 1 or 3 mg/kg FeTM-4-PyP^{5+} 30 min before 5 min of global ischemia in gerbils also showed a dose-dependent effect in improving neurological score, locomotor activity, and passive avoidance test scores at 24 h and decreased CA1 neuronal death measured 4 days post-ischemia [26]. Thus, two new FePs—FeTE-2-$PyP^{5+}$and FeTnHex-2-PyP^{5+}—were synthesized by replacing Mn with Fe and compared with their respective MnPs [86]. In one experiment that assessed the effect of both compounds on the aerobic growth of SOD-deficient *E. coli*, 1 μm FeTE-2-PyP^{5+} had an effect equal to 20 μM MnTE-2-PyP^{5+}. However, 10 mg/kg FeTE-2-PyP^{5+} given twice a day for 1 week was more toxic in mice compared to MnTE-2-PyP^{5+}. The toxicity issues of FePs are still an issue to be explored given involvement of either free Fe^{2+} or involvement of the Fe site in the Fenton chemistry formation of $^{\bullet}OH$ species [11, 86]. Interestingly, in mice treated with a single intraperitoneal injection of 5 mg/kg FeTnHex-2-PyP^{5+} or MnTnHex-2-PyP^{5+}, all 4 of the MnTnHex-2-PyP^{5+} mice died within 2 h after injection, but two of the four FeTnHex-2-PyP^{5+} mice survived for 1 week. Further, 10 mg/kg FeTnHex-2-PyP^{5+} injected intraperitoneally did not lead to hypotension, while the same dose of MnTnHex-2-PyP^{5+} caused lethal hypotension. Such data suggest that FeTnHex-2-PyP^{5+} may be preferred as it does not cause hypotension, but it may not have a beneficial vascular effect potentially limiting its application. The origin of hypotension is still not understood. The differential efficacy of MnPs and FePs requires further exploration.

A very lipophilic MnP compound, Mn(III) meso-tetrakis(*N*-n-octylpyridium-2-yl) porphyrin (MnTnOc-2-PyP^{5+}), should also be mentioned here. It had a beneficial effect in an in vitro experiment showing >1000-fold higher efficacy than its hydrophilic analogue, MnTE-2-PyP^{5+}, in attenuating OGD-induced lactate LDH release [92]. MnTE-2-PyP^{5+} is 4.5 log units less lipophilic than MnTnOct-2-PyP^{5+} [85].

24.4 Conclusion

MnPs are clearly neuroprotective against insults to the central nervous system such as cerebral ischemia and reperfusion injury (Table 24.1). Both hydrophilic and lipophilic compounds can improve post-injury neurological scores and decrease tissue damage even when given at 6 h after reperfusion. Further investigation is needed to define the dosing regime in each particular disease model including optimal dosing regimens, particularly in the context of clinically relevant long-term outcome

Table 24.1 Summary of the effects of three MnPs in preclinical studies of CNS injuries

	$MnTE\text{-}2\text{-}PyP^{5+}$	$MnTDE\text{-}2\text{-}ImP^{5+}$	$MnTnHex\text{-}2\text{-}PyP^{5+}$
Focal cerebral ischemia Transient MCAO	R-ICV-P M-ICV/IV-P	R-ICV-P M-IV-P	R-IV/Sq-P
Focal cerebral ischemia Permanent MCAO	?	?	?
Global cerebral ischemia Transient 2VO plus hypotension	R-ICV-N	?	?
Subarachnoid hemorrhage	?	?	M-IP-P
Closed head injury	M-IV-N	?	?
Spinal cord injury	?	M-IT-P M-IV-N	?

MCAO middle cerebral artery occlusion, *2VO* bilateral common carotid artery occlusion. Data were expressed in the format of species-administrative route-result. Species: *R* rats, *M* mice; Administrative route: *IV* intravenous, *ICV* intracerebroventricular, *Sq* subcutaneous, *IP* intraperitoneal, *IT* intrathecal; Results: *P* positive, *N* negative, *?* not yet tested

efficacy analysis. Inflammatory cytokines and reactive species are essential to direct stem cell migration in the process of neurogenesis, and suppression of their function may slow the tissue repair process, although this is undocumented. Additional studies are needed to determine whether sustained treatment after traumatic brain injury, global cerebral ischemia, or intracerebral hemorrhage is effective in contrast to single bolus doses, which have proven ineffective. Although those concerns must be addressed, MnP-based therapy is promising.

References

1. Aird KM, Allensworth JL, Batinic-Haberle I, Lyerly HK, Dewhirst MW, Devi GR. ErbB1/2 tyrosine kinase inhibitor mediates oxidative stress-induced apoptosis in inflammatory breast cancer cells. Breast Cancer Res Treat. 2012;132:109–19.
2. Allensworth JL, Aird KM, Aldrich AJ, Batinic-Haberle I, Devi GR. XIAP inhibition and generation of reactive oxygen species enhances TRAIL sensitivity in inflammatory breast cancer cells. Mol Cancer Ther. 2012;11:1518–27.
3. Ansari MA, Roberts KN, Scheff SW. A time course of contusion-induced oxidative stress and synaptic proteins in cortex in a rat model of TBI. J Neurotrauma. 2008;25:513–26.
4. Assenza G, Zappasodi F, Squitti R, Altamura C, Ventriglia M, Ercolani M, Quattrocchi CC, Lupoi D, Passarelli F, Vernieri F, Rossini PM, Tecchio F. Neuronal functionality assessed by magnetoencephalography is related to oxidative stress system in acute ischemic stroke. Neuroimage. 2009;44:1267–73.
5. Awasthi D, Church DF, Torbati D, Carey ME, Pryor WA. Oxidative stress following traumatic brain injury in rats. Surg Neurol. 1997;47:575–81. discussion 581-572.
6. Ayer RE, Zhang JH. Oxidative stress in subarachnoid haemorrhage: significance in acute brain injury and vasospasm. Acta Neurochir Suppl. 2008;104:33–41.
7. Batinic-Haberle I. Manganese porphyrins and related compounds as mimics of superoxide dismutase. Methods Enzymol. 2002;349:223–33.

8. Batinic-Haberle I, Liochev SI, Spasojevic I, Fridovich I. A potent superoxide dismutase mimic: manganese beta-octabromo-meso-tetrakis-(N-methylpyridinium-4-yl) porphyrin. Arch Biochem Biophys. 1997;343:225–33.
9. Batinic-Haberle I, Rajic Z, Benov L. A combination of two antioxidants (an SOD mimic and ascorbate) produces a pro-oxidative effect forcing Escherichia coli to adapt via induction of oxyR regulon. Anticancer Agents Med Chem. 2011;11:329–40.
10. Batinic-Haberle I, Reboucas JS, Spasojevic I. Superoxide dismutase mimics: chemistry, pharmacology, and therapeutic potential. Antioxid Redox Signal. 2010;13:877–918.
11. Batinic-Haberle I, Spasojevic I. Complex chemistry and biology of redox-active compounds, commonly known as SOD mimics, affect their therapeutic effects. Antioxid Redox Signal. 2014;20:2323–5.
12. Batinic-Haberle I, Tovmasyan A, Roberts ER, Vujaskovic Z, Leong KW, Spasojevic I. SOD therapeutics: latest insights into their structure-activity relationships and impact on the cellular redox-based signaling pathways. Antioxid Redox Signal. 2014;20:2372–415.
13. Batinic-Haberle I, Tovmasyan A, Spasojevic I. An educational overview of the chemistry, biochemistry and therapeutic aspects of Mn porphyrins – from superoxide dismutation to H2O2-driven pathways. 2015. doi:10.1016/j.redox.2015.01.017. Redox Biol.
14. Bayir H, Kagan VE, Tyurina YY, Tyurin V, Ruppel RA, Adelson PD, Graham SH, Janesko K, Clark RS, Kochanek PM. Assessment of antioxidant reserves and oxidative stress in cerebrospinal fluid after severe traumatic brain injury in infants and children. Pediatr Res. 2002;51:571–8.
15. Benov L, Fridovich I. A superoxide dismutase mimic protects sodA sodB Escherichia coli against aerobic heating and stationary-phase death. Arch Biochem Biophys. 1995;322:291–4.
16. Bowler RP, Sheng H, Enghild JJ, Pearlstein RD, Warner DS, Crapo JD. A catalytic antioxidant (AEOL 10150) attenuates expression of inflammatory genes in stroke. Free Radic Biol Med. 2002;33:1141–52.
17. Candelario-Jalil E, Alvarez D, Merino N, Leon OS. Delayed treatment with nimesulide reduces measures of oxidative stress following global ischemic brain injury in gerbils. Neurosci Res. 2003;47:245–53.
18. Cernak I, Savic VJ, Kotur J, Prokic V, Veljovic M, Grbovic D. Characterization of plasma magnesium concentration and oxidative stress following graded traumatic brain injury in humans. J Neurotrauma. 2000;17:53–68.
19. Chan PH. Reactive oxygen radicals in signaling and damage in the ischemic brain. J Cereb Blood Flow Metab. 2001;21:2–14.
20. Chan PH. Role of oxidants in ischemic brain damage. Stroke. 1996;27:1124–9.
21. Chan PH, Epstein CJ, Li Y, Huang TT, Carlson E, Kinouchi H, Yang G, Kamii H, Mikawa S, Kondo T, et al. Transgenic mice and knockout mutants in the study of oxidative stress in brain injury. J Neurotrauma. 1995;12:815–24.
22. Chan PH, Kamii H, Yang G, Gafni J, Epstein CJ, Carlson E, Reola L. Brain infarction is not reduced in SOD-1 transgenic mice after a permanent focal cerebral ischemia. Neuroreport. 1993;5:293–6.
23. Chan PH, Kawase M, Murakami K, Chen SF, Li Y, Calagui B, Reola L, Carlson E, Epstein CJ. Overexpression of SOD1 in transgenic rats protects vulnerable neurons against ischemic damage after global cerebral ischemia and reperfusion. J Neurosci. 1998;18:8292–9.
24. Cornelius C, Crupi R, Calabrese V, Graziano A, Milone P, Pennisi G, Radak Z, Calabrese EJ, Cuzzocrea S. Traumatic brain injury: oxidative stress and neuroprotection. Antioxid Redox Signal. 2013;19:836–53.
25. Darwish RS, Amiridze N, Aarabi B. Nitrotyrosine as an oxidative stress marker: evidence for involvement in neurologic outcome in human traumatic brain injury. J Trauma. 2007;63:439–42.
26. Dhar A, Kaundal RK, Sharma SS. Neuroprotective effects of FeTMPyP: a peroxynitrite decomposition catalyst in global cerebral ischemia model in gerbils. Pharmacol Res. 2006; 54:311–6.

27. Endo H, Nito C, Kamada H, Yu F, Chan PH. Reduction in oxidative stress by superoxide dismutase overexpression attenuates acute brain injury after subarachnoid hemorrhage via activation of Akt/glycogen synthase kinase-3beta survival signaling. J Cereb Blood Flow Metab. 2007;27:975–82.
28. Fatima G, Sharma VP, Das SK, Mahdi AA. Oxidative stress and antioxidative parameters in patients with spinal cord injury: implications in the pathogenesis of disease. Spinal Cord. 2014.
29. Francis JW, Ren J, Warren L, Brown Jr RH, Finklestein SP. Postischemic infusion of Cu/Zn superoxide dismutase or SOD:Tet451 reduces cerebral infarction following focal ischemia/reperfusion in rats. Exp Neurol. 1997;146:435–43.
30. Fujimura M, Morita-Fujimura Y, Kawase M, Copin JC, Calagui B, Epstein CJ, Chan PH. Manganese superoxide dismutase mediates the early release of mitochondrial cytochrome C and subsequent DNA fragmentation after permanent focal cerebral ischemia in mice. J Neurosci. 1999;19:3414–22.
31. Gauter-Fleckenstein B, Fleckenstein K, Owzar K, Jiang C, Reboucas JS, Batinic-Haberle I, Vujaskovic Z. Early and late administration of MnTE-2-PyP5+ in mitigation and treatment of radiation-induced lung damage. Free Radic Biol Med. 2010;48:1034–43.
32. Guo F, Song W, Jiang T, Liu L, Wang F, Zhong H, Yin H, Wang Q, Xiong L. Electroacupuncture pretreatment inhibits NADPH oxidase-mediated oxidative stress in diabetic mice with cerebral ischemia. Brain Res. 2014;1573:84–91.
33. Hall NC, Packard BA, Hall CL, de Courten-Myers G, Wagner KR. Protein oxidation and enzyme susceptibility in white and gray matter with in vitro oxidative stress: relevance to brain injury from intracerebral hemorrhage. Cell Mol Biol (Noisy-le-Grand). 2000;46:673–83.
34. Hayashi T, Hamakawa K, Nagotani S, Jin G, Li F, Deguchi K, Sehara Y, Zhang H, Nagano I, Shoji M, Abe K. HMG CoA reductase inhibitors reduce ischemic brain injury of Wistar rats through decreasing oxidative stress on neurons. Brain Res. 2005;1037:52–8.
35. Jaramillo MC, Briehl MM Batinic-Haberle I, Tome ME. Inhibition of the electron transport chain via the pro-oxidant activity of the manganese porphyrin, MnTE-2-PyP5+ modulates bioenergetics and enhances the response to chemotherapy in lymphoma cells. Free Radic Biol Med. 2015 (In Press).
36. Jensen MP, Riley DP. Peroxynitrite decomposition activity of iron porphyrin complexes. Inorg Chem. 2002;41:4788–97.
37. Jeong HJ, Kim DW, Kim MJ, Woo SJ, Kim HR, Kim SM, Jo HS, Hwang HS, Kim DS, Cho SW, Won MH, Han KH, Park JS, Eum WS, Choi SY. Protective effects of transduced Tat-DJ-1 protein against oxidative stress and ischemic brain injury. Exp Mol Med. 2012;44:586–93.
38. Jia Z, Zhu H, Li J, Wang X, Misra H, Li Y. Oxidative stress in spinal cord injury and antioxidant-based intervention. Spinal Cord. 2012;50:264–74.
39. Kawase M, Murakami K, Fujimura M, Morita-Fujimura Y, Gasche Y, Kondo T, Scott RW, Chan PH. Exacerbation of delayed cell injury after transient global ischemia in mutant mice with CuZn superoxide dismutase deficiency. Stroke. 1999;30:1962–8.
40. Khan I, Batinic-Haberle I, Benov LT. Effect of potent redox-modulating manganese porphyrin, MnTM-2-PyP, on the Na(+)/H(+) exchangers NHE-1 and NHE-3 in the diabetic rat. Redox Rep. 2009;14:236–42.
41. Kim GW, Kondo T, Noshita N, Chan PH. Manganese superoxide dismutase deficiency exacerbates cerebral infarction after focal cerebral ischemia/reperfusion in mice: implications for the production and role of superoxide radicals. Stroke. 2002;33:809–15.
42. Leinenweber SB, Sheng H, Lynch JR, Wang H, Batinic-Haberle I, Laskowitz DT, Crapo JD, Pearlstein RD, Warner DS. Effects of a manganese (III) porphyrin catalytic antioxidant in a mouse closed head injury model. Eur J Pharmacol. 2006;531:126–32.
43. Liu D, Shan Y, Valluru L, Bao F. Mn (III) tetrakis (4-benzoic acid) porphyrin scavenges reactive species, reduces oxidative stress, and improves functional recovery after experimental spinal cord injury in rats: comparison with methylprednisolone. BMC Neurosci. 2013;14:23.
44. Luo J, Shi R. Diffusive oxidative stress following acute spinal cord injury in guinea pigs and its inhibition by polyethylene glycol. Neurosci Lett. 2004;359:167–70.

45. Mackensen GB, Patel M, Sheng H, Calvi CL, Batinic-Haberle I, Day BJ, Liang LP, Fridovich I, Crapo JD, Pearlstein RD, Warner DS. Neuroprotection from delayed postischemic administration of a metalloporphyrin catalytic antioxidant. J Neurosci. 2001;21:4582–92.
46. Matz PG, Fujimura M, Lewen A, Morita-Fujimura Y, Chan PH. Increased cytochrome c-mediated DNA fragmentation and cell death in manganese-superoxide dismutase-deficient mice after exposure to subarachnoid hemolysate. Stroke. 2001;32:506–15.
47. Mehta SL, Lin Y, Chen W, Yu F, Cao L, He Q, Chan PH, Li PA. Manganese superoxide dismutase deficiency exacerbates ischemic brain damage under hyperglycemic conditions by altering autophagy. Transl Stroke Res. 2011;2:42–50.
48. Meng X, Wang M, Wang X, Sun G, Ye J, Xu H, Sun X. Suppression of NADPH oxidase- and mitochondrion-derived superoxide by Notoginsenoside R1 protects against cerebral ischemia-reperfusion injury through estrogen receptor-dependent activation of Akt/Nrf2 pathways. Free Radic Res. 2014;48:823–38.
49. Mu X, Azbill RD, Springer JE. Riluzole improves measures of oxidative stress following traumatic spinal cord injury. Brain Res. 2000;870:66–72.
50. Murakami K, Kondo T, Kawase M, Li Y, Sato S, Chen SF, Chan PH. Mitochondrial susceptibility to oxidative stress exacerbates cerebral infarction that follows permanent focal cerebral ischemia in mutant mice with manganese superoxide dismutase deficiency. J Neurosci. 1998;18:205–13.
51. Murotomi K, Takagi N, Takeo S, Tanonaka K. NADPH oxidase-mediated oxidative damage to proteins in the postsynaptic density after transient cerebral ischemia and reperfusion. Mol Cell Neurosci. 2011;46:681–8.
52. Noshita N, Sugawara T, Fujimura M, Morita-Fujimura Y, Chan PH. Manganese superoxide dismutase affects Cytochrome c release and Caspase-9 activation after transient focal cerebral ischemia in mice. J Cereb Blood Flow Metab. 2001;21:557–67.
53. Odeh AM, Craik JD, Ezzeddine R, Tovmasyan A, Batinic-Haberle I, Benov LT. Targeting mitochondria by Zn(II)N-alkylpyridylporphyrins: the impact of compound sub-mitochondrial partition on cell respiration and overall photodynamic efficacy. PLoS One. 2014;9, e108238.
54. Petronilho F, Feier G, de Souza B, Guglielmi C, Constantino LS, Walz R, Quevedo J, Dal-Pizzol F. Oxidative stress in brain according to traumatic brain injury intensity. J Surg Res. 2010;164:316–20.
55. Piganelli JD, Flores SC, Cruz C, Koepp J, Batinic-Haberle I, Crapo J, Day B, Kachadourian R, Young R, Bradley B, Haskins K. A metalloporphyrin-based superoxide dismutase mimic inhibits adoptive transfer of autoimmune diabetes by a diabetogenic T-cell clone. Diabetes. 2002;51:347–55.
56. Pineda JA, Aono M, Sheng H, Lynch J, Wellons JC, Laskowitz DT, Pearlstein RD, Bowler R, Crapo J, Warner DS. Extracellular superoxide dismutase overexpression improves behavioral outcome from closed head injury in the mouse. J Neurotrauma. 2001;18:625–34.
57. Rabbani ZN, Batinic-Haberle I, Anscher MS, Huang J, Day BJ, Alexander E, Dewhirst MW, Vujaskovic Z. Long-term administration of a small molecular weight catalytic metalloporphyrin antioxidant, AEOL 10150, protects lungs from radiation-induced injury. Int J Radiat Oncol Biol Phys. 2007;67:573–80.
58. Rabbani ZN, Salahuddin FK, Yarmolenko P, Batinic-Haberle I, Thrasher BA, Gauter-Fleckenstein B, Dewhirst MW, Anscher MS, Vujaskovic Z. Low molecular weight catalytic metalloporphyrin antioxidant AEOL 10150 protects lungs from fractionated radiation. Free Radic Res. 2007;41:1273–82.
59. Rabbani ZN, Spasojevic I, Zhang X, Moeller BJ, Haberle S, Vasquez-Vivar J, Dewhirst MW, Vujaskovic Z, Batinic-Haberle I. Antiangiogenic action of redox-modulating Mn(III) meso-tetrakis(N-ethylpyridinium-2-yl)porphyrin, MnTE-2-PyP(5+), via suppression of oxidative stress in a mouse model of breast tumor. Free Radic Biol Med. 2009;47:992–1004.
60. Reboucas JS, Spasojevic I, Tjahjono DH, Richaud A, Mendez F, Benov L, Batinic-Haberle I. Redox modulation of oxidative stress by Mn porphyrin-based therapeutics: the effect of charge distribution. Dalton Trans. 2008;7(9):1233–42.

61. Rodriguez-Rodriguez A, Egea-Guerrero JJ, Murillo-Cabezas F, Carrillo-Vico A. Oxidative stress in traumatic brain injury. Curr Med Chem. 2014;21:1201–11.
62. Ross AD, Sheng H, Warner DS, Piantadosi CA, Batinic-Haberle I, Day BJ, Crapo JD. Hemodynamic effects of metalloporphyrin catalytic antioxidants: structure-activity relationships and species specificity. Free Radic Biol Med. 2002;33:1657–69.
63. Saito A, Hayashi T, Okuno S, Nishi T, Chan PH. Modulation of proline-rich akt substrate survival signaling pathways by oxidative stress in mouse brains after transient focal cerebral ischemia. Stroke. 2006;37:513–7.
64. Saito A, Hayashi T, Okuno S, Nishi T, Chan PH. Oxidative stress is associated with XIAP and Smac/DIABLO signaling pathways in mouse brains after transient focal cerebral ischemia. Stroke. 2004;35:1443–8.
65. Sheng H, Bart RD, Oury TD, Pearlstein RD, Crapo JD, Warner DS. Mice overexpressing extracellular superoxide dismutase have increased resistance to focal cerebral ischemia. Neuroscience. 1999;88:185–91.
66. Sheng H, Brady TC, Pearlstein RD, Crapo JD, Warner DS. Extracellular superoxide dismutase deficiency worsens outcome from focal cerebral ischemia in the mouse. Neurosci Lett. 1999;267:13–6.
67. Sheng H, Chaparro RE, Sasaki T, Izutsu M, Pearlstein RD, Tovmasyan A, Warner DS. Metalloporphyrins as therapeutic catalytic oxidoreductants in central nervous system disorders. Antioxid Redox Signal. 2014;20:2437–64.
68. Sheng H, Enghild JJ, Bowler R, Patel M, Batinic-Haberle I, Calvi CL, Day BJ, Pearlstein RD, Crapo JD, Warner DS. Effects of metalloporphyrin catalytic antioxidants in experimental brain ischemia. Free Radic Biol Med. 2002;33:947–61.
69. Sheng H, Kudo M, Mackensen GB, Pearlstein RD, Crapo JD, Warner DS. Mice overexpressing extracellular superoxide dismutase have increased resistance to global cerebral ischemia. Exp Neurol. 2000;163:392–8.
70. Sheng H, Spasojevic I, Tse HM, Jung JY, Hong J, Zhang Z, Piganelli JD, Batinic-Haberle I, Warner DS. Neuroprotective efficacy from a lipophilic redox-modulating Mn(III) N-Hexylpyridylporphyrin, MnTnHex-2-PyP: rodent models of ischemic stroke and subarachnoid hemorrhage. J Pharmacol Exp Ther. 2011;338:906–16.
71. Sheng H, Spasojevic I, Warner DS, Batinic-Haberle I. Mouse spinal cord compression injury is ameliorated by intrathecal cationic manganese(III) porphyrin catalytic antioxidant therapy. Neurosci Lett. 2004;366:220–5.
72. Sheng H, Yang W, Fukuda S, Tse HM, Paschen W, Johnson K, Batinic-Haberle I, Crapo JD, Pearlstein RD, Piganelli J, Warner DS. Long-term neuroprotection from a potent redox-modulating metalloporphyrin in the rat. Free Radic Biol Med. 2009;47:917–23.
73. Shimanovich R, Groves JT. Mechanisms of peroxynitrite decomposition catalyzed by FeTMPS, a bioactive sulfonated iron porphyrin. Arch Biochem Biophys. 2001;387:307–17.
74. Shohami E, Beit-Yannai E, Horowitz M, Kohen R. Oxidative stress in closed-head injury: brain antioxidant capacity as an indicator of functional outcome. J Cereb Blood Flow Metab. 1997;17:1007–19.
75. Siesjo BK, Agardh CD, Bengtsson F. Free radicals and brain damage. Cerebrovasc Brain Metab Rev. 1989;1:165–211.
76. Sivonova M, Kaplan P, Durackova Z, Dobrota D, Drgova A, Tatarkova Z, Pavlikova M, Halasova E, Lehotsky J. Time course of peripheral oxidative stress as consequence of global ischaemic brain injury in rats. Cell Mol Neurobiol. 2008;28:431–41.
77. Spasojevic I, Chen Y, Noel TJ, Fan P, Zhang L, Reboucas JS, Clair DKS, Batinic-Haberle I. Pharmacokinetics of the potent redox-modulating manganese porphyrin, MnTE-2-PyP(5+), in plasma and major organs of B6C3F1 mice. Free Radic Biol Med. 2008;45:943–9.
78. Spasojevic I, Chen Y, Noel TJ, Yu Y, Cole MP, Zhang L, Zhao Y, Clair DKS, Batinic-Haberle I. Mn porphyrin-based superoxide dismutase (SOD) mimic, MnIIITE-2-PyP5+, targets mouse heart mitochondria. Free Radic Biol Med. 2007;42:1193–200.
79. Spasojevic I, Kos I, Benov LT, Rajic Z, Fels D, Dedeugd C, Ye X, Vujaskovic Z, Reboucas JS, Leong KW, Dewhirst MW, Batinic-Haberle I. Bioavailability of metalloporphyrin-based SOD

mimics is greatly influenced by a single charge residing on a Mn site. Free Radic Res. 2011;45:188–200.
80. Sugawara T, Lewen A, Gasche Y, Yu F, Chan PH. Overexpression of SOD1 protects vulnerable motor neurons after spinal cord injury by attenuating mitochondrial cytochrome c release. FASEB J. 2002;16:1997–9.
81. Takenaga M, Ohta Y, Tokura Y, Hamaguchi A, Nakamura M, Okano H, Igarashi R. Lecithinized superoxide dismutase (PC-SOD) improved spinal cord injury-induced motor dysfunction through suppression of oxidative stress and enhancement of neurotrophic factor production. J Control Release. 2006;110:283–9.
82. Thiyagarajan M, Kaul CL, Sharma SS. Neuroprotective efficacy and therapeutic time window of peroxynitrite decomposition catalysts in focal cerebral ischemia in rats. Br J Pharmacol. 2004;142:899–911.
83. Tovmasyan A, Maia CG, Weitner T, Sampaio RS, Lieb D, Ghazaryan R, Ivanovic-Burmazovic I, Reboucas JS, Benov L, Batinic-Haberle I. A comprehensive evaluation of the catalase-like activity of different classes of redox-active therapeutic. Free Radic Biol Med. 2015;86.
84. Tovmasyan A, Reboucas JS, Benov L. Simple biological systems for assessing the activity of superoxide dismutase mimics. Antioxid Redox Signal. 2014;20:2416–36.
85. Tovmasyan A, Sheng H, Weitner T, Arulpragasam A, Lu M, Warner DS, Vujaskovic Z, Spasojevic I, Batinic-Haberle I. Design, mechanism of action, bioavailability and therapeutic effects of mn porphyrin-based redox modulators. Med Princ Pract. 2013;22:103–30.
86. Tovmasyan A, Weitner T, Sheng H, Lu M, Rajic Z, Warner DS, Spasojevic I, Reboucas JS, Benov L, Batinic-Haberle I. Differential coordination demands in Fe versus Mn water-soluble cationic metalloporphyrins translate into remarkably different aqueous redox chemistry and biology. Inorg Chem. 2013;52:5677–91.
87. Vujaskovic Z, Batinic-Haberle I, Rabbani ZN, Feng QF, Kang SK, Spasojevic I, Samulski TV, Fridovich I, Dewhirst MW, Anscher MS. A small molecular weight catalytic metalloporphyrin antioxidant with superoxide dismutase (SOD) mimetic properties protects lungs from radiation-induced injury. Free Radic Biol Med. 2002;33:857–63.
88. Wakai T, Sakata H, Narasimhan P, Yoshioka H, Kinouchi H, Chan PH. Transplantation of neural stem cells that overexpress SOD1 enhances amelioration of intracerebral hemorrhage in mice. J Cereb Blood Flow Metab. 2014;34:441–9.
89. Warner DS, Sheng H, Batinic-Haberle I. Oxidants, antioxidants and the ischemic brain. J Exp Biol. 2004;207:3221–31.
90. Weitner T, Kos I, Sheng H, Tovmasyan A, Reboucas JS, Fan P, Warner DS, Vujaskovic Z, Batinic-Haberle I, Spasojevic I. Comprehensive pharmacokinetic studies and oral bioavailability of two Mn porphyrin-based SOD mimics, MnTE-2-PyP5+ and MnTnHex-2-PyP5+. Free Radic Biol Med. 2013;58:73–80.
91. Wellons 3rd JC, Sheng H, Laskowitz DT, Mackensen GB, Pearlstein RD, Warner DS. A comparison of strain-related susceptibility in two murine recovery models of global cerebral ischemia. Brain Res. 2000;868:14–21.
92. Wise-Faberowski L, Warner DS, Spasojevic I, Batinic-Haberle I. Effect of lipophilicity of Mn (III) ortho N-alkylpyridyl- and diortho N,N′-diethylimidazolylporphyrins in two in-vitro models of oxygen and glucose deprivation-induced neuronal death. Free Radic Res. 2009;43:329–39.
93. Xu Y, Liachenko SM, Tang P, Chan PH. Faster recovery of cerebral perfusion in SOD1-overexpressed rats after cardiac arrest and resuscitation. Stroke. 2009;40:2512–8.
94. Yu F, Sugawara T, Nishi T, Liu J, Chan PH. Overexpression of SOD1 in transgenic rats attenuates nuclear translocation of endonuclease G and apoptosis after spinal cord injury. J Neurotrauma. 2006;23:595–603.
95. Zhan Y, Chen C, Suzuki H, Hu Q, Zhi X, Zhang JH. Hydrogen gas ameliorates oxidative stress in early brain injury after subarachnoid hemorrhage in rats. Crit Care Med. 2012;40:1291–6.

Chapter 25
The Contribution of Nitroxidative Stress to Pathophysiological Pain and Opioid Analgesic Failure

Ashley M. Symons-Liguori, Kali Janes, William L. Neumann, and Daniela Salvemini

Abbreviations

COX	Cyclooxygenase
DA	Dopamine
EAAC1	Excitatory amino acid carrier 1
GABA	Gamma aminobutyric acid
GLAST/EAAT1	Glutamate aspartate transporter/excitatory amino acid transporter 1
GLT-1/EAAT2	Glutamate transporter 1/excitatory amino acid transporter 2
GS	Glutamine synthase
HDAC	Histone deacetylase
MENK	Met-enkephalin
NaV	Voltage-gated sodium channel
NE	Norepinephrine
NO	Nitric oxide
NOS	Nitric oxide synthase
NOX	NADPH oxidase
NSAID	Nonsteroidal anti-inflammatory drug
PAG	Periaqueductal grey
PN	Peroxynitrite
PNDC	Peroxynitrite decomposition catalyst

A.M. Symons-Liguori (✉) • K. Janes • D. Salvemini (✉)
Department of Pharmacological and Physiological Science, Saint Louis University School of Medicine, 1402 South Grand Blvd, St. Louis, MO 63104, USA
e-mail: salvemd@slu.edu; salvemd@slu.edu; salvemd@slu.edu

W.L. Neumann
Department of Pharmaceutical Sciences, School of Pharmacy, Southern Illinois University Edwardsville, 200 University Park, Edwardsville, IL 62026, USA
e-mail: wneuman@siue.edu

I. Batinić-Haberle et al. (eds.), *Redox-Active Therapeutics*,
Oxidative Stress in Applied Basic Research and Clinical Practice,
DOI 10.1007/978-3-319-30705-3_25

PNI	Peripheral nerve injury
PTM	Post-translational modification
RVM	Rostral ventromedial medulla
SC	Spinal cord
PK	Protein kinase
SO	Superoxide
SOD	Superoxide dismutase
SODm	Superoxide dismutase mimetic
TH	Tyrosine hydroxylase
TRP	Transient receptor potential
XO	Xanthine oxidase

25.1 Introduction

Physiological pain is an unpleasant protective response that signifies potential or actual tissue damage. As such, acute pain plays a vital role in survival and demands an inherent plasticity in order to attune physiological responses to the threat, degree, or persistence of injury. In some cases, acute pain responses can undergo a transition to chronic pain that results in a continuous state of non-productive pain that can severely hamper an individual's quality-of-life [1]. Changes in cellular oxidative status serve an important function in detecting tissue insult and arbitrating or regulating appropriate and inappropriate stress responses; to this end, it is known that the overproduction of oxidant molecules can signal a genomic antioxidant response and additionally produce molecular oxidative modifications that play a critical role in the initiation and persistence of pathophysiological pain and opioid analgesic failure (opioid tolerance and paradoxical hyperalgesia) [2–6]. The key molecules involved in pathological pain responses are superoxide radical anion ($O_2^{\bullet-}$, SO), nitric oxide (NO), and their reaction product peroxynitrite ($ONOO^-$, PN) [7]. These molecules can be produced by cellular enzymatic and mitochondrial sources and be potentiated as a result of the feedback-inactivation of scavenger systems by PN (e.g., the nitrative inactivation of mitochondrial superoxide dismutase, MnSOD). Collectively, processes resulting in and from the overproduction of SO, NO, and PN can be described by the term "nitroxidative stress." In this chapter, we will discuss the role of nitroxidative stress in pathophysiological pain by first examining the tools used to uncover the actions of these mediators and then dissecting the sources of nitroxidative stress in nociceptive processing. Finally, we will explore the contribution of nitroxidative modifications to important classical pro- and anti-nociceptive signaling systems in pain, including glutamate, opioid, vanilloid, catecholamine, and prostaglandin systems, and the transcriptional regulation of these pathways. The aim of this chapter is to foster a greater understanding of nitroxidative mediators in the induction and maintenance of persistent pathophysiological pain.

25.2 Pharmacological Strategies Targeting Nitroxidative Stress

Because reactive species are inherently short-lived and therefore difficult to measure directly, pharmacological scavengers or decomposition catalysts have provided the greatest insight into the role of nitroxidant molecules in pain and their potential as therapeutic targets. The earliest studies investigating the role of nitroxidant molecules in pain utilized scavengers such as phenyl *N*-tert-butylnitrone (PBN) and 4-hydroxy-2,2,6,6-tetramethylpiperidine-1-oxyl (TEMPOL) and unanimously implicated nitroxidant molecules in a variety of inflammatory [8], neurogenic [9–12], visceral [13], orofacial [14], and neuropathic [10, 15–17] pain states. However, it was quickly discovered that PBN and TEMPOL nonspecifically scavenge a number of reactive species including NO, SO, and in particular PN-derived species, such as the carbonate radical anion and hydroxyl radical, depending upon their oxidative status (reduced nitroxides form hydroxylamine or oxidized nitroxides form oxoammonium cations) [18–21]. Accordingly, these molecules failed to provide informative data about specific molecule involvement. To further the burgeoning understanding of the umbrella term "oxidative stress" as a category of distinct signaling molecules with specific pro-nociceptive actions, the development of more precise tools for probing the action of nitroxidative mediators in nociception was required.

Synthetic metal-based and non-metal SO dismutase enzyme mimetics (SODm) or "synzymes" [20, 22, 23] have been reported to facilitate the disproportionation of SO to hydrogen peroxide and molecular oxygen in paired single-electron transfer reactions [29]. Current SODm compounds that have offered the greatest preclinical insights fall into two categories: (1) functionalized manganese(II) polyazamacrocycles (e.g., SC-72325) and (2) iron(III) and manganese(III) porphyrins. Among Mn porphyrins, the *ortho* isomeric MnTE-2-PyP^{5+} and MnTnBuOE-2-PyP^{5+} are the most frequently and successfully studied compounds and are entering clinical trials. The most frequently studied Fe porphyrin is FeTM-4-PyP^{5+} (Fig. 25.1) [20, 21, 24]. Functionalized manganese (II) compounds offer some selectivity toward various oxidant species through the low reduction potential of the metal center, which cannot be further reduced by many one-electron reductants (e.g., ascorbate, thiols), but can be oxidized by SO to Mn(III) and then rapidly reduced back to Mn(II) by SO [4]. Mn(II) polyazamacrocycles have been reported as specific SODms, but we now know that they are most likely SO-selective agents that can react at slower rates with PN [20]. These compounds are also shown to dismute NO into nitroxyl, HNO, and nitrosonium ion [25, 26]. As such, they have limited utility in sorting the contribution of SO from PN in various pathological states given their kinetic preference for reacting with SO. Indeed, Mn(II)-based SODms have demonstrated efficacy in inflammatory [27], osteoarthritis, and chronic articular pain models [28] and successfully identified SO as a critical mediator in these pain states [20].

The Mn(III) and Fe(III) metalloporphyrins are excellent SODms that operate through a Mn(III)-resting/Mn(II) redox couple. In addition to SODm activity, they also behave as PN decomposition catalysts (PNDCs) in the presence of biological

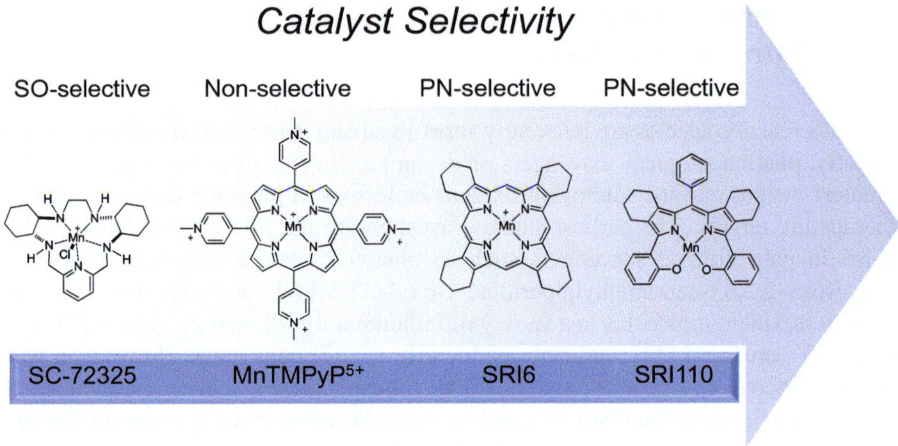

Fig. 25.1 Structures of catalysts used to reduce nitroxidative stress. From *left* to *right*, manganese decomposition catalyst structures ordered from superoxide-selective to superoxide-sparing. The PN-selective SRI compounds possess substitutions that lower the oxidation potential of the manganese center in order to preclude SO reaction and favor PN decomposition as well as bioavailability

reductants [30]. These types of complexes are often considered to be SODms with PNDC activity and are thus powerful but nonselective antioxidants (Fig. 25.1) [20, 21, 24]. The nonselective metalloporphyrin compounds have shown efficacy in sphingosine-1-phosphate-induced [31] and carrageenan-induced [32] hyperalgesia, neuropathic pain associated with ischemia-reperfusion injury [33] and chemotherapy-induced neuropathy [34], and opioid tolerance and hyperalgesia [35–37]. Importantly, the potential clinical utility of these early metalloporphyrins remains limited by poor lipid solubility and oral bioavailability [4].

Since the metalloporphyrin framework is an ideal scaffold for designing oxidation catalysts with electronically tuned redox potentials, it was realized that metalloporphyrins could also be developed as PN-selective, SO-sparing tools through modifications of the ligand system. One interesting example is the widely used "SODm" Mn(III) tetrakis-(4-benzoic acid)porphyrin (MnTBAP). This compound was recently shown to lack SODm activity, indicating that its biological activity is most likely due to its weak PNDC properties [38, 39]. Furthermore, data indicating anti-nociceptive synergy between PNDC compounds and clinically employed analgesics (e.g., COX-1/COX-2 inhibitors [40] and opioids [35]) spurred the development of SO-sparing metalloporphyrins with an eye towards clinical utility. Specifically, changes were made in order to generate compounds with in vivo oral bioavailability, tolerability (reduced toxicity), and SO-sparing properties. Utilization of manganese as a catalytic center circumvents the toxic Fenton chemistry that is seen with iron complexes [41]. In order to avoid the highly cationic nature of ortho and para Mn porphyrins—which renders these compounds unacceptable for clinical use due to poor lipophilicity and ion channel interaction-associated acute toxicity—lipophilic cyclohexenyl functionality was incorporated to shield the hydrophilic metal center during passive transcellular transport [4]. These compounds therefore have enhanced membrane permeability, reduced overall positive charge, and possessed a metal-centered reduction potential that is well below the useful range for dismutation of superoxide (+0.30 V is optimal for high SOD activity) [42, 43]. Applying the aforementioned ligand substitutions to the bis(hydroxyphenyl)dipyromethene and porphyrin classes has produced the clinically usable compounds SRI110 [44] and SRI6 [4] (Fig. 25.1). These next-generation PNDCs have already demonstrated anti-nociceptive qualities in models of neuropathic pain associated with chemotherapy-induced peripheral neuropathy (CIPN) [34] and are perhaps the first PNDCs poised to bridge the preclinical use of metalloporphyrins with the clinical targeting of PN in chronic pain.

25.3 Generation of Nitroxidant Mediators in Pain

As discussed, a variety of pharmacological tools have intimated a role for SO, NO, and PN in several pathophysiological pain states. Post-translational modification (PTM) of proteins is a key pro-nociceptive mechanism of nitroxidative stress [45], and specifically the relevance of PN nitration among the alternative actions of PN has been controversially highlighted in the literature [46]. Importantly, the consequence of NO nitrosylation [47] is separate from the ability of PN to nitrate tyrosine residues [48–50] and consequently modify protein function. The latter appears to be a mechanism specific to pathological rather than physiological pain. For example, in pathophysiological pain, an abundance of NO in the presence of unmanaged mitochondrial SO production (e.g., due to the nitrative inactivation of MnSOD) is thought to favor nitration rather than nitrosylation of amino acid residues. Evaluating the relative abundance of specific molecules is therefore important in understanding nitroxidative mechanisms in pain. Unlike other PTM mechanisms like phosphorylation, a “denitrase” capable of reversing nitrative PTMs has only been fleetingly described [51, 52] and to date remains uninvestigated in pain. At present, it appears that targeting reactive species directly is the best approach to alleviate nitroxidative stress in the context of chronic pain. As such, unraveling the enzymatic sources of SO and PN in nociceptive processing is of paramount importance [3, 23].

25.3.1 Superoxide

Superoxide ($O_2^{\bullet -}$, SO) has a longstanding role in the etiology of chronic pain states and the induction of central sensitization [4, 5]. By measuring the SO dismutation product H_2O_2 or utilizing SODm compounds to scavenge SO, a host of literature has implicated SO generation in a variety of inflammatory and neuropathic pain states [12, 15, 53, 54]. While SO (through the formation of hydrogen peroxide) plays a vital role in regulating the oxidation status of thiol protein residues [55] and can therefore alter protein function directly, the role of SO in pain is largely regarded as a rate-limiting precursor to PN formation [4, 5].

SO is classically produced as a byproduct of mitochondrial metabolism. Specifically, electron transport through complexes I and III of the respiratory chain generates a basal level of SO [56, 57] and represents a coupling of cellular energy demand to oxidative stress. Pharmacological disruption of electron transport in dorsal horn spinal neurons shifts the nociceptive threshold and leads to mechanical hypersensitivity [58]. Bioenergetic dysfunction and unregulated mitochondrial SO production has been implicated in fibromyalgia [59], capsaicin-induced secondary hyperalgesia [11], persistent pain associated with CIPN [60] and diabetic neuropathy [61, 62], loss of opioid analgesia [63], and opioid-induced hyperalgesia [64].

Alternatively, cytosolic production of SO can occur via the enzymes NADPH oxidase (NOX) and xanthine oxidase (XO). The membrane-bound NOX generates SO through the transfer of electrons across the plasma membrane coupled to a reaction process wherein NADPH donates an electron to oxygen. NOX activity is initiated and regulated through PTMs and requires the assembly of several components that can include the membranous subunits, $p22^{phox}$ and the catalytic core $gp91^{phox}$, as well as several cytosolic components, including $p47^{phox}$, $p40^{phox}$, $p67^{phox}$, and rac1/2 [65]. Importantly, the enhanced availability of SO reinforces the expression of the $gp91^{phox}$ [66] and other NOX subunits [67]. The protein components of NOX are expressed in neurons, astrocytes, and microglia [68–70] and its induction is implicated in central sensitization associated with inflammatory hyperalgesia [71], peripheral nerve injury-induced neuropathic pain [72], and the development of paclitaxel-induced CIPN [73]. Accordingly, the regulation of the NOX homologs NOX1, NOX2, and NOX4 are growing targets in the field of chronic pain [74]. NOX2 is thought to generate SO in reactive microglia during peripheral nerve injury (PNI)-induced neuropathic pain [72, 75], and indeed, activation of spinal NOX2 leads to mechanical hypersensitivity and is involved in the induction phase of neuropathic pain resulting from chronic constriction injury [76]. Spinally expressed NOX1 [77] and NOX2 [35] have also been demonstrated to play a pivotal role in sensitizing mechanisms associated with the development of opioid analgesic tolerance [78]. NOX4 is functional in a subset of nonpeptidergic nociceptors and myelinated dorsal root ganglion (DRG) neurons and plays a pro-nociceptive role in the maintenance of PNI-induced neuropathic pain [79]; silencing NOX4 by restoring miR23b microRNA function ameliorates neuropathic pain by protecting GABAergic neurons from reactive oxygen species-mediated apoptosis [80]. Thus, there is a clear role for NOX-derived SO in the etiology of several pain conditions.

The generation of SO as a byproduct of XO-catalyzed reactions is not the exclusive purpose of the enzyme, and accordingly, fewer anti-nociceptive strategies have targeted this enzyme with the intention of SO attenuation. Importantly, XO inhibitors are clinically utilized for the treatment of gout and gout-associated pain. While some work has suggested that diminishing SO generation could be one mechanism by which XO inhibitors function to relieve pain [81] and indeed, allopurinol has demonstrated efficacy in a model of post-ischemic pain [82], XO is not currently targeted as a fundamental mechanism of SO production in pathological pain.

As discussed above, several mechanisms contribute to the generation of SO under normal and pathological conditions. Accordingly, the SO dismutase (SOD) enzymatic system is dedicated to the disproportionation of SO in the mitochondrial matrix via SOD2/MnSOD [83] and SO in the cytosolic and mitochondrial intermembrane spaces via SOD1/CuZnSOD [84]. In pathophysiological pain states, the functional inactivation of the mitochondrial SOD system can occur via the Mn-catalyzed nitration of the tyrosine residue Tyr34 by PN [85]. The functional inactivation of MnSOD provides a critical self-perpetuating mechanism for the continued production of SO as a substrate for PN formation in the development and maintenance of central sensitization [11, 12, 27, 36, 42, 86]. Inactivation of MnSOD has also been implicated in the loss of opioid analgesia [63] and the development of opioid-induced hyperalgesia [35, 78], further suggesting a role for this mechanism in nociceptive plasticity. As such, the mitochondrial SOD system is an important target for regulating the generation of SO as a precursor to PN during pain.

25.3.2 *Nitric Oxide*

Nitric oxide (NO) has a long and well-investigated history in neuronal signaling as a neurotransmitter [87–89] and as a retrograde signaling molecule [90] that functions to activate guanylyl cyclase to result in the production of cyclic GMP and protein kinase G (PKG) activation [91, 92]. NO has been intensively investigated for its pro-nociceptive functions as an activator of cGMP [93] and as a nitrosylating agent [94], but more recent data has focused on NO as a rate-limiting step in the formation of PN [4, 5, 95]. Whereas the induction of canonical NO signaling has important functions in protective physiological pain, the function of NO subserving the formation of PN is only implicated in pathophysiological pain states [27, 40]. In therapeutic development for pathological pain, it is vital to target mechanisms that do not interrupt the normal pain response to damaging stimuli; as such, this chapter will focus on NO as a rate-limiting precursor to PN. Many reviews are available on the role of canonical NO signaling in physiological pain [96, 97].

NO is a diffusible gas mediator synthesized from L-arginine by nitric oxide synthase (NOS), of which there exist three isoforms [98]. Endothelial NOS (eNOS) is most well-known for its expression in the cardiovascular system as a regulator of vascular tone. eNOS is a membrane-bound enzyme that is constitutively expressed; however, it requires the interaction of calcium and calmodulin for its activation [99]. Neuronal NOS (nNOS) is also constitutively expressed and also requires calcium for its activation [100], but is primarily found in the cytosolic compartment of neurons—specifically, anchored to the post-synaptic density [101]. nNOS activation is therefore coupled to neuronal activity, as the availability of cytosolic calcium results in its activity. In abnormal pain states, neuronal hyperexcitability and excessive activation of postsynaptic *N*-methyl-D-aspartate (NMDA) receptors can result in increased calcium influx and hyperactivation of nNOS in a calcium/calmodulin-dependent manner [102]. Genetic studies have implicated nNOS in the development of neuropathic pain and specific nNOS inhibitors show preclinical utility in PNI and CIPN models [103–106]. Activation of nNOS contributes to the PN-mediated downstream phosphorylation of NMDA receptors in a mechanism of central sensitization [10] and is a principal source for PN formation in the DRG during diabetic neuropathy [107]. Lastly, inducible NOS (iNOS) is a cytosolic isoform that is widely expressed in many cells of the immune system and glia. iNOS expression first requires transcriptional activation, and once translated, the iNOS isoform is constitutively active—that is, iNOS activity is independent of calcium [108]. The induction of iNOS is one of the most provocative mechanisms of PN formation in glia, although as a gas mediator, glial-derived NO can easily diffuse across membranes to directly impact neuronal function. The contribution of glial iNOS to PN generation has been shown to be significant in neuropathic pain [95] and some studies have suggested that iNOS induction is more critical to the maintenance phase rather than the induction phase of neuropathic pain states [103]. Orally bioavailable iNOS inhibitors have been developed preclinically in inflammatory and neuropathic models of pain [109], but the effect of these agents on PN formation remains to be investigated.

It is now evident that in spite of the nomenclature, all three NOS isoforms are relevant to aberrant nociception [102, 110]. However, while NOS activation is a prerequisite to the formation of PN, targeting NOS isoforms in order to abrogate PN formation may be limited by the off-target functions of NO in the cardiovascular system, the immune system, and physiological nociception.

25.3.3 Peroxynitrite

As we have emphasized in this chapter, peroxynitrite ($ONOO^-$, PN) is a highly reactive oxidant and nitrating agent that largely accounts for the biological actions of SO and NO in pathological pain. PN is generated by a diffusion–controlled coupling of SO and NO, and as such, the proximal availability of SO and NO regulates the production of PN. While there is not a single PN-synthesizing enzyme, some studies have suggested that a phenomena known as NOS uncoupling could result in the production of both SO and NO as a mechanism of direct PN generation in vivo. In diabetic patients, the production of PN is hypothesized to produce NOS uncoupling as a positive feedback mechanism on PN generation [111]. However, the physiological occurrence and relevance of NOS uncoupling in pain remains uninvestigated [112].

The coordinated proximal induction of SO and NO has been documented in endothelial cells through the formation of membranous microdomain "redox signalosomes" in response to pro-inflammatory stimuli (e.g., TNF-α) [113]; in these microdomains, eNOS and the catalytic subunits of NOX are tethered in close proximity to reinforce the interaction of SO and NO to form PN. The relevance of redox signalosome microdomains in non-endothelial cells and NOS enzymes that are non-membrane bound (i.e., iNOS and nNOS) is an area ripe for investigation. It is known that eNOS residues that allow its association with caveolin-1 in microdomains is conserved across all three NOS isoforms [114]. Moreover, we have discussed the membrane tethering of nNOS in neurons, but it is not known whether nNOS or iNOS associate with NOX subunits in lipid microdomains in neurons or glia. Such a mechanism could fulfill an important gap in our knowledge about the cooperative induction of NOS and NOX to form PN in response to inflammatory or stress-associated stimuli in the CNS. In Fig. 25.2, we have summarized our discussion of the known and hypothesized mechanisms of PN generation in neurons and glia.

25.4 Nitrative Modifications of Neurotransmitter Systems in Pathological Pain

During pathophysiological pain, changes in neuroplasticity occur at multiple levels of the neuraxis in order to enhance the perception of pain; we have referred to this neuroplasticity as sensitization. Nociceptive sensitization is a theoretical mechanism underlying hyperexcitability that can occur peripherally in afferent neurons or centrally at spinal or supraspinal sites. Herein, we will discuss how nitroxidative stress underpins systems-level changes in endogenous pro-nociceptive and anti-nociceptive pathways (Fig. 25.3). Specifically, we will discuss how nitroxidative stress elicits hyperexcitability in pain neurons by modifying the synaptic half-life, efficacy, or functionality of neurotransmitters, altering ion conductance and membrane potential, or interfering with intracellular signaling. We will highlight the possibility of nitration events (see Table 25.1 for summary) in systems such as lipidergic and peptidergic neurotransmitter systems, and how attenuating PN in these systems may lead to relief of pain. It is important to note that this discussion is limited by a dearth of pain-related studies of nitration in pain-related neurotransmitter systems, and this research is urgently needed in order to better realize the therapeutic utility of targeting PN-driven nociception.

25.4.1 Part 1: Pro-nociceptive Systems

Glutamatergic Signaling

Abnormal pain states and loss of pharmacological analgesia feature dysfunctional glutamatergic signaling [115–117]. Alterations in glutamate function are a fundamental mechanism of spinal central sensitization [118] and supraspinal descending facilitation [119] in pathophysiological pain. Nitrative alterations in the glutamatergic system therefore provide an archetypal example of the relevance of nitroxidative signaling modification in pathophysiological pain [32].

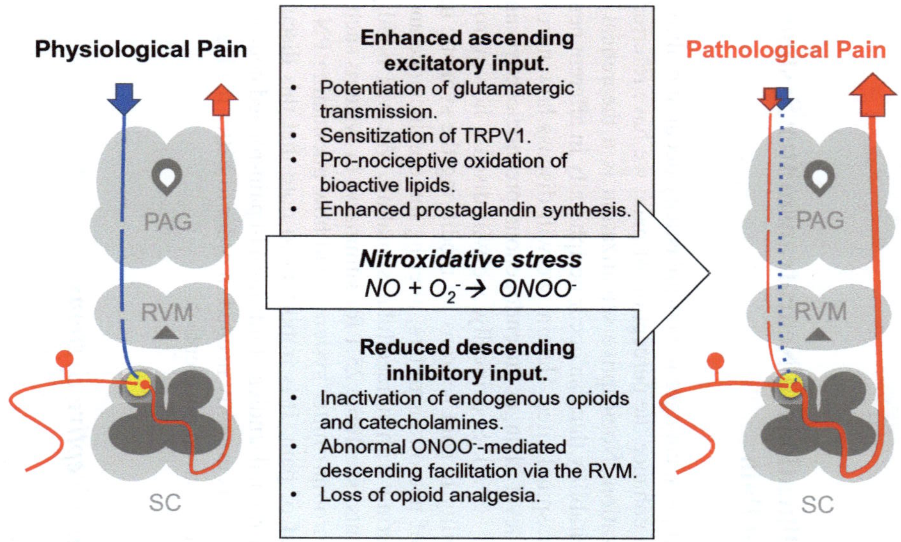

Fig. 25.2 Nitroxidative modulation of the pain pathway in pathological states. In physiological pain, the superficial laminae of the dorsal horn of the spinal cord function as an integrative site for nociceptive input and processing. The primary afferent neuron synapses in the dorsal horn and is conveyed to higher order areas via the spinothalamic tract, known as the ascending pathway. During pathological pain (e.g., chronic pain, opioid analgesic failure), nitroxidative stress can exert its pro-nociceptive effects by directly increasing ascending input or alternatively by modifying the descending input to the dorsal horn to provide less inhibition or facilitation. In opioid tolerance and hyperalgesia, it is thought that these mechanisms overlap to produce a net increase in nociceptive signaling that modifies and/or precludes the analgesic efficacy of opioid therapeutics (*PAG* periaqueductal grey, *RVM* rostral ventromedial medulla, *SC* spinal cord)

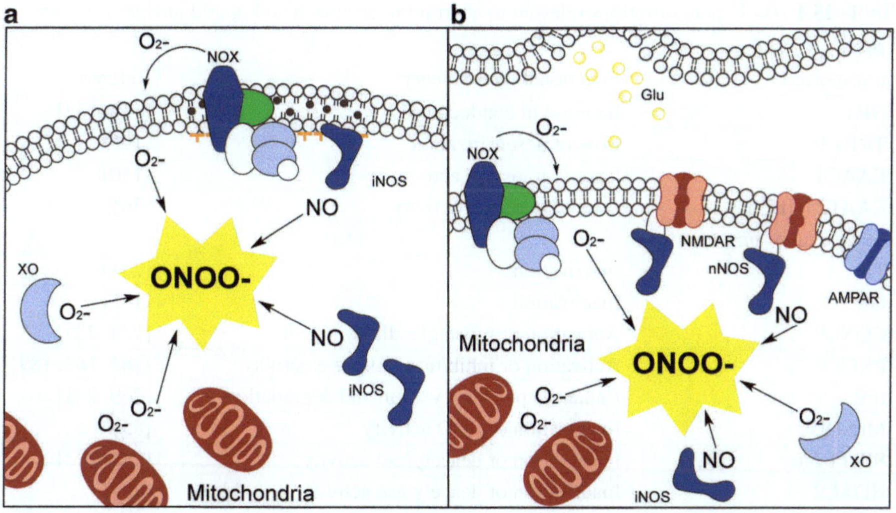

Fig. 25.3 Generation of nitroxidative stress in the nervous system. The production of nitroxidant mediators in glia (**a**) and neurons (**b**). Mitochondrial respiration is a primary source of intracellular SO and is a key mechanism in sustaining the production of PN in both neurons and glia. Membrane-bound NOX isoforms can oxidize intracellular NADPH to produce extracellular SO that can then be transported into the cell or alternatively diffuse paracellularly. In glia, NO can be generated through the transcriptional induction of the constitutively active, calcium-independent iNOS enzyme. In neurons, constitutively expressed but calcium-dependent nNOS is tethered to the post-synaptic density by PSD-95 in order to produce NO in a calcium-dependent and therefore activity-dependent fashion. Accordingly, PN is readily generated in metabolically active cells and, specifically, activated neurons and glia (*NOX* NADPH oxidase, *XO* xanthine oxidase, *iNOS* inducible nitric oxide synthase, *nNOS* neuronal nitric oxide synthase, *Glu* glutamate)

Glutamate is a small molecule neurotransmitter that is the primary excitatory mediator in nociception and is a signaling partner to the slow peptide transmitter substance P in peptidergic nociceptors [120]. Glutamate is released presynaptically to function at post-synaptic depolarizing ionotropic NMDA, kainate (KA), and α-amino-3-hydroxy-5-methyl-4-isoxazolepropionic acid (AMPA) receptors as well as at pre/post-synaptic metabotropic glutamate receptors [121]. Given the excitatory nature of glutamate binding, the glutamatergic synapse is tightly regulated in order to protect neurons from excitotoxicity [122–124]. Following neurotransmission, glutamate is removed from the synaptic cleft by sodium-dependent high-affinity membranous glutamate transporters (GTs) in both neurons and glia [125–128] and enzymatically decomposed to glutamine by glutamine synthase (GS). Glial glutamate reuptake transporters GLAST (EAAT1) and GLT-1 (EAAT2) [129, 130] are responsible for >90 % of total glutamate transport in the superficial dorsal horn [131], and the functional inhibition of these transporters by nitration [132–139] plays a role in the etiology of inflammatory hyperalgesia [32], neuropathic pain [73], and opioid analgesic tolerance [36]. Attenuating glial GT nitration directly with PNDCs [32, 34] or indirectly by reducing PN formation with the spinal administration of A_3 adenosine receptor agonists [73] ameliorates behavioral indicators of pathological pain.

Table 25.1 Biological nitrations relevant to aberrant nociception and opioid analgesic failure

Receptors and transporters	Functional consequence	References
NR1	Increase in conductance	[156–158]
TRPV1[a]	Loss of desensitization	[195]
EAAC1	Loss of transport activity	[150]
EAAT2 (GLT-1)[b]	Loss of transport activity	[36]
Intracellular enzymes		
TH	Inactivation	[264]
GS[b]	Inactivation	[135]
COX-2	Activation of prostaglandin synthesis	[224, 225]
PKC	Activation or inhibition of kinase activity	[162, 163, 189]
IkB	Enhances phosphorylation and degradation	[269, 271]
MnSOD[b]	Inactivation of SOD activity	[85]
SIRT1/3/6	Inactivation of deacetylase activity	[291, 292]
HDAC2	Inactivation of deacetylase activity	[287]
Ligands and mediators		
Met-enkephalin[b]	Loss of opioid receptor activity	[37]
Leu-enkephalin	Protective in neutrophils	[236, 237]
NE	Loss of activity	[257]
Linoleic acid	Gain of TRPA1 agonist properties	[205]

[a]Computational hypothesis

[b]Known contribution to opioid analgesic failure (tolerance and/or paradoxical hyperalgesia)

Similarly, the neuronal GT, EAAC1 [140–145], can be inactivated via nitration and thus further contribute to the synaptic availability of glutamate. Importantly, neuronal EAAC1 is also largely responsible for the uptake of cysteine [146–148], a rate-limiting substrate for the synthesis of the thiol-protective antioxidant glutathione [149]. PN nitrative inactivation of EAAC1 leads to a reduction in neuronal oxidant-handling capacity [139, 150] and reinforces glutamate-driven pro-nociception [151], such that anti-PN strategies are likely to produce dual inhibition of synaptic nitroxidative stress by restoring thiol-protective redox systems (i.e., glutathione).

In addition to the inactivation of transporters that limit the synaptic half-life of glutamate, nitration of glutamine synthase produces a loss-of-function that eliminates GS-driven glutamate decomposition [134, 135]. GS inactivation results in neuronal excitability [152, 153] and astrocyte hypertrophy associated with enhanced glutamate release [154, 155]. Similar to GLT-1 and GLAST, nitrative inhibition of GS in pathological pain states [32, 36] can be attenuated directly using peroxynitrite decomposition catalysts (PNDCs) [32, 34] or indirectly by inhibition of PN generation via A_3 adenosine receptor agonists [73] and is associated with the reversal of nociceptive behaviors in inflammatory, neuropathic, and chemotherapy-induced pain states.

Finally, nitration events can modify postsynaptic glutamate receptors [156–158] as a mechanism of nociceptive sensitization [118, 119]. The NR1 subunit of the NMDA receptor (NMDAR) is a critical regulator of NMDAR function [159] and is subject to several PTMs including nitration and phosphorylation. Interestingly, PN can elicit both of these mechanisms of NMDAR regulation in pathological pain states. For example, tyrosine nitration of NR1 by PN results in the potentiation of synaptic current and calcium conductance [156–158] and is thought to be a vital pro-nociceptive mechanism in neuropathic pain. NMDAR function is also positively regulated by PKC-mediated phosphorylation of NR1 [160, 161] and PN has been illustrated to produce the nitrative activation of PKC [162, 163] and the PN-dependent phosphorylation of NR1 [164]. Accordingly, a free radical scavenger reduces spinal NR1 phosphorylation in neuropathic and inflammatory hyperalgesia [10], underscoring the contribution of nitroxidative species to NMDAR activity in central sensitization. Adenosine-targeted pharmacological strategies have also been useful in attenuating the effects of PN. In hypoxia-reoxygenation injury, the adenosine 2A receptor agonist attenuates NR1 phosphorylation-mediated excitotoxicity [165]. Furthermore, our group has utilized selective adenosine 3 receptor agonists to preclude the nitration of GLT-1 and GS in chemotherapy-induced neuropathic pain [73]. These studies implicate the adenosine axis as a pharmacological receptor-mediated mechanism by which nitration events can be influenced.

Transient Receptor Potential Channels

The transient receptor potential (TRP) family of nonselective cation channels plays a vital role in the molecular integration of multiple endogenous and exogenous sensory stimuli including but not limited to noxious heat or cold, pH, stretch, pressure, oxidants, and noxious chemicals [166, 167]. The TRP vanilloid channel 1 (TRPV1) is of particular relevance to pain [168–171] as these receptors are found on small and medium primary afferent nociceptors and in CNS areas responsible for nociceptive processing [172–174]; acting as critical mediators of inflammatory thermal hyperalgesia [175–177]. Because TRPV1 mediates the influx of sodium and calcium ions in neurons, its activation promotes glutamatergic signaling [178–182] and is thought to promote long-term potentiation [183], which we have discussed as an important contributor to central sensitization. Accordingly, TRPV1 has been targeted in the management of various pain states with a degree of success [184–186].

The interplay of TPRV1 and PN in pathological nociception is an area that warrants further investigation. Importantly, a host of literature implicates oxidation events in the activity, expression, and sensitivity of TRPV1 receptors: SO stimulates TRPV1 activity in inflammation [187, 188] and this may occur through a PN-mediated mechanism involving PKC nitrative activation [162, 163, 189] and the subsequent phosphorylative activation of TRPV1 [190, 191]. Further, in vitro studies demonstrate that NOX activity and oxidant donors induce TRPV1 channel expression and activity [192, 193]. While the physiological nitration of TRPV1 by PN has not been conclusively demonstrated to date, studies have identified the tyrosine residue Tyr671 on TM6 of TRPV1 as essential to the characteristic slow desensitization of the channel—a mechanism that permits sensory neuron adaptation to noxious stimuli over time [194]. Substitutions of Tyr671 result in the loss of TRPV1 desensitization [195], leading to the hypothesis that Tyr671 nitration may modify TRPV1 desensitization properties. Conversely, the activation of TRPV1 in skeletal muscle during muscular hypertrophy is dependent on the formation of PN via the concurrent activation nNOS and NOX4 [196]. Furthermore, oxidation of cysteine residues in the cytoplasmic termini of TRPV1 produces receptor sensitization [197, 198], such that oxidative or nitroxidative modifications of TRPV1 can lead to the persistent activation of TRPV1-expressing nociceptors in pathophysiological pain [197].

TRPV1 activation also appears to provide positive feedback on PN generation in pain. TRPV1 sensitization drives the production of nitroxidative species in CGRP-positive neurons [199] and trigeminal ganglion neurons [200], and furthermore, TRPV1 drives NOX expression in activated microglia [201] and DRG neurons during inflammatory pain [71]. Recent evidence suggests that TRPV1-mediated increases in oxidant and nitrative mediators contribute to the maintenance of inflammation by promoting the expression of the tumor necrosis factor (TNF) receptor in DRG neurons [202].

Finally, the ability of specific oxidized lipids to act as TRP ligands has been hypothesized to play a role in the development of hyperalgesia and allodynia. The oxidized linoleic acid metabolites (9- and 13-hydroxyoctadecadienoic acid) act as TRPV1 agonists in the periphery and spinal cord to promote hyperalgesia [203, 204], and importantly, the nitrative product nitrooleic acid also leads to nociceptor activation through the TRPA1 channel [205]. Indeed, oxidized linoleic acid metabolites are thought to drive persistent pain following thermal injury [206]. Given the extensive documentation of nitroxidative stress in the periphery and spinal cord during inflammatory hyperalgesia [27, 86, 207], it is plausible that nitroxidant mediators also regulate TRPV1 activity through the generation of oxidized linoleic acid and the increased formation of oxidized metabolites. Taken together, the cycle of TRPV1-mediated nitroxidative stress and nitroxidant activation of the TRPV1 system provides a provocative target in persistent pain states; however, a greater understanding of the specific contributions of PN, SO, and other oxidants to this system is required.

Cyclooxygenases and Prostaglandin Signaling

Since the advent of NSAID pain relievers, cyclooxygenase (COX) enzymes have held a time-honored role in nociceptive processing; these enzymes include the constitutive COX-1 and inducible COX-2 [208, 209]. COX is responsible for the generation of prostanoids through the conversion of free arachidonic acid to prostaglandin H2 (PGH_2), which can serve as a precursor for downstream prostaglandins such as PGE_2, prostacyclin, and thromboxanes [210]. The function of PGE_2 at its cognate receptors ($EP_{1\text{-}4}$) in pain has been documented at the peripheral level, and most famously in inflammatory pain where it plays a vital role in the activation and sensitization of peripheral nociceptors [211–213]. Prostaglandin signaling has also been documented to play a role at the spinal level in the dorsal horn in somatic pain and in the late phase of carrageenan-induced peripheral inflammatory pain [214–216] and is implicated in preclinical neuropathic pain and clinical neuropathic pain resultant from nerve injury and painful neuroma in humans [217]. Accordingly, successful antinociceptive strategies have inhibited COX-1 and/or COX-2 function and/or prostaglandin generation. The impact of PN on COX activity and prostaglandin generation is therefore of paramount relevance to nociception.

COX enzymes are "receptor targets" for the multifaceted action of NO in nociception and as such are regulated in its presence [209, 218–221]. Whereas we have previously mentioned that this chapter will discuss NO as a substrate for PN due to the role of NO in protective pain states, the interaction of iNOS and the inducible COX-2 isoform represents a unique and important mechanism in the maintenance of chronic and pathological pain states. A physical interaction between iNOS and COX-2 exists wherein iNOS binds COX-2 and facilitates an increase in the catalytic activity of COX-2 through S-nitrosylation [222]. Excessive prostaglandin synthesis by COX-2 is thought to play a priming role in the acute-to-chronic pain transition both peripherally and spinally [223], such that the pathological induction of these enzyme isoforms could serve as a "bridge" between normal protective pain and pathophysiological chronic pain.

It is now acknowledged that PN can also produce the activation of COX enzymes in pathological states, although this interaction is less understood. For example, PN can produce the oxidative modification of key amino acids residues in the COX polypeptide backbone in order to result in enhanced prostaglandin biosynthesis [224, 225]. To this end, PN-mediated increases in prostaglandin synthesis via COX-1 and COX-2 contribute to the development of peripheral sensitization in inflammatory pain [40]. In other biological systems, the PN-mediated nitrative inactivation of prostacyclin (PGI_2) synthase shifts the balance of prostanoid metabolism to favor PGH_2 and subsequently PGE_2 [226, 227]. Taken together, these data illustrate a clear mechanism for PN-mediated favor of prostaglandin signaling in nociception and may represent a primary avenue by which PN-scavenging strategies attenuate pain.

25.4.2 Part 2: Anti-nociceptive Systems

Opioid signaling

Perhaps most famous in the realm of analgesia is the endogenous and exogenous modification of opioid signaling. In pathological pain states, a variety of hypotheses exist regarding endogenous anti-nociceptive opioid signaling (e.g., endorphins and enkephalins [228, 229]) and the function of pro-nociceptive opioidergic signaling (i.e., dynorphin [230, 231]) in descending facilitation [232, 233]. Importantly, the endogenous anti-nociceptive endorphins and enkephalins possess tyrosine residues that can be prone to nitration [234]. The nitration of leu-enkephalin with PN at physiological pH has been described in liquid and gas chromatography studies [235, 236] and nitrated enkephalins have been isolated from human neutrophils [237]. Because the tyrosine in met-enkephalin (MENK) is the key biologically active pharmacophore required for binding with all opioid receptors [238, 239], it is hypothesized that nitration [240] reduces or eliminates binding and anti-nociceptive properties at μ- and δ-opioid receptors. It is known that the supraspinal (i.e., rostral ventromedial medulla, RVM) activity of PN suppresses descending inhibitory input, as intra-RVM administration of PNDCs attenuates inflammatory hyperalgesia [37] and opioid tolerance [63]. Indeed, the PN-mediated nitration of MENK to 3-nitrotyrosine-methionine-sulfoxide (NSO-MENK) results in a functional inactivation of the peptide. NSO-MENK has no appreciable binding to opioid receptors and furthermore lacks analgesic qualities and anti-nociceptive efficacy in neuropathic pain relative to MENK [37]. A clear role exists for nitroxidative modification of the descending opioidergic system in pathological pain states, but largely remains to be investigated.

In counterpoint, we have referenced at length the driving role of nitroxidative stress in the development of opioid tolerance and opioid paradoxical hyperalgesia. There exists an intricate and only partially understood link between opioid tolerance and hyperalgesia, and certainly these properties of the opioidergic system have been targeted in an attempt to increase the clinical efficacy of opioids in pathological pain. Our laboratory has extensively characterized the overlapping pro-nociceptive and pro-tolerance role of PN in the opioidergic system. At the spinal level, induction of pathological SO derived from NOX and mitochondrial activity drives PN accumulation in a spinal mechanism that underpins tolerance [35, 64, 78]. At a supraspinal level (RVM), the nitrative inactivation of MnSOD contributes to opioid tolerance, and co-administration of a PNDC prevents both MnSOD nitration and the development of tolerance [63]. Finally, PN formation plays a critical role in neuroimmune activation associated with tolerance and nitration of glial proteins [241, 242]. There is a definitive role for the activation of glia in tolerance mechanisms [117], and accordingly, it is thought that PN could drive the reactive state of these cells. The diversity of mechanisms by which PN enacts opioid plasticity in rodent models lends support to the use of PNDC or SODm compounds in tandem with opioids as a mechanism of ameliorating tolerance mechanisms and improving analgesic efficacy.

Catecholamine Signaling

Catecholamines, including dopamine (DA) and norepinephrine (NE), play a vital role in the descending modulation of spinal nociception and pathological pain [243–249]. Dopaminergic signaling is thought to mediate tonic pain suppression via the mesolimbic dopamine circuit [250, 251] and in ventrolateral orbital cortex-evoked descending inhibition of nociception [252]; the loss of these inhibitory pathways has been implicated in neuropathic pain [253] and may permit the persistence of pain states. Similarly, norepinephrine comprises a vital inhibitory system that principally originates in the locus coeruleus and descends to stimulate spinal α2 receptors in a mode of descending inhibition of nociceptive transmission [254, 255]. Because the catecholamine structure lends itself to nitration, the nitrative modifications of DA, epinephrine, and NE have been described chemically [256], but have not been directly investigated in pain to date. In sepsis, catecholamine inactivation by PN mediates the hyporeactivity of patients to NE-mediated vasoconstriction; in these instances, NE sensitivity is restored with administration of a SODm [257]. It may be hypothesized that nitrative inactivation of NE and/or DA could underlie persistent pain in pathological states.

In addition to the direct nitration of catecholamines, the nitrative inactivation of tyrosine hydroxylase—the enzyme responsible for converting L-tyrosine into L-DOPA, a precursor to both DA and NE [258]—has been extensively documented in neurodegenerative literature. Nitrative inactivation of TH in the striatum and subsequent loss of DA transmission is observed in models of Parkinsonian neurodegeneration [259, 260]. Furthermore, chronic estradiol-17β elicits the nitrative inactivation of TH associated with decreases in NE production in areas of the CNS [261]. Critically, it remains controversial whether PN mediates TH inactivation via sulfhydryl oxidation [262, 263] or tyrosine nitration [264], as both of these modifications are known to occur in the presence of PN. A greater understanding of the interaction between PN and the catecholamine systems will benefit our ability to target these systems via PN-attenuating strategies in pain.

25.5 Nitroxidative Regulation of Transcription in Pathological Pain

25.5.1 NF-κB Transcriptional Regulation

Nuclear factor kappa-light-chain-enhancer of activated B cells (NF-κB) is a transcriptional regulatory complex that serves as a cytosolic "stress response" sensor in virtually all vertebrate cell types. The complex NF-κB detects a variety of noxious or damage-associated stimuli including inflammatory mediators, bacterial or viral antigens, and changes in cellular redox status [265]. There is a demonstrated relationship between PN and NF-κB activation in a variety of pain states: NF-κB activation occurs in persistent inflammatory pain and in the development and processing of pathological pain [266]. The pathological activation of the NF-κB complex in glia contributes to reactive hypertrophic states and the production of inflammatory mediators during pain [117]. In human intervertebral disc cells, PN stimulation or treatment with SIN-1, a concomitant SO/NO donor, produces the translocation of NF-κB and gene expression of pro-inflammatory IL-1β and IL-6, implicating PN in degenerative back pain [267]. Furthermore, paw injection of PN produces activation of NF-κB in paw tissues and subsequent changes in COX expression in inflammatory pain [40]. NF-κB therefore has a reputable role in pathophysiological nociception due to the tight correlation between nociceptive mechanisms and cellular stress; however, current research techniques struggle to uncouple the role of NF-κB in pain from its broader role in other insults such as ischemia, excitotoxicity, and cell death.

In unstimulated cells, the NF-κB dimer is sequestered in the cytoplasm by its cytosolic inhibitor, IκB. The canonical activation of NF-κB involves TNF, TLR, or T-cell receptor activation of the IκB serine/threonine kinase (IKK) complex, resulting in phosphorylation and proteolysis of IκB and the nuclear translocation of activated NF-κB [265]. Importantly, IκB can also be phosphorylated on Tyr42 to impede its degradation and persistently inactivate NF-κB [268]. It has been suggested that PN can activate NF-κB signaling directly through tyrosine nitration, and this may occur via the documented ability of PN impair regulatory tyrosine phosphorylation by alternatively nitrating the residue [269, 270]. Indeed, PN produces an increase in nitration of IκB at Tyr42, a dose-dependent decrease in IkB phosphorylation, and concomitant increases in NF-κB activation in human peripheral blood monocytes as a mechanism of pro-inflammatory activation [271]. These data suggest that PN-mediated IκB Tyr42 nitration facilitates the persistent activation of NF-κB. In counterpoint, it has been suggested that NF-κB p65 can also be nitrated in a NO-dependent manner on residues Tyr66 and Tyr152 to result in the inactivation of NF-κB signaling in the P19 embryonic carcinoma cell line [272]. Additional work in a lung carcinoma cell line implicates this mechanism only in the absence of inflammatory activation, suggesting that the persistent activation of NF-κB by PN is context-dependent [273]. In the circumstance of aberrant nociception, it is likely that neuroinflammation and pathological production of PN coincide to produce a favorable context for PN-mediated activation of NF-κB [6]. Indeed, in vivo, PN has been implicated in the direct and robust activation of NF-κB signaling during pain and neurological insult: administration of PNDCs reduces NF-κB expression and associated production of inflammatory mediators in inflammatory pain [40] and spinal cord injury [274, 275]. While in-depth studies are required to better understand the unique interplay between PN generation and NF-κB signaling, these studies highlight the potential for PN-targeted therapies to impact transcriptional regulation in pain.

25.5.2 Histone Modification

In addition to the modulation of cytosolic sensors that mediate transcriptional events, PN is a membrane-permeant molecule and can cross the nuclear envelope in order to mediate oxidation events. While highly reactive molecules such as the hydroxyl radical and even PN are known to play a role in oxidative DNA damage and cell death, PN can in fact impact transcription by modifying the coiling and therefore transcriptional availability of DNA on histones. Histone acetylation and deacetylation events play a remarkably intricate molecular role in the transition between neuronal excitation and potentiation [276]; similarly, it is thought that epigenetic modifications can play a role in the transition from acute to chronic pain states [277]. Persistent inflammatory pain and neuropathic pain are in part maintained by histone deacetylase (HDAC)-mediated epigenetic suppression of glutamate decarboxylase (Gad2) transcription—encoding GAD65, a GABA synthesis enzyme—in the nucleus raphe magnus [278]. Indeed, class II-selective HDAC inhibitors show analgesic efficacy in inflammatory hyperalgesia [279] and class I selective-HDAC inhibitors appear to have therapeutic value in neuropathic pain through epigenetic regulation of targets including cytokines, metabotropic glutamate receptors, GABA synthesis, and relevant ion channels such as NaV1.8 [278, 280–282]. Conversely, studies have indicated that some HDAC inhibitors can promote nociception and enhance opioid-induced hyperalgesia [283]. Epigenetics is an increasingly provocative topic in pathophysiological pain states and we are only beginning to understand the complex, specific mechanisms of epigenetic regulation in protective versus aberrant nociception.

The nitration of HDACs and histones by PN has been identified in painful autoimmune disorders such as rheumatoid arthritis [284] and lupus [285, 286]. PN can mediate the specific nitrative inactivation of HDAC2, a class I HDAC, through nitration of Tyr253 [287]. Interestingly, HDAC2 can also be phosphorylated and positively modulated by tyrosine kinases [288], such that it may be hypothesized that in this system as well, tyrosine is a competitive site for nitration versus phosphorylation. Given the role of class I HDACs in anti-nociception as previously discussed, it is plausible that the PN-mediated inactivation of HDAC2 contributes to anti-nociception. Furthermore, evidence exists for the nitrative inactivation of sirtuin HDACs in pain; in particular, the SIRT6 deacetylase plays a role involved in nociception [289] presumably through its role in synaptic function [290]. PN nitration functionally inhibits the deacetylase activity of SIRT6 [291]. Additionally, spinal PN generated as a consequence of peripheral inflammatory pain has been demonstrated to nitrate SIRT1 and 3 as a mechanism to decrease transcriptional production and activation of MnSOD and catalase in a model of acute inflammatory pain [292]. These data offer a unique insight into the way in which PN nitration is functionally relevant to transcriptional regulation in pain.

25.6 Conclusions and Future Outlook

The mechanism by which nitroxidative stress drives nociception in pathophysiological pain is only beginning to be understood as a specific cellular process that can be therapeutically targeted. In this chapter, we have lead the reader from the infancy of nitroxidant generation to the formation of PN as a critical signaling molecule, through the multiplicative actions of PN as a nitrating agent and an oxidant in the nociceptive system, and frequently, to a positive feedback loop that perpetuates the actions of PN both cellularly and in persistent pain. In conclusion, we purport that the clinical targeting nitroxidant mediators has and continues to be limited by two factors: first, the tolerability and bioavailability of SODm and PNDC compounds has been unacceptable until the recent concerted effort to develop the SRI-series of compounds. Secondly and most importantly, the literature on the role of nitroxidative stress in pathophysiological pain has been disconnected and vague at best. It is the goal of this chapter to provide a unified perspective on nitroxidative stress and furthermore to posit specific, targetable actions of nitroxidative stress (i.e., nitration) to make it more accessible for preclinical and clinical development. Given the inadequacies of the present therapeutic repertoire for chronic pain, a new generation of more drug-like PNDCs and SODms offers a unique opportunity to improve the clinical standard of pain management.

Acknowledgment We would like to acknowledge our funding support from NIH/NIDA RO1 DA024074 and the NIH/NCI RO1 CA169519.

References

1. McCarberg BH, Nicholson BD, Todd KH, Palmer T, Penles L. The impact of pain on quality of life and the unmet needs of pain management: results from pain sufferers and physicians participating in an Internet survey. Am J Ther. 2008;15(4):312–20.
2. Salvemini D. Peroxynitrite and opiate antinociceptive tolerance: a painful reality. Arch Biochem Biophys. 2009;484(2):238–44.
3. Salvemini D, Neumann WL. Peroxynitrite: a strategic linchpin of opioid analgesic tolerance. Trends Pharmacol Sci. 2009;30(4):194–202.
4. Rausaria S, Ghaffari MM, Kamadulski A, Rodgers K, Bryant L, Chen Z, et al. Retooling manganese(III) porphyrin-based peroxynitrite decomposition catalysts for selectivity and oral activity: a potential new strategy for treating chronic pain. J Med Chem. 2011;54(24):8658–69.
5. Little JW, Doyle T, Salvemini D. Reactive nitroxidative species and nociceptive processing: determining the roles for nitric oxide, superoxide, and peroxynitrite in pain. Amino Acids. 2012;42(1):75–94.
6. Janes K, Neumann WL, Salvemini D. Anti-superoxide and anti-peroxynitrite strategies in pain suppression. Biochim Biophys Acta. 2012;1822(5):815–21.
7. Beckman JS, Beckman TW, Chen J, Marshall PA, Freeman BA. Apparent hydroxyl radical production by peroxynitrite: implications for endothelial injury from nitric oxide and superoxide. Proc Natl Acad Sci U S A. 1990;87(4):1620–4.
8. Khattab MM. TEMPOL, a membrane-permeable radical scavenger, attenuates peroxynitrite- and superoxide anion-enhanced carrageenan-induced paw edema and hyperalgesia: a key role for superoxide anion. Eur J Pharmacol. 2006;548(1-3):167–73.
9. Lee I, Kim HK, Kim JH, Chung K, Chung JM. The role of reactive oxygen species in capsaicin-induced mechanical hyperalgesia and in the activities of dorsal horn neurons. Pain. 2007;133(1-3):9–17.
10. Gao X, Kim HK, Chung JM, Chung K. Reactive oxygen species (ROS) are involved in enhancement of NMDA-receptor phosphorylation in animal models of pain. Pain. 2007; 131(3):262–71.
11. Schwartz ES, Kim HY, Wang J, Lee I, Klann E, Chung JM, et al. Persistent pain is dependent on spinal mitochondrial antioxidant levels. J Neurosci. 2009;29(1):159–68.
12. Schwartz ES, Lee I, Chung K, Chung JM. Oxidative stress in the spinal cord is an important contributor in capsaicin-induced mechanical secondary hyperalgesia in mice. Pain. 2008; 138(3):514–24.
13. Wang J, Cochran V, Abdi S, Chung JM, Chung K, Kim HK. Phenyl N-t-butylnitrone, a reactive oxygen species scavenger, reduces zymosan-induced visceral pain in rats. Neurosci Lett. 2008;439(2):216–9.
14. Tang N, Ong WY, Yeo JF, Farooqui AA. Anti-allodynic effect of intracerebroventricularly administered antioxidant and free radical scavenger in a mouse model of orofacial pain. J Orofac Pain. 2009;23(2):167–73.
15. Park ES, Gao X, Chung JM, Chung K. Levels of mitochondrial reactive oxygen species increase in rat neuropathic spinal dorsal horn neurons. Neurosci Lett. 2006;391(3):108–11.

16. Siniscalco D, Fuccio C, Giordano C, Ferraraccio F, Palazzo E, Luongo L, et al. Role of reactive oxygen species and spinal cord apoptotic genes in the development of neuropathic pain. Pharmacol Res. 2007;55(2):158–66.
17. Tal M. A novel antioxidant alleviates heat hyperalgesia in rats with an experimental painful peripheral neuropathy. Neuroreport. 1996;7(8):1382–4.
18. Muscoli C, Cuzzocrea S, Riley DP, Zweier JL, Thiemermann C, Wang ZQ, et al. On the selectivity of superoxide dismutase mimetics and its importance in pharmacological studies. Br J Pharmacol. 2003;140(3):445–60.
19. Salvemini D, Doyle TM, Cuzzocrea S. Superoxide, peroxynitrite and oxidative/nitrative stress in inflammation. Biochem Soc Trans. 2006;34(Pt 5):965–70.
20. Salvemini D, Riley DP, Cuzzocrea S. SOD mimetics are coming of age. Nat Rev Drug Discov. 2002;1(5):367–74.
21. Batinic-Haberle I, Reboucas JS, Spasojevich I. Superoxide dismutase mimics: chemistry, pharmacology and therapeutic potential. Antioxid Redox Signal. 2010;13(6):877–918.
22. Salvemini D, Wang ZQ, Zweier JL, Samouilov A, Macarthur H, Misko TP, et al. A nonpeptidyl mimic of superoxide dismutase with therapeutic activity in rats. Science. 1999;286(5438):304–6.
23. Salvemini D, Neumann W. Targeting peroxynitrite driven nitroxidative stress with synzymes: A novel therapeutic approach in chronic pain management. Life Sci. 2010;86(15-16):604–14.
24. Salvemini D, Riley DP. Nonpeptidyl mimetics of superoxide dismutase in clinical therapies for diseases. Cell Mol Life Sci. 2000;57(11):1489–92.
25. Filipovic MR, Duerr K, Mojovic M, Simeunovic V, Zimmermann R, Niketic V, et al. NO dismutase activity of seven-coordinate manganese(II) pentaazamacrocyclic complexes. Angew Chem Int Ed Engl. 2008;47(45):8735–9.
26. Filipovic MR, Koh AC, Arbault S, Niketic V, Debus A, Schleicher U, et al. Striking inflammation from both sides: manganese(II) pentaazamacrocyclic SOD mimics act also as nitric oxide dismutases: a single-cell study. Angew Chem Int Ed Engl. 2010;49(25):4228–32.
27. Wang ZQ, Porreca F, Cuzzocrea S, Galen K, Lightfoot R, Masini E, et al. A newly identified role for superoxide in inflammatory pain. J Pharmacol Exp Ther. 2004;309(3):869–78.
28. Di Cesare Mannelli L, Bani D, Bencini A, Brandi ML, Calosi L, Cantore M, et al. Therapeutic effects of the superoxide dismutase mimetic compound MnIIMe2DO2A on experimental articular pain in rats. Mediators Inflamm. 2013;2013:905360.
29. Day BJ, Fridovich I, Crapo JD. Manganic porphyrins possess catalase activity and protect endothelial cells against hydrogen peroxide-mediated injury. Arch Biochem Biophys. 1997;347(2):256–62.
30. Lee JB, Hunt JA, Groves JT. Manganese porphyrins as redox-coupled peroxynitrite reductases. J Am Chem Soc. 1998;120(24):6053–61.
31. Doyle T, Finley A, Chen Z, Salvemini D. Role for peroxynitrite in sphingosine-1-phosphate-induced hyperalgesia in rats. Pain. 2011;152(3):643–8.
32. Chen Z, Muscoli C, Doyle T, Bryant L, Cuzzocrea S, Mollace V, et al. NMDA-receptor activation and nitroxidative regulation of the glutamatergic pathway during nociceptive processing. Pain. 2010;149(1):100–6.
33. Kwak KH, Jung H, Park JM, Yeo JS, Kim H, Lee HC, et al. A peroxynitrite decomposition catalyst prevents mechanical allodynia and NMDA receptor activation in the hind-paw ischemia reperfusion injury rats. Exp Ther Med. 2014;7(2):508–12.
34. Doyle T, Chen Z, Muscoli C, Bryant L, Esposito E, Cuzzocrea S, et al. Targeting the overproduction of peroxynitrite for the prevention and reversal of paclitaxel-induced neuropathic pain. J Neurosci. 2012;32(18):6149–60.
35. Doyle T, Esposito E, Bryant L, Cuzzocrea S, Salvemini D. NADPH-oxidase 2 activation promotes opioid-induced antinociceptive tolerance in mice. Neuroscience. 2013;241:1–9.
36. Muscoli C, Cuzzocrea S, Ndengele MM, Mollace V, Porreca F, Fabrizi F, et al. Therapeutic manipulation of peroxynitrite attenuates the development of opiate-induced antinociceptive tolerance in mice. J Clin Invest. 2007;117(11):3530–9.
37. Little JW, Chen Z, Doyle T, Porreca F, Ghaffari M, Bryant L, et al. Supraspinal peroxynitrite modulates pain signaling by suppressing the endogenous opioid pathway. The Journal of neuroscience : the official journal of the Society for Neuroscience. 2012;32(32):10797–808.

38. Batinic-Haberle I, Cuzzocrea S, Reboucas JS, Ferrer-Sueta G, Mazzon E, Di Paola R, et al. Pure MnTBAP selectively scavenges peroxynitrite over superoxide: comparison of pure and commercial MnTBAP samples to MnTE-2-PyP in two models of oxidative stress injury, an SOD-specific Escherichia coli model and carrageenan-induced pleurisy. Free Radic Biol Med. 2009;46(2):192–201.
39. Reboucas JS, Spasojevic I, Batinic-Haberle I. Pure manganese(III) 5,10,15,20-tetrakis(4-benzoic acid)porphyrin (MnTBAP) is not a superoxide dismutase mimic in aqueous systems: a case of structure-activity relationship as a watchdog mechanism in experimental therapeutics and biology. J Biol Inorg Chem. 2008;13(2):289–302.
40. Ndengele MM, Cuzzocrea S, Esposito E, Mazzon E, Di Paola R, Matuschak GM, et al. Cyclooxygenases 1 and 2 contribute to peroxynitrite-mediated inflammatory pain hypersensitivity. FASEB J. 2008;22(9):3154–64.
41. Cheton PL, Archibald FS. Manganese complexes and the generation and scavenging of hydroxyl free radicals. Free Radic Biol Med. 1988;5(5-6):325–33.
42. Batinic-Haberle I, Ndengele MM, Cuzzocrea S, Reboucas JS, Spasojevic I, Salvemini D. Lipophilicity is a critical parameter that dominates the efficacy of metalloporphyrins in blocking the development of morphine antinociceptive tolerance through peroxynitrite-mediated pathways. Free Radic Biol Med. 2009;46(2):212–9.
43. Reboucas JS, DeFreitas-Silva G, Spasojevic I, Idemori YM, Benov L, Batinic-Haberle I. Impact of electrostatics in redox modulation of oxidative stress by Mn porphyrins: protection of SOD-deficient Escherichia coli via alternative mechanism where Mn porphyrin acts as a Mn carrier. Free Radic Biol Med. 2008;45(2):201–10.
44. Rausaria S, Kamadulski A, Rath NP, Bryant L, Chen Z, Salvemini D, et al. Manganese(III) complexes of bis(hydroxyphenyl)dipyrromethenes are potent orally active peroxynitrite scavengers. J Am Chem Soc. 2011;133(12):4200–3.
45. Hess DT, Matsumoto A, Kim SO, Marshall HE, Stamler JS. Protein S-nitrosylation: purview and parameters. Nat Rev Mol Cell Biol. 2005;6(2):150–66.
46. Ischiropoulos H. Protein tyrosine nitration--an update. Arch Biochem Biophys. 2009;484(2):117–21.
47. Sawa T, Akaike T, Maeda H. Tyrosine nitration by peroxynitrite formed from nitric oxide and superoxide generated by xanthine oxidase. J Biol Chem. 2000;275(42):32467–74.
48. Radi R, Cassina A, Hodara R, Quijano C, Castro L. Peroxynitrite reactions and formation in mitochondria. Free Radic Biol Med. 2002;33(11):1451–64.
49. Radi R. Nitric oxide, oxidants, and protein tyrosine nitration. Proc Natl Acad Sci U S A. 2004;101(12):4003–8.
50. Szabo C, Ischiropoulos H, Radi R. Peroxynitrite: biochemistry, pathophysiology and development of therapeutics. Nat Rev Drug Discov. 2007;6(8):662–80.
51. Irie Y, Saeki M, Kamisaki Y, Martin E, Murad F. Histone H1.2 is a substrate for denitrase, an activity that reduces nitrotyrosine immunoreactivity in proteins. Proc Natl Acad Sci U S A. 2003;100(10):5634–9.
52. Kamisaki Y, Wada K, Bian K, Balabanli B, Davis K, Martin E, et al. An activity in rat tissues that modifies nitrotyrosine-containing proteins. Proc Natl Acad Sci U S A. 1998;95(20):11584–9.
53. McCord JM, Fridovich I. Superoxide dismutase. An enzymic function for erythrocuprein (hemocuprein). J Biol Chem. 1969;244(22):6049–55.
54. Viggiano A, Monda M, Viggiano A, Viggiano D, Viggiano E, Chiefari M, et al. Trigeminal pain transmission requires reactive oxygen species production. Brain Res. 2005;1050(1-2):72–8.
55. Cardey B, Foley S, Enescu M. Mechanism of thiol oxidation by the superoxide radical. J Phys Chem A. 2007;111(50):13046–52.
56. Drose S, Brandt U. Molecular mechanisms of superoxide production by the mitochondrial respiratory chain. Adv Exp Med Biol. 2012;748:145–69.
57. Bleier L, Wittig I, Heide H, Steger M, Brandt U, Drose S. Generator-specific targets of mitochondrial reactive oxygen species. Free Radic Biol Med. 2015;78:1–10.
58. Kim HY, Chung JM, Chung K. Increased production of mitochondrial superoxide in the spinal cord induces pain behaviors in mice: the effect of mitochondrial electron transport complex inhibitors. Neurosci Lett. 2008;447(1):87–91.

59. Cordero MD, De Miguel M, Moreno Fernandez AM, Carmona Lopez IM, Garrido Maraver J, Cotan D, et al. Mitochondrial dysfunction and mitophagy activation in blood mononuclear cells of fibromyalgia patients: implications in the pathogenesis of the disease. Arthritis Res Ther. 2010;12(1):R17.
60. Janes K, Doyle T, Bryant L, Esposito E, Cuzzocrea S, Ryerse J, et al. Bioenergetic deficits in peripheral nerve sensory axons during chemotherapy-induced neuropathic pain resulting from peroxynitrite-mediated post-translational nitration of mitochondrial superoxide dismutase. Pain. 2013;154(11):2432–40.
61. Zherebitskaya E, Akude E, Smith DR, Fernyhough P. Development of selective axonopathy in adult sensory neurons isolated from diabetic rats: role of glucose-induced oxidative stress. Diabetes. 2009;58(6):1356–64.
62. Stavniichuk R, Shevalye H, Lupachyk S, Obrosov A, Groves JT, Obrosova IG, et al. Peroxynitrite and protein nitration in the pathogenesis of diabetic peripheral neuropathy. Diabetes Metab Res Rev. 2014;30(8):669–78.
63. Doyle T, Bryant L, Batinic-Haberle I, Little J, Cuzzocrea S, Masini E, et al. Supraspinal inactivation of mitochondrial superoxide dismutase is a source of peroxynitrite in the development of morphine antinociceptive tolerance. Neuroscience. 2009;164(2):702–10.
64. Little JW, Cuzzocrea S, Bryant L, Esposito E, Doyle T, Rausaria S, et al. Spinal mitochondrial-derived peroxynitrite enhances neuroimmune activation during morphine hyperalgesia and antinociceptive tolerance. Pain. 2013;154(7):978–86.
65. Babior BM, Lambeth JD, Nauseef W. The neutrophil NADPH oxidase. Arch Biochem Biophys. 2002;397(2):342–4.
66. Muzaffar S, Shukla N, Angelini GD, Jeremy JY. Superoxide auto-augments superoxide formation and upregulates gp91(phox) expression in porcine pulmonary artery endothelial cells: inhibition by iloprost. Eur J Pharmacol. 2006;538(1-3):108–14.
67. Puntambekar P, Mukherjea D, Jajoo S, Ramkumar V. Essential role of Rac1/NADPH oxidase in nerve growth factor induction of TRPV1 expression. J Neurochem. 2005;95(6):1689–703.
68. Abramov AY, Canevari L, Duchen MR. Beta-amyloid peptides induce mitochondrial dysfunction and oxidative stress in astrocytes and death of neurons through activation of NADPH oxidase. J Neurosci. 2004;24(2):565–75.
69. Green SP, Cairns B, Rae J, Errett-Baroncini C, Hongo JA, Erickson RW, et al. Induction of gp91-phox, a component of the phagocyte NADPH oxidase, in microglial cells during central nervous system inflammation. J Cereb Blood Flow Metab. 2001;21(4):374–84.
70. Tejada-Simon MV, Serrano F, Villasana LE, Kanterewicz BI, Wu GY, Quinn MT, et al. Synaptic localization of a functional NADPH oxidase in the mouse hippocampus. Mol Cell Neurosci. 2005;29(1):97–106.
71. Ibi M, Matsuno K, Shiba D, Katsuyama M, Iwata K, Kakehi T, et al. Reactive oxygen species derived from NOX1/NADPH oxidase enhance inflammatory pain. J Neurosci. 2008;28(38): 9486–94.
72. Kim D, You B, Jo EK, Han SK, Simon MI, Lee SJ. NADPH oxidase 2-derived reactive oxygen species in spinal cord microglia contribute to peripheral nerve injury-induced neuropathic pain. Proc Natl Acad Sci U S A. 2010;107(33):14851–6.
73. Janes K, Esposito E, Doyle T, Cuzzocrea S, Tosh DK, Jacobson KA, et al. A3 adenosine receptor agonist prevents the development of paclitaxel-induced neuropathic pain by modulating spinal glial-restricted redox-dependent signaling pathways. Pain. 2014;155(12):2560–7.
74. Kallenborn-Gerhardt W, Schroder K, Geisslinger G, Schmidtko A. NOXious signaling in pain processing. Pharmacol Ther. 2013;137(3):309–17.
75. Lim H, Kim D, Lee SJ. Toll-like receptor 2 mediates peripheral nerve injury-induced NADPH oxidase 2 expression in spinal cord microglia. J Biol Chem. 2013;288(11):7572–9.
76. Choi SR, Roh DH, Yoon SY, Kang SY, Moon JY, Kwon SG, et al. Spinal sigma-1 receptors activate NADPH oxidase 2 leading to the induction of pain hypersensitivity in mice and mechanical allodynia in neuropathic rats. Pharmacol Res. 2013;74:56–67.
77. Ibi M, Matsuno K, Matsumoto M, Sasaki M, Nakagawa T, Katsuyama M, et al. Involvement of NOX1/NADPH oxidase in morphine-induced analgesia and tolerance. The Journal of neuroscience : the official journal of the Society for Neuroscience. 2011;31(49):18094–103.

78. Doyle T, Bryant L, Muscoli C, Cuzzocrea S, Esposito E, Chen Z, et al. Spinal NADPH oxidase is a source of superoxide in the development of morphine-induced hyperalgesia and antinociceptive tolerance. Neurosci Lett. 2010;483(2):85–9.
79. Kallenborn-Gerhardt W, Schroder K, Del Turco D, Lu R, Kynast K, Kosowski J, et al. NADPH oxidase-4 maintains neuropathic pain after peripheral nerve injury. J Neurosci. 2012;32(30):10136–45.
80. Im YB, Jee MK, Jung JS, Choi JI, Jang JH, Kang SK. miR23b ameliorates neuropathic pain in spinal cord by silencing NADPH oxidase 4. Antioxid Redox Signal. 2012;16(10):1046–60.
81. Connor M. Allopurinol for pain relief: more than just crystal clearance? Br J Pharmacol. 2009;156(1):4–6.
82. Kwak KH, Han CG, Lee SH, Jeon Y, Park SS, Kim SO, et al. Reactive oxygen species in rats with chronic post-ischemia pain. Acta Anaesthesiol Scand. 2009;53(5):648–56.
83. Melov S. Mitochondrial oxidative stress. Physiologic consequences and potential for a role in aging. Ann N Y Acad Sci. 2000;908:219–25.
84. Ho YS, Magnenat JL, Gargano M, Cao J. The nature of antioxidant defense mechanisms: a lesson from transgenic studies. Environ Health Perspect. 1998;106 Suppl 5:1219–28.
85. MacMillan-Crow LA, Thompson JA. Tyrosine modifications and inactivation of active site manganese superoxide dismutase mutant (Y34F) by peroxynitrite. Arch Biochem Biophys. 1999;366(1):82–8.
86. Muscoli C, Mollace V, Wheatley J, Masini E, Ndengele M, Wang ZQ, et al. Superoxide-mediated nitration of spinal manganese superoxide dismutase: a novel pathway in N-methyl-D-aspartate-mediated hyperalgesia. Pain. 2004;111(1-2):96–103.
87. Bredt DS, Snyder SH. Nitric oxide, a novel neuronal messenger. Neuron. 1992;8(1):3–11.
88. Dawson TM, Dawson VL, Snyder SH. A novel neuronal messenger molecule in brain: the free radical, nitric oxide. Ann Neurol. 1992;32(3):297–311.
89. Meller ST, Gebhart GF. Nitric oxide (NO) and nociceptive processing in the spinal cord. Pain. 1993;52(2):127–36.
90. O'Dell TJ, Hawkins RD, Kandel ER, Arancio O. Tests of the roles of two diffusible substances in long-term potentiation: evidence for nitric oxide as a possible early retrograde messenger. Proc Natl Acad Sci U S A. 1991;88(24):11285–9.
91. Ignarro LJ. Haem-dependent activation of guanylate cyclase and cyclic GMP formation by endogenous nitric oxide: a unique transduction mechanism for transcellular signaling. Pharmacol Toxicol. 1990;67(1):1–7.
92. Ignarro LJ. Heme-dependent activation of guanylate cyclase by nitric oxide: a novel signal transduction mechanism. Blood Vessels. 1991;28(1-3):67–73.
93. Schmidtko A, Tegeder I, Geisslinger G. No NO, no pain? The role of nitric oxide and cGMP in spinal pain processing. Trends Neurosci. 2009;32(6):339–46.
94. Tegeder I, Scheving R, Wittig I, Geisslinger G. SNO-ing at the nociceptive synapse? Pharmacol Rev. 2011;63(2):366–89.
95. Tanabe M, Nagatani Y, Saitoh K, Takasu K, Ono H. Pharmacological assessments of nitric oxide synthase isoforms and downstream diversity of NO signaling in the maintenance of thermal and mechanical hypersensitivity after peripheral nerve injury in mice. Neuropharmacology. 2009;56(3):702–8.
96. Riedel W, Neeck G. Nociception, pain, and antinociception: current concepts. Z Rheumatol. 2001;60(6):404–15.
97. Cury Y, Picolo G, Gutierrez VP, Ferreira SH. Pain and analgesia: The dual effect of nitric oxide in the nociceptive system. Nitric Oxide. 2011;25(3):243–54.
98. Schmidt HH, Pollock JS, Nakane M, Forstermann U, Murad F. Ca^{2+}/calmodulin-regulated nitric oxide synthases. Cell Calcium. 1992;13(6-7):427–34.
99. Forstermann U, Closs EI, Pollock JS, Nakane M, Schwarz P, Gath I, et al. Nitric oxide synthase isozymes. Characterization, purification, molecular cloning, and functions. Hypertension. 1994;23(6 Pt 2):1121–31.
100. Bredt DS, Hwang PM, Snyder SH. Localization of nitric oxide synthase indicating a neural role for nitric oxide. Nature. 1990;347(6295):768–70.

101. Cao J, Viholainen JI, Dart C, Warwick HK, Leyland ML, Courtney MJ. The PSD95-nNOS interface: a target for inhibition of excitotoxic p38 stress-activated protein kinase activation and cell death. J Cell Biol. 2005;168(1):117–26.
102. Ahlawat A, Rana A, Goyal N, Sharma S. Potential role of nitric oxide synthase isoforms in pathophysiology of neuropathic pain. Inflammopharmacology. 2014;22(5):269–78.
103. Keilhoff G, Schroder H, Peters B, Becker A. Time-course of neuropathic pain in mice deficient in neuronal or inducible nitric oxide synthase. Neurosci Res. 2013;77(4):215–21.
104. Mukherjee P, Cinelli MA, Kang S, Silverman RB. Development of nitric oxide synthase inhibitors for neurodegeneration and neuropathic pain. Chem Soc Rev. 2014;43(19):6814–38.
105. Annedi SC, Maddaford SP, Mladenova G, Ramnauth J, Rakhit S, Andrews JS, et al. Discovery of N-(3-(1-methyl-1,2,3,6-tetrahydropyridin-4-yl)-1H-indol-6-yl) thiophene-2-carboximidamide as a selective inhibitor of human neuronal nitric oxide synthase (nNOS) for the treatment of pain. J Med Chem. 2011;54(20):7408–16.
106. Mihara Y, Egashira N, Sada H, Kawashiri T, Ushio S, Yano T, et al. Involvement of spinal NR2B-containing NMDA receptors in oxaliplatin-induced mechanical allodynia in rats. Mol Pain. 2011;7:8.
107. Vareniuk I, Pacher P, Pavlov IA, Drel VR, Obrosova IG. Peripheral neuropathy in mice with neuronal nitric oxide synthase gene deficiency. Int J Mol Med. 2009;23(5):571–80.
108. Knowles RG, Moncada S. Nitric oxide synthases in mammals. Biochem J. 1994;298(Pt 2):249–58.
109. Bonnefous C, Payne JE, Roppe J, Zhuang H, Chen X, Symons KT, et al. Discovery of inducible nitric oxide synthase (iNOS) inhibitor development candidate KD7332, part 1: Identification of a novel, potent, and selective series of quinolinone iNOS dimerization inhibitors that are orally active in rodent pain models. J Med Chem. 2009;52(9):3047–62.
110. Kuboyama K, Tsuda M, Tsutsui M, Toyohara Y, Tozaki-Saitoh H, Shimokawa H, et al. Reduced spinal microglial activation and neuropathic pain after nerve injury in mice lacking all three nitric oxide synthases. Mol Pain. 2011;7:50.
111. Cassuto J, Dou H, Czikora I, Szabo A, Patel VS, Kamath V, et al. Peroxynitrite disrupts endothelial caveolae leading to eNOS uncoupling and diminished flow-mediated dilation in coronary arterioles of diabetic patients. Diabetes. 2014;63(4):1381–93.
112. Luo S, Lei H, Qin H, Xia Y. Molecular mechanisms of endothelial NO synthase uncoupling. Curr Pharm Des. 2014;20(22):3548–53.
113. Zhang C, Li PL. Membrane raft redox signalosomes in endothelial cells. Free Radic Res. 2010;44(8):831–42.
114. Felley-Bosco E, Bender F, Quest AF. Caveolin-1-mediated post-transcriptional regulation of inducible nitric oxide synthase in human colon carcinoma cells. Biol Res. 2002;35(2):169–76.
115. Carozzi VA, Canta A, Chiorazzi A, Cavaletti G. Chemotherapy-induced peripheral neuropathy: What do we know about mechanisms? Neurosci Lett. 2015;596:90–107.
116. Osikowicz M, Mika J, Przewlocka B. The glutamatergic system as a target for neuropathic pain relief. Exp Physiol. 2013;98(2):372–84.
117. Hutchinson MR, Shavit Y, Grace PM, Rice KC, Maier SF, Watkins LR. Exploring the neuroimmunopharmacology of opioids: an integrative review of mechanisms of central immune signaling and their implications for opioid analgesia. Pharmacol Rev. 2011;63(3):772–810.
118. Latremoliere A, Woolf CJ. Central sensitization: a generator of pain hypersensitivity by central neural plasticity. J Pain. 2009;10(9):895–926.
119. Ossipov MH, Lai J, Malan Jr TP, Porreca F. Spinal and supraspinal mechanisms of neuropathic pain. Ann N Y Acad Sci. 2000;909:12–24.
120. De Felipe C, Herrero JF, O'Brien JA, Palmer JA, Doyle CA, Smith AJ, et al. Altered nociception, analgesia and aggression in mice lacking the receptor for substance P. Nature. 1998;392(6674):394–7.
121. Bleakman D, Alt A, Nisenbaum ES. Glutamate receptors and pain. Semin Cell Dev Biol. 2006;17(5):592–604.
122. Hertz L. Functional interactions between neurons and astrocytes I. Turnover and metabolism of putative amino acid transmitters. Prog Neurobiol. 1979;13(3):277–323.

123. Choi BH. Radial glia of developing human fetal spinal cord: Golgi, immunohistochemical and electron microscopic study. Brain Res. 1981;227(2):249–67.
124. Tanaka K, Watase K, Manabe T, Yamada K, Watanabe M, Takahashi K, et al. Epilepsy and exacerbation of brain injury in mice lacking the glutamate transporter GLT-1. Science. 1997;276(5319):1699–702.
125. Brustovetsky T, Purl K, Young A, Shimizu K, Dubinsky JM. Dearth of glutamate transporters contributes to striatal excitotoxicity. Exp Neurol. 2004;189(2):222–30.
126. Jabaudon D, Scanziani M, Gahwiler BH, Gerber U. Acute decrease in net glutamate uptake during energy deprivation. Proc Natl Acad Sci U S A. 2000;97(10):5610–5.
127. Mennerick S, Shen W, Xu W, Benz A, Tanaka K, Shimamoto K, et al. Substrate turnover by transporters curtails synaptic glutamate transients. J Neurosci. 1999;19(21):9242–51.
128. Semba J, Wakuta MS. Regional differences in the effects of glutamate uptake inhibitor L-trans-pyrrolidine-2,4-dicarboxylic acid on extracellular amino acids and dopamine in rat brain: an in vivo microdialysis study. Gen Pharmacol. 1998;31(3):399–404.
129. Mennerick S, Zorumski CF. Glial contributions to excitatory neurotransmission in cultured hippocampal cells. Nature. 1994;368(6466):59–62.
130. Rothstein JD, Martin L, Levey AI, Dykes-Hoberg M, Jin L, Wu D, et al. Localization of neuronal and glial glutamate transporters. Neuron. 1994;13(3):713–25.
131. Danbolt NC. Glutamate uptake. Prog Neurobiol. 2001;65(1):1–105.
132. Minana MD, Kosenko E, Marcaida G, Hermenegildo C, Montoliu C, Grisolia S, et al. Modulation of glutamine synthesis in cultured astrocytes by nitric oxide. Cell Mol Neurobiol. 1997;17(4):433–45.
133. Gorg B, Qvartskhava N, Bidmon HJ, Palomero-Gallagher N, Kircheis G, Zilles K, et al. Oxidative stress markers in the brain of patients with cirrhosis and hepatic encephalopathy. Hepatology. 2010;52(1):256–65.
134. Gorg B, Qvartskhava N, Voss P, Grune T, Haussinger D, Schliess F. Reversible inhibition of mammalian glutamine synthetase by tyrosine nitration. FEBS Lett. 2007;581(1):84–90.
135. Gorg B, Wettstein M, Metzger S, Schliess F, Haussinger D. Lipopolysaccharide-induced tyrosine nitration and inactivation of hepatic glutamine synthetase in the rat. Hepatology. 2005;41(5):1065–73.
136. Schliess F, Gorg B, Fischer R, Desjardins P, Bidmon HJ, Herrmann A, et al. Ammonia induces MK-801-sensitive nitration and phosphorylation of protein tyrosine residues in rat astrocytes. FASEB J. 2002;16(7):739–41.
137. Bidmon HJ, Gorg B, Palomero-Gallagher N, Schleicher A, Haussinger D, Speckmann EJ, et al. Glutamine synthetase becomes nitrated and its activity is reduced during repetitive seizure activity in the pentylentetrazole model of epilepsy. Epilepsia. 2008;49(10):1733–48.
138. Trotti D, Rolfs A, Danbolt NC, Brown Jr RH, Hediger MA. SOD1 mutants linked to amyotrophic lateral sclerosis selectively inactivate a glial glutamate transporter. Nat Neurosci. 1999;2(9):848.
139. Trotti D, Rossi D, Gjesdal O, Levy LM, Racagni G, Danbolt NC, et al. Peroxynitrite inhibits glutamate transporter subtypes. J Biol Chem. 1996;271(11):5976–9.
140. Arriza JL, Kavanaugh MP, Fairman WA, Wu YN, Murdoch GH, North RA, et al. Cloning and expression of a human neutral amino acid transporter with structural similarity to the glutamate transporter gene family. J Biol Chem. 1993;268(21):15329–32.
141. Fairman WA, Vandenberg RJ, Arriza JL, Kavanaugh MP, Amara SG. An excitatory amino-acid transporter with properties of a ligand-gated chloride channel. Nature. 1995;375(6532):599–603.
142. Kanai Y, Hediger MA. Primary structure and functional characterization of a high-affinity glutamate transporter. Nature. 1992;360(6403):467–71.
143. Pines G, Danbolt NC, Bjoras M, Zhang Y, Bendahan A, Eide L, et al. Cloning and expression of a rat brain L-glutamate transporter. Nature. 1992;360(6403):464–7.
144. Robinson MB, Dowd LA. Heterogeneity and functional properties of subtypes of sodium-dependent glutamate transporters in the mammalian central nervous system. Adv Pharmacol. 1997;37:69–115.

145. Storck T, Schulte S, Hofmann K, Stoffel W. Structure, expression, and functional analysis of a Na(+)-dependent glutamate/aspartate transporter from rat brain. Proc Natl Acad Sci U S A. 1992;89(22):10955–9.
146. Chen Y, Swanson RA. The glutamate transporters EAAT2 and EAAT3 mediate cysteine uptake in cortical neuron cultures. J Neurochem. 2003;84(6):1332–9.
147. Shanker G, Allen JW, Mutkus LA, Aschner M. The uptake of cysteine in cultured primary astrocytes and neurons. Brain Res. 2001;902(2):156–63.
148. Himi T, Ikeda M, Yasuhara T, Nishida M, Morita I. Role of neuronal glutamate transporter in the cysteine uptake and intracellular glutathione levels in cultured cortical neurons. J Neural Transm. 2003;110(12):1337–48.
149. Dringen R, Pfeiffer B, Hamprecht B. Synthesis of the antioxidant glutathione in neurons: supply by astrocytes of CysGly as precursor for neuronal glutathione. J Neurosci. 1999; 19(2):562–9.
150. Aoyama K, Matsumura N, Watabe M, Nakaki T. Oxidative stress on EAAC1 is involved in MPTP-induced glutathione depletion and motor dysfunction. Eur J Neurosci. 2008;27(1):20–30.
151. Tao YX, Gu J, Stephens Jr RL. Role of spinal cord glutamate transporter during normal sensory transmission and pathological pain states. Mol Pain. 2005;1:30.
152. Muscoli C, Visalli V, Colica C, Nistico R, Palma E, Costa N, et al. The effect of inflammatory stimuli on NMDA-related activation of glutamine synthase in human cultured astroglial cells. Neurosci Lett. 2005;373(3):184–8.
153. Suarez I, Bodega G, Fernandez B. Glutamine synthetase in brain: effect of ammonia. Neurochem Int. 2002;41(2-3):123–42.
154. Kimelberg HK, Goderie SK, Higman S, Pang S, Waniewski RA. Swelling-induced release of glutamate, aspartate, and taurine from astrocyte cultures. J Neurosci. 1990;10(5):1583–91.
155. Rose C, Kresse W, Kettenmann H. Acute insult of ammonia leads to calcium-dependent glutamate release from cultured astrocytes, an effect of pH. J Biol Chem. 2005;280(22):20937–44.
156. Mishra OP, Delivoria-Papadopoulos M. Cellular mechanisms of hypoxic injury in the developing brain. Brain Res Bull. 1999;48(3):233–8.
157. Zanelli SA, Ashraf QM, Delivoria-Papadopoulos M, Mishra OP. Peroxynitrite-induced modification of the N-methyl-D-aspartate receptor in the cerebral cortex of the guinea pig fetus at term. Neurosci Lett. 2000;296(1):5–8.
158. Zanelli SA, Ashraf QM, Mishra OP. Nitration is a mechanism of regulation of the NMDA receptor function during hypoxia. Neuroscience. 2002;112(4):869–77.
159. South SM, Kohno T, Kaspar BK, Hegarty D, Vissel B, Drake CT, et al. A conditional deletion of the NR1 subunit of the NMDA receptor in adult spinal cord dorsal horn reduces NMDA currents and injury-induced pain. J Neurosci. 2003;23(12):5031–40.
160. Lin SL, Tsai RY, Shen CH, Lin FH, Wang JJ, Hsin ST, et al. Co-administration of ultra-low dose naloxone attenuates morphine tolerance in rats via attenuation of NMDA receptor neurotransmission and suppression of neuroinflammation in the spinal cords. Pharmacol Biochem Behav. 2010;96(2):236–45.
161. Mao J, Price DD, Phillips LL, Lu J, Mayer DJ. Increases in protein kinase C gamma immunoreactivity in the spinal cord of rats associated with tolerance to the analgesic effects of morphine. Brain Res. 1995;677(2):257–67.
162. Robles-Flores M, Melendez L, Garcia W, Mendoza-Hernandez G, Lam TT, Castaneda-Patlan C, et al. Posttranslational modifications on protein kinase c isozymes. Effects of epinephrine and phorbol esters. Biochim Biophys Acta. 2008;1783(5):695–712.
163. Balafanova Z, Bolli R, Zhang J, Zheng Y, Pass JM, Bhatnagar A, et al. Nitric oxide (NO) induces nitration of protein kinase Cepsilon (PKCepsilon), facilitating PKCepsilon translocation via enhanced PKCepsilon -RACK2 interactions: a novel mechanism of no-triggered activation of PKCepsilon. J Biol Chem. 2002;277(17):15021–7.
164. Brenner GJ, Ji RR, Shaffer S, Woolf CJ. Peripheral noxious stimulation induces phosphorylation of the NMDA receptor NR1 subunit at the PKC-dependent site, serine-896, in spinal cord dorsal horn neurons. Eur J Neurosci. 2004;20(2):375–84.

165. Yang ZJ, Wang B, Kwansa H, Heitmiller KD, Hong G, Carter EL, et al. Adenosine A2A receptor contributes to ischemic brain damage in newborn piglet. J Cereb Blood Flow Metab. 2013;33(10):1612–20.
166. Palkar R, Lippoldt EK, McKemy DD. The molecular and cellular basis of thermosensation in mammals. Curr Opin Neurobiol. 2015;34C:14–9.
167. Islam MS, editor. Transient receptor potential channels. Dordrecht: Springer; 2011.
168. Tominaga M, Caterina MJ, Malmberg AB, Rosen TA, Gilbert H, Skinner K, et al. The cloned capsaicin receptor integrates multiple pain-producing stimuli. Neuron. 1998;21(3):531–43.
169. Szallasi A, Blumberg PM. Vanilloid (Capsaicin) receptors and mechanisms. Pharmacol Rev. 1999;51(2):159–212.
170. Caterina MJ, Schumacher MA, Tominaga M, Rosen TA, Levine JD, Julius D. The capsaicin receptor: a heat-activated ion channel in the pain pathway. Nature. 1997;389(6653):816–24.
171. Pingle SC, Matta JA, Ahern GP. Capsaicin receptor: TRPV1 a promiscuous TRP channel. Handb Exp Pharmacol. 2007;179:155–71.
172. Valtschanoff JG, Rustioni A, Guo A, Hwang SJ. Vanilloid receptor VR1 is both presynaptic and postsynaptic in the superficial laminae of the rat dorsal horn. J Comp Neurol. 2001; 436(2):225–35.
173. Carlton SM, Coggeshall RE. Peripheral capsaicin receptors increase in the inflamed rat hindpaw: a possible mechanism for peripheral sensitization. Neurosci Lett. 2001;310(1):53–6.
174. Mezey E, Toth ZE, Cortright DN, Arzubi MK, Krause JE, Elde R, et al. Distribution of mRNA for vanilloid receptor subtype 1 (VR1), and VR1-like immunoreactivity, in the central nervous system of the rat and human. Proc Natl Acad Sci U S A. 2000;97(7):3655–60.
175. Cui Y, Chen Y, Zhi JL, Guo RX, Feng JQ, Chen PX. Activation of p38 mitogen-activated protein kinase in spinal microglia mediates morphine antinociceptive tolerance. Brain Res. 2006;1069(1):235–43.
176. Davis JB, Gray J, Gunthorpe MJ, Hatcher JP, Davey PT, Overend P, et al. Vanilloid receptor-1 is essential for inflammatory thermal hyperalgesia. Nature. 2000;405(6783):183–7.
177. Caterina MJ, Leffler A, Malmberg AB, Martin WJ, Trafton J, Petersen-Zeitz KR, et al. Impaired nociception and pain sensation in mice lacking the capsaicin receptor. Science. 2000;288(5464):306–13.
178. Palazzo E, de Novellis V, Marabese I, Cuomo D, Rossi F, Berrino L, et al. Interaction between vanilloid and glutamate receptors in the central modulation of nociception. Eur J Pharmacol. 2002;439(1-3):69–75.
179. Jin YH, Bailey TW, Li BY, Schild JH, Andresen MC. Purinergic and vanilloid receptor activation releases glutamate from separate cranial afferent terminals in nucleus tractus solitarius. J Neurosci. 2004;24(20):4709–17.
180. Marinelli S, Vaughan CW, Christie MJ, Connor M. Capsaicin activation of glutamatergic synaptic transmission in the rat locus coeruleus in vitro. J Physiol. 2002;543(Pt 2):531–40.
181. Musella A, De Chiara V, Rossi S, Prosperetti C, Bernardi G, Maccarrone M, et al. TRPV1 channels facilitate glutamate transmission in the striatum. Mol Cell Neurosci. 2009;40(1): 89–97.
182. Yang K, Kumamoto E, Furue H, Yoshimura M. Capsaicin facilitates excitatory but not inhibitory synaptic transmission in substantia gelatinosa of the rat spinal cord. Neurosci Lett. 1998;255(3):135–8.
183. Li HB, Mao RR, Zhang JC, Yang Y, Cao J, Xu L. Antistress effect of TRPV1 channel on synaptic plasticity and spatial memory. Biol Psychiatry. 2008;64(4):286–92.
184. Wong GY, Gavva NR. Therapeutic potential of vanilloid receptor TRPV1 agonists and antagonists as analgesics: Recent advances and setbacks. Brain Res Rev. 2009;60(1):267–77.
185. Schumacher MA. Transient receptor potential channels in pain and inflammation: therapeutic opportunities. Pain Pract. 2010;10(3):185–200.
186. Palazzo E, Luongo L, de Novellis V, Berrino L, Rossi F, Maione S. Moving towards supraspinal TRPV1 receptors for chronic pain relief. Mol Pain. 2010;6:66.
187. Schultz HD, Ustinova EE. Capsaicin receptors mediate free radical-induced activation of cardiac afferent endings. Cardiovasc Res. 1998;38(2):348–55.

188. Ustinova EE, Schultz HD. Activation of cardiac vagal afferents by oxygen-derived free radicals in rats. Circ Res. 1994;74(5):895–903.
189. Knapp LT, Kanterewicz BI, Hayes EL, Klann E. Peroxynitrite-induced tyrosine nitration and inhibition of protein kinase C. Biochem Biophys Res Commun. 2001;286(4):764–70.
190. Cerutti PA. Mechanisms of action of oxidant carcinogens. Cancer Detect Prev. 1989;14(2):281–4.
191. Gopalakrishna R, Jaken S. Protein kinase C signaling and oxidative stress. Free Radic Biol Med. 2000;28(9):1349–61.
192. Schilling T, Eder C. Importance of the non-selective cation channel TRPV1 for microglial reactive oxygen species generation. J Neuroimmunol. 2009;216(1-2):118–21.
193. Westlund KN, Kochukov MY, Lu Y, McNearney TA. Impact of central and peripheral TRPV1 and ROS levels on proinflammatory mediators and nociceptive behavior. Mol Pain. 2010;6:46.
194. Tominaga M, Tominaga T. Structure and function of TRPV1. Pflugers Arch. 2005; 451(1):143–50.
195. Mohapatra DP, Wang SY, Wang GK, Nau C. A tyrosine residue in TM6 of the Vanilloid Receptor TRPV1 involved in desensitization and calcium permeability of capsaicin-activated currents. Mol Cell Neurosci. 2003;23(2):314–24.
196. Ito N, Ruegg UT, Kudo A, Miyagoe-Suzuki Y, Takeda S. Activation of calcium signaling through Trpv1 by nNOS and peroxynitrite as a key trigger of skeletal muscle hypertrophy. Nat Med. 2013;19(1):101–6.
197. Chuang HH, Lin S. Oxidative challenges sensitize the capsaicin receptor by covalent cysteine modification. Proc Natl Acad Sci U S A. 2009;106(47):20097–102.
198. Susankova K, Tousova K, Vyklicky L, Teisinger J, Vlachova V. Reducing and oxidizing agents sensitize heat-activated vanilloid receptor (TRPV1) current. Mol Pharmacol. 2006;70(1):383–94.
199. Gazzieri D, Trevisani M, Springer J, Harrison S, Cottrell GS, Andre E, et al. Substance P released by TRPV1-expressing neurons produces reactive oxygen species that mediate ethanol-induced gastric injury. Free Radic Biol Med. 2007;43(4):581–9.
200. Sato H, Shibata M, Shimizu T, Shibata S, Toriumi H, Ebine T, et al. Differential cellular localization of antioxidant enzymes in the trigeminal ganglion. Neuroscience. 2013;248: 345–58.
201. Schilling T, Eder C. Stimulus-dependent requirement of ion channels for microglial NADPH oxidase-mediated production of reactive oxygen species. J Neuroimmunol. 2010;225(1-2): 190–4.
202. Ma F, Zhang L, Westlund KN. Reactive oxygen species mediate TNFR1 increase after TRPV1 activation in mouse DRG neurons. Mol Pain. 2009;5:31.
203. Patwardhan AM, Akopian AN, Ruparel NB, Diogenes A, Weintraub ST, Uhlson C, et al. Heat generates oxidized linoleic acid metabolites that activate TRPV1 and produce pain in rodents. J Clin Invest. 2010;120(5):1617–26.
204. Patwardhan AM, Scotland PE, Akopian AN, Hargreaves KM. Activation of TRPV1 in the spinal cord by oxidized linoleic acid metabolites contributes to inflammatory hyperalgesia. Proc Natl Acad Sci U S A. 2009;106(44):18820–4.
205. Taylor-Clark TE, Ghatta S, Bettner W, Undem BJ. Nitrooleic acid, an endogenous product of nitrative stress, activates nociceptive sensory nerves via the direct activation of TRPA1. Mol Pharmacol. 2009;75(4):820–9.
206. Green DP, Ruparel S, Roman L, Henry MA, Hargreaves KM. Role of endogenous TRPV1 agonists in a postburn pain model of partial-thickness injury. Pain. 2013;154(11):2512–20.
207. Salvemini D, Wang ZQ, Wyatt PS, Bourdon DM, Marino MH, Manning PT, et al. Nitric oxide: a key mediator in the early and late phase of carrageenan-induced rat paw inflammation. Br J Pharmacol. 1996;118(4):829–38.
208. Cuzzocrea S, Salvemini D. Molecular mechanisms involved in the reciprocal regulation of cyclooxygenase and nitric oxide synthase enzymes. Kidney Int. 2007;71(4):290–7.

209. Mollace V, Muscoli C, Masini E, Cuzzocrea S, Salvemini D. Modulation of prostaglandin biosynthesis by nitric oxide and nitric oxide donors. Pharmacol Rev. 2005;57(2):217–52.
210. Ricciotti E, FitzGerald GA. Prostaglandins and inflammation. Arterioscler Thromb Vasc Biol. 2011;31(5):986–1000.
211. Gold MS, Levine JD, Correa AM. Modulation of TTX-R INa by PKC and PKA and their role in PGE2-induced sensitization of rat sensory neurons in vitro. J Neurosci. 1998;18(24): 10345–55.
212. Sachs D, Villarreal C, Cunha F, Parada C, Ferreira S. The role of PKA and PKCepsilon pathways in prostaglandin E2-mediated hypernociception. Br J Pharmacol. 2009;156(5):826–34.
213. Kawabata A. Prostaglandin E2 and pain—an update. Biol Pharm Bull. 2011;34(8):1170–3.
214. Minami T, Nakano H, Kobayashi T, Sugimoto Y, Ushikubi F, Ichikawa A, et al. Characterization of EP receptor subtypes responsible for prostaglandin E2-induced pain responses by use of EP1 and EP3 receptor knockout mice. Br J Pharmacol. 2001;133(3):438–44.
215. Nakayama Y, Omote K, Kawamata T, Namiki A. Role of prostaglandin receptor subtype EP1 in prostaglandin E2-induced nociceptive transmission in the rat spinal dorsal horn. Brain Res. 2004;1010(1-2):62–8.
216. Nakayama Y, Omote K, Namiki A. Role of prostaglandin receptor EP1 in the spinal dorsal horn in carrageenan-induced inflammatory pain. Anesthesiology. 2002;97(5):1254–62.
217. Durrenberger PF, Facer P, Casula MA, Yiangou Y, Gray RA, Chessell IP, et al. Prostanoid receptor EP1 and Cox-2 in injured human nerves and a rat model of nerve injury: a time-course study. BMC Neurol. 2006;6:1.
218. Salvemini D, Misko TP, Masferrer JL, Seibert K, Currie MG, Needleman P. Nitric oxide activates cyclooxygenase enzymes. Proc Natl Acad Sci U S A. 1993;90(15):7240–4.
219. Yang T, Zhang A, Pasumarthy A, Zhang L, Warnock Z, Schnermann JB. Nitric oxide stimulates COX-2 expression in cultured collecting duct cells through MAP kinases and superoxide but not cGMP. Am J Physiol. 2006;291(4):F891–5.
220. Cheng HF, Zhang MZ, Harris RC. Nitric oxide stimulates cyclooxygenase-2 in cultured cTAL cells through a p38-dependent pathway. Am J Physiol. 2006;290(6):F1391–7.
221. Salvemini D, Kim SF, Mollace V. Reciprocal regulation of the nitric oxide and cyclooxygenase pathway in pathophysiology: relevance and clinical implications. Am J Physiol Regul Integr Comp Physiol. 2013;304(7):R473–87.
222. Kim SF, Huri DA, Snyder SH. Inducible nitric oxide synthase binds, S-nitrosylates, and activates cyclooxygenase-2. Science. 2005;310(5756):1966–70.
223. Ferrari LF, Bogen O, Reichling DB, Levine JD. Accounting for the delay in the transition from acute to chronic pain: axonal and nuclear mechanisms. J Neurosci. 2015;35(2):495–507.
224. Markey CM, Alward A, Weller PE, Marnett LJ. Quantitative studies of hydroperoxide reduction by prostaglandin H synthase. Reducing substrate specificity and the relationship of peroxidase to cyclooxygenase activities. J Biol Chem. 1987;262(13):6266–79.
225. Landino LM, Crews BC, Timmons MD, Morrow JD, Marnett LJ. Peroxynitrite, the coupling product of nitric oxide and superoxide, activates prostaglandin biosynthesis. Proc Natl Acad Sci U S A. 1996;93(26):15069–74.
226. Zou M, Jendral M, Ullrich V. Prostaglandin endoperoxide-dependent vasospasm in bovine coronary arteries after nitration of prostacyclin synthase. Br J Pharmacol. 1999;126(6):1283–92.
227. Bachschmid M, Thurau S, Zou MH, Ullrich V. Endothelial cell activation by endotoxin involves superoxide/NO-mediated nitration of prostacyclin synthase and thromboxane receptor stimulation. FASEB J. 2003;17(8):914–6.
228. Roques BP, Fournie-Zaluski MC, Wurm M. Inhibiting the breakdown of endogenous opioids and cannabinoids to alleviate pain. Nat Rev Drug Discov. 2012;11(4):292–310.
229. Akil H, Watson SJ, Young E, Lewis ME, Khachaturian H, Walker JM. Endogenous opioids: biology and function. Annu Rev Neurosci. 1984;7:223–55.
230. Mika J, Obara I, Przewlocka B. The role of nociceptin and dynorphin in chronic pain: implications of neuro-glial interaction. Neuropeptides. 2011;45(4):247–61.
231. Laughlin TM, Larson AA, Wilcox GL. Mechanisms of induction of persistent nociception by dynorphin. J Pharmacol Exp Ther. 2001;299(1):6–11.

232. Gardell LR, Vanderah TW, Gardell SE, Wang R, Ossipov MH, Lai J, et al. Enhanced evoked excitatory transmitter release in experimental neuropathy requires descending facilitation. J Neurosci. 2003;23(23):8370–9.
233. Vera-Portocarrero LP, Zhang ET, King T, Ossipov MH, Vanderah TW, Lai J, et al. Spinal NK-1 receptor expressing neurons mediate opioid-induced hyperalgesia and antinociceptive tolerance via activation of descending pathways. Pain. 2007;129(1-2):35–45.
234. Przewlocki R, Przewlocka B. Opioids in chronic pain. Eur J Pharmacol. 2001;429(1-3): 79–91.
235. Yi D, Smythe GA, Blount BC, Duncan MW. Peroxynitrite-mediated nitration of peptides: characterization of the products by electrospray and combined gas chromatography-mass spectrometry. Arch Biochem Biophys. 1997;344(2):253–9.
236. Fontana M, Pecci L, Schinina ME, Montefoschi G, Rosei MA. Oxidative and nitrative modifications of enkephalins by reactive nitrogen species. Free Radic Res. 2006;40(7):697–706.
237. Capuozzo E, Pecci L, Giovannitti F, Baseggio Conrado A, Fontana M. Oxidative and nitrative modifications of enkephalins by human neutrophils: effect of nitroenkephalin on leukocyte functional responses. Amino Acids. 2012;43(2):875–84.
238. Loew GH, Burt SK. Energy conformation study of Met-enkephalin and its D-Ala2 analogue and their resemblance to rigid opiates. Proc Natl Acad Sci U S A. 1978;75(1):7–11.
239. Schwyzer R. Molecular mechanism of opioid receptor selection. Biochemistry. 1986;25(20): 6335–42.
240. Yokoyama K, Uhlin U, Stubbe J. Site-specific incorporation of 3-nitrotyrosine as a probe of pKa perturbation of redox-active tyrosines in ribonucleotide reductase. J Am Chem Soc. 2010;132(24):8385–97.
241. Ndengele MM, Cuzzocrea S, Masini E, Vinci MC, Esposito E, Muscoli C, et al. Spinal ceramide modulates the development of morphine antinociceptive tolerance via peroxynitrite-mediated nitroxidative stress and neuroimmune activation. J Pharmacol Exp Ther. 2009; 329(1):64–75.
242. Muscoli C, Doyle T, Dagostino C, Bryant L, Chen Z, Watkins LR, et al. Counter-regulation of opioid analgesia by glial-derived bioactive sphingolipids. J Neurosci. 2010;30(46):15400–8.
243. Urban MO, Coutinho SV, Gebhart GF. Involvement of excitatory amino acid receptors and nitric oxide in the rostral ventromedial medulla in modulating secondary hyperalgesia produced by mustard oil. Pain. 1999;81(1-2):45–55.
244. Porreca F, Ossipov MH, Gebhart GF. Chronic pain and medullary descending facilitation. Trends Neurosci. 2002;25(6):319–25.
245. Ren K, Dubner R. Neuron-glia crosstalk gets serious: role in pain hypersensitivity. Curr Opin Anaesthesiol. 2008;21(5):570–9.
246. Urban MO, Gebhart GF. Central mechanisms in pain. Med Clin North Am. 1999;83(3):585–96.
247. Wei F, Guo W, Zou S, Ren K, Dubner R. Supraspinal glial-neuronal interactions contribute to descending pain facilitation. J Neurosci. 2008;28(42):10482–95.
248. Millan MJ. Descending control of pain. Prog Neurobiol. 2002;66(6):355–474.
249. Vanegas H. To the descending pain-control system in rats, inflammation-induced primary and secondary hyperalgesia are two different things. Neurosci Lett. 2004;361(1-3):225–8.
250. Manning BH, Morgan MJ, Franklin KB. Morphine analgesia in the formalin test: evidence for forebrain and midbrain sites of action. Neuroscience. 1994;63(1):289–94.
251. Altier N, Stewart J. The role of dopamine in the nucleus accumbens in analgesia. Life Sci. 1999;65(22):2269–87.
252. Sheng HY, Qu CL, Huo FQ, Du JQ, Tang JS. D2-like but not D1-like dopamine receptors are involved in the ventrolateral orbital cortex-induced antinociception: a GABAergic modulation mechanism. Exp Neurol. 2009;215(1):128–34.
253. Cobacho N, de la Calle JL, Paino CL. Dopaminergic modulation of neuropathic pain: analgesia in rats by a D2-type receptor agonist. Brain Res Bull. 2014;106:62–71.
254. Pertovaara A. Noradrenergic pain modulation. Prog Neurobiol. 2006;80(2):53–83.
255. Pertovaara A. The noradrenergic pain regulation system: a potential target for pain therapy. Eur J Pharmacol. 2013;716(1-3):2–7.

256. d'Ischia M, Costantini C. Nitric oxide-induced nitration of catecholamine neurotransmitters: a key to neuronal degeneration? Bioorg Med Chem. 1995;3(7):923–7.
257. Macarthur H, Westfall TC, Riley DP, Misko TP, Salvemini D. Inactivation of catecholamines by superoxide gives new insights on the pathogenesis of septic shock. Proc Natl Acad Sci U S A. 2000;97(17):9753–8.
258. Nagatsu T, Levitt M, Udenfriend S. tyrosine hydroxylase. The initial step in norepinephrine biosynthesis. J Biol Chem. 1964;239:2910–7.
259. Ara J, Przedborski S, Naini AB, Jackson-Lewis V, Trifiletti RR, Horwitz J, et al. Inactivation of tyrosine hydroxylase by nitration following exposure to peroxynitrite and 1-methyl-4-phenyl-1,2,3,6-tetrahydropyridine (MPTP). Proc Natl Acad Sci U S A. 1998;95(13):7659–63.
260. Rubio-Osornio M, Montes S, Perez-Severiano F, Aguilera P, Floriano-Sanchez E, Monroy-Noyola A, et al. Copper reduces striatal protein nitration and tyrosine hydroxylase inactivation induced by MPP+ in rats. Neurochem Int. 2009;54(7):447–51.
261. Kasturi BS, MohanKumar SM, Sirivelu MP, Shin AC, Mohankumar PS. Chronic estradiol-17beta exposure suppresses hypothalamic norepinephrine release and the steroid-induced luteinizing hormone surge: role of nitration of tyrosine hydroxylase. Brain Res. 2013;1493:90–8.
262. Kuhn DM, Aretha CW, Geddes TJ. Peroxynitrite inactivation of tyrosine hydroxylase: mediation by sulfhydryl oxidation, not tyrosine nitration. J Neurosci. 1999;19(23):10289–94.
263. Kuhn DM, Sadidi M, Liu X, Kreipke C, Geddes T, Borges C, et al. Peroxynitrite-induced nitration of tyrosine hydroxylase: identification of tyrosines 423, 428, and 432 as sites of modification by matrix-assisted laser desorption ionization time-of-flight mass spectrometry and tyrosine-scanning mutagenesis. J Biol Chem. 2002;277(16):14336–42.
264. Blanchard-Fillion B, Souza JM, Friel T, Jiang GC, Vrana K, Sharov V, et al. Nitration and inactivation of tyrosine hydroxylase by peroxynitrite. J Biol Chem. 2001;276(49):46017–23.
265. Gilmore TD. Introduction to NF-kappaB: players, pathways, perspectives. Oncogene. 2006;25(51):6680–4.
266. Niederberger E, Geisslinger G. The IKK-NF-kappaB pathway: a source for novel molecular drug targets in pain therapy? FASEB J. 2008;22(10):3432–42.
267. Poveda L, Hottiger M, Boos N, Wuertz K. Peroxynitrite induces gene expression in intervertebral disc cells. Spine. 2009;34(11):1127–33.
268. Livolsi A, Busuttil V, Imbert V, Abraham RT, Peyron JF. Tyrosine phosphorylation-dependent activation of NF-kappa B. Requirement for p56 LCK and ZAP-70 protein tyrosine kinases. Eur J Biochem. 2001;268(5):1508–15.
269. Gow AJ, Duran D, Malcolm S, Ischiropoulos H. Effects of peroxynitrite-induced protein modifications on tyrosine phosphorylation and degradation. FEBS Lett. 1996;385(1-2):63–6.
270. Brito C, Naviliat M, Tiscornia AC, Vuillier F, Gualco G, Dighiero G, et al. Peroxynitrite inhibits T lymphocyte activation and proliferation by promoting impairment of tyrosine phosphorylation and peroxynitrite-driven apoptotic death. J Immunol. 1999;162(6):3356–66.
271. Matata BM, Galinanes M. Peroxynitrite is an essential component of cytokines production mechanism in human monocytes through modulation of nuclear factor-kappa B DNA binding activity. J Biol Chem. 2002;277(3):2330–5.
272. Park SW, Huq MD, Hu X, Wei LN. Tyrosine nitration on p65: a novel mechanism to rapidly inactivate nuclear factor-kappaB. Mol Cell Proteomics. 2005;4(3):300–9.
273. Loukili N, Rosenblatt-Velin N, Rolli J, Levrand S, Feihl F, Waeber B, et al. Oxidants positively or negatively regulate nuclear factor kappaB in a context-dependent manner. J Biol Chem. 2010;285(21):15746–52.
274. Genovese T, Mazzon E, Esposito E, Di Paola R, Murthy K, Neville L, et al. Effects of a metalloporphyrinic peroxynitrite decomposition catalyst, ww-85, in a mouse model of spinal cord injury. Free Radic Res. 2009;43(7):631–45.
275. Genovese T, Mazzon E, Esposito E, Muia C, Di Paola R, Bramanti P, et al. Beneficial effects of FeTSPP, a peroxynitrite decomposition catalyst, in a mouse model of spinal cord injury. Free Radic Biol Med. 2007;43(5):763–80.

276. Cortes-Mendoza J, Diaz de Leon-Guerrero S, Pedraza-Alva G, Perez-Martinez L. Shaping synaptic plasticity: the role of activity-mediated epigenetic regulation on gene transcription. Int J Dev Neurosci. 2013;31(6):359–69.
277. Buchheit T, Van de Ven T, Shaw A. Epigenetics and the transition from acute to chronic pain. Pain Med. 2012;13(11):1474–90.
278. Zhang Z, Cai YQ, Zou F, Bie B, Pan ZZ. Epigenetic suppression of GAD65 expression mediates persistent pain. Nat Med. 2011;17(11):1448–55.
279. Bai G, Wei D, Zou S, Ren K, Dubner R. Inhibition of class II histone deacetylases in the spinal cord attenuates inflammatory hyperalgesia. Mol Pain. 2010;6:51.
280. Kukkar A, Singh N, Jaggi AS. Attenuation of neuropathic pain by sodium butyrate in an experimental model of chronic constriction injury in rats. J Formos Med Assoc. 2014;113(12): 921–8.
281. Denk F, Huang W, Sidders B, Bithell A, Crow M, Grist J, et al. HDAC inhibitors attenuate the development of hypersensitivity in models of neuropathic pain. Pain. 2013;154(9):1668–79.
282. Matsushita Y, Araki K, Omotuyi O, Mukae T, Ueda H. HDAC inhibitors restore C-fibre sensitivity in experimental neuropathic pain model. Br J Pharmacol. 2013;170(5):991–8.
283. Liang DY, Sun Y, Shi XY, Sahbaie P, Clark JD. Epigenetic regulation of spinal cord gene expression controls opioid-induced hyperalgesia. Mol Pain. 2014;10:59.
284. Khan MA, Dixit K, Uddin M, Malik A, Alam K. Role of peroxynitrite-modified H2A histone in the induction and progression of rheumatoid arthritis. Scand J Rheumatol. 2012;41(6):426–33.
285. Khan MA, Dixit K, Moinuddin, Arif Z, Alam K. Studies on peroxynitrite-modified H1 histone: implications in systemic lupus erythematosus. Biochimie. 2014;97:104–13.
286. Ahmad R, Ahsan H. Role of peroxynitrite-modified biomolecules in the etiopathogenesis of systemic lupus erythematosus. Clin Exp Med. 2014;14(1):1–11.
287. Osoata GO, Yamamura S, Ito M, Vuppusetty C, Adcock IM, Barnes PJ, et al. Nitration of distinct tyrosine residues causes inactivation of histone deacetylase 2. Biochem Biophys Res Commun. 2009;384(3):366–71.
288. Gonzalez-Zuniga M, Contreras PS, Estrada LD, Chamorro D, Villagra A, Zanlungo S, et al. c-Abl stabilizes HDAC2 levels by tyrosine phosphorylation repressing neuronal gene expression in Alzheimer's disease. Mol Cell. 2014;56(1):163–73.
289. Tran L, Schulkin J, Ligon CO, Greenwood-Van MB. Epigenetic modulation of chronic anxiety and pain by histone deacetylation. Mol Psychiatry. 2015;20(10):1219–31.
290. Cardinale A, de Stefano MC, Mollinari C, Racaniello M, Garaci E, Merlo D. Biochemical characterization of sirtuin 6 in the brain and its involvement in oxidative stress response. Neurochem Res. 2015;40(1):59–69.
291. Hu S, Liu H, Ha Y, Luo X, Motamedi M, Gupta MP, et al. Posttranslational modification of Sirt6 activity by peroxynitrite. Free Radic Biol Med. 2014;79C:176–85.
292. Muscoli C, Lauro F, Ilari S, Pucci B, Dagostino C, Gliozzi M, et al. Modulation of sirtuins during acute inflammatory pain: the role of ROS. FASEB J. 2013;27:887.6.

Chapter 26
Amyotrophic Lateral Sclerosis: Present Understanding of the Role of SOD

Kristina Ramdial, Fabian H. Rossi, Maria Clara Franco, and Alvaro G. Estevez

Over the last 20 years, the focus on superoxide dismutase research has been to identify the gain-of-function associated with the mutations that cause amyotrophic lateral sclerosis (ALS). The more than 160 mutations in 80 different positions in the enzyme produce the same phenotype—ALS. Expression of the mutant enzyme in rodents leads to a phenotype that closely resembles the human disease. In contrast, gene deletion or overexpression of wild-type human SOD in rodents does not produce major phenotypic alterations, which led to the conclusion that the mutations produce a toxic gain-of-function. Some mutations have no alterations of the enzymatic activity, while others have a reduction or even ablation of the enzymatic activity when tested in vitro in the presence of metal chelators. In spite of the experimental efforts and the many hypotheses that have been proposed to explain the mutant SOD's gain-of-function, the mechanism for the selective motor neuron toxicity remains elusive and highly controversial. However, there are some facts that have generally been accepted. Mutant SOD-induced disease in rodent models of ALS produces aggregation, oxidative stress, mitochondrial dysfunction, astrocyte activation, neuroinflammation, and selective motor neuron death. It is also accepted that in spite of the recent identification of several other genes responsible for familial ALS, it is the mutant SOD transgenic rodent models that better replicate the human disease.

K. Ramdial • M.C. Franco • A.G. Estevez (✉)
Burnett School of Biomedical Sciences, College of Medicine, University of Central Florida, 6900 Lake Nona Blvd, Orlando, FL 32827, USA
e-mail: Kristina.ramdial@ucf.edu; franco@mail.ucf.edu; aest@mail.ucf.edu

F.H. Rossi
Orlando Veteran Affairs Medical Center, 13800 Veterans Way, Orlando, FL 32827, USA
e-mail: fabian.rossi@va.gov

I. Batinić-Haberle et al. (eds.), *Redox-Active Therapeutics*, Oxidative Stress in Applied Basic Research and Clinical Practice, DOI 10.1007/978-3-319-30705-3_26

Evidence supporting and negating every proposed mechanism has been reported. Recently, it was reported that mutant SOD has "prion" properties [1, 2]. However, there is no direct evidence that aggregation of SOD causes the disease. The evidence, suggesting a pathological role of aggregation, is based on correlations without demonstration of any causal relation. Moreover, there is evidence that SOD aggregation can even be protective by preventing the action of soluble toxic forms of SOD [3]. Additionally, treatment with a copper chelator is protective in an animal model of ALS and paradoxically increases the concentrations of the fully metallated mutant SOD in the spinal cord [4]. Cross-breeding of transgenic mice for mutant SOD with transgenic mice overexpressing human wild-type SOD results in enhanced pathology characterized by faster onset and progression of the disease [5–8]. Some authors attribute the enhanced toxicity to increased aggregation [7–9]. Conversely, the enhanced toxicity found in animals transgenic for both wild-type and mutant SOD has also been attributed to the stabilization and increased solubility of mutant SOD by wild-type SOD [6, 10, 11].

We and others have provided evidence for an oxidative mechanism-mediating SOD toxicity. A common property for all mutant SOD tested is their lower affinity for zinc [12–15]. Zn-deficient SOD has a high tendency to form monomers, which aggregate easily (Fig. 26.1). The formation of dimers with fully metallated SOD increases the Zn-deficient SOD heterodimer stability, which enhances solubility and increases the enzyme toxicity (Fig. 26.1) [3]. Additionally, the removal of critical cysteine residues from the SOD decreases the mutant SOD's tendency to aggregate and also increases the Zn-deficient wild-type and mutant SOD toxicity (Fig. 26.1) [3, 16]. The copper in Zn-deficient SOD can be reduced not only by superoxide, but also by biological reductants such as ascorbic acid, and it has an increased capacity to catalyze the nitration of proteins (Fig. 26.2) [13, 17, 18]. The presence of nitrotyrosine, a foot print left by the oxidant, peroxynitrite, in the spinal cord of patients and animal models of ALS, is well-established [19–23]. In addition, intracellular release of Zn-deficient SOD into motor neurons triggers apoptosis by a mechanism that requires the nitration of proteins [3, 17]. Similarly, motor neurons from mouse and rat overexpressing mutant SOD die by apoptosis when incubated with nitric oxide at physiological-relevant concentrations by a mechanism that requires the production of peroxynitrite and tyrosine nitration (Fig. 26.2) [3, 24]. Although the role of nitric oxide in the pathology of ALS has become controversial due to the use of an incomplete knockout for neuronal nitric oxide synthase [25, 26], the use of the complete knockout for the enzyme was impaired due to the number of abnormalities present in the animals [27]. However, the expression of different isoforms of nitric oxide synthase, particularly the neuronal isoform in the spinal motor neurons of ALS patients, is well-established [19, 28, 29].

A critical protein that nitration turns toxic is the protein chaperone heat shock protein 90 (Hsp90). Hsp90 is an abundant and ubiquitous protein chaperone essential for cell survival [30–33]. Nitrated Hsp90 is found in the spinal motor neurons from ALS patients and animal models [34]. The mechanism by which nitrated Hsp90 stimulates apoptosis involves the activation of purine

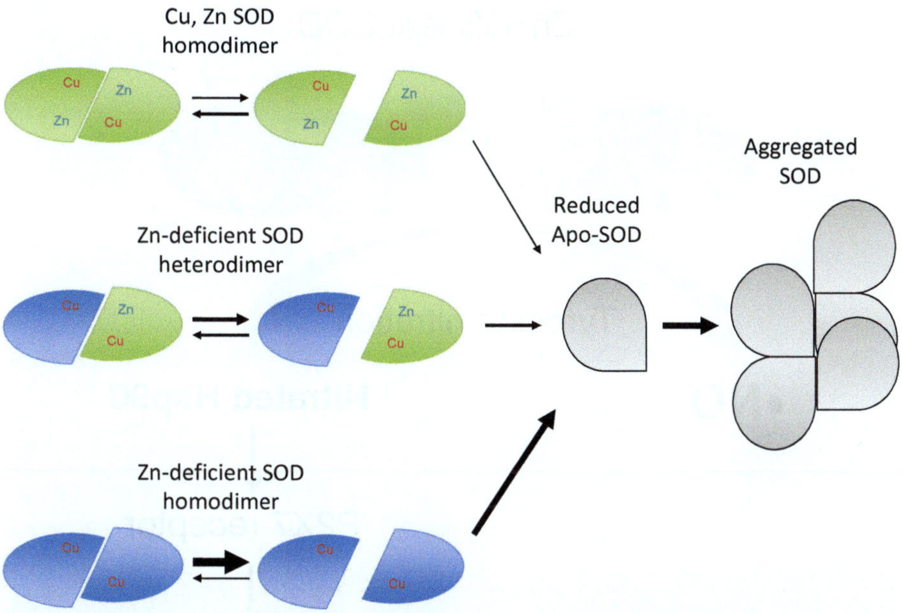

Fig. 26.1 SOD dimer interactions. SOD dimers have a half-life of approximately 17 min at 37 °C [36]. In the absence of reducing agents, SOD remains a dimer with a small proportion in the form of an exchange intermediate, independently of the metal content. The presence of reducing agents slightly but significantly increases the amount of fully metallated SOD in the form of the exchange intermediate, but no monomers can be detected. In contrast, incubation with reducing agents render most Zn-deficient to become a monomer with a large proportion in the form of the exchange intermediate. The incubation of Zn-deficient SOD with Cu, ZnSOD dramatically decreases the amount of monomers in the presence of reducing agents. Under these conditions, most of the enzyme is in the form of the exchange intermediate without significant changes on the amount of dimers. In addition, the presence of fully metallated SOD significantly decreased the susceptibility of Zn-deficient SOD disulfide bonds to be reduced [3]. In summary, Cu, ZnSOD increases the stability and solubility of Zn-deficient SOD

receptor P2X7, followed by the activation of the Fas cell death pathway (Fig. 26.2) [24, 34]. Inactivation of the Fas ligand extends the survival of an ALS animal model [35]. Collectively, these results provide a model for the toxicity of SOD starting with the change in a property of the protein, namely low zinc affinity, all the way to the activation of the cellular mechanism involved in the stimulation of the cell death (Fig. 26.1).

In summary, although knowledge has been accumulated on the possible causes of mutant SOD-induced motor neuron death over the last 2 decades, more research is necessary to translate this knowledge into an effective treatment for ALS patients. Despite the shortcomings in the use of the rodent models of ALS carrying mutant SOD, transgenic mouse and rat models for the mutant SOD are still the best in vivo models of the human disease.

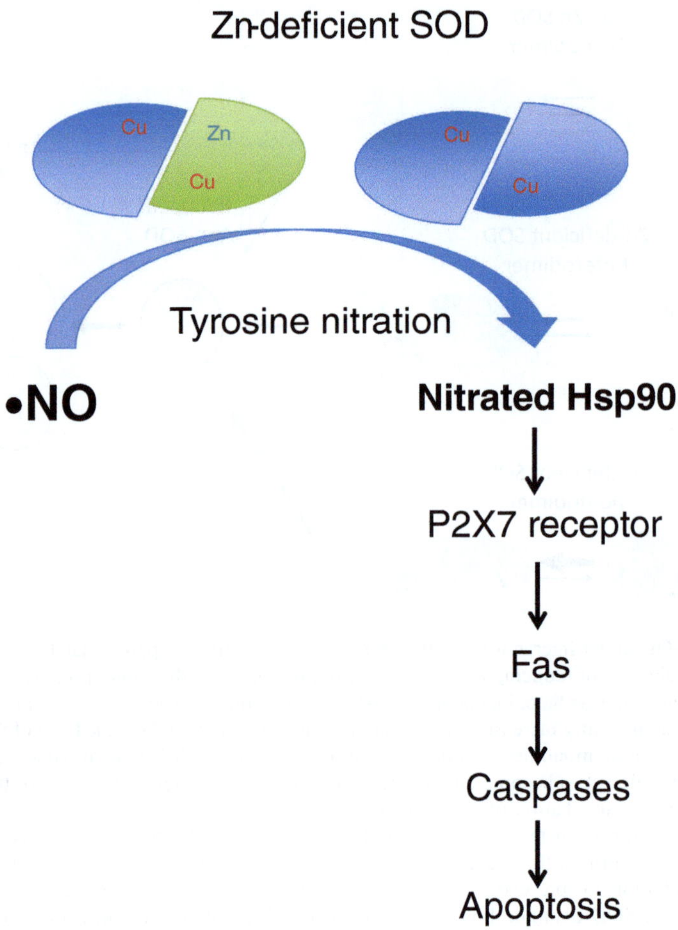

Fig. 26.2 Zn-deficient SOD stimulates nitration and motor neuron apoptosis. Zn-deficient SOD has enhanced catalysis of tyrosine by peroxynitrite [13, 18]. Zn-deficient SOD and mutant SOD require nitric oxide and tyrosine nitration to stimulate motor neuron apoptosis [3, 17, 24]. Hsp90 is nitrated in spinal cord motor neurons from ALS patients and in a mouse model of ALS. Nitration of one of two residues in Hsp90 is necessary and sufficient to stimulate motor neuron apoptosis by activation of the purine receptor P2X7. Downstream of P2X7 receptor activation, Fas ligand is translocated to the plasma membrane where it stimulates the Fas death pathway [24, 34]

Acknowledgment The research was possible thanks to NIH Grants NS36761 (to A.G.E.)

References

1. Chia R, Tattum MH, Jones S, Collinge J, Fisher EM, Jackson GS. Superoxide dismutase 1 and tgSOD1 mouse spinal cord seed fibrils, suggesting a propagative cell death mechanism in amyotrophic lateral sclerosis. PLoS One. 2010;5, e10627.
2. Grad LI, Fernando SM, Cashman NR. From molecule to molecule and cell to cell: prion-like mechanisms in amyotrophic lateral sclerosis. Neurobiol Dis. 2015;77:257–65.
3. Sahawneh MA, Ricart KC, Roberts BR, Bomben VC, Basso M, Ye Y, Sahawneh J, Franco MC, Beckman JS, Estevez AG. Cu, Zn superoxide dismutase (SOD) increases toxicity of mutant and Zn-deficient superoxide dismutase by enhancing protein stability. J Biol Chem. 2010;285:33885–97.
4. Roberts BR, Lim NK, McAllum EJ, Donnelly PS, Hare DJ, Doble PA, Turner BJ, Price KA, Lim SC, Paterson BM, Hickey JL, Rhoads TW, Williams JR, Kanninen KM, Hung LW, Liddell JR, Grubman A, Monty JF, Llanos RM, Kramer DR, Mercer JF, Bush AI, Masters CL, Duce JA, Li QX, Beckman JS, Barnham KJ, White AR, Crouch PJ. Oral treatment with Cu(II)(atsm) increases mutant SOD1 in vivo but protects motor neurons and improves the phenotype of a transgenic mouse model of amyotrophic lateral sclerosis. J Neurosci. 2014;34:8021–31.
5. Jaarsma D, Haasdijk ED, Grashorn JA, Hawkins R, van Duijn W, Verspaget HW, London J, Holstege JC. Human Cu/Zn superoxide dismutase (SOD1) overexpression in mice causes mitochondrial vacuolization, axonal degeneration, and premature motoneuron death and accelerates motoneuron disease in mice expressing a familial amyotrophic lateral sclerosis mutant SOD1. Neurobiol Dis. 2000;7:623–43.
6. Fukada K, Nagano S, Satoh M, Tohyama C, Nakanishi T, Shimizu A, Yanagihara T, Sakoda S. Stabilization of mutant Cu/Zn superoxide dismutase (SOD1) protein by coexpressed wild SOD1 protein accelerates the disease progression in familial amyotrophic lateral sclerosis mice. Eur J Neurosci. 2001;14:2032–6.
7. Deng HX, Shi Y, Furukawa Y, Zhai H, Fu R, Liu E, Gorrie GH, Khan MS, Hung WY, Bigio EH, Lukas T, Dal Canto MC, O'Halloran TV, Siddique T. Conversion to the amyotrophic lateral sclerosis phenotype is associated with intermolecular linked insoluble aggregates of SOD1 in mitochondria. Proc Natl Acad Sci U S A. 2006;103:7142–7.
8. Wang L, Deng HX, Grisotti G, Zhai H, Siddique T, Roos RP. Wild-type SOD1 overexpression accelerates disease onset of a G85R SOD1 mouse. Hum Mol Genet. 2009;18:1642–51.
9. Furukawa Y, Fu R, Deng HX, Siddique T, O'Halloran TV. Disulfide cross-linked protein represents a significant fraction of ALS-associated Cu, Zn-superoxide dismutase aggregates in spinal cords of model mice. Proc Natl Acad Sci U S A. 2006;103:7148–53.
10. Witan H, Kern A, Koziollek-Drechsler I, Wade R, Behl C, Clement AM. Heterodimer formation of wild-type and amyotrophic lateral sclerosis-causing mutant Cu/Zn-superoxide dismutase induces toxicity independent of protein aggregation. Hum Mol Genet. 2008;17:1373–85.
11. Witan H, Gorlovoy P, Kaya AM, Koziollek-Drechsler I, Neumann H, Behl C, Clement AM. Wild-type Cu/Zn superoxide dismutase (SOD1) does not facilitate, but impedes the formation of protein aggregates of amyotrophic lateral sclerosis causing mutant SOD1. Neurobiol Dis. 2009;36:331–42.
12. Lyons TJ, Liu HB, Goto JJ, Nersissian A, Roe JA, Graden JA, Cafe C, Ellerby LM, Bredesen DE, Gralla EB, Valentine JS. Mutations in copper-zinc superoxide dismutase that cause amyotrophic lateral sclerosis alter the zinc binding site and the redox behavior of the protein. Proc Natl Acad Sci U S A. 1996;93:12240–4.
13. Crow JP, Sampson JB, Zhuang Y, Thompson JA, Beckman JS. Decreased zinc affinity of amyotrophic lateral sclerosis-associated superoxide dismutase mutants leads to enhanced catalysis of tyrosine nitration by peroxynitrite. J Neurochem. 1997;69:1936–44.

14. Beckman JS, Esetvez AG, Barbeito L, Crow JP. CCS knockout mice establish an alternative source of copper for SOD in ALS. Free Radic Biol Med. 2002;33:1433–5.
15. Cao X, Antonyuk SV, Seetharaman SV, Whitson LJ, Taylor AB, Holloway SP, Strange RW, Doucette PA, Valentine JS, Tiwari A, Hayward LJ, Padua S, Cohlberg JA, Hasnain SS, Hart PJ. Structures of the G85R variant of SOD1 in familial amyotrophic lateral sclerosis. J Biol Chem. 2008;283:16169–77.
16. Leinweber B, Barofsky E, Barofsky DF, Ermilov V, Nylin K, Beckman JS. Aggregation of ALS mutant superoxide dismutase expressed in Escherichia coli. Free Radic Biol Med. 2004;36:911–8.
17. Estevez AG, Crow JP, Sampson JB, Reiter C, Zhuang Y, Richardson GJ, Tarpey MM, Barbeito L, Beckman JS. Induction of nitric oxide-dependent apoptosis in motor neurons by zinc-deficient superoxide dismutase. Science. 1999;286:2498–500.
18. Crow JP, Ye YZ, Strong M, Kirk M, Barnes S, Beckman JS. Superoxide dismutase catalyzes nitration of tyrosines by peroxynitrite in the rod and head domains of neurofilament-L. J Neurochem. 1997;69:1945–53.
19. Chou S, Wang H, Komai KJ. Colocalization of NOS and SOD1 in neurofilament accumulation within motor neurons of amyotrophic lateral sclerosis: an immunohistochemical study. J Chem Neuroanat. 1996;10:249–58.
20. Abe K, Pan LH, Watanabe M, Kato T, Itoyama Y. Induction of nitrotyrosine-like immunoreactivity in the lower motor neuron of amyotrophic lateral sclerosis. Neurosci Lett. 1995;199:152–4.
21. Beal MF, Ferrante LJ, Browne SE, Matthews RT, Kowall NW, Brown RH. Increased 3-nitrotyrosine in both sporadic and familial amyotrophic lateral sclerosis. Ann Neurol. 1997;42:644–54.
22. Ferrante RJ, Browne SE, Shinobu LA, Bowling AC, Baik MJ, MacGarvey U, Kowall NW, Brown RH, Beal MF. Evidence of increased oxidative damage in both sporadic and familial amyotrophic lateral sclerosis. J Neurochem. 1997;69:2064–74.
23. Ferrante RJ, Shinobu LA, Schulz JB, Mathews RT, Thomas CE, Kowall NW, Gurney ME, Beal MF. Increased 3-nitrotyrosine and oxidative damage in mice with a human copper/zinc superoxide dismutase mutation. Ann Neurol. 1997;42:326–34.
24. Raoul C, Estevez AG, Nishimune H, Cleveland DW, deLapeyriere O, Henderson CE, Haase G, Pettmann B. Motoneuron death triggered by a specific pathway downstream of Fas. potentiation by ALS-linked SOD1 mutations. Neuron. 2002;35:1067–83.
25. Facchinetti F, Sasaki M, Cutting FB, Zhai P, MacDonald JE, Reif D, Beal MF, Huang PL, Dawson TM, Gurney ME, Dawson VL. Lack of involvement of neuronal nitric oxide synthase in the pathogenesis of a transgenic mouse model of familial amyotrophic lateral sclerosis. Neuroscience. 1999;90:1483–92.
26. Huang PL. Mouse models of nitric oxide synthase deficiency. J Am Soc Nephrol. 2000;11:S120–3.
27. Gyurko R, Leupen S, Huang PL. Deletion of exon 6 of the neuronal nitric oxide synthase gene in mice results in hypogonadism and infertility. Endocrinology. 2002;143:2767–74.
28. Anneser JM, Cookson MR, Ince PG, Shaw PJ, Borasio GD. Glial cells of the spinal cord and subcortical white matter up-regulate neuronal nitric oxide synthase in sporadic amyotrophic lateral sclerosis. Exp Neurol. 2001;171:418–21.
29. Catania MV, Aronica E, Yankaya B, Troost D. Increased expression of neuronal nitric oxide synthase spliced variants in reactive astrocytes of amyotrophic lateral sclerosis human spinal cord. J Neurosci. 2001;21:RC148.
30. Richter K, Haslbeck M, Buchner J. The heat shock response: life on the verge of death. Mol Cell. 2010;40:253–66.
31. Li J, Buchner J. Structure, function and regulation of the hsp90 machinery. Biomed J. 2013;36:106–17.
32. Garrido C, Gurbuxani S, Ravagnan L, Kroemer G. Heat shock proteins: endogenous modulators of apoptotic cell death. Biochem Biophys Res Commun. 2001;286:433–42.

33. Didelot C, Schmitt E, Brunet M, Maingret L, Parcellier A, Garrido C. Heat shock proteins: endogenous modulators of apoptotic cell death. In: Gaestel M, editor. Molecular chaperones in health and disease. Berlin: Springer; 2006. p. 171–98.
34. Franco MC, Ye Y, Refakis CA, Feldman JL, Stokes AL, Basso M, de Mera Melero Fernandez RM, Sparrow NA, Calingasan NY, Kiaei M, Rhoads TW, Ma TC, Grumet M, Barnes S, Beal MF, Beckman JS, Mehl R, Estevez AG. Nitration of Hsp90 induces cell death. Proc Natl Acad Sci U S A. 2013;110:E1102–11.
35. Petri S, Kiaei M, Wille E, Calingasan NY, Flint Beal M. Loss of Fas ligand-function improves survival in G93A-transgenic ALS mice. J Neurol Sci. 2006;251:44–9.
36. Roberts BR, Tainer JA, Getzoff ED, Malencik DA, Anderson SR, Bomben VC, Meyers KR, Karplus PA, Beckman JS. Structural characterization of zinc-deficient human superoxide dismutase and implications for ALS. J Mol Biol. 2007;373:877–90.

Chapter 27
Redox Regulation and Misfolding of SOD1: Therapeutic Strategies for Amyotrophic Lateral Sclerosis

Wouter Hubens and Ayako Okado-Matsumoto

Abbreviations

ALS	Amyotrophic lateral sclerosis
CSF	Cerebrospinal fluid
FALS	Familial ALS
HD	Huntington's disease
SALS	Sporadic ALS
SOD1	Cu, Zn-superoxide dismutase
TG2	Transglutaminase 2
WT	Wild type

27.1 Introduction

Amyotrophic lateral sclerosis (ALS) is the most common fatal neurodegenerative disease with an incidence rate of 1–2 per 100,000. The disease is characterized by death of upper and lower motor neurons in the central nervous system. After diagnosis, average life expectancy is 2–5 years, which is mainly attributed to paralysis of the respiratory muscles. Currently, there is no cure available and the only FDA-approved drug (Riluzole) increases life expectancy by a couple of months. This highlights the importance of finding new therapeutic targets.

W. Hubens • A. Okado-Matsumoto (✉)
Department of Biology, Faculty of Science, Toho University,
Miyama 2-2-1, Funabashi, Chiba 274-8510, Japan
e-mail: w.hubens@student.maastrichtuniversity.nl; ayako@bio.sci.toho-u.ac.jp

I. Batinić-Haberle et al. (eds.), *Redox-Active Therapeutics*,
Oxidative Stress in Applied Basic Research and Clinical Practice,
DOI 10.1007/978-3-319-30705-3_27

ALS was first described in 1869 by Jean-Martin Charcot, but despite extensive research the past 150 years, the exact pathophysiology remains to be clarified. 10 % of ALS cases are of hereditary origin, and in 1993, it was discovered that in a subset of these familial ALS (FALS) patients the disease is caused by mutations in Cu, Zn-superoxide dismutase (SOD1) [1]. To date, 178 FALS-related mutations have been reported in the SOD1 gene (http://alsod.iop.kcl.ac.uk). Due to dramatic improvements in sequencing, and an increase of whole-genome association studies, many other familial and sporadic ALS mutations have been discovered. Mutations in TAR DNA Binding Protein (TARDBP), fused in sarcoma (fus), ubiquilin 2 (UBQLN2), and C9orf72, are most frequently mutated in ALS. TARDBP mutations result in TPD-43 inclusions which are a major hallmark of sporadic ALS. These have been reviewed elsewhere. Since this book's primary focus is on oxidative stress in applied basic research and clinical practice, this chapter gives insight in therapeutic strategies related to SOD1.

27.2 SOD1: Enzymatic Function and Structure

SOD1 is a highly conserved enzyme and is extremely stable. The human SOD1 has a homodimeric quaternary structure resisting even temperatures up to 80 °C [2]. Each monomer of 16 kDa consists of an eight-stranded beta barrel with two large loops (Fig. 27.1), of which the "metal-binding" loop binds copper and zinc [3]. Further contributing to stability of SOD1 is a unique intrasubunit disulfide bond between C57 and C146 (Fig. 27.1). SOD1 maintains physiological nM levels of superoxide, $O_2^{\bullet-}$, through catalyzing its dismutation, i.e., oxidation to O_2 and reduction to H_2O_2. Under physiological conditions, SOD1 along with other isoforms (SOD2, MnSOD, and extracellular SOD3) presents our first line of endogenous antioxidative systems. One of the major sites of $O_2^{\bullet-}$ production under physiological conditions is mitochondrial respiration [4, 5].

The importance of SOD1 is reflected by its abundant expression, comprising almost 1 % of all cellular proteins [6]. Wild-type (WT) SOD1 can be found in the cytosol and in the intermembrane space of mitochondria [7, 8]. Mutant SOD1, however, also has vesicular locations in the lysosomes, ER, and the trans-Golgi network [9]. Due to mutations, it interacts with chromogranin and gets secreted via the Golgi network [10].

Initially, it was contemplated that mutations in SOD1 would somehow destabilize the enzyme and cause ALS in a loss-of-function matter, because of its vital role in dismuting toxic $O_2^{\bullet-}$. However, transgenic mice with mutant $SOD1^{G93A}$ or $SOD1^{G37R}$ exhibited ALS-like symptoms despite having increased enzymatic activity [11]. Furthermore, SOD1 knockout models showed no signs of ALS [12]. It is therefore now postulated that mutant SOD1 have an as-of-yet-unknown toxic gain-of-function.

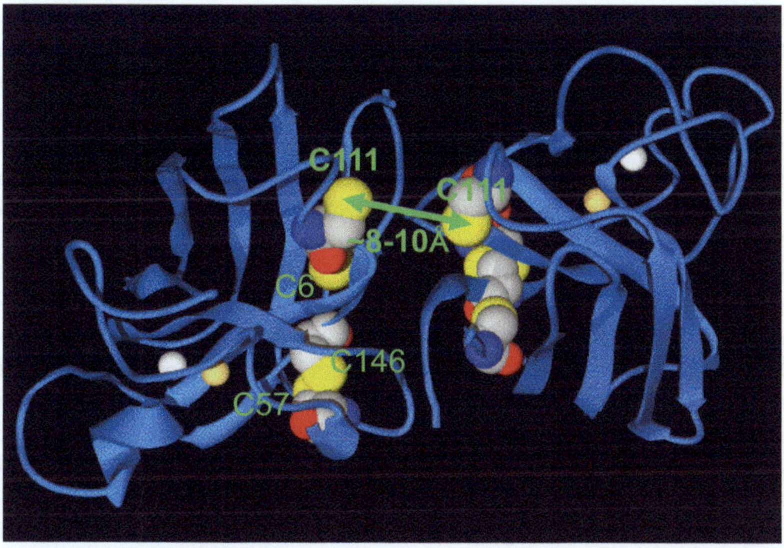

Fig. 27.1 The crystal structure of human SOD1. The crystal structure of human SOD1 (Protein data bank #1SPD) is shown. Human SOD1 consists of a homodimer. Each monomer has 8 β-strands and binds zinc (*small white*) and copper (*small yellow*) in the metal-binding loops. Each monomer has four cysteines (*CPK coloring*), of which C57 and C146 form an intramolecular disfulfide bond. Of the remaining two cysteines, C6 is buried deep within the protein, while C111 is near the dimer interface accessible for modifications. The two monomeric subunits of SOD1 are in very close proximity of each other with only 8–10 Å space between the two C111

27.3 Pro-Oxidant Gain of Function

One of the hypotheses for SOD1 toxicity suggests that SOD1 mutations contribute to oxidative stress. Oxidative stress denotes an imbalance between oxidants and antioxidants in favor of the oxidants, leading to a disruption of redox signaling and to molecular damage [13]. Oxidative stress is a well-known hallmark of ALS, as shown by elevated concentrations of protein carbonyls in the spinal cord and motor cortex of ALS patients [14, 15]. Additionally, many of the environmental risk factors for sporadic ALS (SALS) increase oxidative stress [16]. The level of 3-Nitrotyrosine, which is used as a marker for oxidative/nitrosative stress, is elevated in both sporadic and FALS cases [17, 18].

The nature of pro-oxidant SOD1 remains elusive. Two mutations (A4V and G93A) exhibit altered reactivity. An increased catalysation of oxidative reactions was observed, which induced apoptosis in a neuronal cell line [19]. This apoptosis-inducing effect was diminished with Copper chelator treatment, suggesting oxidative reactions catalyzed by mutant SOD can play a role in ALS.

Most mutant SOD1s also exhibit a 5- to 50-fold lower affinity for Zn(II) and thus have a larger fraction of Zn(II) free SOD (Cu,E,SOD1) compared to WT SOD1 [20, 21]. Loss of Zn affects SOD1 folding, leaving the copper site more accessible. It is hypothesized that this induces toxicity by allowing oxidative reactions to occur through a copper- and peroxynitrite-dependent mechanism. Superoxide can react with nitric oxide to form the powerful oxidant peroxynitrite. Peroxynitrite in turn reacts with SOD1 to nitronium-like intermediates which can nitrate tyrosine residues, slowly injuring cells. When replete Cu,E,SOD1 with zinc, this toxicity is no longer observed [21–24]. The importance of zinc is also reflected in data of dietary depletion and supplementation of zinc in a SOD1^{G93A} mouse model. Zinc deprivation increases disease severity, whereas zinc supplementation provides protection [25]. A rational for this was found in 2011 by Rhoads's research group who utilized the newly developed electrospray mass spectrometry method to measure 1-metal SOD content in vivo. SOD^{G93A} rats had a twofold increase of 1-metal SOD1 in ALS-affected spinal cord regions [26]. However, Rhoads's group could not distinguish between Cu,E,SOD1 and copper-depleted-zinc-repleted SOD1 (E,Zn,SOD1). An increased fraction of E,Zn,SOD1 is also observed in spinal cords of ALS mice [27, 28] and copper treatment experiments slightly alleviate ALS symptoms [28–30]. Combining these findings suggest a role for both 1-metal SOD1 in ALS.

Subramaniam et al. challenged the pro-oxidative hypothesis by creating a mutant SOD1 mouse model in which copper chaperone for SOD1 (CCS) that loads copper into the active site was deleted. In this model, ALS disease progression was not affected, which refutes copper-mediated toxicity as mentioned above [31]. However, despite CCS deletion, 15–20 % of SOD1 activity remained. Instead of disputing the role of copper, this experiment suggested that mutant SOD1 has an alternate method of copper binding [32]. Others also challenged the hypothesis by creating a copper-binding-site-null mouse model where all four histidines of the active site for copper binding are mutated [33]. Although this study shows copper in the active site is not a necessity for SOD1 toxicity, it does not exclude ectopic binding.

Several researchers have reported that mutant SOD1 can ectopically bind copper, giving a rationale why copper chelators can alleviate ALS symptoms in a SOD1^{H46R} mouse model, in which there is reduced copper binding in the active site [32, 34–36]. Using immobilized metal-affinity chromatography, a technique that separates proteins based on their interaction with an immobilized copper [37], it was shown that mutant SOD1 mice have a SOD1 fraction, which has a very high affinity for copper (HAC), whereas WT SOD1 lack this fraction [35, 38]. This fraction gains a pro-oxidative action presumably based on copper-related oxidative events, contributing to oxidative stress. A similar redox-active HAC fraction could be created from WT SOD1 after oxidation. Intriguingly, these fractions are monomerized SOD1, while WT SOD1 is normally present as a homodimer. Monomerization of SOD1 is the basis of the second toxic gain-of-function hypothesis [39].

27.4 Misfolding of Human SOD1

One of the pathological features of FALS is the aggregation of SOD1 in the cytosol [40]. It is therefore hypothesized that SOD1 aggregation may be a mechanism of toxic gain-of-function in mutant SOD1 [39, 41] (Fig. 27.2). Before SOD1 gains its

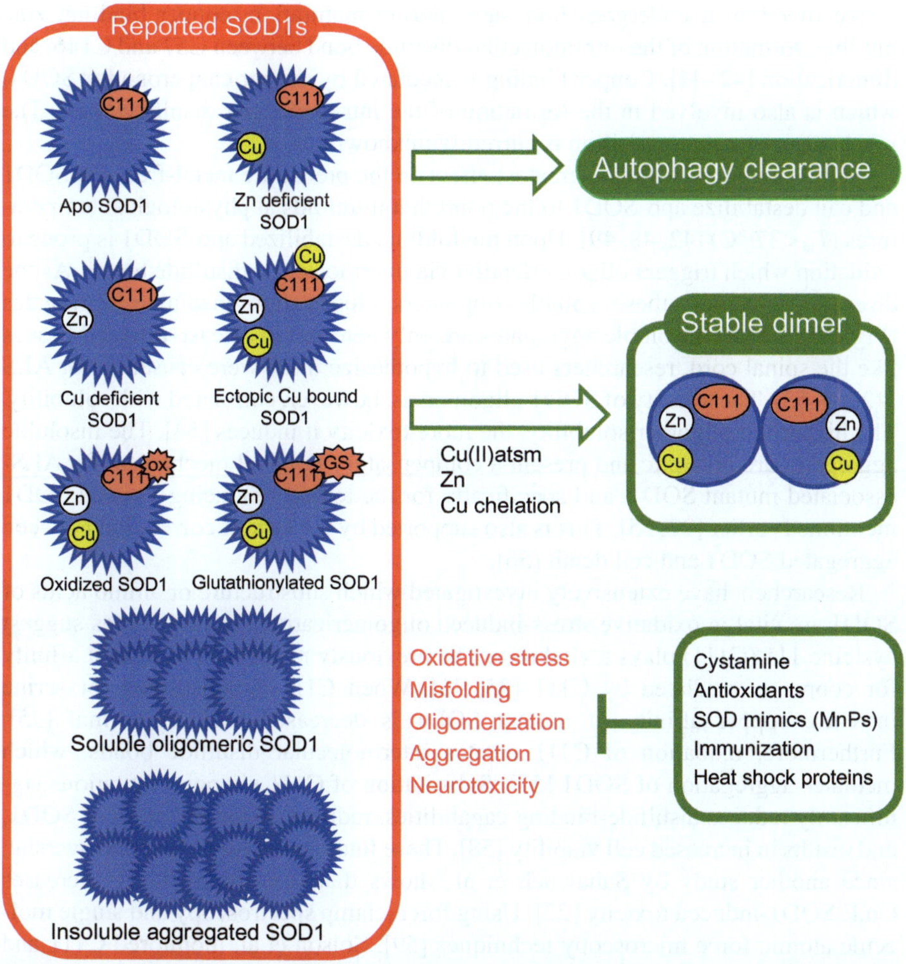

Fig. 27.2 SOD1 toxic gain-of-function hypotheses and therapeutic strategies. According to the pro-oxidant hypothesis, presence of 1-metal SOD1 creates oxidative stress. Possible treatments for restoring metal content include the addition of Cu(II)amts, copper chelation, and zinc repletion, respectively [25, 28–30, 127–129]. Under oxidative stress C111 is oxidized (C111ox), which results in misfolding of SOD1. Misfolding of SOD1 is key element of the second toxic gain-of-function hypothesis. Oxidative stress can be reduced with antioxidant therapy and with SOD mimics (MnPs) [119–121]. Modifying C111 can give protection against oxidation and thereby reduce monomerization and misfolding [101, 106]. However, glutathionylation on C111 induces dimer dissociation and misfolding [98–100]. Unfolded SOD1 oligomerizes by hydrogen bonding, hydrophobic contacts, disulfide bonding, and crosslinking [52, 53, 107, 154]. Cystamine can reduce crosslinking and also aid in lowering oxidative stress [107]. Passive or active immunization can trigger the immune system to clear SOD1 oligomers [133–135]. Lastly, addition of Hsp104, a protein dissagrase, can break down oligomers. Soluble oligomers can also aggregate into insoluble plaques. These are believed to be compensatory and can partly be removed via autophagy [22, 155, 156]. *Zn* zinc, *Cu* copper, *MnP* cationic Mn(III) *N*-substituted pyridylporphyrins, *Cu(II)amts* diacetylbis(*N*(4)-methylthiosemicarbazonato)copper(II)

active function, it undergoes four steps during maturation: copper binding, zinc binding, formation of the intramolecular disulfide bond between C57 and C146, and dimerization [42–44]. Copper binding is mediated by copper chaperone for SOD1, which is also involved in the formation of the intramolecular disulfide bond. The mechanism of zinc acquisition is currently unknown [44–47].

SOD1 mutations have the greatest effect on the premature metal-free apo-SOD1 and can destabilize apo-SOD1 to the point that it unfolds at physiological temperatures (T_m<37 °C) [42, 48, 49]. Upon misfolding, destabilized apo-SOD1 is prone to oxidation which triggers oligomerization via intermolecular disulfide bonds. As the disease progresses, these soluble oligomers cluster into insoluble aggregates (Fig. 27.2). Since insoluble aggregates are only detected in disease-affected tissues like the spinal cord, researchers used to hypothesize these were causative for ALS [42, 50–53]. The toxicity of SOD1 oligomers is, however, correlated with solubility. The higher the oligomer solubility, the more toxicity it induces [54]. The insoluble aggregates are nontoxic and present a compensatory removal mechanism for ALS-associated mutant SOD1, and specifically for the toxicity-inducing 1-metal SOD1 mentioned earlier [22, 55]. This is also supported by the lack of correlation between aggregated SOD1 and cell death [56].

Researchers have extensively investigated which substructure or amino acids of SOD1 are vital in oxidative stress-induced oligomerization. Some findings suggest cysteine 111 (C111) plays a vital role. The previously mentioned increased affinity for copper is mediated by C111 [35, 36]. When C111 is substituted for serine in vitro, copper affinity of mutant SOD1 is decreased back to normal [35]. Furthermore, oxidation of C111 creates intermolecular disulfide bonds, which mediates aggregation of SOD1 [57]. Substitution of C111 via point mutations significantly reduces disulfide-binding capabilities, reduces oligomerization of SOD1, and results in increased cell viability [58]. These findings are however controversial, since another study by Sahawneh et al. shows that substituting C111 increases Cu,E,SOD1-induced toxicity [22]. Using force-clamp spectroscopy and single molecule atomic force microscopy techniques [59], Solson et al. monitored C111 and noticed that mutant SOD are more prone to oligomerize due to altered thiol chemistry at C111 [60]. Modifying C111 might therefore be a therapeutic strategy for FALS if the 1-metal SOD1 toxicity is taken into account (discussed further below).

Molecular evidence for a common pathogenic mechanism between SALS and FALS remains elusive, but due to clinical similarities, it is highly plausible that a common pathway is involved. Gruzman et al. who modified SOD1 via a biotinylated crosslinker observed a covalently crosslinked SOD species in spinal cord extracts of FALS and SALS patients. This crosslinked SOD species was present neither in extracts from normal tissue, nor in extracts from patients with other neurodegenerative diseases [61]. This implies that SOD1 plays a role in both sporadic and FALS.

WT SOD1 could be modified by oxidative stress giving it a similar misfolded state as mutant SOD1. Inducing oxidative stress with hydrogen peroxide causes misfolding of WT SOD1 [62–64]. Since SALS usually has a late onset with first clinical signs visible at an average age of 45–50, and aging is associated with an increase in oxidative stress [65, 66], it is plausible that the accumulated oxidative stress over the years induces misfolding of WT SOD1 in SALS [63, 67]. This notion is supported by the discovery of an over-oxidized form of SOD1 in a subset of SALS patients, which exhibits toxic properties [68]. In yeast cells modified to express SOD1 tagged with green fluorescent protein (SOD1-GFP), two fractions could be observed, one with a low molecular weight and one with a high molecular weight (HMW). The HMW fraction corresponded to SOD1 oligomers and the fluorescent signal of this fraction was increased almost 40 times in aged cells (stationary phase-30 days) compared to young cells (stationary phase-3 days) [67]. This indicates that oxidative stress from aging could be enough to induce misfolding, oligomerization, and subsequent aggregation of SOD1. Authors used native yeast SOD1; this oligomerization was due to the oxidative modification of C146 and H71. Since yeast lacks the free cysteine, C6 and C111 (having A6 and S111 instead) [69], the role of C111 in this age-associated oxidative stress-mediated aggregation of SOD1 remains to be elucidated.

In 2007, the C4F6 antibody was created which detected conformational changes of mutant SOD1, but not of WT SOD [70]. This antibody selectively stained SOD1 in ALS-affected tissues like spinal cords [71]. Remarkably, it also managed to detect misfolded SOD1 from SALS and a fraction of WT SOD1 after exposure to oxidative stress [72]. This confirms that SOD1 has an altered conformation in SALS, induced by oxidative stress, which is similar to the misfolded conformation in FALS. Interestingly, C111 plays a key role in oxidative stress-induced conformational changes. The C4F6 antibody no longer detected oxidative stress-exposed SOD1 when C111 was substituted for serine (C111S). Further studies of the C4F6 antibody revealed that its epitope is at the D90–D96 region of SOD1 [73, 74]. Mutations and oxidative modification of SOD1 destabilize the beta loops IV and VII [75], exposing the D90–D96 epitope region allowing C4F6 binding. Exposure of this C4F6 epitope region positively correlates with SOD1-induced toxicity measured by microglial activity [74].

Overexpression of WT SOD1 in mutant SOD1 mice models of ALS increases disease severity rather then alleviate symptoms. WT SOD1 interacts with mutant SOD1 as heterodimer, which is generally more stable. Sahawaneh et al. attribute this observed increase of toxicity to stabilization of WT SOD1 with mutant Cu,E SOD1 [22, 76–79]. Thus, soluble oligomeric SOD1 and stable misfolded SOD1 with mutation and/or oxidative modification could be therapeutic targets for ALS.

27.5 Altered Signaling Pathway by Mutant SOD1

Another proposed mechanism of SOD1 motoneuron degeneration is that mutations lead to increased susceptibility to activate the neuron-specific Fas receptor death pathway [80]. The Fas receptor signals through two synergetic pathways, the classic Fadd/caspase-8 pathway and the neuron-specific Daxx/Ask1/p38/nNOS pathway. Nitric oxide (NO), one of the end-products of the Daxx/Ask1/p38/nNOS pathway, induces expression of the ligand for Fas receptor (FasL). In mutant SOD1 neurons but not in healthy neurons, this unexpectedly leads to chronic activation of the Fas receptor, causing a Fas/NO amplification feedback loop. This feedback mechanism is supported by an observed increase in some of the pathway intermediates (like Ask and p38) during the pre-symptomatic stage of ALS [81–83]. Chronic low-level activation of this feedback loop results in motoneuron cell death and may explain the relatively slow progression of ALS [84].

Heat shock protein 90 (Hsp90) plays a role in the Fas-induced apoptosis. Due to the feedback loop present in mutant SOD1 neurons, the concentration of NO gradually increases and can react with superoxide to produce peroxynitrite. Peroxynitrite reacts at a currently unknown rate with Hsp90 and modifies one of two tyrosine residues. It remains to be elucidated whether this modification occurs via radicals or that it is a catalyzed reaction. Modified Hsp90 exhibits a toxic gain of function. Normally, Hsp90 has an anti-apoptotic function, but modified Hsp90 stimulates the P2X7 receptor, which results in mobilization of FasL to the plasma membrane, which in turn activates the Fadd/caspase-8 apoptotic pathway [85]. Removing the feedback loop by FasL knockout reduced loss of motor neurons and increased survival in ALS mice [86]. These findings suggest that therapy aimed at the death receptors and the associated Daxx/Ask1/p38/nNOS pathway could be beneficial early in the disease onset.

The role of anti-apoptotic heat shock proteins (Hsps) in ALS motoneuron degeneration has been the focus of many researchers and findings have been extensively reviewed elsewhere [8, 87]. Mutant SOD1 does not only influence Hsps indirectly as described above, but can also directly interact with them, to prevent entering into mitochondrial intermembrane space [88]. After aggregation with SOD1, Hsps lose their function, which has been assigned to mutant SOD1 toxicity. The ongoing phase II clinical trial with arimcolomol, a co-inducer of Hsps, shows therapeutic benefits in ALS [89, 90].

27.6 Therapeutic Strategies for ALS

Based on the main hypotheses of redox regulation and SOD1 misfolding, there is a diverse amount of therapeutic strategies available (Fig. 27.2). One strategy focuses on C111, since it can play a key role in misfolding and oligomerization as well as in prooxidation. Other strategies include focusing on the metal state of SOD1 or decreasing SOD1 oligomers. Current advances in these strategies are discussed below.

27.6.1 Modification of Cysteine 111

In healthy human brain tissue, almost half of the C111 site of SOD1 is modified, of which the most prevalent modification is cysteinylation [91]. In vitro experiments indicated that cysteinylation on C111 protects against H_2O_2-induced oxidation of SOD1. Despite the large excess of cysteine used during in vitro experiments, only one C111 was cysteinylated per SOD1 dimer. Since the distance between two

C111 in the homodimeric formation is only 8–10 Å (Fig. 27.1) [36], cysteinylation of either C111 physically blocks the binding place on the other monomer. This suggests that SOD1 cysteinylation does not completely protect SOD1 against oxidation since one C111 remains free. Unfortunately, this research group only looked at the chemical properties of SOD1 modification via mass spectrometry, so no conclusions can be made whether or not C111 cysteinylation can prevent oxidative stress-mediated aggregation of SOD1. They hypothesize that cysteinylation can greatly ameliorate SOD1 dimer destabilization and minimize (but not completely prevent) formation of SOD1 aggregates. Further research into cysteinylation of C111 looks promising and much can still be unraveled. Cystine, an oxidized dimer of cysteine, can naturally cysteinylate SOD1. In healthy people, cystine concentrations double during aging (52 μM at 26 compared to 105 μM at 60) [92]. This indicates that cysteinylated SOD1 might be an adaptive mechanism to age-related oxidative stress. Further study of the post-translational modification pattern of ALS tissue can determine if there is altered cysteinylation of C111, which (if altered) contributes to development and progression of ALS.

Another naturally occurring modification of C111 is glutathionylation. Glutathione (GSH) is an antioxidant that protects against oxidative damage and gets upregulated in response to oxidative stress [93]. Oxidative stress was a hallmark of ALS [14–18, 94], and GSH concentration is expectedly increased in the cerebrospinal fluid of ALS patients [95]. GSH can bind to C111 by sulfide-glutathionylation to prevent permanent oxidation, as part of a coping mechanism to oxidative stress [96, 97]. However, glutathionylation on C111 changes the tertiary structure of SOD1, modulating stability of the enzyme, and induces dimer dissociation [98, 99]. This is due to the narrow 8–10 Å, space between C111 residues (Fig. 27.1). S-glutathionylation of C111 is therefore detrimental for SOD1 misfolding, indicating that not all antioxidant modifications of C111 are good for therapy. Intriguingly, C4F6 reactivity is increased in SOD1 modified by GSH, suggesting that GSH contributes to misfolding of SOD1 in ALS [100].

Commercially available human SOD1 isolated from erythrocytes shows a distinct absorption at 325 nm and has a mass 30 units more than expected from hSOD1, which could either be a persulfide (Cys-S-SH) or trisulfide bond (Cys-S-S-S-Cys) between subunits [101–103]. De Beus et al. replicated a sulfhydryl modification with sodium sulfide, inducing the formation of a persulfide group at C111. This resulted in a more stable WT SOD1 with higher resistance to SOD1 aggregation in vitro [101]. However, the modification in hSOD1 turned out to be from a trisulfide bond between C111's of the monomers, and SOD1 was detected as a holo dimer in these experimental conditions [102]. Other SOD1 tested in this study were detected as apo monomers suggesting that a trisulfide bond between C111's does not occur naturally in humans, but does have the potential to stabilize SOD1, preventing dissociation to monomers. The observed trisulfide bond in erythrocytes might be explained by increased physiological concentration of hydrogen sulfide (H_2S) or anion hydrosulfide (HS^-). Reactive sulfur species can modify cysteine thiols and thus potentially crosslink C111. Although a truly accurate method for detecting H_2S concentration is lacking, the concentration of H_2S is believed to be much higher in blood compared to brain tissue [104, 105].

Similarly, crosslinking C111 using a maleimide crosslinker (DTME or BMOE) increased thermostability of SOD1 [106]. This study did not investigate aggregation properties of SOD1, but it is hypothesized that SOD1 becomes more resistant to oxidative stress-induced misfolding due to decreased availability of C111.

These in vitro studies indicate that modifying C111 is beneficial in preventing SOD1 misfolding. The next step will be to investigate which Cys111 modification is the safest ex vivo and results in the highest resistance to oxidative stress-induced SOD1 misfolding, accompanied with increased neuronal survival.

27.6.2 Transglutaminase 2

Oono et al. published a paper linking transglutaminase 2 (TG2) with ALS [107]. TG2 is a calcium-dependent endogenous crosslinker that covalently binds glutamine and lysine residues. TG2 is ubiquitously expressed in intra- and extracellular spaces of the central nervous system and has previously been associated with many neurodegenerative diseases [108]. Pathogenic proteins such as α-synuclein (Parkinson's disease) [109], β-amyloid (Alzheimer's disease) [110], and huntingtin (Huntington's disease, HD) [111] could all be crosslinked by TG2. Elevated concentrations of TG2 are observed in the brain and cerebrospinal fluid (CSF) of HD patients and targeting TG2 has therapeutic effects in animal models of HD [112, 113]. Similarly, an increase of TG2 is detected in CSF and in the brain of ALS patients, but its role in the pathogenesis of ALS was only recently clarified [114]. SOD1 has many glutamine and lysine residues prone to TG2-mediated crosslinking (Fig. 27.3). With co-immunoprecipitation and western blot experiments, it became clear that TG2 interacts with misfolded SOD1 (from mutant SOD1 but also apo WT SOD1) and promotes covalently crosslinked oligomerization. TG2-mediated SOD1 oligomerization induces microglial activation and neuroinflammation associated with neurotoxicity. These results provide a novel role of TG2 in SOD1 toxicity and demonstrate that TG2 is a therapeutic target for ALS. Cystamine, an inhibitor of TG2, significantly delayed ALS progression and reduced SOD1 oligomer formation and microglial activation [107]. Since cystamine has multiple functions (including increasing antioxidant levels [115]), a synergistic effect might be expected.

27.6.3 Antioxidant Approach

A different approach is not to modify C111, but to reduce oxidative stress with antioxidants, which may suppress the chance of C111 oxidation. A free radical scavenger edavarone ameliorated ALS symptoms in a phase II Clinical Trial [116] and results of the phase III trial are still pending (NCT00330681, NCT00424463, NCT00415519). Of a different antioxidant, dexpramipexole, the phase III (NCT00600873) results were recently posted, but unfortunately no significant beneficial outcome on survival or disease progression was reported.

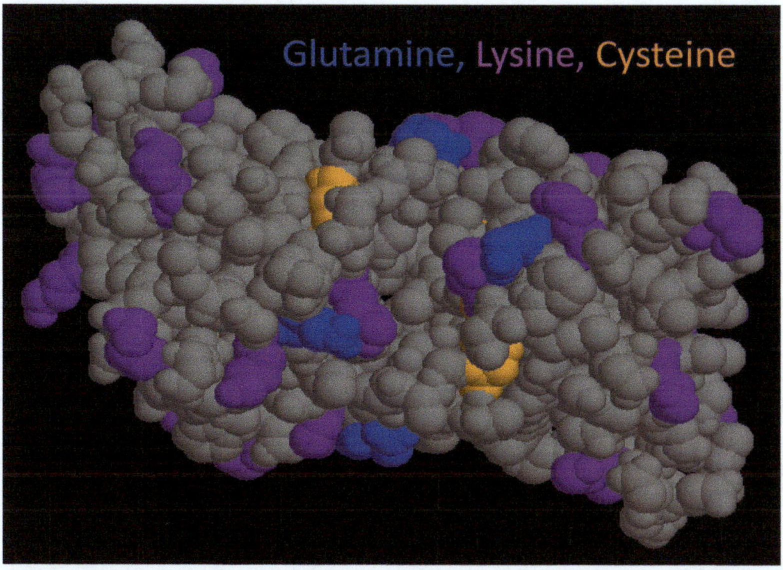

Fig. 27.3 Glutamine, lysine, and cysteine residues on the surface of human SOD1. Glutamine (*blue*), lysine residues (*purple*), and cysteine (*yellow*) on the surface of human SOD1 (Protein data bank #1SPD) are shown. Other amino acid residues were shown in *gray*. Transglutaminase catalyzes covalently crosslinking between glutamine and lysine. Transglutaminase 2 catalyzed FALS-associated mutant SOD1 and apo WT SOD1 [107]

Extensive reviews have been published on porhyrins-based SOD mimics [117, 118]. The most potent SOD mimics are cationic Mn(III) *N*-substituted pyridylporphyrins (MnPs), of which the most frequently studied are the MnTE-2-PyP^{5+}, MnTnHex-2-PyP^{5+}, and MnTDE-2-ImP^{5+}. Beside their ability to reduce $O_2^{\bullet -}$, they can also remove peroxynitrite. In animal experiments, MnTDE-2-ImP^{5+} alleviated ALS symptoms and could markedly extend survival of mice, even if administered during disease onset [119–121]. A phase I toxicity Clinical Trial with ALS patients was performed to assess the safety of MnTDE-2-ImP^{5+} (AEOL 10150) and was deemed not to be toxic at doses far above therapeutic concentrations [122, 123]. MNPs have been explored for many applications including stroke, cancer, and radioprotection. Thus far, no additional clinical trials have been conducted to investigate therapeutic effects with regard to ALS. Recently, it was reported that MnTE-2-PyP^{5+} potentiates glucocorticoid-induced apoptosis in lymphoma cells [124]. In doing so, MnTE-2-PyP^{5+} acts as a pro-oxidant in presence of GSH and H_2O_2 and glutathionylates critical survival proteins including nuclear factor κB (NF-κB) [125]. Intriguingly, the beneficial mechanism of MnPs in cancer treatment might also be applicable for ALS. GSH concentrations are increased in ALS [95] and neuronal cell death is induced by microglia via the NF-κB pathway [126]. Under these conditions, MnTE-2-PyP^{5+} may function as a pro-oxidant at microglia and inhibit NF-κB-induced cell death. This could possibly explain the observed positive effects in ALS animal models [119–121]. Therefore, the research on MnPs in ALS therapy is warranted.

The role of SOD metal content in its SOD1-mediated oxidative stress was discussed earlier in this review [38, 94]. Based on the hypothesis that 1-metal-deficient SOD1 is the toxic intermediate, treatment options could approach this toxicity either via metal chelation or metal addition to apo-SOD1 or holo-SOD1, respectively. At first, therapy was aimed at lowering copper content by copper chelator treatment, which slightly alleviates ALS symptoms [127–129]. However, recent data suggests this is a less effective approach for ALS treatment. In contrast, addition of diacetylbis(*N*(4)-methylthiosemicarbazonato)copper(II) (Cu(II)atsm), an orally bioavailable, blood-brain barrier permeable copper complex, exhibits significant therapeutic effects in ALS mice [28–30]. Survival and locomotor function improved in a dose-dependent matter, consistent with an observed decrease in protein nitration and increased antioxidant activity. Commencing treatment during symptomatic stages of ALS also improved locomotor function and survival albeit less potent as in preventive treatment. The potential to alleviate symptoms even during disease onset makes CuII(atsm) an interesting novel treatment option for ALS, which should be explored in clinical trials.

27.7 Immunization

One of the latest SOD1-targeted therapies is immunization. There are many antibodies created against misfolded SOD1, each with different epitope regions, including: A5C3, D3H5, SEDI, and C4F6 [70, 73, 130–132]. Passive immunization with D3H5 antibodies delayed ALS onset and increased life span in SOD1^{G93A} mice. Passive immunization with A5C3, however, had no effect on disease onset, indicating that not all epitopes of the misfolded SOD1 are associated with SOD1 misfold toxicity. Finding the suitable epitope is therefore essential for successful treatment of ALS [133]. Recently, a novel therapeutic approach mediated by adeno-associated viruses was applied to mice. Authors used a single-chain fragment variable antibody (scFvA) composed of light and heavy chains of D3H5. A single intrathecal injection of a scFvA-containing virus caused sustained expression of scFvA in the spinal cord and delayed disease onset and increased life span up to 28 % [134].

Active immunization against the SEDI epitope was effective as well; delayed disease onset in mice and increased survival up to almost 40 days. This survival correlated with a decrease in misfolded/aggregated SOD1, providing evidence that reduction of misfolded SOD1 is a viable therapy for ALS [135]. Unfortunately, at the time of this study the epitope binding of C4F6 was unknown, so the researchers decided not to test the effect of active immunization against this epitope. Since the exposure of the C4F6 epitope directly correlates with microglial activation [74], an interesting follow-up will be to perform passive or active immunization with C4F6 antibodies.

In contrast to immunotherapy applied in Alzheimer's disease (AD), which failed due to an excessive inflammatory Th1 response, proposed ALS immunizations elicit SOD1 clearance via a Th2 response [135, 136]. This combined with the positive results gained from in vivo experiments are a good sign that SOD1-targeted immunotherapies may soon enter clinical trials.

27.8 Other ALS-Associated Genes and Advances in Targeted Therapies

ALS is a complex disease with a wide variation of causes other than SOD1 mutations and their toxic SOD1 intermediates. Mutations in TAR DNA binding protein (TARDBP), fused in sarcoma (FUS), ubiquilin 2, and C9orf72 are also causative for ALS. The relation between these and ALS are reviewed extensively elsewhere [137–141]. TARDBP and FUS regulate transcription and alternative splicing via direct interaction with DNA, RNA, and mRNA. In ALS, TPD-43 is prone to aggregate into TPD-43 inclusions, which are a major hallmark of sporadic ALS. These inclusions are currently missing from ALS mouse models, which are based on SOD1 mutations. Since our knowledge about TDP-43 function and its regulatory pathways is still limited, the exact mechanism by which mutations in TARDBP induce ALS remains to be elucidated. Nonetheless, several therapeutic strategies are in development. These include protein–protein interaction modulators that modulate the interaction between TPD-43 and other proteins (like SOD1, FUS, and Hsp), targeting TDP-43-RNA interactions and lastly developing compounds which decrease TPD-43 concentrations and/or TPD-43 aggregates. Rational for these strategies and related drug candidates has been reviewed by Budini et al. [142]. Ubiquilin 2 associates with ubiquitin ligases and proteosomes to form an ubiquitin proteasome system (UPS), which mediates protein degradation. Mutations of the gene result in inclusions of ubiquilin 2 in ALS-affected neurons. One of the targets for degradation by UPS is TDP-43, which explains why they are often found together inside inclusions. It is hypothesized perturbations in protein clearance play an important role in neurodegeneration [143, 144]. New therapeutic strategies are aimed at restoring perturbed protein clearance using activators for autophagy, the UPS system, or the chaperone network [145].

The function of C9orf72 is unknown, but C9orf72 gene expansions have great impact on ALS. Within the C9orf72 gene is a hexanucleotide repeat (GGGGCC), which is normally repeated <25 times. Almost 50 % of European FALS patients and 5–10 % of sporadic ALS patients have this repeat more than 1000 times [146]. At the end of 2014, a research group managed to create an inducible mouse model based on this hexanucleotide repeat expansion and observed ubiquitin positive inclusions [147]. This model supports a RNA toxic gain-of-function mechanism of the increased C9orf72 repeats. Further studies utilizing this newly created mouse model can give valuable insights in ALS disease progression and contribute to therapeutic intervention. One therapeutic strategy involves silencing RNA expression of C9orf72. Hypermethylation of CpG islands is known to protect the genome and stably silences gene expression. This effect was also observed in mutant C9orf72 when the promoter was hypermethylated in ALS-derived lymphoblasts [148]. Clinical comparisons of methylation status of C9orf72 in ALS patients further suggest that hypermethylation of the gene is neuroprotective [149]. Combining these findings with the newly created animal model can open the door for creating drugs against C9orf72-mediated ALS.

27.9 Related Therapies in Neurodegenerative Diseases

There is a slight overlap in general disease mechanism between ALS and other neurodegenerative diseases like AD and Parkinson's disease (PD). All these diseases have aggregation of oligomeric proteins and some form of oxidative stress in common.

In AD, similar to SOD1-associated ALS, oligomeric protein (amyloid β) is associated with neurotoxicity rather than amyloid aggregates [150]. One of the most recent therapeutic drugs that are in clinical trial is 8-hydroxyquinolines (8-HQ). These reduce oxidative stress by inhibition of metal-mediated reactive oxygen species. They also seem to inhibit oligomerization of amyloid β by stabilization into small dimeric fractions [151]. Especially the combination of regulating metal homeostasis and dimeric stabilization is a promising approach. Since there is still a struggle between metal content-induced oxidative stress and oligomerization as toxic intermediates in SOD1-associated ALS, an approach like this is highly promising. Even more, considering the success this strategy has in clinical trials of AD patients.

One interesting therapeutic approach covers all aggregation-related neurodegenerative diseases. Hsp104, a protein disaggregase isolated from yeast, transduces energy from ATP hydrolysis to reverse oligomerization and aggregation. Its application to protein misfolding diseases was reviewed by Vashist et al. in 2010 [152]. Recent technological advances has made it possible to modify Hsp104, altering its activity and efficiency. Subtle modifications give increased protective activity against α-synuclein (PD), TDP-43 (ALS, AD and fronto-temporal lobar degeneration), FUS (ALS), and amyloid β 42 (AD) [153]. Since this strategy has a broader focus than previously mentioned strategies which mainly focus on one disease mediator, it may have a higher efficacy to alleviate ALS symptoms.

27.10 Conclusion

ALS is a complex disease with several causative factors. A subset of the disease is a consequence of toxic-gain-of-function of SOD1, of which the exact mechanism remains to be elucidated. SOD1 oligomerization and the metal content of SOD1 are the main two hypotheses and focus for current therapeutic strategies. Since evidence for both hypotheses is strong, it is plausible the disease is a combination of oligomerization and SOD1 metal content. It is therefore vital to keep both hypotheses in mind when trying to create new therapeutic strategies. Preventing SOD1 misfolding by stabilizing cysteine 111 without altering metal-free SOD1 toxicity, restoring SOD1 metal content, reducing oxidative stress with antioxidants, and immunization against specific misfolded SOD1 are all promising therapeutic strategies. Immunization and restoring SOD1 metal content already show promise in mice models and await clinical trials. Perhaps combination treatments that target both hypotheses will show the greatest therapeutic affect and can ameliorate ALS symptoms completely in this particular subset of ALS. Other causes for ALS and therapies were discussed briefly, of which advances in reducing general protein aggregation with Hsp104, restoring perturbed protein clearance, and altering gene expression are currently exciting highlights.

References

1. Rosen DR, Siddique T, Patterson D, Figlewicz DA, Sapp P, Hentati A, et al. Mutations in Cu/Zn superoxide dismutase gene are associated with familial amyotrophic lateral sclerosis. Nature. 1993;362(6415):59–62.
2. Forman HJ, Fridovich I. On the stability of bovine superoxide dismutase. The effects of metals. J Biol Chem. 1973;248(8):2645–9.
3. Tainer JA, Getzoff ED, Beem KM, Richardson JS, Richardson DC. Determination and analysis of the 2 A-structure of copper, zinc superoxide dismutase. J Mol Biol. 1982;160(2):181–217.
4. Raha S, Robinson BH. Mitochondria, oxygen free radicals, disease and ageing. Trends Biochem Sci. 2000;25(10):502–8.
5. Lippard SJ, Berg JM. Principles of bioinorganic chemistry. Mill Valley.: University Science Books; 1994. xvii, 411 p. p.
6. Pardo CA, Xu Z, Borchelt DR, Price DL, Sisodia SS, Cleveland DW. Superoxide dismutase is an abundant component in cell bodies, dendrites, and axons of motor neurons and in a subset of other neurons. Proc Natl Acad Sci U S A. 1995;92(4):954–8.
7. Okado-Matsumoto A, Fridovich I. Subcellular distribution of superoxide dismutases (SOD) in rat liver: Cu,Zn-SOD in mitochondria. J Biol Chem. 2001;276(42):38388–93.
8. Ido A, Fukuyama H, Urushitani M. Protein misdirection inside and outside motor neurons in Amyotrophic Lateral Sclerosis (ALS): a possible clue for therapeutic strategies. Int J Mol Sci. 2011;12(10):6980–7003.
9. Urushitani M, Ezzi SA, Matsuo A, Tooyama I, Julien JP. The endoplasmic reticulum-Golgi pathway is a target for translocation and aggregation of mutant superoxide dismutase linked to ALS. FASEB J. 2008;22(7):2476–87.
10. Urushitani M, Sik A, Sakurai T, Nukina N, Takahashi R, Julien JP. Chromogranin-mediated secretion of mutant superoxide dismutase proteins linked to amyotrophic lateral sclerosis. Nat Neurosci. 2006;9(1):108–18.
11. Cleveland DW. From Charcot to SOD1: mechanisms of selective motor neuron death in ALS. Neuron. 1999;24(3):515–20.
12. Reaume AG, Elliott JL, Hoffman EK, Kowall NW, Ferrante RJ, Siwek DF, et al. Motor neurons in Cu/Zn superoxide dismutase-deficient mice develop normally but exhibit enhanced cell death after axonal injury. Nat Genet. 1996;13(1):43–7.
13. Sies H, Jones D. Oxidative stress. In: Fink G, editor. Encyclopedia of stress. Amsterdam: Elsevier; 2007. p. 45–8.
14. Shaw PJ, Ince PG, Falkous G, Mantle D. Oxidative damage to protein in sporadic motor neuron disease spinal cord. Ann Neurol. 1995;38(4):691–5.
15. Ferrante RJ, Browne SE, Shinobu LA, Bowling AC, Baik MJ, MacGarvey U, et al. Evidence of increased oxidative damage in both sporadic and familial amyotrophic lateral sclerosis. J Neurochem. 1997;69(5):2064–74.
16. D'Amico E, Factor-Litvak P, Santella RM, Mitsumoto H. Clinical perspective on oxidative stress in sporadic amyotrophic lateral sclerosis. Free Radic Biol Med. 2013;65:509–27.
17. Abe K, Pan LH, Watanabe M, Konno H, Kato T, Itoyama Y. Upregulation of protein-tyrosine nitration in the anterior horn cells of amyotrophic lateral sclerosis. Neurol Res. 1997;19(2):124–8.

18. Beal MF, Ferrante RJ, Browne SE, Matthews RT, Kowall NW, Brown Jr RH. Increased 3-nitrotyrosine in both sporadic and familial amyotrophic lateral sclerosis. Ann Neurol. 1997;42(4):644–54.
19. Wiedau-Pazos M, Goto JJ, Rabizadeh S, Gralla EB, Roe JA, Lee MK, et al. Altered reactivity of superoxide dismutase in familial amyotrophic lateral sclerosis. Science. 1996;271(5248):515–8.
20. Lyons TJ, Liu H, Goto JJ, Nersissian A, Roe JA, Graden JA, et al. Mutations in copper-zinc superoxide dismutase that cause amyotrophic lateral sclerosis alter the zinc binding site and the redox behavior of the protein. Proc Natl Acad Sci U S A. 1996;93(22):12240–4.
21. Crow JP, Sampson JB, Zhuang Y, Thompson JA, Beckman JS. Decreased zinc affinity of amyotrophic lateral sclerosis-associated superoxide dismutase mutants leads to enhanced catalysis of tyrosine nitration by peroxynitrite. J Neurochem. 1997;69(5):1936–44.
22. Sahawneh MA, Ricart KC, Roberts BR, Bomben VC, Basso M, Ye Y, et al. Cu, Zn-superoxide dismutase increases toxicity of mutant and zinc-deficient superoxide dismutase by enhancing protein stability. J Biol Chem. 2010;285(44):33885–97.
23. Estevez AG, Crow JP, Sampson JB, Reiter C, Zhuang Y, Richardson GJ, et al. Induction of nitric oxide-dependent apoptosis in motor neurons by zinc-deficient superoxide dismutase. Science. 1999;286(5449):2498–500.
24. Beckman JS, Carson M, Smith CD, Koppenol WH. ALS, SOD and peroxynitrite. Nature. 1993;364(6438):584.
25. Ermilova IP, Ermilov VB, Levy M, Ho E, Pereira C, Beckman JS. Protection by dietary zinc in ALS mutant G93A SOD transgenic mice. Neurosci Lett. 2005;379(1):42–6.
26. Rhoads TW, Lopez NI, Zollinger DR, Morre JT, Arbogast BL, Maier CS, et al. Measuring copper and zinc superoxide dismutase from spinal cord tissue using electrospray mass spectrometry. Anal Biochem. 2011;415(1):52–8.
27. Lelie HL, Liba A, Bourassa MW, Chattopadhyay M, Chan PK, Gralla EB, et al. Copper and zinc metallation status of copper-zinc superoxide dismutase from amyotrophic lateral sclerosis transgenic mice. J Biol Chem. 2011;286(4):2795–806.
28. Roberts BR, Lim NK, McAllum EJ, Donnelly PS, Hare DJ, Doble PA, et al. Oral treatment with Cu(II)(atsm) increases mutant SOD1 in vivo but protects motor neurons and improves the phenotype of a transgenic mouse model of amyotrophic lateral sclerosis. J Neurosci. 2014;34(23):8021–31.
29. Soon CP, Donnelly PS, Turner BJ, Hung LW, Crouch PJ, Sherratt NA, et al. Diacetylbis(N(4)-methylthiosemicarbazonato) copper(II) (CuII(atsm)) protects against peroxynitrite-induced nitrosative damage and prolongs survival in amyotrophic lateral sclerosis mouse model. J Biol Chem. 2011;286(51):44035–44.
30. McAllum EJ, Lim NK, Hickey JL, Paterson BM, Donnelly PS, Li QX, et al. Therapeutic effects of CuII(atsm) in the SOD1-G37R mouse model of amyotrophic lateral sclerosis. Amyotroph Lateral Scler Frontotemporal Degener. 2013;14(7-8):586–90.
31. Subramaniam JR, Lyons WE, Liu J, Bartnikas TB, Rothstein J, Price DL, et al. Mutant SOD1 causes motor neuron disease independent of copper chaperone-mediated copper loading. Nat Neurosci. 2002;5(4):301–7.
32. Beckman JS, Esetvez AG, Barbeito L, Crow JP. CCS knockout mice establish an alternative source of copper for SOD in ALS. Free Radic Biol Med. 2002;33(10):1433–5.
33. Wang J, Slunt H, Gonzales V, Fromholt D, Coonfield M, Copeland NG, et al. Copper-binding-site-null SOD1 causes ALS in transgenic mice: aggregates of non-native SOD1 delineate a common feature. Hum Mol Genet. 2003;12(21):2753–64.
34. Bush AI. Is ALS, caused by an altered oxidative activity of mutant superoxide dismutase? Nat Neurosci. 2002;5(10):919. author reply 20.
35. Watanabe S, Nagano S, Duce J, Kiaei M, Li QX, Tucker SM, et al. Increased affinity for copper mediated by cysteine 111 in forms of mutant superoxide dismutase 1 linked to amyotrophic lateral sclerosis. Free Radic Biol Med. 2007;42(10):1534–42.
36. Liu H, Zhu H, Eggers DK, Nersissian AM, Faull KF, Goto JJ, et al. Copper(2+) binding to the surface residue cysteine 111 of His46Arg human copper-zinc superoxide dismutase, a familial amyotrophic lateral sclerosis mutant. Biochemistry. 2000;39(28):8125–32.

37. Porath J, Carlsson J, Olsson I, Belfrage G. Metal chelate affinity chromatography, a new approach to protein fractionation. Nature. 1975;258(5536):598–9.
38. Kishigami H, Nagano S, Bush AI, Sakoda S. Monomerized Cu, Zn-superoxide dismutase induces oxidative stress through aberrant Cu binding. Free Radic Biol Med. 2010;48(7):945–52.
39. Johnston JA, Dalton MJ, Gurney ME, Kopito RR. Formation of high molecular weight complexes of mutant Cu, Zn-superoxide dismutase in a mouse model for familial amyotrophic lateral sclerosis. Proc Natl Acad Sci U S A. 2000;97(23):12571–6.
40. Kim J, Lee H, Lee JH, Kwon DY, Genovesio A, Fenistein D, et al. Dimerization, oligomerization, and aggregation of human amyotrophic lateral sclerosis copper/zinc superoxide dismutase 1 protein mutant forms in live cells. J Biol Chem. 2014;289(21):15094–103.
41. Bruijn LI, Houseweart MK, Kato S, Anderson KL, Anderson SD, Ohama E, et al. Aggregation and motor neuron toxicity of an ALS-linked SOD1 mutant independent from wild-type SOD1. Science. 1998;281(5384):1851–4.
42. Furukawa Y, O'Halloran TV. Amyotrophic lateral sclerosis mutations have the greatest destabilizing effect on the apo- and reduced form of SOD1, leading to unfolding and oxidative aggregation. J Biol Chem. 2005;280(17):17266–74.
43. Rakhit R, Chakrabartty A. Structure, folding, and misfolding of Cu, Zn superoxide dismutase in amyotrophic lateral sclerosis. Biochim Biophys Acta. 2006;1762(11-12):1025–37.
44. Furukawa Y, O'Halloran TV. Posttranslational modifications in Cu, Zn-superoxide dismutase and mutations associated with amyotrophic lateral sclerosis. Antioxid Redox Signal. 2006;8(5-6):847–67.
45. Banci L, Bertini I, Cantini F, Kozyreva T, Massagni C, Palumaa P, et al. Human superoxide dismutase 1 (hSOD1) maturation through interaction with human copper chaperone for SOD1 (hCCS). Proc Natl Acad Sci U S A. 2012;109(34):13555–60.
46. Son M, Elliott JL. Mitochondrial defects in transgenic mice expressing Cu, Zn superoxide dismutase mutations, the role of copper chaperone for SOD1. J Neurol Sci. 2014;336(1-2):1–7.
47. Banci L, Barbieri L, Bertini I, Luchinat E, Secci E, Zhao Y, et al. Atomic-resolution monitoring of protein maturation in live human cells by NMR. Nat Chem Biol. 2013;9(5):297–9.
48. Tiwari A, Hayward LJ. Familial amyotrophic lateral sclerosis mutants of copper/zinc superoxide dismutase are susceptible to disulfide reduction. J Biol Chem. 2003;278(8):5984–92.
49. Bourassa MW, Brown HH, Borchelt DR, Vogt S, Miller LM. Metal-deficient aggregates and diminished copper found in cells expressing SOD1 mutations that cause ALS. Front Aging Neurosci. 2014;6:110.
50. Rakhit R, Cunningham P, Furtos-Matei A, Dahan S, Qi XF, Crow JP, et al. Oxidation-induced misfolding and aggregation of superoxide dismutase and its implications for amyotrophic lateral sclerosis. J Biol Chem. 2002;277(49):47551–6.
51. Banci L, Bertini I, Boca M, Girotto S, Martinelli M, Valentine JS, et al. SOD1 and amyotrophic lateral sclerosis: mutations and oligomerization. PLoS One. 2008;3(2), e1677.
52. Furukawa Y, Fu R, Deng HX, Siddique T, O'Halloran TV. Disulfide cross-linked protein represents a significant fraction of ALS-associated Cu, Zn-superoxide dismutase aggregates in spinal cords of model mice. Proc Natl Acad Sci U S A. 2006;103(18):7148–53.
53. Niwa J, Yamada S, Ishigaki S, Sone J, Takahashi M, Katsuno M, et al. Disulfide bond mediates aggregation, toxicity, and ubiquitylation of familial amyotrophic lateral sclerosis-linked mutant SOD1. J Biol Chem. 2007;282(38):28087–95.
54. Brotherton TE, Li Y, Glass JD. Cellular toxicity of mutant SOD1 protein is linked to an easily soluble, non-aggregated form in vitro. Neurobiol Dis. 2013;49:49–56.
55. Trumbull KA, Beckman JS. A role for copper in the toxicity of zinc-deficient superoxide dismutase to motor neurons in amyotrophic lateral sclerosis. Antioxid Redox Signal. 2009;11(7):1627–39.
56. Lee JP, Gerin C, Bindokas VP, Miller R, Ghadge G, Roos RP. No correlation between aggregates of Cu/Zn superoxide dismutase and cell death in familial amyotrophic lateral sclerosis. J Neurochem. 2002;82(5):1229–38.

57. Chen X, Shang H, Qiu X, Fujiwara N, Cui L, Li XM, et al. Oxidative modification of cysteine 111 promotes disulfide bond-independent aggregation of SOD1. Neurochem Res. 2012;37(4):835–45.
58. Cozzolino M, Amori I, Pesaresi MG, Ferri A, Nencini M, Carri MT. Cysteine 111 affects aggregation and cytotoxicity of mutant Cu, Zn-superoxide dismutase associated with familial amyotrophic lateral sclerosis. J Biol Chem. 2008;283(2):866–74.
59. Alegre-Cebollada J, Kosuri P, Rivas-Pardo JA, Fernandez JM. Direct observation of disulfide isomerization in a single protein. Nat Chem. 2011;3(11):882–7.
60. Solsona C, Kahn TB, Badilla CL, Alvarez-Zaldiernas C, Blasi J, Fernandez JM, et al. Altered Thiol Chemistry in Human Amyotrophic Lateral Sclerosis-linked Mutants of Superoxide Dismutase 1. J Biol Chem. 2014.
61. Gruzman A, Wood WL, Alpert E, Prasad MD, Miller RG, Rothstein JD, et al. Common molecular signature in SOD1 for both sporadic and familial amyotrophic lateral sclerosis. Proc Natl Acad Sci U S A. 2007;104(30):12524–9.
62. Fujiwara N, Nakano M, Kato S, Yoshihara D, Ookawara T, Eguchi H, et al. Oxidative modification to cysteine sulfonic acid of Cys111 in human copper-zinc superoxide dismutase. J Biol Chem. 2007;282(49):35933–44.
63. Kabashi E, Valdmanis PN, Dion P, Rouleau GA. Oxidized/misfolded superoxide dismutase-1: the cause of all amyotrophic lateral sclerosis? Ann Neurol. 2007;62(6):553–9.
64. Ezzi SA, Urushitani M, Julien JP. Wild-type superoxide dismutase acquires binding and toxic properties of ALS-linked mutant forms through oxidation. J Neurochem. 2007;102(1):170–8.
65. Gemma C, Vila J, Bachstetter A, Bickford PC. Oxidative stress and the aging brain: from theory to prevention. In: Riddle DR, editor. Brain aging: models, methods, and mechanisms. Boca Raton: Frontiers in Neuroscience; 2007.
66. Berlett BS, Stadtman ER. Protein oxidation in aging, disease, and oxidative stress. J Biol Chem. 1997;272(33):20313–6.
67. Martins D, English AM. SOD1 oxidation and formation of soluble aggregates in yeast: relevance to sporadic ALS development. Redox Biol. 2014;2:632–9.
68. Guareschi S, Cova E, Cereda C, Ceroni M, Donetti E, Bosco DA, et al. An over-oxidized form of superoxide dismutase found in sporadic amyotrophic lateral sclerosis with bulbar onset shares a toxic mechanism with mutant SOD1. Proc Natl Acad Sci U S A. 2012;109(13):5074–9.
69. Fink RC, Scandalios JG. Molecular evolution and structure--function relationships of the superoxide dismutase gene families in angiosperms and their relationship to other eukaryotic and prokaryotic superoxide dismutases. Arch Biochem Biophys. 2002;399(1):19–36.
70. Urushitani M, Ezzi SA, Julien JP. Therapeutic effects of immunization with mutant superoxide dismutase in mice models of amyotrophic lateral sclerosis. Proc Natl Acad Sci U S A. 2007;104(7):2495–500.
71. Brotherton TE, Li Y, Cooper D, Gearing M, Julien JP, Rothstein JD, et al. Localization of a toxic form of superoxide dismutase 1 protein to pathologically affected tissues in familial ALS. Proc Natl Acad Sci U S A. 2012;109(14):5505–10.
72. Bosco DA, Morfini G, Karabacak NM, Song Y, Gros-Louis F, Pasinelli P, et al. Wild-type and mutant SOD1 share an aberrant conformation and a common pathogenic pathway in ALS. Nat Neurosci. 2010;13(11):1396–403.
73. Ayers JI, Xu G, Pletnikova O, Troncoso JC, Hart PJ, Borchelt DR. Conformational specificity of the C4F6 SOD1 antibody; low frequency of reactivity in sporadic ALS cases. Acta Neuropathol Commun. 2014;2:55.
74. Rotunno MS, Auclair JR, Maniatis S, Shaffer SA, Agar J, Bosco DA. Identification of a misfolded region in superoxide dismutase 1 that is exposed in amyotrophic lateral sclerosis. J Biol Chem. 2014;289(41):28527–38.
75. Molnar KS, Karabacak NM, Johnson JL, Wang Q, Tiwari A, Hayward LJ, et al. A common property of amyotrophic lateral sclerosis-associated variants: destabilization of the copper/zinc superoxide dismutase electrostatic loop. J Biol Chem. 2009;284(45):30965–73.

76. Fukada K, Nagano S, Satoh M, Tohyama C, Nakanishi T, Shimizu A, et al. Stabilization of mutant Cu/Zn superoxide dismutase (SOD1) protein by coexpressed wild SOD1 protein accelerates the disease progression in familial amyotrophic lateral sclerosis mice. Eur J Neurosci. 2001;14(12):2032–6.
77. Wang L, Deng HX, Grisotti G, Zhai H, Siddique T, Roos RP. Wild-type SOD1 overexpression accelerates disease onset of a G85R SOD1 mouse. Hum Mol Genet. 2009;18(9):1642–51.
78. Witan H, Kern A, Koziollek-Drechsler I, Wade R, Behl C, Clement AM. Heterodimer formation of wild-type and amyotrophic lateral sclerosis-causing mutant Cu/Zn-superoxide dismutase induces toxicity independent of protein aggregation. Hum Mol Genet. 2008;17(10):1373–85.
79. Weichert A, Besemer AS, Liebl M, Hellmann N, Koziollek-Drechsler I, Ip P, et al. Wild-type Cu/Zn superoxide dismutase stabilizes mutant variants by heterodimerization. Neurobiol Dis. 2014;62:479–88.
80. Raoul C, Estevez AG, Nishimune H, Cleveland DW. deLapeyriere O, Henderson CE, et al. Motoneuron death triggered by a specific pathway downstream of Fas. potentiation by ALS-linked SOD1 mutations. Neuron. 2002;35(6):1067–83.
81. Tortarolo M, Veglianese P, Calvaresi N, Botturi A, Rossi C, Giorgini A, et al. Persistent activation of p38 mitogen-activated protein kinase in a mouse model of familial amyotrophic lateral sclerosis correlates with disease progression. Mol Cell Neurosci. 2003;23(2):180–92.
82. Hu JH, Chernoff K, Pelech S, Krieger C. Protein kinase and protein phosphatase expression in the central nervous system of G93A mSOD over-expressing mice. J Neurochem. 2003;85(2):422–31.
83. Wengenack TM, Holasek SS, Montano CM, Gregor D, Curran GL, Poduslo JF. Activation of programmed cell death markers in ventral horn motor neurons during early presymptomatic stages of amyotrophic lateral sclerosis in a transgenic mouse model. Brain Res. 2004;1027(1-2):73–86.
84. Raoul C, Buhler E, Sadeghi C, Jacquier A, Aebischer P, Pettmann B, et al. Chronic activation in presymptomatic amyotrophic lateral sclerosis (ALS) mice of a feedback loop involving Fas, Daxx, and FasL. Proc Natl Acad Sci U S A. 2006;103(15):6007–12.
85. Franco MC, Ye Y, Refakis CA, Feldman JL, Stokes AL, Basso M, et al. Nitration of Hsp90 induces cell death. Proc Natl Acad Sci U S A. 2013;110(12):E1102–11.
86. Petri S, Kiaei M, Wille E, Calingasan NY, Flint BM. Loss of Fas ligand-function improves survival in G93A-transgenic ALS mice. J Neurol Sci. 2006;251(1-2):44–9.
87. Niforou K, Cheimonidou C, Trougakos IP. Molecular chaperones and proteostasis regulation during redox imbalance. Redox Biol. 2014;2:323–32.
88. Okado-Matsumoto A, Fridovich I. Amyotrophic lateral sclerosis: a proposed mechanism. Proc Natl Acad Sci U S A. 2002;99(13):9010–4.
89. Kieran D, Kalmar B, Dick JR, Riddoch-Contreras J, Burnstock G, Greensmith L. Treatment with arimoclomol, a coinducer of heat shock proteins, delays disease progression in ALS mice. Nat Med. 2004;10(4):402–5.
90. Kalmar B, Lu CH, Greensmith L. The role of heat shock proteins in amyotrophic lateral sclerosis: the therapeutic potential of Arimoclomol. Pharmacol Ther. 2014;141(1):40–54.
91. Auclair JR, Johnson JL, Liu Q, Salisbury JP, Rotunno MS, Petsko GA, et al. Post-translational modification by cysteine protects Cu/Zn-superoxide dismutase from oxidative damage. Biochemistry. 2013;52(36):6137–44.
92. Johnson JM, Strobel FH, Reed M, Pohl J, Jones DP. A rapid LC-FTMS method for the analysis of cysteine, cystine and cysteine/cystine steady-state redox potential in human plasma. Clin Chim Acta. 2008;396(1-2):43–8.
93. Dalle-Donne I, Rossi R, Giustarini D, Colombo R, Milzani A. S-glutathionylation in protein redox regulation. Free Radic Biol Med. 2007;43(6):883–98.
94. Barber SC, Mead RJ, Shaw PJ. Oxidative stress in ALS: a mechanism of neurodegeneration and a therapeutic target. Biochim Biophys Acta. 2006;1762(11-12):1051–67.

95. Tohgi H, Abe T, Yamazaki K, Murata T, Ishizaki E, Isobe C. Increase in oxidized NO products and reduction in oxidized glutathione in cerebrospinal fluid from patients with sporadic form of amyotrophic lateral sclerosis. Neurosci Lett. 1999;260(3):204–6.
96. Townsend DM. S-glutathionylation: indicator of cell stress and regulator of the unfolded protein response. Mol Interv. 2007;7(6):313–24.
97. Wilcox KC, Zhou L, Jordon JK, Huang Y, Yu Y, Redler RL, et al. Modifications of superoxide dismutase (SOD1) in human erythrocytes: a possible role in amyotrophic lateral sclerosis. J Biol Chem. 2009;284(20):13940–7.
98. Redler RL, Wilcox KC, Proctor EA, Fee L, Caplow M, Dokholyan NV. Glutathionylation at Cys-111 induces dissociation of wild type and FALS mutant SOD1 dimers. Biochemistry. 2011;50(32):7057–66.
99. McAlary L, Yerbury JJ, Aquilina JA. Glutathionylation potentiates benign superoxide dismutase 1 variants to the toxic forms associated with amyotrophic lateral sclerosis. Sci Rep. 2013;3:3275.
100. Redler RL, Fee L, Fay JM, Caplow M, Dokholyan NV. Non-native soluble oligomers of Cu/Zn superoxide dismutase (SOD1) contain a conformational epitope linked to cytotoxicity in amyotrophic lateral sclerosis (ALS). Biochemistry. 2014;53(14):2423–32.
101. de Beus MD, Chung J, Colon W. Modification of cysteine 111 in Cu/Zn superoxide dismutase results in altered spectroscopic and biophysical properties. Protein Sci. 2004;13(5):1347–55.
102. Okado-Matsumoto A, Guan Z, Fridovich I. Modification of cysteine 111 in human Cu,Zn-superoxide dismutase. Free Radic Biol Med. 2006;41(12):1837–46.
103. Calabrese L, Federici G, Bannister WH, Bannister JV, Rotilio G, Finazzi-Agro A. Labile sulfur in human Superoxide dismutase. Eur J Biochem. 1975;56(1):305–9.
104. Li L, Rose P, Moore PK. Hydrogen sulfide and cell signaling. Annu Rev Pharmacol Toxicol. 2011;51:169–87.
105. Zhao J, Qian Y, Zhang L, Ding S, Deng X, He C, et al. A fluorescent probe for rapid detection of hydrogen sulfide in blood plasma and brain tissues in mice. Chem Sci. 2012;3(10):2920–3.
106. Auclair JR, Boggio KJ, Petsko GA, Ringe D, Agar JN. Strategies for stabilizing superoxide dismutase (SOD1), the protein destabilized in the most common form of familial amyotrophic lateral sclerosis. Proc Natl Acad Sci U S A. 2010;107(50):21394–9.
107. Oono M, Okado-Matsumoto A, Shodai A, Ido A, Ohta Y, Abe K, et al. Transglutaminase 2 accelerates neuroinflammation in amyotrophic lateral sclerosis through interaction with misfolded superoxide dismutase 1. J Neurochem. 2014;128(3):403–18.
108. Ruan Q, Johnson GV. Transglutaminase 2 in neurodegenerative disorders. Front Biosci. 2007;12:891–904.
109. Andringa G, Lam KY, Chegary M, Wang X, Chase TN, Bennett MC. Tissue transglutaminase catalyzes the formation of alpha-synuclein crosslinks in Parkinson's disease. FASEB J. 2004;18(7):932–4.
110. Ho GJ, Gregory EJ, Smirnova IV, Zoubine MN, Festoff BW. Cross-linking of beta-amyloid protein precursor catalyzed by tissue transglutaminase. FEBS Lett. 1994;349(1):151–4.
111. Karpuj MV, Garren H, Slunt H, Price DL, Gusella J, Becher MW, et al. Transglutaminase aggregates huntingtin into nonamyloidogenic polymers, and its enzymatic activity increases in Huntington's disease brain nuclei. Proc Natl Acad Sci U S A. 1999;96(13):7388–93.
112. Lesort M, Chun W, Johnson GV, Ferrante RJ. Tissue transglutaminase is increased in Huntington's disease brain. J Neurochem. 1999;73(5):2018–27.
113. Karpuj MV, Becher MW, Steinman L. Evidence for a role for transglutaminase in Huntington's disease and the potential therapeutic implications. Neurochem Int. 2002;40(1):31–6.
114. Fujita K, Honda M, Hayashi R, Ogawa K, Ando M, Yamauchi M, et al. Transglutaminase activity in serum and cerebrospinal fluid in sporadic amyotrophic lateral sclerosis: a possible use as an indicator of extent of the motor neuron loss. J Neurol Sci. 1998;158(1):53–7.
115. Lesort M, Lee M, Tucholski J, Johnson GV. Cystamine inhibits caspase activity. Implications for the treatment of polyglutamine disorders. J Biol Chem. 2003;278(6):3825–30.

116. Yoshino H, Kimura A. Investigation of the therapeutic effects of edaravone, a free radical scavenger, on amyotrophic lateral sclerosis (Phase II study). Amyotroph Lateral Scler. 2006;7(4):241–5.
117. Batinic-Haberle I, Rajic Z, Tovmasyan A, Reboucas JS, Ye X, Leong KW, et al. Diverse functions of cationic Mn(III) N-substituted pyridylporphyrins, recognized as SOD mimics. Free Radic Biol Med. 2011;51(5):1035–53.
118. Batinic-Haberle I, Tovmasyan A, Roberts ER, Vujaskovic Z, Leong KW, Spasojevic I. SOD therapeutics: latest insights into their structure-activity relationships and impact on the cellular redox-based signaling pathways. Antioxid Redox Signal. 2014;20(15):2372–415.
119. Crow JP, Calingasan NY, Chen J, Hill JL, Beal MF. Manganese porphyrin given at symptom onset markedly extends survival of ALS mice. Ann Neurol. 2005;58(2):258–65.
120. Crow JP. Catalytic antioxidants to treat amyotropic lateral sclerosis. Expert Opin Investig Drugs. 2006;15(11):1383–93.
121. Petri S, Kiaei M, Kipiani K, Chen J, Calingasan NY, Crow JP, et al. Additive neuroprotective effects of a histone deacetylase inhibitor and a catalytic antioxidant in a transgenic mouse model of amyotrophic lateral sclerosis. Neurobiol Dis. 2006;22(1):40–9.
122. Benatar M. Lost in translation: treatment trials in the SOD1 mouse and in human ALS. Neurobiol Dis. 2007;26(1):1–13.
123. Orrell RW. AEOL-10150 (Aeolus). Curr Opin Investig Drugs. 2006;7(1):70–80.
124. Jaramillo MC, Frye JB, Crapo JD, Briehl MM, Tome ME. Increased manganese superoxide dismutase expression or treatment with manganese porphyrin potentiates dexamethasone-induced apoptosis in lymphoma cells. Cancer Res. 2009;69(13):5450–7.
125. Jaramillo MC, Briehl MM, Crapo JD, Batinic-Haberle I, Tome ME. Manganese porphyrin, MnTE-2-PyP5+, Acts as a pro-oxidant to potentiate glucocorticoid-induced apoptosis in lymphoma cells. Free Radic Biol Med. 2012;52(8):1272–84.
126. Frakes AE, Ferraiuolo L, Haidet-Phillips AM, Schmelzer L, Braun L, Miranda CJ, et al. Microglia induce motor neuron death via the classical NF-kappaB pathway in amyotrophic lateral sclerosis. Neuron. 2014;81(5):1009–23.
127. Tokuda E, Ono S, Ishige K, Watanabe S, Okawa E, Ito Y, et al. Ammonium tetrathiomolybdate delays onset, prolongs survival, and slows progression of disease in a mouse model for amyotrophic lateral sclerosis. Exp Neurol. 2008;213(1):122–8.
128. Nagano S, Fujii Y, Yamamoto T, Taniyama M, Fukada K, Yanagihara T, et al. The efficacy of trientine or ascorbate alone compared to that of the combined treatment with these two agents in familial amyotrophic lateral sclerosis model mice. Exp Neurol. 2003;179(2):176–80.
129. Hottinger AF, Fine EG, Gurney ME, Zurn AD, Aebischer P. The copper chelator d-penicillamine delays onset of disease and extends survival in a transgenic mouse model of familial amyotrophic lateral sclerosis. Eur J Neurosci. 1997;9(7):1548–51.
130. Rakhit R, Robertson J, Vande Velde C, Horne P, Ruth DM, Griffin J, et al. An immunological epitope selective for pathological monomer-misfolded SOD1 in ALS. Nat Med. 2007;13(6):754–9.
131. Kerman A, Liu HN, Croul S, Bilbao J, Rogaeva E, Zinman L, et al. Amyotrophic lateral sclerosis is a non-amyloid disease in which extensive misfolding of SOD1 is unique to the familial form. Acta Neuropathol. 2010;119(3):335–44.
132. Prudencio M, Borchelt DR. Superoxide dismutase 1 encoding mutations linked to ALS adopts a spectrum of misfolded states. Mol Neurodegener. 2011;6:77.
133. Gros-Louis F, Soucy G, Lariviere R, Julien JP. Intracerebroventricular infusion of monoclonal antibody or its derived Fab fragment against misfolded forms of SOD1 mutant delays mortality in a mouse model of ALS. J Neurochem. 2010;113(5):1188–99.
134. Patel P, Kriz J, Gravel M, Soucy G, Bareil C, Gravel C, et al. Adeno-associated virus-mediated delivery of a recombinant single-chain antibody against misfolded superoxide dismutase for treatment of amyotrophic lateral sclerosis. Mol Ther. 2014;22(3):498–510.
135. Liu HN, Tjostheim S, Dasilva K, Taylor D, Zhao B, Rakhit R, et al. Targeting of monomer/misfolded SOD1 as a therapeutic strategy for amyotrophic lateral sclerosis. J Neurosci. 2012;32(26):8791–9.

136. Gilman S, Koller M, Black RS, Jenkins L, Griffith SG, Fox NC, et al. Clinical effects of Abeta immunization (AN1792) in patients with AD in an interrupted trial. Neurology. 2005;64(9):1553–62.
137. Al-Chalabi A, Jones A, Troakes C, King A, Al-Sarraj S, van den Berg LH. The genetics and neuropathology of amyotrophic lateral sclerosis. Acta Neuropathol. 2012;124(3):339–52.
138. Iguchi Y, Katsuno M, Ikenaka K, Ishigaki S, Sobue G. Amyotrophic lateral sclerosis: an update on recent genetic insights. J Neurol. 2013;260(11):2917–27.
139. Sreedharan J, Brown Jr RH. Amyotrophic lateral sclerosis: Problems and prospects. Ann Neurol. 2013;74(3):309–16.
140. Leblond CS, Kaneb HM, Dion PA, Rouleau GA. Dissection of genetic factors associated with amyotrophic lateral sclerosis. Exp Neurol. 2014;262 Pt B:91–101.
141. Ince PG, Highley JR, Kirby J, Wharton SB, Takahashi H, Strong MJ, et al. Molecular pathology and genetic advances in amyotrophic lateral sclerosis: an emerging molecular pathway and the significance of glial pathology. Acta Neuropathol. 2011;122(6):657–71.
142. Budini M, Baralle FE, Buratti E. Targeting TDP-43 in neurodegenerative diseases. Expert Opin Ther Targets. 2014;18(6):617–32.
143. Zhang KY, Yang S, Warraich ST, Blair IP. Ubiquilin 2: a component of the ubiquitin-proteasome system with an emerging role in neurodegeneration. Int J Biochem Cell Biol. 2014;50:123–6.
144. Scotter EL, Vance C, Nishimura AL, Lee YB, Chen HJ, Urwin H, et al. Differential roles of the ubiquitin proteasome system and autophagy in the clearance of soluble and aggregated TDP-43 species. J Cell Sci. 2014;127(Pt 6):1263–78.
145. Scotter EL, Chen HJ, Shaw CE. TDP-43 proteinopathy and als: insights into disease mechanisms and therapeutic targets. Neurotherapeutics. 2015;12(2):352–63.
146. Beck J, Poulter M, Hensman D, Rohrer JD, Mahoney CJ, Adamson G, et al. Large C9orf72 hexanucleotide repeat expansions are seen in multiple neurodegenerative syndromes and are more frequent than expected in the UK population. Am J Hum Genet. 2013;92(3):345–53.
147. Hukema RK, Riemslagh FW, Melhem S, van der Linde HC, Severijnen L, Edbauer D, et al. A new inducible transgenic mouse model for C9orf72-associated GGGGCC repeat expansion supports a gain-of-function mechanism in C9orf72 associated ALS and FTD. Acta Neuropathol Commun. 2014;2(1):166.
148. Liu EY, Russ J, Wu K, Neal D, Suh E, McNally AG, et al. C9orf72 hypermethylation protects against repeat expansion-associated pathology in ALS/FTD. Acta Neuropathol. 2014;128(4):525–41.
149. McMillan CT, Russ J, Wood EM, Irwin DJ, Grossman M, McCluskey L, et al. C9orf72 promoter hypermethylation is neuroprotective: Neuroimaging and neuropathologic evidence. Neurology. 2015.
150. Jan A, Adolfsson O, Allaman I, Buccarello AL, Magistretti PJ, Pfeifer A, et al. Abeta42 neurotoxicity is mediated by ongoing nucleated polymerization process rather than by discrete Abeta42 species. J Biol Chem. 2011;286(10):8585–96.
151. Ryan TM, Roberts BR, McColl G, Hare DJ, Doble PA, Li QX, et al. Stabilization of nontoxic Abeta-oligomers: insights into the mechanism of action of hydroxyquinolines in Alzheimer's disease. J Neurosci. 2015;35(7):2871–84.
152. Vashist S, Cushman M, Shorter J. Applying Hsp104 to protein-misfolding disorders. Biochem Cell Biol. 2010;88(1):1–13.
153. Jackrel ME, Shorter J. Engineering enhanced protein disaggregases for neurodegenerative disease. Prion. 2015;9(2):90–109.
154. Chiti F, Dobson CM. Amyloid formation by globular proteins under native conditions. Nat Chem Biol. 2009;5(1):15–22.
155. Kim J, Kim TY, Cho KS, Kim HN, Koh JY. Autophagy activation and neuroprotection by progesterone in the G93A-SOD1 transgenic mouse model of amyotrophic lateral sclerosis. Neurobiol Dis. 2013;59:80–5.
156. Morimoto N, Nagai M, Ohta Y, Miyazaki K, Kurata T, Morimoto M, et al. Increased autophagy in transgenic mice with a G93A mutant SOD1 gene. Brain Res. 2007;1167:112–7.

Chapter 28
Redox-Based Therapeutics for Prevention, Mitigation, and Treatment of Lung Injury Secondary to Radiation Exposure

Isabel L. Jackson and Zeljko Vujaskovic

28.1 Introduction

Ionizing radiation plays an important role in medical diagnostics and cancer treatment. Approximately, two-thirds of all cancer patients will receive radiation therapy as part of treatment regimens to control local disease, prevent recurrence or metastasis, and improve the odds for surgical resection. The accompanying risk for development of normal tissue injury, however, greatly limits the maximum effective radiation dose that can be delivered to achieve local tumor control and improve treatment outcomes. Normal tissue complications for which patients are at risk range from acute toxicities, leading to hospitalizations, treatment interruptions, and poor clinical outcomes, to chronic injury resulting in reduced quality of life and, in extreme cases, organ failure.

In addition to treatment-related complications, exposure to supratherapeutic doses of ionizing radiation may occur as a result of clinical overdose, during use of industrial sources or transport of radioactive sources, in nuclear power plant accidents, or in events involving large-scale radioactivity-producing weapons or radiation dispersal devices ("dirty bombs").

At this time, no U.S. Food and Drug Administration (FDA)-approved treatments are available to reduce the probability for pulmonary complications following radiation therapy. The lack of postexposure treatment options also presents a major impediment to survival following exposure to supratherapeutic doses such as those that might occur in a nuclear accident or attack. In part, the development of targeted

I.L. Jackson (✉) • Z. Vujaskovic
Division of Translational Radiation Sciences, Department of Radiation Oncology, University of Maryland School of Medicine, Baltimore, MD 21202, USA
e-mail: ijackson@som.umaryland.edu; zvujaskovic@som.umaryland.edu

I. Batinić-Haberle et al. (eds.), *Redox-Active Therapeutics*,
Oxidative Stress in Applied Basic Research and Clinical Practice,
DOI 10.1007/978-3-319-30705-3_28

therapeutic approaches to prevent, mitigate, and/or treat the side effects of radiation exposure has been hindered by the complexity surrounding the pathogenesis of radiation-induced normal tissue injury [111].

It is well-known that the processes leading to functional recovery or failure of irradiated tissues are multifactorial and, as such, a single therapeutic target is unlikely to be sufficient to dampen the proapoptotic, proinflammatory, and profibrotic processes that underlie the development of injury. However, at the heart of each of these processes lies chronic oxidative stress, a disruption in the balance between pro- and antioxidants, which hijacks the cellular machinery responsible for regulatory control over tissue response to insults and leads to destruction of the physiologic processes that orchestrate wound healing (Fig. 28.1). A number of therapeutic strategies to prevent, mitigate, and/or treat

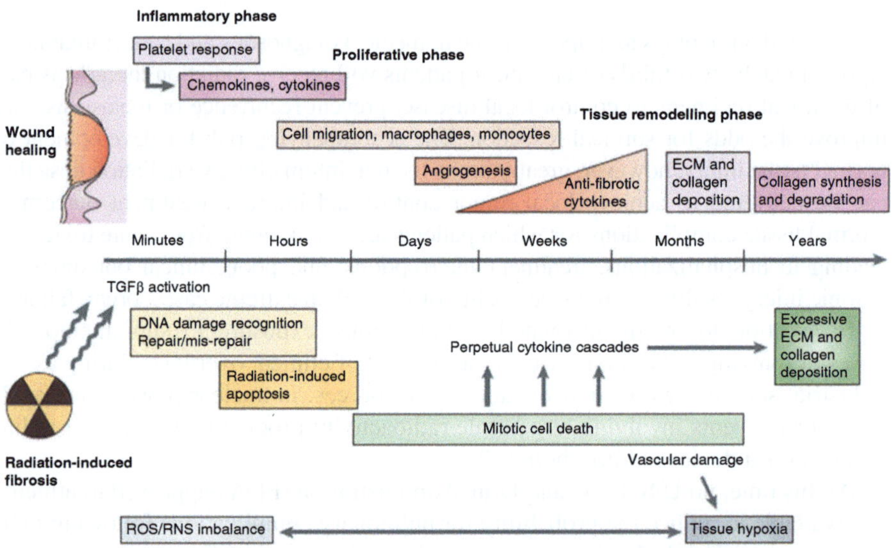

Fig. 28.1 Radiation fibrosis: a wound that does not heal. The pathogenesis of radiation fibrosis begins at the time of insult and progressing through a so-called "latent" period that may last months to years before manifesting clinically. Unlike the normal wound healing response which is tightly controlled, radiation fibrosis results from excessive tissue remodeling that occurs over a prolonged period of time. The mechanisms that drive the tissue response to radiation damage not towards resolution but rather acute pathogenesis are complex and not fully understood. Following the initial ionizing event, secondary production of reactive species alters the environmental milieu leading to vascular leakage and edema, inflammatory cell recruitment, tissue hypoxia, and excessive collagen deposition. Therapeutic targeting of chronic oxidative stress is an attractive approach to mitigate the pathogenic biochemical, molecular, and cellular events that lead to radiation fibrosis. *TGFβ* transforming growth factor beta, *ECM* extracellular matrix, *ROS/RNS* reactive oxygen/nitrogen species. *Reproduced from Bentzen SM. Preventing or reducing late side effects of radiation therapy: radiobiology meets molecular pathology. Nature Reviews. 2006 Sep; 6(9):702-713 with permission from Nature Publishing Group[1]*

radiation-induced normal tissue injury have focused on neutralizing chronic oxidative stress and restoring homeostasis to mitigate the injurious process and promote wound healing.

28.2 Clinical Relevance and Pathogenesis of Radiation-Induced Lung Injury

The lung is one of the most radiosensitive organs. In patients undergoing thoracic radiotherapy, pulmonary complications can severely hinder the ability to achieve local tumor control. Functional deficits resulting from radiation exposure can have significant effects on long-term quality of life among survivors. Apart from clinical radiotherapy, radiation lung damage plays a major contributory role in multiorgan dysfunction syndrome and organ failure following partial- or total-body irradiation after successful treatment of acute radiation syndromes (hematopoietic and gastrointestinal) [2, 3]. Radiation lung damage is generally characterized as acute pneumonitis occurring 1–6 months after exposure, with a peak incidence between 3 and 4 months, or as chronic fibrosis that may occur months to years later.

Clinical manifestations of lung damage include cough, dyspnea, exhaustion on exertion, impaired gas exchange, and, in cases in which the entire thorax has been exposed, possible organ failure. Molecular features of lung damage include vascular hyperpermeability and edema, type II pneumocyte hyperplasia, mononuclear inflammatory cell infiltrates, increased alveolar wall thickness, and fibrous plaques [4]. Although clinical symptoms may not manifest until several weeks to months after exposure, radiographic injury may be seen much earlier because of the ongoing biochemical, molecular, and cellular changes that occur at the subclinical level beginning immediately upon exposure [4]. Medical management involves corticosteroid treatment for pneumonitis at the time of symptoms and, on occasion, antibiotics and oxygen supplementation [5]. No FDA-approved strategy is currently available to prevent, mitigate, or treat radiation pneumonitis/fibrosis.

Radiation pneumonitis/fibrosis has been studied experimentally in rodent [6–9], canine [10], and nonhuman primate [11] models, as well as other species the mechanisms underlying pulmonary radiobiology have been primarily characterized in the C57BL/6J (B6) mouse strain as a result of its designation as “fibrosis prone” in the late 1980s and advances in genetic manipulation in the 1990s. However, recent studies have shown significant differences in the pulmonary radiotolerance of B6 mice and that of other strains, such as the C57L/J, with a right-shift in the dose–response curve for the B6 strain when compared with that of the C57L/J strain, rhesus macaques, and humans [12]. In addition to similarities in dose–response, the C57L/J strain appears to be more analogous to the human pulmonary response to radiation in temporal onset and pathogenesis [13–16]. As a result, the relevance of the B6 strain in enhancing understanding of mechanisms associated with radiation pneumonitis and, to a lesser extent, fibrosis has been questioned. Studies are ongoing to evaluate the predictive value of the various strains, but the CBA/J and C57L/J strains currently appear to hold the most promise.

28.3 Role of Redox Biology in Radiation Pneumonitis/ Fibrosis

Although the initial ionizing event is short-lived, lasting only a few milliseconds, the biological consequences of radiation exposure may take days, weeks, or even years to manifest clinically. At the time of exposure, energy is deposited along the radiation track, disrupting the constitutive structure of the molecular framework and leading to both isolated and clustered DNA lesions [17, 18].

Isolated DNA lesions are efficiently repaired through the base excision repair pathway or nonhomologous end joining. However, it is well-documented that clustered DNA damage sites and complex double-strand breaks (DSB) present greater challenges to DNA repair machinery, resulting in less reparability, genetic instability, and lethality through apoptosis or mitotic catastrophe [17]. The number of clustered DNA lesions and complex DSBs increases with increasing ionization density. As a result, the level of complexity of clustered DNA lesions, a hallmark of ionizing radiation, plays an important role in determining the biologic consequences of exposure [17].

At the time of exposure, irradiated cells, sensing the danger posed, set off myriad alarm signals that activate resting antigen-presenting cells to initiate an immune response [19, 20]. The majority of these alarm signals are thought to be free radical intermediates, but are also likely to include cytokines, growth factors, and even free DNA, which activate a complex biological response to orchestrate cellular repair and regenerate damaged tissue [21]. However, the number of alarm signals activated multiplies in a dose-dependent manner (corresponding to increasing severity of injury), recruiting more and more responders of constitutive or inflammatory cell origin to the site of injury to participate in repair processes. If not resolved, the propagation and ultimate exhaustion of repair and recovery mechanisms leads to bioenergetic collapse of the tissue and fulminant organ failure [22].

The complex biological processes involved in radiation-induced normal tissue injury are propagated at least in part by oxidative stress secondary to the initial ionizing event [17, 18]. Redox-sensitive biomolecules play a role in both physiological and pathological conditions through their ability to mobilize energy resources, act as intracellular or extracellular messengers, and control gene expression [23]. In irradiated tissues, cellular macromolecules are modified directly by ionization or indirectly by the radiolytic products of water (i.e., hydroxyl radical, $^{\bullet}OH$; electron, e^-; superoxide, $O_2^{\bullet-}$; hydrogen peroxide, H_2O_2) [24]. Prompt damage to critical cellular biomolecules is amplified through secondary production of intracellular reactive species, RS (reactive oxygen ROS, reactive nitrogen RNS, reactive sulfur species RSS, etc.) via chemical cascades (e.g., $O_2^{\bullet-}$; H_2O_2; and peroxynitrite, $ONOO^-$, formation), mitochondrial leakage, and activation of enzymatic sources (e.g., NADPH oxidases, nitric oxide synthases) (Fig. 28.2) [24]. Oxidative/nitroxidative modifications of DNA, lipids, and proteins set off a cascade of downstream phenomena, including activation of transcription factors, growth factor signaling, inflammatory cell recruitment, cytokine production, mitochondrial dysfunction, and

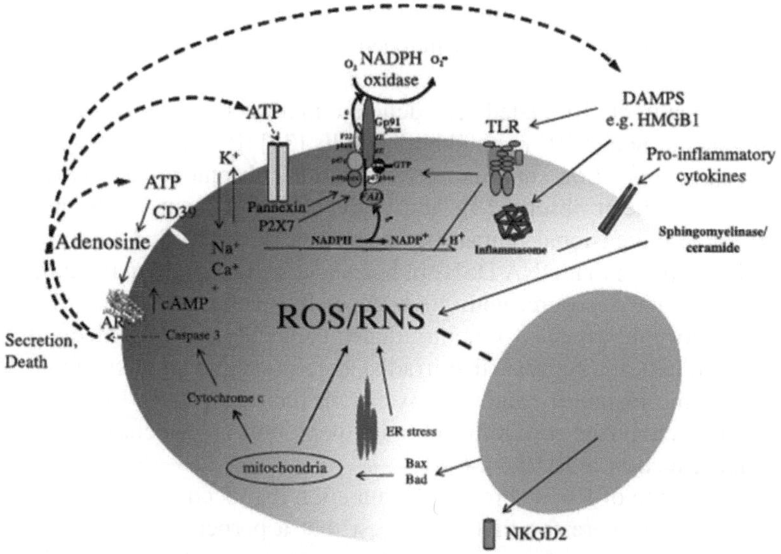

Fig. 28.2 Sources of reactive oxygen/nitrogen species after irradiation. The pro-inflammatory, pro-oxidant microenvironment is a key driver of radiation-induced lung injury. *ATP* adenosine 5′ triphosphate, *CD39* cluster of differentiation 39, *DAMPs* damage associated molecular pattern molecules, *ER* endoplasmic reticulum, *HMGB1* high mobility group protein B1, *NADPH* nicotinamide adenine dinucleotide phosphate, *NKGD2* natural killer group 2 MemberD, *ROS/RNS* reactive oxygen/nitrogen species, *TLR* toll-like receptor. *Reproduced from Schaue D, Kachikwu EL, and McBride W. Cytokines in Radiobiological Responses: A Review. Radiation Research 178(6):505-23 with permission from Radiation Research [25]*

vascular changes, which underlie the pathogenesis of radiation damage. These processes lead to an environment of chronic oxidative stress that overwhelms endogenous antioxidant defense mechanisms and disrupts the physiological processes that normally contribute to wound healing [26].

One of the major sources of cellular ROS is mitochondrial respiration leading to adenosine triphosphate production, ion channel activation, or purinergic signaling [18]. Mitochondria-derived ROS regulate cellular homeostasis and, in times of oxidative stress or low physiologic oxygen concentrations, may shift metabolic pathways away from oxidative phosphorylation toward glycolytic metabolism and fatty acid oxidation to provide cells with the energy needed to undergo repair and regeneration. Radiation induces mitochondrial dysfunction in vitro, including alterations in mitochondrial membrane potential, complexes I–IV activities, and O_2 consumption, which lead to genomic instability [27]. The damaging effects of radiation on mitochondria are not unexpected, given the lack of introns and histone proteins to protect mitochondrial DNA and its relatively inefficient DNA repair mechanisms [28]. When the antioxidant defense system is overwhelmed, the regenerative cycle of mitochondrial ROS formation and release may result in potentially cytotoxic levels of ROS, leading to further deterioration of the functional cell population and organ failure [29].

Mitochondrial ROS are postulated to be "sensors" of radiation damage because of their prominent role in radiation-induced bystander effect (RIBE) signaling and response [30]. Bystander effects are important in propagating cell damage outside of the radiation track. Irradiated cells deficient in mitochondrial DNA are less efficient at generating a RIBE in neighboring cells [31]. Bystander effects, including DNA damage induction, micronuclei formation, chromosomal instability, and apoptosis, have been observed in non-irradiated neighboring cells up to 1 mm away in three-dimensional human tissue models [32, 33]. Sedelnikova and colleagues have shown that DNA DSBs in bystander cells reach a maximum between 12 and 48 h after exposure, with 40–60 % of bystander cells affected [32]. The reduced clonogenic capacity of nonirradiated cells cultured with media from [34] Co irradiated cultures suggests that irradiated cells secrete an unknown factor that causes genomic instability and cytotoxicity in their neighbors [35]. In addition, impaired mitochondrial respiration through depletion of cytochrome c leads to a significant reduction in RIBEs [36].

Another source of RS is chronic inflammation that occurs after radiation exposure. Rubin et al. were the first to suggest that a perpetual cytokine cascade is activated within hours of exposure, perpetuating tissue damage and leading to delayed effects (e.g., fibrosis) in late-responding tissues [37]. Schaue et al. provided an excellent overview of the role of cytokines in radiobiologic responses in a review published in *Radiation Research* [38].

Apart from DNA damage response, the cellular response to radiation is mostly proinflammatory and includes activation of genes, such as c-fos, c-myc, and c-jun, and beta-actin, TNF-α (tumor necrosis factor α), COX2 (cyclooxygenase 2), and ICAM1 (intracellular adhesion molecule 1) [25]. Inflammatory mediators play a role in endothelial cell activation and vascular leakage. As inflammatory cells are recruited to areas of tissue injury, they participate in further amplification of oxidative stress that supersedes the ability of the antioxidant machinery to neutralize damaging ROS/RNS [39].

28.4 Redox-Based Therapeutics to Prevent, Mitigate, and/or Treat Radiation-Induced Lung Injury

As redox signaling and oxidative stress have emerged as key players in the development of radiation-induced normal tissue injury, academic and industry scientists have focused on designing broad-spectrum therapeutic interventions that neutralize reactive species, upregulate antioxidant defense, or both (Table 28.1).

The ability of redox-based therapeutics to protect cells against radiation damage and improve the likelihood of cell survival was first shown by Patt et al. in 1949 [61]. Since then, a number of clinical trials have shown that redox-based therapeutics are capable of reversing radiation-induced fibrosis involving the skin and underlying areas [62–65] as well as sternocostal and mandibular osteoradionecrosis [66, 67]. Such therapeutics have led to symptom improvement in patients with proctitis/enteritis [68] and genitourinary dysfunction [69, 70].

Table 28.1 Redox-based therapeutics to mitigate radiation-induced lung injury

Product/Company	References	Results
Amifostine (Ethyol, WR2721)/ MedImmune, Inc.	Antonadou et al. (2001) [40]	Significantly reduced ≥ grade 2 acute and late lung toxicity in patients receiving RT ($n=146$)
	Antonadou et al. (2003) [41]	Significantly reduced ≥ acute and late lung toxicity in NSCLC patients receiving RCT ($n=73$)
	Komaki et al. (2004) [42]	Reduced incidence of RP in patients receiving RCT
	Vujaskovic et al. (2002) [43, [44]	Reduced functional and histologic damage in irradiated lungs of rats; reduced plasma levels of TGFβ1 and expression/activation of TGFβ1 in tissue
Pentoxyifylline (Ptx) ± tocopherol (vitamin E)	Ozturk et al. (2004) [45]	Reduced severity of radiographic changes; improved CO diffusion; and regional perfusion in patients with lung or breast cancer receiving RT ($n=40$)
	Kaya et al. (2014) [46]	Vitamin E + Ptx prevented radiation fibrosis in rat lungs after thoracic RT
AEOL 10150/	Rabbani et al. (2007) [47, 48]	AEOL 10150 significantly mitigated respiratory dysfunction and reduced structural damage and collagen deposition in lungs of rats after single-dose or fractionated radiation
Aeolus Pharmaceuticals, Inc.	Garofalo et al. (2014) [49]	Improved survival of rhesus macaques by 28.6 % after lethal WTLI; reduced quantitative radiation injury and dexamethasone support in treated animals
SDG/LignaMed	Christofidou-Solomidou et al. (2011) [50]	Dietary flaxseed improved survival and mitigated pneumonitis in murine lungs after WTLI
	Christofidou-Solomidou et al. (2012) [51]	Attenuated lung inflammation and improved survival in mice after WTLI

(continued)

Table 28.1 (continued)

Product/Company	References	Results
Genistein	Day et al. (2008) [52]	Genistein prevented development of collagen-rich plaques in the lungs of C57BL/6J mice after TBI
	Calveley et al. (2010) [53]	Genistein delayed but did not prevent lethality from pneumonitis resulting from thoracic irradiation; decreased proinflammatory and profibrotic cytokines; and reduced fibrosis in rat lungs
	Mahmood et al. (2011) [54]	Genistein + Euk-207 prevented respiratory rate increases and decreased hydroxyproline content in lungs after thoracic RT in rats; decreased micronuclei formation acutely
	Mahmood et al. (2013) [1]	Genistein + Euk-207 significantly mitigated functional lung damage after thoracic RT in rats
	Medina et al. (2014) [55]	BIO 300 (genistein oral nanosuspension; Humanetics Corp.) mitigated pneumonitis/fibrosis and improved survival after WTLI
MnSalens (Euk-207; Euk-189)	Gao et al. (2012) [56]	Euk-207 significantly improved morbidity associated with pneumonitis after TBI and mitigated fibrosis in rat lungs after WTLI
MnPorphyrins	Vujaskovic et al. (2002) [57]	MnTE-2-PyP^{5+} (AEOL 10113, FBC-007) significantly reduced the severity of late-radiation injury in the lungs of rats after hemithoracic irradiation
	Gauter-Fleckenstein et al. (2008) [58]	MnTE-2-PyP^{5+}, and to a lesser extent MnTnHex-2-PyP^{5+}, mitigated functional damage and reduced collagen deposition in the lungs of rats after HTI
	Gauter-Fleckenstein et al. (2010) [59]	MnTE-2-PyP^{5+} administered at the time of symptomatic injury was capable of reversing fibrosis and improving lung function in rats after HTI
	Gauter-Fleckenstein et al. (2014) [60]	MnTnHex-2-PyP^{5+}, given at a 120-fold lower dose than MnTE-2-PyP^{5+}, significantly reduced functional injury and histopathologic damage in rat lungs after HTI

RT radiation therapy, *NSCLC* non-small cell lung cancer, *RCT* radiochemotherapy, *RP* radiation pneumonitis, *WTLI* whole-thorax lung irradiation, *TBI* total-body irradiation, *HTI* hemithoracic irradiation

The approach to medical management of radiation pneumonitis/fibrosis has not changed significantly since the 1970s and mainly includes administration of corticosteroids and anti-inflammatories, with supplemental oxygen in the most severe cases [5, 71]. Amifostine (Ethyol; WR2721; MedImmune, Inc., Gaithersburg, MD), a sulfhydral-containing pro-drug that acts as a free radical scavenger when dephosphorylated, is the only FDA-approved radiomitigator to date, with the approved indication of prevention of xerostomia and nephrotoxicity. Although preclinical data have been promising, Amifostine has not been conclusively shown to prevent and/or mitigate radiation pneumonitis and fibrosis in the clinic. Because of Amifostine's potential side effects (nausea, vomiting, hypotension), the required intravenous administration, and minimal-to-moderate efficacy for radiation pneumonitis and fibrosis, new therapeutic interventions to mitigate and/or treat these indications are warranted [72].

28.4.1 SOD-Mimics

Scientists began investigating the ability of superoxide dismutase (SOD) to confer radiobiologic protection not long after McCord and Fridovich first characterized its enzymatic activity in 1968 [73]. SODs are important in preserving homeostasis through catalyzing the one-electron reduction and oxidation of superoxide anion ($O_2^{\bullet-}$) to hydrogen peroxide (H_2O_2) and oxygen (O_2), respectively. H_2O_2 is further eliminated via either reduction to water (H_2O) by glutathione peroxidase or other systems such as peroxyredoxins or is dismutated, i.e., reduced two-electronically to water and oxygen by catalase. Petkau was the first to demonstrate that administration of Cu/Zn-SOD (bovine) prior to and after radiation exposure could significantly increase the lethal dose for 50 % of animals at 30 days (LD50/30), from 7.34 to 11.4 Gy [74]. Symonyan and Nalbandyan later observed that this effect was more pronounced when Cu/Zn-SOD was delivered immediately after irradiation rather than before, leading them to investigate the effects of X-rays on SOD enzymatic activity in vitro and in vivo. Those studies demonstrated that irradiation led to reduction of enzymatic copper and loss of activity [75]. Investigators hypothesized that by restoring antioxidant defense capabilities, the damaging effects of radiation at the molecular and cellular levels could be mitigated, thereby reducing normal tissue functional deficits and morbidity.

Despite acknowledgment that SOD can protect cells from the damaging effects of ionizing radiation (and other diseases characterized by oxidative stress), its therapeutic potential has yet to be realized clinically. This is in part because of the unfavorable pharmacologic properties of the enzyme as well as difficulty in striking the proper oxidant–antioxidant balance: too little leads to oxidative damage to DNA and proteins; too much increases lipid peroxidation and exacerbates the injury [76]. This very narrow window of efficacy on the bell-shaped curve between helpful and harmful poses significant challenges to identifying the appropriate therapeutic dosing regimen [76].

Scientists have attempted to overcome the short half-life of SOD through pegylation, liposomal delivery, and development of SOD mimics with various degrees of lipophilicity. Those that have shown the greatest efficacy include the MnSalens, MnSOD plasmid/liposomes (MnSOD PL), liposomal SOD (LipSOD), and Mn(III) porphyrins. Among Mn porphyrins, the most frequently explored are Mn(III) *meso-tetrakis*(*N,N'*-diethylimidazolium-2-yl)porphyrin, MnTDE-2-ImP^{5+} (AEOL10150), Mn(III) *meso-tetrakis*(*N*-ethylpyridinium-2-yl) porphyrin, MnTE-2-PyP^{5+} (BMX-010, AEOL10113), Mn(III) *meso-tetrakis(*N-n-hexylpyridinium-2-yl)porphyrin, MnTnHex-2-PyP^{5+} and Mn(III) *meso-tetrakis*(*N*-n-butoxyethylpyridinium-2-yl)porphyrin, MnTnBuOE-2-PyP^{5+} (BMX-001) [77, 78].

The therapeutic efficacy of LipSOD for radiation-induced fibrosis of the skin and underlying tissues has been extensively tested alone and in combination with anti-inflammatories in phase I/II clinical trials. Those trials demonstrated that LipSOD (6 injections of 5 mg over 3 weeks) was capable of reversing severe radiation-induced fibrosis [64, 79]. In cases of prefibrotic skin disease, complete regression was observed. Of a total of 50 patients receiving treatment, a reduction in the average size of the fibrotic lesion was observed in one-third and 82 % saw a significant softening of the area [64]. The success of LipSOD in radiation-induced skin fibrosis in clinical cancer therapy provides a rationale for evaluating SOD-like compounds in other normal tissues commonly affected by radiation, such as the lung and esophagus. Preclinical studies have shown that both Cu/Zn-SOD (SOD1) and MnSOD (SOD2) mitigate a number of radiation-induced pathologies in clinical and preclinical trials, including skin fibrosis [80], esophagitis [81], and lung damage [82].

Sections 28.4.1.1 and 28.4.1.2 describe successes and failures experienced with the most relevant SOD-like compounds undergoing investigation as radiomitigators of lung injury.

SOD Mimics

SOD mimics were among the first compounds developed soon after the discovery of the functional role of the SOD protein [34]. As mentioned, the therapeutic potential of the SOD protein is hindered by its large molecular weight, short half-life, antigenicity, and high cost of manufacture [83]. In contrast, low-molecular-weight SOD mimics have greater cellular permeability and retention than the bulky SOD proteins, thereby improving their half-life and localization to sites of oxidative stress [83]. These manganese-containing catalytic antioxidants are capable of performing a variety of redox reactions, many of which are as yet unknown [78]. It has been demonstrated that the ability of SOD mimics to catalyze the $O_2^{\bullet-}$ dismutation parallels their ability to reduce peroxynitrite and parallels their therapeutic potential. Among those which are emerging as major ones are oxidative modifications of thiols of signaling proteins such as NF-κB (nuclear factor-κB) and complexes I and III of mitochondrial respiration [78, 84]. Due to the ability to accept (be reduced) and give away electron (be oxidized) as exemplified in catalysis of $O_2^{\bullet-}$ dismutation, Mn porphyrins can impart their therapeutic effects via both anti- and pro-oxidative actions. The most prominent pro-oxidative action is the oxidative modification of NF-kB thiols with subsequent inhibition of its transcription and in turn suppression of excessive inflammation [78].

Since the late 1990s, SOD mimics have undergone exhaustive testing to assess their efficacy as mitigators of radiation-induced lung injury, with a focus on their ability to dampen inflammation through modulation of key redox-sensitive transcription factors, including HIF-1α (hypoxia inducible factor α), NF-κB, AP-1 (activator protein 1), and SP-1 (specificity protein 1) [34]. The best-characterized SOD mimics, the ortho isomeric MnTDE-2-ImP^{5+} (AEOL 10150), MnTE-2-PyP^{5+} (AEOL 10113, BMX-010), and MnTnHex-2-PyP^{5+}, have been shown to mitigate pneumonitis and fibrosis after radiation by suppression of the four major mediators of radiation injury: oxidative stress [85, 86], apoptosis [86, 87], inflammation [58, 59], and fibrosis [58, 59].

Vujaskovic and colleagues were the first to demonstrate that MnTE-2-PyP^{5+} could significantly mitigate functional damage and reduce collagen content in the lungs of rats following high-dose hemithoracic irradiation [57]. This was associated with a significant reduction in plasma levels of transforming growth factor β, TGF-β, a biologic predictor of pulmonary normal tissue response to radiation [88]. This was followed by studies by Gauter-Fleckenstein et al., who compared the efficacy of MnTE-2-PyP^{5+} (3 or 6 mg/kg) with the more lipophilic MnTnHex-2-PyP^{5+} (0.05, 0.3, 0.6, or 1.0 mg/kg) when administered daily for 2 weeks after hemithoracic irradiation in rats [58]. Both compounds mitigated lung damage, as assessed by respiratory function and histologic evaluation, and affected expression of HIF-1α, vascular endothelial growth factor, VEGF, macrophage activation, and suppressed oxidative stress (as measured by 8-hydroxydeoxyguanosine, 8-OHdG, marker of DNA oxidative damage) [58, 60]. The investigators hypothesized that MnTnHex-2-PyP^{5+}, with ~7000-fold greater lipophilicity than MnTE-2-PyP^{5+}, would better mitigate lung damage because of greater depth of penetration and increased tissue retention [89]. While MnTE-2-PyP^{5+} afforded radioprotection at 6 mg/kg, MnTnHex-2-PyP^{5+} was similarly efficacious at 120-fold lower concentration [60]. Both compounds were successful in reducing TGF-β and carbonic anhydrase IX, CAIX (an endogenous marker of hypoxia), in irradiated lungs [58]. Proteomic profiling of treated and untreated irradiated lungs showed reduced expression of biologic markers of protein degradation and apoptosis, indicating that MnTE-2-PyP^{5+} mitigates radiation pneumonitis/fibrosis by preventing cell loss and preserving pulmonary structural integrity following hemithoracic radiotherapy [86]. Follow-up studies to determine the therapeutic window found that MnTE-2-PyP^{5+} could mitigate lung damage even when administered starting 8 weeks after exposure, suggesting that it either stopped or reversed the proinflammatory and profibrotic processes leading to severe functional and histologic damage. This was associated with a decrease in HIF-1α, TGF-β, and VEGF. This is consistent with previous studies demonstrating activated macrophages produce $O_2^{\bullet -}$, TGF-β, and VEGF, which can be mitigated by MnTE-2-PyP^{5+} in vitro [90].

The SOD mimic most advanced along the development pipeline, AEOL 10150 (Aeolus Pharmaceuticals; Mission Viejo, CA), has been shown to protect both rodent and nonhuman primate lungs from irradiation [47–49]. The compound is currently under development as a medical countermeasure (MCM) to reduce morbidity and improve survival from the delayed effects of acute radiation exposure on lung (i.e., pneumonitis/fibrosis). In pneumonitis-prone CBA/J mice, AEOL 10150 delivered daily starting 24 h after whole-thorax lung irradiation (consistent with the current United States Government concept of operations document) at a dose of 25 mg/kg (but not 5–10 mg/kg) was effective in improving survival >30 % when administered for a total of 28 days [91]. In a pilot study in nonhuman primates, administration of AEOL 10150 starting 24 h after lethal whole-thorax radiation exposure resulted in a 28.6 % improvement in survival. This was associated with a reduction in quantitative radiologic injury (pneumonitis, fibrosis, pleural effusions), less dexamethasone support, and improved peripheral capillary oxygen saturation among AEOL 10150-treated animals compared to nontreated irradiated controls [49]. Larger studies found similar results with a doubling of 180-day survival among AEOL 10150-treated nonhuman primates compared to nontreated irradiated controls (personal communication with Dr. Thomas J. MacVittie).

In mechanistic studies, AEOL 10150 reduced proapoptotic signaling and apoptosis of lung parenchymal cells [87] and decreased oxidative stress associated with metabolic dysfunction [85] in fibrosis-prone C57BL/6J mice. Ongoing studies using a systems biology approach and state-of-the-art matrix-assisted laser desorption ionization (MALDI) mass spectrometric imaging are being performed to assess temporal changes in irradiated nonhuman primate and murine lungs and mitigation by AEOL 10150 and other MCMs to better define the mechanisms associated with injury and improve therapeutic targeting [22, 92–95].

MnSOD Plasmid/Liposomes

In the latter part of the 1990s, Engelhardt and colleagues demonstrated that MnSOD plasmid/liposome (PL) gene therapy to increase SOD expression could significantly reduce lung epithelial cell apoptosis in vitro at 72 h postexposure [96]. It was also shown that cellular protection was dependent on mitochondrial localization of MnSOD-PL both in vitro and in vivo [97]. MnSOD PL was shown to reduce levels of proinflammatory (interleukin-1, IL-1, tumor necrosis factor-α, TNF-α) and profibrotic (TGF-β) cytokines and mitigate radiation pneumonitis/fibrosis in wild-type [82, 98] and SOD2-deficient mice [99]. Of note, MnSOD PL did not alter radiation-induced cytotoxicity in orthotopic thoracic tumors, a critical requirement for clinical translation [100].

In addition to providing protection against pulmonary injury, MnSOD PL was shown to decrease lipid peroxidation of the esophagus [101] and protect multilineage squamous epithelial progenitor cells [102], leading to mitigation of esophageal stricture and esophagitis after single or fractionated doses of radiation [81, 103, 104].

Although MnSOD PL has not entered into clinical trials for radiation pneumonitis/fibrosis, a phase I dose-escalation study to assess the safety and feasibility of oral administration of MnSOD PL in patients with locally advanced non-small cell lung cancer receiving concurrent radiochemotherapy was completed in 2011 [105]. The study authors concluded that oral administration at all dose levels (0.3, 3, and 30 mg) was safe and recommended 30 mg as the starting dose for phase II investigations to assess the efficacy of MnSOD PL in reducing the incidence of grade 3 or 4 esophageal toxicity. Other endpoints include clinical response to combination chemoradiotherapy and MnSOD PL.

28.4.2 Genistein

Genistein is a soy isoflavone that belongs to a family of redox-active polyphenols which can act both as pro- and antioxidants similar to SOD mimics. Genistein acts as a tyrosine kinase inhibitor, has been shown by multiple investigators to reduce micronuclei formation, decrease functional injury, and delay morbidity/mortality when administered after radiation exposure [52–54]. However, one of the limitations of the efficacy of genistein is its low bioavailability because of poor dissolution in water. To address this limitation, Landauer and colleagues at the Armed Forces Radiation Research Institute developed an orally available nanoparticle suspension to improve water solubility and increase bioavailability by >300-fold [106]. The genistein nanosuspension (BIO 300) was subsequently licensed to Humanetics Corporation (Minneapolis, MN), which, through financial backing from the Biomedical Advanced Research and Development Authority, evaluated the efficacy of BIO 300 as a mitigator of radiation pneumonitis/fibrosis in the lungs of C57L/J mice. Those studies found that BIO 300 (oral gavage) could significantly improve survival when treatment was started 24 h after whole-thorax lung irradiation and given daily thereafter for a total of 4–6 weeks [55].

28.4.3 Activators of Antioxidant Defense Enzymes

A therapeutic area that has been relatively unexplored is development of compounds that upregulate natural antioxidant defense enzymes [107]. New compounds, such as TMC (Cureveda; Halethorpe, MD) and secoisolariciresinol diglucoside (SDG, LignaMed; Philadelphia, PA), are under development to target oxidative stress and restore redox balance through modulation of nuclear factor (erythroid-derived 2)-like 2 (Nrf2)-mediated cytoprotection. The Nrf2/Keap1 pathway maintains the redox balance of the cell by regulating phase II detoxifying and antioxidant enzymes, including heme-oxygenase 1 (HO-1), metallothioneins, NAD(P)H dehydrogenase quinone 1 (NQO1), aldo-keto reductases, glutathione reductase and glutathione S-transferases, and others. Studies have shown that Nrf2-deficient C57BL/6J mice have increased susceptibility to radiation-induced lung damage [108]. In addition to increased oxidative burden,

Nrf2-deficient mice display elevated levels of plasminogen activator inhibitor-1, a TGF-β1 target gene that is coincident with an earlier onset of lung damage than in wild-type mice [108].

Christofidou-Solomidou and colleagues have shown that dietary flaxseed mitigates oxidative damage, reduces pulmonary inflammation/fibrosis, and improves survival in C57/BL6 mice [50, 51, 109, 110]. The bioactive lignan component of flaxseed, SDG, has potent antioxidant properties, including the ability to inhibit lipid peroxidation [51]. Cytoprotective proteins including transcription factor Nrf-2 and its regulated enzymes HO-1 and NQO1 are shown to be upregulated in the irradiated lungs of flaxseed/SDG-treated mice [51]. An increased likelihood of survival associated with improved pulmonary hemodynamics and blood oxygenation levels, decreased inflammation, and reduced collagen deposition/fibrosis is seen when flaxseed is provided in the diet ad libitum starting as late as 6 weeks after thoracic irradiation in C57/BL6 mice [50].

Targeting upregulation of Nrf2 gene expression and cytoprotection may have the ability to reverse the cellular redox imbalance that facilitates chronic oxidative stress, DNA damage and apoptosis, inflammation, and fibrosis leading to acute and chronic lung damage.

28.5 Conclusions

At this point in time, there is no FDA-approved therapy to mitigate and/or treat radiation-induced lung injury. Redox-based therapeutics are attractive as radiomitigators due to their ability to neutralize chronic oxidative stress that drives the development of radiation pneumonitis/fibrosis. A number of compounds have shown therapeutic efficacy in small and large animal models using endpoints that include mitigation of functional injury and radiographic changes and reduction of major morbidity/mortality. Thus, the development and optimization of redox therapeutics offers a promising approach for the treatment of radiation-induced lung injury.

References

1. Mahmood J, Jelveh S, Zaidi A, Doctrow SR, Hill RP. Mitigation of radiation-induced lung injury with EUK-207 and genistein: effects in adolescent rats. Radiat Res. 2013;179:125–34.
2. Baranov AE, Selidovkin GD, Butturini A, Gale RP. Hematopoietic recovery after 10-Gy acute total body radiation. Blood. 1994;83:596–9.
3. Uozaki H, Fukayama M, Nakagawa K, et al. The pathology of multi-organ involvement: two autopsy cases from the Tokai-mura criticality accident. Br J Radiol Supp. 2005;27:13–6.
4. Coggle JE, Lambert BE, Moores SR. Radiation effects in the lung. Environ Health Perspect. 1986;70:261–91.

5. Gross NJ. Pulmonary effects of radiation therapy. Ann Intern Med. 1977;86:81–92.
6. Sharplin J, Franko AJ. A quantitative histological study of strain-dependent differences in the effects of irradiation on mouse lung during the intermediate and late phases. Radiat Res. 1989;119:15–31.
7. Sharplin J, Franko AJ. A quantitative histological study of strain-dependent differences in the effects of irradiation on mouse lung during the early phase. Radiat Res. 1989;119:1–14.
8. Travis EL, Down JD. Repair in mouse lung after split doses of X rays. Radiat Res. 1981;87:166–74.
9. Travis EL, Down JD, Holmes SJ, Hobson B. Radiation pneumonitis and fibrosis in mouse lung assayed by respiratory frequency and histology. Radiat Res. 1980;84:133–43.
10. Poulson JM, Vujaskovic Z, Gillette SM, Chaney EL, Gillette EL. Volume and dose-response effects for severe symptomatic pneumonitis after fractionated irradiation of canine lung. Int J Radiat Biol. 2000;76:463–8.
11. Garofalo M, Bennett A, Farese AM, et al. The delayed pulmonary syndrome following acute high-dose irradiation: a rhesus macaque model. Health Phys. 2014;106:56–72.
12. Jackson IL, Xu PT, Nguyen G, et al. Characterization of the dose response relationship for lung injury following acute radiation exposure in three well-established murine strains: developing an interspecies bridge to link animal models with human lung. Health Phys. 2014;106:48–55.
13. Jackson IL, Goswami C, Katz BP, Vujaskovic Z. Gene expression profiling identifies differences in early tissue respones to radiation among murine strains with distinct phenotypic expression of lung injury. In: 58th Annual meeting of the Radiation Research Society. Rio Grande, Puerto Rico; 2012.
14. Jackson IL, Vujaskovic Z, Down JD. Revisiting strain-related differences in radiation sensitivity of the mouse lung: recognizing and avoiding the confounding effects of pleural effusions. Radiat Res. 2010;173:10–20.
15. Jackson IL, Vujaskovic Z, Down JD. A further comparison of pathologies after thoracic irradiation among different mouse strains: finding the best preclinical model for evaluating therapies directed against radiation-induced lung damage. Radiat Res. 2011;175:510–18.
16. Jackson IL, Xu P, Hadley C, et al. A preclinical rodent model of radiation-induced lung injury for medical countermeasure screening in accordance with the FDA animal rule. Health Phys. 2012;103:463–73.
17. O'Neill P, Wardman P. Radiation chemistry comes before radiation biology. Int J Radiat Biol. 2009;85:9–25.
18. Klammer H, Mladenov E, Li F, Iliakis G. Bystander effects as manifestation of intercellular communication of DNA damage and of the cellular oxidative status. Cancer Lett. 2015;356:58–71.
19. Matzinger P. Essay 1: the Danger model in its historical context. Scand J Immunol. 2001;54:4–9.
20. Gallucci S, Matzinger P. Danger signals: SOS to the immune system. Curr Opin Immunol. 2001;13:114–9.
21. McBride WH, Chiang CS, Olson JL, et al. A sense of danger from radiation. Radiat Res. 2004;162:1–19.
22. Carter C, Jones J, Jackson I, et al. Using MALDI MSI to enabling biomarker identification and medical countermeasure development for radiation induced lung injury. In: Annual Meeting of the Radiation Research Society; 2014 September 21, 2014; Las Vegas; 2014
23. Catala A. Lipid peroxidation of membrane phospholipids generates hydroxy-alkenals and oxidized phospholipids active in physiological and/or pathological conditions. Chem Phys Lipids. 2009;157:1–11.
24. Reisz JA, Bansal N, Qian J, Zhao W, Furdui CM. Effects of ionizing radiation on biological molecules-mechanisms of damage and emerging methods of detection. Antioxid Redox Signal. 2014;21:260–92.

25. Schaue D, McBride WH. Links between innate immunity and normal tissue radiobiology. Radiat Res. 2010;173:406–17.
26. Zhou D, Shao L, Spitz DR. Reactive oxygen species in normal and tumor stem cells. Adv Cancer Res. 2014;122:1–67.
27. Dayal D, Martin SM, Owens KM, et al. Mitochondrial complex II dysfunction can contribute significantly to genomic instability after exposure to ionizing radiation. Radiat Res. 2009;172:737–45.
28. Kim HR, Won SJ, Fabian C, Kang MG, Szardenings M, Mitochondrial SMG, DNA. Aberrations and pathophysiological implications in hematopoietic diseases, chronic inflammatory diseases, and cancers. Ann Lab Med. 2015;35:1–14.
29. Zorov DB, Juhaszova M, Sollott SJ. Mitochondrial reactive oxygen species (ROS) and ROS-induced ROS release. Physiol Rev. 2014;94:909–50.
30. Hei TK, Zhou H, Ivanov VN, et al. Mechanism of radiation-induced bystander effects: a unifying model. J Pharm Pharmacol. 2008;60:943–50.
31. Chen S, Zhao Y, Han W, et al. Mitochondria-dependent signalling pathway are involved in the early process of radiation-induced bystander effects. Br J Cancer. 2008;98:1839–44.
32. Sedelnikova OA, Nakamura A, Kovalchuk O, et al. DNA double-strand breaks form in bystander cells after microbeam irradiation of three-dimensional human tissue models. Cancer Res. 2007;67:4295–302.
33. Belyakov OV, Mitchell SA, Parikh D, et al. Biological effects in unirradiated human tissue induced by radiation damage up to 1 mm away. Proc Natl Acad Sci U S A. 2005;102:14203–8.
34. Batinic-Haberle I, Rajic Z, Tovmasyan A, et al. Diverse functions of cationic Mn(III) N-substituted pyridylporphyrins, recognized as SOD mimics. Free Radic Biol Med. 2011;51:1035–53.
35. Mothersill C, Seymour C. Medium from irradiated human epithelial cells but not human fibroblasts reduces the clonogenic survival of unirradiated cells. Int J Radiat Biol. 1997;71:421–7.
36. Yang G, Wu L, Chen S, et al. Mitochondrial dysfunction resulting from loss of cytochrome c impairs radiation-induced bystander effect. Br J Cancer. 2009;100:1912–6.
37. Rubin P, Johnston CJ, Williams JP, McDonald S, Finkelstein JN. A perpetual cascade of cytokines postirradiation leads to pulmonary fibrosis. Int J Radiat Oncol Biol Phys. 1995;33:99–109.
38. Schaue D, Kachikwu EL, McBride WH. Cytokines in radiobiological responses: a review. Radiat Res. 2012;178:505–23.
39. Fleckenstein K, Zgonjanin L, Chen L, et al. Temporal onset of hypoxia and oxidative stress after pulmonary irradiation. Int J Radiat Oncol Biol Phys. 2007;68:196–204.
40. Antonadou D, Coliarakis N, Synodinou M, et al. Randomized phase III trial of radiation treatment +/- amifostine in patients with advanced-stage lung cancer. Int J Radiat Oncol Biol Phys. 2001;51:915–22.
41. Antonadou D, Petridis A, Synodinou M, et al. Amifostine reduces radiochemotherapy-induced toxicities in patients with locally advanced non-small cell lung cancer. Semin Oncol. 2003;30:2–9.
42. Komaki R, Lee JS, Milas L, et al. Effects of amifostine on acute toxicity from concurrent chemotherapy and radiotherapy for inoperable non-small-cell lung cancer: report of a randomized comparative trial. Int J Radiat Oncol Biol Phys. 2004;58:1369–77.
43. Vujaskovic Z, Feng QF, Rabbani ZN, Anscher MS, Samulski TV, Brizel DM. Radioprotection of lungs by amifostine is associated with reduction in profibrogenic cytokine activity. Radiat Res. 2002;157:656–60.
44. Vujaskovic Z, Thrasher BA, Jackson IL, Brizel MB, Brizel DM. Radioprotective effects of amifostine on acute and chronic esophageal injury in rodents. Int J Radiat Oncol Biol Phys. 2007;69:534–40.

45. Ozturk B, Egehan I, Atavci S, Kitapci M. Pentoxifylline in prevention of radiation-induced lung toxicity in patients with breast and lung cancer: a double-blind randomized trial. Int J Radiat Oncol Biol Phys. 2004;58:213–9.
46. Kaya V, Yazkan R, Yildirim M, et al. The relation of radiation-induced pulmonary fibrosis with stress and the efficiency of antioxidant treatment: an experimental study. Med Sci Monit. 2014;20:290–6.
47. Rabbani ZN, Batinic-Haberle I, Anscher MS, et al. Long-term administration of a small molecular weight catalytic metalloporphyrin antioxidant, AEOL 10150, protects lungs from radiation-induced injury. Int J Radiat Oncol Biol Phys. 2007;67:573–80.
48. Rabbani ZN, Salahuddin FK, Yarmolenko P, et al. Low molecular weight catalytic metalloporphyrin antioxidant AEOL 10150 protects lungs from fractionated radiation. Free Radic Res. 2007;41:1273–82.
49. Garofalo MC, Ward AA, Farese AM, et al. A pilot study in rhesus macaques to assess the treatment efficacy of a small molecular weight catalytic metalloporphyrin antioxidant (AEOL 10150) in mitigating radiation-induced lung damage. Health Phys. 2014;106:73–83.
50. Christofidou-Solomidou M, Tyagi S, Tan KS, et al. Dietary flaxseed administered post thoracic radiation treatment improves survival and mitigates radiation-induced pneumonopathy in mice. BMC Cancer. 2011;11:269.
51. Christofidou-Solomidou M, Tyagi S, Pietrofesa R, et al. Radioprotective role in lung of the flaxseed lignan complex enriched in the phenolic secoisolariciresinol diglucoside (SDG). Radiat Res. 2012;178:568–80.
52. Day RM, Barshishat-Kupper M, Mog SR, et al. Genistein protects against biomarkers of delayed lung sequelae in mice surviving high-dose total body irradiation. J Radiat Res. 2008;49:361–72.
53. Calveley VL, Jelveh S, Langan A, et al. Genistein can mitigate the effect of radiation on rat lung tissue. Radiat Res. 2010;173:602–11.
54. Mahmood J, Jelveh S, Calveley V, Zaidi A, Doctrow SR, Hill RP. Mitigation of radiation-induced lung injury by genistein and EUK-207. Int J Radiat Biol. 2011;87:889–901.
55. Medina J, Buck C, Pavlovic R, et al. Optimal dose and duration of BIO 300 for mitigating radiation induced lung injury in C57L/J mice. In: Annual Meeting of the Radiation Research Society; 2014 September 21, 2014; Las Vegas; 2014
56. Gao F, Fish BL, Szabo A, et al. Short-term treatment with a SOD/catalase mimetic, EUK-207, mitigates pneumonitis and fibrosis after single-dose total-body or whole-thoracic irradiation. Radiat Res. 2012;178:468–80.
57. Vujaskovic Z, Batinic-Haberle I, Rabbani ZN, et al. A small molecular weight catalytic metalloporphyrin antioxidant with superoxide dismutase (SOD) mimetic properties protects lungs from radiation-induced injury. Free Radic Biol Med. 2002;33:857–63.
58. Gauter-Fleckenstein B, Fleckenstein K, Owzar K, Jiang C, Batinic-Haberle I, Vujaskovic Z. Comparison of two Mn porphyrin-based mimics of superoxide dismutase in pulmonary radioprotection. Free Radic Biol Med. 2008;44:982–9.
59. Gauter-Fleckenstein B, Fleckenstein K, Owzar K, et al. Early and late administration of MnTE-2-PyP5+ in mitigation and treatment of radiation-induced lung damage. Free Radic Biol Med. 2010;48:1034–43.
60. Gauter-Fleckenstein B, Reboucas JS, Fleckenstein K, et al. Robust rat pulmonary radioprotection by a lipophilic Mn N-alkylpyridylporphyrin, MnTnHex-2-PyP(5+). Redox Biol. 2014;2:400–10.
61. Patt HM, Tyree EB, Straube RL, Smith DE. Cysteine protection against X irradiation. Science. 1949;110:213–4.
62. Delanian S, Balla-Mekias S, Lefaix JL. Striking regression of chronic radiotherapy damage in a clinical trial of combined pentoxifylline and tocopherol. J Clin Oncol. 1999;17:3283–90.

63. Lefaix JL, Delanian S, Vozenin MC, Leplat JJ, Tricaud Y, Martin M. Striking regression of subcutaneous fibrosis induced by high doses of gamma rays using a combination of pentoxifylline and alpha-tocopherol: an experimental study. Int J Radiat Oncol Biol Phys. 1999;43:839–47.
64. Baillet F, Housset M, Michelson AM, Puget K. Treatment of radiofibrosis with liposomal superoxide dismutase. Preliminary results of 50 cases. Free Radic Res Commun. 1986;1:387–94.
65. Delanian S, Baillet F, Huart J, Lefaix JL, Maulard C, Housset M. Successful treatment of radiation-induced fibrosis using liposomal Cu/Zn superoxide dismutase: clinical trial. Radiother Oncol. 1994;32:12–20.
66. Delanian S, Lefaix JL. Complete healing of severe osteoradionecrosis with treatment combining pentoxifylline, tocopherol and clodronate. Br J Radiol. 2002;75:467–9.
67. Delanian S, Chatel C, Porcher R, Depondt J, Lefaix JL. Complete restoration of refractory mandibular osteoradionecrosis by prolonged treatment with a pentoxifylline-tocopherol-clodronate combination (PENTOCLO): a phase II trial. Int J Radiat Oncol Biol Phys. 2011;80:832–9.
68. Hille A, Christiansen H, Pradier O, et al. Effect of pentoxifylline and tocopherol on radiation proctitis/enteritis. Strahlentherapie und Onkologie : Organ der Deutschen Rontgengesellschaft [et al.] 2005;181:606-14.
69. Gothard L, Cornes P, Brooker S, et al. Phase II study of vitamin E and pentoxifylline in patients with late side effects of pelvic radiotherapy. Radiother Oncol. 2005;75:334–41.
70. Delanian S, Lefaix JL. Current management for late normal tissue injury: radiation-induced fibrosis and necrosis. Semin Radiat Oncol. 2007;17:99–107.
71. Sekine I, Sumi M, Ito Y, et al. Retrospective analysis of steroid therapy for radiation-induced lung injury in lung cancer patients. Radiother Oncol. 2006;80:93–7.
72. Citrin D, Cotrim AP, Hyodo F, Baum BJ, Krishna MC, Mitchell JB. Radioprotectors and mitigators of radiation-induced normal tissue injury. Oncologist. 2010;15:360–71.
73. McCord JM, Fridovich I. Superoxide dismutase. An enzymic function for erythrocuprein (hemocuprein). J Biol Chem. 1969;244:6049–55.
74. Petkau A. Radiation protection by superoxide dismutase. Photochem Photobiol. 1978;28:765–74.
75. Symonyan MA, Nalbandyan RM. The effect of x-rays on properties of superoxide dismutase. Biochem Biophys Res Commun. 1979;90:1207–13.
76. McCord JM, Fridovich I. The utility of superoxide dismutase in studying free radical reactions. I. Radicals generated by the interaction of sulfite, dimethyl sulfoxide, and oxygen. J Biol Chem. 1969;244:6056–63.
77. Miriyala S, Spasojevic I, Tovmasyan A, et al. Manganese superoxide dismutase, MnSOD and its mimics. Biochim Biophys Acta. 1822;2012:794–814.
78. Batinic-Haberle I, Tovmasyan A, Roberts ER, Vujaskovic Z, Leong KW, Spasojevic I. SOD therapeutics: latest insights into their structure-activity relationships and impact on the cellular redox-based signaling pathways. Antioxid Redox Signal. 2014;20:2372–415.
79. Emerit J, Michelson AM, Robert HG, et al. [Superoxide dismutase treatment of 2 cases of radiation-induced sclerosis]. Sem Hop. 1983;59:277–81.
80. Lefaix JL, Delanian S, Leplat JJ, et al. Successful treatment of radiation-induced fibrosis using Cu/Zn-SOD and Mn-SOD: an experimental study. Int J Radiat Oncol Biol Phys. 1996;35:305–12.
81. Stickle RL, Epperly MW, Klein E, Bray JA, Greenberger JS. Prevention of irradiation-induced esophagitis by plasmid/liposome delivery of the human manganese superoxide dismutase transgene. Radiat Oncol Investig. 1999;7:204–17.
82. Epperly MW, Travis EL, Sikora C, Greenberger JS. Manganese [correction of Magnesium] superoxide dismutase (MnSOD) plasmid/liposome pulmonary radioprotective gene therapy: modulation of irradiation-induced mRNA for IL-I, TNF-alpha, and TGF-beta correlates

with delay of organizing alveolitis/fibrosis. Biol Blood Marrow Transplant. 1999;5:204–14.
83. Day BJ. Catalytic antioxidants: a radical approach to new therapeutics. Drug Discov Today. 2004;9:557–66.
84. Jaramillo MC, Briehl MM, Batinic-Haberle I, Tome ME. Manganese (III) meso-tetrakis N-ethylpyridinium-2-yl porphyrin acts as a pro-oxidant to inhibit electron transport chain proteins, modulate bioenergetics, and enhance the response to chemotherapy in lymphoma cells. Free Radic Biol Med. 2015.
85. Jackson IL, Zhang X, Hadley C, et al. Temporal expression of hypoxia-regulated genes is associated with early changes in redox status in irradiated lung. Free Radic Biol Med. 2012;53:337–46.
86. Yakovlev VA, Rabender CS, Sankala H, et al. Proteomic analysis of radiation-induced changes in rat lung: Modulation by the superoxide dismutase mimetic MnTE-2-PyP(5+). Int J Radiat Oncol Biol Phys. 2010;78:547–54.
87. Zhang Y, Zhang X, Rabbani ZN, Jackson IL, Vujaskovic Z. Oxidative stress mediates radiation lung injury by inducing apoptosis. Int J Radiat Oncol Biol Phys. 2012;83:740–8.
88. Fu XL, Huang H, Bentel G, et al. Predicting the risk of symptomatic radiation-induced lung injury using both the physical and biologic parameters V(30) and transforming growth factor beta. Int J Radiat Oncol Biol Phys. 2001;50:899–908.
89. Weitner T, Kos I, Sheng H, et al. Comprehensive pharmacokinetic studies and oral bioavailability of two Mn porphyrin-based SOD mimics, MnTE-2-PyP(5+) and MnTnHex-2-PyP(5+). Free Radic Biol Med. 2013.
90. Jackson IL, Chen L, Batinic-Haberle I, Vujaskovic Z. Superoxide dismutase mimetic reduces hypoxia-induced O2*-, TGF-beta, and VEGF production by macrophages. Free Radic Res. 2007;41:8–14.
91. Murigi FN, Hung C, Salimi S, et al. Dose optimization study of AEOL 10150 in mitigation of radiation induced lung injury in CBA/J mice. In: Annual Meeting of the Radiation Research Society; 2014; Las Vegas; 2014.
92. Carter CL, Jones JW, Jackson IL, et al. MALDI-MSI Lipidomic Investigation into the Delayed Effect of Acute Radiation Exposure: The Lung Syndrome and the Efficacy of a Medical Countermeasure. In: American Society of Mass Spectrometry Annual Meeting; 2014; Baltimore; 2014.
93. Carter CL, MacVittie TJ, Kane MA. MALDI-MSI: Biomarker discovery for radiation exposures. In: Three encyclopedia, systems biology; 2014
94. Jones JW, Li F, Carter CL, et al. identification of lipid biomarkers from mouse lung tissue via the use of UPC2 tandem mass spectrometry. In: American Society of Mass Spectrometry Annual Meeting; 2014; Baltimore; 2014.
95. Jones JW, Li F, Carter CL, et al. Targeted and discovery-based mass spectrometry metabolomics for biomarker identification and validation in mouse and non-human primate radiation models. Las Vegas: Radiation Research Society; 2014.
96. Zwacka RM, Dudus L, Epperly MW, Greenberger JS, Engelhardt JF. Redox gene therapy protects human IB-3 lung epithelial cells against ionizing radiation-induced apoptosis. Hum Gene Ther. 1998;9:1381–6.
97. Epperly MW, Gretton JE, Sikora CA, et al. Mitochondrial localization of superoxide dismutase is required for decreasing radiation-induced cellular damage. Radiat Res. 2003;160:568–78.
98. Epperly M, Bray J, Kraeger S, et al. Prevention of late effects of irradiation lung damage by manganese superoxide dismutase gene therapy. Gene Ther. 1998;5:196–208.
99. Epperly MW, Epstein CJ, Travis EL, Greenberger JS. Decreased pulmonary radiation resistance of manganese superoxide dismutase (MnSOD)-deficient mice is corrected by human manganese superoxide dismutase-Plasmid/Liposome (SOD2-PL) intratracheal gene therapy. Radiat Res. 2000;154:365–74.

100. Epperly MW, Defilippi S, Sikora C, Gretton J, Kalend A, Greenberger JS. Intratracheal injection of manganese superoxide dismutase (MnSOD) plasmid/liposomes protects normal lung but not orthotopic tumors from irradiation. Gene Ther. 2000;7:1011–8.
101. Epperly MW, Tyurina YY, Nie S, et al. MnSOD-plasmid liposome gene therapy decreases ionizing irradiation-induced lipid peroxidation of the esophagus. In Vivo. 2005;19:997–1004.
102. Niu Y, Shen H, Epperly M, et al. Protection of esophageal multi-lineage progenitors of squamous epithelium (stem cells) from ionizing irradiation by manganese superoxide dismutase-plasmid/liposome (MnSOD-PL) gene therapy. In Vivo. 2005;19:965–74.
103. Epperly MW, Gretton JA, DeFilippi SJ, et al. Modulation of radiation-induced cytokine elevation associated with esophagitis and esophageal stricture by manganese superoxide dismutase-plasmid/liposome (SOD2-PL) gene therapy. Radiat Res. 2001;155:2–14.
104. Epperly MW, Kagan VE, Sikora CA, et al. Manganese superoxide dismutase-plasmid/liposome (MnSOD-PL) administration protects mice from esophagitis associated with fractionated radiation. Int J Cancer. 2001;96:221–31.
105. Tarhini AA, Belani CP, Luketich JD, et al. A phase I study of concurrent chemotherapy (paclitaxel and carboplatin) and thoracic radiotherapy with swallowed manganese superoxide dismutase plasmid liposome protection in patients with locally advanced stage III non-small-cell lung cancer. Hum Gene Ther. 2011;22:336–42.
106. Ha CT, Li XH, Fu D, Xiao M, Landauer MR. Genistein nanoparticles protect mouse hematopoietic system and prevent proinflammatory factors after gamma irradiation. Radiat Res. 2013;180:316–25.
107. McCord JM. Therapeutic control of free radicals. Drug Discov Today. 2004;9:781–2.
108. Travis EL, Rachakonda G, Zhou X, et al. NRF2 deficiency reduces life span of mice administered thoracic irradiation. Free Radic Biol Med. 2011;51:1175–83.
109. Lee JC, Krochak R, Blouin A, et al. Dietary flaxseed prevents radiation-induced oxidative lung damage, inflammation and fibrosis in a mouse model of thoracic radiation injury. Cancer Biol Ther. 2009;8:47–53.
110. Pietrofesa R, Turowski J, Tyagi S, et al. Radiation mitigating properties of the lignan component in flaxseed. BMC Cancer. 2013;13:179.
111. Bentzen SM. Preventing or reducing late side effects of radiation therapy: radiobiology meets molecular pathology. Nat Rev Cancer. 2006;6:702–13.

Chapter 29
Using Metalloporphyrins to Preserve β Cell Mass and Inhibit Immune Responses in Diabetes

Gina M. Coudriet, Dana M. Previte, and Jon D. Piganelli

29.1 Introduction

As of 2012, 29.1 million Americans are living with diabetes [1], creating a significant healthcare concern for the United States. Diabetes is a disease highly driven by inflammation and reactive oxygen species (ROS). Therefore, agents that affect secondary inflammation as a result of oxidative stress like metalloporphyrins are a desirable approach to alleviate the inflammation associated with disease. Here, we will discuss the role ROS play in the immunopathology of diabetes, the work that has elucidated the efficacy of metalloporphyrins in the disease, and their future utility in the clinic.

29.2 Diabetes Mellitus

Diabetes mellitus encompasses a group of diseases whereby blood sugar levels are uncontrolled due to inadequate insulin production, secretion, and/or signaling. Progression to diabetes occurs when the β cell is damaged and can no longer perform appropriate glucose sensing and insulin secretion, resulting in hyperglycemia and glucotoxicity. Common symptoms of diabetes include increased urination, thirst, and hunger. Diabetes is a chronic disease and the consequences of poorly controlled blood glucose levels, if left untreated, include metabolic ketoacidosis and death [1].

Gina M. Coudriet and Dana M. Previte contributed equally to this work.

G.M. Coudriet • D.M. Previte • J.D. Piganelli (✉)
Department of Surgery, University of Pittsburgh School of Medicine,
4401 Penn Ave., Rangos Research Center 6125, Pittsburgh, PA 15224, USA
e-mail: jdp51@pitt.edu

I. Batinić-Haberle et al. (eds.), *Redox-Active Therapeutics*,
Oxidative Stress in Applied Basic Research and Clinical Practice,
DOI 10.1007/978-3-319-30705-3_29

Even if blood sugar levels can be somewhat maintained, exposure to high glucose can damage target organs leading to secondary complications such as neuropathy, nephropathy, retinopathy, and cardiovascular disease [2]. These secondary complications remain a major health concern for patients.

Specifically, type 1 diabetes (T1D) is an autoimmune disease accounting for approximately 10 % of all diabetes patients. T1D ensues from self-reactive T cells escaping tolerogenic mechanisms and mediating targeted destruction of pancreatic β cells [3]. This destruction results in reduced β cell mass and uncontrolled blood glucose levels from diminished insulin production. While the initiating events that trigger immune activation against the β cell are not fully elucidated, it is well-accepted that genetic and environmental factors may be responsible for promoting the break in tolerance. These factors include: viral infection [4], family history [5], pancreatic ROS and inflammation [6], and chemical exposure [7].

By the time T1D patients present with hyperglycemia, 80–90 % of the β cell mass has already been lost [8, 9]. Therefore, in order to maintain blood glucose levels, patients must administer exogenous insulin for the remainder of their lives. While this therapy is efficacious for most, there are a percentage of patients who cannot manage blood glucose with insulin alone. Islet transplantation has become a potential therapy for these patients. However, with transplantation, patients must be placed on rigorous immunosuppressive regimens to minimize islet graft rejection. In order to study autoimmune-mediated diabetes in the laboratory, the NOD, or non-obese diabetic, is the primary mouse model used, as it closely mimics human disease onset [10]. Knockout and transgenic NOD mice are regularly exploited for studying T1D in vivo.

Type 2 diabetes (T2D) is the more common form of the disease, accounting for about 90% of cases. Although the genetic susceptibility is more complex, obesity is a major risk factor for progression to T2D [11]. In contrast with T1D, patients suffering from T2D experience insulin resistance through a loss of insulin sensitivity. Lacking an autoimmune component, T2D often manifests with islet hypertrophy and hyperinsulinemia [12]; however, insulin signaling is greatly impaired, ultimately resulting in poorly controlled blood glucose levels. Treatment options for obese T2D patients begin with weight loss, in conjunction with insulin-sensitizing drugs. However, poor control can result in eventual insulin therapy and potential β cell loss by apoptosis. Mice with a genetic predisposition for obesity are available for studying T2D; however, the high fat diet (HFD) model of obesity is widely studied as it closely mimics what is seen in the human population [13].

ROS and reactive nitrogen species (RNS) contribute to β cell damage in both T1D [14, 15] and T2D [16], yet the underlying mechanisms differ. In T1D, β cell damage occurs from autoimmune-mediated proinflammatory cytokine production which fuels oxidative stress. Superoxide anion, hydrogen peroxide, hydroxyl radicals, and nitric oxide (NO) are all produced as a result of the inflammation, and these molecules are sufficient for inducing β cell death [17]. In T2D, glucose toxicity is the driving force behind generation of ROS and resultant β cell dysfunction [17]. With this in mind, it is logical to think that antioxidant therapies would be advantageous in both of these disease settings.

In order to combat the chronic inflammation and oxidative stress displayed in diabetes, manganese metalloporphyrin antioxidants are seemingly viable therapeutics. MnP abbreviation is used to describe MnTE-2-PyP^{5+}/AEOL10113, BMX-010, except when indicated that it relates to MnTDE-2-ImP^{5+}/AEOL10150. Both compounds have similar redox-properties, yet differ with respect to size and shape which differentially affects their bioavailabilities [18]. Not only can these agents scavenge free radicals and other reactive species such as superoxide and peroxynitrite [19, 20], but they also possess a high bioavailability; particularly relevant is their preferable mitochondrial relative to cytosolic accumulation. Their actions are frequently catalytic and display oxidoreductase properties [21, 22]. These characteristics allow for preservation of β cell integrity and modulation of autoimmunity in T1D, which have been extensively studied. While work has yet to be published examining the potential of MnP in T2D, preliminary studies suggest its efficacy.

29.3 The Oxidative Environment of the β Cell

The primary role of the pancreatic β cell is to sense blood glucose levels, and in response, produce and secrete insulin to maintain glucose homeostasis. Insulin production and secretion requires continual obligation for protein translation, modification, and export by β cells. To power these processes, β cells must oxidize glucose via oxidative phosphorylation to generate the required ATP. Oxidative phosphorylation is responsible for mitochondrial-derived ROS production (hydrogen peroxide and superoxide) [23]. Acute doses of these ROS and RNS are required for insulin secretion and aid β cell function [24–26]. However, the overall demand on the β cell in the pre-diabetic environment results in increased stress, and if uncontrolled, excessive ROS and RNS production can tip the balance of oxidative homeostasis. This ensuing, chronic oxidative stress is detrimental to proper insulin signaling and can drive β cell apoptosis [15, 26, 27].

With increased susceptibility to oxidative stress due to high rates of glucose oxidation, one might assume that β cells would express high levels of antioxidant enzymes and effectively repair DNA following oxidative damage. Notably, β cells express very low levels of catalase, glutathione peroxidase, and superoxide dismutase [28] and demonstrate poor DNA repair following oxidative damage [29]. Therefore, since the sole function of the β cell, glucose-stimulated insulin secretion follows a metabolic pathway concomitant with ROS production, it has been hypothesized that the β cell sets itself up for failure in terms of oxidative damage [30]. To circumvent this defect, studies where antioxidant enzymes were overexpressed in β cells were conducted. Results showed that overexpression of these enzymes protected β cells from cytokine and ROS-induced apoptosis and limited NF-κB activation, thereby reducing proinflammatory cytokine production by the β cell itself [31, 32]. These studies indicate the potential for antioxidant therapy to preserve β cell mass and function.

Nicotinamide adenine dinucleotide phosphate (NADPH) oxidases, or NOX, are also key mediators of ROS in β cells [33]. The NOX enzymes are responsible for transferring electrons across membranes, leading to the generation of superoxide and hydrogen peroxide [33]. β cells have been shown to express several NOX isoforms including NOX-1, 2, and 4, NOXA1, and NOXO1 [16, 33–35]. While NOX-1 and 2 are located at the plasma membrane, NOX-4 is expressed in the mitochondria [36]. NOX are activated following glucose sensing by β cells, and the ROS generated aid in glucose-stimulated insulin secretion [24, 26]. Moreover, NOX-generated ROS can serve as a positive feedback loop, enabling more ROS production [37]. If uncontrolled, NOX are yet another oxidative stress-inducing mechanism working against the β cell in diabetes. Increased serum levels of free fatty acids and proinflammatory cytokines, like those seen in diabetic patients, are sufficient for upregulating NOX expression in β cells, leading to increased ROS generation [38]. This upregulation, coupled with low antioxidant enzyme expression, further renders the β cell susceptible to oxidative damage.

Studies have been conducted to specifically examine if NOX inhibition preserves β cell function; however, results have been contradictory. Work by Thayer et al. utilized islets from the transgenic NOD.NCF-1^{m1J} mouse model [39]. These mice contain a point mutation that results in truncation of the p47^{phox} subunit and prevents functional NOX assembly and ROS production [39, 40]. Specifically in β cells, this results in loss of functional NOX-1 and 2. Interestingly, these NOX-deficient islets, when treated with inflammatory cytokines, produced similar levels of NO and had similar viability, as compared to wild-type control islets [39]. However, in another study, treatment of human and murine islets with the NOX-1 inhibitor ML171 provided protection from inflammatory cytokines [41].

Overall, constant oxidative stress due to glucose sensing and insulin secretion, coupled with low expression of oxidative stress defense mechanisms, render the β cell highly susceptible to apoptosis. A therapeutic with the ability to globally scavenge free radicals and enhance the antioxidant defense of the β cell would therefore be beneficial for use in preserving β cell viability and function in diabetic patients.

29.4 Preserving β Cells with Metalloporphyrins

Despite vast advances in medicine, oxidative stress and ROS-mediated damage still remain hurdles in maintaining β cell function and viability [17, 42]. In order to overcome these issues, multiple studies have been performed to examine if MnP treatment is able to dissipate ROS and preserve β cell mass and function.

β cells require high rates of glucose oxidation in response to glucose sensing and for the production of insulin [17, 30]. Increased glucose oxidation results in high levels of mitochondrial-derived ROS, leading to oxidation-induced β cell dysfunction and death [27, 43]. Aconitase is an important enzyme responsible for catalyzing the reaction of citrate to isocitrate in the tricarboxylic acid cycle (TCA). Additionally, aconitase's iron sulfur cluster is highly susceptible to the direct inactivation by superoxide produced as a result of increased glucose oxidation by the mitochondria [43, 44]. In a healthy environment, ROS serve as a negative feedback loop to modulate mitochondrial metabolism in order to avoid overproduction of ROS. Conversely, increased aconitase inactivation, due to an overabundance of superoxide, leads to a build-up of glucose, resulting in a feed-forward loop of oxidative stress in diabetes [45]. Recent studies investigating various metabolic pathways showed that MnP treatment sustained aconitase activity and resulted in increased glucose oxidation, most likely through the scavenging of mitochondrial-derived ROS [46]. Therefore, it was concluded that MnP treatment enabled more efficient TCA cycle function, respiratory chain coupling, and ultimately mitochondrial respiration, while minimizing the deleterious effects of the ROS produced [46]. This treatment may prove beneficial for β cells due to their high susceptibility to oxidative stress and their dependence on glucose oxidation for driving their physiological functions. While this is promising, future studies are necessary to elucidate this potential in β cells, specifically.

Insulin injections are the main therapeutic for maintaining blood glucose in diabetic patients. Currently, pancreatic islet transplantation is a therapeutic approach for those who cannot successfully maintain blood glucose levels by injection alone. With this approach, islets must be successfully isolated from the donor, purified, and transplanted into recipient patients. The nature of this labor-intensive process is unfavorable to the already fragile β cell, as it induces hypoxia and ischemia-reperfusion injury [47]. Preservation of the graft is of utmost importance; however, several hurdles still remain, including preventing β cell apoptosis, limiting oxidative stress, and maintaining functionality. Several studies have explored the potential for MnP in overcoming these hurdles. Treatment of human donor islets with MnP during and after isolation enhanced islet cell mass 2 and 7 days post-isolation [48] (Fig. 29.1a). MnP treatment also resulted in increased pro-insulin staining of islet preparations following isolation, indicating preserved insulin production. While functional glucose responsiveness did not differ between MnP-treated and untreated islets in vitro, islets treated with MnP prior to transplantation better normalized blood glucose levels of diabetic mice in vivo following transplantation [48]. Additionally, MnP afforded protection against streptozotocin (STZ)-induced β cell damage. STZ is a chemical agent that causes oxidative stress-induced β cell death, thus mimicking the stress that occurs from islet isolation. MnP protected human islets from STZ damage in vitro [49] (Fig. 29.1b, c). In vivo, mice were similarly protected from STZ-induced diabetes onset by the systemic administration of MnP. Additionally, islets pretreated with MnP improved β cell function in syngeneic, allogeneic, and xenogeneic murine islet transplants [49].

Follow-up studies from Bottino et al. provided a mechanistic explanation for MnP preservation of islet mass and function. They determined that MnP treatment resulted in decreased islet NF-κB nuclear translocation and reduced poly (ADP-ribose) polymerase (PARP) cleavage, indicative of islet survival [50]. Specifically, NF-κB-DNA binding was inhibited, thereby decreasing expression of NF-κB-dependent monocyte chemoattractant protein-1 (MCP-1) and interleukin-6 (IL-6) [50]. While other NF-κB-regulated genes such as interleukin-1β (IL-1β), tumor

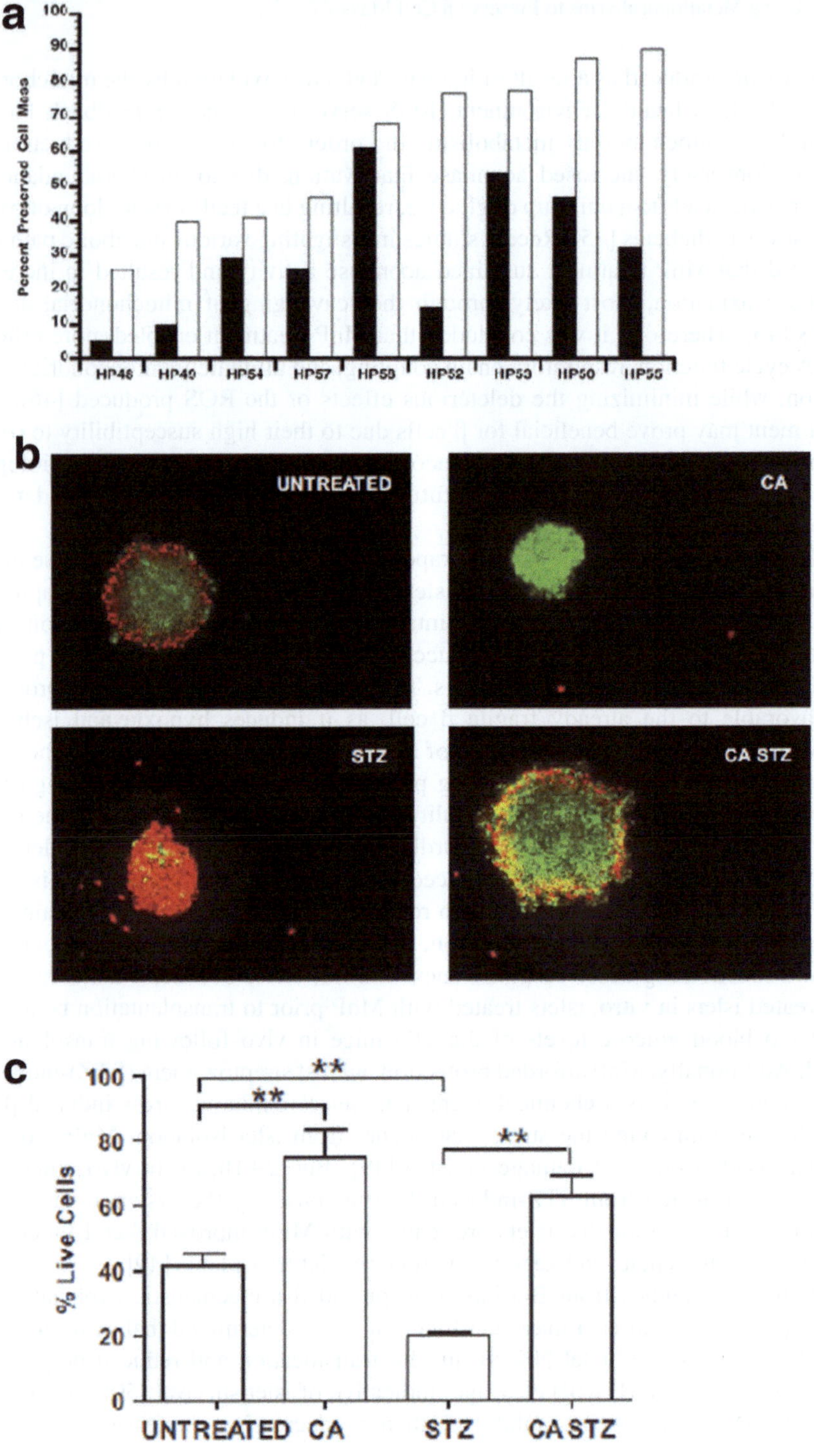

Fig. 29.1 MnP protects human islets from streptozotocin (STZ)-induced cell death. (**a**) Percentage of preserved human β cell mass following 5-day culture in control (*black bars*) and MnP-treated (*white bars*) groups ($n=9$) ($p=0.02$). Data adapted from Bottino et al. [48]. (**b**) Human islets

necrosis factor-α (TNF-α), and macrophage inflammatory protein 1α (MIP-1α) were not examined, the data would suggest that these inflammatory mediators would also be dampened with MnP treatment.

These studies demonstrate that oxidative and inflammatory damage is detrimental to islets and blocking these pathways via MnP treatment is a viable option for maintaining graft mass and function. These studies have led to recent clinical trials in Canada, where donor pancreata are being treated with MnP during perfusion, digestion, and culture, prior to transplantation. At present, the results of the studies have yet to be published.

29.5 Immunopathology of T1D

As mentioned previously, T1D is an autoimmune disease with an orchestrated immunopathology involving both the innate and adaptive arms of the immune system. Macrophages and T cells explicitly contribute to β cell destruction via direct and indirect mechanisms. It is well-documented that ROS play significant roles in activation of both macrophages and T cells in T1D models [6]. Therefore, MnP treatment remains a reasonable therapeutic to minimize ROS generation and aberrant immune responses. The various roles ROS play in the activation of macrophages and T cells, and how MnP has been used to dampen these responses and delay T1D (Fig. 29.2), will be discussed.

29.5.1 Macrophages

Under normal conditions, macrophages are present in the pancreas as a means of immune surveillance. Following genetic and environmental triggers, β cells release antigens, activating tissue-resident macrophages. Once activated, macrophages serve two important roles in driving the early stages of T1D: (1) Producing proinflammatory cytokines and ROS that result in β cell death, and (2) Serving as antigen-presenting cells (APC) to activate antigen-specific T cells [6].

In pre-diabetic NOD mice, macrophages are observed during the early stages of insulitis, [51, 52] concomitant with islet MCP-1 expression [53]. Macrophages are also considered the early and major producer of TNF-α, which can directly mediate β cell destruction [54]. They also produce large amounts of IL-1β, driving more direct inducible nitric oxide synthase (iNOS)-mediated β cell death [55]. Macrophage depletion studies in animal models of T1D display inhibition of disease onset [39, 56].

Fig. 29.1 (continued) (60 islets/group) were cultured in media alone, 68 μM MnP (CA), 11 mM STZ, or 68 μM MnP (CA) + 11 mM STZ. Fluorescence microscopy of islets after an 8-h incubation, stained with acridine orange (*green*/live) and ethidium bromide (*red*/dead). (**c**) Quantification of live cells following treatment, demonstrating that MnP (CA) treatment protects islets from chemical-induced death ($n \geq 3$). Data are presented as mean ± SEM. Significance was tested using one-way ANOVA (**$p < -0.05$). Data adapted from Sklavos et al. [49]

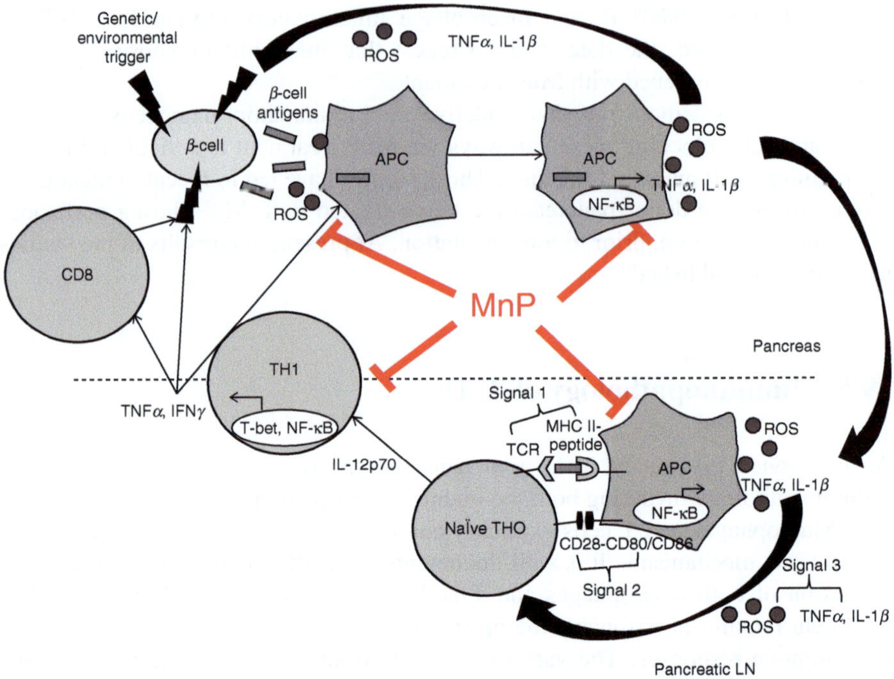

Fig. 29.2 Reactive oxygen species (ROS) play a significant role in the immunopathology of type 1 diabetes (T1D). An initiating environmental and/or genetic insult(s) to the pancreatic β cell induces the release of β cell antigens and ROS. Phagocytosis of these antigens in the presence of ROS leads to macrophage activation via NF-κB. Following activation, macrophages ferry antigen to the pancreatic lymph nodes (LN) where they stimulate antigen-specific CD8⁺ and CD4⁺ T cells (TH0), with the help of ROS (third signal). The activated T cells (TH1) traffic to the pancreas where they can directly or indirectly contribute to β cell death. Studies from our laboratory and others have indicated that redox modulation by MnP treatment is capable of inhibiting T1D immunopathology at the indicated stages (shown in *red*). Figure adapted from Delmastro et al. [6]

In vitro, co-culture experiments show that β cell death is increased in the presence of activated macrophages [57]. Collectively, these studies underscore the critical role of macrophages in early-onset and progression to diabetes.

Similar to β cells, macrophages also express a functional NOX, which during activation generates superoxide and hydrogen peroxide-free radicals [58, 59]. These free radicals create an oxidative environment in the islet, inducing even more oxidative stress in the β cell. ROS can also act as signaling molecules that activate various cellular pathways including the mitogen-activated protein kinases (MAPK), NF-κB, and cell cycle pathways. To further elucidate the role of NOX in T1D progression, NOD.NCF-1^{mlJ} were studied [39, 40]. Ablation of NOX in macrophages resulted in decreased respiratory burst, a skewed cytokine response [60], and a dampened Toll-like receptor 3 (TLR3) response [61], emphasizing the pivotal role NOX plays in mediating macrophage activation and immune responses. Similar results were shown in vivo following adoptive transfer of diabetogenic T cells into NOD.NCF-1^{mlJ} animals [62].

Given that the pathology of T1D is highly driven by ROS and inflammation, M1 polarized (proinflammatory) macrophages, induced by interferon-γ (IFN-γ) and TNF-α, prevail. M1 macrophages not only act as antigen-presenting cells (APC), but also as effectors by their expression of ROS (superoxide), RNS (nitric oxide), cytokines (TNF-α and IL-1β), and chemokines (CCR5, CXCR3, and CCR8) [63, 64]. These effector molecules are important not only for driving T cell activation and trafficking, but also for inducing β cell death directly. The importance of NOX in driving M1 macrophage differentiation is highlighted by Padgett et al. Here, NOX-deficient intra-islet macrophages displayed a diminished M1 phenotype, with an increase of arginase positive M2 polarized (anti-inflammatory) macrophages, due to reduced STAT1 phosphorylation [62].

NF-κB is one of the hallmark transcription factors promoting inflammation. Under non-inflammatory conditions, binding to its inhibitor, IκB sequesters NF-κB in the cytoplasm. Following activation signals, like ROS, IκB releases NF-κB and is targeted to the proteasome for degradation. NF-κB then translocates to the nucleus, binds targeted DNA regions, and drives proinflammatory gene expression [65]. NF-κB itself is a redox-dependent transcription factor, in that the cysteine 62 residue of the p50 subunit must be reduced for optimal DNA binding [66]. Macrophages isolated from the diabetes-prone NOD mouse exhibit exacerbated NF-κB activity and proinflammatory cytokine production, as compared to those from immune-competent strains [67]. The propensity for NF-κB activation by NOD macrophages also results in an enhanced ability to stimulate T cells [68].

These studies illustrate the importance of innate inflammation, highly potentiated by macrophages, and NF-κB activation of proinflammatory cytokines. Additionally, the role of NOX and continual ROS production facilitates mediators of β cell death and progression to T1D onset.

29.6 Inhibition of the Innate Immune Response by Metalloporphyrins

As discussed, macrophages are responsible for presenting β cell antigens to pathogenic autoreactive T cells. They also are the first cell type to infiltrate the islet [51, 52], rendering them key to the pathogenesis of T1D. If treatment with MnP can ameliorate the initial inflammatory events driven by macrophages, then perhaps disease pathology could be dramatically lessened. Work by Tse et al. addressed this issue with extensive studies on MnP treatment of LPS-stimulated bone marrow-derived macrophages [69].

MnP treatment of LPS-stimulated macrophages in vitro dampened superoxide and nitrite production and inhibited production of the inflammatory cytokines TNF-α and IL-1β (Fig. 29.3a, b). This suppression was neither attributed to alterations in the MAPK pathway, nor effects on phosphorylation/degradation of IκB, nor p65 nuclear translocation. In fact, MnP prevented the binding of NF-κB to a DNA consensus sequence in the TNF-α promoter (Fig. 29.3c). Since MnP possess oxido-reductase capabilities, creating an oxidizing environment, it can impair NF-κB

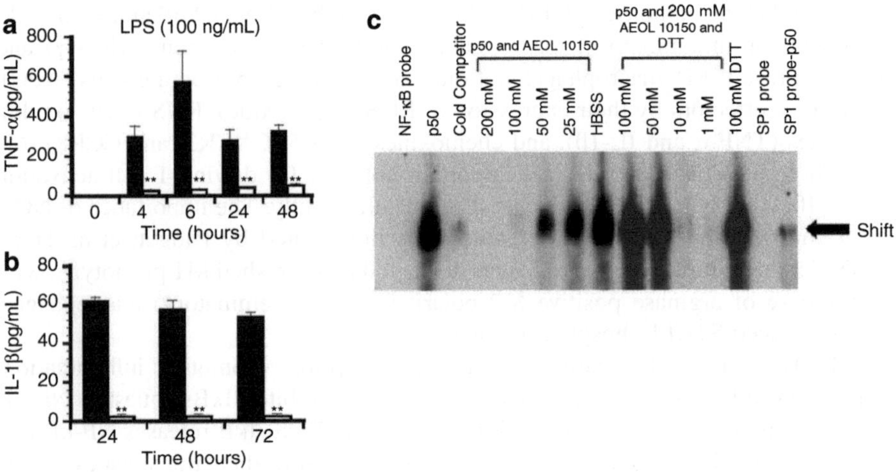

Fig. 29.3 Redox modulation of macrophages with MnP blocks proinflammatory cytokine production by inhibiting NF-κB DNA binding. (**a**) TNF-α and (**b**) IL-1β secretion was measured in culture supernatants from LPS-stimulated (100 ng/mL) macrophages (control; *black bars*) and those stimulated following MnP (MnTE-2-PyP^{5+}) treatment (*white bars*) (**$p<$–0.001). (**c**) Electrophoretic mobility shift assay (EMSA) results from purified recombinant NF-κB p50 treated with and without MnP (MnTDE-2-ImP^{5+}/AEOL 10150) to measure p50-DNA binding capacity. Results are representative of five independent experiments. Data were adapted from Tse et al. [69]

activity, as it is a redox-sensitive transcription factor. The underlying mechanism was attributed to the ability of MnP to oxidize the cysteine 62 residue of NF-κB, thereby inhibiting DNA binding and transcription of proinflammatory genes. This was the first mechanistic explanation of how MnPs control inflammation and demonstrated its potential for dampening inflammation in various diseases.

These data provided evidence for the applicability of MnP in inflammatory-mediated diseases, including diabetes. In T1D, β cells are highly susceptible to the ROS and cytokines secreted by activated macrophages, due to the islet's poor antioxidant defense. Therefore, the effect MnP has on dampening macrophage activation, while protecting β cell mass from bystander destruction, is twofold: (1) promoting β cell viability by scavenging free radicals, and (2) reducing NF-κB-mediated cytokine production. This mechanism is also pertinent in the pathology of T2D as obesity drives a chronic state of macrophage-driven innate immunity and hyperglycemia, mediating enhanced NF-κB activation.

29.6.1 T Cells

T cell activation is an intricate process that requires three specific signals in order to protect from aberrant immune activation. First, antigen must be presented in the context of major histocompatibility complex (MHC) on the APC to the T cell via its T cell receptor (TCR). Next, co-stimulatory molecules on both the APC (i.e., CD80,

CD86) and T cell (i.e., CD28) must engage [70]. With only MHC:TCR engagement and no co-stimulation, T cell anergy, or hyporesponsiveness, will result. Last, signaling molecules like cytokines and ROS are released by macrophages to drive T cell clonal expansion and effector function [71].

In the context of T1D, activated macrophages and dendritic cells (DC) within the β cell phagocytose β cell antigens and ferry them to the pancreatic draining lymph nodes (Fig. 29.2). In the draining lymph node, activated macrophages and DCs present β cell antigens to self-reactive $CD4^+$ and $CD8^+$ T cells. Once activated, these T cells traffic to the pancreas and induce β cell destruction by either direct killing ($CD8^+$ T cells) or production of inflammatory cytokines like IFN-γ ($CD4^+$ T cells) [6].

While ROS production is often associated with oxidative stress, it has been widely accepted that moderate levels of ROS are required for optimal T cell activation [72]. These levels ensure proper TCR activation and downstream signaling, NF-κB activation, as well as effector cytokine production. If the scale is tipped towards high levels of ROS, oxidative damage and apoptosis will ensue [73, 74]. Conversely, if ROS levels are too low, improper TCR signaling and suboptimal clonal expansion will result. As a source of ROS, both $CD8^+$ cytotoxic T cells and $CD4^+$ T helper cells are known to express a functional phagocyte-type NOX [75]. The T cell-expressed NOX is activated following TCR and cytokine stimulation and results in the production of hydrogen peroxide and superoxide [75–77]. Interestingly, experiments performed with αCD3/αCD28 stimulation demonstrated that ROS via the APC are not required for optimal T cell activation [75, 76]. Those produced by the T cells themselves are sufficient. These results were corroborated in crisscross assays utilizing T cells and APCs from NOD.WT and NCF-1^{m1J} mice [40]. Using this mouse model, it was concluded that ROS are required in some capacity, either from the APC or T cell, for priming the T cell response exhibited in diabetes, and defects can lead to altering the overall T helper cell phenotype, delaying disease onset.

ROS produced by APC or T cells serve several roles in T cell activation. First, activated T cells must clonally expand 100–1000-fold, and the ROS produced during MHC:TCR engagement are important mitogenic signals that push T cells out of quiescence and into cell cycle [72]. Second, as T cells transition from naïve to effector, many signaling pathways must be activated to support these changes. ROS generated during activation can serve as secondary signaling molecules to drive these cascades including the MAPK [76], PI3K/Akt [78, 79], and NF-κB [80] pathways. As a specific example, linker for activation of T cells (LAT) is an important adaptor protein for the TCR complex and is required for amplifying and mediating downstream TCR signaling [81]. LAT conformation and localization is highly dependent upon intracellular glutathione levels, highlighting the influence of cellular redox balance on T cell signaling [82, 83].

T cell metabolism has recently gained increased interest in the field. Understanding how immune cells utilize different metabolites like glucose and fatty acids has provided further insight into the regulation of various effector processes. Mitochondria are known to be pertinent ROS generators, and oxidation of glucose, fatty acids, and glutamine via the mitochondria results in ROS production [43]. Driving metabolism through the mitochondria and electron transport chain results in superoxide production and alterations in redox balance. In T cells, mitochondrial-derived superoxide plays an important role in regulating changes in T cell metabolism and activation. Studies by Sena et al. showed that mitochondrial-derived ROS via the electron transport chain are necessary and sufficient for driving T cell activation [84]. Through the use of various electron transport chain complex inhibitors, they verified that superoxide generated by complex III was most critical.

Given the pivotal role of ROS in T cell activation and T1D pathogenesis, it is not surprising that spontaneous T1D onset is significantly delayed in NOD.NCF-1^{mlJ} mice, in part, due to defective T cell responses [40]. Similarly, T1D onset was delayed in the adoptive transfer of diabetogenic $CD4^+$ T cell clones into NOD.NCF-1^{mlJ} recipients [39], highlighting the requirement for ROS during T cell activation and in the context of T1D. These genetic studies further support the hypothesis that NOX- and ROS-targeted therapies are relevant for immune modulation and may prove beneficial in autoimmune diseases like T1D.

29.7 Metalloporphyrins and T Cell Inhibition

As both $CD8^+$ and $CD4^+$ T cells play important roles in mediating T1D, our laboratory has performed extensive in vitro and in vivo work, including various T1D and transplantation models, in order to understand how MnP treatment modulates responses by both T cell subsets [46, 85–87]. Each of these models serves as a potential mimic for delaying T1D onset and preserving transplanted allogenic β cell grafts in human T1D patients.

Specifically in $CD8^+$ T cells, Sklavos et al. showed that proliferation and IFN-γ and TNF-α production were significantly inhibited by MnP treatment in a mixed lymphocyte reaction (MLR) [87]. Cytotoxicity of MnP-treated and untreated $CD8^+$ T cells was tested in an antigen-specific cytotoxicity assay. MnP treatment resulted in a 50% reduction in targeted killing by CTLs at both 100:1 and 50:1 effector:target ratios (Fig. 29.4a). MnP-treated OT-I $CD8^+$ T cells during in vitro stimulation with SIINFEKL peptide also exhibited similar reductions in IFN-γ and TNF-α production, proliferation, and targeted killing. In both models, the reduced cytotoxicity by MnP treatment was in part due to reduced expression of both perforin and granzyme B [87]. Taken together, the results from both the MLR and OT-I models indicated that MnP's ability to inhibit $CD8^+$ T cell activation and effector function was not specific to any one model, supporting its applicability to suppress diabetogenic and anti-graft T cell responses. Here, MnP's effects are relevant to islet transplantation in autoimmune prone individuals, as it would suppress both anti-graft and autoimmune responses.

Delmastro et al. investigated the effects of MnP on activation and effector function of $CD4^+$ diabetogenic T cells. These studies used the NOD.BDC-2.5.TCR.Tg mouse expressing the rearranged TCR of the diabetogenic BDC-2.5 T cell clone that respond to the β cell antigen, chromogranin [88]. Splenocytes isolated from these mice were stimulated in vitro with their cognate peptide, mimetope, with or

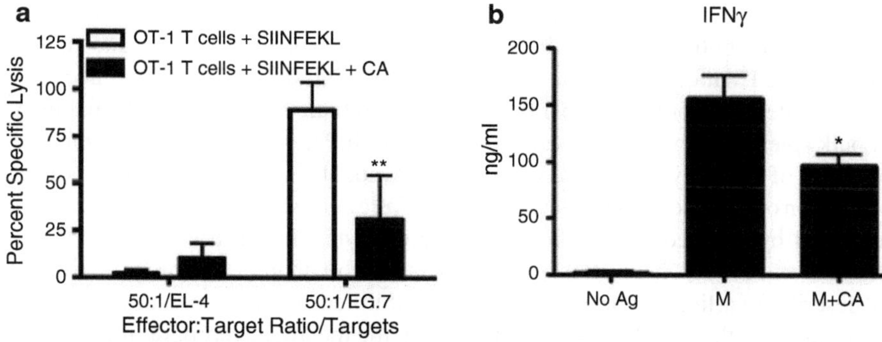

Fig. 29.4 CD8+ and CD4+ effector responses are inhibited upon treatment with MnP. (**a**) OT-I CD8+ T cells were stimulated with their cognate peptide SIINFEKL, with or without MnP (CA) for 96 h. Lytic capacity was measured following incubation with fluorescently labeled targets EL-4 (non-OVA expressing; control) or EG.7 (OVA-expressing) (**p<–0.05). Data were adapted from Sklavos et al. [87]. (**b**) BDC-2.5.TCR.Tg splenocytes were stimulated with no antigen, mimetope (M), or mimetope+MnP (M+CA) for 96 h. IFN-γ production by CD4+ T cells was measured in culture supernatants by ELISA (*p<–0.05). Data adapted from Delmastro et al. [85]

without MnP treatment. In vitro results indicated that MnP treatment inhibited CD4+ T cell activation and effector function by reducing IFN-γ production (Fig. 29.4b), proliferation, and CD69 (marker of activation) and Tbet expression [85]. These results correlated with MnP treatment inhibiting the cleavage of the T cell surface molecule Lymphocyte Activation Gene 3 (LAG-3). LAG-3 is an inhibitory receptor, and upon ligand engagement, mediates signaling that results in dampened T cell activation [89, 90]. In order for optimal T cell activation, LAG-3 must be cleaved from the T cell surface by the metalloproteases ADAM10 and ADAM17 [91]. ADAM17, like many metalloproteases, is a redox-dependent enzyme in that oxidation of specific cysteine residues must occur to change its confirmation from its latent to its activated form [92–94]. As MnP has been shown to scavenge ROS during immune cell activation, Delmastro et al. determined that MnP treatment inhibited ADAM17 activation, thereby maintaining LAG-3 expression on the T cell surface, leading to CD4+ T cell inhibition and delayed T1D onset [85].

In vivo studies by Delmastro et al. further showed that MnP treatment of diabetogenic splenocytes diminished their potential for inducing T1D in a murine adoptive transfer model. Severe combined immunodeficiency mice on the NOD background (NOD.*scid*), lacking endogenous adaptive immunity, received diabetogenic BDC-2.5.TCR.Tg splenocytes. Throughout the course of the study, serum was taken from the MnP treated and control animals, to measure levels of soluble LAG-3 (sLAG-3). Results indicated that elevated levels of sLAG-3 (suggesting CD4+ T cell activation) correlated with progression to diabetes in untreated animals [85]. Alternatively, sLAG-3 levels in MnP treated mice remained steady and all animals were euglycemic. Collectively, these results not only demonstrate that MnP treatment in vivo inhibits T1D onset, but also that sLAG-3 may serve as an early biomarker of T1D progression.

Recent work from our laboratory has demonstrated that MnP treatment can modulate the metabolic profile of diabetogenic T cells both in vivo [46] and in vitro (unpublished data). During in vivo studies, NOD.BDC-2.5.TCR.Tg mice were either treated with or without MnP for 7 days, and then various metabolic pathways of bulk splenocytes were assessed. Not only did MnP treatment inhibit diabetogenic potential of splenocytes in an adoptive transfer model, but it also resulted in reduced utilization of glucose via aerobic glycolysis. While aerobic glycolysis was decreased, indicated by reduced lactate production, splenocytes demonstrated more efficient glucose oxidation. These effects were attributed to enhanced aconitase activation, as described earlier. While this result was exhibited in naïve splenocytes, we are also exploring the effects of MnP on inhibiting the metabolic reprogramming exhibited by T cells during activation (unpublished data). Our data suggest that redox modulation may inhibit this reprogramming, indicating another layer of immunosuppressive potential for MnP.

Not only does MnP protect the β cell and modulate the innate immune response, it also has dramatic effects on the adaptive immunity involved in T1D pathogenesis. MnP's ability to modulate the behavior of specific cell types involved in all arms of the immune system makes it a powerful therapeutic option not only in T1D, but also in other complex immune-mediated diseases, including cancer [95].

29.8 Metalloporphyrins Delay T1D Onset

As mentioned previously, the NOD mouse model is used to recapitulate T1D as these mice develop spontaneous autoimmune diabetes. In addition, the adoptive transfer of BDC-2.5. splenocytes into NOD.*scid* mice leads to rapid immune-mediated β cell destruction and diabetes onset within 7–14 days. To understand the role that ROS play in the NOD spontaneous T1D model, the NOD.NCF-1^{m1J} model was studied [40]. These mice experienced a delay in spontaneous diabetes progression, in large part due to suboptimal immune activation. These results further supported targeting NOX and ROS as a potential therapeutic in T1D.

Early studies in T1D were performed using an adoptive transfer model of the diabetogenic BDC2.5 T cell clone into NOD.*scid* animals pretreated with or without MnP [86]. MnP intraperitoneal injections were continued several days during the experiment, and then ceased. Of the mice treated with MnP, 50 % developed diabetes by 28 days post-transfer, while 100 % of the untreated animals progressed to full-blown disease (Fig. 29.5a). Most notably, histology results showed decreased pancreatic immune cell infiltrate with MnP treatment. Additional in vitro studies indicated that MnP treatment reduced IFN-γ production by BDC-2.5 T cells and inhibited respiratory burst of peritoneal macrophages upon LPS stimulation [86]. Together, these results demonstrated the immunomodulatory capacity of MnP treatment in T1D and its potential therapeutic use in autoimmunity.

More recent studies in T1D with MnP have included the use of 14-day sustain release MnP pellets implanted subcutaneously at the nape of the neck in NOD mice

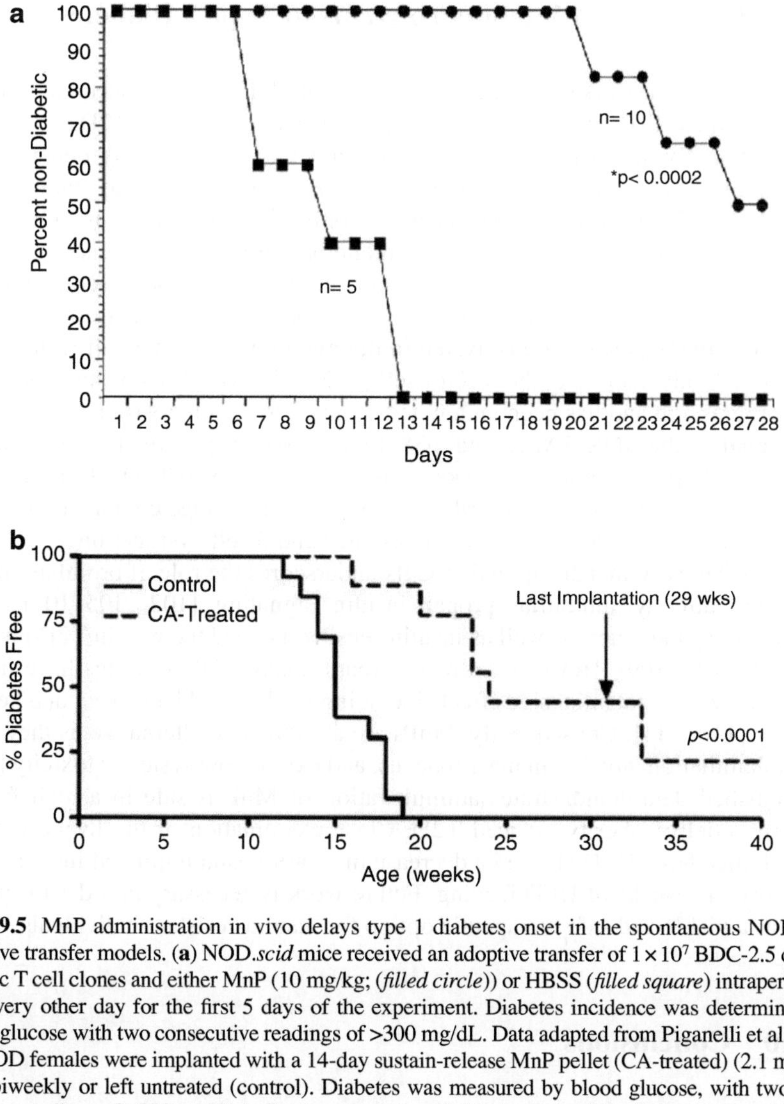

Fig. 29.5 MnP administration in vivo delays type 1 diabetes onset in the spontaneous NOD and adoptive transfer models. (**a**) NOD.*scid* mice received an adoptive transfer of 1×10^7 BDC-2.5 diabetogenic T cell clones and either MnP (10 mg/kg; (*filled circle*)) or HBSS (*filled square*) intraperitoneally every other day for the first 5 days of the experiment. Diabetes incidence was determined by blood glucose with two consecutive readings of >300 mg/dL. Data adapted from Piganelli et al. [86]. (**b**) NOD females were implanted with a 14-day sustain-release MnP pellet (CA-treated) (2.1 mg/kg/day) biweekly or left untreated (control). Diabetes was measured by blood glucose, with two consecutive readings of >300 mg/dL indicating overt disease. Data adapted from Delmastro et al. [85]

[85] (Fig. 29.5b). This administration allows for systemic MnP treatment and drastically decreases the number of administrations. In this study, all untreated NOD mice developed T1D by 20 weeks of age [85]. By 30 weeks of age, only 50 % of animals treated with MnP developed diabetes. These data, together with the data from the adoptive transfer models, further support targeting ROS for preventing T1D. Nevertheless, the protective effects of MnP end when treatment ends, as 25 % more animals progressed to diabetes following MnP treatment conclusion [85]. Overall, MnP treatment does delay diabetes onset; yet, it does not provide durable tolerance. Therefore, the need for other therapies in conjunction with MnP treatment may be necessary.

29.9 Potential of Metalloporphyrins in Type 2 Diabetes

The majority of the work on the therapeutic role that MnP plays in diabetes has focused on autoimmune T1D. However, MnPs have therapeutic potential in T2D as well.

Traditionally, the pathology and progression to T2D is heavily dependent on macrophages fueling the innate immune response [96, 97], whereas studies on the role of T cells in insulin resistance are only recently emerging. Obesity is well-characterized as a state of chronic inflammation, and the presence of adipose tissue correlates with an increase in macrophage infiltration and concomitant IL-6, TNF-α, and MCP-1 expression [98–101]. Additionally, oxidative stress occurs and stress-sensitive pathways are activated in the hyperglycemic state affecting all tissues of the body, including the β cell itself [102]. The NF-κB pathway is pivotal in T2D progression and has been undoubtedly linked to insulin resistance [103, 104]. Additionally, the JNK/SAPK and p38 MAPK pathways are also activated in response to hyperglycemia and are key in the pathobiology of T2D [102]. Sustained activation of these pathways contributes to not only late stage complication of diabetes, but also can exacerbate insulin resistance and β cell dysfunction.

Extensive work in T2D animal models underscores the role of proinflammatory cytokines directly inhibiting proper insulin signaling [102, 105–107]. Anti-inflammatory therapies, as well as insulin sensitizers, lead the way in T2D pharmaceuticals [108, 109]. However, in recent years, many of these drugs have incited controversy over potential side effects including stroke, bladder cancer, and hepatotoxicity [110, 111]. Consequently, MnPs are an attractive alternative as they regulate inflammation, notably in macrophages, and exhibit low systemic toxicity [112]. Unpublished data demonstrates administration of MnP is safe in a high fat diet (HFD) model of obesity-induced T2D. Close examination of the livers of MnP-treated mice fed a HFD showed a decrease in steatosis and improved insulin sensitivity over 12 weeks of HFD feeding. Future work is necessary in order to support MnP as a viable and safe therapeutic option for insulin-resistant individuals.

29.10 Conclusions

Diabetes mellitus is a group of diseases where pancreatic β cells are either destroyed by the immune system (T1D) or become insulin-resistant (T2D). It has been well-documented that the pathologies of both types are highly driven by oxidative stress and ROS leading to secondary inflammation [14–16]. Under normal conditions, low levels of ROS are required for mediating dynamic glucose sensing and insulin production in the β cell, as a means of regulating blood glucose. However, the pancreatic β cell demonstrates a high susceptibility to oxidative stress. This is exemplified by their physiological glucose oxidation function, which results in high ROS output, coupled with low expression of antioxidant enzymes and inefficient DNA repair. Overexpression of antioxidant enzymes in β cells alleviates oxidative stress and reduces apoptosis, indicating redox modulation as a potential therapeutic strategy for preserving β cell function and viability.

Specifically in T1D, the autoimmune threat to the β cell creates another layer of complexity with regard to treatment. ROS is required for tissue-resident macrophage activation as well as stimulating ROS production from these cells. In addition, ROS are required for optimal T cell activation following antigen encounter. Therefore, tipping the redox balance towards a more reduced state is favorable for suppressing aberrant immune responses, like those in autoimmunity and T1D.

Since alleviating ROS production results in enhanced β cell function and immunosuppression, therapies using manganese metalloporphyrins would circumvent both of the main issues faced in T1D. In addition, we postulate that due to the inflammatory nature of T2D, MnPs would likely limit inflammation and restore insulin sensitivity. While these agents are not yet approved for use in diabetic patients, currently, clinical trials in Canada are being conducted to further study the efficacy of MnPs in preserving β cells following donor islet-isolation and transplantation. With these trials underway, MnPs are one step closer to becoming a regular therapeutic option for patients suffering from diabetes.

References

1. Prevention, C.f.D.C.a., National Diabetes Statistics Report: estimates of diabetes and its burden in the United States, 2014., U.D.o.H.a.H. Services, Editor. 2014: Atlanta.
2. Federation ID. Complication of diabetes. 2015. http://www.idf.org/complications-diabetes.
3. Bluestone JA, Herold K, Eisenbarth G. Genetics, pathogenesis and clinical interventions in type 1 diabetes. Nature. 2010;464(7293):1293–300.
4. Knip M, Simell O. Environmental triggers of type 1 diabetes. Cold Spring Harb Perspect Med. 2012;2(7):a007690.
5. Achenbach P, et al. Natural history of type 1 diabetes. Diabetes. 2005;54 Suppl 2:S25–31.
6. Delmastro MM, Piganelli JD. Oxidative stress and redox modulation potential in type 1 diabetes. Clin Dev Immunol. 2011;2011:593863.
7. Bodin J, Stene LC, Nygaard UC. Can exposure to environmental chemicals increase the risk of diabetes Type 1 development? BioMed Res Int. 2015;2015:19.
8. Willcox A, et al. Analysis of islet inflammation in human type 1 diabetes. Clin Exp Immunol. 2009;155(2):173–81.
9. Foulis AK, Stewart JA. The pancreas in recent-onset type 1 (insulin-dependent) diabetes mellitus: insulin content of islets, insulitis and associated changes in the exocrine acinar tissue. Diabetologia. 1984;26(6):456–61.
10. Atkinson MA, Leiter EH. The NOD mouse model of type 1 diabetes: as good as it gets? Nat Med. 1999;5(6):601–4.
11. Gregor MF, Hotamisligil GS. Inflammatory mechanisms in obesity. Annu Rev Immunol. 2011;29:415–45.
12. Butler AE, et al. Beta-cell deficit and increased beta-cell apoptosis in humans with type 2 diabetes. Diabetes. 2003;52(1):102–10.
13. Lutz TA, Woods SC. Overview of animal models of obesity. Curr Protoc Pharmacol. 2012;Chapter 5:Unit5 61.
14. Haskins K, et al. Oxidative stress in type 1 diabetes. Ann N Y Acad Sci. 2003;1005:43–54.

15. Rabinovitch A. Free radicals as mediators of pancreatic islet beta-cell injury in autoimmune diabetes. J Lab Clin Med. 1992;119(5):455–6.
16. Nakayama M, et al. Increased expression of NAD(P)H oxidase in islets of animal models of Type 2 diabetes and its improvement by an AT1 receptor antagonist. Biochem Biophys Res Commun. 2005;332(4):927–33.
17. Lenzen S. Oxidative stress: the vulnerable beta-cell. Biochem Soc Trans. 2008;36(Pt 3):343–7.
18. Batinic-Haberle I, et al. New class of potent catalysts of O_2-dismutation. Mn(III) ortho-methoxyethylpyridyl- and di-ortho-methoxyethylimidazolylporphyrins. Dalton Trans. 2004; 11:1696–702.
19. Batinic-Haberle I, et al. SOD therapeutics: latest insights into their structure-activity relationships and impact on the cellular redox-based signaling pathways. Antioxid Redox Signal. 2014;20(15):2372–415.
20. Batinic-Haberle I, Tovmasyan A, Spasojevic I. An educational overview of the chemistry, biochemistry and therapeutic aspects of Mn porphyrins - From superoxide dismutation to HO-driven pathways. Redox Biol. 2015;5:43–65.
21. Batinic-Haberle I, et al. Design of Mn porphyrins for treating oxidative stress injuries and their redox-based regulation of cellular transcriptional activities. Amino Acids. 2012;42(1):95–113.
22. Delmastro-Greenwood MM, Tse HM, Piganelli JD. Effects of metalloporphyrins on reducing inflammation and autoimmunity. Antioxid Redox Signal. 2013;20(15):2465–77.
23. Antico Arciuch VG, et al. Mitochondrial regulation of cell cycle and proliferation. Antioxid Redox Signal. 2012;16(10):1150–80.
24. Morgan D, et al. Association of NAD(P)H oxidase with glucose-induced insulin secretion by pancreatic beta-cells. Endocrinology. 2009;150(5):2197–201.
25. Newsholme P, et al. Insights into the critical role of NADPH oxidase(s) in the normal and dysregulated pancreatic beta cell. Diabetologia. 2009;52(12):2489–98.
26. Pi J, et al. Reactive oxygen species as a signal in glucose-stimulated insulin secretion. Diabetes. 2007;56(7):1783–91.
27. Robertson RP, Harmon JS. Diabetes, glucose toxicity, and oxidative stress: A case of double jeopardy for the pancreatic islet beta cell. Free Radic Biol Med. 2006;41(2):177–84.
28. Lenzen S, Drinkgern J, Tiedge M. Low antioxidant enzyme gene expression in pancreatic islets compared with various other mouse tissues. Free Radic Biol Med. 1996;20(3):463–6.
29. Modak MA, Parab PB, Ghaskadbi SS. Pancreatic islets are very poor in rectifying oxidative DNA damage. Pancreas. 2009;38(1):23–9.
30. Fridlyand LE, Philipson LH. Does the glucose-dependent insulin secretion mechanism itself cause oxidative stress in pancreatic beta-cells? Diabetes. 2004;53(8):1942–8.
31. Azevedo-Martins AK, et al. Improvement of the mitochondrial antioxidant defense status prevents cytokine-induced nuclear factor-kappaB activation in insulin-producing cells. Diabetes. 2003;52(1):93–101.
32. Gurgul E, et al. Mitochondrial catalase overexpression protects insulin-producing cells against toxicity of reactive oxygen species and proinflammatory cytokines. Diabetes. 2004;53(9):2271–80.
33. Taylor-Fishwick DA. NOX, NOX who is there? The contribution of NADPH oxidase one to beta cell dysfunction. Front Endocrinol (Lausanne). 2013;4:40.
34. Oliveira HR, et al. Pancreatic beta-cells express phagocyte-like NAD(P)H oxidase. Diabetes. 2003;52(6):1457–63.
35. Uchizono Y, et al. Expression of isoforms of NADPH oxidase components in rat pancreatic islets. Life Sci. 2006;80(2):133–9.
36. Drummond GR, et al. Combating oxidative stress in vascular disease: NADPH oxidases as therapeutic targets. Nat Rev Drug Discov. 2011;10(6):453–71.
37. Weaver JR, Taylor-Fishwick DA. Regulation of NOX-1 expression in beta cells: a positive feedback loop involving the Src-kinase signaling pathway. Mol Cell Endocrinol. 2013;369(1-2): 35–41.

38. Morgan D, et al. Glucose, palmitate and pro-inflammatory cytokines modulate production and activity of a phagocyte-like NADPH oxidase in rat pancreatic islets and a clonal beta cell line. Diabetologia. 2007;50(2):359–69.
39. Thayer TC, et al. Superoxide production by macrophages and T cells is critical for the induction of autoreactivity and type 1 diabetes. Diabetes. 2011;60(8):2144–51.
40. Tse HM, et al. NADPH oxidase deficiency regulates Th lineage commitment and modulates autoimmunity. J Immunol. 2010;185(9):5247–58.
41. Weaver JR, Grzesik W, Taylor-Fishwick DA. Inhibition of NADPH oxidase-1 preserves beta cell function. Diabetologia. 2015;58(1):113–21.
42. West IC. Radicals and oxidative stress in diabetes. Diabet Med. 2000;17(3):171–80.
43. Sivitz WI, Yorek MA. Mitochondrial dysfunction in diabetes: from molecular mechanisms to functional significance and therapeutic opportunities. Antioxid Redox Signal. 2010;12(4):537–77.
44. Gardner PR. Superoxide-driven aconitase FE-S center cycling. Biosci Rep. 1997;17(1):33–42.
45. Raza H, et al. Impaired mitochondrial respiratory functions and oxidative stress in streptozotocin-induced diabetic rats. Int J Mol Sci. 2011;12(5):3133–47.
46. Delmastro-Greenwood MM, et al. Mn porphyrin regulation of aerobic glycolysis: implications on the activation of diabetogenic immune cells. Antioxid Redox Signal. 2013;19(16): 1902–15.
47. Rosenberg L, et al. Structural and functional changes resulting from islet isolation lead to islet cell death. Surgery. 1999;126(2):393–8.
48. Bottino R, et al. Preservation of human islet cell functional mass by anti-oxidative action of a novel SOD mimic compound. Diabetes. 2002;51(8):2561–7.
49. Sklavos MM, et al. Redox modulation protects islets from transplant-related injury. Diabetes. 2010;59(7):1731–8.
50. Bottino R, et al. Response of human islets to isolation stress and the effect of antioxidant treatment. Diabetes. 2004;53(10):2559–68.
51. Kolb-Bachofen V, Kolb H. A role for macrophages in the pathogenesis of type 1 diabetes. Autoimmunity. 1989;3(2):145–54.
52. O'Reilly LA, et al. Characterization of pancreatic islet cell infiltrates in NOD mice: effect of cell transfer and transgene expression. Eur J Immunol. 1991;21(5):1171–80.
53. Chen MC, et al. Monocyte chemoattractant protein-1 is expressed in pancreatic islets from prediabetic NOD mice and in interleukin-1 beta-exposed human and rat islet cells. Diabetologia. 2001;44(3):325–32.
54. Dahlen E, et al. Dendritic cells and macrophages are the first and major producers of TNF-alpha in pancreatic islets in the nonobese diabetic mouse. J Immunol. 1998;160(7):3585–93.
55. Arnush M, et al. IL-1 produced and released endogenously within human islets inhibits beta cell function. J Clin Invest. 1998;102(3):516–26.
56. Calderon B, Suri A, Unanue ER. In CD4+ T-cell-induced diabetes, macrophages are the final effector cells that mediate islet beta-cell killing: studies from an acute model. Am J Pathol. 2006;169(6):2137–47.
57. Burkart V, Kolb H. Macrophages in islet destruction in autoimmune diabetes mellitus. Immunobiology. 1996;195(4-5):601–13.
58. Jiang F, Zhang Y, Dusting GJ. NADPH oxidase-mediated redox signaling: roles in cellular stress response, stress tolerance, and tissue repair. Pharmacol Rev. 2011;63(1):218–42.
59. Vignais PV. The superoxide-generating NADPH oxidase: structural aspects and activation mechanism. Cell Mol Life Sci. 2002;59(9):1428–59.
60. Tse HM, et al. Disruption of innate-mediated proinflammatory cytokine and reactive oxygen species third signal leads to antigen-specific hyporesponsiveness. J Immunol. 2007;178(2): 908–17.
61. Seleme MC, et al. Dysregulated TLR3-dependent signaling and innate immune activation in superoxide-deficient macrophages from nonobese diabetic mice. Free Radic Biol Med. 2012;52(9):2047–56.

62. Padgett LE, et al. Loss of NADPH oxidase-derived superoxide skews macrophage phenotypes to delay type 1 diabetes. Diabetes. 2015;64(3):937–46.
63. Cantor J, Haskins K. Recruitment and activation of macrophages by pathogenic CD4 T cells in type 1 diabetes: evidence for involvement of CCR8 and CCL1. J Immunol. 2007;179(9):5760–7.
64. Cantor J, Haskins K. Effector function of diabetogenic CD4 Th1 T cell clones: a central role for TNF-alpha. J Immunol. 2005;175(11):7738–45.
65. Vallabhapurapu S, Karin M. Regulation and function of NF-kappaB transcription factors in the immune system. Annu Rev Immunol. 2009;27:693–733.
66. Matthews JR, et al. Thioredoxin regulates the DNA binding activity of NF-kappa B by reduction of a disulphide bond involving cysteine 62. Nucleic Acids Res. 1992;20(15):3821–30.
67. Sen P, et al. NF-kappa B hyperactivation has differential effects on the APC function of nonobese diabetic mouse macrophages. J Immunol. 2003;170(4):1770–80.
68. Poligone B, et al. Elevated NF-kappaB activation in nonobese diabetic mouse dendritic cells results in enhanced APC function. J Immunol. 2002;168(1):188–96.
69. Tse HM, Milton MJ, Piganelli JD. Mechanistic analysis of the immunomodulatory effects of a catalytic antioxidant on antigen-presenting cells: implication for their use in targeting oxidation-reduction reactions in innate immunity. Free Radic Biol Med. 2004;36(2):233–47.
70. Murphy K, et al. Janeway's immunobiology, vol. xix. 8th ed. New York: Garland Science; 2012. 868 p.
71. Curtsinger JM, et al. Inflammatory cytokines provide a third signal for activation of naive CD4+ and CD8+ T cells. J Immunol. 1999;162(6):3256–62.
72. Kesarwani P, et al. Redox regulation of T-cell function: from molecular mechanisms to significance in human health and disease. Antioxid Redox Signal. 2013;18(12):1497–534.
73. Hildeman DA, et al. Reactive oxygen species regulate activation-induced T cell apoptosis. Immunity. 1999;10(6):735–44.
74. Williams MS, Henkart PA. Role of reactive oxygen intermediates in TCR-induced death of T cell blasts and hybridomas. J Immunol. 1996;157(6):2395–402.
75. Jackson SH, et al. T cells express a phagocyte-type NADPH oxidase that is activated after T cell receptor stimulation. Nat Immunol. 2004;5(8):818–27.
76. Devadas S, et al. Discrete generation of superoxide and hydrogen peroxide by T cell receptor stimulation: selective regulation of mitogen-activated protein kinase activation and fas ligand expression. J Exp Med. 2002;195(1):59–70.
77. Purushothaman D, Sarin A. Cytokine-dependent regulation of NADPH oxidase activity and the consequences for activated T cell homeostasis. J Exp Med. 2009;206(7):1515–23.
78. Wani R, et al. Isoform-specific regulation of Akt by PDGF-induced reactive oxygen species. Proc Natl Acad Sci U S A. 2011;108(26):10550–5.
79. Wani R, et al. Oxidation of Akt2 kinase promotes cell migration and regulates G1-S transition in the cell cycle. Cell Cycle. 2011;10(19):3263–8.
80. Suzuki YJ, Forman HJ, Sevanian A. Oxidants as stimulators of signal transduction. Free Radic Biol Med. 1997;22(1-2):269–85.
81. Zhang W, et al. Functional analysis of LAT in TCR-mediated signaling pathways using a LAT-deficient Jurkat cell line. Int Immunol. 1999;11(6):943–50.
82. Gringhuis SI, et al. Displacement of linker for activation of T cells from the plasma membrane due to redox balance alterations results in hyporesponsiveness of synovial fluid T lymphocytes in rheumatoid arthritis. J Immunol. 2000;164(4):2170–9.
83. Gringhuis SI, et al. Effect of redox balance alterations on cellular localization of LAT and downstream T-cell receptor signaling pathways. Mol Cell Biol. 2002;22(2):400–11.
84. Sena LA, et al. Mitochondria are required for antigen-specific T cell activation through reactive oxygen species signaling. Immunity. 2013;38(2):225–36.
85. Delmastro MM, et al. Modulation of redox balance leaves murine diabetogenic TH1 T cells "LAG-3-ing" behind. Diabetes. 2012;61(7):1760–8.
86. Piganelli JD, et al. A metalloporphyrin-based superoxide dismutase mimic inhibits adoptive transfer of autoimmune diabetes by a diabetogenic T-cell clone. Diabetes. 2002;51(2):347–55.

87. Sklavos MM, Tse HM, Piganelli JD. Redox modulation inhibits CD8 T cell effector function. Free Radic Biol Med. 2008;45(10):1477–86.
88. Dobbs C, Haskins K. Comparison of a T cell clone and of T cells from a TCR transgenic mouse: TCR transgenic T cells specific for self-antigen are atypical. J Immunol. 2001;166(4): 2495–504.
89. Workman CJ, et al. Phenotypic analysis of the murine CD4-related glycoprotein, CD223 (LAG-3). Eur J Immunol. 2002;32(8):2255–63.
90. Workman CJ, Vignali DA. The CD4-related molecule, LAG-3 (CD223), regulates the expansion of activated T cells. Eur J Immunol. 2003;33(4):970–9.
91. Li N, et al. Metalloproteases regulate T-cell proliferation and effector function via LAG-3. EMBO J. 2007;26(2):494–504.
92. Edwards DR, Handsley MM, Pennington CJ. The ADAM metalloproteinases. Mol Aspects Med. 2008;29(5):258–89.
93. Van Wart HE, Birkedal-Hansen H. The cysteine switch: a principle of regulation of metalloproteinase activity with potential applicability to the entire matrix metalloproteinase gene family. Proc Natl Acad Sci U S A. 1990;87(14):5578–82.
94. Wang Y, et al. Regulation of mature ADAM17 by redox agents for L-selectin shedding. J Immunol. 2009;182(4):2449–57.
95. Oberley-Deegan RE, et al. The antioxidant, MnTE-2-PyP, prevents side-effects incurred by prostate cancer irradiation. PLoS One. 2012;7(9), e44178.
96. Wellen KE, Hotamisligil GS. Obesity-induced inflammatory changes in adipose tissue. J Clin Invest. 2003;112(12):1785–8.
97. Weisberg SP, et al. Obesity is associated with macrophage accumulation in adipose tissue. J Clin Invest. 2003;112(12):1796–808.
98. Tataranni PA, Ortega E. A burning question: does an adipokine-induced activation of the immune system mediate the effect of overnutrition on type 2 diabetes? Diabetes. 2005; 54(4):917–27.
99. Pickup JC, Crook MA. Is type II diabetes mellitus a disease of the innate immune system? Diabetologia. 1998;41(10):1241–8.
100. Pickup JC. Inflammation and activated innate immunity in the pathogenesis of type 2 diabetes. Diabetes Care. 2004;27(3):813–23.
101. Kanda H, et al. MCP-1 contributes to macrophage infiltration into adipose tissue, insulin resistance, and hepatic steatosis in obesity. J Clin Invest. 2006;116(6):1494–505.
102. Evans JL, et al. Are oxidative stress-activated signaling pathways mediators of insulin resistance and beta-cell dysfunction? Diabetes. 2003;52(1):1–8.
103. Arkan MC, et al. IKK-beta links inflammation to obesity-induced insulin resistance. Nat Med. 2005;11(2):191–8.
104. Cai D, et al. Local and systemic insulin resistance resulting from hepatic activation of IKK-beta and NF-kappaB. Nat Med. 2005;11(2):183–90.
105. Hirosumi J, et al. A central role for JNK in obesity and insulin resistance. Nature. 2002;420(6913):333–6.
106. Hotamisligil GS. Inflammatory pathways and insulin action. Int J Obes Relat Metab Disord. 2003;27 Suppl 3:S53–5.
107. Tuncman G, et al. Functional in vivo interactions between JNK1 and JNK2 isoforms in obesity and insulin resistance. Proc Natl Acad Sci U S A. 2006;103(28):10741–6.
108. Tran L, et al. Pharmacologic treatment of Type 2 diabetes: injectable medications. Ann Pharmacother. 2015;49(6):700–14.
109. Tran L, et al. Pharmacologic treatment of Type 2 diabetes: oral medications. Ann Pharmacother. 2015;49(5):540–56.
110. Burcharth J. The epidemiology and risk factors for recurrence after inguinal hernia surgery. Dan Med J. 2014;61(5):B4846.
111. Marks DH. Drug utilization, safety and clinical use of Actos and Avandia. Int J Risk Saf Med. 2013;25(1):39–51.
112. Gad SC, et al. A nonclinical safety assessment of MnTE-2-PyP, a manganese porphyrin. Int J Toxicol. 2013;32(4):274–87.

Chapter 30
Redox-Active Metal Complexes in Trypanosomatids

Cynthia Demicheli, Frédéric Frézard, and Nicholas P. Farrell

Abbreviations

ABC	ATP-binding cassette
Cys	Cysteine
Cys-Gly	Cysteinyl-glycine
GSH	Glutathione
iNOS	Inducible nitric oxide synthase
NO	Nitric oxide
MRPA	Multidrug resistance-associated protein A
ROS	Reactive oxygen species
TR	Trypanothione reductase
$T(SH)_2$	Reduced trypanothione
WHO	World Health Organization

C. Demicheli (✉)
Departamento de Química, Instituto de Ciências Exatas, Universidade Federal de Minas Gerais, Av. Antônio Carlos 6627, Pampulha, 31270-901 Belo Horizonte, MG, Brazil
e-mail: demichel@netuno.lcc.ufmg.br

F. Frézard
Departamento de Fisiologia e Biofísica, Instituto de Ciências Biológicas, Universidade Federal de Minas Gerais, Av. Antônio Carlos 6627, Pampulha, 31270-901 Belo Horizonte, MG, Brazil
e-mail: frezard@icb.ufmg.br

N.P. Farrell
Department of Chemistry, Virginia Commonwealth University, 1001 W. Main Street, Richmond, VA 23220, USA
e-mail: npfarrell@vcu.edu

I. Batinić-Haberle et al. (eds.), *Redox-Active Therapeutics*, Oxidative Stress in Applied Basic Research and Clinical Practice, DOI 10.1007/978-3-319-30705-3_30

30.1 Historical Use of Metal Complexes Against Trypanosomatids

Trypanosomatids are protozoan parasites that cause various diseases in human, such as leishmaniasis, Chagas disease, and sleeping sickness. According to World Health Organization (WHO), trypanosomiasis and leishmaniasis are the most challenging among the neglected tropical diseases [1, 2].

At the beginning of the last century, Paul Ehrlich, considered the father of modern chemotherapy, discovered the anti-trypanosomal arsenic-based drug, atoxyl, for treatment of sleeping sickness [3]. However, atoxyl caused side effects, mainly blindness. In 1934, Ernst Friedheim, in the search for a safer drug, designed and developed melarsoprol which saved three millions of lives in the 1940s [4]. This arsenical drug is still in use for sleeping sickness, but is limited to cases of advanced infections with *Trypanosoma brucei gambiense* and *rhodesiense*. More recently, trivalent arsenicals in the form of As_2O_3 started to be used clinically in the treatment of acute promyelocytic leukemia [5].

Antimony-based drugs have also a long history of use in the treatment of infectious diseases. The importance of antimony in the early medicine is well-documented due to the debate created around their utilization in this period [6]. The first antimony compounds prepared for medicinal applications were introduced in the sixteenth and seventeenth centuries, with emphasis on antimony(III) potassium tartrate (tartar emetic). Tartar emetic was first obtained placing sour wine in glasses of antimony metal and its use was prescribed for many diseases, especially the lung diseases. During the late nineteenth century, tartar emetic was used for fever and pneumonia and its use declined slowly until the beginning of twentieth century.

At the beginning of the last century, the Brazilian physician Gaspar Vianna, pioneer researcher in the treatment of leishmaniasis, reported the efficacy of tartar emetic for clinical treatment of muco-cutaneous leishmaniasis [7]. In India, Brahmachari discovered in 1912 the organoantimonial compound urea stibamine, the first effective drug against visceral leishmaniasis, and was a nominee for the Nobel Prize in 1929 for this finding [8].

From the 1940s, these antimonials were substituted by the less toxic pentavalent antimony (Sb(V)) complexes, meglumine antimoniate and sodium stibogluconate (Fig. 30.1), in the therapeutics of leishmaniases. Typically, these Sb(V) drugs are given at 20 mg of Sb/kg per day intramuscularly or intravenously for 20–30 days [9]. Even though pentavalent antimonials are still the first-line drugs against all forms of leishmaniasis in most developing countries, their mechanism of action is still not fully understood and their use in the clinical setting is limited by their side effects and the emergence of resistance to antimony [9].

Recent advances in understanding the molecular and cellular biology of leishmaniasis as well as the cellular accumulation pathways, mechanisms of resistance, and target identification allow a more systematic rationale-based approach for development of new anti-parasitic drugs [10–12]. In this context, complexes with other metals including gold showed promising pharmacological activities.

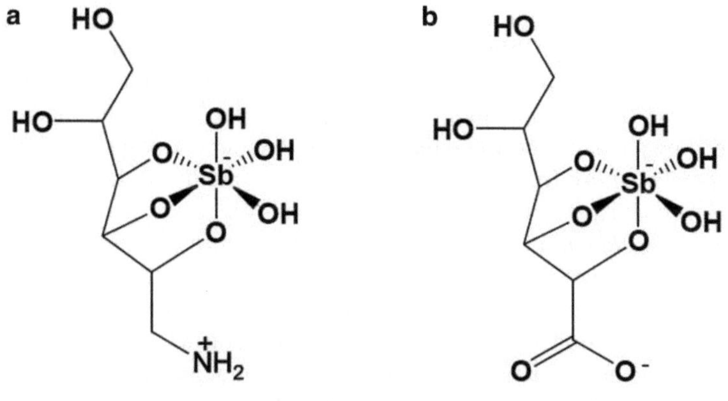

Fig. 30.1 Structural formula proposed for meglumine antimoniate (**a**) and stibogluconate (**b**). Adapted from Frézard et al. [9]

Most of the progress regarding drug development and mechanism of action was achieved using cell and animal models of leishmaniasis [13]. The in vitro models include extracellular log-phase leishmania promastigotes and established axenic leishmania amastigotes and the intracellular amastigotes in primary mouse peritoneal macrophages or monocyte transformed macrophages as host cells. The in vivo models use mostly inbred strains of mice infected with leishmania species causing either cutaneous or visceral leishmaniasis.

Interestingly, in the case of these metal-based compounds, evidence was obtained that those exert their antiparasitic action by causing oxidative stress. This chapter will describe in details the current knowledge on their mechanism of action, with emphasis on antimonial drugs and gold complexes for leishmaniasis.

30.2 Importance of Redox Systems in the Host–Leishmania Parasite Relationship

The protozoan leishmania parasite has a relatively simple life cycle with two principal stages: the flagellated mobile promastigote living in the gut of the sandfly vector and the intracellular amastigote within phagolysosomal vesicles of the vertebrate host macrophage. After recognition of *Leishmania* spp., macrophages are activated, triggering phagocytosis and various cellular processes to destroy the parasite. These cellular processes include production of phagolysosomal degradation enzymes, oxidative burst generation, and nitric oxide (NO) production. The oxidative burst provoked by the enzyme NADPH oxidase is a result of the increase in oxygen consumption as a consequence of the phagocytosis process. After macrophage activation, increased concentrations of various cytokines such as IFN-γ and TNF-α enhance NADPH oxidase activity and subsequently production of reactive oxygen species (ROS), such as superoxide radical. The production of superoxide radical leads to the spontaneous or enzymatic formation of hydrogen peroxide, hydroxyl radical, hypochlorite, and peroxynitrite. The increased NO and NO-metabolite levels in activated macrophages are the result of inducible nitric oxide synthase (iNOS) activation. Parasite persistence within the macrophages is determined by a balance between the ability of the immune response to sufficiently activate *Leishmania*-infected macrophages and the ability of the parasite to resist cytotoxic mechanisms of macrophage activation [14].

Although *Leishmania* species are susceptible in vitro to exogenous superoxide radical, hydrogen peroxide, nitric oxide, and peroxynitrite, they manage to survive the endogenous oxidative burst during phagocytosis and the subsequent elevated nitric oxide production in the macrophage. The parasite adopts various defense mechanisms against oxidative stress: the lipophosphoglycan membrane decreases superoxide radical production by inhibiting NADPH oxidase assembly and the parasite also protects itself through antioxidant enzymes [15]. Among the various parasite defense mechanisms against host attack, thiol metabolism appears as a first-line defense. One major system involved in the redox homeostasis in trypanosomatids is the trypanothione ($T(SH)_2$)/trypanothione reductase (TR)system. The ($T(SH)_2$)/TR system, which keeps $T(SH)_2$ under the reduced state, replaces the nearly ubiquitous glutathione/GR system, protects trypanosomatids from oxidative damage, and delivers the reducing equivalents for DNA synthesis [16].

It has been reported that $T(SH)_2$ is capable of reducing NO (generated by the host cell) and iron into harmless stable dinitrosyl iron complex with 600 time greater affinity than mammalian glutathione (GSH) reductase system [17]. This is the mechanism by which the parasite protects itself from such lethal environment.

30.3 Pentavalent Antimonial Drugs Against Leishmaniasis

The metabolism and mechanism of action of pentavalent antimonials against leishmania parasites are not fully understood [9]. However, the data available so far converge towards the central role of redox processes.

30.3.1 Drug Activation Though Metal Reduction

It is generally assumed that Sb(V) behaves as a prodrug, undergoing reduction within the organism into the more toxic and active trivalent form. This model is supported by several reports that Sb(V) is reduced into Sb(III) in the vertebrate host and that reduction could also happen in the parasite [18–22]. However, the exact microenvironment where the reduction occurs is still unclear.

Leishmania promastigote forms are insensitive to Sb(V). On the other hand, amastigote-like cultured parasites are sensitive to pentavalent antimonials, suggesting the occurrence of intracellular Sb(V) reduction in this life-cycle stage [23].

However, in the latter case, thiol availability from the culture medium and low pH as well as high temperature may also be determinant factors in Sb(V) to Sb(III) reduction and consequently high sensitivity to Sb(V) [24, 25].

The higher sensitivity of intramacrophagic amastigotes to Sb(V) compared to axenic amastigote-like favors the hypothesis of host-associated Sb(V) to Sb(III) reduction [23].

Thiols have been reported as potential reducing agents in this conversion [24, 26]. Reduction of Sb(V) to Sb(III) was found to occur spontaneously in the presence of the following thiols: (1) GSH, which is the main thiol in the cytosol of mammalian cells; (2) cysteine (Cys) and cysteinyl-glycine (Cys-Gly) found predominantly within lysosomes; and (3) the bis(glutathione)-spermidine conjugate $T(SH)_2$, which is the main thiol within the parasite [27].

The observations that Cys, Cys-Gly, and $T(SH)_2$ are more effective reducing agents than GSH and that this reaction is favored in acidic pH [18] led to the hypothesis that Sb(V) may be reduced in vivo by $T(SH)_2$ within leishmania parasites and by Cys or Cys-Gly within the acidic compartments of mammalian cells. Both promastigotes and intracellular amastigotes maintain intracellular pH values close to neutral (~7) even in the presence of extracellular acidic pH (~4–5). This observation reinforces the hypothesis that thiol-mediated Sb(V) to Sb(III) reduction takes place preferentially in the host cell.

Other studies have suggested the participation of a parasite-specific enzymes in the process of reduction of Sb(V) to Sb(III) [28, 29]. However, these parasite-specific enzymes are expressed in both life-cycle stages and therefore it is difficult to accommodate the different toxicity of pentavalent antimony toward promastigotes and amastigotes.

Recently, the pentavalent antimonial drug meglumine antimoniate was found to contain up to 30 % of Sb(III), indicating that the mode of action of this drug could be mediated by this residual amount of Sb(III) [30]. Additionally, the availability of Sb(III) increased at low pH values, suggesting that this drug may act as molecular carrier releasing the active $Sb(OH)_3$ form specifically in the acidic intracellular compartment where leishmania parasite resides.

30.3.2 Antimony-Induced Redox Imbalance

Sb(III) is classified as a borderline metal ion within Pearson's hard-soft acid-base theory and has a high affinity towards nitrogen- and sulfhydryl-containing ligands. The anti-leishmanial mechanisms of Sb(III) are probably related to its interaction with sulfhydryl-containing biomolecules, including thiols, peptides, proteins, and enzymes.

Thus, Sb(III) was found to form stable complex with the major intracellular thiols, GSH and $T(SH)_2$, in the form of a 1:3 and 1:1 Sb-thiol complexes, respectively [24, 26, 31]. Once Sb(III) is in the cell and conjugated to $T(SH)_2$, the complex can be sequestered inside a vacuole or extruded by ATP-binding cassette (ABC) transporters [32, 33].

It has been reported that Sb(III) interferes with the thiol metabolism, not only by inducing efflux of intracellular $T(SH)_2$ and GSH from intact leishmania cells, but also by inhibiting TR through specific interaction with the redox-active catalytic site [34], resulting in lower $[T(SH)_2]/[[T(S)_2]$ ratio [35]. Both actions synergistically contribute to lowering the parasite-neutralizing capacity of reactive species coming from the host. It is worth noting that similar effects in thiol metabolism were observed with Sb(V) on axenic amastigotes in these studies. Now, it is also clearly established that Sb(III) triggered apoptotic cell death associated with ROS [36].

Whether the oxidative stress resulting from metal interference in thiol metabolism is sufficient to promote cell death or additional event involving metal interaction with other molecular target(s) takes place still has to be determined. In that sense, recent studies have identified zinc-finger protein as potential target of Sb(III) [37, 38]. The zinc finger domain is characterized by the coordination of a zinc atom by several amino acid residues, including cysteine and histidine. Zinc finger proteins sharing the CCHC motif have been identified in trypanosomatids and are likely to be involved in DNA replication, structure, and repair [39]. On the other hand, CCCH zinc finger domains, which are found mainly in RNA-binding proteins with regulatory functions at all stages of mRNA metabolism [40], are suspected to play a crucial role in the biology of kinetoplastid protozoa, because of the unusual emphasis on post-transcriptional control of gene expression in this group of organisms [41]. Interestingly, Sb(III) promotes Zn(II) ejection more effectively from CCCH zinc finger peptides than CCHC peptides [38], suggesting that the action of antimonial drugs could be related mainly to interaction of Sb(III) with CCCH zinc finger proteins.

Despite the strong evidence that Sb(III) mediates the antileishmanial action of pentavalent antimonials, some studies have suggested intrinsic pharmacological actions of Sb(V) [42–44].

Demicheli and co-workers have reported the formation of a complex between adenine ribonucleoside and Sb(V). This was the first report of a physiologically relevant biomolecule capable of forming stable complexes with Sb(V). Both 1:1 and 1:2 Sb(V)-ribonucleoside complexes were evidenced [19, 42, 45–47]. The large NMR resonance changes for H2′signal suggested that –OH groups in the ribosome are the binding sites for Sb(V) probably via ring chelation at C2′ and C3′.

Stibogluconate was found to be a potent inhibitor of protein tyrosine phosphatases, which leads to an increase in cytokine responses [44]. Another recent study revealed that this drug induced generation of ROS and NO via phosphoinositide 3-kinase and mitogen-activated protein kinase activation in *Leishmania donovani*-infected macrophages [48, 49], indicating that Sb(V) can stimulate the innate arm of the immune system. Meglumine antimoniate was also reported to increase the phagocytic capacity of monocytes and neutrophils and enhance superoxide anion production by phagocytes, which represent the first line of defense against the parasite [50].

As summarized in Fig. 30.2, these data taken altogether suggest that pentavalent antimonials affect the parasite viability through both Sb(III)-induced imbalance of thiol metabolism in parasite and Sb(V)-induced stimulation of macrophage microbicidal activity.

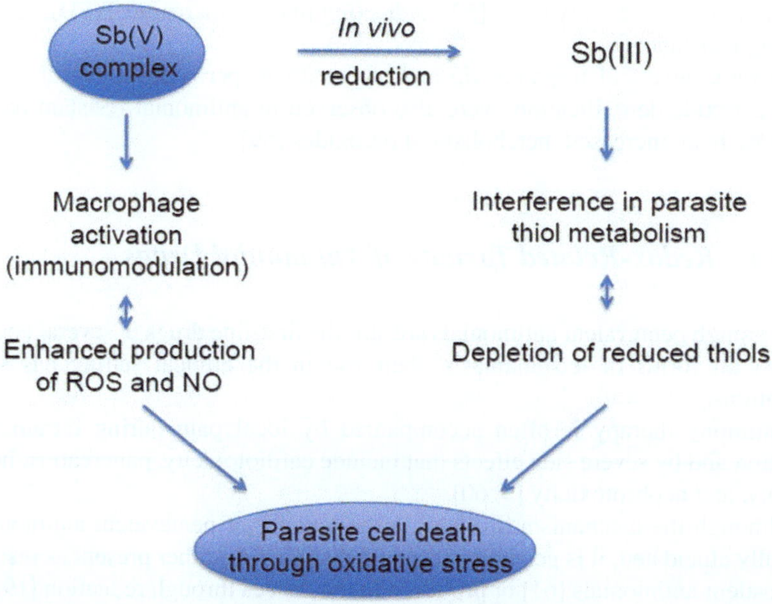

Fig. 30.2 Model for the mechanism of action of pentavalent antimonial drugs against leishmaniasis

30.3.3 Redox Changes in Antimony-Resistant Parasites

Resistance of leishmania parasites to antimony has been extensively studied in both laboratory-selected resistant leishmania lines and field isolates [51, 52, 53], allowing important insights into the molecular and functional factors that modulate parasite sensitivity to the drug.

Most antimony-resistance associated genes are involved in metabolic process related to oxidative stress, cell redox homeostasis, and thiol biosynthesis [54], which is not surprising considering the mode of action of antimonial drugs.

A change often encountered in both laboratory-selected and field-isolated resistant leishmania parasites is overexpression of rate-limiting enzymes of thiol biosynthesis, such as ornithine decarboxylase and γ-glutamylcysteine synthetase, which causes an overproduction of intracellular thiols and extra protection against the oxidative stress upon drug exposure [55–58]. The higher rate of thiol synthesis may also result in enhanced rates of efflux through ABC transporters contributing to the resistance phenotype [33]. Overexpression of the multidrug resistance-associated protein A transporter responsible for sequestration of Sb-thiol conjugates in intracellular vesicles is another frequently observed change that contributes to the resistance phenotype in *Leishmania* [32]. Thiol depletion of these strains reestablished their susceptibility to Sb [57], indicating that the increased $T(SH)_2$ levels are causing resistance.

Elevated levels of tryparedoxin and tryparedoxin peroxidase, key enzymes in hydroperoxide detoxification, were also observed in antimonial resistant parasites resulting in an increased metabolism of peroxides [59].

30.3.4 Redox-Related Toxicity of Antimonial Drugs

Even though pentavalent antimonials are still the first-line drugs in several countries against all forms of leishmaniasis, their use in the clinical setting has several limitations.

Antimony therapy is often accompanied by local pain during intramuscular injection and by severe side effects that include cardiotoxicity, pancreatitis, hepatotoxicity, and nephrotoxicity [9, 60].

Although the mechanism involved in the toxicity of pentavalent antimonials is not fully elucidated, it is generally accepted that Sb(III), either present as residue in pentavalent antimonials [61] or produced in the tissues through reduction [19], may be responsible for their side effects and antileishmanial action.

Studies of the mechanism of cytotoxicity of the trivalent tartar emetic drug suggest that Sb(III) compromises thiol homeostasis through depletion of intracellular glutathione and inhibition of glutathione reductase [22]. Then, Sb(III) enhances oxidative stress and leads to apoptosis through increase of ROS [22, 62–64].

Further evidence that pentavalent antimonials exert toxicity through induction of oxidative stress was obtained by Kato et al. [65], who demonstrated the protective effect of the antioxidant ascorbic acid during antimonial chemotherapy in a murine model of visceral leishmaniasis.

30.4 Gold Complexes Against Leishmaniasis

Auranofin (Fig. 30.3) is a US Food and Drug Administration-approved drug used therapeutically for rheumatoid arthritis. It is the prototypical gold drug with remarkable broad-spectrum medicinal properties, inspiring the development of other Au(I) and Au(III) compounds [66]. More recently, the awarding of orphan-drug status to auranofin for the possible treatment of amebiasis (caused by *Entamoeba histolytica*, an intestinal protozoan parasite) has significant global health implications in developing countries [67]. Transcriptional profiling and thioredoxin reductase assays suggested that auranofin targets the *E. histolytica* thioredoxin reductase, preventing the reduction of thioredoxin and enhancing sensitivity of trophozoites to ROS-mediated killing. Auranofin also inhibits viral load in simian virus [68, 69]. In combination

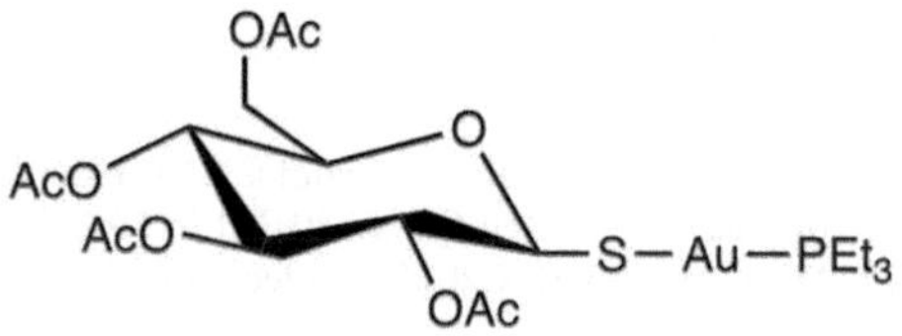

Fig. 30.3 Structure of Auranofin

with auranofin, administration of buthionine sulfoximine, an inhibitor of glutathione synthesis, under a highly intensified antiretroviral treatment was followed, after therapy suspension, by a significant decrease of viral RNA and DNA in peripheral blood as compared to pre-therapy levels.

Auranofin has also shown antiparasitic (malaria, leishmaniasis) activity, very likely arising from inhibition of parasitic enzymes involved in the control of the redox metabolism [11, 66]. The validation of TR as a key enzyme of *Leishmania infantum* polyamine-dependent redox metabolism and a target for antileishmanial drugs suggested that thiophilic agents besides Sb(III) could be effective enzyme inhibitors and potential antileishmanial agents. The X-ray crystal structure of the auranofin–trypanothione reductase–NADPH complex resolved at 3.5 Å resolution showed gold bound to the two active site cysteine residues of TR [70]. The thio-sugar moiety of auranofin is located at the trypanothione binding site, suggesting that auranofin may inhibit TR through a dual mechanism. Enzymatic assays revealed that auranofin causes a pronounced enzyme inhibition and the drug kills the promastigote stage of *L. infantum* at micromolar concentrations.

An integrated in vitro and in vivo screening platform incorporating multiple leishmania life cycles and species probed a focused library of pharmaceutically active compounds for identification and prioritization of *bona fide* cytotoxic chemotypes toward leishmania parasites. Auranofin was confirmed as a potent cytotoxic antileishmanial agent and inducer of apoptotic-like death in promastigotes. Significantly, the antileishmanial activity of auranofin transferred to cell-based amastigote assays as well as in vivo murine models [71].

Auranofin may represent a prototype drug that can be used to identify signaling pathways within the parasite and host cell critical for parasite growth and survival. Indeed, a structurally diverse group of Au(I)/Au(III) compounds behave as highly effective inhibitors of *Leishmania infantum* TR, some being even more effective than antimonials [72]. Simultaneous consideration of TR inhibition and antiproliferative potency has identified appropriate candidates for further evaluation. The current results suggest a foundation for potential exploitation of gold-based complexes as chemical tools or the basis of therapeutics for leishmaniasis.

Acknowledgment We acknowledge the Brazilian agencies CNPq, FAPEMIG and CAPES for financial support. We thank the support of NSF-CHE-1413189 and Sciences Without Borders CAPES PVES 154/2012.

References

1. Nussbaum K, Honek J, Cadmus CM, Efferth T. Trypanosomatid parasites causing neglected diseases. Curr Med Chem. 2010;17:1594–617.
2. WHO [Internet]. C2015. Neglected tropical diseases. http://www.who.int/neglected_diseases/diseases/en/.
3. Riethmiller S. From atoxyl to salvarsan: searching for the magic bullet. Chemotherapy. 2005;51:234–42.
4. Chersterman C. Dr. Ernest A. H. Friedheim. A tribute on his eightieth birthday. Trans R Soc Trop Med Hyg. 1979;73:597–8.
5. Shen ZX, Chen GQ, Ni JH, Li XS, Xiong SM, Qiu QY, et al. Use of arsenic trioxide (As_2O_3) in the treatment of acute promyelocytic leucemia (APL): II. Clinical efficacy and pharmacokinetics in relapsed patients. Blood. 1997;89:3354–60.
6. Duffin J, René P. Anti-moine; anti-biotique: The public fortunes of the secret properties of antimony potassium tartrate (tartar emetic). J Hist Med Allied Sci. 1991;46:440–56.
7. Vianna G. Tratamento da leishmaniose tegumentar por injeções intravenosas de tártaro emético. 7 Congresso Brasileiro de Medicina Tropical de São Paulo, São Paulo, Brasil. 1912;4:426-8.
8. Peters W. The treatment of kala-azar. New approaches to an old problem. Indian J Med Res. 1981;73:1–18.
9. Frézard F, Demicheli C, Ribeiro RR. Pentavalent antimonials: new perspectives for old drugs. Molecules. 2009;14:2317–36.
10. Fricker SP, Mosi RM, Cameron BR, Baird I, Zhu Y, Anastassov V, et al. Metal compounds for the treatment of parasitic diseases. J Inorg Biochem. 2008;102:1839–45.
11. Navarro M, Gabiani C, Chiara M, Messore L, Gambino D. Metal-based drugs for malaria, trypanosomiasis and leishmaniasis: recent achievements and perspectives. Drug Discov Today. 2010;15:1070–8.
12. Setzer WN. Trypanosomatid disease drug discovery and target identification. Future Med Chem. 2013;5:1703–4.
13. Croft SL, Seifert K, Yardley V. Current scenario of drug development for leishmaniasis. Indian J Med Res. 2006;123:399–410.
14. Mukbel RM, Patten Jr C, Gibson K, Ghosh M, Petersen C, Jones DE. Macrophage killing of Leishmania amazonensis amastigotes requires both nitric oxide and superoxide. Am J Trop Med Hyg. 2007;76:669–75.
15. Van Assche T, Deschacht M, da Luz RA, Maes L, Cos P. Leishmania-macrophage interactions: insights into the redox biology. Free Radic Biol Med. 2011;51:337–51.
16. Krauth-Siegel RL, Comini MA. Redox control in trypanosomatids, parasitic protozoa with trypanothione-based thiol metabolism. Biochim Biophys Acta. 2008;1780:1236–48.
17. Bocedi A, Dawood KF, Fabrini R, Federici G, Gradoni L, Pedersen JZ, et al. Trypanothione efficiently intercepts nitric oxide as a harmless iron complex in trypanosomatid parasites. FASEB J. 2010;24:1035–42.
18. Ferreira CS, Martins PS, Demicheli C, Brochu C, Ouellette M, Frézard F. Thiol-induced reduction of antimony(V) into antimony(III): a comparative study with trypanothione, cysteinylglycine, cysteine and glutathione. Biometals. 2003;16:441–3.
19. Goodwin LC, Page JE. A study of the excretion of organic antimonials using a polarographic procedure. Biochem J. 1943;22:236–40.
20. Hansen C, Hansen EW, Hansen HR, Gammelgaard B, Sturup S. Reduction of Sb(V) in a human macrophage cell line measured by HPLC-ICP-MS. Biol Trace Elem Res. 2011;144:234–43.
21. Shaked-Mishan P, Ulrich N, Ephros M, Zilberstein D. Novel intracellular Sb(V) reducing activity correlates with antimony susceptibility in *Leishmania donovani*. J Biol Chem. 2001;276:3971–6.

22. Wyllie S, Fairlamb AH. Differential toxicity of antimonial compounds and their effects on glutathione homeostasis in a human leukaemia monocyte cell line. Biochem Pharmacol. 2006;71:257–67.
23. Sereno D, Cavaleyra M, Zemzoumi K, Maquaire S, Ouaissi A, Lemesre JL. Axenically grown amastigotes of Leishmania infantum used as an in vitro model to investigate the pentavalent antimony mode of action. Antimicrob Agents Chemother. 1998;42:3097–102.
24. Frézard F, Demicheli C, Ferreira CS, Costa MAP. Glutathione-induced conversion of pentavalent antimony to trivalent antimony in meglumine antimoniate. Antimicrob Agents Chemother. 2001;45:913–6.
25. Carrio J, de Colmenares M, Riera C, Gallego M, Arboix M, Portus M. *Leishmania infantum*: stage-specific activity of pentavalent antimony related with the assay conditions. Exp Parasitol. 2000;95:209–14.
26. Yan SC, Li F, Ding KY, Sun H. Reduction of pentavalent antimony by trypanothione and formation of a binary and ternary complex of antimony(III) and trypanothione. J Biol Inorg Chem. 2003;8:689–97.
27. Fairlamb AH, Cerami A. Metabolism and functions of trypanothione in the Kinetoplastida. Annu Rev Microbiol. 1992;46:695–729.
28. Denton H, McGregor JC, Coombs GH. Reduction of anti-leishmanial pentavalent antimonial drugs by a parasite-specific thiol-dependent reductase, TDR1. Biochem J. 2004;381:405–12.
29. Zhou Y, Messier N, Ouellette M, Rosen BP, Mukhopadhyay R. *Leishmania major* LmACR2 is a pentavalent antimony reductase that confers sensitivity to the drug Pentostam. J Biol Chem. 2004;279:37445–51.
30. Salaün P, Frézard F. Unexpectedly high levels of antimony (III) in the pentavalent antimonial drug Glucantime: insights from a new voltammetric approach. Anal Bioanal Chem. 2013;405:5201–14.
31. Sun H, Yan SC, Cheng WS. Interaction of antimony tartrate with the tripeptide glutathione. Eur J Biochem. 2000;267:5450–7.
32. Légaré D, Richard D, Mukhopadhyay R, Stierhof YD, Rosen BP, Haimeur A, et al. The Leishmania ATP-binding cassette protein PGPA is an intracellular metal-thiol transporter ATPase. J Biol Chem. 2001;276:26301–7.
33. Mukhopadhyay R, Dey S, Xu N, Gage D, Lightbody J, Ouellette M, et al. Trypanothione overproduction and resistance to antimonials and arsenicals in *Leishmania*. Proc Natl Acad Sci U S A. 1996;93:10383–7.
34. Baiocco P, Colotti G, Franceschini S, Ilari A. Molecular basis of antimony treatment in leishmaniasis. J Med Chem. 2009;52:2603–12.
35. Wyllie S, Cunningham ML, Fairlamb AH. Dual action of antimonial drugs on thiol redox metabolism in the human pathogen *Leishmania donovani*. J Biol Chem. 2004;279:39925–93.
36. Moreira W, Leprohon P, Ouellette M. Tolerance to drug-induced cell death favours the acquisition of multidrug resistance in *Leishmania*. Cell Death Dis. 2011;2, e201.
37. Demicheli C, Frézard F, Mangrum JB, Farrell NP. Interaction of trivalent antimony with a CCHC zinc finger domain: potential relevance to the mechanism of action of antimonial drugs. Chem Commun. 2008;39:4828–30.
38. Frézard F, Silva H, Pimenta AM, Farrell N, Demicheli C. Greater binding affinity of trivalent antimony to a CCCH zinc finger domain compared to a CCHC domain of kinetoplastid proteins. Metallomics. 2012;4:433–40.
39. Webb JR, McMaster WR. Molecular cloning and expression of a *Leishmania major* gene encoding a single-stranded DNA-binding protein containing nine “CCHC” zinc finger motifs. J Biol Chem. 1998;268:13994–4002.
40. Lai WS, Kennington EA, Blackshear PJ. Interactions of CCCH zinc finger proteins with mRNA: non-binding tristetraprolinmutants exert an inhibitory effect on degradation of AU-rich element-containing mRNAs. J Biol Chem. 2002;277:9606–13.
41. Clayton C, Shapira M. Post-transcriptional regulation of gene expression in trypanosomes and leishmanias. Mol Biochem Parasitol. 2007;156:93–101.
42. Demicheli C, Frézard F, Lecouvey M, Garnier-Suillerot A. Antimony(V) complex formation with adenine nucleosides in aqueous solution. Biochim Biophys Acta. 2002;1570:192–8.

43. Lucumi A, Robledo S, Gama V, Saravia NG. Sensitivity of *Leishmania viannia panamensis* to pentavalent antimony is correlated with the formation of cleavable DNA-protein complexes. Antimicrob Agents Chemother. 1998;42:1990–5.
44. Pathak MK, Yi T. Sodium stibogluconate is a potent inhibitor of protein tyrosine phosphatases and augments cytokine responses in hemopoietic cell lines. J Immunol. 2001;167:3391–7.
45. Chai Y, Yan S, Wong ILK, Chow LMC, Sun H. Complexation of antimony [Sb(V)] with guanosine 5′-monophosphate and guanosine 5′-diphospho-D-mannose: formation of both mono and bis-adducts. J Inorg Biochem. 2005;99:2257–63.
46. Demicheli C, Santos LS, Ferreira CS, Bouchemal N, Hantz E, Eberlin MN, et al. Synthesis and characterization of Sb(V)–adenosine and Sb(V)–guanosine complexes in aqueous solution. Inorg Chim Acta. 2006;359:159–67.
47. Hansen HR, Pergantis SA. Mass spectrometric identification and characterization of antimony complexes with ribose-containing biomolecules and an RNA oligomer. Anal Bioanal Chem. 2006;385:821–33.
48. Ghosh M, Roy K, Roy S. Immunomodulatory effects of antileishmanial drugs. J Antimicrob Chemother. 2013;68:2834–8.
49. Mookerjee Basu J, Mookerjee A, Sen P, Bhaumik S, Sen P, Banerjee S, et al. Sodium antimony gluconate induces generation of reactive oxygen species and nitric oxide via phosphoinositide 3-kinase and mitogen-activated protein kinase activation in Leishmania donovani-infected macrophages. Antimicrob Agents Chemother. 2006;50:1788–97.
50. Muniz-Junqueira MI, Paula-Coelho VN. Meglumine antimonate directly increases phagocytosis, superoxide anion and TNF-α production, but only via TNF-α it indirectly increases nitric oxide production by phagocytes of healthy individuals, in vitro. Int Immunopharmacol. 2008;8:1633–8.
51. Croft SL, Sundar S, Fairlamb AH. Drug resistance in leishmaniasis. Clin Microbiol Rev. 2006;19:111–26.
52. Decuypere S, Vanaerschot M, Brunker K, Imamura H, Muller S, Khanal B, et al. Molecular mechanisms of drug resistance in natural Leishmania populations vary with genetic background. PLoS Negl Trop Dis. 2012;6, e1514.
53. Ouellette M, Drummelsmith J, Papadopoulou B. Leishmaniasis: drugs in the clinic, resistance and new developments. Drug Resist Updat. 2004;7:257–66.
54. Frézard F, Monte-Neto R, Reis PG. Antimony transport mechanisms in resistant leishmania parasites. Biophys Rev. 2014;6:119–32.
55. Grondin K, Haimeur A, Mukhopadhyay R, Rosen BP, Ouellette M. Co-amplification of the gamma-glutamylcysteine synthetase gene gsh1 and of the ABC transporter gene pgpA in arsenite-resistant *Leishmania tarentolae*. EMBO J. 1997;16:3057–65.
56. Haimeur A, Guimond C, Pilote S, Mukhopadhyay R, Rosen BP, Poulin R, et al. Elevated levels of polyamines and trypanothione resulting from overexpression of the ornithine decarboxylase gene in arsenite-resistant *Leishmania*. Mol Microbiol. 1999;34:726–35.
57. Mandal G, Wyllie S, Singh N, Sundar S, Fairlamb AH, Chatterjee M. Increased levels of thiols protect antimony unresponsive *Leishmania donovani* field isolates against reactive oxygen species generated by trivalent antimony. Parasitology. 2007;134:1679–87.
58. Mukherjee A, Padmanabhan PK, Singh S, Roy G, Girard I, Chatterjee M, et al. Role of ABC transporter MRPA, gamma-glutamylcysteine synthetase and ornithine decarboxylase in natural antimony-resistant isolates of *Leishmania donovani*. J Antimicrob Chemother. 2007;59:204–11.
59. Wyllie S, Mandal G, Singh N, Sundar S, Fairlamb AH, Chatterjee M. Elevated levels of tryparedoxin peroxidase in antimony unresponsive *Leishmania donovani* field isolates. Mol Biochem Parasitol. 2010;173:162–4.
60. Marsden PD. Pentavalent antimonials: old drugs for new diseases. Rev Soc Bras Med Trop. 1985;18:187–98.
61. Dzamitika SA, Falcão CA, de Oliveira FB, Marbeuf C, Garnier-Suillerot A, Demicheli C et al. Role of residual Sb(III) in meglumine antimoniate cytotoxicity and MRP1-mediated resistance. Chem Biol Interact. 2006;160:217–24.

62. Lecureur V, Le Thiec A, Le Meur A, Amiot L, Drenou B, Bernard M, et al. Potassium antimonyl tartrate induces caspase- and reactive oxygen species-dependent apoptosis in lymphoid tumoral cells. Br J Haematol. 2002;119:608–15.
63. Lösler S, Schlief S, Kneifel C, Thiel E, Schrezenmeier H, Rojewski MT. Antimony trioxide- and arsenic-trioxide-induced apoptosis in myelogenic and lymphatic cell lines, recruitment of caspases, and loss of mitochondrial membrane potential are enhanced by modulators of the cellular glutathione redox system. Ann Hematol. 2009;88:1047–58.
64. Timerstein MA, Plews PI, Walker CV, Woolery MD, Wey HE, Toraason MA. Antimony induces oxidative-stress and toxicity in cultured cardiac myocytes. Toxicol Appl Pharmacol. 1995;130:41–7.
65. Kato KC, Morais-Teixeira E, Reis PG, Silva-Barcellos NM, Salaün P, Campos PP, et al. Hepatotoxicity of pentavalent antimonial drug: possible role of residual Sb(III) and protective effect of ascorbic acid. Antimicrob Agents Chemother. 2014;58:481–8.
66. Berners-Price SJ, Filipovska A. Gold compounds as therapeutic agents for human diseases. Metallomics. 2011;3:863–73.
67. Debnath A, Parsonage D, Andrade RM, He C, Cobo ER, Hirata K, et al. A high-throughput drug screen for *Entamoeba histolytica* identifies a new lead and target. Nat Med. 2012;18:956–62.
68. Lewis MG, DaFonseca S, Chomont N, Palamara AT, Tardugno M, Mai A et al. Gold drug auranofin restricts the viral reservoir in the monkey AIDS model and induces containment of viral load following ART suspension. AIDS. 2011;25:1347–56.
69. De Luca A, Hartinger CG, Dyson PJ, Lo Bello M, Casini A. A new target for gold(I) compounds: glutathione-S-transferase inhibition by auranofin. J Inorg Biochem. 2013; 119:38–42.
70. Ilari A, Baiocco P, Messori L, Fiorillo A, Boffi A, Gramiccia M, et al. A gold-containing drug against parasitic polyamine metabolism: the X-ray structure of trypanothione reductase from *Leishmania infantum* in complex with auranofin reveals a dual mechanism of enzyme inhibition. Amino Acids. 2012;42:803–11.
71. Sharlow ER, Leimgruber S, Murray S, Lira A, Sciotti RJ, Hickman M, et al. Auranofin is an apoptosis-simulating agent with in vitro and in vivo anti-leishmanial activity. ACS Chem Biol. 2014;9:663–72.
72. Colotti G, Ilari A, Fiorillo A, Baiocco P, Cinellu MA, Maiore L, et al. Metal-based compounds as prospective antileishmanial agents: inhibition of trypanothione reductase by selected gold complexes. ChemMedChem. 2013;8:1634–7.

ERRATUM TO

Chapter 15
Advances in Breast Cancer Therapy Using Nitric Oxide and Nitroxyl Donor Agents

Debashree Basudhar, Katrina M. Miranda, David A. Wink, and Lisa A. Ridnour

I. Batinić-Haberle et al. (eds.), *Redox-Active Therapeutics*,
Oxidative Stress in Applied Basic Research and Clinical Practice,
DOI 10.1007/978-3-319-30705-3_15

DOI 10.1007/978-3-319-30705-3_31

On Page 378, in line no-4, soluble guanylate cyclase is incorrectly captured. The correct format is as follows: soluble guanylate cyclase (SGC)

On Page 378, line no-4, cGMP is incorrectly captured. The correct format should be: cyclic guanosive monophospase (cGMP)

On Page 379, under section 15.2 line no-6 it is given as "HER2" is incorrectly captured. The correct format is "human epidermal growth factor (HER2)"

On page 379, under section 15.2.1, line 26, it is given as "metal center" is incorrectly captured. The correct format should be "metal centers"

On Page 380, under section 15.2.1, line-36"O_2 Concentration" is incorrectly captured. The correct format is "O_2 Concentrations"

The updated original online version for this chapter can be found at http://dx.doi.org/10.1007/978-3-319-30705-3_15

I. Batinić-Haberle et al. (eds.), *Redox-Active Therapeutics*,
Oxidative Stress in Applied Basic Research and Clinical Practice,
DOI 10.1007/978-3-319-30705-3_31

On Page 380, under section 15.2.1, line 43 “function and” is incorrectly captured. The correct format is “function to”.

On Page 380, under section 15.2.1, line 51 it is (Cys 520 in human) is incorrectly captured. The correct format should be (Cys 520 in humans).

On Page 380, under section 15.2.1, line 54,it is given as 100 to 300 is incorrectly captured. The correct format should be “100 to 500”.

On Page 380, under section 15.2.1, line 58, it is given as “metastasis [76]” is incorrectly captured. The correct format should be “metastasis [76, 77].

On Page 384, the numbering style is 15.3.3 which was incorrectly captured. The correct format is 15.3.2.1.

On Page 385, the numbering style is 15.3.4 was incorrectly captured. The correct format is 15.3.2.2

On Page 386, the numbering is given as 15.3.5: The correct format should be changed to 15.3.2.3

On Page 387, the numbering is given as 15.3.6: The correct format is 15.3.3

On Page 387 under section:15.3.3 , HNO Donors and Their Potential Usage the text has been inserted:

“The role of HNO as a key pharmacological agent in the treatment of alcoholism, cardiovascular disease, and cancer has generated much interest. However, HNO chemical biology has been limited due to its irreversible dimerization, thus necessitating use of donor compounds for production of HNO in situ”

On Page 387 above “Angeli’s salt (AS, Na2N2O3)” a new subheading needs to be added:

15.3.3.1 Diazeniumdiolate based HNO donors

On Page 387, under section: 15.3.3, line-10, it is incorrectly captured as “a structurally”. The correct format is “the structurally”

On Page 387, under section 15.3.3, line-11, it is given as “NO and HNO”. The correct format should be “NO and/or HNO”.

On Page 387, under section 15.3.3, line-10 “class of compounds known as” should be deleted. On Page 389, the numbering is 15.3.7:

The correct numbering should be changed to 15.3.3.2

On Page 390, the numbering is 15.3.8: The correct numbering should be 15.3.4

On page 391, the under 15.4 Conclusion below the line-11, “as novel therapeutic agents” the next paragraph is to be added:

Acknowledgement

This work was supported in part by the Intramural Research Program of the NIH, Cancer and Inflammation Program.

Index

I. Batinić-Haberle et al. (eds.), *Redox-Active Therapeutics*,
Oxidative Stress in Applied Basic Research and Clinical Practice,
DOI 10.1007/978-3-319-30705-3

Q

MIX
Papier aus verantwortungsvollen Quellen
Paper from responsible sources
FSC® C105338

If you have any concerns about our products,
you can contact us on
ProductSafety@springernature.com

In case Publisher is established outside the EU,
the EU authorized representative is:
Springer Nature Customer Service Center GmbH
Europaplatz 3, 69115 Heidelberg, Germany

Printed by Libri Plureos GmbH
in Hamburg, Germany